NUMBERS EXPRESSED IN SCIENTIFIC NOTATION

1 000 000	$= 10 \times 10 \times 10 \times 10 \times 10 \times 10$	$= 10^6$
100 000	$= 10 \times 10 \times 10 \times 10 \times 10$	$= 10^5$
10 000	$= 10 \times 10 \times 10 \times 10$	$= 10^4$
1000	$= 10 \times 10 \times 10$	$= 10^3$
100	$= 10 \times 10$	$= 10^2$
10	$= 10$	$= 10^1$
1	$= 1$	$= 10^0$
0.1	$= 1/10$	$= 10^{-1}$
0.01	$= 1/100 = 1/10^2$	$= 10^{-2}$
0.001	$= 1/1000 = 1/10^3$	$= 10^{-3}$
0.0001	$= 1/10\,000 = 1/10^4$	$= 10^{-4}$
0.000 01	$= 1/100\,000 = 1/10^5$	$= 10^{-5}$
0.000 001	$= 1/1\,000\,000 = 1/10^6$	$= 10^{-6}$

PHYSICAL DATA

Speed of light in a vacuum	$= 2.9979 \times 10^8$ m/s
Speed of sound (20°C, 1 atm)	$= 343$ m/s
Standard atmospheric pressure	$= 1.01 \times 10^5$ Pa
1 astronomical unit (A.U.), (average earth-sun distance)	$= 1.50 \times 10^{11}$ m
Average earth-moon distance	$= 3.84 \times 10^8$ m
Equatorial radius of the sun	$= 6.96 \times 10^8$ m
Equatorial radius of Jupiter	$= 7.14 \times 10^7$ m
Equatorial radius of the earth	$= 6.37 \times 10^6$ m
Equatorial radius of the moon	$= 1.74 \times 10^6$ m
Average radius of hydrogen atom	$= 5 \times 10^{-11}$ m
Mass of the sun	$= 1.99 \times 10^{30}$ kg
Mass of Jupiter	$= 1.90 \times 10^{27}$ kg
Mass of the earth	$= 5.98 \times 10^{24}$ kg
Mass of the moon	$= 7.36 \times 10^{22}$ kg
Proton mass	$= 1.6726 \times 10^{-27}$ kg
Neutron mass	$= 1.6749 \times 10^{-27}$ kg
Electron mass	$= 9.1 \times 10^{-31}$ kg
Electron charge	$= 1.602 \times 10^{-19}$ C

STANDARD ABBREVIATIONS

A	ampere	g	gram	M	molarity
amu	atomic mass unit	h	hour	min	minute
atm	atmosphere	hp	horsepower	mph	mile per hour
Btu	British thermal unit	Hz	Hertz	N	newton
C	coulomb	in.	inch	Pa	pascal
°C	degree Celsius	J	joule	psi	pound per square inch
cal	calorie	K	kelvin	s	second
eV	electron volt	kg	kilogram	V	volt
°F	degree Fahrenheit	lb	pound	W	watt
ft	foot	m	meter	Ω	ohm

CONCEPTUAL Physical Science

SECOND EDITION

Paul G. Hewitt
City College of San Francisco

John Suchocki
Leeward Community College

Leslie A. Hewitt

Addison Wesley Longman

An imprint of Addison Wesley Longman, Inc.

Originally published by HarperCollins College Publishers

Menlo Park, California • Reading, Massachusetts • New York • Harlow, England
Don Mills, Ontario • Sydney • Mexico City • Madrid • Amsterdam

About the cover: The image is Kilauea Volcano on the island of Hawaii, the most active volcano on Earth. Brad Lewis describes his photo:

"In March of 1992, I journeyed to Pu'u O'o Vent to experience the beginning of an eruptive episode and began photographing when the spatter cone was less than 12 hours old. Shortly after sunrise, a rainbow formed above the forest, and the elements came together. Nowhere else is there a more visual reminder that the Earth is alive and continually recreating itself."

Acquisitions Editor: Sami Iwata
Publisher: Robin J. Heyden
Market Development Manager: Andy Fisher
Marketing Manager: Gay Meixel
Publishing Associates: Bridget Biscotti-Bradley, Erika Buck
Developmental Editor: Irene Nunes
Production Coordination: Joan Marsh
Production Service: Julie Kranhold, Ex Libris
Photo Research: Tracey A. David
Text and Cover Designer: Juan Vargas
Title Logo: Ernie Brown
Composition and Prepress: Thompson Type
Cover Printer: Coral Graphics
Printer and Binder: Von Hoffmann Press

For permission to use copyrighted material, grateful acknowledgment is made to the copyright holders on pp.773–774, which are hereby made part of this copyright page.

Conceptual Physical Sciences, Second Edition

Library of Congress Cataloging-in-Publication Data

Hewitt, Paul G.
Conceptual physical science / Paul G. Hewitt, John Suchocki, Leslie A. Hewitt.—2nd ed.
p. cm.
Includes index.
ISBN 0-321-00191-5
1. Physical sciences. I. Suchocki, John. II. Hewitt, Leslie A. III. Title
Q158.5.H48 1999
500.2—dc21

1 2 3 4 5 6 7 8 9 10—VHP—02 01 00 99 98-31269
ISBN 0-321-03540-2 CIP

To Marjorie Hewitt Suchocki
sister, mom, and aunt

and to the memory of James Kyle Hewitt
son, cousin, and brother

Brief Contents

Detailed Contents

Conceptual Physical Science Photo Album

Conceptual Physical Science is a very personalized book, a family undertaking shown in many photographs throughout the book. Photos of Marjorie Hewitt Suchocki and James Hewitt, to whom this book is dedicated, are both found in Chapter 12. James, who died at the age of 24 in a car crash, is shown at the age of 10 in Figure 12.39, page 301. Marjorie, who is alive and well, teaching theology and philosophy in the graduate programs of the Claremont Colleges, is shown in Figure 12.3 on page 287.

Charlie Spiegel, our mentor to whom the first edition was dedicated, and who passed away in 1997, opens the book facing page 1. Great grand-daughter Sarah Stafford holding the chickie sits on Charlie's lap.

All eight part openers feature children who are friends and family. Part I, page 11, is tiny friend Andrea Wu (who shows up again on page 66); Part II on page 133 is Terrance Jones, son of niece Corine Jones; Part III on page 183 is Hayli Holmes; Part IV on page 231 is author Paul's grandson (and Leslie's and John's nephew) Alexander Hewitt; Part V, page 313, is author Paul's first grandson, Manuel Hewitt (son of James); Part VI on page 361 is John and Tracy's first son, Ian Suchocki (pronounced Su-hock'-ee, with a silent c) ; Part VII on page 531 is Ian's brother, Reece Suchocki. Concluding Part VIII, page 683, is W.J. Akil Marshall, son of artist and UH Hilo colleague, Mike Marshall.

Author Paul is shown with friend and high-school physics teacher Pablo Robinson, who is sandwiched between beds of nails in Figure 5.3 on page 108. Pablo is the author of the lab manuals that accompany the 7th and 8th Editions of Hewitt's physics text, *Conceptual Physics.* Some of those labs appear in the *Conceptual Physical Science* lab manual. Dear friend Marshall Ellenstein, Chicago's finest physics teacher, walks barefoot on broken glass on page 128. Paul's friend since childhood, Paul Ryan, dips his hand in molten lead in Figure 6.30 on page 154. In a much less violent activity, friend Ken Ford demonstrates noise-canceling earphones in Figure 10.23 on page 245. Close friend Tim Gardner safely demonstrates Bernoulli's Principle in Figure 5.33 on page 123, and stands among the telescopes atop Mauna Kea in Hawaii in the opening photo to Chapter 29, page 706. In the same chapter in Figure 28.26, page 722, is little Melissa, daughter of friends Dennis and Tai McNelis. Physics buddy John Hubisz stands next to a decayed building to illustrate entropy in Figure 7.26, page 178. Brother David Hewitt (no, not a twin) and his wife Barbara work the water pump shown in Figure 5.27, page 119. Niece Stephanie Hewitt, of Costa Rica, holds up a redwood tree on page 367.

Author John walks barefoot across red-hot coals in Figure 7.2, page 160 (his first fire walk— an assignment from senior author Uncle Paul!). Wife Tracy and son Ian illustrate Newton's Third Law in Figure 2.25 on page 48. Tracy and son Reece appear on page 487, and Ian and Reece appear together on page 488. Figure 10.43 on page 254 is part of the wedding party of John and Tracy, including, left to right, John's brother-in-law Butch Orr, sister Cathy Orr, bride and groom, sister Joan Lucas, mom Marjorie Hewitt Suchocki, Tracy's parents Sharon and David Hopwood, friends Kellie Dippel and Mark Werkmeister, and Uncle Paul. John's niece Alexandra Lucas is shown on page 368, and nephew Graham Orr drinks water on page 372.

Author Leslie, at her dad's insistence, appears in three figures. The first is Leslie at the age of 16 in Figure 13.1, page 315 (the black and white photo that has been in the last six editions of *Conceptual Physics*). The second is more recent, Figure 25.4 on page 621, and the third is with her dad in Figure 22.40 on page 556. Leslie's mom fearlessly holds her hand above expanding steam in Figure 7.7, page 162, and on the same page in Figure 7.6, Leslie's brother Paul demonstrates adiabatic expansion. Paul again demonstrates adiabatic compresssion by pumping air in his tire on page 170. Paul's wife, Ludmila, looks through Polaroid filters in Figure 11.43 on page 281, and is shown with son Alexander in Figure 11.18 on page 268. Alexander's dog Hanz is shown in Figure 6.20, page 147. Last but certainly not least of Leslie's entourage is her husband, Bob Abrams, who helped considerably with Part VII, shown in Figure 22.9 on page 538, and Figure 24.29 on page 606. Leslie's and Bob's first child is shown before birth in the sonogram in Figure 10.14 on page 240.

Author Paul's dear friends include former teaching assistants at CCSF. The first to appear in the book is Tenny Lim, seen pulling back her arrow in Figure 3.13 on page 62. Tenny went on from CCSF to a degree in mechanical engineering and is presently a design engineer at JPL in Pasadena. A more recent photo of Tenny, with yet another "arrow," a probe that will be injected into Comet Tempel 1 to analyze its nucleus, is shown on page 703. The second teaching assistant shown, in Figure 7.16 on page 166, is Helen Yan, now a physics instructor at CCSF, who did the hand lettering for the part openers. Helen poses again with the same black-hole box she posed with more than fifteen year ago for the black and white photo that appeared in three editions of Conceptual Physics. The third shown in the book, Figure 10.10 on page 237, is Lillian Lee, who was Paul's TA back in 1978-79. Twenty years later she remains a close friend and assisted in the production of the ancillaries to this edition. More recent CCSF teaching assistants are shown in the air track photo of Figure 3.10 on page 59. They are, left to right, Alex Diaz and Glenda Gin. In the same photo are colleagues Annette Rappleyea, who assisted with test bank materials, and Will Maynez, who designed and built the air track. His assistants are to his right; Ray Choi and Kumiko Furukawa.

Many of the photographs in this book were taken by artist Meidor Hu, who stands electrified in Figure 8.10, page 191. Meidor is shown again in Figure 10.22 on page 245. Of the several photos Meidor took are one of her brother Tin Hoy in Figure 9.9, page 215, her sister Mei Tuck with friend Gabe Vitelli on page 174, her Uncle Chiu Man Wu in Figure 6.19, page 146, and Meidor's bunny in Figure 11.11 on page 264.

The inclusion of these people who are so dear to the authors makes *Conceptual Physical Science* all the more our labor of love.

To the Students

Physical Science is about the rules of the physical world--physics, chemistry, geology, and astronomy. Just as you can't enjoy a ball game, computer game, or party game until you know its rules, so it is with nature. Nature's rules are beautifully elegant and can be neatly described mathematically. That's why many physical sciences texts are treated as applied mathematics. But too much emphasis on computation misses something essential--*comprehension*--a gut feeling for the concepts. This book is *conceptual*, focusing on concepts in down-to-earth English rather than in mathematical language. You'll see the mathematical structure in frequent equations, but you'll find them *guides to thinking* rather than recipes for computation.

We enjoy physical science, and you will too--because you'll understand it. Just as a person who knows the rules of botany best appreciates plants, and a person who knows the intricacies of music best appreciates music, you'll better appreciate the physical world about you when you learn its rules.

Enjoy your physical science!

Paul G. Hewitt
John Suchocki
Leslie A. Hewitt

To the Instructor

Conceptual Physical Science, Second Edition, with its important ancillaries, provides a first introduction to physics, chemistry, earth science, and astronomy, melded in a manner to captivate student interest. It builds a conceptual base in physics and chemistry, which is then applied to earth science and astronomy. For the nonscience student, it is a base from which to view nature more perceptively—to see that surprisingly few relationships make up its rules. For the science student, it is this and a springboard to involvement in other sciences such as biology and health related fields.

As in the first edition, the book begins with physics—the most basic of the sciences because it reaches up to chemistry, which in turn reaches to the earth sciences, and ultimately up to the life sciences. Physics is about the laws of motion, energy, electricity, heat, sound, and light. The three chapters on heat in the first edition have been tightened to make two chapters in this edition. Likewise with the three previous chapters on the atom and radioactivity, which now make up two chapters. Like the first edition, the physics is treated conceptually, meaning that its focus is on qualitative comprehension more than on mathematical expression. Physics in this text is not treated as applied mathematics. We minimize mathematical language and mathematical problems that are roadblocks to many students. Although a flip though the pages will show that the equations are there, they are presented as guides to thinking rather than recipes for algebraic manipulation. Their derivations are addressed in the footnotes. The treatment of physics concludes with the realm of the atom—a bridge to chemistry.

The chemistry chapters in this edition have been completely reworked, with new art by the chemistry co-author. The historical flavor of the first edition has been toned down in favor of chemistry concepts as they relate to the chemist's submicroscopic perspective. Two new chapters have been added, the first an expanded discussion of chemical bonding as it relates to the periodic table, and the second an expanded discussion of acid/base and redox reactions. Expansion of the organic chapter includes discussions of natural and synthetic polymers. As with the first edition, emphasis is on visual models. Because most physical science students have neither the time nor resources to adequately delve into a quantum mechanical approach to chemistry, electron configurations are treated via the easy-to-visualize shell model and chemical bonding is treated in terms of overlapping atomic shells and Coulomb's law. Throughout Part 6, chemistry is related to the student's familiar world—the fluorine in their toothpaste, the Teflon on their frying pans, and how washing soda serves to soften water—with many environmental aspects of chemistry addressed.

Like the first edition, Earth Science encompasses the science of geology and the science of meteorology. Geology, the study of planet Earth, includes the formation of rocks and minerals, the study of the internal dynamics that have and continue to influence Earth's surface, and the study of water that nourishes it. New sections on the oceans have been added to the chapter on surface properties, and two new chapters are added—one that integrates oceans and the atmosphere, and a new chapter on weather. Like the rest of the book, the treatment is conceptual, focusing on the central concepts of geology with emphasis on processes. The treatment is quite visual, as the many figures attest. The theme of the Earth Science chapters is that change is an ongoing process, where the present is the key to the past.

The applications of physics, chemistry, and geology applied to other massive bodies in the universe culminate with astronomy. Astronomy is about "out there," where space and time differ from the students' everyday notions. As in the previous edition, both special and general relativity make up most of concluding Chapter 30. So our tour of physical science begins with the physics of atoms, then proceeds to the chemistry of molecules, then to the geology of aggregates of molecules, and finally to the aggregates of matter in the cosmos—astronomy.

This second edition has a greater number of exercises and problems at the end of each chapters. Another new feature throughout the book is the "Link-To" Boxes, that show the connections of subject matter to real world applications and phenomena.

Pedagogy

At the end of each chapter are Review Questions, Exercises, and in many chapters, Home Projects and Problems. All of the important ideas from each chapter are framed in the relatively easy-to-answer Review Questions, grouped by chapter sections. They are, as the name implies, a review of chapter material. Their purpose is simply to provide a structured way to review the chapter. They are not meant to challenge the student's intellect, for in the vast majority of cases, the answers can be simply looked up. The Exercises, on the other hand, play a different role. Some of these are designed to prompt the application of physical science to everyday situations, while others are more sophisticated and call for considerable critical thinking.

The Problems are mainly simple computations that aid in learning concepts. There are fewer Problems than Exercises to decrease the likelihood of students focusing on number crunching rather than conceptual reasoning. Exercises call for critical thinking. Although building confidence in math is a worthy goal, it is not what this book is about.

Students can find the answers to the odd-numbered Exercises and Problems in the back part of the *Conceptual Physical Science Practice Book.* Complete answers to all Exercises and Problems are in the Instructor's Manual.

Units of measurement are not emphasized in this text. When used, they are almost exclusively expressed in SI (exceptions include such units as calories, grams per centimeter cubed, and light years). Mathematical derivations are avoided in the main body of the text and appear in footnotes or in the appendixes.

Linking physical science to real world applications and phenomena is now enhanced with a new feature in this second edition—the "Link-to" boxes. Most every chapter has one or more of these boxes.

Ancillary Materials

More than enough material is included for a one-year course, which allows for a variety of course designs to fit your taste. These are suggested in the **Instructor's Manual** (0-321-03536-4), which you'll find to be different from most instructor's manuals. It contains many lecture ideas and topics not treated in the textbook, as well as teaching tips and suggested step-by-step lectures and demonstrations. It has full-page answers to all Exercises and Problems in the text.

Answers to the odd-numbered Exercises and Problems are available to students in the student supplement, **Conceptual Physical Science Practice Book** (0-321-03531-

3). This very important and greatly expanded book, our most creative work, guides your students to a sometimes computational way of developing concepts. It spans a wide use of analogies and intriguing situations, all with a "user friendly" tone.

The **Next-Time Questions** book (0-321-03539-9) has 143 insightful full-page questions with Hewitt cartoons, with answers on the back of each page. Use these as overhead transparencies, or for posting. There are Next-Time Questions for every chapter.

The expanded **Test Bank** book (0-321-03535-6) has more than 2400 multiple choice questions as well as short answer and essay questions. The questions are categorized according to level of difficulty. The **TestGen-EQ Computerized Testing Software** (Win: 0-321-03533-X; Mac: 0-321-03534-8) contains everything in the Test Bank and allows you to edit and change the order of the questions, add new questions, and print different versions of a test. Available for PC or Macintosh.

The **Conceptual Physical Science Laboratory Manual** (0-321-03531-3) is written by the authors. In addition to interesting laboratory experiments, it includes a range of activities similar to the home projects in the textbook. These guide students to experience phenomena before they quantify the same phenomena in a follow-up laboratory experiment. Answers to questions in the lab manual are in the Instructor's Manual.

Transparency Acetates (0-3212-03532-1) features more than 100 important figures from the text, which are available to qualified adopters from your Addison Wesley Longman representative.

Last, but not least, is a **video** on home projects in physics and chemistry (0-321-05194-7). Consult your Addison Wesley Longman representative about other videos, including the 34-video set of Hewitt's *Conceptual Physics Alive!*

Go to it! Your conceptual physical science course really can be the most interesting, informative, and worthwhile science course available to your students.

Acknowledgments

For input to this edition we are grateful to Bob Abrams (Leslie's husband), Robert Baruffaldi, Debra Brice, Milton Cha, Jim Court, Richard Crowe, George Curtis, Suk Hwang, Dan Johnson, Chris Lock, Jack Ott, Michael Reese, Bill Sakai, Russell Schnell, Jane Sneed, Walter Steiger, Phil Wolf, Larry Woolf, and Mike Young. We remain grateful to Charlie Spiegel who helped so much in the first edition, and whose death prevented his participation in this new edition (the first edition was dedicated to Charlie). We thank Ken Ford and Josip Slisko for the many valued suggestions to the eighth edition of *Conceptual Physics,* which have carried over to this book.

Thanks go to Tracy Suchocki for critically reviewing the chemistry chapters, and for her loving support as the wife of chemistry co-author John. We also thank Tracy's mother, Sharon Hopwood for her photography and for her assistance in the final proofing of the chemistry chapters. Thanks go to brother and uncle David Hewitt for his photography. We thank Meidor Hu for her photographs. We are grateful to Ernie Brown for designing the lettering on the cover, and to Helen Yan for hand lettering the part openers.

For physics input to the first-edition, which carries over to this edition, we remain thankful to friends Howie Brand, Paul Doherty, Marshall Ellenstein, John Hubisz, and Pablo Robinson. Appreciation to Peter Brancazio, Brooklyn College, Nick Brown, Cal Poly San Luis Obispo, Arnie Feldman, University of Hawaii at Manoa, and David Willey, University of Pennsylvania, Johnstown, for helpful physics suggestions. For chemistry suggestions we remain grateful to Leeward Community College colleagues Bob Asato, Manny Cabral, George Shiroma, and Pearl Takeuchi, and Ted Brattstrom, Pearl

City High School. Special thanks to Albert Sneden and Everette May of Virginia Commonwealth University for their support and guidance during the early development of the chemistry chapters. For Earth science suggestions we remain grateful to Karen Grove and Lisa White, both of the Department of Geosciences at San Francisco State University. For many highly valued Earth-science suggestions for this edition we are indebted to Jim Court, City College of San Francisco, and Mike Young, Westfield State College. For astronomy suggestions we again thank Richard Crowe at the University of Hawaii at Hilo. We remain grateful to amateur astronomer Forrest Luke of Lelihua High School in Hawaii for valued ideas implemented in the first edition, and to author Paul Tipler for suggestions concerning relativity.

For help with the production of ancillaries we thank Lillian Lee.

We thank the following reviewers for their feedback: M. Lynn James, University of Northern Colorado; Courtney Willis, University of Northern Colorado; Joanne Rhodes, Webster University; Richard Schmidt, Lakeland Community College; Karen Wetz, Manatee Community College; Jaime Nieman, Fiorello H. La Guardia Community College; Mark Yeager, Mira Costa College; John Martin, Wright State University; Kenneth Ladner, Western New Mexico University; Russell Roy, Santa Fe Community College; Daniel P. Smith, Taylor University; Paul Tiskus, Rhode Island College; Alvin K. Benson, Brigham Young University; Claude E. Bolze, Tulsa Community College; Edward B. Hanrahan, Hampton University; Linda Roach, Northwestern State University; Thomas Willard, Florida Southern College; Diane M. Bunce, Catholic University of America; Michael Young, Westfield State College; Cheng-ming Fou, University of Delaware; Eric Harms, Brevard Community College; and Janan Hayes, Merced College.

Enormous thanks to very talented Irene Nunes for her first-rate edits in greatly improving our manuscripts, which are now much more readable. And last but not least, we thank Addison-Wesley Science Editor Sami Iwata and her staff for their professional care.

Paul G. Hewitt

John Suchocki

Leslie Hewitt

Wow, Great Grandpa Charlie! Before this chickie exhausted its inner space resources and poked out of its shell, it must have thought it was at its last moments. But what seemed like its end was a new beginning. Are we like chickies, ready to poke through to a new environment and new beginning -- like humanizing outer space maybe?

About Science

Much of science is organized common sense about the physical world. Science as a body of knowledge about nature is the result of the collective efforts, experimental findings, insights, and wisdom of the human race. Above all, science is the process of discovering and explaining the order of nature and how its parts connect to one another. The earliest preludes to science predate recorded history, when people first discovered regularities and relationships in nature, such as star patterns in the night sky and weather patterns. From these regularities and relationships, people learned to make predictions that gave them some control over their surroundings.

Rational thinking, the premise of science, gained headway in Greece in the third and fourth centuries B.C. and spread throughout the Mediterranean world. Advancement toward scientific methods came to a halt in Europe when the Roman Empire fell in the fifth century A.D. Barbarian hordes destroyed much in their path as they overran Europe and ushered in the Dark Ages. During this time the Chinese and Polynesians continued charting the stars and the planets, while Arab nations developed mathematics and learned how to make such materials as glass, paper, metals, and various chemicals. The Greek philosophy of rational thinking was reintroduced to Europe by Islamic influences that penetrated into Spain during the tenth, eleventh, and twelfth centuries. Universities emerged in Europe in the thirteenth century when scholars developed the philosophy that reason is compatible with religious faith. The fifteenth century saw the advent of the printing press, which furthered the progress of scientific thought by allowing information to be stored and communicated much more efficiently.

The sixteenth-century Polish astronomer Copernicus caused great controversy when he published a book proposing that the Sun was stationary and the Earth revolved it. These ideas conflicted with the popular view that the Earth was the center of the universe. They also conflicted with Church teachings and were banned for 200 years.

Modern science began in the sixteenth century when the Italian physicist Galileo Galilei used experiments, rather than speculation, to study nature's behavior (we'll say more about Galileo in following chapters). He was arrested for popularizing the Copernican theory and for his other contributions to scientific thought. Yet a century later, advocates of Copernican ideas were accepted by the intellectual establishment.

This cycle of opposition to acceptance of new ideas happens age after age. In the early 1800s, geologists met with violent condemnation because their views of how the Earth was formed differed from the Genesis account of creation. Later in the same century these views of geology were accepted, but theories of evolution were condemned and the teaching of them forbidden. Every age has its groups of intellectual rebels who are persecuted, condemned, or suppressed at the time but who later seem harmless and often essential to the elevation of human conditions. "At every crossway on the road that leads to the future, each progressive spirit is opposed by a thousand men appointed to guard the past."[1]

Mathematics and Scientific Measurements

Science and human conditions advanced dramatically after the discovery, some four centuries ago, that nature can be analyzed and described mathematically. Expressed in mathematical terms, the ideas of science are unambiguous. They don't have the double meanings that so often confuse the discussion of ideas expressed in words. When findings about nature are expressed mathematically, they are often easier to verify or disprove by experiment. The methods of mathematics and experimentation have led to the enormous success of science.[2]

Measurements are a hallmark of good science. How much you know about something is often related to how well you can measure it. This was well put by the famous physicist Lord Kelvin in the nineteenth century: "I often say that when you can measure something and express it in numbers, you know something about it. When you

---

[1]From Count Maurice Maeterlinck's "Our Social Duty."

[2]The mathematical structure of physical science is evident in the equations you will encounter throughout this book. You will see that equations are simply shortcut expressions of relationships that can also be expressed in words. The focus of this book, however, is on understanding concepts—in English, and so equations are treated as compact statements that help guide thinking. The usefulness of equations as recipes for mathematical problem solving is of secondary importance. A premature emphasis on mathematical problem solving can obscure science—hence our emphasis on concept building first and problem solving second. You'll see many more conceptual exercises than problems in the chapter endmatter. *Conceptual Physical Science* puts comprehension comfortably before computation.

cannot measure it, when you cannot express it in numbers, your knowledge is of a meager and unsatisfactory kind. It may be the beginning of knowledge, but you have scarcely in your thoughts advanced to the stage of science, whatever it may be."

Mathematics and scientific measurements are the tools of good science (as you'll likely see in the lab part of this course). These tools are not something new, but go back to ancient times. In the third century B.C., for example, fairly accurate measurements were made of the sizes of the Earth, Moon, Sun, and the distances between them. We'll see how in Chapter 28.

The Scientific Method

In the sixteenth century, Galileo and the English philosopher Francis Bacon, working independently, developed a formal method for doing science—a procedure known as the **scientific method**. Based on rational thinking and experimentation, this method works as follows:

1. Recognize a question or a problem.
2. Make an educated guess—a **hypothesis**—as to the answer.
3. Predict the consequences that should be observable if the hypothesis is correct and that should be absent if the hypothesis is not correct.
4. Perform experiments to see if the predicted consequences are observed.
5. Formulate the simplest general rule that organizes the three ingredients—hypothesis, predicted effects, and experimental findings.

A scientific hypothesis is an educated guess that is only presumed to be factual until demonstrated by experiments. When a hypothesis has been tested over and over again and has not been contradicted, it may become known as either a **law** or a *principle*.

Scientists use the word *theory* in a way that differs from its usage in everyday speech. In everyday speech a theory is no different from a hypothesis—a supposition that has not been verified. A scientific **theory**, however, is a synthesis of a large body of information that encompasses well-tested and verified hypotheses about certain aspects of the natural world. Physicists, as we shall learn, speak of the quark theory of the atomic nucleus, chemists speak of the theory of metallic bonding in metals, geologists subscribe to the theory of plate tectonics, and astronomers speak of the theory of the Big Bang.

The theories of science are not fixed, but rather evolve as they go through stages of redefinition and refinement. During the last hundred years, for example, the theory of the atom has been repeatedly refined as new evidence on atomic behavior has been gathered. Similarly, chemists have refined their view of the way atoms bond, geologists have refined the plate tectonics theory, and astronomers armed with new data from the Hubble telescope are presently sharpening their view of the universe. The refinement of theories is a strength of science.

Although the scientific method is powerful, good science isn't always done this way. Many scientific advances have involved trial and error, experimentation without guessing, or just plain accidental discovery. Trained observation, however, is essential for noticing questions in the first place and for making sense of evidence. The success of science has to do more with an attitude common to scientists, rather than to any particular method. This attitude is one of inquiry, experimentation, and humility before the facts.

The Scientific Attitude

In science, a **fact** is generally a close agreement among competent observers studying the same phenomena. A fact is not something that is unchanging and absolute, as is commonly thought. For example, whereas it was once a recognized fact that the universe is unchanging and permanent, today it is an equally recognized fact that the universe is expanding and evolving. It is a fact that an amputated limb of a salamander may grow back. That the severed limb of a human may also be able to regenerate with the appropriate biochemical manipulations is not a fact, but a hypothesis requiring much experimentation. The motivation for uncovering new facts in science is often closely related to the applications these new facts might bring. If we could regenerate severed arms, for instance, might we also be able to regenerate damaged heart tissue?

If a scientist believes a certain hypothesis is true but finds contradicting evidence, the scientific attitude holds that the hypothesis must be either discarded or revised. For example, the greatly respected Greek philosopher Aristotle (384–322 B.C.) claimed that an object falls at a speed proportional to its weight. This false idea was held to be true for nearly 2000 years because of Aristotle's compelling authority. Galileo allegedly demonstrated the falseness of Aristotle's claim with one experiment—showing that heavy and light objects dropped from the Leaning Tower of Pisa fell at nearly equal speeds. In the scientific spirit, a single verifiable experiment outweighs any authority, regardless of reputation or the number of advocates. In modern science, argument by appeal to authority has little value.

Competent scientists must be open to changing their minds. They change their minds, however, only when confronted with solid experimental evidence or when a conceptually simpler hypothesis forces them to a new point of view. More important than defending beliefs is improving them. Better hypotheses are made by those who are honest in the face of fact.

Scientists must accept their experimental findings even when they would like them to be different. They must strive to distinguish between what they see and what they wish to see, for scientists, like most people, have a vast capacity for fooling themselves.[3] People have always tended to adopt general rules, beliefs, creeds, ideas, and hypotheses without thoroughly questioning their validity and to retain them long after they have been shown to be meaningless, false, or at least questionable. The most widespread assumptions are often the least questioned. Too often, when an idea is adopted, particular attention is given to cases that seem to support it, while cases that seem to refute it are distorted, belittled, or ignored.

Away from their profession, scientists may be no more honest or ethical than most other people. But in their profession they work in an arena that puts a high premium on honesty. The cardinal rule in science is that all hypotheses must be testable—they must be susceptible, at least in principle, to being proved *wrong*. In science, it is more important that there be a means of proving an idea wrong than that there be a means of proving it right. At first this may seem strange, for when we wonder about most things, we concern ourselves with ways of finding out whether they are true. This emphasis on determining possible wrongness distinguishes science from nonscience. If you want to distinguish whether a hypothesis is scientific or not, look to see if there is a test for proving it wrong. If there is no test for its possible wrongness, then the hypothesis is not scientific.

---

[3]In your education it is not enough to be aware that other people may try to fool you: It is more important to be aware of your own tendency to fool yourself.

Consider the biologist Charles Darwin's hypothesis that life forms evolve from simpler to more complex forms. This could be proved wrong if paleontologists found that more complex forms of life appeared before their simpler counterparts. Einstein hypothesized that light is bent by gravity. This might be proved wrong if starlight that grazed the Sun were undeflected from its normal path. As it turns out, less complex life forms are found to precede their more complex counterparts, and starlight is found to bend as it passes close to the Sun (as seen during a solar eclipse), which support the claims. If and when a hypothesis or scientific claim is confirmed, it is regarded as useful and a stepping stone to additional knowledge.

Consider on the other hand the hypothesis that "intelligent life exists on other planets somewhere in the universe." At present, this hypothesis is not scientific. Reasonable or not, it is *speculation*. Although it can be proved correct by the verification of a single instance of intelligent life existing elsewhere in the universe, there is no way to prove the hypothesis wrong if no life is ever found. If we search the far reaches of the universe for eons and find no life, we would not prove that it doesn't exist around the next corner. A hypothesis that is capable of being proved right but not capable of being proved wrong is not a scientific hypothesis. Many such statements are quite reasonable and useful, but they lie outside the domain of science.

None of us has the time, energy, or resources to test every idea, and so most of the time we take somebody's word. How do we know whose word to take? To reduce the likelihood of error, scientists accept the word only of those whose ideas, theories, and findings are testable—if not in practice, at least in principle. Speculations that cannot be tested are regarded as unscientific. This approach has the long-run effect of compelling honesty—findings widely publicized among fellow scientists are generally subjected to further testing. Sooner or later, mistakes (or deception) are found out; wishful thinking is exposed. A discredited scientist does not get a second chance in the community of scientists. Honesty, so important to the progress of science, thus becomes a matter of self-interest to scientists. There is relatively little bluffing in a game where all bets are called. In fields of study where correct and incorrect are not so easily established, the pressure to be honest is considerably less.

Question

Which of these hypotheses are scientific?

a. Better stock market decisions are made when the planets Venus, Earth, and Mars are aligned.
b. Atoms are the smallest particles of matter that exist.
c. Space is permeated with an essence that is undetectable.
d. Albert Einstein is the greatest physicist of the twentieth century.

Answer

Both *a* and *b* are scientific because there is a test for falseness. Hypothesis *a* can be proved wrong by going to the library and researching the performance of the stock market during times when these planets were aligned. Hypothesis *b* is not only *capable* of being proved wrong but in fact *has* been proved wrong. Hypotheses *c* and *d* have no test for possible falseness and are therefore unscientific. If Einstein was not the greatest physicist, how could we know?

Some pseudoscientists and other pretenders of knowledge will not even consider a test for the possible wrongness of their statements. It is important to note that because the name Einstein is generally held in high esteem, hypothesis d is a favorite of pseudoscientists. So we should not be surprised that the name of Einstein, like that of Jesus and other highly respected sources, is cited often by charlatans who wish to bring respect to themselves and their points of view. In all fields it is prudent to be skeptical of those who wish to credit themselves by calling upon the authority of others.

Link to Pseudoscience

In pre-science times any attempt to harness nature meant forcing nature against her will. Nature had to be subjugated, usually with some form of magic or by means that were above nature—supernatural. Science does just the opposite, and works within nature's laws. The methods of science have displaced the methods of magic—but not entirely. The old ways persist full force in primitive cultures, and they survive in technologically advanced cultures, too, often disguised as science. This is fake science—**pseudoscience.** The hallmark of a pseudoscience, or "alternative science," is that it lacks the key ingredients of (1) evidence and (2) having a test for wrongness. In the realm of pseudoscience, skepticism and tests for possible wrongness are downplayed or flatly ignored.

There are various ways to view our place in the universe, and mysticism is one of them. Astrology is an ancient form of magic that supposes a mystical connection between individuals and the universe—that human affairs are influenced by the positions and movements of stars and other celestial bodies. This nonscientific view is quite appealing with its implication that, however insignificant we may feel at times, we are in fact intimately connected to the workings of the cosmos. Astrology as ancient magic is one thing, but astrology in the guise of science is another. When it poses as a science related to astronomy, then it becomes pseudoscience. Some astrologers today present their craft in a scientific guise. When they use up-to-date astronomical information and computers that chart the movements of heavenly bodies, they're in the realm of science, but when they use this data to concoct astrological revelations, they have crossed over into full-fledged pseudoscience.

Pseudoscience, like science, makes predictions. The predictions of a dowser, who locates underground water with a dousing stick, have a very high rate of success—nearly 100%. Whenever the dowser goes through his or her ritual and points to a spot on the ground, the well digger is sure to find water. Dowsing works. Of course, the dowser can hardly miss because there is some groundwater beneath the surface at nearly every spot on Earth. (The real test of a dowser would be finding a place where water wouldn't be found!)

A shaman who studies the oscillations of a pendulum suspended over the abdomen of a pregnant woman can predict the sex of the fetus with an accuracy of 50%. This means if the shaman tries this magic many times on many fetuses, half the predictions will be right and half wrong—the predictability of ordinary guessing. The best that can be said for the shaman is the 50% success rate is a lot better than that of astrologers, palm readers, or other pseudoscientists who predict the future.

An example of pseudoscience that has zero success is energy-multiplying machines. These machines, which are alleged to deliver more energy than they take in, are "still on the drawing boards and needing funds for development." They are touted by quacks who sell shares to an ignorant public who succumb to the pie-in-the-sky promises of success. Then there are palm readers, who claim to have sacred knowledge of the "old ways" that transcends science. And others who claim possession of a "secret method" for astral travel, crystal power, memory recovery, or channeling. Pseudoscientists are everywhere, are usually successful in recruiting apprentices for money or labor, and can be very convincing even to seemingly reasonable people. Their books greatly outnumber books on science in bookstores. Psuedoscience, or junk science as it is often called, is thriving.

We humans have learned much in the nearly 500 years since the onset of science. Gaining this knowledge and overthrowing superstition was by enormous human effort and painstaking experimentation. We should rejoice in what we've learned. We have come a long way in comprehending nature and in liberating ourselves from ignorance, and we should be proud of this. We no longer have to die whenever an infectious disease strikes. We no longer live in fear of demons. We no longer pour molten lead in the boots of women accused of witchery, as was done for nearly three centuries during medieval times. Today we have no need to pretend that superstition is anything but superstition, or that junk notions are anything but junk notions—whether dispensed by shamans, street-corner quacks, or hacks who write health books.

Yet there is reason to fear that what people of one time fight for, a following generation surrenders. The grip that belief in magic and superstition had on people took centuries to overcome. Yet today the same magic and superstition are perceived as enchanting to a growing number of people. James Randi reports in his book *Flim-Flam!* that more than 20,000 practicing astrologers in the United States service millions of credulous believers. Science writer Martin Gardner reports that a greater percentage of Americans today believe in astrology and occult phenomena than did people in medieval Europe. Few newspapers carry a daily science column, but nearly all provide daily horoscopes. And then there are the flourishing television psychics who gain adherents daily.

Many people believe that the human condition is slipping backward because of growing technology. More likely, however, we'll slip backward because science and technology will bow to the magic and superstitions of the past. Watch for their spokespeople. Junk science is a big and successful business.

Science, Art, and Religion

The search for order and meaning in the world has taken different forms: One is science, another is art, and another is religion. These three domains differ from one another in important ways, although they often overlap. Science is principally engaged with discovering and recording natural phenomena, the arts are an expression of human experience as they pertain to the senses, and religion addresses the source, purpose, and meaning of it all.

Science and the arts are comparable. In art we find what is possible in human experience. We can learn about emotions ranging from anguish to love, even if we haven't yet experienced them. The arts do not necessarily give us those experiences but describe them to us and suggest possibilities. Science similarly tells us about possibilities, but those of the natural world rather than those of human experience. It provides us with a way of connecting things, of seeing relationships between and among them, and of making sense of the myriad of natural events around us. Science broadens our perspective of the natural environment of which we are a part. A knowledge of both the arts and the sciences makes for a wholeness that affects the way we view the world and the decisions we make about it and ourselves. A truly educated person is knowledgeable in both the arts and the sciences.

Science and religion deal with very different domains. Science is concerned with the physical realm; religion is concerned with the spiritual realm. Simply put, science asks *how*; religion asks *why*. The practices of science and religion are also different. Whereas scientists experiment to find nature's secrets, religious practitioners worship God and work to build human community. In these respects, science and religion are as different as apples and oranges and do not contradict each other. Science and religion are two different yet complementary fields of human activity.

When we study the nature of light later in this book, we shall treat light first as a wave and then as a particle. To the person who knows only a little about science, waves and particles are contradictory; light can be only one or the other, and we have to choose between them. But to the enlightened person, waves and particles complement each other and provide a deeper understanding of light. In a similar way, it is mainly people who are either uninformed or misinformed about the deeper natures of both science and religion who feel that they must choose between believing in religion and believing in science. Unless one has a shallow understanding of either or both, there is no contradiction in being religious and being scientific in one's thinking.[4]

Science and Technology

Science and technology are also different from each other. Whereas science has to do with establishing theories that organize and make sense of observed phenomena, technology develops the tools for putting the findings of science to use.

Technology is a double-edged sword that can be both helpful and harmful. We have the technology, for example, to extract fossil fuels from the ground and then burn them to produce energy. Such energy production from fossil fuels has benefited our society in countless ways. On the flip side, the burning of fossil fuels endangers the environment. It is tempting to blame technology for many of today's problems, such as pollution, resource depletion, and even overpopulation. These problems, however, are not the fault of technology any more than a shotgun wound is the fault of the shotgun. It is humans who use the technology, and humans who are responsible for how it is used.

Remarkably, we already possess the technology to solve almost all of our problems. The twenty-first century will likely see a switch from fossil fuels to more sustainable energy sources, such as photovoltaics, solar thermal electric generation, and biomass conversion. Whereas the paper of this book comes from trees, the paper of the future may well come from fast-growing weeds. More and more we are recycling waste products. In some parts of the world, progress is being made on stemming the human population explosion that aggravates almost every problem faced by humans today.

The greatest obstacle to solving today's problems lies more in social inertia than in a lack of technology. Technology is our tool. What we do with this tool is up to us. The promise of technology is a cleaner and healthier world. Wise applications of technology *can* lead to a better world.

Physics, Chemistry, Geology, and Astronomy

The study of science today branches into the study of living things and nonliving things: the life sciences and the physical sciences. The life sciences branch into such areas as biology, zoology, and botany. The physical sciences branch into such areas as physics, chemistry, geology, and astronomy—the areas addressed in this book.

We begin with physics because it is basic to the other physical sciences. Physics is about basic concepts—motion, force, energy, matter, heat, sound, light, and the insides of atoms. Chemistry builds on physics and tells us how atoms combine to form molecules, and how molecules combine to make the materials around us. Applied to the Earth and its processes, physics and chemistry make up the science of geology; applied to the other planets and to the stars, they all make up the science of astronomy.

Biology is more complex than physics and chemistry, for it involves matter that is alive. Underneath biology is chemistry, and underneath chemistry is physics. The concepts of physics reach up to these more complicated sciences. So physics provides the foundation for both physical science and life science. Thus an understanding of science in general begins with an understanding of physics, which begins this book. We treat

[4]Of course, this doesn't apply to certain fundamentalists—Christian, Muslim, or otherwise—who steadfastly assert that one cannot embrace both their brand of religion and science.

physics conceptually, as we do the other topics, so that you can enjoy understanding your knowledge.

In Perspective

Only a few centuries ago, the most talented and most skilled artists, architects, and artisans of the world directed their genius and effort to the construction of the great cathedrals, synagogues, temples, and mosques. Some of these architectural structures took centuries to build, which means that nobody witnessed both the beginning and the end of construction. Even the architects and early builders who lived to a ripe old age never saw the finished results of their labors. Entire lifetimes were spent in the shadows of construction that must have seemed without beginning or end. This enormous focus of

Risk Assessment

Technology supplies numerous benefits—but with risks. When the benefits of a technological innovation are seen to outweigh its risks, the technology is accepted and applied. Xrays, for example, continue to be used for medical diagnosis despite their potential for causing cancer. But when risks outweigh benefits, technology is used sparingly or not at all, as can be mandated by government. When negative consequences associated with certain medicines, for example, outweigh their benefits, the medicines are prohibited.

Risk can vary from one group to another. Aspirin is useful for adults, but for young children it can cause a potentially lethal condition known as Reye's Syndrome. Dumping raw sewage into the local river may pose little risk for a town located upstream, but for towns downstream the untreated sewage is a health hazard. Storing radioactive wastes underground may pose little risk for us today, but for future generations the risks are great if there is leakage into groundwater. Technologies involving different risks for different people, as well as different benefits, raise questions that are often hotly debated. Which medications should be sold over the counter and how should they be labeled? Is the cost of retrofitting a wastewater treatment plant worth the benefit? What should we do with the radioactive wastes generated from nuclear power plants as well as from hospitals employing nuclear medicine? Many risk analyses need to be considered when making decisions affecting a wider segment of society.

People seem to have a hard time accepting the impossibility of zero risk. Airplanes cannot be made perfectly safe. Processed foods cannot be completely free of toxicity, for all foods are toxic to some degree. You cannot go to the beach without risking skin cancer no matter how much sunscreen you apply. You cannot avoid radioactivity, for it's in the air you breathe and the foods you eat, and has been that way since before humans first walked the Earth. Even the cleanest rain contains radioactive carbon-14, not to mention the same in our bodies. Between each heart beat in the human body, there have always been about 10,000 naturally occurring radioactive decays. You might hide yourself in the hills, eat the most natural of foods, practice obsessive hygiene and still die from cancer caused by the radioactivity emanating from your own body. The probability of eventual death is 100%. Nobody is exempt.

Science is the determining of the most probable. As the tools of science improve, then assessment of the most probable gets closer to being on target. Acceptance of risk, on the other hand, is a societal issue. Placing zero risk as a societal goal is not only impractical, but selfish. Any society striving toward a policy of zero risk would consume its present and future economic resources. Isn't it more noble to accept nonzero risk and minimize risk as much as possible within the limits of practicality? A society that accepts no risks receives no benefits.

human energy was inspired by a vision that went beyond worldly concerns—a vision of the cosmos. To the people of that time, the structures they erected were their "spaceships of faith," firmly anchored but pointing to the cosmos.

Today the efforts of many of our most skilled scientists, engineers, artists, and artisans are directed to building the spaceships that already orbit the Earth and others that will voyage beyond. The time required to build these spaceships is extremely brief compared with the time spent building the stone and marble structures of the past. Many people working on today's spaceships were alive before the first jetliner aircraft carried passengers. Where will younger lives lead in a comparable time?

We seem to be at the dawn of a major change in human growth, for, as little Sarah suggests in the photo at the beginning of this book, we may be like the hatching chicken who has exhausted the resources of its inner-egg environment and is about to break through to a whole new range of possibilities. The Earth is our cradle and has served us well. But cradles, however comfortable, are one day outgrown. So with the inspiration that in many ways is similar to the inspiration of those who built the early cathedrals, synagogues, temples, and mosques, we aim for the cosmos.

We live in an exciting time!

Question

Which of the following activities involves the utmost human expression of passion, talent, and intelligence?

a. painting and sculpture b. literature c. music d. religion
e. science

Answer

All of them! The human value of science, however, may be the least understood by most individuals in our society. The reasons are varied, ranging from the common notion that science is incomprehensible to people of average intellectual ability to the extreme view that science is a dehumanizing force in our society. Most of the misconceptions about science probably stem from the confusion between the *abuses* of science and science itself.

Science is an enchanting human activity shared by a wide variety of people who, with present-day tools and know-how, are reaching further and finding out more about themselves and their environment than people in the past were ever able to do. The more you know about science, the more passionate you feel toward your surroundings. There is physics and chemistry in everything you see, hear, smell, taste, and touch!

Summary of Terms

- **Scientific method** An orderly method for gaining, organizing, and applying new knowledge.
- **Hypothesis** An educated guess; a reasonable explanation of an observation or experimental result that is not fully accepted as factual until tested over and over again by experiment.
- **Law** A general hypothesis or statement about the relationship of natural quantities that has been tested over and over again and has not been contradicted. Also known as a *principle*.
- **Theory** A synthesis of a large body of information that encompasses well-tested and verified hypotheses about certain aspects of the natural world.
- **Fact** A phenomenon about which competent observers who have made a series of observations are in agreement.
- **Pseudoscience** A fake science, erroneously regarded as scientific, that is characterized by a lack of (1) evidence and (2) a test for wrongness.

PART I

Mechanics

CHAPTER

1

Motion

Motion is everywhere—from the vibrations of atoms in matter to the swirl of galaxies in the universe. A systematic study of motion goes back to Aristotle, the leading philosopher of the fourth century B.C. In his view, every object in the universe had a proper place determined by its "nature," and any object not in its proper place would "strive" to get there. An unsupported lump of clay, for instance, being of earth, properly fell to the ground; a puff of smoke, being of the air, properly rose. According to Aristotle, the *distance* of an object from its natural place was the fundamentally important factor governing motion.

Galileo, a leading scientist of the seventeenth century A.D., broke with this traditional concept by realizing that *time* was an important missing factor. According to Galileo, motion was best described as a change in distance over a change in time. To measure motion, Galileo used *rates.* A rate tells how fast something happens—in other words, how much something changes in a certain amount of time. Rates that measure motion are *speed, velocity,* and *acceleration.*

Aristotle (384–322 B.C.)

Greek philosopher, scientist, and educator, Aristotle was the son of a physician who personally served the king of Macedonia. At 17, Aristotle entered the Academy of Plato, where he worked and studied for 20 years until Plato's death. He then became the tutor of young Alexander the Great. Eight years later he formed his own school. Aristotle's aim was to systematize existing knowledge, and to this end he made critical observations, collected specimens, and gathered together, summarized, and classified almost all existing knowledge of the physical world. His systematic approach became the method from which Western science later arose. After his death, his voluminous notebooks were preserved in caves near his home and were later sold to the library at Alexandria.

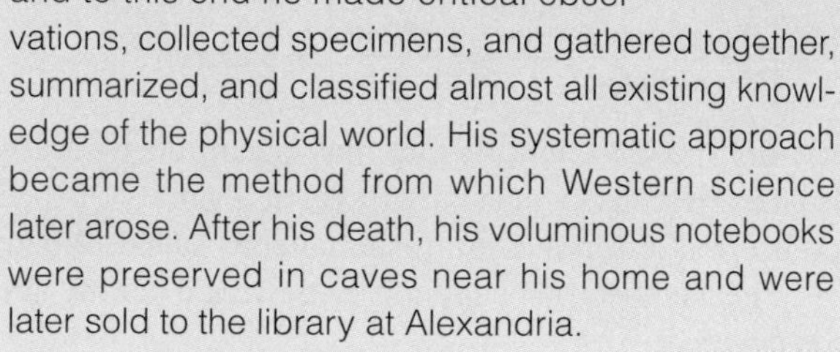

Scholarly activity ceased in most of Europe through the Dark Ages, and the works of Aristotle were forgotten. Scholarship continued in the Byzantine and Islamic empires, and various texts of Aristotle's teachings were reintroduced to Europe during the eleventh and twelfth centuries and translated into Latin. The Church, at that time the dominant political and cultural force in Western Europe, first prohibited the works of Aristotle and then accepted and incorporated them into Christian doctrine. Any attack on Aristotle was an attack on the Church itself. It was in this climate that Galileo effectively challenged Aristotle's ideas on motion, and ushered in a new method of knowing—experimentation.

1.1 Speed

Any object in motion moves a certain distance in a given time. An automobile, for example, may travel so many kilometers in an hour. **Speed** is a measure of how fast something is moving. It is the rate at which distance is covered and is always measured in terms of a unit of distance divided by a unit of time:

$$\textbf{Speed} = \frac{\textbf{distance}}{\textbf{time}}$$

Fig. 1.1
A cheetah is the champion runner over distances less than 500 meters and can achieve peak speeds of 100 km/h.

Any distance unit and time unit can be used. For long distances, the units kilometers per hour (km/h) or miles per hour (mi/h or mph) are commonly used. For shorter distances, meters per second (m/s) are often useful. The slash symbol (/) is read as *per* and means *divided by.* Table 1.1 shows speeds in various units.[1]

The speed that something has at any instant is **instantaneous speed**. It is the speed shown by the speedometer

[1]Conversion is based on 1 h = 3600 s, 1 mi = 1609.344 m.

of a car. When we say that a car's speed at some instant is 60 kilometers per hour, we are specifying its instantaneous speed, and we mean that if the car continued moving at exactly that speed for an hour, it would travel 60 kilometers. If it continued at exactly that speed for half an hour, it would go half the distance, 30 kilometers. If it continued for 1 minute, it would go 1 kilometer.

A car or any other moving body rarely travels at constant speed. So we speak of the **average speed**:

$$\textbf{Average speed} = \frac{\textbf{total distance covered}}{\textbf{time interval}}$$

If we cover a distance of 80 kilometers in a time interval of 1 hour, for example, our average speed is 80 kilometers per hour. Likewise, if we travel 320 kilometers in 4 hours, our average speed is

$$\textbf{Average speed} = \frac{\textbf{total distance covered}}{\textbf{time interval}} = \frac{\textbf{320 km}}{\textbf{4 h}} = \textbf{80 km/h}$$

Table 1.1
Approximate Speeds in Different Units

25 km/h = 16 mi/h = 7 m/s
60 km/h = 37 mi/h = 17 m/s
75 km/h = 47 mi/h = 21 m/s
100 km/h = 62 mi/h = 28 m/s
120 km/h = 75 mi/h = 33 m/s

Average speed doesn't show the speed variations that may occur during a given time interval, including possible stops. Because we usually travel at a variety of speeds on most trips, average speed is often different from instantaneous speed. Whether we talk about average speed or instantaneous speed, however, we are talking about the rate at which a distance is covered.

If we know average speed and time of travel, distance traveled is easy to find. A simple rearrangement of the preceding definition gives

$$\textbf{Total distance covered} = \textbf{average speed} \times \textbf{time interval}$$

If your average speed is 80 kilometers per hour on a 4-hour trip, for example, you cover a total distance of 320 kilometers.

Questions

1. What is the average speed of a cheetah that sprints (a) 100 m in 4 s (seconds)? (b) 50 m in 2 s?
2. A car moved at an average speed of 60 km/h for 1 h. (a) How far did it go? At this rate, how far would it go (b) in 4 h? (c) in 10 h?
3. During this 1-h trip, is it possible for the car to have an average speed of 60 km/h and never exceed a reading of 60 km/h on the speedometer?

Fig. 1.2
A speedometer gives readings in both miles per hour and kilometers per hour.

Speed Is Relative

In a strict sense, everything moves—even things that appear to be at rest. To understand why this is so, think about the Earth revolving around the Sun. In its orbit, the Earth travels at a speed of about 30 kilometers per second. This means that everything on Earth is also moving at 30 kilometers per second relative to the Sun. Thus a book that is at rest relative to the table it lies on is moving at about 30 kilometers per second relative to the Sun.

Galileo Galilei (1564–1642)

Galileo was born in Pisa in the same year Shakespeare was born and Michelangelo died. Galileo studied medicine at the University of Pisa and then changed to mathematics. He developed an early interest in the mechanics of motion and was soon at odds with his contemporaries, who held to Aristotelian ideas on falling bodies. Galileo left Pisa to teach at the University of Padua and became an advocate of the new Copernican theory of the solar system. He was one of the first to build a telescope and the first to direct it to the nighttime sky, an activity that led to his discovering mountains on the Moon and the moons of Jupiter.

Because Galileo published his findings in Italian instead of in the Latin expected of so reputable a scholar, and because of the recent invention of the printing press, his ideas reached a wide readership. Galileo soon ran afoul of the Church and was warned not to teach and hold to Copernican views. He restrained himself publicly for nearly 15 years and then defiantly published his observations and conclusions, which were counter to Church doctrine. The outcome was a trial in which he was found guilty, and he was forced to renounce his discoveries. By then an old man broken in health and spirit, he was sentenced to perpetual house arrest. Nevertheless, he completed his studies on motion, and his writings were smuggled from Italy and published in Holland.

From damage to his eyes caused by years of looking at the Sun through a telescope, Galileo became blind at the age of 74. He died 4 years later.

Whenever we discuss the speed of something, we are describing its motion relative to something else. When we say an express train travels at 200 kilometers per hour, we mean relative to the track. When we say a space shuttle moves at 8 kilometers per second, we mean relative to the Earth below. Unless stated otherwise, when we discuss the speed of things in our environment, we mean relative to the surface of the Earth.

Answers

(Are you reading this before trying to answer the questions on your own? If so, do you also exercise your body by watching others do push-ups? Exercise your brain: When you encounter the questions throughout this book, try your best to answer them before you read the answer.)

1. In both cases the answer is 25 m/s:

$$\textbf{Average speed} = \frac{\textbf{distance covered}}{\textbf{time interval}} = \frac{\textbf{100 m}}{\textbf{4 s}} = \textbf{25 m/s}$$

$$\textbf{Average speed} = \frac{\textbf{distance covered}}{\textbf{time interval}} = \frac{\textbf{50 m}}{\textbf{2 s}} = \textbf{25 m/s}$$

2. The distance traveled is average speed x time of travel:
   a. Distance = 60 km/h x 1 h = 60 km
   b. Distance = 60 km/h x 4 h = 240 km
   c. Distance = 60 km/h x 10 h = 600 km
3. No, not if the trip started from rest and ended at rest, because there would then be intervals during which the instantaneous speed was less than 60 km/h. Unless there is compensation in the form of intervals during which the instantaneous speed was greater than 60 km/h, it would not be possible for the average speed to be 60 km/h. In practice, average speeds are usually appreciably less than peak instantaneous speeds.

1.2 Velocity

When we describe speed plus the *direction* of motion, we are specifying **velocity**. Loosely speaking, we can use the words *speed* and *velocity* interchangeably. Strictly speaking, however, there is a distinction between the two. When we say that something travels at 60 kilometers per hour, we are specifying its speed, but when we say that something travels at 60 kilometers per hour to the north, we are specifying its velocity. A race car driver is concerned with his speed—how fast he is moving; an airplane pilot is concerned with her velocity—how fast and in what direction she is moving.

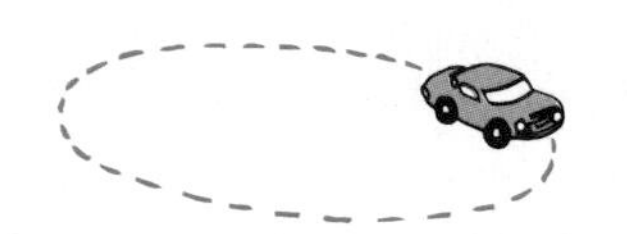

Fig. 1.3
The car on the circular track may have a constant speed, but its velocity is changing every instant. Why?

When something moves at constant velocity or constant speed, then *equal distances* are covered in equal intervals of time. Constant velocity and constant speed, however, can be very different from one another. Constant velocity means constant speed *with no change in direction.* A car that rounds a curve at a constant speed does not have a constant velocity—its velocity is nonconstant because the direction of travel is continuously changing. Velocity can be constant only when motion is along a straight-line path.

Questions

1. "She moves at a constant speed in a constant direction." Express the same thought in fewer words.
2. The speedometer of a car moving to the east reads 100 km/h. The car passes another car moving to the west at 100 km/h. Do both cars have the same speed? Do they have the same velocity?
3. During a certain period of time, the speedometer of a car reads a constant 60 km/h. Does this indicate a constant speed? A constant velocity?

Velocity Vectors

Pictures are often more descriptive than words. A pictorial way to represent velocity is by an arrow. When the length of the arrow represents the speed of a moving object and the direction of the arrow represents the direction of motion, the arrow is called a **velocity vector**.

Any quantity that is described by both a *magnitude* (how much) and a *direction* (which way) is a *vector* quantity and can be represented by a vector (arrow). The mag-

Answers

1. "She moves at constant velocity."
2. Both cars have the same speed, but they have opposite velocities because they are moving in opposite directions.
3. The constant speedometer reading indicates a constant speed but not a constant velocity, because the car may not be moving along a straight-line path—in which case its velocity is changing (Figure 1.3). A speedometer shows the magnitude of velocity but gives no information about direction.

Fig. 1.4
This vector, scaled so that 1 cm equals 20 km/h, represents a velocity of 60 km/h to the right.

nitude of a velocity vector is what we call speed. The velocity vector in Figure 1.4 is scaled so that each centimeter of its length represents a speed of 20 kilometers per hour; it is 3 centimeters long and points to the right, and therefore it represents a velocity of 60 kilometers per hour to the right.

Vectors are useful when adding and subtracting velocities. Consider an airplane flying due north at 100 kilometers per hour relative to the surrounding air. Suppose there is a tailwind (wind from behind) that moves due north at 20 kilometers per hour. This example is represented with the vectors on the left in Figure 1.5. Here the velocity vectors are scaled so that 1 centimeter represents 20 kilometers per hour. Thus the 100-kilometer-per-hour velocity of the airplane is shown by the 5-centimeter-long vector, and the 20-kilometer-per-hour tailwind is shown by the 1-centimeter-long vector. You can see (with or without the vectors) that the resulting velocity of the plane is 120 kilometers per hour north relative to the ground. Without the tailwind, the airplane would travel 100 kilometers in 1 hour relative to the ground. With the tailwind, it travels 120 kilometers over the ground in 1 hour. When we combine parallel vectors in the same direction, the vectors add.

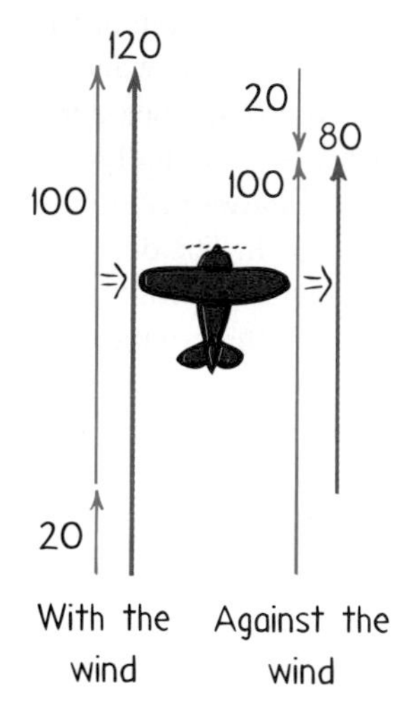

Fig. 1.5
The velocity of an airplane relative to the ground depends on its velocity relative to still air and the wind.

Now suppose the airplane makes a U-turn and flies south into the wind (wind head-on). The resulting velocity of the plane relative to the ground is 100 kilometers per hour minus 20 kilometers per hour, which equals 80 kilometers per hour. Flying against a 20-kilometer-per-hour headwind, the airplane travels only 80 kilometers relative to the ground in 1 hour (Figure 1.5, right). When we combine parallel vectors that are in opposite directions, the vectors subtract.

■ Question

A motor boat is moving 10 km/h relative to the water. If the boat travels in a river that flows at 10 km/h, what is its velocity relative to the shore when it heads directly upstream? When it heads directly downstream?

So combining vectors that act along parallel directions is simple: When in the same direction, they add; When in opposite directions, they subtract. The algebraic sum of two or more vectors is called their **resultant**. To find the resultant of two vectors that are not parallel to each other, we use the *parallelogram rule.* Construct a parallelogram with the two vectors as adjacent sides, with tails together; the diagonal of the parallelogram is the resultant of the two vectors. This is shown in Figure 1.6, where the parallelogram is a rectangle. (In this chapter the only vectors we'll treat are those that are either parallel or at right angles to each other. A more general treatment of vectors is given in the *Practice Book,* and in Appendix C at the end of this book.)

■ Answer
When the boat heads upstream, its velocity is zero relative to the land (a velocity of +10 added to a velocity of −10 equals zero). When the boat heads directly downstream, its velocity is 20 km/h in the direction of river flow (a velocity of +10 added to a velocity of +10 equals +20 in the same direction).

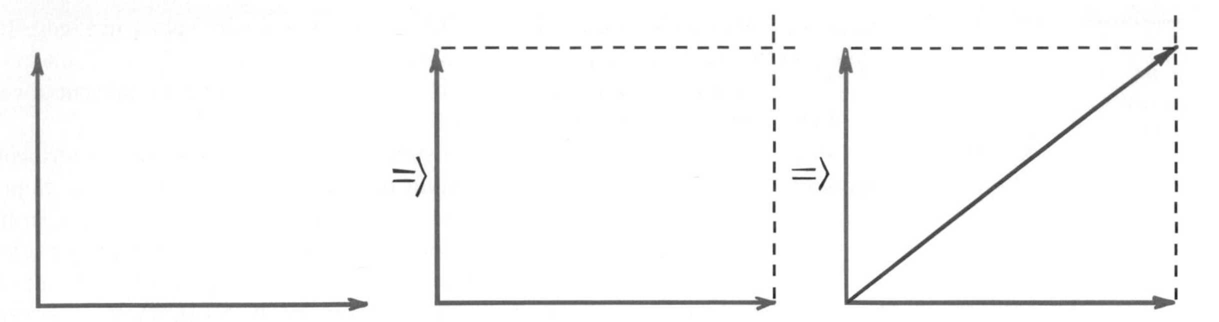

Fig. 1.6
The pair of vectors at right angles to each other make two sides of a rectangle, the diagonal of which is their resultant.

Suppose an airplane flies in a crosswind. Will the plane's speed relative to the ground be affected? The answer is yes, for the airplane will be blown off course and its speed relative to the ground will increase. We can see this with vectors. Consider a slow-moving airplane that is flying 80 kilometers per hour north when suddenly it is caught in an easterly crosswind of 60 kilometers per hour. Figure 1.7 shows vectors for the airplane velocity and wind velocity. The scale here is 1 cm:20 km/h. The diagonal of the constructed rectangle measures 5 centimeters, which represents 100 kilometers per hour. So the airplane moves at 100 kilometers per hour relative to the ground, in a direction between north and northeast.[2]

There is a special case of the parallelogram that often occurs. When two vectors have the same length (same magnitude) and are at right angles to each other, the parallelogram is a square. We see this in Figure 1.8. Since for any square the length of a diagonal is $\sqrt{2}$, or 1.414, times the length of any side, the resultant is $\sqrt{2}$ times one of

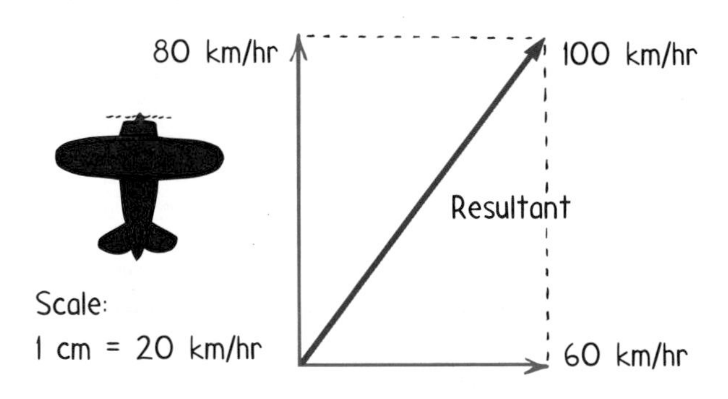

Fig. 1.7
An airplane flying at a speed of 80 km/h relative to the surrounding air in a 60-km/h crosswind has a resultant speed of 100 km/h relative to the ground.

[2]Whenever vectors are at right angles to each other, their resultant can be found by the Pythagorean Theorem: The square of the hypotenuse of a right triangle is equal to the sum of the squares of the other two sides. Note that two right triangles are contained in the parallelogram (rectangle in this case) in Figure 1.7. From either one of these triangles we get:

$$\begin{aligned}\text{Resultant}^2 &= (60\text{ km/h})^2 + (80\text{ km/h})^2\\ &= (3600\text{ km}^2/\text{h}^2) + (6400\text{ km}^2/\text{h}^2)\\ &= (10{,}000\text{ km}^2/\text{h}^2)\end{aligned}$$

The square root of $(10{,}000\text{ km}^2/\text{h}^2)$ is 100 km/h, as expected.

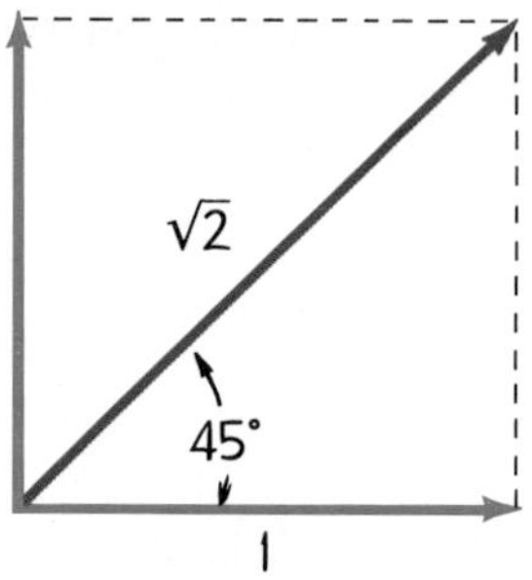

Fig. 1.8
The diagonal of a square is $\sqrt{2}$ times the length of one of its sides.

the vectors. For example, the resultant of two equal vectors of magnitude 100 acting at right angles to each other is 141.4.

■ Question

A boat travels between the two banks of a river at a speed of 10 km/h relative to the water (at right angles to the river flow), and the water flows at 10 km/h. What's the velocity of the boat relative to its starting point on shore?

The kind of motion we have considered so far has primarily been steady motion—velocity that has not undergone change. Much of the motion around us, however, undergoes change. This is accelerated motion.

1.3 Acceleration

When the velocity of a moving object changes, we say the object accelerates. We can change the velocity of something by changing its speed, by changing its direction, or by changing both its speed *and* its direction. We define the rate of change in velocity as **acceleration**:

$$\textbf{Acceleration} = \frac{\textbf{change of velocity}}{\textbf{time interval}}$$

We are all familiar with acceleration in an automobile. When driving, we call it "pickup"; we experience it when we tend to lurch toward the rear of the car as the driver steps on the gas (which is why the gas pedal is appropriately called the "accelerator"). The key idea that defines acceleration is *change.* Suppose we are driving and in 1 second steadily increase our speed from 30 kilometers per hour to 35 kilometers per hour, and then to 40 kilometers per hour in the next second, to 45 kilometers per hour in the next second, and so on. Each second we change our velocity by 5 kilometers per hour. This change of velocity is what we mean by acceleration:

$$\textbf{Acceleration} = \frac{\textbf{change of velocity}}{\textbf{time interval}} = \frac{\textbf{5 km/h}}{\textbf{1 s}} = \textbf{5 km/h/s}$$

In this case the acceleration is 5 kilometers per hour per second (abbreviated 5 km/h/s). Note that a unit for time enters twice: once for the unit of velocity and again

---

■ Answer
When the boat heads cross-stream, its velocity relative to its starting point is $\sqrt{2}$ times 10 km/hr, or 14.14 km/h, 45 degrees downstream (in accord with the diagram in Figure 1.8).

for the interval of time in which the velocity is changing. Also note that acceleration is not just the total change in velocity; it is the *time rate of change,* or *change per second,* of velocity.

Fig. 1.9
We say that an object undergoes acceleration when there is a *change* in its velocity.

Question

1. A particular car can go from rest to 90 km/h in 10 s. What is its acceleration?
2. In 2.5 s, a car increases its speed from 60 km/h to 65 km/h while a bicycle goes from rest to 5 km/h. Which undergoes the greater acceleration? What is the acceleration of each?

The term *acceleration* applies to decreases as well as increases in velocity. We say the brakes of a car, for example, produce a large retarding acceleration; that is, there is a large decrease per second in the velocity of the car. We often call this either *deceleration* or *negative acceleration.* We experience deceleration when we tend to lurch toward the front of the car when the driver hits the brakes.

We also accelerate whenever we move in a curved path, even if we are moving at constant speed, because our direction and hence our velocity are changing.[3] We experience this acceleration as we tend to lurch toward the outer part of the curve. We distinguish speed and velocity for this reason and define *acceleration* as the rate at which *velocity* changes, thereby encompassing changes both in speed and in direction.

Anyone who has ever had to stand up during a bus ride has experienced the difference between velocity and acceleration. Except for the effects of a bumpy road, you

---

Answers

1. Its acceleration is 90 km/h ÷ 10 s = 9 km/h/s. Strictly speaking, this is its average acceleration, for there may have been some variation in its rate of picking up speed.
2. The accelerations of both are the same: 2 km/h/s.

$$\text{Acceleration}_{\text{car}} = \frac{\text{change of velocity}}{\text{time interval}} = \frac{65\text{ km/h} - 60\text{ km/h}}{2.5\text{ s}} = \frac{5\text{ km/h}}{2.5\text{ s}} = 2\text{ km/h/s}$$

$$\text{Acceleration}_{\text{bike}} = \frac{5\text{ km/h} - 0\text{ km/h}}{2.5\text{ s}} = \frac{5\text{ km/h}}{2.5\text{ s}} = 2\text{ km/h/s}$$

Although the velocities involved are quite different, the rates of change of velocity—in other words, the accelerations—are equal.

---

[3]When only the direction of velocity changes, we speak of *centripetal* acceleration. A body moving in a circular path, for example, accelerates toward the center of the circle. The magnitude of centripetal acceleration is the square of the speed divided by the radius of curvature: $a = v^2/r$. For brevity, we'll not treat centripetal acceleration in this book.

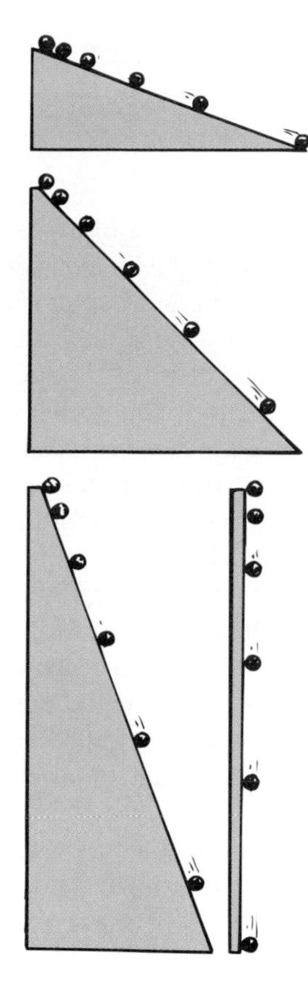

Fig. 1.10
The greater the slope of the incline, the greater the acceleration of the ball. What is its acceleration if the incline is vertical?

can easily stand without falling inside a bus that moves at constant velocity, no matter how fast it is going. You can flip a coin and catch it exactly as if the bus were at rest. It is only when the bus accelerates—speeds up, slows down, or turns—that you experience difficulty.

In much of this book we shall be concerned primarily with motion along a straight line. When straight-line motion is being considered, it is common to use *speed* and *velocity* interchangeably. When the direction is not changing, acceleration may be expressed as the rate at which *speed* changes:

$$\textbf{Acceleration (along a straight line)} = \frac{\textbf{change in speed}}{\textbf{time interval}}$$

Acceleration, like velocity, is a vector quantity. Hence its direction as well as its magnitude should be specified. For motion along a straight-line path, acceleration is also along the straight-line path—in the direction of velocity when speed increases and in the direction opposite the velocity direction when speed decreases.

Galileo developed the concept of acceleration in his experiments on inclined planes. His main interest was not objects sliding down a hill (real-world example of an inclined plane) but rather objects falling freely after being dropped from a great height. But because he lacked suitable timing devices, he used inclined planes to slow down motion so that he could investigate it more carefully.

Galileo found that a ball rolling down an inclined plane picks up the same amount of speed in successive seconds; that is, the ball rolls with uniform, or constant, acceleration. For example, a ball rolling down a plane inclined at a certain angle might pick up a speed of 2 meters per second for each second it rolls. This gain in speed per second is its acceleration: 2 meters per second per second. Its instantaneous velocity at 1-second intervals, at this acceleration, is then 0, 2, 4, 6, 8, 10, . . . meters per second. We can see that the instantaneous speed or velocity of the ball at any given time after being released from rest is simply equal to its acceleration multiplied by the length of time it has been accelerating:[4]

$$\textbf{Velocity acquired} = \textbf{acceleration} \times \textbf{time}$$

If we substitute the acceleration of the ball in this relationship, we can see that at the end of 1 second, the ball is traveling at 2 meters per second; at the end of 2 seconds, it is traveling at 4 meters per second; at the end of 10 seconds, it is traveling at 20 meters per second; and so on.

Galileo found greater accelerations for steeper inclines. The ball attains its maximum acceleration when the incline is tipped vertically. Then the acceleration is the same as that of a falling object (Figure 1.10). Regardless of weight or size, Galileo discovered that when air resistance is small enough to be neglected, all material objects fall with the same constant acceleration.

Galileo did what Aristotle failed to do—investigate a part of nature while ignoring the complexities of air resistance. For with no air resistance, the simple rule in the formula was revealed.

---

[4]Note that this relationship follows from the definition of acceleration. From $a = (change\ in\ v)/t$, simple rearrangement (multiplying both sides of the equation by t) gives $change\ in\ v = at$.

1.4 Free-Fall

How Fast?

Fig. 1.11
Consider a falling rock equipped with a speedometer. In each succeeding second of fall, the rock's speed increases by the same amount: 10 m/s. (Table 1.2 shows the speed we would read at various seconds of fall.)

Things fall because of the force of gravity. When a falling object is free of all restraints—no friction, air or otherwise—and falls under the influence of gravity alone, it is in **free-fall**. (We'll consider the effects of air resistance on falling in Chapter 2). Table 1.2 shows the instantaneous speed of a freely falling object at 1-second intervals. The important thing to note in these numbers is the way the speed changes. *During each second of fall, the object gains a speed of 10 meters per second.* This gain per second is the acceleration. Free-fall acceleration is approximately 10 meters per second each second, or, in shorthand notation, 10 m/s/s = 10 m/s^2 (read as 10 meters per second squared). Again, note that the unit of time, the second, enters twice—once for the unit of speed and again for the time interval during which the speed changes.

In the case of freely falling objects, it is customary to use the letter g to represent the acceleration (because the acceleration is due to *gravity*). Although the value of g varies slightly in different parts of the world, its average value is 9.8 meters per second each second, or, in shorter notation, 9.8 m/s^2. We round this off to 10 m/s^2 in our present discussion and in Table 1.2 to establish the ideas involved more clearly (multiples of 10 are more obvious than multiples of 9.8). Where precision is important, the value 9.8 m/s^2 should be used.

Note in Table 1.2 that the values listed for the instantaneous speed of an object falling from rest are those calculated from the equation Galileo deduced with his inclined planes:

$$\textbf{Velocity acquired} = \textbf{acceleration} \times \textbf{time}$$

For an object falling from rest,[5] the instantaneous velocity v at any time t can be expressed in shorthand notation as

$$v = gt$$

The letter v symbolizes both speed and velocity. To see that this equation makes good sense, take a moment to check it with Table 1.2.

Table 1.2
Free-Fall from Rest

Time of Fall (s)	Instantaneous Speed (m/s)
0	0
1	10
2	20
3	30
4	40
5	50
.	.
.	.
.	.
t	$10t$

[5]If instead of being dropped from rest the object is thrown downward at speed v_0, the speed v after any elapsed time t is $v = v_0 + gt$. We will not be concerned with this added complication here, and will instead learn as much as we can from the most simple cases. That will be a lot!

■ Question

What would the speedometer on the falling rock in Figure 1.11 read 3.5 s, 6 s, and 100 s after the rock drops from rest?

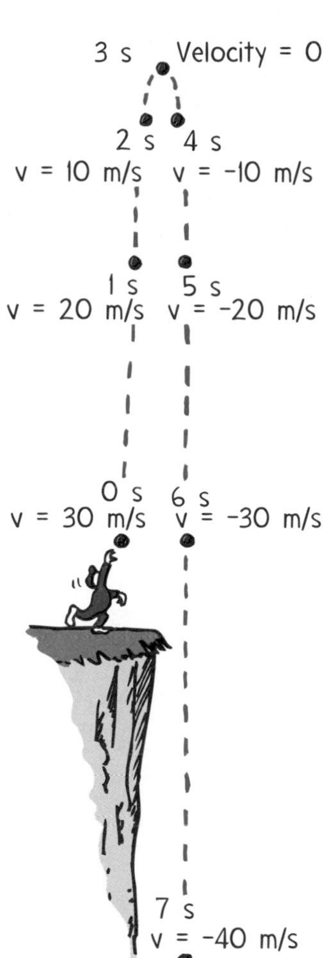

Fig. 1.12
The rate at which the velocity changes each second is constant.

So far we've considered objects moving straight downward, toward the ground. How about an object thrown straight upward? Once released, it continues to move upward for a while and then comes back down. At the highest point, when it is changing its direction of motion from upward to downward, its instantaneous speed is zero. Then it starts downward just as if it had been dropped from rest at that height. What causes the upward motion to reverse itself?

To answer this question correctly, it is important to bear in mind that gravity is always pulling downward on every object, *even an object moving in the upward direction.* During the upward part of this motion, the object slows from its initial upward velocity to zero velocity. How much does its speed decrease each second? Because of gravity, the speed of the upward-moving object decreases at the rate of 10 meters per second each second—the same acceleration it experiences on the way down. So, as Figure 1.12 shows, the instantaneous speed at points of equal elevation in the path is the same whether the object is moving upward or downward. The velocities are opposite, of course, because they are in opposite directions. During each second, the speed or the velocity changes by 10 meters per second. The acceleration is 10 meters per second squared the whole time, whether the object is moving upward or downward.

How Far?

How *fast* an object falls is altogether different from how *far* it falls. With his inclined planes, Galileo found that the distance a uniformly accelerating object travels is proportional to the *square of the time taken to cover that distance.* The details of this relationship are in Appendix B, and here we only state the results. The distance traveled by a uniformly accelerating object starting from rest is

$$\textbf{Distance traveled} = \tfrac{1}{2}\,\textbf{(acceleration} \times \textbf{time} \times \textbf{time)}$$

In addition to applying to motion in the horizontal direction, such as cars on the highway, this relationship also applies to the distance something falls. For a freely falling object that starts from rest, the relationship is, in shorthand notation,[6]

$$d = \tfrac{1}{2} g t^2$$

where d is the distance the object falls, g is the acceleration caused by gravity, and t is the time of fall in seconds. If we use 10 m/s^2 for the value of g, the distance fallen for various times is as shown in Table 1.3.

■ **Answer**
35 m/s, 60 m/s, and 1000 m/s, respectively. You can reason this from Table 1.2 or use the equation $v = gt$, where g is 10 m/s^2.

[6] d = average velocity × time (equation from Section 1.1)

$$= \frac{\text{initial velocity} + \text{final velocity}}{2} \times \text{time}$$

$$= \frac{0 + gt}{2} \times t = \tfrac{1}{2} g t^2$$

Fig. 1.13
Suppose that a falling rock is equipped with an odometer. The readings indicating distance fallen increase with time according to the relationship $d = \frac{1}{2}gt^2$ and are shown in Table 1.3.

Note from Table 1.3 that an object falls a distance of only 5 meters during the first second of fall, although its speed at 1 second is 10 meters per second. This may be confusing, for you may think that the object should fall 10 meters. In order to fall 10 meters in the first second, however, the object would have to fall at an *average* speed of 10 meters per second for the entire second. It starts its fall at a speed of 0 meters per second, and its speed is 10 meters per second only in the last instant of the 1-second interval. Its average speed during this interval is the average of its initial speed (0 meters per second) and final speed (10 meters per second). To find the average value of these or any other two numbers, simply add the two numbers and divide by 2, as shown in the footnote. This calculation yields an average speed of 5 meters per second, which over a time interval of 1 second gives a distance of 5 meters. As the object continues to fall, it covers ever-increasing distances as its speed continuously increases.

Table 1.3
Distance Fallen in Free-Fall

Time of Fall (s)	Distance of Fall (m)
0	0
1	5
2	20
3	45
4	80
5	125
.	.
.	.
.	.
t	$\frac{1}{2}10t^2$

Question

A cat steps off a ledge and drops to the ground in $\frac{1}{2}$ second.
a. What is its speed on striking the ground?
b. What is its average speed during the $\frac{1}{2}$ s?
c. How high is the ledge above the ground?

Use the approximate value 10 m/s² for g.

A common observation is that all objects do not fall with equal accelerations. A dry leaf, a feather, or a sheet of paper may flutter to the ground, while a stone or a coin falls rapidly. The fact that air resistance is responsible for those different accelerations

---

Answer

a. Speed: $v = gt = 10\text{ m/s}^2 \times \frac{1}{2}\text{ s} = 5\text{ m/s}$

b. Average speed: $\bar{v} = \dfrac{\text{initial } v + \text{final } v}{2} = \dfrac{0\text{ m/s} + 5\text{ m/s}}{2} = 2.5\text{ m/s}$

We put a bar over the symbol to denote average speed—$\bar{v}$.
c. The height of the ledge is equal to the distance the cat falls in the half second: $d = \bar{v}t = 2.5\text{ m/s} \times \frac{1}{2}\text{ s} = 1.25\text{ m}$

or, equivalently,

$$d = \frac{1}{2}gt^2$$
$$= \frac{1}{2} \times 10\text{ m/s}^2 \times (\tfrac{1}{2}\text{ s})^2 = \frac{1}{2} \times 10\text{ m/s}^2 \times \frac{1}{4}\text{ s}^2 = 1.25\text{ m}$$

Notice that we can find the distance by either of these equivalent relationships. Doing it both ways is a good check!

Link to Sports—Hang Time

Some people—basketball players, track and field athletes, ballet dancers—are gifted with great jumping ability. The greatest jumpers seem to hang in the air in defiance of gravity. Ask your friends to estimate the "hang time" of the great jumpers—the amount of time a jumper has his or her feet off the ground. Two or three seconds? Several seconds? Nope. Surprisingly, the hang time of the greatest jumpers is nearly always less than 1 second! The seemingly longer time is one of many illusions we have about nature.

Very few athletes can jump higher than 2 feet. Basketball star Spud Webb won the NBA slam dunk contest back in 1986 and at this writing is still the champ, with his vertical jump of 4 feet (1.25 meters). It is important to distinguish between how high one can reach and how high one can jump. A very tall basketball player, for example, doesn't have to jump very high to outdistance shorter rivals. Spud, surprisingly, is one of the shorter basketball players (5 feet 7 inches), making his jumping ability all the more outstanding.

We distinguish between the height of a horizontal bar a jumper clears and the actual vertical displacement of the jumper's center of gravity (jumpers clearing a high bar can contort their bodies to clear bars that their centers of gravity pass beneath).

Jumping ability is best measured by a standing vertical jump. Stand facing a wall and, with feet flat on the floor and arms extended upward, make a mark on the wall at the top of your reach. Then make your jump and at the peak make another mark. The distance between these two marks measures your vertical leap. If its more than 2 feet (0.6 meter), you're exceptional.

Now let's look at the physics. As you leap upward, your push (jumping force) against the floor lasts only as long as your feet are in contact with the floor. The greater the push that propels you upward, the greater your launch speed and the higher the jump. It is important to note that as soon as your feet leave the ground, whatever upward speed you attain immediately decreases at the steady rate of g—10 m/s^2. Maximum height is attained when your upward speed decreases to zero. Then you begin to fall, gaining speed at exactly the same rate, g. Time rising equals time falling: hang time is the sum of time up and time down. While airborne, no amount of leg or arm pumping or other bodily motions can change your hang time.

The relationship between time up or down t and vertical height d is given by

$$d = \frac{1}{2} g t^2$$

We can rearrange this expression to read

$$t = \sqrt{\frac{2d}{g}}$$

Let's use Spud's jumping height of 1.25 meters for d and 9.8 m/s^2 for g. Solving for t, half the hang time, we get

$$t = \sqrt{\frac{2d}{g}} = \sqrt{\frac{2(1.25)}{9.8}} = 0.5 \text{ s}$$

Double this (because this is the time for one way of an up-and-down round-trip) and we see that Spud's record-breaking hang time is 1 second.[7]

We've only been talking about vertical motion. How about running jumps? We'll learn in Chapter 4 that hang time depends only on a jumper's vertical speed at launch and is independent of horizontal speed. While airborne, the jumper moves horizontally at a constant speed while only vertical speed undergoes acceleration. Interesting physics!

[7] A general rule is that height jumped in *feet* is equal to four times hang time squared. Here we use $g = 32$ ft/s^2. We let T equal hang time (twice t). Then $d = (g/2)(T/2)^2 = g/2(T^2/4) = (g/8)(T^2) = (32/8)(T^2) = 4T^2$.

Fig. 1.14
A falling feather and a coin have equal accelerations in a vacuum.

can be shown very nicely with a closed glass tube containing light and heavy objects—a feather and a coin, for example. In the presence of air, the feather and coin fall with quite different accelerations. When the air in the tube is removed by a vacuum pump, however, the feather and coin fall with the same acceleration (Figure 1.14). Although air resistance appreciably alters the motion of lightweight things such as falling feathers, it doesn't appreciably affect the motion of heavier objects, such as stones and baseballs, traveling at low speeds. The relationships $v = gt$ and $d = \frac{1}{2} gt^2$ can be used to a very good approximation for most objects falling from rest in air .

So we see how much speed something acquires is different from how far something falls. And we see that acceleration—how quickly how fast changes—is not velocity, nor is it a change in velocity. Acceleration is the rate at which velocity changes.

Please remember that it took people nearly 2000 years from the time of Aristotle to reach a clear understanding of motion, so be patient with yourself if you find that you require a few hours to achieve as much!

Summary of Terms

- **Speed** Distance traveled divided by time.
- **Instantaneous speed** Speed at any given instant.
- **Average speed** Total distance traveled divided by time.
- **Velocity** The speed of an object and specification of its direction of motion.
- **Velocity vector** An arrow drawn to scale that represents the magnitude and direction of a given velocity.
- **Resultant** The single vector that results when two or more vectors are combined.
- **Acceleration** The rate at which velocity changes with time; the change in velocity may be in magnitude or direction or both.
- **Free-fall** A state of fall under the influence of only gravity—free from air resistance.

Review Questions

Each chapter in this book concludes with a set of review questions and exercises, and most chapters also contain computational problems. The **Review Questions** are designed to help you fix ideas in your mind and catch the essentials of the chapter material. You'll notice that answers to the questions can be found in the chapters.

The **Exercises** stress thinking rather than mere recall, and answering them requires an understanding of the definitions, principles, and relationships of the chapter material. In many cases the intention of particular exercises is to help you to apply the ideas of physics to familiar situations. Unless you cover only a few chapters in your course, you will likely be expected to tackle only a few exercises for each chapter. The large number of exercises is to allow your instructor a wide choice of assignments.

Problems are found only for those chapters that feature concepts most clearly understood with numerical values and straightforward calculations. The problems are relatively few in number so as to avoid an undue emphasis on problem solving that could obscure the primary goal of *Conceptual Physical Science*—to develop a gut feel for the concepts of science in your everyday language. Calculations are to enhance the learning of concepts, not the other way around.

Speed

1. What two units of measurement are necessary for describing speed?
2. Distinguish average speed from instantaneous speed.
3. What is the average speed of a horse that gallops a distance of 15 km in a time interval of $\frac{1}{2}$ hour?
4. How far does a horse travel if it gallops at an average speed of 25 kilometers per hour for $\frac{1}{2}$ hour?

Velocity

5. Distinguish between speed and velocity.
6. If a car moves with a constant velocity, does it also move with a constant speed? Explain.

$$\text{Speed} = \frac{\text{distance}}{\text{time}}$$

Time = 1 hour

San Francisco

Livermore

$$\text{Speed} = \frac{80\ \text{km}}{1\ \text{h}} = 80\ \text{km/h}$$

(a)

Velocity = {speed *and* direction}

San Francisco

E

Velocity = 300 km/h, east

(b)

Acceleration = {Rate of *change* in velocity} *due to* {change in speed and/or direction}

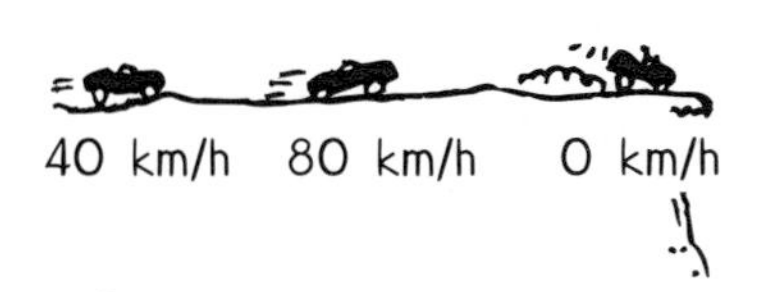

Change in speed but *not* direction

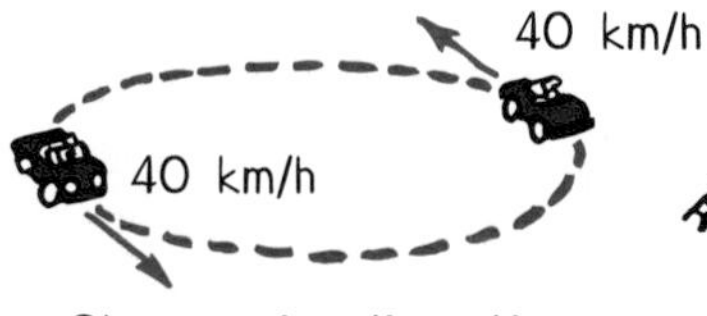

Change in direction but *not* speed

Change in speed *and* direction

(c)

$$\text{Acceleration} = \frac{\text{change in velocity}}{\text{time}}$$

Time = 0, velocity = 0

Time = 1 s, velocity = 10 m/s

Time = 2 s, velocity = 20 m/s

$$\text{Acceleration} = \frac{20\ \text{m/s}}{2\ \text{s}}$$

$$a = 10\ \frac{\text{m/s}}{\text{s}}$$

$$a = 10\ \text{m/ss}$$

$$a = 10\ \text{m/s}^2$$

(d)

Fig. 1.15
Motion analysis.

7. If a car moves with a constant speed, can you say that it also moves with a constant velocity? Give an example to support your answer.
8. What two pieces of information are necessary in order to define a vector quantity?
9. Why do we say velocity is a vector quantity and speed is not?

Acceleration

10. Distinguish between velocity and acceleration.
11. What is the acceleration of a car that increases its velocity from 0 to 100 km/h in 10 s?
12. What is the acceleration of a car that maintains a constant velocity of 100 km/h for 10 s? (Why do many students who correctly answer the preceding question get this one wrong?)
13. When are you most aware of motion in a moving vehicle—when it is moving steadily in a straight line or when it is accelerating? If a car could move with absolutely constant velocity (no bumps at all), would you be aware of motion?
14. Acceleration is generally defined as the time rate of change of velocity. When can it be defined as the time rate of change of speed?

Free Fall

15. What is meant by a "freely falling" object?
16. What is the velocity acquired by a freely falling object 5 s after being dropped from rest? What is it 6 s after?
17. The acceleration of free-fall is about 10 m/s2. Why does the seconds unit appear twice?
18. What relationship between distance traveled and time of travel did Galileo discover with his inclined planes?
19. What is the distance covered by a freely falling object 5 s after it is dropped from rest? What is it 6 s after?
20. Consider these measurements: 10 m, 10 m/s, and 10 m/s^2. Which is a measure of distance, which of speed, and which of acceleration?

Exercises

1. Charlie drove his car around the block at constant velocity. True or false?
2. Is a fine for speeding based on one's average speed or instantaneous speed? Explain.
3. Can an automobile having a velocity toward the north have an acceleration toward the south? Explain.
4. Can an object reverse its direction of travel while maintaining a constant acceleration? If so, give an example. If not, explain why not.
5. If you were standing in a bus moving at constant velocity, would you have to lean in some special way to compensate for the bus's motion? What if the bus were accelerating? Explain.
6. Harry says acceleration is how fast you go. Carol says acceleration is how fast you get fast. They look to you for confirmation. Who's correct?
7. On which of these hills does the ball roll with increasing speed and decreasing acceleration along the path? (Use

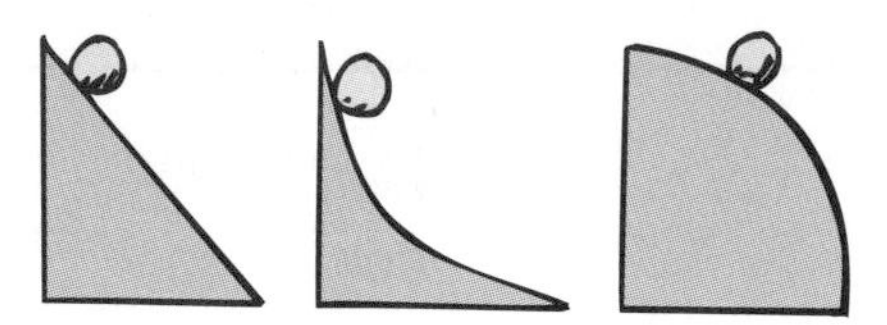

this example if you wish to explain to someone the difference between speed and acceleration.)
8. Why does vertically falling rain make slanted streaks on the side windows of a moving automobile? If the streaks make an angle of 45°, what does this tell you about the relative speed of the car and the falling rain?
9. What is the velocity relative to the ground of an airplane that has an airspeed of 120 km/h when it is in a 90-km/h crosswind?
10. What is the acceleration of a car that moves at a steady velocity of 100 km/h for 100 s? Explain your answer and state why this question is an exercise in careful reading as well as physics.
11. For a freely falling object dropped from rest, what is its acceleration at the end of the fifth second of fall? The tenth second? Defend your answer.
12. Extend Tables 1.2 and 1.3 (which give values of from 0 to 5 s) to 0 to 10 s, assuming no air resistance.
13. Suppose a freely falling object were somehow equipped with a speedometer. By how much would its speed reading increase each second?
14. Suppose that the freely falling object in the preceding exercise were also equipped with an odometer. Would the readings indicate equal or unequal distances covered in successive seconds? Explain.
15. When a ball player throws a ball straight up, by how much does the ball's speed decrease each second during the ascent? In the absence of air resistance, by how much does the speed increase each second the ball is descending? How much time, from the starting position, is required for rising compared to falling?
16. Someone standing at the edge of a cliff (as in Figure 1.12) throws a ball straight up at a certain speed and another ball straight down with the same initial speed. If air resistance is negligible, which ball has the greater speed when it strikes the ground?
17. If you drop an object, its acceleration toward the ground is 10 m/s^2. If you throw it down instead, is its acceleration after throwing greater than 10 m/s^2? Why or why not?

18. In the preceding exercise, can you think of a reason the acceleration would be less than 10 m/s^2?
19. Two balls, A and B, are released simultaneously from rest at the left end of the equal-length tracks A and B as shown. Which ball reaches the end of its track first?

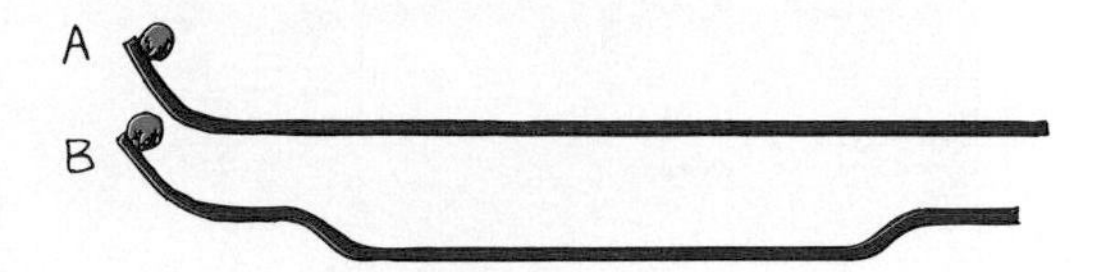

20. Referring to the tracks above, (a) does ball B roll faster along the lower part of track B than ball A rolls along track A? (b) Is the speed gained by ball B going down the extra dip the same as the speed it loses going up near the right-hand end—and doesn't this mean that the speed of balls A and B are the same at the ends of both tracks? (c) On track B, won't the average speed dipping down and up be greater than the average speed of ball A during the same time? (d) So overall, which ball has the greater average speed? (Do you wish to change your answer to the previous exercise?)

Problems

1. The ocean's level is currently rising at a rate of about 1.5 mm per year. At this rate, in how many years will the level be 3 m (meters) higher than it is now?
2. A reconnaissance plane flies 600 km away from its base at 200 km/h, then flies back to its base at 300 km/h. What is its average speed?
3. What is the acceleration of a vehicle that changes its speed from 100 km/h to a dead stop in 10 s?
4. A ball is thrown straight up with an initial speed of 30 m/s. How high does it go, and how long is it in the air (neglecting air resistance)?
5. A ball is thrown straight up with enough speed so that it is in the air several seconds. (a) What is the velocity of the ball when it gets to its highest point? (b) What is its velocity 1 s before it reaches its highest point? (c) What is the change in its velocity during this 1-s interval? (d) What is its velocity 1 s after it reaches its highest point? (e) What is the change in velocity during this 1-s interval? (f) What is the change in velocity during the 2-s interval? (Caution: velocity, not speed!) (g) What is the acceleration of the ball during any of these time intervals and at the moment the ball has zero velocity?
6. What is the instantaneous velocity of a freely falling object 10 s after it is released from rest? What is its average velocity during this 10-s interval? How far does it go during this time?
7. A climber near the summit of a vertical cliff accidentally knocks loose a large rock. She sees it shatter at the bottom of the cliff 8 s later. What was the speed of impact? How far did the rock fall?
8. A car goes from $v = 0$ to $v = 50$ m/s in 10 s. If you wish to find the distance traveled using the equation $d = \frac{1}{2}at^2$, what value should you use for a?
9. Consider a planet where the acceleration due to gravity is 20 m/s^2 and air resistance is nil. If a rock initially at rest is dropped on this planet, how fast is it falling after 1 s? What distance is covered in that 1-s interval? If it were dropped on the Earth, how fast would the same rock be falling after 1 s (ignore air resistance) and how far would it fall?
10. Very few athletes can jump more than 2 ft (0.6 m) high. Use $d = \frac{1}{2}gt^2$ to solve for the time one spends moving upward in a 2-ft vertical jump. Then double it for the "hang time"—the time one's feet are off the ground.

CHAPTER

2

Newton's Laws of Motion

Aristotle believed that natural laws could be understood by logical reasoning. One of his assertions was that heavy objects necessarily fall faster than lighter objects. Another was that moving objects must necessarily have forces exerted on them to keep them moving. These ideas were completely turned around 2,000 years later by Galileo, who stated that experiment was superior to logic in uncovering natural laws. The idea that heavy things fall faster than lighter things was demolished by Galileo in his famous Leaning Tower of Pisa experiment, where he supposedly dropped objects of different weights and showed that, except for the effects of air resistance, they fell to the ground together. In his inclined-plane experiments he showed that moving things, once moving, continued in motion *without* the application of forces. Galileo called the property of objects to behave this way *inertia.*

Galileo's work set the stage for Isaac Newton, who was born shortly after Galileo died in 1642. By the time Newton was 23, he had developed his famous three laws of motion that completed the overthrow of Aristotelian ideas. These three laws first appeared in one of the most important books of all time, Newton's *Philosophiae Naturalis Principia Mathematica.*[1] The first law is a restatement of Galileo's concept of inertia; the second law relates acceleration to the cause that produces it, force; and the third is the law of action and reaction.

2.1 Newton's First Law of Motion

Fig. 2.1
Galileo's famous demonstration.

Newton refined Galileo's idea of **inertia**—the tendency of things to resist changes in motion—and made it his first law, appropriately called the **law of inertia**. From Newton's *Principia:*

> **Law 1:** ***Every material object continues in its state of rest, or of uniform motion in a straight line, unless it is compelled to change that state by forces impressed upon it.***

The key word in this law is *continues:* An object *continues* to do whatever it happens to be doing unless a force is exerted upon it. If it is at rest, it *continues* in a state of rest. If it is moving, it *continues* to move without turning or changing its speed. In short, the law says that an object does not accelerate of itself; acceleration must be imposed against the object's tendency to retain its state of motion. Things at rest tend to stay at rest—things moving tend to continue moving.

We use the principle of inertia when we stamp our feet to remove snow or shake a garment to remove dust. Probably the most celebrated classroom demonstration of inertia is the classic tablecloth-and-dishes stunt. When the tablecloth is yanked properly and suddenly, the dishes remain intact on the table. Things at rest tend to remain at rest.

If you perform this stunt in the dining car of a train moving at constant velocity, you'll produce the same result. Or, toss a ball back and forth in a train at rest and again in a train moving at constant velocity and you'll see the ball behaves the same way both times. Likewise, if the ride is smooth, a waiter in a moving train pours coffee into your cup in the same way as if the train were at rest. The law of inertia holds true whether you are at rest or moving at constant velocity.[2]

Fig. 2.2
Inertia in action.

Question

When we jump straight up, why do we land in our footsteps rather than at a location equal to the distance the Earth moves during our jump?

Answer
We return to our footsteps because before the jump we are already moving along with the moving Earth, and in accord with Newton's first law, we continue in that state of motion during the jump.

[1]The Mathematical Principles of Natural Philosophy.

[2]We say that a frame of reference that is either at rest or moving at constant velocity is an *inertial frame of reference.* Newton's first law holds in an inertial frame of reference. In a frame of reference that accelerates, such as a rotating merry-go-round, tossing a ball back and forth or pouring coffee is very different for different speeds. A rotating frame of reference is an accelerated frame of reference—a noninertial frame, in which the law of inertia doesn't hold.

Mass

Every object possesses inertia; how much depends on the amount of matter in the object—the more matter, the more inertia. In speaking of how much matter something contains, we use the term *mass.* The greater the mass of an object, the greater the amount of matter in the object and the greater its inertia. **Mass** is a measure of the inertia of an object.

Loosely speaking, mass corresponds to our intuitive notion of weight. We say something contains a lot of matter if it weighs a lot. That's because we are accustomed to measuring the amount of matter by gravitational attraction to the Earth. Mass is more fundamental than weight, however; it is a fundamental quantity that completely escapes the notice of most people. There are times, however, when weight corresponds to our unconscious notion of inertia. For example, if you are trying to determine which of two small objects is the heavier, you might shake them back and forth in your hands or move them in some way instead of lifting them. In doing so, you are judging which of the two is more difficult to accelerate—in other words, which is the more resistant to a *change* in motion. You are making a comparison of the inertias of the objects.

It is easy to confuse mass and weight, mainly because they are directly proportional to each other. If the mass of an object is doubled, its weight is also doubled; if the mass is halved, its weight is halved. There is a distinction between the two, however, as the following definitions show:

> **Mass:** ***The quantity of matter in an object. More specifically, mass is the measurement of the inertia, or sluggishness, an object exhibits in response to any effort made to start it, stop it, or change in any other way its state of motion.***
>
> **Weight:** ***The gravitational force exerted on an object by the nearest most-massive body (locally, by the Earth).***

In the United States, the quantity of matter in an object is commonly described by the gravitational pull between it and the Earth—in other words, its *weight*—usually expressed in the unit called the *pound.* In most of the world, however, the quantity of matter in an object is commonly expressed in a mass unit, the **kilogram**. At the surface of the Earth, a brick having a mass of 1 kilogram weighs 2.2 pounds. In the metric system of units, the unit of weight (or any other force) is the **newton**, which is equal to a little less than a quarter pound (like the weight of a quarter-pounder hamburger *after* it is cooked). A 1-kilogram brick weighs 9.8 newtons (9.8 N). Away from the Earth's surface, where the influence of gravity is less, the same brick weighs less. It would also weigh less on the surface of planets with less gravity than the Earth. On the Moon's surface, for example, where the gravitational force on things is only $\frac{1}{6}$ as strong as on Earth, 1 kilogram weighs about 1.6 newtons (or 0.37 pounds). On more massive planets it would weigh more. But, the mass of the brick is the same everywhere. The brick offers the same resistance to speeding up or slowing down whether it's on the Earth, Moon, or any body attracting it. In a drifting spaceship where a scale with a brick on it reads zero, the brick still has mass. Even though it doesn't press down on the scale, the brick has the same resistance to a change in motion as it has on Earth. Just as much force would have to be exerted by an astronaut in the spaceship to shake the brick back and forth as would be required to shake it back and forth while on Earth. You'd have to provide the same amount of push to accelerate a huge truck up to 60 miles per hour on a level surface on the Moon as on Earth. The difficulty of lifting it against gravity (weight), however, is something else. Mass and weight are different from each other.

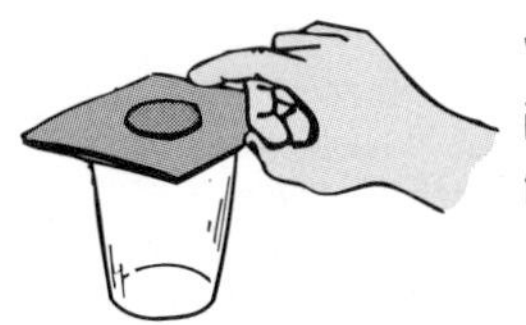

Fig. 2.3
Examples of inertia.

A nice demonstration that shows the difference between mass and weight is the massive ball suspended on the string, shown in Figure 2.3. The top string breaks when the lower string is pulled with a gradual increase in force, but the bottom string breaks when the string is jerked. Which of these cases illustrates the weight of the ball, and which illustrates the mass of the ball? Note that only the top string bears the weight of the ball. So when the lower string is gradually pulled, the tension supplied by the pull is also transmitted to the top string. So total tension, due to the ball's weight, is greater in the top string. The top string breaks when the breaking point is reached. But when the bottom string is jerked, the mass of the ball, its tendency to remain at rest, is responsible for the bottom string breaking.

■ Question

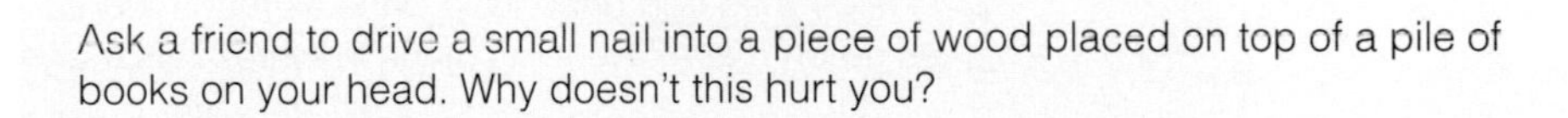

Ask a friend to drive a small nail into a piece of wood placed on top of a pile of books on your head. Why doesn't this hurt you?

It is also easy to confuse mass and volume. When we think of a massive object, we often think of a big object. An object's size, however, is not necessarily a good way to judge its mass. Which is easier to get moving: a car battery or an empty cardboard box of the same size? These two objects have identical volumes but very different masses. So, we see that mass is neither weight nor volume.

■ Answer

The relatively large mass of the books and block atop your head resists being moved. The force that is successful in driving the nail will not be as successful in accelerating the massive books and block, which don't move very much when the nail is struck. Can you see the similarity of this to the suspended massive ball demonstration, where the supporting string doesn't break when the bottom string is jerked?

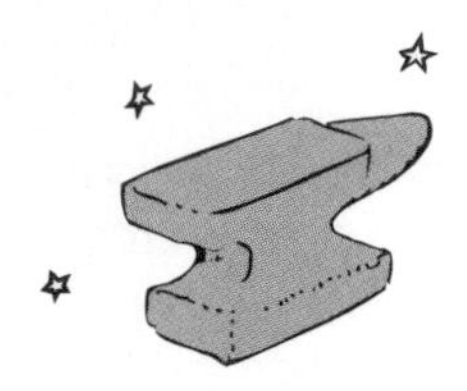

Fig. 2.4
An anvil in outer space between the Earth and the Moon may be weightless, but it is not massless.

Fig. 2.5
The astronaut in space finds it just as difficult to shake the "weightless" anvil as it would be on Earth. If the anvil is more massive than the astronaut, which shakes more—the anvil or the astronaut?

Although Galileo introduced the idea of inertia, it was Newton who grasped its significance. The law of inertia defines natural motion and tells us what kinds of motion are the result of applied forces. We shall see in the next section that *changes* in motion result from applied forces. Whereas Aristotle maintained that the forward motion of an arrow through the air requires a steady force, Newton's law of inertia instead tells us that the behavior of the arrow is natural; constant speed along a straight line (or, simply, constant velocity) requires no force. Aristotle and his followers held that the circular motions of heavenly bodies are natural, requiring no applied forces. The law of inertia, however, clearly states that in the absence of forces of some kind, the planets would not move in the divine circles of ancient and medieval astronomy but would move instead in straight-line paths off into space. Newton maintained that the curved motion of the planets was evidence of some kind of force being applied to them. We shall see in the next chapter that his search for this force led to the law of gravity.

Questions

1. A hockey puck sliding across the ice finally comes to rest. How would Aristotle interpret this behavior? How would Galileo and Newton interpret it? How would you interpret it?
2. Does a 2-kg iron brick have twice as much *inertia* as a 1-kg iron brick? Twice as much *mass?* Twice as much *volume?* Twice as much *weight?*
3. Would it be easier to lift a huge truck on the Earth or on the Moon?

Answers

1. Aristotle would say that the puck slides to a stop because it seeks its proper and natural state, one of rest. Galileo and Newton would say that, once in motion, the puck would continue in motion and what prevents continued motion is not the puck's nature or its proper rest state, but the friction between puck and ice (this friction is small compared to the friction between a puck and a wooden floor, which is why the puck slides so much farther on ice). Only you can answer the last question.
2. The answers to all parts are yes. A 2-kg iron brick contains twice as many iron atoms and therefore twice the amount of matter. Twice the amount of matter means twice the inertia and twice the mass. In the same location, it also has twice the weight. And because both bricks have the same density (the same mass/volume), the 2-kg brick also has twice the volume.
3. A huge truck would be easier to lift on the Moon because gravity is less on the Moon. When you *lift* an object, you are contending with the force of gravity (its weight). Although its mass is the same on the Earth, the Moon, or anywhere else, the truck's weight on the Moon is only $\frac{1}{6}$ its weight on the Earth, and so only $\frac{1}{6}$ as much effort is required to lift it on the Moon. Moving it *horizontally,* however, when weight isn't a factor, is a different story. We shall see that when mass is the only factor, equal forces will produce equal accelerations whether the object is on the Earth or the Moon.

Isaac Newton (1642–1727)

Isaac Newton was born prematurely and barely survived on Christmas Day 1642, the same year Galileo died. Newton's birthplace was his mother's farmhouse in Woolsthorpe, England. His father died several months before his birth, and he grew up under the care of his mother and grandmother. As a child he showed no particular signs of brightness, and at the age of 14½ he was taken out of school to work on his mother's farm. As a farmer he was a failure. He preferred to read books he borrowed from a neighboring pharmacist. An uncle sensed the scholarly potential in young Isaac and prompted him to study at the University of Cambridge, which he did for 5 years, graduating without particular distinction.

A plague swept through London, and Newton retreated to his mother's farm—this time to continue his studies. At the farm, at the age of 23, he laid the foundations for the work that was to make him immortal. Story has it that seeing an apple fall to the ground led him to consider the force of gravity extending to the Moon and beyond, and he formulated the law of universal gravitation (which he later proved); he invented the calculus, an indispensable mathematical tool in science; he extended Galileo's work and formulated the three fundamental laws of motion; and he formulated a theory of the nature of light and showed with prisms that white light is composed of all colors of the rainbow. It was his experiments with prisms that first made him famous.

When the plague subsided, Newton returned to Cambridge and soon established a reputation for himself as a first-rate mathematician. His mathematics teacher resigned in his favor and Newton was appointed the Lucasian professor of mathematics, a post he held for 28 years. In 1672, he was elected to the Royal Society, where he exhibited the world's first reflector telescope.

Newton was 42 when he began his now famous *Principia Mathematica Philosophiae Naturalis.* He wrote it in Latin and completed it in 18 months. It appeared in print in 1687 and wasn't printed in English until 1729, 2 years after his death. When asked how he was able to make so many discoveries, Newton replied that he found his solutions to problems not by sudden insight but by thinking very long and hard about them until he worked them out. He also said that he had stood on the shoulders of giants, acknowledging others like Galileo.

At the age of 46, his energies turned somewhat from science when he was elected a member of Parliament. He attended the sessions in Parliament for 2 years and never gave a speech. One day he rose and the House fell silent to hear the great man. Newton's "speech" was very brief; he simply requested that a window be closed because of a draft.

A further turn from his work in science was his appointment as warden and then as master of the mint, to the dismay of counterfeiters who flourished at that time. He maintained his membership in the Royal Society and was elected president, and re-elected each year for the rest of his life. At the age of 62, he wrote *Opticks,* which summarized his work on light. Nine years later he wrote a second edition of his *Principia.*

Although Newton's hair turned gray at 30, it remained full, long, and wavy all his life, and unlike others in his time he did not wear a wig. He was a modest man, very sensitive to criticism, and never married. He remained healthy in body and mind into old age. At 80, he still had all his teeth, his eyesight and hearing were sharp, and his mind was alert. In his lifetime he was regarded by his countrymen as the greatest scientist who ever lived. In 1705, he was knighted by Queen Anne. Newton died at the age of 85 and was buried in Westminster Abbey along with England's kings and heroes.

2.2 Newton's Second Law of Motion

Fig. 2.6
The greater the mass, the greater the force needed for a given acceleration.

Every day we see things that do not continue in a constant state of motion: objects initially at rest may later move; moving objects may follow paths that are not straight lines; things in motion may stop. Most of the motion we observe undergoes changes and is the result of one or more applied forces. The overall net force acting on an object, whether it be from a single source or a combination of sources, produces acceleration. The relationship of acceleration to force and inertia is given in Newton's second law:

> **Law 2:** ***The acceleration of an object is directly proportional to the net force acting on the object, is in the direction of the net force, and is inversely proportional to the mass of the object.***

In summarized form, this is

$$\textbf{Acceleration} \sim \frac{\textbf{net force}}{\textbf{mass}}$$

In symbol notation, this is simply

$$a \sim \frac{F_{\text{net}}}{m}$$

The wiggly line ~ is a symbol meaning "is proportional to." We say that acceleration a is directly proportional to the net force F and inversely proportional to the mass m. By this we mean that if F increases, a increases by the same factor (if F doubles, a doubles); if m increases, a decreases by the same factor (if m doubles, a is cut in half). With appropriate units of F, m, and a, the proportionality may be expressed as an exact equation:

$$a = \frac{F_{\text{net}}}{m}$$

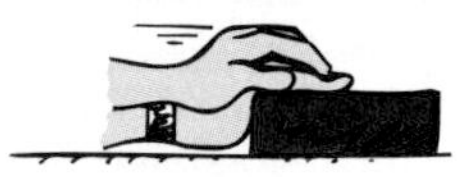

Fig. 2.7
Acceleration is directly proportional to force.

An object is accelerated in the direction of the net force acting on it. Applied in the direction of the object's motion, a net force increases the object's speed. Applied in the opposite direction, it decreases the object's speed. Applied at right angles, it deflects the object. Any other direction of application will result in a combination of speed change and deflection. *The acceleration of an object is always in the direction of the net force.*

A **force**, in the simplest sense, is a push or a pull. Its source may be gravitational, electrical, magnetic, or simply muscular effort. In the second law, Newton gives a precise idea of force by relating it to the acceleration it produces. He says in effect that *force is anything that can accelerate an object.*

We say *net* force because oftentimes more than a single force acts on an object. Let's examine more carefully what we mean by *net.* If you and a friend pull on an object in the same direction and with forces of equal magnitude, the two forces add to produce a net force twice as great as your single force. If, however, you each pull with equal forces but in opposite directions, the two forces on the object cancel each other. One of the forces can be considered to be the negative of the other, and they add algebraically to zero. The net force is zero.

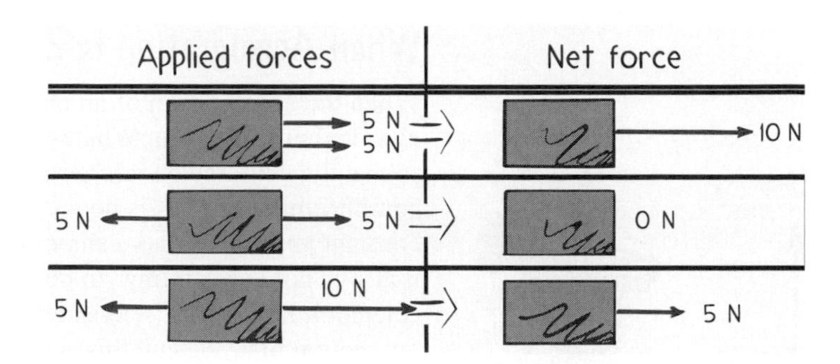

Fig. 2.8
Net force.

Force of hand accelerates the brick

The same force accelerates 2 bricks 1/2 as much

3 bricks, 1/3 as much acceleration

Fig. 2.9
Acceleration is inversely proportional to mass.

Suppose you pull on an object with a force of 10 newtons and your friend pulls in the opposite direction with a force of 5 newtons. The result is the same as if you pulled alone with a force of 5 newtons. The resulting 5 newtons is the net force. If the forces are in the same direction, they are added; if in opposite directions, they are subtracted (Figure 2.8). It is the net force that accelerates the object.

Because force involves both magnitude and direction, it is a *vector quantity.* When two or more forces are exerted on an object, they combine by vector rules. When the forces are parallel, either in the same direction or in opposite directions, they add algebraically. When two or more forces are exerted at angles to one another so they are neither in the same direction nor in opposite directions, they combine via the parallelogram rule described in Section 1.2. To simplify our study of physics, we do not treat forces at angles in this chapter, and leave such to Appendix C.

So, we see that the acceleration of an object depends on both the net force exerted on the object and the mass of the object.[3]

Question

In Chapter 1, acceleration was defined as the time rate of change of velocity: a = (change in v)/time. Are we in this chapter saying that acceleration is instead the ratio of force to mass: $a = F/m$? Which is it?

Answer
Acceleration is defined as the time rate of change of velocity and this change is *produced by* a force. How much force/mass (the cause) acts determines the rate of change in velocity in a given time interval (the effect). So whereas we *defined acceleration* in Chapter 1, in this chapter we define *the terms that cause acceleration.*

---

[3]Mass can be operationally defined as the proportionality constant between force and acceleration in Newton's second law, rearranged to read $m = F/a$. A 1-unit mass is that which requires 1 unit of force to produce 1 unit of acceleration. So 1 kg is the amount of matter that 1 N of force will accelerate 1 m/s^2. (In British units, 1 slug is the amount of matter that 1 pound of force will accelerate 1 ft/s^2). We shall see later that mass is simply a form of concentrated energy.

When Acceleration Is Zero—Equilibrium

When the acceleration of an object is zero, we say the object is in **mechanical equilibrium**. Whatever forces may act on it are canceled out, so that the net force on it is zero. A book lying motionless on a table is in equilibrium because it is not accelerating. And the same book sliding at constant velocity across a smooth surface is also in equilibrium because it also is not accelerating. In both cases the acceleration is zero. Let's consider the case without motion first. We call this *static* equilibrium.

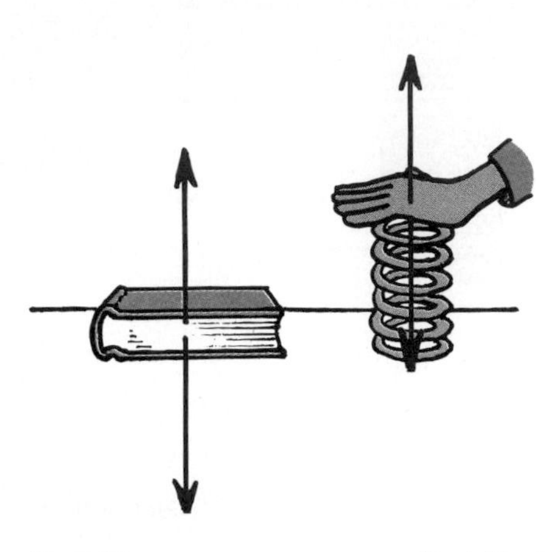

Fig. 2.10
(a) The table pushes up on the book with a support force that is equal in magnitude to the magnitude of the downward force of gravity on the book. (b) The spring pushes up on your hand (support force) with as much force as you push down on the spring.

A book lying motionless on a table has zero acceleration, and so according to Newton's second law, the net force acting on the book must be zero. Weight acts downward, which means that another force must act upward on the book. This other force is the *support force* exerted by the table (often called the *normal force*).[4] The table exerts on the book an upward support force that is equal in magnitude to the downward force of gravity exerted on the book—its weight. The table pushes upward the same way a spring pushes upward when you compress it. When a book similarly pushes down on a table, the book compresses atoms in the table, which behave as microscopic springs, producing the support force (Figure 2.10).

Fig. 2.11
The sum of the upward-pulling forces exerted by the rings must equal her weight. Then we see the vector sum of all three forces equals zero ($\Sigma F = 0$). She is then in equilibrium.

When you step on a bathroom scale, the downward pull of gravity and the upward support force of the floor compress a spring that is calibrated to give your weight. In effect, the scale shows the support force exerted on you by the scale/floor. Because you are not accelerating, the net force on you is zero, which means the support force and your weight are equal in magnitude.

This idea that zero net force acts on things in static equilibrium is often quite useful. For example, if you see that a force is exerted on something that doesn't move, then you know there is another force acting—one that is equal in magnitude and opposite in direction. The net force on a stationary object is always zero.

■ Question

Suppose you stand on two bathroom scales with your weight evenly divided between them. What will each scale read? How about if you stand with more of your weight on one foot than on the other?

■ Answer

The readings on the two scales add up to your weight. This is because the sum of the scale readings, which equals the support force exerted by the floor, must counteract your weight so that the net force on you will be zero. If you stand equally on each scale, each will read half your weight. If you lean more on one scale than the other, more than half your weight will be read on that scale but less on the other, with the result that they will still add up to your weight.

---

[4]This force acts at right angles to the surface of the table. When we say "normal to," we are saying "at right angles to," which is why this force is called a normal force.

Now let's move onto the case of an object *moving* but *not accelerating*. Any such object is said to be in *dynamic* equilibrium. Consider a hockey puck that slides across the ice at constant velocity. It slides without accelerating—in other words, it is in dynamic equilibrium—because the net force on it is zero. A bowling ball rolling along a bowling alley is in dynamic equilibrium—until it interacts with the pins. When you push a crate at constant velocity across a factory floor, the crate is in dynamic equilibrium. In this case the force you apply is balanced by the force of friction between crate and floor. The net force is zero; hence the acceleration is zero.

Fig. 2.12
The crate slides to the right because of an applied force of 75 N. A friction force of 75 N opposes the sliding motion and results in a zero net force on the crate. As a result, the crate slides at constant velocity (zero acceleration).

It's important to emphasize that zero acceleration does not mean zero velocity. Zero acceleration means that the object will maintain the velocity it happens to have, neither speeding up nor slowing down nor changing direction. Most things that are moved in our environment must be pushed or pulled to overcome friction.

When surfaces slide or tend to slide over one another, a force of **friction** acts. The amount of friction force depends on the kinds of material and on how much they are pressed together, and results from the mutual contact of irregularities in the surfaces.[5] The irregularities act as obstructions to motion. Even surfaces that appear to be very smooth have irregular surfaces when viewed microscopically.

The direction of a friction force is always in a direction opposing motion. Thus, if an object is to move at constant velocity, a force must be applied that is equal to the opposing force of friction so that the two forces exactly cancel each other. The zero net force then results in zero acceleration.

■ **Question**

A jumbo jet cruises at a constant velocity of 1000 km/h when the thrusting force of its engines is a constant 100,000 N. What is the acceleration of the jet? What is the force of air friction (air resistance, or air drag) on the jet?

When Acceleration Is g—Free-Fall

Although Galileo founded both the concept of inertia and the concept of acceleration, and was the first to measure the acceleration of falling objects, he could not explain why objects of various masses fall with equal accelerations. Newton's second law provides the explanation.

A falling object accelerates toward the Earth because of gravitational attraction (a force between object and Earth that we'll treat in Chapter 4). We call this force the

---

■ **Answer**

The acceleration is zero because the velocity is constant. Therefore, from Newton's second law, the net force on the jet is zero, which means that the frictional force of air resistance must just equal the thrusting force of 100,000 N and act in the opposite direction. So the air drag on the jet is 100,000 N. (Here *air resistance* and *air drag* are other words for air friction. And note that we don't need to know the velocity of the jet to answer this question. We need only know that it is constant, our clue that acceleration and therefore net force are zero.)

---

[5]Even though it may not seem so yet, most of the concepts in physics are not really complicated. Friction is an exception, however —it is a very complicated phenomenon. The findings are empirical (gained from a wide range of experiments) and the predictions approximate (also based on experiment).

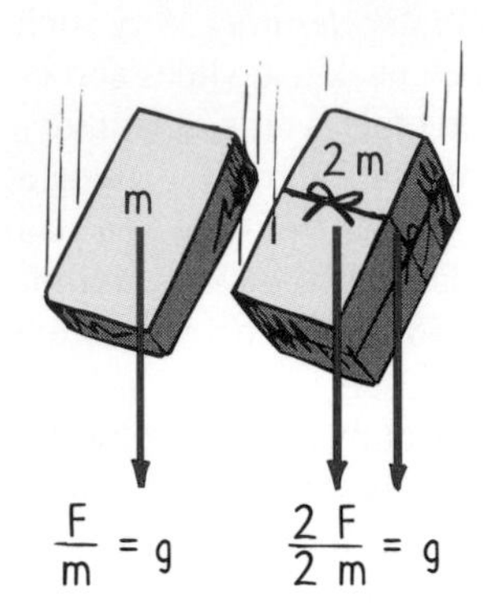

Fig. 2.13
The ratio of gravitational force F to mass m is the same for all objects in the same locality; hence, in the absence of air drag, their accelerations all have the same value: g.

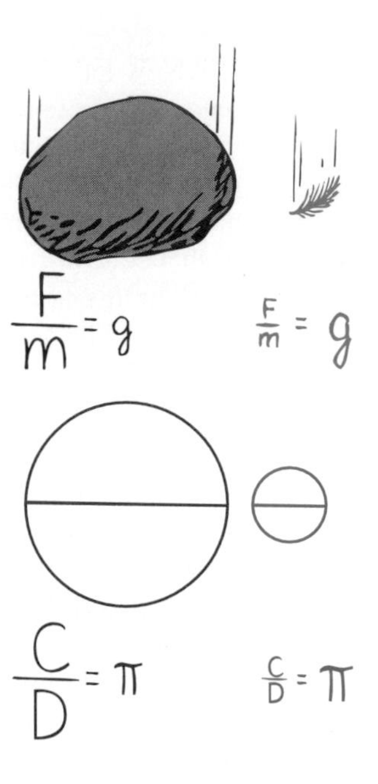

Fig. 2.14
The ratio of gravitational force F to mass m is the same for the large rock and the small feather; similarly, the ratio of circumference C to diameter D is the same for the large and the small circle.

weight of the object.[6] When this is the only force acting on the object—that is, when friction such as air resistance is negligible—we say that the object is in *free-fall.*

The attractive force between a more massive object and the Earth is greater than that between a less massive object and the Earth. The double brick in Figure 2.13, for example, has twice the attraction as the single brick. Why then, as Aristotle supposed, doesn't the double brick fall twice as fast? The answer is that the acceleration of an object depends not only on the force acting on the object—in this case, the weight—but also on the object's resistance to motion, its inertia: from Newton's second law, $a = F/m$. Whereas a force produces an acceleration, inertia is a *resistance* to acceleration. So twice the force exerted on twice the inertia produces the same acceleration as half the force exerted on half the inertia. Both objects accelerate equally.

The ratio of weight to mass for freely falling objects equals a constant—g. We use the symbol g, rather than a, to denote that acceleration is due to gravity alone. This is similar to the constant ratio of circumference to diameter for circles, which equals the constant π. The ratio of weight to mass is the same for both heavy and light objects, just as the ratio of circumference to diameter is the same for both large and small circles (Figure 2.14).

Although Galileo first defined acceleration (change in velocity/time interval), he was not aware of the other expression for acceleration, that given by Newton's second law. So he wasn't able to say *why* objects of different masses dropped from the Leaning Tower of Pisa have equal accelerations. Newton's second law tells us why the acceleration of free-fall is independent of an object's mass. For example, a boulder 100 times heavier than a pebble falls at the same acceleration as the pebble because the boulder has 100 times as much resistance to changes in its motion. In other words, a force 100 times greater but with 100 times as much mass to move experiences no gain in accelera-

---

[6]As we learned in Section 2.1, weight and mass are directly proportional to each other. The constant of proportionality is g: $w \sim m$, $W = gm$, or more conventionally, $W = mg$. So the weight of a 1-kg mass $= mg = (1\ \text{kg})(9.8\ \text{m/s}^2) = 9.8\ \text{N}$.

tion over an object with 100 times less force and 100 times less mass. The force/mass ratio is the same for all freely falling objects.

Question

In a vacuum, a coin and a feather fall side by side. Would it be correct to say that in a vacuum equal forces of gravity act on both bodies?

When Acceleration Is Less Than g—Nonfree-Fall

Although a feather and a coin fall equally fast in a vacuum, they fall quite differently in air. How do Newton's laws apply to objects falling in air? The answer is that Newton's laws apply to *all* objects, whether freely falling or falling in the presence of resistive forces. The accelerations, however, are quite different for the two cases. The important thing to keep in mind is the idea of *net force.* In a vacuum or in cases where air resistance can be neglected, the net force acting on the falling object is the object's weight because it is the only force. In the presence of air resistance, however, the net force is less than the weight—it is the object's weight minus air drag.

The air drag experienced by a falling object depends on two things. First, it depends on the frontal area of the falling object—that is, on the amount of air the object must plow through as it falls. The greater the frontal area of the object, the greater the air drag. Second, it depends on the speed of the falling object; the greater the speed, the greater the number of air molecules an object encounters per second and the greater the force of molecular impact—in other words, the greater the air drag. So air drag depends on the size and the speed of a falling object.

R

mg

Fig. 2.15
When the weight mg is greater than the force of air resistance R, the falling sack accelerates. As speed increases, so does R. When $R = mg$, acceleration reaches zero and the sack reaches its terminal speed.

In some cases air drag greatly affects falling; in other cases it doesn't. Air drag is important for a falling feather. Because it has so much frontal area relative to its low weight, a feather doesn't have to fall very fast before the upward-acting air drag cancels the downward-acting weight. The net force on the feather is then zero, and so acceleration terminates. When acceleration terminates, we say the feather has reached its **terminal speed**. (If we are concerned with direction, down for falling objects, we say the object has reached its **terminal velocity**.)

Consider skydiving. As a falling skydiver gains speed, air drag builds up until finally it equals the skydiver's weight. If and when this happens, the *net* force on the diver becomes zero and she no longer accelerates; she has reached her terminal speed. For a feather, terminal speed is a few centimeters per second, whereas for a skydiver it is about 200 kilometers per hour. A skydiver may vary speed by varying body orientation. Head-first or feet-first is a way of encountering less air and thus less air drag and attaining a high terminal speed. A lower terminal speed is attained by spreading oneself out like a flying squirrel. Minimum terminal speed is attained when the parachute is opened because most of the air drag is due to increased frontal area.

Answer

No, no, no, a thousand times no! These objects accelerate equally not because the forces of gravity on them are equal (they aren't!) but because the ratios of their weights to masses are equal. Although air resistance is not present in a vacuum, gravity is (you'd know this if you stuck your hand into a vacuum chamber and a massive truck rolled over it). If you answered yes to this question, let this be an alarm to be more careful when you think physics!

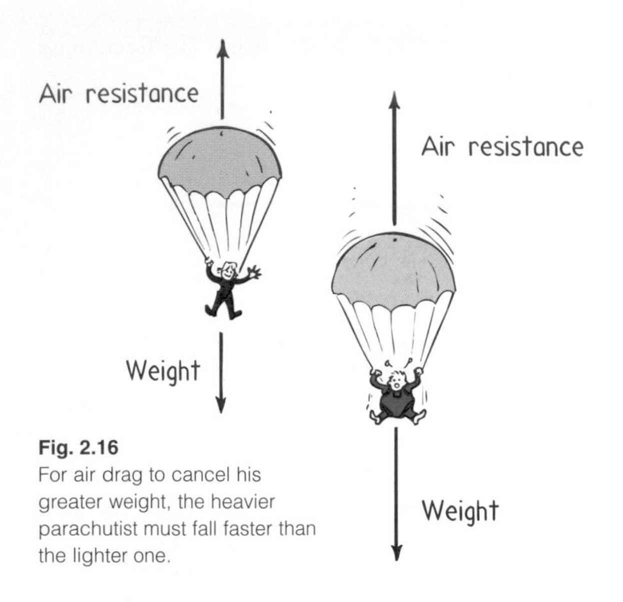

Fig. 2.16
For air drag to cancel his greater weight, the heavier parachutist must fall faster than the lighter one.

Consider a man and woman parachuting together from the same altitude (Figure 2.16). Suppose the man is twice as heavy as the woman and that their same-sized chutes are initially opened. Same-size chutes means that at equal speeds the air resistance is the same on each. Who gets to the ground first—the heavy man or the lighter woman? The answer is, the person who falls faster gets to the ground first—that is, the person with the greater terminal speed. At first we might think that because the chutes are the same size, the terminal speeds are the same and therefore the two persons reach the ground together. This doesn't happen, however, because air drag also depends on speed. The woman will reach her terminal speed when air drag against her chute equals her weight. When this happens, the same amount of air drag against the chute of the man will not yet equal his weight. He must fall faster than she does in order for air drag to match his greater weight.[7] Terminal speed is greater for the heavier person, with the result that the heavier person reaches the ground first.

■ **Question**

A skydiver jumps from a high-flying helicopter. As she falls faster and faster through the air, does her acceleration increase, decrease, or remain the same?

Consider a pair of tennis balls, one a hollow regular ball and the other filled with iron pellets. If you hold them above your head and drop them simultaneously you'll see that they strike the ground at the same time. If you drop them from a greater height, however, say from the top of a building, you'll see the heavier ball strikes the ground first. Why? In the first case the balls do not gain much speed in their short fall. Any air drag encountered is small relative to their weights, even for the regular ball. So the tiny

---

■ **Answer**

Acceleration decreases because the net force acting on her decreases. Net force is equal to her weight minus her air drag, and being that air drag increases with increasing speed, net force and hence acceleration decrease. By Newton's second law,

$$a = \frac{F_{net}}{m} = \frac{mg - R}{m}$$

where mg is her weight and R is the air drag she encounters. As R increases, a decreases. Note that if she falls fast enough that $R = mg$, then $a = 0$, and she falls at constant speed.

---

[7]Terminal speed for the man will be about 41% greater than terminal speed for the half-as-heavy woman. This is because the retarding force of air resistance is proportional to speed squared. So

$$\frac{(\text{man's speed})^2}{(\text{woman's speed})^2} = (1.41)^2 = 2.$$

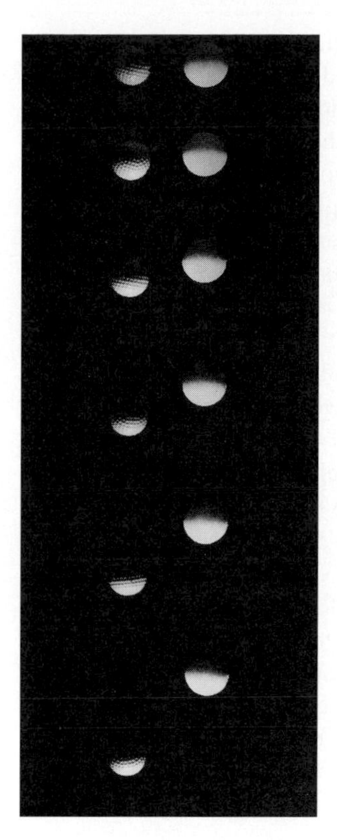

Fig. 2.17
A stroboscopic study of a golf ball (left) and a Styrofoam ball (right) falling in air. Air drag is negligible for the heavier golf ball, and its acceleration is nearly equal to g. Air drag is not negligible for the lighter Styrofoam ball, which reaches its terminal speed sooner, and therefore lags behind the golf ball.

difference in their arrival time is not noticed. But when dropped from a greater height, the greater speeds are met with greater air drag. At any given speed, each ball encounters the same air drag because the two are the same size. This same air drag may be a lot relative to the weight of the regular ball, but it's only a little relative to the weight of the heavier ball (like the parachutists in Figure 2.16). For example, 1 newton of air drag acting on a 2-newton object cuts the acceleration to $g/2$, but 1 newton of air drag on a 200-newton object diminishes acceleration only slightly. So even with equal air drags, the accelerations of the two balls differ.

There is a moral to be learned here. Whenever you consider the acceleration of something, use the equation of Newton's second law to guide your thinking: The acceleration is equal to the ratio of *net* force to mass. For the falling tennis balls, the net force on the regular ball is appreciably reduced as air drag builds up, while the net force on the iron-filled ball is only slightly reduced. Acceleration decreases as net force decreases, which in turn decreases as air drag increases. If and when the air drag builds up to equal the weight of the falling object, the net force becomes zero and acceleration terminates.

2.3 Newton's Third Law of Motion

Drop a sheet of tissue paper in front of the heavyweight boxing champion of the world and challenge him to hit it in midair with 50 pounds (222 newtons) of force. Sorry, the champ can't do it. In fact, his best punch couldn't even come close. Why is this?

So far we have treated force in its simplest sense—as a push or pull. In a broader sense, a force is not a thing in itself but makes up an *interaction* between one thing and another. We pull on a cart, and it accelerates. A hammer hits a stake and drives it into the ground. One object interacts with another object. Which exerts the force and which receives the force? Newton's answer to this is that neither force has to be identified as "exerter" or "receiver"; he concluded that both objects must be treated equally. For example, when we pull the cart, the cart pulls us, as evidenced perhaps by the tightening of the rope wrapped around our hand. This pair of forces, our pull exerted on the cart and the cart's pull exerted on us, make up the single interaction between us and the cart. In the interaction between hammer and stake, the hammer exerts a force on the stake but is itself brought to a halt in the process. That's because the stake exerts a force on the hammer. Such observations led Newton to his **third law**—the **law of action and reaction**.

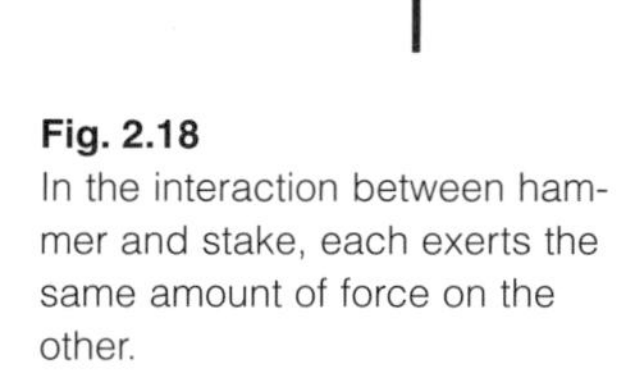

Fig. 2.18
In the interaction between hammer and stake, each exerts the same amount of force on the other.

Law 3: *Whenever one object exerts a force on a second object, the second object exerts an equal and opposite force on the first.*

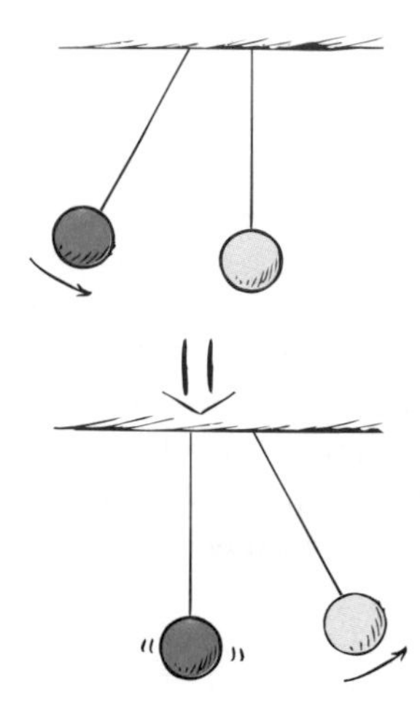

Fig. 2.19
The impact force between the blue and yellow balls moves the yellow ball and stops the blue one.

Newton's third law is often stated thus: "To every action there is always opposed an equal reaction." In any interaction, there is an action and reaction pair of forces that are equal in magnitude and opposite in direction. Neither force exists without the other—forces always come in *pairs,* one action and the other reaction. The action and reaction pair of forces make up one interaction between two things.

You interact with the floor when you walk on it. Your push against the floor is coupled to the floor's push against you. The pair of forces occur simultaneously. Likewise, the tires of a car push against the road while the road pushes back on the tires—the tires and road push against each other. In swimming you interact with the water: You push it backward while it pushes you forward—you and the water push against each other. In each case, there is a pair of forces, one action and the other reaction, that make up one interaction. Which force we call *action* and which we call *reaction* doesn't matter. The point is that neither exists without the other.

Being that action and reaction forces are equal and opposite, why don't they cancel to zero so that nothing ever accelerates? To answer this question, we must consider the *system* involved in an interaction. Consider the force pair between the apple and orange in Figure 2.21. We'll define the system as being only the orange. A force is exerted on the orange, and so it accelerates. The fact that the orange simultaneously exerts a force on the apple, which is external to the system, may affect the apple but does not affect the orange. The interaction is between the system (the orange) and something external (the apple), and so the action and reaction forces don't cancel. However, if we consider the system to be both orange and apple together, the force pair

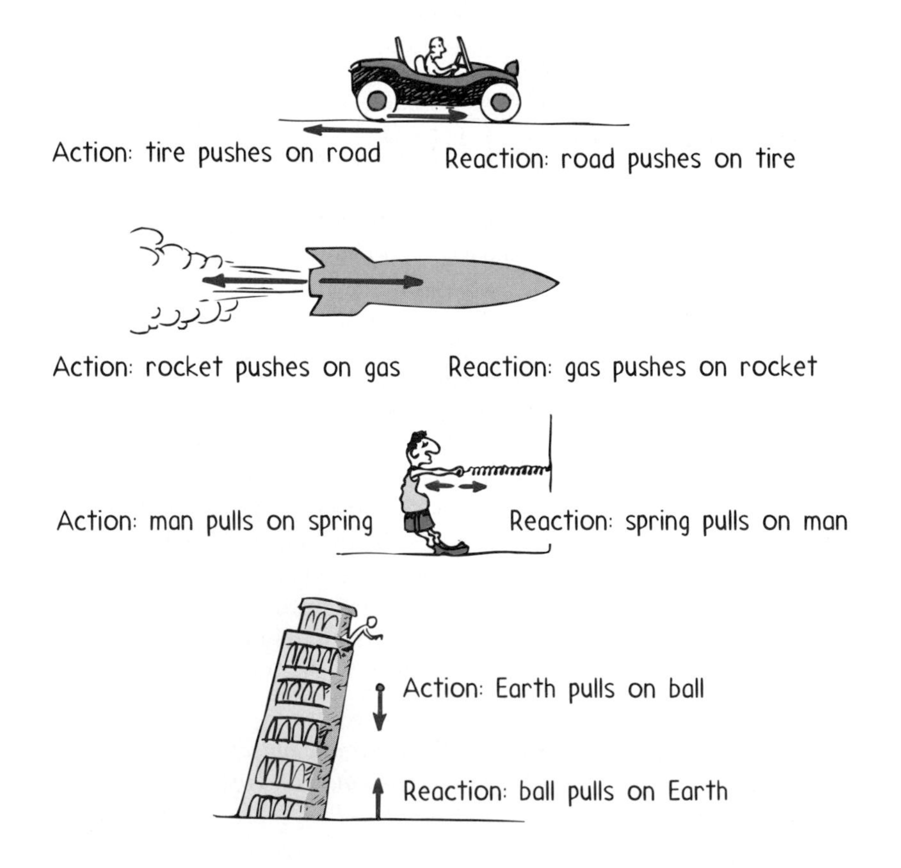

Fig. 2.20
Action and reaction forces. Note that when the action is "A exerts force on B," the reaction is "B exerts force on A."

Fig. 2.21
When an apple pulls on an orange, the orange accelerates—period! The fact that the orange pulls back on the apple affects the apple—not the orange. The force exerted by the apple on the orange is not canceled by the force exerted by the orange on the apple.

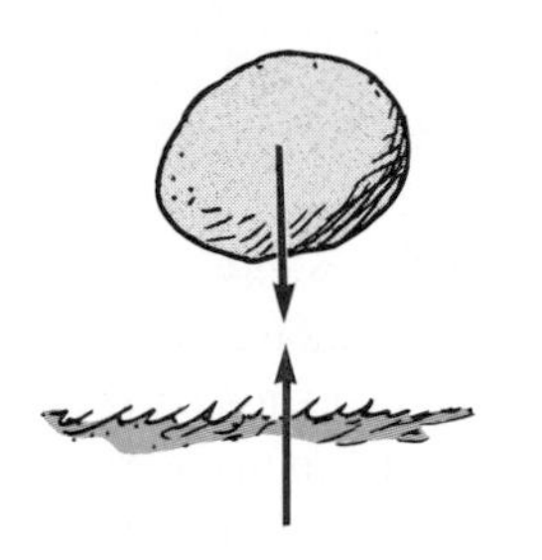

Fig. 2.22
The gravitational interaction between the boulder and the Earth consists of a pair of equal and opposite forces; one is the Earth pulling down on the boulder, the other is the boulder pulling up on the Earth.

is internal to the system. In this case the forces do cancel each other. Any acceleration of the system is due to other forces, such as the external force of friction between the apple's feet and the floor.

Similarly, the many force pairs between molecules in a golf ball may hold the ball together into a cohesive solid, but they play no role at all in accelerating the ball. A force external to the ball is needed to accelerate it. If action and reaction forces are internal to a system, they cancel each other and contribute nothing to accelerating the system. Any acceleration of a system is due to a force or forces external to the system. Internal forces cancel out. If this is confusing, it may be well to note that Newton had difficulties with the third law himself.

Questions

1. On a cold, rainy day, your car battery is dead and you must push the car to get it started. Why can't you push the car by remaining comfortably inside and pushing against the dashboard?
2. Does a speeding missile possess force?
3. We know that the Earth pulls on the Moon. Does it follow that the Moon also pulls on the Earth?
4. Can you identify the action and reaction forces in the case of an object falling in a vacuum?
5. Why does a book sitting on a table never accelerate "spontaneously" in response to the trillions of interatomic forces acting within it?

Answers

1. The system to be accelerated is the car. If you remain inside, the force pair consisting of you and the dashboard produce actions and reactions within the system. These forces cancel out as far as any motion of the car is concerned. For the car to accelerate, there must be an interaction between it and something external to it—you pushing outside against the road, for example.
2. No, a force is not something an object *has,* like mass, but is part of an interaction between one object and another. A speeding missile may be capable of exerting a force on another object when interaction occurs, but it does not possess force as a thing in itself. As we shall see in the following chapter, a speeding missile possesses momentum and kinetic energy.
3. Yes, the two pulls make up an action-reaction pair of forces associated with the gravitational interaction between Earth and Moon. We can say that (1) the Earth pulls on the Moon, and (2) the Moon likewise pulls on the Earth, but it is more insightful to think of this as a single interaction—the Earth and Moon simultaneously pull on each other, with forces of equal magnitude.
4. To identify a pair of action-reaction forces in any situation, first identify the pair of interacting objects involved. Something is interacting with something. In this case the whole Earth is interacting (gravitationally) with the falling object. So the Earth pulls downward on the object (call it action), while the object pulls upward on the Earth (reaction).
5. Every one of these interatomic forces is part of an action-reaction pair within the book. These forces add up to zero, no matter how many of them there are. This is what makes Newton's *first* law apply to the book. The book has zero acceleration unless an *external* force acts on it.

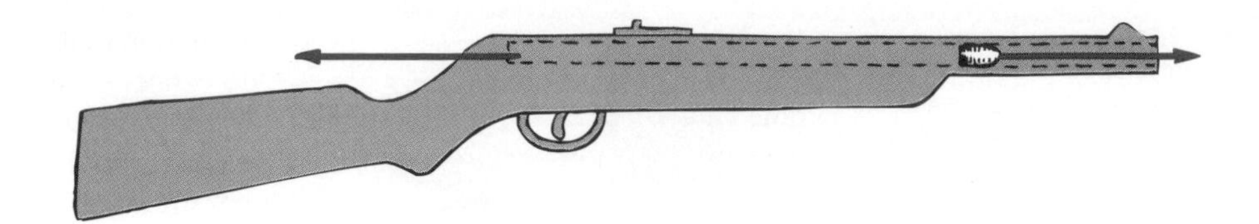

Fig. 2.23
The force exerted by the bullet on the recoiling rifle is just as great as the force exerted by the rifle on the bullet. Why does the bullet accelerate more than the rifle?

When a rifle is fired, there is an interaction between rifle and bullet (Figure 2.23). A pair of forces act on both the rifle and the bullet. The force exerted by the rifle on the bullet is as great as the reaction force exerted by the bullet on the rifle; hence, the rifle "kicks." Being that the forces are equal in magnitude, why doesn't the rifle recoil with the same speed as the bullet? In analyzing changes in motion, Newton's second law reminds us that we must also consider the masses involved. Suppose we let F represent both the action force and the reaction force, m the mass of the bullet, and $\mathcal{M}$ the mass of the more massive rifle. The accelerations of the bullet and the rifle are found by taking the ratio of force to mass. The acceleration of the bullet is

$$\frac{F}{m} = \mathbf{a}$$

while the acceleration of the recoiling rifle is

$$\frac{F}{\mathbf{m}} = a$$

We see why the change in motion of the bullet is so huge relative to the change in motion of the rifle: a given force divided by a small mass produces a large acceleration, whereas the same force divided by a large mass produces a small acceleration.

If we extend the idea of a rifle recoiling from the bullet it fires, we can understand rocket propulsion. Consider a machine gun recoiling each time a bullet is fired. If the machine gun is fastened such that it is free to slide on a vertical wire (Figure 2.24a), it accelerates upward as bullets are fired downward. A rocket accelerates the same way as it continuously "recoils" from the ejected exhaust gas. Each molecule of exhaust gas is like a tiny bullet shot from the rocket (Figure 2.24b).

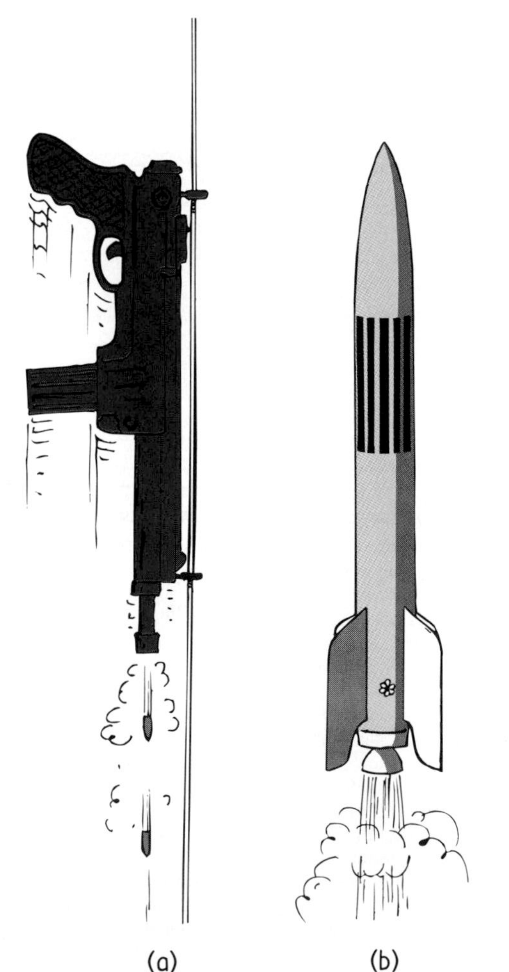

Fig. 2.24
(a) The machine gun recoils from the bullets it fires and climbs upward.
(b) The rocket recoils from the "molecular bullets" it fires and climbs upward.

A common misconception is that a rocket is propelled by the impact of exhaust gases against the atmosphere. In fact, in the early 1900s, before the advent of rockets, some people thought that sending a rocket to the Moon was impossible because of the absence of an atmosphere for the rocket to push against. But this is like saying a gun won't kick unless the bullet has air to push against. Not true! Both the rocket and the recoiling gun accelerate not because of any pushes on the air but because of the reaction forces exerted by the "bullets" they fire—air or no air. A rocket works better, in fact, above the atmosphere where there is no air resistance to restrict its speed.

Using Newton's third law, we can understand how a helicopter gets its lifting force. The whirling blades are shaped to force air particles down (action), and the air forces the blades up (reaction). This upward reaction force is called *lift.* When lift equals the weight of the craft, the helicopter hovers in midair. When lift is greater, the helicopter climbs upward.

The same principle applies to birds and airplanes. Birds fly by pushing air downward. The air in turn pushes the bird upward. When the bird is soaring, the wing must be shaped so that moving air particles are deflected downward. Slightly tilted wings that deflect oncoming air downward produce the lift on an airplane. Air must be pushed downward continuously to maintain lift. This supply of air is obtained by the forward motion of the aircraft, which results from propellers or jets that push air backward. When the propellers of jets push air backward, the air in turn pushes the propellers or jets forward. We shall learn in Chapter 4 that the curved surface of a wing is an airfoil, which enhances the lifting force.

■ Questions

1. A car accelerates along a level road. What is the force that moves the car?
2. A high-speed bus and an innocent bug have a head-on collision. The force exerted by the bus on the bug splatters the bug over the windshield. Is the magnitude of the corresponding force exerted by the bug on the windshield greater than, less than, or the same as the force exerted by the windshield on the bug? Is the resulting deceleration of the bus greater than, less than, or the same as that of the bug?

We see Newton's third law at work everywhere. A fish pushes the water backward with its fins, and the water in turn pushes the fish forward. The wind pushes against the branches of a tree, and the branches in turn push back on the wind and we have whistling sounds. Forces are interactions between different things. Every contact requires at least

---

■ Answers

1. The force between the road and tires push the car along—really! Except for air drag when moving, only the road exerts a horizontal force on the car. How? The rotating tires push backward on the road (action), and the road simultaneously pushes forward on the tires (reaction). How about that!
2. The magnitudes of the two forces are the same, for the forces constitute an action-reaction pair that makes up the interaction between bus and bug. The accelerations, however, are very different because the masses involved are different. The bug undergoes an enormous and lethal acceleration, whereas the bus undergoes only a very tiny deceleration—so tiny that the very slight slowing of the bus is unnoticed by its passengers. But if the bug were instead a more massive body—another bus, for example—the slowing down would be quite evident!

Link to Zoology—Geese Flight

Ever wonder why geese fly in a V formation? Simple physics! A bird's wings deflect air downward (action), and the air pushes the wings upward (reaction). This is how birds attain lift. Interestingly, the downward-moving air at the wing tips swirls upward. This region of updraft is nicely employed by an-

other bird, which gets added lift by positioning itself in this updraft. This bird, in turn, creates an updraft for a following bird, and so on. Birds take turns at the lead position, which requires more effort. Now you have a better idea of why geese fly in a V formation.

Fig. 2.25
You cannot touch without being touched—Newton's third law.

a twoness; there is no way that an object can exert a force on nothing or that nothing can exert a force on something. Forces, whether large shoves or slight nudges, always occur in pairs, each of which is opposite to the other. Thus, we cannot touch without being touched.

Newton's laws are the essence of classical physics and provide the foundation for understanding more modern concepts. Whatever the insights of modern physics developed in this century, it's safe to say that Newton's three laws of motion and, as we'll see, his law of universal gravitation, will remain central concepts of physics in general. The simplicity of Newton's laws makes them not only beautiful but also enormously useful. They are all that is needed to get rockets to the Moon and beyond.

Summary of Terms

- **Inertia** The tendency of things to resist changes in motion.
- **Law of inertia (Newton's first law)** Every material object continues in its state of rest, or of uniform motion in a straight line, unless it is compelled to change that state by forces impressed upon it.
- **Mass** The quantity of matter in an object. More specifically, it is the measurement of the inertia or sluggishness that an object exhibits in response to any effort made to start it, stop it, deflect it, or change in any way its state of motion.
- **Weight** The gravitational force exerted on an object by the nearest most-massive body (locally, by the Earth).
- **Kilogram** The fundamental SI unit of mass. One kilogram (symbol kg) is this amount of mass in 1 liter (L) of water at 4°C.
- **Newton** The SI unit of force. One newton (symbol N) is the force that will give an object of mass 1 kg an acceleration of 1 m/s^2.
- **Newton's second law** The acceleration of an object is directly proportional to the net force acting on the object, is in the direction of the net force, and is inversely proportional to the mass of the object.

- **Force** Any influence that can cause an object to be accelerated, measured in newtons in the metric system and in pounds in the British system.
- **Mechanical equilibrium** The state of an object or system of objects for which any impressed forces cancel to zero and no acceleration occurs.
- **Friction** The resistive force that opposes the motion or attempted motion of an object past another with which it is in contact, or through a fluid.
- **Terminal speed** The constant speed of a falling object where acceleration terminates because air resistance balances the weight. When direction is specified, we speak of terminal velocity.
- **Law of action and reaction (Newton's third law)** Whenever one object exerts a force on a second object, the second object exerts an equal and opposite force on the first.

Review Questions

Newton's First Law of Motion

1. Is inertia the *reason* objects maintain their states of motion or the *name* given to this property of matter?
2. Clearly distinguish among *mass, weight,* and *volume.*
3. Which is more fundamental, *mass* or *weight*? (Does an object that has mass also have weight?)
4. Does a 2-kg iron brick have twice as much inertia as a 1-kg block of wood? Twice as much volume? (Why do these two questions have different answers?)
5. What kind of path would the planets follow if suddenly no force acted on them?

Newton's Second Law of Motion

6. If we say that one quantity is *proportional* to another quantity, does this mean they are *equal* to each other? Explain briefly, using mass and weight as an example.
7. A cart is simultaneously pulled to the left with a force of 100 N and to the right with a force of 30 N. What is the net force on the cart?
8. Why do we say force is a vector quantity?
9. If the net force acting on a sliding block is tripled, by how much does the acceleration increase?
10. If the mass of a sliding block is tripled while a constant net force is applied, by how much does the acceleration decrease?
11. If the mass of a sliding block is tripled at the same time the net force on it is tripled, how does the resulting acceleration compare with the original acceleration?
12. What is the net force on something in mechanical equilibrium?
13. Consider a book that weighs 15 N and is at rest on a flat table. How many newtons of support force does the table provide? What is the net force on the book?
14. Consider a woman weighing 500 N who stands with her weight evenly divided on a pair of bathroom scales. What is the reading on each scale? If she shifts her weight so one scale reads 300 N, what does the other scale read?
15. What is the acceleration of an object that moves at constant velocity? What is the net force on the object?
16. In what direction does the force of friction act when a body slides?
17. If you push horizontally with a force of 50 N on a crate and make it slide at constant velocity, how much friction acts on the crate? If you increase your force, does the crate accelerate? Explain.
18. What is the net force on a 10-N freely falling object?
19. Why doesn't a heavy object accelerate more than a light object when both are freely falling?
20. What is the net force on a 10-N falling object when it encounters 4 N of air resistance? 10 N of air resistance?
21. What two principal factors affect the force of air resistance on a falling object?
22. What is the acceleration of a falling object that has reached its terminal velocity?
23. Why does a heavy parachutist fall faster than a lighter parachutist who wears the same size parachute?

Newton's Third Law of Motion

24. First we said a force was a push or pull; now we say it is an interaction. Which is it, a push or pull, or an interaction? And what does it mean to say *interaction*?
25. Consider hitting a baseball with a bat. If we call the force exerted on the ball by the bat the *action* force, identify the *reaction* force.
26. When do action and reaction pairs of forces cancel each another and when do they not?
27. If the forces that act on a bullet and on the recoiling gun from which it is fired are equal in magnitude, why do the bullet and gun have very different accelerations?
28. How does a helicopter get its lifting force?

Home Projects

1. Drop a sheet of paper and a coin at the same time. Which reaches the ground first? Why? Now crumple the paper into a small, tight wad and again drop it with the coin. Which reaches the ground first this time? Explain the difference. Will they fall together if dropped from a second-, third-, or fourth-story window? Try it and explain your observations.
2. Drop a book and a sheet of paper and note that the book has a greater acceleration *g*. Place the paper beneath the book, and it is forced against the book as both fall, with the result that both fall at *g*. How do the accelerations compare if you place the paper on top of the book and then

drop both? You may be surprised, so try it and see. Then explain your observation.

3. Drop two balls of different weight from the same height, and at low speeds they practically fall together. Will they roll together down an inclined plane? If each is suspended from an equal length of string, making a pair of pendulums, and displaced through the same angle, will they swing back and forth in unison? Try it and see; then explain using Newton's laws.
4. The net force acting on an object and the resulting acceleration are always in the same direction. You can demonstrate this with a spool. If the spool is pulled horizontally to the right, in which direction will it roll?

5. Hold your hand like a flat wing outside the window of a moving automobile. Then slightly tilt the front edge upward and notice the lifting effect. Can you see Newton's laws at work here?

Exercises

Please do not be intimidated by the large number of exercises in this and other meatier chapters. If your course work is to cover many chapters, your instructor will likely assign only a few exercises from each.

1. In an orbiting space shuttle, you are handed two identical boxes, one filled with sand and the other with feathers. How can you tell which is which without opening them?
2. Your empty hand is not hurt when it bangs lightly against a wall. Why is it hurt if it bangs the same wall while carrying a heavy load? Which of Newton's laws is most applicable here?
3. Why is a massive cleaver more effective for chopping vegetables than an equally sharp knife?
4. When a junked car is crushed into a compact cube, does its mass change? Its weight? Explain.
5. The gravitational force on the Moon is only $\frac{1}{6}$ the gravitational force on the Earth. What is the weight of a 10-kg object on the Moon and on the Earth? What is its mass on the Moon and on the Earth?
6. What is your mass in kilograms? Your weight in newtons?
7. Each of the bones in the chain forming your spine is separated from its neighbors by disks of elastic tissue. What happens when you jump heavily on your feet from an elevated position? Can you think of a reason you are a little taller in the morning than at night? (*Hint:* Think about the hammerhead in Figure 2.3.)
8. Before the time of Galileo and Newton, it was thought by many learned scholars that a stone dropped from the top of a tall mast of a moving ship would fall straight down and hit the deck at a point behind the mast. The distance between the point of impact and the base of the mast would be equal to how far the ship had moved forward while the stone was falling. In light of your understanding of Newton's laws, what do you think about this?
9. To pull a wagon across a lawn with constant velocity, you have to exert a steady force. Reconcile this fact with Newton's first law, which says that motion with constant velocity requires no force.
10. When your car moves along the highway at constant velocity, the net force on it is zero. Why, then, do you continue running your engine?
11. If an object has no acceleration, can you conclude that no forces are exerted on it? Explain.
12. A rocket becomes progressively easier to accelerate as it travels through space. Why? (*Hint:* About 90% of the mass of a newly launched rocket is fuel.)
13. As you stand on a floor, does the floor exert an upward force against your feet? How much force does it exert? Why are you not moved upward by this force?
14. As you are leaping into the air, how does the force that you exert on the ground compare with your weight?
15. A common saying goes, "It's not the fall that hurts you; it's the sudden stop." Translate this into Newton's laws of motion.
16. The girl hangs at rest from the ends of the rope as shown. How does the reading on the scale compare with her weight?

17. Harry the painter swings year after year from his bosun's chair. His weight is 500 N, and the rope, unknown to him, has a breaking point of 300 N. Why doesn't the rope break when he is supported as shown in the sketch on the left? One day Harry is painting near a flagpole, and, for a change, he ties the free end of the rope to the flagpole instead of to his chair as shown in the sketch on the right. Why did Harry end up taking his vacation early?

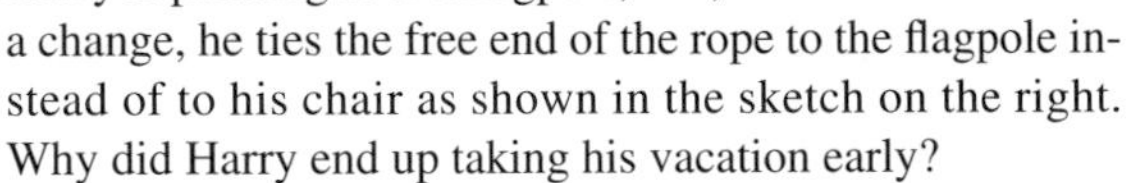

18. Consider the two forces acting on the person standing at rest: the downward pull of gravity and the upward support of the floor. Are these forces equal and opposite? Do they form an action-reaction pair? Why or why not?

19. Two 100-N weights are attached to a spring scale as shown. Does the scale read 0, 100, or 200 N, or give some other reading? (*Hint:* Would it read any differently if one of the ropes were tied to the wall instead of to the hanging 100-N weight?)

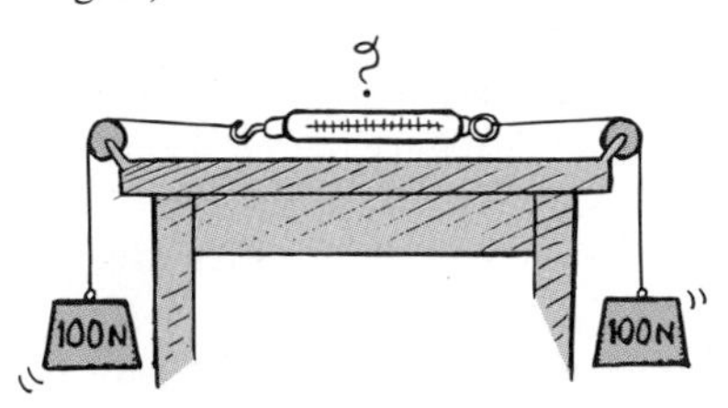

20. What is the net force on an apple that weighs 1 N when you hold it at rest in your hand? When you release it?

21. You hold an apple in your hand. (a) Identify all the forces acting on the apple and their reaction forces. (b) When you drop the apple, identify all the forces acting on it as it falls and the corresponding reaction forces.

22. A "shooting star" is usually a grain of sand from outer space that burns up and gives off light as it enters the atmosphere. What exactly causes this burning?

23. Aristotle claimed the speed of a falling object depends on its weight. We now know that all objects in free-fall, whatever their weight, undergo the same gain in speed. Why does weight not affect acceleration?

24. Does a stick of dynamite contain force?

25. Can a dog wag its tail without the tail in turn "wagging the dog" ? (Consider a dog with a relatively massive tail.)

26. When the athlete holds the barbell overhead, the reaction force is the weight of the barbell on his hand. How does this force vary as the barbell is accelerated upward? Downward?

27. Why can you exert greater force on the pedals of a bicycle if you pull up on the handlebars?

28. If the Earth exerts a gravitational force of 1000 N on an orbiting communications satellite, how much force does the satellite exert on the Earth?

29. The strong man will push the two initially stationary freight cars of equal mass apart before he himself drops straight to the ground. Is it possible for him to give one car a greater speed than the other? Why or why not?

30. Your weight is the result of a gravitational force exerted by the Earth on your body. What is the corresponding reaction force?

31. Suppose two carts, one twice as massive as the other, fly apart when a compressed spring that joins them is released. How fast does the heavier cart roll relative to the lighter cart?

32. If you exert a horizontal force of 200 N to slide a crate across a factory floor at constant velocity, how much friction is exerted by the floor on the crate? Is the force of friction equal in magnitude to your 200-N push? Does the force of friction make up the reaction force to your push? Why not?

33. If a Mack truck and Honda Civic have a head-on collision, upon which vehicle is the impact force greater? Which vehicle undergoes the greater change in motion? Explain.

34. Standing on frictionless ice, two people of equal mass attempt a tug-of-war with a 12-m rope. When they pull on the rope, each slides toward the other. How do their accelerations compare, and how far does each slide before they meet?

35. Suppose in the preceding exercise that one person has twice the mass of the other. How far does each slide before they meet?

36. Which team wins in a tug of war: the team that pulls harder on the rope or the team that pushes harder against the ground? Explain.

37. Why is it that a cat that falls from the top of a 50-story building will hit the ground no faster than if it fell from the 20th story?

38. Free-fall is motion in which gravity is the only force acting. (a) Is a sky diver who has reached terminal speed in free-fall? (b) Is a satellite circling the Earth above the atmosphere in free-fall?

39. How does the weight of a falling body compare with the air resistance it encounters just before it reaches terminal velocity? What about just after?

40. You tell your friend that the acceleration of a sky diver decreases as falling progresses. Your friend then asks if this means the skydiver is slowing down. What is your response?
41. As an object falls faster and faster through the air, where air resistance is a factor, does its acceleration increase, decrease, or remain constant?
42. If Galileo really did drop two balls from the top of the Leaning Tower of Pisa, air resistance was not really negligible. Assuming the two balls were the same size but one was much heavier than the other, which struck the ground first? Why?
43. If you simultaneously drop a pair of tennis balls from the top of a building, they will strike the ground at the same time. If one of the balls is filled with lead pellets, will it fall faster and hit the ground first? Which of the two will encounter more air resistance? Defend your answers.
44. What is the acceleration of a rock at the top of its trajectory when thrown straight upward? (Is your answer consistent with Newton's second law?)
45. Suzie Skydiver with her parachute has a mass of 50 kg.
 a. Before opening her chute, what force of air resistance will she encounter when she reaches terminal velocity?
 b. What force of air resistance will she encounter when she reaches terminal velocity after the chute is open?
 c. Discuss why your answers are the same or different.
46. Make up three multiple-choice questions, one for each of Newton's laws, that check a classmate's understanding of these laws.

Problems

1. A 400-kg bear grasping a vertical tree slides down at constant velocity. What is the friction force that acts on the bear?
2. Find the net force produced by a 30-N force and a 20-N force in each of the following cases:
 a. Both forces act in the same direction.
 b. The two act in opposite directions.
 c. The two act at right angles to each other.
3. When two horizontal forces are exerted on a cart, 600 N forward and 400 N backward, the cart undergoes acceleration. What additional force is needed to produce nonaccelerated motion?
4. A horizontal force of 100 N is required to push a box across a floor at a constant speed.
 a. What is the net force acting on the box?
 b. What is the force of friction acting on the box?
5. If a mass of 1 kg is accelerated 1 m/s^2 by a force of 1 N, what would be the acceleration of a mass of 2 kg acted on by a force of 2 N?
6. How much acceleration does a 747 jumbo jet of mass 30,000 kg experience in takeoff when the thrust for each of four engines is 30,000 N?
7. You slide a 2-kg sled across a horizontal surface by pushing it with a horizontal force of 20 N. The force of friction on the sled is 12 N. What is the acceleration of the sled?
8. What horizontal force must be applied to produce an acceleration of g for a 1-kg puck on a horizontal friction-free air table?
9. A firefighter of mass 80 kg slides down a vertical pole with an acceleration of 4 m/s^2. What is the friction force that acts on the firefighter?
10. Suzy Skydiver and her parachute have a mass of 50 kg. How much air drag will be acting on her when she accelerates at $g/2$?

CHAPTER

3

Momentum and Energy

Moving objects have a quantity that objects at rest don't have. More than a hundred years ago, this quantity was called *impedo.* A boulder at rest had no impedo, while the same boulder rolling down a steep incline had much impedo. The faster an object moved, the greater its impedo. Any change in impedo depended on an applied force and on how long the force acted. Apply a force to a cart at rest and you give the cart impedo. Apply a long force and you give the cart more impedo. But what do we mean by "how long?" Does "how long" refer to time or to distance? When this distinction was made, the term *impedo* gave way to two new terms—*momentum* and *kinetic energy.* This chapter is about both these important concepts.

3.1 Momentum

Fig. 3.1
The boulder, unfortunately, has more momentum than the runner.

We all know that a heavy truck is harder to stop than a small car moving at the same speed. We state this fact by saying that the truck has more momentum than the car. By **momentum** we mean inertia in motion or, to be more specific, the product of the mass of an object and its velocity:

Momentum = mass × velocity

or, in shorthand notation,

Momentum = *mv*

When direction is not an important factor, we can say momentum = mass × speed, which we still abbreviate *mv*.

We can see from the definition that a moving object can have a large momentum if either its mass or its speed is large, or if both are. A truck has more momentum than a car moving at the same speed because the truck has a greater mass. A huge ship moving at a low speed has a large momentum because of its great mass, and a fast-moving small bullet has a large momentum because of its high speed. A massive truck with no brakes rolling down a steep hill has a large momentum, whereas the same truck at rest has no momentum at all—because the *v* part of *mv* is zero.

3.2 Impulse Changes Momentum

To change the momentum of something requires force, and very importantly, how long the force acts. Here we'll take "how long" to be *time.* Apply a force briefly to a stalled automobile, and you produce a small change in its momentum. Apply the same force over an extended time interval, and a greater change in momentum results. We name the product of force and this time interval **impulse**.

The **relationship between impulse and momentum** can be seen by rearranging Newton's second law ($a = F/m$). The time-interval part of impulse is "buried" in the term for acceleration (change in velocity/time interval). Rearrangement of Newton's second law gives[1]

Force × time interval = change in (mass × velocity)

We can express all terms in this relationship in shorthand notation by introducing the delta symbol Δ (Greek letter *D),* which stands for "change in" (or "difference in"):

$$Ft = \Delta mv$$

which reads, "force multiplied by the time-interval during which it acts equals change in momentum."

---

[1]If we equate the cause of acceleration ($a = \frac{F}{m}$) to what acceleration is (a = change in v/change in t), we get

$$F/m = \frac{\textbf{(change in } v\textbf{)}}{\textbf{(change in } t\textbf{)}}$$

If we let (change in t) be simply t, then simple rearrangement gives

$$Ft = \textbf{change in } (mv) \quad \textbf{or} \quad Ft = \Delta mv$$

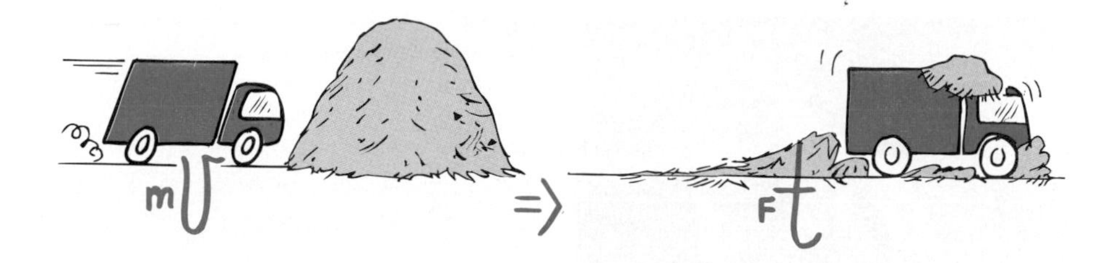

Fig. 3.2
A change in momentum that takes place over a long time interval requires a small force.

This rearrangement of Newton's second law explains why "follow-through" is important in increasing the momentum of things. You apply the largest force you can, and you apply this force for the longest possible time to impart maximum velocity to something. We also see why the long barrel of a cannon increases the velocity of an emerging cannonball: The force of exploding gunpowder in a long barrel acts on the cannonball for a longer time. The longer the time interval over which the force acts, the greater the increase in momentum.

Fig. 3.3
A change in momentum that takes place over a short time interval requires a large force.

Correspondingly, if we decrease momentum over a long time, a smaller force results. A truck out of control is better off hitting a haystack than a brick wall (Figure 3.2). When the truck hits the haystack, the time of impact may be extended 100 times. Then the force of impact is reduced by 100. So whenever you wish the force of impact to be small, extend the time of impact. The safety net used by acrobats provides an obvious example of applying a small-impact force over a long time to reduce the momentum of fall. We can see from the impulse-momentum equation why a boxer rides or rolls with the punch to reduce the force of impact (Figure 3.4).

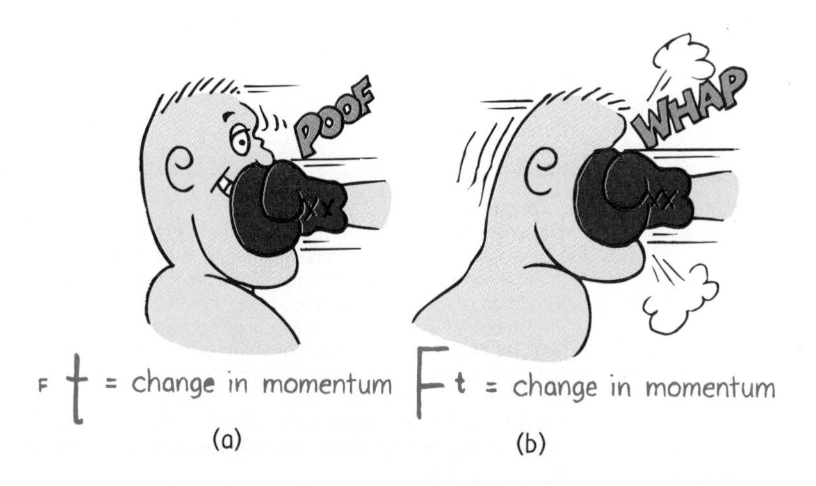

Fig. 3.4
In both cases the boxer's jaw provides an impulse that reduces the momentum of the punch. (a) The boxer is moving away when the glove hits, thereby extending the time of contact. This means the force is less than if the boxer had not moved. (b) The boxer is moving into the glove, thereby lessening the time of contact. This means that the force is greater than if the boxer had not moved.

Fig. 3.5
The karate expert imparts a large impulse to the bricks in a short time and produces a considerable force.

The idea of short time of contact explains how a karate expert can sever a stack of bricks with the blow of his bare hand (Figure 3.5). He brings his arm and hand swiftly against the bricks with considerable momentum. This momentum is quickly reduced when he delivers an impulse to the bricks. The impulse is the force of his hand against the bricks multiplied by the time his hand makes contact with the bricks. By swift execution he makes the time of contact very brief and correspondingly makes the force of impact huge. If his hand is made to bounce upon impact, the force is even greater.

Questions

1. If you stand on a skateboard and toss a ball forward and horizontally, you'll roll backward with the same amount of momentum given to the ball. Will you roll backward if you instead go through the motions of tossing the ball but don't release it?
2. If the boxer in Figure 3.4 is able to make the duration of impact three times as long by riding with the punch, by how much is the force of impact reduced?
3. If the boxer instead moves into the punch so that the duration of impact is decreased by half, by how much is the force of impact increased?
4. A boxer being hit with a punch contrives to extend time for best results, whereas a karate expert delivers a force in a short time for best results. Isn't there a contradiction here?

The impulse-momentum relationship is put to a thrilling test in bungee jumping. When your fall is brought to a halt by the cord, you'll be glad the cord stretches, for whatever momentum you gain in your fall, the cord must supply the same impulse to

---

Answers

1. No, you'll not roll back without immediately rolling forward to produce no net rolling. In terms of momentum, if no net momentum is imparted to the ball, no net momentum is imparted to you. Or in third-law fashion, if no net force acts on the ball, no net force acts on you.
2. The force of impact will be three times less than if he doesn't pull back.
3. The force of impact will be two times greater than if he holds his head still. This force is further increased because of the additional impulse produced when his momentum of approach is stopped. The increased impulse and short time of impact result in forces that account for many knockouts.
4. There is no contradiction because the best results for each are quite different. The best result for the boxer is the smallest possible force, accomplished by maximizing the time of contact, and the best result for the karate expert is the greatest possible force, accomplished by minimizing the time of contact.

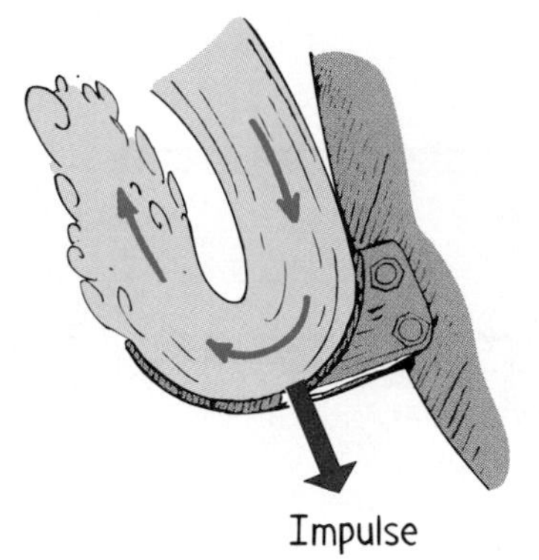

bring you to a halt—hopefully above ground level. Note how $Ft = \Delta mv$ applies here. The mv you wish to change is the momentum you have gained before the cord does its thing. The Ft is the impulse the cord must supply to reduce your mv to zero. Because the cord takes so long stretching, a large Δt ensures a correspondingly small force F on you. (Would you like to try such a jump with a nonstretchable cord? Ouch!) Bungee cords typically stretch to about twice their length, depending on momentum.

Bouncing

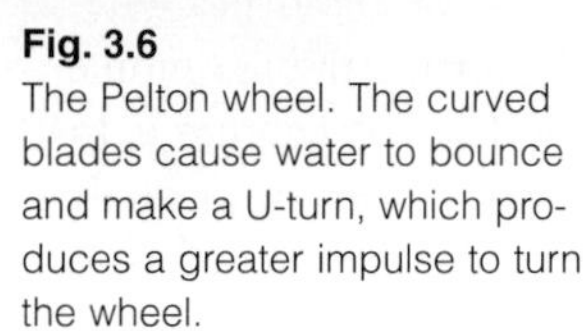

Fig. 3.6
The Pelton wheel. The curved blades cause water to bounce and make a U-turn, which produces a greater impulse to turn the wheel.

You know that if a flower pot falls from a shelf onto your head, you may be in trouble. And whether you know it or not, if it bounces from your head, you're certainly in trouble. Impulses are greater when bouncing takes place. This is because the impulse required to bring something to a stop and then, in effect, "throw it back again" (the bouncing part) is greater than the impulse required merely to bring something to a stop. Suppose you catch the falling pot with your hands. That is, you provide an impulse to catch it and reduce its momentum to zero. If you were to then throw the pot upward, you would have to provide additional impulse. So it would take more impulse to catch it and throw it back up than merely to catch it. The same greater impulse is supplied by your head if the pot bounces from it.

An interesting application of the greater impulse that occurs when bouncing takes place was employed with great success in California during the gold rush days. The water wheels used in gold-mining operations were inefficient. A man named Lester A. Pelton saw that the problem had to do with their flat paddles. He designed curved-shape paddles that would cause the incident water to make a U-turn upon impact—to "bounce." In this way the impulse exerted on the water wheels was greatly increased. Pelton patented his idea and made more money from his invention, the Pelton wheel, than most of the gold miners made from gold.

3.3 Conservation of Momentum

Newton's second law tells us that if we want to accelerate an object, we must apply a force. The force must be an external force. If you want to accelerate a car, you must push from outside. Inside forces don't count—sitting inside an automobile pushing against the dashboard, with the dashboard pushing back, has no effect in accelerating the automobile. Likewise, if we want to change the momentum of an object, the force must be external. In the impulse-momentum concept, $Ft = \Delta mv$, internal forces don't count. If no external force is present, then no change in momentum is possible.

Consider a rifle being fired. The force that drives the bullet and the force that makes the rifle recoil are equal and opposite (Newton's third law). To the system composed of the rifle and the bullet, these two forces are internal ones. No external net force acts on the rifle-bullet system, and so the total momentum of the system undergoes no net

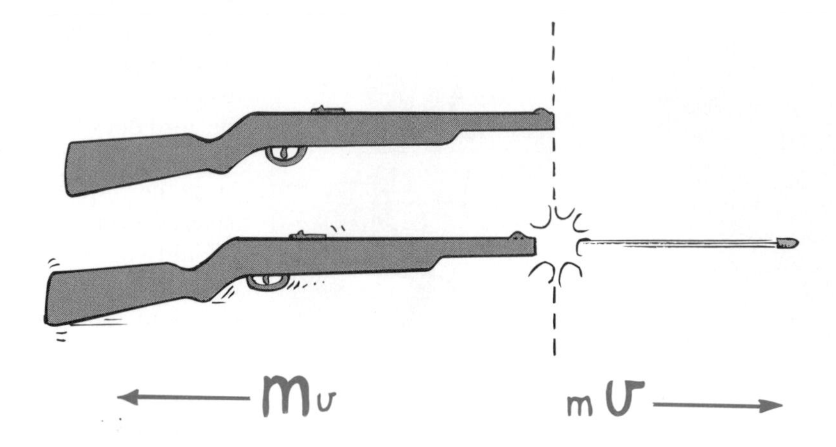

Fig. 3.7
Before firing, the net momentum of the rifle-bullet system is zero. After firing, the net momentum of the system is still zero because the leftward momentum of the rifle is equal and opposite to the rightward momentum of the bullet.

change (Figure 3.7). Before firing, the momentum is zero; after firing, the *net* momentum is still zero. Like velocity, momentum is a *vector quantity.* The momentum gained by the bullet is equal in magnitude to the momentum gained by the rifle.[2] Because their directions are opposite, they cancel. No momentum is gained and no momentum is lost.

Whenever a physical quantity remains unchanged during a process, that quantity is said to be *conserved.* We say momentum is conserved. The **conservation of momentum** is especially useful in collisions, where the forces involved are internal ones. In any collision, we can say

Net momentum before collision = net momentum after collision

When a moving billiard ball hits a resting one head-on, the first ball comes to rest and the second ball moves away with the initial velocity of the first ball. We call this an **elastic collision**; the colliding objects rebound without lasting deformation or the generation of heat. In this collision, momentum is transferred from the first ball to the second. Momentum is conserved. The collisions between molecules in a gas are perfectly elastic. Billiard balls approximate perfectly elastic collisions (Figure 3.8).

Momentum is conserved even when the colliding objects are not elastic. An **inelastic collision** is characterized by deformation, the generation of heat, or both. Sometimes an inelastic collision results in the coupling of colliding objects. Consider, for example, a freight car moving along a track and colliding with another freight car initially at rest (Figure 3.9). If the freight cars are of equal mass and are coupled by the collision, can we predict the velocity of the coupled cars after impact?

Suppose the the mass of each car is 10,000 kg and the orange car is initially moving at 10 m/s. From the conservation of momentum,

$$(\text{total } mv)_{\text{before}} = (\text{total } mv)_{\text{after}}$$

$$(10{,}000 \times 10)_{\text{before}} = (20{,}000 \times v)_{\text{after}}$$

By simple algebra, $v = 5$ m/s. This makes sense because twice as much mass is moving after the collision and so the velocity must be half as much as the velocity before collision. The two sides of the equation are then equal.

[2]Here we neglect gases ejected from exploded gunpowder, which can be significant. In 1992, actor John Eric Hexum accidentally killed himself when he held a pistol loaded with a blank bullet to his head and fired it. Although no slug emerged from the gun, exhaust gases did, which killed him. So strictly speaking, momentum of a bullet + gunpowder gases is equal to the opposite momentum of a recoiling rifle.

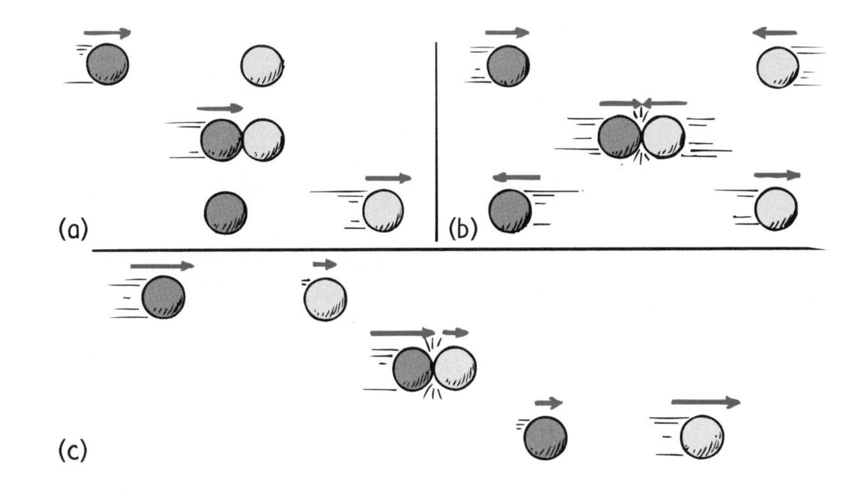

Fig. 3.8
Elastic collisions of equally massive balls. In each case, the balls exchange roles, which means that each acquires the momentum the other one had. (a) A green ball strikes a yellow ball initially at rest. (b) A head-on collision. (c) A collision between balls moving in the same direction.

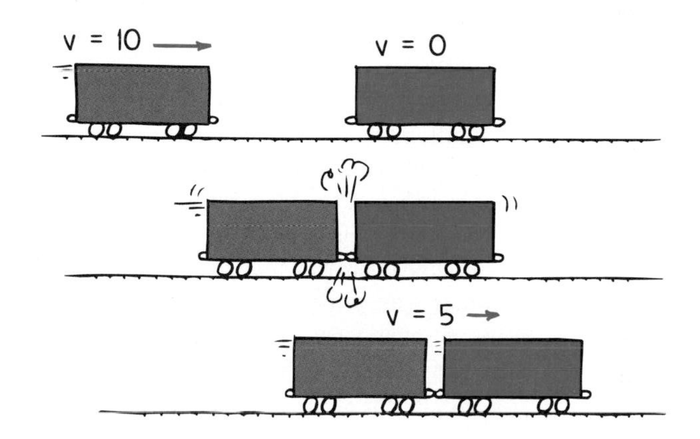

Fig. 3.9
Inelastic collision. After the collision, the two cars together have the same momentum that the freight car on the left had before the collision.

Fig. 3.10
An air track. Blasts of air from tiny holes provide a friction-free air cushion for the carts to glide upon.

■ **Question**

Consider the air track in Figure 3.10. Suppose a gliding cart that has a mass of 0.5 kg bumps into and sticks to a stationary cart that has a mass of 1.5 kg. If the speed of the gliding cart before impact is 4 m/s, how fast do the coupled carts glide after collision?

So we see that changes in an object's motion depend both on an applied force and on how long that force acts. When "how long" means time, we call the quantity "force × time" *impulse.* But "how long" can mean distance also. When we consider the quantity force × *distance,* we are talking about something entirely different—the concept of *energy.*

3.4 Energy

Energy is perhaps the concept most central to all of science. The combination of energy and matter makes up the universe: Matter is substance, and energy is the mover of substance. The idea of matter is easy to grasp. Matter is stuff we can see, smell, or feel. It has mass and occupies space. Energy, on the other hand, is abstract. We cannot see, smell, or feel most forms of energy. Surprisingly, the idea of energy was unknown to Isaac Newton, and its existence was still being debated in the 1850s. Although energy is familiar to us today, it is difficult to define because it is not only a "thing" but both a thing and a process—as if it were both a noun and a verb. Persons, places, and things all have energy, but we usually observe energy only when it undergoes change. Put another way, energy is nature's way of keeping score. We sense energy only when the score changes—either when there is a transformation from one form of energy to another (as when radiant energy from the Sun changes to thermal energy when it vibrates cells in our skin) or when there is a transfer of energy from one point to another (as when water from a lake high in the mountains rushes through an electric generator at a lower location). Interestingly, just as water is composed of tiny lumps (H_2O molecules) energy changes also occur in tiny lumps—*quanta,* which we'll study in Part 4. For the larger-scale energy changes we consider in this chapter, we'll ignore the lumpy nature of energy. (Even matter is condensed, bottled-up energy, as set forth in Einstein's famous formula, $E = mc^2$, which we'll return to in the last part of this book.)

We continue our study of energy by learning about a relatively simple and related concept: work.

■ **Answer**

According to momentum conservation,

$$(\text{total } mv)_{\text{before}} = (\text{total } mv)_{\text{after}}$$

$$(0.5\text{ kg})\,(4\text{ m/s}) = (0.5\text{ kg} + 1.5\text{ kg})v$$

$$v = \frac{(0.5\text{ kg})(4\text{ m/s})}{(0.5\text{ kg} + 1.5\text{ kg})} = \frac{2.0\text{ kg·m/s}}{2.0\text{ kg}} = 1\text{ m/s}$$

This makes sense, for four times as much mass is moving after the collision, and so this larger mass glides more slowly. In order to keep the momentum of the system unchanged, four times the mass glides $\frac{1}{4}$ as fast.

3.5 Work

Fig. 3.11
Work is done in lifting the barbell. If the weight lifter were taller, he would have to expend proportionally more energy to press the barbell over his head.

We have seen that force × time is impulse, which changes momentum. Force × distance, however, changes energy. We call the quantity "force × distance" work. When we lift a load against Earth's gravity, we do work on the load. The heavier the load or the higher we lift it, the more work we do. Two things enter into every case where work is done: (1) the exertion of a force and (2) the movement of something by that force. We define the work done on an object as the product of the applied force and the distance through which the object is moved:[3]

Work = force × distance

If we lift two loads one story up, we do twice as much work as in lifting one load the same distance, because the *force* needed to lift twice the weight is twice as much. Similarly, if we lift a load two stories high instead of one story high, we do twice as much work because the *distance* is twice as much.

A weight lifter who holds a barbell weighing 1000 newtons overhead does no work on the barbell. He may get really tired doing so, but if the barbell is not moved by the force he exerts, he does no work *on the barbell.* Work may be done on his muscles as they stretch and contract, which is force times distance on a biological scale, but this work is not done *on the barbell. Lifting* the barbell, however, is a different story. When the weight lifter raises the barbell from the floor, he does work on the barbell.

The unit of work, which combines the unit of force (N) with the unit of distance (m), is the newton-meter (N·m), also called the *joule* (J) (rhymes with *cool*). One joule of work is done when a force of 1 newton is exerted over a distance of 1 meter, as in lifting an apple over your head. For larger values we speak of kilojoules (kJ), thousands of joules, or megajoules (MJ), millions of joules. The weight lifter in Figure 3.11 does work in kilojoules. The energy released by 1 kilogram of gasoline is rated in megajoules.

Fig. 3.12
He may expend energy when he pushes on the wall, but if it doesn't move, no work is performed on the wall.

Work is done in lifting the heavy ram of a pile driver, and as a result the ram acquires the property of being able to do work on a piling beneath it when it falls. When work is done by an archer in drawing a bow, the bent bow has the ability to do work on the arrow. When work is done to wind a spring mechanism, the spring acquires the ability to do work on various gears to run a clock, ring a bell, or sound an alarm. In each case, something has been acquired by an object. This something enables the object to do work. This something may be a compression of atoms in the material of the object; it may be a physical separation of attracting bodies; it may be a rearrangement of electric charges in the molecules of a substance. This "something" that enables an object to do work is **energy**.[4] Like work, energy is measured in joules.

[3] The force in this definition is the force parallel to the distance moved. If the force acts at an angle, then work is the product of the component of force that acts in the direction of motion and the distance moved.

3.6 Forms of Mechanical Energy

Energy appears in many forms, which we shall discuss in the following chapters. For now we focus on *mechanical energy*—the form of energy due either to (1) the *position* of interacting bodies—*potential energy,* or (2) to their motion—*kinetic energy.* Mechanical energy may be in the form of either potential energy, kinetic energy, or both.

Fig. 3.13
The potential energy of Tenny's drawn bow equals the work (average force × distance) she did in drawing the arrow into position. When the bow is released, most of the potential energy will become the kinetic energy of the arrow. The rest heats the bow.

Potential Energy

An object may store energy because of its position. Such energy stored and held in readiness is called **potential energy** (PE) because in the stored state it has the potential to do work. For example, a stretched or compressed spring has the potential for doing work. When a bow is drawn, energy is stored in the bow. When a rubber band is stretched, it is capable of doing work—especially if it is part of a slingshot.

The chemical energy in fuels is potential energy, for it is energy of position from a microscopic point of view. Such energy characterizes fossil fuels, electric batteries, and the food we eat. This energy is available when atoms are rearranged, that is, when the positions of the atoms are changed. Any substance that can do work through chemical action possesses potential energy.

The easiest visualized form of potential energy occurs when work is done to elevate objects against Earth's gravity. The potential energy due to elevated position is called *gravitational potential energy.* Water in an elevated reservoir and the elevated ram of a pile driver have gravitational potential energy. The amount of gravitational potential energy possessed by an

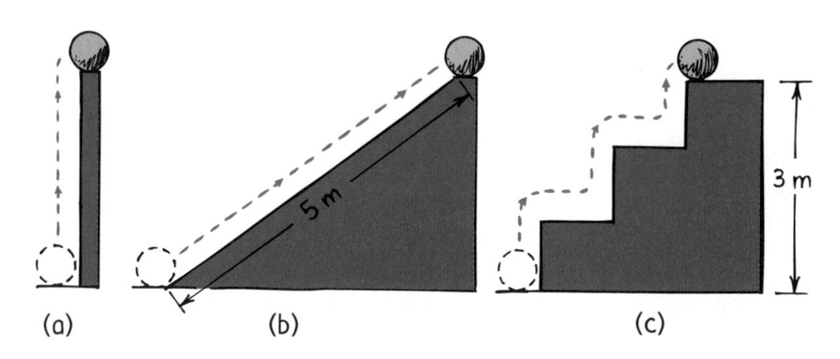

Fig. 3.14
The potential energy of the 10-N ball is the same (30 J) in all three cases because the work done in elevating it 3 m is the same whether it is (a) lifted straight up with 10 N of force, (b) pushed with 6 N of force up the 5-m incline, or (c) lifted with 10 N up each 1-m stair. No work is done in moving it horizontally (neglecting friction).

---

[4]Strictly speaking, that which enables an object to do work is its available energy, for not all the energy in an object can be transformed to work; some unavoidably dissipates as heat.

elevated object is equal to the work done against gravity in lifting it. The work done equals the force required to move it upward times the vertical distance it is moved ($W = Fd$). The upward force equals the weight mg of the object. So the work done in lifting it through a height h is given by the product mgh:

Gravitational potential energy = weight × height

$$\text{PE} = mgh$$

Note that the height h is the distance above some reference level, such as the ground or the floor of a building. The potential energy mgh is relative to that level and depends only on mg and the height h. You can see in Figure 3.14 that the potential energy of the ball at the top of the structure depends on vertical displacement only and not on the path taken to get the ball to the top.

1. How much work is done in lifting the 100-N block of ice a vertical distance of 3 m?
2. How much work is done in pushing the same block of ice up the 6-m-long ramp? The force needed is only 50 N (which is why inclines are used).
3. What is the increase in the block's potential energy in each case?

Kinetic Energy

If we push on an object, we can set it in motion. A moving object, by virtue of its motion, is capable of doing work. We call energy of motion **kinetic energy** (KE).

The kinetic energy of an object depends on that object's mass and speed. It is equal to half the mass multiplied by the square of the speed:

Kinetic energy = $\frac{1}{2}$ mass × speed²

$$\text{KE} = \frac{1}{2}mv^2$$

A car moving along a road has kinetic energy. A twice-as-heavy car moving at the same speed has twice the kinetic energy. That's because a twice-as-heavy car has twice the mass. Note how kinetic energy depends on not just plain speed, but speed multiplied by itself—*speed squared.* If you double the speed of a moving car, you increase its kinetic energy by *four* ($2^2 = 4$). Or drive three times as fast and you have *nine* times the kinetic energy ($3^2 = 9$). The fact that kinetic energy depends on the square of the speed means that small changes in speed produce large changes in kinetic energy.

Note the kinetic and potential energy conversions in Figures 3.16, 3.17, and 3.18. Got the idea? Now let's relate this to work.

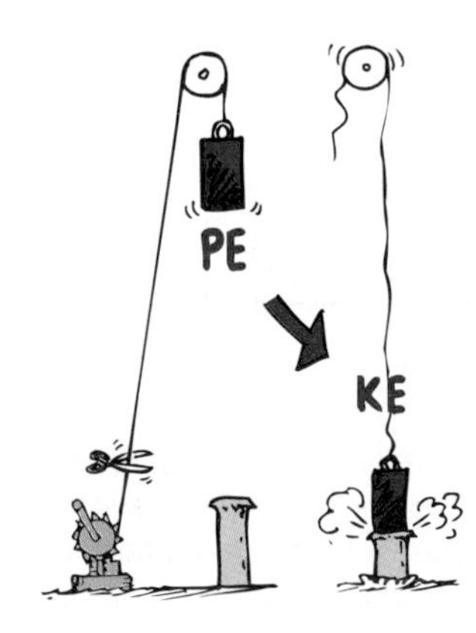

Fig. 3.15
The potential energy of the elevated ram is converted to kinetic energy when the ram is released.

Answers
1. 300 J
2. 300 J
3. Either way increases the block's potential energy by 300 J. The ramp simply makes this work easier to perform.

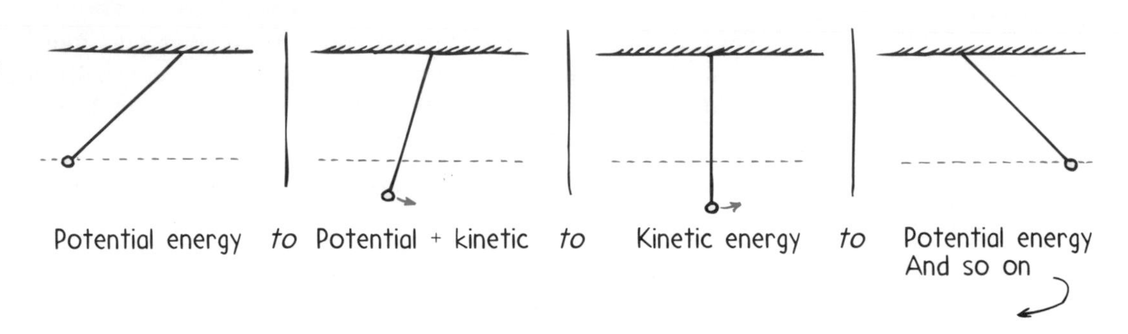

Fig. 3.16
Energy transitions in a pendulum.

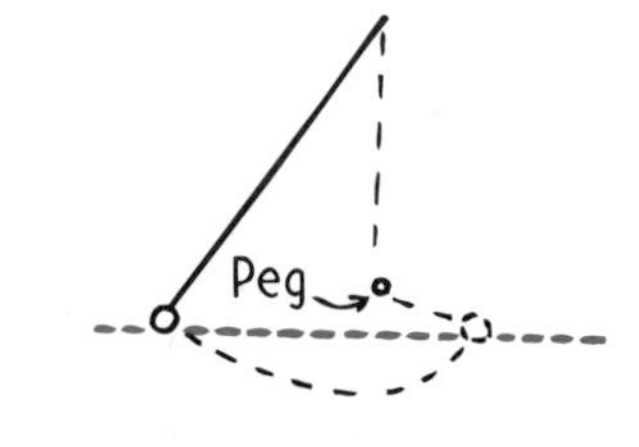

Fig. 3.17
The pendulum bob will swing to its original height whether or not the peg is present. Why?

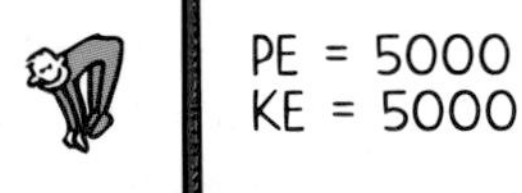

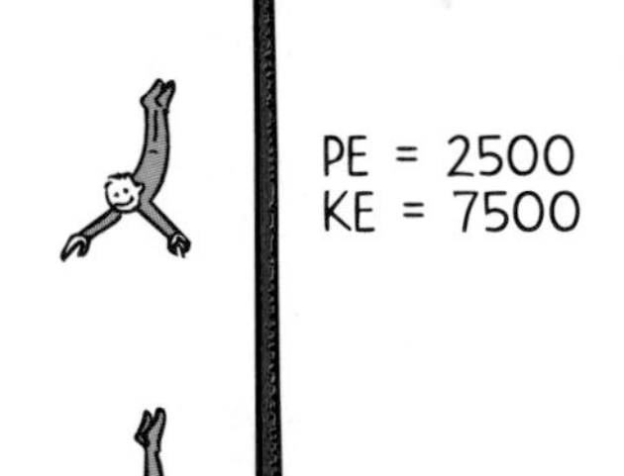

Fig. 3.18
A circus diver at the top of a pole has a potential energy of 10,000 J. As he dives, his potential energy transforms to kinetic energy. Note that at successive positions one-fourth, one-half, three-fourths, and all the way down, the total energy is constant. (Adapted from K. F. Kuhn and J. S. Faughn, *Physics in Your World*, Philadelphia: Saunders, 1980.)

3.7 Work-Energy Theorem

The kinetic energy of an object moving at a given speed is equal either to the work done in bringing the object from rest to that speed or to the work the object can do in being brought to rest. The work done is equal to the change in kinetic energy. This relationship is an important one in physics, and is called the **work-energy theorem**. We abbreviate "change in" with the delta symbol, Δ, and say

$$\textbf{Work} = \Delta KE$$

Work equals change in kinetic energy. The work in this equation is the *net* work—that is, the work based on the net force.

The work-energy theorem is not confined to changes in kinetic energy but applies to potential energy changes as well. More generally we say

$$\textbf{Work} = \Delta E$$

where ΔE is any change in mechanical energy—either kinetic or potential.

The work-energy theorem is a central concept of mechanics, emphasizing the role of energy *changes.* If there is no change in an object's energy, then we know no work was done on the object. Recall our previous example of the weight lifter and the barbell. Work was done on the barbell while its potential energy was being changed. But when it was held stationary, no further work was being done *on the barbell* as evidenced by no further change in its energy. Similarly, push against a crate on a factory floor, and if it slides, then you're doing work on it. If it doesn't, then you are not doing work on the crate. Put rollers beneath the crate and push again. Now your push does work on the crate. The amount of work done on the crate is matched by its gain in kinetic energy.

The work-energy theorem applies as well to decreasing speed. The more kinetic energy something has, the more work is required to stop it. Twice as much kinetic en-

Kinetic Energy and Momentum Compared

Momentum and kinetic energy are both properties of moving things, but they are different from each other. Like velocity, momentum is a vector quantity and is therefore directional and capable of being canceled entirely. But kinetic energy is a nonvector (scalar) quantity, like mass, and can never be canceled. The momenta of two firecrackers approaching each other may cancel, but when they explode, there is no way their energies can cancel. Energies transform to other forms; momenta do not. Another difference is the velocity dependence of the two. Whereas momentum depends on velocity (mv), kinetic energy depends on the square of velocity ($\frac{1}{2}mv^2$). An object that moves with twice the velocity of another object of the same mass has twice the momentum but four times the kinetic energy. So when a car going twice as fast crashes, it crashes with four times the energy.

If the distinction between momentum and kinetic energy is not really clear to you, you're in good company. Failure to make this distinction, when impedo was in vogue, resulted in disagreements and arguments between the best British and French physicists for two centuries.

ergy means twice as much work. We do work when we apply the brakes to slow a car. This work is the friction force supplied by the brakes multiplied by the distance over which the friction force acts.

It is interesting to note that the friction supplied by the brakes is the same whether the car moves slowly or quickly. Friction doesn't depend on speed. The variable is the *distance* of braking. Note what this means: A car moving twice as fast as another takes four times ($2^2 = 4$) as much work to stop and therefore takes four times as much distance to stop. Accident investigators are well aware that an automobile going 100 kilometers per hour has four times the kinetic energy it would have at 50 kilometers per hour. This means a car going 100 kilometers per hour will skid four times as far when its brakes are applied as it would going 50 kilometers per hour. Speed is squared for kinetic energy.

Questions

1. When the brakes of a car going 90 km/h lock, how much farther will it skid than when the brakes lock at 30 km/h?
2. Can an object have energy?
3. Can an object have work?

Answers

1. Nine times farther. The car has nine times as much energy when it travels three times as fast: $\frac{1}{2}m(3v)^2 = \frac{1}{2}m9v^2 = 9(\frac{1}{2}mv^2)$. The friction force will ordinarily be the same in either case; therefore, to do nine times the work requires nine times as much sliding distance.
2. Yes, but only in a relative sense. For example, an elevated object may possess PE relative to the ground but none relative to a point at the same elevation. Similarly, the KE of an object is relative to a frame of reference, usually taken to be the Earth's surface. We shall see later that material objects have energy of being—the congealed energy that makes up mass—$E = mc^2$. Read on!
3. No, unlike momentum or energy, work is not something an object *has.* Work is something an object *does* to some other object. An object can *do* work only if it has energy.

The Swinging Wonder

Momentum conservation is nicely demonstrated with the swinging wonder, the novel device shown in the photo. When a single ball is raised and allowed to swing into the array of other identical balls, a single ball from the other side pops out. When two balls are similarly raised and released, presto—two balls on the other side pop out. The number of balls incident on the array is always the same as the number of balls that emerge. We can see that *momentum before = momentum after.* That is, $mv = mv$, or $2mv = 2mv$, or $3mv = 3mv$, and so on. The intriguing question arises: When a single ball is raised, released, and makes impact, why cannot two balls emerge with half the speed? Or if two balls make impact, why cannot one ball emerge with twice the speed? If either of these cases occurred, the momentum before would still be equal to the momentum after: $mv = 2m(\frac{1}{2}v)$; or $2mv = m(2v)$. Intriguingly, this never happens. Nor can it happen.

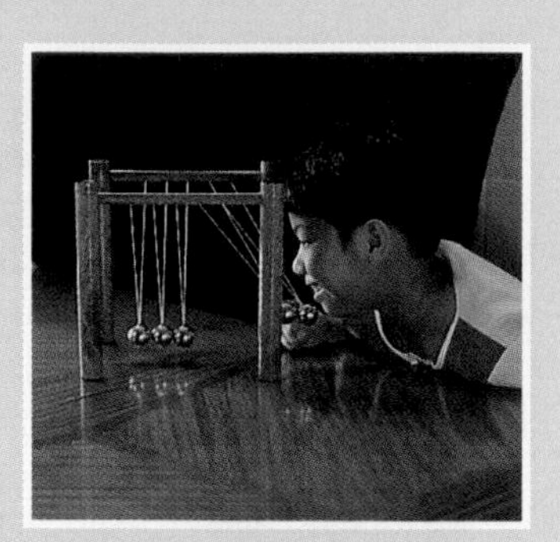

Why? Because in addition to momentum, something else must be conserved in this interaction—energy. Because the collisions are quite elastic, with very little energy transforming to heat and sound, to a good approximation the kinetic energy before equals the kinetic energy after. That is, $\frac{1}{2}mv^2_{before} = \frac{1}{2}mv^2_{after}$. Consider dropping two balls with one emerging at twice the speed. Then will $\frac{1}{2}2mv^2 = \frac{1}{2}m(2v)^2$? The answer is no! If this were to occur, there would be more energy after the collision than before (we'll leave it to you to figure how much more). Give this some thought and you'll see there is a reason why, for identical balls, the number of balls that make impact will always equal the number of balls that emerge.

In any collision, elastic or inelastic, momentum before collision equals momentum after collision. In the special case of a perfectly elastic collision, where no energy is transformed to other forms, kinetic energy before collision equals kinetic energy after collision.

Why is this device called the swinging wonder? Because the unequal-balls situation and its impossibility has left many people wondering—and wondering—and wondering. But you know the reason the number of incident and emerging balls must be the same. It's nice to know some physics!

Kinetic energy underlies other seemingly different forms of energy, such as heat, sound, and light. Random molecular motion is sensed as heat: When fast-moving molecules bump into the molecules in the surface of your skin, they transfer kinetic energy to your molecules much as balls in a game of pool or billiards. Sound consists of molecules vibrating in rhythmic patterns: Shake a group of molecules in one place and in cascade fashion they disturb neighboring molecules that in turn disturb others, preserving the rhythm of shaking throughout the medium. Electrons in motion make electric currents. Even light energy originates from the motion of electrons within atoms. Kinetic energy is far-reaching.

3.8 Conservation of Energy

More important than being able to state *what energy is* is understanding how it behaves—*how it transforms.* We can better understand processes or changes that occur in nature if we analyze them in terms of energy either transforming from one form to another or transferring from one place to another.

As we draw back the arrow in a bow, we do work in bending the bow; we give the arrow and bow potential energy. When released, most of this potential energy is transferred to the arrow as kinetic energy (the rest slightly warms the bow). The arrow in turn transfers this energy to its target, perhaps a bale of hay. The distance the arrow penetrates into the hay multiplied by the average force of impact—in other words, the work done on the hay—doesn't quite match the kinetic energy of the arrow. The energy score doesn't balance. But if we investigate further, we find that both arrow and hay are a bit warmer. By how much? By the energy difference. Energy changes from one form to another. Taking heat energy into account, we find energy transforms without net loss or net gain. Quite remarkable!

The study of various forms of energy and their transformations from one form into another has led to one of the greatest generalizations in physics—the law of **conservation of energy**:

> **Energy cannot be created or destroyed; it may be transformed from one form into another, but the total amount of energy never changes.**

When we consider any system in its entirety, whether it is as simple as a swinging pendulum or as complex as an exploding galaxy, there is one quantity that doesn't change: energy. It may change form or it may be transferred from one place to another, but, as far as we can tell, the total energy score stays the same. This energy score takes into account the fact that the atoms that make up matter are themselves concentrated bundles of energy. When the nuclei (cores) of atoms rearrange themselves, enormous amounts of energy can be released. Enormous gravitational forces in the deep, hot interior of the Sun crush the hydrogen nuclei together to form helium. This welding together of atomic cores is called *thermonuclear fusion.* This process releases radiant energy, some of which reaches the Earth as sunshine. Part of this energy falls on plants, and part of this in turn later becomes coal. Another part supports life in the food chain that begins with plants, and part of this energy later becomes oil. Part of the energy from the Sun goes into the evaporation of water from the ocean, and part of this returns to the Earth as rain that may be trapped behind a dam. By virtue of its position, the water in a dam has energy that may be used to power a generating plant below, where it will be transformed to electric energy. The energy travels through wires to homes, where it is used for lighting, heating, cooking, and operating electric gadgets. How nice that energy is conserved as it transforms from one form to another!

■ Question

Does an automobile consume more fuel when its air conditioner is on? When its lights are on? When its radio is on while it is sitting in the parking lot?

■ Answer

The answer to all three questions is yes, for the energy these devices consume ultimately comes from the fuel. Even the energy from the battery must be given back to the battery by the alternator, which is turned by the engine, which runs from the energy of the fuel. There's no free lunch!

3.9 Power

Energy can be transformed quickly or it can be transformed slowly. The rate at which energy is changed, or—to say the same thing another way—the rate at which work is done, is called **power**. We say power is equal to the amount of work done per time it takes to do it:

$$\textbf{Power} = \frac{\textbf{work done}}{\textbf{time interval}}$$

Fig. 3.19
The three main engines of a space shuttle can develop 33,000 MW of power when fuel is burned at the enormous rate of 3400 kg/s. This fuel-consumption rate is like emptying an average-size swimming pool in 20 s.

An engine of great power can do work rapidly. An automobile engine that has twice the power of another does not necessarily do twice as much work or go twice as fast as the less powerful engine. Twice the power means it will do the same amount of work in half the time or twice the work in the same time. The main advantage of a powerful automobile engine is the acceleration it can produce. It can get the automobile up to a given speed in less time than less powerful engines.

Here's another way to look at power: a liter (L) of gasoline can do a certain amount of work, but the power produced when we burn it can be any amount, depending on how *fast* it is burned. The liter may power a lawnmower for a half hour or a jet engine at 3600 times the power for a half second.

The unit of power is the joule per second (J/s), also known as the watt (in honor of James Watt, the eighteenth-century developer of the steam engine). One watt (W) of power is expended when 1 joule of work is done in 1 second. One kilowatt (kW) equals 1000 watts. One megawatt (MW) equals 1 million watts. In the United States we customarily rate engines in units of horsepower and electricity in kilowatts, but either may be used. In the metric system of units, automobiles are rated in kilowatts. (One horsepower is the same as three-fourths of a kilowatt, and so an engine rated at 134 horsepower is a 100-kW engine.)

3.10 Machines

A *machine* is a device for multiplying forces or simply changing the direction of forces. The operation of every machine ever built, from the simplest to the most complex, is governed by the conservation of energy concept. Consider one of the simplest machines, the *lever,* shown in Figure 3.20. Push one end down, and a load at the other end is pushed upward. That is, at the same time that work is done on one end of the lever, work is done on a load at the other end. If the heat from friction forces is small enough to neglect, the work input is equal to the work output:

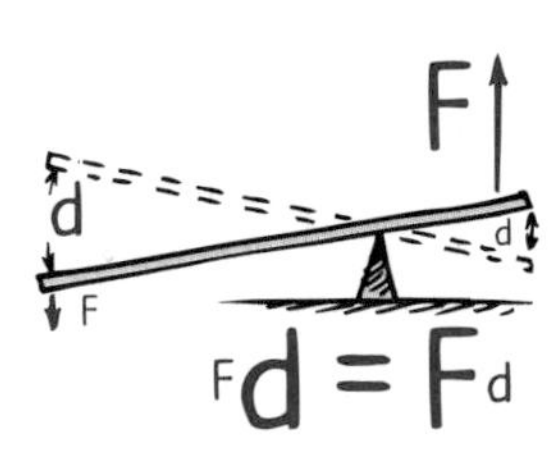

Fig. 3.20
The lever. The work (force × distance) done at one end is equal to the work done on a load at the other end.

$$\textbf{Work input} = \textbf{work output}$$

Since work equals force times distance, we can say that input force × input distance = output force × output distance.

$$(\textbf{Force} \times \textbf{distance})_{\textbf{input}} = (\textbf{force} \times \textbf{distance})_{\textbf{output}}$$

Fig. 3.21
Input force × input distance = output force × output distance.

If the pivot point of the lever, called the *fulcrum,* is relatively close to the load, then a small input force produces a large output force. This is because the input force is exerted through a large distance and the load is moved over a correspondingly short distance. In this way, a lever can multiply forces. But no machine can multiply work or multiply energy. That's a conservation of energy no-no!

Consider the ideal example illustrated in Figure 3.21. The child uses the principle of the lever as she jacks up the front end of an automobile. By exerting a small force through a large distance, she is able to provide a large force acting through a small distance. Every time she pushes the jack handle down 25 centimeters, the car rises only a hundredth as far but with 100 times the force.

A block and tackle, which is a system of pulleys, is another simple machine that multiplies force at the expense of distance. One can exert a relatively small force through a relatively large distance and lift a heavy load through a relatively short distance. In the ideal pulley system shown in Figure 3.22, the man pulls 7 meters of rope with a force of 50 newtons and lifts 500 newtons through a vertical distance of 0.7 meter. The work the man does in pulling the rope is numerically equal to the increased potential energy of the 500-newton block.

Any machine that multiplies force does so at the expense of distance. Likewise, any machine that multiplies distance, such as your forearm and elbow, does so at the expense of force. No machine or device can put out more energy than is put into it. No machine can create energy; it can only transform it from one form to another.

Efficiency

The three examples just discussed are *ideal* machines, which means that 100 percent of the work input is transformed to work output. An ideal machine operates at 100 percent efficiency. In practice, this doesn't happen, and we can never expect it to happen. In any energy transformation, some energy is always dissipated to molecular kinetic energy—heat. This makes the machine warmer and also means that its efficiency is less than 100 percent.

Efficiency can be expressed by the ratio

$$\textbf{Efficiency} = \frac{\textbf{work done}}{\textbf{energy used}}$$

Fig. 3.22
Input force × input distance = output force × output distance.

Inefficiency exists whenever energy is transformed from one form to another. Even a lever rocks about its fulcrum and converts a small fraction of the input energy into heat. We may input 100 joules of work but only get out 98 joules of useful energy. The lever is then 98 percent efficient, and we waste 2 joules of work input on heat. If the girl in Figure 3.21 puts in 100 joules of work and increases the potential energy of the car by 60 joules, the jack is 60 percent efficient; 40 joules of work has been used to do the work in overcoming the friction force, and this work appears as heat. In a pulley system, a larger fraction of input energy goes into heat. If we do 100 joules of work, the forces of friction acting through the distances through which the pulleys turn and rub about their axles may dissipate 60 joules of energy as heat. So the work output is only 40 joules, and the pulley system has an efficiency of 40 percent. The lower the efficiency of a machine, the greater the amount of energy wasted as heat.

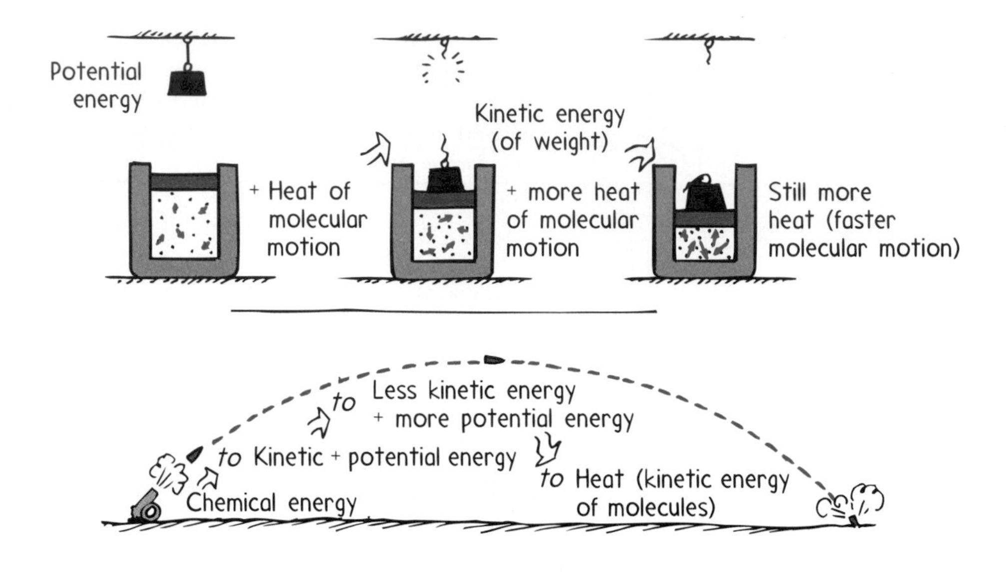

Fig. 3.23
Energy transitions. The graveyard of kinetic energy is strewn with heat.

Question

Consider an imaginary miracle car that has a 100 percent efficient engine and burns fuel that has an energy content of 40 megajoules per liter. If the air drag plus frictional forces on the car traveling at highway speed is 500 N, what is the maximum distance the car can go on 1 L of fuel?

An automobile engine is a machine that transforms chemical energy stored in fuel into mechanical energy. We'd like all this energy converted into mechanical energy; that is, we'd like an engine that is 100 percent efficient. This is never possible, however, because some of the energy goes out in the hot exhaust gases, and nearly half is wasted in the friction of the moving engine parts. In addition to these inefficiencies, the fuel doesn't burn completely, and so a certain amount of fuel energy goes unused.

Look at the inefficiency that accompanies transformations of energy this way: In any transformation there is a dilution of available *usable energy.* The amount of usable energy decreases with each transformation until nothing is left but thermal energy at low temperature. When we study thermodynamics, we'll see that energy in the form of heat is useless for doing work unless it can be transformed to heat energy at a lower temperature. Once it reaches the lowest practical temperature, that of our environment, it cannot be used. The environment around us is the graveyard of useful energy.

Answer

From the definition work = force × distance, simple rearrangement gives distance = work/force. If all 40,000,000 J of energy in 1 L is used to do the work of overcoming the air drag and frictional forces, the distance covered is

$$\textbf{Distance} = \frac{\textbf{work}}{\textbf{force}} = \frac{\textbf{40,000,000 J/L}}{\textbf{500 N}} = \textbf{80,000 m/L} = \textbf{80 km/L}$$

The important point here is that even with a perfect engine, there is an upper limit of fuel economy dictated by the conservation of energy.

3.11 Sources of Energy

Except for nuclear power, the source of practically all our energy is the Sun. This includes the energy we obtain from the combustion of petroleum, coal, natural gas, and wood, for these materials are the result of photosynthesis, a biological process that incorporates the Sun's radiant energy into plant tissue. There are many other ways of using the Sun as a source of energy. Sunlight, for example, can be directly transformed into electricity by way of photovoltaic cells, like those found in solar-powered calculators. Solar radiation can also be used indirectly to generate electricity. Sunlight evaporates water, which later falls as rain; rainwater flows into rivers and turns water wheels or modern generator turbines as it returns to the sea. Or using mirrors, solar radiation can be concentrated to heat water into steam, which can also be used to turn generator turbines. Furthermore, wind is solar energy that has already been converted into mechanical energy. The mechanical energy of wind can be used to turn generator turbines within specially equipped windmills. A problem with these sources of power is gathering it, for the energy is dilute. Concentrated energy sources such as fossil and nuclear fuels are the choice contenders for large-scale power production.

The most concentrated form of usable energy is in uranium and plutonium—nuclear fuels. While nuclear-fission power plants account for nearly 80 percent of electrical power in France, and growing, less than 20 percent of electricity in the United States comes from nuclear power—and this percentage is scheduled to diminish in coming years. Public fear about anything nuclear is high in America and relatively absent in France and other countries. The reasons for this are varied, and worthy of debate (a physical science class activity?). It is interesting to note that, as we shall see in Part 5, the Earth's interior is kept hot because of nuclear power, which has been with us since time zero.

Geothermal energy is an offspring of the heat generated by radioactivity in the Earth's interior. Geothermal energy is predominantly limited to areas of volcanic activity, such as Iceland, New Zealand, Japan, and Hawaii, where heated water near the Earth's surface is tapped to provide steam for running turbo-generators. In locations where heat from volcanic activity is near the ground surface and groundwater is absent, another method holds promise for producing electricity cheaply and cleanly: dry-rock geothermal power, where cavities are made in deep, dry, hot rock into which water is introduced. When the water turns to steam, it is piped to a turbine at the surface. Then it is returned to the cavity for re-use.

Except for geothermal power, methods for obtaining energy have serious environmental consequences. Although nuclear fission isn't a polluter of the atmosphere, it remains controversial because of the nuclear wastes generated. The combustion of fossil fuels, on the other hand, leads to increased atmospheric concentrations of carbon dioxide, sulfur dioxide, and other pollutants. Methods using solar energy are limited in that they require proper atmospheric conditions.

As impressive as solar energy may seem, it is dilute. Solar energy is the least concentrated of the contenders for large-scale power production. Nevertheless, solar energy is a practical source at smaller scales, such as powering orbiting satellites and the like.

As the world population increases, so does our need for energy. Common sense dictates that as new sources are being developed, we should continue to optimize present sources and use what we consume efficiently and wisely.

Summary of Terms

- **Momentum** The product of the mass of an object and its velocity.
- **Impulse** The product of the force acting on an object and the time during which it acts.
- **Relationship** between impulse and momentum Impulse is equal to the change in the momentum of the object that the impulse acts on. In symbol notation,

$$Ft = \Delta mv$$

- **Conservation of momentum** When no external net force acts on an object or a system of objects, no change of momentum takes place. Hence, the momentum before an event involving only internal forces is equal to the momentum after the event:

$$mv_{\text{(before event)}} = mv_{\text{(after event)}}$$

- **Elastic collision** A collision in which colliding objects rebound without lasting deformation or the generation of heat.
- **Inelastic collision** A collision in which the colliding objects become distorted, generate heat, and possibly stick together.
- **Work** The product of the force and the distance through which the force moves:

$$W = Fd$$

- **Energy** The property of a system that enables it to do work.
- **Potential energy** The stored energy that a body possesses because of its position.
- **Kinetic energy** Energy of motion, described by the relationship

$$\text{Kinetic energy} = \tfrac{1}{2}mv^2$$

- **Work-energy theorem** The work done on an object is equal to the energy gained by the object.

$$\text{Work} = \Delta E$$

- **Conservation of energy** Energy cannot be created or destroyed; it may be transformed from one form into another, but the total amount of energy never changes. In an ideal machine, where no energy is transformed into heat, $\text{work}_{\text{input}} = \text{work}_{\text{output}}$ and $(Fd)_{\text{input}} = (Fd)_{\text{output}}$.
- **Power** The time rate of work:

$$\text{Power} = \frac{\text{work done}}{\text{time interval}}$$

Efficiency The percent of the work put into a machine that is converted into useful work output.

Review Questions

Momentum

1. Which has a greater momentum, a heavy truck at rest or a moving skateboard?

Impulse Changes Momentum

2. How does impulse differ from force?
3. What are the two ways in which the impulse exerted on something can be increased?
4. Is the impulse-momentum relationship related to Newton's second law?
5. Why is it incorrect to say that impulse equals momentum?
6. For the same force, which cannon imparts the greater speed to a cannonball—a long cannon or a short one? Explain.
7. Why is it a good idea to have your hand extended forward when you are getting ready to catch a fast-moving baseball with your bare hand?
8. Consider a baseball that is caught and thrown at the same speed. Which case illustrates the greatest change in momentum: the baseball (1) being caught, (2) being thrown, or (3) being caught and then thrown back?
9. In the preceding question, in which case is the greatest impulse required?

Conservation of Momentum

10. Can you produce a net impulse on an automobile by sitting inside and pushing on the dashboard? Can the internal forces within a basketball produce an impulse on the ball to change its momentum? Explain.
11. Is it correct to say that, if no net impulse is exerted on an object, there is no change in the object's momentum?
12. What does it mean to say that a quantity is *conserved*?
13. Distinguish between an *elastic* collision and an *inelastic* collision. Why is momentum conserved for both types of collisions?
14. Railroad car A rolls at a certain speed and collides elastically with car B of the same mass. After the collision, car A is at rest. How does the speed of B after the collision compare with the initial speed of A?
15. If the equally massive cars of the previous question stick together after colliding inelastically, how does their speed after the collision compare with the initial speed of A?

Energy

16. When is energy most evident?

Work

17. A force sets an object in motion. When the force is multiplied by the time interval of its application, we call the quantity *impulse,* which changes the *momentum* of that object. What do we call the quantity *force* × *distance,* and what quantity does it change?
18. Which requires more work—lifting a 50-kg sack a vertical distance of 2 m or lifting a 25-kg sack a vertical distance of 4 m?

Forms of Mechanical Energy

19. In what units are work and energy measured?
20. A car is lifted a certain distance in a service station and therefore has potential energy with respect to the floor. If it were lifted twice as high, how much potential energy would it have?
21. Two cars are lifted to the same elevation in a service station. If one car is twice as heavy as the other, how do their potential energies compare?
22. How many joules of kinetic energy does a 1-kg book have when it is tossed across the room at a speed of 2 m/s? How much energy is imparted to the wall it accidentally encounters?
23. Which has the greater kinetic energy—a car traveling at 30 km/h or a half-as-heavy car traveling at 60 km/h?

Work-Energy Theorem

24. What is the evidence for determining whether or not work is done on an object?
25. A car moving at some initial speed accelerates until its speed is four times the initial speed. By how much does its kinetic energy increase? The brakes do a certain amount of work to stop the car. How much work must the brakes do to stop the four-times-as-fast car?

Conservation of Energy

26. Ideally, how much kinetic energy can be given to an arrow shot horizontally from a bow that has a potential energy of 40 J?
27. An apple hanging from a limb has potential energy. If it falls, what becomes of this energy just before the apple hits the ground? After it hits the ground?

Power

28. True or false: One watt is the unit of power equivalent to 1 joule per second.
29. How many watts of power are expended when a force of 1 N moves a book 2 m in a time interval of 1 s?

Machines

30. Can a machine multiply input force? Input distance? Input energy? (If your three answers are the same, seek help, for the last question is especially important.)
31. A force of 50 N applied to the end of a lever moves that end a certain distance. If the other end of the lever is moved half as far, how much force does it exert?
32. Is it possible to design a machine that has an efficiency greater than 100 percent? Discuss.

Sources of Energy

33. Explain how the energy that operates an electric toothbrush is actually the energy of sunlight.

Home Project

When you get a bit ahead in your studies, cut classes some afternoon and visit your local pool hall or billiards parlor and bone up on momentum conservation. When balls move at angles to one another, the vector nature of momentum comes into play. Note that no matter how complicated the collision of balls, two things are always true: (1) The momentum vector of the cue ball before impact (this vector lies along the line of action—the direction in which the cue ball moves) is the same as the resultant momentum vector for all balls after impact, and (2) the components of momenta perpendicular to this line of action cancel to zero after impact. The zero momentum perpendicular to the line of action is the same before and after collision; the momentum along the line of action is also the same before and after collision.

The vector nature of momentum and its conservation are seen more clearly when rotational skidding—English—is not imparted to the cue ball. When English is imparted by striking the cue ball off center, rotational momentum, which is also conserved, somewhat complicates analysis. Regardless of how the cue ball is struck, however, in the absence of external forces, both linear and rotational momentum are always conserved. Pool or billiards offers a first-rate exhibition of momentum conservation in action.

Exercises

1. Why might a wine glass survive a fall onto a carpeted floor but not onto a concrete floor?
2. To bring a supertanker to a stop at the dock, its engines are typically cut off about 25 kg from port. Why?
3. In terms of impulse and momentum, why are nylon ropes, which stretch considerably under stress, favored by mountain climbers?
4. It is generally much more difficult to stop a heavy truck than a skateboard moving at the same speed. State a case

where the moving skateboard could require more stopping force. (Consider impulses over different time intervals.)

5. Throw an egg against a wall and the egg breaks. Throw an egg at the same speed into a sagging sheet and it doesn't break. Why?
6. Why is a punch more forceful with a bare fist than with a boxing glove?
7. A boxer can punch a heavy bag for more than an hour without tiring, but tires after only a few minutes when boxing with an opponent. Why? (*Hint:* When punches are aimed at the bag, what supplies the impulse to stop them? When punches are aimed at the opponent, what or who supplies the impulse to stop the punches that miss?)
8. Railroad cars are loosely coupled so that there is a noticeable delay between the moment the first car begins to move and the moment the last car begins to move. Discuss the advisability of this loose coupling and slack between cars from an impulse-momentum point of view.

9. A fully dressed person is at rest in the middle of a pond on perfectly frictionless ice and must get to shore. How can this be accomplished?
10. Answer Exercise 31 in Chapter 2 in terms of momentum conservation.
11. Why is it difficult for a fire fighter to hold a hose that ejects large amounts of water at high speed?
12. Your friend says the conservation of momentum is violated when you step off a chair and gain momentum as you fall. What do you say?
13. If a Mack truck and a Honda Civic have a head-on collision, which vehicle experiences the greater force of impact? The greater impulse? The greater change in its momentum? The greater acceleration?
14. Will a head-on collision between two cars be more damaging to the occupants if the cars stick together upon impact or if they rebound?
15. Suppose there are three astronauts outside a spaceship, and two of them decide to play catch, using the third as their "ball." All the astronauts weigh the same on Earth and are equally strong. The first astronaut throws the "ball" toward the second one, and the game begins. Describe the motion of the astronauts as the game proceeds. How long will the game last?

16. Can momenta cancel? Can energies cancel? Why are your answers different from each other?
17. When a rifle that has a long barrel is fired, the force of expanding gases acts on the bullet for a longer distance than in a short-barreled rifle. What effect does this have on the velocity of the emerging bullet? (Do you see why long-range cannons have such long barrels?)
18. Drop an object and it falls, gaining speed as time goes by. Identify (a) the work being done on the falling object, and (b) the impulse that changes its momentum.
19. The momentum of a skydiver increases as speed increases when falling. Does the momentum of the Earth simultaneously change? How much and in what direction? How does the system made up of the Earth and the skydiver change?
20. To determine the potential energy of Tenny's drawn bow (Figure 3.13), would it be an underestimate or an overestimate to multiply the force with which she holds the arrow in its drawn position by the distance she pulled it? (Why do we say the work done is the *average* force $\times$ distance?)
21. You and a flight attendant toss a ball back and forth in an airplane in flight. Does the KE of the ball depend on the speed of the airplane? Carefully explain.
22. Can something have energy without having momentum? Explain. Can something have momentum without having energy? Defend your answer.
23. If a golf ball and a Ping Pong ball both move with the same kinetic energy, can you say which has the greater speed? Explain in terms of KE. Similarly, in a gaseous mixture where massive molecules and light molecules all have the same average KE, can you say which have the greater speed?
24. You have a friend who says that, after a golf ball collides with a bowling ball at rest, although the speed gained by the bowling ball is very small, its momentum exceeds the initial momentum of the golf ball. Your friend further asserts this is related to the "negative" momentum of the recoiling golf ball after collision. Another friend says this is hogwash—that momentum conservation would be violated. Which friend do you agree with?
25. At what point in its motion is the KE of a pendulum bob a maximum? At what point is the bob's PE a maximum? When its KE is half its maximum value, how much PE does it have?
26. A physics instructor demonstrates energy conservation by releasing a heavy pendulum bob, as shown in the sketch, allowing it to swing to and fro. What would happen if in his exuberance he gave the bob a slight shove as it left his nose? Why?

27. No work is done on an object unless there is a force, or some component of a force, along its direction of motion. This being so, does the string that supports a pendulum bob do work on the bob as it swings to and fro? Explain. Does the force of gravity do work on the bob? Explain.

28. Discuss the design of the roller coaster shown in the sketch in terms of the conservation of energy.

29. Consider the identical balls released from rest on tracks A and B as shown. When the balls reach the right ends of the tracks, which will have the greater speed? Why is this question easier to answer than the similar question asked in Exercise 19 back in Chapter 1?

30. Does a car burn more gasoline when its lights are on? When the radio is playing? Does the overall consumption of gasoline depend on whether or not the engine is running? Defend your answer.

31. You tell your friend that no machine can possibly put out more energy than is put into it, and your friend states that a nuclear reactor puts out more energy than is put into it. What do you say?

32. This may seem like an easy question for a physics type to answer: With what force does a rock that weighs 10 N strike the ground if dropped from a rest position 10 m high? This question does not have a straightforward numerical answer. Why?

33. In the hydraulic machine shown, the large piston is raised 1 cm when the small piston is pushed down 10 cm. If the small piston is pushed down with a force of 100 N, how much force is the large piston capable of exerting?

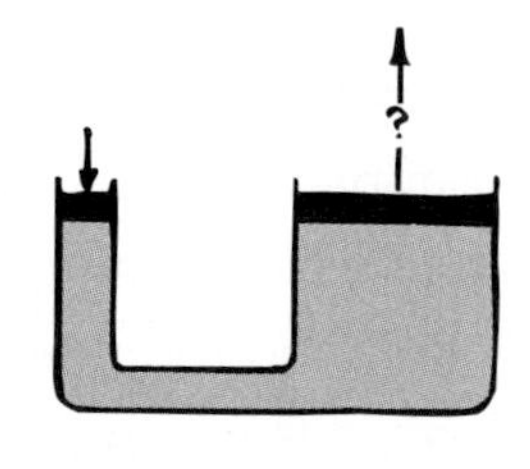

34. Consider the inelastic collision between the two freight cars in Figure 3.9. The momentum before and after the collision is the same. The KE, however, is less after the collision than before. How much less, and what becomes of this energy?

35. If an automobile had a 100 percent efficient engine, would it be warm to your touch? Would its exhaust heat the surrounding air? Would it make any noise? Would it vibrate? Would any of its fuel go unused?

36. Which requires more work to stop—a light truck or a heavy truck moving with the same momentum?

Problems

1. A railroad engine weighs four times as much as a freight car. If the engine coasts at 5 km per hour into a freight car that is initially at rest, how fast do the two coast after they couple together?
2. A 5-kg fish swimming at 1 m/s swallows an absent-minded 1-kg fish at rest. (a) What is the speed of the larger fish after lunch? (b) What would be its speed if the smaller fish were swimming toward it at 4 m/s?
3. This question is typical on some driver's license exams: A car moving at 50 km/h skids 15 m with locked brakes. How far will the car skid with locked brakes at 150 km/h?
4. How many kilometers per liter will a car obtain if its engine is 25 percent efficient and it encounters an average retarding force of 1000 N? Assume that the energy content of gasoline is 40 MJ/L.
5. A car that has a mass of 1000 kg moves at 20 m/s. What braking force is needed to bring the car to a halt in 10 s?
6. What is the efficiency of a pulley system that raises a 1000-N load 1 m when 3000 J of effort is involved?
7. What is the efficiency of the body when a cyclist expends 1000 W of power to deliver mechanical energy to her bicycle at the rate of 100 W?
8. Your monthly electric bill is probably expressed in kilowatt-hours (KW h), a unit of energy delivered by the flow of 1 kW of electricity for 1 h. How many joules of energy do you get when you buy 1 kW h?
9. The decrease of PE for a freely falling object equals its gain in KE, in accordance with the law of conservation of energy. (a) By simple algebra, find an equation for an object's speed v after falling a vertical distance h. Do this by equating KE to its change of PE. (b) Then figure out how much higher a freely falling object must fall to have twice the speed when it hits the ground.
10. Using the definitions of momentum, $p = mv$, and kinetic energy, $\text{KE} = (\frac{1}{2})mv^2$, show by algebraic manipulation that you can write $\text{KE} = \frac{p^2}{2m}$. This equation tells us that if two objects have the same momentum, the one of less mass has the greater kinetic energy (see Exercise 36).

CHAPTER

4

Gravity and Satellite Motion

It would be erroneous to say that Isaac Newton discovered gravity, for that discovery goes all the way back to when Earth dwellers first felt the consequences of tripping and falling. What Newton discovered was that gravity is universal—that it's not a phenomenon unique to the Earth, as his contemporaries had supposed.

From the time of Aristotle, the circular motions of heavenly bodies were regarded as natural. The ancients believed that the stars, planets, and Moon moved in divine circles, free from any impressed forces. Newton, however, recognized that a force must be acting on the planets; otherwise, their paths would be straight lines. And whereas others of his time, influenced by Aristotle, said that any such force would be directed along the planets' motion, Newton reasoned it must be perpendicular to their motion, directed toward the center of their curved paths—toward the Sun. This was the force of gravity, the same force that pulls apples off trees. Newton's stroke of intuition, that the force between the Earth and apples is the same force that pulls moons and planets and everything else in our universe, was a revolutionary break with the prevailing notion that there were two sets of natural laws, one for earthly events and another altogether for motions in the heavens.

4.1 The Law of Universal Gravitation

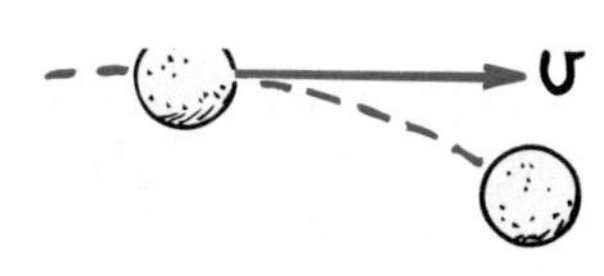

Fig. 4.1
If the Moon did not fall, it would follow the straight-line path. Because of its attraction to the Earth, it falls along a curved path.

According to popular legend, Newton was sitting under an apple tree when he got the idea that gravity extends beyond the Earth. Perhaps he looked up through tree branches toward the origin of a falling apple and noticed the Moon. In any event, he had the insight to see that the force between the Earth and a falling apple is the same force that pulls the Moon in an orbital path around the Earth, a path similar to a planet's path around the Sun.

To test this hypothesis, Newton compared the fall of an apple with the fall of the Moon. He realized that the Moon falls in the sense that *it falls away from the straight line it would follow if there were no forces acting on it.* Because of its horizontal speed, it "falls around" the round Earth. By simple geometry, the distance the Moon falls per second could be compared to the same distance from the Earth that an apple or anything else that far away would fall in 1 second. Newton's calculations didn't check. Disappointed but recognizing that brute fact must always win over beautiful hypothesis, he placed his papers in a drawer, where they remained for nearly 20 years. During this period he founded and developed the field of geometric optics for which he first became famous.

Newton's interest in mechanics was rekindled with the appearance of a spectacular comet in 1680 and another 2 years later. He returned to the Moon problem at the prodding of his astronomer friend Edmond Halley, for whom the second comet was later named. Newton made corrections in the experimental data used in his earlier method and obtained excellent results. Only then did he publish what is one of the most far-reaching generalizations of the human mind: the **law of universal gravitation**.[1]

Everything in the universe pulls on everything else in a beautifully simple way that involves only mass and distance. According to Newton, every mass attracts every other mass with a force that, for any two masses, is directly proportional to the product of the masses and inversely proportional to the square of the distance separating them. This statement can be expressed symbolically as

$$F \sim \frac{m_1 m_2}{d^2}$$

where m_1 and m_2 are the masses and d is the distance between their centers. Thus the greater the masses m_1 and m_2, the *greater* the force of attraction between them—in direct proportion to the masses.[2] The greater the distance of separation *d,* the *weaker* the

---

[1]This is a dramatic example of the painstaking effort and cross-checking that go into the formulation of a scientific theory. Contrast Newton's approach with the failure to "do one's homework," the hasty judgments, and the absence of cross-checking that so often characterize the pronouncements of people advocating less-than-scientific theories.

[2]Thus far we have treated mass as a measure of inertia, which is called *inertial mass.* Now we see mass as a measure of gravitational force, which in this context is called *gravitational mass.* It is experimentally established that the two are equal, and, as a matter of principle, the equivalence of inertial and gravitational mass is the foundation of Einstein's general theory of relativity.

force of attraction—in inverse proportion to the square of the distance between their centers of mass.

The Universal Gravitational Constant, G

The proportionality form of the law of universal gravitation can be expressed as an equation when the constant of proportionality G, called the *universal gravitational constant,* is introduced:

$$F = G\frac{m_1 m_2}{d^2}$$

In words, the force of gravity between two objects is found by multiplying their masses, dividing by the square of the distance between their centers, and then multiplying this result by the constant G. The magnitude of G is the same as the magnitude of the force between two 1-kilogram masses 1 meter apart: 0.0000000000667 newton. This small magnitude of G indicates an extremely weak force. In standard units, G has this same numerical value, expressed in scientific notation as[3]

$$G = 6.67 \times 10^{-11}\ \text{N·m}^2/\text{kg}^2$$

The constant G was first measured in the eighteenth century, long after the time of Newton. The measurement was made by the English physicist Henry Cavendish, who used an extremely sensitive torsion balance to measure the tiny force between two lead masses. A simpler method was later developed by Philipp von Jolly, who attached a spherical flask of mercury to one arm of a sensitive balance (Figure 4.2). After the balance was put in equilibrium, a 6-ton lead sphere was rolled beneath the mercury flask. The gravitational force between the two masses was equal to the weight that had to be placed on the opposite end of the balance to restore equilibrium. All the quantities m_1, m_2, F, and d were known, and so the ratio G was easily calculated:

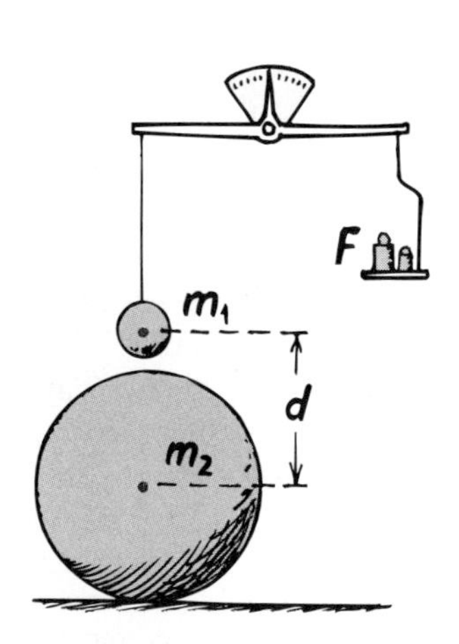

Fig. 4.2
Jolly's method of measuring G. The spherical flask m_1 and the lead sphere m_2 attract each other with a force F equal to the weights needed to restore the balance to its equilibrium position.

$$\frac{F}{m_1 m_2/d^2} = 6.67 \times 10^{-11}\ \text{N/kg}^2/\text{m}^2 = 6.67 \times 10^{-11}\ \text{N·m}^2/\text{kg}^2$$

The value of G tells us that the force of gravity is a very weak force. It is the weakest of the presently known four fundamental forces. (The other three are the electromagnetic force and two kinds of nuclear forces.) We sense the force of gravitation only when masses like that of the Earth are involved. The force of attraction between you and a large ship on which you stand, for example, is too weak for ordinary measurement. The force of attraction between you and the Earth, however, can be measured. It is your weight.

Your weight depends on your distance from the center of the Earth. At the top of a mountain, your mass is no different from what it is anywhere else, but your weight is slightly less than at ground level because your distance from the center of the Earth is greater.

Once the value of G was known, the mass of the Earth was easily calculated. The force that the Earth exerts on a 1-kilogram mass placed at its surface is 9.8 newtons.

---

[3]The numerical value of G depends entirely on the units we choose for mass, distance, and time. The international system of choice uses kilogram for mass, meter for distance, and second for time. Scientific notation is discussed in Appendix A.

The distance between the 1-kilogram mass and the center of mass of the Earth is the Earth's radius, 6.4×10^6 meters. Therefore, from $F = G(m^1\, m^2/d^2)$, where m^1 is the mass of the Earth,

$$9.8 \text{ N} = 6.67 \times 10^{-11} \text{ N·m}^2/\text{kg}^2 \frac{1 \text{ kg} \times m_1}{(6.4 \times 10^6 \text{ m})^2}$$

from which the mass of the Earth is calculated to be $m_1 = 6 \times 10^{24}$ kilograms.

■ Question

If there is an attractive force between all objects, why do we not feel ourselves gravitating toward massive buildings in our vicinity?

Gravity and Distance: The Inverse-Square Law

We can better understand how gravity is diluted with distance by considering how paint from a spray can spreads with increasing distance from the can (Figure 4.3). Suppose we position the can at the center of a sphere that has a radius of 1 meter and press the nozzle. A burst of paint spray travels that 1 meter and coats the inside of the sphere with a square patch of paint 1 millimeter thick. How thick would the patch be if the experiment were done in a sphere having twice the radius? In this case, the same amount of paint would travel in a straight line for 2 meters. It would spread to a patch twice as tall and twice as wide, which means it would be spread over an area four times as big, and so its thickness would be only $\frac{1}{4}$ millimeter. Can you see from the figure that, for a sphere of radius 3 meters, the thickness of the paint patch would be only $\frac{1}{9}$ millimeter? Can you see that the thickness of the paint decreases as the square of the distance increases? This is known as the **inverse-square law**. The inverse-square law holds for gravity and for all other phenomena wherein the effect from a localized source spreads uniformly throughout the surrounding space: the electric field about an isolated electron, light from a match, radiation from a piece of uranium, and sound from a cricket.

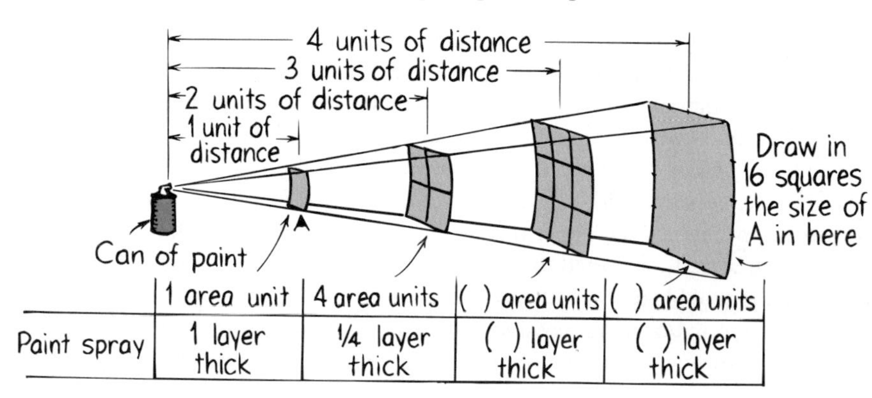

Fig. 4.3
The inverse-square law. Paint spray travels radially away from the nozzle of the can in straight lines. Like gravity, the "strength" of the spray obeys the inverse-square law.

■ **Answer**

Gravity pulls us to massive buildings and to everything else in the universe. Physicist Paul A. M. Dirac, winner of the 1933 Nobel Prize, put it this way: "Pick a flower on Earth and you move the farthest star!" *How much* we are influenced by buildings or *how much* interaction there is between flowers and stars is another story. The forces between us and buildings are relatively small and therefore not noticeable because the masses are small relative to the mass of the Earth. These tiny forces are overwhelmed by the Earth's overpowering attraction.

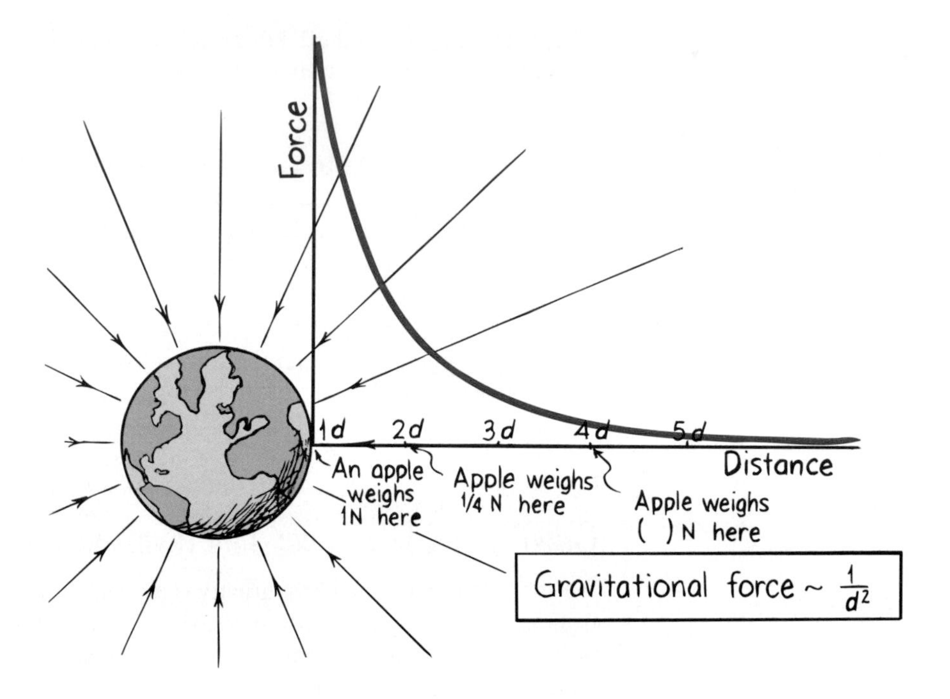

Fig. 4.4
If an apple weighs 1 N at the Earth's surface, it weighs only $\frac{1}{4}$ N twice as far from the center of the Earth. At three times the distance, it weighs only $\frac{1}{9}$ N. What would it weigh at four times the distance? Five times? Gravitational force versus distance is plotted in color.

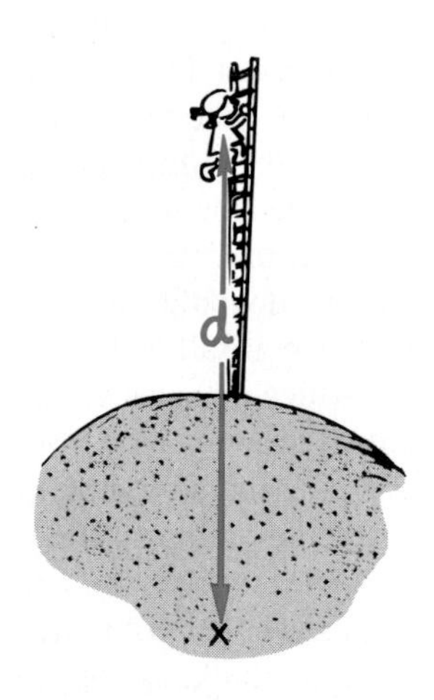

Fig. 4.5
According to Newton's equation, the child's weight (not her mass) decreases as she increases her distance from the Earth's center (not surface).

It is important to emphasize that the distance term *d* in Newton's equation for gravity is the distance between the centers of masses of the objects attracted to each other. Note in Figure 4.4 that an apple that normally weighs 1 newton at the Earth's surface weighs only $\frac{1}{4}$ as much when it is twice the distance from the Earth's center. The greater the distance from the Earth's center, the less the weight of an object. A child that weighs 300 newtons at sea level weighs only 299 newtons atop Mt. Everest. The Earth's gravitational force approaches, but never reaches, zero—no matter how far away a body gets. Even if you were transported to the far reaches of the universe, the gravitational influence of home would still be with you. It may be overwhelmed by the gravitational influences of nearer and/or more massive bodies, but it would still be there. The gravitational influence of every material object, however small or however far, is exerted through all of space.

Movie and video coverage of astronauts in Earth orbit gives the impression they are free from gravity. Not so, for as we shall soon discuss, they are hurling around the Earth in a continuous state of free fall. If they experienced no gravity, they'd fly away in straight-line paths in accord with the law of inertia.

Questions

1. By how much does the gravitational force between two objects decrease when the distance between their centers is doubled? Tripled? Increased tenfold?
2. Consider an apple that is at the top of a tree and is being pulled by the Earth's gravity with a force of 1 N. If the tree were twice as tall, would the force of gravity be only $\frac{1}{4}$ as strong? Defend your answer.

Link to Geology—Low-Density Floating Mountains

The Earth's gravitation is less atop a mountain because of the greater distance to the Earth's center, in accord with the law of universal gravitation. Interestingly, gravitation at the Earth's surface also varies slightly due to varying densities of underlying rock—valued information to geologists and oil prospectors.

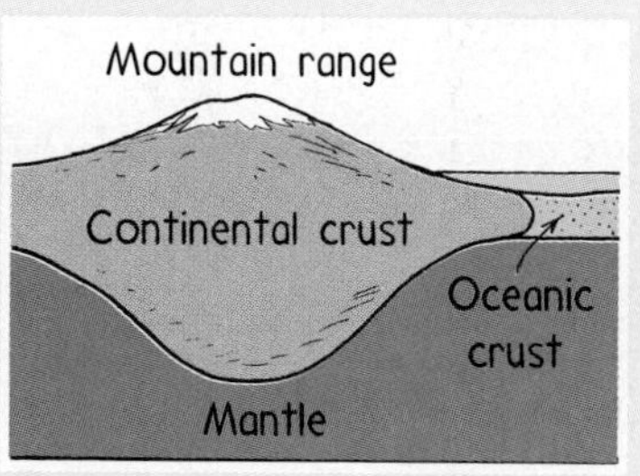

An additional reason for weaker gravitation atop a mountain has to do with the density of mountain material. Mountains are less dense than the semiliquid part of the Earth they float on—the mantle. Ice floats on water for the same reason—it is less dense than water. So just as the bottom of icebergs extend far beneath the surface of water, the bottom of mountains extend far into the mantle. The result is a greater distance between the top of mountains and the greater-density parts of the Earth beneath. This increased "gap" further reduces the strength of gravitation at the top of mountains.

4.2 Tides

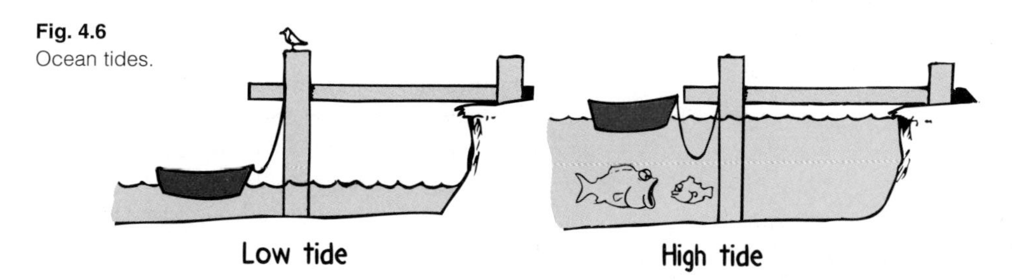

Fig. 4.6
Ocean tides.

Seafaring people have always known there is a connection between the ocean tides and the Moon. Newton was the first to show that tides are caused by *differences* in the gravitational pull exerted by the Moon on the Earth. The Moon's gravitational force is strongest on the side of the Earth nearest the Moon and weakest on the side farthest from the Moon.

To understand why differences in the strength of the Moon's gravitational pull produce tides, consider a spherical ball of Jell-O. If you exert the same force in the same direction on every part of the ball, it remains a sphere as it accelerates. But if you pull harder on one side than on the other, there is a difference in accelerations, with the result that the ball elongates (Figure 4.7). That's what is happening to this big ball we're living on. The different strengths of the Moon's pull on opposite sides of the Earth cause the Earth to stretch, and this stretching produces the ocean bulges that extend nearly 1 meter above the average surface level of the ocean. The Earth spins once per day, and so a given point on the Earth passes beneath both of these bulges each day.

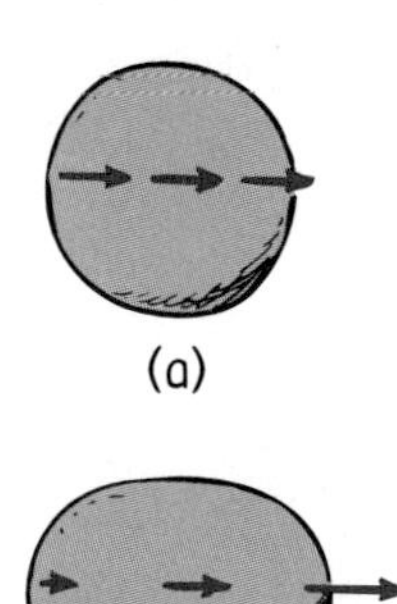

Fig. 4.7
(a) A ball of Jell-O remains spherical when different parts of it are pulled equally in the same direction. (b) When one side is pulled more than the other, however, the shape is elongated.

Answers

1. It decreases to one-fourth, one-ninth, and one-hundredth of its initial value.
2. No, because the twice-as-tall apple tree is not twice as far from the Earth's center. The taller tree would have to have a height equal to the radius of the Earth (6370 km) before the weight of the apple at its top reduced to $\frac{1}{4}$ N. Before its weight decreases by 1%, an apple or any object must be raised 32 km—nearly four times the height of Mt. Everest. So as a practical matter, we disregard the effects of everyday changes in elevation.

Force Fields

We know the Earth and Moon pull on each other. This type of interaction is called *action at a distance*, because the two bodies interact with each other without touching. We can look at this in a different way, however: We can regard the Moon as in being in contact with the *gravitational field* of the Earth. A **gravitational field** is the space surrounding a body in which another body experiences an attractive force. A gravitational field is an example of a **force field**; any body in the field experiences a force. It is common to think of rockets and distant space probes as being influenced by the gravitational field right where they are in space rather than by Earth and other planets or stars acting from a distance. The field concept plays an in-between role in our thinking about the forces between bodies.

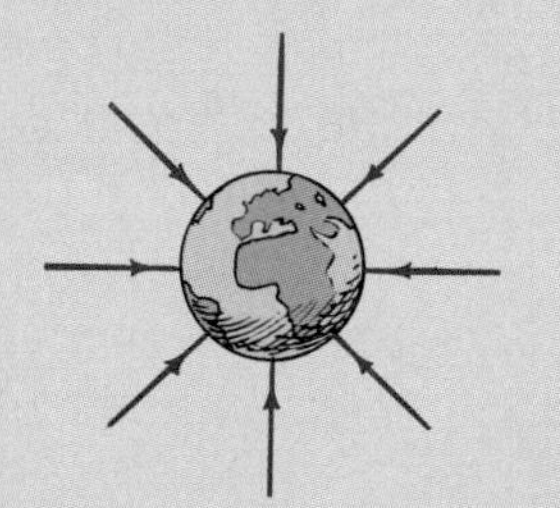

A more familiar force field is the *magnetic field* of a magnet. Iron filings sprinkled over a sheet of paper on top of the magnet reveal the shape of the magnet's magnetic field. The pattern of filings shows the strength and direction of the magnetic field at different locations around the magnet. Where the filings are close together, the field is strong. The direction of the filings show the direction of the field at each point. Planet Earth is a giant magnet, and like all magnets is surrounded by a magnetic field. Evidence of the field is easily seen by the orientation of a magnetic compass.

The pattern of the Earth's gravitational field can be represented by field lines. Like the iron filings around a magnet, the field lines are closer together where the gravitational field is stronger. At each point on a field line, the direction of the field at that point is along the line. Arrows show the field direction. A particle, astronaut, spaceship, or any other body in the vicinity of the Earth is accelerated in the direction of the field line at that location. The strength of the Earth's gravitational field, like the strength of its force on objects, follows the inverse-square law. It is strongest near the Earth's surface and weakens with increased distance from the Earth.

Another example of a force field is the one that surrounds electrical charges—the electric field, which we shall study in Chapter 8. In Chapter 9 we'll learn how magnets align with the magnetic fields of the Earth to become compasses. And then in Chapter 28 we'll learn how the Moon similarly aligns with the Earth's gravitational field, which is why only one side of the Moon faces us. Force fields are far-reaching.

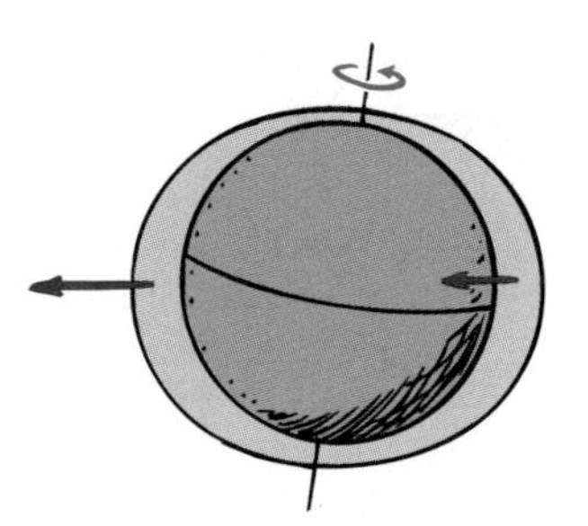

Fig. 4.8
Two tidal bulges remain relatively fixed with respect to the Moon while the Earth spins daily beneath them.

This produces two sets of ocean tides per day—two high and two low. It turns out that while the Earth spins, the Moon moves in its orbit and appears at the same position in our sky every 24 hours and 50 minutes. For this reason, the two-high-tide cycle is actually at 24-hour-and-50 minute intervals. This means tides do not occur at the same time every day.

Fig. 4.9
When the attractions of the Sun and the Moon are lined up with each other, spring tides occur.

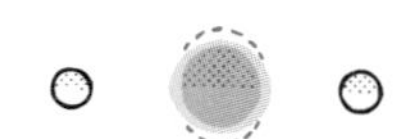

Fig. 4.10
When the attractions of the Sun and the Moon are about 90° apart (at the time of a half moon), neap tides occur.

The Sun also contributes to ocean tides but is only about half as effective as the Moon—even though the Sun pulls 180 times more on the Earth than the Moon pulls. Why aren't tides due to the Sun 180 times greater than lunar tides? Because the *difference* in gravitational pulls on opposite sides of the Earth is very small for the Sun (only about 0.017 percent, compared to 6.7 percent across the Earth by the Moon).

When the Sun, Earth, and Moon are all lined up, as shown in Figure 4.9, the tides due to the Sun and the Moon coincide and we have higher-than-average high tides and lower-than-average low tides. These are called *spring tides*. (Spring tides have nothing to do with the spring season.) Spring tides occur at the times of a new or full Moon. When the Moon is half way between new and full, as shown in Figure 4.10, the solar and lunar tides partly cancel each other. Then high tides are lower than average and low tides higher than average. These are called *neap tides*.

Because much of the Earth is molten, we have Earth tides, though they are less pronounced than ocean tides. There are also atmospheric tides, which regulate the cosmic rays that reach the Earth's surface. Our treatment of tides is quite simplified, for the tilt of the Earth's axis, interfering land masses, friction with the ocean bottom, and other factors complicate tidal motions.

Newton deduced that the difference in gravitational pulls between two bodies decreases as the *cube* of the distance between the centers of the bodies—twice as far away produces $\frac{1}{8}$ the pull; three times as far, only $\frac{1}{27}$ the pull, and so on. (The *Conceptual Physical Science Practice Book* treats this cube relationship further.) Only relatively close distances result in appreciable tides, and so the nearby Moon "out-tides" the enormously more massive but farther-away Sun. Tide height also depends on the size of the body having tides. Although the Moon produces a considerable tide in the Earth's oceans, which are thousands of kilometers apart, it produces scarcely any in a lake. That's because no part of the lake is significantly closer to the Moon than any other part, so there is no significant *difference* in the Moon's pulls on the lake. Similarly for the fluids in your body. Any tides in these fluids caused by the Moon are negligible. You're not tall enough for tides. What microtides the Moon may produce in your body are only about $\frac{1}{200}$ of the tides produced by a 1-kilogram melon held 1 meter above your head!

Fig. 4.11
The tidal force difference due to a 1-kg body 1 m over the head of a person of average height is about 60 trillionths N/kg. For our Moon overhead, it is about 0.3 trillionths N/kg. So holding a melon over your head produces about 200 times as much tidal effect in your body as the Moon.

Link to Power Production—Tidal Power

The ocean tides have been producing electricity since 1968 in Brittany, France. A short distance inland on a river that empties into the ocean, a large dam has been built in the town of La Rance. The dam gets its power from the rising and falling of the tides. First the water is higher on one side of the dam, and its level is maintained at 3 meters higher than the level on the lower side. Water flows through a tube to the lower side, turning 24 huge turbines in the process. When the tide changes, the flow of water is in the reverse direction, again turning the turbines.

The dam produces more than 200 MW of power for a city of 300,000 people and also supports a highway that allows traffic to cross the water.

4.3 Weight and Weightlessness

When you step on a spring balance, such as a bathroom scale, you compress a spring inside. When the pointer stops, the strong electrical forces between molecules inside the spring material balance the gravitational attraction between you and the Earth—with the result that nothing moves. You and the scale are in static equilibrium. The pointer is calibrated to show your weight. If you stood on a bathroom scale in an accelerating elevator, you'd find variations in your weight. If the elevator accelerated upward, the springs inside the scale would be more compressed and your weight reading would increase. If the elevator accelerated downward, the springs inside the scale would be less compressed and your weight reading would decrease. If the elevator cable broke and the elevator fell freely, the reading on the scale would go to zero. According to the reading, you would be weightless. Would you really be weightless? We can answer this question only if we agree on what we mean by *weight.*

Recall that in Chapter 2 we defined weight as the gravitational force exerted on an object by the nearest most massive body. According to this definition, you have weight whether or not you are falling because you are still gravitationally attracted to the Earth

Fig. 4.12
The sensation of weight (your apparent weight) equals the force with which you press against the supporting floor or scale. If the scale accelerates up or down, your apparent weight varies (while your weight *mg* remains the same).

during the fall. So your weight and the weight you experience, your *apparent weight*, can be very different from each other. We define **apparent weight** as the force an object exerts against a supporting floor or a weighing scale. According to this definition, you are as heavy as you feel; so in an elevator that accelerates downward, the supporting force of the floor is less and your apparent weight is less. If the elevator is in free-fall, your apparent weight is zero (Figure 4.12). Even in this weightless condition, however, there is still a gravitational force acting on you, causing your downward acceleration. But gravity now is not felt as weight because there is no support force.

Fig. 4.13
Both are "weightless."

Consider an astronaut in orbit. She feels weightless because she is not supported by anything. There would be no compression in the springs of a bathroom scale placed beneath her feet because the bathroom scale is falling as fast as she is. (We'll see shortly that a satellite is an object in free-fall; its sideways motion is great enough to ensure it falls around the Earth rather than into it. In the case we're discussing here, the astronaut and scale are both satellites.) If she drops a couple of objects in her vicinity, she drops with them and they remain in her vicinity, unlike what happens on the ground. Local effects of gravity seem to be eliminated. The body organs respond as though gravity forces were absent, and this gives the sensation of weightlessness. The astronaut experiences the same sensation in orbit that she would feel in a falling elevator—a state of free-fall.

On the other hand, if the astronaut were in deep space far removed from any attracting objects, she'd experience the sensation of weight if her spaceship were being accelerated. Like the girl in the upward-accelerating elevator, she'd press against a scale or a supporting surface.

The space station depicted by NASA artists in Figure 4.14 provides a weightless environment. The shuttle, station facility, and astronauts all accelerate equally toward the Earth. The acceleration is somewhat less than $1g$ because of their altitude. Although they and their surroundings are accelerating toward the Earth in free-fall, they experience no acceleration at all relative to the space station.

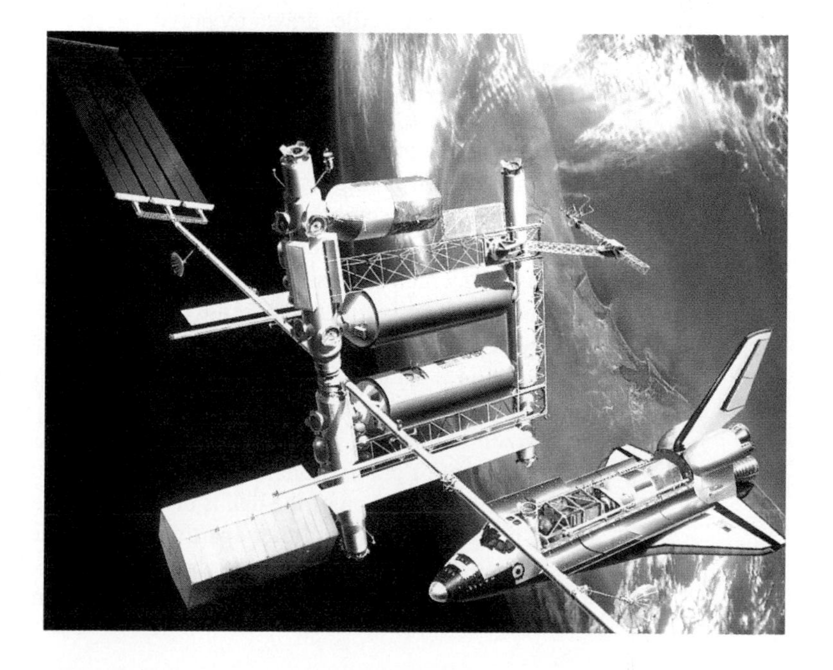

Fig. 4.14
The half-dozen or so inhabitants in this proposed laboratory and docking facility will continuously experience weightlessness. They are in free-fall around the Earth. Is there a force of gravity between them and the Earth?

4.4 Projectile Motion

Without gravity, you could toss a rock at an angle skyward and it would follow a straight-line path. But because of gravity the path curves. A tossed rock, a cannonball, or any other object that is projected by some means and continues in motion under the influence of gravity is called a **projectile**. The path of a projectile is called a *trajectory*. To the early cannoneers, the trajectories of projectiles seemed hopelessly complex. Now we see that these trajectories are surprisingly simple when we look at the horizontal and vertical components of motion separately.

Fig. 4.15
(a) Drop a ball, and it accelerates downward and covers a greater vertical distance each second. (b) Roll a ball along a level surface, and its velocity is constant because no component of gravitational force acts horizontally.

The horizontal component of motion for a projectile is no more complex than the horizontal motion of a bowling ball rolling freely along a level bowling alley. If the retarding effect of friction can be ignored, the ball moves at constant velocity. Because there is no force acting horizontally on the ball, it rolls of its own inertia and covers equal distances in equal time intervals. In other words, it rolls without accelerating. The *horizontal* component of a projectile's motion is just like the bowling ball's motion along the alley (Figure 4.15b).

The *vertical* component of motion for a projectile following a curved path is just like the motion described in Chapter 1 for a freely falling object. Like a ball dropped in mid-air, the projectile, drawn by gravity, accelerates downward (Figure 4.15a).

The trajectory of a projectile is a combination of horizontal and vertical motion. The horizontal component is completely independent of the vertical component. Unless air drag or some other horizontal force acts, the constant horizontal velocity component is not affected by the vertical force of gravity. Gravity affects only the vertical component. It is important to understand that *each component acts independently of the other*. Their combined effects produce the trajectories of projectiles.

These ideas are nicely illustrated in the simulated multiple-flash exposure in Figure 4.16, which shows equally timed successive positions for a ball rolled off a horizontal table. Investigate it carefully, for there's a lot of good physics there. The trajectory of the ball is best analyzed by considering the horizontal and vertical components of motion separately. There are two important things to notice. The first is that the ball's horizontal component of motion doesn't change over time. The ball travels the same horizontal distance in the equal times between each flash. That's because there is no component of gravitational force acting horizontally. Gravity acts only downward, so the only acceleration of the ball is downward. The second thing to note is that the vertical positions become farther apart with time. The distances traveled vertically are the same as if the ball were simply dropped. Figure 4.17 is an actual strobe-light photograph of two balls that start to move from the same height at the same time.

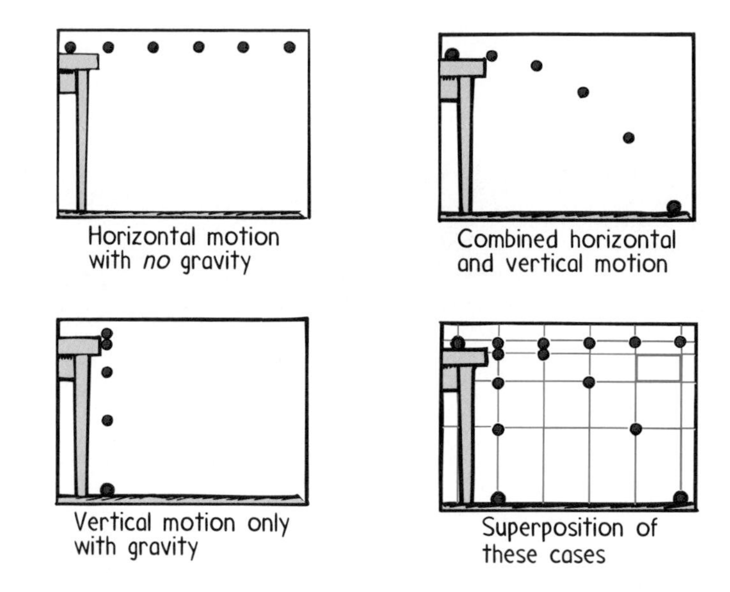

Fig. 4.16
Simulated photographs of a moving ball illuminated with a strobe light.

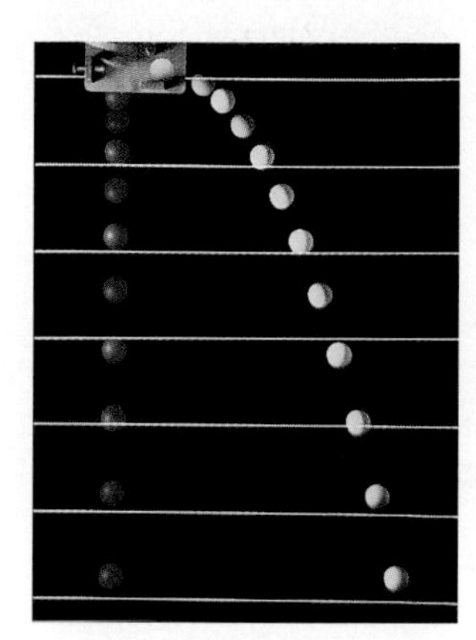

Fig. 4.17
A strobe-light photograph of two golf balls released simultaneously from a mechanism that allows one ball to drop freely while the other is projected horizontally.

The path traced by a projectile that accelerates only in the vertical direction while moving at a constant horizontal velocity is a *parabola*. When air resistance can be neglected, which is usually the case for slow-moving projectiles or for those that are very heavy relative to the forces of air resistance, projectile trajectories are parabolic.

Question

At the instant a horizontally held rifle is fired over a level range, a bullet held at the side of the rifle is released and drops to the ground. Which bullet, the one fired downrange or the one dropped from rest, strikes the ground first?

Consider a cannonball shot at an upward angle (Figure 4.18). Pretend for a moment that there is no gravity; according to the law of inertia, the cannonball would follow the straight-line path shown by the dashed line. But there *is* gravity, so this doesn't happen. What really happens is that the cannonball continuously falls beneath the imaginary dashed line until it finally strikes the ground. Get this: The vertical distance it falls beneath any point on the dashed line is the same vertical distance it would fall if it

---

Answer
Both bullets fall the same vertical distance with the same acceleration g due to gravity and therefore strike the ground at the same time. Can you see that this is consistent with our analysis of Figures 4.16 and 4.17? We can reason this another way by asking which bullet would strike the ground first if the rifle were pointed at an upward angle. In this case the dropped bullet would hit the ground first. Now consider the case where the rifle is pointed downward—the fired bullet hits first. There must be some angle at which there is a dead heat—where both hit at the same time. Can you see that this angle is the one the rifle makes when it is pointing neither upward nor downward—in other words, when it is horizontal?

Fig. 4.18
With no gravity, the projectile would follow a straight-line path (dashed line). But because of gravity, it falls beneath this line the same vertical distance it would fall if released from rest. Compare the distances fallen with those give in Table 1-3 in Chapter 1. (With $g = 9.8\ \text{m/s}^2$, these distances are more accurately 4.9 m, 19.6 m, and 44.1 m.)

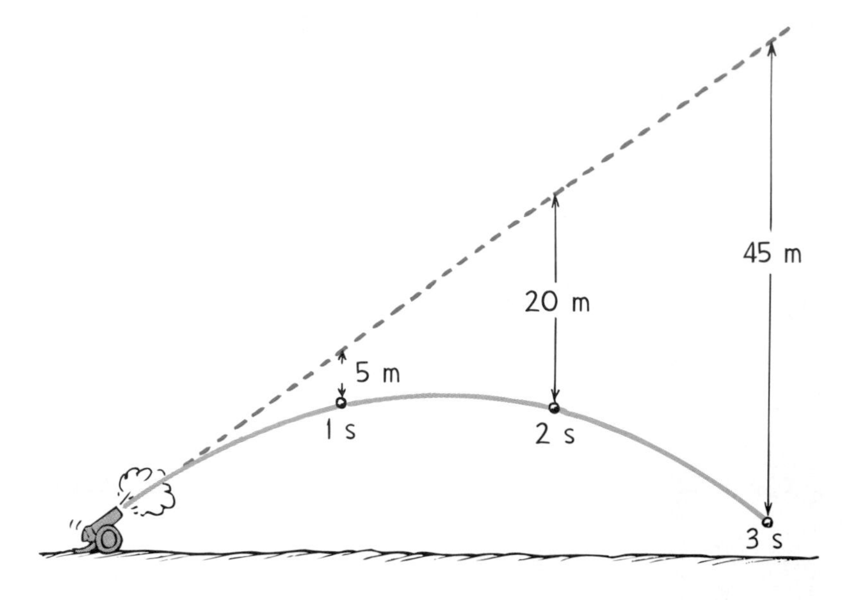

were dropped from rest and had been falling for the same amount of time. This distance, as introduced in Chapter 1, is given by $d = \frac{1}{2}gt^2$, where t is the elapsed time.

We can put it another way: Shoot a projectile skyward at some angle and pretend there is no gravity. After so many seconds t, the projectile should be at a certain point along a straight-line path. But because of gravity, it isn't. Where is it? The answer is that it's directly below this point. How far below? The answer in meters is $5t^2$ (or, more precisely, $4.9t^2$). How about that!

Questions

1. Why does the cannonball in Figure 4.18 move equal horizontal distances in equal time intervals?
2. Suppose the cannonball in Figure 4.18 were fired faster. How many meters below the dashed line would it be at the end of 5 s?
3. If the horizontal component of the cannonball's velocity were 20 m/s, how far downrange would the cannonball be at the end of 5 s?

The sketch of the girl tossing a ball (Figure 4.19) shows a velocity vector and its horizontal and vertical components. Only the vertical component changes with time. When

---

Answers

1. Because no horizontal force acts, there's no horizontal acceleration. Hence the horizontal component of the cannonball's motion is constant.
2. Using $g = 10\ \text{m/s}^2$, we calculate the vertical distance beneath the dashed line at the end of 5 s to be 125 m [$d = 5t^2 = 5(5)^2 = 5(25) = 125$]. (If we use $g = 9.8\ \text{m/s}^2$, then this distance is 122.5 m). It is interesting to note that this distance doesn't depend on the angle of the cannon. If air resistance is neglected, any projectile will fall $5t^2$ meters below where it would have reached if there were no gravity.
3. In the absence of air resistance, the cannonball would travel a horizontal distance of 100 m [$d = vt = (20\ \text{m/s})(5\ \text{s}) = 100\ \text{m}$]. Note that because gravity acts only vertically and there is no acceleration in the horizontal direction, the cannonball travels equal horizontal distances in equal times. This distance is simply its horizontal component of velocity multiplied by the time (and not $5t^2$, which applies only to vertical motion under the acceleration of gravity).

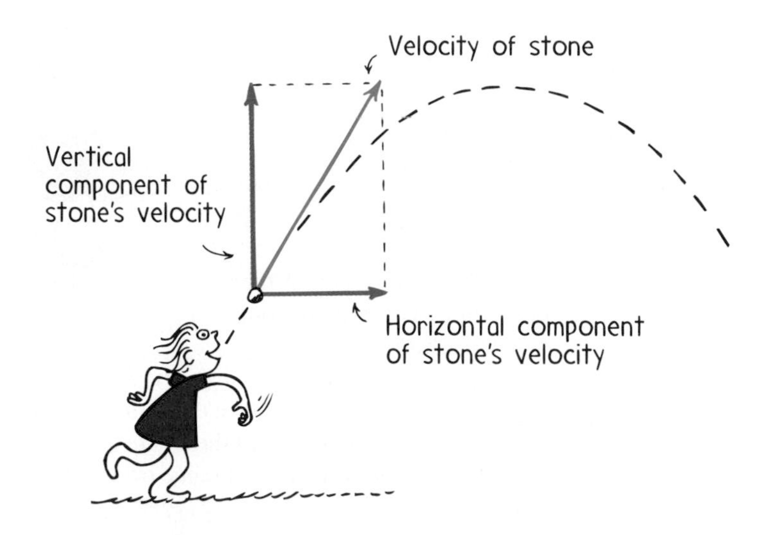

Fig. 4.19
Vertical and horizontal components of a projectile's velocity.

rising, it decreases with time as the projectile goes *against* gravity. When descending, the vertical component increases with time as the projectile travels *with* gravity. The horizontal vector component remains constant because there is no horizontal component of gravity to change it.

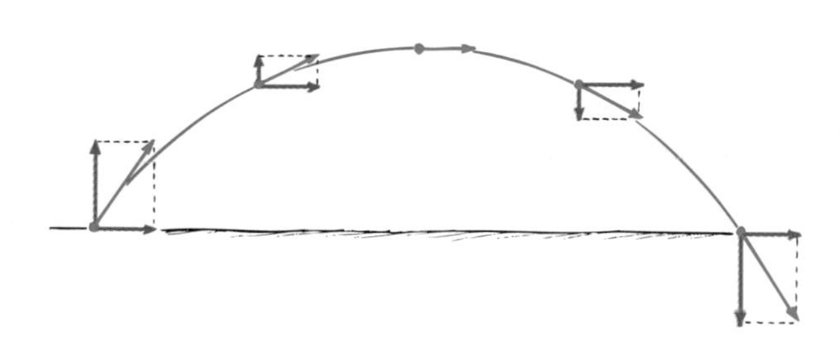

Fig. 4.20
The velocity of a projectile at various points along its path. Note that the vertical component changes and the horizontal component is the same everywhere.

In Figure 4.20, we see sample vectors along a projectile's trajectory. Notice that the horizontal component is everywhere the same, and only the vertical component changes. Note also that the actual velocity is represented by the vector that forms the diagonal of the rectangle formed by the vector components. At the top of the trajectory the vertical component is zero, so the actual velocity there is just equal to the horizontal component of velocity. Everywhere else the magnitude of velocity is greater (just as the diagonal of a rectangle is greater than either of its sides).

Figure 4.21 shows the trajectories of several projectiles in the absence of air resistance, all having the same initial speed but different launching angles. Notice that these projectiles reach different *altitudes* (heights above the ground). They also have different *horizontal ranges* (distances traveled horizontally). The remarkable thing to note

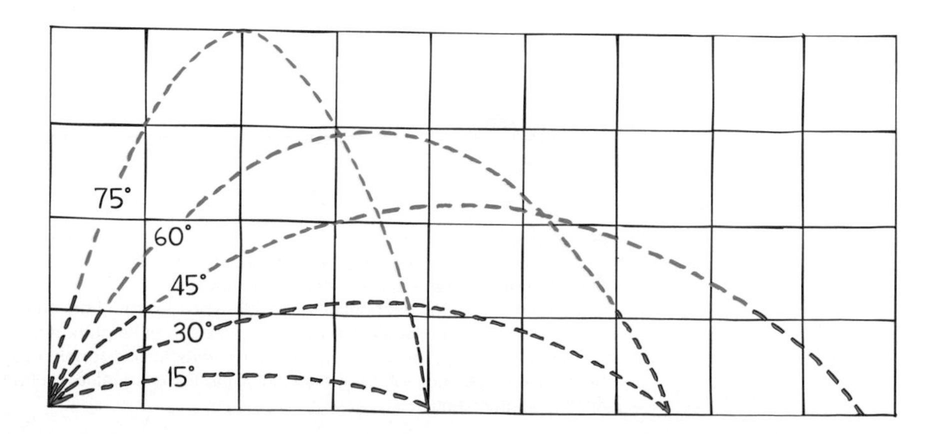

Fig. 4.21
Ranges of a projectile shot at the same speed but different projection angles.

Fig. 4.22
If the batted ball's speed were the same for all angles and if air resistance weren't a factor, maximum range would be attained when the ball is batted at an angle of 45°. Actually, a more forceful ball-bat contact is greater at a lower angle. This fact combined with air resistance makes the favored angle for maximum range on the baseball diamond about 35°.

from Figure 4.21 is that the same range is obtained from two launching angles—any two angles that add up to 90 degrees! An object thrown into the air at an angle of 60 degrees, for example, will have the same range as if it were thrown at the same speed at an angle of 30 degrees. For the smaller angle, however, the object remains in the air for a shorter time. The greatest range occurs when the launching angle is 45 degrees.

Questions

1. A projectile is shot at an angle into the air. If air resistance is negligible, what is the acceleration of its vertical component of motion? Of its horizontal component of motion?
2. At what part of its trajectory does a projectile have minimum speed?

We have emphasized the special case of projectile motion without air resistance. When there is air resistance, the range of the projectile is shorter and its altitude less (Figure 4.23).

When air resistance is negligible, a projectile rises to its maximum height in the same time it takes to fall from that height to the ground (Figure 4.24). This is because

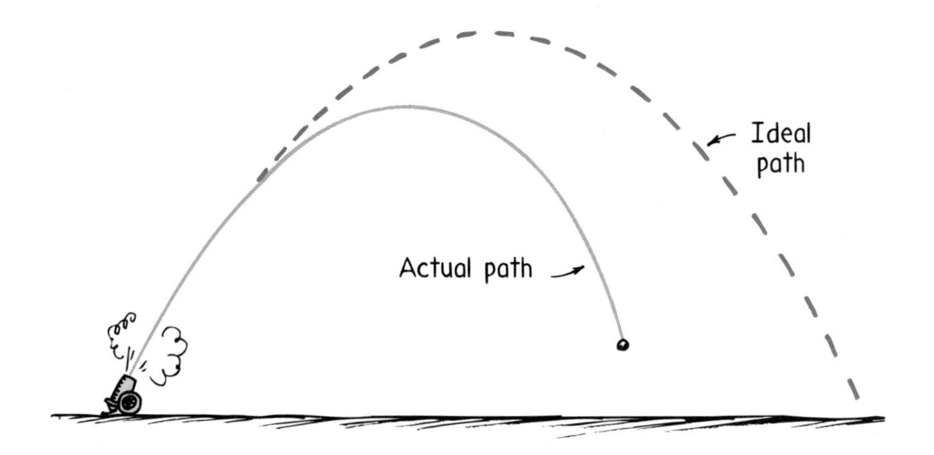

Fig. 4.23
In the presence of air resistance, the trajectory of a high-speed projectile falls short of a parabolic path.

Answers

1. The acceleration in the vertical direction is *g* because the force of gravity is along the vertical direction. (Recall from Chapter 2 that acceleration is always in the direction of the force that acts on an object.) The acceleration is zero in the horizontal direction because no horizontal force acts on the projectile.
2. The speed of a projectile is minimum at the top of its path. If it is launched vertically, its speed at the top is zero. If it is projected at an angle, the vertical component of speed is zero at the top, leaving only the horizontal component. So the speed at the top is equal to the horizontal component of the projectile's velocity at any point.

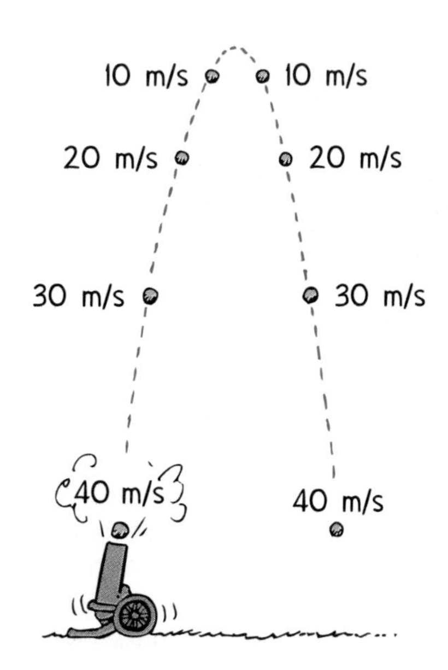

Fig. 4.24
Without air resistance, speed, lost while going up equals speed gained while coming down; so time going up equals time coming down.

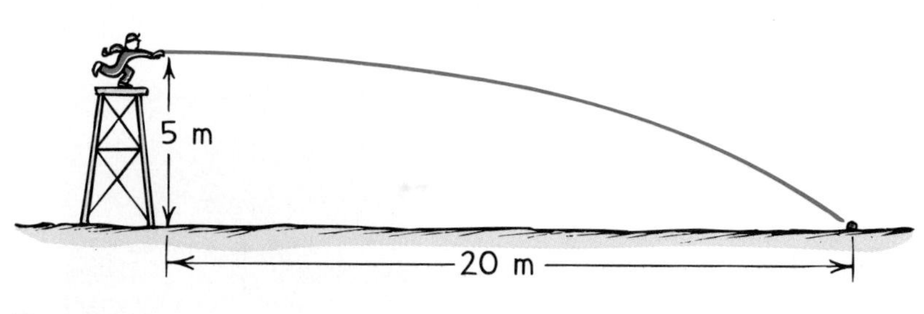

Fig. 4.25
How fast is the ball thrown?

its deceleration by gravity while going up is the same as its acceleration by gravity while coming down. The speed it loses while going up is therefore the same as the speed it gains while coming down. So the projectile arrives at the ground with the same speed it had when it was projected from the ground.

Question

A boy on a tall platform throws a ball 20 m downrange, as shown in Figure 4.25. What is his pitching speed?

Baseball games normally take place on level ground. For the short-range projectile motion of the playing field, the Earth can be considered flat because the ball's flight is not affected by the Earth's curvature. For very long range projectiles, however, the curvature of the Earth's surface must be taken into account. We'll now see that if an object is projected fast enough, it will fall all the way around the Earth and become an Earth satellite.

4.5 Satellite Motion

Consider the girl throwing the stone in Figure 4.26. If gravity did not act on the stone, it would follow the straight-line path shown by the dashed line. But there is gravity, so the ball falls below this straight-line path. In fact, 1 second after the stone leaves her hand it will have fallen a vertical distance of 5 meters below the dashed line—whatever the pitching speed. It is important to understand this, for it is the crux of satellite motion.

Answer

The ball is thrown horizontally, so the pitching speed is horizontal distance divided by time. A horizontal distance of 20 m is given, but the time is not stated. But note that the ball has a vertical drop of 5 m, which we have learned takes 1 s. So while the ball moves horizontally at constant speed, it falls for 1 s. This means that pitching speed $v = d/t = (20\text{ m})/(1\text{ s}) = 20\text{ m/s}$. It is interesting to note that consideration of the equation for constant speed, $v = d/t$, guides thinking about the crucial factor in this problem—the time.

Link to Sports—Hangtime Revisited

Recall our discussion of hang time back in Chapter 1. We stated that the time one is airborne during a jump is independent of horizontal speed. We see now why this is so—the horizontal and vertical components of motion are independent of each other. We can apply the rules of projectile motion to jumping. Whatever maneuverings are employed to attain maximum launching velocity, once the feet leave the ground, the athlete can be considered a projectile. Barring air resistance effects, no amount of arm pumping or other bodily motions can change the time spent in the air—the hang time. Hang time depends only on the vertical component of lift-off velocity. This lift-off component can be increased by running and bounding against the ground with greater force, which is why a running jump is often higher than a standing jump. With a greater lift-off velocity, hang time is increased, but hang time from a standing jump on a skateboard at rest and from a skateboard that's moving is the same.

Can time in the air be increased by the transfer of kinetic energy to potential energy? Not unless there is a mechanism for doing so, such as a rope or pole. Tarzan could grab a vine while running and swing higher than he could jump. Similarly, a pole vaulter transfers the kinetic energy of running into bending the pole and increasing its elastic potential energy, which is then transferred into gravitational potential energy. But without a means of converting kinetic energy into potential energy, horizontal motion has no effect on hang time.

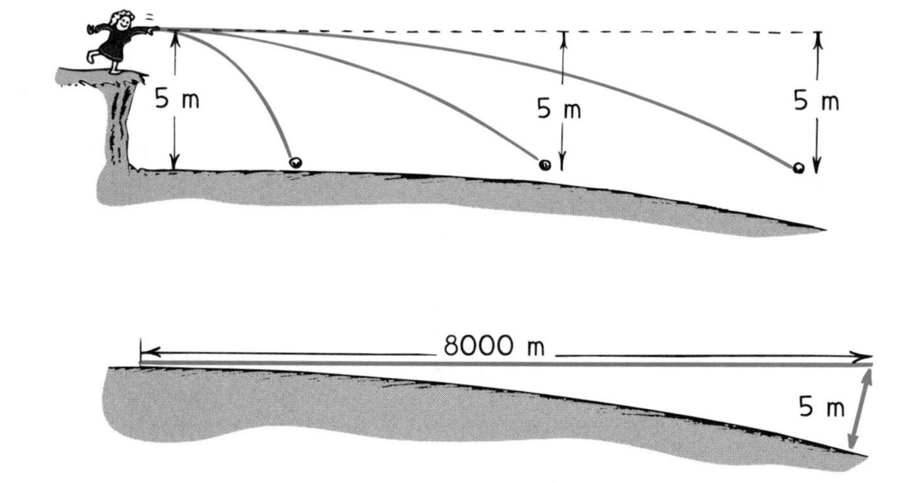

Fig. 4.26
Throw a stone at any speed, and 1 s later it will have fallen 5 m below where it would have been without gravity.

Fig. 4.27
Earth's curvature—not to scale!

An Earth satellite is simply a projectile that falls *around* the Earth rather than *into* it. The speed of the satellite must be great enough to ensure that its falling distance matches the Earth's curvature.[4] A geometrical fact about the Earth is that its surface drops a vertical distance of 5 meters for every 8000 meters tangent to the surface (Fig-

[4]The conventional definition of *to fall* is to "to get closer to the Earth"; satellites, such as the Moon, do not do this. In science we find many cases where the technical definition differs from the conventional one. For example, we say "the Sun rises" and "the Moon sets," but technically they do not.

ure 4.27).[5] This means that if you were floating in a calm ocean, you would be able to see only the top of a 5-meter mast on a ship 8 kilometers away. So if a baseball could be thrown fast enough to travel a horizontal distance of 8 kilometers during the time (1 second) it takes to fall 5 meters, then it would follow the curvature of the Earth. A little thought will show that this speed is 8 kilometers per second. If this doesn't seem fast, convert it to kilometers per hour and you get an impressive 29,000 kilometers per hour (or 18,000 miles per hour)!

At this speed, atmospheric friction would burn the baseball or even a piece of iron to a crisp. This is the fate of grains of sand and other meteorites that graze the Earth's atmosphere and burn up, appearing as "falling stars." That is why satellites such as the space shuttles are launched to altitudes of 150 kilometers and higher. A common misconception is that satellites orbiting at high altitudes are free from gravity. Nothing could be further from the truth. The force of gravity on a satellite 150 kilometers above the Earth's surface is nearly as great as at the surface. The high altitude is to put the satellite beyond the Earth's atmosphere, not beyond the Earth's gravity.

Satellite motion was understood by Newton, who reasoned that the Moon is simply a projectile circling the Earth under the attraction of gravity. This concept is illustrated in a drawing by Newton, shown in Figure 4.30. He compared the motion of the Moon to that of a cannonball fired from the top of a high mountain. Nearly 300 years before the advent of artificial satellites, Newton imagined a cannonball fired from a mountaintop above the Earth's atmosphere, where air resistance would not impede motion. Fired with a small horizontal speed, the cannonball would follow a curved path and soon hit the Earth. Fired faster, it would follow a path less curved and would hit

Fig. 4.28
If the speed of the stone and the curvature of its trajectory are great enough, the stone may become a satellite.

Fig. 4.29
The space shuttle is a projectile in a constant state of free-fall. Because of its tangential velocity, it falls around the Earth rather than vertically into it.

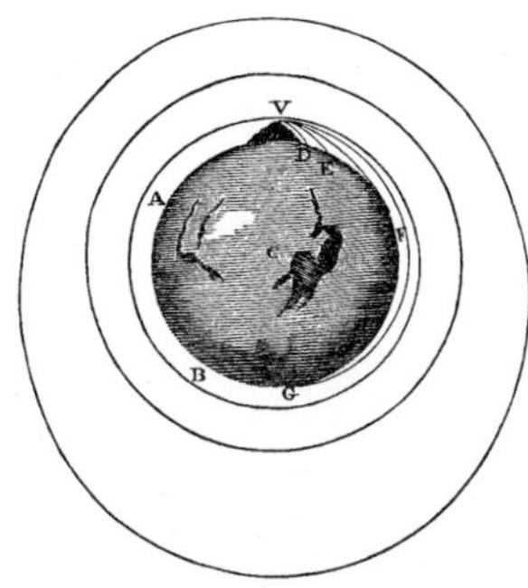

Fig. 4.30
"The greater the velocity . . . with which [a stone] is projected, the farther it goes before it falls to the Earth. We may therefore suppose the velocity to be so increased, that it would describe an arc of 1, 2, 5, 10, 100, 1000 miles before it arrived at the Earth, till at last, exceeding the limits of the Earth, it should pass into space without touching."—Isaac Newton, *System of the World*.

[5]A tangent to a circle or to the Earth's surface is a straight line that touches the circle or surface at only one place. Any tangent is parallel to the circle at the point of contact.

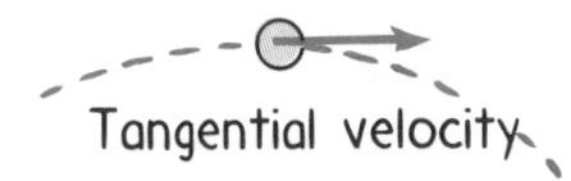

Fig. 4.31
The tangential velocity about the Sun allows the Earth to fall around the Sun rather than into it. If this tangential velocity were reduced to zero, what would be the fate of the Earth?

the Earth farther away. If the cannonball were fired fast enough, Newton reasoned, the curved path would become a circle and the cannonball would circle indefinitely. It would be in orbit.

Both cannonball and Moon have "sideways" velocity, or *tangential velocity*—the velocity parallel to the Earth's surface—sufficient to ensure motion *around* the Earth rather than *into* it. If there is no resistance to reduce its speed, the Moon "falls" around and around the Earth indefinitely. Similarly with the planets that continuously fall around the Sun in closed paths. Why don't they crash into the Sun? They don't because of their tangential velocities. What would happen if their tangential velocities were reduced to zero? The answer is simple enough: Their motion would be straight toward the Sun and they would indeed crash into it. Any objects in the solar system having insufficient tangential velocities long ago crashed into the Sun. What remains is the harmony we observe.

Circular Orbits

So the tangential velocity a projectile needs to orbit the Earth is 8 kilometers per second. That's because in the 1 second the projectile travels 8 kilometers horizontally, it falls a vertical distance of 5 meters—the same vertical distance the Earth curves for each 8-kilometer tangent. So an 8-kilometers-per-second cannonball fired horizontally from Newton's mountain would follow the Earth's curvature and coast around the Earth again and again (provided the cannoneer and the cannon got out of the way). Fired slower, the cannonball would strike the Earth's surface; fired faster, it would overshoot a circular orbit, as we discuss shortly. Newton calculated the speed for circular orbit, and being that such a cannon-muzzle velocity was clearly impossible, he did not foresee people launching satellites (he did not foresee multistage rockets).

Note that in circular orbit the speed of a satellite is not changed by gravity; only the direction changes. We can understand this by comparing a satellite in circular orbit with a bowling ball rolling along a bowling alley. Why doesn't the gravity that acts on the bowling ball change its speed? Because gravity is not pulling forward or backward; gravity pulls straight downward. The bowling ball has no component of gravitational force along the direction of the alley (Figure 4.32).

The same is true for a satellite in circular orbit. In circular orbit a satellite is always moving in a direction perpendicular to the force of gravity that acts on it. With no component of force in the direction of motion, no change in speed occurs—only change in direction. A satellite in circular orbit around the Earth coasts parallel to the surface of the Earth at constant speed—a special form of free-fall.

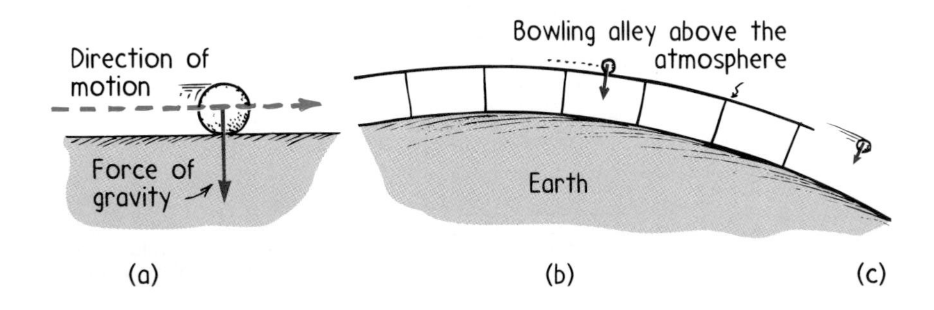

Fig. 4.32
(a) The force of gravity on the bowling ball is at 90° to the ball's direction of motion and so has no component to pull the ball forward or backward. As a result, the ball rolls at constant speed. (b) The same is true when the bowling alley is larger and remains "level" with the curvature of the Earth. (c) If the ball moves at 8 km/s with no air resistance, does it need the alley?

Questions

Consider a ball rolling along a bowling alley that completely circles the Earth and is elevated high enough so air resistance can be neglected.

1. Why would the force of gravity not change the speed of the ball?
2. If a section of the alley were cut away to leave a large gap, how fast must the ball travel to clear the gap and continue its motion? At this speed, what would be the maximum gap for unchanged motion?

For a satellite close to the Earth, the period (time for a complete orbit) is about 90 minutes. For higher altitudes, the orbital speed is less and the period longer. For example, communication satellites located in orbit 5.5 Earth-radii above the Earth have a period of 24 hours. This period matches the period of daily Earth rotation. For an orbit around the equator, these satellites always stay above the same point on the ground (which means that dish antennas on the Earth can be directed toward them continuously). The Moon is even farther away and has a period of 27.3 days. The higher the orbit of a satellite, the lower its speed and the longer its period.[6]

Questions

1. The beauties of physics is that there are usually different ways to view and explain a given phenomenon. Is the following explanation valid? Satellites remain in orbit instead of falling to the Earth because they are beyond the main pull of the Earth's gravity.
2. Satellites in close circular orbit fall about 5 m during each second of orbit. Why doesn't this distance accumulate and send satellites crashing into the Earth's surface?

Answers

1. A change in speed requires a component of force along the direction of travel. In this case the alley and gravitational force on the ball are everywhere perpendicular to each other, so there is no force component in the direction of motion.
2. To clear the gap without bumping into the edge of the alley, the ball must have orbital speed. Then its curved path matches that of the alley's surface. In this case the alley can be completely removed—it will have in effect a 360° gap—because the ball would be in Earth orbit anyway!

Answers

1. No, no, a thousand times no! If any moving object were beyond the pull of gravity, it would move in a straight line and would not curve around the Earth. Satellites remain in orbit because they are being pulled by gravity, not because they are beyond it. For the altitudes of most Earth satellites, the force of gravity is only a few percent weaker than at the Earth's surface.
2. In each second, the satellite falls about 5 m below the straight-line tangent it would have taken if there were no gravity. The Earth's surface also curves 5 m beneath a straight-line 8-km tangent. The process of falling with the curvature of the Earth continues from tangent line to tangent line, so the curved path of the satellite and the curve of the Earth's surface "match" all the way around the Earth. Satellites in fact do crash to the Earth's surface from time to time, but this is principally because they encounter upper-air resistance that decreases their orbital speed.

---

[6]The speed of a satellite in circular orbit is given by $v = \sqrt{GM/d}$ and the period of satellite motion is given by $T = 2\pi\sqrt{d^3/GM}$, where G is the universal gravitational constant, M is the mass of the Earth (or whatever body the satellite orbits), and d is the altitude of the satellite measured from the center of the Earth or parent body.

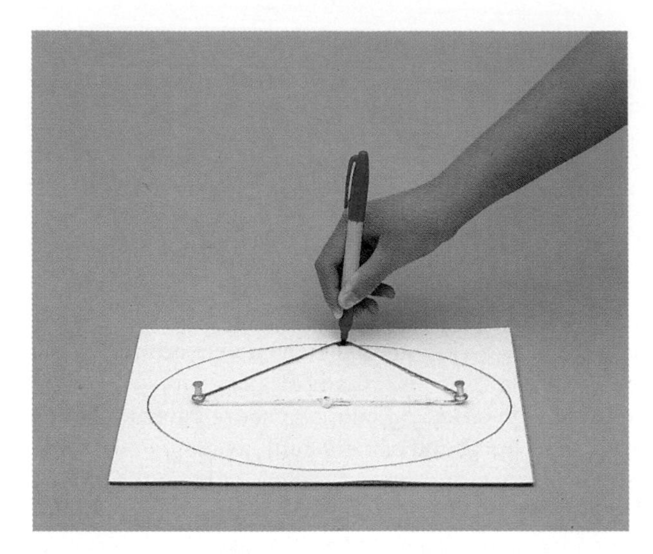

Fig. 4.33
A simple method for drawing an ellipse.

Elliptical Orbits

If a projectile just above the drag of the atmosphere is given a horizontal speed somewhat greater than 8 kilometers per second, it will overshoot a circular path and trace an oval path called an **ellipse**.

An ellipse is a specific curve; it is the closed path taken by a point that moves in such a way that the sum of its distances from two fixed points (called *foci*) is constant. For a satellite orbiting a planet, one focus is at the center of the planet; the other focus is somewhere in space. An ellipse can be constructed by using a pair of tacks, one at each focus, a loop of string, and a pencil (Figure 4.33). The closer the foci are to each other, the closer the ellipse is to a circle. When both foci are at the same location, the ellipse is a circle. So we see that a circle is a special case of an ellipse.

Unlike the constant speed of a satellite in a circular orbit, speed varies in an elliptical orbit. When the initial speed is greater than 8 kilometers per second, the satellite overshoots a circular path and moves away from the Earth, against the force of gravity. It therefore loses speed. Like a rock thrown into the air, it slows to a point where it no longer recedes and then begins to fall back toward the Earth. In a way similar to Figure 4.24, the speed it loses in receding is regained as it falls back toward the Earth, and it finally crosses its original path with the same speed it had initially (Figure 4.35). The procedure repeats over and over, and an ellipse is traced each cycle.

Fig. 4.34
The shadows cast by the ball are all ellipses, one for each lamp in the room. The point at which the ball makes contact with the table is the common focus of the three ellipses.

It's interesting to note that the curved path of a projectile is actually a tiny segment of an ellipse that extends within and just beyond the center of the Earth (Figure 4.36a). In Figure 4.36b, we see several paths of cannonballs fired from Newton's mountain. All ellipses have the center of the Earth as one focus. As muzzle velocity is increased, the ellipses are less eccentric

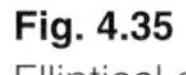

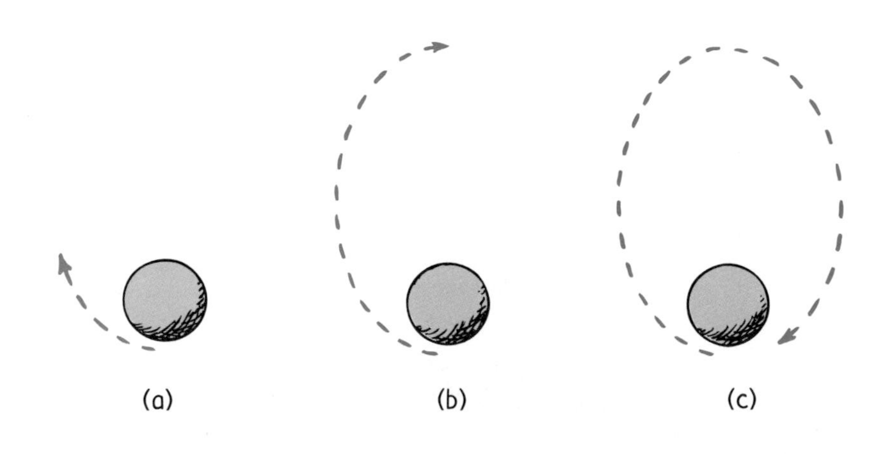

Fig. 4.35
Elliptical orbit. (a) An Earth satellite that has a speed somewhat greater than 8 km/s overshoots a circular orbit and travels away from the Earth. (b) The pull of Earth's gravity slows the satellite to a point where it no longer moves farther from the Earth (c). It then falls toward the Earth, gaining the speed it lost in receding. When it falls to the point depicted in (a), the cycle begins again.

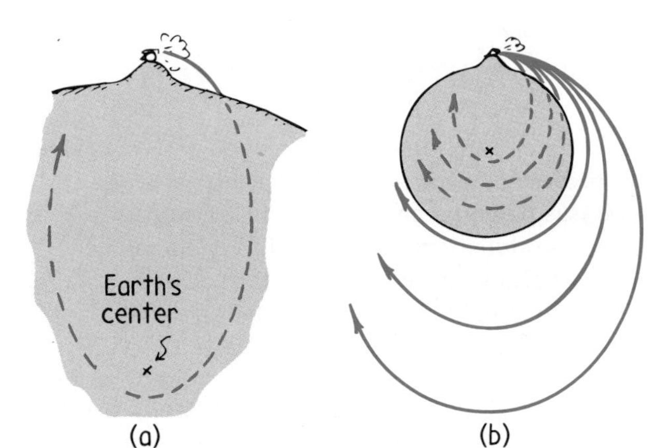

Fig. 4.36
(a) The parabolic path of the cannonball is part of an ellipse that extends within the Earth. The Earth's center is the far focus. (b) All paths of the cannonball are ellipses. For less than orbital speeds, the center of the Earth is the far focus; for circular orbit, both foci are the Earth's center; for greater speeds, the Earth's center becomes the near focus.

(more nearly circular), and when muzzle velocity reaches 8 kilometers per second, the ellipse rounds into a circle and no longer intercepts the Earth's surface. The cannonball then coasts in circular orbit. At greater muzzle velocities, the orbiting cannonball traces the familiar external ellipse.

Comets, which we'll learn more about in Chapter 28, follow highly eccentric elliptical paths about the Sun. Their orbits extend far beyond the planets and their orbital planes may be very different from the orbital planes of the planets.

Putting a satellite into Earth orbit requires control over the speed and direction of the rocket that carries the satellite above the atmosphere. A rocket initially fired vertically is intentionally tipped from the vertical course; then, once above the drag of the atmosphere, it is aimed *horizontally*, whereupon the satellite is given a final thrust to a speed of 8 kilometers per second or more. This is shown in Figure 4.37, where for the sake of simplicity the satellite is an entire single-stage rocket. We see that, with the proper tangential velocity, the rocket satellite falls around the Earth rather than into it.

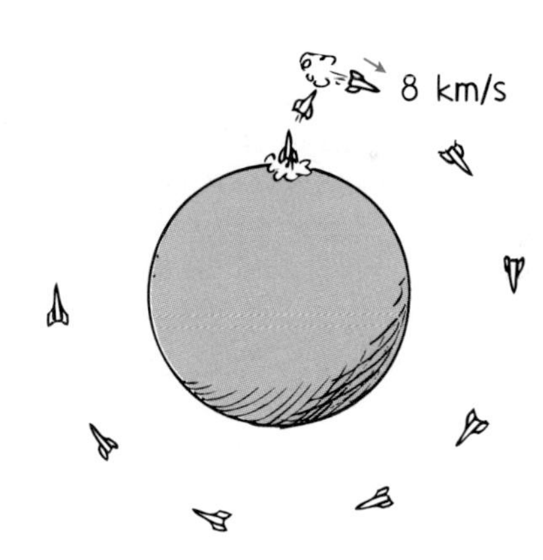

Fig. 4.37
The initial vertical thrust of the rocket pushes it up above the atmosphere. Another thrust to a horizontal speed of at least 8 km/s is required if the rocket is to fall around, rather into, the Earth.

Question

The orbital path of a satellite is shown in the sketch. In which of the marked positions A through D does the satellite have the highest speed? Lowest speed?

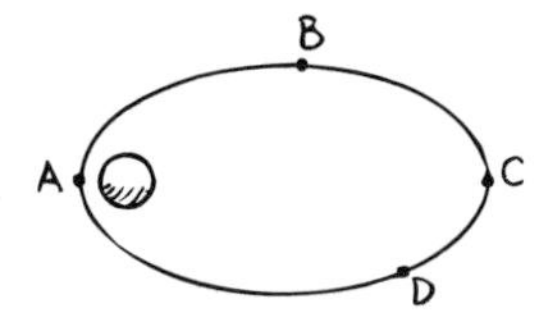

Answer
The satellite has its greatest speed as it whips around A and its lowest speed at C. After passing C, it gains speed as it falls back to A to repeat its cycle.

4.6 Energy Conservation and Satellite Motion

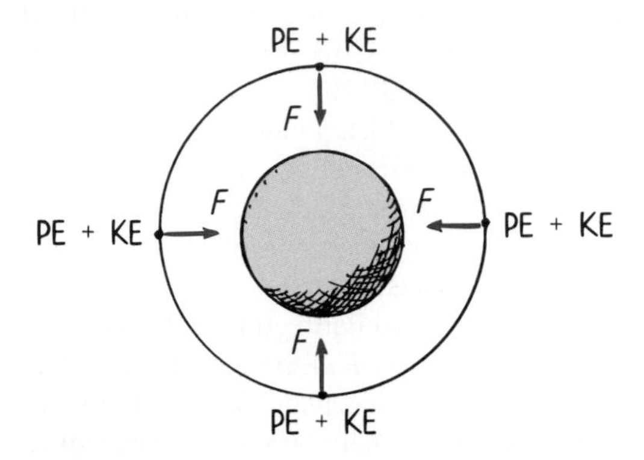

Fig. 4.38
The force of gravity on a satellite is always toward the center of the body it orbits. For a satellite in circular orbit, no component of force acts along the direction of motion. Hence the speed and the KE do not change.

Recall from Chapter 3 that an object in motion possesses kinetic energy (KE) because of its motion, and an object above the Earth's surface possesses potential energy (PE) because of its position. Everywhere in its orbit, a satellite has both PE with respect to the body it orbits and KE. In accordance with the law of conservation of energy, the sum of the KE and PE is a constant all through the orbit.

In circular orbit, the distance between the Earth's center and the satellite does not change, which means the PE of the satellite is the same everywhere in the orbit. By the conservation of energy, therefore, the KE must also be constant. So a satellite in circular orbit coasts at an unchanging PE, KE, and speed (Figure 4.38).

In elliptical orbit the situation is different, as Figure 4.39 shows. Both the satellite's speed and its distance from the Earth's center vary. The satellite's PE is greatest when the satellite is farthest away (at the *apogee*) and least when it is closest (at the *perigee*). The satellite's KE is least when its PE is greatest, and KE is greatest when PE is least. At every point in the orbit, the *sum* of KE and PE is the same.

At all points along the elliptical orbit, there is a component of gravitational force in the direction of motion (with two exceptions, apogee and perigee, where the force is perpendicular only). The component of force in the direction of motion changes the speed of the satellite. Or, by the work-energy theorem we can say (component of force in the direction of motion) $\times$ (distance moved) $= \Delta$KE. Either way, when the satellite gains altitude and moves against this component, its speed and its KE decrease. The decrease continues to the apogee. Once past the apogee, the satellite moves in the same direction as the force component, and the speed and KE increase. The increase continues until the satellite whips past the perigee and repeats the cycle.

Fig. 4.39
The sum of KE and PE for a satellite is constant at all points along its orbit.

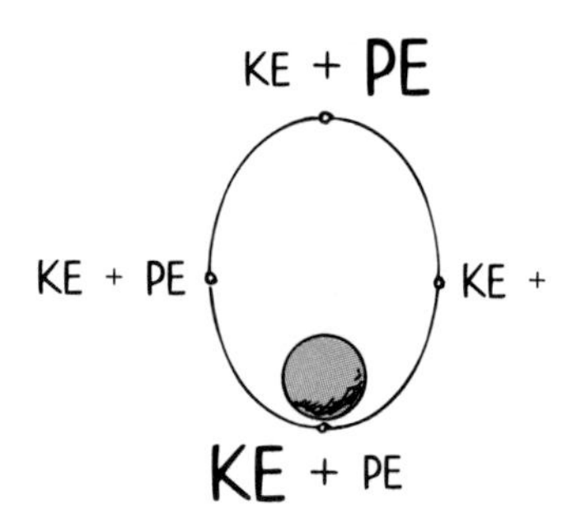

Fig. 4.40
In an elliptical orbit, a component of force exists along the direction of the satellite's motion. This component changes the satellite's speed and thus its KE. (The perpendicular component changes only the direction.)

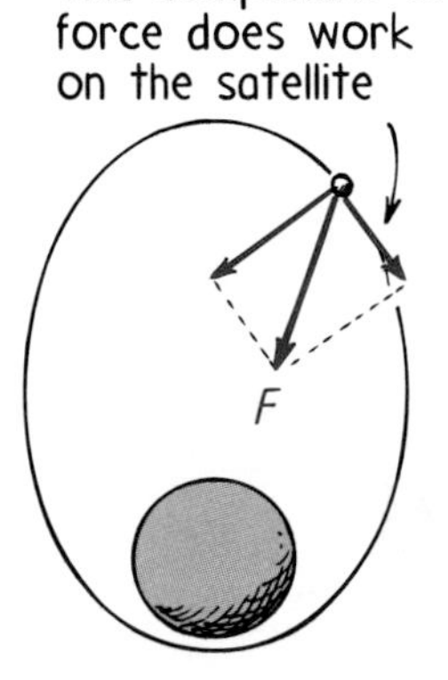

■ **Question**

Why does the force of gravity change the speed of a satellite when it is in an elliptical orbit but not when it is in a circular orbit?

4.7 Escape Speed

We know that a cannonball fired horizontally at 8 kilometers per second from Newton's mountain would find itself in orbit. But what would happen if the cannonball were instead fired at the same speed *vertically*? It would rise to some maximum height, reverse direction, and then fall back to the Earth. The old saying "What goes up must come down" would hold true, just as surely as a stone tossed skyward will be returned by gravity (unless, as we shall see, its speed is too great).

In today's spacefaring age, it is more accurate to say "What goes up *may* come down," for there is a critical speed at which a projectile is able to outrun gravity and escape the Earth. This critical speed is called **escape speed** or, if direction is involved, *escape velocity*. From the surface of the Earth, escape speed is 11.2 kilometers per second.[7] Launch a projectile at any speed greater than that and it will leave the Earth, traveling slower and slower, never stopping due to Earth gravity. Gravitational interaction with the Earth becomes weaker and weaker with increased distance, its speed becomes less and less, though both are never reduced to zero. The projectile outruns the Earth's influence. Although it never escapes the Earth's gravitational field, it does manage to escape the Earth itself.

The escape speeds of other bodies in the solar system are shown in Table 4-1. Note that the escape speed from the Sun is 620 kilometers per second at the surface of the Sun. Even at a distance equaling that of the Earth's orbit, the escape speed from the Sun is 42.5 kilometers per second, considerably more than the escape speed from the Earth. An object projected from the Earth at a speed greater than 11.2 kilometers per second

---

■ **Answer**

At any point on its path, the direction of motion of a satellite is always tangent to the path. If a component of force exists along this tangent, the acceleration of the satellite involves a change in speed as well as direction. In circular orbit, the gravitational force is always perpendicular to the direction of motion, just as every part of the circumference of a circle is perpendicular to the radius. So there is no component of gravitational force along the tangent, and only the direction of motion changes, not the speed. But when the satellite moves in directions that are not perpendicular to the force of gravity, as in an elliptical path, there is a component of force along the direction of motion that changes the satellite's speed. From a work-energy point of view, a component of force along the distance the satellite moves does work to change its KE.

---

[7]From an energy point of view, as a projectile continues outward, its PE increases and its KE decreases. By energy conservation, the PE of a 1-kg mass infinitely far from the Earth is 60 million joules. So to put a satellite that far from the Earth's surface requires an initial KE of at least 60 MJ/kg. This corresponds to a speed of 11.2 km/s, whatever the mass involved.

Escape speed, from any planet or any body, is given by $v = \sqrt{2GM/d}$, where G is the universal gravitational constant, M is the mass of the attracting body, and d is the distance from its center. (At the surface of the body, d is the body's radius.)

Fig. 4.41
Pioneer 10 was launched from Earth in 1972. Twelve years later it escaped the solar system, the first spaceship to do so, and is now wandering in interstellar space.

but less than 42.5 kilometers per second will escape the Earth but not the Sun. Rather than recede forever, it will take up an orbit around the Sun.

Interestingly, escape speed might well be called *maximum falling speed.* Any object, however far from the Earth, released from rest and allowed to fall to the Earth under the influence only of the Earth's gravity would not exceed a speed of 11.2 kilometers per second. As a closer-to-home example, suppose you toss a ball upward at 30 meters per second to a friend atop a building and the ball just barely reaches your friend. When the friend drops the ball back to you, it will return at 30 meters per second. By the same token, if it takes a speed of 11.2 kilometers per second to launch a package to a friend beyond Pluto, you'll catch it at the same speed if it is dropped back to you.

The first probe to escape the solar system, *Pioneer 10,* was launched from the Earth in 1972 at a speed of only 15 kilometers per second. The escape was accomplished by directing the probe into the path of oncoming Jupiter. It was whipped about by Jupiter's great gravitation, picking up speed in the process—similar to the increase in the speed of a ball encountering an oncoming bat when it departs from the bat. The probe's departure speed from Jupiter was increased enough to exceed the Sun's escape speed at the distance of Jupiter. *Pioneer 10* passed the orbit of Pluto in 1984. Unless it collides with another body, it will wander indefinitely through interstellar space. Like a note in a bottle cast into the sea, *Pioneer 10,* and since then *Voyager 1* and *Voyager 2* also, contain information about the Earth that might be of interest to extraterrestrials, in hopes that the information will one day wash up and be found on some distant "seashore."

It is important to point out that the escape speed from a body is the initial speed given by a brief thrust, after which there is no force to assist motion. Given enough time, one could escape the Earth at any sustained speed more than zero. For example, suppose a rocket is launched to a destination such as the Moon. If the rocket engines burn out when still close to the Earth, the rocket needs a minimum speed of 11.2 kilometers per second. But if the engines can be sustained longer, the rocket could go to the Moon without ever attaining 11.2 kilometers per second.

Table 4-1
Escape Speeds at the Surface of Bodies in the Solar System

Astronomical Body	Mass (Earth-masses)	Radius (Earth-radii)	Escape speed (km/s)
Sun	330,000	109	620
Sun(at a distance of the Earth's orbit)	23,500	42.5	
Jupiter	318	11	60.2
Saturn	95.2	9.2	36.0
Neptune	17.3	3.5	24.9
Uranus	14.5	3.7	22.3
Earth	1.00	1.00	11.2
Venus	0.82	0.95	10.4
Mars	0.11	0.53	5.0
Mercury	0.055	0.38	4.3
Moon	0.0123	0.27	2.4

Interestingly, the accuracy with which an non-crewed rocket reaches its destination is not accomplished by staying on a preplanned path or by getting back on that path if it strays off course. No attempt is made to return the rocket to its original path. Instead, the control center in effect asks, "Where is it now with respect to where it ought to go? What is the best way to get there from here, given its present situation?" With the aid of high-speed computers, the answers to these questions are used in finding a new path. Corrective thrusters put the rocket on this new path. This process is repeated over and over again all the way to the goal.[8]

Summary of Terms

- **Law of universal gravitation** Every body in the universe attracts every other body with a force that, for two bodies, is directly proportional to the product of their masses and inversely proportional to the square of the distance separating them:

$$F = \frac{Gm^1 m^2}{d^2}$$

- **Inverse-square law** A law relating the intensity of an effect to the inverse square of the distance from the cause:

$$\text{Intensity} \sim \frac{1}{\text{distance}^2}$$

- **Apparent weight** The force you exert against a supporting floor or a weighing scale, wherein you are as heavy as you feel.
- **Projectile** Any object that moves through the air or space under the influence of gravity.
- **Ellipse** An oval path. The sum of the distances from any point on the path to two points inside called foci is a constant.
- **Escape speed** The speed that a projectile, space probe, or similar object must reach in order to escape the gravitational influence of the Earth or celestial body to which it is attracted.

Review Questions

The Law of Universal Gravitation

1. In Newton's insight, what did a falling apple have in common with the Moon?
2. How can the Moon "fall" without getting closer to the Earth?
3. How does the force of gravity between two objects depend on their masses?
4. How does the force of gravity depend on the distance between two objects?
5. How was *G* first measured?
6. What is the magnitude of the gravitational force between you and the Earth?
7. What is the magnitude of the gravitational force between the Earth and a 1-kg mass? (*Hint:* You don't have to use the law of universal gravitation to figure this out.)
8. What happens to the force of gravity between a pair of objects when the distance between them is doubled?
9. Hanging from a tree is an apple that weighs 1 N. Twice as high above the ground on the same tree is another apple of the same mass. Why is its weight practically the same as that of the other apple and not $\frac{1}{4}$ N?
10. Can you escape from the Earth's gravity by getting above the atmosphere? By going to the Moon? Defend your answers.

Tides

11. Why do both the Sun and the Moon exert a greater gravitational force on one side of the Earth than on the other?
12. Do tides depend more on the strength of gravitational pull or on the *difference* in strengths? Explain.
13. Why are all tides greatest at the time of a full or new Moon?
14. (Fill in the blank.) Whereas gravitational force depends on the inverse square of distance, tidal force, the difference in gravitational forces per unit mass, depends on the inverse ____________ of distance.

Weight and Weightlessness

15. Would the springs inside a bathroom scale be more compressed or less compressed if you weighed yourself in an elevator that was accelerating upward? Downward?
16. Would the springs inside a bathroom scale be more compressed or less compressed if you weighed yourself in an elevator that moved upward at constant velocity? Downward at constant velocity?

---

[8] Is there a lesson to be learned here? Suppose you find that you are off course. You may, like the rocket, find it more fruitful to take a course that leads to your goal as best plotted from your present position and circumstances, rather than trying to get back on the course you plotted from a previous position and under, perhaps, different circumstances.

Projectile Motion

17. A rock is thrown upward at an angle. What happens to the horizontal component of its velocity as it rises? As it falls?

18. Why does the horizontal component of a projectile's motion remain constant?

19. Why does the vertical component of a projectile's motion undergo change?

20. How does the vertical distance a projectile falls below an otherwise straight-line path compare with the vertical distance it would fall from rest in the same time?

21. What angle from the horizontal gives the greatest range for a projectile launched from ground level?

22. A projectile is launched upward at an angle of 75 degrees from the horizontal and strikes the ground a certain distance down range. For what other angle of launch at the same speed does this projectile land just as far away?

23. A projectile is launched vertically at 100 m/s. If air resistance can be neglected, at what speed does it return to its initial level?

24. How does air resistance affect the range of projectiles?

Satellite Motion

25. During a 1-s interval, how far does a baseball fall beneath a straight-line tangent to its motion?

26. During a 1-s interval, how far does an Earth satellite in close orbit fall beneath a straight line tangent to its motion?

27. What is *tangential velocity*?

28. How can a projectile "fall around the Earth"?

29. Why don't satellites like the space shuttle fall to Earth like tossed baseballs?

30. Why doesn't the force of gravity change the speed of a satellite in circular orbit?

31. How long does it take a satellite in close orbit about the Earth to complete one revolution?

32. For orbits of greater altitude, is the period greater or less than your answer to the previous question?

33. Why does the force of gravity change the speed of a satellite moving in an elliptical orbit?

34. At what part of an elliptical orbit does a satellite have the greatest speed? The least speed?

Energy Conservation and Satellite Motion

35. Why is kinetic energy constant for a satellite in circular orbit?

36. Why does kinetic energy vary for a satellite in an elliptical orbit?

37. With respect to the apogee and perigee of an elliptical orbit, where is the gravitational potential greatest? Least?

38. Is the sum of kinetic and potential energies constant for satellites in circular orbits? For those in elliptical orbits?

Escape Speed

39. What is the minimum speed for orbiting the Earth in close orbit? The maximum speed? What happens above this speed?

40. How was *Pioneer 10* able to escape the solar system with an initial speed less than escape speed?

Home Project

Hold your hands outstretched, one twice as far from your eyes as the other. Which hand looks bigger? Most people see the hands as being about the same size, while some see the nearer hand as slightly bigger. Almost nobody sees the nearer hand as four times as big. But by the inverse-square law, the nearer hand should appear twice as tall and twice as wide and therefore occupy four times as much of your visual field as the farther hand. Your belief that your hands are the same size is so strong that you likely overrule this information. Now if you overlap your hands slightly and view them with one eye closed, you'll see the nearer hand as clearly bigger. This raises an interesting question: What other illusions do you have that are not so easily checked?

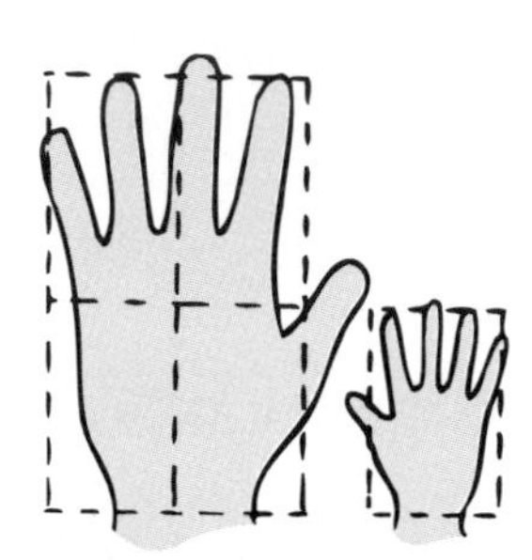

Exercises

1. Comment on whether or not this label on a consumer product should be cause for concern. *CAUTION: The mass of this product pulls on every other mass in the universe, with an attracting force that is proportional to the product of the masses and inversely proportional to the square of the distance between them.*

2. Gravitational force acts on all bodies in proportion to their masses. Why, then, doesn't a heavy body fall faster than a light body?

3. Which weighs more, a sheet of aluminum foil or the same sheet crumpled into a tight wad?

4. Which planets, those closer to the Sun than the Earth or those farther from the Sun than the Earth, have a period greater than 1 Earth-year?

5. What are the magnitude and direction of the gravitational force that acts on a man who weighs 700 N at the surface of the Earth?

6. Somewhere between the Earth and the Moon, the gravitational pull exerted by these two bodies on a space pod would cancel. Is this location nearer the Earth or nearer the Moon?

7. The Earth and the Moon are attracted to each other by gravitational force. Does the more massive Earth attract the less massive Moon with a force that is greater than, smaller than, or the same as the force with which the Moon attracts the Earth? (With an elastic band stretched between your thumb and forefinger, which is pulled more strongly by the band, your thumb or your forefinger?)

8. If the mass of the Earth somehow increased, with all other factors remaining the same, would your weight also increase? (*Hint:* Let the equation for gravitational force guide your thinking.)

9. A small light bulb located 1 m in front of a 1-m^2 opening illuminates a wall behind the opening. If the wall is 1 m behind the opening (2 m from the light source), the illuminated area covers 4 m^2. How many square meters is illuminated if the wall is 3 m from the light bulb? 5 m? 10 m?

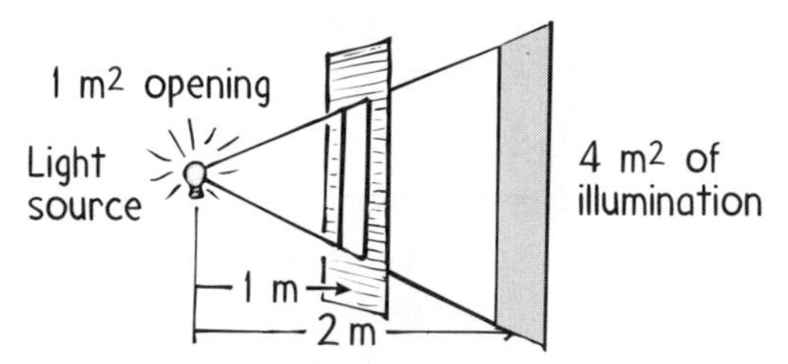

10. The planet Jupiter is about 300 times as massive as the Earth, so it might seem that a body on the surface of Jupiter would weigh about 300 times as much as on the Earth. But it so happens that a body on the surface of Jupiter weighs slightly less than 3 times as much as it would on the surface of the Earth. Can you explain this? (*Hint*: Let the terms in the equation for gravitational force guide your thinking.)

11. From the data in the preceding exercise, estimate the radius of Jupiter relative to that of the Earth.

12. Why do passengers on high-altitude jet planes feel the sensation of weight while passengers in an orbiting space shuttle do not?

13. If the Earth made one rotation on its axis each 90 min instead of one each 24 h, would you press against the Earth's surface if you were at the equator? At the poles? In the middle of the United States? Explain.

14. If you were in a car that drove off a cliff, why would you be weightless? Would gravity still be acting on you?

15. If you were in a freely falling elevator and dropped a pencil, you'd see the pencil hovering. Is the pencil falling? Explain.

16. Explain why the following reasoning is wrong. "The Sun attracts all bodies on the Earth. At midnight, when the Sun is directly below, the direction in which it pulls on an object located on Earth is the same as the direction in which the Earth pulls on the object; at noon, when the Sun is directly overhead, the direction in which it pulls on the object is the opposite of the direction in which the Earth pulls on the object. Therefore, all objects should be heavier at midnight than at noon." (*Hint*: Relate this to the preceding two exercises.)

17. If the mass of the Earth increased, your weight would correspondingly increase, but if the mass of the Sun increased, your weight would not be affected. Why?

18. Most people today know that the ocean tides are caused by the gravitational influence of the Moon. And most people therefore think that the gravitational pull of the Moon on the Earth is greater than the gravitational pull of the Sun on the Earth. What do you think?

19. Would ocean tides exist if the gravitational pull of the Moon (and Sun) was somehow equal on all parts of the planet? Explain.

20. Why aren't high ocean tides exactly 12 hours apart?

21. With respect to spring and neap ocean tides, when are the lowest low tides? That is, when is it best for digging clams?

22. Whenever a high tide is unusually high, will the following low tide be unusually low? Defend your answer in terms of "conservation of water." (If you slosh water in a tub so that it is extra deep at one end, will the other end be extra low?)

23. The Mediterranean Sea has very little sediment churned up and suspended in its waters, mainly because of the absence of any substantial tides. Why do you suppose the Mediterranean Sea has practically no tides? Similarly, are there tides in the Black Sea? Great Salt Lake? Your county reservoir? A glass of water? Explain.

24. The human body is composed mostly of water. Why does the gravitational pull of the Moon overhead cause appreciably less biological tides in the fluids in your body than a 1-kg melon held over your head?

25. If the Moon didn't exist, would the Earth still have ocean tides? If so, how often?

26. Which requires more fuel—a rocket going from the Earth to the Moon or one coming back? Why?

27. True or false:

a. When air resistance does not affect its motion, a projectile covers equal horizontal distances in equal time intervals.

b. When air resistance does not affect the motion of a projectile, the horizontal and vertical components of its velocity remain constant.

c. Changes in the vertical component of velocity for a projectile are proportional to changes in the horizontal component of velocity.

28. If there were no gravity, a projectile would follow a straight-line path. Because of gravity, a projectile falls beneath the straight-line path it would otherwise follow. How many meters does it fall below this line if it has been traveling for 1 s? For 2 s? Do your answers depend on the angle at which the projectile is launched?

29. If a cannonball is fired from a tall mountain, gravity changes its speed all along its trajectory. If it is fired fast enough to go into circular orbit, however, gravity does not change its speed. Explain.

30. It being true that the Moon is gravitationally attracted to the Earth, why doesn't it simply crash into the Earth?

31. Does the speed of a falling object depend on its mass? Does the speed of a satellite in orbit depend on its mass? Defend your answers.

32. If you have ever watched the launching of an Earth satellite, you may have noticed that the rocket departs from a vertical course and continues its climb at an angle. Why?

33. Why are satellites usually sent into orbit by firing in an easterly direction (the direction in which the Earth spins)?

34. Of all the United States, why is Hawaii the most efficient launching site for nonpolar satellites? (*Hint:* Look at the spinning Earth from above either pole and compare it to a spinning turntable.)
35. In the sketch below, a ball gains KE when rolling down a hill because work is done by the component of weight (F) that acts in the direction of motion. Sketch in the similar component of the gravitational force that does work to change the KE of the satellite on the right.

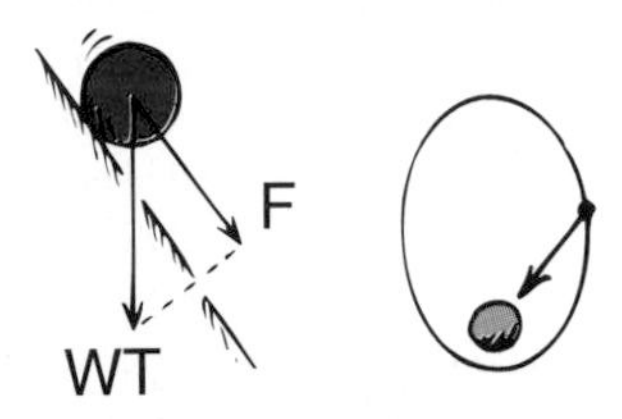

36. Why is work done by the force of gravity on a satellite in an elliptical orbit but not one in a circular orbit?
37. What is the shape of a satellite's orbit when its velocity is everywhere perpendicular to the force of gravity?
38. Would the speed of a satellite in close circular orbit about Jupiter be greater than, equal to, or less than 8 km/s?
39. If a flight mechanic drops a wrench from a high-flying jumbo jet, it crashes to Earth. If an astronaut on the orbiting space shuttle drops a wrench, does it crash to Earth also? Defend your answer.
40. The orbiting space shuttle travels at 8 km/s relative to the Earth. Suppose it projects a capsule rearward at 8 km/s relative to the shuttle. Describe the path of the capsule relative to the Earth.
41. The orbital speed of the Earth about the Sun is 30 km/s. If the Earth were suddenly stopped in its tracks, it would fall in a radial straight-line path into the Sun. Suppose you wish to dispose of radioactive wastes by sending a rocketload of them to the Sun. How fast and in what direction relative to the Earth's orbit should the rocket be fired?
42. If you stopped an Earth satellite dead in its tracks, it would crash into the Earth. Why, then, don't the communications satellites that "hover motionless" above the same spot on Earth crash into the Earth?
43. Escape speed from the surface of the Earth is 11.2 km/s. How could a space vehicle escape from the Earth while traveling at half this speed or less?
44. At which of the indicated positions—A, B, C, or D—does the satellite experience the greatest gravitational force? The greatest speed? The greatest velocity? The greatest momentum? The greatest kinetic energy? The greatest gravitational potential energy? The greatest total energy? The greatest angular momentum? The greatest acceleration?

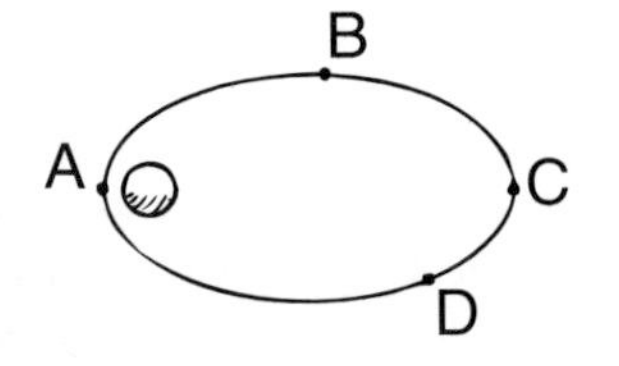

Problems

1. The value of g at the Earth's surface is about 9.8 m/s2. What is its value at a distance from the Earth's center that is four times the Earth's radius?
2. Many people mistakenly believe that astronauts orbiting the Earth are "above gravity." Calculate g for space shuttle territory, 200 kilometers above the Earth's surface. Earth's mass is 6×10^{24} kg, and its radius is 6.38×10^{6} m (6380 km). Your answer is what percentage of 9.8 m/s^2?

3. A 3-kg newborn baby at the Earth's surface is gravitationally attracted to the Earth with a force of about 30 N. (a) Calculate the force of gravity with which the baby on the Earth is attracted to the planet Mars when Mars is closest to the Earth. (The mass of Mars is 6.4×10^{23} kg, and its closest distance is 5.6×10^{10} m). (b) Calculate the force of gravity between the baby and the physician who delivers it. Assume the physician has a mass of 100 kg and is 0.5 m from the baby. (c) How do the forces compare?
4. Calculate the force of gravity between the Earth (mass = 6×10^{24} kg) and the Sun (mass = 2×10^{30} kg, distance = 1.5×10^{11} m).
5. Suppose the force of gravity between the Sun and the Earth vanishes. Instead, a steel cable between the two bodies keeps the Earth in orbit. Estimate the thickness (diameter) of such a cable. You need to know that the tensile strength of steel cable is about 5.0×10^{8} N/m^2. That is, a steel cable that has a cross-sectional area of 1 m^2 will support a force of 5.0×10^{8} N.

6. The Moon is about 3.8×10^{5} km from the Earth. Find its average orbital speed about the Earth.
7. Calculate the speed in meters per second at which the Earth revolves about the Sun. Assume the orbit is nearly circular.

CHAPTER

Fluid Mechanics

Liquids and gases have the ability to flow; hence they are called *fluids.* Because they are both fluids, they obey similar mechanical laws. How is it that iron boats don't sink in water and helium balloons don't sink from the sky? Why is it impossible to breathe through a snorkel when you're under more than a meter of water? Why do your ears pop when riding an elevator? How do hydrofoils and airplanes attain lift? To discuss fluids properly, we need to understand two concepts—*density* and *pressure.*

5.1 Density

Fig. 5.1
When the volume of the bread is reduced, its density increases.

An important property of materials—whether in the solid, liquid, or gaseous phases—is the measure of compactness: **density**. We think of density as the "lightness" or "heaviness" of materials having the same volume. It is a measure of how much mass is squeezed into a given space, the amount of matter per unit volume:

$$\textbf{Density} = \frac{\textbf{mass}}{\textbf{volume}}$$

Here mass is measured in either grams or kilograms and volume in either cubic centimeters (cm^3) or cubic meters (m^3).[1] The mass of 1 cubic centimeter of water at 4°C is 1 gram, and so water has a density of 1 gram per cubic centimeter. A given volume of mercury, density 13.6 grams per cubic centimeter, is therefore 13.6 times as massive as an equal volume of water. Osmium, a hard, bluish-white metallic element, is the densest solid substance on the Earth. The densities of a few materials are given in Table 5.1.

A quantity known as *weight density,* commonly used when discussing liquid pressure, is weight per unit volume:[2]

$$\textbf{Weight density} = \frac{\textbf{weight}}{\textbf{volume}}$$

Questions

1. Which has the greater density—1 kg of water or 10 kg of water?
2. Which has the greater density—5 kg of lead or 10 kg of aluminum?
3. Which has the greater density—a whole candy bar or half a candy bar?

5.2 Pressure

Imagine placing a book on a bathroom scale. Whether you place it on its back, on its side, or balanced on a corner, it exerts the same force. The weight reading is the same. Now place the book on the palm of your hand in these three different positions and you sense a difference—the *pressure* of the book depends on the area over which the force

---

Answers
1. The density of any amount of water is the same: 1 g/cm^3 or, equivalently, 1000 kg/m^3, which means that the density of 1 kg of water is exactly the same as the density of 10 kg, or any other amount, of water.
2. Density is a *ratio* of weight or mass per volume, and this ratio is greater for any amount of lead than for any amount of aluminum—see Table 5.1.
3. Both have the same density.

---

[1]A cubic meter is a sizable volume and contains a million cubic centimeters, so there is a million grams of water in a cubic meter (or, equivalently, a thousand kilograms of water in a cubic meter). Hence, 1 g/cm3 5 1000 kg/m3.

[2]Weight density is common to British units, in which 1 cubic foot of fresh water (almost 7.5 gallons) weighs 62.4 pounds. So fresh water has a weight density of 62.4 lb/ft3. Salt water is a bit denser, 64 lb/ft3.

Table 5.1
Densities of some materials

Material	Grams per cubic centimeter	Kilograms per cubic meter
Liquids		
Mercury	13.6	1,360
Glycerin	1.26	1,260
Seawater	1.03	1,025
Water at 4°C	1.00	1,000
Benzene	0.90	899
Ethyl alcohol	0.81	806
Solids		
Osmium	22.5	22,480
Platinum	21.5	21,450
Gold	19.3	19,320
Uranium	19.0	19,050
Lead	11.3	11,344
Silver	10.5	10,500
Copper	8.9	8,920
Brass	8.6	8,560
Iron	7.8	7,800
Tin	7.3	7,280
Aluminum	2.7	2,702
Ice	0.92	917
Gases (atmosphere pressure at sea level)		
Dry air		
0°C	0.00129	1.29
10°C	0.00125	1.25
20°C	0.00121	1.21
30°C0.00116	1.16	
Hydrogen at 0°C	0.00090	0.090
Helium at 0°C	0.00178	0.178

Fig. 5.2
Although the weight of both books is the same, the upright book exerts greater pressure against the table.

is distributed. There is a difference between force and pressure. **Pressure** is defined as the force exerted over a unit of area, such as a square meter or square foot:[3]

$$\textbf{Pressure} = \frac{\textbf{force}}{\textbf{area}}$$

A dramatic illustration of pressure is shown in Figure 5.3. One of your authors applies appreciable force when he breaks the cement block with a sledge hammer. Yet his teaching colleague, who is sandwiched between two beds of sharp nails, is unharmed. This is because the transmitted force of the hammer impact is distributed over more than 200 nails making contact with his body. The combined surface area of the nails results in a tolerable pressure that does not puncture the skin. Force and pressure are different from each other.

[3]Pressure may be measured in any unit of force divided by any unit of area. The standard international (SI) unit of pressure, the newton per square meter, is called the *pascal* (Pa), after the seventeenth-century theologian and scientist Blaise Pascal. A pressure of 1 Pa is very small and approximately equals the pressure exerted by a dollar bill resting flat on a table. Science types more often use kilopascals (1 kPa = 1000 Pa).

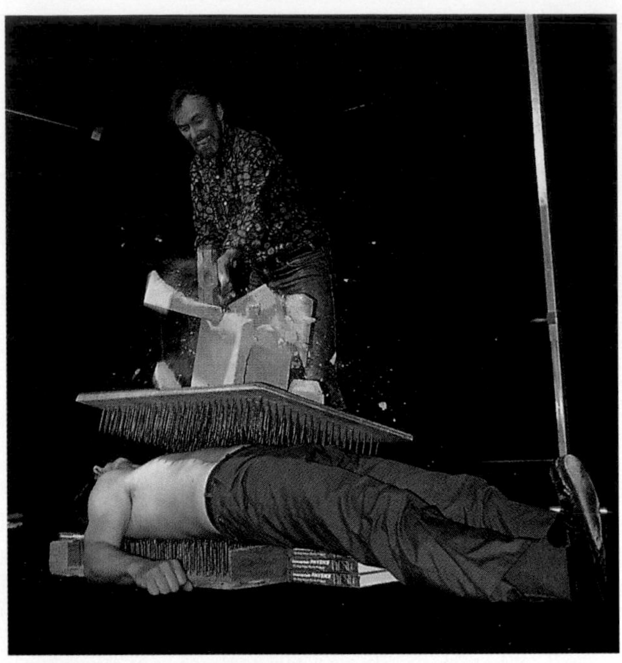

Fig. 5.3
Paul Hewitt applies a force to fellow physics teacher Paul Robinson, bravely sandwiched between beds of sharp nails. The driving force per nail is not enough to puncture the skin. From an inertia point of view, is Robinson better off with a substantially massive block or a less massive one? From an energy point of view, is he better off if the block breaks or if it stays in one piece as the hammer strikes?

Pressure in a Liquid

When you swim under water, you can feel the water pressure acting against your eardrums. The deeper you swim, the greater the pressure. What causes this pressure? It is simply the weight of the water (and air) above pushing against you. If you swim twice as deep, there is twice the weight of water above, and twice the water pressure. At three times the depth, there is three times the water pressure, and so on. The pressure of the air above is always transmitted down through the water and adds to the water pressure. However, because air pressure near the Earth's surface is nearly constant, the only thing that changes with depth is the water pressure. That's what we consider here.

The pressure exerted by the liquid depends on density as well as depth. If you were submerged in a liquid more dense than water, the pressure would be proportionally greater.

Liquid pressure = weight density × depth

It is important to note that pressure does not depend on amount of liquid. You feel the same pressure a meter deep in a small pool as you do a meter deep in the middle of the ocean. This is illustrated by the connecting vases shown in Figure 5.4. If the pressure at the bottom of a large vase were greater than the pressure at the bottom of a neighboring, narrower vase, the greater pressure would force water sideways and then up the narrower vase to a higher level. This doesn't happen, however, and we conclude that pressure is dependent upon depth, not volume. That a liquid seeks its own level can be demonstrated by filling a garden hose with water and holding the two ends upright. The water levels is the same in the two sides even when the two ends are held far apart. Pressure is depth-dependent and not volume-dependent, so we see there is a reason why water seeks its own level.

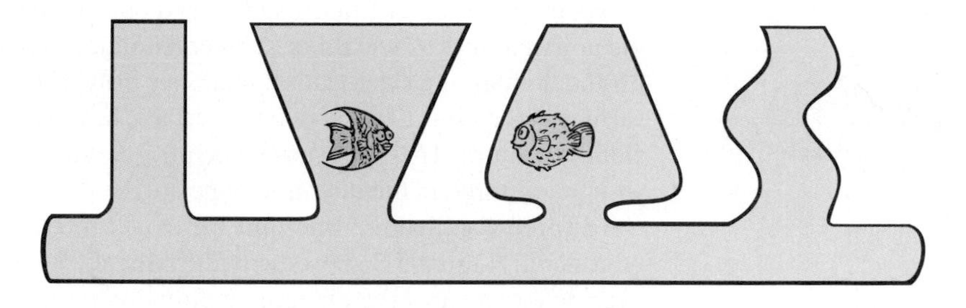

Fig. 5.4
The pressure exerted by a liquid is the same at any given depth below the surface, regardless of the shape of the containing vessel.

Another interesting fact about liquid pressure is that it is exerted equally in all directions. For example, if we are submerged in water, no matter which way we tilt our heads we feel the same amount of water pressure on our ears. Because a liquid can flow, the pressure isn't only downward. We know pressure acts upward when we try to push a beach ball beneath the water surface. The bottom of a boat is certainly pushed upward by water pressure. And we know water pressure acts sideways when we see water spurting sideways from a leak in an upright can. Pressure in a liquid at any point is exerted in equal amounts in all directions.

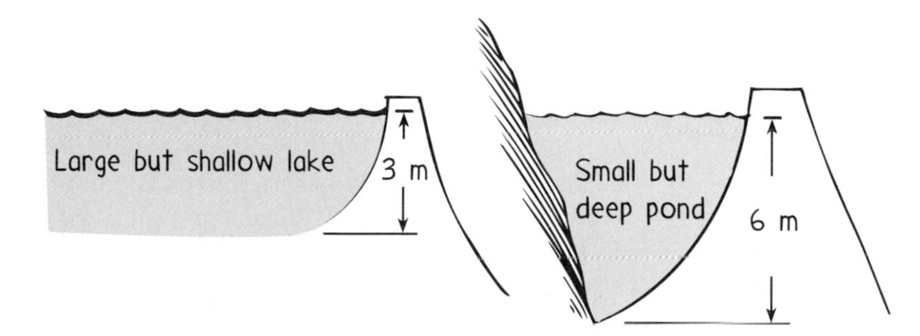

Fig. 5.5
The average water pressure acting against the dams depends on the average depth of the water and not on the volume of water held back. The average pressure exerted by the large, shallow lake is only one-half that exerted by the small, deep pond.

When liquid presses against a surface, there is a net force directed perpendicular to the surface (Figure 5.6). If there is a hole in the surface, the liquid spurts at right angles to the surface before curving downward due to gravity (Figure 5.7). At greater depths the pressure is greater and the speed of the escaping liquid is greater.[4]

Fig. 5.6
The forces that produce pressure against a surface add up to a net force that is perpendicular to the surface.

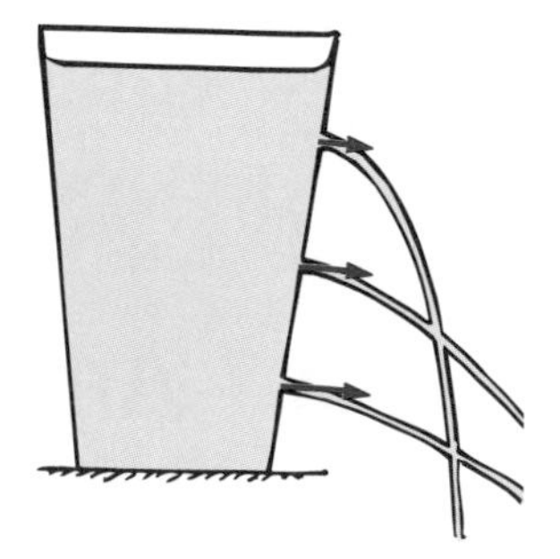

Fig. 5.7
Water pressure pushes perpendicularly against the sides of a container and increases with increasing depth.

[4]The speed of liquid out of the hole is $\sqrt{2gh}$, where h is the depth below the free surface. Interestingly, this is the same speed the water or anything else would have if freely falling the same distance h.

5.3 Buoyancy in a Liquid

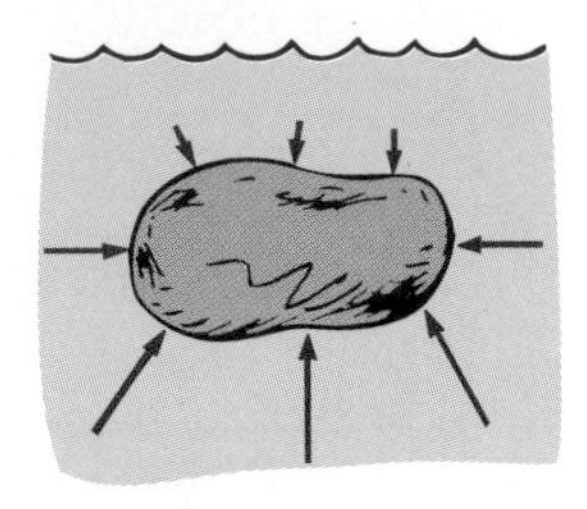

Fig. 5.8
The greater pressure against the bottom of a submerged object produces an upward buoyant force.

Anyone who has ever lifted a submerged object out of water is familiar with *buoyancy,* the apparent loss of weight of submerged objects. For example, lifting a large boulder off the bottom of a riverbed is a relatively easy task as long as the boulder is below the surface. When it is lifted above the surface, however, the force required to lift it is considerably more. This is because when the boulder is submerged the water exerts on it an upward force in the direction opposite the direction of gravitational force. This upward force is called the **buoyant force** and is a consequence of pressure increasing with depth. Figure 5.8 shows why the buoyant force acts upward. Pressure is exerted everywhere against the object in a direction perpendicular to its surface. The arrows represent the magnitude and directions of forces at different places. Forces that produce pressures against opposite sides cancel one another because they are at the same depth. Pressure is greatest against the bottom of the boulder because the bottom is at the greatest depth. The upward forces against the bottom are greater than the downward forces against the top, so the forces do not cancel and there is a net force upward. This net force is the buoyant force.

If the weight of the submerged object is greater than the buoyant force, the object sinks. If the weight is equal to the buoyant force, the object remains at any level, like a fish. If the buoyant force is greater than the weight of the completely submerged object, the object rises to the surface and floats.

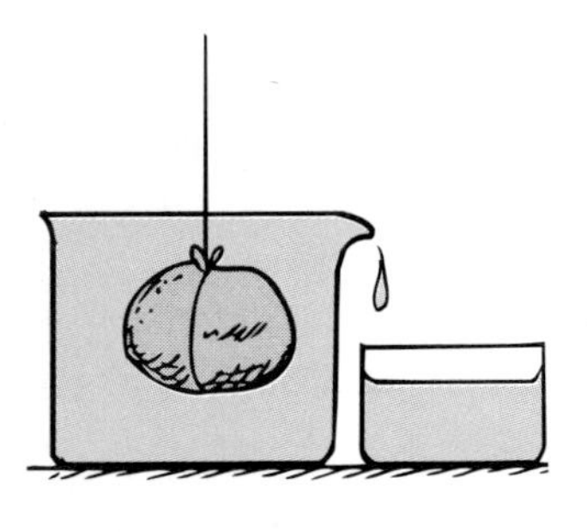

Fig. 5.9
When a stone is submerged, it displaces a volume of water equal to the volume of the stone.

Understanding buoyancy requires understanding the meaning of the expression "volume of water displaced." If a stone is placed in a container that is brimful of water, some water will overflow (Figure 5.9). Water is *displaced* by the stone. A little thought tells you that the *volume of the stone*—that is, the amount of space it takes up or the number of its cubic centimeters—is equal to the *volume of water displaced.* Place any object in a container partly filled with water, and the level of the surface rises (Figure 5.10). By how much? By exactly the same amount as if we had added a volume of water equal to the volume of the immersed object. This is a good method for determining the volume of irregularly shaped objects: *A completely submerged object always displaces a volume of liquid equal to its own volume.*

Fig. 5.10
The increase in water level is the same as that which would occur if, instead of putting the stone in the container, we had poured in a volume of water equal to the stone's volume. 7

5.4 Archimedes' Principle

The relationship between buoyancy and displaced liquid was first discovered in the third century B.C. by the Greek scientist Archimedes. It is stated as follows:

An immersed body is buoyed up by a force equal to the weight of the fluid it displaces.

Fig. 5.11
A liter of water occupies a volume of 1000 cm^3, has a mass of 1 kg, and weighs 9.8 N. Its density may therefore be expressed as 1 kg/L and its weight density as 9.8 N/L. (Seawater is slightly denser, about 10 N/L).

This relationship is called **Archimedes' principle** and is true of all fluids, both liquids and gases. If an immersed body displaces 1 kilogram of fluid, the buoyant force acting on the body is equal to the weight of 1 kilogram.[5] By *immersed,* we mean either *completely* or *partially submerged.* If we immerse a sealed 1-liter container halfway into a tub of water, the container displaces a half-liter of water and is buoyed up by a force equal to the weight of a half-liter of water. If we immerse the container completely (submerge it), it is buoyed up by a force equal to the weight of a full liter or 1 kilogram of water (which is 9.8 newtons). Unless the container is compressed, the buoyant force is 9.8 newtons at *any* depth, as long as the container is completely submerged. This is because at any depth the container can displace no greater volume of water than its own volume. And the weight of this volume of water (not the weight of the submerged object!) is equal to the buoyant force.

If, upon being immersed, a 25-newton object displaces 20 newtons of fluid, the buoyant force pushing upward on the object is 20 newtons and its apparent weight is 5 newtons. Note that in Figure 5.12 the 3-newton block has an apparent weight of 1 newton when submerged. The apparent weight of a submerged object is its weight in air minus the buoyant force.

Fig. 5.12
Objects weigh more in air than in water. When submerged, this 3-N block appears to weigh only 1 N. The "missing" weight is equal to the weight of water displaced, 2 N, which equals the buoyant force.

Questions

1. Does Archimedes' principle tell us that if an immersed object displaces 10 N of fluid, the buoyant force on the object is 10 N?
2. A 1-L container completely filled with lead has a mass of 11.3 kg and is submerged in water. What is the buoyant force acting on it?
3. A boulder is thrown into a deep lake. As it sinks deeper and deeper, does the buoyant force on it increase, decrease, or remain constant?

Answers

1. Yes. Looking at it another way, the immersed object pushes aside a volume of fluid that weighs 10 N. The displaced fluid reacts by pushing back on the immersed object with a force of 10 N.
2. The buoyant force is equal to the weight of 1 kg (9.8 N) because the volume of water displaced is 1 L, which has a mass of 1 kg and weighs 9.8 N. The 11.3 kg is irrelevant; 1 L of *anything* submerged in water displaces 1 L and is buoyed up with a force of 9.8 N, the weight of 1 kg.
3. The buoyant force does not change because the boulder displaces the same volume of water at any depth. Because water is practically incompressible, its density is very nearly the same at all depths; hence, the weight of water displaced, or the buoyant force, is practically the same at all depths.

---

[5]Remember from Chapter 2 that a kilogram is not a unit of force but a unit of mass. So, strictly speaking, the buoyant force is not 1 kg, but the *weight* of 1 kg, which is 9.8 N. We could as well say that the buoyant force is 1 *kilogram weight,* not simply 1 kg.

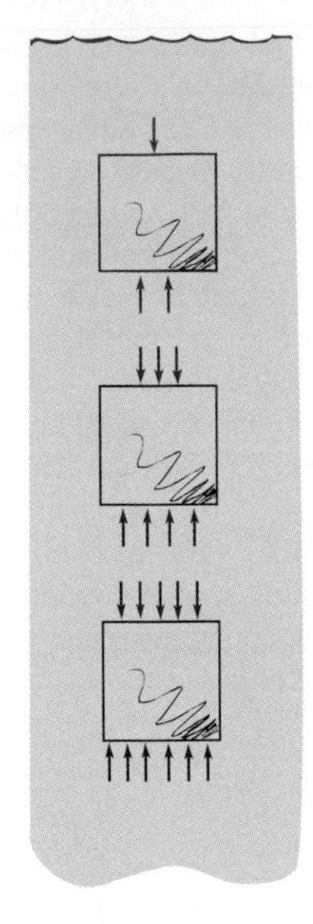

Fig. 5.13
The difference in the upward and downward force acting on the submerged block is the same at any depth.

Perhaps your instructor will summarize Archimedes' principle by way of a numerical example showing that the difference between the upward-acting and downward-acting forces is proportional to the pressure difference on a submerged cube. And you'll see that the difference in forces is numerically identical to the weight of fluid displaced. It makes no difference how deep the cube is placed, for although the pressure is greater with increasing depth, the *difference* between the pressure exerted upward against the bottom of the cube and the pressure exerted downward against the top of the cube is the same at any depth (Figure 5.13). Whatever the shape of the submerged body, the buoyant force is equal to the weight of fluid displaced.

5.5 Flotation

Iron is nearly eight times as dense as water and therefore sinks, but an iron ship floats. Why? Consider a solid 1-ton block of iron. When submerged, it doesn't displace 1 ton of water because it's eight times more compact than water. It displaces only $\frac{1}{8}$ ton of water—certainly not enough to keep it from sinking. Suppose we reshape the same iron block into a bowl, as shown in Figure 5.14. It still weighs 1 ton, but now when we place it in the water, it displaces a greater volume of water than before because its own volume has been increased. The deeper it is immersed, the more water it displaces and the greater the buoyant force acting on it. When the buoyant force equals 1 ton, it sinks no farther.

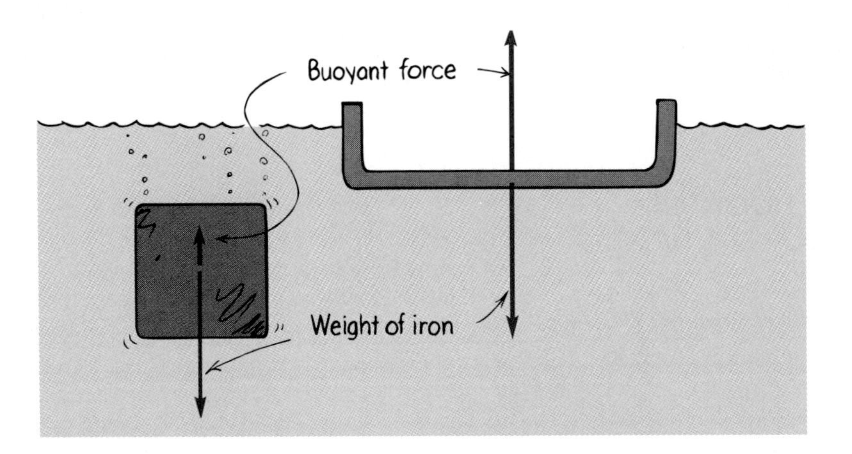

Fig. 5.14
An iron block sinks, but the same quantity of iron shaped like a bowl floats.

When the iron bowl displaces a weight of water equal to its own weight, it floats. This is sometimes called the **principle of flotation**:

A floating object displaces a weight of fluid equal to its own weight.

Every ship, every submarine, and every dirigible must be designed to displace a weight of fluid equal to its own weight. Thus a 10,000-ton ship must be built wide enough to displace 10,000 tons of water before it sinks too deep in the water. The same holds true for vessels in air, which is also a fluid. A dirigible or huge balloon that weighs 100 tons displaces at least 100 tons of air. If it displaces more, it rises; if it displaces less, it falls. If it displaces exactly its weight, it hovers at constant altitude.

Fig. 5.16
A floating object displaces a weight of fluid equal to its own weight.

Because the buoyant force acting on a body equals the weight of the fluid the body displaces, denser fluids exert a greater buoyant force on a body than do less dense fluids. A ship therefore floats higher in salt water than in fresh water because salt water is slightly denser than fresh water. In the same way, a solid chunk of iron floats in mercury even though it sinks in water.

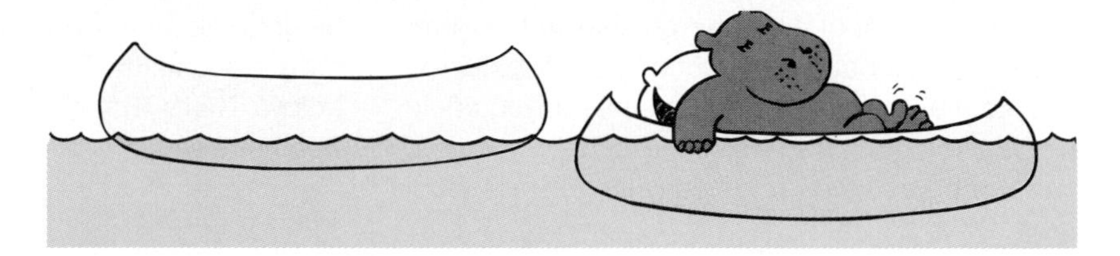

Fig. 5.15
The weight of a floating object equals the weight of the water displaced by the submerged part.

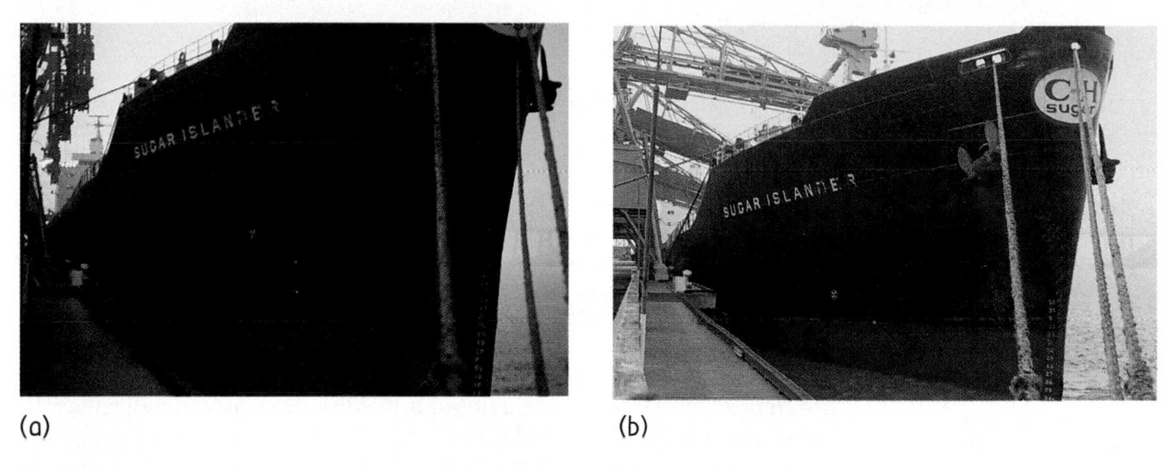

Fig. 5.17
The same ship empty (a) and loaded (b). How does the weight of its load compare with the weight of extra water displaced?

Link to History—Archimedes and the Gold Crown

According to popular legend, Archimedes (287–212 B.C.) was given the task of determining whether a crown made for King Hieron II was of pure gold or whether it contained some cheaper metals. Archimedes' problem was to determine the density of the crown without destroying it. He could weigh the gold, but determining its volume was a problem. Story has it that he came to the solution while bathing in the public baths of Syracuse and immediately rushed naked through the streets shouting "Eureka, Eureka" ("I have found it, I have found it").

What Archimedes had discovered was a simple and accurate way of finding the volume of an irregular object—the displacement method of determining volumes. Once he knew both the weight and volume of the crown, he could calculate its density. Then that density could be compared with the density of gold. Archimedes' insight preceded Newton's law of motion, from which Archimedes' principle can be derived, by almost 2000 years.

Link to Geology—Floating Mountains

Mountains float on the Earth's semimolten mantle just as icebergs float in water. Both mountains and icebergs are less dense than the material they float upon. Just as most of an iceberg is below the water surface (90%), most of a mountain (about 85%) extends into the dense semiliquid mantle. If you could shave off the top of an iceberg, the iceberg would be lighter and be buoyed up to nearly its original height before its top was shaved. Similarly, when mountains erode and wear away, they become lighter and are pushed up from below to float to nearly their original heights. So when a kilometer of mountain erodes away, about 0.85 kilometer of mountain pops up from below. That's why it takes so long for mountains to weather away.

■ Questions

Fill in the blanks:

1. The volume of a submerged body is equal to the __________ of the fluid displaced.
2. The weight of a floating body is equal to the __________ of the fluid displaced.

5.6 Pressure in a Gas

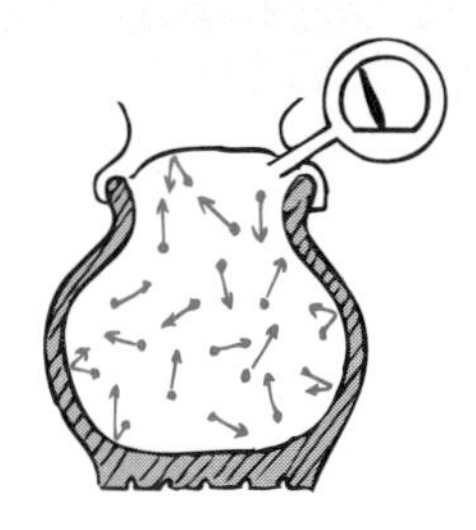

There are similarities and there are differences between gases and liquids. The primary difference between a gas and a liquid is the distance between molecules. In a gas, the molecules are far apart and free from the cohesive forces that dominate their motions when in the liquid and solid phases. The motions of gas molecules are less restricted. A gas expands, fills all the space available to it, and exerts a pressure against its container. Only when the quantity of gas is very large, such as the Earth's atmosphere or a star, do gravitational forces limit the size or determine the shape of the mass of gas.

Fig. 5.18
When the density of gas in the tire is increased, pressure is increased.

Boyle's Law

We know that the air pressure inside the inflated tires of an automobile is considerably greater than the atmospheric pressure outside. The density of air inside is also more than that of the air outside. To understand the relationship between pressure and density, think of the molecules of air (primarily nitrogen and oxygen) inside the tire, where they behave like tiny billiard balls, perpetually moving helter skelter and banging against the inner walls. Their impacts on the inner surface of the tire produce a jittery force that appears to our coarse senses as a steady push. This pushing force averaged over a unit of area provides the pressure of the enclosed air.

■ Answers

1. Volume
2. Weight

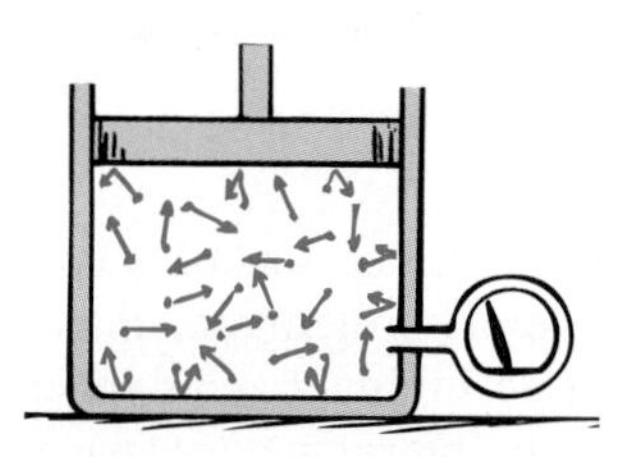

Suppose we now add more air until there are twice as many molecules in the same volume (Figure 5.18). The air density is now doubled. If the molecules move at the same average speed—or, equivalently, if they have the same temperature—then the number of collisions is doubled. This means the pressure is doubled. So pressure is proportional to density.

In the tire we doubled the density by doubling the amount of air. We can also double the density of a *fixed* amount of air by compressing it to half its volume. Consider the cylinder and movable piston in Figure 5.19. If the piston is pushed downward so that the volume is half the original volume, the density of molecules is doubled, and the pressure is correspondingly doubled. Decrease the volume to a third its original value, and the pressure increases by three, and so forth (provided the temperature remains the same).

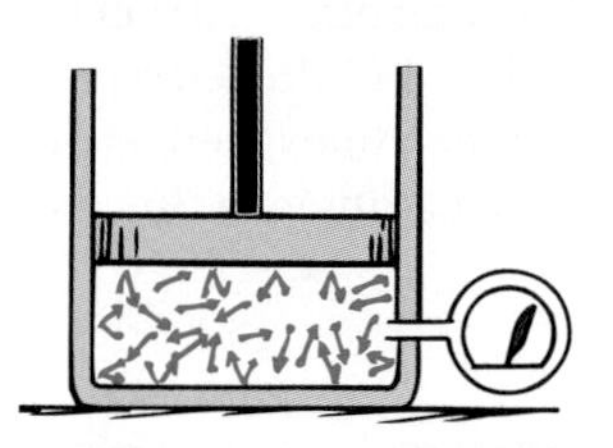

Fig. 5.19
When the volume of gas is decreased, density and therefore pressure are increased.

Notice in these examples with the piston that the product of pressure and volume remains constant. For example, a doubled pressure multiplied by a halved volume gives the same value as a tripled pressure multiplied by a one-third volume. In general, the product of pressure and volume for a given mass of gas is a constant as long as the temperature does not change. "Pressure × volume" for a quantity of gas at one time is equal to any "different pressure × different volume" at any other time. In shorthand notation,

$$P_1V_1 = P_2V_2$$

where P_1 and V_1 represent the original pressure and volume, respectively, and P_2 and V_2 the second pressure and volume. This relationship is called **Boyle's Law,** after Robert Law, the seventeenth-century physicist who is credited with its discovery.[6]

Boyle's law applies to ideal gases. An *ideal gas* is one in which the disturbing effects of the forces between molecules and the finite size of the individual molecules can be neglected. Air and other gases under normal pressures approach ideal-gas conditions.

Questions

1. A piston in an airtight pump is withdrawn so that the volume of the air chamber is increased by a factor of 3. What is the change in pressure?
2. A scuba diver beneath the water surface breathes compressed air. Why is it dangerous for the diver to hold her breath while ascending?

Answers

1. The pressure in the piston chamber is reduced to one-third its original value. This principle underlies a mechanical vacuum pump.
2. In accordance with Boyle's law, if the diver holds her breath while ascending (while surrounding pressure decreases), her lungs tend to inflate—ouch! A first lesson in scuba diving is not to hold your breath when ascending. To do so can be fatal.

[6]Humor aside, Boyle's law is named after Robert Boyle. A general law that takes temperature changes into account is $P_1V_1/T_1 = P_2V_2/T_2$, where T_1 and T_2 represent the initial and final *absolute* temperatures, measured in SI units called kelvins (Chapter 8).

5.7 Atmospheric Pressure

We live at the bottom of an ocean of air. Unlike the situation with oceans of water, however, there is no sharp surface at the "top" of the atmosphere. Furthermore, unlike water, the atmosphere becomes less dense with altitude. Air gets thinner and thinner the higher one goes—eventually thinning to near emptiness in interplanetary space. About 50 percent of the atmosphere is below an altitude of 5.6 kilometers (18,000 ft), 75 percent is below 11 kilometers (36,000 ft), 90 percent is below 18 kilometers, and 99 percent is below about 30 kilometers. Although the top of the atmosphere is not well defined, for practical purposes we consider it as being at an altitude of 30 or so kilometers.

Fig. 5.20
The famous "Magdeburg hemispheres" experiment, demonstrating atmosphere pressure. Two teams of horses couldn't pull the evacuated hemisphere apart. Were the hemispheres sucked together or pushed together? By what?

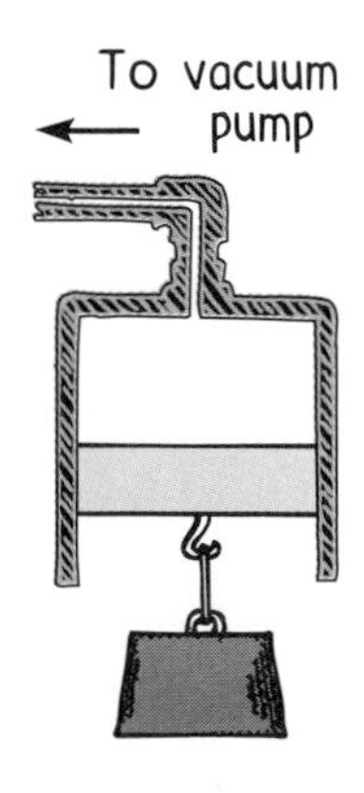

Fig. 5.21
Is the piston that supports the load pulled up or pushed up?

Like water in the ocean, the atmosphere exerts a pressure. One of the most celebrated experiments demonstrating atmospheric pressure was conducted in 1654 by Otto von Gueicke, burgermeister of Magdeburg and inventor of the vacuum pump. Von Gueicke placed together two copper hemispheres about $\frac{1}{2}$ meter in diameter to form a sphere, as shown in Figure 5.20. He set a gasket made of a ring of leather soaked in oil and wax between them to make an airtight joint. When he evacuated the sphere with his vacuum pump, two teams of eight horses each were unable to pull them apart.

When the air pressure inside a cylinder like that shown in Figure 5.21 is reduced, there is an upward force on the piston. This force is large enough to lift a heavy weight. If the inside diameter of the cylinder is 12 centimeters or greater, a person can be lifted by this force.

What do the experiments of Figures 5.20 and 5.21 demonstrate? Do they show that air exerts pressure or that there is a "force of suction"? If we say there is a force of suction, then we assume that a vacuum can exert a force. But what is a vacuum? It is an absence of matter; it is a condition of nothingness. How can nothing exert a force? The hemispheres are not sucked together, nor is the piston holding the weight sucked upward. The hemispheres and the piston are being pushed against by the pressure of the atmosphere.

Fig. 5.22
You don't notice the weight of a bag of water while you're submerged in water. Similarly, you don't notice the weight of air while you are submerged in an "ocean" of air.

Just as water pressure is caused by the weight of water, **atmospheric pressure** is caused by the weight of air. We humans have adapted so completely to the invisible air surrounding us that we sometimes forget it has weight. Perhaps a fish "forgets" about the weight of water in the same way. The reason we don't feel this weight crushing against our bodies is that the pressure inside our bodies equals that of the surrounding air. There is no net force for us to sense.

At sea level, 1 cubic meter of air at 20°C has a mass of about 1.2 kilograms. Estimate the number of cubic meters in your room, multiply by 1.2 kilograms per cubic meter, and you'll have the mass of air in your room. Multiply by 9.8 and you'll have its weight in newtons. Don't be surprised if it's heavier than your kid sister. If your kid sister doesn't believe air has weight, maybe it's because she's always surrounded by it. Hand her a plastic bag of water and she'll tell you it has weight. But hand her the same bag of water while she's submerged in a swimming pool, and she won't feel the weight. Similarly with air; we don't notice its weight because we're submerged in air. The weight of air produces atmospheric pressure.

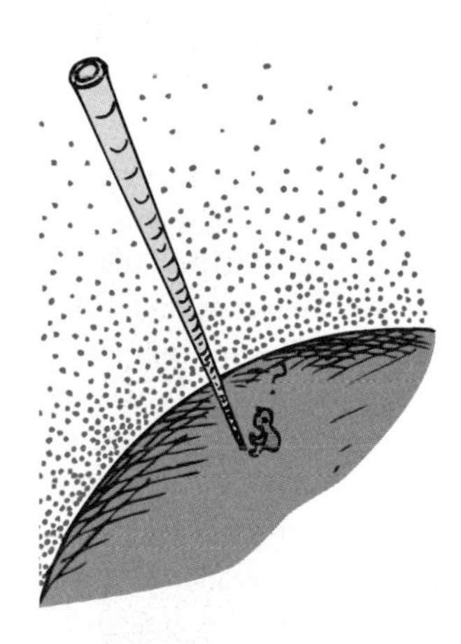

Fig. 5.23
The mass of air that would occupy a bamboo pole that extends 30 km up—to the "top" of the atmosphere—is about 1 kg. This air weighs about 10 N.

Unlike the constant density of water in a lake, the density of air in the atmosphere decreases with altitude. At 10 kilometers, 1 cubic meter of air has a mass of only about 0.4 kilogram. To compensate for the corresponding low atmospheric pressure at high altitudes, airplanes are pressurized; the additional air needed to fully pressurize a 747 jumbo jet, for example, is more than 1000 kilograms. Air is heavy if you have enough of it.

Consider the mass of air in an upright 30-kilometer-tall bamboo pole that has an inside cross-sectional area of 1 square centimeter. If the density of air inside the pole matches the density of air outside, the inside air has a mass of about 1 kilogram. Thc weight of this much air is about 10 newtons. So the air pressure at the bottom of the pole is about 10 newtons per square centimeter (10 N/cm^2). Of course, the same is true without the bamboo pole. There are 10,000 square centimeters in 1 square meter, so a column of air 1 square meter in cross section that extends all the way up through the atmosphere has a mass of about 10,000 kilograms. The weight of this air is about 100,000 newtons (10^5 N). This weight produces a pressure of 100,000 newtons per square meter—or, equivalently, 100,000 pascals, or 100 kilopascals. To be more exact, the average atmospheric pressure at sea level is 101.3 kilopascals (101.3 kPa).[7]

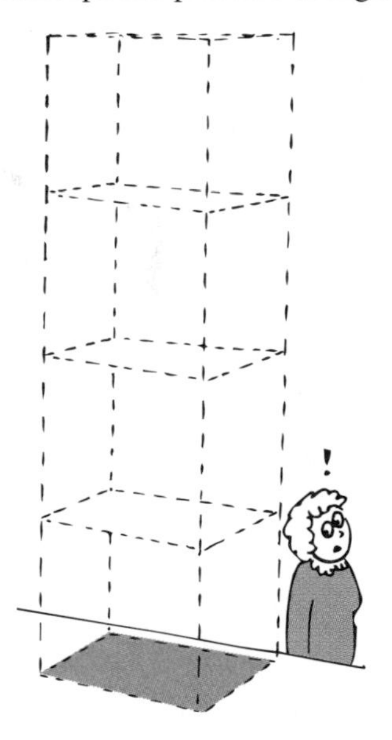

Fig. 5.24
The weight of the air bearing down on a 1=m^2 surface at sea level is about 100,000 N. In other words, atmospheric pressure is about 10^5 N/m^2, or about 100 kPa.

The pressure of the atmosphere is not uniform. Besides altitude variations, there are variations at any one

[7]The average atmospheric pressure at sea level is often called 1 atmosphere (atm). In British units, the average atmospheric pressure at sea level is 14.7 lb/in^2 (psi).

locality due to moving fronts and storms. Measurement of changing air pressure is important to meteorologists in predicting weather.

Questions

1. About how many kilograms of air fill a classroom that has a 200-m^2 floor area and a 4-m-high ceiling? (See Table 5.1 for air densities, and assume a chilly 10°C temperature).
2. Why doesn't the pressure of the atmosphere break windows?

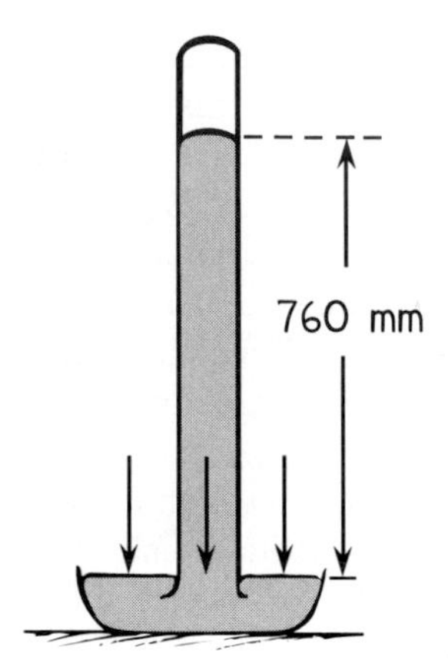

Fig. 5.25
A simple mercury barometer.

Barometers

Instruments used for measuring the pressure of the atmosphere are called **barometers.** A simple mercury barometer is illustrated in Figure 5.25. A glass tube, longer than 760 millimeters (76 centimeters) and closed at one end, is filled with mercury and tipped upside down into a dish of mercury. The mercury in the tube runs out of the submerged open bottom until the difference in the mercury levels in the tube and the dish is 760 millimeters. Except for some trace amounts of mercury vapor, the empty space trapped above is a pure vacuum. The vertical height of the mercury column remains constant even when the tube is tilted, unless the top of the tube is less than 76 centimeters above the level in the dish—in which case the mercury completely fills the tube.

Why does mercury behave this way? The explanation is similar to the reason a simple see-saw balances when the weights of the people at its two ends are equal. The barometer "balances" when the weight of liquid in the tube exerts the same pressure on the mercury dish beneath as that exerted by the atmosphere outside. Whatever the width of the tube, a 760-millimeter column of mercury weighs the same as the air that would fill a vertical 30-kilometer tube of the same width. If the atmospheric pressure increases, then the atmosphere pushes down harder on the mercury in the dish and it takes a column of mercury higher than 760 millimeters to exert an equal, balancing pressure. The mercury is literally pushed up into the tube of a barometer by atmospheric pressure.

Fig. 5.26
Strictly speaking, they do not suck the soda up the straws. They instead reduce air pressure in the straws and allow the weight of the atmosphere to press the liquid up into the straws. Could they drink a soda this way on the Moon?

Could water be used to make a barometer? The answer is yes, but the glass tube would have to be much longer—13.6 times as long, to be exact. You may recognize this number as the magnitude of the density of mercury. Because mercury is 13.6 times more dense than water, a volume of water 13.6 times that of mercury is needed to provide the same weight as the mercury in the tube. So the tube would have to be at least 13.6 times taller than the mercury column. A water barometer would have to be 13.6 $\times$ 0.76 meter—10.3 meters high—too tall to be practical.

What happens in a barometer is similar to what happens when you drink through a straw. By sucking on the straw in a can of soda, you reduce the air pressure in the straw. Atmospheric pressure on the surface of the soda pushes liquid up into the reduced-

---

Answers

1. The volume of air is 200 m^2 $\times$ 4 m = 800 m^3; each cubic meter of air has a mass of about 1.25 kg, and so 800 m^3 $\times$ 1.25 kg/m^3 = 1000 kg.
2. Atmospheric pressure is exerted on both sides of a window, so no net force is exerted on the window. If for some reason the pressure is reduced or increased on one side only, then watch out! It can happen when a tornado passes nearby.

Fig. 5.27
The atmosphere pushes water from below up a pipe that is evacuated of air by the pumping action.

pressure region. Strictly speaking, the liquid is not sucked up; it is pushed up by the pressure of the atmosphere. If the atmosphere is prevented from pushing on the surface of the drink, as in the party-trick bottle with the straw through the air-tight cork stopper, one can suck and suck and get no drink.

If you understand these ideas, you can understand why there is a 10.3-meter limit on the height water can be lifted with a vacuum pump. The old fashioned farm-type pump, shown in Figure 5.27, operates by producing a partial vacuum in a pipe that extends down into the water below. The atmospheric pressure exerted on the surrounding surface of the water simply pushes the water up into the region of reduced pressure inside the pipe. Can you see that even with a perfect vacuum, the maximum height to which water can be lifted in this way is 10.3 meters?

■ Question

What is the maximum height to which water can be drunk through a straw?

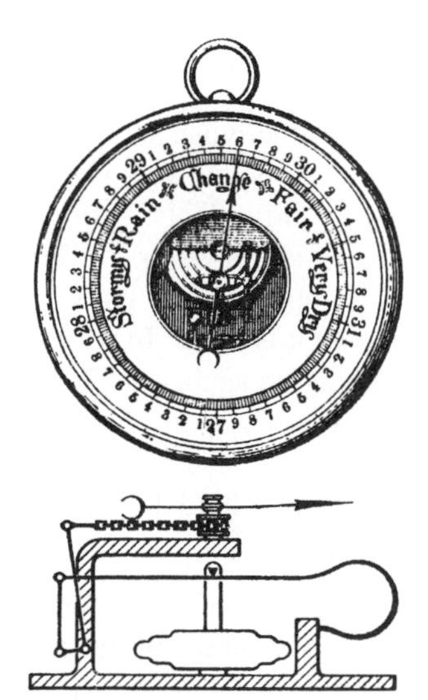

Fig. 5.28
An aneroid barometer (top) and its cross section (bottom).

A small portable instrument that measures atmospheric pressure is the *aneroid barometer* (Figure 5.2). It uses a metal box that is partially exhausted of air and has a slightly flexible lid that bends in or out with changes in atmospheric pressure. Motion of the lid is indicated on a scale by a mechanical spring-and-lever system. Because atmospheric pressure decreases with increasing altitude, a barometer can be used to determine elevation. An aneroid barometer calibrated for altitude is called an altimeter (altitude meter). Some of these instruments are sensitive enough to indicate a change in elevation as you walk up a flight of stairs.[8]

The pressure ("vacuum") inside a television picture tube is about 1 ten-thousandth pascal (10^{-4} Pa). At an altitude of about 500 kilometers—artificial satellite territory—gas pressure is about one ten-thousandth of this (10^{-8} Pa). This is a pretty good vacuum by earthbound standards. Still greater vacuums exist in the wakes of satellites orbiting at this distance, and they can reach 10^{-13} pascals. This is called a "hard vacuum."

■ **Answer**
However strong your lungs may be, or whatever device you use to make a vacuum in the straw, at sea level the water cannot be pushed up by the atmosphere higher than 10.3 m.

---

[8]Evidence of a noticeable pressure difference over a 1-m or less difference in elevation can be seen in any small helium-filled balloon. The atmosphere really does push harder against the lower bottom than against the higher top!

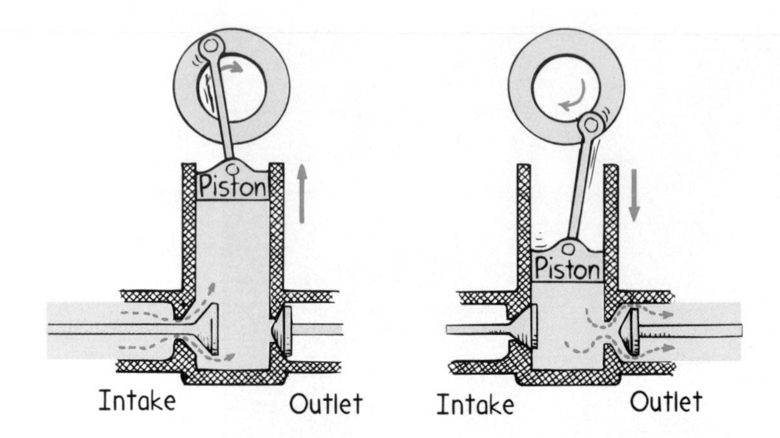

Fig. 5.29
A mechanical vacuum pump. When the piston is lifted, the intake valve opens and air moves in to fill the empty space. When the piston is moved downward, the outlet valve opens and the air is pushed out. What changes would you make to convert this pump to an air compressor?

Vacuums on the Earth are produced by pumps, which operate by the tendency of a gas to fill its container. If a space with low gas pressure is provided, gas flows from the region of higher pressure to the one of lower pressure. A vacuum pump simply provides a region of lower pressure into which normally fast-moving gas molecules randomly move. The air pressure is repeatedly lowered by piston and valve action (Figure 5.29). The best vacuums attainable with mechanical pumps are about 1 pascal. Better vacuums, down to 10^{-8} pascal, are attainable with vapor diffusion pumps. Vacuums greater than that are very difficult to attain. Technologists requiring hard vacuums are looking at the prospects of orbiting laboratories in space.

5.8 Buoyancy in a Gas

Fig. 5.30
All bodies are buoyed up by a force equal to the weight of the air they displace. Why, then, don't all objects float like this balloon?

A crab lives at the bottom of its ocean of water and looks upward at jellyfish and other lighter-than-water objects drifting above it. Similarly, we live at the bottom of our ocean of air and look upward at balloons and other lighter-than-air objects drifting above us. A balloon is suspended in air and a jellyfish is suspended in water for the same reason: Each is buoyed upward by a displaced weight of fluid equal to its own weight. In one case the displaced fluid is air, and in the other case it is water. In water, immersed objects are buoyed upward because the pressure acting up against the bottom of them exceeds the pressure acting down against the top. Likewise, air pressure acting up against an object immersed in air is greater than the pressure above pushing down. The buoyancy in both cases is numerically equal to the weight of fluid displaced. Archimedes' principle holds for air just as it does for water:

An object surrounded by air is buoyed up by a force equal to the weight of the air displaced.

We know that a cubic meter of air at normal atmospheric pressure and room temperature has a mass of about 1.2 kilograms, so its weight is about 12 newtons. Therefore any 1-cubic-meter object in air is buoyed up with a force of 12 newtons. If the mass of the 1-cubic-meter object is greater than 1.2 kilograms (so that its weight is greater than 12 newtons), it falls to the ground when released. If this size object has a mass less than 1.2 kilograms, the buoyant force is greater than its weight and it rises in the air. Any object that has a mass less than the mass of an equal volume of air rises in the air. Another way to say this is that any object less dense than air rises in air. Gas-filled balloons that rise in air are less dense than air.

The gas used in a balloon prevents the atmosphere from collapsing the balloon. No gas at all would mean no weight, which is good, but nothing to prevent collapse, which is bad. Hydrogen is the least heavy gas but is seldom used in balloons because it is highly flammable. In sport balloons, the gas is heated air. In balloons intended to reach very high altitudes or to stay up a long time, helium is usually used. Its density is low enough that the combined weight of helium, balloon, and cargo is less than the weight of the air it displaces. Low-density gas is used in a balloon for the same reason cork is used in a swimmer's life preserver. The cork possesses no strange tendency to be drawn toward the surface of water, and the gas possess no strange tendency to rise. Cork and gases are buoyed upward like anything else. They are simply light enough for the buoyancy to be significant.

Will a lighter-than-air balloon rise indefinitely? Just how high will it rise? We can state the answer in several ways. A balloon will rise only so long as it displaces a volume of air that weighs more than the balloon. Because air becomes less dense with altitude, the same volume of air (that equal to the balloon's volume) weighs less and less as the balloon rises. When the weight of displaced air equals the weight of the balloon, the balloon's upward acceleration ceases. We can also say that when the buoyant force on the balloon equals its weight, the balloon ceases rising. Equivalently, when the density of the balloon equals the density of the surrounding air, the balloon will cease rising. Helium-filled toy balloons usually break when released in the air when expansion stretches the rubber until it ruptures.

Questions

1. Is there a buoyant force acting on you? If there is, why are you not buoyed up by it?
2. (This one calls for your best thinking!) How does buoyancy change as a helium-filled balloon ascends?

Large dirigible airships are designed so that when loaded they will slowly rise in air; that is, their total weight is a little less than the weight of air displaced. When in motion, the ship may be raised or lowered by means of horizontal "elevators."

Thus far we have treated pressure only as it applies to stationary fluids. Motion produces an additional influence.

Answers

1. There is a buoyant force acting on you, and you are buoyed up by it. You don't notice it only because your weight is so much greater.
2. If the balloon is free to expand as it rises, the increase in volume is counteracted by a decrease in the density of higher-altitude air. So, interestingly, the greater volume of displaced air doesn't weigh more, and buoyancy stays the same. If a balloon is not free to expand, buoyancy will decrease as a balloon rises because of the less-dense displaced air. Usually balloons expand when they initially rise, and if they don't finally rupture, fabric stretching reaches a maximum and balloons settle where buoyancy matches weight.

5.9 Bernoulli's Principle

True or false: Atmospheric pressure increases in a gale, tornado, or hurricane. If you answered true, sorry; the statement is false. High-speed winds may blow the roof off your house, but the pressure within the winds is actually less than that of still air of the same density inside the house. As strange as it may first seem, when the speed of a fluid increases, the internal pressure decreases. This is true whether the fluid is a gas or a liquid.

Fig. 5.31
The pressure in the spout drops when the plug is removed.

Daniel Bernoulli, a Swiss scientist of the eighteenth century, studied the relationship between fluid speed and pressure. When a fluid flows through a narrow constriction, its speed increases. This is easily noticed by the increased speed of the water coming out of the nozzle of a garden hose when you narrow the opening of the nozzle. The fluid must speed up in the constricted region if the flow is to be continuous.

Bernoulli wondered how the fluid got this extra speed. He reasoned that it is acquired at the expense of a lowered internal pressure. His discovery, now called **Bernoulli's principle,** states

When the speed of a fluid increases, pressure in the fluid decreases.

Bernoulli's principle is a consequence of the conservation of energy. In a steady flow of fluid, there are three kinds of energy; kinetic energy due to motion, gravitational potential energy due to elevation, and work done by pressure forces. In a steady fluid flow where no energy is added or taken away, the sum of these forms of energy remains constant.[9] If the elevation of the flowing fluid doesn't change, then an increase in speed means a decrease in pressure, and vice versa.

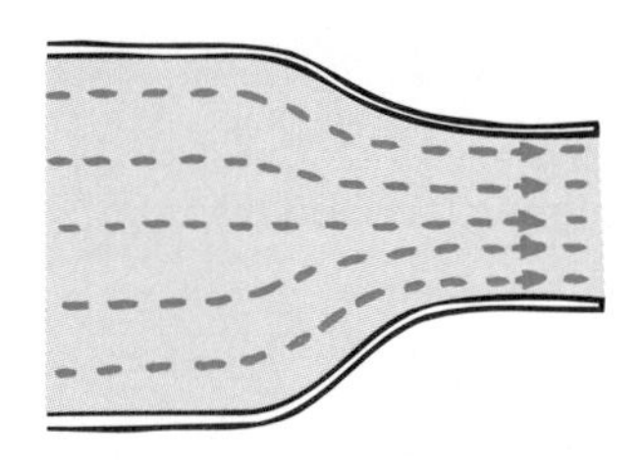

Fig. 5.32
Water speeds up when it flows into the narrower part of the pipe. The close-together streamlines there indicate increased speed and decreased internal pressure.

The decrease of fluid pressure with increasing speed may at first seem surprising, particularly if we fail to distinguish between the pressure *in* the fluid and the pressure exerted *by* the fluid on something that interferes with its flow. The pressure within fast-moving water in a fire hose is relatively low, whereas the pressure that the water can exert on anything in its path may be huge.

In steady flow, one small bit of fluid follows along the same path as a bit of fluid in front of it. The motion of a fluid in steady flow follows streamlines, which are represented by dashed lines in Figure 5.31 and later figures. *Streamlines* are the smooth paths, or trajectories, of the bits of fluid. The lines are closer together in narrower regions, where the flow speed is greater and the pressure within the fluid less.

Bernoulli's principle holds only for steady flow. If the speed is too great, the flow may become turbulent and follow a changing, curling path known as an *eddy***.** This type of flow exerts friction on the fluid and causes some of its energy to be transformed to heat. Then Bernoulli's principle does not hold.

---

[9] In mathematical form: $\frac{1}{2}mv^2 + mgy + pV = \text{constant}$; where m is the mass of a unit volume V, v its speed, p its pressure, y its elevation, and g the acceleration due to gravity. If mass m is expressed in terms of density (Greek symbol "rho"), $\rho = m/V$, and each term is divided by V, Bernoulli's equation takes the form: $\frac{1}{2}\rho v^2 + \rho gy + p = \text{constant}$. Then all three terms have units of pressure. If y does not change, then an increase in v means a decrease in ρ, and vice versa.

Applications of Bernoulli's Principle

Fig. 5.33
The paper rises when Tim blows air across its top surface.

Hold a sheet of paper in front of your mouth, as shown in Figure 5.33. When you blow across the top surface, the paper rises. This is because the pressure of the moving air against the top of the paper is less than the pressure of the air at rest against the lower surface.

We began our discussion of Bernoulli's principle by stating that atmospheric pressure decreases in a strong wind, tornado, or hurricane. As it turns out, an unvented building that has airtight, closed windows is in more danger of losing its roof in a storm than a well-vented building. This is because during the storm the air pressure inside may be appreciably greater than the reduced atmospheric pressure outside, and the roof is more likely to be pushed off by the higher-pressure air in the building than blown off by the wind. When the wind is blowing over a peaked roof as shown in Figure 5.34, the effect is even more pronounced. The crowding of the streamlines shows this. The difference between outside and inside pressure need not be very much. A small pressure difference over a large area can be formidable. So if you're ever caught in an unvented building in a tornado or hurricane, consider opening the windows a bit so that the pressures inside and outside are more nearly equal.[10]

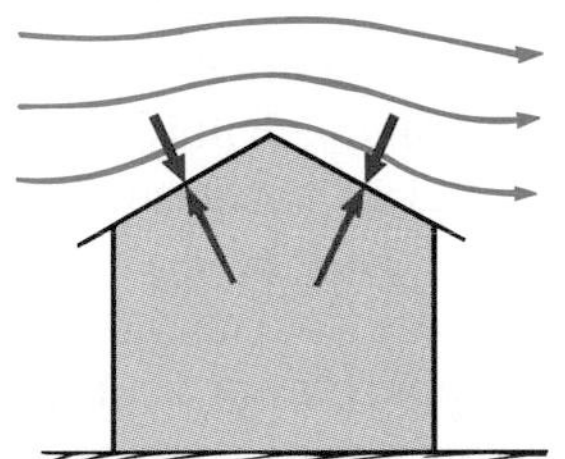

Fig. 5.34
Air pressure above the roof is less than air pressure beneath the roof.

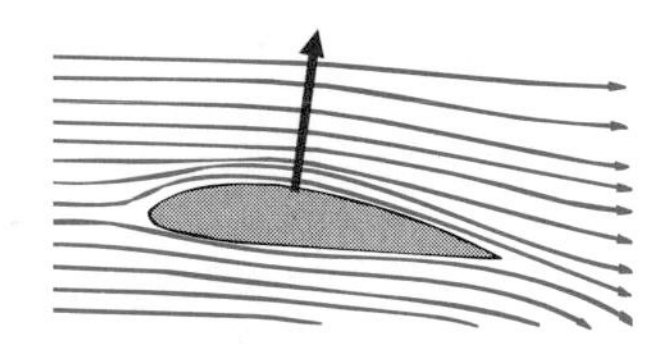

Fig. 5.35
The arrow represents the net upward force (lift) that results from less air pressure above the wing than below the wing.

Fig. 5.36
Where is air pressure greater—on the top or bottom surface of the hang glider?

If we think of a blown-off roof as an airplane wing, we can better understand the lifting force that supports a heavy airliner. In both cases a greater pressure below pushes the roof and wing into a region of lesser pressure above. When a wing moves through the air, the pressure difference is established by causing air to move faster over the top of the wing than across the bottom of the wing. This difference in speed is achieved by having the wing tilt slightly upward in front (where the tilt is called "angle of attack"). Some wings are flatter on the bottom and more curved on the top to accentuate the difference in speeds for the air going over and under the wing.

[10] A word of caution: In a building that has venting adequate to tolerate the sudden pressure drop without the need to open windows, opening a window may actually increase damage.

The net upward pressure on a wing multiplied by the surface area of the wing gives the net lifting force. Lift is greater when there is a large wing area and when the plane is traveling fast. Gliders have a very large wing area relative to their weight so that they do not have to be going very fast for sufficient lift. At the other extreme, fighter planes designed for high speed have small wing areas relative to their weight. Consequently, they must take off and land at high speeds.

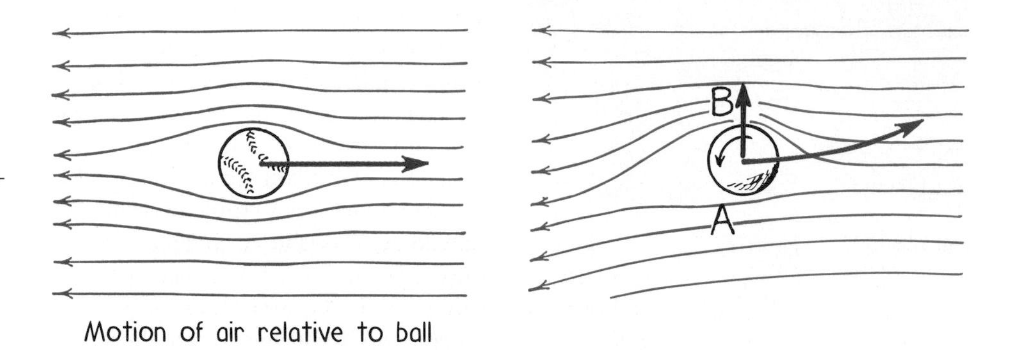

Fig. 5.37
Left: The streamlines are the same on either side of a non-spinning baseball. Right: A spinning ball produces a crowding of streamlines. The resulting "lift" (red arrow) causes the ball to curve as shown by the blue

Sports buffs know that a baseball pitcher can throw a ball in such a way that it curves off to one side of its trajectory. This is accomplished by imparting a large spin to the ball. Similarly, a tennis player can hit a ball that curves. A thin layer of air is dragged around the spinning ball by friction, which is enhanced by the baseball's threads or the tennis ball's fuzz. The moving layer produces a crowding of streamlines on one side. Note in Figure 5.37(b) that the streamlines are more crowded at B than at A for the direction of spin shown. Air pressure is therefore greater at A, and the ball curves as shown.

Bernoulli's principle accounts for the fact that passing ships run the risk of a sideways collision. Water flowing between the ships travels faster than water flowing past the outer sides, which means the streamlines are more compressed between the ships than outside. Water pressure acting against the hulls is therefore reduced between the ships. Unless the ships are steered to compensate for this, the greater pressure against the outer sides of the ships then forces them together. Figure 5.38 shows how this can be demonstrated in your kitchen sink or bathtub.

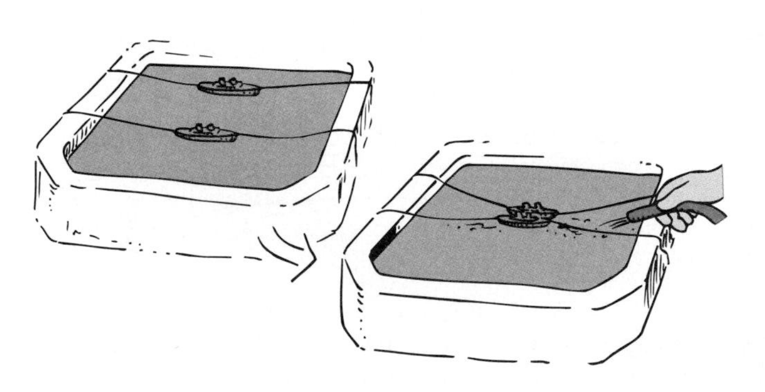

Fig. 5.38
Try this in your sink. Loosely moor a pair of toy boats side by side. Then direct a stream of water between them. The boats draw together and collide. Why?

Question

If you take a barometer outside on a windy day, will it give a lower reading than if you keep it indoors?

Answer

Because atmospheric pressure is reduced where the wind is blowing, the barometer should give a lower reading. (The effect is very small, however, and so you might not detect it unless the wind is very strong or your barometer is very sensitive.)

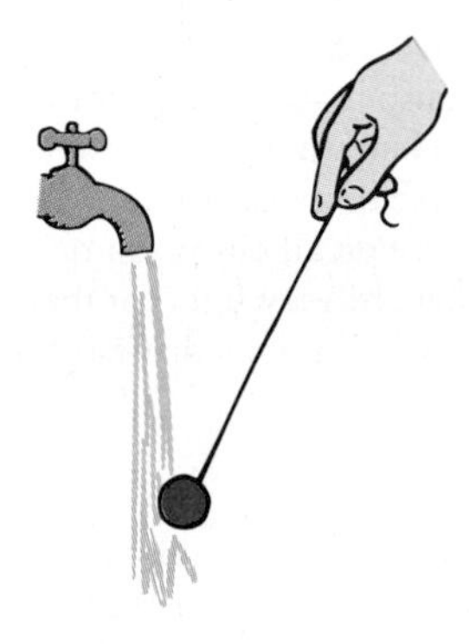

Fig. 5.39
Pressure is greater in the stationary fluid (air) than in the moving fluid (water). Therefore the atmosphere pushes the ball into the region of reduced pressure.

There's another way you can demonstrate Bernoulli's principle in your kitchen sink (Figure 5.39). Tape a Ping-Pong ball to a string and allow the ball to swing into a stream of running water. You'll see that it remains in the stream even when tugged slightly to the side, as shown. This is because the pressure of the stationary air next to the ball is greater than the pressure of the moving water. The ball is pushed into the region of reduced pressure by the atmosphere.

A similar thing happens to a bathroom shower curtain when the shower water is turned on full blast. The pressure in the shower stall is reduced, and the relatively greater pressure outside the curtain pushes it inward. Although convection produced by temperature differences may play an even greater role, think of Daniel Bernoulli the next time you're taking a shower and the curtain swings in against your legs!

Fig. 5.40
The curved shape of an umbrella can be disadvantageous on a windy day.

Summary of Terms

- **Density** The amount of matter per unit volume:

$$\text{Density} = \frac{\text{mass}}{\text{volume}}$$

Weight density is weight per unit volume.

- **Pressure** The ratio of force to the area over which that force is distributed:

$$\text{Pressure} = \frac{\text{force}}{\text{area}}$$

Liquid pressure = weight density × depth

- **Buoyant force** The net upward force that a fluid exerts on an immersed object.
- **Archimedes' principle** An immersed body is buoyed up by a force equal to the weight of the fluid it displaces.
- **Principle of flotation** A floating object displaces a weight of fluid equal to its own weight.
- **Boyle's law** The product of pressure and volume is constant for a given mass of confined gas so long as temperature remains unchanged:
$P_1V_1 = P_2V_2$
- **Atmospheric pressure** The pressure exerted against bodies immersed in the atmosphere resulting from the weight of air pressing down from above. At sea level, atmospheric pressure is about 101 kPa.
- **Barometer** Any device that measures atmospheric pressure.
- **Bernoulli's principle** When the speed of a fluid increases, pressure in the fluid decreases.

Review Questions

Density

1. Distinguish between the "heaviness" of a material and its density.
2. Distinguish between *mass density* and *weight density.* What are the mass density and weight density of water?

Pressure

3. Distinguish between *force* and *pressure.*
4. How does the pressure exerted by a liquid change with depth in the liquid? How does the pressure exerted by a liquid change as the density of the liquid changes?
5. (a) If you swim twice as deep in water, how much more water pressure is exerted on your ears? (b) If you swim in salt water, is the pressure greater than in fresh water at the same depth? Why or why not?
6. How does the water pressure 1 m below the surface of a small pond compare with the water pressure 1 m below the surface of a huge lake?
7. If you punch a hole in a container filled with water, in what direction does the water initially flow outward from the container?

Buoyancy in a Liquid

8. Why does buoyant force act upward on an object submerged in water?
9. How does the volume of a completely submerged object compare with the volume of water displaced?

Archimedes' Principle

10. How does the buoyant force that acts on a fish compare with the weight of the fish?

11. If a 1-L container is immersed halfway in water, what is the volume of water displaced? What is the buoyant force on the container?
12. How does the buoyant force on a fully submerged object compare with the weight of water displaced?
13. What is the mass in kilograms of 1 L of water? What is its weight in newtons?
14. Does the buoyant force on a fully submerged object depend on the weight of the object or on the weight of the fluid displaced by the object? Does the force depend on the weight of the object or on its volume? Defend your answer.

Flotation

15. What weight of water is displaced by a 100-ton ship? What is the buoyant force that acts on this ship?
16. Does the buoyant force on a floating object depend on the weight of the object or on the weight of the fluid displaced by the object? Or are these two weights the same for the special case of floating? Defend your answer.

Pressure in a Gas

17. Describe the main differences between liquids and gases.
18. By how much does the density of air increase when it is compressed to half its volume?
19. What happens to the air pressure inside a balloon when the balloon is squeezed to half its volume at constant temperature?
20. What is an ideal gas?

Atmospheric Pressure

21. What is the cause of atmospheric pressure?
22. What is the mass in kilograms of a cubic meter of air at room temperature (20°C)?
23. What is the approximate mass in kilograms of a column of air that has a cross-sectional area of 1 cm^2 and extends from sea level to the upper atmosphere? What is the weight in newtons of this amount of air?
24. What is the SI unit of atmospheric pressure?
25. How does the downward pressure of the 76-cm column of mercury in a barometer compare with the air pressure at the bottom of the atmosphere?
26. How does the weight of mercury in a barometer tube compare with the weight of an equal cross section of air from sea level to the top of the atmosphere?
27. Why would a water barometer have to be 13.6 times taller than a mercury barometer?
28. When you drink liquid through a straw, is it more accurate to say the liquid is pushed up the straw rather than sucked up? What exactly does the pushing? Defend your answer.
29. Why will a vacuum pump not operate for a well that is more than 10.3 m deep?

Buoyancy in a Gas

30. A balloon that weighs 1 N is suspended in air, drifting neither up nor down. How much buoyant force acts on it? What happens if the buoyant force decreases? Increases?
31. Does the air exert a buoyant force on all objects or only on objects such as balloons that are very light for their size? Why does the air support only things that have a very low density?

Bernoulli's Principle

32. What happens to the internal pressure in a fluid flowing in a horizontal pipe when the fluid's speed increases?
33. What are streamlines? Is pressure greater or less in regions where streamlines are crowded?
34. (a) What does Bernoulli's principle have to do with the flight of airplanes? (b) What does the tilt of the wing have to do with the flight of airplanes?
35. Why does a spinning ball curve in its flight?
36. Does atmospheric pressure increase or decrease in a strong wind?

Home Projects

1. Try to float an egg in water. Then dissolve salt in the water until the egg floats. How does the density of an egg compare to that of tap water? To salt water?
2. Punch a couple of holes in the bottom of a water-filled container, and water will spurt out because of water pressure. Now drop the container, and as it freely falls note that the water no longer spurts out! If your friends don't understand this, could you figure it out and then explain it to them?

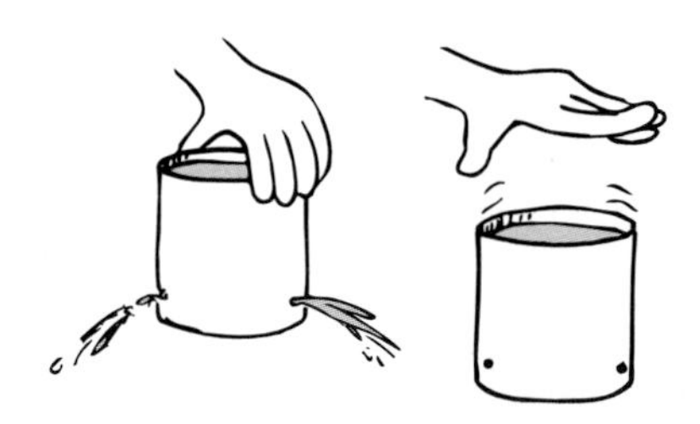

3. Place a wet Ping-Pong ball in a can of water held high above your head. Then drop the can on a rigid floor. As the can falls, surface tension pulls the ball beneath the surface. What then happens as the can comes to an abrupt stop is worth watching!

4. Try this in the bathtub or when you're washing dishes. Lower a drinking glass, mouth downward, over a small floating object. What do you observe? How deep does the glass have to be pushed in order to compress the enclosed air to half its volume? (You won't be able to do this in your bathtub unless it's 10.3 m deep!)
5. You can find the pressure exerted by the tires of your car on the road and compare it with the air pressure in the tires. For this project, you need to get the weight of your car from the manual or a dealer, then divide by 4 to get the approximate weight held up by one tire. You can closely approximate the area of contact of a tire with the road by tracing the edges of tire contact on a sheet of paper marked with 1-in. squares beneath the tire. After you get the pressure of the tire on the road, compare it with the air pressure in the tire. Are they nearly equal? Which one is greater?

6. You ordinarily pour water from a full glass to an empty glass by simply placing the full glass above the empty glass and tipping. Have you ever poured air from one glass to the other? The procedure is similar. Lower two glasses in water, mouths downward. Let one fill with water by tilting its mouth upward. Then hold the water-filled glass mouth downward above the air-filled glass. Slowly tilt the lower glass and let the air escape, filling the upper glass. You will be pouring air from one glass to another!

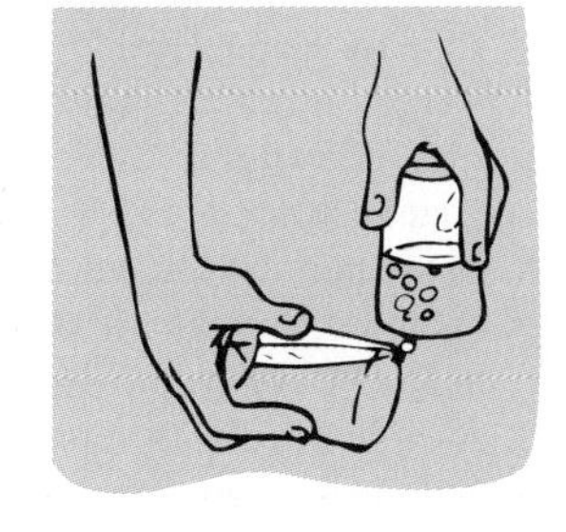

7. Raise a filled glass of water above the waterline, but with its mouth beneath the surface. Why does the water not run out? How tall would a glass have to be before water began to run out? (You won't be able to do this indoors unless you have a 10.3-m ceiling.)

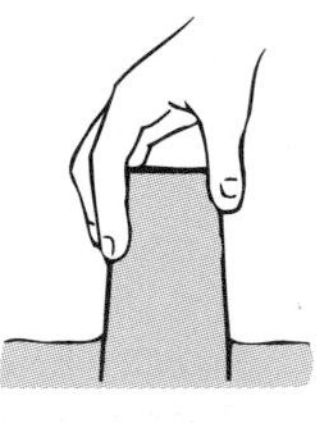

8. Place a card over the open top of a glass filled to the brim with water, and invert it. Why does the card stay intact? Try it sideways.

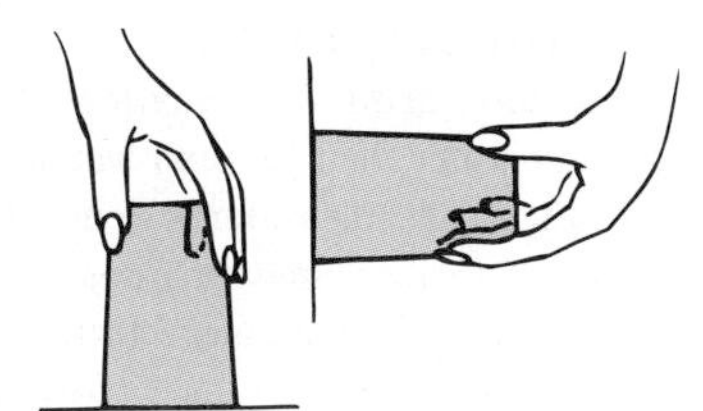

9. Invert a water-filled pop bottle or small-necked jar. Notice that the water doesn't simply fall out, but gurgles out of the container. Air pressure won't let it get out until some air has pushed its way up inside the bottle to occupy the space above the liquid. How would an inverted, water-filled bottle empty if this were done on the Moon?
10. Heat a small amount of water to boiling in an aluminum soda-pop can and invert it quickly into a dish of cold water. Surprisingly dramatic!
11. Make a small hole near the bottom of an open tin can. Fill it with water, which proceeds to spurt from the hole. Cover the top of the can firmly with the palm of your hand and the flow stops. Explain.

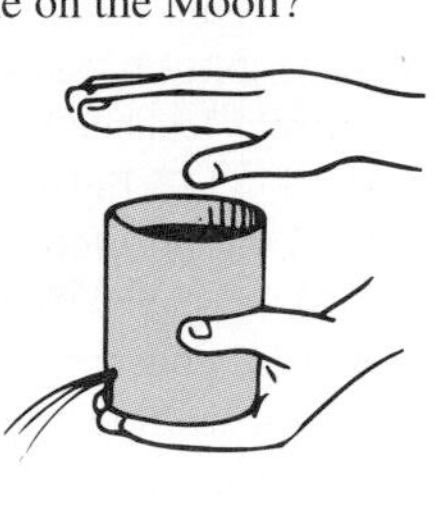

12. Lower a narrow glass tube or drinking straw into water and place your finger over the top of the tube. Lift the tube from the water and then lift your finger from the top of the tube. What happens? (You'll do this often in chemistry experiments.)

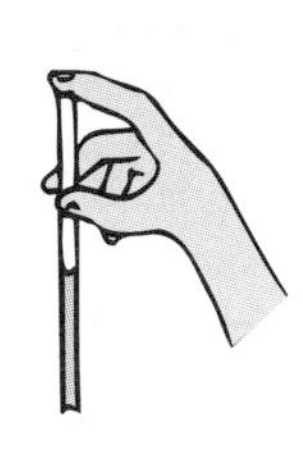

13. Fold the ends of a filing card down so that you make a little bridge. Stand it on the table and blow through the arch as shown. No matter how hard you blow, you will not succeed in blowing the card off the table (unless you blow against the side of it). Try this with your nonphysics friends. Then explain it to them!

14. Push a pin through a small card and place it in the hole of a thread spool. Try to blow the card from the spool by blowing through the hole. Try it in all directions.
15. Hold a spoon in a stream of water as shown and feel the effect of the differences in pressure.

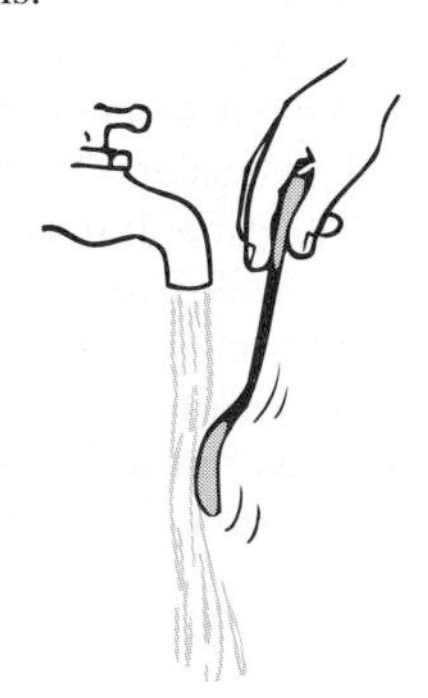

Exercises

1. You know that a sharp knife cuts better than a dull knife. Why is this so?
2. The photo shows physics teacher Marshall Ellenstein walking barefoot on broken glass. What physics concept is he demonstrating, and why is he careful that the broken pieces of glass are small and numerous? (The band-aids on his feet are for humor!)

3. Which do you suppose exerts more pressure on the ground—an elephant or a woman standing on spike heels? (Which will be more likely to make dents in a linoleum floor?) Can you approximate a rough calculation for each?
4. If water faucets upstairs and downstairs are turned fully on, will more water per second come out of the upstairs faucet or the downstairs one?
5. The sketch shows the reservoir that supplies water to a farm. It is made of wood and reinforced with metal hoops. (a) Why is it elevated? (b) Why are the hoops closer together near the bottom?

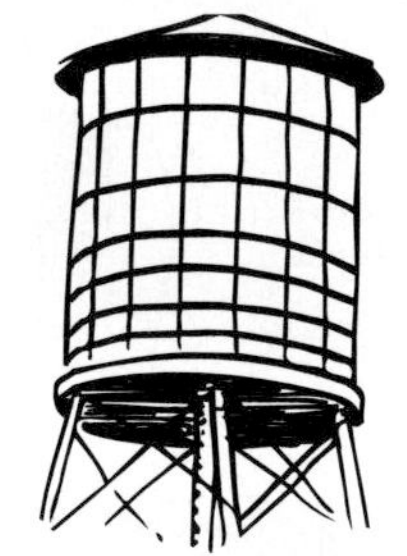

6. There is a legend of a Dutch boy who bravely held back the whole Atlantic Ocean by plugging a hole in a dike with his finger. Is this possible and reasonable?
7. Why does water "seek its own level"?
8. Suppose you wish to lay a level foundation for a home on hilly, bushy terrain. How can you use a garden hose filled with water to determine equal elevations for distant points?
9. A block of aluminum that has a volume of 10 cm^3 is placed in a beaker of water filled to the brim. Water overflows. The same is done in another beaker with a 10-cm^3 block of lead. Does the lead displace more, less, or the same amount of water?
10. A block of aluminum that has a mass of 1 kg is placed in a beaker of water filled to the brim. Water overflows. The same is done in another beaker with a 1-kg block of lead. Does the lead displace more, less, or the same amount of water?
11. A block of aluminum that weighs 10 N is placed in a beaker of water filled to the brim. Water overflows. The same is done in another beaker with a 10-N block of lead. Does the lead displace more, less, or the same amount of water? (Why are your answers to this exercise and Exercise 10 different from your answer to Exercise 9?)
12. If liquid pressure were the same at all depths, would there be a buoyant force on an object submerged in the liquid? Explain.
13. How much force is needed to push a nearly weightless but rigid 1-L carton beneath a surface of water?
14. The density of ice is 0.9 g/cm^3, which is 90 percent of the density of water. What proportion of a floating chunk of ice is above water level?
15. The Himalayan Mountains are slightly less dense than the semimolten material upon which they "float." Do you suppose that, like floating icebergs, they are deeper than they are high?
16. Why is a high mountain composed mainly of lead impossible for planet Earth?
17. Why is it inaccurate to say that heavy objects sink and light ones float? Give exaggerated examples to support your answer.
18. Compared with an empty ship, would a ship loaded with a cargo of Styrofoam sink deeper into the water or rise higher? Defend your answer.
19. A barge filled with scrap iron is in a canal lock. If the iron is thrown overboard, does the water level at the side of the lock rise, fall, or remain unchanged? Explain.
20. Would the water level in a canal lock go up or down if a battleship in the lock sank?
21. A balloon is weighted so that it is barely able to float in water. If it is pushed beneath the surface, will it come back to the surface, stay at the depth to which it is pushed, or sink? Explain. (*Hint:* What change in density, if any, does the balloon undergo?)
22. The density of a rock doesn't change when the rock is submerged in water, but your density changes when you are submerged. Explain.
23. In answering the question of why bodies float higher in salt water than in fresh water, a friend replies that the reason is that salt water is denser than fresh water. (Does your friend often answer questions by reciting only factual statements that relate to the answers but don't provide any concrete reasons?) How would you answer the same question?

24. Suppose you wear two life preservers that are identical in size, first a light one filled with Styrofoam and then a very heavy one filled with lead pellets. If you wear these life preservers in the water, upon which will the buoyant force be greater? Upon which will the buoyant force be ineffective? Why are your answers different?

25. When an ice cube in a glass of water melts, does the water level in the glass rise, fall, or remain unchanged?

26. Count the tires on a large tractor trailer that is unloading food at your local supermarket, and you may be surprised to count 18 or so tires. Why so many tires? (*Hint:* Consider Home Project 5.)

27. Why do your ears pop when you ascend to higher altitudes?

28. Two teams of eight horses each were unable to pull the Magdeburg hemispheres apart (Figure 5.20). Why? Suppose two teams of nine horses each could pull them apart. Then would one team of nine horses succeed if the other team were replaced with a strong tree? Defend your answer.

29. Before boarding an airplane, you buy a roll of camera film or any item packaged in an airtight foil package, and while in flight you notice that it is puffed up. Explain why this happens.

30. We can understand how pressure in water depends on depth by considering a stack of bricks. The pressure at the bottom of the first brick corresponds to the weight of the entire stack. Halfway up the stack, the pressure is half because the weight of the bricks above is half. To explain atmospheric pressure, we should consider compressible bricks, like foam rubber. Why?

31. At which location would it be slightly more difficult to draw soda through a straw—at sea level or on top of a very high mountain? Explain.

32. The pressure exerted against the ground by an elephant's weight distributed evenly over its four feet is less than 1 atmosphere. Why, then, would you be crushed beneath the foot of an elephant but are unharmed by the pressure of the atmosphere?

33. A friend says that the buoyant force exerted by the atmosphere on an elephant is significantly greater than the buoyant force exerted by the atmosphere on a small helium-filled balloon. What do you say?

34. Why is it so difficult to breathe when snorkeling at a depth of 1 m and practically impossible at a 2-m depth? Why can't a diver simply breathe through a hose that extends to the surface?

35. A block of wood and a block of iron on weighing scales each weigh 1 ton. Which has the greater mass?

36. Why does the weight of an object in air differ from its weight in a vacuum? Cite an example where this would be an important consideration.

37. Two identical balloons filled with air to the same volume are suspended on the ends of a stick that is horizontally balanced. One of the balloons is then punctured. Is the balance upset? If so, which way does the stick tip?

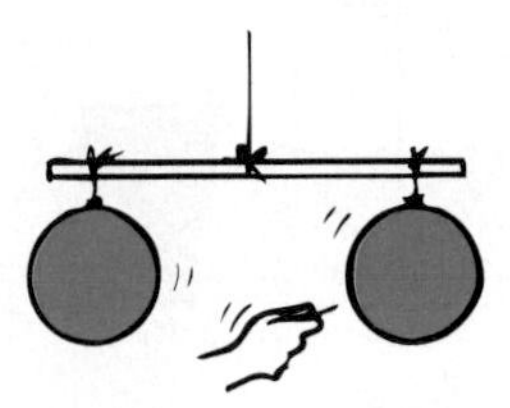

38. Two balloons that have the same weight and volume are filled with equal amounts of helium. One is rigid and the other is free to expand as the pressure outside decreases. When released, which will rise higher? Explain.

39. Estimate the buoyant force that acts on you. (To do this, you can estimate your volume by knowing your weight and by assuming that your weight density is a bit less than that of water.)

40. Estimate the *force* that the atmosphere exerts on you, and then compare this force with the weight of something familiar in your environment. To do this, estimate your surface area from the area of your clothes and multiply by 10^5 N/m^2. Why is your answer so different from your answer to the previous question?

41. At sea level the force exerted by the atmosphere against the outside of a 10-m^2 store window is about a million newtons. Why does this force not shatter the window? Why might the window shatter in a strong wind?

42. In a department store, an airstream from a hose connected to the exhaust of a vacuum cleaner blows upward at an angle and supports a beach ball in midair. Does the air blow under or over the ball to provide support?

43. What provides the lift to keep a Frisbee in flight?

44. Why is it that when passing an oncoming truck on the highway, your car tends to lurch toward the truck?

45. Why is it that the canvas roof of a convertible automobile bulges upward when the car is traveling at high speeds?

46. Why is it that the windows of older trains sometimes break when a high-speed train passes by on the next track?

47. A steady wind blows over the waves of an ocean. Why does the wind increase the size of the humps and troughs of the waves?

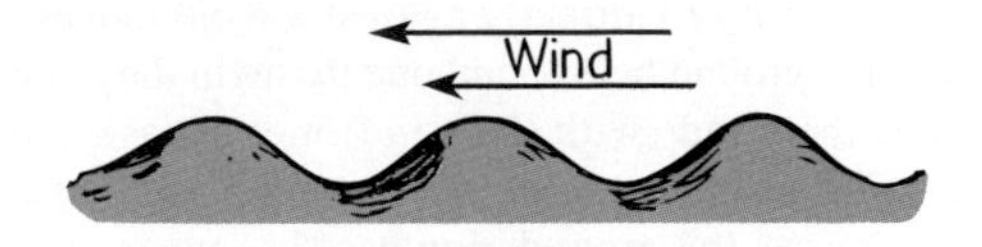

48. In answering the question of why a flag flaps in the wind, a friend replies that it flaps because of Bernoulli's principle. (Not really a convincing explanation, is it?) How would you answer the same question?

49. Wharves are made with pilings that permit the free passage of water. Why would a solid-wall wharf be disadvantageous to ships attempting to pull alongside?

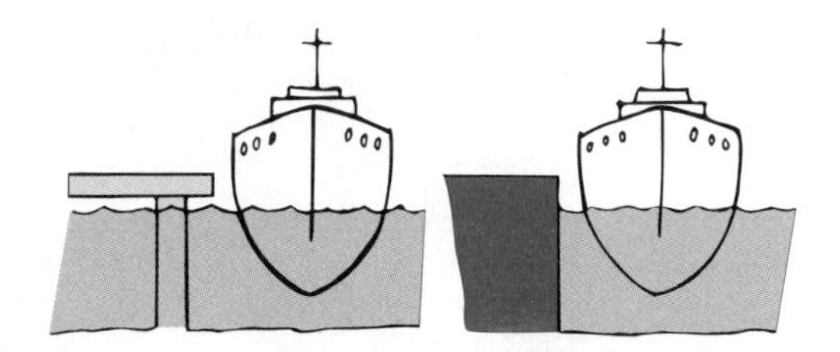

50. It is often said that fast-moving air has lower pressure than air at rest or slower-moving air. Can you make an argument that the opposite is the case—that fast-moving air is the *result,* not the *cause* of, lower pressure?

Problems

1. A rectangular barge 5 m long and 2 m wide floats in fresh water. (a) Find how much deeper it floats when its load is a 400-kg horse. (b) If the barge can be pushed only 15 cm deeper into the water before water overflows to sink it, how many 400-kg horses can it carry?

2. A merchant in Katmandu sells you a solid gold 1-kg statue for a very reasonable price. When you get home, you wonder whether or not you got a bargain, and so you lower the statue into a measuring cup and measure its volume. What volume will verify that it's pure gold?

3. When a 2.0-kg object is suspended by a scale in water, the scale reads 1.5 kg. What is the density of the object?

4. An ice cube measures 10 cm on a side and floats in water, with 1 cm extending above the water surface. If you shaved off the 1-cm part, how many centimeters of the remaining ice would extend above the water surface?

5. A vacationer floats lazily in the ocean with 90% of his body below the surface. The density of the ocean water is 1025 kg/m^3. What is the vacationer's average density?

6. On a perfect fall day, you are hovering at low altitude in a hot-air balloon, accelerated neither upward nor downward. The total weight of the balloon, including its load and the hot air in it, is 20,000 N. (a) What is the weight of the displaced air? (b) What is the volume of the displaced air?

Part I—Sample Exam Questions

Choose the BEST answer to each of the following.

1. A pebble falls from rest atop a vertical cliff on the Moon's surface, where the acceleration due to gravity is 1.6 m/s2. What will be its speed when it has fallen for 3 seconds?
(a) 2.4 m/s (b) 3.2 m/s (c) 4.8 m/s (d) 7.2 m/s (e) none of these

2. Consider drops of water that leak from a dripping faucet. As the drops fall they
(a) get closer together (b) get farther apart
(c) remain at a relatively fixed distance from one another

3. The acceleration of a bowling ball rolling along a smooth horizontal bowling alley is
(a) zero (b) about 10 m/s^2 (c) constant

4. If an object moves along a straight-line path, then it must be
(a) accelerating (b) acted on by a force (c) both of these (d) none of these

5. An airplane flying directly against a wind moves slower across the ground below, and one flying in the same direction as the wind (with the wind) moves faster. When a plane flies in a cross-wind (wind coming from the side) its speed across the ground, compared to when there is no wind at all, is
(a) greater (b) less (c) no different

6. You stand at the top of a cliff and throw a rock downward, and another rock horizontally at the same speed(d) The rock that stays in the air for the longest time is the one thrown
(a) downward (b) horizontally (c) both take the same time

7. A heavy rock and a light rock in free fall have the same acceleration. The reason the heavy rock does not have more acceleration is because
(a) the force of gravity on each is the same
(b) there is no air resistance
(c) the inertia of both rocks is the same
(d) all of these (e) none of these

8. A ball rolls down a curved ramp as shown. As its speed increases, its rate of gaining speed
(a) increases (b) decreases (c) remains unchanged

9. A pair of tennis balls fall through the air from a tall building. One is regular and the other is filled with iron pellets. The ball to reach the ground first is the
(a) regular ball (b) iron-filled ball (c) both reach the ground in the same time

10. The same pair of tennis balls (regular and iron filled) fall from a tall building. Air resistance just before they hit the ground is greater for the
(a) regular ball (b) iron-filled ball
(c) is the same for both

11. A karate chop delivers a force of 3000 N to a board that breaks. The force that acts on the hand during this event is
(a) less than 3000 N (b) 3000 N (c) more than 3000 N
(d) not enough information to say

12. Apply the equation $Ft = \Delta mv$ to the case of a person falling on a wooden floor. If v is the speed of the person as she strikes the floor, then m is
(a) the mass of the person (b) the mass of the floor
(c) both of these (d) none of these

13. A high-speed truck collides head on with a flying bug that spatters on the windshield(d) Consider these statements:
- The amount of force is the same on the bug and truck.
- The amount of impulse is the same on the bug and truck.
- The change in momentum is the same for the bug and truck.
- The amount of acceleration is the same for the bug and truck.

(a) All statements are true (b) Three statements are true
(c) Two statements are true (d) One statement is true
(e) No statements are true

14. When wind blows against a windmill connected to a generator to produce electricity, the speed of the wind after interacting with the windmill is
(a) reduced (b) unchanged (c) increased (d) there's no physics principle to guide an answer to this question!

15. When the brakes are applied for a car traveling at a certain speed, it skids a certain distance to a stop. If the car were traveling twice as fast, skidding distance would be
(a) twice (b) four times (c) eight times (d) there's no physics principle to guide an answer to this question!

16. When a bullet is fired from a rifle, the force that accelerates the bullet is equal in magnitude to the force that makes the rifle recoil. But compared with the rifle, the bullet has a greater
(a) inertia (b) potential energy (c) kinetic energy
(d) momentum

17. The reason for the answer to the preceding question has to do with the fact that the force on the bullet acts over
(a) the same time (b) a longer time (c) a longer distance
(d) none of these

18. Which pulls with the greater force on the Earth's oceans?
(a) Moon (b) Sun (c) both the same

19. The space shuttle orbiting the Earth is above the Earth's
(a) atmosphere (b) gravitational field (c) both
(d) neither

20. If an astronaut in an orbiting space shuttle wishes to drop an object to the Earth below, she should launch the object at a speed of 8 km/s
(a) straight downward toward Earth (b) forward, in the direction of orbit (c) backward (d) none of these

21. Squeeze an air-filled balloon to half size and the pressure inside
(a) remains the same (b) halves (c) doubles

22. As a weighed air-filled balloon sinks deeper and deeper in water, the buoyant force on it
(a) increases (b) decreases (c) remains about the same

Answers: 1 c, 2 b, 3 a, 4 d, 5 a, 6 b, 7 e, 8 b, 9 b, 10 b, 11 b, 12 a, 13 d, 14 a, 15 b, 16 c, 17 c, 18 b, 19 a, 20 c, 21 c, 22 (b)

PART

Heat

Although the temperature of these sparks exceeds 2000°C, the heat they impart in striking my skin is very small-- which illustrates that temperature and heat are different concepts. Learning to distinguish between closely related concepts is the essence of *Conceptual Physical Science*.

CHAPTER

6

Thermal Energy

All matter is composed of continuously jiggling atoms or molecules. Whether these particles combine to form solids, liquids, or gases depends on how fast the particles are moving. By virtue of their motion, the particles in matter possess kinetic energy. The average kinetic energy of the individual particles is directly related to a property you can sense: how hot something is. Whenever something becomes warmer, the kinetic energy of its particles increases. When struck with a hammer, a penny becomes warm because the hammer's blow causes the penny's atoms to jostle faster. Put a flame to a liquid and the liquid becomes warmer. Rapidly compress air in a tire pump and the air becomes warmer. When a solid, liquid, or gas gets warmer, its atoms or molecules move faster. The particles have more *thermal motion* and therefore more kinetic energy. We say the substance made up of these particles has more *thermal energy*.

6.1 Temperature

Fig. 6.1
Can we trust our sense of hot and cold? Will both fingers feel the same temperature when they are put in the warm water?

The quantity that tells how warm or cold an object is with respect to some standard is called **temperature**. We express the temperature of matter by a number that corresponds to the degree of hotness on some chosen scale. A common thermometer measures temperature by means of the expansion and contraction of a liquid, usually mercury or colored alcohol.

The most common thermometer in the world is the *Celsius thermometer*, named in honor of the Swedish astronomer Anders Celsius (1701–1744), who first suggested the scale of 100 degrees between the freezing point and boiling point of water. The number 0 is assigned to the temperature at which water freezes and the number 100 to the temperature at which water boils (at standard atmospheric pressure), with 100 equal parts called *degrees* between. In the United States, the number 32 is assigned to the temperature at which water freezes and the number 212 to the temperature at which water boils. This scale is used in the Fahrenheit thermometer, named after its originator, the German physicist G. D. Fahrenheit (1686–1736). The Fahrenheit scale will become obsolete if and when the United States goes metric.[1]

Arithmetic formulas are used for converting from one temperature scale to the other and are popular in classroom exams. Because such arithmetic exercises are not really physics, and because the probability of your having the occasion to do this task elsewhere is small, we shall not be concerned with it here. Besides, this conversion can be very closely approximated by simply reading the corresponding temperature from the side-by-side scales in Figure 6.2.[2]

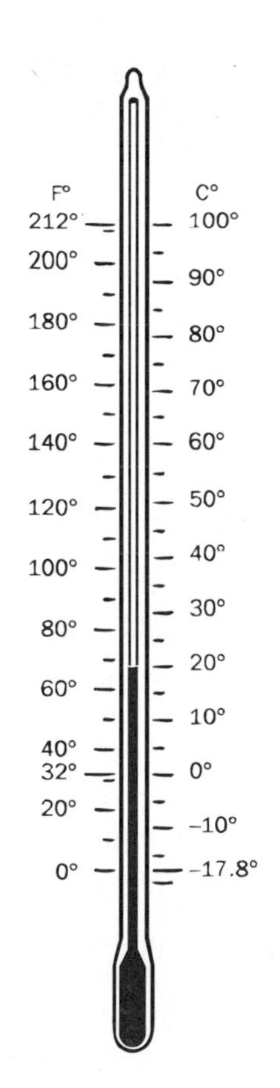

Fig. 6.2
Fahrenheit and Celsius scales on a thermometer.

Temperature is related to the random motion of the atoms and molecules in a substance. (For brevity, from now on in this chapter, we'll simply say *molecules* to mean *atoms and molecules*.)[3] Temperature is proportional to the average kinetic energy of molecular translational motion (that is, molecular motion along either a straight or a curved path).

It is interesting to note that what a thermometer actually registers is its own temperature. When a thermometer is in thermal contact with something whose temperature we wish to know, energy flows between the two until their temperatures are equal; at this point, thermal equilibrium is established. When we look at the temperature of the thermometer, we are learning the temperature of something. A thermometer should be small enough that it doesn't appreciably alter the temperature being measured. If you are measuring the room air temperature, your thermometer is small enough. But if you are measuring the temperature of a drop of water, contact between the drop and the thermometer may change the drop's temperature.

---

[1]Changing any long-established custom is difficult, and the Fahrenheit scale does have some advantages in everyday use. For example, its degrees are smaller ($1\ \text{F}^\circ = \frac{5}{9}\ \text{C}^\circ$), which gives greater accuracy when reporting the weather in whole-number temperature readings. Then, too, people somehow attribute a special significance to numbers increasing by an extra digit, so that when the temperature of a hot day is reported to reach 100°F, the idea of heat is conveyed more dramatically than by saying it is 38°C. Like so much of the rest of the British system of measure, the Fahrenheit scale is geared to human beings.

[2]Okay, if you really want to know, the formulas for temperature conversion are $\text{C}^\circ = \frac{5}{9}(\text{F}^\circ - 32^\circ)$ and $\text{F}^\circ = \frac{9}{5}\ \text{C}^\circ + 32^\circ$.

[3]As we shall see in Chapter 17, a molecule is a submicroscopic unit of matter composed of a group of atoms. Atoms make up molecules, not the other way around.

■ **Question**

True or false: Temperature is a measure of the total kinetic energy in a substance.

6.2 Absolute Zero

In principle, there is no upper limit of temperature. As thermal motion increases, a solid object melts to a liquid and the liquid then vaporizes. As the temperature of the vapor is further increased, molecules break into atoms and the atoms lose some or all of their electrons, thereby forming a cloud of electrically charged particles—a *plasma*. This situation exists in stars, where the temperature is many millions of degrees Celsius.

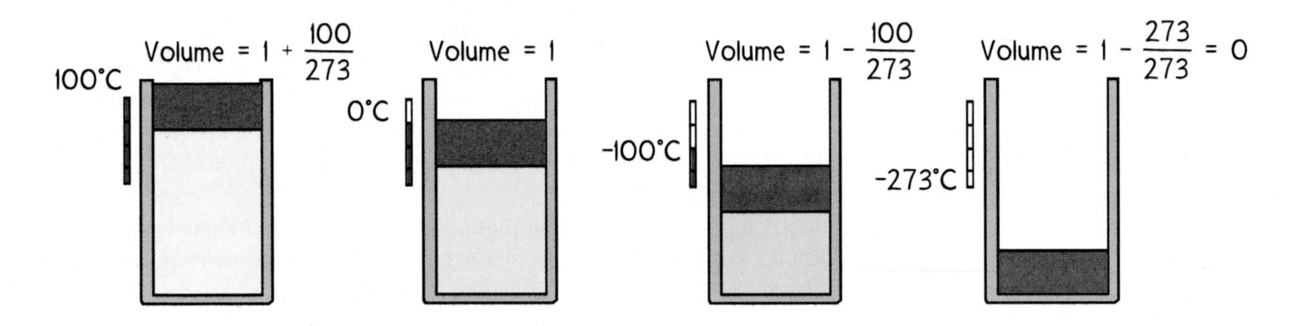

Fig. 6.3
When pressure is held constant, the volume of a gas changes by $\frac{1}{273}$ of its volume at 0C° with each 1C° change in temperature. At 100°C, the volume is $\frac{100}{273}$ greater than it is at 0°C. When the temperature is reduced to −100°C, the volume is reduced by $\frac{100}{273}$. At −273°C, the volume of the gas would be reduced by $\frac{273}{273}$ and so would be zero.

In contrast, there is a definite limit at the other end of the temperature scale. Gases expand when heated and contract when cooled. Nineteenth-century experiments found that all gases, regardless of their initial pressure or volume, change by $\frac{1}{273}$ of their volume at 0°C for each Celsius degree change in temperature, provided the pressure is held constant. So if a gas at 0°C were cooled by 273 C°, it would contract $\frac{273}{273}$ volumes and be reduced to zero volume. Clearly, we cannot have a substance with zero volume.

It was also found that the pressure of any gas in any container of *fixed* volume changes by $\frac{1}{273}$ for each Celsius degree change in temperature. So a gas in a container of fixed volume cooled 273 C° below zero would have no pressure whatsoever. In practice, every gas liquefies before it gets this cold. Nevertheless, these decreases by $\frac{1}{273}$ increments suggested the idea of a lowest temperature: −273°C. So there is a lower limit of temperature. When molecules lose all available kinetic energy, they reach the

---

■ **Answer**

False. Temperature is a measure of the *average* (not the *total*!) kinetic energy of the molecules in a substance. For example, there is twice as much total molecular kinetic energy in 2 L of boiling water as in 1 L—but the temperatures of the two amounts of water are the same because the *average* kinetic energy per molecule in each is the same.

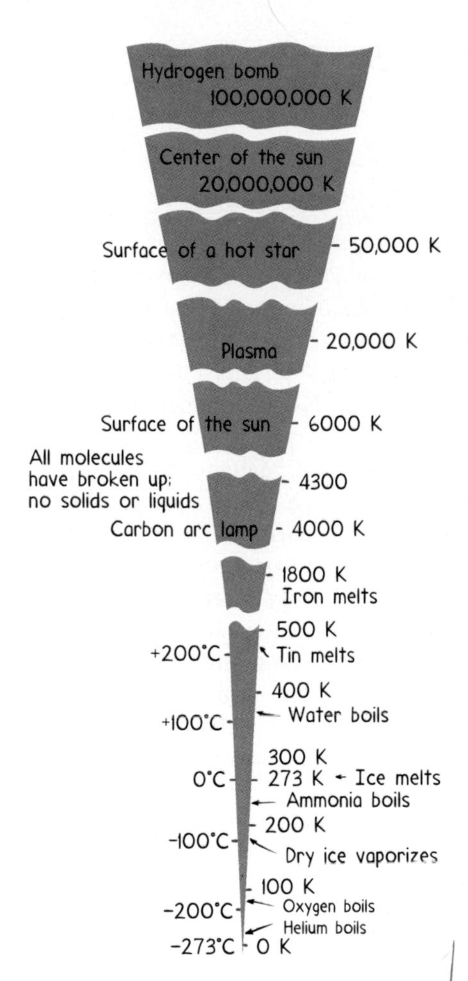

Fig. 6.4
Some absolute temperatures.

absolute zero of temperature.[4] At absolute zero, no more energy can be extracted from a substance, and no further lowering of its temperature is possible.

On the absolute temperature scale, also called the Kelvin scale,[5] absolute zero is 0 K (short for "0 kelvin,"; note that the word "degrees" is not used with Kelvin temperatures). There are no negative numbers in the Kelvin scale, and degrees on it are calibrated with the same-sized divisions as on the Celsius scale. The melting point of ice thus is 273 K, and the boiling point of water is 373 K.

Questions

1. Which is larger, a Celsius degree or a kelvin?
2. A piece of metal has a temperature of 0°C. If someone says a second, identical piece of metal is twice as hot, what does that person mean? What is the temperature of the second piece?

6.3 Thermal Energy

We all know there is a vast amount of energy locked in all materials—this book, for example. Its pages are composed of molecules in constant motion. They have kinetic energy. Because of their interactions with neighboring molecules, they also have potential energy. The pages can be easily burned, so we know they store chemical energy, which is really electric potential energy at the molecular level. We know there are vast amounts of energy associated with atomic nuclei. Then there is the "energy of being," described by the celebrated equation $E = mc^2$ (mass energy). Energy within a substance is found in these and other forms, which, when taken

Answers

1. Neither. They are equal.
2. To say a piece of metal is twice as hot means it has twice the thermal energy. This means it has twice the absolute temperature. Because the first piece is at 0°C = 273 K, the second piece is at a temperature of 546 K. In Celsius, this is 273°C. Get it?

---

[4]Even at absolute zero, molecules still have a small kinetic energy, called the *zero-point energy*. Helium, for example, has enough motion at absolute zero to keep it from freezing. The explanation for this involves quantum theory.

[5]Named after the famous British physicist Lord Scale, who coined the word *thermodynamics* and was the first to suggest this thermodynamic temperature scale. (Lord Scale? Yes, like Washington Street is named after Mr. Street!)

together, are called **thermal energy**.[6] Changes in temperature indicate changes in thermal energy.

6.4 Heat

If you touch a hot stove, thermal energy enters your hand because the stove is warmer than your hand. When you touch a piece of ice, on the other hand, thermal energy passes out of your hand and into the colder ice. The direction of thermal energy flow is always from a warmer thing to a neighboring cooler thing. A physicist defines **heat** as the thermal energy transferred from one thing to another due to a temperature difference between the two things.

According to this definition, matter does not *contain* heat. Matter contains thermal energy, not heat. Heat is *thermal energy in transit*. Once heat has been transferred to an object or substance, it ceases to be heat. It becomes thermal energy.

For substances in thermal contact, thermal energy flows from the higher-temperature substance to the lower-temperature one but not necessarily from a substance that has more thermal energy to one that has less thermal energy. For instance, there is more thermal energy in a large bowl of warm water than in a red-hot thumbtack; if the tack is immersed in the water, thermal energy doesn't flow from the warm water to the tack. Instead, it flows from the hot tack to the cooler water. Thermal energy never flows of itself from a low-temperature substance to a higher-temperature substance.

Questions

1. Suppose you apply a flame to 1 L of water for a certain length of time and the temperature of the water rises by 2 C°. If you apply the same flame for the same length of time to 2 L of water, by how much does its temperature rise?
2. If a fast-moving marble collides with some slow-moving marbles, does the fast marble usually speed up or slow down? Which lose(s) kinetic energy and which gain(s) kinetic energy, the initially fast-moving marble or the initially slow ones? How do these questions relate to the direction of heat flow?

Fig. 6.5
The temperature of the sparks is very high, about 2000°C. That's a lot of thermal energy per molecule of spark. Because there are only a few molecules per spark, however, the total amount of thermal energy in the sparks is safely small. Temperature is one thing; transfer of thermal energy is another.

Answers

1. Its temperature rises by only 1 C° because there are twice as many molecules in 2 L of water and each molecule receives only half as much energy on the average. So the average kinetic energy, and thus the temperature, increase by half as much. See Figure 6.7.
2. A fast-moving marble slows when it hits slower-moving marbles. It gives up some of its kinetic energy to the slower ones. Likewise with the flow of thermal energy. Molecules having more kinetic energy in contact with molecules having less kinetic energy give up some of their excess kinetic energy to the less energetic ones. The direction of energy transfer is from hot to cold. For both marbles and molecules, however, the total amount of energy before and after contact is the same.

[6] In place of the term *thermal energy,* physicists use the term *internal energy*—the grand total of all energies inside a substance.

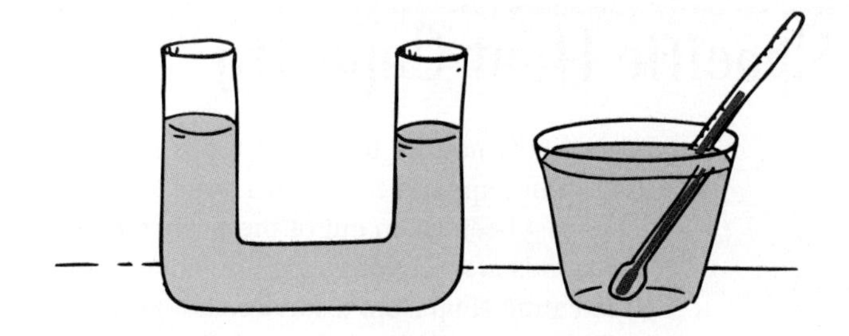

Fig. 6.6
Just as water in the two arms of the U-tube seeks a common level (where the pressures at any given depth are the same), the thermometer and its immediate surroundings reach a common temperature (at which the average molecular KE is the same in both). If the water is warmer than the thermometer, for instance, thermal energy transfers from the water to the thermometer until the molecules in both have the same average KE.

6.5 Heat Units

Heat is a form of energy and can be measured in joules. But a more common unit of heat is the **calorie**, defined as the amount of heat required to change the temperature of 1 gram of water by 1 Celsius degree.[7]

The energy ratings of foods and fuels are determined by burning them and measuring the energy released. (Digestion is really "burning"—at a slow rate). The heat unit used to label foods is the kilocalorie, which is 1000 calories (the heat required to change the temperature of 1 kilogram of water by 1 C°). To distinguish this unit from the smaller calorie, the food unit is sometimes called a *Calorie* with a capital C.

It is important to remember that the calorie and Calorie are units of energy. These names are historical carryovers from the early idea that heat was an invisible fluid called *caloric*, a view that persisted into the nineteenth century. We now know that heat is a form of energy, and so doesn't need its own unit. Someday the calorie may give way to the SI unit, the joule, as the common unit for measuring heat. (The relationship between calories and joules is 1 calorie = 4.184 joules.)

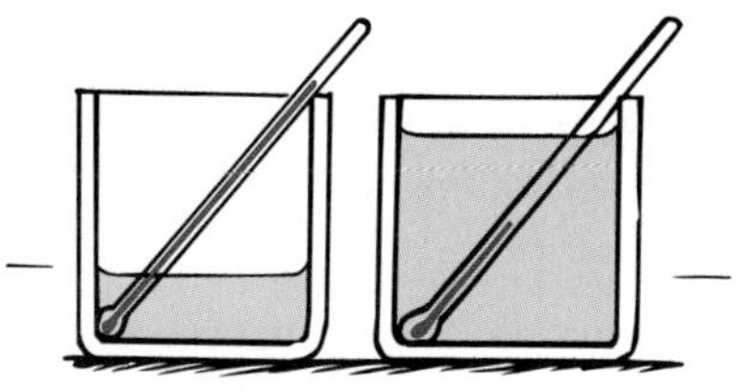

Hot stove

Fig. 6.7
Although the same quantity of heat is added to both pots, the temperature increases more in the pot containing the smaller amount of water.

Fig. 6.8
To the weight watcher, the peanut contains 10 Calories; to the physicist, it releases 10,000 calories (41,840 joules) of energy when burned or digested

[7]Another common unit of heat is the British thermal unit (Btu), defined as the amount of heat required to change the temperature of 1 lb of water by 1 F°.

6.6 Specific Heat Capacity

Fig. 6.9
The filling of hot apple pie may be too hot to eat even though the crust is not.7

You've likely noticed that some foods remain hotter much longer than others. Like the filling of hot apple pie can burn your tongue while the crust does not, even when the pie has just been taken out of the oven. Or a piece of toast may be comfortably eaten a few seconds after coming from the hot toaster, whereas you must wait several minutes before eating soup from a stove as hot as the toaster.

Different substances have different capacities for storing thermal energy. If we heat a pot of water on a stove, we might find that it has to be heated for 15 minutes in order to raise its temperature from room temperature to boiling. If we put an equal mass of iron on the same flame, we would find it rising through the same temperature range in only about 2 minutes. For silver, the time would be less than a minute. Different materials require different quantities of heat to raise the temperature of a given mass by a specified number of degrees. This is because different materials absorb energy in different ways. The added energy may increase the jiggling motion of molecules, which raises the temperature, or it may increase the amount of internal vibration or rotation within the molecules and therefore go into potential energy, which does not raise the temperature. Generally there is a combination of both.

From the definition of the calorie, we know that it takes 1 calorie of energy to raise the temperature of 1 gram of water by 1 Celsius degree. It takes only about one-eighth as much energy to raise the temperature of a gram of iron by the same amount. In other words, water absorbs more heat than iron for the same change in temperature. We say water has a higher **specific heat capacity** (sometimes simply called *specific heat*)[8]:

> **The specific heat capacity of any substance is defined as the quantity of heat required to change the temperature of a unit mass of the substance by 1 degree.**

We can think of specific heat capacity as thermal inertia. Recall that inertia is a term used in mechanics to signify the resistance of an object to a change in its state of motion. Specific heat capacity is like a thermal inertia in that it signifies how much resistance a substance has to a change in temperature.

Question

Which has a higher specific heat capacity, water or sand? In other words, which warms up more slowly in the sunlight or cools more slowly at night?

Answer
Water has the higher specific heat capacity. In the same sunlight, the temperature of water increases more slowly than the temperature of sand. And water will cool slower at night. Sand's low specific heat capacity, as evidenced by how quickly it warms in the morning sun and how quickly it cools at night, affects local climates.

[8]If we know the specific heat capacity c, the formula for the quantity of heat Q involved when a mass m of a substance undergoes a change in temperature, T is $Q = mc\Delta T$. In words, heat transferred = mass × specific heat capacity × temperature change.

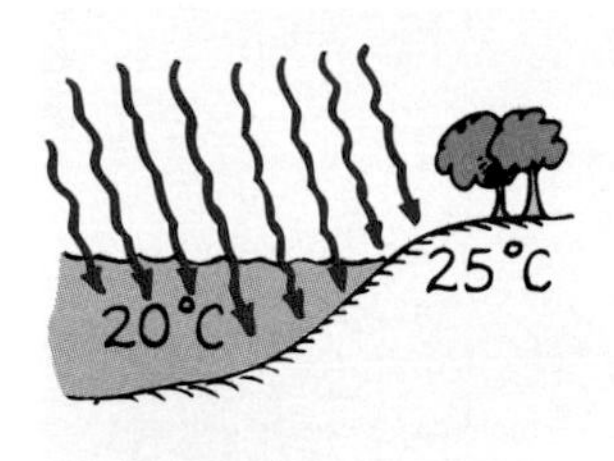

Fig. 6.10
Because water has a high specific heat capacity and is transparent, it takes more energy to warm the water than to warm the land. Solar energy incident on the land is concentrated at the surface, but that hitting the water extends beneath the surface and so is "diluted."

Water has a much higher capacity for storing energy than all but a few uncommon materials. A relatively small amount of water absorbs a great deal of heat for a correspondingly small temperature rise. Because of this, water is a very useful cooling agent and is used in the cooling system of automobiles and other engines. If a liquid of lower specific heat capacity were used in cooling systems, its temperature would rise higher for a comparable amount of heat absorbed. Water also takes a long time to cool, a fact that explains the wide use of hot-water bottles by old timers on cold winter nights. (Better blankets, including electric ones, have since taken their place.)

This tendency on the part of water to resist changes in temperature improves the climate in many places. The next time you're looking at a world globe, notice the high latitude of Europe. If water did not have a high specific heat capacity, the countries of Europe would be as cold as the northeastern regions of Canada, for both Europe and Canada get about the same amount of sunlight per square kilometer. The Atlantic current known as the Gulf Stream carries warm water northeast from the Caribbean. It holds much of its thermal energy long enough to reach the North Atlantic off the coast of Europe, where it then cools. The energy released, 1 calorie per Celsius degree for each gram of water that cools, is carried by westerly winds over the European continent. A similar effect occurs in the United States. The winds in the latitudes of North America are westerly. On the West Coast, air moves from the Pacific Ocean to the land. Because of water's high specific heat capacity, ocean temperatures do not vary much from summer to winter. The water is warmer than the air in the winter and cooler than the air in the summer. In winter, the water warms the air, which then moves over it and warms the coastal regions of North America. In summer, the water cools the air and the coastal regions are cooled. The East Coast does not benefit from the moderating effects of water because air moves from the land to the Atlantic Ocean. Land, with a lower specific heat capacity, gets hot in the summer but cools rapidly in the winter. Because of water's high specific heat capacity and westerly winds, the West Coast city of San Francisco is warmer in the winter and cooler in the summer than the East Coast city of Washington, D.C., which is at about the same latitude.

Because they are more or less surrounded by water, islands and peninsulas do not have the extremes of temperatures observed in the interior of a continent. The high summer and low winter temperatures common in Manitoba and the Dakotas, for example, are largely due to the absence of large bodies of water. Europeans, islanders, and people living near ocean air currents should be glad that water has such a high specific heat capacity. San Franciscans are!

6.7 Thermal Expansion

When the temperature of a substance is increased, its molecules jiggle faster and on average move farther apart. The result is that the substance expands. With few exceptions, all phases of matter—solids, liquids, gases, and plasmas—generally expand when heated and contract when cooled. In many cases these changes in volume are not very noticeable, but careful observation will usually detect them. Telephone wires are longer and sag more on a hot summer day than on a cold winter day. Metal lids on glass jars can often be loosened by heating them under hot water. If one part of a piece of

Fig. 6.11
One end of the bridge is fixed, but the end shown rides on rockers to allow for thermal expansion.

Fig. 6.12
This gap in the roadway of a bridge is called an expansion joint; it allows the bridge to expand and contract. (Was this picture taken on a warm or a cold day?)

glass is heated or cooled more rapidly than adjacent parts, the expansion or contraction that results may break the glass. This is especially true with thick glass. Pyrex glass is an exception because it is specially formulated to expand very little with increasing temperature.

Thermal expansion must be allowed for in structures and devices of all kinds. A dentist uses filling material that has the same rate of expansion as teeth. In automobile engines made with aluminum pistons in steel cylinders, the diameter of the pistons is

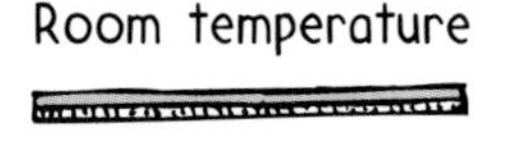

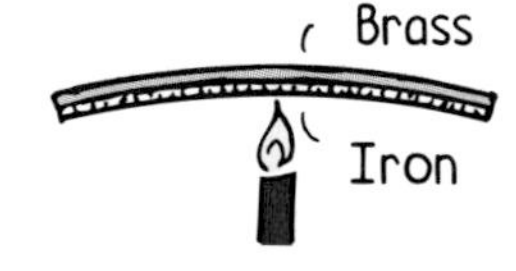

Fig. 6.13
A bimetallic strip. Brass expands more when heated than iron does and contracts more when cooled. Because of this behavior, the strip bends as shown.

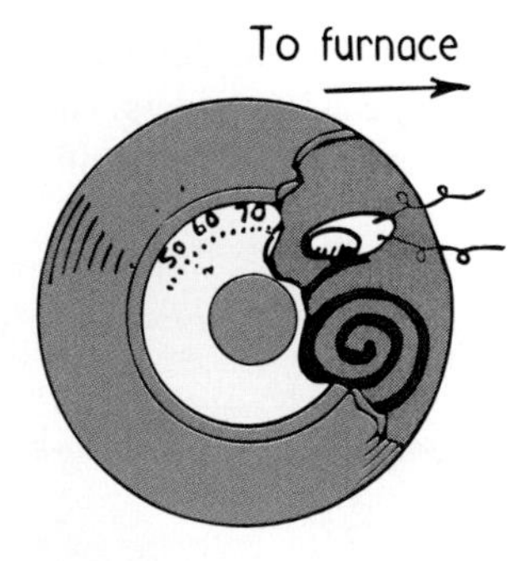

Fig. 6.14
A thermostat. When the bimetallic coil expands, the drop of liquid mercury rolls away from the electrical contacts and breaks the electrical circuit. When the coil contracts, the mercury rolls against the contacts and completes the circuit.

less than the diameter of the cylinders to allow for the much greater expansion rate of aluminum. A civil engineer uses reinforcing steel of the same expansion rate as concrete. A long steel bridge commonly has one end fixed while the other rests on rockers (Figure 6.11), and the roadway is segmented with tongue-and-groove gaps called *expansion joints* (Figure 6.12). Similarly, concrete roadways and sidewalks are intersected by gaps, sometimes filled with tar, so that the concrete can expand freely in summer and contract in winter.

Different substances expand at different rates. When two strips of different metals, say one of brass and the other of iron, are welded or riveted together, the greater expansion of one results in the bending shown in Figure 6.13. Such a compound thin bar is called a *bimetallic strip.* The movement of the strip may be used to turn a pointer, regulate a valve, or close a switch.

A practical application of differential thermal expansion is the thermostat (Figure 6.14). The back-and-forth bending of the bimetallic coil opens and closes an electric circuit. When the room becomes too cold, the coil bends toward the brass side and in so doing activates an electrical switch that turns on the furnace. When the room becomes too warm, the coil bends toward the iron side, which breaks the electrical circuit and so turns off the heating unit. Bimetallic strips are used in oven thermometers, refrigerators, electric toasters, and various other devices.

Liquids expand appreciably with increases in temperature. In most cases the expansion of liquids is greater than the expansion of solids. The gasoline overflowing a car's tank on a hot day is evidence of this. If the tank and contents expanded at the same rate, no overflow would occur.

Question

A Concorde supersonic airplane in flight is 20 cm longer than when parked on the ground. Offer an explanation.

Expansion of Water

Increasing the temperature of a liquid usually results in expansion. Water near its freezing temperature is an exception, however, for the following reason. Ice has a crystalline structure, with open-structured crystals. In the solid phase, H_2O molecules arranged in this open structure occupy a greater volume than they do in the liquid phase (Figure 6.15). Consequently, ice is less dense than water. When ice melts, not all the six-sided crystals collapse. Some remain in the ice-water mixture, making up a microscopic slush that slightly "bloats" the water (increases its volume slightly). The result is that liquid water at 0°C is less dense than slightly warmer water. As the temperature of the icy water at 0°C is increased, more of the remaining ice crystals collapse, and so the volume of the water *decreases*. So we see that cold liquid water *contracts* when its temperature is raised. This contraction continues until the water reaches a temperature of 4°C. With further increase in temperature, the water undergoes a net expansion that continues all the way to the boiling point, 100°C (Figure 6.16).

We say *net* expansion because the near-freezing water has been expanding as a result of greater thermal motion while *simultaneously* contracting due to crystal collapse.

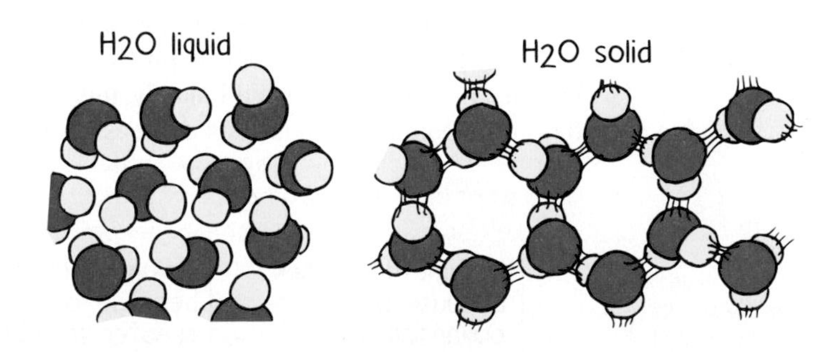

Fig. 6.15
Crystals of ice have an open-structured, hexagonal arrangement, and it is this openness that accounts for expansion as liquid water freezes. Ice is therefore less dense than water.

Answer
At cruising speed (faster than the speed of sound), air friction against the Concorde raises its temperature dramatically, resulting in this significant thermal expansion.

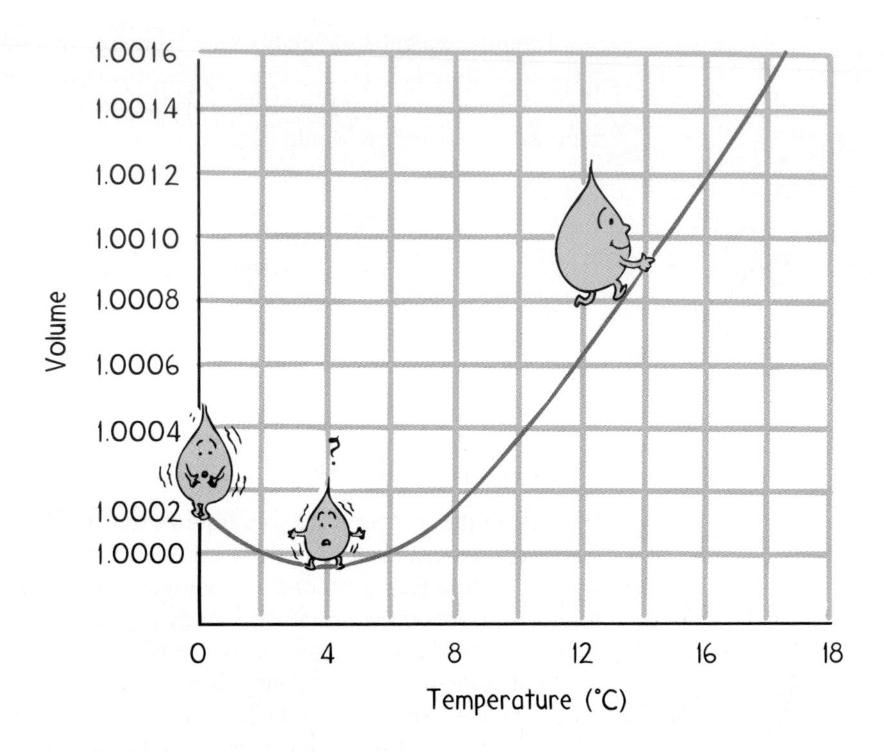

Fig. 6.16
The contraction and then expansion of water with increasing temperature.

Expansion simply overtakes contraction at 4°C. Ice crystals are pretty well gone by about 10°C, and expansion thereafter has free reign. This odd behavior is shown graphically in Figure 6.17.

So water has its smallest volume, and thus its greatest density, at 4°C. Just below 0°C, when the water becomes solid ice, its volume is considerably increased—and its density lower. Note that the volume of *ice* at 0°C is not shown in Figure 6.16. (If it were plotted to the same exaggerated scale, the graph would extend far beyond the top of the page.) After water has turned to ice, further cooling causes it to contract.

This behavior of water is of great importance in nature. If water were most dense at its freezing point, as is true of most other liquids, then the coldest water in a pond would settle to the bottom. The pond would freeze from the bottom up, destroying organisms in winter months. Fortunately, this doesn't happen. The densest water that settles at the bottom of a pond is 4 degrees above the freezing temperature. Water at the

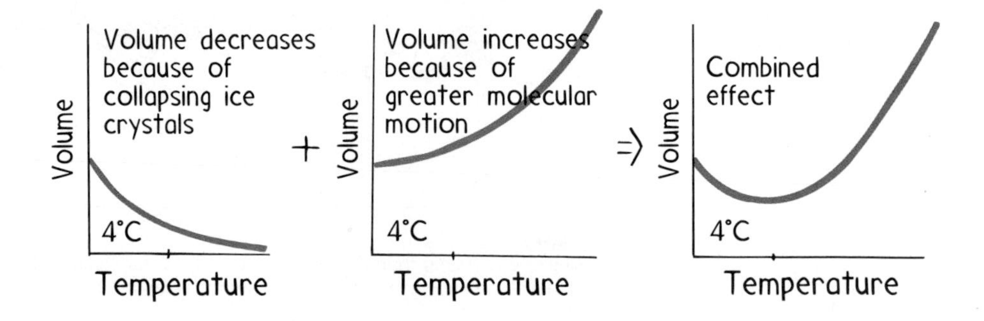

Fig. 6.17
As the temperature of water initially at 0°C rises, the simultaneous collapse of ice crystals and increase in molecular motion produce the overall effect of water being most dense at 4°C.

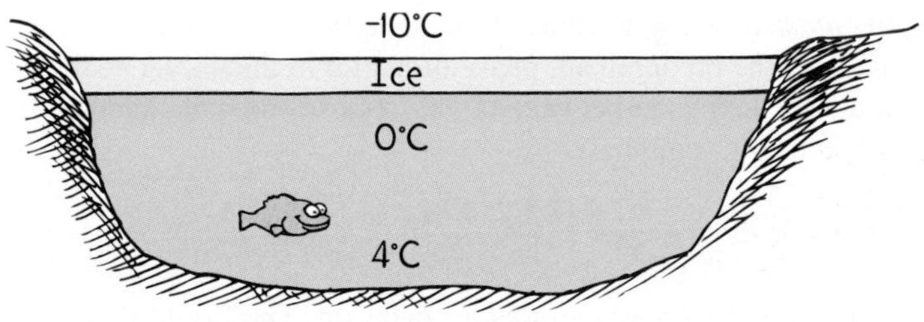

Fig. 6.18
As water cools, it sinks until the entire pond is at 4°C. Then, as water at the surface is cooled further, it floats on top and can freeze. Once ice is formed, temperatures lower than 4°C can extend down into the pond.

freezing point, 0°C, is less dense and "floats." That's why ice forms at the surface, with the organisms that require a liquid environment happily swimming below in a "warm" 4°C water (Figure 618).

Consider a pond initially at, say, 10°C. When the surface air is colder than the water, cooling occurs primarily at the surface. For the surface water to cool to 0°C, it must first cool to 4°C. But water cooled to 4°C is denser and sinks to the bottom. Can you see why it takes so long for a pond to freeze over? Before water at the top can be cooled to 0°C, all the water in the pond must first be cooled to 4°C. Only when this condition is met can the surface water be cooled to 3°, 2°, 1°, and 0°C without sinking. Then ice can form.

Continued cooling of the pond results in the freezing of the water just below the ice. This means that a pond freezes from the surface downward. In a cold winter the ice will be thicker than in a milder winter. Very deep bodies of water are not ice-covered even in the coldest of winters. This is because all the water must be cooled to 4°C before lower temperatures can be reached. For deep water, the winter is not long enough for this to occur. If only some of the water is 4°C, it lies on the bottom. Because of water's high specific heat and poor ability to conduct heat, the bottom of deep bodies of water in cold regions is a constant 4°C year-round. Fish should be glad that this is so.

■ **Question**

What was the precise temperature at the bottom of Lake Michigan on New Year's Eve in 1901?

6.8 Change of Phase

Matter exists in four common phases.[9] Ice, for example, is the *solid* phase of H_2O. Add thermal energy, and you add motion to the rigid molecular structure, which breaks down to form H_2O in the *liquid* phase, water. Add more thermal energy, and the liquid changes to the *gaseous* phase. Add still more energy, and the molecules break into ions

■ **Answer**

The temperature at the bottom of any body of water that has 4°C water in it is 4°C at the bottom, for the same reason that rocks are at the bottom. Both 4°C water and rocks are more dense than water at any other temperature. Water is a poor heat conductor, and so if the body of water is deep and in a region of long winters and short summers, as is the case for Lake Michigan, the water at the bottom is 4°C year-round.

[9]Some texts use the word *state* in place of *phase*. In this context, matter exists in four common *states*. Either term may be used.

and electrons, giving the *plasma* phase. Plasma, an incandescent gas found in fluorescent and other vapor lamps, is the predominant phase of matter in the universe, making up the Sun, stars, and much of the matter between them. Let's see how changes in temperature and pressure affect changes of phase.

Evaporation

Fig. 6.19
When wet, the cloth covering on the canteen promotes cooling. As the faster-moving water molecules evaporate from the wet cloth, its temperature decreases and cools the metal, which in turn cools the water within. Water in the canteen can be appreciably cooler than air temperature.

Water in an open container eventually evaporates, or dries up. The liquid that disappears becomes water vapor in the air. **Evaporation** is a change of phase from liquid to gas that takes place at the surface of a liquid. Molecules in the liquid phase move about in all directions and bump into one another while moving at different speeds. In bumping, some gain kinetic energy while others lose kinetic energy. Molecules at the surface that gain kinetic energy by being bumped from below may acquire enough energy to break free of the liquid. If so, they can leave the surface and fly into the space above the liquid. In this way they become molecules of gas.

The increased kinetic energy of molecules bumped free comes from molecules remaining in the liquid. This is "billiard-ball physics": When balls bump into one another and some gain kinetic energy, the others lose the same amount. Molecules going from liquid to gas are the energy gainers, while the energy losers remain in the liquid. As a result, the average kinetic energy of the molecules remaining in the liquid is lowered—evaporation is a cooling process.

Unless heated by the surroundings, the temperature of a container of water decreases as evaporation proceeds. As the water cools, so does the rate of evaporation. A higher rate of evaporation can be maintained if the water is in contact with a relatively warm surface, such as your skin. Body heat then flows from you into the water. In this way the water maintains a higher temperature and evaporation continues at a relatively high rate. This is why you feel cool as you dry off after getting wet—you are losing your body heat to the energy-requiring process of evaporation.

When our bodies tend to overheat, our sweat glands produce perspiration. This is part of nature's thermostat, for the evaporation of perspiration cools us and helps us maintain a stable body temperature. Many animals do not have sweat glands and must cool themselves by other means (Figures 6.20 and 6.21).

■ Question

Would evaporation be a cooling process if each molecule at the surface of a liquid had the same kinetic energy before and after bumping into other molecules?

■ Answer
No. If there were no transfer of kinetic energy during molecular bumping, there would be no change in temperature. A liquid cools only when there is a decrease in the average kinetic energy of its molecules. This decrease occurs when some molecules (like billiard balls) gain speed at the expense of others that lose speed. Those that leave (evaporate) are gainers while losers remain behind to effectively lower the temperature of the water.

Link to Entomology—Life at the Extremes

Some deserts reach surface temperatures of 60°C (140°F). Too hot for life? Not for a species of ant (*Cataglyphis*) that thrives at this searing temperature and can withstand higher temperatures than any other creatures in the desert. At this extremely high temperature, the desert ants can forage for food without the presence of the lizards who otherwise prey upon them. How they are able to do this is currently being researched. They scavenge the desert surface for corpses of those who did not find cover from the hot Sun, touching the hot sand as little as possible while often sprinting on four legs with two high in the air. Although their foraging paths zig zag over the desert floor, their return paths are almost straight lines to their nest holes. They attain speeds of 100 body lengths per second. During an average 6 day life, most of these ants retrieve 15–20 times their weight in food.

From deserts to glaciers, a variety of creatures have invented ways to survive the harshest corners of the world. A species of worm thrives in the glacial ice in the Arctic. There are insects in the Antarctic ice that pump their bodies full of antifreeze to ward off becoming frozen solid. Some fish beneath the ice are able to do the same. Then there are bacteria that thrive in boiling hot springs as a result of heat-resistant proteins in their bodies.

An understanding of how creatures survive at the extremes of temperature can provide clues for practical solutions to the physical challenges of humans. Astronauts who venture from our nest, for example, will need all the techniques available for coping with unfamiliar environments.

Fig. 6.20
Hanz has no sweat glands (except between the toes). He cools himself by panting. In this way evaporation occurs in the mouth and within the bronchial tract.

Fig. 6.21
Pigs have no sweat glands and therefore cannot cool by the evaporation of perspiration. Instead, they wallow in the mud to cool themselves.

The rate of evaporation is greater at higher temperatures because there are a greater proportion of molecules having sufficient kinetic energy to escape the liquid. We shall learn in Chapter 18 that the hydrogen bonds between water molecules are broken when evaporation occurs. Water evaporates at lower temperatures, too, but at a lower rate. A puddle of water, for example, may evaporate to dryness on a cool day.

Even frozen water "evaporates." This form of evaporation, in which molecules jump directly from a solid to a gaseous phase, is called **sublimation**. Because water molecules are so tightly held in the solid phase, frozen water does not evaporate (sublime) as readily as liquid water does. Sublimation, however, does account for the loss of significant portions of snow and ice, especially high on sunny, dry mountain tops.

Condensation

Fig. 6.22
Heat is given up by steam when it condenses inside the radiator.

The opposite of evaporation is **condensation**—the changing of a gas to a liquid. When gas molecules near the surface of a liquid are attracted to the liquid, they strike the surface with increased kinetic energy and become part of the liquid. This kinetic energy is absorbed by the liquid, with the result that temperature is increased. Condensation is a warming process.

A dramatic example of the warming that results from condensation is the energy given up by steam when it condenses—a painful experience if it condenses on you. That's why a burn from 100°C steam is much more damaging than a burn from 100°C boiling water; the steam gives up considerable energy when it condenses to a liquid and wets the skin. This energy release by condensation is utilized in steam-heating systems.

Steam is water vapor at a high temperature, usually 100°C or more. Cooler water vapor also gives up energy when it condenses. In taking a shower, for example, you're warmed by condensation of vapor in the shower region—even vapor from a cold shower—if you remain in the moist shower area. You quickly sense the difference if you step outside. Away from the moisture, net evaporation takes place quickly and you feel chilly. When you remain in the shower stall, even with the water off, the warming effect of condensation counteracts the cooling effect of evaporation. If as much moisture condenses as evaporates, you feel no change in body temperature. If condensation exceeds evaporation, you are warmed. If evaporation exceeds condensation, you are cooled. So now you know why you can dry yourself with a towel much more comfortably if you remain in the shower area. To dry yourself thoroughly, you can finish the job in a less-moist area.

Fig. 6.23
If you're chilly outside the shower stall, step back inside and be warmed by the condensation of the excess water vapor there.

Spend a July afternoon in dry Tucson or Phoenix, where evaporation is appreciably greater than condensation. The result of this pronounced evaporation is a much cooler feeling than what you would experience on a same-temperature July afternoon in New York City or New Orleans. In these humid locations, condensation noticeably counteracts evaporation, and you feel the warming effect as vapor in the air condenses on your skin. You are literally being "tattooed" by the impact of H_2O molecules in the air slamming into you. Put more mildly, you are warmed by the condensation of vapor in the air upon your skin. We shall explore condensation in the atmosphere when we study meteorology in Chapter 26.

■ Question

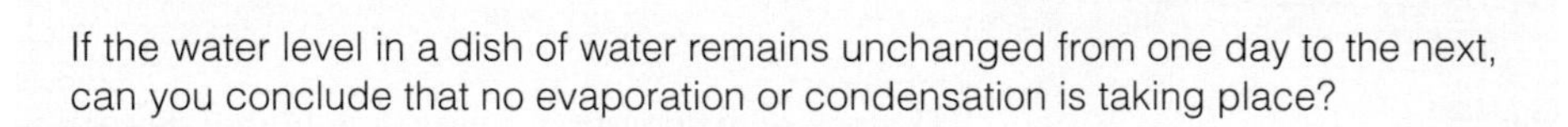

If the water level in a dish of water remains unchanged from one day to the next, can you conclude that no evaporation or condensation is taking place?

■ **Answer**

Not at all, for there is much activity taking place at the molecular level. Both evaporation and condensation occur continuously. The fact that the water level remains constant indicates equal rates of evaporation and condensation.

Boiling

Evaporation occurs beneath the surface of a liquid in the process called **boiling**. Bubbles of vapor form in the liquid and are buoyed to the surface, where they escape. Bubbles can form only when the pressure of the vapor within them is great enough to resist the pressure exerted by the surrounding water, which includes atmospheric pressure. This condition of the pressure inside the bubble equaling or exceeding that of the surrounding water occurs at the boiling temperature of the liquid. At lower temperatures, the vapor pressure in the bubbles is not enough, and the surrounding pressure collapses any bubbles that form.

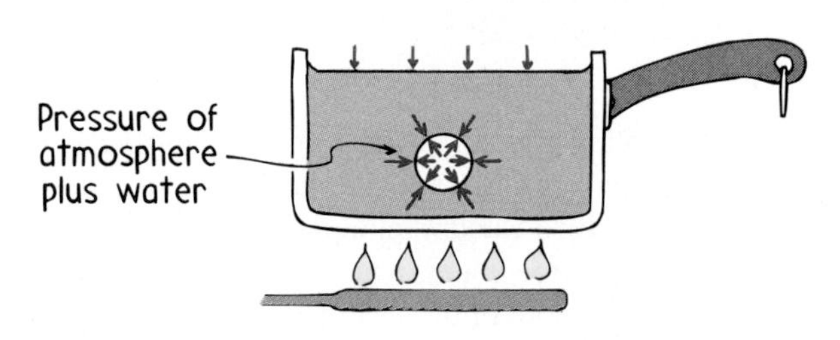

Fig. 6.24
The motion of vapor molecules in the bubble of steam (much enlarged) creates a gas pressure (called the *vapor pressure*) that counteracts the atmospheric and water pressure against the bubble.

At what point boiling begins depends not only on temperature but also on pressure. As atmospheric pressure increases, the vapor molecules inside any bubbles that form must move faster in order to exert increased pressure within the bubble to counteract the additional atmospheric pressure. So increasing the pressure on the surface of a liquid raises its boiling temperature. Conversely, lowered atmospheric pressure (as at high altitudes) decreases the boiling temperature. In Denver, Colorado, the "mile-high city," for example, water boils at 95°C, instead of the 100°C boiling temperature characteristic of sea level. If you try to cook food in boiling water that is cooler than 100°C, you must wait a longer time for proper cooking. A three-minute boiled egg in Denver is yukky. If the temperature of the boiling water were very low, food would not cook at all.

The role of pressure is evident in a pressure cooker (Figure 6.25). Vapor pressure builds up inside and prevents boiling, which results in a higher water temperature. It is important to note that it is the high temperature of the water that cooks the food, not the boiling process itself. At high altitudes, water boils at a lower temperature.

Fig. 6.25
The tight lid of a pressure cooker holds pressurized vapor above the water surface, and this inhibits boiling. In this way the boiling temperature of the water is increased to above 100°C.

Boiling, like evaporation, is a cooling process. At first thought, this may seem surprising—perhaps because we usually associate boiling with heating. However, heating water is one thing; boiling it is another. When 100°C water at atmospheric pressure is boiling, it is in thermal equilibrium. It is being cooled by boiling as fast as it is being heated by energy from the heat source (Figure 6.26). If cooling did not take place, continued

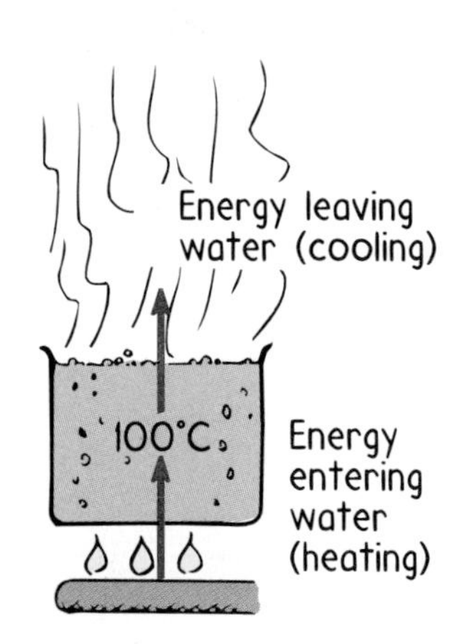

Fig. 6.26
Heating warms the water from below, and boiling cools it from above.

application of heat to a pot of boiling water would raise its temperature. The reason a pressure cooker reaches higher temperatures is because boiling is forestalled, which in effect prevents cooling.

■ Question

Being that boiling is a cooling process, would it be a good idea to cool your hot, sticky hands by dipping them into boiling water?

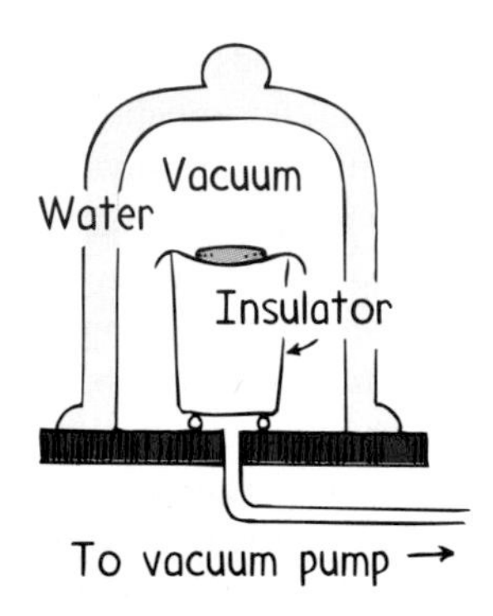

Fig. 6.27
Apparatus to demonstrate that in a vacuum water freezes and boils at the same time. A gram or two of water is placed in a dish that is insulated from the base by a polystyrene cup.

We can dramatically show the cooling effect of evaporation and boiling by placing a shallow dish of room-temperature water in a vacuum jar (Figure 6.27). When the pressure in the jar is slowly reduced by a vacuum pump, the water starts to boil. The boiling process takes heat away from the water, which consequently cools. As the pressure is further reduced, more and more of the slower-moving molecules boil away. Continued boiling results in a lowering of temperature until the freezing point of approximately 0°C is reached. Continued cooling by boiling causes ice to form over the surface of the bubbling water. Boiling and freezing take place at the same time! Frozen bubbles of boiling water are a remarkable sight.

Spray some drops of coffee into a vacuum chamber, and they, too, boil until they freeze. Even after they are frozen, the water molecules continue to evaporate into the vacuum until little crystals of coffee solids are left. This is how freeze-dried coffee is made. The low temperature of this process tends to keep the chemical structure of the coffee solids from changing. When hot water is added, more of the original flavor of the coffee is retained.

Melting and Freezing

Melting occurs when a substance changes from a solid to a liquid. To visualize what happens during this phase change, imagine you are holding hands with someone and that each of you starts jumping around randomly. The more violently you jump, the more difficult keeping hold is. If you jump violently enough, keeping hold becomes impossible. Something like this happens to the molecules of a solid when it is heated. As heat is absorbed, the molecules vibrate more and more violently. If enough heat is absorbed, the attractive forces between the molecules are no longer able to hold them together. The solid melts.

Freezing occurs when a liquid changes to a solid—the converse of melting. As energy is withdrawn from a liquid, molecular motion diminishes until finally the molecules, on average, are moving so slowly that the attractive forces between them are able to cause cohesion. The molecules then vibrate about fixed positions and form a solid. The liquid freezes.

■ **Answer**

No, no, no! When we say boiling is a cooling process, we mean that the water (not your hands!) is being cooled relative to the higher temperature it would attain otherwise. Because of the cooling effect of the boiling, the water remains at 100°C instead of getting hotter. A dip in 100°C water would be most uncomfortable for your hands!

Snowballs

The open structure of H_2O molecules in the solid phase has some interesting applications. Pressure exerted on the ice crystals lowers their melting point. Simply put, the crystals are crushed to the liquid phase. The effect is small, for the melting point is only lowered 0.007 C° for each atmosphere of added pressure. Nonetheless, when the pressure is removed, the molecules crystallize and refreezing occurs. This phenomenon of melting under pressure and freezing again when the pressure is reduced is called *regelation*. It is one of the properties of water that make it different from other materials.

When we make a snowball, we are using this principle. We compress the snow with our hands and cause a slight melting of the ice crystals; when we remove the pressure, the ice refreezes and binds the snow together. Making snowballs is difficult in very cold weather because the pressure we can apply may not be enough to melt the snow.

At atmospheric pressure, ice forms at 0°C. At this temperature, water molecules on average are moving so slowly that crystals can form—unless some other substance is dissolved in the water. Then the freezing point is lower because the "foreign" molecules impede crystal formation. The result of this impedance is that slower-than-normal molecular motion is required for the formation of the ice crystals. As ice crystals do form, the impedance is intensified because the proportion of "foreign" molecules among the still-liquid water molecules increases. Connections become more and more difficult. Only when the water molecules move slowly enough for attractive forces to play an unusually large part can freezing be completed.

In general, adding anything to water lowers its freezing temperature. Antifreeze is a practical application of this process.

6.9 Energy and Changes of Phase

Whenever a substance changes phase, a transfer of energy occurs. To melt ice, for example, heat must be applied—changing ice to water requires thermal energy. Likewise, in boiling water, thermal energy is required to change liquid water to steam. The steam in turn can be turned into a glowing incandescent gas, a plasma, by the addition of still more thermal energy. The energy required to melt ice is the energy required to break the bonds between ice crystals, and the energy that turns liquid water to steam is the energy required to break the molecular bonds between liquid water molecules.

Conversely, energy must be extracted from a substance to change its phase in the direction from plasma to gas to liquid to solid (Figure 6.28).

Energy is absorbed when change of phase is in this direction
Solid ⇄ Liquid ⇄ Gas
Energy is released when change of phase is in this direction

Fig. 6.28
Energy changes with change of phase.

The cooling cycle of a refrigerator neatly employs the concepts shown in Figure 6.28. A liquid that has a low boiling point, often a chlorofluorocarbon[10], is pumped into the cooling unit, where it turns into a gas and in doing so draws heat from the things stored there. The gas with its added energy is directed outside the cooling unit to coils located in the back, appropriately called condensation coils, where heat is given off to the air as the gas condenses to a liquid. A motor pumps the fluid through the system, where it is made to undergo the cyclic process of vaporization and condensation. The next time you're near a refrigerator, place your hand near the condensation coils in the back, and you'll feel the heat that has been extracted from inside.

An air conditioner employs the same principle and simply pumps heat energy from one part of the unit to another. When the roles of vaporization and condensation are reversed, the air conditioner becomes a heater.

So we see that a solid must absorb energy to melt and a liquid must absorb energy to vaporize. Conversely, a gas must release energy to liquefy, and a liquid must release energy to solidify.

Questions

1. To impart a hickory flavor to a roasted turkey, a cook places a pot of water containing hickory chips in an oven with the turkey. Why does the turkey take longer than expected to cook?
2. When H_2O in the vapor state condenses, is the surrounding air warmed or cooled?

The general behavior of many substances as they change phase can be illustrated by a description of the changes of phase of H_2O. Consider a 1-gram piece of ice at −50°C in a closed container, put on a stove to heat. A thermometer in the container reveals a slow increase in temperature up to 0°C, as shown in Figure 6.29. At 0°C, the temperature stops rising, even though heat is still being added. Now the added heat is being used to break crystal bonds in the ice, effectively increasing their potential energies rather than their kinetic energies. (Remember, only increased *kinetic* energy means increased temperature.) So any thermal energy going into the water now doesn't make the water molecules move faster; they simply break apart from one another, and the temperature remains constant. Melting 1 gram of ice requires 80 calories. Only when all the ice melts does the temperature begin to rise again. Then each calorie absorbed

Answers

1. Much of the heat from the oven is consumed in changing the phase of the water from liquid to gas. As long as water remains in the liquid phase, the temperature of the oven will not rise much higher than the boiling temperature of water—100°C.
2. The surrounding air is warmed because thermal energy is released by the gaseous H_2O molecules as they turn into liquid H_2O molecules.

[10]Present-day research is directed to making thermoelectric devices wherein electrons take the place of a fluid. Electric currents undergo expansion (cooling) and compression (heating) when moving between materials having different electron configurations. Watch for motorless refrigerators in the future!

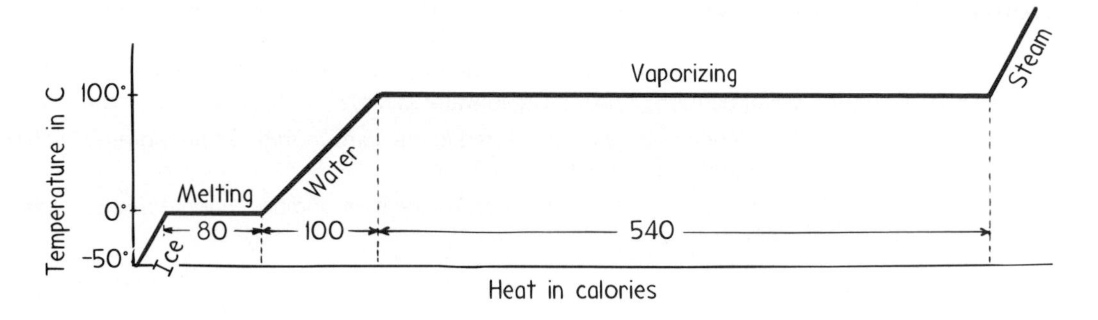

Fig. 6.29
A graph showing the energy involved in the heating and change of phase of 1 g of H_2O.

by the water increases its temperature by 1 degree Celsius until the boiling temperature, 100°C, is reached. At this point, the 100°C temperature remains constant as heat is added while all the water is vaporized to steam. The gram of liquid water must absorb 540 calories of thermal energy in order to vaporize. Finally, when all the water has become steam at 100°C, the temperature begins to rise once again and continues to rise as long as heat is added. The plasma phase, not shown in Figure 6.29, is similarly achieved at temperatures above 4000°C.

The 540 calories required to vaporize a gram of water is a relatively large amount of energy—much more than the amount required to bring a gram of ice at absolute zero (−273°C) to boiling water at 100°C! Although the molecules in steam at 100°C and boiling water at 100°C have the same average kinetic energy, steam has an extra 540 calories per gram potential energy because the molecules are relatively free of one another and not bound together as in the liquid phase. This 540 calories per gram of steam is released to the surroundings when the steam condenses to water. Then the potential energy of the far-apart steam molecules transforms to heat. This is like the potential energy of two attracting magnets separated from each other; when released their potential energy is converted into kinetic energy, and then into heat as the magnets strike each other.

So we see that the energies required to melt ice (80 calories per gram) and boil water (540 calories per gram) are the same as the amounts released when the phase changes are in the opposite direction. The processes are reversible.

The amount of energy required to change any substance from solid to liquid (and vice versa) is called the **heat of fusion** for the substance. For water, this is 80 calories per gram. The amount of energy required to change any substance from liquid to gas (and vice versa) is called the **heat of vaporization** for the substance. For water this is a whopping 540 calories per gram.[11] We shall see in Chapter 18 that these relatively high values are due to the strong forces between water molecules—hydrogen bonds.

These concepts are not limited to water. Whenever any solid is warmed to its melting point, slow continued addition of heat melts the solid. An addition of heat to a liquid at its vaporization temperature vaporizes the liquid. In Part 7 we'll consider the concept of *partial melting* for rocks composed of different minerals with different melting points. Depending on temperature, some minerals in the rock melt while others remain solid. But that comes later.

---

[11]In SI units, the heat of vaporization of water is 2.26 MJ/kg, and the heat of fusion of water is 0.335 MJ/kg.

Questions

1. How much energy is transferred to the surroundings when a gram of steam at 100°C condenses to water at 100°C?
2. How much energy is transferred to the surroundings when a gram of boiling water at 100°C cools to liquid water at 0°C?
3. How much energy is transferred to the surroundings when a gram of liquid water at 0°C freezes to ice at 0°C?
4. How much energy is transferred to the surroundings when a gram of steam at 100°C turns to ice at 0°C?

Fig. 6.30
Paul Ryan tests the hotness of molten lead by dragging his wetted finger through it.

Some people who live in cold climates take advantage of water's high heat of fusion by placing large tubs of water in their basements. Outside, the temperature may drop to well below freezing. Downstairs in the basement, however, as the water freezes, millions of calories of thermal energy are released to keep the basement considerably warmer.

Water's high heat of vaporization allows you to briefly touch your *wetted* finger to a hot skillet on a hot stove without harm. You can even touch it a few times in succession *as long as your finger remains wet*. This is because energy that ordinarily would go into burning your finger goes instead into changing the phase of the moisture on your finger to a vapor. This same wetted-finger trick lets you judge the hotness of a hot clothes iron.

Supervisor Paul Ryan of the Department of Public Works in Malden, Massachusetts, has for years used molten lead to seal pipes in certain plumbing operations. He startles onlookers by dragging his finger through the molten lead to judge its hotness (Figure 6.30). He doesn't get burned because he does this only when he is sure the lead is very hot and his finger is thoroughly wet. (Do not try this on your own, for if the lead is not hot enough it will stick to your finger—ouch!)

Answers

1. Figure 6.29 tells you that the vaporization of 1 g of liquid water requires 540 calories, and this same amount of thermal energy must be released in the reverse process. Therefore 1 g of steam at 100°C transfers 540 calories of energy to the surroundings when it condenses to water at the same temperature.
2. Again from Figure 6.29 (or from the definition of the calorie), 1 g of boiling water transfers 100 calories to the surroundings when it cools 100 C° to become liquid water at 0°C.
3. And again from Figure 6.29, 1 g of liquid water at 0°C releases 80 calories to become ice at 0°C.
4. One gram of steam at 100°C transfers to the surroundings a grand total of the preceding values, $540 + 100 + 80 = 720$ calories, as it becomes ice at 0°C.

■ Question

Suppose 4 g of water at 100°C is spread over a large surface so that 1 g evaporates rapidly. If evaporation of the 1 gram takes 540 calories from the remaining 3 g of water and no other heat transfer takes place, what are the temperature and phase of the remaining 3 g once the 1 g is completely evaporated?

■ **Answer**

The remaining 3 g turns to 0°C ice. For 540 calories to come from 3 g, each gram gives up 180 calories. Taking 100 calories from a gram of 100°C water reduces its temperature to 0°C (remember the definition of a calorie), and taking away 80 more calories turns it to ice. This is why hot water spread thinly so that its surface is large compared with its volume quickly turns to ice in a freezing-cold environment.

Summary of Terms

- **Temperature** A measure of the hotness of an object, related to the average kinetic energy per molecule in the object, measured in degrees Celsius, degrees Fahrenheit, or kelvins.
- **Absolute zero** The lowest possible temperature; the temperature at which all particles have their minimum kinetic energy.
- **Thermal energy** (*internal energy*) The total energy (kinetic plus potential) of the particles that make up a substance.
- **Heat** The thermal energy that flows from an object at higher temperature to one at a lower temperature, commonly measured in calories or joules.
- **Calorie** A unit of thermal energy, or heat. One calorie is the thermal energy required to raise the temperature of 1 gram of water 1 Celsius degree (1 cal = 4.184 J). One Calorie (with a capital C) is equal to 1000 calories and is the unit used in describing the energy available from food.
- **Specific heat capacity** The quantity of heat per unit of mass required to raise the temperature of a substance by 1 Celsius degree.
- **Evaporation** The change of phase from liquid to gaseous.
- **Sublimation** The change of phase from solid to gaseous, skipping the liquid phase.
- **Condensation** The change of phase from gaseous to liquid.
- **Boiling** Rapid evaporation that takes place within a liquid as well as at its surface.
- **Melting** The change of phase from solid to liquid.
- **Freezing** The change of phase from liquid to solid.
- **Heat of fusion** The amount of thermal energy required to change a substance from solid to liquid or from liquid to solid.
- **Heat of vaporization** The amount of thermal energy required to change a substance from liquid to gas or from gas to liquid.

Review Questions

1. Why does a penny become warmer when struck by a hammer?
2. What happens to the temperature of air when it is rapidly compressed?

Temperature

3. What are the temperatures for freezing water on the Celsius and Fahrenheit scales? For boiling water at sea level?
4. Is the temperature of an object a measure of the total kinetic energy of molecules in the object or a measure of the average kinetic energy per molecule in the object?
5. What is meant by the statement "a thermometer measures its own temperature"?

Absolute Zero

6. By how much does the pressure of a gas in a rigid vessel decrease when the temperature is decreased by 1 C°?
7. What pressure would you expect in a rigid container of 0°C gas if you cooled it by 273 C°?
8. What is the temperature of melting ice on the Kelvin scale? Of boiling water at atmospheric pressure?
9. Why are there no negative numbers on the Kelvin scale?

Thermal Energy

10. When is thermal energy in a substance evident?

Heat

11. When you touch a cold surface, does "coldness" travel from the surface to your hand or does thermal energy travel from your hand to the cold surface? Explain.
12. Distinguish between temperature and heat.
13. Distinguish between heat and thermal energy.
14. What determines the direction of heat flow?

Heat Units

15. How is the energy value of foods determined?
16. Distinguish between a calorie and a Calorie.
17. Distinguish between a calorie and a joule.

Specific Heat Capacity

18. Which warms up faster when heat is applied—iron or silver?
19. Does a substance that heats up quickly have a high or a low specific heat capacity?
20. How does the specific heat capacity of water compare with the specific heat capacities of other common materials?
21. Northeastern Canada and much of Europe receive about the same amount of sunlight per unit area. Why is Europe generally warmer in the winter?
22. Because of the law of energy conservation, something must warm up if ocean water cools. What is that something?

Thermal Expansion

23. Why is hot water poured into a drinking glass more likely to break the glass if the glass is thick?
24. How is a bimetallic strip used to regulate temperature?
25. Which generally expands more for the same increase in temperature—solids or liquids?

Expansion of Water

26. When the temperature of ice-cold water is increased slightly, does it undergo a net expansion or net contraction?
27. What is the reason for ice being less dense than water?
28. Does "microscopic slush" in water tend to make it more dense or less dense?
29. What happens to the amount of "microscopic slush" in cold water when its temperature is increased?
30. What happens to the kinetic energy of molecular motion in water when its temperature is increased?
31. At what temperature do the combined effects of contraction and expansion produce the smallest volume for water?
32. Why does ice form at the surface of a body of water instead of at the bottom?

Change of Phase

33. What are the four phases of matter?

Evaporation

34. What is evaporation, and why is it a cooling process? What cools during evaporation?
35. Do the molecules in a liquid all have about the same speed or do they all have very different speeds?
36. Why is perspiration a cooling process?
37. Why does a hot dog pant?
38. Give an example of a solid bypassing the liquid phase and changing directly to the gaseous phase. What is the name of this process?

Condensation

39. What is condensation, and why is it a warming process? What warms up during condensation?
40. Why is a steam burn more damaging than a burn from boiling water at the same temperature?
41. Why do we feel uncomfortably warm on a hot, humid day?

Boiling

42. Distinguish between evaporation and boiling.
43. Why does water not boil at 100°C when it is under pressure that is higher than normal atmospheric pressure?
44. Is it the boiling or the higher temperature that cooks food faster in a pressure cooker?
45. We observe that the temperature of boiling water doesn't increase with continued heat input. Explain how this observation is evidence that boiling is a cooling process.
46. What condition permits water to boil at a temperature below 100°C?
47. We can add heat to water and boil it to form steam. Is it a contradiction to say we can boil water to form ice? Explain.

Melting and Freezing

48. Why does increasing the temperature of a solid make it melt?
49. Why does decreasing the temperature of a liquid make it freeze?
50. Why does freezing of water not occur at 0°C when molecules other than H_2O are mixed in with the water molecules?
51. Which freezes at the lower temperature, pure water or salt water?

Energy and Changes of Phase

52. Does a liquid give off or absorb energy when it evaporates? When it solidifies?
53. Does a gas give off or absorb energy when it condenses?

54. Why does the temperature of melting ice not rise when heat is applied?
55. Why does the temperature of boiling water not rise when heat is applied?

Home Projects

1. Place a Pyrex funnel mouth down in a saucepan full of water so that the straight tube of the funnel sticks above the water. Rest one side of the funnel rim on a nail or coin so that water can get under it. Heat the pan on a stove and watch the water as it begins to boil. Where do the bubbles form first? Why? As the bubbles rise, they expand rapidly and push water ahead of them. The funnel confines the water, which is forced up the tube and driven out at the top. Now do you know how a geyser and a coffee percolator work?

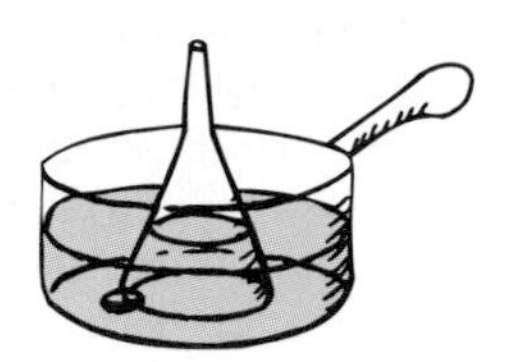

2. Watch the spout of a teakettle of boiling water. Notice that you cannot see the steam that issues from the spout. The cloud that you see farther away from the spout is not steam but rather condensed water droplets—much bigger than the particles that make up water *vapor*. Now hold a candle in the cloud of condensed droplets. Can you explain your observations?
3. Measure the temperature of boiling water and the temperature of a boiling solution of table salt and water.

Exercises

1. Why can't you establish whether you are running a high temperature by touching your own forehead?
2. Which has the greater amount of thermal energy, an iceberg or a cup of hot coffee? Explain.
3. Would a mercury thermometer be feasible if glass and mercury expanded at the same rates for a given change in temperature? Explain.
4. What is *temperature* a measurement of?
5. In a mixture of hydrogen and oxygen gases at the same temperature, which molecules on average move faster? Defend your answer.
6. If you drop a hot rock into a pail of water, the temperature of the rock and the water change until the two are equal. The rock cools and the water warms. Does this hold true if the hot rock is dropped into the Atlantic Ocean? Explain.
7. Why does the pressure of a gas enclosed in a rigid container increase as the temperature increases?
8. If you add the same amount of heat to two different materials of the same mass, will you produce the same increase in temperature? Why or why not?
9. Why does a watermelon stay cool for a longer time than sandwiches when both are removed from a cooler on a hot day?
10. Bermuda is close to North Carolina, but unlike North Carolina it has a tropical climate year-round. Why?
11. To keep your feet warm on a long, cold winter night, which is better to take to bed: a hot brick wrapped in a towel or a hot bottle of water wrapped in a towel?
12. San Francisco is warmer in winter than Washington, D.C. Why isn't it warmer in summer?
13. In addition to the molecular to-and-fro vibrations associated with temperature, some molecules can absorb large amounts of energy in the form of internal vibrations and rotations in the atoms making up the molecules. Would you expect materials composed of such molecules to have a high or a low specific heat? Explain.
14. The desert sand is very hot in the day and very cool at night. What does this tell you about its specific heat?
15. Suppose you use natural gas to heat your home. Who would gain by having the gas warmed before it passes through your gas meter: you or the gas company?
16. A metal ball is just able to pass through a metal ring. When the ball is heated, however, it cannot pass through the ring. What happens if the ring, rather than the ball, is heated? Does the size of the hole increase, stay the same, or decrease?
17. After a machinist very quickly slips a hot, snugly fitting iron ring over a very cold brass cylinder, there is no way the two can be separated intact. Explain.
18. Suppose you cut a small gap in a metal ring. If you heat the ring, will the gap become wider or narrower?
19. When a mercury thermometer is warmed, the mercury level momentarily goes down before it rises. Why?
20. An old remedy for a pair of nested drinking glasses stuck together is to run water at one temperature into the inner glass and then run water at a different temperature over the surface of the outer glass. Which water should be hot and which should be cold?
21. After you use a steel tape to measure the dimensions of a plot of land on a hot day, you come back and re-measure on a cold day. On which day do you determine the larger area?
22. Why can you predict the present temperature at the bottom of Lake Superior but not the present temperature at the bottom of Lake Toba in Sumatra?
23. Suppose that water is used in a thermometer instead of mercury. If the temperature is 4°C and then changes, why can't the thermometer indicate whether the temperature is rising or falling?

24. How does the combined volume of the billions and billions of hexagonal open spaces in the ice crystals in a piece of floating ice compare with the portion of ice that floats above the water line?
25. How would the shape of the curve in Figure 6.16 differ if density were plotted against temperature instead of against volume? Make a rough sketch.
26. Why is it important to protect water pipes so they don't freeze?
27. If cooling occurred at the bottom of a pond instead of at the surface, would a pond freeze from the bottom up? Explain.
28. If water had a lower specific heat, would ponds be more likely to freeze or less likely to freeze?
29. You can determine wind direction by wetting your finger and holding it up in the air. Explain.
30. Can you give two reasons why pouring a cup of hot coffee into the saucer results in faster cooling?
31. Why does blowing over hot soup cool it?
32. Touch one ear with a bit of perfume and the other with a bit of same-temperature water. Which ear feels momentarily cooler, and why?
33. A covered glass of water sits for days with no drop in water level. Can you say that nothing has happened, that no evaporation and no condensation have taken place? Explain.
34. If all the molecules in a liquid had the same speed and some were able to evaporate, would the remaining liquid be cooled? Explain.
35. Porous canvas bags filled with water are used by travelers in hot weather. When the bags are slung on the outside of a fast-moving car, the water inside is cooled considerably. Explain.
36. Why will wrapping a bottle in a wet cloth at a picnic often produce a cooler bottle than placing the bottle in a bucket of cold water?
37. The human body can maintain its customary temperature of 37°C on a day when the temperature is above 40°C. How is this done?
38. Why does the temperature of boiling water remain the same as long as the heating and boiling continue?
39. In a nuclear submarine power plant, the temperature of the water in the reactor is above 100°C. How is this possible?
40. Water spontaneously boils in a vacuum—which means that no heat input is needed. Could you cook an egg in this boiling water? Explain.
41. An inventor friend proposes to design cookware that allows water to boil at a temperature below 100°C so that food can be cooked with less energy. Comment on this idea.
42. Your instructor hands you a closed flask of room-temperature water. When you hold it, the heat from your bare hands causes the water to boil. Quite impressive! How is this accomplished?
43. When you boil potatoes, does the length of time it takes to cook them depend on how vigorously the water is boiling? (Directions for cooking spaghetti call for vigorously boiling water—not to lessen cooking time, but to prevent something from happening. If you don't know what it is, ask a cook.)
44. Why does putting a lid over a pot of water on a stove shorten the time it takes for the water to come to a boil, whereas using the lid after the water is boiling shortens the cooking time only slightly?
45. People who live where snowfall is common will tell you that air temperatures are higher on snowy day than on clear days. Some misinterpret this by stating that snowfall can't occur on very cold days ("it's too cold to snow"). Explain this misinterpretation.
46. What effect does melting ice have on the temperature of the surrounding air?
47. What effect does freezing water have on the temperature of the surrounding air?
48. Why does dew form on the surface of an ice-cold soft-drink can?
49. Why is it that in cold winters a tub of water placed in a farmer's canning cellar helps prevent canned food from freezing?
50. How does keeping an open tub of water inside an Arctic igloo prevent the interior from falling below 0°C?
51. Why does spraying fruit trees with water before a frost help to protect the fruit from freezing?

Problems

1. Suppose a bar 1 m long expands $\frac{1}{2}$ cm when heated. By how much will a bar 100 m long of the same material expand when similarly heated?
2. Steel expands 1 part in 100,000 for each Celsius degree increase in temperature. Suppose the 1.3-km main span of the Golden Gate Bridge had no expansion joints. How much longer would it be for an increase in temperature of 10 C°?
3. Consider a 40,000-km-long steel strip that forms a ring that fits snugly all around the Earth at the equator. Suppose people along its length breathe on it so as to raise its temperature 1 C°. The pipe gets longer and so is no longer snug. How high does it now stand above ground level? (To simplify, consider only the expanded distance of the circular strip's radius—the distance from the Earth's center—and apply the geometry formula that relates circumference C and radius r, $C = 2\pi r$. The result is surprising!)
4. Find the mass of 0°C ice that 10 g of 100°C steam will completely melt.
5. Radioactive minerals in common granite release energy at the rate 0.01 cal/kg·yr. If the granite is thermally isolated so that all this energy heats it, how many years does it take for a 1-kg chunk of granite at 50°C to reach its approximate melting temperature, 700°C? (The specific heat of granite is 200 cal/kg·C°.)

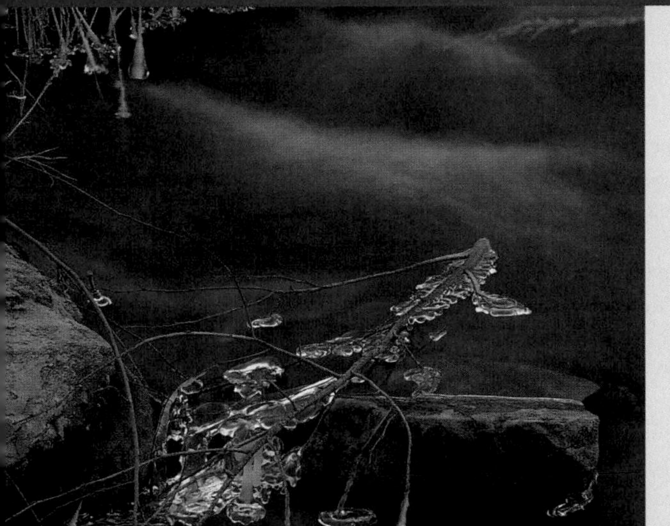

CHAPTER

7

Heat Transfer and Thermodynamics

The study of heat and its transformation to mechanical energy is called **thermodynamics** (which stems from Greek words meaning "movement of heat"). The science of thermodynamics was developed at the beginning of the last century, before the atomic and molecular theory of matter was understood. Because the early workers in thermodynamics knew nothing about electrons, atoms, and other microscopic particles, the models they used invoke macroscopic notions—such as mechanical work, pressure, and temperature—and their roles in energy transformations. The foundation of thermodynamics is the conservation of energy and the fact that heat flows from hot to cold and not the other way around. Thermodynamics provides the basic theory of heat engines, from steam turbines to fusion reactors, and the basic theory of refrigerators and heat pumps.

We'll preface our study of thermodynamics with the three common ways in which heat is transferred from warmer to cooler things: *conduction, convection,* and *radiation.*

7.1 Conduction

Fig. 7.1
The tile floor feels colder than the wooden floor, even though both floors are at the same temperature. This is because tile is a better conductor of heat than wood is, and so heat is more readily conducted out of the foot touching the tile.

Hold one end of an iron nail in a flame and it quickly becomes too hot to hold. The heat enters the nail at the end in the flame and is transmitted along the nail's whole length. The transfer of heat in this manner is called **conduction**. Every atom contains electrons, and metal atoms contain one or more loosely held electrons that are responsible for the conduction of heat (and, as we shall see in Chapter 8, the conduction of electricity also). The loosely held electrons in the iron atoms in the end of the nail held in the flame are jostled by the energy given to them by the fire, and they collide with one another and with iron atoms farther up the nail, which in turn collide with their neighbors, and so on. This bumping process continues until the increased motion has been transmitted to all the atoms and the entire nail has become hot. Heat conduction occurs by collisions between the particles inside the body being heated.

Solids whose atoms or molecules contain one or more loosely held outer electrons conduct heat well. Metals have the "loosest" outer electrons and are therefore the best conductors of heat. Silver is the best, copper next, and, among the common metals, aluminum and then iron. Wool, wood, straw, paper, cork, and Styrofoam are poor conductors of heat. The outer electrons in the molecules of these materials are firmly attached to the molecules. Poor conductors are called *insulators.*

Because wood is a good insulator, it is used for cookware handles. Even when it is hot, you can quickly grasp the wooden handle of a pot with your bare hand without harm. An iron handle of the same temperature would surely burn your hand. Wood is a good insulator even when it's red hot, which is why firewalking co-author John Suchocki can walk barefoot on red-hot wooden coals without burning his feet (Figure 7.2). (CAUTION: Don't try this on your own; even experienced firewalkers sometimes receive bad burns when conditions aren't just right.) The principal factor here is the low conductivity of wood—even red-hot wood. Although its temperature is high, relatively little heat is conducted to the feet, just as little heat is conducted by air when you put your hand briefly in a hot pizza oven. If you touch the metal in the hot oven, OUCH! Similarly, a firewalker who steps on a hot piece of metal or another good conductor will be burned.

Fig. 7.2
John Suchocki walks barefoot on red-hot wooden coals, demonstrating that wood is a poor conductor of heat even when red hot.

Liquids and gases, in general, are poor heat conductors. Air is a very poor conductor, which, as previously mentioned, is why your hand isn't harmed if put briefly in a hot oven. The good insulating properties of such things as wool, fur, and feathers are largely due to the air spaces they contain. Porous substances are likewise good insulators because of their many small air spaces. Be glad that air is a poor conductor; if it weren't, you'd feel quite chilly on a 20°C (68°F) day!

Snow is a poor heat conductor (good insulator)—about the same as dry wood. Hence a blanket of snow keeps the ground warm in winter. Snowflakes are formed of crystals, which collect into feathery masses, imprisoning air and thereby interfering with the escape of heat from the Earth's surface. Traditional Eskimo winter dwellings are shielded from the cold by their snow covering. Animals in the forest find shelter from the cold in snow banks and in holes in the snow. The snow doesn't provide them with heat; it simply slows down the loss of the body heat they generate.

Heat is conducted from a higher to a lower temperature, not the other way around. We often hear people say they wish to keep the cold out of their homes. A better way to put this is to say they wish to prevent the heat from escaping by conduction through the walls and roof. There is no "cold" that flows into a warm home. If the home becomes

Fig. 7.3
Snow patterns on the roof of a house show areas of conduction and insulation. Bare parts show where heat from inside has leaked through the roof and melted the snow.

colder, it is because heat has flowed out. Homes are insulated with rock wool or spun glass to prevent heat from escaping rather than to prevent cold from entering. The insulation does not prevent heat from getting through it, however; it simply slows down the rate at which heat penetrates. Even a well-insulated warm home gradually cools. Insulation merely delays the rate at which heat is conducted from the warmer indoors to the cooler outdoors.

■ **Question**

In desert regions that are hot in the daytime and cold at night, the walls of houses are often made of mud. Why is it important that the mud walls be thick?

7.2 Convection

Fig. 7.4
Convection currents in a gas (air) and a liquid.

Liquids and gases transmit heat mainly by **convection**, which is heat transfer by the motion of the fluid—by currents. Convection occurs in all fluids. Whether we heat water in a pan or air in a room, the process is the same (Figure 7.4). As the fluid is heated from below, the molecules at the bottom begin moving faster and therefore spreading farther apart. This greater distance between molecules means the heated fluid becomes less dense and is buoyed upward. Denser cooler fluid then takes its place at the bottom. In this way, convection currents keep the fluid stirred up as it heats—warmer fluid moving away from the heat source and cooler fluid moving toward the heat source to be warmed.

In the atmosphere, warm air expands and in doing so becomes less dense than the surrounding air and is buoyed upward like a balloon. When the rising air reaches an al-

■ **Answer**
A wall of appropriate thickness keeps the house warm at night by slowing the flow of heat from warmer inside to cooler outside and cool in the daytime by slowing the flow of heat from warmer outside to cooler inside. Such a wall has "thermal inertia."

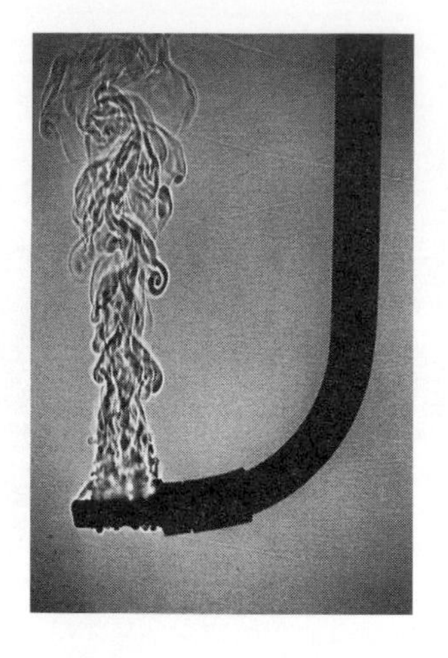

Fig. 7.5
A heater at the tip of a J-tube submerged in water produces convection currents, which are revealed as shadows (caused by deflections of light in water of different temperatures).

titude at which its density matches that of the surrounding air, it no longer rises. This is evident when we see smoke from a fire rise and then settle off as it cools and its density matches that of the surrounding air. Rising warm air expands because less atmospheric pressure squeezes on it as it rises to higher altitudes. As the air expands, it cools. (Do the following experiment right now. With your mouth open, blow on your hand. Your breath is warm. Now repeat but this time pucker your lips to make a small hole so your breath expands as it leaves your mouth. Note that your breath is appreciably cooler! Expanding air cools.) This is the opposite of what occurs when air is compressed. If you've ever compressed air with a tire pump, you probably noticed that both air and pump became quite hot.

A dramatic example of cooling by expansion occurs with steam expanding through the nozzle of a pressure cooker (Figure 7.7). The cooling effect of both expansion and rapid mixing with cooler air allows you to hold your hand comfortably in the jet of condensed vapor. (Caution: If you try this, be sure to place your hand high above the nozzle at first and then lower it to a comfortable distance. If you put your hand at the nozzle where no steam appears, watch out! Steam is invisible and is near the nozzle before it expands and cools. As noted in Chapter 6, the cloud of "steam" you see is actually condensed water vapor, which is much cooler.)

Convection currents stirring the atmosphere result in winds. Some parts of the Earth's surface absorb heat from the Sun more readily than others, and as a result the air near the surface is heated unevenly and convection currents form. This is evident at the seashore, as Figure 7.8 shows. In the daytime, the land warms more easily than the water, and the warmer air above the land rises and is replaced by cooler air moving in from above the water. The result is a sea breeze. At night, the process reverses because the shore cools off more quickly than the water, and then the warmer air is over the sea. Build a fire on the beach and you'll notice that the smoke sweeps inward during the day and seaward at night.

Fig. 7.6
Blow warm air onto your hand from your wide-open mouth. Now reduce the opening between your lips so that the air expands as you blow. Do you notice a difference in air temperature?

Fig. 7.7
The hot steam from the pressure cooker expands and cools as it rises and so is cool to Millie's touch.

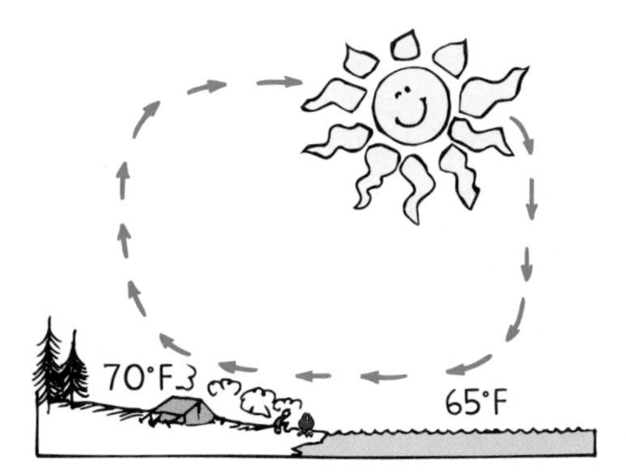

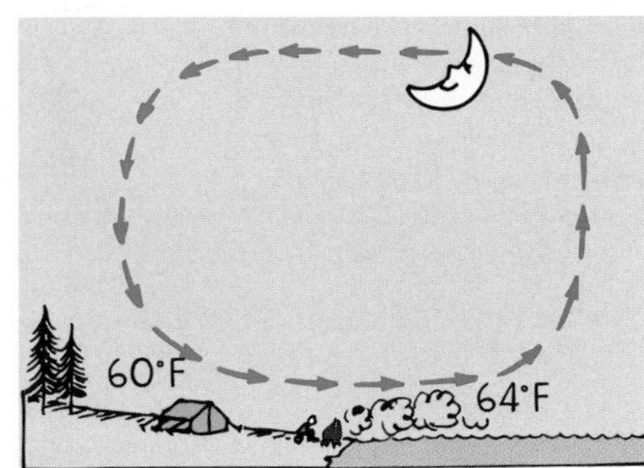

Fig. 7.8
Convection currents produced by unequal heating of land and water. During the day, warm air above the land rises, and cooler air over the water moves in to replace it. At night the direction of air flow is reversed because now the water is warmer than the land.

7.3 Radiation

Heat from the Sun passes through space and then through the atmosphere before it warms the Earth's surface. This heat does not pass through the atmosphere by conduction, for air is a poor conductor. Nor does it pass through by convection, for convection begins only after the Earth is warmed. We also know that neither convection nor conduction is possible in the empty space between our atmosphere and the Sun. We can see that heat must be transmitted some other way—by **radiation**.[1] The energy so transmitted is called *radiant energy.*

Radiant energy is in the form of *electromagnetic waves.* It includes radio waves, microwaves, infrared waves, visible-light waves, ultraviolet waves, X rays, and gamma

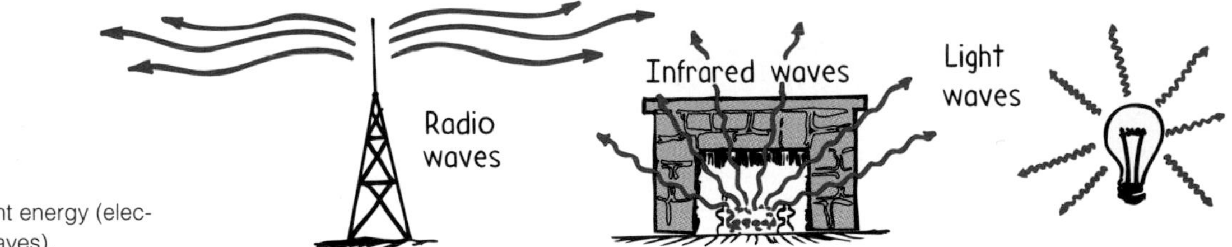

Fig. 7.9
Types of radiant energy (electromagnetic waves).

rays. These types of radiant energy are listed in order of decreasing wavelength, from longest to shortest. The waves of infrared (below-the-red) waves, for instance, are longer than those of visible-light waves. The longest visible wavelengths are for red light, and

[1]Do not confuse radiation with radioactivity, which are reactions that involve the atomic nucleus and are characteristic of nuclear power plants and the like. Radiation here is electromagnetic radiation, "heat" waves of low-frequency light, which we shall treat in detail in Parts 3 and 4.

Fig. 7.10
A wave of long wavelength is produced when the rope is shaken gently (at a low frequency). When the rope is shaken more vigorously (high frequency), a wave of shorter wavelength is produced.

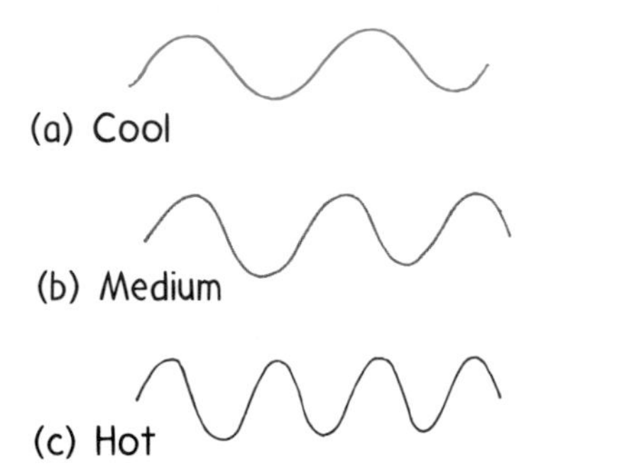

Fig. 7.11
(a) A low-temperature (cool) source emits primarily low-frequency, long-wavelength waves. (b) A medium-temperature source emits primarily medium-frequency, medium-wavelength waves. (c) A high-temperature source (hot) emits primarily high-frequency, short-wavelength waves.

the shortest are for violet light. We'll treat waves further in Chapters 10 and 11, and electromagnetic waves in Chapters 9 and 11. What is important to know at this point is that a short-wavelength wave has a high frequency. Frequency refers to how frequently something vibrates—its rate of vibration. When you vibrate a stretched rope at a certain frequency, the wave produced has the same frequency, as illustrated in Figure 7.10.

Emission of Radiant Energy

All substances at any temperature above absolute zero emit radiant energy. The peak frequency f of the radiant energy is directly proportional to the absolute temperature T of the emitter (Figure 7.12):

$$f \sim T$$

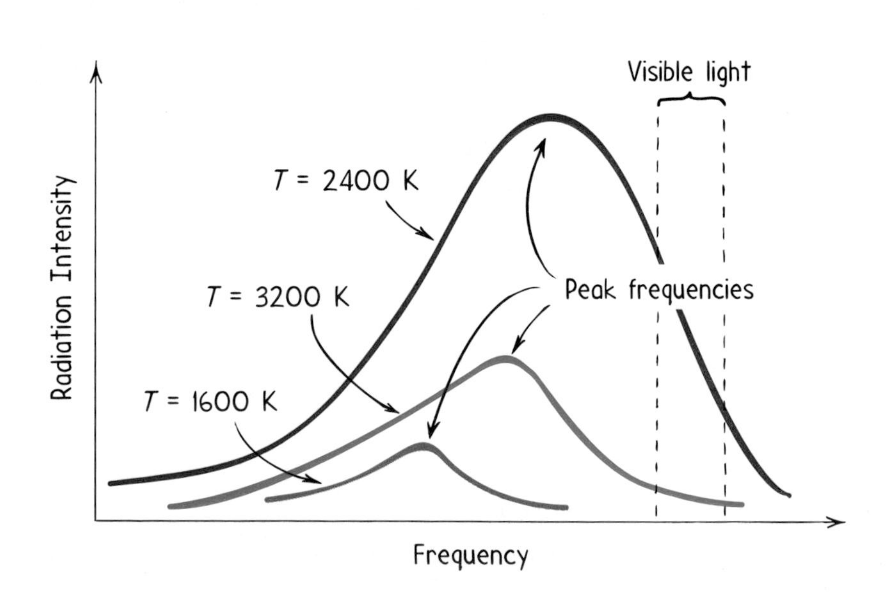

Fig. 7.12
Radiation curves for different temperatures. The peak frequency of radiant energy is directly proportional to the absolute temperature of the emitter.

Fig. 7.13
Both the Sun and the Earth emit the same kind of radiant energy. The Sun's glow is visible to the eye; the Earth's glow is of longer waves and so is not visible to the eye.

The Sun has a very high temperature and therefore emits radiant energy at a high frequency—high enough to be in the visible portion of the electromagnetic spectrum. The Earth, in comparison, is relatively cool, and so the radiant energy it emits has a frequency lower than that of visible light; the radiation emitted by the Earth is in the form of infrared waves. Radiant energy emitted by the Earth is called *terrestrial radiation.*

Most people know that the Sun glows and emits radiant energy, and many educated people know that the source of the Sun's radiant energy involves nuclear reactions in its deep interior. However, relatively few people know that the Earth also "glows" and emits radiant energy of the same nature, whose source also involves nuclear reactions in its interior (we'll treat nuclear reactions in Chapter 14). Visit the depths of any mine and you'll find it's warm down there—year-round. Radioactivity in the Earth's interior warms the Earth. Much of this heat conducts to the surface, where it is radiated as terrestrial radiation. So radiant energy is emitted by both the Sun and the Earth, and differs only in the range of frequencies, and the amount. When we study meteorology in Chapter 26, we'll learn how the atmosphere is transparent to the high-frequency solar radiation but opaque to much of the lower-frequency terrestrial radiation, producing a "greenhouse effect" and, likely, some global warming.

All objects continuously emit radiant energy in a mixture of frequencies (because temperature corresponds to a mixture of molecular kinetic energies). Objects of everyday temperatures emit mostly long-wavelength infrared waves. Shorter-wavelength infrared waves absorbed by our skin produce the sensation of heat, and so it is common to refer to infrared radiation as *heat radiation.*

Common sources that give the sensation of heat are the burning embers in a fireplace, a lamp filament, and the Sun. All of these emit both infrared radiation and visible light. When this radiant energy falls on other objects, it is partly reflected and partly absorbed. The part that is absorbed increases the thermal energy of the objects. If one of the objects happens to be your skin, you feel the radiation as warmth.

Absorption of Radiant Energy

Good absorbers of radiant energy are also good emitters; poor absorbers are poor emitters. For example, a radio antenna constructed to be a good emitter of radio waves is also, by its very design, a good receiver (absorber) of them. A poorly designed transmitting antenna is also a poor receiver.

It is interesting to note that if a good absorber were not also a good emitter, black objects would remain warmer than lighter-colored objects and the two would never reach a common temperature. As we learned in Chapter 6, however, objects in thermal contact, given sufficient time, do reach the same temperature. When time is not sufficient, as when a blacktop pavement or dark automobile body remains hotter than its surroundings on a hot day, thermal equilibrium is not reached. But at nightfall the dark objects cool faster! Given sufficient time, all objects come to thermal equilibrium. So a dark object that absorbs a lot must emit a lot as well.

You can check this out with a pair of metal containers of the same size and shape, one having a white or mirrorlike surface and the other having a blackened surface (Figure 7.14). Fill the containers with hot water and place a thermometer in each. You will find that the black container cools faster. The blackened surface is a better emitter. Coffee or tea stays hot longer in a shiny, mirrorlike pot than in a blackened one. The same

Fig. 7.14
When the containers are filled with hot (or cold) water, the blackened one cools (or warms) faster.

experiment can be done in reverse. This time fill each container with ice water and place the containers in front of a fireplace or outside on a sunny day—wherever there is a good source of radiant energy. You'll find that the black container warms up faster. A good emitter of radiant energy is also a good absorber.

Every surface, hot or cold, both absorbs and emits radiant energy. If the surface absorbs more than it emits, it is a net absorber; if it emits more than it absorbs, it is a net emitter. Whether a surface plays the role of net emitter or net absorber depends on whether its temperature is above or below that of its surroundings. If hotter than its surroundings, the surface will be a net emitter and will cool. If colder than its surroundings, it will be a net absorber and will become warmer.

Reflection of Radiant Energy

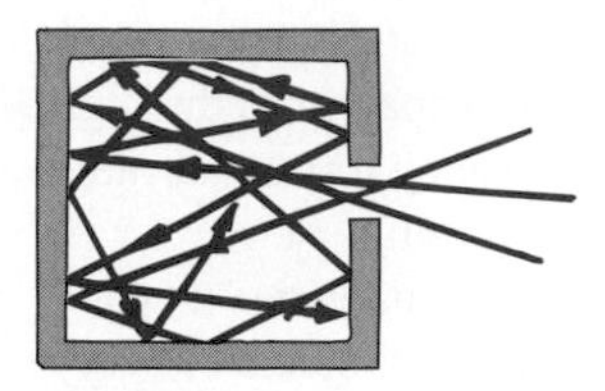

Fig. 7.15
Radiation that enters the opening has little chance of leaving because most of it is absorbed. For this reason, the opening to any cavity looks black to us.

Absorption and reflection are opposite processes. Therefore, a good absorber of radiant energy reflects very little radiant energy, including visible light. A surface that reflects very little or no radiant energy looks dark. Hence a good absorber appears dark. A perfect absorber reflects no radiant energy and appears perfectly black. The pupil of the eye, for example, allows light to enter with no reflection, which is why it appears black. (An exception occurs in flash photography when pupils appear pink, which occurs when very bright light is reflected off the eye's inner pink surface and back through the pupil.)

Look at the open ends of pipes in a stack; the holes appear black. Look at open doorways or windows of distant houses in the daytime, and they too look black. Openings appear black because the light that enters is reflected back and forth on the inside walls many times and is partly absorbed at each reflection; as a result, very little or none remains to come back out the opening and travel to your eyes (Figure 7.15).

Good reflectors, on the other hand, are poor absorbers. Clean snow is a good reflector and therefore does not melt rapidly in sunlight. If the snow is dirty, it absorbs radiant energy from the Sun and melts faster. Dropping black soot from an aircraft onto snow-covered mountains is a technique sometimes used in flood control. Controlled

Fig. 7.16
The hole looks perfectly black and indicates a black interior, when in fact the interior has been painted a bright white.

melting at favorable times rather than a sudden runoff of melted snow is thereby accomplished.

Light-colored buildings stay cooler in summer because they reflect much of the oncoming radiant energy. Light-colored buildings are also poor emitters, and so they retain more of their internal energy and stay warm. Paint your house a light color.

Questions

1. If a good absorber of radiant energy were a poor emitter, how would its temperature compare with the temperature of its surroundings?
2. Is it more efficient to paint a heating radiator black or silver?

Cooling at Night by Radiation

Bodies that radiate more energy than they receive become cooler. This happens at night when solar radiation is absent. A lone object left out in the open at night radiates energy into the air and, because of the absence of any warmer bodies in the vicinity, receives very little energy in return. Thus it gives out more energy than it receives and becomes cooler. But if the object is made of a material that is a good heat conductor—such as metal, stone, or concrete—heat conducts from the ground to it, which somewhat stabilizes its temperature. On the other hand, materials such as wood, straw, and grass are poor conductors, and little heat is conducted into them from the ground. These insulating materials are net radiators and get *colder than the air.* It is common for frost to form on these kinds of nonconducting materials even when the temperature of the air does not go down to freezing. Have you ever seen a frost-covered lawn or field on a chilly but above-freezing morning before the Sun is up? The next time you see this, notice that the frost forms only on the grass, straw, or other poor conductors, not on cement, stone, or other good conductors.

Fig. 7.17
Patches of frost crystals betray the hidden entrances to mouse burrows. Each cluster of crystals is frozen mouse breath!

Snow is a good example of an insulator becoming colder than the surrounding air. Because of snow's high reflectivity, during the day it gains

Answers

1. If a good absorber were not also a good emitter, there would be a net absorption of radiant energy and the temperature of the absorber would remain higher than its temperature of the surroundings. Things around us approach a common temperature only because good absorbers are, by their very nature, also good emitters.
2. Most of the heat provided by a heating radiator is accomplished by convection, and so the color is not really that important. For optimum efficiency, however, the radiators should be painted a dull black so that the contribution by radiation is increased.

very little heat energy from the Sun, and at night it loses energy rapidly by radiating infrared radiation into the air. Because snow is such a poor heat conductor, very little heat moves from the ground to it, and so the surface of the snow becomes cooler than the surrounding air. The bottom of the snow, interestingly, remains at about ground temperature. That's why Midwestern wheat farmers like deep snows in winter—to protect their wheat fields from the harsh, cold weather.

Questions

1. Which is likely to be colder, a night when the stars are out or a night with no stars?
2. In winter, why do road surfaces on bridges tend to be more icy than the road surfaces on either side?

Our study of conduction and convection has been in terms of microscopic particle movements—electrons, atoms, and molecules—"billiard-ball" physics. Now we look at macroscopic phenomena in describing thermal processes—mechanical work, pressure, and temperature. This is thermodynamics, characterized by two principal laws.[2]

7.4 First Law of Thermodynamics

Recall from Chapter 6 that more than a hundred years ago heat was thought to be an invisible fluid called *caloric,* which flowed from hot objects to cold objects. Caloric was conserved in its interactions, and this discovery was the forerunner of the law of conservation of energy. In the mid-1880s, it became apparent that there was no such "invisible fluid" and that the flow of heat was nothing more than the flow of energy. The caloric theory of heat was gradually abandoned,[3] and today we view heat as defined in Section 6.4: thermal energy transferred from one place to another because of a temperature difference. Heat is energy in transit. Like other forms of energy, it cannot be either created or destroyed.

Answers

1. The stars being out means there are no clouds in the atmosphere, so that all the heat the Earth radiates keeps traveling off into space. On a cloudy night, net radiation is less because the clouds radiate some energy back to the Earth's surface. Therefore it is likely to be colder on a starry night than on a cloudy one!
2. Energy radiated by road surfaces built on the ground is partly replaced by heat conducted from the warmer ground below. The road surfaces of bridges make very little thermal contact with the ground, however, and so very little energy is conducted from the ground to the bridge road surfaces as they radiate energy away. So the road surface on bridges get colder, with more chance of ice formation.

[2]There is also a zeroth law of thermodynamics (bearing this charming name because it was formulated *after* the first and second law were), which states that two systems each in thermal equilibrium with a third system are in equilibrium with each other. Then there is a third law, which states that no system can have its absolute temperature reduced to zero.

[3]Popular ideas, when proved wrong, are seldom immediately discarded. People tend to identify with the ideas that characterize their time; hence it is often the young who are more prone to discover and accept new ideas and push the human adventure forward.

The Thermos Bottle

A common Thermos bottle, a double-walled glass with a vacuum between silvered walls, nicely summarizes heat transfer. Any hot or cold liquid poured into such a bottle remains at very nearly the same temperature for many hours because the transfer of heat by conduction, convection, and radiation is severely inhibited.

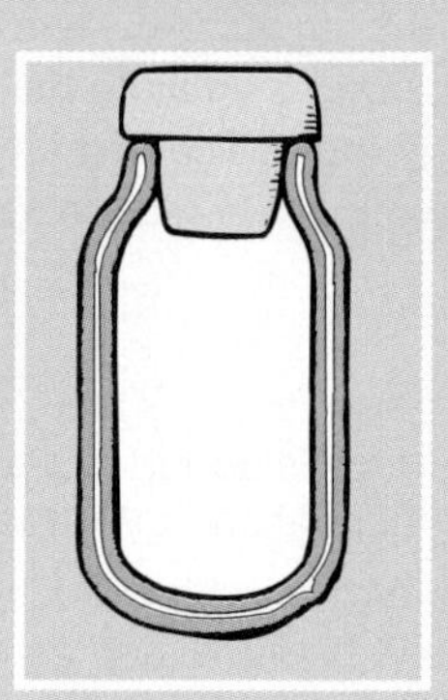

1. Heat transfer by *conduction* through the vacuum is impossible. Some heat escapes by conduction through the glass and stopper, but this is a slow process because glass and plastic or cork are poor conductors.
2. The vacuum also prevents heat loss through the walls by *convection.*
3. Heat loss by *radiation* is prevented by the silvered walls, which reflect heat waves back into the bottle.

When the law of energy conservation is applied to thermal systems, we call it the **first law of thermodynamics**:

> **When heat flows to or from a system, the system gains or loses an amount of thermal energy equal to the amount of heat transferred.**

By *system,* we mean a well-defined group of atoms, molecules, particles, or objects. The system may be the steam in a steam engine or it may be the whole Earth's atmosphere. It can even be the body of a living creature. The important point is that we must be able to define what is contained *within* the system and what is *outside* it. If we add heat energy to the steam in a steam engine, to the Earth's atmosphere, or to the body of a living creature, we increase the thermal energies of these systems. And because of this added energy, these systems may do work on external things. This added energy does one or both of two things: (1) It increases the thermal energy of the system if it remains in the system, or (2) it does work on things external to the system if it leaves the system. So, more specifically, the first law states

> **Heat added to a system = increase in thermal energy of the system + external work done by the system.**

Note that the first law is an overall principle that is not concerned with the inner workings of the system. Whatever the details of molecular behavior in the system, the added heat serves only two functions: It increases the thermal energy of the system or it enables the system to do external work (or both). Our ability to describe and predict the behavior of systems that may be too complicated to analyze in terms of atomic and molecular processes is one of the beauties of thermodynamics. Thermodynamics bridges the microscopic and macroscopic worlds.

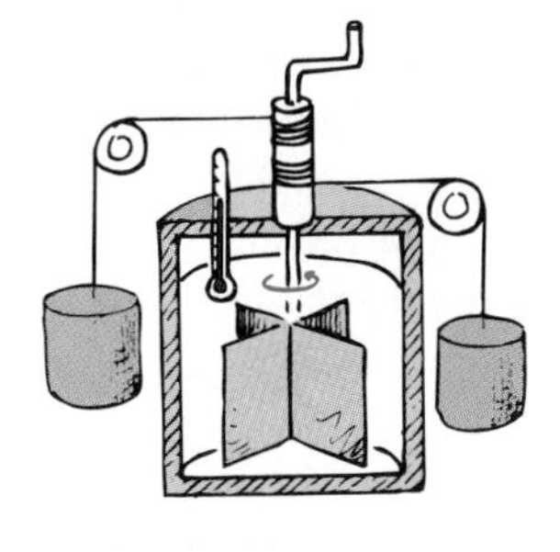

Fig. 7.18
Paddle-wheel apparatus used to compare thermal energy with mechanical energy. As the weights fall, they give up potential energy (mechanical), which is converted to thermal energy. This equivalence between mechanical and thermal energy was first demonstrated by James Joule, for whom the unit of energy is named.

Put an airtight can of air on a hot stove and heat it up. Because the can is of fixed volume, no work is done on it (because work involves movement by a force), and so all the heat that goes into the can increases the thermal energy of the enclosed air. So its temperature rises. If the can is fitted with a movable piston, then the heated air can do work as it expands and pushes the piston outward. Can you see that the temperature of the enclosed air must now be less than if no work were done on the piston? If heat is added to a system that does no external work, then the amount of heat added is equal to the increase in the thermal energy of the system. If the system does external work, then the increase in thermal energy is correspondingly less.

As another example, consider a given amount of heat supplied to a steam engine. The added heat goes partly to increase the thermal energy of the steam and partly to do mechanical work the engine is designed to do. The sum of the increase in thermal energy and the work done is equal to the heat input. In no way can energy output in any form exceed energy input. The first law of thermodynamics is simply the thermal version of the law of conservation of energy.

Questions

1. If 100 J of energy is added to a system that does no external work, by how much is the thermal energy of that system raised?
2. If 100 J of energy is added to a system that does 40 J of external work, by how much is the thermal energy of that system raised?

Fig. 7.19
Do work on the pump by pressing down on the piston and you compress the air inside. What happens to the temperature of the enclosed air? What happens to its temperature if it expands and pushes the piston outward?

Adding heat to a system so that the system can do mechanical work is only one application of the first law of thermodynamics. If instead of adding heat to a system, we do mechanical work on it, the first law tells us what we can expect: an increase in thermal energy. A bicycle pump provides a good example. When we pump on the handle, the pump becomes hot. Why? Because we are primarily doing mechanical work on the system and raising its thermal energy. If the process happens so quickly that very little heat is conducted out of the system, then most of the work input goes into increasing the thermal energy, and the system becomes hotter.

Adiabatic Processes

Compressing or expanding a gas while no heat enters or leaves the system is said to be an **adiabatic process** (from the Greek for "impassable"). Adiabatic conditions can be achieved either by thermally insulating a system from its surroundings (with Styrofoam, for example) or by performing the process so rapidly that heat has no time to enter or leave. In an adiabatic process, therefore, because no heat enters or leaves the system, the "heat added" part of the first law of thermodynamics must be zero. Then under adiabatic conditions, changes in thermal energy are equal to the work done on or by the system.[4] If we do work on a system—by compressing it, for example—its thermal energy increases; we raise its temperature. If work is done by the system—as it expands against its surroundings, for example—its thermal energy decreases; it cools.

Answers
1. 100 J.
2. 60 J. We see from the first law that 100 J − 40 J = 60 J.

---

[4] ΔHeat = Δthermal energy + work
0 = Δthermal energy + work
Then we can say
−Work = Δthermal energy

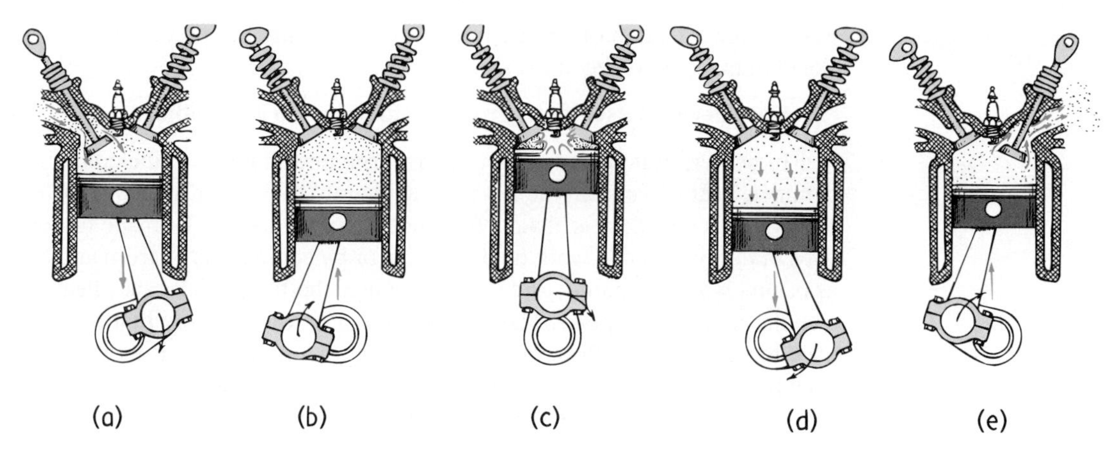

Fig. 7.20
A four-cycle internal-combustion engine. (a) A fuel-air mixture from the carburetor fills the cylinder as the piston moves down. (b) The piston moves upward and compresses the mixture—adiabatically, because no heat is transferred in or out. (c) The spark plug fires, ignites the mixture, and raises it to a high temperature. (d) Adiabatic expansion pushes the piston downward, the power stroke. (e) The burned gases are pushed out the exhaust pipe. Then the intake valve opens and the cycle repeats.

A common example of an adiabatic process is the compression and expansion of the gases in the cylinders of an automobile engine (Figure 7.20). Compression and expansion occur in only hundredths of a second, too short a time for any appreciable amount of energy to leave the combustion chamber. For very high compressions, like those that occur in diesel engines, the temperatures achieved are sufficient to ignite a fuel mixture without the use of spark plugs. Diesel engines have no spark plugs.

So when work is done on a gas by adiabatically compressing it, the gas gains thermal energy and becomes warmer. We note this by the warmth of a bicycle pump when air is compressed. When a gas adiabatically expands, it does work on its surroundings and gives up thermal energy and becomes cooler. Expanding air cools.

Adiabatic cooling is central to cloud formation, temperature inversions, climate, and tornadoes. Adiabatic processes occur not only in air but in ocean waters and in molten parts of the Earth's interior. We'll return to adiabatic processes when we study meteorology in Chapter 24.

7.5 Second Law of Thermodynamics

Suppose we place a hot brick next to a cold brick in a thermally insulated region. We know that the hot brick cools as it gives heat to the cold brick, which warms. The two bodies arrive at a common temperature: thermal equilibrium. No energy is lost, in accordance with the first law of thermodynamics. Now suppose the hot brick extracts heat from the cold brick and becomes hotter. Would this violate the first law of thermodynamics? Not if the cold brick becomes correspondingly colder so that the combined energy of the two bricks remains constant. If this were to happen, it would not violate the first law, but it would violate the **second law of thermodynamics**, which distinguishes the direction of energy transformation in natural processes. This law can be stated in many ways, but most simply it is this:

Heat never of itself flows from a cold object to a hot object.

In winter, heat flows from inside a warm, heated home to the cold air outside. In summer, heat flows from the hot air outside into the cooler interior. The direction of spontaneous heat flow is always from hot to cold. Heat can be made to flow the other way, but only by doing work on the system or by adding energy from another source—as occurs with heat pumps and air conditioners, both of which cause heat to flow from a cooler to a warmer place. The huge amount of internal energy in the ocean cannot be used to light a single flashlight lamp without external effort. Energy cannot of itself flow from the lower-temperature ocean to the higher-temperature lamp filament. Without external effort, the direction of heat flow is from hot to cold.

Heat Engines

It is easy to change work completely into heat—simply rub your hands together briskly. The heat created by your work adds to the internal energy of your hands, making them warmer. Or push a crate at constant speed along a floor. All the work you do in overcoming friction is completely converted to heat, which warms the crate and the floor. However, the reverse process, changing heat completely into work, can never occur. The best we can do is convert *some* heat to mechanical work. The first heat engine to do this was the steam engine, invented in about 1700.

A **heat engine** is any device that changes internal energy into mechanical work. The basic idea behind a heat engine—whether steam, internal combustion, or jet—is that mechanical work can be obtained only when heat flows from a high-temperature region to a low-temperature region. In every heat engine, only *some of the heat* can be transformed into work.

In considering heat engines, we usually talk about *reservoirs* instead of regions: Heat flows out of a high-temperature reservoir and into a low-temperature one. Every heat engine (1) gains heat from a reservoir of higher temperature, increasing the engine's internal energy; (2) converts some of this energy into mechanical work; and (3) expels the remaining energy as heat to a lower-temperature reservoir, usually called a *sink* (Figure 7.21). In a gasoline engine, for example, (1) the products of burning fuel in the combustion chamber provide the high-temperature reservoir, (2) the hot gases do mechanical work on the piston, and (3) heat is expelled to the environment via the cooling system and the exhaust.

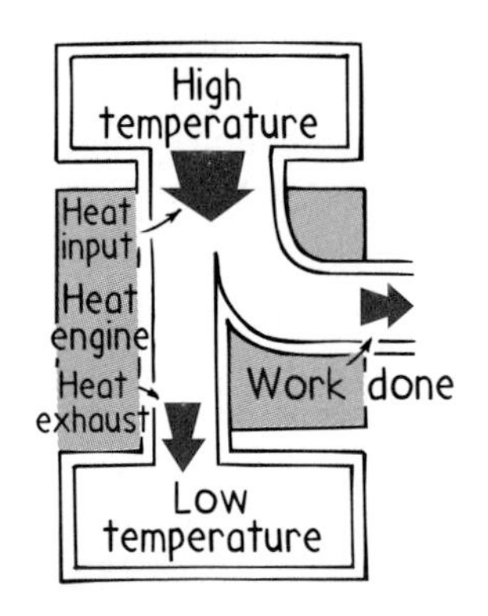

Fig. 7.21
When heat in a heat engine flows from the high-temperature reservoir to the low-temperature sink, part of the heat can be turned into work. (If work is put into a heat engine, the flow of heat may be from the low-temperature sink to the high-temperature reservoir, as in a refrigerator or air conditioner.)

The second law tells us that no heat engine can convert all the heat supplied into mechanical energy. Only some of the heat can be transformed into work, with the remainder expelled in the process. Applied to heat engines, the second law may be stated:

When work is done by a heat engine running between two temperatures T_{hot} and T_{cold}, only some of the heat at T_{hot} can be converted to work; the rest is expelled at T_{cold}.

Every heat engine discards some heat, which may be desirable or undesirable. Hot air expelled in a laundry on a cold winter day may be quite desirable, while the same hot air on a hot summer day is something else. When expelled heat is undesirable, we call it *thermal pollution.*

Before scientists understood the second law, many people thought that a very-low-friction heat engine could convert nearly all the input heat energy to useful work. But

Link to Power Production—OTEC

The OTEC, acronym for ocean thermal energy conversion, is a nonpolluting power plant developed off the Kona Coast in Hawaii. It produces electric power by exploiting the temperature difference between warm surface waters and cold deep waters. More specifically, it is a heat engine that runs on the temperature difference between solar-lit 26°C tropical surface waters and deep, dark 4°C water. Because of the small temperature difference, the efficiency is low.

The low efficiency requires that large amounts of water be circulated, but that requirement can have some advantages. Cold, deep water, saturated with nitrogen and other nutrients locked away from the surface food chain in darkness for centuries, is circulated to the surface. This upwelling of nutrient-rich water exposed to sunlight should produce an explosion of phytoplankton growth comparable to that obtained when fertilizers are added to land crops. So the promise of OTEC is not only pollution-free electric energy, but also an increased food supply from the sea.

The operation of OTEC is similar to that of a conventional, steam-powered plant. Warm seawater is admitted into a partially evacuated chamber where it boils (recall that the boiling point of water depends on pressure). The expanding "steam," actually a vapor, turns a turbogenerator, after which it is condensed by cold water pumped up from the depths. In an alternative closed-cycle system, a working fluid such as ammonia is vaporized and condensed through heat exchange with warm and cold seawater, providing across the turbine a greater pressure difference that allows more straightforward scale-up to commercial-sized plants.

Like wind power and hydropower, OTEC power comes ultimately from sunlight. It needs no other fuel. Although funding for OTEC was recently cut in Washington, watch for OTEC in the new century!

not so. In 1824, the French engineer Sadi Carnot analyzed the functioning of a heat engine and made a fundamental discovery. He showed that the greatest fraction of energy input that can be converted to useful work, even under ideal conditions, depends on the temperature difference between the hot reservoir and the cold sink. His equation is[5]

$$\textbf{Ideal efficiency} = \frac{T_{hot} - T_{cold}}{T_{hot}}$$

Ideal efficiency depends only on the temperature difference between input and exhaust. Whenever ratios of temperatures are involved, the absolute temperature scale must be used. So T_{hot} and T_{cold} are expressed in kelvins. For example, when the hot reservoir in a steam turbine is 400 K (127°C) and the sink is 300 K (27°C), the ideal efficiency is

$$\frac{400 - 300}{400} = \frac{1}{4}$$

This means that even under ideal conditions, only 25 percent of the heat provided by the steam can be converted into work, while the remaining 75 percent is expelled as waste heat. This is why steam is superheated to high temperatures in steam engines and power plants. The higher the steam temperature driving a motor or turbogenerator, the higher the efficiency of power production. (Increasing operating temperature in our ex-

---

[5]The general definition of efficiency is work output/heat input. To see how this can be written in the form $(T_{hot} - T_{cold},)/T_{hot}$, start with energy conservation: heat input = work output + heat exhaust (output) (Figure 7.21). So work output = heat input − heat output, and so the definition of efficiency can be written (heat input − heat output)/(heat input). In the ideal case, it can be shown that the ratio (heat out)/(heat in) = T_{cold}/T_{hot}. Then we can say ideal efficiency = $(T_{hot} - T_{cold})/T_{hot}$.

Thermodynamics Dramatized!

Put a small amount of water in an aluminum soda-pop can and heat it on a stove until steam issues from the opening. When this occurs, air has been driven out and replaced by steam. Then with a pair of tongs quickly invert the can into a pan of water. Crunch! The can is crushed by atmospheric pressure! Why? When the molecules of steam encounter water from the pan, condensation occurs, leaving a vacuumlike very low pressure in the can, whereupon the surrounding atmospheric pressure crunches the can. Here we see, dramatically, how pressure is reduced by condensation. Can you now better see the role of condensation in the turbine of Figure 7.22?

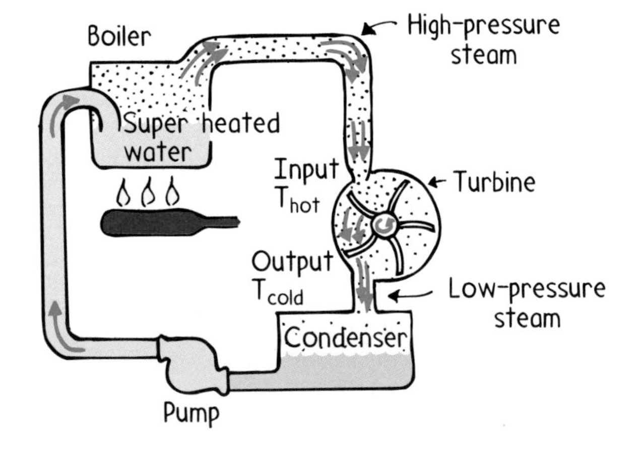

Fig. 7.22
A simplified steam turbine. The turbine turns because pressure exerted by the high-temperature steam on the front side of the turbine blades is greater than that exerted by the low-temperature steam on the back side of the blades. Without a pressure difference, the turbine would not turn and deliver energy to an external load (an electric generator, for example). The presence of steam pressure on the back side of the blades, even in the absence of friction, prevents a perfectly efficient engine.

ample to 600 K, for instance, yields an efficiency (600 − 300)/600 $= \frac{1}{2}$, twice as high as the efficiency at 400 K.)

We can see the role of reservoir-sink temperature difference in the operation of the steam-turbine engine in Figure 7.22. The hot reservoir is steam from the boiler, and the cold sink is the exhaust region in the condenser. The hot steam exerts pressure and does work on the turbine blades when it pushes on their front sides. This is nice. But steam pressure is not confined to the front sides of the blades; steam pressure is also exerted on the backs of the blades—countereffective and not so nice. Pressure on the back sides is reduced as steam condenses and cools after giving much of its energy to the blades. Even if friction were absent, the turbine's net work output would be the difference between the work done on the blades by the hot steam and the work done by the blades on the cooler steam in exhausting it. We know that with confined steam, temperature and pressure go hand in hand; increase temperature and you increase pressure—decrease temperature and you decrease pressure. So the pressure difference necessary for the operation of a heat engine is directly related to the temperature difference between reservoir and sink. The greater the temperature difference, the greater the efficiency.

Carnot's equation states the upper limit of efficiency for all heat engines. The higher the operating temperature (relative to exhaust temperature) of any heat engine, whether in an ordinary automobile, a nuclear-powered ship, or a jet aircraft, the higher the efficiency of that engine. In practice, friction is always present in all engines, and efficiency is always less than ideal.[6] So whereas friction is solely responsible for the inefficiencies of many devices, in the case of heat engines the overriding concept is the second law of thermodynamics; only some of the heat input can be converted to work—even without friction.

Questions

1. What is the ideal efficiency of an engine whose hot reservoir and sink are both at 400 K?
2. What is the ideal efficiency of a machine having a hot reservoir at 400 K and a cold sink at absolute zero, 0 K?

7.6 Order Tends to Disorder

The first law of thermodynamics states that energy can be neither created nor destroyed. The second law qualifies this by adding that the form energy takes in transformations "deteriorates" to less useful forms. Energy becomes more diffuse and ultimately degenerates into waste. Another way to say this is that organized energy (concentrated and therefore usable energy) degenerates into disorganized energy (nonusable). The energy of gasoline is organized and usable. When gasoline burns in a car engine, part of its energy does useful work, part heats the engine and surroundings, and part goes

Answers

1. Zero: $(400 - 400)/400 = 0$. No work output is possible for any heat engine unless a temperature difference exists between reservoir and sink.
2. $(400 - 0)/400 = 1$; only in this idealized case is an ideal efficiency of 100% possible.

---

[6]The *ideal* efficiency of an ordinary automobile engine is more than 50 percent, but the *actual* efficiency is about 25%. Engines that operated at higher temperatures (relative to sink temperatures) would be more efficient, but the melting point of engine materials limits the upper temperatures at which they can operate. Higher efficiencies await engines made with new materials that have melting points higher than those of today's material. (See the box on ceramic engines.)

Link to Technology—Ceramic Engines

More than 10,000 years ago, someone cooked a hunk of clay, and it hardened into pottery. Today we call clay hardened this way ceramic. Compared to metals, ceramics can be harder, lighter, stiffer, and more resistant to heat and corrosion. Ceramics are used for coating metals to extend their lives. They're also of considerable interest to engine technology. The ideal efficiency of an engine goes up with higher operating temperatures. Today's metal automobile engines are relatively inefficient because their operating temperatures must be kept lower than the melting point of the metals. Costly radiators are installed to get rid of valuable heat that otherwise would raise efficiency. Hence the interest in ceramic engines.

Being brittle, ceramics shatter and crack when dropped or hit. Unlike metals, ceramics cannot bend and deform to absorb impacts. Intense research is presently under way to combat the problem of brittleness, with some success. Ceramic engines are already employed in Isuzu's Ceramic Research Institute in Yokohama, Japan, where a high-powered sedan labeled CERAMIC features a small engine and no radiator. This car uses heat instead of rejecting it. Likewise with the prototype turbine ceramic automotive engines developed in the United States by the Department of Energy. In place of pistons, the United States' versions feature two gas turbines made of silicon nitride. With other ceramic parts sharing the same high-temperature area, the cars will run at a hot 1600 K, with higher efficiency and cleaner emissions. Keep your eye out for ceramic engines sometime in the future—maybe not right away because the ceramic engine will likely require a long, continued, and coordinated international effort.

Fig. 7.23
Push a heavy crate across a rough floor and all your work goes into heating the floor and crate. Work against friction turns into thermal energy, which cannot do any work on the crate.

out the exhaust. Useful energy degenerates to nonuseful forms and is unavailable for doing the same work again, such as driving another car. Heat, diffused into the environment as thermal energy, is the graveyard of useful energy.

The quality of energy is lowered with each transformation as energy of an organized form tends to disorganized forms. With this broader perspective, the second law can be stated another way:

Natural systems tend to proceed toward a state of greater disorder.

Consider a system consisting of a stack of pennies on a table, all heads up. Somebody walks by and accidentally bumps into the table and the pennies topple to the floor, certainly not all heads up. Order becomes disorder. Molecules of gas all moving in harmony make up an orderly state—and also an unlikely state. On the other hand, molecules of gas moving in haphazard directions and speeds make up a disorderly and more probable state. If you remove the lid of a bottle containing some gas, the gas molecules escape into the room and make up a more disorderly state (Figure 7.24). Relative order becomes disorder. You would not expect the reverse to happen; that is, you would not expect the molecules to spontaneously order themselves back into the bottle and thereby return to the more ordered containment. Processes in which disorder goes to order without any external help don't occur in nature.

Fig. 7.24
Molecules of gas go from the bottle (a more-ordered state) to the air (a less-ordered state), and not vice versa.

Disordered energy can be changed to ordered energy, but only at the expense of some organizational effort or work input. For example, water in a refrigerator freezes and becomes more ordered because work is put into the refrigeration cycle; gas can be ordered into a small region if a compressor supplied with outside energy does work. Without some outside energy input, however, processes in which the net effect is an increase in order don't occur in nature.

7.7 Entropy

The idea of lowering the "quality" of energy is embodied in the idea of **entropy**, the measure of the *amount of disorder* in a system.[7] The second law states that, in the long run, entropy always increases. Gas molecules escaping from a bottle move from a relatively orderly state to a disorderly state. Disorder increases; entropy increases. Whenever a physical system is allowed to distribute its energy freely, it always does so in a manner such that entropy increases while the amount of energy in the system available for doing work decreases.

Fig. 7.25
Why is the motto of this contractor—"Increasing entropy is our business"—so appropriate?

Consider the old riddle, "How do you unscramble an egg?" The answer is simple: "Feed it to a chicken." But even then, you won't get all your original egg back—egg-making has its inefficiencies, too. All living organisms, from bacteria to trees to human beings, extract energy from their surroundings and use it to increase their own organization. In living organisms, entropy decreases, but any entropy decrease is maintained by an entropy increase elsewhere. Life forms plus their waste products always have a net increase in entropy.[8] Energy from the outside must be transported into the living system to support life. When it isn't, the organism soon dies and tends toward disorder.

---

[7]Entropy can be expressed as a mathematical equation, stating that the increase in entropy ΔS in an ideal thermodynamic system is equal to the amount of heat added to a system ΔQ divided by the temperature T of the system: $\Delta S = \Delta Q/T$.

[8]Interestingly, the American writer Ralph Waldo Emerson, who lived during the time the second law of thermodynamics was the new science topic of the day, philosophically speculated that not everything becomes more disordered with time and cited the example of human thought. Ideas about the nature of things grow increasingly refined and better organized as they pass through the minds of succeeding generations. Human thought is evolving toward more order.

The first law of thermodynamics is a universal law of nature for which no exceptions have been observed. The second law, however, is a probabilistic statement. Given enough time, even the most improbable states may occur; in other words, entropy may sometimes decrease. For example, the haphazard motions of air molecules could momentarily become harmonious in one corner of a room, just as a barrelful of pennies dumped on the floor could all come up heads. These situations are possible—but not probable. The second law tells us the most probable course of events, not the only possible one.

The laws of thermodynamics are often stated this way: You can't win (because you can't get any more energy out of a system than you put into it), you can't break even (because you can't even get as much energy out as you put in), and you can't get out of the game (entropy in the universe is always increasing).

Fig. 7.26
Entropy

Summary of Terms

- **Thermodynamics** The study of heat and its transformation to mechanical energy.
- **Conduction** The transfer of heat energy by collisions between the particles in a substance (especially a solid).
- **Convection** The transfer of heat energy in a gas or liquid by means of currents in the heated fluid. The fluid moves, carrying energy with it.
- **Radiation** The transfer of energy by means of electromagnetic waves.
- **First law of thermodynamics** A restatement of the law of energy conservation, usually as it applies to systems involving changes in temperature: The heat added to a system equals the increase in the thermal energy of the system plus the external work the system does on its environment
- **Adiabatic process** A process, usually of expansion or compression, wherein no heat enters or leaves a system.
- **Second law of thermodynamics** Heat never spontaneously flows from a cold object to a hot one. Also, no machine can be completely efficient in converting energy to work; some input energy is dissipated as waste heat at lower temperature. And finally, all systems tend to become more and more disordered as time goes by.
- **Heat engine** A device that uses heat as input and supplies mechanical work as output or one that uses work as input and moves heat "uphill" from a cooler to a warmer place.
- **Entropy** A measure of the disorder of a system. Whenever energy freely transforms from one form to another, the direction of transformation is toward a state of greater disorder and therefore greater entropy.

Review Questions

1. What are the three ways in which heat is transferred?

Conduction

2. What is the role of "loose" electrons in heat conductors?
3. Distinguish between a heat conductor and a heat insulator.
4. Why does room-temperature tile feel cooler to the bare feet than a wooden floor at the same temperature?
5. Why are materials such as wood, fur, feathers, and even snow good insulators?
6. How does a blanket keep you warm on a cold night, even though it is not really a source of thermal energy?
7. In what sense do we say there is no such thing as cold?

Convection

8. How is heat transferred from one place to another by convection?
9. How does buoyancy relate to convection?
10. What happens to the pressure of air as the air rises? What happens to its volume? Its temperature?
11. In Figure 7.7, why is Millie's hand not burned?
12. Why does the direction of coastal winds change from day to night?

Radiation

13. What is radiant energy?
14. How do the wavelengths of radiant energy vary with the temperature of the radiating source?
15. How does the frequency of radiant energy vary with the temperature of the radiating source?
16. All objects continuously emit radiant energy. Why then doesn't the temperature of all objects continuously decrease?
17. All objects continuously absorb energy from their surroundings. Why then doesn't the temperature of all objects continuously increase?
18. Which body glows, the Sun, the Earth, or both? Explain.
19. Which normally cools faster, a black pot of hot water or a silvered pot of hot water? Explain.
20. Which normally warms up faster, a black pot of cold water or a silvered pot of cold water? Explain.
21. What determines whether an object is a net emitter or a net absorber of radiant energy?
22. Why does a good absorber of radiant energy appear black to the human eye?
23. Is a good absorber of radiation a good emitter or a poor emitter?
24. Why does the pupil of the eye appear black?
25. What happens to the temperature of something that radiates energy without absorbing the same amount in return?
26. Does a good conductor in contact with a relatively warm ground become significantly colder than the ground when it radiates energy? Why or why not?
27. Does a good insulator in contact with a relatively warm ground become significantly colder than the ground when it radiates energy? Why or why not?

First Law of Thermodynamics

28. How does the law of the conservation of energy relate to the first law of thermodynamics?
29. What happens to the thermal energy of a system when mechanical work is done on the system? What happens to the temperature of the system?
30. What condition is necessary in order for a process to be adiabatic?
31. If work is done *on* a system, does the thermal energy of the system increase or decrease? If work is done *by* a system, does the thermal energy of the system increase or decrease?
32. What happens to the temperature of rapidly expanding air? Of rapidly compressed air?
33. Why do diesel engines need no spark plugs?

Second Law of Thermodynamics

34. How does the second law of thermodynamics relate to the direction of heat flow?
35. What three processes occur in every heat engine?
36. What is thermal pollution?
37. If all friction could be removed from a heat engine, would it be 100% efficient? Explain.
38. Why is it advantageous to use steam as hot as possible in a steam-driven turbine?

Order Tends to Disorder

39. What does it mean to say that when energy is transformed, it becomes less useful?
40. With respect to orderly and disorderly states, what do natural systems tend to do? Can a disorderly state ever transform to an orderly state? Explain.

Entropy

41. What is the physicist's term for *measure of messiness*?
42. Distinguish between the first and second laws of thermodynamics in terms of whether or not exceptions occur.
43. Water put in the freezer compartment of a refrigerator goes to a state of less molecular disorder when it freezes. Is this an exception to the entropy principle? Explain.

Home Projects

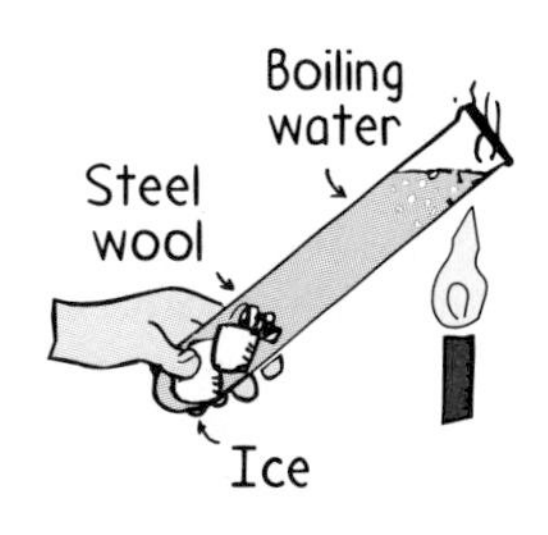

1. Hold the bottom end of a test tube full of cold water in your hand. Heat the top part in a flame until it boils. The fact that you can still hold the bottom shows that water is a poor conductor of heat. This is even more dramatic when you wedge chunks of ice at the bottom with some steel wool; the water above can be brought to a boil without melting the ice.
2. If you live where there is snow, do as Benjamin Franklin did nearly two centuries ago and lay samples of light and dark cloth on the snow. Note the difference in the rate of melting beneath the cloths.

3. Wrap a piece of paper around a thick metal bar and place it in a flame for a minute or so. Note that the paper does not catch fire. Can you figure out why? (Paper generally doesn't ignite until its temperature reaches 233°C.)

4. How much energy is in a nut? Burn it and find out. Remove a nut (pecan or walnut halves work best) from its hard shell and pierce it with a bent paper clip that holds it above a flameproof surface. Above this, secure a beaker of water so you can measure the temperature change of the water when the nut burns. Use about 10 cm^3 (10 ml) of water and a Celsius thermometer. Ignite the nut with a match, and then record water temperature the moment that the flame goes out. The number of calories released by the burning nut can be calculated by the formula $Q = mc\Delta T$, where Q is in calories, m is the mass of the water, c its specific heat capacity (1 cal/g·C°) and ΔT the change in temperature. (Remember from Section 6.5 that the energy in food is expressed in terms of the dietetic Calorie, which is 1000 of the calories you'll measure. To find the number of dietetic Calories, divide your result by 1000. If you used half a nut, double your answer for the energy in a whole nut!)

Exercises

1. Wrap a fur coat around a thermometer. Will the temperature rise?
2. If 70°F air feels warm and comfortable to us, why does 70°F water feel cool when we swim in it?
3. At what common temperature do both a block of wood and a block of metal both feel neither hot nor cold to the touch?
4. Many tongues have been injured by licking a piece of metal on a very cold day. Why would no harm result if a piece of wood were licked on the same day?
5. In defrosting frozen food, why is defrosting faster if the food is placed on a slab of black aluminum rather than on a wooden countertop?
6. Visit a snow-covered cemetery and note that the snow does not slope upward against the gravestones but instead forms depressions as shown. Can you think of a reason for this?

7. If you were caught in freezing weather with only a candle for a heat source, would you be warmer in an Eskimo igloo or in a wooden shack?
8. You can bring water in a paper cup to a boil by placing it in a hot flame. Why doesn't the cup burn?
9. Why can you comfortably hold your fingers close beside a candle flame but not very close above the flame?
10. You can safely hold your bare hand in a hot pizza oven for a few seconds, but if you momentarily touch the metal insides, you'll burn yourself. Why?
11. Wood has a very low thermal conductivity. Does it still have a low conductivity if it is hot? Could you safely grab the wooden handle of a pan from a hot oven with your bare hand? Although the pan handle is hot, does much heat conduct from it to your hand if you do it quickly? Could you do the same with an iron handle? Explain.
12. Wood has a very low thermal conductivity. Does it still have a low conductivity if it is very hot—that is, in the stage of smoldering red hot coals? Could you safely walk across a bed of red-hot wooden coals with bare feet? Although the coals are hot, does much heat conduct from them to your feet if you step quickly? Could you do the same on red-hot iron coals? Explain.
13. The snow made by snow-making machines at ski areas consists of a mixture of compressed air and water blown through a nozzle. The temperature of the mixture may initially be well above the freezing temperature of water, and yet crystals of snow are formed as the mixture is ejected from the nozzle. Explain.
14. A steam radiator warms a room by what two kinds of heat transfer? What is the most efficient color for a radiator?
15. Turn a common incandescent lamp on and off quickly while holding your hand a few inches from the bulb. You feel its heat but find when you touch the bulb that it is not hot. Explain.
16. A good absorber of radiation appears black because little or no light is reflected. Why does a good *emitter* of heat radiation appear black at room temperature?
17. A number of bodies at different temperatures placed in a closed room ultimately come to the same temperature. Would this thermal equilibrium be possible if good absorbers were poor emitters and if poor absorbers were good emitters? Explain.
18. From the rules that a good absorber of radiation is a good radiator and a good reflector is a poor absorber, state a rule relating the reflecting and radiating properties of a surface.
19. Suppose at a restaurant you are served coffee before you are ready to drink it. For it to be as hot as possible when you are ready for it, would you be wiser to add cream to it right away or when you are ready to drink it?
20. Even though metal is a good thermal conductor, frost can be seen on parked cars in the early morning even when the air temperature is above freezing. What does this have to do with the rubber tires of the car?
21. As we humans consume more energy from fossil and other nonrenewable fuels, the overall temperature of the Earth tends to rise. Regardless of the increase in energy, however, the temperature does not rise indefinitely. By what process is an indefinite rise prevented? Explain.

22. A friend said the temperature inside a certain oven is 500 and the temperature inside a certain star is 50,000. You're unsure about whether your friend meant Celsius degrees or kelvins. How much difference does it make in each case?

23. The temperature of the Sun's interior is about 10^7 degrees. Does it matter whether this is degrees Celsius or kelvins? Explain.

24. Helium has the special property that its internal energy is directly proportional to its absolute temperature. Consider a flask of helium at 10°C. If it is heated until it has twice the internal energy, what will its temperature be?

25. If you shake a can of liquid back and forth, does the temperature of the liquid increase? (Try it and see.)

26. When you pump up a tire with a bicycle pump, the cylinder of the pump becomes hot. Give two reasons why.

27. Is it possible to wholly convert a given amount of mechanical energy into heat? Is it possible to wholly convert a given amount of heat into mechanical energy? Cite examples to illustrate your answers.

28. Imagine a gigantic dry-cleaner's bag full of air floating like a balloon with a string hanging from it high above the ground. What would be the effect on its temperature if it were yanked suddenly to ground?

29. The combined molecular kinetic energies of molecules in a very large container of cold water are greater than the combined molecular kinetic energies in a cup of hot tea. Pretend you partially immerse the teacup in the cold water and that the tea absorbs 10 units of energy from the water and becomes hotter, while the water becomes cooler. Would this energy transfer violate the first law of thermodynamics? The second law of thermodynamics? Explain.

30. What is the ultimate source of energy in coal, oil, and wood?

31. Why is *thermal pollution* a relative term?

32. Is it possible to construct a heat engine that produces no thermal pollution? Defend your answer.

33. What happens to the efficiency of a heat engine when the temperature of the reservoir into which heat energy is rejected is lowered?

34. Will the efficiency of a car engine increase, decrease, or remain the same if driven on a very cold day? Will the efficiency change if the muffler is removed? How? Defend your answers.

35. Under what conditions would a heat engine be 100% efficient?

36. Suppose one wishes to cool a kitchen by leaving the refrigerator door open and closing the kitchen door and windows. What will happen to the room temperature? Why?

37. Could you warm a kitchen by leaving the hot oven door open? Explain.

38. An electric fan in a room not only doesn't decrease the temperature of air but actually *increases* it. Why? And how then are you cooled by a fan on a hot day?

39. What percentage of the electrical energy transformed by a common light bulb becomes heat energy?

40. In buildings that are being electrically heated, is it wasteful to turn all the lights on? Is turning all the lights on wasteful if the building is being cooled by air conditioning?

41. Comment on this statement: The second law of thermodynamics is one of the most fundamental laws of nature, and yet it is not an exact law at all.

42. Using both the first and the second law of thermodynamics, defend the statement that 100% of the electrical energy that goes into lighting a lamp is converted to thermal energy.

43. As a chicken grows from an egg, it becomes more ordered with time. Does this violate the principle of entropy? Explain.

44. (a) If you spent 10 minutes repeatedly shaking and throwing down a pair of coins, would you expect to see two heads come up at least once? (b) If you spent an hour shaking a handful of ten coins and throwing them down, would you expect to see all ten come up heads at least once? (c) If you stirred a box of 10,000 coins and dumped them repeatedly on the floor all day long, would you expect to see all 10,000 come up heads at least once?

45. In your bedroom are probably some 10^{27} air molecules. If they all happened to congregate on the opposite side of the room, you could suffocate. But this is unlikely. Is such a circumstance less likely, more likely, or of the same probability if there are many times fewer molecules in the room?

Problems

1. On a chilly 10°C day, a friend who likes cold weather says she wishes it were twice as cold. Taking this literally, what temperature would this be?

2. What is the ideal efficiency of an automobile engine in which gasoline is heated to 2700 K and the outdoor air is 300 K?

3. What is the ideal efficiency of an OTEC power plant that operates on the temperature difference between deep 4°C water and 25°C surface water?

4. Construct a table of all the possible combination of numbers that can come up when you throw two dice. A friend says, "Yes, I know that seven is the most likely total number when two dice are thrown, but *why*?" Based on your table, answer your friend, and explain that in thermodynamics the situations likely to be observed are those that can be formed in the greatest number of different ways.

Part II—Sample Exam Questions

Choose the BEST answer to each of the following.

1. When scientists discuss kinetic energy per molecule, the concept being discussed is
(a) heat (b) temperature (c) thermal energy
(d) entropy

2. In a mixture of hydrogen gas, oxygen gas, and nitrogen gas, the molecules with the greatest average speed are those of
(a) hydrogen (b) oxygen (c) nitrogen (d) that all have the same speed

3. The reason that the white-hot sparks that strike your face from a 4th-of-July-type sparkler don't harm you is because
(a) they have a low temperature (b) the energy per molecule is low (c) the energy per molecule is high, but little energy is transferred because of relatively few molecules in the spark

4. As a piece of metal with a hole in it cools, the diameter of the hole
(a) increases (b) decreases (c) remains the same size

5. If water had a higher specific heat capacity, ponds in cold weather would be
(a) less likely to freeze (b) more likely to freeze
(c) neither more nor less likely to freeze

6. Consider a sample of water at 2 °(C) If the temperature is increased slightly, say by 1 degree, the volume of water
(a) increases (b) decreases (c) remains unchanged

7. The temperature of water at the bottom of a deep ice-covered lake is
(a) slightly below freezing temperature (b) itself at the temperature of freezing (c) somewhat above freezing temperature

8. The principle reason one can walk barefoot on red-hot wooden coals without burning the feet has to do with
(a) low temperature of the coals (b) low thermal conductivity of the coals (c) mind-over-matter techniques

9. If the slower-moving molecules in a liquid were more likely to undergo evaporation, then evaporation would make the remaining liquid
(a) warmer (b) cooler
(c) no warmer or cooler than without evaporation

10. Soon after you remove a soda pop can from a cold refrigerator it becomes wet This wetness is primarily due to
(a) conduction (b) convection (c) radiation
(d) evaporation (e) condensation

11. Planet Earth loses heat primarily by
(a) conduction (b) convection (c) radiation
(d) all of these

12. Melting ice actually
(a) tends to warm the surroundings
(b) tends to cool the surroundings
(c) has no effect on the temperature of the surroundings

13. Consider a flask of 5-°C helium If it is heated until it has twice the internal energy, its temperature will be
(a) 10 °(C) (b) 273 °(C) (c) 278 °(C) (d) 283 °(C)
(e) 556 °(C)

14. The interior of Earth is kept hot by
(a) enormous pressure (b) great insulation (c) radioactivity (d) its own natural heat

15. On a hot day in your hot kitchen you operate your refrigerator with its door open in an attempt to cool the room. Doing this will
(a) cool the room, but inefficiently
(b) actually warm the room
(c) have no effect on changing the room temperature

Answers: 1b, 2a, 3c, 4b, 5a, 6b, 7c, 8b, 9a, 10e, 11c, 12b, 13d, 14c, 15(b)

PART

III

Electricity and Magnetism

Just think, this magnet outpulls the whole world when it lifts these nails. The pull between the nails and the earth I call a **gravitational force**, and the pull between the nails and the magnet I call a **magnetic force**. I can *name* these forces, but I don't yet *understand* them. My learning begins by realizing there's a big difference in *knowing the names* of things and really *understanding* those things.

CHAPTER

8

Electricity

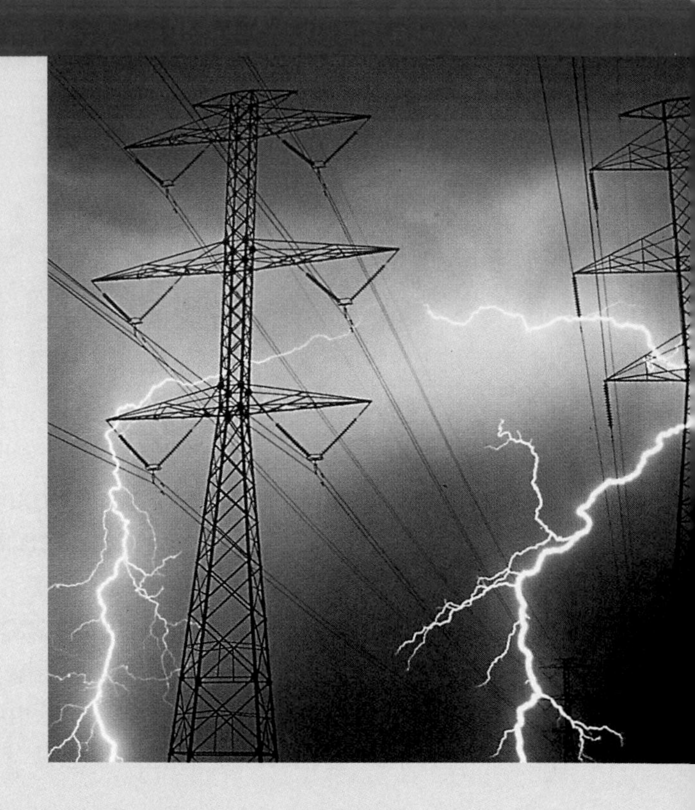

Electricity underlies just about everything around us. It's in the lightning from the sky, it's in the spark beneath our feet when we scuff across a rug, and it's what holds atoms together to form molecules. The control of electricity is evident in technological devices of many kinds, from lamps to computers. In this technological age it is important to have an understanding of the basics of electricity and of how these basics can be manipulated to produce a prosperity unknown before recent times.

This chapter begins with electric charges, the forces between them, the aura that surrounds them, and their behavior in materials. We learn that electric current is the flow of electric charges produced by an electrical pressure called *voltage.* In the next chapter we see that electric currents produce magnetism, and how both can be controlled to operate meters and motors. An understanding of electricity requires a step-by-step approach, as one concept builds upon the next, and so on. So put extra care in studying this material. If you're hasty it can be difficult, confusing, and frustrating. But with careful effort, it can be comprehensible and rewarding. Onward!

8.1 Electric Force and Charge

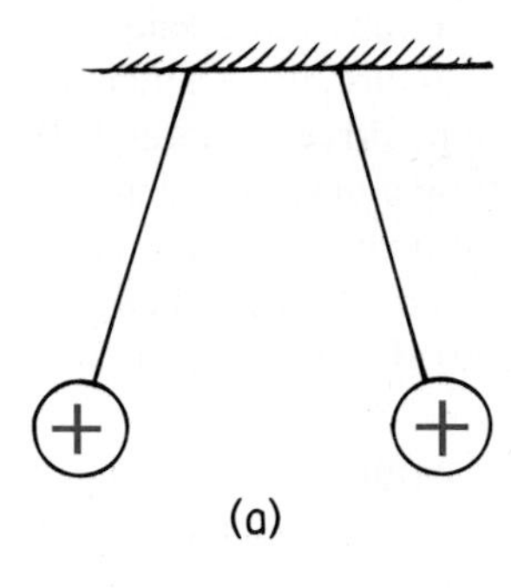

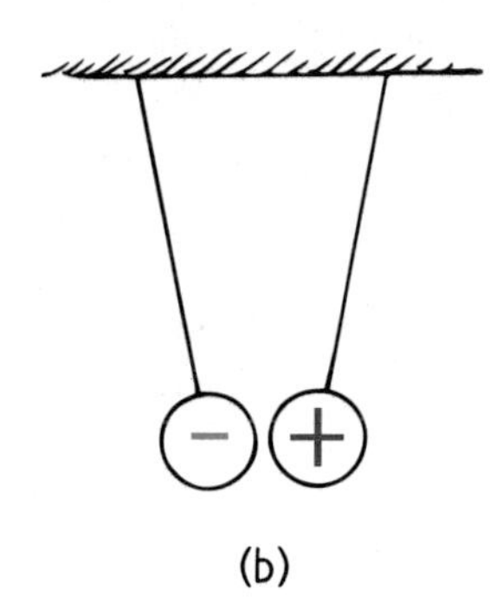

Fig. 8.1
(a) Like charges repel.
(b) Unlike charges attract.

What if there were a universal force that, like gravity, varies inversely as the square of distance but is billions upon billions of times stronger? If there were such a force and if it were attractive like gravity, the universe would be pulled together into a tight ball, with all matter pulled as close together as it could get. But suppose this force were a repelling force, with every bit of matter repelling every other bit. What then? The universe would be an ever-expanding gaseous cloud. Suppose, however, that the universe consisted of two kinds of particles—positives and negatives, say. Suppose positives repelled positives but attracted negatives and negatives repelled negatives but attracted positives. In other words, like kinds repel and unlike kinds attract (Figure 8.1). Suppose there were equal numbers of each so that this strong force was perfectly balanced! What would the universe be like? The answer is simple: It would be like the one we are living in. For there are such particles, and there is such a force. We call it the *electric force.*

The terms *positive* and *negative* refer to electric *charge,* the fundamental quantity that underlies all electrical phenomena. The positively charged particles are protons, and the negatively charged particles are electrons. The attractive force between these particles causes them to lump together into incredibly small units, which we call atoms. (Atoms also contain neutral particles called neutrons.) Most of the interesting details about atoms are presented in Chapter 13 and subsequent chapters. To understand the basic principles of electricity, however, it is important to be aware of some fundamental facts about atoms:

1. Every atom is composed of a positively charged *nucleus* surrounded by negatively charged electrons.
2. The electrons of all atoms are identical. Each has the same quantity of negative charge and the same mass.
3. Protons and neutrons compose the nucleus. (The common form of hydrogen that has no neutron is the only exception.) Protons are about 1800 times more massive than electrons but carry an amount of positive charge equal to the negative charge of electrons. Neutrons have slightly more mass than protons and have no net charge.

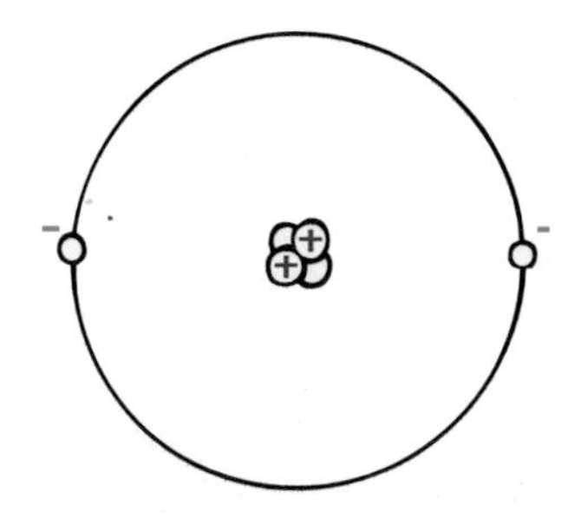

Fig. 8.2
Model of a helium atom. The atomic nucleus is made up of two protons and two neutrons. The positively charged protons attract two negative electrons. What is the net charge of this atom?

Normally, an atom has as many electrons as protons. When an atom loses one or more electrons, it has a positive net charge, and when it gains one or more electrons, it has a negative net charge. A charged atom is called an **ion.** A *positive ion* has a net positive charge. A *negative ion,* with one or more extra electrons, has a net negative charge.

Material objects are made of atoms, which means they are composed of electrons and protons (and neutrons). Objects ordinarily have equal numbers of electrons and protons and are therefore electrically neutral. But if there is a slight imbalance in the numbers, the object is electrically charged. An imbalance comes about when electrons are added to or removed from an object. Although electrons closest to the atomic nucleus, the innermost electrons, are bound very tightly to the oppositely charged atomic nucleus, the electrons farthest from the nucleus, the outermost electrons, are bound very loosely and can be easily dislodged. How much work is required to tear an electron away from an atom varies from one substance to another. The electrons are held

Fig. 8.3
Electrons are transferred from the fur to the rod. The rod is then negatively charged. Is the fur charged? How much compared to the rod? Positively or negatively?

more firmly in rubber and plastic than in your hair, for example. Thus, when a comb is rubbed through your hair, electrons transfer from the hair to the comb. The comb then has an excess of electrons and is said to be *negatively charged.* Your hair, in turn, has a deficiency of electrons and is said to be positively charged. To take another example, if you rub a glass or plastic rod with silk, the rod becomes positively charged. Silk has a greater affinity for electrons than does glass or plastic. Electrons are therefore rubbed off the rod and onto the silk.

So we see that an object having unequal numbers of electrons and protons is electrically charged. If it has more electrons than protons, it is negatively charged. If it has fewer electrons than protons, it is positively charged.

It is important to note that when we charge something, no electrons are created or destroyed. They are simply transferred from one material to another. Charge is *conserved.* In every event, whether large-scale or at the atomic and nuclear level, the principle of *conservation of charge* has always been found to apply. No case of the creation or destruction of net electric charge has ever been found. Conservation of charge ranks with conservation of energy and momentum as a significant fundamental principle in physics.

Question

If you walk across a rug and scuff electrons from your feet, are you negatively or positively charged?

Link to Electronics Technology—Electrostatic Charge

Static charge can be dangerous. Two hundred years ago, young boys called powder monkeys ran below the decks of warships to bring sacks of black gunpowder to the cannons above. Maritime law required that this task be done barefoot. Why? Because it was important that no static charge build up on the powder on their bodies as they ran to and fro. Bare feet scuffed the decks much less than shoes and assured no charge buildup that might result in an igniting spark and an explosion.

Today static charge is again a danger in many industries—not because buildings may blow up but because delicate electronic circuits may be destroyed by static charge. Some circuit components are sensitive enough to be “fried” by static electric sparks. Electronics technicians often wear clothing of special fabrics with ground wires between their sleeves and their socks. Some wear special wrist bands that are connected to a grounded surface so that static charge will not build up—when moving a chair, for example. The smaller the electronic circuit, the more hazardous are sparks that may short-circuit the circuit elements.

Answer
You have fewer electrons after you scuff your feet, and so you are positively charged (and the rug is negatively charged).

Coulomb's Law

The electric force, like the gravitational force between two masses, is inversely proportional to the square of the distance between charged particles. It increases as the particle-to-particle distance decreases and decreases as that distance increases. This relationship was discovered by Charles Coulomb in the eighteenth century and is called **Coulomb's law**. It states that for two charged particles or objects that are much smaller than the distance between them, the force between the two varies directly as the product of their charges and inversely as the square of the separation distance. The force acts along a straight line from one charged particle to the other. Coulomb's law can be expressed as

$$F = k\frac{q_1 q_2}{d^2}$$

where k is the proportionality constant, q_1 represents the quantity of charge of one particle, q_2 represents the quantity of charge of the other particle, and d is the distance between the charged particles.

The unit of charge is the **coulomb**, abbreviated C. It turns out that a charge of 1 coulomb is the charge associated with 6.25 billion billion electrons. This might seem like a great number of electrons, but it represents only the amount of charge that passes through a common 100-watt light bulb in a little over a second.

The proportionality constant k in Coulomb's law is similar to G in Newton's law of gravity. Instead of being a very small number like G (6.67×10^{-11}), k is a very large number, approximately

$$k = 9{,}000{,}000{,}000\ \mathrm{N{\cdot}m^2/C^2}$$

or, in scientific notation, $k = 9 \times 10^9\ \mathrm{N{\cdot}m^2/C^2}$. The unit $\mathrm{N{\cdot}m^2/C^2}$ is not central to our interest here; it simply converts the right-hand side of the equation to the unit of force, the newton (N). What is important is the large magnitude of k. If, for example, a pair of like charged particles each carrying a charge of 1 coulomb were 1 meter apart, the force of repulsion between them would be 9 billion newtons.[1] That would be about ten times

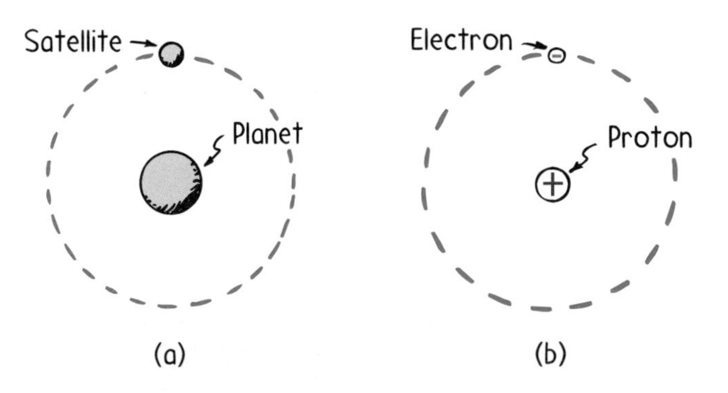

Fig. 8.4
(a) A gravitational force holds the satellite in orbit about the planet, and (b) an electric force holds the electron in orbit about the proton. In both cases there is no contact between the bodies. We say that the orbiting bodies interact with the *force fields* of the planet and proton and are everywhere in contact with these fields. Thus the force that one electrically charged body exerts on another can be described as the interaction between one body and the field set up by the other.

---

[1] Contrast this enormous value to the gravitational force of attraction between two unit masses (kilograms) 1 m apart: 6.67×10^{-11} N. This is an extremely small force. For the gravitational force to be 1 N, the masses at 1 m apart would have to have a mass of nearly 123,000 kg each! Gravitational forces between ordinary objects are exceedingly small, and electrical forces (noncancelled) between ordinary objects are exceedingly huge. We don't sense them because the positives and negatives normally balance out, and even for highly charged objects, the imbalance of electrons to protons is normally less than one part in a trillion trillion.

the weight of a battleship! Obviously, such amounts of net charge do not usually exist in our everyday environment.

So Newton's law of gravitation for masses is similar to Coulomb's law for electrically charged bodies.[2] Whereas the gravitational force of attraction between particles such as an electron and a proton is extremely small, the electric force between these particles is relatively enormous. Other than the big difference in strength, the most important difference between gravitational and electric forces is that electric forces may be either attractive or repulsive, whereas gravitational forces are only attractive.

Questions

1. The proton in the nucleus of the hydrogen atom attracts the electron that orbits it. Relative to the force exerted by the proton on the electron, does the electron attract the proton with less force, more force, or the same amount of force?
2. If a proton at a particular distance from a charged particle is repelled with a given force, by how much does the repelling force decrease when the proton is three times farther away from the particle? Five times farther?
3. Is the charge on the particle repelling the proton in Question 2 positive or negative?

Charge Polarization

Charge an inflated balloon by rubbing it on your hair. Then place the balloon against a wall and it sticks. This is because the charge on the balloon alters the charge distribution in the atoms in the wall, effectively inducing an opposite charge in the wall. The atoms cannot move from their relatively fixed positions, but their "centers of charge" are moved. The positive part of each atom is attracted toward the balloon, and the negative part is repelled. This has the effect of distorting the atoms (Figure 8.5), which are said to be **electrically polarized**. We treat electrical polarization further in Chapter 17. (We know that after you rub the balloon on your hair, the balloon sticks to the wall. In a humorous vein, if you put your head to the wall, will that stick, too?)

Answers

1. The same amount of force, in accord with Newton's third law—basic mechanics! Recall that a force is an interaction between two things, in this case between the proton and the electron. They pull on each other—equally but in opposite directions.
2. It decreases to $\frac{1}{9}$ its original value, to $\frac{1}{25}$.
3. Positive.

---

[2]The similarities between these two forces have made some people think they may be different aspects of the same thing. Albert Einstein was one of these people, and he spent the latter part of his life searching with little success for a "unified field theory." More recently, the electrical force has been unified with one of the two nuclear forces, the *weak force,* which plays a role in radioactive decay. Physicists are still looking for a way to unify electrical and gravitational forces.

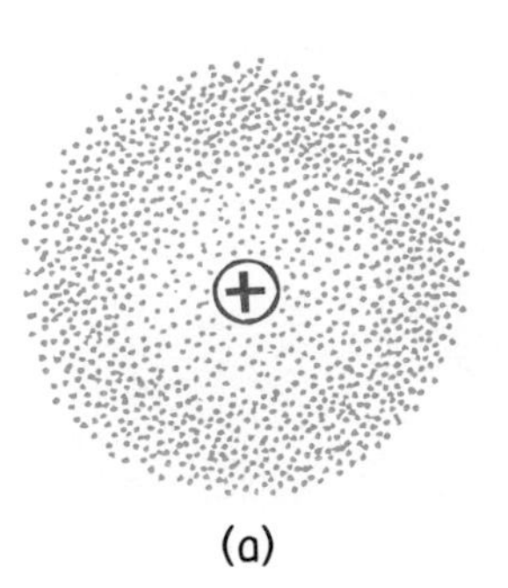

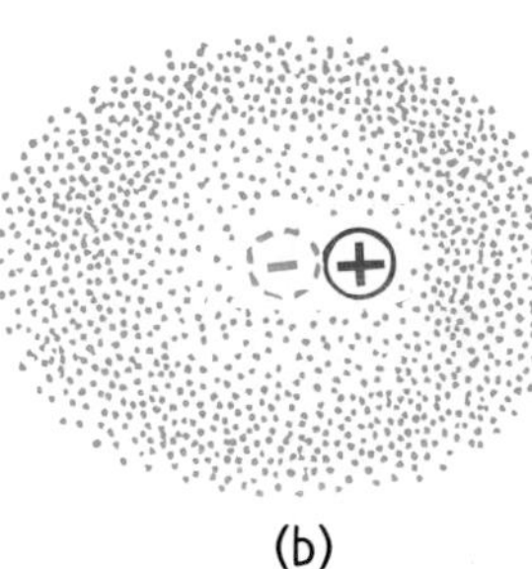

Fig. 8.5
(a) The center of the negative "cloud" of electrons coincides with the center of the positive nucleus in an atom. (b) When an external negative charge is brought nearby to the left, as on a charged balloon, the electron cloud is distorted so the centers of negative and positive charge no longer coincide. The atom is electrically polarized.

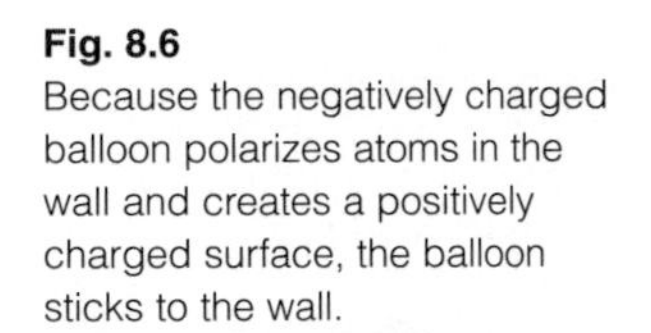

Fig. 8.6
Because the negatively charged balloon polarizes atoms in the wall and creates a positively charged surface, the balloon sticks to the wall.

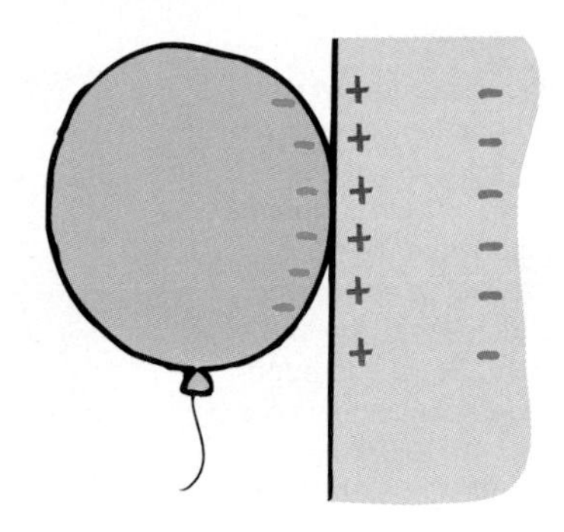

8.2 Electric Fields

Electric forces, like gravitational forces, act between things that are not in contact with each other. For both electricity and gravity, there exists a *force field* that influences distanced charges and masses, respectively. Just as the space between any two masses is filled with a gravitational field, the space around every electric charge is filled with an **electric field**—a kind of aura that extends through space.[3] About a proton, the electric field extends radially away from the proton. About an electron, the field is in the opposite direction, shown by the field lines in Figure 8.7. Some electric field configurations are shown in Figure 8.8, and photographs of field patterns are shown in Figure 8.9. In the next chapter we'll see how bits of iron similarly align with magnetic fields.

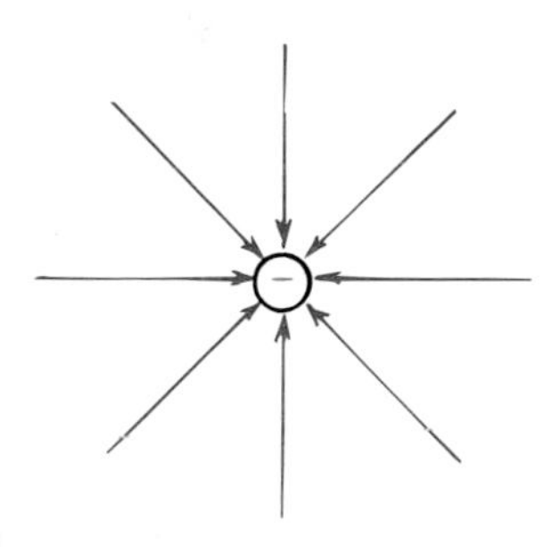

Fig. 8.7
Electric field representation about a negative charge.

Place a charged particle in an electric field and the particle experiences an electric force. The direction of force on a positive charge is the same as the direction of the electric field.

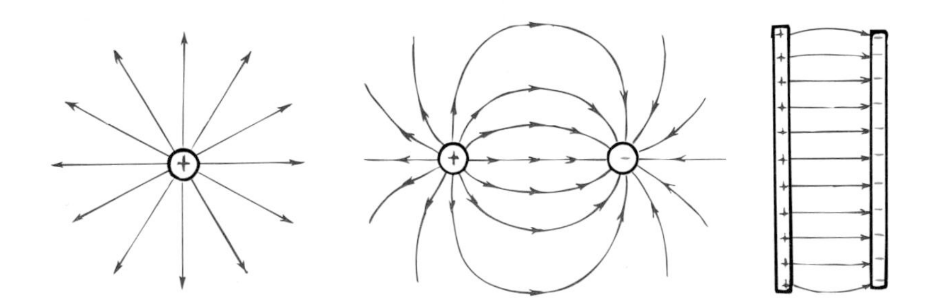

Fig. 8.8
Some electric field configurations. (a) Lines of force emanating from a single positively charged particle. (b) Lines of force for a pair of equal but oppositely charged particles. Note that the lines emanate from the positive particle and terminate on the negative particle. (c) Uniform lines of force between two oppositely charged parallel plates.

[3] An electric field has both magnitude (strength) and direction. The magnitude of the field at any point is simply the force per unit of charge. If a charge q experiences a force F at some point in space, then the electric field E at that point is

$$E = \frac{F}{q}$$

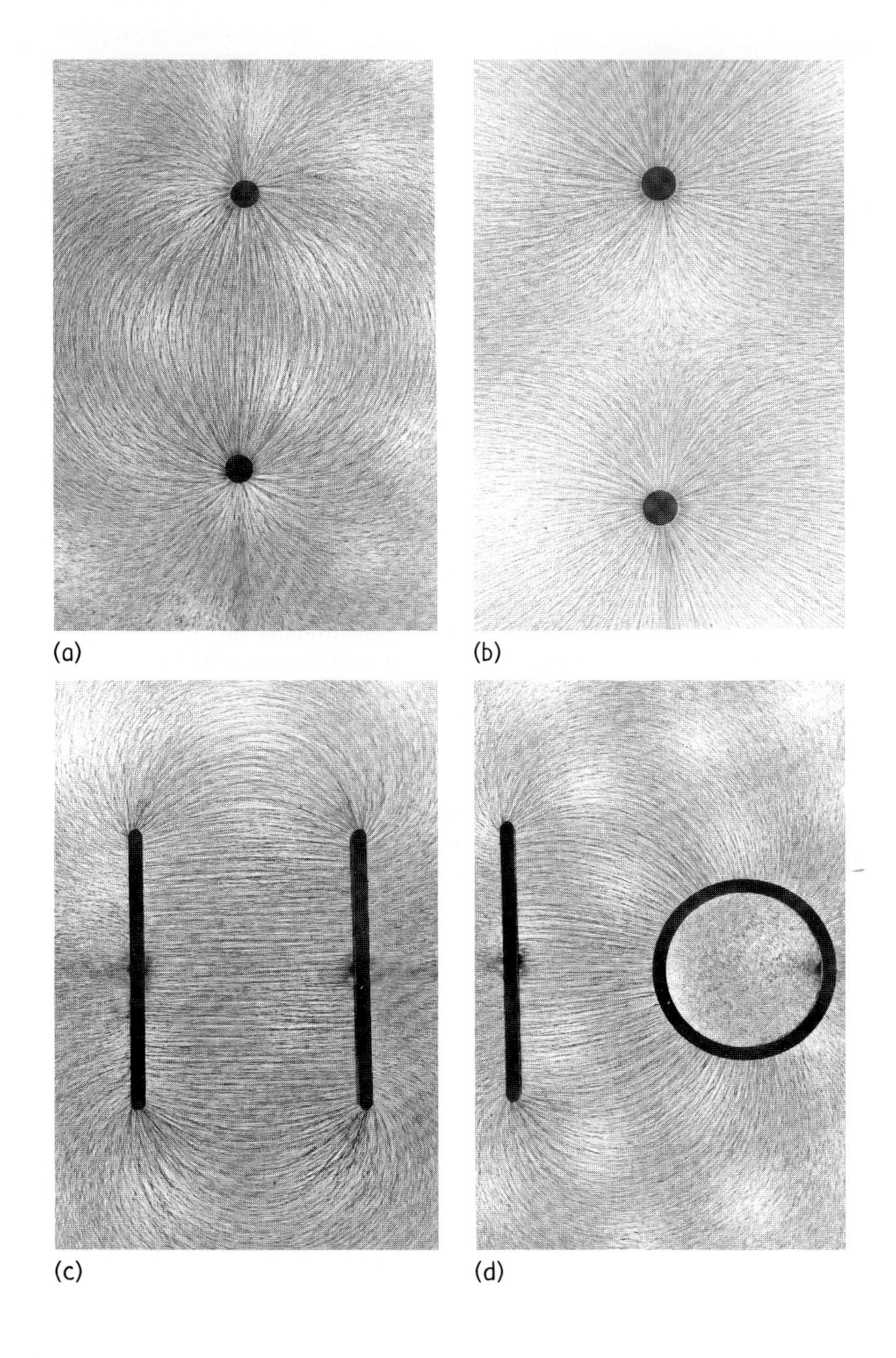

Fig. 8.9
The electric field due to a pair of charged conductors is shown by bits of thread suspended in an oil bath surrounding the conductors. Note that the threads line up end-to-end along the direction of the electrical field. (a) Equal and oppositely charged conductors (like Figure 8.8b). (b) Equal like-charged conductors. (c) Oppositely charged plates. (d) Oppositely charged cylinder and plate.

Perhaps your instructor will demonstrate the effects of the electric field that surrounds the charged dome of a Van de Graaff generator (Figure 8.10). Charged objects in the field of the dome are either attracted or repelled, depending on their sign of charge.

Fig. 8.10
Both Meidor and the spherical dome of the Van de Graaff generator are electrically charged. Why does her hair stand out?

Question

In Figure 8.10, Meidor, like the dome of the Van de Graaff generator, is charged (because she was in contact with the dome when charging took place). Why does her hair stand out?

8.3 Electric Potential

When we studied energy in Chapter 3, we learned that an object has gravitational potential energy because of its location in a gravitational field. Similarly, a charged object has **electrical potential energy** by virtue of its location in an electric field. Just as work is required to lift a massive object against the gravitational field of the Earth, work is also required to push a charged particle against the electric field of a charged body. This work changes the electrical potential energy of the charged particle.[4] The work done in compressing a spring increases the potential energy of the spring, and the work done in pushing a charged particle closer to a charged sphere increases the potential energy of the charged particle (Figure 8.12). If the particle is released, it accelerates

---

Answer
Meidor's hair is also charged. Each hair is repelled by others around it—evidence that *like charges repel.* Even a small charge produces an electric force greater than hair weight. Fortunately, the electric force is not great enough to make her arms stand out!

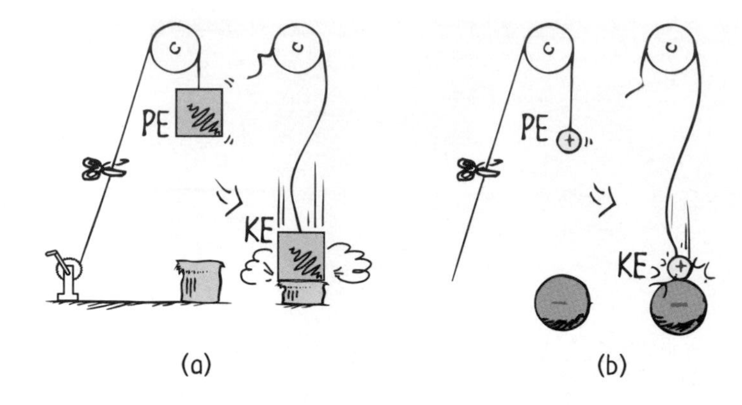

Fig. 8.11
(a) The PE (gravitational potential energy) of a mass is held in a gravitational field. (b) The PE of a charged particle is held in an electric field. When the mass and particle are released, how does the KE (kinetic energy) acquired by each compare with the decrease in PE?

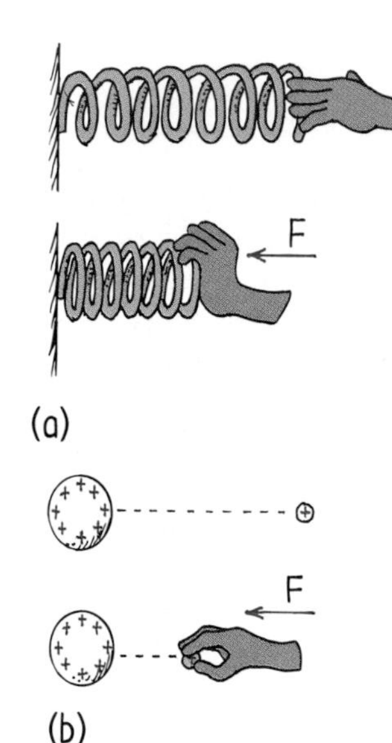

Fig. 8.12
(a) The spring has more mechanical PE when compressed. (b) The charged particle similarly has more electrical PE when pushed closer to the charged sphere. In both cases the increased PE is the result of work input.

in a direction away from the sphere and its electrical potential energy changes to kinetic energy.

If we push a particle that carries twice the charge, we do twice as much work, and the double charge in the same location has twice the electrical potential energy as before; three times the charge, and we have three times as much potential energy, and so on. Rather than dealing with the total electrical potential energy of a charged body, it is convenient to consider the electrical potential energy *per charge.* We simply divide the amount of potential energy in any case by the amount of charge. The concept of electrical potential energy per charge is called **electric potential**:

$$\textbf{Electric potential} = \frac{\textbf{electric potential energy}}{\textbf{amount of charge}}$$

The unit of measurement for electric potential is the **volt**, so electric potential is often called *voltage.* A potential of 1 volt (V) equals 1 joule (J) of energy per 1 coulomb (C) of charge.

$$\textbf{1 volt} = \frac{\textbf{1 joule}}{\textbf{coulomb}}$$

Thus a 1.5-volt battery gives 1.5 joules of energy to every 1 coulomb of charge passing through it. The terms *electric potential* and *voltage* are both common, so either may be used. In this book these terms are used interchangeably.

The significance of electric potential (voltage) is that a definite value for it can be assigned to a location. We can speak about the electric potentials at different locations in an electric field whether or not charges occupy those locations. Likewise with electric potentials at various locations in an electric circuit. Later in this chapter we shall see that the location of the positive

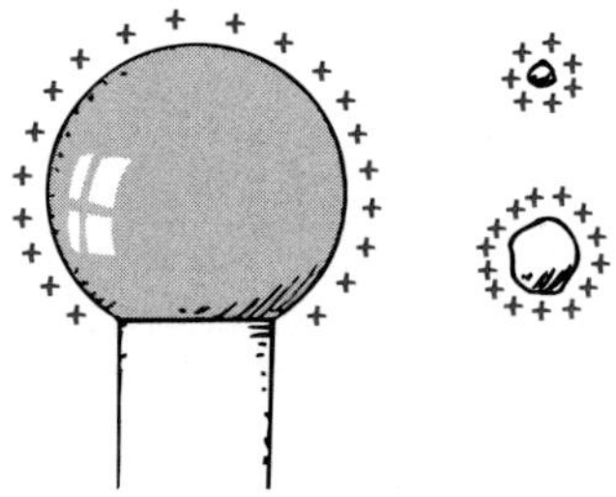

Fig. 8.13
Of the two charged bodies on the right, the one carrying the greater charge has the higher electrical PE in the field of the charged dome on the left, but the *electric potential* of any amount of charge at the same location is the same. Why?

[4]This work is positive if it increases the electrical potential energy of the charged particle and negative if it decreases it. In a follow-up course, you can distinguish between positive particles and negative particles moving with or against electric fields—which if covered here would contribute to "information overload."

Fig. 8.14
Although the electric potential (voltage) of the charged balloon is high, the electrical PE is low because of the small amount of charge.

terminal of a 12-volt battery is maintained at a voltage 12 volts higher than the location of the negative terminal. When a conducting medium connects this voltage difference, charges in the medium move between these two locations.

Questions

1. If there were twice as many coulombs in each of the charged bodies near the charged sphere in Figure 8.13, would the electrical potential energy of the bodies relative to the charged sphere be the same or twice as great? Would the electric potential of each body be the same or twice as great?
2. What does it mean to say that your car has a 12-volt battery?

Rub a balloon on your hair, and the balloon becomes negatively charged—perhaps to several thousand volts! That would be several thousand joules of energy if the charge were 1 coulomb. However, 1 coulomb is a fairly respectable amount of charge. The charge on a balloon rubbed on hair is more typically much less than a millionth of a coulomb. Therefore, the amount of energy associated with the charged balloon is very, very small. A high electric potential means a lot of energy only if a lot of charge is involved.[5] There is an important difference between electrical potential energy and electric potential.

Now we see what happens when one electric potential is applied to one end of a piece of metal wire and a different electric potential is applied to the other end. Any difference in electric potential is called a **potential difference** (or *voltage difference*) and acts like an "electric pressure" that produces an electric current—a flow of electric charge.

8.4 Voltage Sources

When the ends of a conductor of heat are at different temperatures, heat energy flows from the higher-temperature part to the lower-temperature part. The flow ceases when both ends reach the same temperature. Similarly, when the ends of an electrical con-

Answers

1. The result of twice as many coulombs is that the charged body then has twice as much *electrical potential energy* because it takes twice as much work to put the charged body at that point in the electric field. But the *electric potential* would be the same because twice the energy divided by twice the charge gives the same potential as one unit of energy divided by one unit of charge. Electric potential is not the same thing as electrical potential energy. Be sure you understand this before you continue.
2. It means that one of the battery terminals is 12 V higher in electric potential than the other one. Soon we'll see that it also means that when a circuit is connected between these terminals, each coulomb of charge in the resulting current will be given 12 J of energy as it passes through the battery (and 12 J of energy "spent" in the circuit—a 12-V voltage drop).

[5]This is very similar to the harmless high-temperature sparks emitted by a fireworks sparkler. Temperature is the ratio of energy/molecule. A high temperature means a lot of energy only if a lot of molecules are involved.

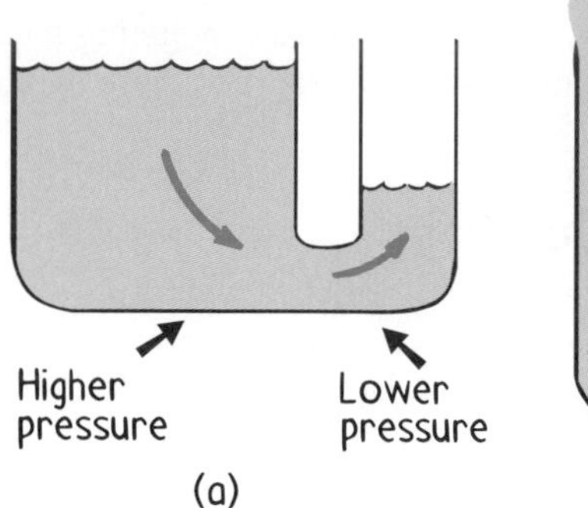

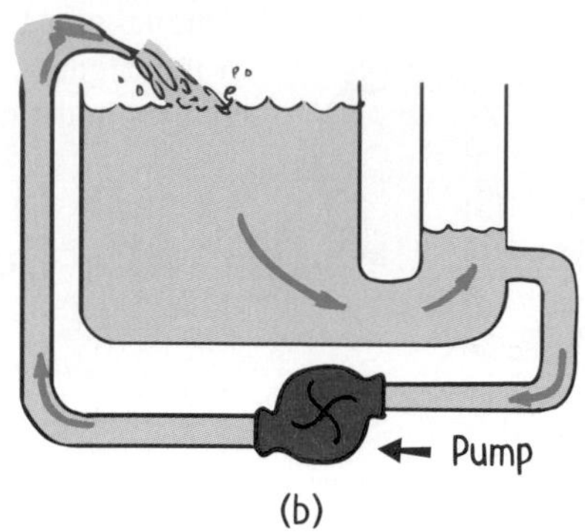

Fig. 8.15
(a) Water flows from the reservoir of higher pressure to the reservoir of lower pressure. The flow ceases when the difference in pressure ceases.
(b) Water continues to flow because a difference in pressure is maintained with the pump.

Fig. 8.16
An unusual source of voltage. The electric potential between the head and tail of the electric eel (*Electrophorus electricus*) can be up to 600 V.

ductor are at different electric potentials, charges in the conductor flow from the higher-potential end to the lower-potential end. The flow of charge persists until both ends reach the same potential. Without a difference in electric potential, no flow of charge occurs. Connect one end of a wire to a charged Van de Graaff generator, for example, and the other end to the ground, and a surge of charge flows through the wire. The flow is brief, however, for the sphere quickly reaches the same potential as the ground.

To attain a sustained flow of charge in a conductor, some arrangement must be provided to maintain a difference in potential while charge flows from one end to the other. The situation is analogous to the flow of water from a higher reservoir to a lower one (Figure 8.15a). Water flows in a pipe that connects the reservoirs only as long as a difference in water level exists. The flow of water in the pipe, like the flow of charge in the wire that connects the Van de Graaff generator to the ground, will cease when the pressures at the two ends are equal (we imply this when we say that water seeks its own level). A continuous flow is possible if the difference in water levels—and hence the difference in water pressures—are maintained with a pump (Figure 8.15b).

A sustained electric current requires a suitable pumping device to provide a difference in electric potential—in other words, to provide a voltage difference. Chemical batteries or generators are "electrical pumps" that maintain a steady flow of charge. These devices do work to pull negative charges away from positive ones. In chemical batteries, this work is done by the chemical disintegration of zinc or lead in acid, and the energy originally stored in chemical bonds is converted to electric potential energy.[6] Generators separate charge by electromagnetic induction, a process described in the next chapter. The work done by whatever means in separating opposite charges is available at the terminals of the battery or generator. This energy per charge provides the difference in potential (the voltage difference) that provides the "electrical pressure" to move electrons through a circuit joined to these terminals.

A common automobile battery provides an "electrical pressure" of 12 volts to a circuit connected across its terminals. Then 12 joules of energy is supplied to each coulomb of charge that is made to flow in the circuit. When it completes the circuit, the charge is without this energy, which has been "dropped off" in one or more electrical devices. We say there is a 12-volt *voltage drop* in the circuit.

---

[6]The chemical nature of batteries is described in Chapter 20.

Link to History of Technology—110 Volts

In the early days of electricity, high voltages burned out electric light filaments, and so low voltages were more practical. The hundreds of power plants built in the United States prior to 1900 adopted 110 volts (or 115 or 120 volts) as their standard. Tradition has it that 110 volts was settled on because it made bulbs of the day glow as brightly as a gas lamp. By the time electricity became widely available in Europe, engineers had figured out how to make light bulbs that would not burn out so fast at higher voltages. Because power transmission is more efficient at higher voltages, Europe adopted 220 volts as their standard. The United States stayed with 110 volts (today officially 120 volts) because of the installed base of 110-volt equipment.

8.5 Electric Current

Just as water current is the flow of H_2O molecules, **electric current** is the flow of charged particles. In circuits of metal wires, electrons are the flowing charged particles. This is because one or more electrons from each metal atom are free to move throughout the atomic lattice. These charge carriers are called *conduction electrons*. Protons, on the other hand, do not move because they are bound inside the nuclei of atoms that are more or less locked in fixed positions. In fluids, however, positive ions, negative ions, and electrons may compose the flow of electric charge.

The *rate* of electrical flow is measured in amperes (abbreviation A). An **ampere** is the rate of flow of 1 coulomb of charge per second. (Recall that 1 coulomb, the standard unit of charge, is the electric charge of 6.25 billion billion electrons.) In a wire that carries 5 amperes, for example, 5 coulombs of charge passes any cross section in the wire each second. In a wire that carries 10 amperes, twice as many coulombs pass any cross section each second.

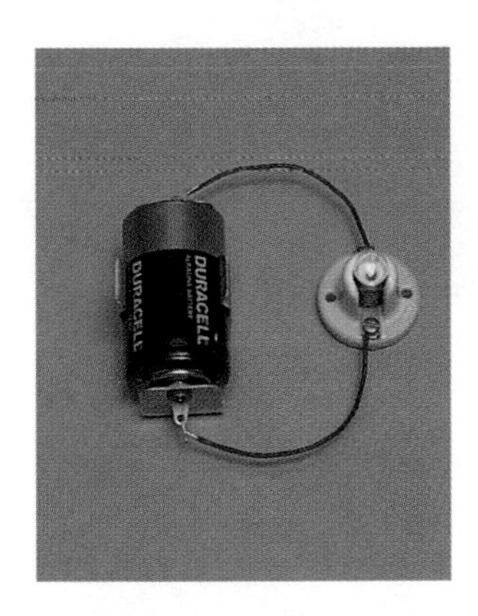

Fig. 8.17
Each coulomb of charge made to flow in the circuit that connects the ends of this 1.5-V flashlight cell is energized with 1.5 J.

It is interesting to note that a current-carrying wire is not electrically charged. Under ordinary conditions, negative conduction electrons swarm through the atomic lattice made up of positively charged atomic nuclei. So there are as many electrons as protons in the wire. Whether a wire carries a current or not, the net charge of the wire is normally zero at every moment.

There is often some confusion about charge flowing *through* a circuit and voltage placed, or impressed, *across* a circuit. We can distinguish between these ideas by considering a long pipe filled with water. Water flows *through* the pipe if there is a difference in pressure *across,* or between, its ends. Water flows from the high-pressure end to the low-pressure end. Only the water flows, not the pressure. Similarly, electric charge flows because of a difference in electrical pressure (voltage difference). You say that charges flow *through* a circuit because of an applied voltage *across* the circuit.[7] You don't say that voltage flows through a circuit. Voltage doesn't go anywhere, for it is the charges that move. Voltage produces current (if there is a complete circuit).

[7]We often think of current flowing through a circuit, but don't say this around somebody who is picky about grammar, for the expression "current flows" is redundant. More properly, charge flows—which *is* current.

Direct Current and Alternating Current

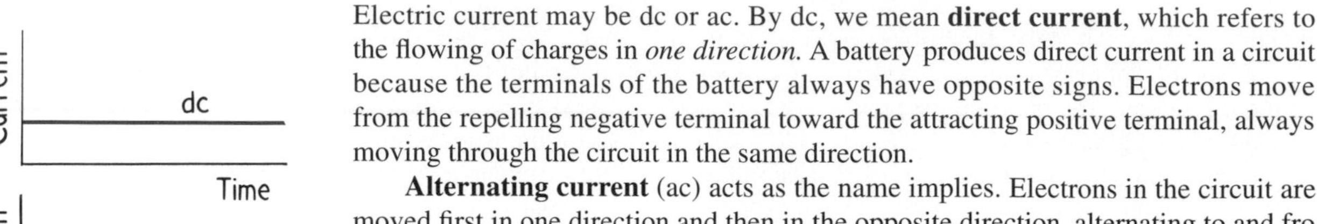

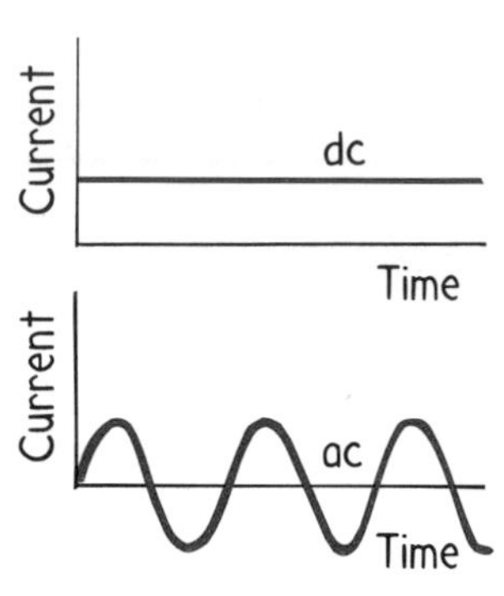

Fig. 8.18
Time graphs of dc and ac.

Electric current may be dc or ac. By dc, we mean **direct current**, which refers to the flowing of charges in *one direction.* A battery produces direct current in a circuit because the terminals of the battery always have opposite signs. Electrons move from the repelling negative terminal toward the attracting positive terminal, always moving through the circuit in the same direction.

Alternating current (ac) acts as the name implies. Electrons in the circuit are moved first in one direction and then in the opposite direction, alternating to and fro about relatively fixed positions. This can be accomplished by periodically switching the sign at the terminals. Nearly all commercial ac circuits involve currents that alternate back and forth at a frequency of 60 cycles per second. This is 60-hertz current (a cycle per second is called a *hertz* Hz). In some countries, 25-, 30-, or 50-hertz current is used. Throughout the world, most residential and commercial circuits are ac because electric energy in the form of ac can easily be stepped up to high voltage to be transmitted great distances with small heat losses, then stepped down to convenient voltages where the energy is used. Why this is so is quite interesting and is touched on in the next chapter.

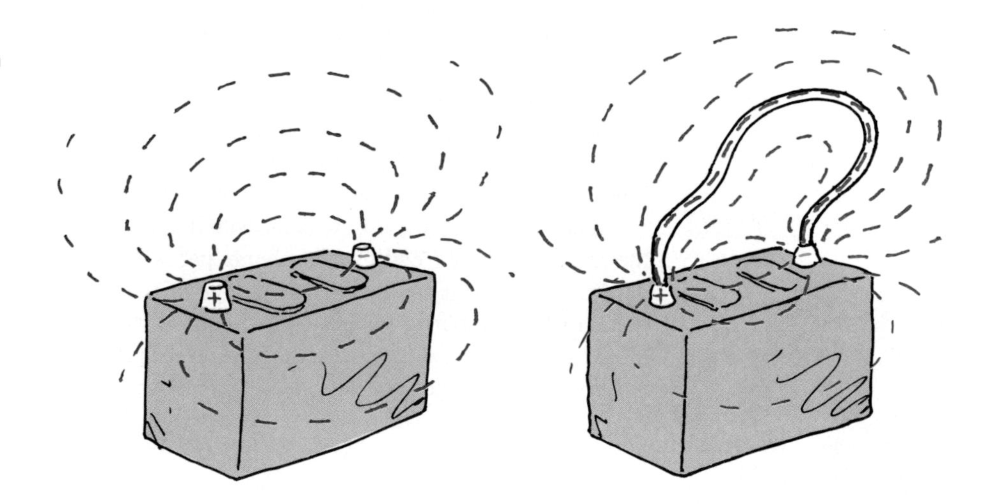

Fig. 8.19
The electric field lines between the terminals of a battery are directed through a conductor that joins the terminals. A thick metal wire is shown here, but the path from one terminal to the other is usually an electric circuit. (If you touch this conducting wire, you won't be shocked, but the wire will heat quickly and may burn your hand!)

8.6 Electrical Resistance

A battery or generator of some kind is the prime mover of charge and source of voltage in an electric circuit. How much current there is in the circuit depends not only on the voltage but also on the **electrical resistance** the conductor offers to the flow of charge. This is similar to the rate of water flow in a pipe, which depends not only on the pressure behind the water but also on the resistance offered by the pipe. The resistance of a wire depends on the conductivity of the material and also on its thickness and length. Electrical resistance is less in thick wires. The longer the wire, of course, the greater the resistance. In addition, electrical resistance depends on temperature. The greater the jostling about of atoms within the conductor (in other words, the

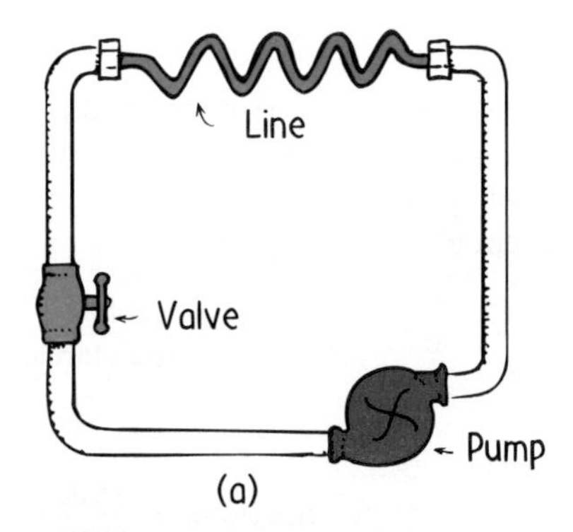

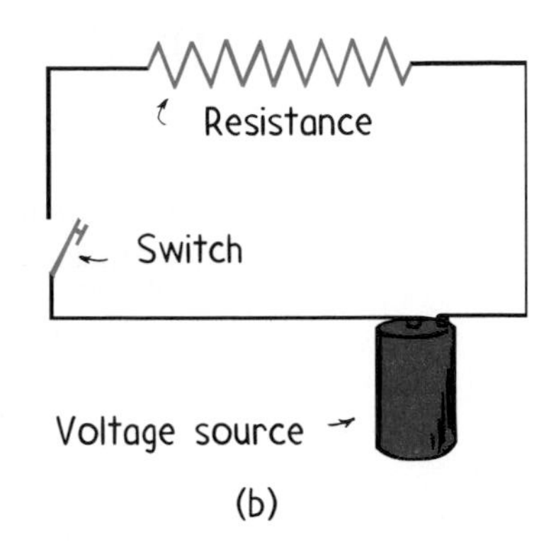

Fig. 8.20
(a) In a hydraulic circuit, a narrow pipe (green) offers resistance to water flow. (b) In an electric circuit, a lamp or other device (shown by the zig-zag symbol for resistance) offers resistance to electron flow.

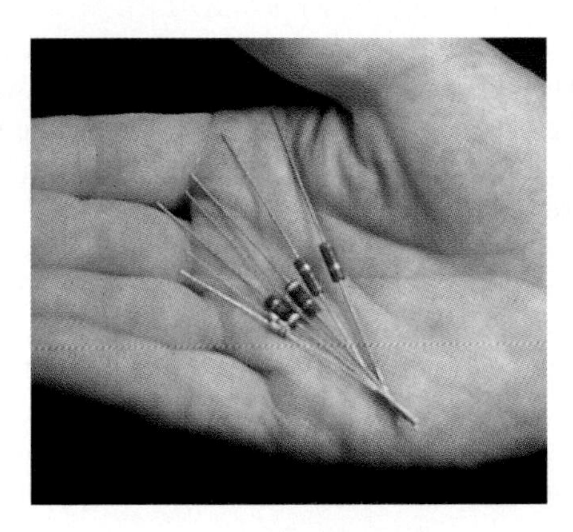

Fig. 8.21
Resistors. The symbol of resistance in an electric circuit is ww

higher the temperature), the greater resistance most conductors offer to the flow of charge.[8] The resistance of some materials reaches zero at very low temperatures. These are superconductors.

Electrical resistance is measured in units called *ohms.* The Greek letter *omega,* Ω, is commonly used as the symbol for the ohm. This unit is named after Georg Simon Ohm, a German physicist who in 1826 discovered a simple and very important relationship among voltage, current, and resistance.

Link to Technology—Superconductors

In ordinary conductors such as house wiring, the moving electrons that constitute an electric current often collide with atomic nuclei in the wire, transferring their kinetic energy to the wire as heat. Early in this century certain metals in a bath of 4 K liquid helium lost all electrical resistance. The electrons in these conductors traveled pathways that avoided these collisions, permitting them to flow indefinitely. These materials are called **superconductors**, having zero electrical resistance to the flow of charge. In superconductivity no current is lost and no heat is generated.

Until recently it was generally thought that zero electrical resistance could be brought about only in certain metals and only at temperatures near absolute zero. Then in 1986 superconductivity was achieved at 30 K, which spurred hopes of finding superconductivity above 77 K, the point at which nitrogen liquefies. Nitrogen is easier to handle than the liquid helium needed for colder environments. The historic leap came in the following year with a yttrium compound that lost its resistance at 90 K. Various ceramic oxides have since been found to be superconducting at temperatures above 100 K.

Once electric current is established in a superconductor, the charges flow indefinitely, even without an electric field. Steady currents have been observed to persist for years in certain superconductors without apparent loss in energy. There is presently enormous interest in the physics community as to exactly why certain materials acquire superconducting properties. Explanations generally have to do with the wave nature of matter (quantum mechanics) and are being vigorously researched.

---

[8]Carbon is an interesting exception. As temperature increases, more carbon atoms shake loose an electron. This increases the electric current. So the resistance of carbon lowers with increasing temperature. This and (primarily) its high melting point are why carbon is used in arc lamps.

Ohm's Law

The relationship among voltage, current, and resistance is summarized by a statement called **Ohm's law**. Ohm discovered that the amount of current in a circuit is directly proportional to the voltage established across the circuit and inversely proportional to the resistance of the circuit:

$$\textbf{Current} = \frac{\textbf{voltage}}{\textbf{resistance}}$$

Or, in units form,

$$\textbf{Amperes} = \frac{\textbf{volts}}{\textbf{ohms}}$$

So for a given circuit of constant resistance, current and voltage are proportional to each other.[9] This means we get twice the current for twice the voltage. The greater the voltage, the greater the current. But if the resistance is doubled for a circuit, the current is cut in half. The greater the resistance, the smaller the current. Ohm's law makes good sense.

Ohm's law tells us that a difference in potential of 1 volt established across a circuit that has a resistance of 1 ohm produces a current of 1 ampere. If 12 volts is impressed across the same circuit, the current is 12 amperes. The resistance of a typical lamp cord is much less than 1 ohm, and a typical light bulb has a resistance of more than 100 ohms. An iron or electric toaster has a resistance of 15 to 20 ohms. The current inside these and all other electrical devices is regulated by circuit elements called *resistors,* whose resistance may be a few ohms or millions of ohms.

Questions

1. How much current flows through a lamp that has a resistance of 60 Ω when 12 V is impressed across it?
2. What is the resistance of an electric frying pan that draws a current of 12 A when connected to a 120-V circuit?

Electric Shock

Which causes electric shock in the human body—current or voltage? The damaging effects of shock are the result of current through the body. From Ohm's law we can see that this current depends on the voltage applied and also on the body's electrical resis-

Answers

1. 0.2 A. This is calculated from Ohm's law: 0.2 A = 12 V/60 Ω.
2. 10 Ω. Rearrange Ohm's law to read:

$$\textbf{Resistance} = \frac{\textbf{voltage}}{\textbf{current}} = \frac{\textbf{120 V}}{\textbf{12 A}} = \textbf{10 }\Omega$$

[9] Many texts use V for voltage, I for current, and R for resistance and express Ohm's law as $V = IR$. It then follows that $I = V/R$, or $R = V/I$, so if any two variables are known, the third can be found. Units are abbreviated V for volts, A for amperes, and Ω for ohms.

Table 8.1
Effect of Electric Current on the Body

Current (A)	Effect
0.001	Can be felt
0.005	Is painful
0.010	Causes involuntary muscle contractions (spasms)
0.015	Causes loss of muscle control
0.070	Goes through the heart; serious damage, probably fatal for if current lasts for more than 1 s

tance. The resistance of the human body ranges from about 100 ohms if the body is soaked with salt water to about 500,000 ohms if the skin is very dry. If we touch the two electrodes of a battery with dry fingers, completing the circuit from one hand to the other, we can expect to offer a resistance of about 100,000 ohms. We usually cannot feel 12 volts if we do this, though 24 volts just barely tingles. If our skin is moist, however, 24 volts can be quite uncomfortable. Table 8-1 describes the effects of different amounts of current on the human body.

Questions

1. At 100,000 Ω, how much current flows through your body if you touch the terminals of a 12-V battery?
2. If your skin is very moist—so that your resistance is only 1000 Ω—and you touch the terminals of a 12-V battery, how much current do you receive?

For you to receive a shock, there must be a *difference* in electric potential between one part of your body and another part. Most of the charge making up the current will pass along the path of least electrical resistance connecting these two points. Suppose you fell from a bridge and managed to grab onto a high-voltage power line, halting your fall. So long as you touch nothing else of different potential, you receive no shock. Even if the wire is a few thousand volts above ground potential and even if you hang by it with two hands, no appreciable amount of charge flows from one hand to the other. This is because there is no appreciable difference in electric potential between your hands. If, however, you reach over with one hand and grab onto a wire of different potential . . . zap! We have all seen birds perched on high-voltage wires. Every part of their bodies is at the same high potential as the wire, and so they feel no ill effects.

Most electric plugs and sockets today are wired with three connections. The two flat prongs on a plug are for the current-carrying double wire inside the socket, one part

---

Answers:

1. $\dfrac{12\ \text{V}}{100{,}000\ \Omega} = 0.00012\ \text{A}$

2. $\dfrac{12\ \text{V}}{1000\ \Omega}$ 5 0.012 A (Ouch!)

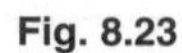

Fig. 8.22
The bird can stand harmlessly on one wire of high potential, but it had better not reach over and grab a neighboring wire! Why not?

Fig. 8.23
The round prong connects the body of the appliance directly to ground (the Earth). Any charge that builds up on an appliance is therefore conducted to the ground—preventing accidental shock.

of which is "live" (energized) and the other neutral, while the round prong connects to a wire in the electrical system that is grounded—connected directly to the ground (Figure 8.23). The electrical appliance at the other end of the plug is therefore connected to all three wires. If the live wire in the plugged-in appliance accidentally comes in contact with the metal surface of the appliance, and you touch the appliance, you could receive a dangerous shock. This won't occur when the appliance casing is grounded via the ground wire, which assures that the appliance casing is always at zero ground potential.

Question

Which causes electric shock, current or voltage?

Link to Safety—Electric Shock

Many people are killed each year by current from common 120-volt electric circuits. If you touch a faulty 120-volt light fixture with your hand while your feet are on the ground, there may be a 120-volt "electrical pressure" (potential difference) between your hand and the ground. Resistance to current is usually greatest between your feet and the ground, and so the current is usually not enough to do serious harm. But if your feet and the ground are wet, there is a low-resistance electrical path between you and the ground. The 120 volts across this lowered resistance may produce a current greater than your body can stand.

Pure water is not a good conductor, but the ions normally found in water make it a fair conductor. More dissolved materials, especially small amounts of salt, reduce the resistance even more. There is usually a layer of salt left from perspiration on your skin, which when wet lowers your skin resistance to a few hundred ohms or less. Handling electrical devices while taking a bath is a definite no-no.

Injury by electric shock comes in three forms: (1) burning of tissues by heating, (2) muscle contraction, and (3) disruption of cardiac rhythm. These conditions are caused by too much power delivered in critical body volumes for too long a time.

Electric shock can upset the nerve center that controls breathing. In rescuing shock victims, the first thing to do is clear them from the electric supply with a dry wooden stick or some other nonconductor so that you don't get electrocuted yourself. Then apply artificial respiration. It is important to continue artificial respiration, for there have been cases of victims of lightning who did not breathe for several hours, were eventually revived and who completely regained good health.

Answer
Electric shock *occurs* when current is produced in the body, which is *caused* by an impressed voltage.

8.7 Electric Circuits

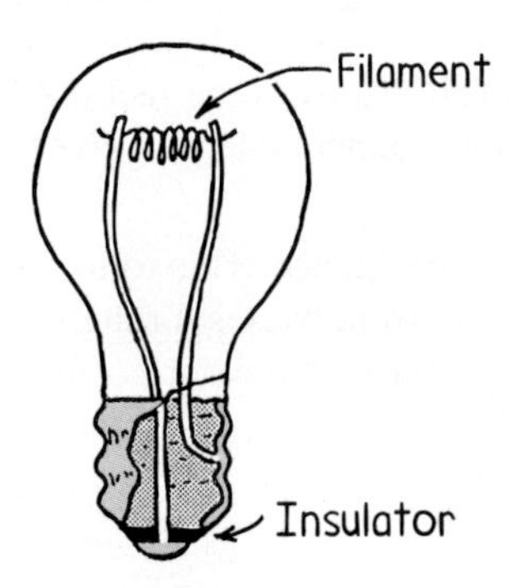

Fig. 8.24
The conduction electrons that surge to and fro in the filament of the lamp do not come from the voltage source. They are in the filament to begin with. The voltage source simply provides them with surges of energy.

Any path along which electrons can flow is a *circuit.* For a continuous flow of electrons, there must be a complete circuit with no gaps. A gap is usually provided by an electric switch that can be opened or closed to either cut off or allow energy flow.

Most circuits contain more than one device that receives electric energy from the circuit. These devices are commonly connected in a circuit in one of two ways, *series* or *parallel.* When connected in series, the devices and the wires connecting them form a single pathway for electron flow between the terminals of the battery, generator, or wall socket. When connected in parallel, the devices and wires connecting them form branches, each of which is a separate path for the flow of electrons. Series and parallel connections each have their own distinctive characteristics.

Series Circuits

A simple **series circuit** is shown in Figure 8.25, where three lamps are connected in series with a battery. When the switch is thrown shut, the same current exists almost immediately in all three lamps. The current does not "pile up" in any lamp but flows *through* each lamp. Electrons that make up this current leave the negative terminal of the battery, pass through each resistive filament of the lamps in turn, and then return to the positive terminal of the battery (the electrons then pass through the battery and out the negative terminal again, which means that the amount of current passing through the battery is the same as the amount passing through the lamps). This is the only path of the electrons through the circuit. A break anywhere in the path results in an open circuit, and the flow of electrons ceases. Burning out of one of the lamp filaments or simply opening the switch causes such a break. The circuit shown in Figure 8.25 illustrates the following important characteristics of series connections:

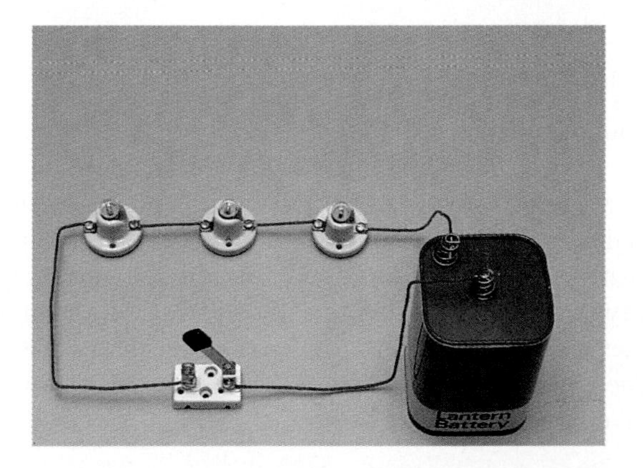

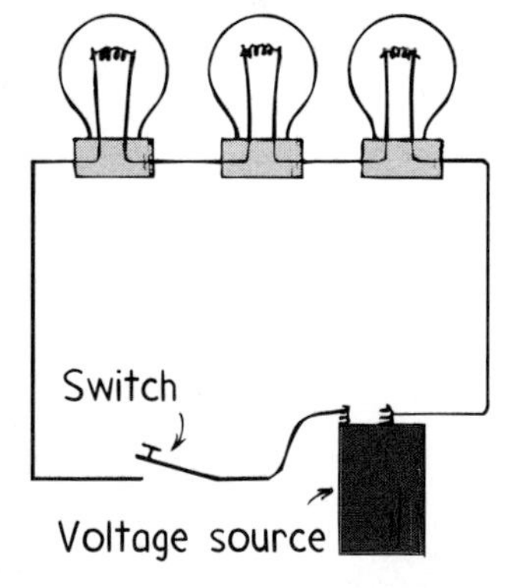

Fig. 8.25
A simple series circuit. The 6-V battery provides 2 V across each lamp.

1. Electric current has but a single pathway through a series circuit. This means that the current is the same through each electrical device in the circuit.
2. This current is resisted by the resistance of the first device, the resistance of the second, and that of the third also, and so the total resistance to current in the circuit is the sum of the individual resistances along the circuit path (assuming the resistance of the connecting wires is negligible).

3. The current in the circuit is numerically equal to the voltage supplied by the source divided by the total resistance of the circuit. This is in accord with Ohm's law.
4. The total voltage impressed across a series circuit divides among the electrical devices in the circuit so that the sum of the "voltage drops" across each device is equal to the total voltage supplied by the source. This follows from the fact that the amount of energy given to the total current is equal to the sum of energies given to each device.
5. The voltage drop across each device is proportional to its resistance. This follows from the fact that more energy is wasted as heat when a current passes through a high-resistance device than when the same current passes through a device offering little resistance.

Question

1. What happens to current in other lamps if one lamp in a series circuit burns out?
2. What happens to the light intensity of each lamp in a series circuit when more lamps are added to the circuit?

It's easy to see the main disadvantage of a series circuit: If one device fails, current in the whole circuit ceases. Some cheap Christmas tree lights are connected in series. When one bulb burns out, it's fun and games (or frustration) trying to find which one to replace.

Most circuits are wired so that it is possible to operate several electrical devices, each independently of the others. In your home, for example, a lamp can be turned on or off without affecting the operation of other lamps or electrical devices on the same circuit. This is because these devices are connected not in series, but in parallel.

Parallel Circuits

A simple **parallel circuit** is shown in Figure 8.26. Three lamps are connected to the same two points A and B. Electrical devices connected to the same two points of an electrical circuit are said to be *connected in parallel.* Electrons leaving the negative terminal of the battery need travel through only *one* lamp filament before returning to the positive terminal of the battery. In this case current branches into three separate pathways from A to B. A break in any one path does not interrupt the flow of charge in the other paths. Each device operates independently of the other devices.

The circuit shown in Figure 8.26 illustrates the following major characteristics of parallel connections:

Answers

1. If one lamp filament burns out, the path connecting the terminals of the voltage source breaks and current ceases. All lamps go out.
2. The addition of more lamps in a series circuit results in a greater circuit resistance. This decreases the current in the circuit and therefore in each lamp, which causes the lamps to dim. The same amount of energy is divided among more lamps, which means less energy per lamp—less voltage drop across each lamp.

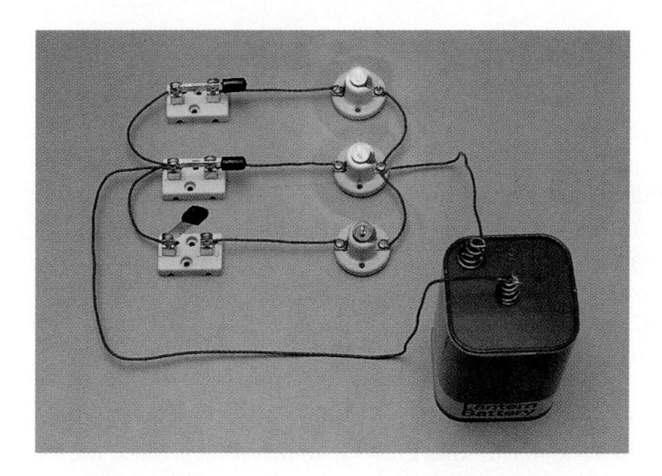

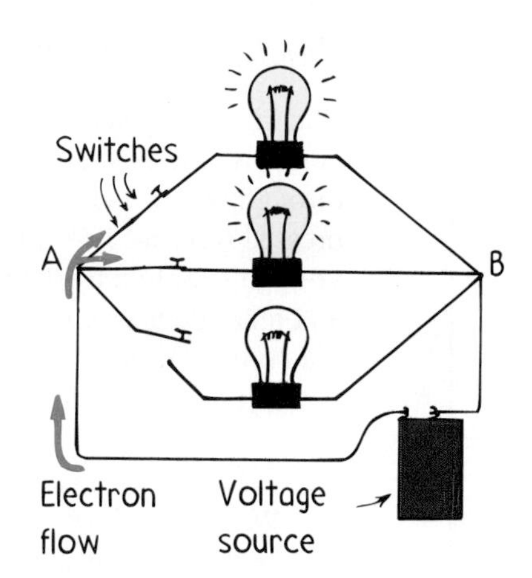

Fig. 8.26
A simple parallel circuit. A 6-V battery provides 6 V across each lamp.

1. Each device connects the same two points A and B of the circuit. The voltage is therefore the same across each device.
2. The total current in the circuit divides among the parallel branches. Because the voltage across each branch is the same, the amount of current in each branch is inversely proportional to the resistance of the branch.
3. The total current in the circuit equals the sum of the currents in its parallel branches.
4. As the number of parallel branches is increased, the overall resistance of the circuit is decreased (just as more check-out cashiers at a supermarket decreases people-flow resistance). With each added parallel path, the overall circuit resistance is lowered. This means the overall resistance of the circuit is less than the resistance of any one of the branches.

Questions

1. What happens to the current in other lamps if one lamp in a parallel circuit burns out?
2. What happens to the light intensity of each lamp in a parallel circuit when more lamps are added?

Answers

1. If one lamp burns out, the others are unaffected. The current in each branch, according to Ohm's law, is equal to voltage/resistance, and because neither voltage nor resistance is affected in the other branches, the current in those branches is unaffected. The total current in the overall circuit (the current through the battery), however, is lowered by an amount equal to the current drawn by the lamp in question before it burned out. But the current in any other single branch is unchanged.
2. The light intensity of each lamp is unchanged as other lamps are introduced (or removed). Only the total resistance and total current in the circuit change, which is to say the current through the battery changes. (We neglect any resistance in the wires and in the battery, which is normally low compared with the resistance of the lamps.) As lamps are introduced, more paths are available between the battery terminals, which effectively decreases total circuit resistance. This decreased resistance is accompanied by an increased current, the same increase that feeds energy to the lamps as they are introduced. Although changes of resistance and current occur for the circuit as a whole, no changes occur in any individual branch in the circuit.

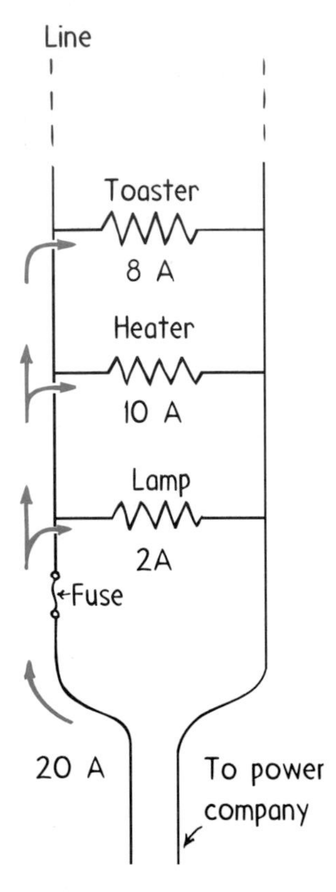

Fig. 8.27
Circuit diagram for appliances connected to a household circuit.

Parallel Circuits and Overloading

Electricity is usually fed into a home by way of two wires called *lines.* These lines are very low in resistance and are connected to wall outlets in each room—sometimes through two or more separate circuits. About 110–120 volts of electric potential is impressed on these lines by a transformer in the neighborhood that steps down the higher voltage supplied by the power utility. This voltage is then applied across appliances and other devices connected in parallel by plugs to the house circuit. As more devices are connected to a circuit, more pathways for current result in lowering of the combined resistance of the circuit. Therefore a greater amount of current exists in the circuit, and this can be a problem. Circuits that carry more than a safe amount of current are said to be *overloaded.*

We can see how overloading occurs by considering the circuit in Figure 8.27. The supply line is connected to an electric toaster that draws 8 amperes, an electric heater that draws 10 amperes, and an electric lamp that draws 2 amperes. When only the toaster is operating and drawing 8 amperes, the total line current is 8 amperes. When the heater is also operating, the total line current increases to 18 amperes (8 amperes to the toaster and 10 amperes to the heater). If you turn on the lamp, the line current increases to 20 amperes. Connecting any more devices increases the current still more. Connecting too many devices into the same circuit results in wire overheating that may cause a fire.

To prevent overloading in circuits, fuses are connected in series along the supply line. In this way the entire line current must pass through the fuse. The fuse shown in Figure 8.28 is constructed with a wire ribbon that heats up and melts at a given current. If the fuse is rated at 20 amperes, it will pass 20 amperes but no more. A current above 20 amperes melts the fuse, which "blows out" and breaks the circuit. Before a blown fuse is replaced, the cause of overloading should be determined and remedied. Often, insulation that separates the wires in a circuit wears away and allows the wires to touch each other. This greatly reduces the resistance in the circuit, which causes the current to overload, and is called a *short circuit.*

In modern buildings, fuses have been largely replaced by circuit breakers, which use magnets or bimetallic strips to open a switch when the current is too great. Utility companies use circuit breakers to protect their lines all the way back to the generators.

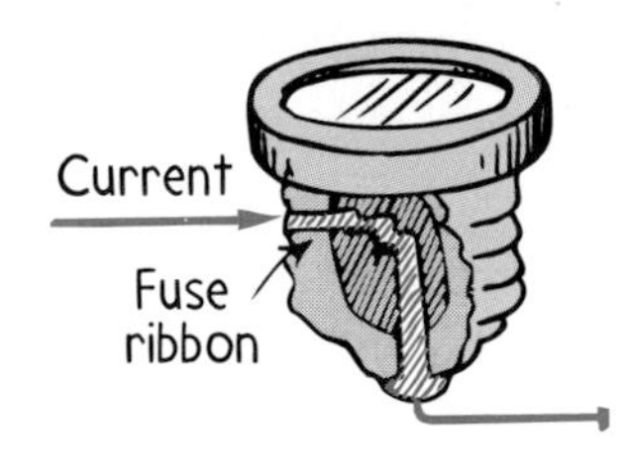

Fig. 8.28
A safety fuse.

8.8 Electric Power

The moving charges in an electric current do work. This work, for example, can heat a circuit or turn a motor. The rate at which work is done—that is, the rate at which electric energy is converted into another form, such as mechanical energy, heat, or light—is called **electric power**. Electric power is equal to the product of current and voltage.[10]

---

[10]Recall from Chapter 3 that power = work/time. Because work and energy are both measured in the same unit (joules), we can also say power = energy/time = (voltage × charge)/time = voltage × charge/time = voltage × current. Note that the units for mechanical power and electrical power check:

$$\textbf{Power} = \frac{\cancel{\textbf{charge}}}{\textbf{time}} \times \frac{\textbf{energy}}{\cancel{\textbf{charge}}} = \frac{\textbf{energy}}{\textbf{time}}$$

Fig. 8.29
The power and voltage on the light bulb reads "100 W 120 V." How many amperes flow through the bulb?

$$\textbf{Power} = \textbf{current} \times \textbf{voltage}$$

If the voltage is expressed in volts and the current in amperes, then the power is expressed in watts. So, in units form,

$$\textbf{Watts} = \textbf{amperes} \times \textbf{volts}$$

If a lamp rated at 120 watts operates on a 120-volt line, you can see that it draws a current of 1 ampere (120 watts = 1 ampere × 120 volts). A 60-watt lamp draws $\frac{1}{2}$ ampere on a 120-volt line. This relationship becomes a practical matter when you wish to know the cost of electrical energy, which is usually a small fraction of a dollar per kilowatt-hour, depending on locality. A kilowatt is 1000 watts, and a kilowatt-hour represents the amount of energy consumed in 1 hour at the rate of 1 kilowatt.[11] Therefore, in a locality where electrical energy costs 25 cents per kilowatt-hour, a 100-watt electric light bulb can be run for 10 hours at a cost of 25 cents, or a half nickel for each hour. A toaster or iron, which draws much more current and therefore much more energy, costs about ten times as much to operate.

So we've learned some electricity—beginning with the knowledge that protons and electrons of atoms are the sources of electric charge. We've learned that charges attract or repel one another—likes repel, unlikes attract. We've seen that electrons in the wires of a circuit flow away from the negative terminal of a battery and toward the positive terminal to produce electric current. And what have we done with electric current? Much more than was done with currents of water that powered industry in previous centuries. Understanding and control of electric currents has transformed the world, where almost everything about us now depends on the flow of electrons—from microchips in electrical devices to the motors of industry. How long will it be before electric automobiles are commonplace?

Questions

1. If a 120-V line to a socket is limited to 15 A by a safety fuse, will it operate a 1200-W hair dryer?

2. At 30¢/kWh, what does it cost to operate the 1200-W hair dryer for 1 h?

Answers

1. From the expression watts = amperes × volts, we see that the current drawn by the hair dryer is 1200 W/120 V = 10 A, and so the hair dryer will operate when connected to the circuit. But because two hair dryers on the same circuit will draw a total of 20 A, the fuse will blow.
2. 36¢ (1200 W = 1.2 kW; 1.2 kW × 1 h × 30¢/1 kWh = 36¢).

---

[11]Power = energy/time (Section 3.9), and simple rearrangement gives energy = power × time; thus energy can be expressed in the unit *kilowatt-hours* (kWh).

Summary of Terms

- **Ion** A charged atom—one with an excess or deficiency of electrons.
- **Coulomb's law** For any two electrically charged bodies, the relationship among the electric force the bodies exert on each other, the charge on the two bodies, and the distance between them:

$$F = k\frac{q_1 q_2}{d^2}$$

  If the charges are alike in sign, the force is repelling; if the charges are unlike, the force is attractive.
- **Coulomb** The SI unit of electrical charge. One coulomb (symbol C) is equal in magnitude to the total charge of 6.25×10^{18} electrons.
- **Electrically polarized** Term applied to an atom or molecule in which the charges are aligned so that one side has a slight excess of positive charge and the other side a slight excess of negative charge.
- **Electric field** Defined as force per unit charge, it can be considered an "aura" surrounding charged objects and is a storehouse of electric energy. About a charged point, the strength of the electric field decreases with distance according to the inverse-square law.
- **Electrical potential energy** The energy a charge possesses by virtue of its location in an electric field.
- **Electric potential** The electric potential energy per amount of charge, measured in volts:

$$\text{Electric potential} = \frac{\text{electric potential energy}}{\text{amount of charge}}$$

- **Volt** The unit of electric potential, a potential of 1 volt equals 1 joule of energy per 1 coulomb of charge; 1 V = 1 J/C.
- **Potential difference** (synonymous with *voltage difference*) The difference in electric potential between two points, measured in volts. Electrical potential difference can be compared with the difference in water pressure between two containers of water: When connected by a pipe, water flows from the container having the higher pressure to the one having the lower pressure—until the two pressures are equal. Similarly, when two points having different electric potential are connected by a conductor, charge flows from the point having the greater potential to the one having the lesser potential so long as a potential difference exists.
- **Electric current** The flow of electric charge that transports energy from one place to another. Measured in amperes.
- **Ampere** The unit of electric current; 1 ampere = 1 coulomb per second (the flow of 6.25×10^{18} electrons per second); 1 A = 1 C/s.
- **Direct current (dc)** Electrically charged particles flowing in one direction only.
- **Alternating current (ac)** Electrically charged particles that repeatedly reverse direction, vibrating about relatively fixed positions. In the United States the vibrational rate is 60 Hz.
- **Electrical resistance** The property of a material that resists the flow of charged particles through it; measured in ohms (Ω).
- **Ohm's law** The current in a circuit varies in direct proportion to the voltage across the circuit and inversely with the circuit's resistance:

$$\text{Current} = \frac{\text{voltage}}{\text{resistance}}$$

  A potential difference of 1 V across a resistance of 1 Ω produces a current of 1 A.
- **Series circuit** An electric circuit in which electrical devices are connected in such a way that the same electric current exists in all of them.
- **Parallel circuit** An electric circuit in which electrical devices are connected in such a way that the same voltage acts across each one and any single one completes the circuit independently of all the others.
- **Electric power** The rate of energy transfer, or the rate of doing work; the amount of energy transferred per unit time, which electrically can be measured by

$$\text{Power} = \text{current} \times \text{voltage}$$

  Measured in watts (or kilowatts), where 1 A × 1 V = 1 W.

Review Questions

Electric Force and Charge

1. Which is stronger, the electric force between an electron and a proton or the gravitational force between these particles? Is the difference between these forces large or small?
2. Why does gravitational force predominate over electric force for astronomical bodies?
3. Which part of an atom is *positively* charged and which part is *negatively* charged?
4. How does the charge of one electron compare with that of another electron?
5. How do the masses of electrons compare with the masses of protons? Neutrons?
6. How does the number of protons in the atomic nucleus normally compare with the number of electrons that orbit the nucleus?
7. What kind of charge does an object acquire when electrons are stripped from it?
8. What is meant by saying charge is *conserved*?
9. How does a *coulomb* of charge compare with the charge of a *single* electron?

10. How is Coulomb's law similar to Newton's law of gravitation? How is it different?
11. How does the magnitude of electric force between a pair of charged particles change when the particles are moved twice as far apart? Three times as far apart?
12. How does an electrically *polarized* object differ from an electrically *charged* object?

Electric Fields

13. Give two examples of common force fields.
14. How is the direction of an electric field defined?

Electric Potential

15. Distinguish between *electric potential energy* and *electric potential.*
16. A balloon may easily be charged to several thousand volts. Does that mean it has several thousand joules of energy? Explain.
17. Why does Meidor's hair stand out in Figure 8.10?
18. What condition is necessary for heat energy to flow from one end of a metal bar to the other? For electric charge to flow?
19. For electric circuits, what is analogous to the statement "water seeks its own level"?
20. What condition is necessary for a sustained flow of electric charge through a conducting medium?

Voltage Sources

21. How much energy is given to each coulomb of charge passing through a 6-V battery?
22. Does electric charge flow *across* a circuit or *through* a circuit? Does voltage *flow* across a circuit or is it *impressed* across a circuit? Explain.
23. Does voltage produce current or does current produce voltage? Which is the cause and which is the effect?

Electric Current

24. Why do electrons rather than protons make up the flow of charge in a metal wire?
25. Distinguish between *dc* and *ac* .
26. Does a battery produce dc or ac? Does the generator at a power station produce dc or ac?

Electrical Resistance

27. Which has the greater resistance, a thick wire or a thin wire of the same length?
28. What is the effect on current through a circuit of steady resistance when the voltage is doubled? What if both voltage and resistance are doubled?
29. How much current flows through a radio speaker that has a resistance of 8 Ω when 12 V is impressed across the speaker?
30. Which has the greater electrical resistance, wet skin or dry skin?
31. High voltage by itself does not produce electric shock. What does?
32. What is the function of the third prong on the plug of an electric appliance?
33. What is the source of the electrons that produce an electric shock when you touch a charged conductor?

Electric Circuits

34. What is an electric circuit, and what is the role of a gap in one?
35. In a circuit consisting of two lamps connected in series, if the current through one lamp is 1 A, what is the current through the other lamp?
36. If 6 V is impressed across the above circuit and the voltage across the first lamp is 2 V, what is the voltage across the second lamp?
37. In a circuit consisting of two lamps connected in parallel, if there is 6 V across one lamp, what is the voltage across the other lamp?
38. If the current through each of the two branches of a parallel circuit is the same, what does this tell you about the resistance of the two branches?
39. How does the total current through the branches of a parallel circuit compare with the current through the voltage source?
40. As more lines are opened at a fast-food restaurant, the resistance to the motion of people trying to get served is reduced. How is this similar to what happens when more branches are added to a parallel circuit?
41. Are household circuits normally wired in series or in parallel?
42. How does the amount of current in a home circuit differ from the amount of current in a reading lamp?
43. Why will too many electrical devices operating at one time often blow a fuse?

Electric Power

44. What is the relationship among electric power, current, and voltage?
45. Which draws more current, a 40-W bulb or a 100-W bulb?

Home Projects

1. Demonstrate charging by friction and discharging from points with a friend who stands at the far end of a carpeted room. Scuff your shoes across the rug until your noses are close together. This can be a delightfully tingling experience, depending on how dry the air is (and how pointed your noses are).

2. Briskly rub a comb on your hair or a woolen garment and bring it near a small but smooth stream of running water. Is the steam of water charged? (Before you say yes, note the behavior of the stream when an opposite charge is brought nearby.)

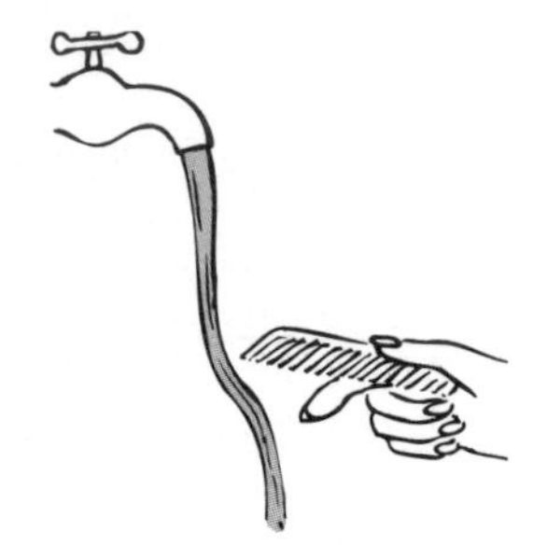

3. An electric cell is made by placing two plates made of different materials that have different affinities for electrons in a conducting solution. (A battery is actually a series of cells.) You can make a simple 1.5-V cell by placing a strip of copper and a strip of zinc in a tumbler of salt water. The voltage of a cell depends on the materials used and the solution they are placed in, not on the size of the plates.

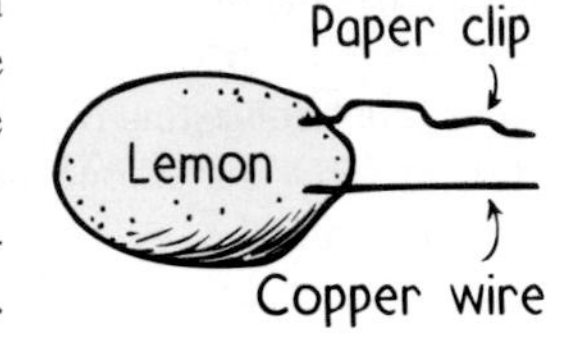

An easy cell to construct is the citrus cell. Stick a paper clip and a piece of copper wire into a lemon. Hold the ends of the wire close together, but not touching, and place the ends on your tongue. The slight tingle you feel and the metallic taste you experience result from a slight current of electricity pushed by the citrus cell through the wires when your moist tongue closes the circuit.

4. Examine the electric meter in your house. It is probably in the basement or on the outside of the house. You will see that in addition to the clocklike dials in the meter, there is a circular aluminum disk that spins between the poles of magnets when electric current goes into the house. The more electric current, the faster the disk turns. The speed of the disk is directly proportional to the number of watts used; for example, it spins five times as fast for 500 W as for 100 W.

You can use the meter to determine how many watts an electrical device uses. First, see that all electrical devices in your home are disconnected (you may leave electric clocks connected, for the 2 watts they use will hardly be noticeable). The disk will be practically stationary. Then connect a 100-W bulb and note how many seconds it takes for the disk to make five complete revolutions. The black spot painted on the edge of the disk makes this easy. Disconnect the 100-W bulb and plug in a device of unknown wattage. Again, count the seconds for five revolutions. If it takes the same time, it's a 100-W device; if it takes twice the time, it's a 50-W device; half the time, a 200-W device; and so forth. In this way you can estimate the power consumption of devices fairly accurately.

Exercises

1. Why do clothes often cling together after tumbling in a clothes dryer?
2. When one material is rubbed against another, electrons jump readily from one to the other but protons do not. Why is this? (Think in atomic terms.)
3. In a crystal of salt, there are electrons and positive ions. How does the net charge of the electrons compare with the net charge of the ions? Explain.
4. The 5 thousand billion billion freely moving electrons in a penny repel one another. Why don't they fly out of the penny?
5. If electrons were positive and protons negative, would Coulomb's law be written the same or differently?
6. By what factor does the magnitude of the electric force between a pair of charged particles change when they are brought to half their original distance of separation? To one-quarter their original distance? To four times their original distance? (What law guides your answers?)
7. Two charged pellets are pulled apart to twice their original separation. (a) Which is likely to be larger, the gravitational or the electric force between them? (b) Which changes more when they are pulled apart, the gravitational or the electric force between them?
8. Two equal charges exert equal forces on each other. What if one charge has twice the magnitude of the other? How do the forces they exert on each other compare?
9. An electroscope is a simple device consisting of a metal ball attached by a conductor to two fine gold leaves that are protected from air disturbances in a jar, as shown. When the ball is touched by a charged body, the leaves, which normally hang straight down, spread apart. Why? (Electroscopes are useful not only as charge detectors, but also for measuring the quantity of charge: the more charge transferred to the ball, the more the leaves diverge.)

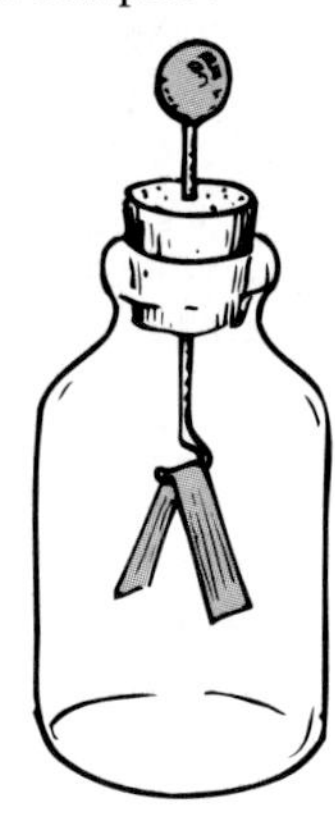

10. The leaves of a charged electroscope collapse in time. At higher altitudes they collapse more readily. Why? (*Hint:* The existence of cosmic rays was first indicated by this observation.)
11. It is relatively easy to strip the outer electrons from a heavy atomic nucleus like that of uranium (which is then a uranium ion) but very difficult to strip the inner electrons. Why do you suppose this is so?

12. Why is a good conductor of electricity also a good conductor of heat?
13. Why are materials such as glass and rubber good insulators?
14. A piece of paper becomes polarized in the presence of, say, a negative charge. The positive side of the paper is attracted to the negative charge, and the negative side of the paper is repelled by the negative charge. How does the thickness of the paper play a role in these forces not canceling out?
15. Measurements show that there is an electric field surrounding the Earth. Its magnitude is about 100 N/C at the Earth's surface and points inward toward the Earth's center. From this information, can you state whether the Earth carries a net negative charge or a net positive charge?
16. Suppose that the strength of the electric field about an isolated charge has a certain value at a distance of 1 m. By what factor does the field strength change at a distance of 2 m from the charge? What law guides your answer?
17. If you put in 10 J of work to push a charged pellet against an electric field, is the pellet's electric potential greater or less than what it was at its original position? When the pellet is released, what is its kinetic energy as it flies past its starting position?
18. If you rub an inflated balloon against your hair and then bring it into the vicinity of uncharged bits of paper, the bits of paper are attracted to the balloon. Then upon touching the balloon, they are repelled. Explain.
19. Why are metal-spiked shoes not a good idea for golfers on a stormy day?
20. You are not harmed by contact with a charged balloon, even though its voltage may be very high. Is the reason similar to why you are not harmed by the greater-than-1000°C sparks from a 4th of July–type sparkler? Defend your answer by comparing electric potential to temperature.
21. A friend says that the reason one's hair stands out while touching a charged van de Graaff generator is simply that the hair strands become charged and are light enough so that the repulsion between strands is visible. Do you agree or disagree?
22. After a comb is rubbed through your hair, the comb is negatively charged and your hair is positively charged. When your hair stands out when you touch a negatively charged Van de Graaff generator, is your hair positively charged? Explain.
23. One example of a water system is a garden hose that waters a garden. Another is the cooling system of an automobile. Which of these examples exhibits behavior more analogous to that of an electric circuit? Why?
24. What happens to the brightness of light emitted by a light bulb when the current through it increases?
25. Your tutor tells you that an *ampere* and a *volt* really measure the same thing and that the different terms only serve to make a simple concept seem confusing. Why should you consider getting a different tutor?
26. A simple lie detector consists of an electric circuit, one part of which is part of your body—such as between your fingers. A sensitive meter shows the current that exists when a small voltage is applied. How does this technique indicate a person is lying? (And when does this technique not tell when someone is lying?)
27. Only a small percentage of the electric energy fed into a common light bulb is transformed into light. What happens to the rest?
28. Why are thick wires rather than thin wires usually used to carry large currents?
29. A copper wire 1 mile long has a resistance of 10 Ω. What is its resistance when it is shortened (a) by cutting it in half, and (b) by doubling it over and using it as "one" wire?
30. Is the current in a light bulb connected to a 220-V source greater or less than when the same bulb is connected to a 110-V source?
31. Which does less damage—plugging a 110-V appliance into a 220-V circuit or plugging a 220-V appliance into a 110-V circuit? Explain.
32. Would you expect to find dc or ac in the filament of a light bulb in your home? How about in an automobile?
33. Are automobile headlights wired in parallel or in series? What is your evidence?
34. A car's headlights consume 40 W on low beam and 50 W on high beam. Is there more or less resistance in the high-beam filament?
35. Is the resistance of a 110-W bulb greater or less than the resistance of a 60-W bulb? Assuming the filaments in the two bulbs are of the same length, which bulb has the thicker filament?
36. The wattage marked on a light bulb is not an inherent property of the bulb but depends on the voltage to which it is connected, usually 110 or 120 V. How many amperes flow through a 60-W bulb connected in a 120-V circuit?
37. The damaging effects of electric shock result from the amount of current through the body. Why, then, do we see signs that read "Danger—High Voltage" rather than "Danger—High Current"?
38. If electrons flow very slowly through a circuit, why does a lamp glow quickly when you turn on a distant switch?
39. If a glowing light bulb is jarred and oxygen leaks inside, the bulb momentarily brightens considerably before burning out. Putting excess current through a light bulb will also burn it out. What physical change occurs when a light bulb burns out?
40. Consider a pair of flashlight bulbs connected to a battery. Will they each glow brighter connected in series or in parallel? Will the battery run down faster if they are connected in series or in parallel?

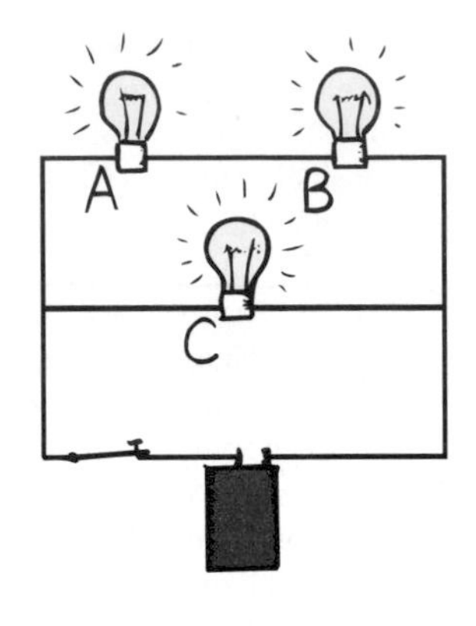

41. In the circuit shown, how do the brightnesses of the identical light bulbs compare? Which light bulb draws the most current? What happens if bulb A is unscrewed? If C is unscrewed?

42. As more and more bulbs are connected in series to a flashlight battery, what happens to the brightness of each bulb? Assuming heating inside the battery is negligible, what happens to the brightness of each bulb when more and more bulbs are connected in parallel?

43. What current changes occur when more devices are introduced in a series circuit? In a parallel circuit? Why are your answers different from each other?

44. It so happens that if too great a load is placed on a battery, the internal resistance of the battery is increased. This lowers the voltage supplied to the external circuit. If too many lamps are connected in parallel across a battery, does their brightness diminish? Explain.

45. Why are devices in household circuits almost never connected in series?

46. If a 60-W bulb and a 100-W bulb are connected in series in a circuit, across which bulb is there the greater voltage drop? How about if they are connected in parallel?

47. A friend says that power companies sell energy, not power. Do you agree or disagree?

48. Make up two multiple-choice questions that distinguish between (a) voltage and current, and (b) series and parallel circuits.

Problems

1. Two charged particles are separated by 6 cm. The attractive force between them is 20 N. Find the force between them when they are separated by 12 cm. (Why can you solve this problem without knowing the magnitudes of the charges?)

2. If the charges attracting each other in the preceding problem are of equal magnitude, what is that magnitude?

3. Two pellets, each carrying a charge of 1 microcoulomb (10^{-6} C), are located 3 cm (0.03 m) apart. What is the electric force between them? What mass would an object have to have to experience this same force in the Earth's gravitational field?

4. Electronic types neglect the gravitational force exerted by the Earth on electrons. To see why, compute the force of the Earth's gravity on an electron and compare it with the force exerted on the electron by an electric field of magnitude 10,000 V/m (a relatively small field). The mass and charge of an electron are given on the inside back cover.

5. Atomic physicists ignore the effect of gravity within an atom. To see why, calculate and compare the gravitational and electric forces between an electron and a proton separated by 10^{-10} m. The needed charges and masses are given on the inside back cover.

6. In 1909, Robert Millikan was the first to measure the charge of an electron in his now-famous oil drop experiment. In the experiment tiny oil drops are sprayed into a uniform electric field that exists between a horizontal pair of oppositely charged plates. The drops are observed with a magnifying eyepiece, and the electric field is adjusted so that the upward force on some negatively charged oil drops is just sufficient to balance the downward force of gravity. That is, when the drop is properly suspended, the upward force qE on it just equals the downward force mg. Millikan accurately measured the charges on many oil drops and found the values to be whole-number multiples of 1.6×10^{-19}C—the charge of the electron. For this he won the Nobel prize.
(a) If a drop of mass 1.1×10^{-14} kg remains stationary in an electric field of 1.68×10^5 N/C, what is the charge of this drop?
(b) How many extra electrons are on this particular oil drop (given the presently known charge of the electron)?

7. Rearrange the equation current = voltage/resistance to express *resistance* in terms of current and voltage. Then solve the following: A certain device in a 120-V circuit has a current rating of 20 A. What is its resistance (how many ohms)?

8. Using the formula power = current × voltage, find the current drawn by a 1200-W hair dryer connected to 120 V. Then, using the method you used in the previous exercise, find the resistance of the hair dryer.

9. The useful life an automobile battery has without being recharged is given in terms of ampere-hours. A typical 12-V battery has a rating of 60 ampere-hours, which means that a current of 60 A can be drawn for 1 h, 30 A can be drawn for 2 h, and so forth. Suppose you forget to turn off the headlights in your parked automobile. If each of the two headlights draws 3 A, how long before your battery is "dead"?

10. A certain light bulb that has a resistance of 95 W is labeled "150 W." Was this bulb designed for use in a 120-V circuit or a 220-V circuit?

11. How much does it cost to operate a 100-W lamp continuously for 1 month if the power utility rate is 20¢/kWh?

12. A sample of water absorbs 24.1 kJ of energy when placed for 1 min in a 500-W microwave oven. What percentage of the oven's energy output is this? In other words, how efficiently does the microwave oven warm water?

CHAPTER

9

Magnetism

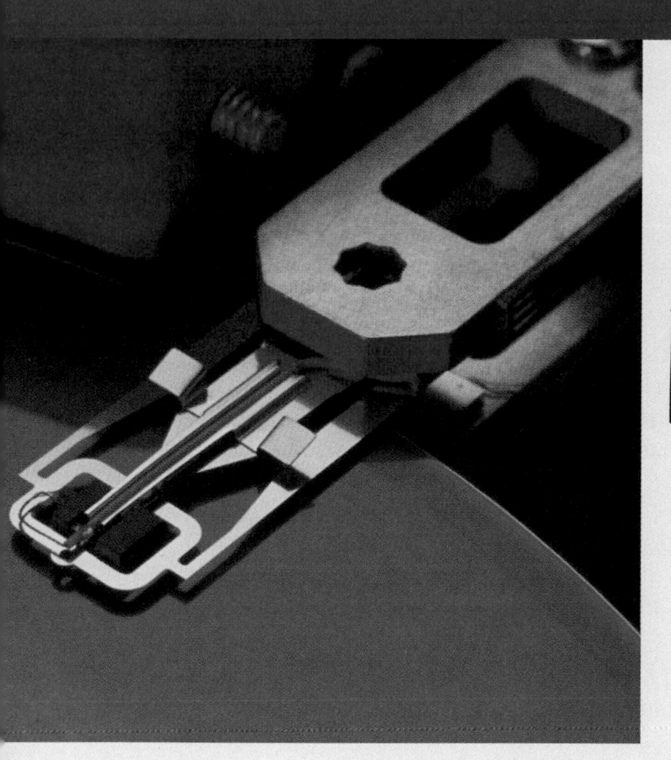

The term *magnetism* comes from the region of Magnesia, a province of Greece, where certain stones were found by the Greeks more than 2000 years ago. These stones, called *lodestones,* had the unusual property of attracting pieces of iron. Magnets were first fashioned into compasses and used for navigation by the Chinese in the twelfth century.

In the sixteenth century, William Gilbert, Queen Elizabeth's physician, made artificial magnets by rubbing pieces of iron against lodestone, and suggested that a compass always points north and south because the Earth has magnetic properties. Later, in 1750, John Michell in England found that magnetic poles obey the inverse-square law, and his results were confirmed by Charles Coulomb. The subjects of magnetism and electricity developed almost independently of each other until 1820, when a Danish physicist named Hans Christian Oersted discovered in a classroom demonstration that an electric current affects a magnetic compass.[1] He saw that magnetism was related to electricity. Shortly thereafter, the French physicist André-Marie Ampere proposed that electric currents are the source of all magnetic phenomena.

9.1 Magnetic Poles

Fig. 9.1
A horseshoe magnet.

Anyone who has played around with magnets knows they exert forces on one another. **Magnetic forces** are similar to electric forces, for they can both attract and repel without touching, depending on which ends of two magnets are held near one another. And similar to what we find with electric forces, the strength of the magnetic interaction depends on the distance between the two magnets. Whereas electric charges produce electric forces, regions called *magnetic poles* give rise to magnetic forces.

Suspend a bar magnet at its center by a piece of string and you've got a compass. One end, called the *north-seeking pole,* points northward, and the opposite end, called the *south-seeking pole,* points southward. More simply, these are called the *north* and *south poles.* All magnets have both a north and a south pole. In a simple bar magnet these poles are located at the two ends. A common horseshoe magnet is simply a bar magnet bent into a U shape. Its poles are also at its two ends.

When the north pole of one magnet is brought near the north pole of another magnet, they repel each other.[2] The same is true of a south pole near a south pole. If opposite poles are brought together, however, attraction occurs. We find that

Like poles repel each other; opposite poles attract.

This rule is similar to the rule for the forces between electric charges, where like charges repel one another and unlike charges attract. But there is a very important difference between magnetic poles and electric charges. Whereas electric charges can be isolated, magnetic poles cannot. Electrons and protons are entities by themselves. A cluster of electrons need not be accompanied by a cluster of protons, and vice versa. But a north magnetic pole never exists without the presence of a south pole, and vice versa. The north and south poles of a magnet are like the head and tail of the same coin.

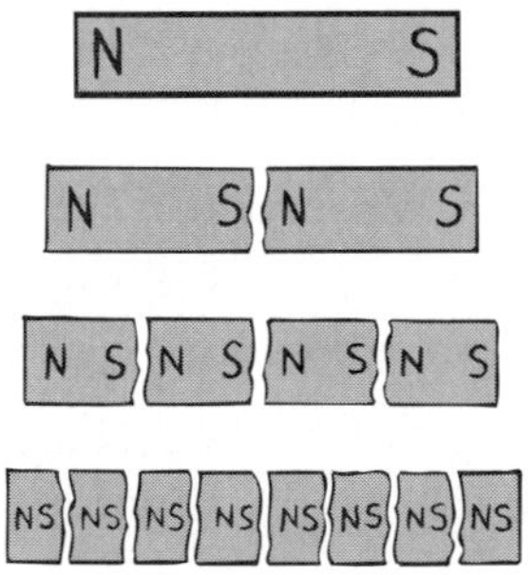

Fig. 9.2
Break a magnet in half and you have two magnets. Break these in half and you have four magnets, each with a north and south pole. Keep breaking the pieces further and further and you find the same results. Magnetic poles always exist in pairs.

If you break a bar magnet in half, each half still behaves as a complete magnet. Break the pieces in half again, and you have four complete magnets. You can continue breaking the pieces in half and never isolate a single pole. Even when your piece is one atom thick, there are two poles. This suggests that atoms themselves are magnets.

[1]We can only speculate about how often such relationships become evident when they "aren't supposed to" and are dismissed as "something wrong with the apparatus." Oersted, however, was keen enough to see that nature was revealing another of its secrets.

[2]The force of interaction between magnetic poles is given by $F \sim p_1 p_2/d^2$, where p_1 and p_2 represent magnetic pole strengths and d represents the separation distance between the poles. Note the similarity of this relationship to Coulomb's law.

■ Question

Does every magnet necessarily have a north and south pole?

9.2 Magnetic Fields

Sprinkle some iron filings on a sheet of paper placed on a magnet and you'll see that the filings trace out an orderly pattern of lines that surround the magnet. The space around the magnet contains a **magnetic field**. The shape of the field is revealed by the filings, which align with the magnetic field lines that spread out from one pole and re-

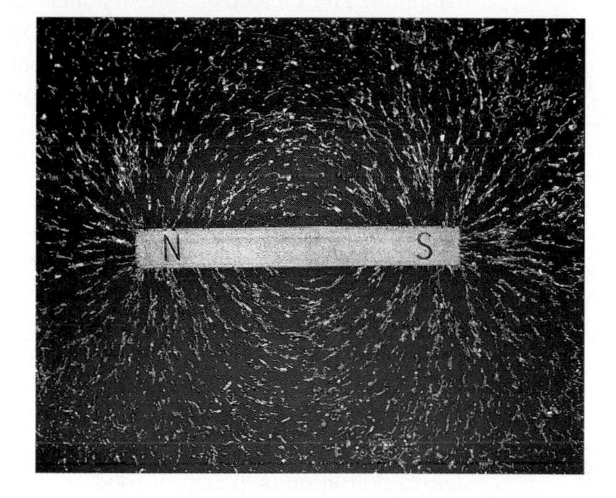

Fig. 9.3
Top view of iron filings sprinkled around a magnet. The filings trace out a pattern of *magnetic field lines* in the space surrounding the magnet. Interestingly, the magnetic field lines continue inside the magnet (not revealed by the filings) and form closed loops.

turn to the other. It is interesting to compare the field patterns in Figures 9.3 and 9.5 with the electric field patterns in Figure 8.9.

The direction of the field outside a magnet is from the north to the south pole. Where the lines are closer together, the field is stronger. The concentration of iron filings at the poles of the magnet in Figure 9.3 shows that the magnetic field strength is greater there. If we place another magnet or a small compass anywhere in the field, its poles line up with the magnetic field.

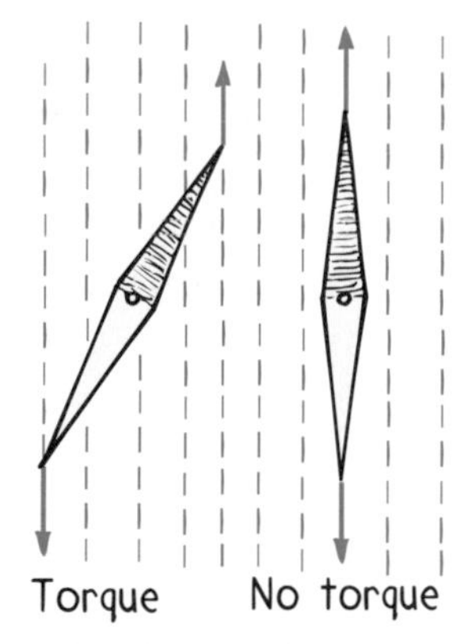

Fig. 9.4
When the compass needle is not aligned with the magnetic field (left), the oppositely directed forces on the needle produce a pair of torques (twisting forces) called a *couple* that twist the needle into alignment (right) .

Magnetism is very much related to electricity. Just as an electric charge is surrounded by an electric field, the same charge is also surrounded by a magnetic field if it is moving. This magnetic field is due to the "distortions" in the electric field caused by motion and was explained by Albert Einstein in 1905 in his special theory of relativity. We won't go into the details except to acknowledge that a magnetic field is a relativistic byproduct of the electric field. Charged particles in motion have associated with

---

■ **Answer**
Yes, just as every coin has two sides, a head and a tail. Some trick magnets may have more than one pair of poles, but nevertheless poles occur in pairs.

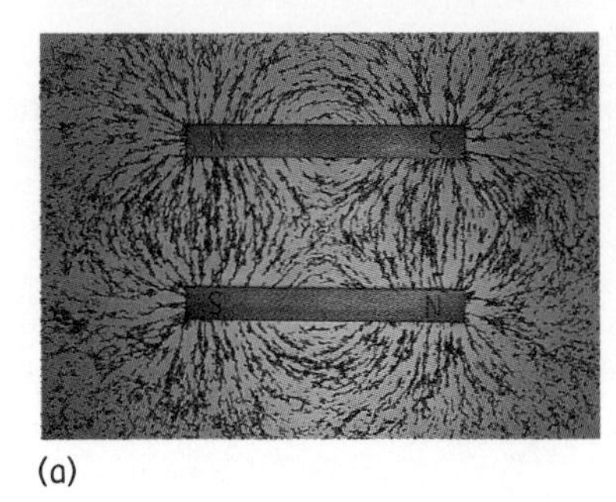

(a)

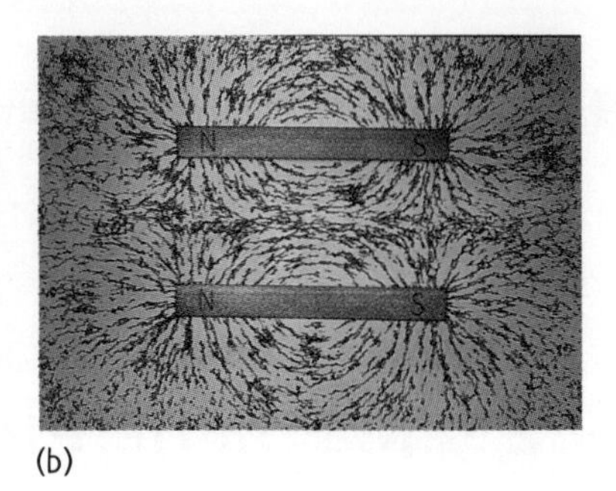

(b)

Fig. 9.5
The magnetic field patterns for a pair of magnets. (a) Opposite poles are nearest each other. (b) Like poles are nearest each other. Compare these fields with those of the electric charges in Figure 8.9a and b.

them both an electric and a magnetic field. A magnetic field is produced by the motion of electric charge.[3]

If the motion of electric charges produces magnetism, where is this motion in a common bar magnet? The answer is, in the electrons of the atoms that make up the magnet. These electrons are in constant motion. Two kinds of electron motion contribute to magnetism: electron spin and electron revolution. Electrons spin about their own axes like tops, and they revolve about the atomic nucleus. In most common magnets, electron spin is the chief contributor to magnetism.

Every spinning electron is a tiny magnet. A pair of electrons spinning in the same direction makes up a stronger magnet. A pair of electrons spinning in opposite directions, however, work against each other. The magnetic fields cancel. This is why most substances are not magnets. In most atoms the various fields cancel one another because the electrons spin in opposite directions. In materials such as iron, nickel, and cobalt, however, the fields do not cancel each other entirely. Each iron atom has four electrons whose spin magnetism is uncanceled. Each iron atom, then, is a tiny magnet. The same is true to a lesser extent for the atoms of nickel and cobalt. Most common magnets are therefore made from alloys containing iron, nickel, and cobalt in various proportions.

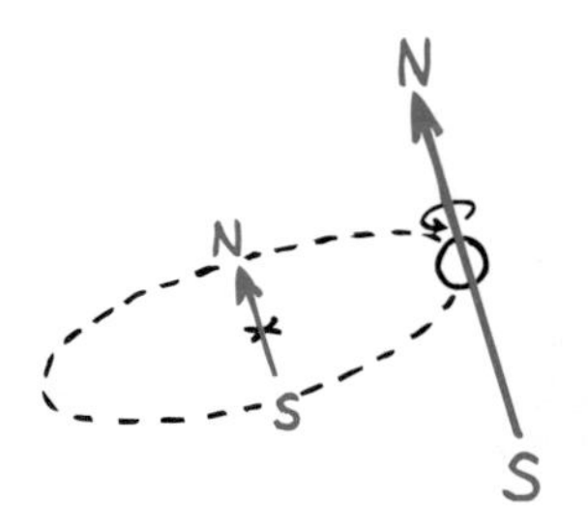

Fig. 9.6
Both the spinning motion and the revolving (orbital) motion of every electron in an atom produce magnetic fields. The field due to spin (large vector) combines with the field due to revolution (small vector) to produce the magnetic field of the atom. The resulting field is greatest for iron atoms.

Most iron objects around you are magnetized to some degree. A filing cabinet, a refrigerator, or even cans of food on your pantry shelf have north and south poles induced by the Earth's magnetic field. Pass a compass from their bottoms to their tops and their poles are easily identified. Turn cans upside down and see how many days it takes for the poles to reverse themselves!

9.3 Magnetic Domains

The magnetic field of an individual iron atom is so strong that interactions among adjacent atoms cause large clusters of them to line up with one another. These clusters of aligned atoms are called **magnetic domains**. Each domain is made up of billions of aligned atoms. The domains are microscopic (Figure 9.7), and there are many of them in a crystal of iron. Like the alignment of iron atoms within domains, domains themselves can align with one another.

[3]Because motion is relative, the magnetic field is relative. For example, when an electron moves by you, there is a definite magnetic field associated with it. But if you move along with the electron, so that there is no motion relative to you, you find no magnetic field associated with the electron. Magnetism is relativistic, as Albert Einstein explained when he published his first paper on special relativity, "On the Electrodynamics of Moving Bodies."

Fig. 9.7
A microscopic view of magnetic domains in a crystal of iron. Each domain consists of billions of aligned iron atoms. The blue arrows pointing in different directions tell us that these domains are not aligned.

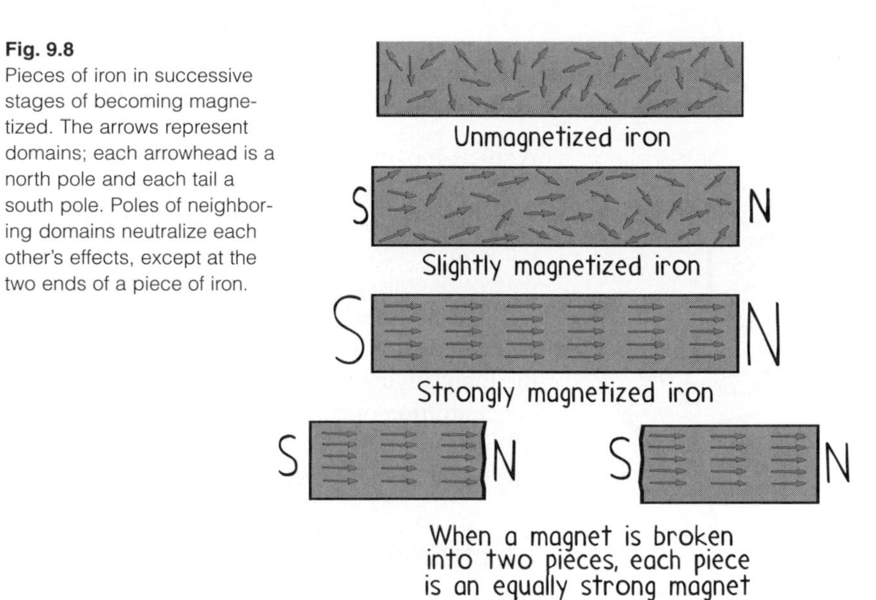

Fig. 9.8
Pieces of iron in successive stages of becoming magnetized. The arrows represent domains; each arrowhead is a north pole and each tail a south pole. Poles of neighboring domains neutralize each other's effects, except at the two ends of a piece of iron.

Not every piece of iron is a magnet because the domains in ordinary iron are not aligned. Consider a common iron nail: The domains in the nail are randomly oriented. They can be induced into alignment, however, when a magnet is brought nearby. (It is interesting to listen with an amplified stethoscope to the clickety-clack of domains undergoing alignment in a piece of iron when a strong magnet approaches.) The domains align themselves much as electrical charges in a piece of paper align themselves in the presence of a charged rod. When you remove the nail from the magnet, ordinary thermal motion causes most or all of the domains in the nail to return to a random arrangement.

Permanent magnets can be made by placing pieces of iron or other magnetic materials in strong magnetic fields. Alloys of iron differ in their ability to become magnetized; soft iron is easier to magnetize than steel. It helps to tap the material to nudge any stubborn domains into alignment. Another way is to stroke the material with a magnet. The stroking motion aligns the domains. If a permanent magnet is dropped or heated outside of the strong magnetic field from which it was made, some of the domains are jostled out of alignment and the magnet becomes weaker.

Fig. 9.9
The iron nails become induced magnets.7

Question

How can a magnet attract a piece of iron that is not magnetized?

Answer
Like the compass needle in Figure 9.4, domains in the unmagnetized piece of iron are induced into alignment by the magnetic field of the magnet. One domain pole is attracted to the magnet and the other domain pole is repelled. Does this mean the net force is zero? No, because the domain pole that is attracted is closer than the pole repelled. Closeness wins, and there is a net attraction. In this way a magnet attracts nonmagnetized pieces of iron (Figure 9.9). (This induction of pole alignment is similar to the way a charged comb attracts bits of uncharged paper.)

9.4 Electric Currents and Magnetic Fields

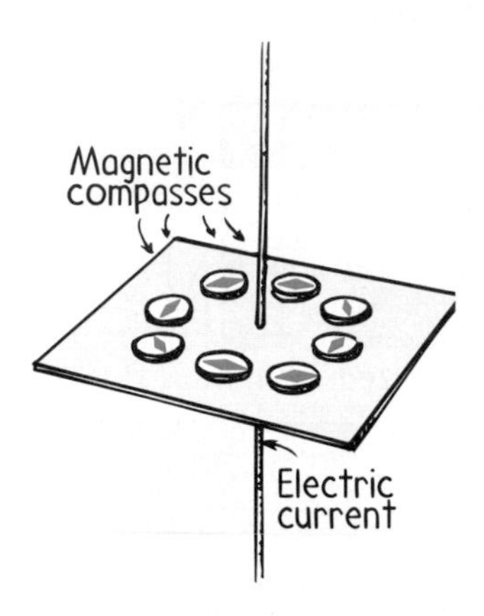

Fig. 9.10
The compasses show the circular shape of the magnetic field surrounding the current-carrying wire.

A single moving charged particle produces a magnetic field. A current of charged particles, then, also produces a magnetic field. The magnetic field that surrounds a current-carrying wire can be demonstrated by arranging an assortment of compasses around the wire (Figure 9.10). The magnetic field about the current-carrying wire makes up a pattern of concentric circles. When the current reverses direction, the compass needles turn around, showing that the direction of the magnetic field changes also.[4]

If the wire is bent into a loop, the magnetic field lines become bunched up inside the loop (Figure 9.11). If the wire is bent into another loop, overlapping the first, the concentration of magnetic field lines inside the loops is doubled. It follows that the magnetic field intensity in this region is increased as the number of loops is increased. The magnetic field intensity is appreciable for a current-carrying coil of many loops.

Electromagnets

If a piece of iron is placed in a current-carrying coil of wire, the magnetic domains in the iron are induced into alignment. This further increases the magnetic field intensity, and we have an **electromagnet**! The strength of an electromagnet is increased by simply increasing the current through the coil. Strong electromagnets are used to control charged-particle beams in high-energy accelerators.

Electromagnets powerful enough to lift automobiles are a common sight in junkyards. The strength of these electromagnets is limited by overheating of the current-carrying coils (due to electrical resistance) and saturation of magnetic domain alignment in the core. The most powerful electromagnets omit the iron core and use superconducting coils through which large electrical currents easily flow.

Superconducting Electromagnets

The relatively new ceramic superconductors (Chapter 8) have the interesting property of expelling magnetic fields. Because magnetic fields cannot penetrate the surface of a superconductor, magnets levitate above them. The reasons for this behavior are presently being researched and seem explainable only in terms of quantum mechanics. One of the hot applications of superconducting electromagnets is the levitation of high-speed trains for transportation. Prototype trains have already been demonstrated in the United States, Japan, and Germany. Watch for the growth of this relatively new technology.

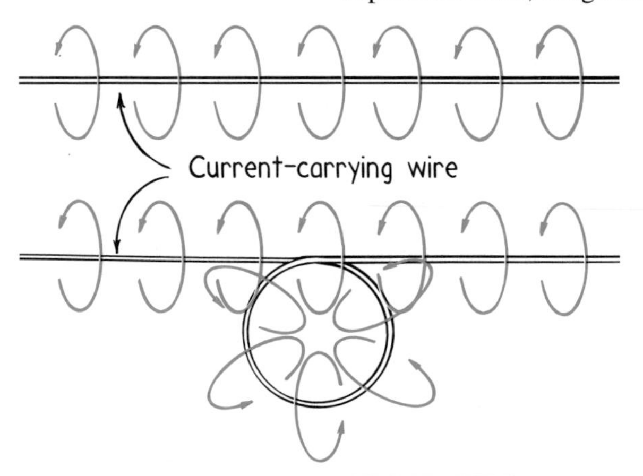

Fig. 9.11
Magnetic field lines about a current-carrying wire crowd up when the wire is bent into a loop.

[4]The Earth's magnetism is thought to be the result of electric currents that accompany thermal convection in the molten parts of the Earth's interior. Scientists have found evidence that the Earth's poles periodically reverse their polarity—more than 20 reversals in the past 5 million years. This is perhaps the result of changes in the direction of electric currents within the Earth. More about this in Chapter 23.

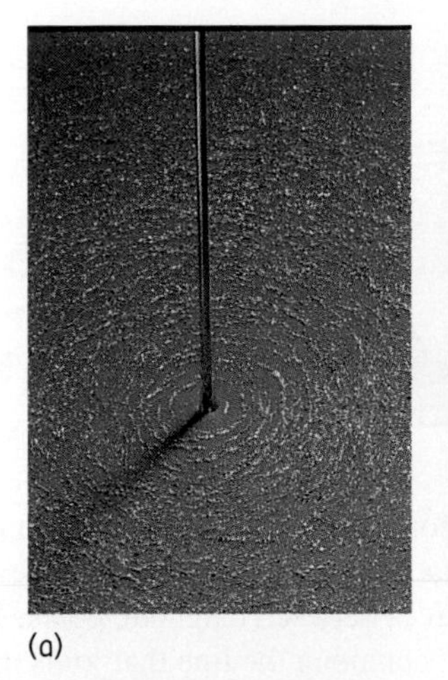
(a)

(b)

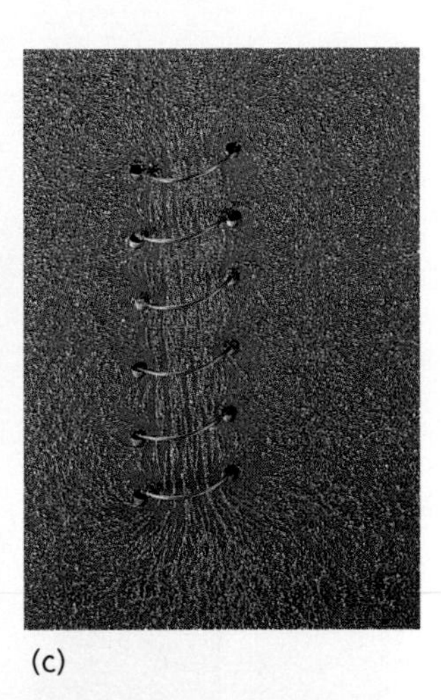
(c)

Fig. 9.12
Iron filings sprinkled on paper reveal the magnetic field configurations about (a) a current-carrying wire, (b) a current-carrying loop, and (c) a current-carrying coil of loops.

Fig. 9.13
A permanent magnet levitates above a superconductor because the magnet's magnetic field cannot penetrate the superconducting material.

Fig. 9.14
Scale model of a prototype magnetically levitated vehicle—a *magplane*. Whereas conventional trains vibrate as they ride on rails at high speeds, magplanes can travel vibration-free at high speeds because they make no physical contact with the guideway they float above.

9.5 Magnetic Forces on Moving Charges

A charged particle at rest does not interact with a magnetic field. But if the charged particle moves in a magnetic field, the magnetic character of a charge in motion becomes evident. It experiences a deflecting force.[5] The force is greatest when the particle moves in a direction perpendicular to the magnetic field lines. At other angles, the force is less and becomes zero when the particle moves parallel to the field lines. In any case the direction of the force is always perpendicular to the magnetic field lines and perpendicular

[5]When particles of electric charge q and velocity v move perpendicularly into a magnetic field of strength B, the force F on each particle is $F = qvB$. For nonperpendicular angles, v in this relationship must be the component of velocity perpendicular to B.

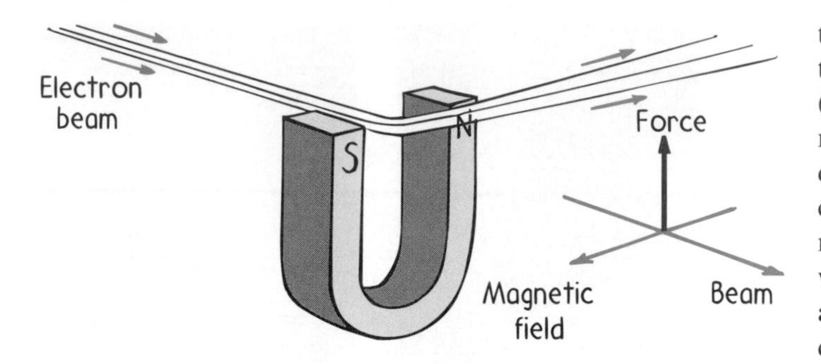

Fig. 9.15
A beam of electrons is deflected by a magnetic field.

to the velocity of the charged particle (Figure 9.15). So a moving charge is deflected when it crosses through a magnetic field, but when it travels parallel to the field, no deflection occurs.

The force that causes this sideways deflection is very different from the forces that occur in other interactions, such as the gravitational forces between masses, the electric forces between charges, and the magnetic forces between magnetic poles. The force that acts on a moving charged particle does not act along the line that joins the sources of interaction, but instead acts perpendicularly to both the magnetic field and the electron beam.

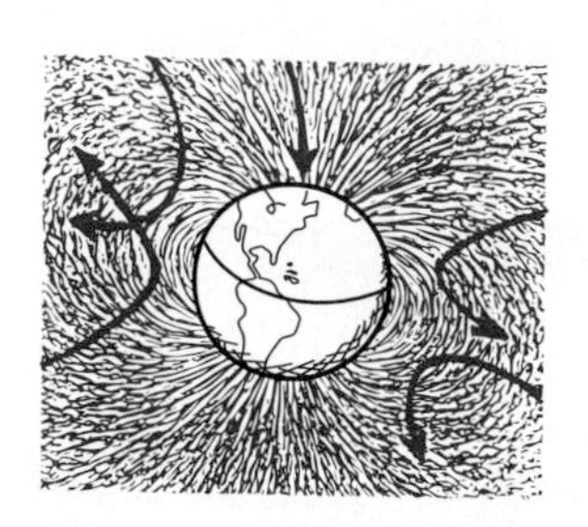

Fig. 9.16
The magnetic field of the Earth deflects many charged particles that make up cosmic radiation.

We are fortunate that charged particles are deflected by magnetic fields. This fact is employed to guide electrons onto the inner surface of a TV tube and provide a picture. More interesting, charged particles from outer space are deflected by the Earth's magnetic field. The intensity of harmful cosmic rays bombarding the Earth's surface would be more intense otherwise.

Magnetic Force on Current-Carrying Wires

Simple logic tells you that if a charged particle moving through a magnetic field experiences a deflecting force, then a current of charged particles moving through a magnetic field experiences a deflecting force also. If the particles are trapped inside a wire when they respond to the deflecting force, the wire is also pushed (Figure 9.17).

If we reverse the direction of current, the deflecting force acts in the opposite direction. The force is strongest when the current is perpendicular to the magnetic field lines. The direction of force is not along the magnetic field lines nor along the direction of current. The force is perpendicular to both field lines and current. It is a sideways force.

Just as a current-carrying wire deflects a magnet such as a compass needle—as discovered by Oersted in a physics classroom in 1820—a magnet deflects a current-carrying wire. The discovery of these complementary links between electricity and

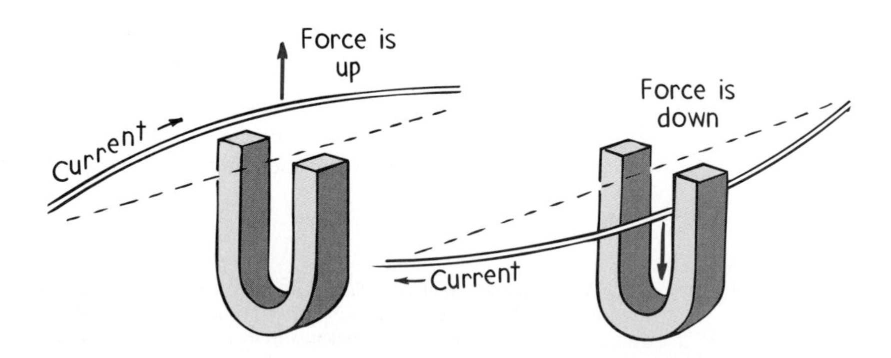

Fig. 9.17
A current-carrying wire experiences a force in a magnetic field. (Can you see that this is a follow-up of what happens in Figure 9.15?)

magnetism created much excitement, for almost immediately people began harnessing the electromagnetic force for useful purposes—with great sensitivity in electric meters and with great force in electric motors.

Electric Meters

The simplest meter capable of detecting an electric current is a magnetic compass. The next level of complexity is a compass in a coil of wires (Figure 9.18). When an electric current exists in the coil, each loop produces its own effect on the needle, and so a very small current can be detected. A current-indicating instrument is called a *galvanometer.*

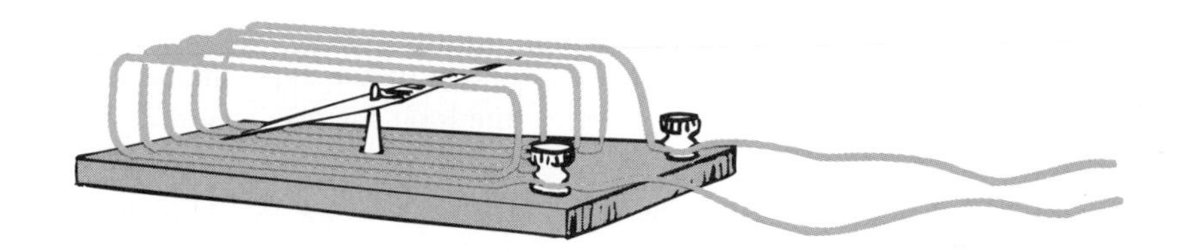

Fig. 9.18
A very simple galvanometer.

A more common galvanometer design is shown in Figure 9.19. It employs more loops of wire and is therefore more sensitive. The coil is free to move, and the magnet is held stationary. The coil turns against a spring, and the greater the current in its windings, the greater its deflection. A needle fixed to the coil shows the amount of deflection. A galvanometer may be calibrated to measure current (amperes), in which case it is called an *ammeter.* Or it may be calibrated to measure electric potential (volts), in which case it is called a *voltmeter.*[6]

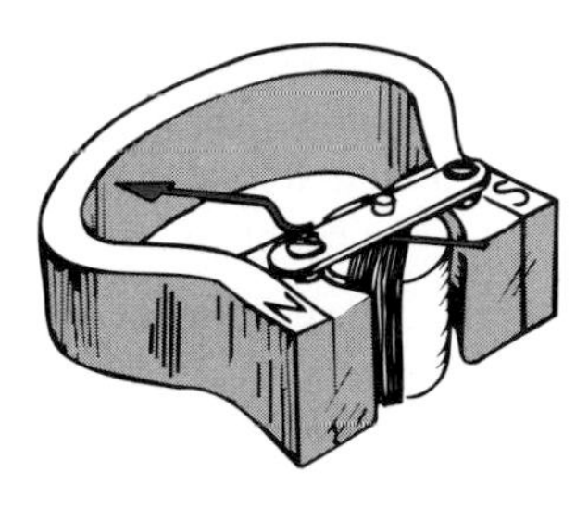

Fig. 9.19
A common galvanometer design.

Electric Motors

If we modify the design of the galvanometer slightly, so that deflection makes a complete rather than a partial rotation, we have an **electric motor**. The principal difference is that in a motor the current is made to change direction every time the coil makes a half rotation. After being forced to turn one half rotation, the coil continues in motion just in time for the current to reverse, whereupon instead of the coil reversing direction, it is forced to continue another half rotation in the same direction. This happens in cyclic fashion to produce continuous rotation, which has been harnessed to run clocks, operate gadgets, and lift heavy loads.

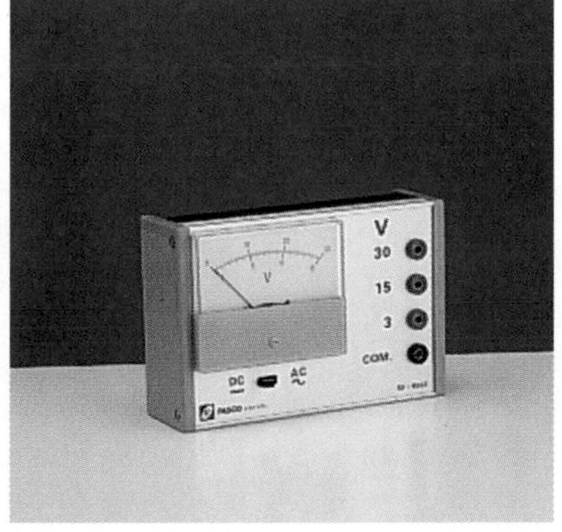

Fig. 9.20
Both the ammeter and the voltmeter are basically galvanometers

[6]Measuring instruments to some degree change what is being measured—ammeters and voltmeters included. Because an ammeter is connected in series with the circuit it measures, its resistance is made very low so that it doesn't appreciably lower the current it measures. A voltmeter is connected in parallel, and its resistance is made very high so that it draws very little current for its operation.

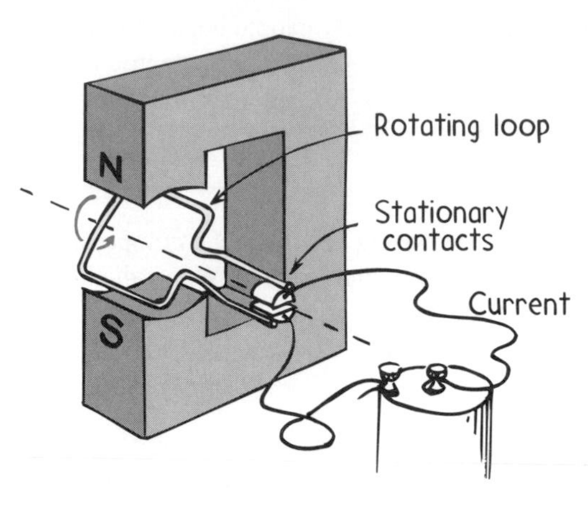

Fig. 9.21
A simplified electric motor.

In Figure 9.21 we see the principle of the electric motor in bare outline. A permanent magnet produces a magnetic field in a region where a rectangular loop of wire is mounted to turn about the dashed axis shown. Any current in the loop has one direction in the upper side of the loop and the opposite direction in the lower side (it has to do this because if charges flow into one end of the loop, they must flow out the other end). If the upper side of the loop is forced to the left by the magnetic field, the lower side is forced to the right, as if it were a galvanometer. But unlike the situation in a galvanometer, in a motor the current is reversed during each half revolution by means of stationary contacts on the shaft. The parts of the wire that rotate and brush against these contacts are called *brushes.* In this way the current in the loop alternates so that the forces on the upper and lower regions do not change directions as the loop rotates. The rotation is continuous as long as current is supplied.

We have described here only a very simple dc motor. Larger motors, dc or ac, are usually made by replacing the permanent magnet by an electromagnet that is energized by the power source. Of course, more than a single loop is used. Many loops of wire are wound about an iron cylinder, called an *armature,* which then rotates when the wire carries current.

The advent of electric motors brought to an end much human and animal toil in many parts of the world. Electric motors have greatly changed the way people live.

Question

What is the major similarity between a galvanometer and a simple electric motor? What is the major difference?

Link to Medical Technology—MRI: Magnetic Resonance Imaging

Magnetic resonance image scanners provide high-resolution pictures of the tissues inside a body. Superconducting coils produce a magnetic field that is up to 60,000 times stronger than the intensity of the Earth's magnetic field, and this field is used to align the protons of hydrogen atoms in the body of the patient.

Like electrons, protons have a "spin" property and therefore align with a magnetic field. Unlike a compass needle, which aligns with the Earth's magnetic field, the protons' axes wobble about the applied magnetic field. Wobbling protons are slammed with a burst of radio waves tuned to push the protons' spin axes sideways, perpendicular to the applied magnetic field. When the radio waves pass and the protons quickly return to their wobbling pattern, they emit faint electromagnetic signals whose frequencies depend slightly on the chemical environment in which the proton resides. The signals are picked up by sensors, which when analyzed by a computer reveal varying densities of hydrogen atoms in the body and their interactions with surrounding tissue. The images clearly distinguish fluid and bone.

It is interesting to note that MRI was formerly called NMR (nuclear magnetic resonance) because hydrogen nuclei resonate with the applied fields. Because of public fear of anything "nuclear," the devices are now called MRI. Tell your friends that every atom in their bodies contains a nucleus!

9.6 Electromagnetic Induction

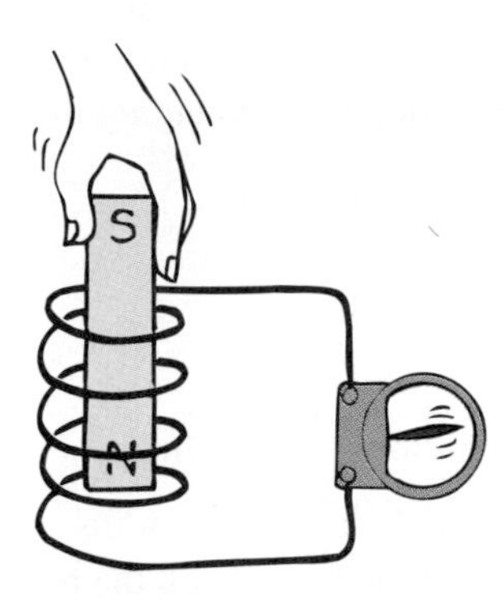

Fig. 9.22
When the magnet is plunged into the coil, charges in the coil are set in motion; voltage is induced in the coil.

In the early 1800s, the only current-producing devices were voltaic cells, which produced small currents by dissolving metals in acids. These were the forerunners of our present-day batteries. The question arose as to whether electricity could be produced from magnetism. The answer was provided in 1831 by two physicists, Michael Faraday in England and Joseph Henry in the United States—each working without knowledge of the other. Their discovery changed the world by making electricity commonplace—powering industries by day and lighting cities at night.

Faraday and Henry both discovered that electric current can be produced in a wire simply by moving a magnet in or out of a coiled part of the wire (Figure 9.22). No battery or other voltage source is needed—only the motion of a magnet in a wire loop. They discovered that voltage is caused, or *induced,* by the relative motion between a wire and a magnetic field. Whether the magnetic field moves near a stationary conductor or the conductor moves in a stationary magnetic field doesn't matter—voltage is induced either way (Figure 9.23).

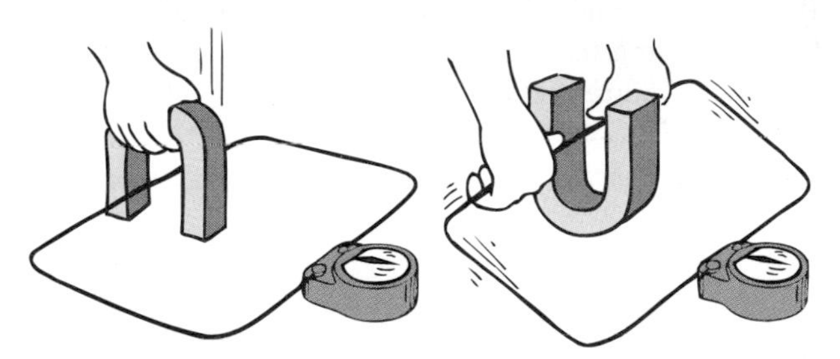

Fig. 9.23
Voltage is induced in the wire loop either when the magnetic field moves past the wire or when the wire moves through the magnetic field.

The greater the number of loops of wire that move in a magnetic field, the greater the induced voltage (Figure 9.24). Pushing a magnet into twice as many loops induces twice as much voltage; pushing into ten times as many loops induces ten times as much voltage; and so on. It may seem that we get something (energy) for nothing by simply increasing the number of loops in a coil of wire. But we don't: We find it is

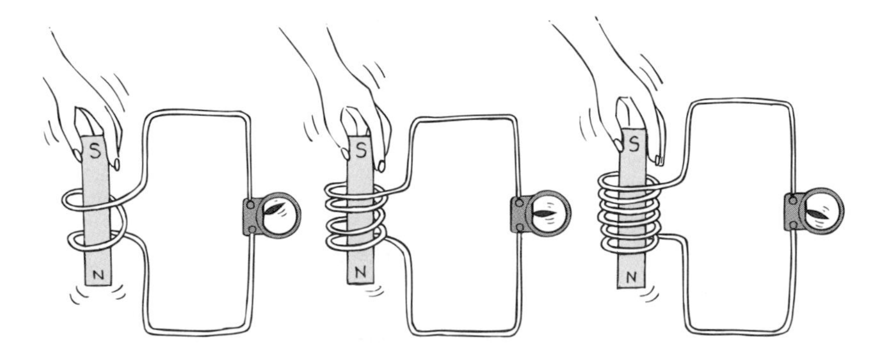

Fig. 9.24
When a magnet is plunged into a coil of twice as many loops as another, twice as much voltage is induced. If the magnet is plunged into a coil with three times as many loops, three times as much voltage is induced.

Answer
A galvanometer and a motor are similar in that they both employ coils positioned in a magnetic field. When a current passes through the coils, forces on the wires rotate the coils. The fundamental difference is that the maximum rotation of the coil in a galvanometer is one-half turn, whereas in a motor the coil (wrapped on an armature) rotates through many complete turns. This is accomplished by alternating the direction of the current with each half turn of the armature.

Fig. 9.25
It is more difficult to push the magnet into a coil made up of many loops because the magnetic field of each current loop resists the motion of the magnet.

more difficult to push the magnet into a coil made up of more loops. This is because the induced voltage makes a current, which makes an electromagnet, which repels the magnet in our hand. So we do more work to induce more voltage (Figure 9.25). The amount of voltage induced depends on how fast the magnetic field lines are entering or leaving the coil. Very slow motion produces hardly any voltage at all. Quick motion induces a greater voltage. This phenomenon of inducing voltage by changing the magnetic field in a coil of wire is called **electromagnetic induction**.

Faraday's Law

Electromagnetic induction is summarized by **Faraday's law**:

> **The induced voltage in a coil is proportional to the number of loops times the rate at which the magnetic field changes within those loops.**

The amount of *current* produced by electromagnetic induction depends on the resistance both in the coil and in the circuit that it connects, as well as on the induced voltage.[7] For example, we can plunge a magnet in and out of a closed rubber loop and in and out of a closed loop of copper. The voltage induced in each is the same, providing the loops are the same size and the magnet moves with the same speed. But the current in each is quite different. The electrons in the rubber sense the same voltage as those in the copper, but their bonding to the fixed atoms prevents the movement of charge that so freely occurs in the copper.

Question

If you push a magnet into a coil, as shown in Figure 9.25, you'll feel a resistance to your push. Why is this resistance greater in a coil with more loops?

We have mentioned two ways in which voltage can be induced in a loop of wire: by moving the loop near a magnet or by moving a magnet near the loop. There is a third way, by changing a current in a nearby loop. All three cases possess the same essential ingredient—a changing magnetic field in the loop.

We see electromagnetic induction all around us. On the road we see it operate when a car drives over buried coils of wire to activate a nearby traffic light; when iron

---

Answer
Simply put, more work is required to provide more energy to be dissipated by more current in the resistor. You can also look at it this way: When you push a magnet into a coil, you cause the coil to become an electromagnet. The more loops on the coil, the stronger the electromagnet that you produce and the stronger it pushes back against the magnet you are moving. (If the coil's electromagnet attracted your magnet instead of repelling it, energy would be created from nothing and the law of energy conservation would be violated. So the coil has to repel your magnet.)

---

[7]Current also depends on the *inductance* of the coil, which measures a coil's tendency to resist a change in current because the magnetism produced by one part of the coil opposes the change of current in other parts. We do not treat this complication in this book.

parts of a car move over the buried coils, the Earth's magnetic field in the coils is changed, which induces voltage to trigger the changing of the traffic lights. Similarly, when one walks through the upright coils in the security system at an airport, any metal you are carrying slightly alters the magnetic field in the coils, induces voltage, and sounds an alarm. When the magnetic strip on the back of a credit card is scanned, changing magnetic fields induce voltages that identify the card. Similarly with the recording head of a tape recorder, when magnetic domains are aligned as blank tape moves past a current-carrying coil. Variations in the current are recorded as variations in domain alignment in the tape. On playback, the variations in magnetic field strength produce changes in the magnetic field that move by the playback coil. The variations in induced voltage match the variations that produced the domain alignment upon recording. The result is beautiful music, if that's what you started with. Of course the voltages induced are tiny, but even they are amplified with transformers, which themselves are based on electromagnetic induction. Electromagnetic induction is everywhere. As we shall see in Chapter 11, it even underlies the electromagnetic waves we call light.

9.7 Generators and Alternating Current

When a magnet is repeatedly plunged into and pulled back out of a coil of wire, the direction of the induced voltage alternates. As the magnetic field strength inside the coil is increased (magnet entering), the induced voltage in the coil is directed one way. When the magnetic field strength diminishes (magnet leaving), the voltage is induced in the opposite direction. The frequency of the alternating voltage induced is equal to the frequency of the changing magnetic field within the loop.

It is more practical to induce voltage by moving a coil rather than by moving a magnet. This can be done by rotating the coil in a stationary magnetic field (Figure 9.26). This arrangement is called a **generator**. The construction of a generator is in principle identical to that of a motor. Only the roles of input and output are reversed. In a motor, electrical energy is the input and mechanical energy the output; in a generator, mechanical energy is the input and electric energy the output. Both devices simply transform energy from one form to another.

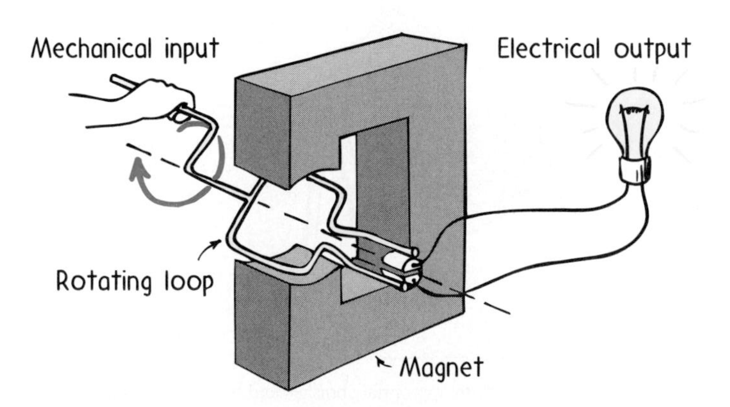

Fig. 9.26
A simple generator. Voltage is induced in the loop when it is rotated in the magnetic field.

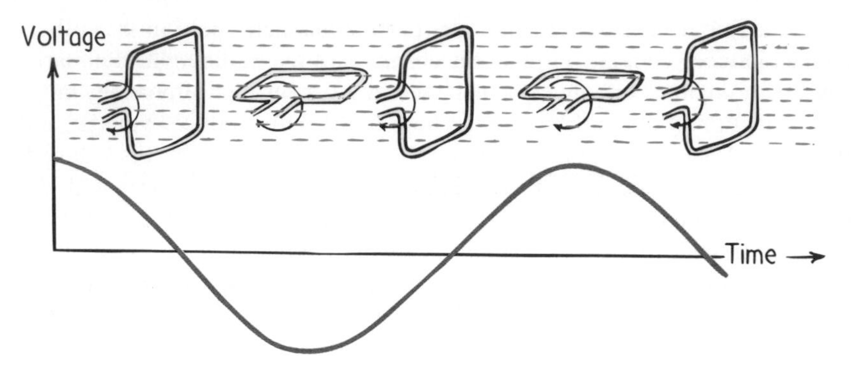

Fig. 9.27
As the loop rotates, the induced voltage (and current) change in magnitude and direction. One complete rotation of the loop produces one complete cycle in voltage (and current).

We can see the electromagnetic induction cycle in Figure 9.27. Note that when the loop of wire is rotated in the magnetic field, there is a change in the number of magnetic field lines enclosed within the loop. When the plane of the loop is perpendicular to the field lines, maximum lines are enclosed. As the loop rotates, it in effect chops the lines, with fewer enclosed. When the plane of the loop is parallel to the lines, none are enclosed. Continued rotation increases and decreases the number of enclosed lines in cyclic fashion, with the greatest rate of change of field lines occurring when the number of enclosed field lines goes through zero. Hence the induced voltage is greatest at these points. Because the voltage induced by the generator alternates, the current produced is an alternating current.[8] The alternating current in our homes is produced by generators standardized so that the current goes through 60 cycles of change each second—60 hertz.

9.8 Power Production

Fifty years after Faraday and Henry discovered electromagnetic induction, Nikola Tesla and George Westinghouse put those findings to practical use and showed the world that electricity could be generated reliably and in sufficient quantities to light entire cities.

Tesla built generators much like those still in use but quite a bit more complicated than the simple model we have discussed. Tesla's generators had armatures consisting of bundles of copper wires that were made to spin within strong magnetic fields by means of a turbine, which in turn was spun by the energy of either falling water or steam. The rotating loops of wire in the armature cut through the magnetic field of the surrounding electromagnets, thereby inducing alternating voltage and current.

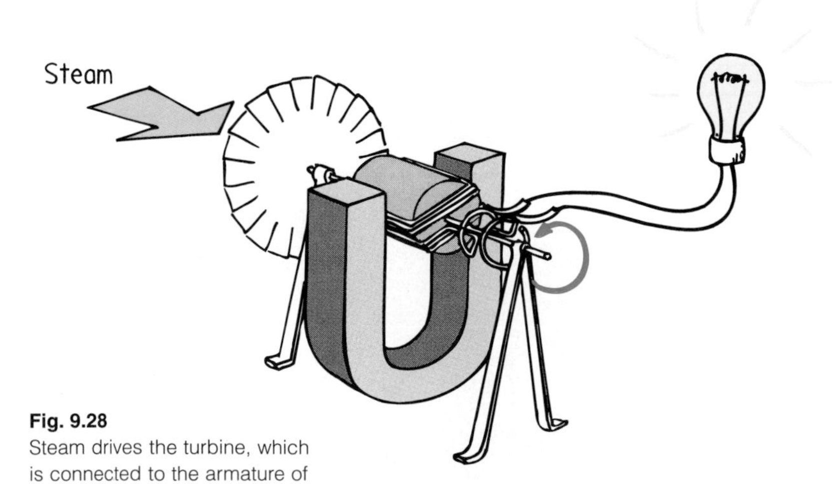

Fig. 9.28
Steam drives the turbine, which is connected to the armature of the generator.

We can look at this process from an atomic point of view. When the wires in the spinning armature cut through the magnetic field, oppositely directed electromagnetic forces act on the negative and positive charges. Electrons respond to this force by momentarily swarming relatively freely in one direction throughout the crystalline cop-

[8]With appropriate brushes and by other means, the ac in the loop(s) can be taken off as dc to make a dc generator.

per lattice; the positive copper ions are forced in the opposite direction. Because the ions are anchored in the lattice, however, they move hardly at all. Only the electrons move, sloshing back and forth in alternating fashion with each rotation of the armature. The energy of this electronic sloshing is tapped at the electrode terminals of the generator.

Generators, of course, don't produce energy—they simply convert energy from some other form to electric energy. As we discussed in Chapter 8, energy from the source, whether fossil or nuclear fuel or wind or water, is converted to mechanical energy to drive the turbine, and the generator converts most of this to electrical energy. The electricity produced simply carries this energy to distant places. Some people think that electricity is a primary source of energy. It is not. It is a form of energy that must have a source.

Boosting or Lowering Voltage—the Transformer

When changes in the magnetic field of a current-carrying coil of wire are intercepted by a second coil of wire, voltage is induced in the second coil. This is the principle of the **transformer**—a simple electromagnetic-induction device consisting of an input coil of wire (the primary) and an output coil of wire (the secondary). The coils need not physically touch each other, but they are commonly wound on a common iron core so that the magnetic field of the primary passes through the secondary. The primary is powered by an ac voltage source, and the secondary is connected to some external circuit. Changes in primary current produce changes in its magnetic field. These changes extend to the secondary, and by electromagnetic induction, voltage is induced in the secondary. If the number of turns of wire is the same in both coils, voltage input and voltage output are the same. Nothing is gained. But if the secondary has more turns than the primary, greater voltage is induced in the secondary. This is a *step-up transformer.* If the secondary has fewer turns than the primary, the ac voltage induced in the secondary is lower than that in the primary. This is a *step-down transformer.* Arrange the secondary so it can be varied with the turn of a knob and you have a dimmer switch.

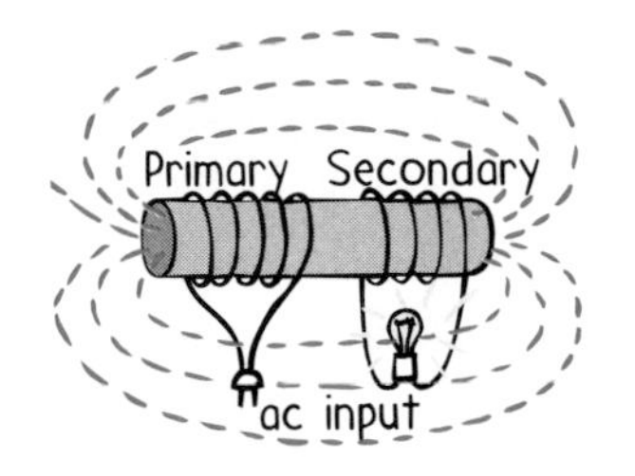

Fig. 9.29
A simple transformer.

The relationship between primary and secondary voltages relative to the number of turns is

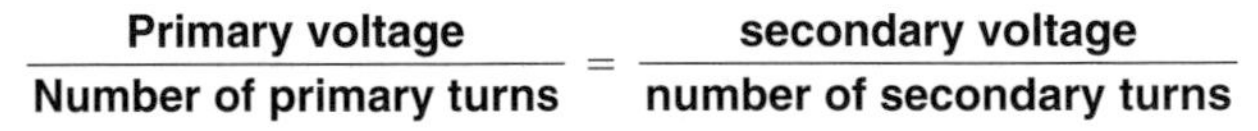

$$\frac{\textbf{Primary voltage}}{\textbf{Number of primary turns}} = \frac{\textbf{secondary voltage}}{\textbf{number of secondary turns}}$$

It might seem we get something for nothing with a step-up transformer, but we don't. When voltage is stepped up, current in the secondary is less than in the primary.

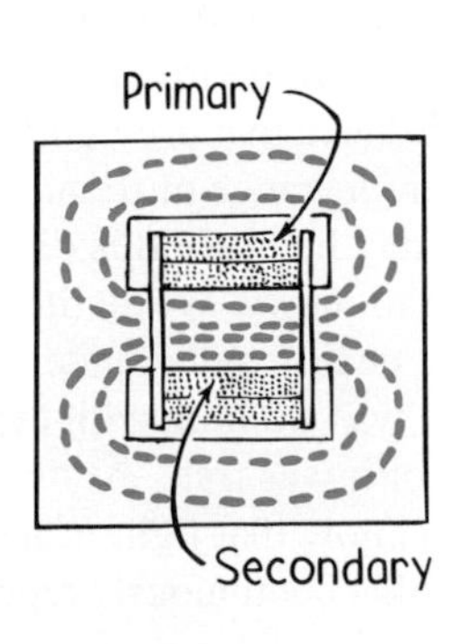

Fig. 9.30
A practical transformer. Both primary and secondary coils are wrapped on the inner part of the iron core (yellow), which guides alternating magnetic field lines (green) produced by ac in the primary. The alternating field induces ac voltage in the secondary. Thus power at one voltage from the primary is transferred to the secondary at a different voltage.

The transformer actually transfers energy from one coil to the other. As we learned in Chapter 3, the rate of transferring energy is *power.* The power used in the secondary is supplied by the primary. The primary gives no more than the secondary uses, in accordance with the law of energy conservation. If slight power losses due to heating of the core are neglected, then

Power into primary = power out of secondary

Electric power is equal to the product of voltage and current, and so we can say

$$(\text{Voltage} \times \text{current})_{\text{primary}} = (\text{voltage} \times \text{current})_{\text{secondary}}$$

The ease with which voltages can be stepped up or down with a transformer is the principal reason that most electric power is ac rather than dc.

9.9 Field Induction

Electromagnetic induction has thus far been discussed as the production of voltages and currents. Actually, the more fundamental *fields* underlie both voltages and currents. The modern view of electromagnetic induction holds that electric and magnetic fields are induced, which in turn give rise to the voltages we have considered. Induction takes place whether or not a conducting wire or any material medium is present. In this more general sense, Faraday's law states

> **An electric field is induced in any region of space in which a magnetic field is changing with time. The magnitude of the induced electric field is proportional to the rate at which the magnetic field changes. The direction of the induced electric field is at right angles to the direction of the changing magnetic field.**

There is a second effect, which is the counterpart to Faraday's law. It is the same as Faraday's law except that the roles of electric and magnetic fields are interchanged. It is one of the many symmetries in nature. This effect was advanced by the British physicist James Clerk Maxwell in about 1860 and is known as **Maxwell's counterpart to Faraday's law**:

> **A magnetic field is induced in any region of space in which an electric field is changing with time. The magnitude of the induced magnetic field is proportional to the rate at which the electric field changes. The direction of the induced magnetic field is at right angles to the direction of the changing electric field.**

Maxwell was the first person to see the link between electromagnetic waves and light. If electric charges are set into vibration in the range of frequencies that match those of light, waves are produced that *are* light! Maxwell discovered that light is simply electromagnetic waves in the range of frequencies to which the eye is sensitive.

On the evening of the day he made this discovery, Maxwell had a date with the young woman he was later to marry. While the couple was walking in a garden, the woman remarked about the beauty and wonder of the stars. Maxwell asked how she would feel to know that she was walking with the only person in the world who knew what starlight really was. For it was true: At that time, James Clerk Maxwell was the only person in the world to know that light of any kind is energy carried in waves of electric and magnetic fields that continuously regenerate each other.

Summary of Terms

- **Magnetic force** (1) Between magnets, the attraction of unlike magnetic poles for each other and the repulsion between like magnetic poles. (2) Between a magnetic field and a moving charged particle, a deflecting force due to the motion of the particle.
- **Magnetic field** The region of magnetic influence around either a magnetic pole or a moving charged particle.
- **Magnetic domains** Clustered regions of aligned magnetic atoms. When these regions are aligned with one another, the substance containing them is a magnet.
- **Electromagnet** A magnet whose field is produced by an electric current. Electromagnets are usually in the form of a wire coil with a piece of iron inside the coil.
- **Electric motor** A device that uses a current-carrying coil forced to rotate in a magnetic field to convert electrical energy to mechanical energy.
- **Electromagnetic induction** The induction of voltage when a magnetic field changes with time. If the magnetic field within a closed loop changes in any way, a voltage is induced in the loop:

$$\text{Voltage induced} \sim \text{number of loops} \times \frac{\text{magnetic field change}}{\text{time}}$$

  This is a statement of Faraday's law. The induction of voltage is the result of a more fundamental phenomenon: the induction of an electric *field,* as defined for the more general case below.
- **Faraday's law** An electric field is induced in any region of space in which a magnetic field is changing with time. The magnitude of the induced electric field is proportional to the rate at which the magnetic field changes. The direction of the induced field is at right angles to the direction of the changing magnetic field.
- **Generator** An electromagnetic induction device that produces electric current by rotating a coil within a stationary magnetic field; converts mechanical energy to electrical energy.
- **Transformer** A device for transferring electric power from one coil of wire to another by means of electromagnetic induction, for the purpose of transforming one value of voltage to another.
- **Maxwell's counterpart to Faraday's law** A magnetic field is induced in any region of space in which an electric field is changing with time. The magnitude of the induced magnetic field is proportional to the rate at which the electric field changes. The direction of the induced magnetic field is at right angles to the changing electric field.

Review Questions

Magnetic Poles

1. Where are the magnetic poles located on a bar magnet?
2. In what way is the rule for the interaction between magnetic poles similar to the rule for the interaction between electric charges?
3. In what way are magnetic poles very different from electric charges?

Magnetic Fields

4. An electric field surrounds an electric charge. What additional field surrounds a moving electric charge?
5. Why is *motion* a key word for magnetism?
6. What two kinds of motion are exhibited by electrons in an atom?

Magnetic Domains

7. What is a magnetic domain?
8. Why is iron magnetic and wood not?
9. Why will dropping an iron magnet on a hard floor make it a weaker magnet?

Electric Currents and Magnetic Fields

10. How does the direction of magnetic field lines about a current-carrying wire differ from the direction of electric field lines about a charge?
11. What happens to the direction of the magnetic field about an electric current when the direction of the current is reversed?
12. Why is the magnetic field strength inside a current-carrying loop of wire greater than the field strength about a straight section of wire?

Magnetic Forces on Moving Charges

13. In what direction relative to a magnetic field does a charged particle move in order to experience maximum deflecting force? Minimum deflecting force?
14. Both gravitational and electric forces act along the direction of the force fields. How is the direction of the magnetic force on a moving charge different?
15. What effect does the magnetic field about the Earth have on cosmic-ray bombardment?
16. Since a magnetic force acts on a moving charged particle, does it make sense that a magnetic force also acts on a current-carrying wire? Defend your answer.
17. What relative direction between a magnetic field and a current-carrying wire results in the greatest force on the wire? Smallest force?
18. What happens to the direction of the force exerted on a wire when the current in the wire is reversed?

19. What is a galvanometer called when calibrated to read current? Voltage?

Electromagnetic Induction

20. When a magnet is thrust into a coil of wire, voltage is induced in the wire, which in turn produces a current in the wire if the wire is connected to something. Is each current-carrying loop of wire in the coil then an electromagnet?
21. What must change in order for electromagnetic induction to occur?
22. What are the three ways that voltage can be induced in a wire?

Generators and Alternating Current

23. How does the frequency of induced voltage compare with how frequently a magnet is plunged in and out of a coil of wire?
24. What is the basic difference between a generator and an electric motor?
25. What is the basic similarity between a generator and an electric motor?
26. Where in the rotation cycle of a simple generator is the greatest rate of change of enclosed field lines? Where, then, is induced voltage a maximum?
27. Why does the voltage induced in a generator alternate?

Power Production

28. What commonly supplies the energy input to a turbine?

Field Induction

29. What is induced by the rapid alternation of a magnetic field?
30. What is induced by the rapid alternation of an electric field?

Home Projects

1. Find the direction and dip (slant from the vertical) of the Earth's magnetic field lines in your locality. Magnetize a large steel needle or straight piece of steel wire by stroking it a couple of dozen times with a strong magnet. Run the needle or wire through a cork in such a way that when the cork floats, the magnet remains horizontal (parallel to the water surface). Float the cork in a plastic or wooden container of water. The needle points toward one of the Earth's magnetic poles. Then press unmagnetized common pins into the sides of the cork. Rest the pins on the rims of a pair of drinking glasses so that the needle or wire points toward the magnetic pole. It should dip in line with the Earth's magnetic field.

2. An iron bar can be easily magnetized by aligning it with the magnetic field lines of the Earth and striking it lightly a few times with a hammer. This works best if the bar is tilted down to match the dip of the Earth's field. The hammering jostles the domains so they can better fall into alignment with the Earth's field. The bar can be demagnetized by striking it when it is in an east-west direction.

Exercises

1. Small magnets such as those used to hold notes on refrigerator doors are often in the shape of a flat disk or square. Where do you think the poles of these magnets are located?
2. A friend tells you that a refrigerator door, beneath its layer of white painted plastic, is made of aluminum. How could you check to see if this is true (without any scraping)?
3. If you place a chunk of iron near the north pole of a magnet, attraction will occur. Why will attraction also occur if you place the iron near the south pole of the magnet?
4. In what sense are all magnets electromagnets?
5. Being that every iron atom is a tiny magnet, why aren't all iron materials magnets?
6. Cans of food in your kitchen pantry are likely magnetized. Why?
7. "An electron always experiences a force in an electric field but not necessarily in a magnetic field." Defend this statement.
8. The central core of the Earth is probably composed of iron and nickel, excellent metals for making permanent magnets. Why is it unlikely that the Earth's core is a permanent magnet?
9. Why does a magnet attract an ordinary nail or paper clip but not a wooden pencil?
10. One way to make a compass is to stick a magnetized needle into a piece of cork and float it in a glass bowl full of water. The needle aligns itself with the magnetic field of the Earth. Since the north pole of this compass is attracted northward, does the needle float toward the northward side of the bowl? Defend your answer.

11. A friend says that when a compass is taken across the equator, it turns around and points in the opposite direction. Another friend says this is not true, that southern-hemisphere types use the south pole of the compass to find direction. You're on; what do you say?
12. Why does a magnet placed in front of a television picture tube distort the picture? (*Note:* Do NOT try this with a color set. If you succeed in magnetizing the metal mask in back of the glass screen, you will have picture distortion even when the magnet is removed!)
13. Magnet A has twice the magnetic field strength of magnet B and at a certain distance pulls on magnet B with a force

of 50 N. With how much force, then, does magnet B pull on magnet A?

14. A strong magnet attracts a paper clip to itself with a certain force. Does the paper clip exert a force on the magnet? If not, why not? If so, does it exert as much force on the magnet as the magnet exerts on it? Defend your answers.
15. When iron naval ships are built, the location of the shipyard and the orientation in the ship while in the shipyard are recorded on a brass plaque permanently fixed to the ship. Why?
16. A cyclotron is a device for accelerating charged particles in ever-increasing circular orbits to high speeds. The charged particles are subjected to both an electric field and a magnetic field. One of these fields increases the speed of the charged particles, and the other field holds them in a circular path. Which field performs which function?

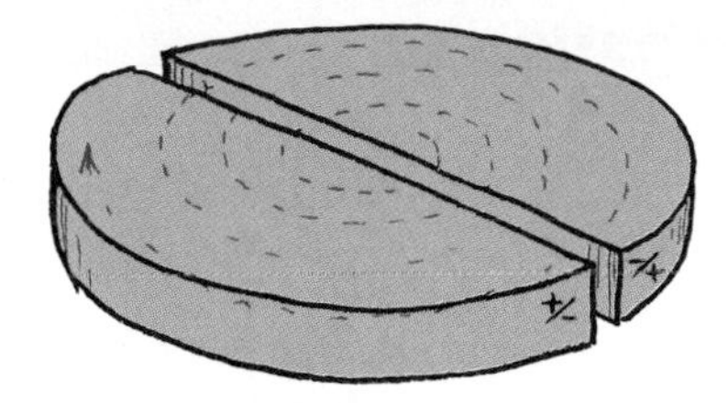

17. A magnetic field can deflect a beam of electrons, but it cannot do work on the electrons to speed them up. Why?
18. Two charged particles are projected into a magnetic field that is perpendicular to their velocities. If the charges are deflected in opposite directions, what does this tell you about them?
19. Will a pair of parallel current-carrying wires exert forces on each other?
20. Why does an iron core increase the magnetic induction of a coil of wire?
21. Why is a generator armature harder to rotate when it is connected to and supplying electric current to a circuit?
22. Will a bicyclist coast farther if the lamp connected to his generator is turned off? Explain.
23. A common pickup for an electric guitar consists of a coil of wire around a permanent magnet. The permanent magnet induces magnetism in the nearby guitar string. When the string is plucked, it vibrates above the coil, thereby changing the magnetic field that passes through the coil. The rhythmic vibrations of the string produce the same rhythmic changes in the magnetic field in the coil, which in turn induce the same rhythmic voltages in the coil, which when amplified and sent to a speaker produce music! Why does this type pickup not work with nylon strings?
24. If your metal car moves over a wide, closed loop of wire embedded in a road surface, is the magnetic field of the Earth within the loop altered? Does this produce a current pulse? Can you think of a practical application for this at a traffic intersection?
25. At the security area of an airport, you walk through a weak magnetic field inside a coil of wire. What happens when a small piece of iron on your person slightly alters the magnetic field in the coil?
26. A certain earthquake detector consists of a little box that contains a massive magnet suspended by sensitive springs. The magnet is surrounded by stationary coils of wire that are fastened to the box, which is firmly anchored to the Earth. Explain how this device works, using two important principles of physics—one studied in Chapter 2 and the other in this chapter.
27. A piece of plastic tape coated with iron oxide is magnetized more in some parts than in others. When the tape is moved past a small coil of wire, what happens in the coil? What is a practical application of this?
28. When a strip of magnetic material, variably magnetized, is embedded in a plastic card that is moved past a small coil of wire, what happens in the coil? What is a practical application of this?
29. A magician places an aluminum ring on a table, underneath which is hidden an electromagnet. When the magician says "abracadabra" (and pushes a switch to produce a current in the coil under the table), the ring jumps into the air. Explain his "trick."
30. How could a light bulb near, yet not touching, an electromagnet be lit? Is ac or dc required? Defend your answer.
31. What are the primary differences and similarities between an electric motor and an electric generator?
32. Does the voltage output increase when a generator is made to spin faster? Explain.
33. If you place a metal ring in a region where a magnetic field is rapidly alternating, the ring may become hot to your touch. Why?
34. The battery in an electric toothbrush is charged by placing it in a plastic holder, but there are no metal-to-metal contacts. Both the toothbrush and the holder have plastic casings. How do you think energy is transferred from the holder to the battery?

35. How could the motor in Figure 9.21 be used as a generator?
36. How could the generator in Figure 9.26 be used as a motor?
37. When a bar magnet is dropped through a vertical length of copper pipe, it falls noticeably more slowly than when dropped through a vertical length of plastic pipe. Why?
38. What is wrong with this scheme? To generate electricity without fuel, arrange a motor to run a generator that will produce electricity that is stepped up with transformers so that the generator can run the motor and simultaneously furnish electricity for other uses.
39. A friend says that changing electric and magnetic fields generate one another and that this gives rise to visible light

when the frequency of change matches the frequencies of light. Do you agree? Explain.

40. Would electromagnetic waves exist if changing magnetic fields could produce electric fields, but changing electric fields could not in turn produce magnetic fields?

Problems

1. The primary coil of a step-up transformer draws 100 W. Find the power provided by the secondary coil.

2. An ideal transformer has 50 turns in its primary and 250 turns in its secondary. For a 12-V ac connected to the primary, find (a) volts ac available at the secondary; (b) current in a 10-Ω device connected to the secondary; (c) power supplied to the primary.

3. A model electric train requires 6 V to operate. If the primary coil of its transformer has 240 windings, how many windings should the secondary have if the primary is connected to a 120-V household circuit?

4. Neon signs require about 12,000 V for their operation. What should be the ratio of the number of loops in the secondary to the number of loops in the primary for a neon-sign transformer that operates off 120-V lines?

Part III —Sample Exam Questions

Choose the BEST answer to each of the following.

1. Protons and electrons
(a) repel each other (b) attract each other (c) do not interact

2. Particle A interacts with particle B, which has twice the charge of particle (A) Compared to the force on particle A, the force on particle B is
(a) four times as much (b) two times as much (c) the same (d) half as much (e) none of these

3. When you touch a negative Van de Graaff generator, your standing hair is
(a) negative also (b) positive

4. Two charged particles held close to each other are released. As the particles move, the velocity of each increases. Therefore the particles have
(a) the same sign of charge (b) opposite signs of charge (c) not enough information is given

5. You can touch and discharge a 10,000-volt Van de Graaff generator with little harm because although the voltage is high, there is relatively little
(a) resistance (b) energy (c) grounding (d) all of these (e) none of these

6. The current through a 12-ohm hair dryer connected to 120-V is
(a) 1 (A) (b) 10 (A) (c) 12 (A) (d) 120 (A) (e) none of these

7. Compared with the amount of electric current in the filament of a lamp, the amount of current in the connecting wire is
(a) definitely less (b) often less (c) actually more (d) the same (e) incredibly, all of these

8. Double the voltage that operates a hair dryer and the current within tends to
(a) halve (b) remain the same (c) double (d) quadruple

9. A woman experiences an electrical shock with a faulty hair dryer. The electrons making the shock come from the
(a) woman's body (b) ground (c) power plant (d) hair dryer (e) electric field in the air

10. As more lamps are connected to a series circuit, the overall current in the power source
(a) increases (b) decreases (c) stays the same

11. As more lamps are connected to a parallel circuit, the overall current in the power source
(a) increases (b) decreases (c) stays the same

12. The headlights, radio, and defroster fan in an automobile are connected in
(a) series (b) parallel

13. The current in a 60-W light bulb connected to a 120-V source is
(a) 0.25 (A) (b) 0.5 (A) (c) 2 (A) (d) 4 (A) (e) more than 4 (A)

14. Change the magnetic field in a closed loop of wire and you induce in the loop a
(a) voltage (b) current (c) electric field (d) all these (e) none of these

15. A step-up transformer increases
(a) power (b) energy (c) both of these (d) neither of these

Answers: 1b, 2c, 3a, 4c, 5b, 7c, 8a, 9a, 10b, 11a, 12b, 13b, 14d, 15d

PART

IV

Waves—Sound and Light

This disc is the pits --billions of them, carefully inscribed in an array that is scanned by a laser beam millions of pits per second. Digitized music! But the beauty of CD recordings is not confined to sound--just look at the brilliant spectrum of colors diffracted by the evenly spaced rows of pits. I find this CD even more beautiful when I know why it's so colorful and why it sounds so great. That's physical science!

CHAPTER

10

Sound Waves

If a tree fell in the middle of a deep forest hundreds of miles away from any living being, would there be a sound? Different people answer this question in different ways. "No," some will say, "sound is subjective and requires a listener. If there is no listener, there will be no sound." "Yes," others will say, "sound is not something in a listener's head, but is an objective thing." Such discussions often are beyond agreement because the participants are arguing not about the nature of sound but about the definition of the word. Either side is correct, depending on which definition is taken. Investigation can proceed only when a definition is agreed on. The physical scientist usually takes the objective position and defines sound as a form of energy that exists, whether or not it is heard, and goes on from there to investigate its nature.

Associated with all sounds are vibrations. Without vibrations there is no sound. The vibration may be a string on a guitar, the reed in a clarinet, or the vocal chords of your larynx when you speak or sing. When they vibrate in air they make the air molecules they touch vibrate also. If the vibrations of air reach your ear, they are transmitted to a part of your brain, and you hear sound.

10.1 Vibrations and Waves

In a general sense, anything that moves back and forth, to and fro, side to side, in and out, or up and down is vibrating. A *vibration* is a wiggle in time. A wiggle in both space and time is a **wave**. A wave extends from one place to another. Light and sound are both vibrations that propagate throughout space as waves, yet they are two very different kinds of waves. Sound is the propagation of vibrations through a material medium—a solid, liquid, or gas. If there is no medium to vibrate, then no sound is possible. Sound cannot travel in a vacuum. But light can, for as we shall learn in the following chapter, light is a vibration of electric and magnetic fields—a vibration of pure energy. Although light can pass though many materials, it needs none. This is evident when it propagates through the vacuum between the Sun and the Earth.

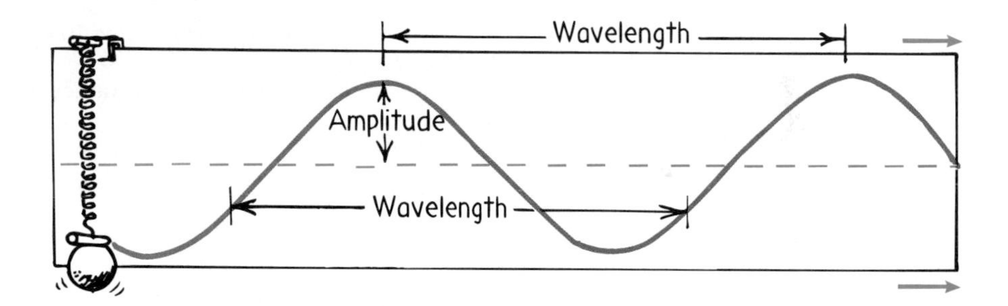

Fig. 10.1
When the bob vibrates up and down, a marking pen traces out a sine curve on paper that is moved horizontally at constant speed.

The relationship between a vibration and a wave is shown in Figure 10.1. A marking pen on a bob attached to a vertical spring vibrates up and down and traces a wave form on a sheet of paper that is touching the pen and moving horizontally at constant speed. The wave form is a *sine curve,* a pictorial representation of a wave. As in a water wave, the high points are called *crests* and the low points *troughs.* The dashed line represents the "home" position, or midpoint, of the vibration. The term **amplitude** refers to the distance from the midpoint to the crest (or trough) of the wave. So the amplitude equals the maximum displacement from equilibrium.

The **wavelength** of a wave is the distance from one crest to the next one, or equivalently, the distance between any two successive identical parts of the wave. The wavelengths of waves at the beach are measured in meters, the wavelengths of ripples in a pond in centimeters, and the wavelengths of light waves in billionths of a meter (nanometers).

How frequently a vibration occurs is described by its **frequency**. The frequency of any vibrating object specifies the number of to-and-fro vibrations (in other words, the number of *cycles*) it makes in a given time (usually 1 second). If one complete to-and-fro vibration (one cycle) occurs in 1 second, the frequency is one vibration per second. If two vibrations occur in 1 second, the frequency is two vibrations per second.

The unit of frequency is the **hertz** (Hz), as we learned in Section 8.5. It is named after Heinrich Hertz, who demonstrated the existence of radio waves in 1886. We call one vibration per second 1 hertz; two vibrations per second is 2 hertz; and so on. Higher

Fig. 10.2
Electrons in the transmitting antenna vibrate 940,000 times each second and produce 940-kHz radio waves.

frequencies are measured in kilohertz (kHz), and still higher frequencies in megahertz (MHz). AM radio waves are usually measured in kilohertz, and FM radio waves are measured in megahertz. A station at 960 kilohertz on the AM radio dial, for example, broadcasts radio waves that have a frequency of 960,000 vibrations per second. A station at 101.7 megahertz on the FM dial broadcasts radio waves that have a frequency of 101,700,000 hertz. These radio-wave frequencies are the frequencies at which electrons are forced to vibrate in the antenna of a radio station's transmitting tower. The source of all waves is something that vibrates. The frequency of the vibrating source and the frequency of the wave it produces are the same.

The **period** of a wave or vibration is the time it takes for a complete vibration. Period can be calculated from frequency, and vice versa. Suppose, for example, that a pendulum makes two vibrations in one second. Its frequency is 2 hertz, and the time needed to complete one vibration—that is, the period—is $\frac{1}{2}$ second. If the vibration frequency is 3 hertz, then the period is $\frac{1}{3}$ second. Frequency and period are the inverse of each other:

Frequency = 1/period

Period = 1/frequency

Questions

1. An electric vibrator completes 60 cycles every second. What are (a) its frequency, and (b) its period?
2. Gusts of wind make the Sears Building in Chicago sway back and forth, completing a cycle every 10 seconds. What are (a) its frequency, and (b) its period?

10.2 Wave Motion

Fig. 10.3
Water waves.

Drop a stone into a quiet pond and waves travel outward in expanding circles. Note that the waves move, not the water. This can be seen by observing a floating leaf the waves encounter. The leaf bobs up and down but doesn't travel with the waves. The waves move, not the water. Likewise with waves of wind over a field of tall grass on a gusty day. Waves travel across the grass, but individual stems of grass do not leave their places; instead, they swing to and fro between definite limits but go nowhere. When you speak, wave motion through the air travels across the room at about 340 meters per second. The air itself doesn't travel across the room at this speed. In these examples, when the wave motion stops, the water, the grass, and the air return to their initial positions. It is characteristic of wave motion that the medium carrying the wave returns to its initial condition after the disturbance has passed.

Answers

1. (a) 60 cycles per second is 60 vibrations per second or 60 Hz; (b) $\frac{1}{60}$ second.
2. (a) 1/10 Hz; (b) 10 s.

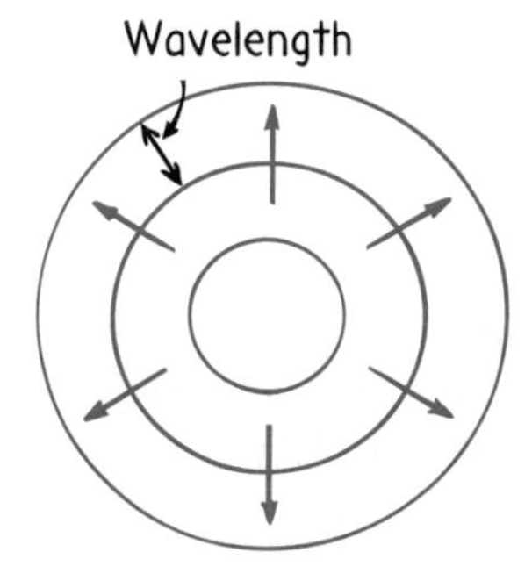

Fig. 10.4
A top view of water waves.

Wave Speed

The speed of periodic wave motion is related to the frequency and wavelength of the waves. We can understand this by considering the simple case of water waves (Figures 10.3 and 10.4). Imagine that we fix our eyes at a stationary point on the surface of water and observe the waves passing by this point. We can measure how much time passes between the arrival of one crest and the arrival of the next one (this time interval is the period) and also observe the distance between crests (the wavelength). We know that speed is defined as distance divided by time. In this case the distance is one wavelength and the time is one period, so wave speed = wavelength/period.

For example, if the wavelength is 10 meters and the time between crests at a point on the surface is 0.5 second, the wave moves 10 meters in 0.5 second and its speed is 10 meters divided by 0.5 second, or 20 meters per second.

Because period is the inverse of frequency, the formula wave speed = wavelength/period can also be written

Wave speed = wavelength × frequency

This relationship holds true for all kinds of waves, whether they are water waves, sound waves, or light waves.

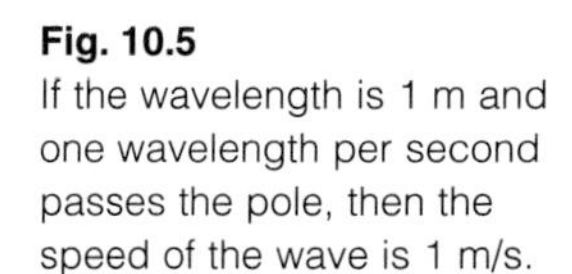

Fig. 10.5
If the wavelength is 1 m and one wavelength per second passes the pole, then the speed of the wave is 1 m/s.

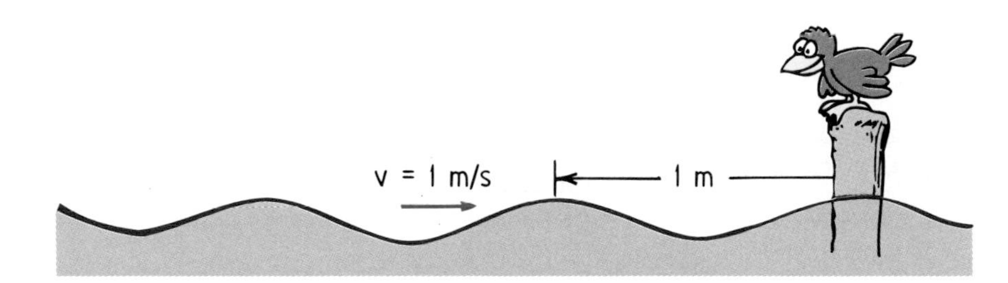

Questions

1. If a train of freight cars, each 10 m long, rolls by you at the rate of three cars each second, what is the speed of the train?
2. If a water wave vibrates up and down three times each second and the distance between wave crests is 2 m, what are (a) the wave's frequency? (b) its wavelength? (c) its wave speed?
3. A 60-Hz vibrator produces air waves that spread out at 340 m/s. What are (a) the frequency of the waves? (b) their period? (c) their speed? (d) their wavelength?

Answers

1. 30 m/s. We can see this in two ways. (1) According to the speed definition from Chapter 1, $v = d/t = 3 \times 10\ \text{m}/1\ \text{s} = 30\ \text{m/s}$, because 30 m of train passes you in 1 s. (2) If we compare the train to wave motion, with wavelength corresponding to 10 m and frequency 3 Hz, then speed = frequency × wavelength = 3 Hz × 10 m = 30 m/s.
2. (a) 3 Hz; (b) 2 m; (c) wave speed = frequency × wavelength = 3/s × 2 m = 6 m/s. (Remember that 3 Hz is 3 vibrations/s; because *vibrations* has no unit, we write 3 Hz = 3/s.) It is customary to express the equation for wave speed as $v = f\lambda$, where v is wave speed, f is wave frequency, and λ (the Greek letter lambda) is wavelength.
3. (a) 60 Hz; (b) $\frac{1}{60}$ s; (c) 340 m/s; (d) 5.7 m.

10.3 Transverse and Longitudinal Waves

Fasten one end of a Slinky to a wall and hold the free end in your hand. Shake it up and down and you produce vibrations that are at right angles to the direction of wave travel. The right-angled, or sideways, motion is called *transverse motion.* This type of wave—in which the direction of wave travel is perpendicular to the direction of the vibrating source—is called a **transverse wave**. Waves in the stretched strings of musical instruments and upon the surfaces of liquids are transverse. We shall see later that electromagnetic waves, some of which are radio waves and light, are also transverse.

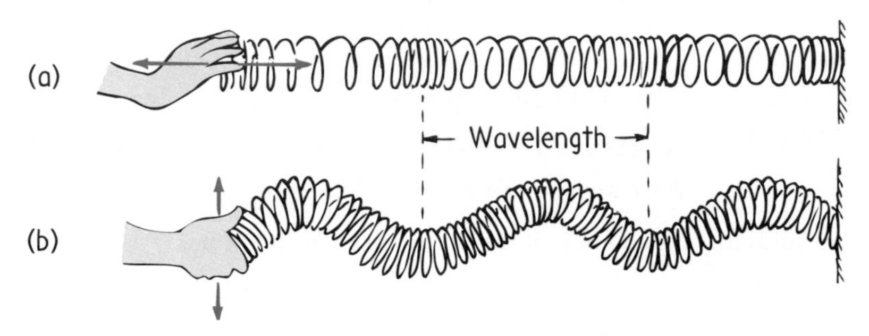

Fig. 10.6
Both waves transfer energy from left to right. (a) When the Slinky is pushed and pulled rapidly along its length, a longitudinal wave is produced. (b) When the end of the Slinky is shaken up and down, a transverse wave is produced.

A **longitudinal wave** is one in which the direction of wave travel is *along* the direction in which the source vibrates. We produce a longitudinal wave with our Slinky when we shake it back and forth along its axis, as shown in Figure 10.6a. The vibrations are then parallel to the direction of energy transfer. Part of the Slinky is compressed, and a wave of *compression* travels along the spring. In between successive compressions is a stretched region, called a *rarefaction.* Both compressions and rarefactions travel in the same direction along the Slinky and together make up the longitudinal wave. Sound waves are longitudinal waves. The wavelength of a longitudinal wave is the distance between successive compressions or, equivalently, the distance between successive rarefactions. In the case of sound, each molecule in the air vibrates to and fro about some equilibrium position as the waves move by.

10.4 Sound Waves

Think of the air molecules in a room as tiny, haphazardly moving Ping-Pong balls. Vibrate a Ping-Pong paddle in their midst and you set them vibrating to and fro in rhythm with your vibrating paddle. As with shaking the Slinky back and forth, in some regions the balls are momentarily bunched up (compressions), and in other regions between the bunched-up ones they are momentarily spread out (rarefactions). The vibrating prongs of a tuning fork do the same to air molecules. Vibrations made up of compressions and rarefactions spread from the tuning fork throughout the air, and we hear these vibrations as sound. So a **sound wave** is a longitudinal wave. It is important to point out again that what travel through the air are the vibrations, *not the air molecules themselves.*

Fig. 10.7
Vibrate a Ping-Pong paddle in the midst of a lot of Ping-Pong balls, and you make the balls vibrate also.

We describe our subjective impression about the frequency of sound by *pitch.* A high-pitch sound like that from a tiny bell has a high vibration frequency. Sound from a large bell has a low pitch because its vibrations are of a low frequency.

The human ear can normally hear pitches corresponding to the range of frequencies from about 20 to 20,000 hertz. As we age, this range shrinks. So by the time you

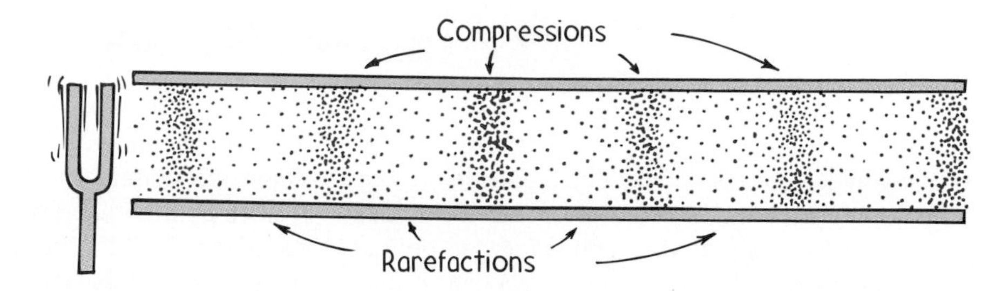

Fig. 10.8
Compressions and rarefactions travel (both at the same speed in the same direction) from the tuning fork through the air in the tube.

can afford to trade in your old radio for an expensive hi-fi set, you may not be able to tell the difference. Sound waves having frequencies below 20 hertz are called *infrasonic waves,* and those having frequencies above 20,000 hertz are called *ultrasonic waves.* We cannot hear infrasonic or ultrasonic sound waves, but dogs and some other animals can.

Most sound is transmitted through air, but any elastic substance—solid, liquid, or gas—can transmit sound.[1] Air is a poor conductor of sound relative to solids and liquids. You can hear the sound of a distant train clearly by placing an ear against the rail.

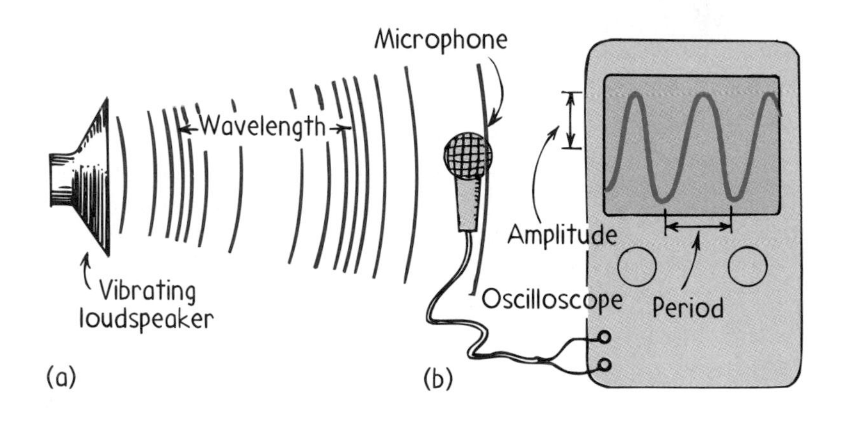

Fig. 10.9
The radio loudspeaker is a paper cone that vibrates in rhythm with an electric signal. The sound produced sets up similar vibrations in the microphone, which are displayed on an oscilloscope. The shape of the wave form on the oscilloscope screen reveals information about the sound.

Fig. 10.10
Waves of compressed and rarefied air, produced by the vibrating cone of the loudspeaker, make up the pleasing sound of music.

When swimming, have a friend at a distance click two rocks together beneath the water surface while you are submerged. Observe how well water conducts the sound. Sound cannot travel in a vacuum because there is nothing to compress and expand. The transmission of sound requires a medium.

Pause to reflect on the physics of sound while you are quietly listening to your radio sometime. The radio loudspeaker is a paper cone that vibrates in rhythm with an electrical signal. Air molecules next to the vibrating cone are set into vibration. These in turn vibrate against

[1]An elastic substance is "springy," has resilience, and can transmit energy with little loss. Steel, for example, is elastic, whereas lead or putty is not.

Link to Technology—Loudspeaker

The loudspeaker of your radio or other sound-producing systems changes electrical signals into sound waves. The electrical signals pass through a coil wound around the neck of a paper cone. This coil acts as an electromagnet, which is located near a permanent magnet. When the current direction is one way, magnetic force pushes the electromagnet toward the permanent magnet, pulling the cone inward. When the current direction reverses, the cone is pushed outward. Vibrations in the electric signal then cause the cone to vibrate. Vibrations of the cone produce sound waves in the air.

neighboring molecules, which in turn do the same, and so on. As a result, rhythmic patterns of compressed and rarefied air emanate from the loudspeaker, showering the whole room with undulating motions. The resulting vibrating air sets your eardrum into vibration, which in turn sends cascades of rhythmic electrical impulses along the cochlea nerve canal and into the brain. And you listen to the sound of music.

Speed of Sound

If we watch a person at a distance chopping wood or hammering, we can easily see that the blow takes place an appreciable time before its sound reaches our ears. Thunder is heard after a flash of lightning is seen. These common experiences show that sound requires time to travel from one place to another. The speed of sound depends on wind conditions, temperature, and humidity. It does not depend on the loudness or the frequency of the sound; all sounds travel at the same speed in a given medium. The speed of sound in dry air at 0°C is about 330 meters per second, nearly 1200 kilometers per hour. Water vapor in the air increases this speed slightly. Sound travels faster through warm air than cold air. This is to be expected because the faster-moving molecules in warm air bump into each other more often and therefore can transmit a pulse in less time.[2] For each degree rise in temperature above 0°C, the speed of sound in air increases by 0.6 meter per second. So in air at a normal room temperature of about 20°C, sound travels at about 340 meters per second. In water, sound speed is about 4 times its speed in air; in steel it's about 15 times.

Questions

1. Do compressions and rarefactions in a sound wave travel in the same direction or in opposite directions from one another?
2. What is the approximate distance of a thunderstorm when you note a 3-s delay between the flash of lightning and the sound of thunder? (Use 340 m/s for the speed of sound.)

Answers

1. They travel in the same direction.
2. In 3 s the sound travels (340 m/s $\times$ 3 s) = 1020 m. Because there is no appreciable delay between the time the lightning bolt is created and the time you see it, the storm is slightly more than 1 km away.

---

[2] The speed of sound in a gas is about $\frac{3}{4}$ the average translational speed of the molecules.

10.5 Reflection of Sound

We call the reflection of sound an *echo.* The fraction of sound energy reflected from a surface is large if the surface is rigid and smooth, and less if the surface is soft and irregular. Sound energy that is not reflected is either transmitted or absorbed.

Sound reflects from a smooth surface the same way light does—the angle of incidence is equal to the angle of reflection (Figure 10.11). Sometimes when sound reflects from the walls, ceiling, and floor of a room, the surfaces are too reflective and the sound becomes garbled. This is due to multiple reflections called *reverberations.* On the other hand, if the reflective surfaces are too absorbent, the sound level is low and the room may sound dull and lifeless. Reflected sound in a room makes it sound lively and full, as you have probably found out while singing in the shower. In the design of an auditorium or concert hall, a balance must be found between reverberation and absorption. The study of sound properties is called *acoustics.*

It is often advantageous to place highly reflective surfaces behind the stage to direct sound out to an audience. In some concert halls, reflecting surfaces are suspended above the stage. The ones in the opera hall in San Francisco are large, shiny, plastic surfaces that also reflect light (Figure 10.12). A listener can look up at these reflectors and see the reflected images of the members of the orchestra (the plastic reflectors are somewhat curved, which increases the field of view). Both sound and light obey the same law of reflection, so if a reflector is oriented so that you can see a particular musical instrument, rest assured that you can hear it also. Sound from the instrument follows the line of sight to the reflector and then to you.

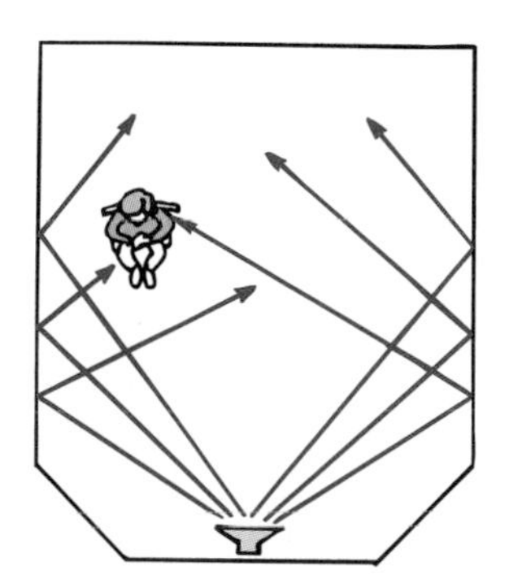

Fig. 10.11
The angle of incident sound is equal to the angle of reflected sound.

Fig. 10.12
The plastic plates above the orchestra reflect both light and sound. Adjusting them is quite simple: What you see is what you hear.

10.6 Refraction of Sound

Sound waves bend when parts of the waves travel at different speeds. This happens in uneven winds or when sound is traveling through air of uneven temperatures. This bending of sound is called **refraction**. On a warm day, air near the ground may be appreciably warmer than air above, and so the speed of sound near the ground increases. Sound waves therefore tend to bend away from the ground, resulting in sound that does not seem to carry well (Figure 10.13).

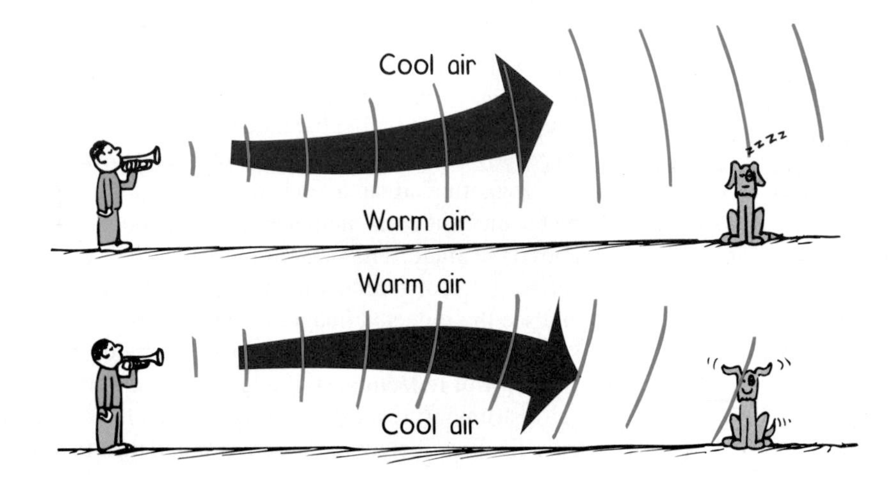

Fig. 10.13
Sound waves are bent in air of uneven temperatures.

The refraction of sound occurs under water, too, where the speed of sound again varies with temperature. This poses a problem for surface vessels that bounce ultrasonic waves off the bottom of the ocean to chart its features, but is a blessing for submarines that wish to escape detection. Because of layers of water at different temperatures (thermal gradients), the refraction of sound leaves gaps, or "blind spots," in the water. This is where submarines hide. If it weren't for refraction, submarines would be easier to detect.

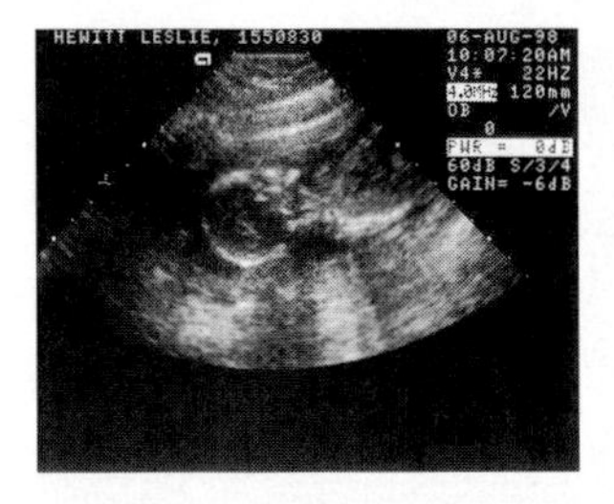

Fig. 10.14
The 14-week-old fetus of M. Hewitt Abrams.

The multiple reflections and refractions of ultrasonic waves are used by physicians in a technique for harmlessly "seeing" inside the body without the use of X rays. When high-frequency sound (ultrasound) enters the body, it is reflected more strongly from the outside of organs than from their interior, and a picture of the outline of the organs is obtained. The ultrasound echo technique may be relatively new to humans, but not to bats, who emit ultrasonic squeaks and locate objects by their echoes, or to dolphins, who do this and more.

■ Question

A depth-sounding vessel surveys the ocean bottom with ultrasonic sound that travels 1530 m/s in seawater. How deep is the water if the time delay of the echo from the ocean floor (which is the time it takes the sound to travel from source to ocean floor back to source) is 2 s?

Link to Zoology—Dolphins and Acoustical Imaging

The primary sense of the dolphin is acoustic, for vision is not a very useful sense in the often murky and dark depths of the ocean. Whereas sound is a passive sense for us, it is an active sense for dolphins, who send out sounds and then perceive their surroundings on the basis of the echoes that come back. The ultrasonic waves emitted by a dolphin enables it to "see" through the bodies of other animals and people. Because skin, muscle, and fat are almost transparent to dolphins, they "see" only a thin outline of the body, but the bones, teeth, and gas-filled cavities are clearly apparent. Physical evidence of cancers, tumors, and heart attacks can all be "seen" by dolphins—as humans have only recently been able to do with ultrasound.

What's more interesting, a dolphin can reproduce the sonic signals that paint the mental image of its surroundings; thus the dolphin probably communicates its experience to other dolphins by communicating the full acoustic image of what is "seen," placing the image directly in the minds of other dolphins. It needs no word or symbol for "fish," for example, but communicates an image of the real thing—perhaps with emphasis highlighted by selective filtering, as we similarly communicate a musical concert to others via various means of sound reproduction. Small wonder that the language of the dolphin is very unlike our own!

Fig. 10.15
A dolphin emits ultrahigh-frequency sound to locate and identify objects in its environment. Distance is sensed by the time delay between sending sound and receiving its echo, and direction is sensed by differences in time for the echo to reach its two ears. A dolphin's main diet is fish and, since hearing in fish is limited to fairly low frequencies, they are not alerted to the fact they are being hunted.

10.7 Forced Vibrations

If you strike an unmounted tuning fork, its sound is rather faint. Do the same when the fork is held against a table after striking it, and the sound is louder. This is because the table is forced to vibrate and, with its larger surface, sets more air in motion. The table can be forced into vibration by a fork of any frequency. This is a case of **forced vibration**. The vibration of a factory floor caused by the running of heavy machinery is an-

---

Answer
The 2-s delay means it takes 1 s for the sound to reach the bottom and another 1 s to return to the detector on the vessel. Sound traveling at 1530 m/s for 1 s tells us the bottom is 1530 m deep.

other example of forced vibration. A more pleasing example is given by the sounding boards of stringed instruments.

Drop a wrench and a baseball bat on a concrete floor, and you can tell the difference in their sounds. This is because each vibrates differently when striking the floor. They are not forced to vibrate at a particular frequency, but instead each vibrates at its own characteristic frequency. When disturbed, any object composed of an elastic material vibrates at its own special set of frequencies, which together form its special sound. We speak of an object's **natural frequency**, which depends on factors such as the elasticity and shape of the object. Bells and tuning forks, of course, vibrate at their own characteristic frequencies. And interestingly enough, most other things, from planets to atoms and almost everything in between, have a springiness to them and vibrate at one or more natural frequencies.

10.8 Resonance

When the frequency of forced vibrations imposed on an object matches the object's natural frequency, a dramatic increase in amplitude occurs. This phenomenon is called **resonance**. Literally, *resonance* means "resounding" or "sounding again." Putty doesn't resonate because it isn't elastic, and a dropped handkerchief is too limp. For something to resonate, it needs a force to pull it back to its starting position and enough energy to keep it vibrating.

A common experience illustrating resonance occurs on a swing. When pumping a swing, we pump in rhythm with the natural frequency of the swing. More important than the force with which we pump is the timing. Even small pumps or small pushes from someone else, if delivered in rhythm with the frequency of the swinging motion, produce large amplitudes.

A common demonstration of resonance is to place a pair of tuning forks having the same natural frequency on wooden sounding boards sitting a meter or so apart. When one of the forks is struck, it sets the other fork into vibration. This is a small-scale version of pushing a friend on a swing—it's the timing that's important. When a series of

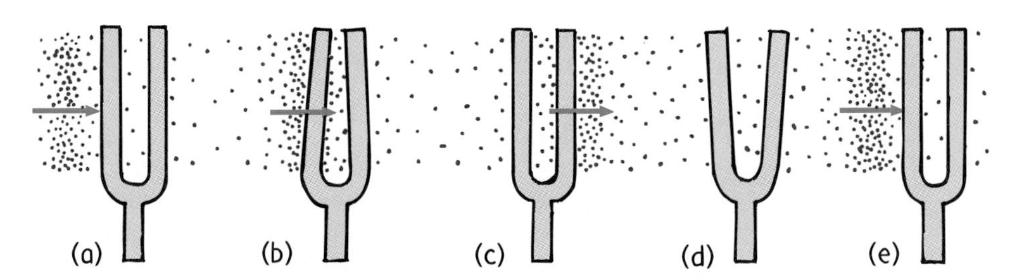

Fig. 10.16
Stages of resonance. (Blue arrows indicate sound waves traveling to the right.) (a) The first compression meets the left prong and gives it a tiny momentary push. (b) The prong bends to the right and then (c) moves leftward to return to its vertical position just at the time a rarefaction arrives at it. (d) Still moving, the prong overshoots leftward into the rarefied region. (e) Just when the prong restores to the vertical position, the next compression arrives to repeat the cycle. Now the prong bends farther to the right (not shown) because it is already moving when the compression pushes it. This effect is clearly audible when the forks are placed on sounding boards.

Fig. 10.17
In 1940, four months after being completed, the Tacoma Narrows Bridge in the state of Washington was destroyed by wind-generated resonance. A mild gale produced an irregular force in resonance with the natural frequency of the bridge, steadily increasing the amplitude of vibration until the bridge collapsed.

sound waves impinge on the fork, each compression gives the prongs a tiny push. Because the frequency of these pushes corresponds to the natural frequency of the fork, the pushes successively increase the amplitude of vibration. This is because the pushes occur at the right time and in the same direction as the instantaneous motion of the fork. The motion of the second fork is called a *sympathetic vibration.*

If the forks do not have the same natural frequency, the timing of pushes is off and resonance does not occur. When you tune your radio set, you are similarly adjusting the natural frequency of the electronics in the set to match one of the many surrounding signals in the air. The set then resonates to one station at a time instead of playing all stations at once.

Resonance occurs whenever successive impulses are applied to a vibrating object in rhythm with its natural frequency. Cavalry troops marching across a footbridge near Manchester, England, in 1831 inadvertently caused the bridge to collapse when they marched in rhythm with the bridge's natural frequency. Since then, it is customary to order troops to "break step" when crossing bridges. A more recent bridge disaster was caused by wind-generated resonance (Figure 10.17).

10.9 Interference

One of the most interesting properties of all waves is **interference**. Consider transverse waves. When the crest of one wave overlaps the crest of another, their individual effects add together. The result is a wave of increased amplitude. This is *constructive interference* (Figure 10.18). When the crest of one wave overlaps the trough of another, their individual effects are reduced. The high part of one wave simply fills in the low part of another. This is called *destructive interference.*

Wave interference is easiest to see in water. In Figure 10.19 we see the interference pattern made when two vibrating objects touch the surface of water. Notice the regions where a crest of one wave overlaps the trough of another to produce regions of

+ = Reinforcement

+ = Cancellation

Fig. 10.18
Constructive and destructive interference in a transverse wave.

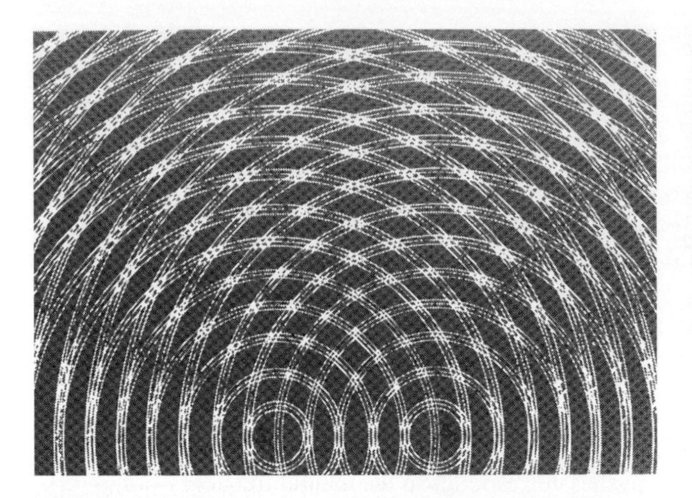

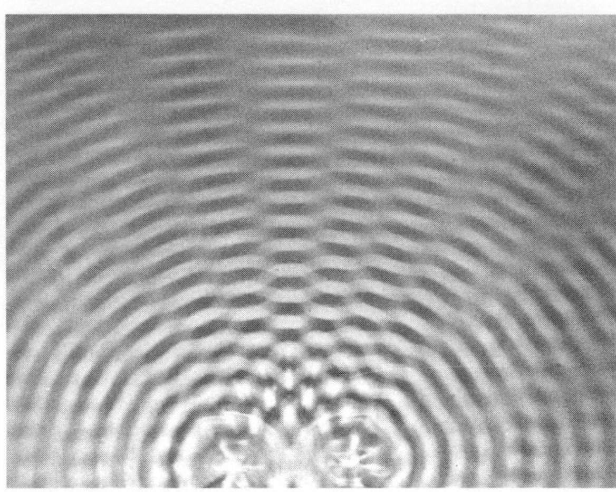

Fig. 10.19
Two sets of overlapping water waves produce an interference pattern.

zero amplitude. At points along these regions, the waves arrive out of step. We say they are *out of phase* with one another.

Interference is characteristic of all wave motion, whether the waves are water waves, sound waves, or light waves. We see a comparison of interference for both transverse and longitudinal waves in Figure 10.20. In either case, when crests overlap crests, increased amplitude results. Or when crests overlap troughs, decreased amplitude results. In the case of sound, the crest of a wave corresponds to a compression and the trough of a wave corresponds to a rarefaction.

Fig. 10.20
Constructive (top two panels) and destructive (bottom two panels) wave interference in transverse and longitudinal waves.

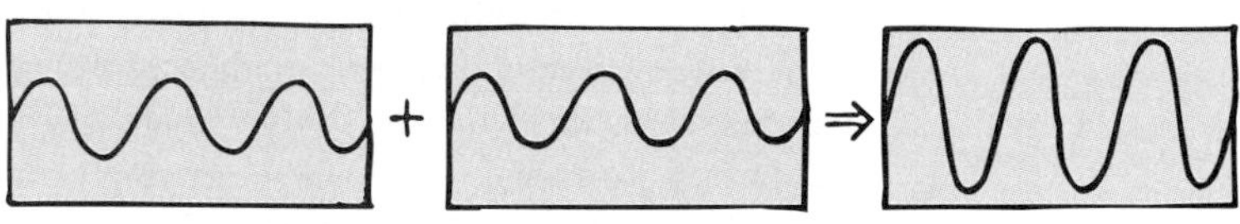

The superposition of two identical transverse waves in phase produces a wave of increased ampitude.

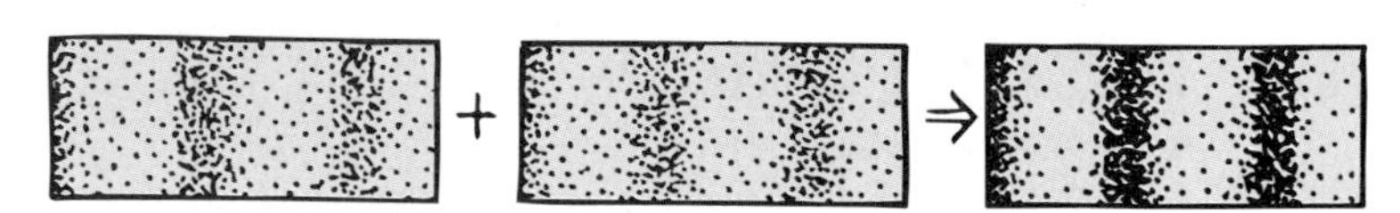

The superposition of two identical longitudinal waves in phase produces a wave of increased intensity.

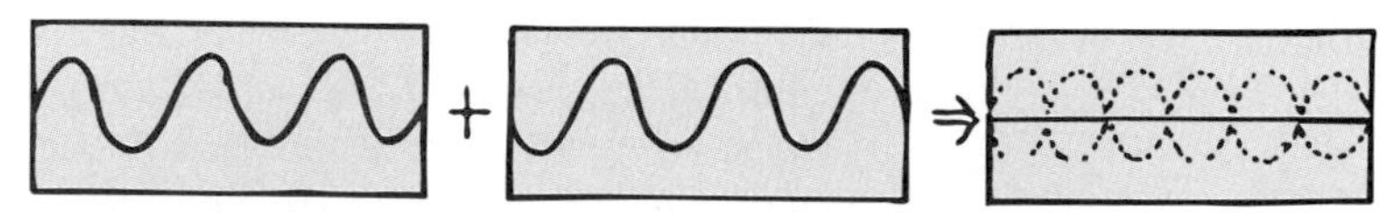

Two identical transverse waves that are out of phase destroy each otherwhen they are superimposed.

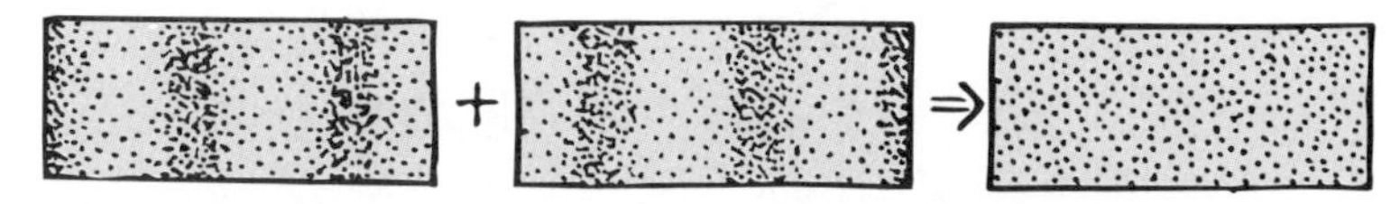

Two identical longitudinal waves that are out of phase destroy each other when they are superimposed.

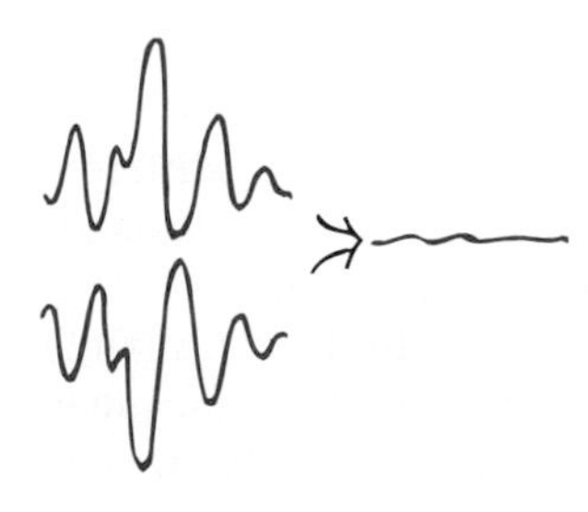

Fig. 10.21
When a mirror image of a sound signal combines with original signal, the sound is canceled.

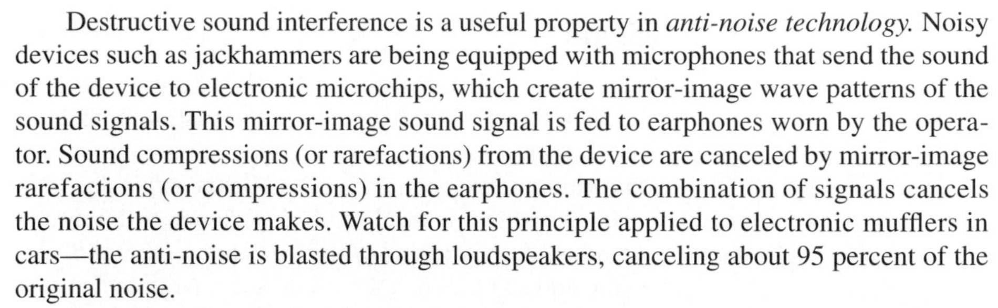

Destructive sound interference is a useful property in *anti-noise technology.* Noisy devices such as jackhammers are being equipped with microphones that send the sound of the device to electronic microchips, which create mirror-image wave patterns of the sound signals. This mirror-image sound signal is fed to earphones worn by the operator. Sound compressions (or rarefactions) from the device are canceled by mirror-image rarefactions (or compressions) in the earphones. The combination of signals cancels the noise the device makes. Watch for this principle applied to electronic mufflers in cars—the anti-noise is blasted through loudspeakers, canceling about 95 percent of the original noise.

Sound interference is dramatically illustrated when monaural sound is played by stereo speakers that are out of phase. Speakers are out of phase when the input wires to one speaker are interchanged (positive and negative wire inputs reversed). For a monaural signal, this means that when one speaker is sending a compression of sound, the other is sending a rarefaction. The sound produced is not as full and not as loud as from speakers properly connected in phase because the longer waves are being canceled by interference. Shorter waves are canceled as the speakers are brought closer together, and when the pair of speakers are brought face to face and touching each other, very little sound is heard! Only the highest frequencies survive cancellation. You must try this to appreciate it.

Fig. 10.22
The positive and negative wire inputs to one of the stereo speakers have been interchanged, resulting in speakers that are out of phase. When the speakers are far apart, monaural sound is not as loud as from properly phased speakers. When they are brought face to face, very little sound is heard. Interference is nearly complete as the compressions of one speaker fill in the rarefactions of the other!

Fig. 10.23
Ken Ford tows gliders in quiet comfort when he wears his noise-canceling earphones.

Beats

A *tone* is a sound of distinct pitch and duration. When two tones of slightly different frequency are sounded together, a fluctuation in the loudness of the combined sounds may be heard; the sound is loud, then faint, then loud, then faint, and so on. This periodic variation in loudness is called **beats** and is due to interference. Strike two slightly mismatched tuning forks, and because one fork vibrates at a different frequency than the other, the vibrations of the forks are momentarily in step, then out of step, then in again, and so on. When the combined waves reach our ears in step—when, say, a compression from one fork overlaps a compression from the other—the sound is a maximum. A moment later, when the forks are out of step, a compression from one fork and a rarefaction from the other reach our ears at the same time, and the sound is a minimum. The sound that reaches our ears throbs between maximum and minimum loudness and produces a tremolo effect.

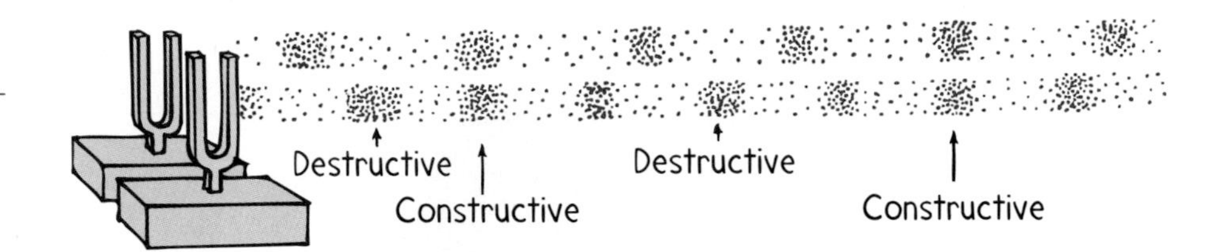

Fig. 10.24
The interference of two sound sources of slightly different frequencies produces beats.

Beats can occur with any kind of wave and provide a practical way to compare frequencies. To tune a piano, for example, a piano tuner listens for beats produced between a standard tuning fork and those of a particular string on the piano. When the frequencies are identical, the beats disappear. The members of an orchestra tune up by listening for beats between their instruments and a standard tone produced by a piano or some other instrument.

Standing Waves

Another interesting effect of interference is *standing waves.* Tie a rope to a wall and shake the free end up and down. The wall is too rigid to shake, so the waves that hit it are reflected back along the rope. By shaking the rope just right, you can cause the incident and reflected waves to interfere and form a **standing wave**, where parts of the rope, called the *nodes,* are stationary. You can hold your fingers on either side of the rope at a node, and the rope will not touch them. Thus nodes are positions of zero rope displacement in a rope containing a standing wave. The distance between any two adjacent nodes is a half wavelength. Note that this distance makes up one loop and that two loops make up a full wavelength. The positions on a standing wave that have the largest displacements are known as *antinodes* and occur halfway between nodes.

Standing waves are produced when two sets of waves of equal amplitude and wavelength pass through each other in opposite directions. Then the waves are steadily in and out of phase with each other and produce stable regions of constructive and destructive interference (Figure 10.25).

It's easy to make standing waves yourself. Tie one end of a rope or, better, one end of a length of rubber tubing to a firm support. Shake the free end of the rope or tubing up and down. When you shake with the right frequency, you set up a standing wave

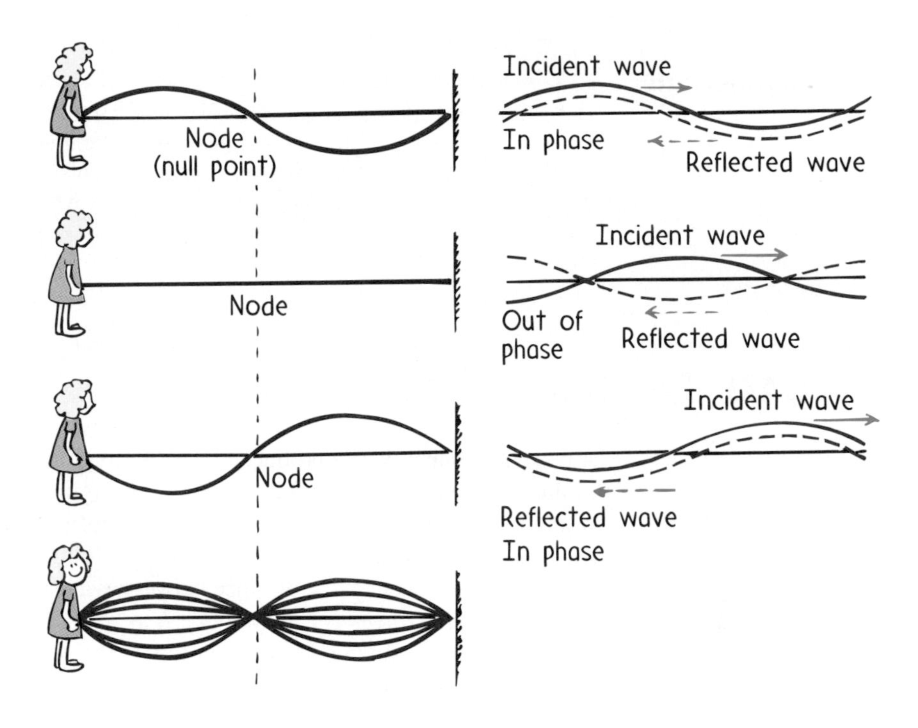

Fig. 10.25
The incident and reflected waves interfere to produce a standing wave.

consisting of one loop (which equals a half wavelength) as shown in Figure 10.26a. Shake with twice the frequency, and you set up a standing wave that consists of two loops and has a wavelength half that of the first wave (Figure 10.26b). Triple the frequency, and a standing wave with one-third the original wavelength, having three loops, results.

Standing waves are set up in the strings of musical instruments when plucked, bowed, or struck. They are set up in the air in an organ pipe, a trumpet, or a clarinet and in the air of a soda-pop bottle when air is blown over the top. Standing waves can be set up in a tub of water or a cup of coffee by sloshing it back and forth with the right frequency. They can be produced with either transverse or longitudinal vibrations.

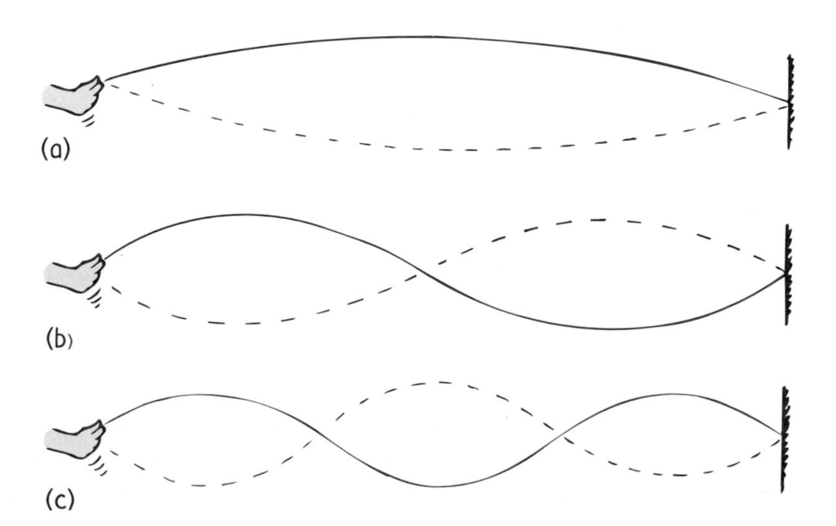

Fig. 10.26
(a) Shake the rope until you set up a standing wave of one loop ($\frac{1}{2}$ wavelength). (b) Shake with twice the frequency and produce a wave having two loops (1 wavelength). (c) Shake with three times the frequency and produce three loops ($\frac{3}{2}$ wavelengths).

Questions

1. Is it possible for one wave to cancel another so that no amplitude remains?
2. Suppose you set up a standing wave of three loops, as shown in Figure 10.26c. If you then shake at twice the frequency, how many loops will your new standing wave have? How many wavelengths?

10.10 Doppler Effect

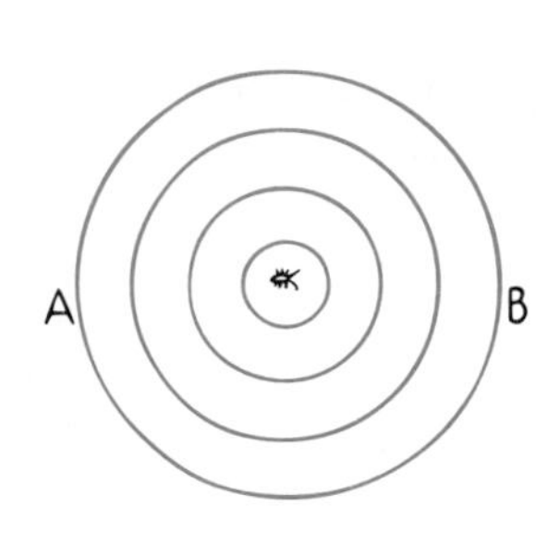

Fig. 10.27
Top view of water waves made by a stationary bug jiggling in still water.

A pattern of water waves produced by a bug jiggling its legs and bobbing up and down in the middle of a quiet puddle is shown in Figure 10.27. The bug is not going anywhere but is merely treading water in a fixed position. The waves it makes are concentric circles because wave speed is the same in all directions. If the bug bobs in the water at a constant frequency, the distance between wave crests (the wavelength) is the same in all directions. Waves encounter point A as frequently as they encounter point B. This means that the frequency of wave motion is the same at points A and B, or anywhere else in the vicinity of the bug. This wave frequency is the same as the bug's bobbing frequency.

Suppose now that the jiggling bug begins moving to the right at a speed less than the wave speed. In effect, the bug chases part of the waves it has produced. The wave pattern is distorted and is no longer made of concentric circles (Figure 10.28). The outermost wave (wave 1 in Figure 10.28) was made when the bug was at the center of that circle. The center of wave 2 was made when the bug was at the center of that circle, and so forth. The centers of the circular waves move in the direction of the swimming bug. Although the bug maintains the same bobbing frequency as before, an observer at B sees the waves coming more often and so measures a *higher* frequency. This is because each successive wave has a shorter distance to travel to B and therefore arrives at that point more frequently than if the bug weren't moving toward it. An observer at A, on the other hand, measures a *lower* frequency because of the longer time between wave-crest arrivals. This occurs because each successive wave travels farther to A as a result of the bug's motion away from A. This change in frequency due to the motion of the source (or receiver) is called the **Doppler effect** (after the Austrian scientist Christian Doppler).

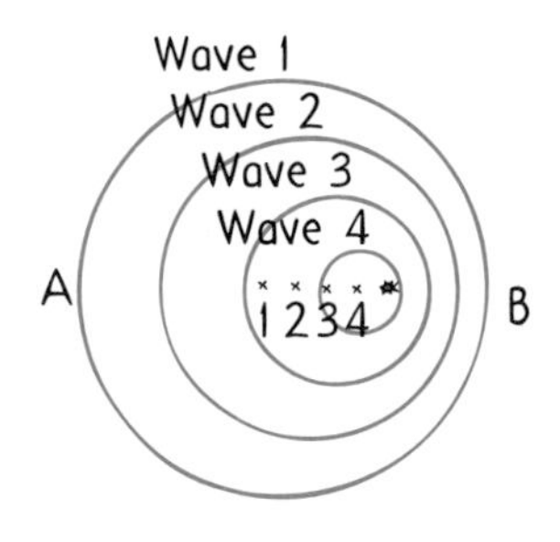

Fig. 10.28
Water waves made by a bug swimming in still water toward point B

Water waves spread over the two-dimensional surface of the water. Sound and light waves, on the other hand, travel in three-dimensional space in all directions like an expanding balloon. Just as circular waves are closer together in front of the swimming bug, spherical sound or light waves ahead of a moving source are closer together and reach a receiver more frequently. The Doppler effect holds for all types of waves.

The Doppler effect is evident when you hear the changing pitch of a car horn as the car drives by. When it approaches, the frequency and hence the pitch are higher

Answers

1. Yes. This is destructive interference. In a standing wave in a rope, for example, parts of the rope—the nodes—have no amplitude because of destructive interference.
2. If you impart twice the frequency to the rope or tubing, you produce a standing wave having twice as many loops—six. Because a full wavelength has two loops, you have three wavelengths in your new standing wave.

Fig. 10.29
The pitch (frequency) of sound increases when the source moves toward you and decreases when the source moves away.

than normal. This is because the source is moving toward you and so the crests of the sound waves hit your ear more frequently. When the car passes and moves away, you hear a drop in pitch because the crests of the waves hit your ear less frequently.

The Doppler effect is also evident when the receiver moves. Move toward a stationary wave source and you encounter its waves more frequently. Move away from the stationary source and you encounter waves less frequently. In general, the Doppler effect results from *relative* motion between any wave source and receiver.

The Doppler effect also occurs with light waves. When a source of light waves approaches a receiver (such as the human eye), there is an increase in the light's measured frequency. When the source recedes from a receiver, there is a decrease in the light's measured frequency. An increase in frequency is called a *blue shift* because the increase is toward the high-frequency, blue end of the color spectrum. A decrease in frequency is called a *red shift,* referring to a shift toward the lower-frequency, red end of the color spectrum. The galaxies, for example, show a red shift in the light they emit. A measurement of this shift permits a calculation of the speeds at which they are receding from the Earth. A rapidly spinning star shows a red shift on the side turning away from us and a relative blue shift on the side turning toward us. This enables astronomers to calculate the star's spin rate.

■ **Question**

When a wave source moves toward you, do you measure an increase or decrease in wave speed?

10.11 Wave Barriers and Bow Waves

When the speed of a moving wave source is as great as the speed of the waves it produces, a *wave barrier* is produced. Consider the bug in our previous example when it swims as fast as the wave speed. Can you see that it keeps up with the waves it produces? Instead of the waves getting ahead of the bug, they pile up and superimpose on one another directly in front of it (Figure 10.30). The bug encounters a wave barrier. Much effort is required of the bug to swim over this barrier before it can swim faster than wave speed.

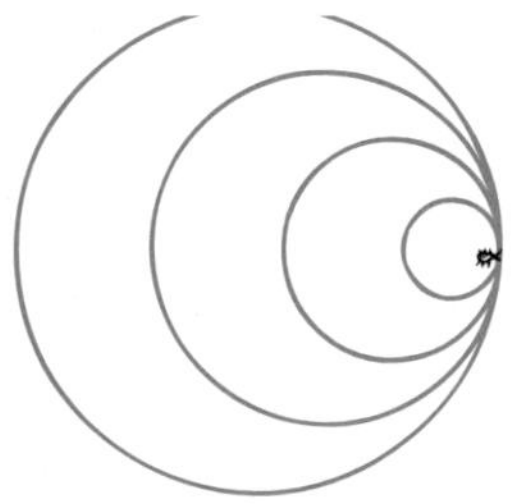

The same thing happens when an aircraft travels at the speed of sound. The sound waves produced by the engines overlap to produce a barrier of compressed air on the leading edges of the wings and other parts of the craft. Considerable thrust is required

Fig. 10.30
Wave pattern made by a bug swimming at wave speed.

■ **Answer**
Neither! It is the *frequency* of the waves that changes for a moving source, not the wave speed. Be clear about the distinction between frequency and speed. How frequently a wave vibrates is altogether different from how fast it moves from one place to another.

Fig. 10.31
This aircraft has just cracked the wave barrier. The cloud is water vapor that has just condensed out of the rapidly expanding air in the rarefied region behind the wall of compressed air.7

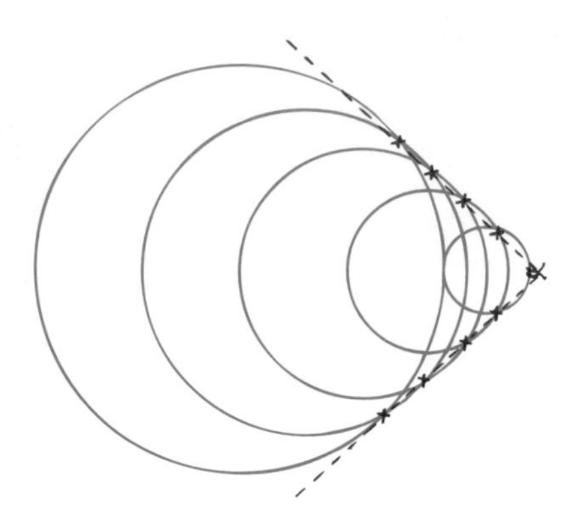

Fig. 10.32
A bow wave, the pattern made by a bug swimming faster than wave speed. The points at which adjacent waves overlap (×) produce the V shape.

for the aircraft to push through this barrier (Figure 10.31). Once through, the craft can fly faster than the speed of sound without similar opposition. It is now *supersonic.*[3] It is like the bug, which once over its wave barrier finds the water ahead relatively smooth and undisturbed.

When the bug swims faster than wave speed, it produces a wave pattern like the one shown in Figure 10.32. It outruns the waves it produces. The overlapping waves form a V shape called a **bow wave**, which appears to be dragging behind the bug. The familiar bow (rhymes with cow) wave generated by a speedboat knifing through the water is produced by the overlapping of many periodic circular waves.

Some wave patterns made by sources moving at various speeds are shown in Figure 10.33. Note that after the speed of the source exceeds wave speed, increased speeds produce a narrower and narrower V shape.[4]

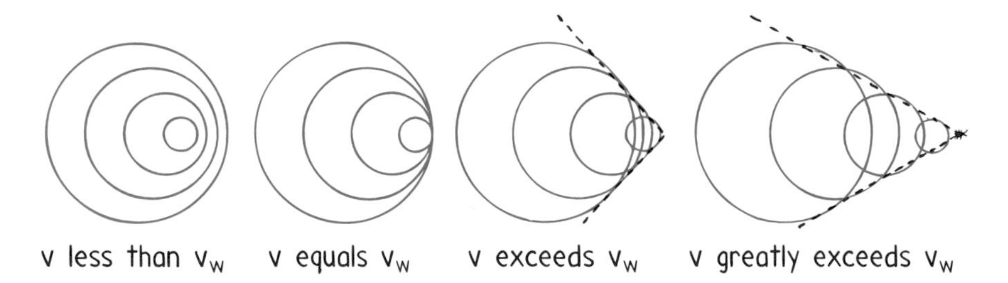

Fig. 10.33
Patterns made by a bug swimming at successively greater speeds. Overlapping at the edges occurs only when the bug swims faster than wave speed.

10.12 Shock Waves and the Sonic Boom

A speedboat knifing through the water generates a two-dimensional bow wave. A supersonic aircraft similarly generates a three-dimensional **shock wave**. Just as a bow wave is produced by overlapping circles that form a V, a shock wave is produced by overlapping spheres that form a cone. And just as the bow wave of a speedboat spreads until it

[3] *Ultra*sonic waves are those that have a frequency higher than the highest frequency the human ear can hear. *Super*sonic waves are those that travel through a medium at a speed higher than the speed of sound in that medium.

[4] Bow waves generated by boats in water are more complex than indicated here. Our idealized treatment serves as an analogy for the production of the less complex shock waves in air.

reaches the shore of a lake, the conical wake generated by a supersonic craft spreads until it reaches the ground.

The bow wave of a speedboat passing by can splash and douse you if you are at the water's edge. In a sense, you can say that you are hit by a "water boom." In the same way, when the conical shell of compressed air that sweeps behind a supersonic aircraft reaches listeners on the ground, the sharp crack they hear is described as a **sonic boom**.

We don't hear a sonic boom from slower-than-sound, or subsonic, aircraft because the sound waves reach our ears one at a time and are perceived as one continuous tone. Only when the craft moves faster than sound do the waves overlap to reach the listener in a single burst. The sudden increase in pressure is much the same in effect as the sudden expansion of air produced by an explosion. Both processes direct a burst of high-pressure air to the listener. The ear is hard pressed to distinguish between the high pressure from an explosion and the high pressure from many overlapping waves.

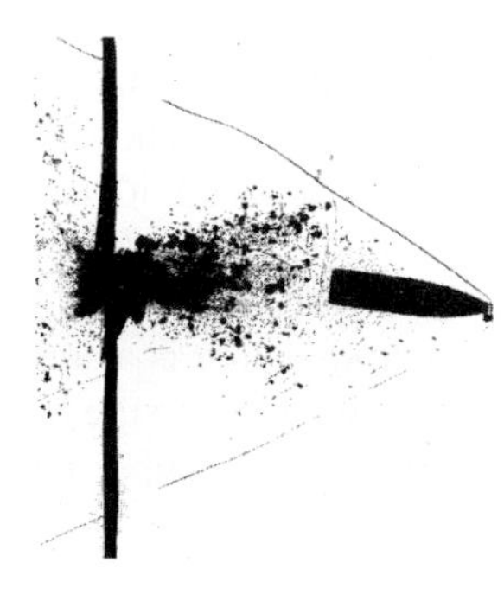

Fig. 10.34
Shock wave of a bullet piercing a sheet of Plexiglas. Light is deflected as it passes through the compressed air that makes up the shock wave, making it visible. Look carefully and see the second shock wave originating at the tail of the bullet.

A water skier is familiar with the fact that next to the high hump of the bow wave is a V-shaped depression. The same is true of a shock wave, which consists of two cones: a high-pressure cone generated at the bow of the supersonic aircraft and a low-pressure cone that follows at the tail of the craft. The edges of these cones are visible in the photograph of the supersonic bullet in Figure 10.34. Between these two cones the air pressure rises sharply to above atmospheric pressure, then falls below atmospheric pressure before sharply returning to normal beyond the inner tail cone (Figure 10.35). This overpressure suddenly followed by underpressure intensifies the sonic boom.

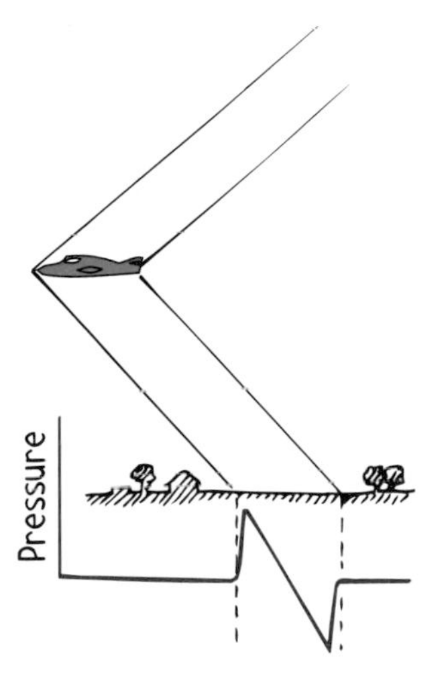

Fig. 10.35
A shock wave and the air pressure differences it causes.

A common misconception is that a sonic boom is produced only at the moment an aircraft breaks through the sound barrier—that is, just as the aircraft surpasses the speed of sound. This is the same as saying that a boat produces a bow wave only at the point where it overtakes its own waves. This is not so. A shock wave and its resulting sonic boom are swept *continuously* behind an aircraft the whole time the craft is traveling faster than sound, just as a bow wave is swept continuously behind a speedboat. In Figure 10.37, listener B is in the process of hearing a sonic

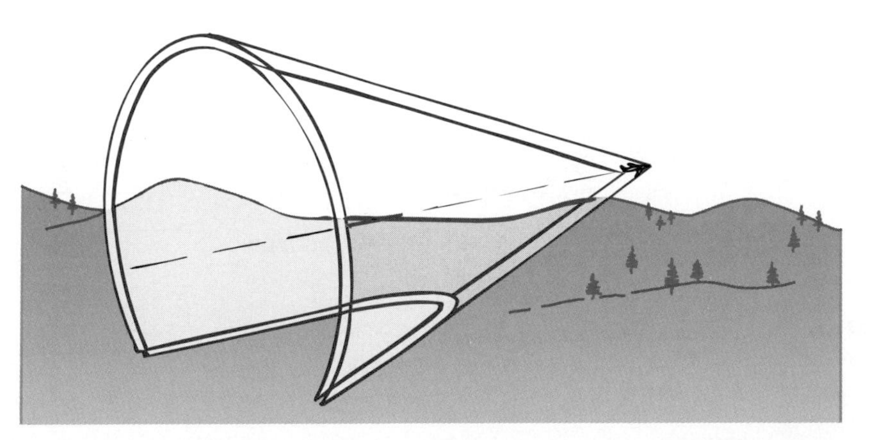

Fig. 10.36
A shock wave is made up of two cones—a high-pressure cone with the apex at the bow of the aircraft and a low-pressure cone with the apex at the tail.

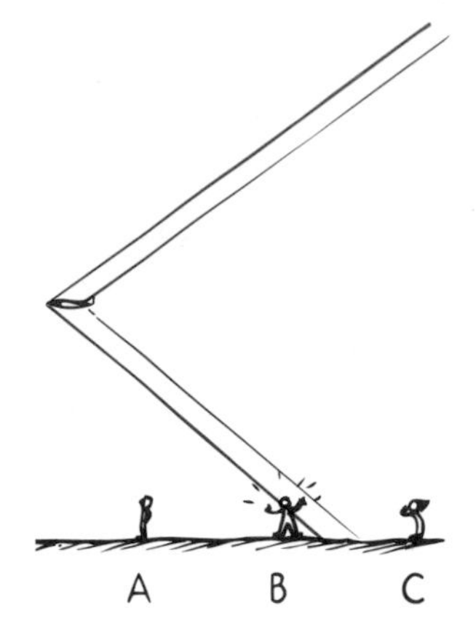

Fig. 10.37
The shock wave has not yet reached listener A, but is now reaching listener B and has already reached listener C.

boom. Listener A has already heard it, and listener C will hear it shortly. The aircraft that generated this shock wave may have broken through the sound barrier hours ago!

It is not necessary that the moving source be "noisy" to produce a shock wave. Once an object—even a silent one—is moving faster than the speed of sound, it *makes* sound. A supersonic bullet passing overhead produces a crack, which is a small sonic boom. If the bullet were larger and disturbed more air in its path, the crack would be more boomlike. When the a lion tamer cracks a whip, the cracking sound is a sonic boom produced by the tip when it travels faster than the speed of sound. Both the bullet and the whip are not in themselves sound sources, but when traveling at supersonic speeds they produce their own sound as they generate shock waves.

10.13 Musical Sounds

Most of the sounds we hear are noises. The impact of a falling object, the slamming of a door, the roaring of a motorcycle, and most of the sounds from traffic in city streets are noises. Noise corresponds to an irregular vibration of the eardrum produced by some irregular vibration originating from some source of noise. If we make a diagram to indicate how air pressure on the eardrum varies with time, the graph corresponding to a noise might look like that shown in Figure 10.38a. The sound of music has a different character, having a variety of tones—or musical "notes." The graph representing a musical sound has a shape that repeats itself over and over again (Figure 10.38b).

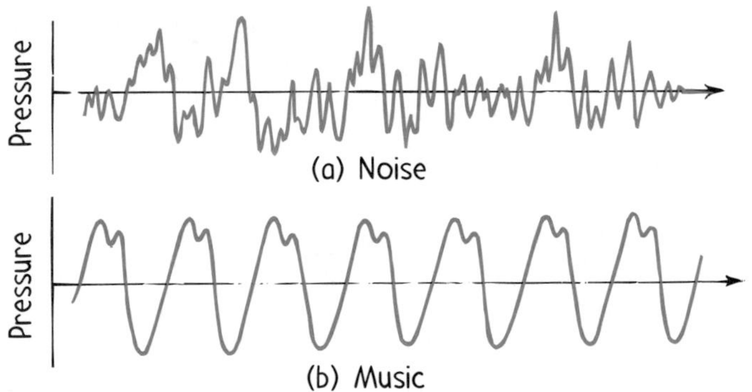

Fig. 10.38
Graphical representation of (a) noise, and (b) music.

We have no trouble distinguishing between the sound from a piano and a like-pitched sound from a clarinet. Each tone has a characteristic sound that differs in *quality,* or *timbre.* Most mu-

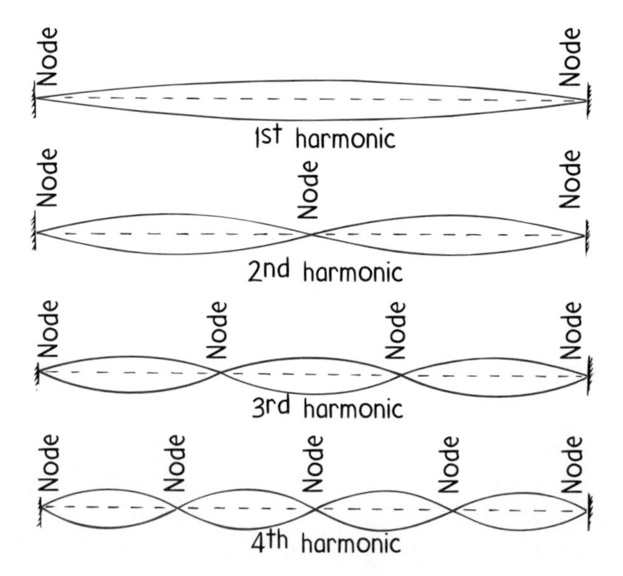

Fig. 10.39
Modes of vibration of a guitar string.

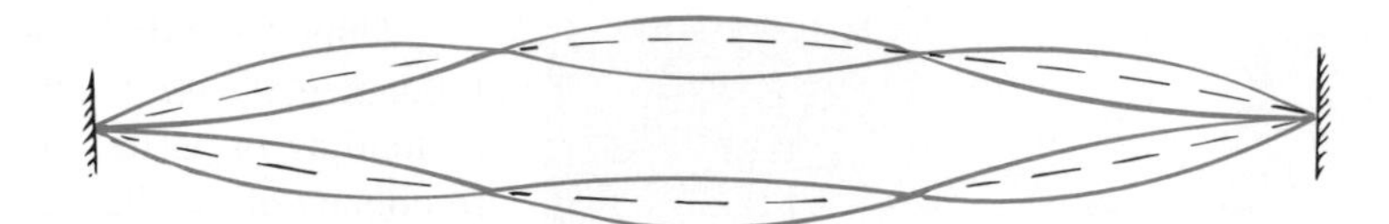

Fig. 10.40
A composite vibration of the fundamental mode and the third harmonic.

sical sounds are composed of a superposition of many frequencies called *partial tones,* or simply *partials.* The lowest frequency of a musical note, called the **fundamental frequency**, determines the pitch of the note. Partial tones that are whole multiples of the fundamental frequency are called **harmonics**. A tone that has twice the frequency

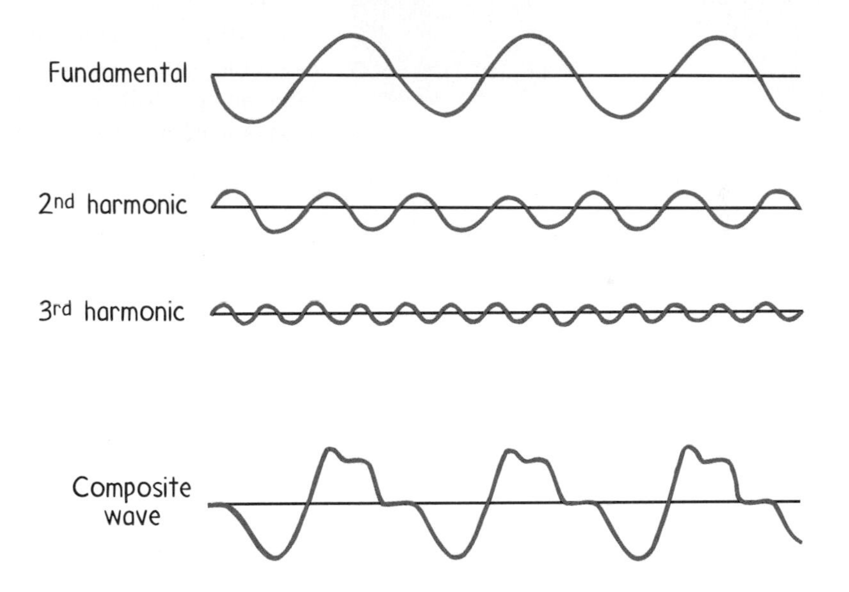

Fig. 10.41
Sine waves combine to produce a composite wave.

of the fundamental is the second harmonic, a tone with three times the fundamental frequency is the third harmonic, and so on (Figure 10.39).[5] It is the variety of partial tones that give a musical note its characteristic quality.

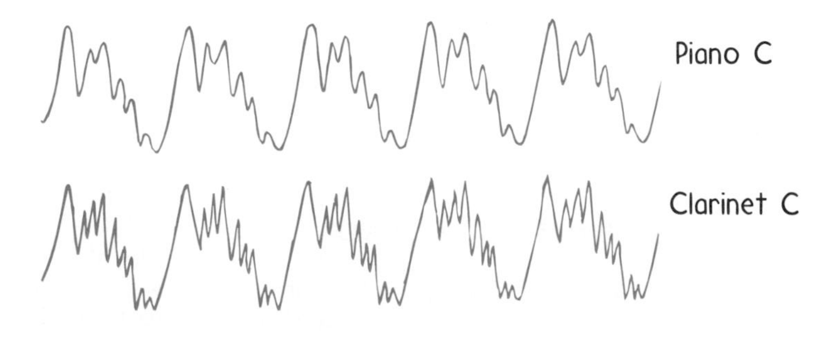

Fig. 10.42
Sounds from the piano and clarinet differ in quality.

[5]Not all partial tones present in a complex tone are integer multiples of the fundamental. Unlike the harmonics of woodwinds and brasses, stringed instruments such as a piano produce "stretched" partial tones that are nearly, but not quite, harmonics. This is an important factor in tuning pianos and happens because the stiffness of the strings adds a little bit of restoring force to the tension.

Fig. 10.43
Does each listener hear the same music?

Thus if we strike middle C on the piano, we produce a fundamental tone that has a pitch of about 262 hertz and also a blending of partial tones of 2, 3, 4, 5, and so on times the frequency of middle C. The number and relative loudness of the partial tones determine the quality of sound associated with the piano. Sound from practically every musical instrument consists of a fundamental plus partials. Pure tones, those having only one frequency, can be produced electronically. Electronic synthesizers, for example, produces pure tones and mixtures of these to give a vast variety of musical sounds.

The sounds produced by a certain note from a piano and a clarinet have different qualities that the ear recognizes because the partials are different. A pair of tones having the same pitch but different qualities have either different partials or a difference in the relative intensity of the partials.

Summary of Terms

- **Wave** A disturbance or vibration propagated from point to point in a medium or in space.
- **Amplitude** For a wave or vibration, the maximum displacement on either side of the equilibrium (midpoint) position.
- **Wavelength** The distance between successive crests, troughs, or identical parts of a wave.
- **Frequency** For a vibrating body or medium, the number of vibrations per unit time. For a wave, the number of crests that pass a particular point per unit time.
- **Hertz** The unit of frequency. One hertz (symbol Hz) equals one vibration per second.
- **Period** The time required for a vibration or a wave to make a complete cycle; equal to 1/frequency.
- **Wave speed** The speed with which waves pass a particular point:

  Wave speed = wavelength × frequency

- **Transverse wave** A wave in which the medium vibrates in a direction perpendicular (transverse) to the direction in which the wave travels. Light is an example of a transverse wave.
- **Longitudinal wave** A wave in which the medium vibrates in a direction parallel (longitudinal) to the direction in which the wave travels. Sound is an example of a longitudinal wave.
- **Sound wave** A longitudinal vibratory disturbance that travels in a medium and can be heard by the human ear when in the approximate frequency range 20–20,000 Hz.
- **Refraction** The bending of a wave as it passes either through a nonuniform medium or from one medium to another, caused by differences in wave speed.
- **Forced vibration** The setting up of vibrations in an object by a vibrating force.
- **Natural frequency** A frequency at which an elastic object naturally tends to vibrate, so that minimum energy is required to produce a forced vibration or to continue vibration at that frequency.
- **Resonance** The response of a body when a forcing frequency matches its natural frequency.
- **Interference** The pattern formed by superposition of different sets of waves that produces mutual reinforcement in some places and cancellation in others.
- **Beats** A series of alternate reinforcements and cancellations produced by the interference of two waves of slightly different frequency, heard as a throbbing effect in sound waves.
- **Standing wave** A stationary wave pattern formed in a medium when two sets of identical waves pass through the medium in opposite directions.
- **Doppler effect** The change in frequency of wave motion resulting from motion of the wave source or receiver.
- **Bow wave** The V-shaped wave made by an object moving across a liquid surface at a speed greater than the wave speed.

- **Shock wave** The cone-shaped wave made by an object moving at supersonic speed through a fluid.
- **Sonic boom** The loud sound resulting from the incidence of a shock wave.
- **Fundamental frequency** The lowest frequency of vibration of a musical note.
- **Harmonic** A frequency that is an integer multiple of the fundamental.

Review Questions

Vibrations and Waves

1. What is a *wiggle in time* called? What is a *wiggle in space and time* called?
2. What is the source of all waves?
3. How do frequency and period relate to each other?

Wave Motion

4. What is it that moves from source to receiver in wave motion?
5. Does the medium in which a wave moves travel along with the wave? Give examples to support your answer.
6. What is the relationship among frequency, wavelength, and wave speed?
7. As the frequency of a wave of constant speed is increased, does the wavelength increase or decrease?

Transverse and Longitudinal Waves

8. In a transverse wave, in what direction are the vibrations relative to the direction of wave travel?
9. In a longitudinal wave, in what direction are the vibrations relative to the direction of wave travel?
10. Distinguish between a compression and a rarefaction.
11. How do the compressions and rarefactions of longitudinal waves compare to the crests and troughs of transverse waves?
12. Do compressions and rarefactions travel in the same or opposite directions from one another in a wave? Cite evidence to support your answer.

Sound Waves

13. How does the paper cone of a radio loudspeaker emit sound?
14. Relative to solids and liquids, how does air rank as a conductor of sound?
15. Why will sound not travel in a vacuum?
16. What factors does the speed of sound depend on? What are some factors that it does *not* depend on?
17. Does sound travel faster in warm air or in cold air? Defend your answer.
18. How does the speed of sound in water compare with the speed of sound in air? How does the speed in steel compare with the speed in air?

Reflection of Sound

19. What is the law of reflection for sound?
20. What is a reverberation?

Refraction of Sound

21. What causes refraction?
22. Does sound tend to bend upward or downward when its speed near the ground is lower than its speed higher up?
23. There is a difference between the way we passively see our surroundings in daylight and the way we actively probe our surroundings with a searchlight in the darkness. Which of these ways of perceiving our surroundings is more like the way a dolphin perceives its environment?

Forced Vibrations

24. Why does a struck tuning fork sound louder when it is held against a table?
25. Give three examples of forced vibration.
26. Does a blob of putty have a natural frequency? Explain.

Resonance

27. Distinguish between forced vibrations and resonance.
28. What is required to make an object resonate?
29. When you listen to a radio, why do you hear only one station at a time instead of all stations at once?
30. Why do troops "break step" when crossing a bridge?

Interference

31. Is it possible for one wave to cancel another? Defend your answer.
32. What kind of waves exhibit interference?
33. Distinguish between constructive interference and destructive interference.
34. What is responsible for "dead spots" in poorly designed theaters or concert halls?
35. What physical phenomenon underlies the production of beats?
36. What does it mean to say one wave is out of phase with another?
37. What causes a standing wave?
38. What is a node? An antinode?
39. In a standing wave, how many wavelengths apart are two adjacent nodes? Two adjacent antinodes?

Doppler Effect

40. In the Doppler effect, does frequency change? Does wavelength change? Does wave speed change?
41. Can the Doppler effect be observed with longitudinal waves, transverse waves, or both?

Wave Barriers and Bow Waves

42. How do the speed of a wave source and the speed of the waves themselves compare when a wave barrier is being produced?

43. How do the speed of a wave source and the speed of the waves themselves compare when a bow wave is being produced?
44. How does the V shape of a bow wave depend on the speed of the wave source?

Shock Waves and the Sonic Boom

45. How does the V shape of a shock wave depend on the speed of the wave source?
46. True or false: A sonic boom occurs only when an aircraft is breaking through the sound barrier.
47. True or false: For an object to produce a sonic boom, it must be a sound source.

Musical Sounds

48. Distinguish between a musical sound and noise.

Home Projects

1. Tie a length of rubber tubing, a spring, or a rope to a fixed support and produce standing waves. See how many nodes you can produce.
2. Test to see which ear has the better hearing by covering one ear and finding how far away your open ear can hear the ticking of a clock; repeat for the other ear. Notice also how the sensitivity of your hearing improves when you cup your ears with your hands.

Exercises

1. What kind of motion should you impart to the nozzle of a garden hose so that the resulting stream of water approximates a sine curve?
2. What kind of motion should you impart to a stretched coiled spring (or Slinky) to provide a transverse wave? A longitudinal wave?
3. If a gas tap is turned on for a few seconds, someone a couple of meters away will hear the gas escaping long before she smells it. What does this indicate about the way in which sound waves travel?
4. If we double the frequency of a vibrating object, what happens to its period?
5. You dip your finger repeatedly into a puddle of water and make waves. What happens to the wavelength if you dip your finger more frequently?
6. How does the frequency of vibration of a small object floating in water compare with the number of waves passing the object each second?
7. In terms of wavelength, how far does a wave travel during one period?
8. Suppose a sound wave and an electromagnetic wave have the same frequency. Which has the longer wavelength?
9. The wave patterns seen in Figure 10.4 are composed of circles. What does this tell you about the speed of waves moving in different directions?
10. A rock is dropped in water, and waves spread over the flat surface of the water. What becomes of the energy in these waves when they die out?
11. What two physics mistakes occur in a science fiction movie that shows a distant explosion in outer space, where you see and hear the explosion at the same time?
12. Why is it undesirable for a radio loudspeaker to have resonant frequencies in the audio range? (Some cheap AM radios have this problem.)
13. Would there be a Doppler effect if the source of sound were stationary and the listener in motion? Why or why not? In which direction should the listener move to hear a higher frequency? A lower frequency?
14. When you blow your horn while driving toward a stationary listener, the listener hears an increase in the horn frequency. Would the listener hear an increase in the horn frequency if he were in another car traveling at the same speed in the same direction as you are? Explain.
15. Is it correct to say that the Doppler effect is the apparent change in the speed of a wave due to motion of the source? (Why is this question a test of reading comprehension as well as a test of physics knowledge?)
16. Astronomers find that light coming from one edge of the Sun has a slightly higher frequency than light from the opposite edge. What do these measurements tell us about the Sun's motion?
17. What can you say about the speed of a boat that makes a bow wave?
18. Does the conical angle of a shock wave open wider, narrow down, or remain constant as a supersonic aircraft increases its speed?
19. Imagine a superfast fish able to swim faster than the speed of sound in water. Would such a fish produce a "water boom"?
20. A weight suspended from a spring bobs up and down over a distance of 20 cm twice each second. What is its frequency? Its period? Its amplitude?
21. Why do flying bees buzz?
22. A cat can hear sound frequencies up to 70,000 Hz. Bats send and receive ultrahigh-frequency squeaks up to 120,000 Hz. Which hears shorter wavelengths, cats or bats?
23. At the stands of a race track, you notice smoke from the starter's gun before you hear it fire. Explain.
24. When a sound wave moves past a point in air, are there changes in the density of air at this point? Explain.
25. At the instant a high-pressure region is created just outside the prongs of a vibrating tuning fork, what is being created between the prongs?
26. Why is it so quiet after a snowfall?

27. A bell jar is a popular lecture demonstration device used to show how a vacuum affects sound. As the air in the jar is evacuated by a vacuum pump, a ringing bell inside is heard less and less. When all the air has been pumped out, you can no longer hear the bell but you can see it. What differences in the properties of sound and light does this indicate?
28. Why is the Moon described as a "silent planet"?
29. As you pour water into a glass, you repeatedly tap the glass with a spoon. As the tapped glass is being filled, does the pitch of the sound increase or decrease? (What should you do to answer this question?)
30. If the speed of sound depended on frequency, how would distant music sound?
31. If the frequency of a sound wave is doubled, what change occurs in its speed? In its wavelength?
32. Why does sound travel faster in warm air than in cooler air?
33. Why does sound travel faster in moist air than in dry air? (*Hint:* At the same temperature, water vapor molecules have the same average kinetic energy as the heavier nitrogen and oxygen molecules in the air. How, then, do the average speeds of H_2O molecules compare with those of N_2 and O_2 molecules?)
34. Would sound refraction occur if the speed of sound were unaffected by wind, temperature, and other conditions? Defend your answer.
35. Why can the tremor of the ground from a distant explosion be felt before the sound of the explosion can be heard?
36. What kinds of wind conditions would make sound more easily heard at long distances? Less easily heard at long distances?
37. Ultrasonic waves have many applications in technology and medicine. One advantage is that high intensities can be used without danger to the ear. Cite another advantage of their short wavelength. (*Hint*: Why do microscopists use blue light rather than white light to see detail?)
38. Why is an echo weaker than the original sound?
39. A rule of thumb for estimating the distance in kilometers between an observer and a lightning stroke is to divide the number of seconds in the interval between the flash and the sound by 3. Is this rule correct?
40. If a single disturbance some unknown distance away sends out both transverse and longitudinal waves that travel with distinctly different speeds in the medium, such as in the ground during earthquakes, how could the origin of the disturbance be located?
41. Why will marchers at the end of a long parade following a band be out of step with marchers near the front?
42. A special device can transmit out-of-phase sound from a noisy jackhammer to earphones worn by its operator. Over the noise of the jackhammer, the operator can easily hear your voice while you are unable to hear his. Explain.
43. If a guitar string vibrates in two loops, where can a tiny piece of folded paper be supported without flying off? How many pieces of folded paper can be supported if the wave form contains three loops?
44. When water is poured into a glass, does the pitch of the sound of the filling glass increase or decrease? (Think about this; then try it and hear for yourself.)

Problems

1. From far away you watch a woman driving nails into her front porch at a regular rate of one stroke per second. You hear the sound of the blows exactly synchronized with the blows you see. And then you hear one more blow after you see her stop hammering. How far away is she?
2. A skipper on a boat notices wave crests passing his anchor chain every 5 s. He estimates the distance between wave crests to be 15 m. He also correctly estimates the speed of the waves. What is this speed?
3. Radio waves travel at the speed of light—300,000 km/s. What is the wavelength of radio waves received at 100.1 MHz on your radio dial?
4. A mosquito flaps its wings 600 vibrations per second (it is this motion that produces the annoying 600-Hz buzz). How far does the sound travel between wing beats? In other words, find the wavelength of the mosquito's sound.
5. An oceanic depth-sounding vessel surveys the ocean bottom with ultrasonic sound that travels 1530 m/s in seawater. How deep is the water if the time delay of the echo from the ocean floor is 6 s?
6. A bat flying in a cave emits a sound and receives its echo 1 s later. How far away is the cave wall?
7. What is the shortest wavelength you can hear?
8. What frequency of sound produces a wavelength of 1 meter in room-temperature air?

CHAPTER

11

Light Waves

Light is the only thing we can really see. But what *is* light? We know that during the day the primary source of light is the Sun and the secondary source is the brightness of the sky. Other common sources are flames, white-hot filaments in light bulbs, and glowing gases in glass tubes. Knowing the various sources of light doesn't tell us what it is, however.

Light originates from the accelerated motion of electrons. It is an electromagnetic phenomenon and only a tiny part of a larger whole—a wide range of electromagnetic waves called the *electromagnetic spectrum.*

We begin our study of light by investigating its electromagnetic properties, how it interacts with materials, and its appearance—color. We see its wave nature in the way it diffracts and interferes and in how it aligns to become polarized.

11.1 The Electromagnetic Spectrum

Fig. 11.1
Shake an electrically charged object to and fro, and you produce an electromagnetic wave.

Shake the end of a stick back and forth in still water, and you produce waves on the water surface. Similarly shake an electrically charged rod to and fro in empty space, and you produce electromagnetic waves in space. This is because the moving charge is an electric current. Recall from Chapter 9 that a magnetic field surrounds an electric current and changes as the charges accelerate. Recall also from Chapter 9 that a changing magnetic field induces an electric field—electromagnetic induction. And what does the changing electric field do? In accordance with Maxwell's counterpart to Faraday's law, it induces a changing magnetic field. The vibrating electric and magnetic fields regenerate each other to make up an **electromagnetic wave**. In a vacuum, all electromagnetic waves move at the same speed.

The classification of electromagnetic waves according to frequency is the **electromagnetic spectrum** (Figure 11.3). Electromagnetic waves that have a frequency as low as 0.01 hertz (Hz) have been detected. Electromagnetic waves that have frequencies on the order of several thousand hertz (kilohertz, kHz) are classified as low-frequency radio waves. A frequency of 1 million hertz (1 megahertz, MHz) lies in the middle of the AM radio band. The very-high-frequency (VHF) television band of waves starts at about 50 megahertz, and FM radio is found from 88 to 108 megahertz. Then come ultrahigh frequencies (UHF), followed by microwaves, beyond which are infrared waves, often called "heat waves." Farther to the right in Figure 11.3 is visible light, which makes up less than a millionth of 1 percent of the electromagnetic spectrum. The lowest frequency of light we can see with our eyes appears red. The highest visible frequencies are nearly twice the frequency of red and appear violet. Still higher frequencies are ultraviolet. These higher-frequency waves are more energetic and cause sunburns. Higher frequencies beyond ultraviolet extend into the X-ray and gamma-ray regions. There is no sharp boundary between these regions, which actually overlap each other. The spectrum is broken up into these arbitrary regions merely for classification.

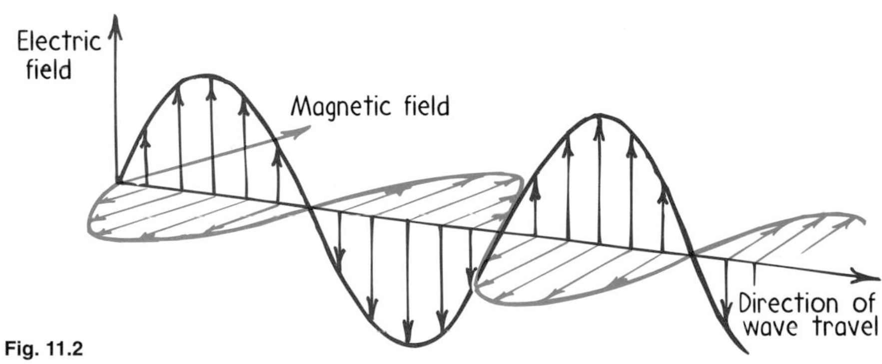

Fig. 11.2
The electric and magnetic fields of an electromagnetic wave are perpendicular to each other and to the direction of motion of the wave.

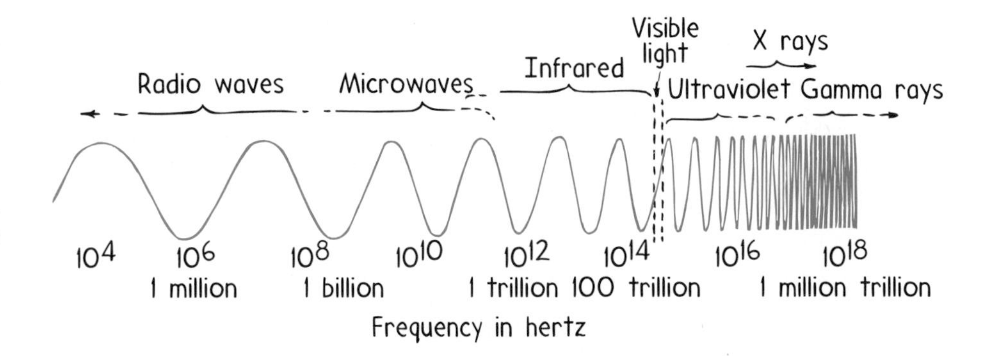

Fig. 11.3
The electromagnetic spectrum is a continuous range of waves extending from radio waves to gamma rays. The descriptive names of the sections are merely a historical classification, for all waves are the same in nature, differing principally in frequency and wavelength; all travel at the same speed.

The frequency of an electromagnetic wave as it vibrates through space is identical to the frequency of the vibrating electric charge that generates it. Different frequencies result in different wavelengths—low frequencies produce long wavelengths, and high frequencies produce short wavelengths. The higher the frequency of the vibrating charge, the shorter the wavelength of radiation.[1]

■ **Question**

Is it correct to say that a radio wave is a low-frequency light wave? Is a radio wave also a sound wave?

11.2 Transparent and Opaque Materials

Light is energy carried in an electromagnetic wave emitted by vibrating electrons in atoms. When light is transmitted through matter, some of the electrons in the matter are forced into vibration. In this way, vibrations in the emitter are transmitted to vibrations in the receiver. This is similar to the way sound is transmitted (Figure 11.4).

Thus the way a receiving material responds when light is incident upon it depends on the frequency of the light and on the natural frequency of the electrons in the material. Visible light vibrates at a very high frequency, some 100 trillion times per second (10^{14} hertz). If a charged object is to respond to these ultrafast vibrations, it must have very, very little inertia. Because the mass of electrons is so tiny, they can vibrate at this rate.

Materials such as glass and water allow light to pass through in straight lines. We say they are **transparent** to light. To understand how light gets through a transparent material, visualize the electrons in the atoms of transparent materials as if they were

Fig. 11.4
Just as a sound wave can force a sound receiver into vibration, a light wave can force electrons in materials into vibration.

■ **Answer**

Both a radio wave and a light wave are electromagnetic waves, which originate in the vibrations of electrons. Because radio waves have lower frequencies than light waves, a radio wave may be considered a low-frequency light wave (and a light wave a high-frequency radio wave). A sound wave, however, is a *mechanical* vibration of matter and is *not* electromagnetic. A sound wave is fundamentally different from an electromagnetic wave. So a radio wave is definitely not a sound wave.

[1]The relationship is $c = f\lambda$, where c is the wave's speed (constant), f is the frequency, and λ is the wavelength. It is common to describe sound waves and radio waves by frequency and light waves by wavelength. In this book, however, we favor the single concept of frequency in describing light.

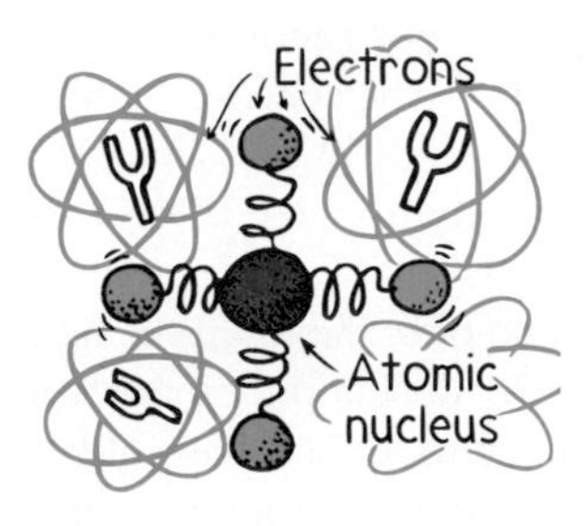

Fig. 11.5
The electrons of atoms have certain natural frequencies of vibration and can be modeled as particles connected to the atomic nucleus by springs. As a result, atoms and molecules behave somewhat like optical tuning forks.

connected to the nucleus by springs (Figure 11.5).[2] When a light wave is incident upon them, the electrons are set into vibration.

As we learned in Section 10.8, materials that are springy (elastic) respond more to vibrations at some frequencies than at other frequencies. Bells ring at a particular frequency, tuning forks vibrate at a particular frequency, and so do the electrons of atoms and molecules. The natural vibration frequencies of an electron depend on how strongly it is attached to its atom or molecule. Different atoms and molecules have different "spring strengths." Electrons in the atoms of glass have a natural vibration frequency in the ultraviolet range. Therefore when ultraviolet waves shine on glass, resonance occurs and the vibration of electrons builds up to large amplitudes, just as pushing someone at the resonant frequency on a swing builds to a large amplitude. The energy any glass atom receives is either reemitted or passed on to neighboring atoms by collisions. Resonating atoms in the glass can hold onto the energy of the ultraviolet light for quite a long time (about 100-millionths of a second). During this time the atom makes about 1 million vibrations, and it collides with neighboring atoms and gives up its energy as heat. Thus glass is not transparent to ultraviolet.

At lower wave frequencies, such as those of visible light, electrons in the glass atoms are forced into vibration, but at less amplitude. The atoms hold the energy for less time, with less chance of collision with neighboring atoms, and less energy transformed to heat. The energy of vibrating electrons is reemitted as light. Glass is transparent to all the frequencies of visible light. The frequency of the reemitted light that is passed from atom to atom is identical to the frequency of the light that produced the vibration in the first place. However, there is a slight time delay between absorption and reemission.

It is this time delay that results in a lower average speed of light through a transparent material (Figure 11.6). Light travels at different average speeds through different materials. We say *average speeds* because the speed of light in a vacuum, whether in interstellar space or in the space between molecules in a piece of glass, is a constant

Fig. 11.6
A wave of visible light incident upon a pane of glass sets up in the glass atoms vibrations that produce a chain of absorptions and reemissions, which pass the light energy through the material and out the other side. Because of the time delay between absorptions and reemissions, the light travels through the glass more slowly than through empty space.

[2]Electrons, of course, are not really connected to the nucleus by springs. We are simply presenting a visual "spring model" of the atom to help us understand the interaction of light with matter. Physicists devise such conceptual models to understand nature, particularly at the submicroscopic level. The worth of a model lies not in whether it is "true" but in whether it is useful. A good model not only is consistent with and explains observations but also predicts what may happen. If predictions of the model are contrary to what happens, the model is usually either refined or abandoned. The simplified model we present here—an atom whose electrons vibrate as if on springs, with a time interval between absorbing energy and reemitting energy—is quite useful for understanding how light passes through transparent material.

[3]The presently accepted value is 299,792 km/s, rounded to 300,000 km/s. (This corresponds to 186,000 mi/s).

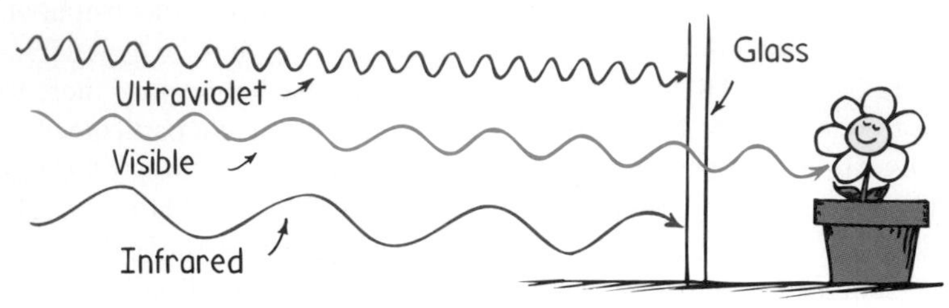

Fig. 11.7
Glass blocks both infrared and ultraviolet waves but is transparent to all the frequencies of visible light.

300,000 kilometers per second. We call this speed of light c.[3] The speed of light in the atmosphere is slightly less than in a vacuum but is usually rounded off as c. In water, light travels at 75 percent of its speed in a vacuum, or $0.75c$. In glass, light travels about $0.67c$, depending on the type of glass. In a diamond, light travels at less than half its speed in a vacuum, only $0.41c$. When light emerges from these materials into the air, it again travels at its original speed c.

Infrared waves, which have frequencies lower than those of visible light, vibrate entire molecules in the glass. This molecular vibration increases the internal energy and temperature of the structure, which is why infrared waves are often called *heat waves.* Glass is transparent to visible light, but not to ultraviolet and infrared light.

Questions

1. Why is glass transparent to visible light but not to ultraviolet and infrared?
2. Suppose that as you walk across a room you make several momentary stops along the way to greet people who are "on your wavelength." How is this analogous to visible light traveling through glass?
3. In what way is it not analogous?

Most things around us are **opaque**—they absorb light without reemitting it. Books, desks, chairs, and people are opaque. Vibrations given by light to their atoms and molecules are turned into random kinetic energy—into internal energy. They become slightly warmer.

Metals are opaque. As we learned in Section 7.15, the outer electrons of atoms in metals are not bound to any particular atom. They are free to wander with very little restraint throughout the material (which is why metal conducts electricity and heat so well). When light shines on metal and sets these free electrons into vibration, their energy does not "spring" from atom to atom in the material but is instead reflected. That's why metals are shiny.

Answers

1. Because the natural vibration frequency for electrons in glass is the same as the frequency of ultraviolet light, resonance occurs when ultraviolet waves shine on glass. The absorbed energy is passed on to other atoms as heat, not reemitted as light, making the glass not transparent at ultraviolet frequencies. In the range of visible light, the forced vibrations of electrons in the glass are at smaller amplitudes—vibrations are more subtle, reemission of light rather than the generation of heat occurs, and the glass is transparent. Lower-frequency infrared light causes whole molecules, rather than electrons, to resonate; again heat is generated and the glass is opaque.
2. Your average speed across the room is less than it would be in an empty room because of the time delays associated with your momentary stops. Likewise, the speed of light in glass is less than in air because of the time delays caused by the light's interactions with atoms along its path.
3. In walking across the room, it is you who begin and complete the walk. This is not analogous to the light case, for according to our model for light passing through a transparent material, the light absorbed by the first electron made to vibrate is not the same light that is reemitted—even though the two, like identical twins, are indistinguishable.

Fig. 11.8
Metals are shiny because light that shines on them forces free electrons into vibration, and these vibrating electrons then emit their "own" light waves as reflection.

The Earth's atmosphere is transparent to some ultraviolet light, all visible light, and some infrared light but is opaque to high-frequency ultraviolet light. The small amount of ultraviolet that does get through is responsible for sunburns. If it all got through, we would be fried to a crisp. Clouds are semitransparent to ultraviolet, which is why you can get a sunburn on a cloudy day. Ultraviolet light is not only harmful to the skin but is also damaging to tar roofs. Now you know why tarred roofs are covered with gravel.

Have you noticed that things look darker when they are wet than when they are dry? Light incident on a dry surface bounces directly to your eye, while light incident on a wet surface bounces around inside the transparent wet region before it reaches your eye. What happens with each bounce? Absorption! So more absorption of light occurs in a wet surface, and the surface looks darker.

11.3 Color

Roses are red and violets are blue; colors intrigue artists and physics types too. To the physicist, the colors of objects are not in the substances of the objects or even in the light they emit or reflect. Color is a physiological experience and is in the eye of the beholder. So when we say that light from a rose is red, in a stricter sense we mean that it *appears* red. Many organisms, including people with defective color vision, do not see the rose as red at all.

The colors we see depend on the frequency of the light we see. Different frequencies of light are perceived as different colors; the lowest frequency we detect appears to most people as the color red, and the highest as violet. Between them range the infinite number of hues that make up the color spectrum of the rainbow. By convention these hues are grouped into the seven colors red, orange, yellow, green, blue, indigo, and violet. These colors blended together appear white. The white light from the Sun is a composite of all the visible frequencies.

Selective Reflection

Except for light sources such as lamps, lasers, and gas discharge tubes (which we treat in Chapter 12), most of the objects around us reflect rather than emit light. They reflect only part of the light incident upon them, the part that gives them their color. A rose, for example, doesn't emit light; it reflects it. If we pass sunlight through a prism and then place a deep-red rose in various parts of the spectrum, the petals appear brown or black in all parts of the spectrum except the red. In the red part of the spectrum, the petals appear red but the green leaves appear black. This shows that the red petals have the ability to reflect red light but not other colors of light; likewise the green leaves have the ability to reflect green light but not other colors. When the rose is held in white light, the petals appear red and the leaves green because the petals reflect the red part of the white light and the leaves reflect the green part. To understand why objects reflect specific colors of light, we must turn our attention to the atom.

Fig. 11.10
The square on the left *reflects* all the colors illuminating it. In sunlight it is white. When illuminated with blue light, it is blue. The square on the right *absorbs* all the colors illuminating it. In sunlight it is warmer than the white square.

Fig. 11.9
Sunlight passing through a prism separates into a color spectrum. The colors of things depend on the colors of the light that illuminate them.

Light is reflected from objects in a manner similar to the way sound is "reflected" from a tuning fork when a nearby tuning fork sets it into vibration. One tuning fork can make another vibrate even when the frequencies are not matched, although at significantly reduced amplitudes. The same is true of atoms and molecules. Electrons can be forced into vibration by the vibrating electric fields of electromagnetic waves. Once vibrating, these electrons send out their own electromagnetic waves just as vibrating acoustical tuning forks send out sound waves.

Usually a material absorbs light of some frequencies and reflects the rest. If a material absorbs most of the visible light that is incident upon it but reflects red, for example, it appears red. That's why the petals of a red rose are red and the leaves green. The atoms of the petals absorb all visible light except red, which they reflect; the leaves absorb all except green, which they reflect. An object that reflects light of all the visible frequencies, as the white part of this page does, is the same color as the light that shines on it. If a material absorbs all the light that shines on it, it reflects none and is black.

Fig. 11.11
The bunny's dark fur absorbs all the radiant energy in incident sunlight and is therefore black. Light fur on other parts of the body reflects light of all frequencies and is therefore white.

Interestingly, the petals of most yellow flowers, like daffodils, reflect red and green as well as yellow. Yellow daffodils reflect a broad band of frequencies. The reflected colors of most objects are not pure single-frequency colors, but are composed of a spread of frequencies.

An object can reflect only those frequencies present in the illuminating light. The appearance of a colored object therefore depends on the kind of light that illuminates it. An incandescent lamp, for instance, emits light that is richer in lower frequencies, enhancing any reds viewed in this light. In a fabric having only a little bit of red in it, the red is more apparent under an incandescent lamp than under a fluorescent lamp. Fluorescent lamps are richer in the higher frequencies, and so blues are enhanced under them. For this reason, it is difficult to tell the true color of objects viewed in artificial light. What a color looks like in daylight is different from what it looks like when illuminated by either kind of lamp (Figure 11.12).

Fig. 11.12
Color depends on the light source.

Selective Transmission

The color of a transparent object depends on the color of the light it transmits. A red piece of glass appears red because it absorbs all the colors that compose white light except red, which it transmits. Similarly, a blue piece of glass appears blue because it

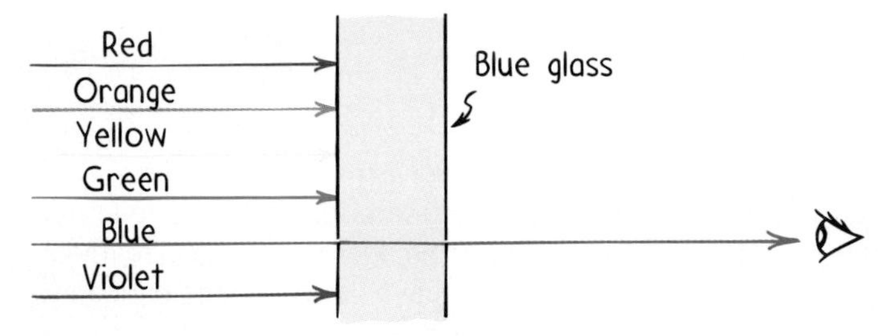

Fig. 11.13
Only energy having the frequency of blue light is transmitted; energy of the other frequencies is absorbed and warms the glass.

transmits primarily blue and absorbs the other colors that illuminate it. The piece of glass contains dyes or *pigments*—fine particles that selectively absorb certain frequencies and selectively transmit others. From an atomic point of view, electrons in the pigment molecules are set into vibration by the illuminating light. Some of the frequencies are absorbed by the pigments, and others are reemitted from molecule to molecule in the glass. The energy of the absorbed frequencies of light increases the kinetic energy of the molecules and the glass is warmed. Ordinary window glass is colorless because it transmits light of all visible frequencies equally well.

Questions

1. When red light shines on a red rose, why do the leaves become warmer than the petals?
2. When green light shines on a red rose, why do the petals look black?
3. If you hold any small source of white light between you and a piece of red glass, you'll see two reflections from the glass: one from the front surface and one from the back surface. What color is each reflection?

Mixing Colored Lights

The fact that white light from the Sun is a composite of all the visible frequencies is easily demonstrated by passing sunlight through a prism and observing the rainbow-colored spectrum. The distribution of solar frequencies (Figure 11.14) is uneven, being most intense in the yellow-green part of the spectrum. Our eyes have evolved to have maximum sensitivity in this range. That is why new fire engines are painted yellow-green, particularly at airports, where visibility is vital. Our sensitivity to yellow-green light is also why at night we see better under the illumination of yellow sodium-vapor lamps than under common tungsten-filament lamps of the same brightness.

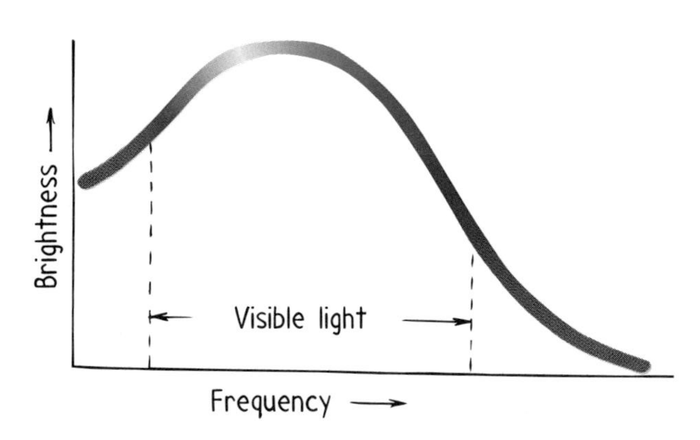

Fig. 11.14
The radiation curve of sunlight is a graph of brightness versus frequency. Sunlight is brightest in the yellow-green region, in the middle of the visible range.

Answers

1. The leaves absorb rather than reflect the red light, and so become warmer.
2. The petals absorb rather than reflect the green light. Because green is the only color illuminating the rose and because green contains no red to be reflected, the rose reflects no color and appears black.
3. The reflection from the front surface is white because the light doesn't go far enough into the colored glass to allow absorption of nonred light. Only red light reaches the back surface because the pigments in the glass absorb all the other colors, and so the back reflection is red.

Link to Ophthalmology—Color Vision

Light from the world around us focuses upon the retina in our eyes, and we see. The *retina* is composed of two kinds of tiny antennae—rods and cones—that resonate to incoming light. As the names imply, rods are rod-shaped and cones cone-shaped. Rods perceive light intensity and cones perceive color. We see color because there are three types of cones—those sensitive to red, those sensitive to green, and those sensitive to blue. Cones are more numerous toward the region of the retina where vision is most distinct—the *fovea.* Rods predominate away from the fovea, toward the periphery of the retina. Primates and a species of ground squirrel are the only mammals that have the three types of cones and experience full-color vision. The retinas of other mammals consist primarily of rods, which are sensitive only to lightness or darkness, like a black-and-white photograph or movie.

Cones require more energy to "fire" an impulse through the nervous system than rods do. If the intensity of light is very low, the things we see have no color. Low-intensity light is seen only with our rods, which is why it is difficult to tell the color of a car by moonlight. Dark-adapted vision is almost entirely due to the rods, while vision in bright light is due to the cones. Stars, for example, look white to us. Yet most stars are actually brightly colored. A time exposure of the stars with a camera reveals reds and red-oranges for the "cooler" stars and blues and blue-violets for the "hotter" stars. The starlight is too weak, however, to fire the color-perceiving cones in the retina. So we see the stars with our rods and perceive them as white or, at best, as only faintly colored. Females have a slightly lower threshold of firing for cones and so can see a bit more color than males. So if she says she sees colored stars and he says she doesn't, she is probably right!

What goes on in the eye is quite complex. Some color sensations depend on intensity, with both rods and cones responding. As intensity increases, orange seems to get yellower and violet bluer—with no change in frequency. Yellow, green, and blue, however, are independent of intensity and are called "psychological primaries." The eye is indeed fascinating.

All the colors combined make white. Interestingly, the perception of white also results from the combination of only red, green, and blue light. We can understand this by dividing the solar radiation curve into three regions, as in Figure 11.15. Three types of cone-shaped receptors in our eyes perceive color. Light in the lowest third of the spectral distribution stimulates the cones sensitive to low frequencies and appears red; light in the middle third stimulates the midfrequency-sensitive cones and appears green; light in the high-frequency third stimulates the higher-frequency-sensitive cones and appears blue. When all three types of cones are stimulated equally, we see white.

Project red, green, and blue lights on a screen. Where they all overlap, white is produced. Where two of the three colors overlap, another color is produced (Figure 11.16). In the language of physicists, colored lights that overlap are said to *add* to each other. So we say that red, green, and blue light *add to produce white light,* and that any two of these colors of light *add to produce another color.* Various amounts of red, green, and blue—the colors to which each of our three types of cones are sensitive—produce any color in the spectrum. For this reason, red, green, and blue are called the **additive primary colors**. A close examination of the picture on most color television tubes will reveal that the picture is an assemblage of tiny spots, each less than a millime-

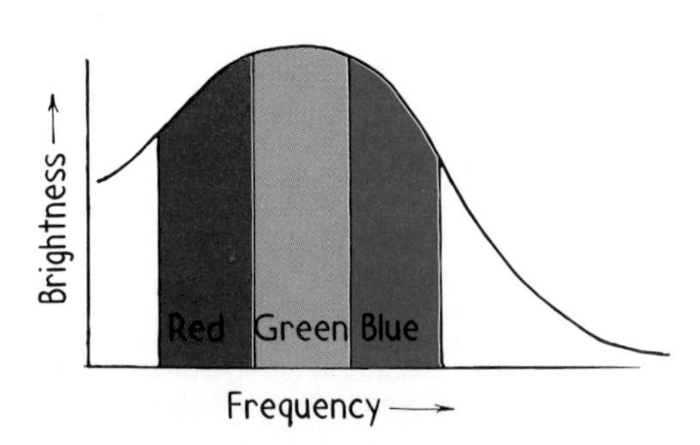

Fig. 11.15
Radiation curve of sunlight divided into three regions—red, green, and blue. These are the additive primary colors.

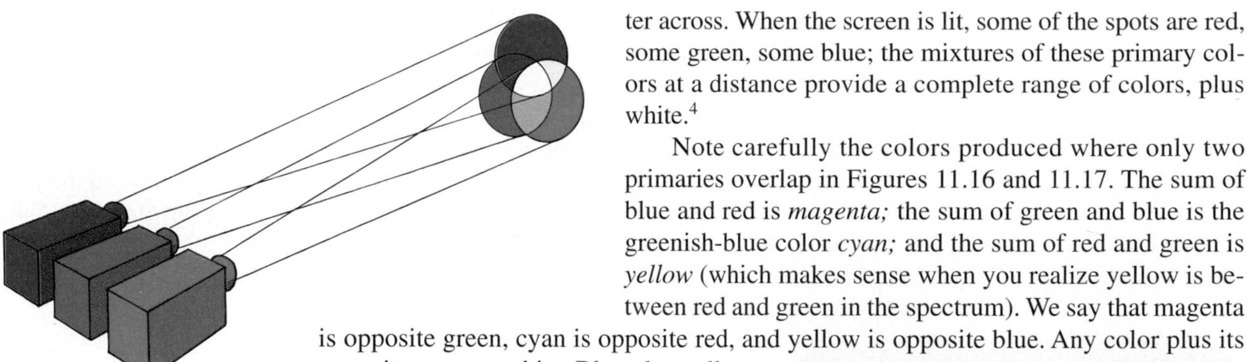

ter across. When the screen is lit, some of the spots are red, some green, some blue; the mixtures of these primary colors at a distance provide a complete range of colors, plus white.[4]

Note carefully the colors produced where only two primaries overlap in Figures 11.16 and 11.17. The sum of blue and red is *magenta;* the sum of green and blue is the greenish-blue color *cyan;* and the sum of red and green is *yellow* (which makes sense when you realize yellow is between red and green in the spectrum). We say that magenta is opposite green, cyan is opposite red, and yellow is opposite blue. Any color plus its opposite appears white. Blue plus yellow, for example, produces white. (This follows logically: Yellow is (red + green), and so blue + (red + green) = white.) From Figures 11.16 and 11.17 we also see that red plus cyan produces white; green plus magenta produces white also. We call any two colors that combine to produce white **complementary colors**.

Fig. 11.16
Color addition by the mixing of colored lights. When three projectors shine red, green, and blue light on a white screen, the overlapping parts produce different colors. White is produced where all three overlap.

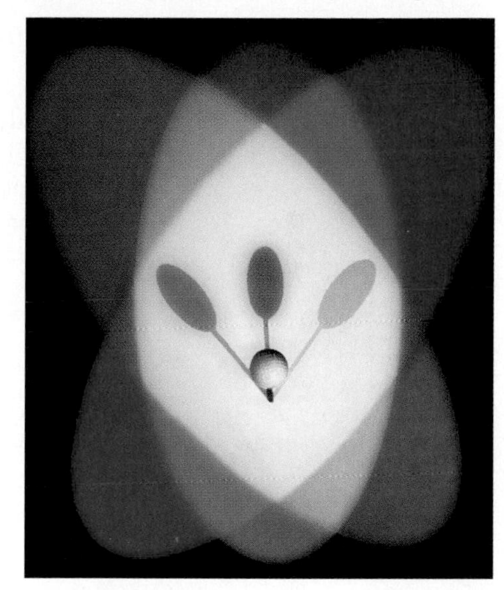

Fig. 11.17
The white golf ball appears white when illuminated with red, green, and blue lights of equal intensities. Why are the shadows of the ball cyan, magenta, and yellow?

Questions

1. From Figure 11.17 find the complements of cyan, of yellow, and of red.
2. Red + cyan = ________.
3. White − cyan = ________.
4. White − red = ________.

Answers
1. Red, blue, cyan.
2. White.
3. Red.
4. Cyan. The cyan color of seawater is the result of red taken away from white sunlight. Because the natural frequency of water molecules is an infrared frequency, infrared light is strongly absorbed by water. To a lesser extent, red light is absorbed by water—enough to give it the greenish-blue, or cyan, color.

[4] The "black" that you see on the darkest scenes on a black-and-white TV tube is simply the color of the tube face itself, which is more a light gray than black. Because our eyes are sensitive to the contrast with the illuminated parts of the screen, we see this gray as black.

The fact that a color and its complement combine to produce white light is nicely used in lighting stage performances. Blue and yellow lights shining on performers, for example, produce the effect of white light—except where one of the two colors is absent, as in the shadows. The shadow of one lamp, say the blue, is illuminated by the yellow lamp and appears yellow. Similarly, the shadow cast by the yellow lamp appears blue. This is a most interesting effect.

We see this effect in Figure 11.17, where red, green, and blue light shine on the golf ball. Note the shadows cast by the ball. The middle shadow is cast by the green spotlight and is not dark because it is illuminated by the red and blue lights, which make magenta. The shadow cast by the blue light appears yellow because it is illuminated by red and green light. Can you see why the shadow cast by the red light appears cyan?

Mixing Colored Pigments

Every artist knows that if you mix red, green, and blue paint, the result is not white but rather a muddy dark brown. Red and green paint certainly do not mix to form yellow, as is the rule for combining colored lights. Mixing pigments in paints and dyes is entirely different from mixing lights. Pigments are tiny particles that absorb specific colors. For example, pigments that produce the color red absorb the complementary color

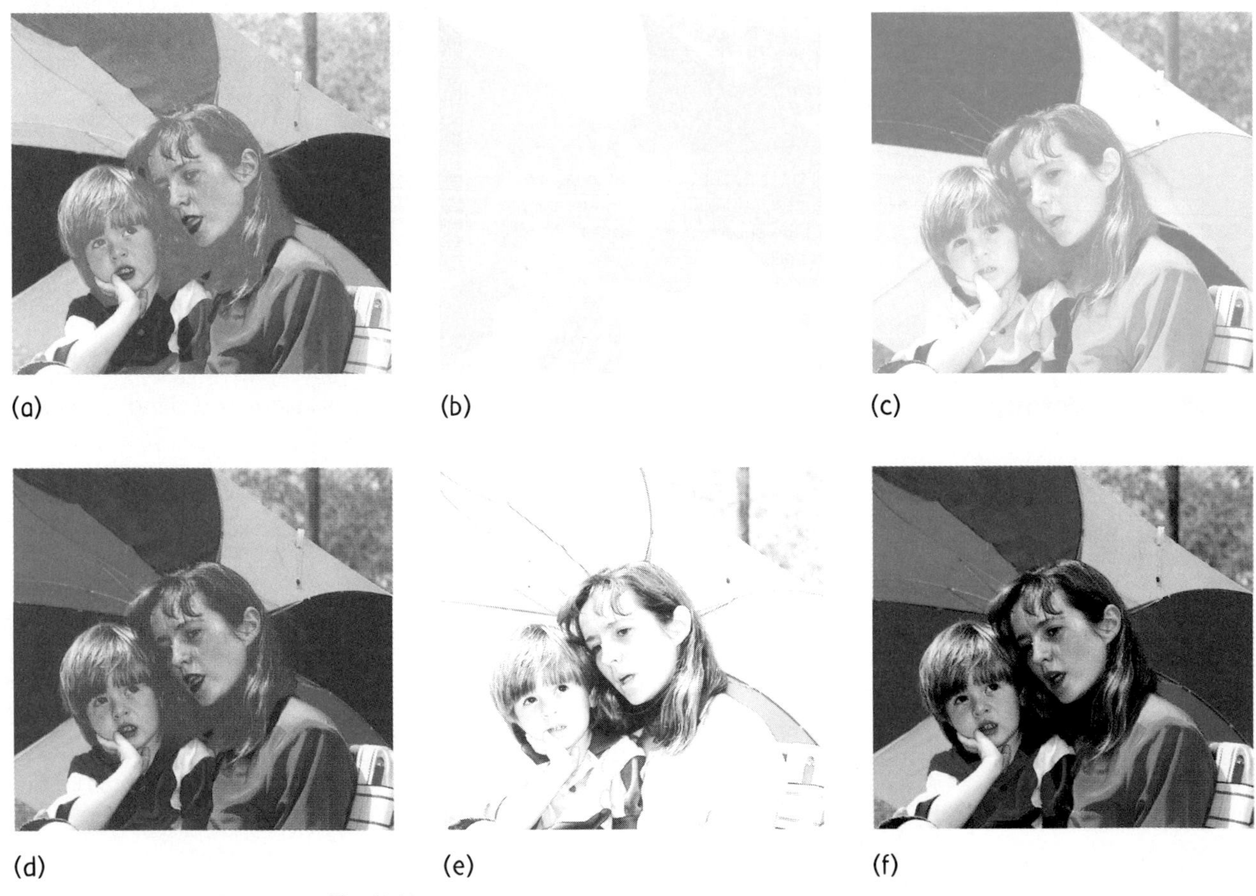

Fig. 11.18
Only four colors of ink are used to print color illustrations and photographs—magenta (a), yellow (b), cyan (c), and black. When magenta, yellow, and cyan are combined, they produce the color shown in (d). Addition of black (e) produces the finished result shown in (f).

Fig. 11.19
Look through a magnifying glass and you will see that the color green on a printed page is made up of blue and yellow dots.

cyan. So something painted red absorbs cyan, which is why it reflects red. In effect, cyan has been *subtracted* from white light. Something painted blue absorbs yellow and so reflects all the colors except yellow. Take yellow away from white and you've got blue. The colors magenta, cyan, and yellow are the **subtractive primaries**. The variety of colors you see in the colored photographs in this or any other book are the result of magenta, cyan, and yellow dots. Light illuminates the book, and light of some frequencies is subtracted from the light reflected. The rules of color subtraction differ from the rules of light addition. We leave this topic to the recommended reading.

Why the Sky Is Blue

Not all colors are the result of the addition or subtraction of light. Some colors, like the blue of the sky, are the result of selective scattering. Consider the analogous case of sound: If a beam of a particular frequency of sound is directed to a tuning fork of similar frequency, the tuning fork is set into vibration and redirects the beam in multiple directions. The tuning fork *scatters* the sound. A similar process occurs with the scattering of light from atoms and particles that are far apart from one another, as they are in the atmosphere.[5]

We know from Section 11.2 that atoms behave like tiny optical tuning forks and reemit light waves that shine on them. Very tiny particles do the same. The tinier the particle, the higher the frequency of light it reemits. This is similar to the way small bells ring with higher notes than larger bells. The nitrogen and oxygen molecules that make up most of the atmosphere are like tiny bells that "ring" with high frequencies when energized by sunlight. Like sound from the bells, the reemitted light is sent in all directions. When light is reemitted in all directions, we say the light is *scattered.*

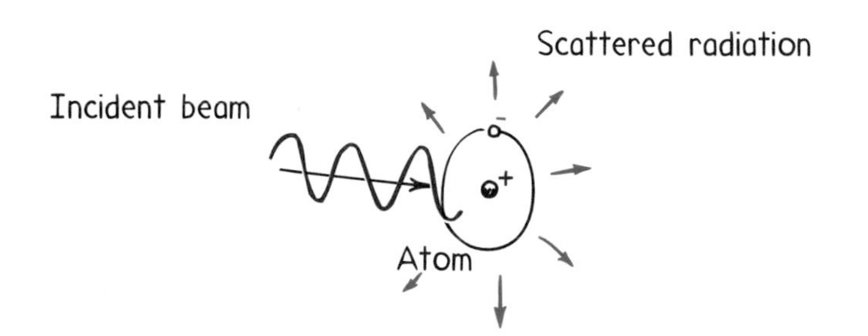

Fig. 11.20
A beam of light falls on an atom and increases the vibrational motion of electrons in the atom. The vibrating electrons reemit the light in various directions. Light is scattered.

Of the visible frequencies of sunlight, violet is scattered the most by nitrogen and oxygen molecules in the atmosphere, followed by blue, green, yellow, orange, and red, in that order. Red is scattered only a tenth as much as violet. Although violet light is scattered more than blue, our eyes are not very sensitive to violet light. Therefore the blue scattered light is what predominates in our vision, and we see a blue sky!

The blue of the sky varies in different places under different conditions. A principal factor is the water-vapor content of the atmosphere. On clear, dry days the sky is a much deeper blue than on clear days with high humidity. Places where the upper air is exceptionally dry, such as Italy and Greece, have beautifully blue skies that have inspired painters for centuries. Where the atmosphere contains a lot of particles of dust and other particles larger than oxygen and nitrogen molecules, light of the lower frequencies is also scattered strongly. This makes the sky less blue, and it takes on a whitish appearance. After a heavy rainstorm when the particles have been washed away, the sky becomes a deeper blue.

The grayish haze in the skies above large cities is the result of particles emitted by car and truck engines and by factories. Even when idling, a typical automobile engine emits more than 100 billion particles per second. Most are invisible but act as tiny cen-

[5]This type of scattering is called *Rayleigh scattering* (after Lord Scattering?) and occurs whenever the scattering particles are much smaller than the wavelength of incident light and resonate at frequencies higher than those of the scattered light.

Fig. 11.21
In clean air, the scattering of high-frequency light gives us a blue sky. When the air is full of particles larger than molecules, lower-frequency light is also scattered. This low-frequency scattered light adds to the high-frequency scattered light, and the result is a whitish sky.

ters to which other particles adhere. These are the primary scatterers of lower-frequency light. With the largest of these particles, absorption rather than scattering takes place, and a brownish haze is produced. Yuk!

Why Sunsets Are Red

Light that isn't scattered is *transmitted.* Because low-frequency red, orange, and yellow light is scattered *least* by the atmosphere, it is better transmitted through the air than the higher-frequency colors. Red, which is scattered the least, and therefore transmitted the most, passes through more atmosphere than any other color. So the thicker the atmosphere through which a beam of sunlight travels, the more time there is to scatter the higher-frequency components of the light. This means the light that best makes it through is red. As Figure 11.22 shows, sunlight travels through more atmosphere at sunset, and that is why sunsets are red.

At noon sunlight travels through the least amount of atmosphere to reach the Earth's surface. Only a small amount of high-frequency light is scattered from the sunlight, enough to make the sun look yellowish. As the day progresses and the sun drops lower in the sky (Figure 11.22), the path through the atmosphere is longer, and more violet and blue are scattered from the sunlight. The removal of violet and blue leaves the transmitted light redder. The sun becomes progressively redder, going from yellow to orange and

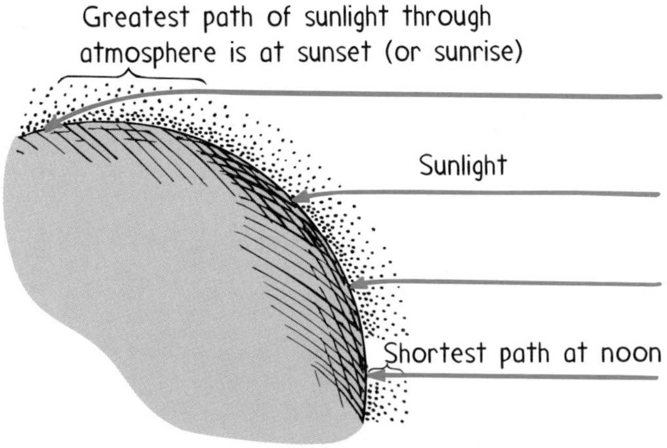

Fig. 11.22
A sunbeam must travel through more atmosphere at sunset than at noon. As a result, more blue is scattered from the beam at sunset than at noon. By the time a beam of initially white light gets to the ground, only light of the lower frequencies survives to produce a red sunset.

finally to a red-orange at sunset. Sunsets and sunrises are unusually colorful following volcanic eruptions because particles larger than atmospheric molecules are then more abundant in the air.

The colors of the sunset are consistent with our rules for color mixing. When blue is subtracted from white light, the complementary color that is left is yellow. When higher-frequency violet is subtracted, the resulting complementary color is orange. When medium-frequency green is subtracted, magenta is left. The combinations of resulting colors vary with atmospheric conditions, which change from day to day and give us a variety of sunsets.

Questions

1. If molecules in the sky scattered low-frequency light more than high-frequency light, what color would the sky be? What color would sunsets be?
2. Distant dark mountains are bluish. What is the source of this blueness? (*Hint:* What is between us and the mountains we see?)
3. Distant snow-covered mountains reflect a lot of light and are bright. Very distant ones look yellowish. Why? (*Hint:* What happens to the reflected white light as it travels from the mountains to us?)

Why Clouds Are White

Clusters of water droplets in a variety of sizes make up clouds. The different-size clusters result in a variety of scattered frequencies: the tiniest, blue; slightly larger clusters, say, green; and still larger clusters, red. The overall result is a white cloud. Electrons close to one another in a cluster vibrate together and in step, which results in a greater intensity of scattered light than from the same number of electrons vibrating separately. Hence clouds are bright!

Larger clusters of droplets absorb much of the light incident upon them, and so the scattered intensity is less. Therefore clouds composed of larger clusters are darker. Further increase in the size of the clusters causes them to fall as raindrops, and we have rain.

The next time you find yourself admiring a crisp blue sky or delighting in the shapes of bright clouds or watching a beautiful sunset, think about all those ultratiny optical tuning forks vibrating away—you'll appreciate these everyday wonders of nature even more!

Answers

1. If low-frequency light were scattered, the noontime sky would appear reddish-orange. At sunset more reds would be scattered by the longer path of the sunlight, and the sunlight would be predominantly blue and violet. So sunsets would appear blue!
2. If we look at distant dark mountains, very little light from them reaches us, and the blueness of the atmosphere between us and them predominates. The blueness we attribute to the mountains is actually the blueness of the low-altitude "sky" between us and the mountains!
3. Bright snow-covered mountains appear yellow because the blue in the white light they reflect is scattered on its way to us. By the time the light gets to us, it is weak in the high frequencies and strong in the low frequencies—hence its yellowish cast. For greater distances, farther away than mountains are usually seen, they would appear orange for the same reason a sunset appears orange.

Why do we see the scattered blue when the background is dark but not when the background is bright? Because the scattered blue is faint. A faint color will show itself against a dark background but not against a bright background. For example, when we look from the Earth's surface at the atmosphere against the darkness of space, the atmosphere is sky blue. But astronauts above who look down through the same atmosphere to the bright surface of the Earth do not see the same blueness.

Fig. 11.23
A cloud is composed of various sizes of water-droplet clusters. The tiniest clusters scatter blue light, slightly larger ones scatter green light, and still larger ones scatter red light. The result is a white cloud.

Fig. 11.24
Water is cyan because it absorbs red light. The froth in the waves is white because, like clouds, it is composed of a variety of tiny clusters of water droplets that scatter all the visible frequencies.

11.4 Diffraction

When you touch your finger to the surface of still water, circular ripples are produced. When you touch the surface with a straight edge, such as a horizontally held meter stick, you produce a plane wave. A series of plane waves can be generated by successively dipping a meter stick into the surface (Figure 11.25). The photographs in Figure 11.26 are top views of a ripple tank in which plane waves are incident upon openings of various sizes. In the left image, where the opening is wide, the waves continue through the opening without change—except at the two ends, where the waves bend. This bending is **diffraction**, which is defined as any bending of light by means other than reflection and refraction. As the width of the opening is narrowed, as in the center image in Figure 11.26, the spreading of waves is more pronounced. When the opening is small relative to the wavelength of the incident wave, as in the right image, diffraction is much more pronounced. We say the waves are *diffracted* as they spread. Diffraction occurs for all kinds of waves, including sound and light waves.

Diffraction is not confined to narrow slits or to openings in general but can be seen for all shadows. On close examination, even the sharpest shadow is blurred slightly at the edge (Figure 11.28).

The amount of diffraction depends on the wavelength of the wave relative to the size of the obstruction that casts the shadow. Long waves are better at filling in shadows, which is why foghorns emit low-frequency sound waves—to fill in any "blind spots." Likewise for radio waves of the standard AM broadcast band, which are very long relative to the size of most objects in their path. The wavelength of AM radio waves ranges from 180 to 550 meters,

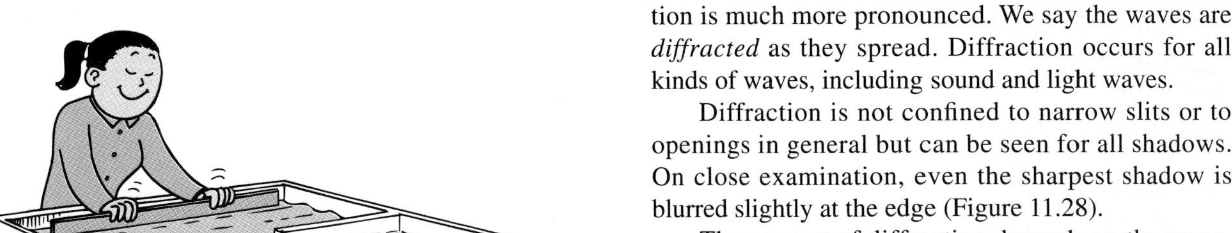

Fig. 11.25
The oscillating meter stick makes plane waves in the tank of water. These waves diffract through the opening.

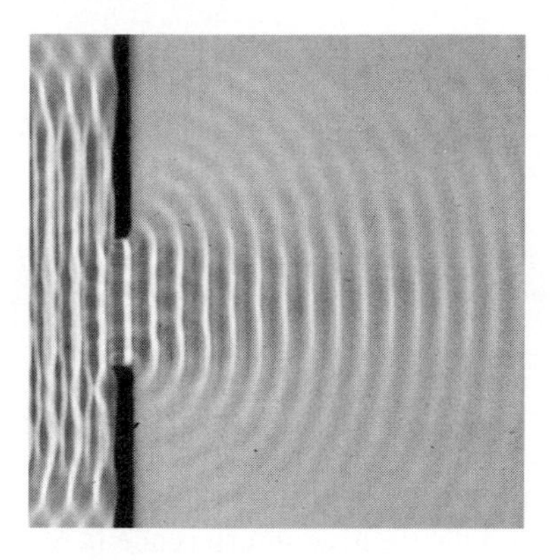

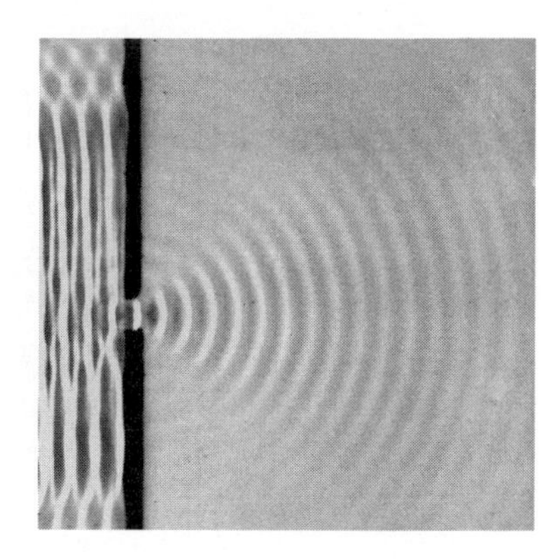

Fig. 11.26
Plane waves passing through openings of various sizes. The smaller the opening, the greater the bending of the waves at the two ends—in other words, the greater the diffraction.

and the waves readily bend around buildings and other objects that might otherwise obstruct them. A long-wavelength radio wave doesn't "see" a relatively small building in its path—but a short-wavelength radio wave does. The radio waves of the FM band range from 2.8 to 3.4 meters and don't bend very well around buildings. This is one of the reasons FM reception is often poor in localities where AM comes in loud and clear. In the case of radio reception, we don't wish to "see" objects in the path of radio waves, and so diffraction is nice.

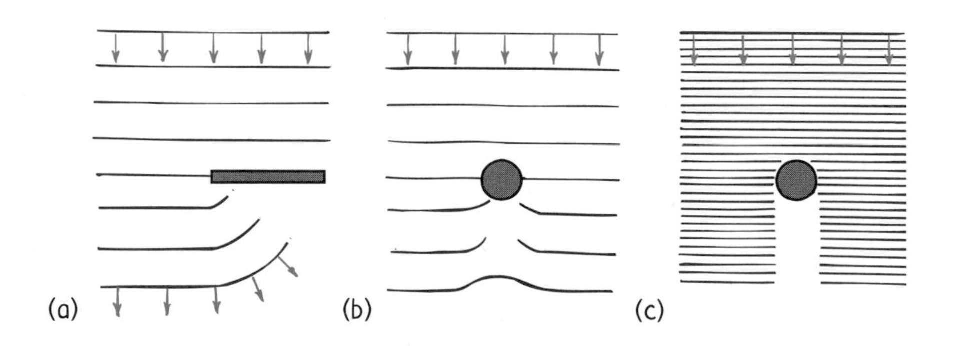

Fig. 11.27
(a) Waves tend to spread into the shadow region. (b) When the wavelength is about the size of the object creating the barrier, the shadow is soon filled in. (c) When the wavelength is short relative to the object's size, a sharp shadow is cast.

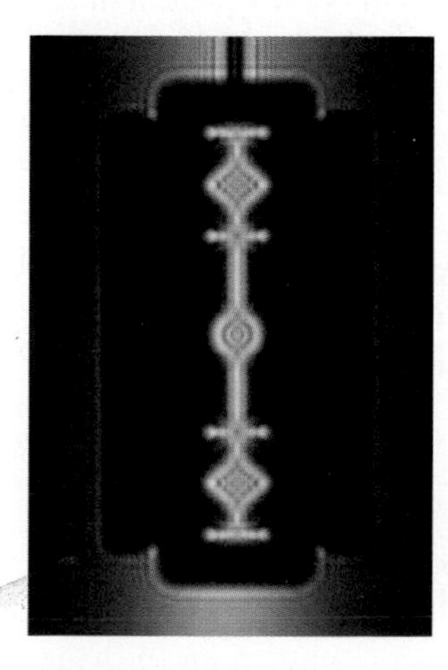

Fig. 11.28
Diffraction fringes are evident in the shadows of monochromatic (single-frequency) laser light.

Diffraction is not so nice for viewing very small objects. If the size of an object is about the same as the wavelength of light, diffraction blurs the image. If the object is smaller than the wavelength of light, no structure can be seen. The entire image is lost due to diffraction. No amount of magnification or perfection of microscope design can defeat this fundamental diffraction limit.

To minimize this problem, microscopists illuminate tiny objects with electron beams rather than light. Relative to light waves, electron beams have extremely short wavelengths. *Electron microscopes* take advantage of the fact that all matter has wave properties: A beam of electrons has a wavelength smaller than those of visible light. In an electron microscope, electric and magnetic fields, rather than optical lenses, are used to focus and magnify images.

The fact that smaller details can be better seen with smaller wavelengths is neatly employed by the dolphin in scanning its environment with ultrasound. The echoes of long-wavelength sound give the dolphin an overall image of objects in its surroundings. To examine more detail, the dolphin emits sound of shorter wavelengths. The dolphin has always done naturally what physicians have only recently been able to do with an ultrasonic imaging devices.

■ **Question**

Why does a microscopist use blue light rather than white light to illuminate objects being viewed?

11.5 Interference

Note that the diffracted light in Figure 11.28 shows alternating dark and light bands of light. These bands, conveniently called *fringes,* are produced by **interference**, which we discussed in the previous chapter. Constructive and destructive interference is reviewed in Figure 11.29. We see that the adding, or *superposition,* of a pair of identical waves in phase with each other produces a wave having the same frequency but twice the amplitude. If the waves are exactly one-half wavelength out of phase, their superposition results in complete cancellation. If they are out of phase by other amounts, partial cancellation occurs.

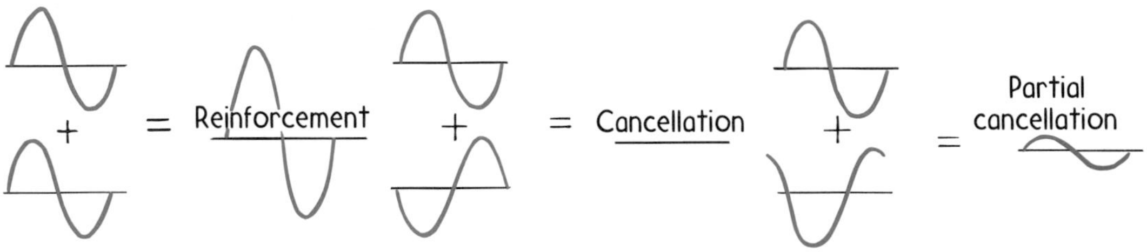

+ = Reinforcement + = Cancellation + = Partial cancellation

Fig. 11.29
Wave interference.

In 1801, the wave nature of light was convincingly demonstrated when the British physicist and physician Thomas Young performed his now-famous interference experiment.[6] Young found that light directed through two closely spaced side-by-side pinholes recombines to produce fringes of brightness and darkness on a screen behind. The bright fringes form when crests overlap crests, while the dark fringes form when crests and troughs overlap. Figure 11.30 shows Young's drawing of the pattern of superimposed waves from the two sources. When his experiment is done with two closely spaced slits instead of pinholes, the fringe patterns are straight lines (Figure 11.31).

We see in Figures 11.32 and 11.33 how the series of bright and dark fringes result from the different path lengths from slits to screen. For the central bright fringe, the paths from the two slits are the same length and so the waves arrive in phase and reinforce each other. The dark fringes on either side of the central fringe result from one path being longer (or shorter) by one-half wavelength, so that the waves arrive half a wavelength out of phase. The other sets of dark fringes occur where the paths differ by odd multiples of one-half wavelength: $\frac{3}{2}$, $\frac{5}{2}$, and so on.

■ **Answer**
There is less diffraction with blue light, and so the microscopist sees more detail (just as a dolphin beautifully investigates fine detail in its environment by the echoes of ultrashort wavelengths of sound).

---

[6]Thomas Young read fluently at the age of 2; by 4, he had read the Bible twice; by 14, he knew eight languages. In his adult life he was a physician and scientist, contributing to an understanding of fluids, work and energy, and the elastic properties of materials. He was the first person to make progress in deciphering Egyptian hieroglyphics. No doubt about it—Thomas Young was a bright guy!

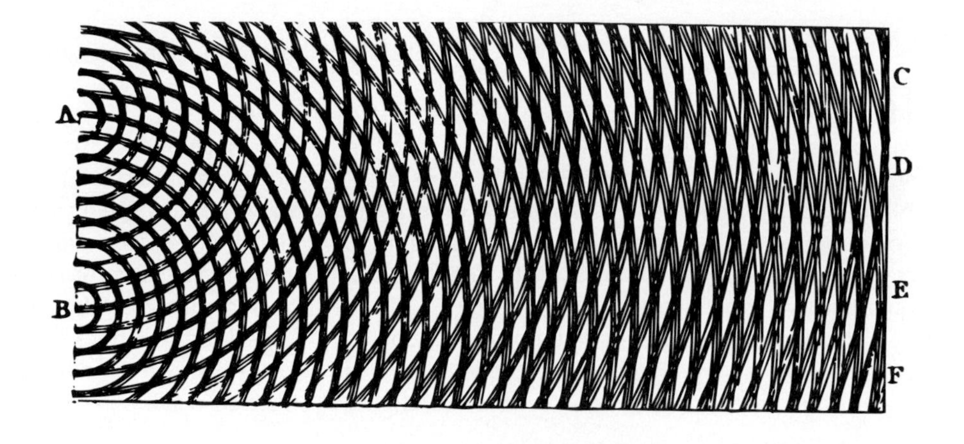

Fig. 11.30
Thomas Young's original drawing of a two-source interference pattern. Letters C, D, E, and F mark regions of destructive interference.

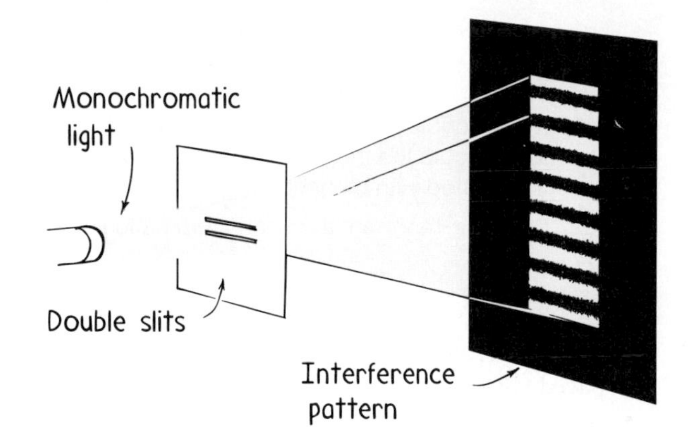

Fig. 11.31
When monochromatic light passes through two closely spaced slits, a striped interference pattern is produced.

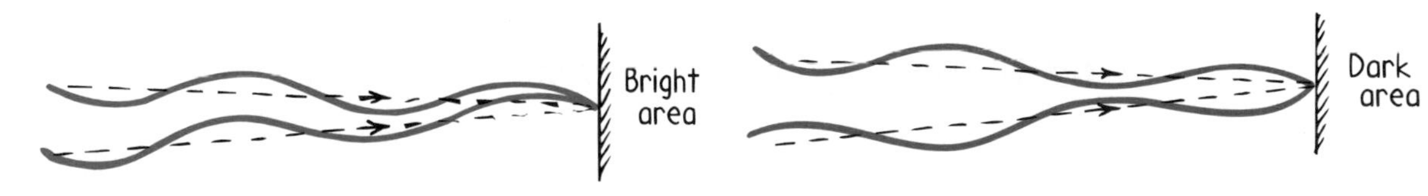

Fig. 11.32
A bright area occurs when waves from both slits arrive in phase; a dark area results from the overlapping of waves that are out of phase.

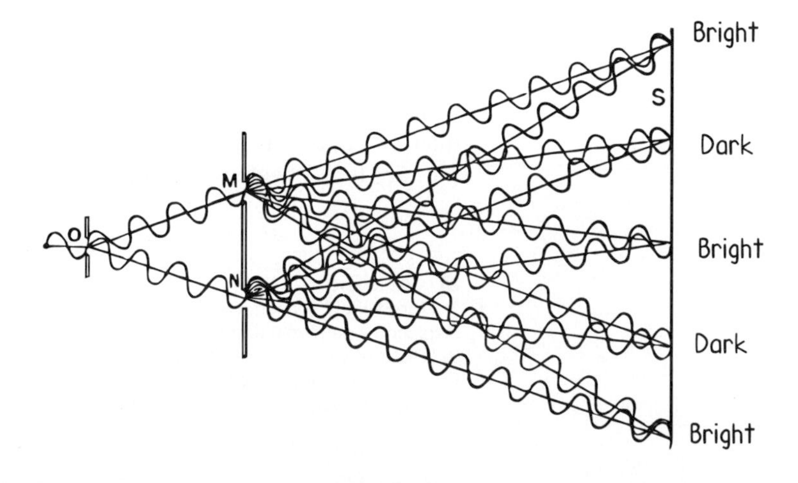

Fig. 11.33
Light from O passes through slits M and N and produces an interference pattern on the screen S.

Link to Optometry—Seeing Star-Shaped Stars

Have you ever wondered why stars are represented with spikes? The stars on the American flag have five spikes, and the Jewish Star of David has six spikes. All through the ages, stars have been drawn with spikes. The reason for this has to do not with the stars, which are point sources of light in the night sky, but rather with poor eyesight.

The surface of our eyes, the cornea, becomes scratched by a variety of causes. These scratches make up a diffraction grating of sorts. A scratched cornea is not a very good diffraction grating, but its effects are evident if you look at a bright point source against a dark background—like a star in the night sky. Instead of seeing a point of light, you see a spiky shape. The spikes even shimmer and twinkle if there are some temperature differences in the atmosphere to produce some refraction. And if you live in a windy desert region where sandstorms are frequent, your cornea will be even more scratched and you'll see more vivid star spikes.

Questions

1. If the double slits were illuminated with monochromatic (single-frequency) red light, would the fringes be more widely or more closely spaced than if illuminated with monochromatic blue light?
2. Why is it important that monochromatic light be used?

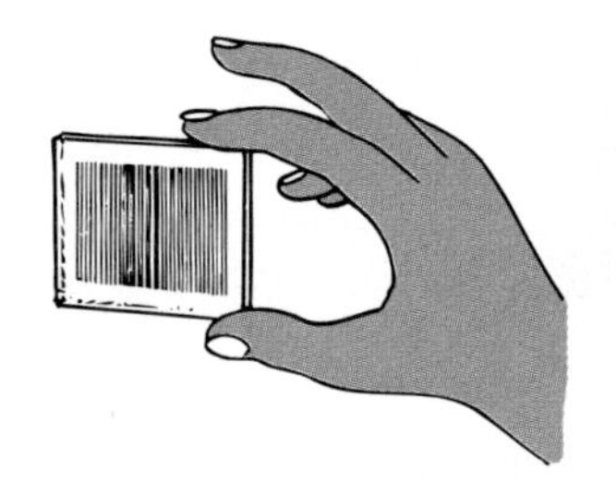

Fig. 11.34
Because of the interference it causes, a diffraction grating disperses light into colors. It may be used in place of a prism in a spectrometer.

Interference patterns are not limited to one or two slits. A multitude of closely spaced slits makes up a *diffraction grating.* These devices, like prisms, disperse white light into colors. These are used in devices called *spectrometers,* which we shall discuss in Chapter 14. The feathers of some birds act as diffraction gratings and disperse colors. Likewise for the microscopic pits on the reflective surface of compact discs.

Interference Colors by Reflection from Thin Films

We have all noticed the beautiful spectrum of colors reflected from a soap bubble or from gasoline on a wet street. These colors are produced by the interference of light waves. This phenomenon is often called *iridescence* and is observed in thin transparent films.

A soap bubble appears iridescent in white light when the thickness of the soap film is about the same as the wavelength of light. Light waves reflected from the outer and inner surfaces of the film to your eye travel different distances. When illuminated by white light, the film may be just the right thickness at one place to cause the destructive interference of, say, yellow light. When yellow light is subtracted from white light, the mixture left appears as the complementary color—blue. At another place, where the

---

Answers

1. More widely spaced. Can you see in Figure 11.33 that a slightly longer—and therefore slightly more displaced—path from entrance slit to screen would result for the longer waves of red light?
2. If light of various wavelengths were diffracted by the slits, dark fringes for one wavelength would be filled in with bright fringes for another, resulting in no distinct fringe pattern. If you haven't seen this, be sure to ask your instructor to demonstrate it.

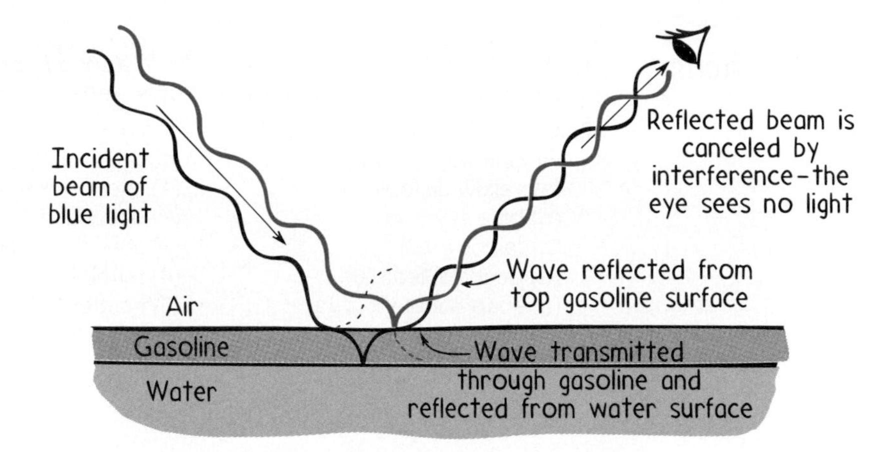

Fig. 11.35
The thin film of gasoline is just the right thickness to cancel the reflections of blue light from the top and bottom surfaces. If the film were thinner, perhaps shorter-wavelength violet would be canceled. (One wave is drawn black to show how, upon reflection, it is out of phase with the blue wave.)

film is thinner, a different color may be canceled by interference, and the light seen is *its* complementary color.

The same thing happens with gasoline on a wet street (Figure 11.35). Light reflects from both the upper, gasoline-air surface and the lower, gasoline-water surface. If the thickness of the gasoline is such to cancel blue, as the figure suggests, then the gasoline surface appears yellow to the eye. Blue subtracted from white leaves yellow. The variety of colors seen in the thin film of gasoline, then, corresponds to different film thicknesses, providing a vivid "contour map" of microscopic differences in surface "elevations."

View the thin film of gasoline at a lower angle and you see different colors. The *apparent thickness* of the film is effectively greater than it was when viewed at the higher angle. This change in apparent thickness is caused by the greater distance the light rays travel in the gasoline at lower angles. Longer waves are canceled in this case, and different colors appear.[7]

Dishes washed in soapy water and poorly rinsed have a thin film of soap on them. Hold such a dish up to a light source so that *interference colors* can be seen. Then turn the dish to a new position, keeping your eye on the same part of the dish, and the color will change. Light reflecting from the bottom surface of the transparent soap film is canceling light reflecting from the top surface of the soap film. Different wavelengths of light are canceled for different angles.

Interference techniques can be used to measure the wavelengths of light and other regions of the electromagnetic spectrum. The principle of interference provides a means of measuring extremely small distances with great accuracy. Instruments called *interferometers,* which use the principle of interference, are the most accurate instruments known for measuring small distances.

---

[7]Phase shifts at some reflecting surfaces also contribute to interference. For simplicity and brevity, our concern with this topic is limited to this footnote. In short, when light passing through one medium is reflected at the surface of a second medium in which the speed of light is less (when there is a greater *index of refraction*), there is a 180° phase shift (that is, half a wavelength), but no phase shift occurs when the second medium is one that transmits light at a higher speed (and there is a lower index of refraction). For example, in a soap bubble, light reflects from the first surface 180° out of phase. Light reflects from the second surface without a phase change. If the thickness of the soap film is very small relative to the wavelength of light—so that the distance through the film is negligible—the parts of the wave reflected from the two surfaces are out of phase and cancel—for all frequencies. This is why parts of a soap film that are extremely thin appear black. Waves of all frequencies are canceled.

Questions

1. What color appears to be reflected from a soap bubble in sunlight when its thickness is such that green light is canceled?
2. In the left column are the colors of certain objects. In the right column are various ways in which colors are produced. Match the right column to the left.

(a) yellow daffodil	(1) interference
(b) blue sky	(2) diffraction
(c) rainbow	(3) selective reflection
(d) peacock feathers	(4) refraction
(e) soap bubble	(5) scattering

11.6 Polarization

Fig. 11.36
A vertically plane-polarized plane wave and a horizontally plane-polarized plane wave

Interference and diffraction provide the best evidence that light is wavelike. As we learned in Chapter 10, waves can be either longitudinal or transverse. Sound waves are longitudinal, which means the vibratory motion of the medium is *along* the direction of wave travel. The fact that light waves exhibit **polarization** demonstrates that they are transverse.

Shake a taut rope either up and down or from side to side as shown in the two parts of Figure 11.36, and you produce a transverse wave along the rope that travels in one plane. We say that such a wave is *plane-polarized,*[8] meaning that as it travels along the rope the wave is confined to a single plane. If we shake the rope up and down, we produce a vertically polarized wave. If we shake it from side to side, we produce a horizontally polarized wave.

A single vibrating electron can emit an electromagnetic wave that is plane-polarized. The plane of polarization matches the vibrational direction of the electron. A vertically accelerating electron, then, emits light that is vertically polarized, and a horizontally accelerating electron emits light that is horizontally polarized (Figure 11.37).

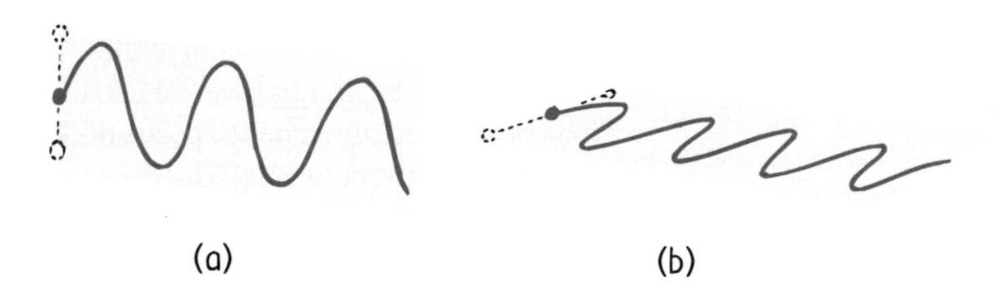

Fig. 11.37
(a) A vertically plane-polarized wave from a charge vibrating vertically. (b) A horizontally plane-polarized wave from a charge vibrating horizontally

Answers

1. The composite of all the visible wavelengths except green results in the complementary color magenta. See Figure 11.17.
2. a–3; b–5; c–4; d–2; e–1.

[8] Light may also be circularly polarized and elliptically polarized, which are also transverse polarizations. But we do not study these cases.

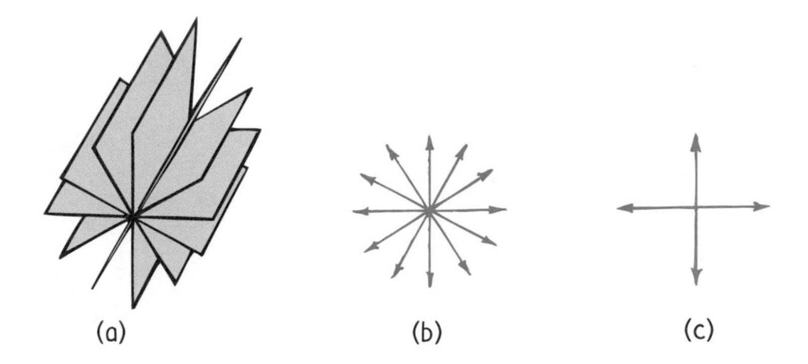

Fig. 11.38
Representations of planes of waves. Configurations (b) and (c) show electric vectors that make up the electric part of electromagnetic waves.

A common light source, such as an incandescent lamp, a fluorescent lamp, or a candle flame, emits light that is unpolarized. This is because there is no preferred vibrational direction for the accelerating electrons emitting the light. The planes of vibration might be as numerous as the accelerating electrons producing them. A few planes are represented in Figure 11.38a. We can represent all these planes by radial lines (Figure 11.38b) or, more simply, by vectors in two mutually perpendicular directions (Figure 11.38c), as if we had resolved all the vectors of configuration c into horizontal and vertical components. This simpler schematic represents unpolarized light. Polarized light would be represented by a single vector.

All transparent crystals that have a noncubic natural shape have the property of transmitting light of one polarization differently from light of another polarization. Certain crystals not only divide unpolarized light into two internal beams polarized at right angles to each other but also strongly absorb one beam while transmitting the other (Figure 11.39). Herapathite is such a crystal. Microscopic crystals of herapathite are embedded between cellulose sheets in uniform alignment and are used in making Polaroid filters. Some Polaroid sheets consist of certain aligned molecules rather than tiny crystals.[9]

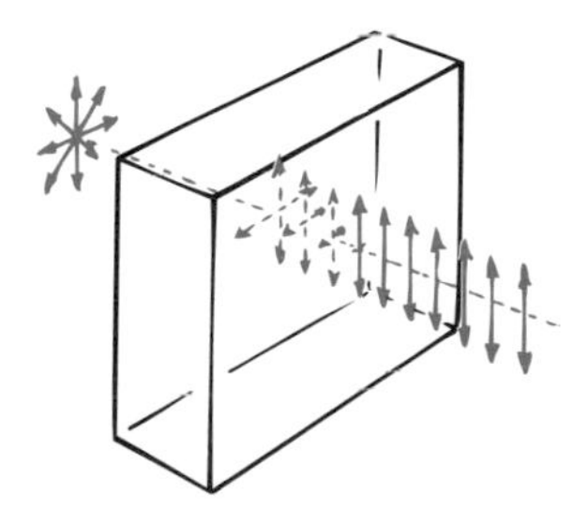

Fig. 11.39
One component of the incident unpolarized light is absorbed, resulting in emerging polarized light.

If you look at unpolarized light through a Polaroid filter, you can rotate the filter in any direction and the light appears unchanged. But if the light is polarized, then as you rotate the filter, you can progressively cut off more and more of the light until it is blocked out. An ideal Polaroid filter transmits 50 percent of incident unpolarized light. That 50 percent is, of course, polarized. When two Polaroid filters are arranged so that their polarization axes are aligned, light is transmitted through both (Figure 11.40a). If their axes are at right angles to each other (in this case we say the filters are *crossed*), no light penetrates the pair (Figure 11.40b). (Actually, some of the shorter wavelengths do get through, but not to any significant degree.) When Polaroid filters are used in pairs like this, the first one is called the *polarizer* and the second one the *analyzer.*

Much of the light reflected from nonmetallic surfaces is polarized. The glare from glass or water is a good example. Except for perpendicular incidence, the reflected ray contains more vibrations parallel to the reflecting surface, whereas the transmitted beam contains more vibrations at right angles to the surface (Figure 11.42). Skipping flat rocks off the surface of a pond is analogous. When the rocks hit parallel to the surface, they easily reflect, but when they hit with their faces at right angles to the surface, they "refract" into the water. The glare from reflecting surfaces can be appreciably diminished with the use of Polaroid sunglasses. The polarization axes of the lenses are vertical because most glare reflects from horizontal surfaces.

---

[9]The molecules are polymeric iodine in a sheet of polyvinyl alcohol or polyvinylene.

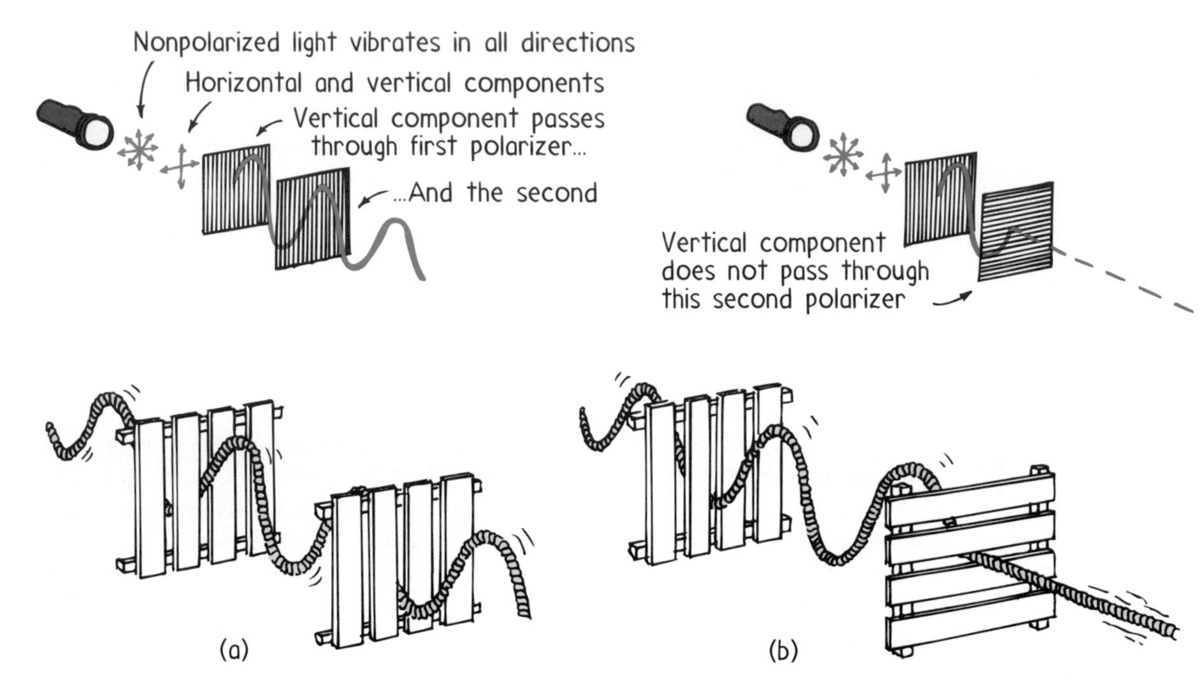

Fig. 11.40
(a) Both light and the vibrations of the rope pass through aligned filters. (b) Neither light nor vibrations of the rope pass through crossed filters. The blue dashed line shows the direction light would travel if the filters weren't crossed.

Fig. 11.41
Polaroid sunglasses block out horizontally vibrating light. When the lenses overlap at right angles, no light gets through.

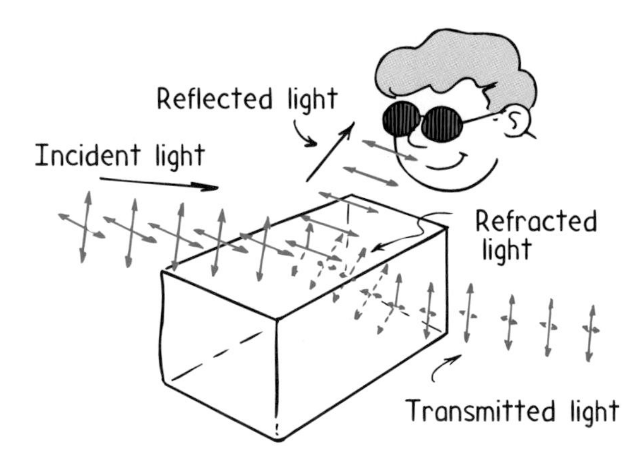

Fig. 11.42
Most glare from nonmetallic surfaces is polarized. Here we see that the components of incident light parallel to the surface are reflected and the components perpendicular to the surface pass through the surface into the medium. Because most of the glare we encounter is from horizontal surfaces, the polarization axes of Polaroid sunglasses are vertical.

Question

Which pair of glasses is best suited for automobile drivers? (The polarization axes are shown by the straight lines.)

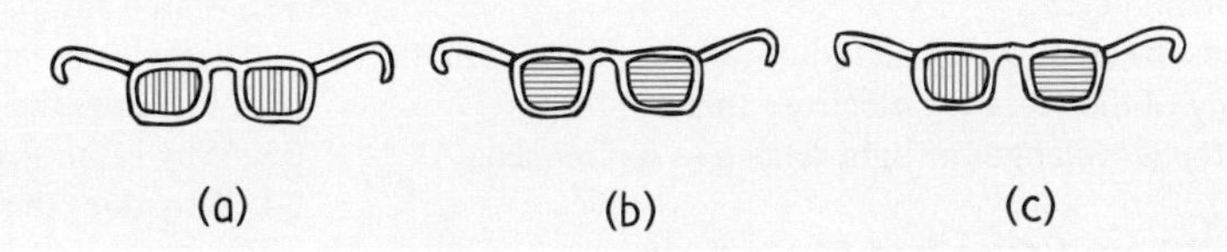

Beautiful colors similar to interference colors can be seen when certain materials are placed between crossed Polaroid filters. Cellophane works wonderfully. Why these colors are produced is another story, one we leave to the recommended reading at the back of the book.

Fig. 11.43
Light is (a) transmitted when the axes of the Polaroid filters are aligned but (b) absorbed when Ludmila rotates one filter so that the axes are at right angles to each other. (c) When she inserts a third Polaroid filter at an angle between the crossed ones, light is again transmitted. Why?

Answer
Pair A is best suited because the vertical axis blocks horizontally polarized light, which composes much of the glare from horizontal surfaces. Pair C can be used to view 3-D movies.

Summary of Terms

- **Electromagnetic wave** A wave emitted by vibrating electrical charges (often electrons) and composed of vibrating electric and magnetic fields that regenerate one another.
- **Electromagnetic spectrum** The range of electromagnetic waves extending in frequency from radio waves to gamma rays.
- **Transparent** The term applied to materials through which light can pass in straight lines.
- **Opaque** The term applied to materials through which light cannot pass.
- **Additive primary colors** The three colors—red, blue, and green—that when added in certain proportions produce any other color in the visible-light part of the electromagnetic spectrum.
- **Complementary colors** Any two colors that when added produce white light.
- **Subtractive primary colors** The three colors of absorbing pigments—magenta, yellow, and cyan—that when mixed in certain proportions reflect any other color in the visible-light part of the electromagnetic spectrum.
- **Diffraction** The bending of light as it passes around an obstacle or through a narrow slit, causing the light to spread and to produce bright and dark fringes.
- **Interference** The result of superposing two or more waves of the same wavelength. Constructive interference results from crest-to-crest reinforcement; destructive interference results from crest-to-trough cancellation.
- **Polarization** The alignment of the transverse electric vectors that make up electromagnetic radiation. Such waves of aligned vibrations are said to be *polarized.*

Review Questions

The Electromagnetic Spectrum

1. Does visible light make up a relatively large part or a relatively small part of the electromagnetic spectrum?
2. What is the principal difference between a radio wave and light?
3. What is the principal difference between light and an X ray?

4. How do the speeds of various electromagnetic waves compare with one another?
5. What color do we perceive for the lowest visible frequencies? The highest?
6. How does the frequency of a radio wave compare with the frequency of the vibrating electrons that produces it?
7. How is the wavelength of light related to its frequency?

Transparent and Opaque Materials

8. The sound coming from one tuning fork can force another to vibrate. What is the analogous effect for light?
9. In what region of the electromagnetic spectrum is the resonant frequency of electrons in glass?
10. What is the fate of the energy in ultraviolet light incident on glass?
11. What is the fate of the energy in visible light incident on glass?
12. Why are your answers for Questions 10 and 11 different?
13. How does the frequency of reemitted light in a transparent material compare with the frequency of the light that stimulates its reemission?
14. How does the average speed of light in glass compare with its speed in a vacuum?

Color

15. What is the relationship between the frequency of light and its color?
16. Which has the higher frequency, red light or blue light?
17. Distinguish between the white of this page and the black of this ink in terms of what happens to the white light that falls on both.
18. How does the color of an object illuminated by candle light differ from the color of the same object illuminated by a fluorescent lamp?
19. What color light is transmitted through a piece of red glass?
20. Which warms more quickly in sunlight, a colorless or a colored piece of glass? Why?
21. What is the evidence for the statement that white light is a composite of all the colors of the visible part of the electromagnetic spectrum?
22. What is the color of the peak frequency of solar radiation?
23. What color light are our eyes most sensitive to?
24. What frequency ranges of the radiation curve do red, green, and blue light occupy?
25. Why are red, green, and blue called the *additive primary colors?*
26. What is the resulting color when equal intensities of blue light and green light are combined? When equal intensities of red light and cyan light are combined?
27. What are the subtractive primary colors? Why are they so called?
28. If you look with a magnifying glass at pictures printed in color in this book or in magazines, you'll notice three colors of ink plus black. What are the names of these three colors?
29. Why are red and cyan called *complementary colors?*
30. Which interacts more with high-pitched sounds, small bells or large bells?
31. Which interacts more with high-frequency light, small particles or large particles?
32. Why does the sky sometimes appear whitish?
33. Why is the sky a deeper blue after a heavy rainstorm?
34. Why does the Sun look reddish at sunrise and sunset but not at noon?
35. What is the evidence for a variety of particle sizes in a cloud?
36. What is the evidence for extra big particles in a rain cloud?

Diffraction

37. Is diffraction more pronounced through a small opening or through a large opening?
38. For an opening of a given size, is diffraction more pronounced for a longer wavelength or a shorter wavelength?
39. What are some of the assets and liabilities of diffraction?

Interference

40. Is interference restricted to only some types of waves or does it occur for all types of waves?
41. What is monochromatic light?
42. What produces iridescence?
43. What causes the variety of colors seen in gasoline splotches on a wet street? Why are these colors not seen on a dry street?
44. What accounts for the variety of colors in a soap bubble?
45. If you look at a soap bubble from different angles so that you're viewing different apparent thicknesses of soap film, do you see different colors? Explain.

Polarization

46. What phenomenon distinguishes between longitudinal and transverse waves?
47. How does the direction of polarization of light compare with the direction of vibration of the electrons that produced it?
48. Why does light pass through a pair of Polaroid filters when the axes are aligned but not when the axes are at right angles to each other?
49. How much unpolarized light does an ideal Polaroid filter transmit?
50. When unpolarized light is incident at a grazing angle upon water, what can you say about the reflected light?

Home Projects

1. Stare at a piece of colored paper for 45 seconds or so. Then look at a white surface. Because the cones in your retina receptive to the color of the paper have become fatigued, you see an afterimage of the complementary color when you look at the white area. This is because the fatigued cones send a weaker signal to the brain. All the colors pro-

duce white, but all the colors minus one produce the color complementary to the missing one.

2. Simulate your own sunset: Add a few drops of milk to a glass of water and look through it at a lit light bulb. The bulb appears to be red or pale orange, while light scattered to the side appears blue.
3. With a razor blade, cut a slit in a card and look at a light source through it. You can vary the size of the opening by bending the card slightly. See the interference fringes? Try it with two closely spaced slits.
4. Next time you're in the bathtub, froth up the soapsuds and notice the colors of highlights from the illuminating light overhead on each tiny bubble. Notice that different bubbles reflect different colors, due to the different thicknesses of soap film. If a friend is bathing with you, compare the different colors you each see reflected from the same bubbles. They will be different—for what you see depends on your point of view!
5. Do this one at your kitchen sink. Dip a dark-colored coffee cup (dark colors make the best background for viewing interference colors) in dishwashing detergent and then hold it sideways and look at the reflected light from the soap film that covers its mouth. Swirling colors appear as the soap runs down to form a wedge that grows thicker at the bottom. The top becomes thinner, so thin that it appears black. This tells us that its thickness is less than one-fourth the thickness of the shortest waves of visible light. Whatever its wavelength, light reflecting from the inner surface reverses phase, rejoins light reflecting from the outer surface, and cancels. The film soon becomes so thin it pops.
6. When you're wearing Polaroid sunglasses, look at the glare from a nonmetallic surface such as a road or body of water. Tip your head from side to side and see how the glare intensity changes as you vary the magnitude of the electric vector component aligned with the polarization axis of the glasses. Also notice the polarization of different parts of the sky when you hold the sunglasses in your hand and rotate them.
7. Place a source of white light on a table in front of you. Then place a sheet of Polaroid in front of the source, a bottle of corn syrup in front of the sheet, and a second sheet of Polaroid in front of the bottle. Look through the Polaroid sheets that sandwich the bottle to view spectacular colors as you rotate one of the sheets.
8. You can also see spectacular interference colors with a polarized-light microscope. Any microscope, including an inexpensive toy one, can be converted to a polarized-light microscope by fitting a piece of Polaroid inside the eyepiece and taping another one onto the stage of the microscope. Mix drops of naphthalene and benzene on a slide and watch the growth of crystals. Rotate the eyepiece and change the colors.
9. Make some slides for a slide projector by sticking some crumpled cellophane onto pieces of slide-sized Polaroid. (Also try strips of cellophane tape overlapped at different angles and experiment with different brands of transparent tape.) Project them onto a large screen or white wall and rotate a second, slightly larger piece of Polaroid in front of the projector lens in rhythm with your favorite music. You'll have your own light show.

Exercises

1. Which waves have the longest wavelengths: light waves, X rays, or radio waves? Which have the highest frequencies?
2. Which requires a physical medium in which to travel: light, sound, or both? Explain.
3. When astronomers observe a supernova explosion in a distant galaxy, they see a sudden, simultaneous rise in visible light and other forms of electromagnetic radiation. Is this evidence to support the idea that the speed of light is independent of frequency? Explain.
4. What is the same about radio waves and light? What is different about them?
5. The intensity of light falls off as the inverse square of the distance from the source. Does this mean that light energy is lost? Explain.
6. What evidence can you cite to support the idea that light can travel in a vacuum?
7. Why would you expect the speed of light to be slightly less in the atmosphere than in a vacuum?
8. If you fire a bullet through a tree, the bullet slows down inside the tree and emerges at a speed less than the speed at which it entered. Does light, then, similarly slow down when it passes through glass? And like the bullet, does it emerge at a lower speed? Defend your answer.
9. Short wavelengths of visible light interact more frequently with the atoms in glass than do longer wavelengths. Does this interaction time tend to speed up or slow down the average speed of light in glass?
10. Is glass transparent or opaque to frequencies of light that match its own natural frequencies? Explain.
11. What determines whether a material is transparent or opaque?
12. You can get a sunburn on a cloudy day, but you can't get a sunburn even on a sunny day if you are behind glass. Explain.
13. Suppose that sunlight falls on both a pair of reading glasses and a pair of dark sunglasses. Which pair of glasses would you expect to be warmer in sunlight? Defend your answer.
14. In a clothing shop lit only by fluorescent lighting, a customer insists on taking garments into the daylight at the doorway to check their color. Is the customer being reasonable? Explain.
15. If sunlight were somehow green instead of white, what color garment would be best to wear on an uncomfortably hot day? On a very cold day?
16. What color does red cloth appear when illuminated by sunlight? By red light from a neon sign? By cyan light?

17. Does a ripe banana appear black when illuminated with red light? With yellow light? With green light? With blue light?
18. When white light is shone on red ink dried on a glass plate, the color transmitted is red but the color reflected is not red. What is it?
19. Stare intently for a minute or so at an American flag. Then turn your view to a white area. What colors do you see in the image of the flag that appears on the wall?
20. Why does a white piece of paper appear white in white light, red in red light, blue in blue light, and so on for every color?
21. A spotlight that has a white-hot filament is coated so that it won't transmit yellow light. What color is the emerging beam of light?
22. How could you use the spotlights at a play to make the yellow clothes of the performers suddenly change to black?
23. Suppose two flashlight beams are shone on a white screen, one beam through a pane of blue glass and the other through a pane of yellow glass. What color appears on the screen where the two beams overlap? Suppose, instead, that the two panes of glass are placed in the beam of a single flashlight. What then?
24. Does a color television work by color addition or by color subtraction? Which dots must be struck by electrons to create the color yellow? Magenta? White?
25. Complete the following equations:
    Yellow light + blue light = ________ light.
    Green light + ________ light = white light.
    Magenta + yellow + cyan = ________ light.
26. Check to see if the following three statements are accurate. Then fill in the last statement. (All colors are combined by addition of light.)
    Red + green + blue = white.
    Red + green + yellow + white = blue.
    Red + blue + magenta + white = green.
    Green + blue + cyan + white = ________.
27. On a photographic print, a friend is seen wearing a red sweater. What color is the sweater on the negative?
28. If the sky were composed of atoms that predominantly scattered orange light rather than blue, what color would sunsets be?
29. Tiny particles, like tiny bells, scatter high-frequency waves more than low-frequency waves. Large particles, like large bells, mostly scatter low frequencies. Intermediate-size particles and bells mostly scatter intermediate frequencies. What does this have to do with the whiteness of clouds?
30. Very big particles, such as droplets of water, absorb more radiation than they scatter. What does this have to do with the darkness of rain clouds?
31. If the atmosphere of the Earth were about 50 times thicker, would snow still seem white or would it be some other color? What color?
32. The atmosphere of Jupiter is more than 1000 km thick. From the surface of this planet, would you expect to see a white Sun?
33. A lunar eclipse occurs when the Moon passes into the Earth's shadow. Rather than being black, the Moon appears a copper red. Is there a connection between the reddish Moon color and the ring of sunsets and sunrises that circle the Earth? Explain.
34. Why do radio waves diffract around buildings but light waves do not?
35. A pattern of fringes is produced when monochromatic light passes through a pair of thin slits. Is such a pattern produced by three parallel thin slits? By thousands of such slits? Give an example to support your answer.
36. The colors of peacocks and hummingbirds are the result not of pigments but of ridges in the surface layers of their feathers. By what physical principle do these ridges produce colors?
37. When dishes are not properly rinsed after washing, different colors are reflected from their surfaces. Explain.
38. If you notice the interference patterns of a thin film of oil or gasoline on water, you'll note that the colors form complete rings. How are these rings similar to the lines of equal elevation on a contour map?
39. Why do Polaroid sunglasses reduce glare, whereas unpolarized sunglasses simply cut down on the total amount of light reaching the eyes?
40. How can you determine the polarization axis for a single sheet of Polaroid?
41. Most of the glare from nonmetallic surfaces is polarized, the axis of polarization being parallel to that of the reflecting surface. Would you expect the polarization axis of Polaroid sunglasses to be horizontal or vertical? Why?
42. How can a single sheet of Polaroid film be used to show that the sky is partially polarized? (Interestingly enough, unlike humans, bees and many insects can discern polarized light; they use this ability for navigation.)
43. What percentage of light is transmitted by two ideal Polaroid filters atop each other with their polarization axes aligned? With their axes at right angles to each other?

Problems

1. The Sun is 1.50×10^{11} m from the Earth. How long does it take for the Sun's light to reach the Earth?
2. In about 1675, the Danish astronomer Olaus Roemer found that light from eclipses of Jupiter's moon took an extra 1000 s to travel 300,000,000 km across the diameter of the Earth's orbit around the Sun. Show how this finding provided the first reasonably accurate measurement for the speed of light.
3. The nearest star beyond the Sun is Alpha Centauri, 4.2×10^{16} m away from the Earth. If we received a radio message from this star today, how long ago would it have been sent?

CHAPTER

12

Properties of Light

Most of the things we see around us do not emit their own light. They are visible because they reemit light reaching their surface from a primary source, such as the Sun or a lamp, or from a secondary source, such as the illuminated sky. When light falls on the surface of a material, the light is usually either reflected, transmitted through the material, or absorbed by the material and turned into heat. We say light is *reflected* when it is returned into the medium from which it came—the process is **reflection**. When light is transmitted from one transparent material into another, we say it is *refracted*—the process is **refraction**. Usually some degree of reflection, refraction, and absorption occurs when light interacts with matter. In this chapter we ignore the light absorbed and converted to heat energy and concentrate on the light that continues to be light after it meets a surface.

12.1 Reflection

When this page is illuminated with sunlight or lamplight, electrons in the atoms of the paper and ink vibrate more energetically in response to the oscillating electric fields of the illuminating light. The energized electrons reemit the light by which you see the page. When the page is illuminated by white light, the paper appears white, which reveals that the electrons reemit all the visible frequencies. Very little absorption occurs. The ink is a different story. Except for a bit of reflection, it absorbs all the visible frequencies and therefore appears black.

Law of Reflection

Anyone who has played pool or billiards knows that when a ball bounces from a surface, the angle of rebound is equal to the angle of incidence. Likewise for light. This is the **law of reflection**, and it holds for all angles:

The angle of refection equals the angle of incidence.

The law of reflection is illustrated in Figure 12.1, with arrows representing light rays. Instead of measuring angles from the reflecting surface, it is customary to measure the angles of each ray from a line perpendicular to the reflecting surface. This imaginary line is called the *normal.* The incident ray, the normal, and the reflected ray all lie in the same plane.

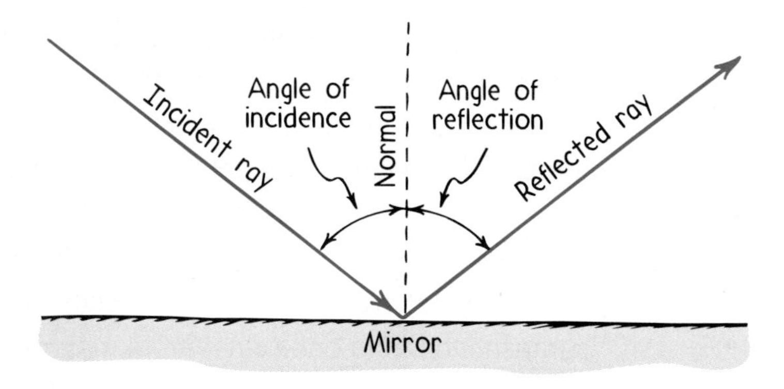

Fig. 12.1
The law of reflection.

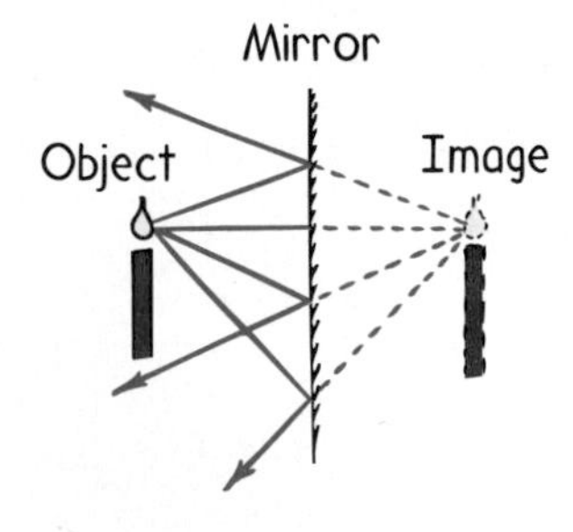

Fig. 12.2
A virtual image is formed behind the mirror and is located at the position where the extended reflected rays (dashed lines) converge.

A practical case of the law of reflection is an ordinary mirror. Suppose a candle flame is placed in front of a mirror. Rays of light are sent from the flame in all directions. Figure 12.2 shows only four of the infinite number of rays leaving one of the infinite number of points on the candle. When these rays encounter the mirror, they are reflected at angles equal to their angles of incidence. The rays diverge from the flame and on reflection diverge from the mirror. These divergent rays appear to emanate from behind the mirror, from a point located where the rays seem to diverge (dashed lines). An observer sees an image of the candle at this point. The image is as far behind the mirror as the object is in front of the mirror, and image and object have the same size. When you view yourself in a mirror, for example, the size of your image is the same as the size your twin would appear if located as far behind the mirror as you are in front—as long as the mirror is flat (we call a flat mirror a *plane mirror*).

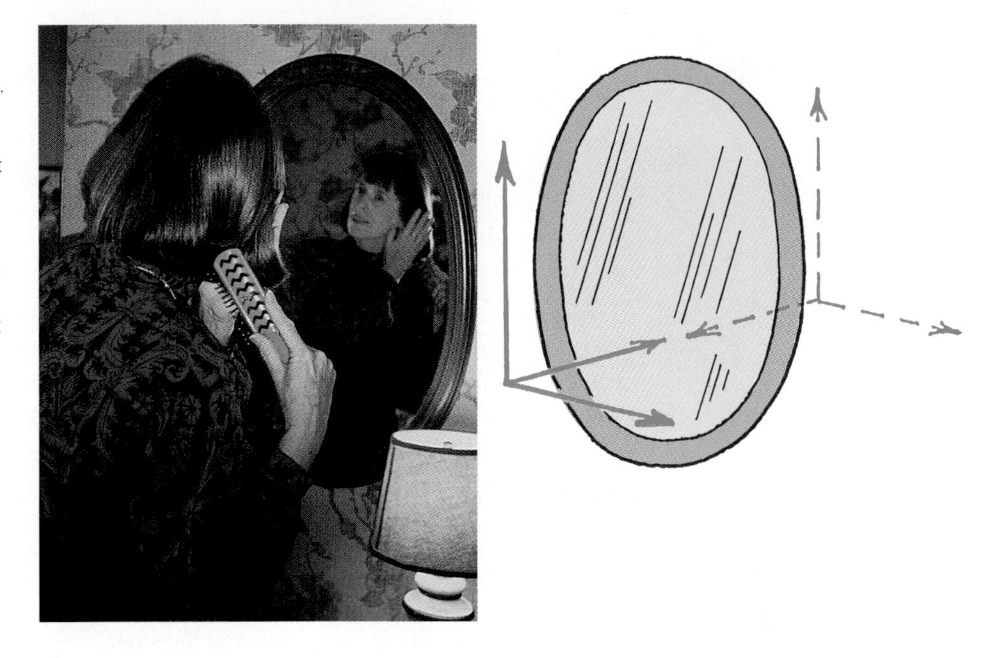

Fig. 12.3
Marjorie's image is as far behind the mirror as she is in front. Note that she and her image have the same color of clothing—evidence that light doesn't change frequency upon reflection. Interestingly, her left-right axis is no more reversed than her up-down axis. The axis that *is* reversed, as shown to the right, is front-back. That's why it seems her left hand faces the right hand of her image.

When the mirror is curved, the sizes and distances of object and image are no longer equal. We shall not get into curved mirrors in this text, except to say that the law of reflection still holds. A curved mirror behaves as a succession of flat mirrors, each at an angular orientation slightly different from the one next to it. At each point, the angle of incidence is equal to the angle of reflection (Figure 12.4). Note that in a curved mirror, unlike in a plane (flat) mirror, the normals (shown by the dashed black lines between the black rays) at different points on the surface are not parallel to one another.

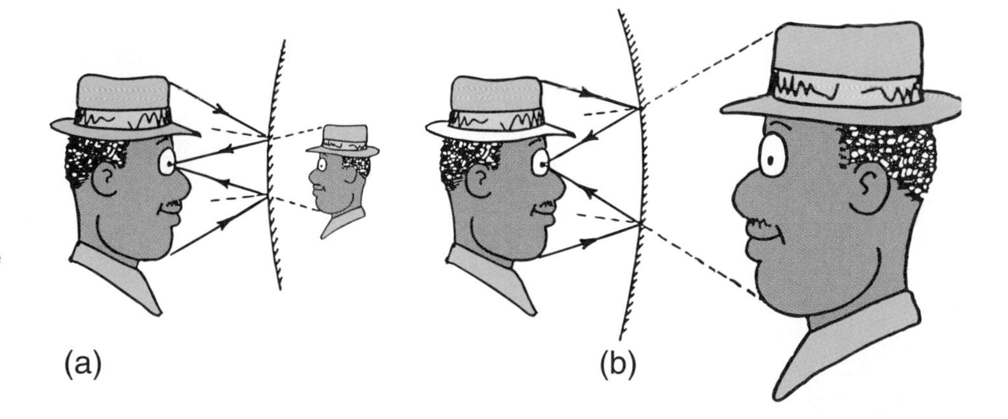

Fig. 12.4
(a) The virtual image formed by a *convex* mirror (a mirror that curves outward) Is smaller and closer to the mirror than the object. (b) When the object is close to a *concave* mirror (a mirror that curves inward like a "cave"), the virtual image is larger and farther away than the object. In either case the law of reflection holds for each ray.

Whether the mirror is flat or curved, the human eye-brain system cannot ordinarily tell the difference between an object and its reflected image. So the illusion that an object exists behind a mirror (or in some cases in front of a concave mirror) is merely due to the fact that the light from the object enters the eye in exactly the same manner, physically, as it would have entered if the object really were at the image location.

Questions

1. What evidence can you cite to support the claim that the frequency of light does not change upon reflection?
2. If you wish to take a picture of your image while standing 5 m in front of a plane mirror, for what distance should you set your camera to provide sharpest focus?

Only part of the light that strikes a surface is reflected. On a pane of clear glass, for example, and for normal incidence (light perpendicular to the surface), only about 4 percent is reflected from each surface of the pane. On a polished aluminum or silver surface, however, about 90 percent of incident light is reflected.

Diffuse Reflection

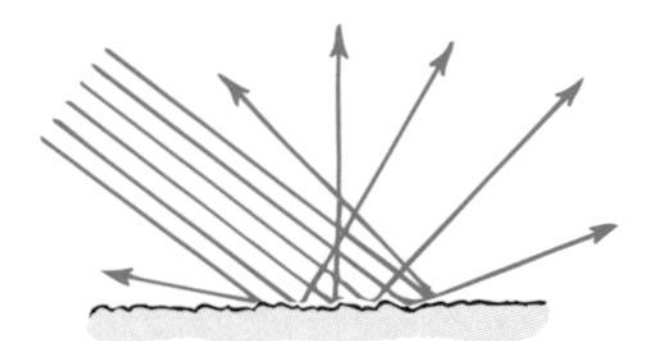

Fig. 12.5
Diffuse reflection. Although each ray obeys the law of reflection, the many different surface angles that light rays encounter in striking a rough surface cause reflection in many directions.

Light incident on a rough surface is reflected in many directions. This is called **diffuse reflection** (Figure 12.5). If a surface is so smooth that the distances between successive elevations on the surface are less than about one-eighth the wavelength of the light, there is very little diffuse reflection and the surface is said to be *polished.* A surface may therefore be polished for radiation of long wavelength but not polished for light of short wavelength. The wire-mesh "dish" shown in Figure 12.6 is very rough for light waves and so hardly mirrorlike. But for long-wavelength radio waves it is "polished" and therefore an excellent reflector.

Fig. 12.6
The open-mesh parabolic dish is a diffuse reflector for short-wavelength visible-light waves but a polished reflector for long-wavelength radio waves.

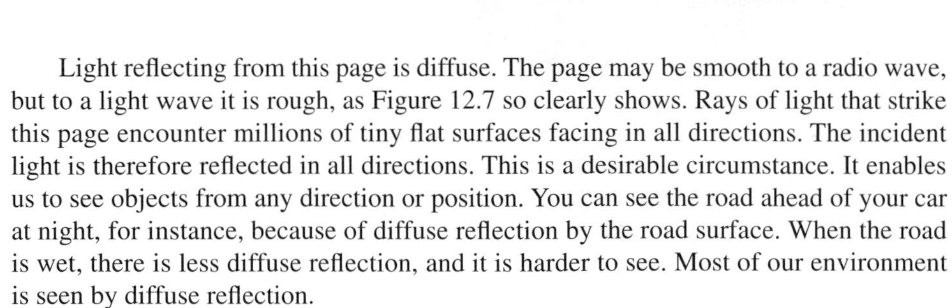

Light reflecting from this page is diffuse. The page may be smooth to a radio wave, but to a light wave it is rough, as Figure 12.7 so clearly shows. Rays of light that strike this page encounter millions of tiny flat surfaces facing in all directions. The incident light is therefore reflected in all directions. This is a desirable circumstance. It enables us to see objects from any direction or position. You can see the road ahead of your car at night, for instance, because of diffuse reflection by the road surface. When the road is wet, there is less diffuse reflection, and it is harder to see. Most of our environment is seen by diffuse reflection.

Fig. 12.7
A magnified view of the surface of ordinary paper.

Answers

1. The color of an image is identical to the color of the object forming the image. Look at yourself in a mirror and the color of, say, your eyes doesn't change. The fact that the color is the same is evidence that the frequency of light doesn't change upon reflection.
2. Set your camera for 10 m; the situation is equivalent to your standing 5 m in front of an open window and viewing your twin standing 5 m in back of the window.

An undesirable circumstance related to diffuse reflection is the ghost image that occurs on a TV set when the TV signal bounces off buildings and other obstructions. For antenna reception, this difference in path lengths for the direct signal and the reflected signal produces a slight time delay. Multiple reflections may produce multiple ghosts.

12.2 Refraction

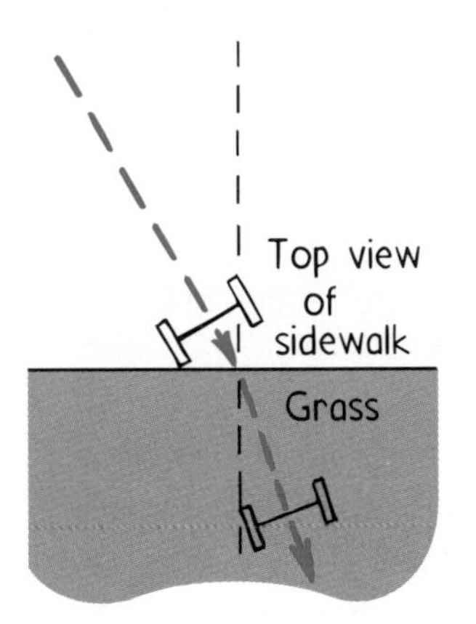

Fig. 12.8
The direction of the rolling wheels changes when one slows down before the other does.

Recall from Chapter 11 that the average speed of light is lower in glass and other transparent materials than in empty space. In other words, light travels at different speeds in different materials.[1] It travels at 300,000 kilometers per second in a vacuum, at a slightly lower speed in air, and at about three-fourths that speed in water. In a diamond, light travels at about 40 percent of its speed in a vacuum. As mentioned at the opening of this chapter, when light bends as it passes obliquely from one medium to another, we call the process *refraction.*

The cause of refraction is the changing of the average speed of light in going from one transparent medium to another. We can understand this by considering the action of a pair of toy cart wheels connected to an axle as the wheels roll from a smooth sidewalk onto a grass lawn. If the wheels meet the grass at some angle, as in Figure 12.8, they are deflected from their straight-line course. Note that on meeting the lawn, where the wheels roll more slowly owing to interaction with the grass, the left wheel slows down first. This is because it meets the grass while the right wheel is still on the smooth sidewalk. The faster-moving right wheel tends to pivot about the slower-moving left wheel because during the same time interval, the right wheel travels farther than the left one. This action bends the direction of the rolling wheels toward the "normal," the black dashed line perpendicular to the grass-sidewalk border in Figure 12.8.

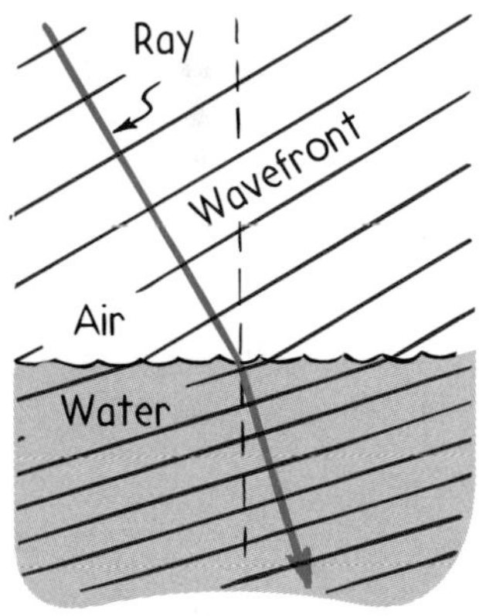

Fig. 12.9
When light is refracted (bent) as it moves from one medium into another, the direction of the light waves changes as one part of each wave slows down before the other part.

A light wave bends in a similar way, as shown in Figure 12.9. Note the direction of light, shown by the blue arrow (the light ray), and also note the *wave fronts* drawn at right angles to the ray. (If the light source were close, the wave fronts would appear as segments of circles; but if we assume the distant Sun is the

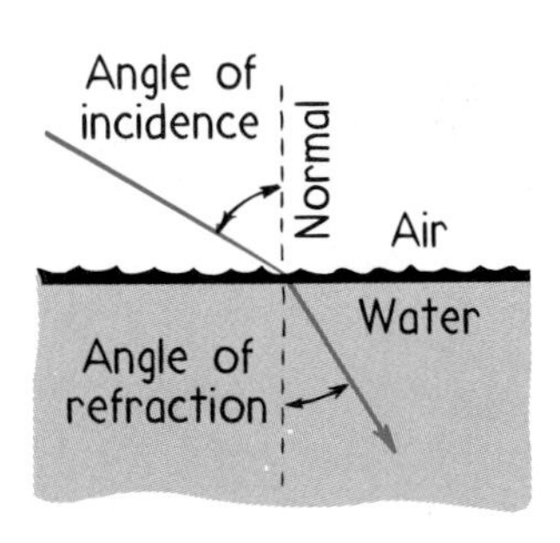

Fig. 12.10
Refraction.

[1]Just how much the speed of light differs from its speed in a vacuum is given by the index of refraction, *n,* of the material:

$$n = \frac{\text{speed of light in vacuum}}{\text{speed of light in material}}$$

For example, the speed of light in diamond is 125,000 km/s, and so for diamond the index of refraction is

$$n = \frac{300{,}000 \text{ km/s}}{125{,}000 \text{ km/s}} = 2.4$$

For a vacuum, $n = 1$.

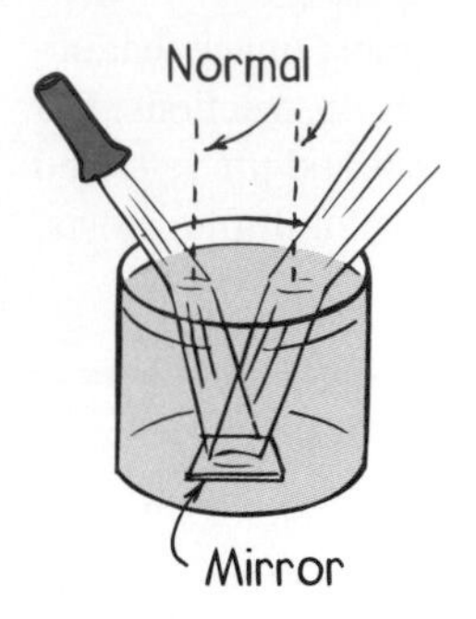

Fig. 12.11
When light slows down in going from one medium to another, as in going from air to water, it refracts toward the normal. When it speeds up in traveling from one medium to another, as in going from water to air, it refracts away from the normal.

source, the wave fronts are practically straight lines.) The wave fronts are everywhere perpendicular to the light rays. In the figure the wave meets the water surface at an angle, and so the left portion of the wave slows down in the water while the part still in the air travels at speed *c*. The light ray must remain perpendicular to the wave front and so bends at the surface, just as the wheels bend to change direction when they roll from the sidewalk into the grass. In both cases the bending is caused by a change in speed.[2]

Figure 12.11 shows a beam of light entering water at the left and exiting at the right. The path would be the same if the light entered from the right and exited at the left. The light paths are reversible for both reflection and refraction. If you see someone's eyes by way of a reflective or refractive device, such as a mirror or a prism, then that person can see you by the device also.

Question

If the speed of light were the same in all media, would refraction still occur when light passes from one medium to another?

The refraction of light is responsible for many illusions; one of them is the apparent bending of a stick partly immersed in water. The submerged part seems closer to the surface than it really is. Likewise when you view a fish in water: The fish appears nearer to the surface and closer than it really is (Figure 12.12). If we look straight down into water, an object submerged 4 meters beneath the surface appears to be only 3 meters deep. Because of refraction, submerged objects appear to be magnified.

Fig. 12.12
Because of refraction, a submerged object appears to be nearer to the surface than it actually is.

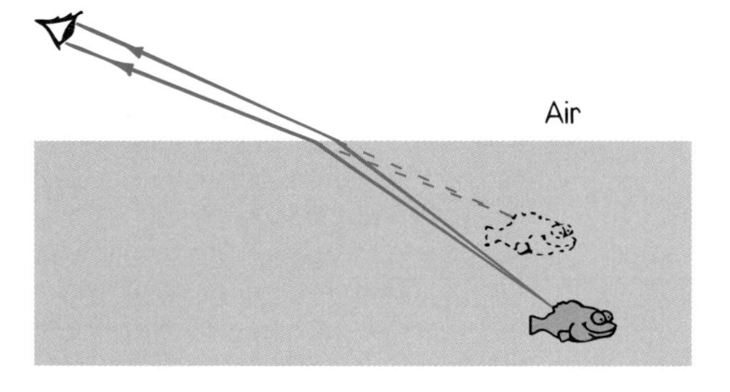

Answer
No.

[2]The quantitative law of refraction, called *Snell's law,* is credited to W. Snell, a seventeenth-century Dutch astronomer and mathematician: $n_1 \sin \phi_1 = n_2 \sin \phi_2$, where n_1 and n_2 are the indices of refraction of the media on either side of the surface, and ϕ_1 and ϕ_2 are the respective angles of incidence and refraction. If three of these values are known, the fourth can be calculated from this relationship.

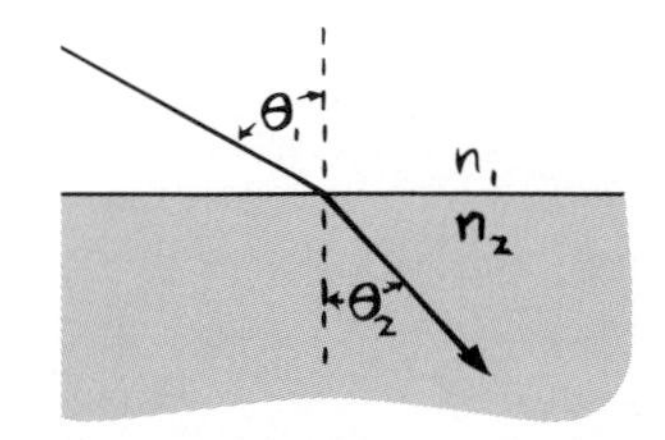

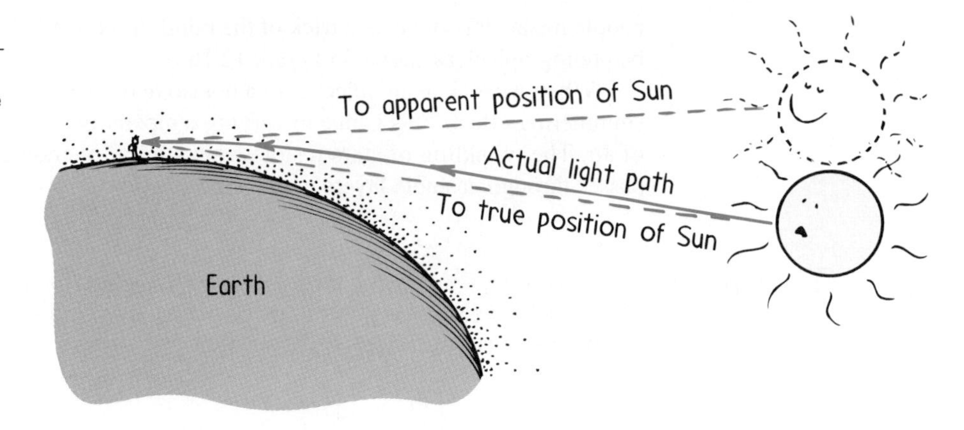

Fig. 12.13
Because of atmospheric refraction, when the Sun is near the horizon it appears higher in the sky.

Fig. 12.14
The Sun's shape is distorted by differential refraction.

Refraction occurs in the Earth's atmosphere. Whenever we watch a sunset, we see the Sun for several minutes after it has sunk below the horizon (Figure 12.13). The Earth's atmosphere is thin at the top and dense at the bottom. Because light travels faster in thin air than in dense air, parts of the wave fronts of sunlight higher up travel faster than parts of the wave fronts closer to ground. Because the density of the atmosphere changes gradually, the light path bends gradually and produces a curved path. So we get a slightly longer period of daylight each day. Furthermore, when the Sun (or Moon) is near the horizon, the rays from the lower edge are bent more than the rays from the upper edge. This produces a shortening of the vertical diameter, causing the Sun to appear elliptical (Figure 12.14).

We are all familiar with the mirage we see while driving on a hot road. The sky appears to be reflected from water on the distant road, but when we get there, the road is dry. Why is this so? The air is very hot just above the road surface and cooler above. Light travels faster through the thinner hot air near the ground than through the denser cool air above. Wave fronts near the ground travel faster than those higher up, and so the light is refracted upward (Figure 12.15). So we see an upside-down view as if reflection were occurring from a water surface. We see a mirage. A mirage is not, as many

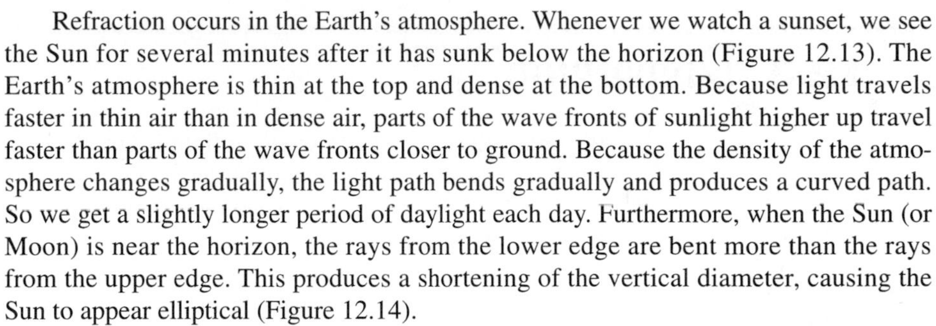

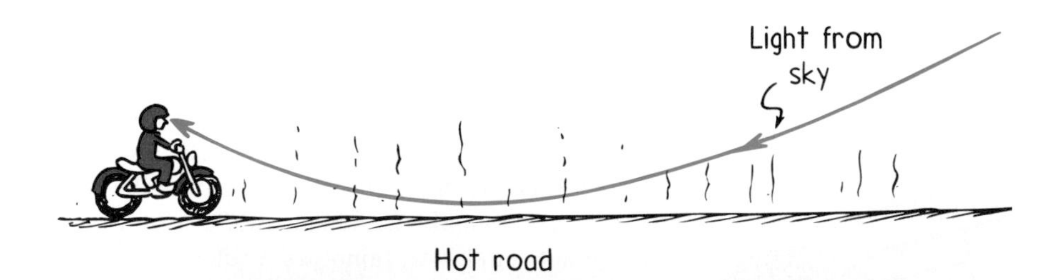

Fig. 12.15
Light from the sky picks up speed in air near the ground because that air is warmer and less dense than the air above. When the light grazes the surface and bends upward, the observer sees a mirage.

Fig. 12.16
A mirage. The apparent wetness of the road is not reflection of the sky by water but rather refraction of sky light through the warmer and less-dense air near the road surface.

people mistakenly believe, a trick of the mind. A mirage is formed by real light and can be photographed, as shown in Figure 12.16.

When we look at an object over a hot stove or over a hot pavement, we see a wavy, shimmering effect. This is due to varying temperatures and therefore varying densities of air. The twinkling of stars results from similar phenomena in the sky, where light passes through unstable layers in the atmosphere.

■ **Question**

If the speed of light were the same in air of various temperatures and densities, would there still be slightly longer daytimes, twinkling stars at night, mirages, and a slightly squashed Sun at sunset?

12.3 Dispersion

We know that the average speed of light is less than c in a transparent medium. How much less depends on the nature of the medium and on the frequency of the light. Recall two things from Section 11.2: (1) Light having frequencies that match resonant electron frequencies in the atoms and molecules of a transparent medium is absorbed, and (2) light having frequencies near the resonant frequencies interacts more often in the absorption/reemission sequence and therefore travels more slowly. Because the resonant frequency of most transparent materials is in the ultraviolet part of the spectrum, the higher frequencies of visible light travel more slowly than the lower frequencies. Violet travels about 1 percent slower in ordinary glass than does red light. The colors between red and violet travel at their own respective speeds.

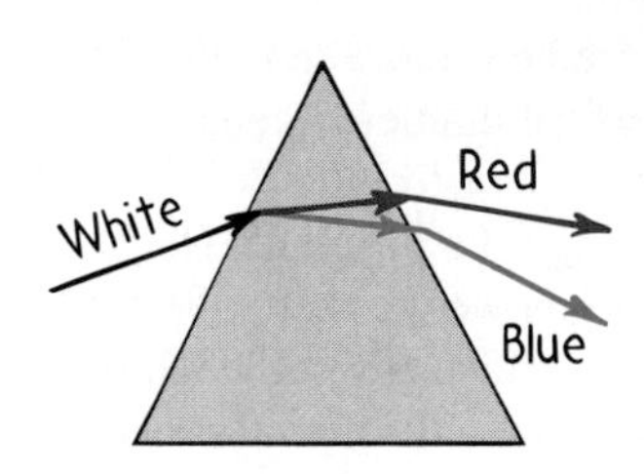

Fig. 12.17
Dispersion by a prism makes the components of white light visible.

Because different frequencies of light travel at different speeds in transparent materials, they refract by different amounts. When white light is refracted twice, as in a prism, the separation of the different colors of light is quite noticeable. This separation of light into colors arranged according to frequency is called *dispersion* (Figure 12.17).

Rainbows

A most spectacular illustration of dispersion is the rainbow. For a rainbow to be seen, the Sun must be shining in one part of the sky and water drops in a cloud or in falling rain must be present in the opposite part of the sky. When our backs are to the Sun, we see the spectrum of colors in a bow. Seen from an airplane near midday, the bow forms a complete circle. All rainbows would be completely round if the ground were not in the way.

The beautiful colors of rainbows are dispersed from the sunlight by millions of tiny spherical droplets that act like prisms. We can better understand this by considering an individual raindrop, as shown in Figure 12.18. Follow the ray of sunlight as it enters the drop near its top surface. Some of the light here is reflected (not shown), and the remainder is refracted into the water. At this first refraction, the light is dispersed into its spectrum colors, violet being deviated the most and red the least. Reaching the

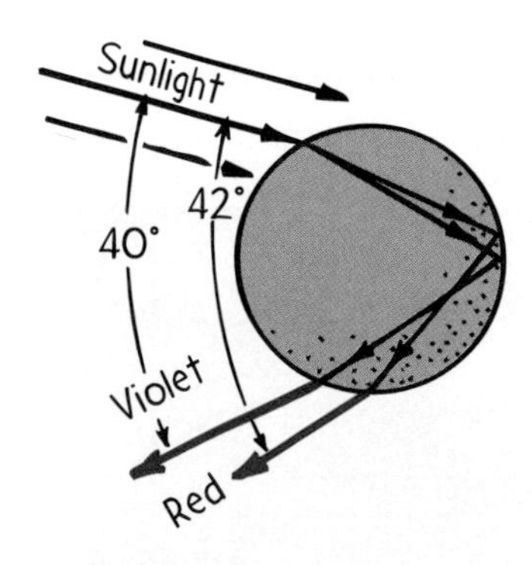

Fig. 12.18
Dispersion of sunlight by a single raindrop.

■ **Answer**

No.

opposite side of the drop, each color is partly refracted out into the air (not shown) and partly reflected back into the water. Arriving at the lower surface of the drop, each color is again partly reflected (not shown) and partly refracted into the air. This refraction at the second surface, like that in a prism, increases the dispersion already produced at the first surface.[3]

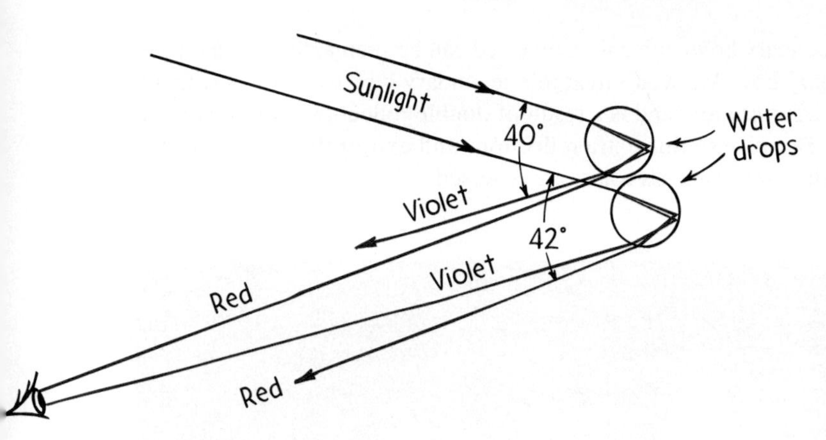

Fig. 12.19
Sunlight incident on two raindrops as shown emerges from them as dispersed light. The observer sees the red light from the upper drop and the violet light from the lower drop. Millions of drops produce the whole spectrum of visible light

Although each drop disperses a full spectrum of colors, an observer is in a position to see only a single color from any one drop (Figure 12.19). If violet light from a single drop reaches the eye of an observer, red light from the same drop is incident elsewhere (toward the observer's feet). To see red light, one must look to a drop higher in the sky. The color red is seen when the angle between a beam of sunlight and the dispersed light is 42 degrees. The color violet is seen when the angle between the sunbeams and dispersed light is 40 degrees.

Why does the light dispersed by the raindrops form a bow? The answer to this involves a bit of geometry. First of all, a rainbow is not the flat, two-dimensional arc it appears to be. It appears flat for the same reason a spherical burst of fireworks high in the sky appears as a disc—because of a lack of distance cues. The rainbow you see is actually a three-dimensional cone with the tip (apex) at your eye. Consider a glass cone, the shape of those paper cones you sometimes see at drinking fountains. If you held the tip of such a glass cone against your eye, what would you see? You'd see the glass as a circle. Likewise with a rainbow. All the drops that disperse the rainbow's light toward *you* lie in the shape of a cone—a cone of different layers with drops that deflect red to your eye on the outside, orange beneath the red, yellow beneath the orange, and so on all the way to violet on the inner conical surface (Figure 12.20). The thicker the region containing water drops, the thicker the conical edge you look through, and the more vivid the rainbow.

Your cone of vision that intersects the cloud of drops that creates your rainbow is different from that of a person next to you. So when a friend says, "Look at the pretty rainbow," you can reply, "Okay, move aside so I can see it, too." Everybody sees his or her own personal rainbow.

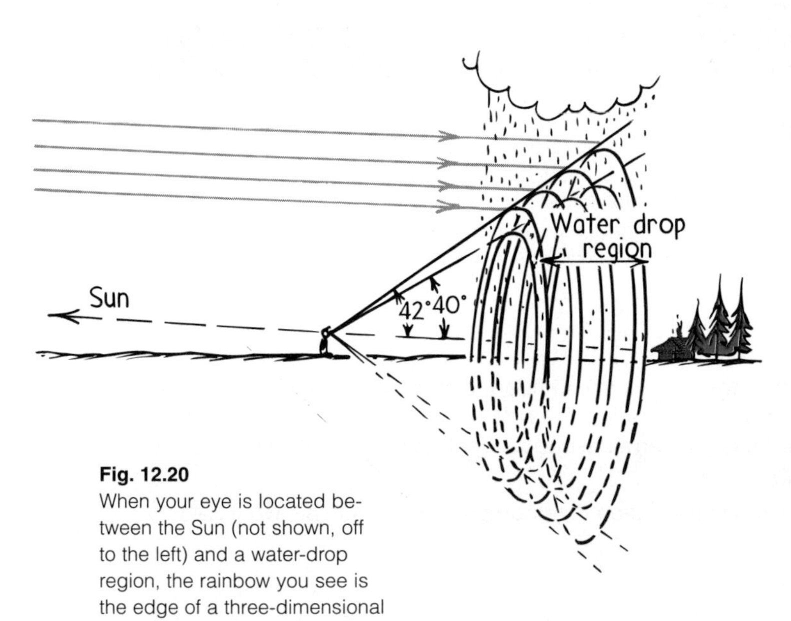

Fig. 12.20
When your eye is located between the Sun (not shown, off to the left) and a water-drop region, the rainbow you see is the edge of a three-dimensional cone that extends through the water-drop region. (Innumerable layers of drops form innumerable two-dimensional arcs like the four suggested here.)

[3] Two refractions and a reflection can result in the angle between the incoming and outgoing rays being anything between 0° and about 42° (0° corresponding to a full 180° reversal of the light). There is a strong concentration of light intensity, however, near the maximum angle of 42°. That is what is shown in Figures 12.18 and 12.19.

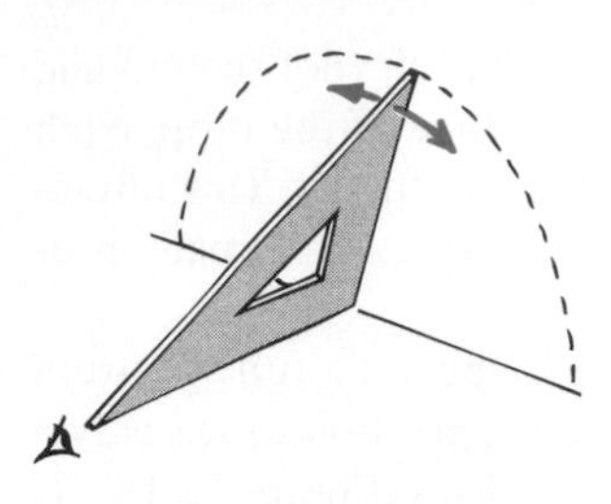

Fig. 12.21
Only raindrops along the dashed line disperse red light to the observer at a 42° angle; hence, the light forms a bow.

Another fact about rainbows: A rainbow always faces you squarely, because of the lack of distance cues mentioned earlier. When you move, your rainbow moves with you. So you can never approach the side of a rainbow or see it end-on as in the exaggerated view of Figure 12.20. You *can't* get to its end. Hence the expression "looking for the pot of gold at the end of the rainbow" means pursuing something you can never reach.

Often a larger, secondary bow with colors reversed can be seen arching at a greater angle around the primary bow. We won't treat this secondary bow except to say that it is formed by similar circumstances and is a result of double reflection within the raindrops (Figure 12.23). Because of this extra reflection (and extra refraction loss), the secondary bow is much dimmer and its colors are reversed.

Fig. 12.22
Two refractions and a reflection in water droplets produce light at all angles up to about 42°, with the intensity concentrated where we see the rainbow at 40° to 42°. No light gets out of the water droplet at angles greater than 42° unless it undergoes two or more reflections inside the drop. So the sky is brighter inside the rainbow than outside it. Notice the weak secondary rainbow to the right of the primary.

Sunlight

Red

Violet

Fig. 12.23
Double reflection in a drop produces a secondary bow.

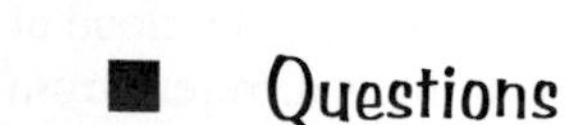

Questions

1. If you point to a wall with your arm extended to make about a 42° angle to the wall, then rotate your arm in a full circle while keeping the 42° angle to the wall, what shape does your arm describe? What shape on the wall does your finger sweep out?
2. If light traveled at the same speed in raindrops as it does in air, would we have rainbows?

Answers
1. Your arm describes a cone, and your finger sweeps out a circle. Likewise with rainbows.
2. No.

12.4 Total Internal Reflection

Some Saturday night when you're taking your bath, fill the tub extra deep and bring a waterproof flashlight into the tub with you. Put the bathroom light out. Shine the submerged light straight up and then slowly tip it, as shown in Figure 12.24. Note how the intensity of the emerging beam diminishes and how more and more light is reflected from the water surface to the bottom of the tub. At a certain angle, called the **critical angle**, you'll notice that the beam no longer emerges into the air above the surface. At this point, the intensity of the emerging beam reduces to zero where the beam grazes the surface. When the flashlight is tipped beyond the critical angle (48 degrees from the normal for water), all the light is reflected back into the tub. This is **total internal reflection**. The light striking the air-water surface obeys the law of reflection: The angle of reflection is equal to the angle of incidence. When the angle of incidence is greater than the critical angle, the only light emerging from the water surface is that diffusely reflected from the bottom of the bathtub.

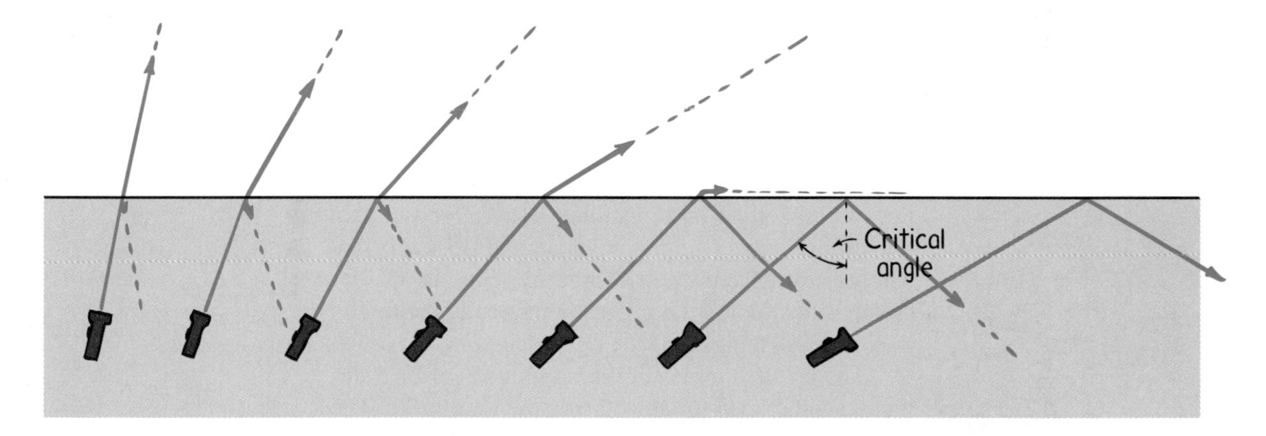

Fig. 12.24
Light emitted in the water is partly refracted and partly reflected at the surface. The blue dashes show the direction of light, and the length of the arrows indicates the proportions refracted and reflected. Beyond the critical angle the beam is totally internally reflected.

Because of total internal reflection, your pet goldfish in the bathtub looks up to see a compressed view of the outside world. The 180 degree view from horizon to opposite horizon is seen through an angle of 96 degrees—twice the critical angle (Figure 12.25). A lens that similarly compresses a wide view, called a *fisheye lens,* is used for special-effect photographs.

The critical angle for glass is about 43 degrees, depending on the type of glass. This means that any light traveling in a slab of glass that hits either surface at any angle greater than 43 degrees is totally internally reflected. This phenomenon is the principal reason for the use of prisms instead of mirrors in many optical instruments. Whereas a mirror reflects only about 90 percent of incident light, glass prisms are more efficient, as shown in Figure 12.26. A tiny amount

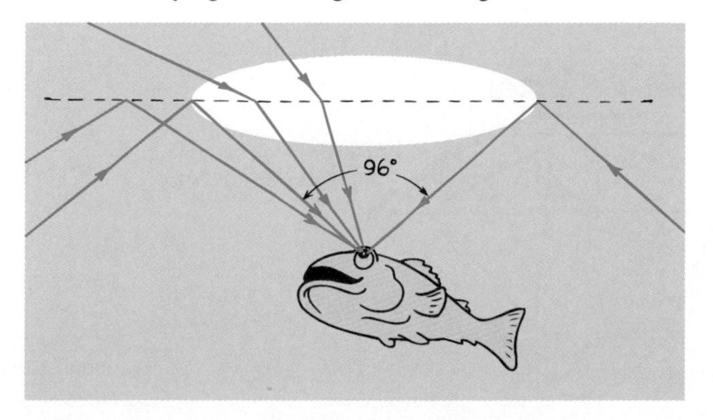

Fig. 12.25
An observer under water sees a circle of light at the still surface. Beyond a cone of 96° (twice the critical angle), the observer sees a reflection of the water interior or bottom.

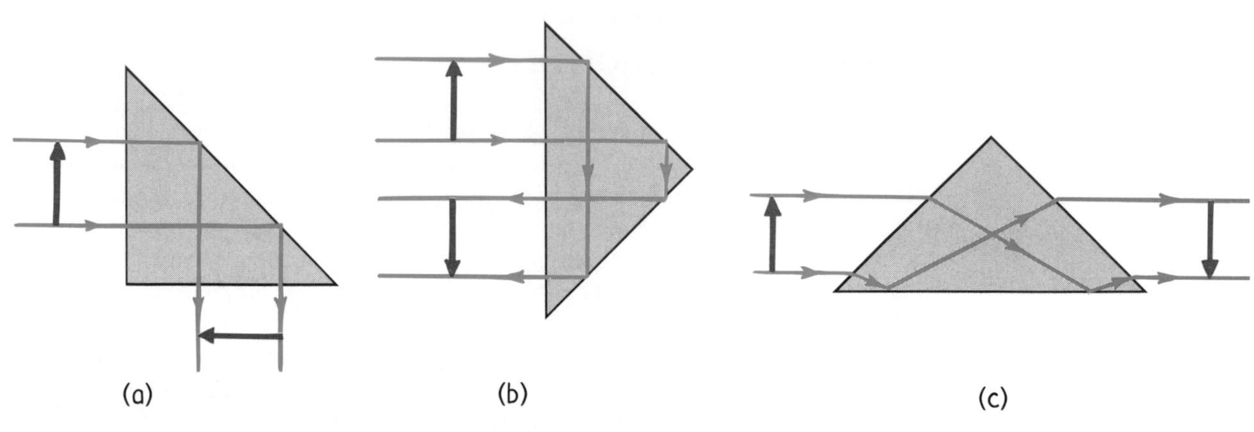

Fig. 12.26
Total internal reflection in a prism. The prism changes the direction of the light beam (a) by 90°, (b) by 180°, and (c) not at all. Note that in each case the orientation of the image is different from the orientation of the object.

of light is lost by reflection as the beam enters the prism, but once inside, reflection on the 45-degree-slanted face is total—100 percent. Moreover, because the reflection is from an interior surface, the light is not marred by any dirt or dust, as might happen with the surface of a mirror.

A pair of prisms each reflecting light through 180 degrees is shown in Figure 12.27. Binoculars use pairs of prisms to lengthen the light path between lenses and thus eliminate the need for long barrels (Figure 12.28). So a compact set of binoculars is as effective as a longer telescope. Another advantage of prisms is that whereas the image of a straight telescope is upside-down, reflection by the prisms in binoculars reinverts the image, so things are seen right-side-up.

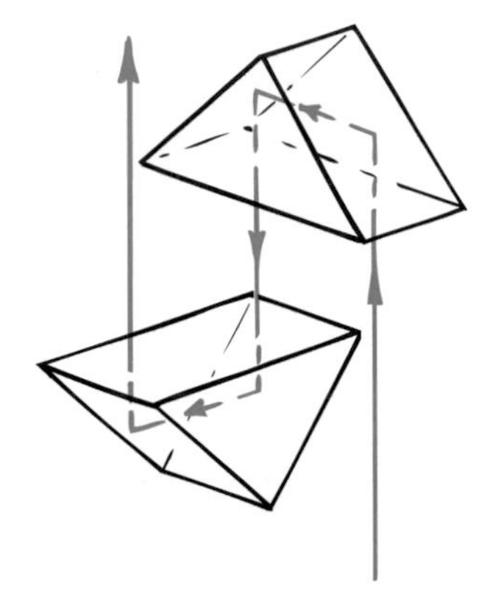

Fig. 12.27
Total internal reflection in a pair of prisms.

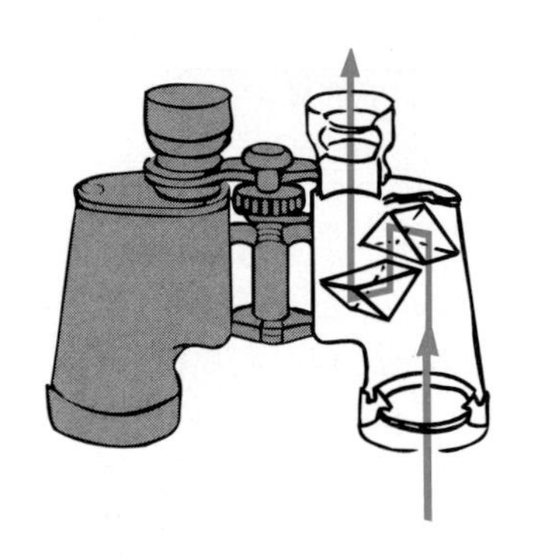

Fig. 12.28
Prism binoculars.

The critical angle for diamond is about 24.5 degrees, smaller than for any other known substance. The critical angle varies slightly for different colors because speed varies slightly for different colors. Once white light enters a diamond gemstone, most is incident on the sloped backsides at angles greater than 24.5 degrees and so is totally internally reflected (Figure 12.29). Because of the great slowdown in speed as light enters a diamond, refraction is pronounced, and because of the frequency-dependence of the speed, there is great dispersion. Further dispersion occurs as the light exits through the many facets at the diamond's face. Hence we see unexpected flashes of a wide array of colors. Interestingly, when these flashes are narrow enough to be seen by only one eye at a time, the diamond "sparkles."

Total internal reflection also underlies the operation of optical fibers, or light pipes (Figure 12.30). An optical fiber "pipes" light from one place to another by a series of total internal

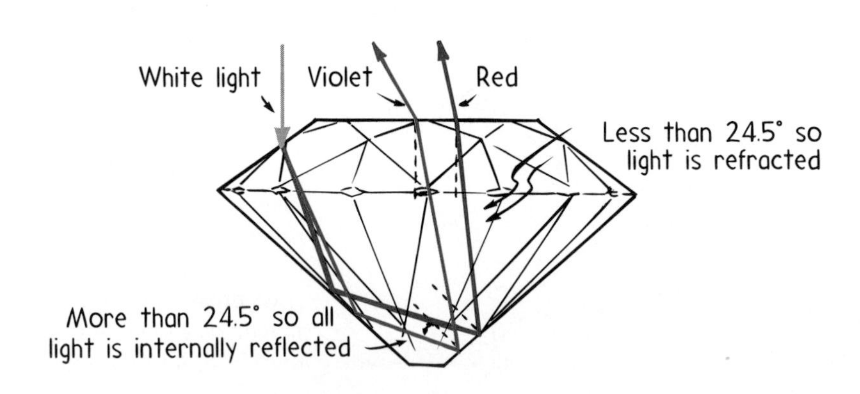

Fig. 12.29
Paths of light in a diamond. Rays that strike the inner surface at angles greater than the critical angle are internally reflected and exit via refraction at the top surface.

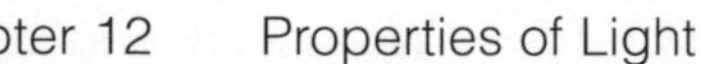

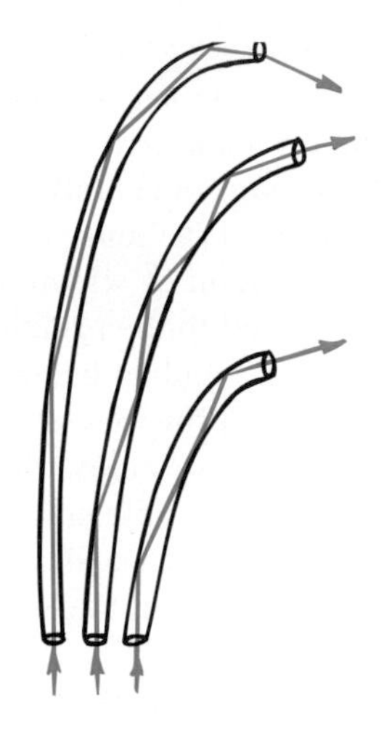

Fig. 12.30
The light is "piped" from below by a succession of total internal reflections until it emerges at the top ends.

reflections, much as a bullet ricochets down a steel pipe. Light rays bounce along the inner walls, following the twists and turns of the fiber. Optical fibers are used in decorative table lamps and to illuminate instrument displays on automobile dashboards from a single bulb. Dentists use them with flashlights to get light where they want it. Bundles of thin flexible glass or plastic fibers are used to see what is going on in inaccessible places, such as the interior of a motor or a patient's stomach. These bundles can be made small enough to snake through blood vessels or through such tubes as the urethra. Light shines down some of the fibers to illuminate the scene and is reflected back along others.

Optical fibers are important in communications because they offer a practical alternative to copper wires and cables. In many places, thin glass fibers now replace thick, bulky, expensive copper cables to carry thousands of simultaneous telephone messages among the major switching centers. In many aircraft, control signals are fed from the pilot to the control surfaces by means of optical fibers. Signals are carried in the modulations of laser light. Unlike electricity, light is indifferent to temperature and fluctuations in surrounding magnetic fields, and so the signal is clearer. Also, it is much less likely to be tapped by eavesdroppers.

12.5 Lenses

A very practical case of refraction occurs in lenses. We can understand a lens by assuming it consists of a set of matched prisms and blocks of glass arranged in the order shown in Figure 12.31. The prisms and blocks refract incoming parallel light rays so that they converge to (or diverge from) a point. The arrangement shown in Figure 12.31a converges the light, and we call such a lens a **converging lens**. Note that it is thicker in the middle. In Figure 12.31b the middle is thinner than the edges, and this lens diverges the light; such a lens is called a **diverging lens**. Note that the prisms in (b) diverge the incident rays in a way that makes them appear to come from a single point in front of the lens.

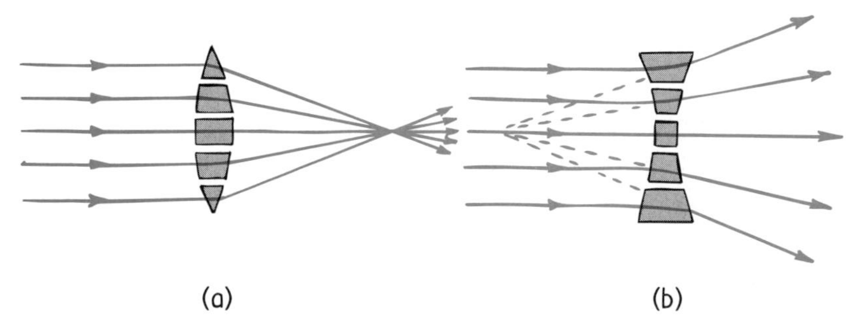

Fig. 12.31
A lens may be thought of as a set of blocks and prisms.
(a) A converging lens.
(b) A diverging lens.

In both lenses the greatest deviation of rays occurs at the outermost prisms, for they have the greatest angle between the two refracting surfaces. No deviation occurs exactly in the middle, for in that region the two surfaces of the glass are parallel to each other. Real lenses are not made of prisms, of course; they are made of a solid piece of glass with surfaces ground usually to a circular curve. In Figure 12.32 we see how smooth lenses refract waves.

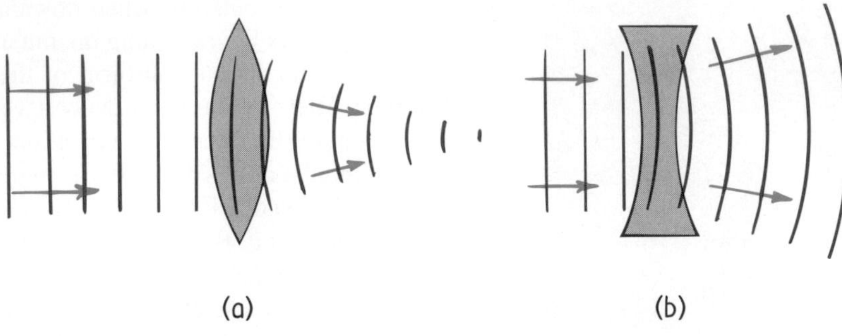

Fig. 12.32
Wave fronts travel more slowly in glass than in air. (a) The waves are retarded more through the center of the lens, and convergence results. (b) The waves are retarded more at the edges, and divergence results.

Some key features of lenses are shown for a converging lens in Figure 12.33. The *principal axis* is the line joining the centers of curvatures of the two surfaces of the lens. The *focal point* is the point at which a beam of light parallel to the principal axis converges. Incident beams not parallel to the principal axis focus at points above or below the focal point. All such possible points make up a *focal plane.* Because a lens has two surfaces, it has two focal points and two focal planes. The *focal length* of the lens is the distance between the center of the lens and either focal point.

In a diverging lens, an incident beam of light parallel to the principal axis is not converged to a point. Instead, it is diverged, with the result that the light appears to come from a point in front of the lens.

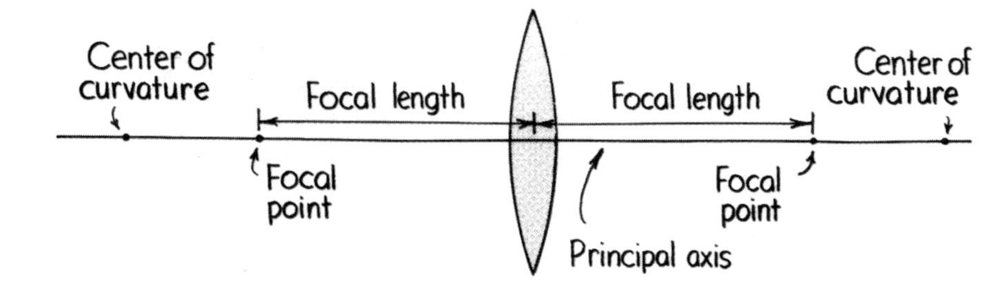

Fig. 12.33
Key features of a converging lens.

Image Formation by a Lens

At this moment, light is reflecting from your face onto this page. Light that reflects from your forehead, for example, strikes every part of the page. Likewise for the light that reflects from your chin. Every part of the page is illuminated with reflected light from your forehead, your nose, your chin, and every other part of your face. You don't see an image of your face on the page because there is too much overlapping of light. But put a barrier with a pinhole in it between your face and the page, and the light that reaches the page from your forehead does not overlap the light from your chin. Likewise for the rest of your face. Without this overlapping, an image of your face is formed on the page. It will be very dim, for very little light reflected from your face gets through the pinhole. To see the image, you'd have to shield the page from other light sources. The same is true of the vase and flowers in Figure 12.35b.

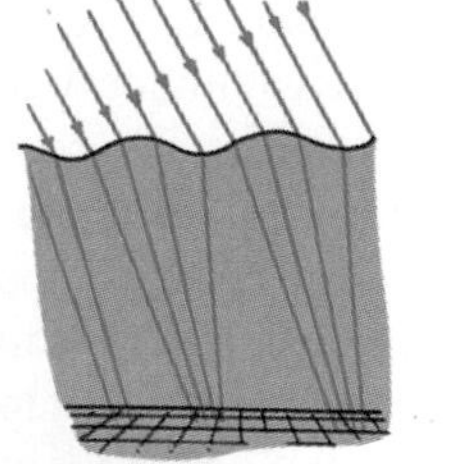

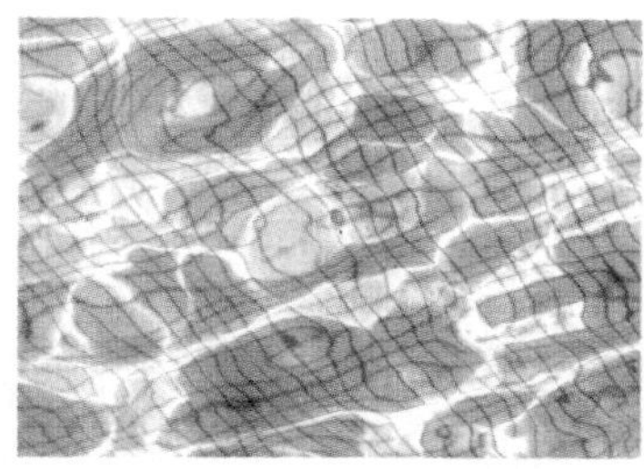

Fig. 12.34
The moving patterns of bright and dark areas at the bottom of the pool result from the uneven surface of the water, which behaves like a blanket of undulating lenses. Just as we see the pool bottom shimmering, a fish looking upward at the Sun would see it shimmering, too. Because of similar irregularities in the atmosphere, we see the stars twinkle.

The first cameras had no lenses and admitted light through a small pinhole.[4] Long exposure times were required because of the small amount of light admitted by the pinhole. A somewhat larger hole would admit more

[4] You can see why the image formed by a camera is upside-down by looking at the sample rays in Figure 12.35b.

Link to Physiology—Your Eye

Light is the only thing we see with the most remarkable optical instrument known—the eye. Light enters through the *cornea,* which does about 70 percent of the necessary bending of the light before it passes through the *pupil* (the aperture—opening—in the iris). Light then passes through the *lens,* which provides the extra bending power needed to focus images of nearby objects on the extremely sensitive *retina* (only quite recently have artificial detectors attained greater sensitivity to light than the human eye). An image of the visual field outside the eye is spread over the retina. The retina is not uniform. There is a spot in the center of our field of view called the *fovea,* or region of most distinct vision. Greater detail is seen here than at the side parts of the eye. There is also a spot in the retina where the nerves carrying all the information exit; this is the *blind spot.*

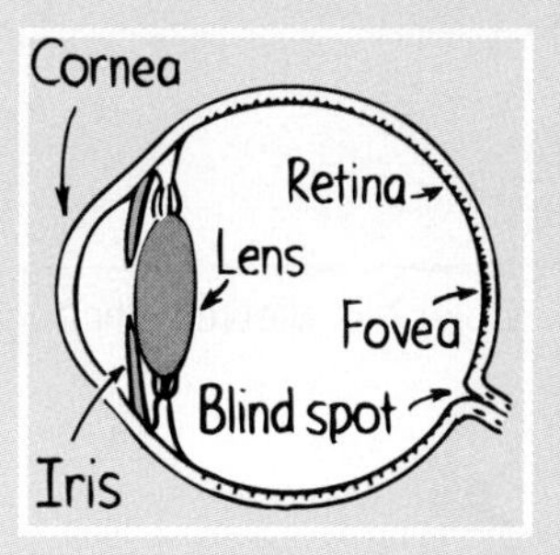

You can demonstrate that you have a blind spot in each eye if you hold this book at arm's length, close your left eye, and look at the round dot and X below with your right eye only. You can see both the dot and the X at this distance. Now move the book slowly toward your face, with your right eye fixed upon the dot, and you'll reach a position about 20–25 centimeters from your eye where the X disappears; when both eyes are open, one eye "fills in" the part to which the other eye is blind. Now repeat with only the left eye open, looking this time at the X, and the dot will disappear. But note that your brain fills in the two intersecting lines. Amazingly, your brain fills in the "expected" view even with one eye closed. Instead of seeing nothing, your brain gratuitously fills in the appropriate background. Repeat this for small objects on various backgrounds. You not only see what's there—you see what's not there!

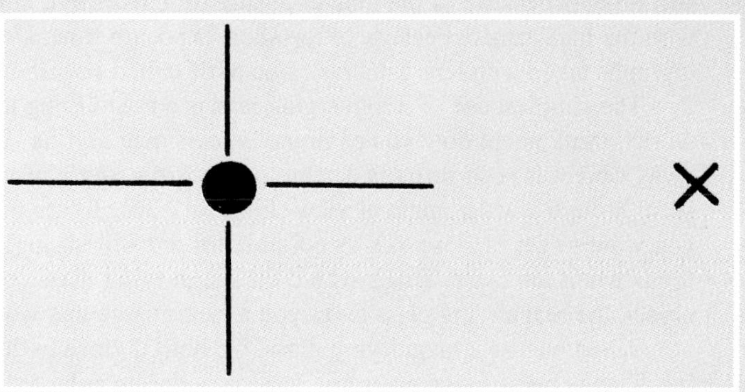

The light receptors in the retina do not connect directly to the optic nerve but are instead interconnected to many other cells. Through these interconnections a certain amount of information is combined and "digested" in the retina. In this way the light signal is "thought about" before it goes to the optic nerve and then to the main body of the brain. So some brain functioning occurs in the eye. Amazingly, the eye does some of our "thinking."

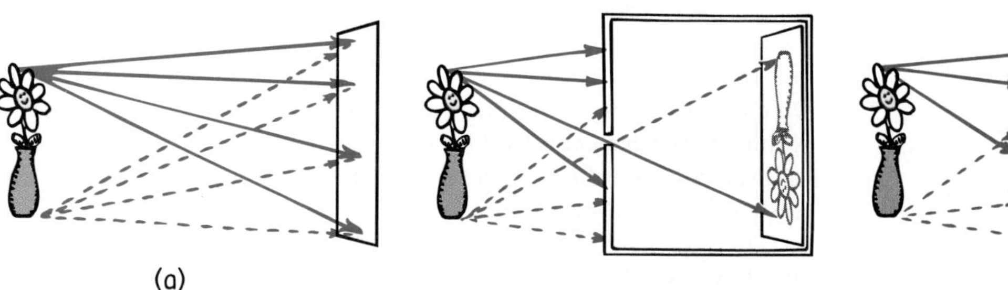

Fig. 12.35
Image formation. (a) No image appears on the wall because rays from all parts of the object overlap all parts of the wall. (b) A single small opening in a barrier prevents overlapping rays from reaching the wall; a dim upside-down image is formed. (c) A lens converges the rays upon the wall without overlapping; more light makes a brighter image.

Fig. 12.36

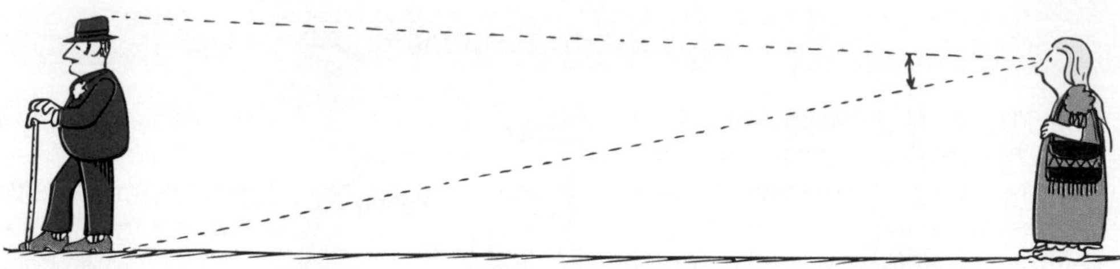

light, but overlapping rays would produce a blurry image. Too large a hole would allow too much overlapping and no image would be discernible. That's where a converging lens comes in (Figure 12.35c). The lens converges light onto the film without the unwanted overlapping of rays. Whereas the first pinhole cameras were useful only for still objects because of the long exposure time required, moving objects can be taken with the lens camera because of the short exposure time. Do you now know why photographs taken with lens cameras came to be called *snapshots*?

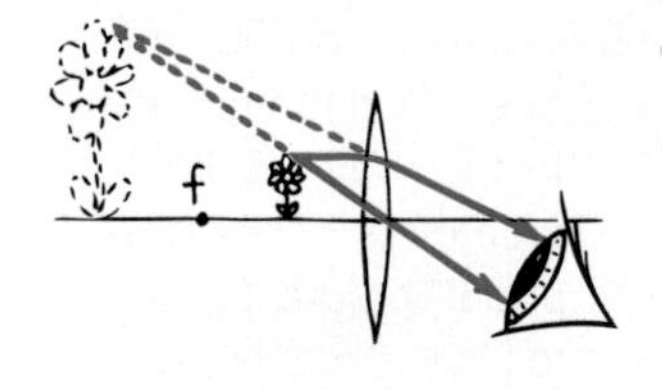

Fig. 12.37
When an object is near a converging lens (inside its focal point *f*), the lens acts as a magnifying glass to produce a virtual image. The image appears larger and farther from the lens than the object.

The simplest use of a converging lens is a magnifying glass. To understand how it works, think about how you examine objects near and far. With unaided vision, a faraway object is seen through a relatively narrow angle of view, and a close object is seen through a wider angle of view (Figure 12.36). To see the details of a small object, you want to get as close to it as possible for the widest-angle view. But your eye can't focus when too close. That's where the magnifying glass comes in. When close to the object, the magnifying glass gives you a clear image that would be blurry otherwise.

When we use a magnifying glass, we hold it close to the object we wish to examine. This is because a converging lens provides an enlarged, right-side-up image only when the object is inside the focal point. If a screen is placed at the image distance, no image appears on it because no light is directed to the image position. The rays that reach our eye, however, behave virtually *as if* they came from the image position, so we call this a **virtual image**.

When the object is far away enough to be outside the focal point of a converging lens, instead of a virtual image a **real image** is formed. Figure 12.38 shows a case in which a converging lens forms a real image on a screen. A real image is upside-down. A similar arrangement is used for projecting slides and motion pictures on a screen and for projecting a real image on the film of a camera. Real images formed with a single lens are always upside-down.

A diverging lens used alone produces a reduced virtual image. It makes no difference how far or how near the object is. When a diverging lens is used alone, the image is always virtual, right-side up, and smaller than the object. A diverging lens is often used as a "finder" on a camera. When you look at the object to be photographed through such a lens, you see a virtual image that approximates the same proportions as the photograph.

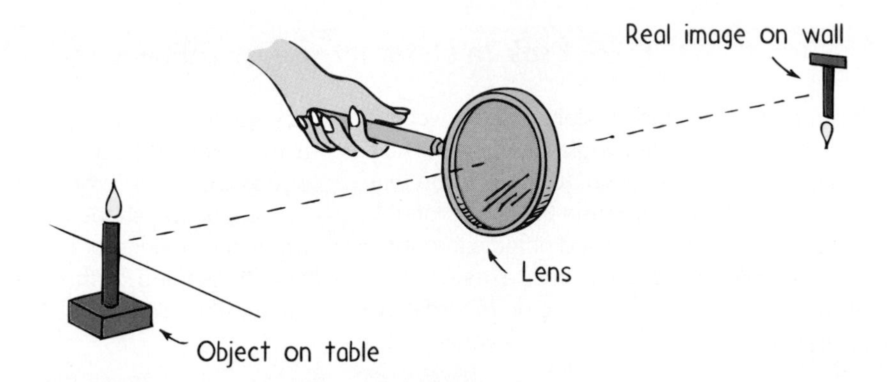

Fig. 12.38
Spherical aberration.

Fig. 12.39
A diverging lens forms a virtual, right-side-up image of Jamie and his cat.

Question

Why is the greater part of the photograph in Figure 12.39 out of focus?

Answer

Both Jamie and his cat and the virtual image of Jamie and his cat are "objects" for the lens of the camera that took this photograph. Because the objects are at different distances from the lens, their respective images are at different distances with respect to the film in the camera. So only one can be brought into focus. The same is true of your eyes. You cannot focus on near and far objects at the same time.

Link to Optometry—Contact Lenses

An option for those with poor sight today is wearing eyeglasses. The advent of eyeglasses probably occurred in Italy in the late 1200s. (Curiously, the telescope wasn't invented until some 300 years later. If, in the meantime, anybody viewed objects through a pair of lenses separated along their axes, such as fixed at the ends of a tube, there is no record of it.) And a present-day option to wearing eyeglasses is contact lenses. Now there is an alternative to both eyeglasses and contacts because today's technology enables eye surgeons to shave the cornea of the eye to a proper shape for normal vision. In tomorrow's world, the wearing of eyeglasses and contact lenses may be a thing of the past. We really do live in a rapidly changing world. And that can be nice.

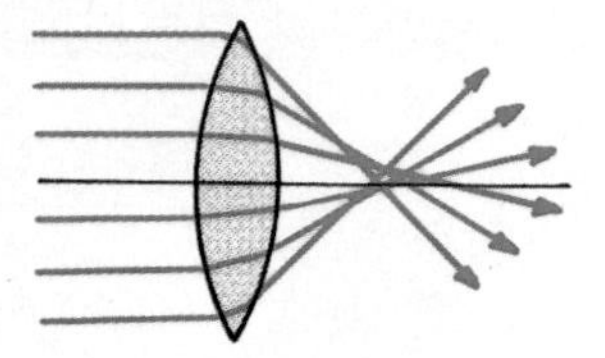

Fig. 12.40 Spherical aberration.

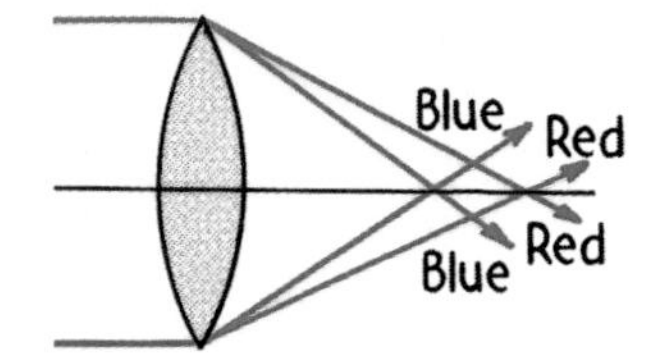

Fig. 12.41 Chromatic aberration.

Lens Defects

No lens forms a perfect image. A distortion in an image is called an *aberration.* By combining lenses in certain ways, aberrations can be minimized. For this reason, most optical instruments use compound lenses, each consisting of several simple lenses, instead of single lenses.

Spherical aberration results when the light passing through the edges of a lens focuses at a slightly different place from where light passing through the center of the lens focuses (Figure 12.40). As a result, the image you see is blurred. This can be remedied by covering the edges of the lens, as with a diaphragm in a camera. In good optical instruments, spherical aberration is corrected by a combination of lenses.

Chromatic aberration is the result of different colors having different speeds and hence different refractions in the lens. In a simple lens (as in a prism), different colors of light do not focus at the same point (Figure 12.41). As a result, some colors in the image you see may be in focus while other colors are out of focus. *Achromatic lenses,* which combine simple lenses of different kinds of glass, correct this defect.

The pupil of the eye changes in size to regulate the amount of light that enters. Vision is sharpest when the pupil is smallest because light then passes through only the center of the eye's lens, where spherical and chromatic aberrations are minimal. Also, the eye then acts more like a pinhole camera, so minimum focusing is required for a sharp image. You see better in bright light because in such light your pupils are smaller.[5]

Astigmatism of the eye is a defect that results when the outer surface of the eye, the cornea, is curved more in one direction than in another, somewhat like the side of a barrel. Because of this defect, the eye does not form sharp images. The remedy is eyeglasses with cylindrical lenses that have more curvature in one direction than in another.

[5]If you wear glasses and ever misplace them, or if you find it difficult to read small print, squint or, even better, hold a pinhole (in a piece of paper or whatever) in front of your eye, close to the page. You'll see the print clearly, and because you're close, it is magnified. Try it and see!

Link to Physiology and Psychology—Lateral Inhibition

The human eye can do what no camera film can do: perceive degrees of brightness that ranges from about 500 million to 1. The difference in brightness between the Sun and Moon, for example, is about 1 million to 1. But, because of an effect called *lateral inhibition,* we don't perceive the actual differences in brightness. The brightest places in our visual field are prevented from outshining the rest, for whenever a receptor cell on our retina sends a strong brightness signal to our brain, that cell also signals neighboring cells to dim their responses. In this way, we even out our visual field, which allows us to discern detail in both very bright areas and dark areas.

Lateral inhibition exaggerates the difference in brightness at the edges of places in our visual field. Edges, by definition, separate one thing from another. So we accentuate differences rather than similarities. This is illustrated in the pair of shaded rectangles below. They look to be different shades of brightness because of the edge that separates them. But cover the edge with your pencil or your finger, and they look equally bright (try it now)! That's because both rectangles *are* equally bright; each rectangle is shaded lighter to darker, moving from left to right. Our eye concentrates on the boundary where the dark edge of the left rectangle joins the light edge of the right rectangle, and our eye-brain system assumes that the rest of the rectangle is the same. We pay attention to the boundary and ignore the rest.

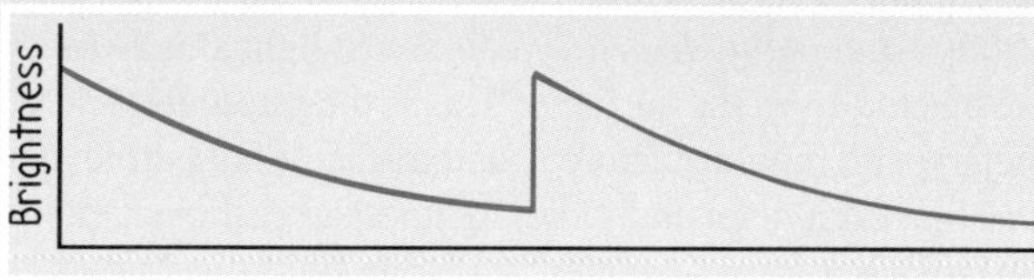

Questions to ponder: Is the way the eye picks out edges and makes assumptions about what lies beyond similar to the way we sometimes make judgments about other cultures and other people? Don't we in the same way tend to exaggerate the differences on the surface while ignoring the similarities and subtle differences within?

Questions

1. If light traveled at the same speed in glass and in air, would glass lenses alter the direction of light rays?
2. Why is there chromatic aberration in light that passes through a lens but none in light that reflects from a mirror?

Answers

1. No.
2. Different frequencies travel at different speeds in a transparent medium and therefore refract at different angles, which produces chromatic aberration. The angles at which light reflects, however, has nothing to do with its frequency. One color reflects the same as any other color. In telescopes, therefore, mirrors are preferable to lenses because with mirrors there is no chromatic aberration.

12.6 Wave-Particle Duality

In ancient times Plato and other Greek philosophers held that light was made up of tiny particles. And in the early 1700s, so did Isaac Newton, who first became famous for his experiments with light. Then a hundred years later the wave nature of light was demonstrated by Thomas Young in the double-slit experiment. This wave view was reinforced in 1862 by James C. Maxwell's finding that light is energy carried in the oscillating electric and magnetic fields of electromagnetic waves. The wave view of light was confirmed experimentally by Heinrich Hertz 25 years later. Then in 1905, Albert Einstein published a Nobel Prize–winning paper that challenged the wave theory of light. Einstein stated that, in its interactions with matter, light is confined not in continuous waves but in tiny particles of energy called *photons.*

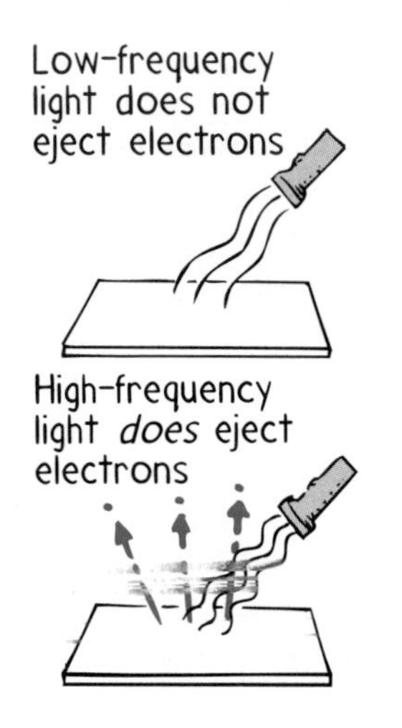

Fig. 12.42
The photoelectric effect depends on frequency.

So science had come full circle in its view of light: particle to wave and back to particle. As we shall see, both views are correct! First, however, let's look at Einstein's particle model of light, which explained a phenomenon that puzzled scientists in the early 1900s—the *photoelectric effect.* When light shines on certain metal surfaces, electrons are ejected from the surfaces. This is the **photoelectric effect**, used in electric eyes, light meters, and motion-picture sound tracks. What perplexed investigators in the early 1900s was that ultraviolet and violet light impart sufficient energy to knock electrons from these surfaces but lower-frequency visible light does not—even when the lower-frequency light is very bright. Ejection of electrons depends only on the frequency of light, and the higher the frequency of light used, the greater the kinetic energy of the ejected electrons. Very dim high-frequency light ejects fewer electrons, but each has the same kinetic energy as the more numerous electrons ejected by brighter light of the same frequency.

Einstein's explanation was that the electrons in the metal are bombarded by "particles of light"—**photons**. Einstein stated that the energy of each photon is proportional to its frequency:[6]

$$E \sim f$$

So Einstein viewed light as a hail of photons, each carrying energy proportional to its frequency. One photon is completely absorbed by each electron ejected from the metal.

All attempts to explain the photoelectric effect by waves failed. A light wave has a broad front, and its energy is spread out along this front. For the light wave to eject a single electron from a metal surface, all the light's energy would somehow have to be concentrated on that one electron. But this is as improbable as an ocean wave hitting a beach and knocking only one single seashell far inland with an energy equal to the energy of the whole wave. Therefore, instead of having us think of light encountering a surface as a continuous train of waves, the photoelectric effect suggests we conceive of light as a succession of particlelike photons. The energy of each photon is proportional to the frequency of the light, and that energy is given completely to a single electron in

---

[6] When the energy of any photon is divided by its frequency, the single number that always results is the proportionality constant called **Planck's constant**, h (6.6×10^{-34} J·s). Planck's constant is a fundamental constant of nature that sets a lower limit on the smallness of things. We can insert this constant in the proportionality $E \sim f$ and express it as an equation:

$$E = hf$$

This equation gives the smallest amount of energy that can be converted to light having frequency f. The light is emitted not continuously (waves) but rather as a stream of photons (particles). Each photon throbs at a frequency f and carries an energy hf.

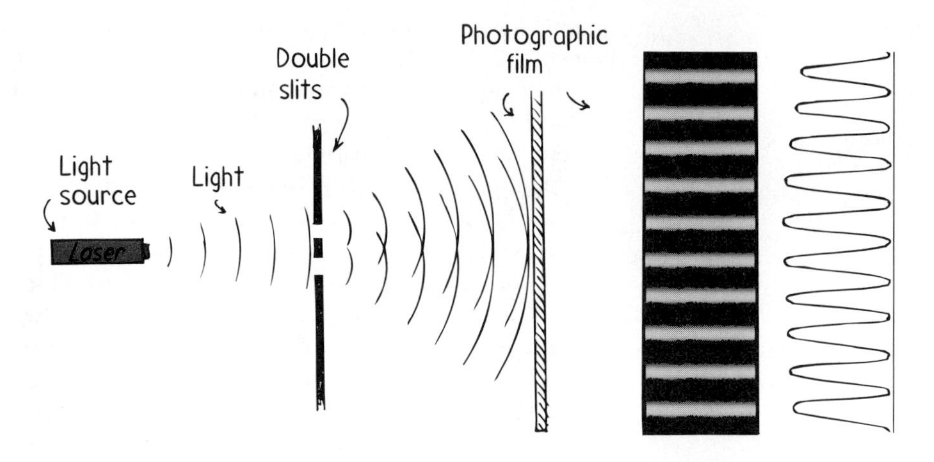

Fig. 12.43
Arrangement for double-slit experiment. The black and white striped rectangle is a photograph of the interference pattern. To the far right is a graphic representation of the pattern.

the metal surface. The number of ejected electrons had to do with the number of photons—the brightness of the light.

Experimental verification of Einstein's explanation of the photoelectric effect was made 11 years later by the American physicist Robert Millikan. Every aspect of Einstein's interpretation was confirmed. The photoelectric effect proves conclusively that light has particle properties. A wave model of light is inconsistent with the photoelectric effect. On the other hand, interference demonstrates convincingly that light has wave properties. A particle model of light is inconsistent with interference.

Recall Thomas Young's double-slit interference experiment, which we earlier discussed in terms of waves. When monochromatic light passes through a pair of closely spaced thin slits, an interference pattern is produced on photographic film (Figure 12.43). Now let's consider the experiment in terms of photons. Suppose we dim our light source so that in effect only one photon at a time reaches the thin slits. If the film behind the slits is exposed to the light for a very short time, the film becomes exposed as simulated in Figure 12.44a. Each spot represents the place where the film has been exposed to a photon. If the light is allowed to expose the film for a longer time, a pattern of fringes begins to emerge as in Figure 12.44b and c. This is quite amazing. Spots on the film are seen to progress photon by photon to form the same interference pattern characterized by waves!

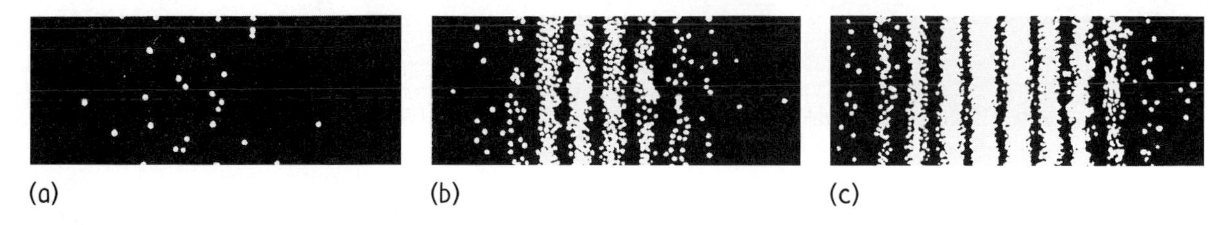

Fig. 12.44
Stages of two-slit interference pattern. The pattern of individually exposed grains progresses from (a) 28 photons to (b) 1000 photons to (c) 10,000 photons. As more photons hit the screen, a pattern of interference fringes appears.

Evidently, light has both a wave nature and a particle nature—a wave-particle duality.[7] This duality is evident in the formation of optical images. We understand the photo-

[7]From a pre-quantum point of view (a *quantum* is the smallest "particle" of something, such as of light, electricity, or any other form of energy), this wave-particle duality is mysterious. This mystery leads some people to believe that photons and other "quanta" have some sort of consciousness, with each photon or quanta having "a mind of its own." The mystery, however, is like beauty. It is in the mind of the beholder rather than in nature itself. We conjure models to understand nature, and when inconsistencies arise, we sharpen or change our models. The wave-particle duality of light doesn't fit a model built on Newtonian ideas. One alternative model is that quanta have minds of their own. Another model is quantum physics. We subscribe to the latter.

graphic image produced by a camera in terms of light waves, which spread from each point of the object, refract as they pass through the lens system, and converge to focus on the photographic film. The path of light from the object through the lens system and to the focal plane can be calculated using methods developed from the wave theory of light.

But now consider carefully the way in which the photographic image is formed. The photographic film consists of an emulsion that contains grains of silver halide crystal, each grain containing about 10^{10} silver atoms. Each photon absorbed gives up its energy to a single grain in the emulsion. This energy activates surrounding crystals in the grain and is used to develop the film. Many photons activating many grains produce the usual photographic exposure. When a photograph is taken with exceedingly feeble light, the image is built up by individual photons that arrive independently and are seemingly random in their distribution. We see this strikingly illustrated in Figure 12.45, which shows how an exposure progresses photon by photon.

What all this means is that light has both wave and particle properties. Simply put, *light behaves as a stream of photons when it interacts with the photographic film or other detectors and behaves as a wave in traveling from a source to the place where it is detected.* Light travels as a wave and hits as a stream of photons. In interference experiments, photons strike the film at places where we would expect to see constructive interference of waves.

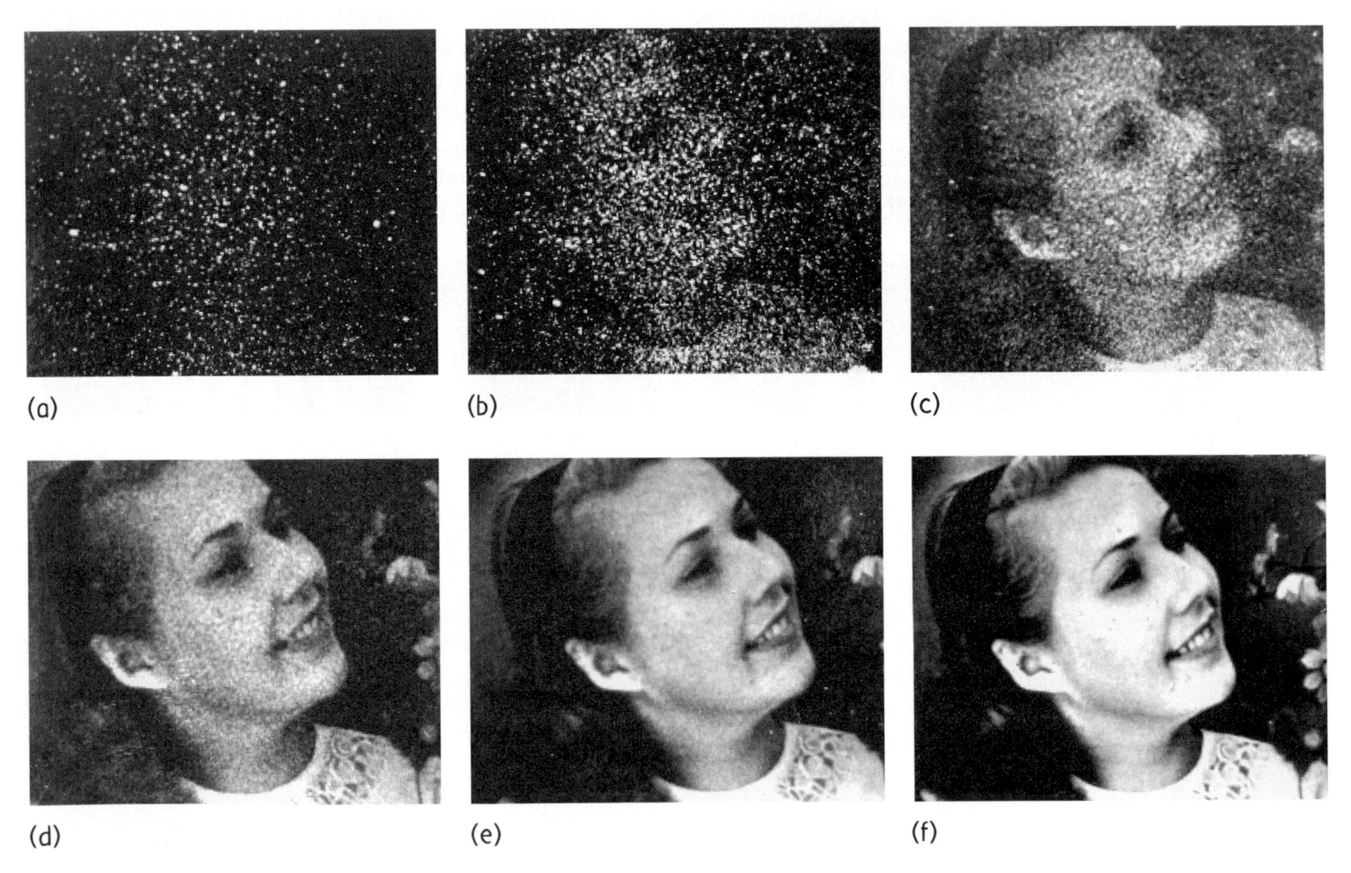

Fig. 12.45
Stages of film exposure reveal the photon-by-photon production of a photograph. The approximate numbers of photons at each stage are (a) 3×10^3; (b) 1.2×10^4; (c) 9.3×10^4; (d) 7.6×10^5; (e) 3.6×10^6; and (f) 2.8×10^7.

The fact that light exhibits both wave and particle behavior is one of the most intriguing surprises that physicists have discovered in this century. The finding that light comes in tiny bunches, tiny *quanta* as they are called, led to a whole new way of looking at nature—wave mechanics, or *quantum mechanics.* An outcome of this new mechanics is that just as light has particle properties, so do particles have wave properties. First electrons were found to have wave properties; a beam of electrons passing through slits exhibits the same type of diffraction pattern as light. Then other particles—even baseballs and orbiting planets—could be described by the new mechanics of waves. Quantum mechanics and Newtonian physics overlap in the macroworld, and both are seen as "correct." But only quantum mechanics, with its emphasis on waves, is wholly accurate in the microworld of the atom. More about this in Part 5. Onward!

Summary of Terms

- **Reflection** The return of light rays from a surface in such a way that the angle at which a given ray is returned is equal to the angle at which it strikes the surface.
- **Refraction** The bending of an oblique ray of light when it passes from one transparent medium to another.
- **Law of reflection** The angle of a reflection equals the angle of incidence. The reflected and incident rays lie in a plane that is normal to the reflecting surface.
- **Diffuse reflection** Reflection in many directions from an irregular surface.
- **Critical angle** The minimum angle of incidence inside a medium at which a light ray is totally reflected.
- **Total internal reflection** The total reflection of light traveling within a denser medium when it strikes the boundary with a less-dense medium at an angle greater than the critical angle.
- **Converging lens** A lens that is thicker in the middle than at the edges and refracts parallel rays passing through it to a focus.
- **Diverging lens** A lens that is thinner in the middle than at the edges, causing parallel rays passing through it to diverge as if from a point.
- **Virtual image** An image formed by light rays that do not converge at the location of the image.
- **Real image** An image formed by light rays that converge at the location of the image.
- **Photoelectric effect** The emission of electrons from a metal surface when light shines on it.
- **Photon** A particle of light, or the basic packet of electromagnetic radiation.

Review Questions

Reflection

1. Distinguish between *reflection* and *refraction.*
2. What does incident light that falls on an object do to the electrons in the atoms of the object?
3. What do the electrons in an illuminated object do when they are made to vibrate with greater energy?
4. What is the law of reflection?
5. Relative to the distance of an object in front of a plane mirror, how far behind the mirror is the image?
6. Does the law of reflection hold for curved mirrors? Explain.
7. Does the law of reflection hold for diffuse reflection? Explain.
8. How can a surface be polished for some waves and not others?

Refraction

9. How does the angle at which light strikes a pane of window glass compare with the angle at which it passes out the other side?
10. How does the angle at which a ray of light strikes a prism compare with the angle at which it passes out the other side?
11. Does light travel faster in thin air or in dense air? What does this difference in speed have to do with the length of a day?
12. What is a mirage?
13. Why do stars twinkle?
15. When a wheel rolls from a smooth sidewalk onto grass, the interaction of the wheel with the blades of grass slows the wheel. What slows light when it passes from air into glass or water?
16. When light passes from one material into another, it may bend toward the normal to the surface or away from the normal. When does it do which?
17. Does refraction make a swimming pool seem deeper or shallower?

Dispersion

18. What happens to light of a certain frequency when it is incident on a material whose natural frequency is the same as the frequency of the light?
19. Which travels more slowly in glass, red light or violet light?

20. If light of different frequencies has different speeds in a material, does it also refract at different angles in the same material? Explain.
21. What prevents rainbows from being seen as complete circles?
22. Does a single raindrop illuminated by sunlight disperse a spectrum of colors?
23. Does a viewer see a single color or a spectrum of colors coming from a single faraway raindrop?
24. Why is a secondary rainbow dimmer than a primary bow?

Total Internal Reflection

25. What is meant by *critical angle*?
26. When is light totally reflected in glass?
27. When is light totally reflected in a diamond?
28. Light normally travels in straight lines, but it "bends" in an optical fiber. Explain.

Lenses

29. Distinguish between a *converging lens* and a *diverging lens.*
30. What is the *focal length* of a lens?
31. Distinguish between a *virtual image* and a *real image.*
32. What kind of lens can be used to produce a real image? A virtual image?
33. Distinguish between *spherical aberration* and *chromatic aberration.*
34. What is astigmatism?

Wave-Particle Duality

35. What evidence can you cite for the wave nature of light? For the particle nature of light?
36. Which are more successful in dislodging electrons from a metal surface, photons of violet light or photons of red light? Why?
37. Why won't a very bright beam of red light impart more energy to an electron than a feeble beam of violet light?
38. Does the brightness of a beam of light primarily depend on the frequency of photons or on the number of photons?
39. Does light behave primarily as a wave or as a particle when it interacts with the crystals of matter in photographic film?
40. Does light travel from one place to another in a wavelike way or a particlelike way?
41. Does light interact with a detector in a wavelike way or a particlelike way?
42. When does light behave as a wave? When does it behave as a particle?

Home Projects

1. Make a pinhole viewer, as illustrated. Cut out one end of a small cardboard box, and cover the end with tissue or wax

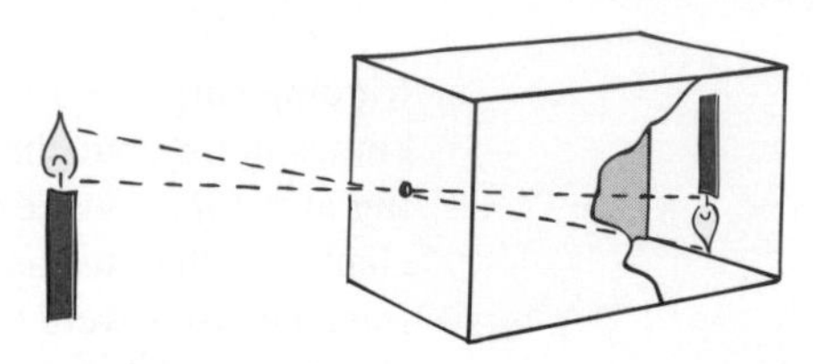

paper. Make a clean-cut pinhole at the other end. (If the cardboard is thick, make the pinhole through a piece of tinfoil placed over a larger opening in the cardboard.)

   Aim the box at a bright object in a darkened room, and you will see an upside-down image on the tissue paper. The tinier the pinhole, the dimmer and sharper the image. If, in a dark room, you replace the tissue paper with unexposed photographic film, cover the back so that it is light-tight, and cover the pinhole with a removable flap, you have a camera. Exposure times differ depending principally on the kind of film and amount of light. Try different exposure times, starting with about 3 s. Also try boxes of various lengths. The lens on a commercial camera is much bigger than the pinhole and therefore admits more light in less time—hence the name *snapshots.*

2. Stand a pair of mirrors on edge with the faces parallel to each other. Place an object such as a coin between the mirrors and look at the reflections in each mirror. Nice?
3. Set up two pocket mirrors at right angles and place a coin between them. You'll see four coins. Change the angle

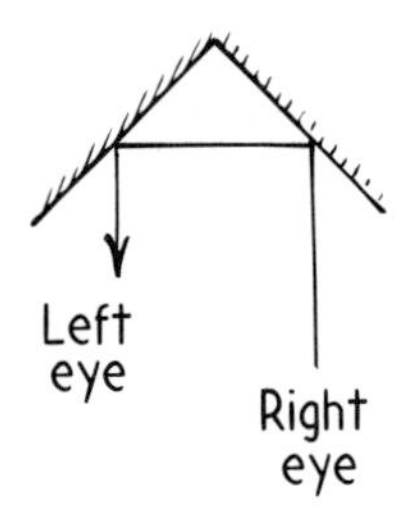

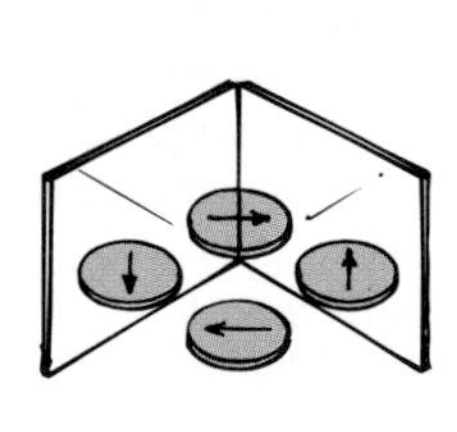

of the mirrors and see how many images of the coin you can see.

4. Look at yourself in a pair of mirrors at right angles to each other. Wink. Notice that you see yourself as others see you. Rotate the mirrors, still at right angles to each other. Does your image rotate also? Now place the mirrors 60 degrees apart so you again see your face. Again rotate the mirrors and see if your image rotates also. Amazing?

Exercises

1. An eye at point P looks into the mirror. Which of the numbered cards is seen reflected in the mirror?

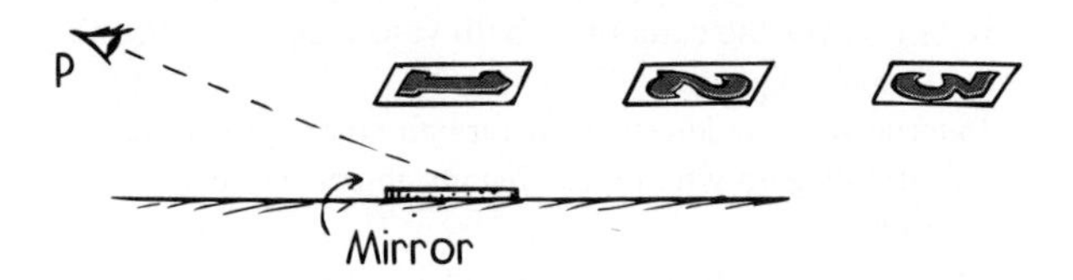

2. Cowboy Joe wishes to shoot his assailant by ricocheting a bullet off a mirrored metal plate. To do so, should he simply aim at the mirrored image of his assailant? Explain.
3. What must be the minimum length of a plane mirror for you to see a full view of yourself?
4. What effect does your distance from the plane mirror have in the preceding answer? (Try it and see!)
5. Hold a pocket mirror at almost arm's length from your face and note the amount of your face you can see. To see more of your face, should you hold the mirror closer or farther, or would you have to have a larger mirror? (Try it and see!)
6. The diagram shows a person and her twin at equal distances on opposite sides of a thin wall. Suppose a window

is to be cut in the wall so that each twin can see a complete view of the other. Show the size and location of the smallest window that can be cut in the wall to do the job. (*Hint:* Draw rays from the top of each twin's head to the other twin's eyes. Do the same from the feet of each to the eyes of the other.)

7. What is wrong with the cartoon of the man looking at himself in the mirror? (Have a friend face a mirror as shown, and you'll see.)
8. Why is the lettering on the front of some vehicles "backwards"?

AMBULANCE

9. A person in a dark room looking through a window can clearly see a person outside in the daylight, whereas the person outside cannot see the person inside. Explain.
10. To reduce the glare of the surroundings, the windows of some department stores slant inward at the bottom, rather than being vertical. How does this reduce glare?
11. We see the bird and its reflection. Why do we not see the bird's feet in the reflection?

12. Show with a simple diagram that when a mirror that has a fixed beam incident on it is rotated through a certain angle, the reflected beam is rotated through an angle twice as large. (This doubling of displacement makes irregularities in ordinary window glass more evident.)
13. Why does reflected light from the Sun or Moon appear as a column in the body of water as shown? How would it appear if the water surface were perfectly smooth?

14. When you look at yourself in the mirror and wave your right hand, your beautiful image waves its left hand. Then why don't the feet of your image wiggle when you shake your head?

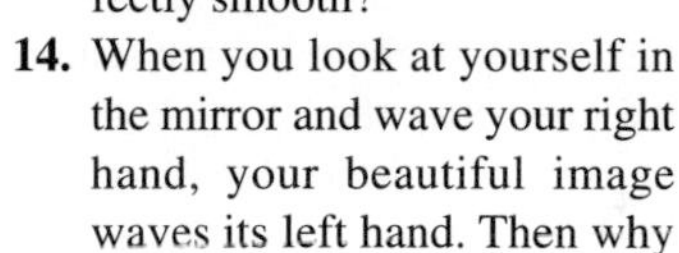

15. A pair of toy cart wheels are rolled obliquely from a smooth surface onto two plots of grass, a rectangular plot and a triangular plot, as shown. The ground is on a slight incline so that after slowing down in the grass, the wheels speed up again when they emerge on the smooth surface. Finish each sketch by showing some positions of the wheels inside the plots and on the other sides, thereby indicating the direction of travel.

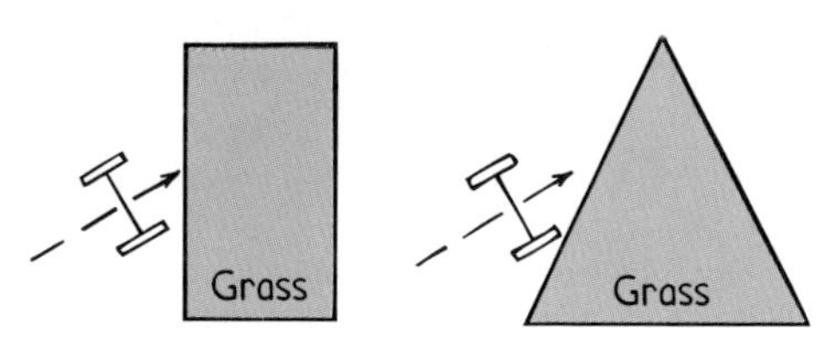

16. If light of all frequencies traveled at the same speed in glass, how would white light appear after passing through a prism?
17. Place a glass test tube in water and you can see the tube. Place it in soybean oil, and you can't see it. What does this tell you about the speed of light in the oil and in the glass?
18. If you were in a boat and spearing a fish you see in the water, would you aim above, below, or directly at the fish to make a direct hit? (Assume the fish is stationary in the water.) If you instead used light from a laser as your "spear," would you aim above, below, or directly at the observed fish? Defend your answers.

19. If the fish in the previous exercise were small and blue, and your laser light were red, what corrections should you make? Explain.
20. Note the different bendings of light in the sketches for a beam of light in water and a piece of square tile dipped in water. Do the bendings of light in these sketches contradict each other? Explain.

21. When a fish looks upward at an angle of 45°, does it see the sky or the reflection of the pond bottom? Defend your answer.
22. If you were to send a beam of laser light to a space station above the atmosphere and just above the horizon, would you aim the laser above, below, or at the visible space station? Defend your answer.
23. What accounts for the large shadows cast by the ends of the thin legs of the water strider? What accounts for the ring of bright light around the shadows?

24. Figure 12.20 shows the rainbow as elliptical rather than circular, to indicate the bow is being viewed from the side. Can one view a rainbow from the side so that it appears as the segment of an ellipse rather than the segment of a circle? Defend your answer.
25. Two observers standing apart from one another do not see the "same" rainbow. Explain.
26. A rainbow viewed from an airplane may form a complete circle. Where will the shadow of the airplane appear? Explain.
27. How is a rainbow similar to the halo sometimes seen around the Moon on a night when ice crystals are in the upper atmosphere?
28. What is responsible for the rainbow-colored fringe commonly seen at the edges of a spot of white light from the beam of a lantern or slide projector?
29. Transparent plastic swimming-pool covers called *solar heat sheets* have thousands of small lenses made up of air-filled bubbles. The lenses in these sheets are advertised as being able to focus heat from the Sun into the water and raise its temperature. Do you think the lenses of such sheets direct any more solar energy into the water than the amount that would enter a pool not covered by one? Defend your answer.
30. Would the average intensity of sunlight measured by a light meter at the bottom of the pool in Figure 12.34 be different if the water were still?
31. What would be the effect of a pinhole camera (Home Project 1) that has two pinholes instead of one? Multiple holes?
32. Can you take a photograph of your image in a plane mirror and focus the camera on both your image and the mirror frame? Explain.
33. In terms of focal length, how far behind the camera lens is the film located when very distant objects are being photographed?
34. Maps of the Moon are actually upside-down. Why?
35. In taking a photograph, what would happen to the image if you cover up the bottom half of the lens?
36. Why do older people who do not wear glasses read books farther away from their eyes than do younger people?
37. Why is chromatic aberration not a problem with mirrors?
38. Rays of light parallel to the principal axis of a converging lens pass through the lens and come to a focus a certain distance from the lens—its focal point. When the lens is under water, is the focal point longer, shorter, or the same distance as in air?
39. Why do goggles allow a swimmer under water to focus more clearly on what is being looking at?
40. If a fish wore goggles above the water surface, why would vision be better for the fish if the goggles were filled with water? Explain.
41. We speak of photons of red light and photons of green light. Can we speak of photons of white light? Why or why not?
42. A beam of red light and a beam of blue light have exactly the same energy. Which beam contains the greater number of photons?
43. Silver bromide (AgBr) is a light-sensitive substance used in some types of photographic film. To be exposed, the film must be illuminated with light having sufficient energy to break apart the AgBr molecules. Why do you suppose this film may be handled without exposure in a darkroom illuminated with red light? How about blue light? How about very bright red light as compared with very dim blue light?
44. Suntanning produces cell damage in the skin. Why is ultraviolet radiation capable of producing this damage but visible radiation is not?
45. Explain briefly how the photoelectric effect is used in the operation of at least two of the following: an electric eye, a photographer's light meter, the sound track of a motion picture.
46. Does the photoelectric effect *prove* that light is made of particles? Do interference experiments *prove* that light is composed of waves? (Is there a distinction between what something *is* and how it *behaves*?)

Problems

1. If you take a photograph of your image in a plane mirror, how many meters away should you set your focus if you are 2 m in front of the mirror?
2. Suppose you walk toward a mirror at 2 m/s. How fast do you and your image approach each other? (The answer is *not* 2 m/s.)
3. When light strikes glass perpendicularly, about 4% is reflected at each surface. How much light is transmitted through a pane of window glass?
4. No glass is perfectly transparent. Mainly because of reflections, about 92% of perpendicular light passes through an average sheet of clear windowpane. The 8% loss is not noticed through a single sheet, but through several sheets it is apparent. How much light is transmitted by two sheets?

Part IV—Sample Exam Questions

Choose the BEST answer to each of the following.

1. A portion of water vibrates up and down two complete cycles in one second as a water wave passes by. The wave's wavelength is 5 meters. What is the wave's speed?
(a) 2m/s (b) 5m/s (c) 10m/s (d) 15m/s (e) none of these
2. A 60-vibration per second wave travels 30 meters in one second. Its frequency is
(a) 30 hertz and it travels at 60m/s (b) 60 hertz and it travels at 30m/s (c) neither of these
3. A mass on the end of a spring bobs up and down one complete cycle every 2 seconds. Its frequency is
(a) 0.5 Hz (b) 2Hz (c) neither of these
4. When a source of sound approaches you, you detect an increase in its
(a) speed (b) wavelength (c) frequency (d) all of these
5. A sonic boom is typically produced when an aircraft flies
(a) from subsonic to supersonic speed (b) faster than the speed of sound (c) neither of these
6. The speed of sound in air depends on
(a) frequency (b) wavelength (c) air temperature (d) all of these (e) none of these
7. A singer holds a high note and shatters a distant crystal wine glass. This phenomenon best demonstrates
(a) forced vibrations (b) the Doppler Effect (c) interference (d) resonance
8. To set a tuning fork of 400 Hz into resonance, it is best to use another of
(a) 200 Hz (b) 400 Hz (c) 800 Hz (d) any of these three
9. True or false: Any radio wave travels faster under all conditions than any sound wave
(a) true (b) false
10. Which of the following does not fit in the same family?
(a) light wave (b) radio wave (c) sound wave (d) microwave (e) X-ray
11. If the resonant frequency of outer electron shells in atoms in a particular material match the frequency of green light, the material will be
(a) transparent to green light (b) opaque to green light
12. If water naturally absorbed blue and violet light rather than infrared, water would appear
(a) green-blue, as it presently appears (b) a more intense green-blue (c) orange-yellowish (d) to have no color at all
13. The sky is blue because air molecules in the sky act as tiny
(a) mirrors that reflect primarily blue light (b) scatterers of high-frequency light (c) incandescent blue-hot sources (d) prisms (e) none of these
14. The average speed of light is greatest in
(a) red glass (b) yellow glass (c) green glass (d) blue glass (e) all the same
15. If different colors of light had the same speed in matter, there would be no
(a) rainbows (b) dispersion by prisms (c) colors from diamonds (d) all of these
16. Lenses work because in different materials light has different
(a) wavelengths (b) frequencies (c) speeds (d) energies (e) none of these
17. A fish outside water will see better if it has goggles that are
(a) tinted green-blue (b) flat (c) filled with water (d) none of these
18. Waves diffract most when their wavelengths are
(a) long (b) short (c) same each way
19. Which photons have the most energy of those listed below?
(a) red (b) white (c) blue (d) all the same
20. Light behaves primarily as a particle when it
(a) travels from one place to another (b) interacts with matter (c) neither

Answers: 1c, 2b, 3a, 4c, 5b, 6c, 7d, 8b, 9a, 10c, 11b, 12c, 13b, 14a, 15d, 16c, 17c, 18a, 19c, 20(b)

PART

V

The Atom

Like everyone, I'm made of atoms. I get lots of them whenever I take a breath of air. Some I exhale right away but many stay for a while to become part of me -- then I may exhale them later. Whenever you take a breath, some of my atoms are in it and become part of you (and likewise, yours become part of me). Atoms cycle and recycle through all of us. Their numbers and smallness are staggering -- there are many more atoms in a breath of air than the entire human population since time zero. So every time we take a breath we inhale atoms that were once a part of every person who ever lived. And atoms that are now part of us will be part of everyone else later on. In this sense, we're all one!

CHAPTER

13

Structure of the Atom

Imagine that you fall off your chair in slow motion and that while falling to the floor, you also shrink in size. What would such an experience be like? What would you see? As you topple from your chair and approach the floor, you brace yourself for impact against the smooth, solid surface. But as you get nearer and nearer to it, becoming smaller and smaller, the smooth floor gives way to a myriad of cracks and crevices that appear as canyons. These are the microscopic irregularities found in all apparently smooth surfaces. Falling into one of these crevices, you note that its solid walls have given way to fuzzy surfaces that throb and pucker. You brace yourself for impact as you approach the hazy surface—closer and closer, smaller and smaller. And then, instead of impact, you penetrate into emptiness. You have entered the world inside the atom—a world as empty of matter as the solar system—a world of mostly empty space and so small that it defies visual description.

All matter, however solid it appears, is made up of tiny building blocks that are themselves mostly empty space. These are atoms—atoms that combine to form molecules, which in turn combine to form the substances of matter.

13.1 The Elements

All manner of things—shoes and ships and sealing wax, cabbages, kings, and anything else we can think of—are composed of atoms. One might think that an incredible number of different kinds of atoms exist to account for the rich variety of substances we find around us. But the number is surprisingly small. The great variety of substances results not from any great variety of atoms but from the many ways a few types of atoms can be combined—just as in a color print three colors can be combined to form almost every conceivable color. To date (1998) we know of 112 distinct atoms. We call these atoms the chemical *elements.* Only 88 elements are found naturally; the others are made in the laboratory with high-energy atomic accelerators and nuclear reactors. These heaviest elements are too unstable (radioactive) to occur naturally in appreciable amounts.

Fig. 13.1
Co-author Leslie, shown here at age 16, is made of stardust—in the sense that the carbon, oxygen, nitrogen, and other atoms that make up her body originated in the deep interiors of ancient stars that have long since exploded.

Hydrogen, which makes up more than 90 percent of the atoms in the universe, was the original element. Elements heavier than hydrogen are manufactured in the deep interiors of stars. There, enormous temperatures and pressures cause hydrogen atoms to fuse into more complex elements. With the exception of some of the hydrogen and trace amounts of other light elements, all the elements in our surroundings are remnants of stars that exploded long before our solar system came into being.

These star remnants are the building blocks of all matter. All matter, however complex, living or nonliving, is some combination of elements. From a pantry having 112 bins, each containing a different element, we have all the materials needed to make up any substance occurring in the universe.

The majority of elements are not found in great abundance, and some are exceedingly rare. Only a dozen or so compose the things we see every day.[1] Living things, for example, consist primarily of four elements: carbon, hydrogen, oxygen, and nitrogen.

Atoms are ageless. The origin of most atoms goes back to the origin of the universe, and nearly all atoms are older than the Sun and Earth. There are atoms in your body that have existed since the beginning of time, recycling throughout the universe among innumerable forms, both nonliving and living. So you don't "own" the atoms that make up your body—you're simply the present caretaker. There will be many to follow.

It is difficult to imagine how small atoms are. They are so small that they can't be seen with visible light because they are smaller than the wavelengths of visible light. Because of diffraction (Chapter 11), we could stack microscope on top of microscope and never "see" an atom. Photographs of atoms, such as that of Figure 13.2, are obtained with a scanning tunneling microscope, a nonlight imaging device that bypasses

[1]Most common substances are formed out of combinations of two or more of these most common elements: hydrogen, H; carbon, C; nitrogen, N; oxygen, O; sodium, Na; magnesium, Mg; aluminum, Al; silicon, Si;

Link to Physiology—A Breath of Air

Atoms are so small that there are about as many atoms of air in your lungs at any moment as there are breaths of air in the Earth's atmosphere. That's because there are about 10^{22} atoms in a liter of air at atmospheric pressure and about 10^{22} liters of air in the atmosphere. Here's what that means. Exhale a deep breath; the number of molecules exhaled approximately equals the number of breathfuls of air in the Earth's atmosphere. It will take about six years for your breath to become uniformly mixed in the atmosphere. Then anyone, anywhere on Earth, who inhales a breath of air takes in, on average, one of the molecules you exhaled. But you exhale many, many breaths, and so other people breathe in many, many molecules that were once in your lungs—that were once a part of you. And of course, vice versa: With each breath you take in, you breathe molecules that were once a part of everyone who ever lived. Considering that exhaled atoms are part of our bodies, it can be truly said that we are literally breathing one another.

light and optics altogether. Even better detail can be seen with newer types of imaging devices that are presently revolutionizing microscopy.

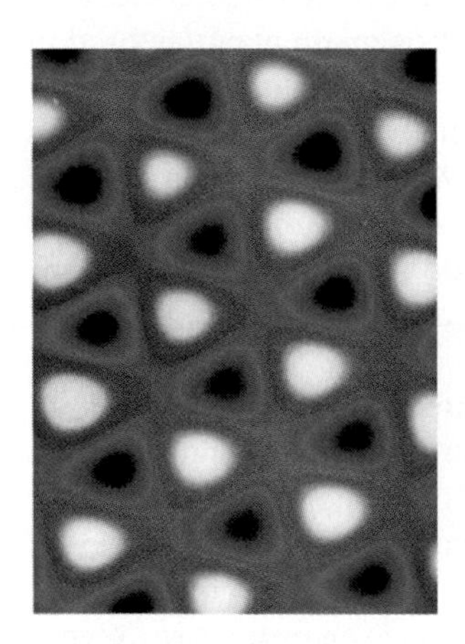

Fig. 13.2
An image of carbon atoms obtained with a scanning tunneling microscope.

The first somewhat direct evidence for the existence of atoms was inadvertently discovered in 1827 by a Scottish botanist, Robert Brown, while he was studying grains of pollen under a microscope. He noticed that the grains were in a constant state of agitation, always moving, always jumping about. At first he thought the grains were some sort of moving life forms. But later he found that inanimate dust particles and grains of soot also moved the same way. This perpetual jiggling of particles—now called *Brownian motion*—results from collisions between visible particles and invisible atoms. Brown's pollen grains were moving because they were constantly being jostled by the atoms that make up the water.

Today we know the atom is made up of smaller, subatomic particles—electrons, protons, and neutrons. We also know that atoms differ from one another only in the number of these subatomic particles they contain. Protons and neutrons are bound together at the atom's center and form a larger particle—the atomic *nucleus.* Most of an atom's mass is concentrated in the nucleus. Surrounding the nucleus are the electrons, which are tinier still. These are the electrons of an electric current. Electrons also dictate chemical behavior, which we treat in subsequent chapters. For now, let's look into the discoveries of these subatomic particles and their role in the atom.

Question

A friend claims there are atoms in his brain that were once in the brain of Albert Einstein. Is your friend's claim likely correct or nonsense?

Answer
Your friend is correct! In addition, there are atoms in your friend's and everyone else's body that were once in Charlie Chaplin and everybody else, too! The configurations of these atoms, however, are now quite different. What's more, the atoms of which you and your friend are composed will live forever in the bodies of all the people on the Earth who are yet to be.

13.2 The Electron

Fig. 13.3
Franklin's kite-flying experiment.

The name *electron* comes from the Greek word for *amber,* the material discovered by early Greeks that was found to exhibit the effects of electrical charging. In 1752, Ben Franklin learned from experiments with the atmosphere that lightning is a flow of electrical energy. This finding told him that electricity was not restricted to solid objects and that it could travel through a gas.

Franklin's experiments later inspired other scientists to produce electric currents through various gases in sealed glass tubes. A glowing ray was produced when a voltage was applied across electrodes at the ends of a sealed gas-filled tube. Experimentation showed that the ray was blocked from reaching the positive electrode (the *anode*) when an object was placed in its path. It was thus reasoned that the ray emerged from the negative electrode (the *cathode*). The apparatus was named the *cathode-ray tube* (Figure 13.4). When electric charges were brought near the tube, the ray was deflected toward the positive charge and away from the negative charge. The ray was also deflected by the presence of a magnet. These findings indicated that the ray was negatively charged. The speed of the ray was found to be considerably less than that of light. Because of these characteristics, it appeared that the ray behaved more like a beam of particles than a beam of light.

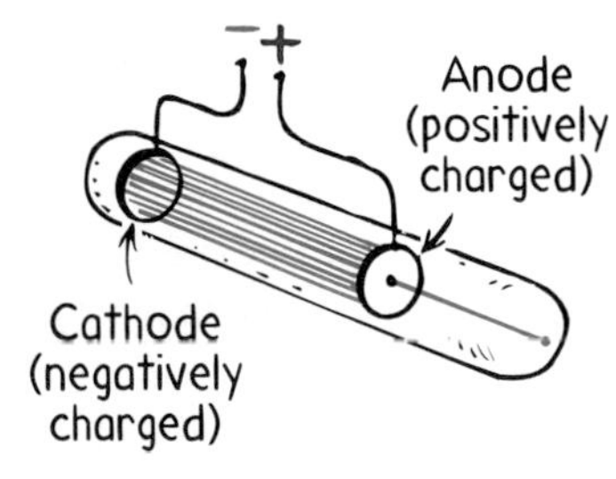

Fig. 13.4
A simple cathode-ray tube. An electric current is produced in the gas when a high voltage is imposed across the ends of the tube.

In 1887, J. J. Thompson measured the deflection angles of cathode-ray particles in a magnetic field. He reasoned that the deflection depended on two things: the mass of the particles and their electrical charge. The greater each particle's mass, the greater the inertia and the less deflection. The greater each particle's charge, the greater the deflection. From his measurements, Thompson was able to calculate the mass-to-charge ratio of the cathode-ray particles. Using only their ratio, however, he could not calculate the mass or the charge of the particles. To calculate the mass, he needed to know the charge, but to calculate the charge, he needed to know the mass.

Robert Millikan addressed this question. He calculated the numerical value of a single unit of electric charge on the basis of an experiment he carried out in 1911. In his experiment, Millikan sprayed tiny oil droplets into an electric field. When the field was strong, some of the droplets moved upward, indicating they carried a very slight negative charge. Millikan adjusted the field so that droplets would hover motionless. He knew that on motionless droplets the downward force of gravity

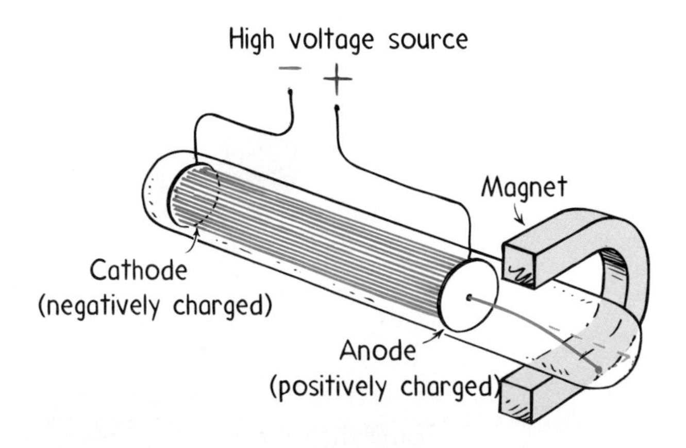

Fig. 13.5
Cathode rays are deflected by magnetic fields

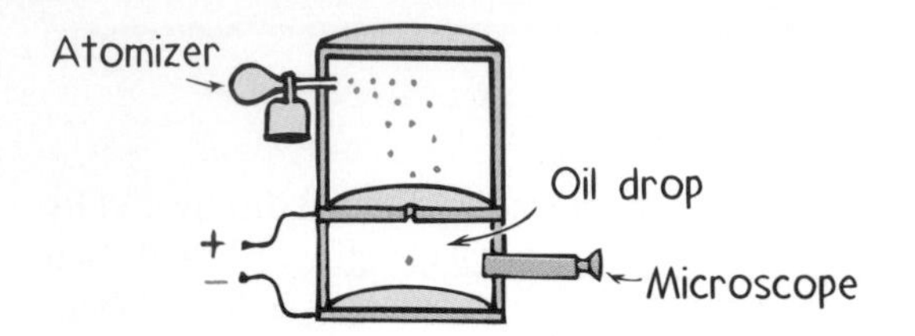

Fig. 13.6
Millikan's oil-drop experiment for determining the charge on the electron. The pull of gravity on the drops is balanced by the upward electrical force.

was exactly balanced by the upward electrical force. Investigation showed that the electrical charges on the droplets were always some multiple of a single very small value, which Millikan proposed to be a fundamental unit of all electrical charge. Using this value and the ratio discovered by Thompson, he calculated the mass of a cathode-ray particle to be considerably less than that of the smallest known atom, hydrogen. This was very surprising to many people for it meant that the atom was no longer the smallest known particle of matter.

The cathode-ray particle is known today as the **electron**, a fundamental component of all atoms. All electrons are identical to one another, each having a negative charge and an incredibly small mass of 9.1×10^{-31} kg. Electrons on the outer edges of atoms dictate the many properties of a material, including chemical reactivity and physical attributes, such as taste, texture, appearance, and color. The cathode ray—a stream of electrons—has found a great number of applications. Most notably, your television set is a cathode-ray tube with one end widened out into a phosphor-coated screen.

It was reasoned that if atoms contained negatively charged particles, some balancing positively charged matter must also exist. From this Thompson put forth what he called a "plum-pudding" model of the atom in which electrons were like plums in a sea of positively charged pudding. Further experimentation, however, soon proved this model wrong.

13.3 The Atomic Nucleus

A more accurate picture of the atom first came to the New Zealand–born British physicist Ernest Rutherford, who in 1909 oversaw the now-famous gold-foil experiment. This significant experiment showed that the atom is mostly empty space, with most of its mass concentrated in the central region—the *atomic nucleus.*

In Rutherford's experiment a beam of positively charged particles (called alpha particles) from a radioactive source was directed through a sheet of very thin gold foil. Because alpha particles are thousands of times more massive than electrons, it was expected that the stream of alpha particles would not be impeded as it passed through the "atomic pudding." This was indeed observed—for the most part. Nearly all alpha particles passed through the gold foil undeflected and produced spots of light when they hit a fluorescent screen surrounding the gold leaf. But some particles were deflected from their straight-line path as they emerged (Figure 13.7). A few alpha particles were widely

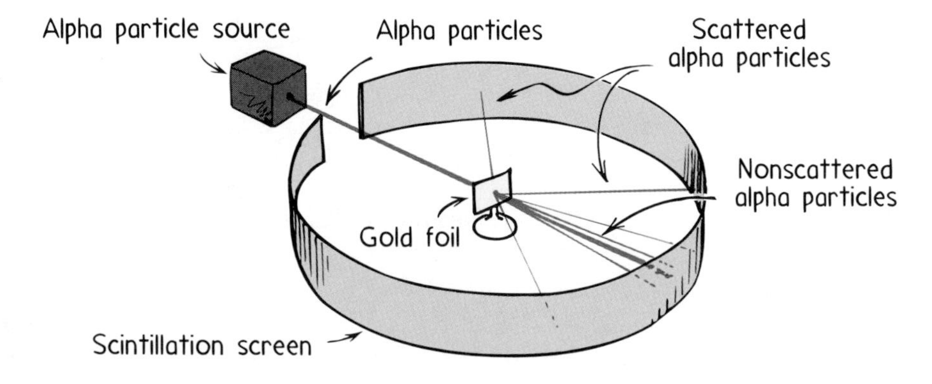

Fig. 13.7
Rutherford's gold-foil experiment. Deflection of alpha particles showed the atom to be mostly empty space with a concentration of mass at its center—the atomic nucleus.

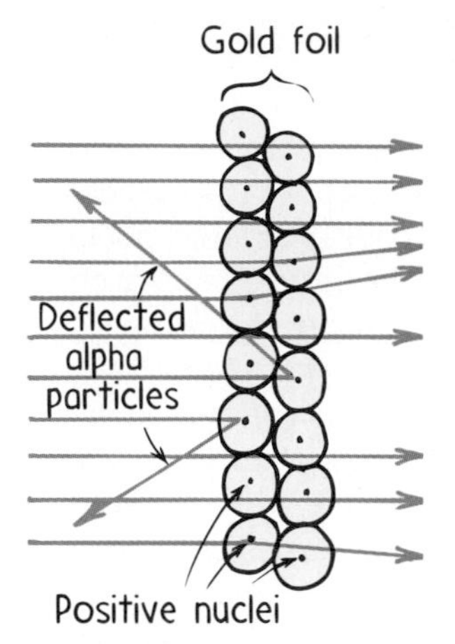

Fig. 13.8
Most alpha particles pass through the gold foil undeflected. Those few that approach a positive nucleus are repelled and scattered through various angles.

deflected, and a small number were even scattered backwards! These alpha particles must have hit something relatively massive, but what? Rutherford reasoned that the undeflected particles traveled through regions of the gold foil that were empty space, while the small number of deflected particles were repelled from extremely dense positively charged centers. Each atom, he concluded, must contain one of these centers, which he named the **atomic nucleus**.

Like the solar system, the atom is mostly empty space. The nucleus and surrounding electrons occupy only a tiny fraction of the atomic volume. The diameter of an atom is generally about 10,000 times greater than the diameter of its nucleus. Hence, if the nucleus were the size of the period at the end of this sentence, the outer edges of the atom would be located some 3.3 meters away.

We and all materials around us are mostly empty space because the atoms of which we are made are mostly empty space. So, why don't atoms simply pass through one another? How is it that we are supported by the floor despite the empty nature of its atoms? Although subatomic particles are small relative to the volume of the atom, the range of their electric fields is several times larger than that volume. On the outer surface of any atom are electrons, which repel the electrons of neighboring atoms. Atoms, therefore, can only get so close to each other before they start repelling (provided they don't combine in a chemical reaction). When the atoms of your hand push against the atoms of a wall, electrical repulsions prevent your hand from passing through the wall. These same electrical repulsions prevent us from falling through the solid floor. They also allow us the sense of touch. Interestingly enough, when you touch someone, your atoms do not meet. Instead, your atoms get close enough so that you sense electrical repulsion forces. There is still a tiny, though imperceptible, gap of space between you.

The nucleus contains nearly all of an atom's mass but occupies only a few quadrillionths of its volume. The nucleus, therefore, is extremely dense. If bare atomic nuclei could be packed against each other into a lump 1 centimeter in diameter (about the size of a large pea), the lump would weigh 133,000,000 tons! Huge electrical forces of repulsion prevent such close packing of atomic nuclei, however, because each nucleus is electrically charged and repels all other nuclei. Only under special circumstances are the nuclei of two or more atoms squashed into contact. When this happens, a violent nuclear reaction may take place. Such reactions happen in the centers of stars and is what makes them shine. This is discussed further in the next chapter.

13.4 Protons and Neutrons

The positive charge of the nucleus of any atom was found to be equal in magnitude to the combined negative charge provided by the electrons. It was thus reasoned, and then experimentally demonstrated, that positively charged particles, **protons**, make up the nucleus. Although the proton was found to be nearly 2000 times more massive than the electron, the proton's charge is equal to that of the electron and carries an opposite sign. The number of protons in the nucleus is electrically balanced by an equal number of electrons whirling about the nucleus. For example, an oxygen atom contains eight electrons and eight protons. A neutral atom has a zero net charge. (Recall from Chapter 8 that when electric charge is not neutralized, the atom is called an *ion;* more about ions in Part 6.)

Elements are classified by the number of protons their atoms contain, which is their **atomic number**. Hydrogen, containing one proton per atom, has atomic number 1; helium, containing two protons per atom, has atomic number 2, and so on in se-

Link to General Science—Physical and Conceptual Models

There are *physical models* and there are *conceptual models*. A physical model replicates an object on some different scale. A toy airplane, for example, is a physical model of a real airplane. The more accurate a physical model, the more it looks like the real thing.

A conceptual model, on the other hand, describes a system, not an object. A thunderstorm, for example, is best described using a conceptual model. Such a model shows how the various components of the system are related to one another. The components of a thunderstorm include humidity, atmospheric pressure, temperature, electric charge, and motion of large masses of air. The more accurate a conceptual model, the better we can predict the behavior of the system.

Like the thunderstorm, the atom is a complex and dynamic system, and it is best described using a conceptual model. We should be careful, however, not to misinterpret the atomic model as being a re-creation of the actual atom. The atomic model serves merely as a *symbolic representation* of atomic behavior.

Imaging devices such as scanning tunneling microscopes provide an exterior view of the atom, but no present-day imaging devices can provide views of the interior. This is where a model comes in. We can gain an understanding of an atom's interior by piecing together a model of the atom—a model shaped by information gained by many experiments. An atomic model is valued not as a means of knowing what the inside of an atom "looks like" but rather as an aide in explaining why atoms behave as they do.

quence to naturally occurring uranium with atomic number 92. The numbers continue through the artificially produced transuranic (beyond uranium) elements, at this writing, up to atomic number 112. The arrangement of elements by their atomic numbers makes up the *periodic table* (Chapter 16).

If we compare the mass-to-charge ratios of different atoms, we see that the atomic nucleus must be made up of more than protons. Nuclear charge and nuclear mass do not always go hand in hand. Helium, for example, has twice the charge of hydrogen but four times the mass. The added mass is due to another particle found in the nucleus, the **neutron**, which has about the same mass as the proton but is without charge.

Although a given type of atom will usually contain a certain number of neutrons in the nucleus, a small percentage will not. For example, most hydrogen atoms contain no neutrons. A small percentage, however, contain one neutron and a smaller percentage two neutrons. Similarly, most iron nuclei contain 30 neutrons, but a small percentage contain 29 neutrons. Atoms of the same element that contain different numbers of neutrons are **isotopes** of each other. The various isotopes of an element all have the same electric charge and so for the most part behave identically. The hydrogens in H_2O, for example, may or may not contain a neutron. The oxygen doesn't "know the difference." We'll return to isotopes in the next chapter.

The total mass of an atom is the **atomic mass number**. This number is the sum of the masses of all an atom's components (electrons, protons, and neutrons) minus a negligible amount of mass that was converted into energy when the components came together to form the atom. Electrons are light relative to protons and neutrons, and so their contribution to the total mass of an atom is negligible. Most elements have a variety of isotopes, each with its own atomic mass number. The atomic mass number for each element listed in the periodic table is the average of the masses of these isotopes based upon relative abundance. For example, about 99 percent of all carbon atoms are the isotope containing six neutrons. The remaining 1 percent is the heavier isotope containing seven neutrons, which raises the average mass of carbon from 12.000 units to 12.011 units.

Protons and neutrons are made of even more fundamental particles, *quarks.* In the same way that Rutherford was able to deduce something about the internal structure of the atom by bombarding it with alpha particles, evidence for the existence of quarks has been obtained by bombarding nuclei with highly energetic electrons. We discuss quarks in the next chapter.

13.5 Bohr's Planetary Model of the Atom

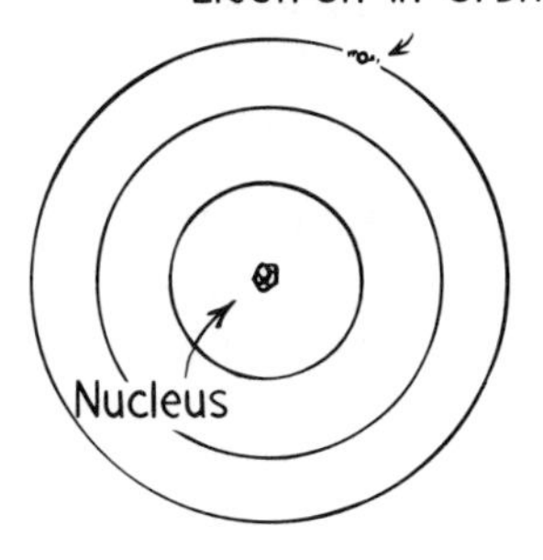

Fig. 13.9
The Bohr model of the atom.

An early conceptual model of the atom is the classic planetary model in which electrons whirl around the small but dense nucleus like planets orbiting the Sun (Figure 13.9). This model, developed by the Danish physicist Niels Bohr in 1911, explains how atoms emit or absorb only particular frequencies of light.

Recall from Chapter 12 the photon model of light, which explains the photoelectric effect. A photon is thought of as a vibrating corpuscle of light. Photons that vibrate at a high frequency are more energetic than photons that vibrate at lower frequencies. A photon of ultraviolet light, for example, is more energetic than a photon of visible light (which is why ultraviolet light produces sunburns). In the visible spectrum, violet and blue photons have more energy than green and red photons. The energy of a photon is directly proportional to its frequency:

$$E \sim f$$

Where does the photon get this energy? Bohr's planetary model of the atom provides an explanation. According to Bohr, electrons orbit the atomic nucleus with different energies. An electron orbiting close to the nucleus has a lower energy than one orbiting farther from the nucleus. Bohr reasoned that a photon of light is emitted from an atom when an electron jumps from a higher-energy outer orbit to a lower-energy inner orbit. Similarly, an electron is boosted from a lower-energy inner orbit to a higher-energy outer orbit when the atom absorbs a photon (Figure 13.10). An atom in this higher energy level is said to be *excited.* When a photon is emitted, the atom de-excites to a lower energy level.

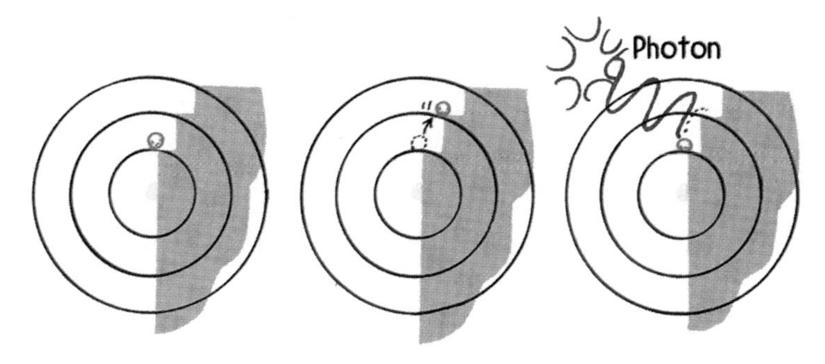

Fig. 13.10
When an electron in an atom is boosted to a higher orbit, the atom is excited. When the electron returns to its original orbit, the atom de-excites and gives off a photon of light.

In this model an electron can have only particular (discrete) amounts of energy. This means that the energy difference between any two orbits has a particular value. An electron making a transition from one of these orbits to the other emits or absorbs light having energy equal to this energy difference. Transitions between orbits farther from the nucleus absorb or emit more energy than transitions between orbits closer in. For example, an electron jump from the third energy level to the second produces a higher-energy photon than does an electron jump from the second energy level to the first.

Atomic Spectra

Long before the advent of the Bohr model, clues to atomic structure were evident in the light emitted by atoms—**atomic spectra**. As discussed in Sections 11.5 and 12.3,

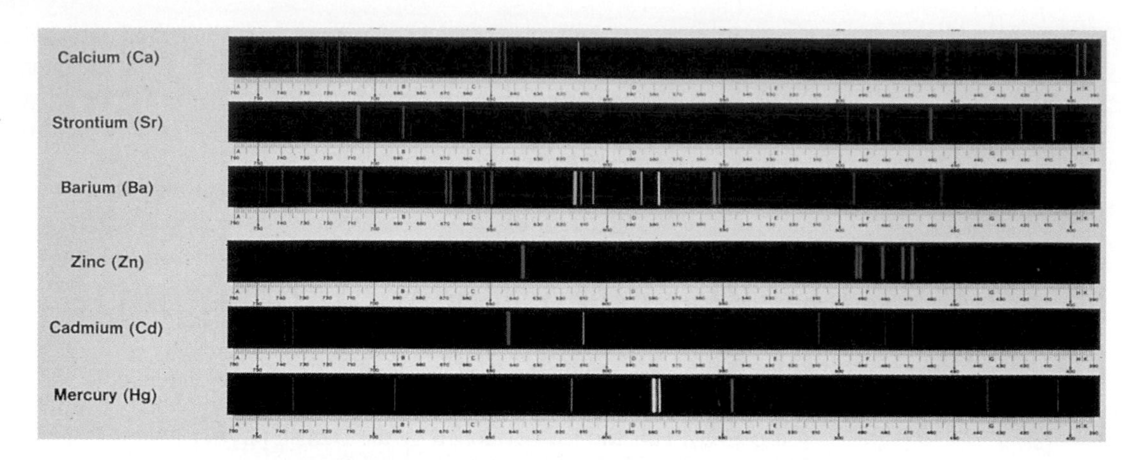

Fig. 13.11 Spectral patterns of some elements.

white light passed through a prism or a diffraction grating is dispersed into a spectrum of colors. It is not only white light that gets dispersed this way. For example, when the yellow light from a sodium lamp is passed though a prism or grating, it is dispersed into two colors of yellow. If the sodium light is first passed through a thin vertical slit before it gets to the prism or grating, separation of the colors is evident in two images of the slit. These are spectral lines representing the element sodium. All elements when excited emit their own frequencies of light, which produce a spectrum characteristic of the element. The atomic spectra of several elements are shown in Figure 13.11. Atomic spectra are the fingerprints of the elements. Every element produces its own characteristic spectrum.

The arrangement of light source, thin slit, and prism or grating is called a spectroscope (Figure 13.12). In the late 1800s, chemists used the spectroscope for chemical analysis, while physicists were busy trying to find order in the confusing arrays of spectral lines. It had long been noted that the lightest element, hydrogen, has a far more orderly spectrum than the other elements. Figure 13.13 shows a portion of the hydrogen spectrum. Note that spacing between successive lines decreases in a regular way. A Swiss schoolteacher, Johann J. Balmer, expressed these line positions by a mathematical formula. The formula worked for hydrogen and even predicted the positions of hydrogen lines not yet measured.

Another regularity in atomic spectra was first observed by Johannes Rydberg. He noticed that the sum of the frequencies of two lines in the spectrum of hydrogen some-

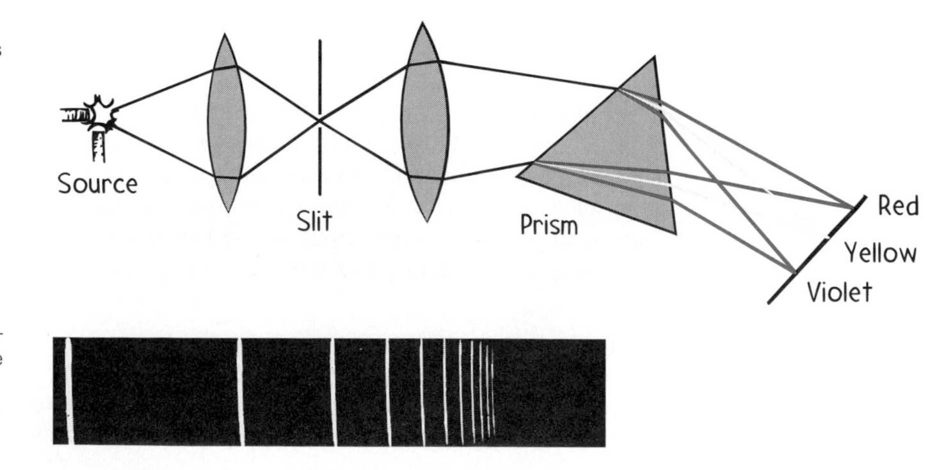

Fig. 13.12 A simple spectroscope. Images of the illuminated slit are cast on a screen and make up a pattern of lines. The spectral pattern is characteristic of the light used to illuminate the slit.

Fig. 13.13 A portion of the hydrogen spectrum. Each line, an image of the slit, represents light of a specific frequency emitted by hydrogen gas when excited.

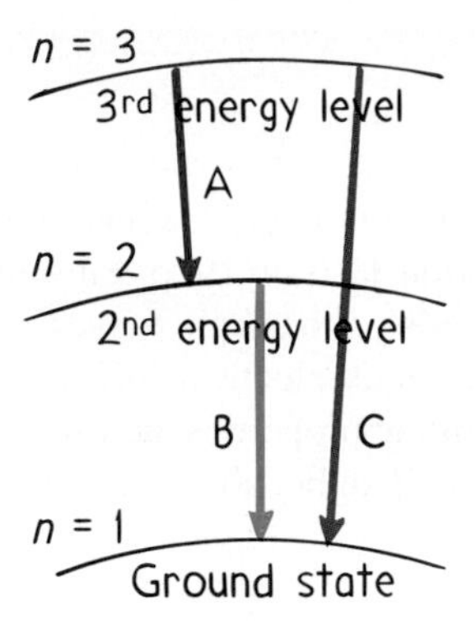

Fig. 13.14
Three energy levels in an atom. An electron jumping from the third level to the second level is shown in red, and one jumping from the second level to the ground state is shown in green. The sum of the energies (and the frequencies) for these two jumps equals the energy (and the frequency) of the single jump from the third level to the ground state, shown in blue.

times equals the frequency of a third line. Rydberg and other investigators could offer no explanation for this—until the Bohr model was proposed. The Bohr model of the atom explains why the sum of two frequencies of light emitted by an atom often equals a third frequency of light emitted by the same atom. If an electron is raised from, say, the orbit closest to the nucleus in Figure 13.14 (this innermost, first orbit is called the *ground state*) to the third energy level (the third-highest orbit), it can return to the first orbit (ground state) by two routes. It can return by a single jump from the third to the first orbit or by a double jump from third to second to first. These two paths produce three spectral lines. Note in Figure 13.14 that energy jump A plus energy jump B is equal to energy jump C. Because frequency is proportional to energy, the frequency for path A plus the frequency for path B equals the frequency for path C.

■ Question

Referring to Figure 13.14, suppose the frequency of light emitted as an electron moves along path A is 5×10^9 Hz, and the frequency as an electron moves along path B is 7×10^9 Hz. What frequency of light is emitted when an electron makes a transition along path C?

The Bohr model accounts for X rays and shows that they are emitted when electrons jump from outermost to innermost orbits. Niels Bohr predicted X-ray frequencies that were later confirmed by experiment. By measuring the frequencies of light emitted by atoms, energy levels can be mapped for the electron orbits of each element. Bohr also accounted for the general chemical properties of an unknown element, hafnium, leading to its discovery. So we see that the planetary model of the atom was a remarkable guide to explaining and predicting atomic behavior.

Despite its successes, Bohr knew his model was only a beginning. It was limited, for it did not explain why an electron was restricted to certain energy levels. There were other theoretical difficulties. An important premise of his model was that an electron does not fall into the positively charged nucleus. Today, however, we find that electrons do occasionally fall into the nucleus and then bounce back out! Electrons therefore do not orbit the nucleus the way planets orbit the Sun. The Bohr model and its revisions are now seen as useful stepping stones to newer conceptual models that help explain most atomic behavior. The newer models involve the wave nature of electrons and of particles in general.

■ Answer
Add the two frequencies from paths A and B together to get the frequency of path C: $(5 \times 10^9 \text{ Hz}) + (7 \times 10^9 \text{ Hz}) = 12 \times 10^9$ Hz, or 1.2×10^{10} Hz.

13.6 Quantum Mechanics

In 1925, the French physicist Louis de Broglie proposed a revolutionary idea regarding the properties of subatomic particles. As we learned in Section 12.6, in 1905, Einstein had published his work on the *photoelectric effect,* which tells us that light waves behave like a hail of particles having enough momentum to knock electrons off metal surfaces. Knowing from Einstein's work that waves have particle properties, de Broglie proposed that particles have wave properties. According to de Broglie, all moving matter possesses a wavelength that is related to its momentum:

$$\textbf{Wavelength} = \frac{h}{\textbf{momentum}}$$

where h is Planck's constant, the extremely small number we met in discussing the photoelectric effect in Section 12.6.[2]

All material things—electrons, protons, atoms, bowling balls, planets, stars, and you—have a wavelength that is related to momentum. For macroscopic objects, however, this wavelength is so tiny that it cannot be measured with any instrument now available. For example, a bullet of mass 0.02 kilogram traveling at the speed of sound (330 meters per second) has a wavelength of 10^{-34} meter, which is one million million million millionth the diameter of a hydrogen atom. The wavelength you have while running down the street is even less! It's a different story, however, for subatomic particles traveling at high speeds. An electron traveling at 3 percent the speed of light (8,994,000 meters per second) has a detectable wavelength of 10^{-12} meter, which is about equal to the diameter of the hydrogen atom.

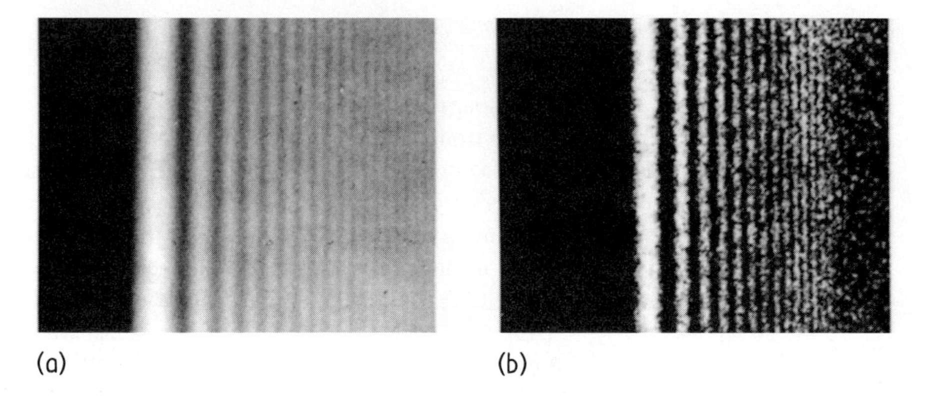

Fig. 13.15
Fringes produced by (a) the diffraction of light and (b) by an electron beam, which reveal the wave nature of both light and particles.

The wave properties of electrons were experimentally confirmed and found to conform to de Broglie's equation. Today the wave nature of electrons and other small but fast-moving particles is routinely detected (Figure 13.15). A practical application is the electron microscope, which focuses electron waves. Because electron waves are much shorter than light waves, electron microscopes are able to show unusually great detail (Figure 13.16). Electron beams are also used for studying atomic arrangements inside solids and for imprinting tiny patterns on semiconductor chips. They may even supplant optical beams as a basis for lithography in the 21st century.

---

[2]Recall that the wavelength of a wave is the distance between successive parts of the wave and is related to frequency and speed by $v = f\lambda$; recall also that momentum = mass × velocity.

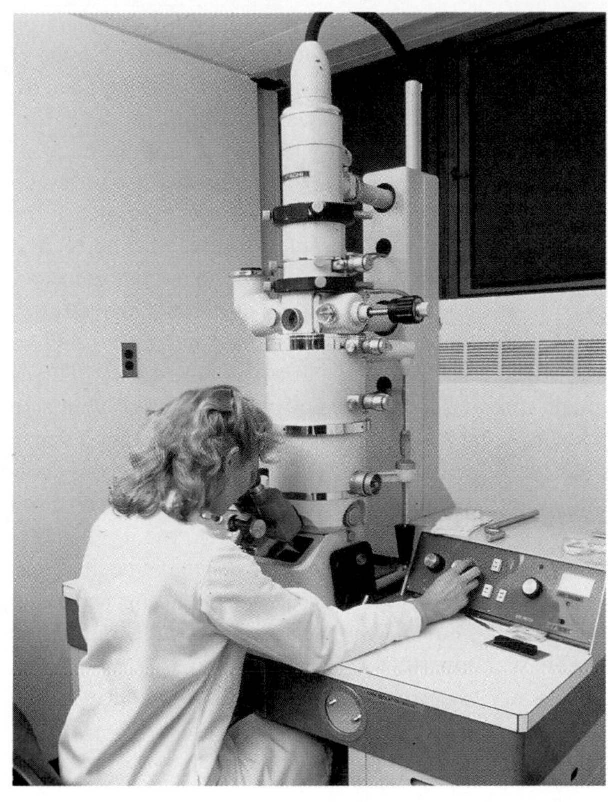

(a)

(b)

Fig. 13.16
(a) An electron microscope makes practical use of the wave nature of electrons. Because the wavelength of electrons is typically thousands of times shorter than the wavelength of visible light, the electron microscope is able to distinguish detail not visible with optical microscopes. (b) Detail of a female mosquito head as seen with a scanning electron microscope at a "low" magnification of 200 times.

■ **Question**

What wavelength does matter have when standing still?

Electron Waves

A satellite can orbit the Sun at any distance—there are no forbidden orbits. Similarly, according to the Bohr model, it seems like an electron should be able to orbit the nucleus at any distance. But this doesn't happen. It can't. Why the electron occupies only discrete levels is understood by considering the electron's wave behavior. Permitted orbits are a natural consequence of the electron wave's closing on itself in phase, as indicated in Figure 13.17a. When this doesn't happen, as indicated in Figure 13.17b, no orbit occurs. In this way, the wave is reinforced in each revolution, similar to the way a standing wave on a drum head or a spherical bell is reinforced by its successive reflections. Such reinforced waves give the appearance of standing still. So we can view an electron as a three-dimensional standing wave surrounding the atomic nucleus. The

---

■ **Answer**
Standing still means zero velocity and zero momentum. When h is divided by zero, the wavelength in the de Broglie equation is infinitely long—or rather, nonexistent. But interestingly enough, all motion is relative. The reason an electron has a wavelength we can detect is because it is moving very fast relative to us. From an electron's point of view, it is not moving at all, and we're the ones that have a wavelength!

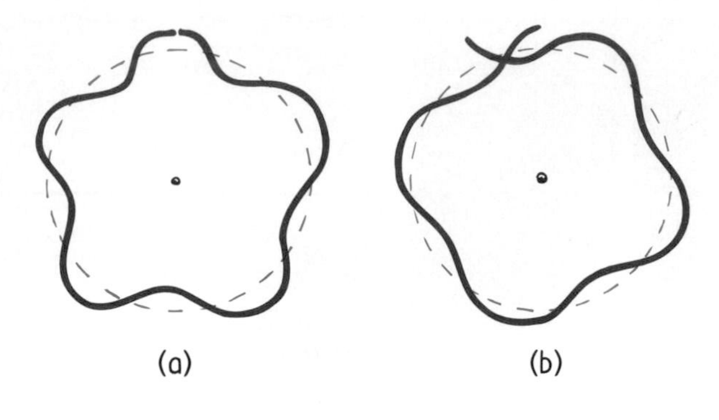

Fig. 13.17
(a) An orbiting electron forms a standing wave only when the circumference of its orbit is equal to a whole-number multiple of the wavelength. (b) When the wave does not close in on itself in phase, it undergoes destructive interference. Hence orbits exist only where waves close in on themselves in phase.

wavelength of this electron equals the circumference of its orbit around the nucleus, as shown in Figure 13.17a.

The circumference of the innermost orbit is equal to one wavelength of the electron wave. The second orbit has a circumference of two electron wavelengths, the third has three, and so forth (Figure 13.18). This is similar to a "chain necklace" made of equal-size paper clips.[3] No matter what size necklace is made, its circumference is equal to some multiple of the length of a single clip. This shows how electrons have only certain quantities of energy. Since only certain sizes of the electron wave are permitted, only certain energy levels for the electron are permitted.

So we see why negative electrons don't spiral closer and closer to the positive nucleus that attracts them. When each electron orbit is viewed as a standing wave, the circumference of the smallest orbit can be no smaller than a single wavelength—no fraction of a wavelength is possible in a constructive standing wave. As long as an electron carries the momentum necessary for wave behavior, atoms don't shrink in on themselves.

Fig. 13.18
The orbits of an electron in an atom have discrete radii because the circumferences of the orbits are whole-number multiples of the electron wavelength. This results in a discrete energy state for each orbit.

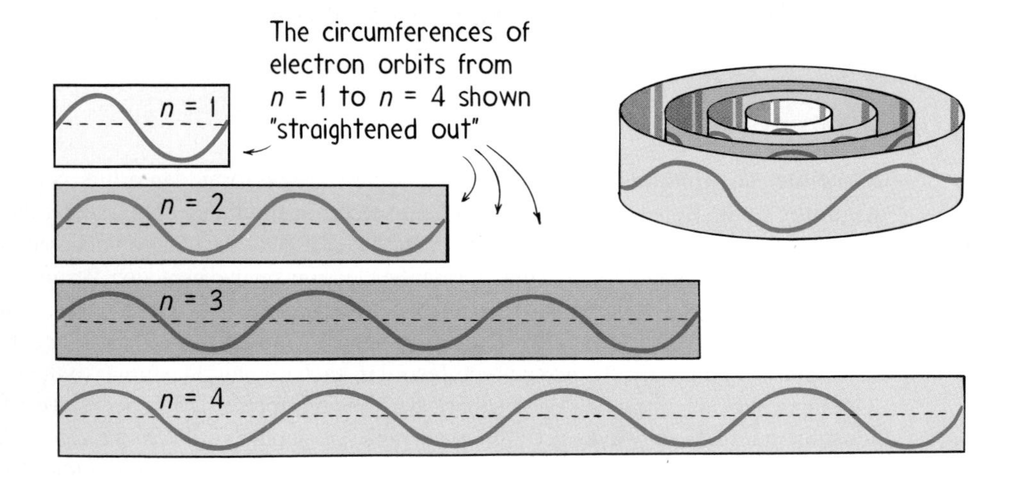

[3]For each orbit the electron has a unique speed, which determines its wavelength. Electron speeds are less and wavelengths longer for orbits of increasing radii; so to make our analogy accurate, we would have to use not only more paper clips to make increasingly longer necklaces but increasingly larger paper clips as well.

13.7 Electron Wave-Cloud Model of the Atom

It is easy to visualize the motion of a planet about the Sun—macrophysics—but very difficult to visualize the motion of an electron about the atomic nucleus—microphysics. Electrons have wave properties; they also exhibit particle properties. So we are still led to ask how does an electron, behaving as a particle, orbit the nucleus? What path does it trace?

One way to answer this question would be to pinpoint the orbiting electron's location over time and plot its path. But here we find a big difference in the methods of macrophysics and microphysics. In noting the motion of planets, we can passively observe the planets' positions in the sky. Observing the planets has no effect on their motion. Observing an electron, however, is a different story. To observe an electron requires probing it somehow—perhaps with electromagnetic radiation, which unfortunately, unpredictably zaps the electron away from its original location. All attempts of investigators to measure specific electron positions in the early 1920s failed. The breakthrough came when atomic processes were viewed in terms of *probabilities* rather than certainties.

In the late 1920s, an Austrian-German physicist, Erwin Schrodinger, and a German physicist, Werner Heisenburg, formulated equations for calculating the *probability* of finding an electron at any given location in an atom. Using their equations, they showed that there were certain regions of space centered about the nucleus where an electron of a given energy state was most likely to be found. One way to picture one of these regions is to plot the various possible locations of the electron as dots. Because some locations are more likely than others, the resulting pattern resembles a cloud having some regions denser than others. These clouds are **probability clouds**.

As investigators suspected, the dimensions of probability clouds correspond to the dimensions of electron waves. For example, a single electron wave centered about the nucleus of an atom has the shape of a sphere that has a diameter of about 10^{-10} meter. This corresponds to the shape and size of the probability cloud of an electron in its lowest energy state.

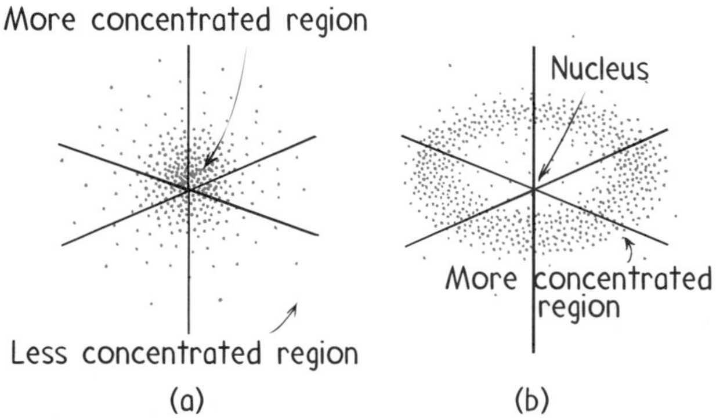

More concentrated region

Less concentrated region

(a)

Nucleus

More concentrated region

(b)

Fig. 13.19
Probability clouds showing probable locations of the electron (a) relatively close to the nucleus, and (b) relatively far from the nucleus.

We see this electron wave-cloud model of the atom as an extension of the planetary model. In the planetary model, the energy of an electron depends upon its distance from the nucleus. This is also true for the wave-cloud model of the atom. An electron in some low-energy state shows both a wave and a probability cloud concentrated close to the nucleus (Figure 13.19a). An electron in a higher-energy state shows both a wave and a probability cloud concentrated farther away from the nucleus (Figure 13.19b).

Heisenburg took the idea of the probability cloud one step further and showed that a probability cloud is a special case of a more general concept, the **uncertainty principle**. He showed that the more you know about an electron's momentum, the less you can know about its position. Likewise, the more you know about its energy, the less you can know when it has that energy. The more certain we are of one value, the more uncertain we are of the other. So, we view the distinct levels of energy for an electron as the electron's *probable locations* rather than exact locations. The probability-cloud model and the uncertainty principle go hand in hand.

13.8 The Quantum Model

Because of its wave properties, an electron in an atom can have only particular quantities of energy. This concept is an important part of a branch of modern physics called **quantum mechanics**, which embraces a conceptual model called the *quantum model.*

With the quantum model, an electron's wave and its probability cloud are recognized as two aspects of the same idea. Rather than viewing an electron occupying a single point about the nucleus at any given moment, the equations of quantum mechanics specify regions of space where an electron having a certain amount of energy is most likely to be found. This is the region in which electrons orbit. We call this region an **atomic orbital**. By convention, orbits are drawn to show where the electron is located 90 percent of the time. This gives the orbital an *imaginary* border. This border is arbitrary, however, for the electron may exist on either side of it.

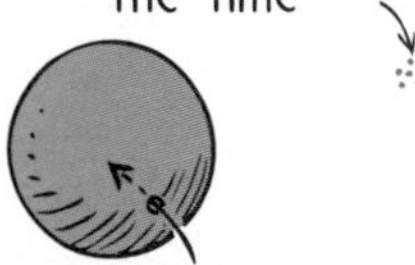

Fig. 13.20
A spherical atomic orbital.

There are a number of different types of atomic orbits, and they differ by shape and/or size. The simplest shape is a sphere, as shown in Figure 13.20. Another type has the shape of an hourglass (Figure 13.21).

The hourglass-shaped orbital illustrates the significance of the wave nature of the electron. Unlike the case with a real hourglass, the two lobes of this orbital are not open to each other. Yet the electron can occupy either lobe because of its wave properties. To understand how this can happen, consider an analogy from the macroscopic world. A guitar player can gently tap a guitar string at its midpoint (the twelfth fret) and pluck it elsewhere at the same time to produce a high-pitched tone called a harmonic. A close inspection of this string reveals that it vibrates everywhere along the string except at the point directly above the twelfth fret. As we learned in Section 10.9, this point is a *node*. Although it has no motion, waves nonetheless travel through it. Thus the guitar string vibrates on both sides of the node when only one side is plucked. Similarly, the point between the two lobes of the hourglass-shaped orbital is a node through which the electron's waves pass.

In summary we see that the planetary model of the atom explains the emission or absorption of light by an atom. Simply put, when an electron jumps from an outer (higher-energy) orbit to an inner (lower-energy) orbit, the atom emits light, and when an electron is boosted from an inner orbit to an outer one, the atom absorbs light. The emission or absorption of light can be explained in a similar fashion using the quantum model. When an electron jumps from an orbital of higher energy to one of lower energy, light is emitted. Likewise, when light is absorbed, an electron is boosted from an orbital of lower energy to one of higher energy. The two models say much the same thing. The difference is that the quantum model incorporates the wave nature of the electron.

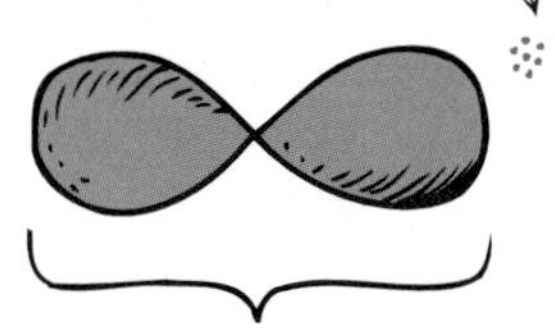

Fig. 13.21
An hourglass-shaped atomic orbital.

The more atomic behavior we observe, the more we are able to refine our model of the atom. A more refined model gives us not a better picture of what atoms "look like" but rather a better idea of how they behave under a variety of conditions. For example, a good atomic model tells us how atoms can combine to form molecules, and how molecules can combine to form various substances. In short, a good model enables us to predict things.

Summary of Terms

- **Electrons** The negatively charged particles in an atom.
- **Atomic nucleus** The core of an atom, consisting of two basic subatomic particles—protons and neutrons.
- **Protons** The positively charged particles in an atomic nucleus.
- **Atomic number** The number that designates the identity of an element, which is the number of protons in the nucleus of an atom; in a neutral atom the atomic number is also the number of electrons in the atom.
- **Neutrons** The electrically neutral particles in an atomic nucleus.
- **Isotopes** Two (or more) atoms whose nuclei contain the same number of protons but different numbers of neutrons.
- **Atomic mass number** The number associated with an atom that is the same as the number of protons plus neutrons in its nucleus.
- **Atomic spectra** The range of frequencies of light emitted by atoms.
- **Probability cloud** The pattern of electron positions plotted over a period of time that shows the likelihood of an electron's position at a given time.
- **Uncertainty principle** It is not possible to measure exactly both the position and the momentum of a particle at the same time, and it is not possible to measure exactly both the energy of a particle and the time at which it has that energy.
- **Quantum mechanics** The field of wave mechanics where atomic and subatomic particles are treated as waves.
- **Atomic orbital** The region of space where electrons of a given energy are likely to be located.

Review Questions

1. Is hydrogen an atom or an element?
2. How many atoms are found naturally in the environment?

The Elements

3. What is the origin of hydrogen?
4. What is the origin of the heavier elements?
5. What are the four most common elements in living things?
6. What is Brownian motion?
7. What part of the atom dictates its chemical properties?

The Electron

8. What is a cathode ray?
9. Why is a cathode ray deflected by a nearby electric charge and a nearby magnet?
10. What did Thompson discover about the electron?
11. What did R. Millikan discover about the electron?

The Atomic Nucleus

12. What did Rutherford discover about the atom?
13. What was the fate of the vast majority of alpha particles that were directed at gold foil in Rutherford's laboratory?
14. What was the fate of a tiny fraction of the alpha particles, a fate that surprised Rutherford?
15. What kind of force prevents atoms from meshing into one another?

Protons and Neutrons

16. How massive is a proton relative to an electron?
17. How much charge does a proton have relative to an electron?
18. What does the atomic number of an element tell you about the element?
19. What does the atomic mass number of an element tell you about the element?
20. If two atoms are isotopes of each other, do they have the same atomic number? The same atomic mass number? Explain your answers.
21. The nucleus of a neutral iron atom contains 26 protons. How many electrons does a neutral iron atom contain?
22. What is the relationship between quarks and protons? Between quarks and neutrons?

Bohr's Planetary Model of the Atom

23. From where does light get its energy?
24. Which light comes from the higher energy transition, red or blue?
25. How does the energy of a photon emitted from an atom compare with the energy change in the atom that emits the photon?
26. What is the relationship between the light emitted by an atom and its electron energy levels?
27. Why do we say the atomic spectra of elements are like atomic fingerprints?
28. What is a spectroscope?
29. What is the connection between the spectra of elements and the atomic structure of the elements?

Quantum Mechanics

30. What was de Broglie's hypothesis about particles and waves?
31. What is the relationship between a particle's wavelength and its momentum?
32. Why are electron orbits at only certain distances from the atomic nucleus?

33. How many electron wavelengths make up the circumference of an electron in its second orbit?

Electron Wave-Cloud Model of the Atom

34. We can passively observe a satellite in Earth orbit, but we cannot passively observe an electron in an atomic orbit. Why?
35. What did Heisenburg discover about probing subatomic particles?
36. How do Bohr's orbits compare with electron probability clouds?
37. What is uncertain in the uncertainty principle?

The Quantum Model

38. How do atomic orbits relate to probability clouds?
39. How do the shapes of atomic orbits relate to wave patterns?
40. How does the quantum model of light emission/absorption differ from the planetary model?
41. What is the function of an atomic model?

Exercises

1. What does light have to do with atomic structure?
2. A cat strolls across your backyard. An hour later a dog with his nose to the ground follows the trail of the cat. Explain this occurrence from a molecular point of view.
3. If no molecules in a body could escape, would the body have any odor?
4. Which are older, the atoms in the body of an elderly person or those in the body of a baby?
5. Where were the atoms that make up a newborn "manufactured"?
6. A class of meteorites called *chondrites* contain a relative abundance of elements identical to the relative abundance observed in the Sun (except for the volatile gases hydrogen and helium). What does this scientific finding suggest about the origin of the solar system?
7. Why are atoms visible under an electron microscope but invisible under even ideal optical microscopes?
8. A teaspoon of oil dropped on the surface of a quiet pond spreads out to cover almost an acre. The oil film has a thickness equal to the diameter of a molecule. If you know the volume of the oil and measure the film area, how can you calculate the diameter of the molecule?
9. In what sense can you truthfully say that you are a part of every person in history? In what sense can you say that you will tangibly contribute to every person on the Earth who will follow?
10. What are the chances that at least one of the atoms exhaled by your very first breath will be in your next breath?
11. Which contributes more to an atom's mass, electrons or protons? Which contributes more to an atom's size?
12. If the particles of a cathode ray were more massive, would the ray be bent more or less in a magnetic field? How about if the particles had greater charges?
13. Why did Rutherford assume that the atomic nucleus was positively charged?
14. How does Rutherford's model of the atom account for the back-scattering of alpha particles directed at the gold leaf?
15. Being that the atom is mostly empty space, why don't atoms simply ooze into one another under pressure?
16. Why is it difficult to compress an atom into a smaller volume despite its empty nature?
17. Would you use a physical model or a conceptual model to describe the following: the brain; the mind; the solar system; the beginning of the universe; a stranger; your best friend; a gold coin; a dollar bill; a car engine; the greenhouse effect; a virus; the spread of sexually transmitted diseases?
18. If two protons and two neutrons are removed from the nucleus of an oxygen atom, the nucleus of which element is created?
19. Why is the light emitted by elements characteristic of the elements?
20. How can a hydrogen atom, which has only one electron, have so many spectral lines?
21. Suppose a certain atom possesses four distinct energy levels. Assuming all transitions between levels are possible, how many spectral lines will this atom exhibit? Which transitions correspond to the highest-energy light emitted? To the lowest-energy light?
22. An electron de-excites from the fourth orbit to the third and then directly to the ground state. Two photons are emitted. How do their combined energies compare with the energy of the single photon emitted by de-excitation from the fourth orbit directly to the ground state?
23. In the process called fluorescence, ultraviolet light falls on certain dyes and causes them to emit visible. When infrared light falls on these materials, visible light is not emitted. Why not?
24. How does the wave model of electrons orbiting the nucleus account for discrete energy values rather than arbitrary energy values?
25. In an atom, why is there no such thing as a stable electron orbit that has a circumference of 2.5 de Broglie wavelengths?
26. How do the number of protons in an atomic nucleus dictate the chemical properties of the element?
27. Two different elements have different atomic numbers but the same atomic mass. Explain.
28. What is similar between electrons in orbit around an atomic nucleus and planets in orbit around the Sun? What is different?

    For the remaining exercises, consult the periodic table.

29. If an atom has 43 electrons, 56 neutrons, and 43 protons, what is its approximate atomic mass? What is the name of this element?

30. Which would be the more valued result: taking one proton from each nucleus in a sample of gold or adding one proton to each gold nucleus? Explain.

31. What element results if you add two protons to the nucleus of a mercury atom?

32. What element results if two protons and two neutrons are ejected from a radium nucleus?

33. If two protons and two neutrons are removed from the nucleus of an oxygen atom, what nucleus remains?

34. You could swallow a capsule of germanium without ill effects. But if a proton were added to each germanium nucleus, you would not want to swallow the capsule. Why?

Problems

This problem set requires some knowledge of exponents (Appendix A).

1. The diameter of an atom is about 10^{-10} m. (a) How many atoms make a line a millionth of a meter (10^{-6} m) long? (b) How many atoms cover a square a millionth of a meter on a side? (c) How many atoms fill a cube a millionth of a meter on a side? (d) If a dollar were attached to each atom, what could you buy with your line of atoms? With your square of atoms? With your cube of atoms?

2. There are approximately 10^{23} H_2O molecules in a thimbleful of water and 10^{46} H_2O molecules in the Earth's oceans. Suppose Columbus threw a thimbleful of water into the ocean and those water molecules have now mixed uniformly with all other water molecules in the oceans. Can you show that if you dip a sample thimbleful of water from anywhere in the ocean, you'll have at least one of the molecules that was in Columbus's thimble? (*Hint:* The ratio of the number of molecules in a thimble to the number of molecules in the ocean equals the ratio of the number of molecules in question to the number of molecules the thimble can hold.)

3. There are approximately 10^{22} molecules in a single medium-sized breath of air and approximately 10^{44} molecules in the atmosphere of the whole world. The number 10^{22} squared is equal to 10^{44}. So how many breaths of air are there in the world's atmosphere? How does this number compare with the number of molecules in a single breath? If all the molecules from Julius Caesar's last dying breath are now thoroughly mixed in the atmosphere, how many of these do we inhale on average with each single breath?

4. Assume that the present world population of 4×10^9 is $\frac{1}{30}$ of the number of people who have ever lived on the Earth. Then the total number of people in the Earth's history is about $30 \times 4 \times 10^9 \cong 10^{11}$. Compare this number with the number of molecules in a single breath (see Problem 3 for this value). Compare the number of molecules of air in your lungs to the total number of people who have ever lived. Which is larger, and by how much? Is it correct to say that the ratio

$$\frac{\text{Number of molecules in single breath}}{\text{Number of people who have ever lived}}$$

is about the same as the ratio

$$\frac{\text{Number of people who have ever lived}}{1} \quad ?$$

CHAPTER

14

The Atomic Nucleus

One of the most misunderstood areas of physics is the atomic nucleus and its processes. Public phobia about anything *nuclear* or anything *radioactive* is greater than the similar phobia about electricity more than a century ago and the phobia about gasoline-powered vehicles nearly a century ago. Just as the phobias about electricity in houses and gasoline in cars carrying children stemmed from ignorance, much of the phobia about anything nuclear stems from a lack of knowledge.

Elements having an unstable nucleus are said to be *radioactive.* Radioactivity has to do with the breakdown of certain atomic nuclei. Radioactive decay is accompanied by a release of energy as particles are ejected from a nucleus and high-frequency electromagnetic radiation is emitted.

A common misconception is that radioactivity is something new in the environment, but it has been around far longer than the human race. It is as much a part of our environment as the Sun and the rain. It has always been in the soil we walk on, the air we breathe, and it is what warms the interior of the Earth and makes it molten. In fact, radioactive decay in the Earth's interior is what heats the water that spurts from a geyser or wells up from a natural hot spring. Even the helium in a child's balloon is nothing more than the products of radioactive decay.

14.1 Alpha, Beta, and Gamma Rays

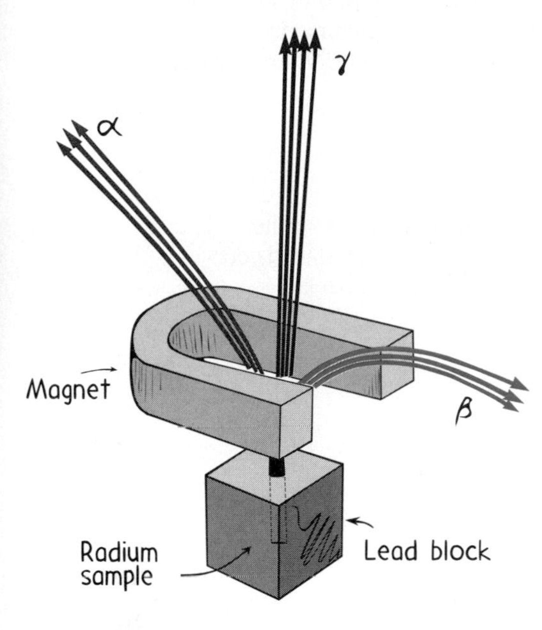

Fig. 14.1
In a magnetic field, alpha rays bend one way, beta rays bend the other way, and gamma rays don't bend at all. The combined beam comes from a radioactive source placed at the bottom of a hole drilled in a lead block.

All elements having an atomic number greater than 82 (lead) are radioactive. These elements, and others, emit three types of radiation, called by the first three letters of the Greek alphabet, α, β, γ—*alpha, beta,* and *gamma.* Alpha rays carry a positive electrical charge, beta rays carry a negative charge, and gamma rays carry no charge. The three rays can be separated by putting a magnetic field across their paths (Figure 14.1).

An **alpha particle** is a combination of two protons and two neutrons (in other words, it is the nucleus of a helium atom). Alpha particles are relatively easy to stop because of their relatively large size and their double positive charge. They do not normally penetrate through light materials such as paper or clothing. Even when traveling through only a few centimeters of air, alpha particles pick up electrons and become nothing more than harmless helium. As a matter of fact, that's where the helium in a child's balloon comes from—practically all the Earth's helium atoms were once energetic alpha particles.

A **beta particle** is an electron ejected from a nucleus. Electrons we have discussed thus far have been those that orbit an atomic nucleus. A beta particle, however, originates inside the nucleus—from a neutron. As we shall soon see, the neutron becomes a proton once it loses the electron that is the beta particle.

A beta particle is normally faster than an alpha particle and carries only a single negative charge. Beta particles are not so easy to stop as alpha particles are, and they are able to penetrate light materials such as paper or clothing. They are not able to penetrate

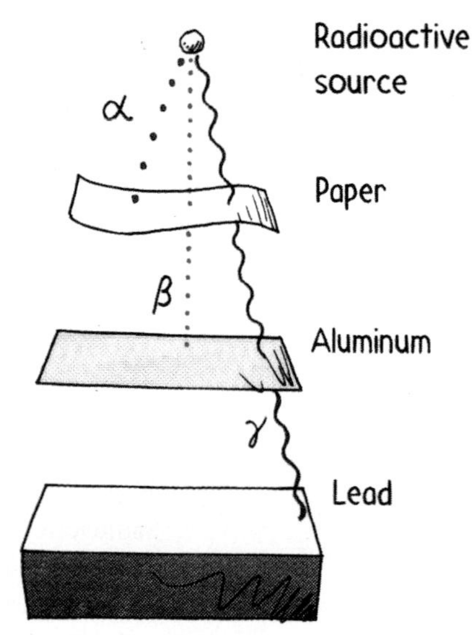

Fig. 14.3
Alpha particles are the least penetrating and can be stopped by a few sheets of paper. Beta particles readily pass through paper but not through a sheet of aluminum. Gamma rays penetrate several centimeters into solid lead.

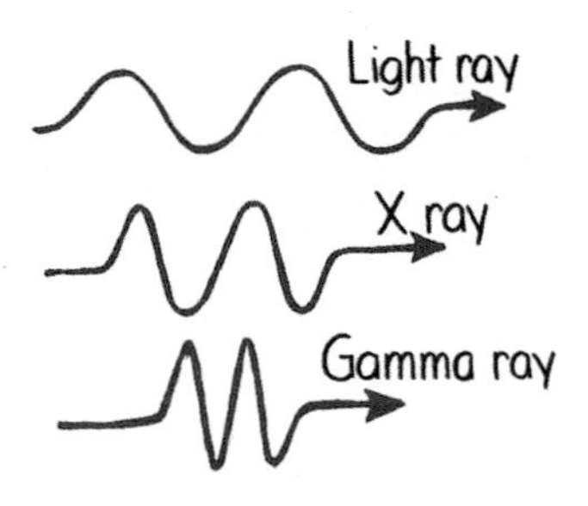

Fig. 14.2
A gamma ray is part of the electromagnetic spectrum. It is simply electromagnetic radiation that is much higher in frequency and energy than light and X rays.

deeply into denser materials, however, such as water or aluminum. Beta particles, once stopped, simply become a part of the material they are in, like any other electron.

Gamma rays are the high-frequency electromagnetic radiation emitted by radioactive elements. Like visible light, a gamma ray is pure energy. The amount of energy in a gamma ray, however, is much greater than the amount in visible light, ultraviolet light, or even X rays. Because they have no mass or electric charge and because of their high energies, gamma rays are able to penetrate through most materials. They cannot penetrate unusually dense materials such as lead, where they are readily absorbed. Delicate molecules inside our cells that get zapped by gamma rays suffer structural damage. Hence, gamma rays are more harmful to our bodies than alpha or beta particles (unless the alphas or betas are ingested).

■ Question

Suppose you are given three radioactive cookies—one an alpha emitter, one a beta emitter, and one a gamma emitter. You must eat one, hold one in your hand, and put the third in your pocket. What can you do to minimize your exposure to radiation?

14.2 Effects of Radiation on Humans

As Figure 14.4 shows, most of the radiation we encounter originates in the natural environment. It is in the ground we stand on and in the bricks and stones of buildings. This natural background radiation was present before humans emerged in the world. If our bodies couldn't tolerate it, we wouldn't be here.

Even the cleanest air we have always breathed is slightly radioactive due to bombardment by cosmic rays. These rays are a combination of high-energy particles and gamma rays that make up the background radiation in space. They originate in the Sun and other stars. At sea level the protective blanket of the atmosphere reduces cosmic radiation, but at higher altitudes cosmic radiation is more intense. For instance, in Denver, the "mile-high city," a person receives more than twice as much cosmic radiation as at sea level. A couple of round-trip flights between New York and San Francisco expose us to as much radiation as we receive in a chest X ray. The air time of airline personnel is limited because of this extra radiation. Cosmic radiation is strongest near the poles, where the Earth's magnetic field does not act as a shield.

Cosmic radiation also affects us indirectly, by transforming nitrogen atoms in the air to radioactive carbon-14, which is incorporated through photosynthesis into the plants we eat. The rocks and minerals of our planet contain significant quantities of radioactive isotopes because most of them contain trace amounts of uranium. Interestingly, people who live in brick, concrete, or stone buildings are exposed to greater amounts of radiation than people who live in wooden buildings.

■ Answer

Ideally you should get as far from all the cookies as possible. But if you must eat one, hold one, and put one in your pocket, hold the alpha because the skin on your hand will shield you. Put the beta in your pocket because your clothing will likely shield you. Eat the gamma because it will penetrate your body in any of these cases anyway.

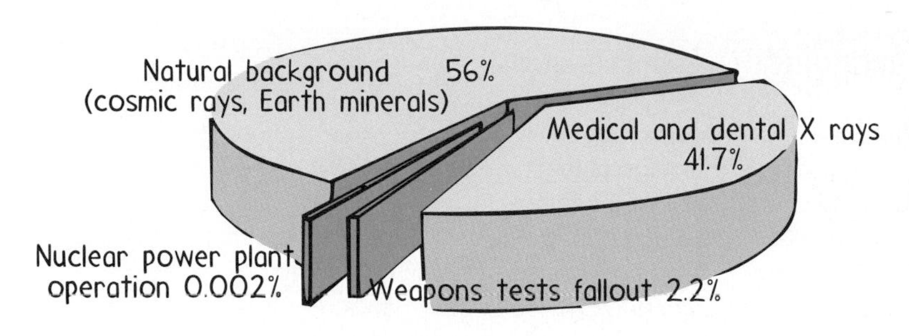

Fig. 14.4
Origins of radiation exposure for an average individual in the United States.

Even the human body is a source of natural radiation, primarily from the potassium we ingest. Our bodies contain about 200 grams of potassium. Of this quantity, about 20 milligrams is the radioactive isotope potassium-40. Between every heartbeat, about 5000 potassium-40 atoms undergo spontaneous radioactive decay.

The leading source of naturally occurring radiation, however, is radon-222, an inert gas arising from uranium deposits. Radon is a heavy gas that tends to accumulate in basements after it seeps up through cracks in the floor. Levels of radon vary from region to region depending upon local geology. You can check the radon level in your home with a radon detector kit. If levels are abnormally high, corrective measures such as sealing the basement foundation and maintaining adequate ventilation are recommended.

Nearly half of our annual exposure to radiation comes from nonnatural sources—primarily medical X rays and radiotherapy. Television sets, fallout from nuclear testing, and the coal and nuclear power industries are also contributors. The coal industry far outranks the nuclear power industry as a source of radiation. The global combustion of coal annually releases about 9000 tons of radioactive thorium and about 4000 tons of radioactive uranium into the atmosphere. Worldwide, the nuclear power industries generate about 10,000 tons of radioactive waste each year. Most all of this waste, however, is contained and *not* released into the environment.

Fig. 14.5
The internationally used symbol to indicate an area where radioactive material is being handled or produced.

When radiation encounters the intricately structured molecules in the watery, ion-rich brine that makes up our body cells, the radiation can create chaos on the atomic scale. Some molecules are broken, and this change alters other molecules, which can be harmful to life processes.

Cells are able to repair most kinds of molecular damage if the radiation is not too intense. A cell can survive an otherwise lethal dose of radiation if the dose is spread over a long period of time to allow intervals for healing. When radiation is sufficient to kill cells, the dead cells can be replaced by new ones. An important exception to this is most nerve cells, which are irreplaceable. Sometimes a radiated cell will survive with a damaged DNA molecule. Defective genetic information will be transmitted to offspring cells when the damaged cell reproduces, and a cell *mutation* will occur. Mutations are usually insignificant, but if significant they will probably result in cells that do not function as well as undamaged ones. In rare cases a mutation will be an improvement. A mutation could also be part of the cause of a cancer that will develop later.

Radioactive Tracers

All the elements have been made radioactive by bombardment with neutrons and other particles. Radioactive materials are extremely useful in scientific research and industry. To check the action of a fertilizer, for example, researchers combine a small amount of radioactive material with the fertilizer and then apply the combination to a few plants. The amount of radioactive fertilizer taken up by the plants can be easily measured with radiation detectors. From such measurements, scientists can inform farmers of the proper amount of fertilizer to use. When used in this way, radioactive isotopes are called *tracers*.

Link to Medical Technology—Positron Emission Tomography (PET)

An innovative technique for studying biochemical reactions in the body is *positron emission tomography,* PET. Behind every PET image is a short-lived positron-emitting isotope such as carbon-11, nitrogen-13, oxygen-15, or fluorine-18. These isotopes, injected intravenously into a patient, emit positrons (positively charged electrons) as they decay. Within a short distance in the tissue, each positron collides with an electron, which results in the annihilation of both particles and the generation of two oppositely directed rays of gamma radiation. Information from the gamma-ray detectors is fed to a computer, which generates a three-dimensional image of the tracer's location over time.

PET technology has been most useful in the study of neurotransmitters—chemicals the nervous system uses to transmit nerve impulses. Using PET technology, scientists can study the rate at which radiolabeled neurotransmitters build up and deplete at various centers of the brain. This is particularly useful for studying the effects of aging, drug addiction and withdrawal, mental illness, and neurogenic disorders such as Parkinson's disease.

Fig. 14.6
Radioactive isotopes are used to check the action of fertilizers in plants and the progress of food in digestion.

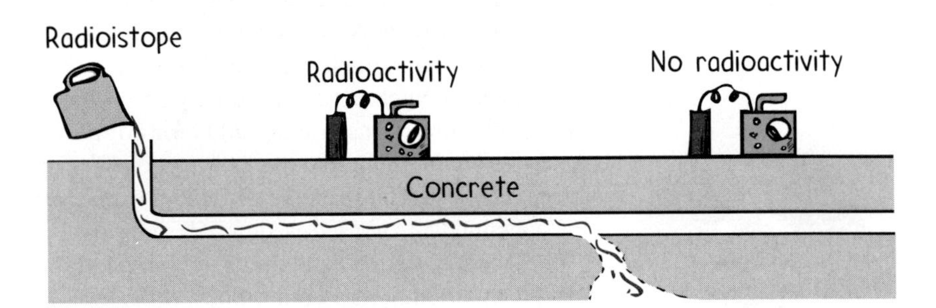

Fig. 14.7
Tracking pipe leaks with radioactive isotopes.

14.3 The Nucleus

As described in Chapter 13, the atomic nucleus occupies only a few quadrillionths of the volume of the atom, leaving most of the atom as empty space. The nucleus is composed of **nucleons**, which is the collective name for the protons and neutrons in any nucleus.

Just as there are energy levels for the orbital electrons of an atom, there are energy levels in the nucleus. Whereas orbiting electrons emit photons when making transitions to lower orbits, similar changes of energy states in the radioactive nucleus result in the emission of gamma-ray photons. This is gamma radiation.

The emission of alpha and beta particles from a nucleus can be understood from the viewpoint of quantum mechanics. Just as orbital electrons form a probability cloud about the nucleus of any atom, there is a similar probability cloud for alpha particles inside a radioactive nucleus (and other particles as well). The probability is extremely high that alpha particles are inside the nucleus, but not 100 percent. When the probability cloud extends outside the nucleus, an alpha particle may also be outside, whereupon it is hurled violently away by electrical repulsion. This is alpha radiation. When electrons are ejected from the nucleus, we have beta radiation.

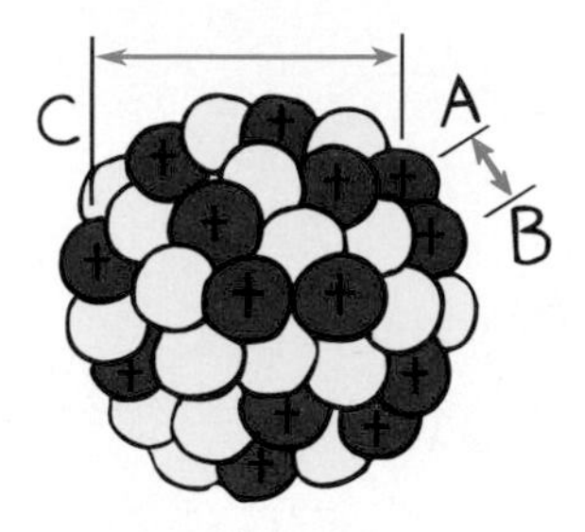

Fig. 14.8
Proton A both attracts (nuclear force) and repels (electrical force) proton B but mainly repels proton C (because the nuclear attraction is weaker at greater distances). The greater the distance between A and C, the greater the role of repulsion and the more unstable the nucleus becomes. Hence larger nuclei are more unstable than smaller nuclei. (The yellow particles are neutrons.)

In addition to alpha, beta, and gamma rays, more than 200 other particles have been detected coming from the nucleus when it is clobbered by energetic particles. We do not think of these so-called elementary particles as being buried within the nucleus and then popping out, just as we do not think of a spark as being buried in a match before it is struck. These particles come into being when the nucleus is disrupted. Regularities in the masses of these particles as well as the particular characteristics of their creation are explained by the existence of a fairly compact family of six subnuclear particles—the *quarks.*[1]

Quarks are the fundamental building blocks of all nucleons. An unusual property of quarks is that they carry fractional electrical charges. One kind, the *up* quark, carries $+\frac{2}{3}$ of the proton charge, and another kind, the *down* quark, has $-\frac{1}{3}$ of the proton charge. The proton consists of the combination *up up down,* and the neutron consists of the combination *up down down.* Each quark has an antiquark, which is a quark carrying the opposite electric charge. As with magnetic poles, no quarks have been isolated and experimentally observed—most investigators think quarks by nature cannot be isolated. We shall not delve further into the nature of quarks in this text.

Particles lighter than protons and neutrons, such as electrons and muons, and still lighter particles called *neutrinos* are members of a class of six particles called *leptons.* Leptons are not composed of quarks. At the present time, the six quarks and six leptons are thought to be the truly *elementary particles,* particles not composed of more basic entities. Investigation of elementary particles is at the frontier of our present knowledge and the area of much current excitement and research.

14.4 Isotopes

The nucleus of a hydrogen atom contains a single proton. A helium nucleus contains two protons, a lithium nucleus has three, and so forth. Every succeeding element in the list of elements has one more proton than the preceding element. In neutral atoms, there are as many protons in the nucleus as there are electrons outside the nucleus. As stated in the previous chapter, the number of protons in the nucleus is the same as the *atomic number.* Hydrogen has atomic number 1; helium, atomic number 2; lithium, 3, and so on.

The number of neutrons in the nucleus of a given element may vary, however. For example, the nucleus of every hydrogen atom contains one proton, but some hydrogen nuclei contain a neutron in addition to the proton. And in rare instances, a hydrogen nucleus may contain *two* neutrons in addition to the proton. As we learned in Section 13.4, atoms that contain like numbers of protons but unlike numbers of neutrons are called **isotopes**. (Some people confuse isotopes with ions. An ion is an electrically charged species—one in which the number of electrons orbiting about the nucleus is different from the number of protons in the nucleus. An isotope is an element containing varying numbers of neutrons in its nucleus.)

The most common isotope of hydrogen is ${}^{1}_{1}H$. The subscript (lower) number refers to the atomic number, and the superscript (upper) number refers to the atomic mass number (approximately but not exactly equal to the *atomic mass*). The double-mass hydrogen isotope ${}^{2}_{1}H$ is called *deuterium.* "Heavy water" is the name usually given to

[1]The name *quark,* inspired by a quotation from *Finnegans Wake* by James Joyce, was chosen in 1963 by Murray Gell-Mann, who first proposed their existence.

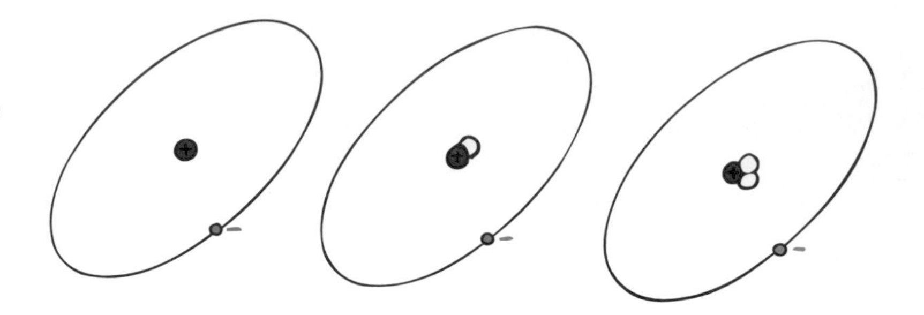

Fig. 14.9
Three isotopes of hydrogen: hydrogen with no neutrons in the nucleus, deuterium with one neutron, and tritium with two neutrons. Each nucleus contains a single proton, around which orbits a single electron, which in turn determines the chemical properties of the atom. The different number of neutrons change the mass of the atom but not its chemical properties.

H_2O in which one or both of the H atoms have been replaced by deuterium atoms. In all naturally occurring hydrogen compounds, such as hydrogen gas and water, there is 1 atom of deuterium to about 6000 atoms of hydrogen. The triple-mass hydrogen isotope $^{3}_{1}H$, which is radioactive and lives long enough to be a known constituent of atmospheric water, is called *tritium.* Tritium is present only in extremely minute amounts—less than 1 per 10^{17} atoms of ordinary hydrogen.

All elements have a variety of isotopes. For instance, three isotopes of uranium naturally occur in the Earth's crust; the most common is $^{238}_{92}U$. In briefer notation we can drop the atomic number and simply say uranium-238 or, even briefer, U-238. Of the 83 elements present in the Earth in significant amounts, 20 have a single stable (nonradioactive) isotope. The others have between 2 and 10 stable isotopes. Counting both radioactive and stable types, more than 2000 distinct isotopes are known.

■ Question

State the numbers of protons and neutrons in $^{1}_{1}H$, $^{14}_{6}C$, and $^{235}_{92}U$.

14.5 Half-Life

The rate of decay of a radioactive isotope is measured in terms of a characteristic time called the **half-life**. This is the time it takes for half of an original quantity of a radioactive element to decay. For example, radium-226 has a half-life of 1620 years. This means that half of a sample of radium will be converted to other elements by the end of 1620 years. In the next 1620 years, half of the remaining radium will decay, leaving only one-fourth the original amount of radium. (After 20 half-lives, the initial quantity of radium-226 will be diminished by a factor of about 1 million.)

Half-lives are remarkably constant and not affected by external conditions. Some radioactive isotopes have half-lives that are less than a millionth of a second, while others have half-lives of more than a billion years. Uranium-238 has a half-life of 4.5 billion years. All uranium eventually decays in a series of steps to the element lead. In 4.5 billion years, half the uranium presently in the Earth today will be lead.

It is not necessary to wait through the duration of a half-life in order to measure it. The half-life of an element can be calculated at any given moment by measuring the

■ **Answer**
The atomic number (subscript) gives the number of protons. The number of neutrons is the atomic mass number (superscript) minus the atomic number. So we see 1 proton and no neutrons in $^{1}_{1}H$, 6 protons and 8 neutrons in $^{14}_{6}C$, and 92 protons and 143 neutrons in $^{235}_{92}U$.

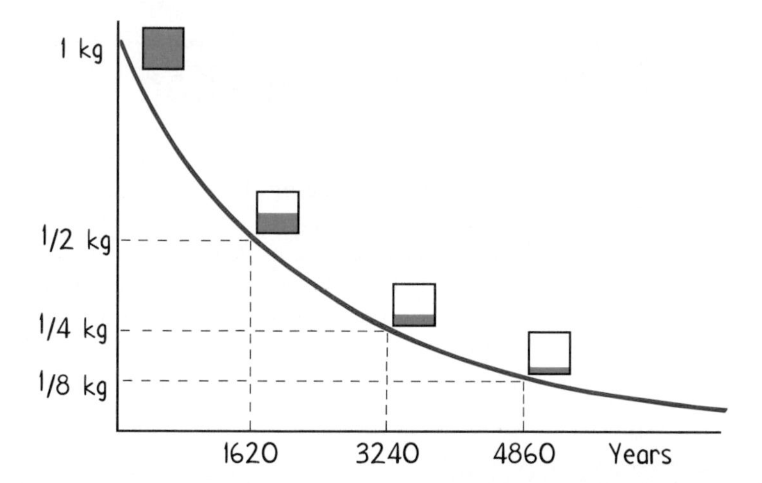

Fig. 14.10
Every 1620 years, the amount of radium decreases by half.

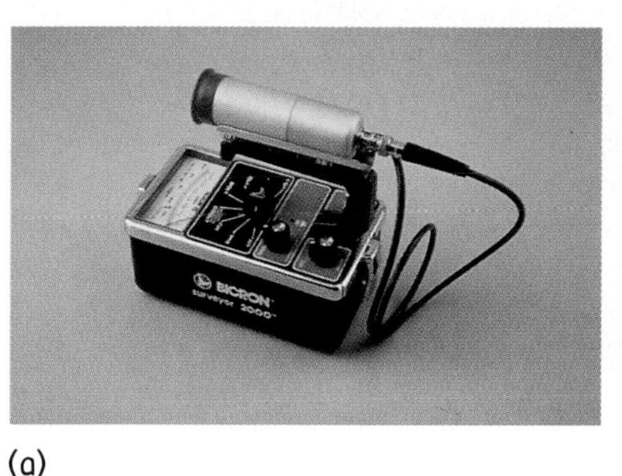

(a) (b)

Fig. 14.11
Some radiation detectors. (a) A Geiger counter detects incoming radiation by the way the radiation ionizes a gas enclosed in the tube. (b) A scintillation counter indicates incoming radiation by flashes of light produced when charged particles or gamma rays pass through the counter.

rate of decay of a known quantity. This is easily done using a radiation detector (Figure 14.11). In general, the shorter the half-life of a substance, the faster it disintegrates and the more radioactivity per amount is detected.

Questions

1. If a sample of a radioactive isotope has a half-life of 1 day, how much of the original sample is left at the end of the second day? The third day?
2. What becomes of isotopes that undergo radioactive decay?
3. Which gives a higher counting rate on a radiation detector, radioactive material that has a short half-life or radioactive material that has a long half-life?

Answers

1. At the end of the first day, one-half of the original sample is left. At the end of the second day, half of this is left. Half of one-half is one-fourth. So one-fourth of the original sample is left at the end of the second day. Can you see that at the end of three days, $\frac{1}{8}$ of the original isotope is left?
2. An isotope that decays becomes a different element altogether.
3. The material with the shorter half-life is more active and so gives a higher counting rate on a radiation detector.

14.6 Natural Transmutation of Elements

When a radioactive nucleus emits an alpha particle, a different element is formed. The changing of one chemical element to another is called **transmutation**. Consider uranium-238, the nucleus of which contains 92 protons and 146 neutrons. When an alpha particle is ejected, the nucleus is reduced by two protons and two neutrons (because an alpha particle is a helium nucleus, which means it consists of two protons and two neutrons). An element is defined by the number of protons in its nucleus, and so the 90 protons and 144 neutrons left behind are no longer uranium, but the nucleus of a different element—*thorium.* This reaction is expressed as

$$^{238}_{92}\text{U} \rightarrow {}^{234}_{90}\text{Th} + {}^{4}_{2}\text{He}$$

The arrow shows that $^{238}_{92}\text{U}$ changes into the two elements written to the right of the arrow. When this transmutation happens, energy is released, partly in the form of gamma radiation, partly in the kinetic energy of the alpha particle ($^{4}_{2}\text{He}$), and partly in the kinetic energy of the thorium atom. In this and all other such equations, the mass numbers at the top balance ($238 = 234 + 4$) and the atomic numbers at the bottom also balance ($92 = 90 + 2$).

Thorium-234, the product of this reaction, is also radioactive. When it decays, it emits a beta particle. Recall that a beta particle is an electron—not an orbital electron but one from the nucleus. A beta particle is emitted by a neutron in the nucleus. You may find it useful to think of a neutron as a combined proton and electron (although it's not really the case) because when the neutron emits an electron, it becomes a proton. A neutron is ordinarily stable when locked in a nucleus, but a free neutron is radioactive and has a half-life of 12 minutes. It decays into a proton by beta emission.[2] So in the case of thorium, which has 90 protons, beta emission leaves the nucleus with one fewer neutron and one more proton. The new nucleus then has 91 protons and is no longer thorium, but the element *protactinium.* Although the atomic number has increased by 1 in this process, the mass number (protons + neutrons) remains the same. The nuclear equation is

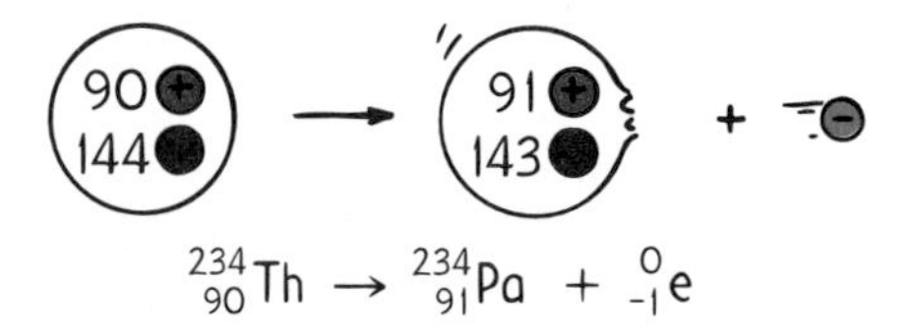

$$^{234}_{90}\text{Th} \rightarrow {}^{234}_{91}\text{Pa} + {}^{0}_{-1}\text{e}$$

[2]Beta emission is always accompanied by the emission of a neutrino (actually an antineutrino), a neutral particle that has a very tiny mass and travels at about the speed of light. The neutrino ("little neutral one") was postulated by Wolfgang Pauli in 1930 and detected in 1956. Neutrinos are hard to detect because they interact very weakly with matter. To capture a neutrino is extremely difficult. Whereas a piece of solid lead a few centimeters thick will stop most gamma rays, it would take a piece of lead about 8 light-years thick to stop half the neutrinos produced in typical nuclear decays. Thousands of them fly through you every second of every day because the universe is filled with them. Only one or two times a year does a neutrino or two interact with the matter of your body. Neutrinos are so numerous in the universe that they might actually make up most of the mass of the universe. Neutrinos may be the "gravitational glue" that holds the universe together.

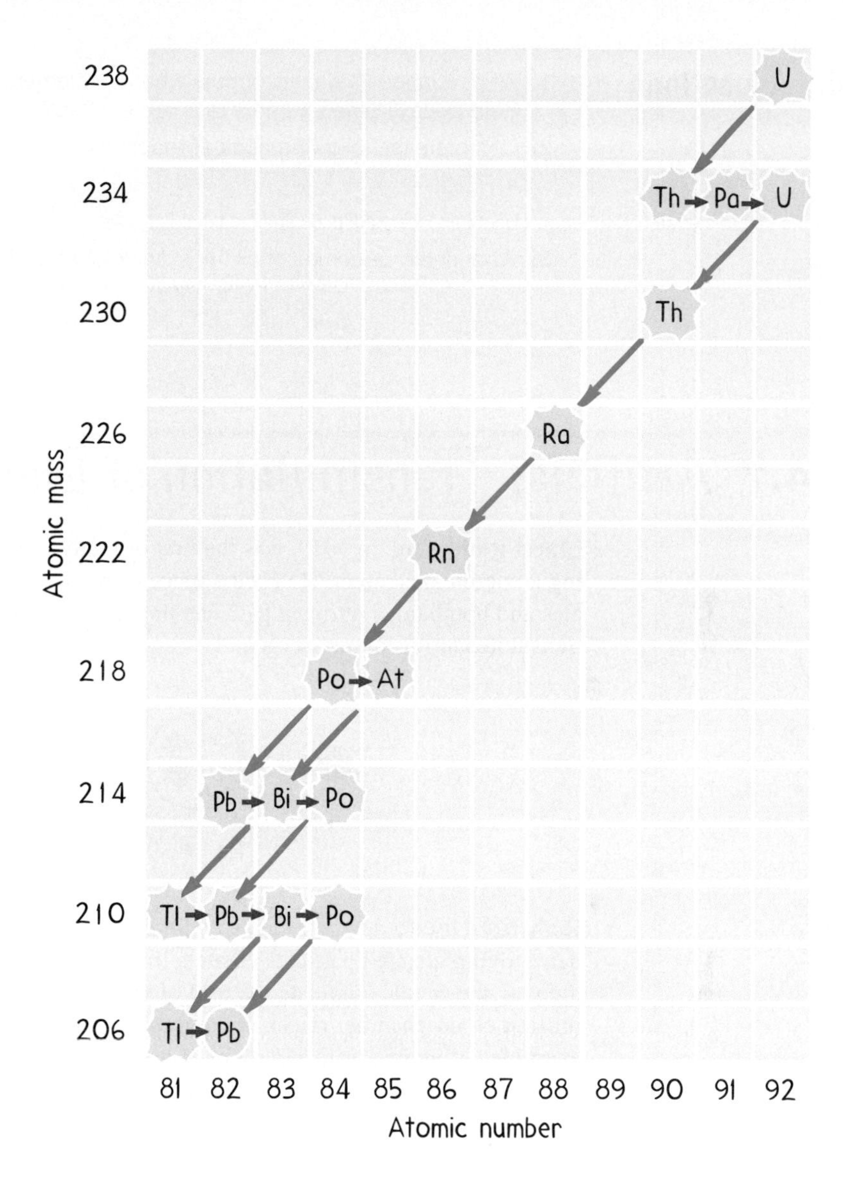

Fig. 14.12
U-238 decays to Pb-206 through a series of alpha and beta decays. Each arrow that slants downward toward the left shows an alpha decay, and each arrow that points to the right shows a beta decay.

We write an electron as $_{-1}^{0}e$. The 0 indicates that an electron's mass is insignificant relative to that of the protons and neutrons that alone contribute to the mass number. The -1 is the charge of the electron. Remember that this electron is a beta particle from the nucleus and not an electron from the electron cloud that surrounds the nucleus.

The successive radioactive decays of $^{238}_{92}U$ to $^{206}_{82}Pb$, an isotope of lead, is shown in Figure 14.12. This is one of several similar radioactive series that occur in nature. Note that radioactive elements can decay backward or forward in the periodic table.[3]

[3]Sometimes a nucleus emits a positron, which is the *antiparticle* of an electron. In this case a proton becomes a neutron, and the atomic number is decreased.

Questions

1. What change in identity occurs when an element undergoes alpha decay? Beta decay? When gamma rays are emitted?
2. Complete the following nuclear reactions:
   a. ${}^{226}_{88}\text{Ra} \rightarrow {}^{?}_{?}\text{X} + {}^{\;\,0}_{-1}e$
   b. ${}^{209}_{84}\text{Po} \rightarrow {}^{205}_{82}\text{Pb} + {}^{?}_{?}?$
3. What finally becomes of all the uranium that undergoes radioactive decay?

14.7 Artificial Transmutation of Elements

Ernest Rutherford, in 1919, was the first of many investigators to succeed in transmuting a chemical element. He used a piece of radioactive ore as a source of alpha particles and bombarded nitrogen gas. The impact of an alpha particle on a nitrogen nucleus transmutes nitrogen into oxygen:

$${}^{4}_{2}\text{He} + {}^{14}_{7}\text{N} \rightarrow {}^{17}_{8}\text{O} + {}^{1}_{1}\text{H}$$

Rutherford used a device called a *cloud chamber* to record this event. In a cloud chamber, moving charged particles create a trail of ions along their path in a way similar to the way ice crystals indicate the trail of jet planes high in the sky. From a quarter of a million cloud-chamber tracks photographed on movie film, Rutherford showed seven examples of atomic transmutations. Analysis of tracks bent by a strong external magnetic field showed that when an alpha particle collided with a nitrogen atom, a proton (shown by H in the preceding reaction) bounced out of the nitrogen nucleus, and the resulting oxygen nucleus recoiled a short distance. The alpha particle disappeared. It was absorbed by the nitrogen nucleus, transforming nitrogen to oxygen.

Answers

1. When an element ejects an alpha particle from its nucleus, the mass number of the resulting atom decreases by 4, and its atomic number decreases by 2. The resulting atom is therefore an element two spaces back in the periodic table of the elements. When an element ejects a beta particle (electron) from its nucleus, the mass of the atom is practically unaffected and so there is no change in mass number, but its atomic number *increases* by 1. The resulting atom is an element one place *forward* in the periodic table. Gamma emission results in no change in either mass number or atomic number.

2. a. ${}^{226}_{88}\text{Ra} \rightarrow {}^{226}_{89}\text{Ac} + {}^{\;\,0}_{-1}e$
   b. ${}^{209}_{84}\text{Po} \rightarrow {}^{205}_{82}\text{Pb} + {}^{4}_{2}\text{He}$

3. All uranium ultimately becomes lead. On the way to becoming lead, it exists as a series of elements, such as indicated in Figure 14.12.

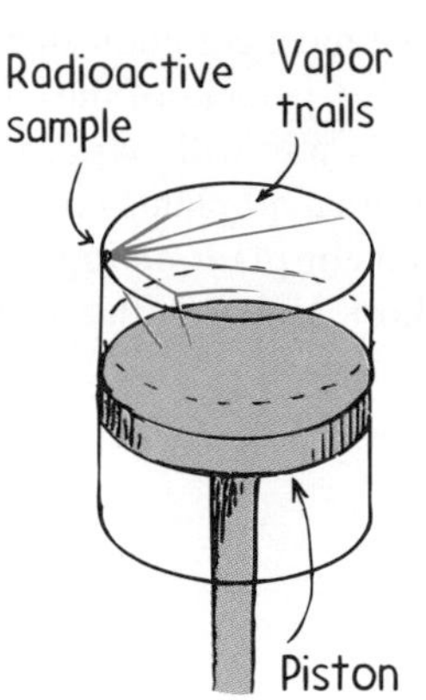

Fig. 14.13
A cloud chamber. Charged particles moving through supersaturated vapor leave trails. When the chamber is in a strong electric or magnetic field, bending of the tracks provides information about the charge, mass, and momentum of the particles.

Fig. 14.14
Tracks of elementary particles in a bubble chamber, a device that is similar to a cloud chamber but more complicated. (The trained eye notes that two particles were destroyed at the point where the spirals emanate.)

Since Rutherford's announcement in 1919, experimenters have carried out many other nuclear reactions, first with natural bombarding projectiles from radioactive ores and then with still more energetic projectiles—protons and electrons hurled by huge particle accelerators. Artificial transmutation is what produces the hitherto unknown synthetic elements from atomic number 93 to 112. All these artificially made elements have short half-lives. If they ever existed naturally when the Earth was formed, they have long since decayed.

14.8 Isotopic Dating

The Earth's atmosphere is continuously bombarded by cosmic rays that cause many atoms in the upper atmosphere to transmute. These transmutations result in many protons and neutrons being "sprayed out" into the environment. Most of the protons quickly capture electrons and become hydrogen atoms in the upper atmosphere. The neutrons, however, keep going for longer distances because they have no charge and therefore do not interact electrically with matter. Eventually, many of them collide with atomic nuclei in the denser lower atmosphere. Any nitrogen atom capturing a neutron becomes an isotope of carbon by emitting a proton:

$$^{1}_{0}n + ^{14}_{7}\mathrm{N} \rightarrow ^{14}_{6}\mathrm{C} + ^{1}_{1}\mathrm{H}$$

This carbon-14 isotope is radioactive and has eight neutrons (the most common stable isotope, carbon-12, has six neutrons). Less than one-millionth of 1 percent of the carbon in the atmosphere is carbon-14. Both carbon-12 and carbon-14 join with oxygen to become carbon dioxide, which is taken in by plants. This means that all plants contain a tiny bit of radioactive carbon-14. All animals eat either plants or plant-eating animals, and therefore all contain a little carbon-14 in them. In short, all living things on the Earth contain some carbon-14.

Carbon-14 is a beta emitter and decays back to nitrogen:

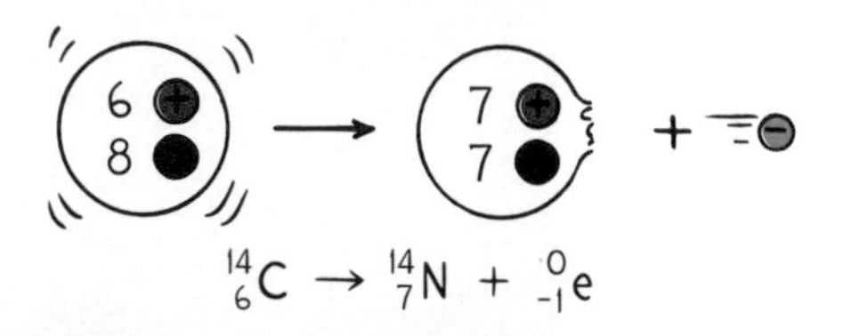

Because plants take in carbon dioxide as long as they live, any carbon-14 lost to decay is immediately replenished with fresh carbon-14 from the atmosphere. In this way, a radioactive equilibrium is reached where there is a ratio of about one carbon-14 atom to every 100 billion carbon-12 atoms. When a plant dies, replenishment stops. Then the percentage of carbon-14 decreases at a constant rate given by its radioactive half-life.[4] The longer a plant is dead, the less carbon-14 it contains. Because animals eat plants, they too contain carbon-14.

The half-life of carbon-14 is about 5730 years. This means that half of the carbon-14 atoms now present in a plant or animal that dies today will decay in the next 5730 years. Half of the remaining carbon-14 atoms will then decay in the following 5730 years, and so forth. The radioactivity of living things gradually decreases at a steady rate after they die.

Fig. 14.15
The radioactive carbon isotopes in the skeleton diminish by one-half every 5730 years.

With this knowledge, investigators such as archeologists are able to calculate the age of carbon-containing artifacts, such as wooden tools or skeletons, by measuring their current level of radioactivity. This process, known as *carbon-14 dating,* enables us to probe as much as 50,000 years into the past.

Carbon dating would be an extremely simple and accurate dating method if the amount of radioactive carbon in the atmosphere had been constant over the ages. But it hasn't been. Fluctuations in the Sun's and Earth's magnetic fields affect cosmic-ray intensities in the Earth's atmosphere, which in turn produce fluctuations in the amount of carbon-14 in the atmosphere at any given time. In addition, changes in the Earth's climate affect the amount of carbon dioxide in the atmosphere. The oceans are great reser-

---

[4]A 1-g sample of contemporary carbon contains about 5×10^{22} atoms, 6.5×10^{10} of which are carbon-14, and has a beta decay rate of about 13.5 decays per minute.

voirs of carbon dioxide. When the oceans are cold, they release less carbon dioxide into the atmosphere than when they are warm. Because of all these fluctuations in the production of carbon-14 through the centuries, carbon dating has an uncertainty of about 15 percent. This means, for example, that the straw of an old adobe brick that is dated to be 500 years old may really be only 425 years old on the low side, or 575 years old on the high side. For many purposes this is an acceptable level of uncertainty. Techniques involving longer-lived radioactive isotopes are used to date more ancient relics.

■ **Question**

Suppose an archeologist extracts 1 g of carbon from an ancient ax handle and finds it to be one-fourth as radioactive as 1 g of carbon extracted from a freshly cut tree branch. About how old is the ax handle?

Uranium Dating

The dating of older, but nonliving, things is accomplished with radioactive *minerals,* such as uranium. The naturally occurring isotopes U-238 and U-235 decay very slowly and ultimately become isotopes of lead—but not the common lead isotope Pb-208. Instead, U-238 decays through several stages to finally become Pb-206 (Figure 14.12). U-223, on the other hand, decays to becomes the isotope Pb-207. Thus any lead 206 and lead -207 that now exist in a uranium-bearing rock were at one time uranium. The older the rock, the higher the percentage of these *remnant isotopes.*

From the half-lives of uranium isotopes and the percentage of lead isotopes in uranium-bearing rock, it is possible to calculate the date at which the rock was formed. Rocks dated in this way have been found to be as much as 3.7 *billion* years old. Samples from the Moon, where there has been an absence of erosion, have been dated at 4.2 billion years, an age that agrees closely with the estimated 4.6 billion year age of the Earth and the rest of the solar system.

We'll return to isotopic dating when we get into geology in Part 7.

14.9 Nuclear Fission

In 1938, two German scientists, Otto Hahn and Fritz Strassmann, made an accidental discovery that was to change the world. While bombarding a sample of uranium with neutrons in the hopes of creating new, heavier elements, they were astonished to find chemical evidence for the production of barium, an element having about half the mass of uranium. Hahn wrote of this news to his former colleague Lise Meitner, who had fled from Nazi Germany to Sweden because of her Jewish ancestry. From the evi-

---

■ **Answer**

Assuming the ratio of $^{14}C/^{12}C$ was the same when the ax was made, the ax handle is two half-lives of ^{14}C; that's 2×5730 years, about 11,460 years old.

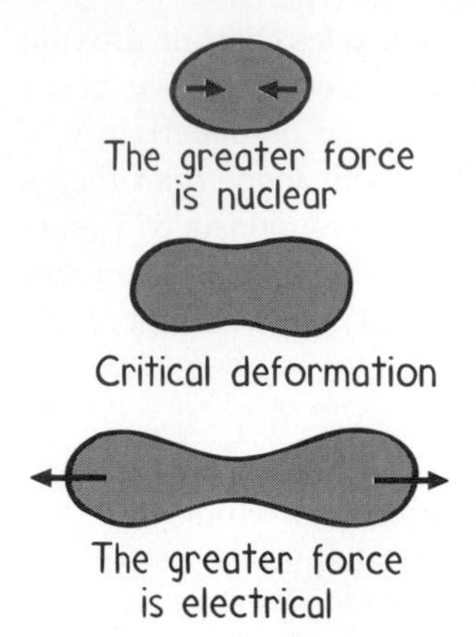

Fig. 14.16
Nuclear deformation may result in repulsive electrical forces overcoming attractive nuclear forces, in which case fission occurs.

dence given to her by Hahn, Meitner concluded that the uranium nucleus, activated by neutron bombardment, had split in half. Soon thereafter, Meitner, working with her nephew Otto Frisch, also a physicist, published a paper in which the term *nuclear fission* was coined.

Within the nucleus of every atom there exists a delicate balance between attractive nuclear forces and repulsive electric forces between protons. In all known nuclei, the nuclear forces dominate. In uranium, however, this domination is tenuous. If a uranium nucleus stretches into an elongated shape (Figure 14.16), the electrical forces may push it into an even more elongated shape. If the elongation passes a certain point, the repulsive electrical forces overwhelm the attractive nuclear forces and the nucleus splits. This splitting of a nucleus is **nuclear fission**.

The energy released by the fission of one U-235 atom is enormous—about 7 million times the energy released by the explosion of one TNT molecule. This energy is mainly in the form of kinetic energy of the fission fragments, which fly apart from one another. A much smaller amount of energy is given to the ejected neutrons and to gamma radiation.

A typical uranium fission reaction is

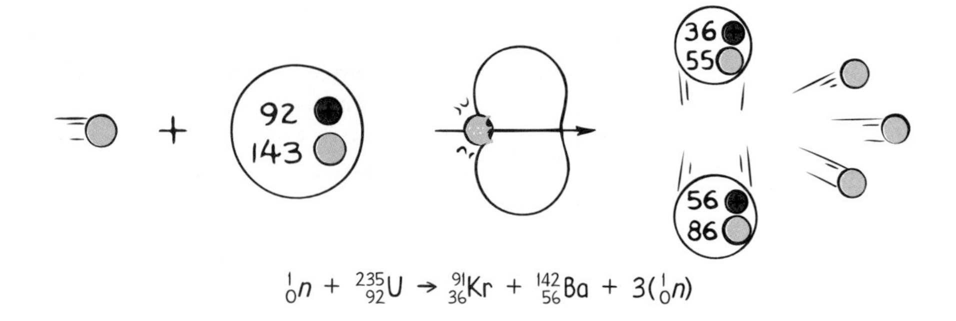

$$ {}^{1}_{0}n + {}^{235}_{92}U \rightarrow {}^{91}_{36}Kr + {}^{142}_{56}Ba + 3({}^{1}_{0}n) $$

Note in this reaction that one neutron (incoming yellow circle on left) starts the fission of the uranium nucleus and that the fission produces three neutrons (outgoing yellow circles on right). These product neutrons can cause three other uranium atoms to fission, releasing a total of nine more neutrons. If each of these neutrons then succeeds in splitting a uranium atom, the next step in the reaction produces 27 neutrons, and so on. Such a sequence is called a **chain reaction**[5] (Figure 4.17).

Chain reactions do not occur to any great extent in naturally occurring uranium ore deposits because not all uranium atoms fission so easily. Fission occurs mainly for the rare isotope U-235, which makes up only 0.7 percent of the uranium in pure uranium metal. When the prevalent isotope U-238 absorbs neutrons from fission, it typically does not undergo fission. So any chain reaction

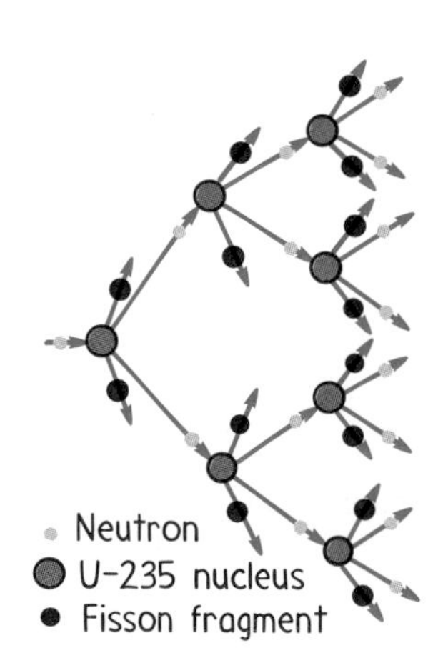

Fig. 14.17
A chain reaction.

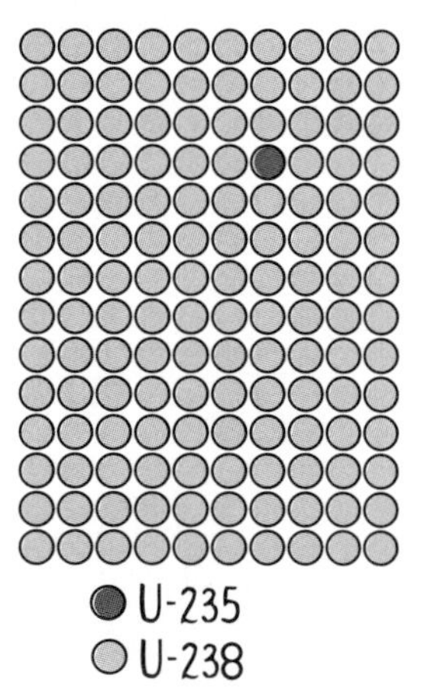

Fig. 14.18
Only 1 part in 140 (0.7 percent) of naturally occurring uranium is U-235.

[5]In the reaction considered here, three neutrons are ejected when fission occurs. In some other reactions, two neutrons may be ejected. On average, fission produces 2.5 neutrons per reaction.

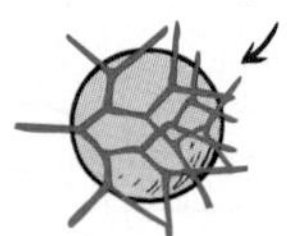

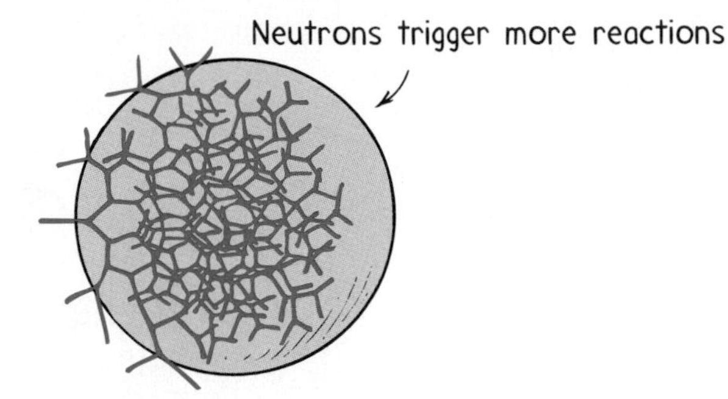

Fig. 14.19
A chain reaction in a small piece of pure U-235 dies out because neutrons leak from the surface too soon. The small piece has a lot of surface area relative to its mass. In a larger piece, more uranium atoms and less surface area are presented to the neutrons.

is snuffed out by the neutron-absorbing U-238, as well as by other neutron-absorbing elements in the rock in which the ore is imbedded.

If a chain reaction occurred in a baseball-size chunk of pure U-235, an enormous explosion would result. If the chain reaction were started in a smaller chunk of pure U-235, however, no explosion would occur. This is because of geometry: The ratio of surface area to mass is larger in a small piece than in a large piece. (Similarly, there is more skin on a kilogram of small potatoes than on a single large, 1-kilogram potato.) In a small piece of uranium, neutrons leak through the surface before an explosion can occur. In a bigger piece, the chain reaction builds up to enormous energies before the neutrons get to the surface and escape (Figure 14.19). For masses greater than a certain amount, called the **critical mass**, an explosion of enormous magnitude may take place.

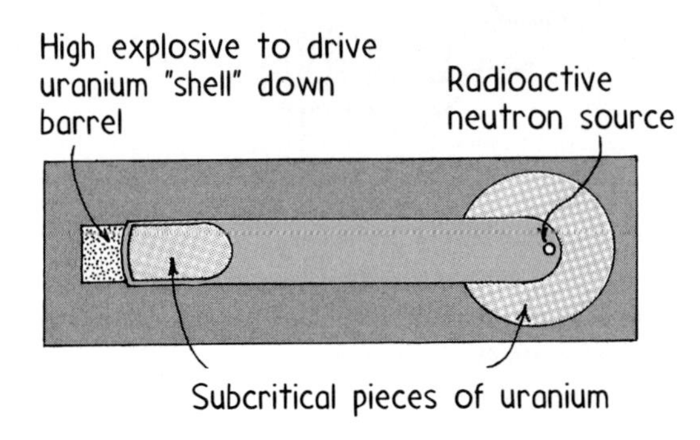

Fig. 14.20
An idealized uranium fission bomb. (In an actual weapon, one piece of uranium is fired toward the other one, which is the target.)

Consider a large quantity of U-235 divided into two units, each having a mass smaller than critical (Figure 14.20). The units are *subcritical.* Neutrons readily reach a surface and escape before a sizable chain reaction builds up. But if the pieces are suddenly driven together, the total surface area decreases. If the timing is right and the combined mass is greater than critical, a violent explosion takes place. Such a device is a nuclear fission bomb.

Constructing a fission bomb is a formidable task. The difficulty is separating enough U-235 from the more abundant U-238. Scientists took more than two years to extract enough U-235 from uranium ore to make the bomb detonated at Hiroshima in 1945. To this day, uranium isotope separation remains a difficult process.

Question

A 1-kg ball of U-235 is critical, but the same ball broken up into small chunks isn't. Explain.

Answer
The small chunks of U-235 have more combined surface area than the ball from which they came (just as the combined surface area of gravel is greater than the surface area of a boulder of the same mass). Neutrons escape via the surface before a sustained chain reaction can build up.

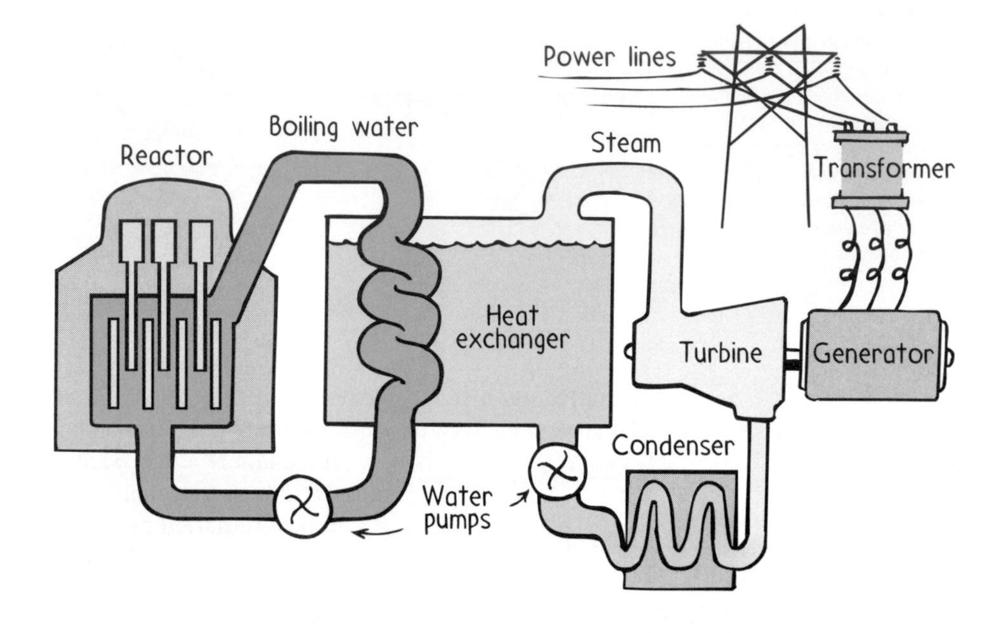

Fig. 14.21
Diagram of a nuclear fission power plant.

Nuclear Fission Reactors

Energy available from of nuclear fission was introduced to the world in the form of nuclear bombs. This violent image still impacts our thinking about nuclear power, making it difficult for many people to recognize its potential usefulness. Currently, about 20 percent of electric energy in the United States is generated by nuclear fission reactors.[6] These reactors are simply nuclear furnaces. Like fossil fuel furnaces, they do nothing more elegant than boil water to produce steam for a turbine (Figure 14.21). The greatest practical difference is the amount of fuel involved. One kilogram of uranium fuel, a chunk smaller than a baseball, yields more energy than 30 freight car loads of coal.

Fig. 14.22
A typical nuclear fission power plant.

A fission reactor contains three components: nuclear fuel, control rods, and liquid (usually water) to transfer the heat created by fission from the reactor to the turbine. The nuclear fuel is primarily U-238 plus about 3 percent U-235. Because the U-235 isotopes are so highly diluted with U-238, an explosion like that of a nuclear bomb is not possible. The reaction rate, which depends on the number of neutrons available to initiate fission of other U-235

[6]Although the United States leads the world in numbers of nuclear power plants, nearly 80 percent of electric power in France is nuclear.

nuclei, is controlled by rods inserted into the reactor. The control rods are made of a neutron-absorbing material, usually cadmium or boron. Water surrounding the nuclear fuel is kept under high pressure to keep it at a high temperature without boiling. Heated by fission, this water then transfers heat to a second, lower-pressure water system that operates a turbine and electric generator. Two separate water systems are used so that no radioactivity reaches the turbine.

One disadvantage of fission power is the generation of waste products that are radioactive. Light atomic nuclei are most stable when composed of equal numbers of protons and neutrons, and it is mainly heavy nuclei that need more neutrons than protons for stability. For example, there are 143 neutrons but only 92 protons in U-235. When uranium fissions into two medium-weight elements, the extra neutrons in their nuclei make them unstable. These fragments are therefore radioactive, and most of them have very short half-lives. Some of them, however, have half-lives of thousands of years. Safely disposing of these waste products as well as materials made radioactive in the production of nuclear fuels requires special storage casks and procedures. Although fission power goes back nearly a half century, the technology of radioactive waste disposal is still in the developmental stage.

The Breeder Reactor

One of the fascinating features of fission power is the *breeding* of fission fuel from nonfissionable U-238. Breeding occurs when small amounts of fissionable isotopes are mixed with U-238 in a reactor. Fission liberates neutrons that convert the relatively abundant nonfissionable U-238 to U-239, which beta-decays to Np-239, which in turn beta-decays to fissionable plutonium—Pu-239. So in addition to the abundant energy produced, fission fuel is bred from relatively abundant U-238 in the process.

Breeding occurs to some extent in all fission reactors, but a reactor specifically designed to breed more fissionable fuel than is put into it is called a **breeder reactor**. Using a breeder reactor is like filling your car's gas tank with water, adding some gasoline, then driving the car and having more gasoline after the trip than at the beginning! The basic principle of the breeder reactor is very attractive, for after a few years of operation a breeder-reactor power plant can produce vast amounts of power while at the same time breeding twice as much fuel as its original fuel.

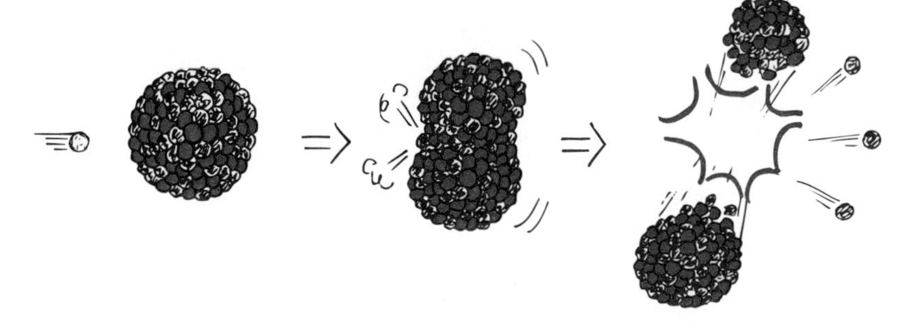

Fig. 14.23
Pu-239, like U-235, undergoes fission when it captures a neutron.

The downside of breeder reactors is the enormous complexity of successful and safe operation. The United States gave up on breeders more than a decade ago, and only France and Germany are still investing in them. Officials in these countries point out that supplies of naturally occurring U-235 are limited. At present rates of consumption, all natural sources of U-235 may be depleted within a century. If countries then

[7]The French, interestingly, do not bury their radioactive waste—they tend and monitor it in underground storage facitities. When a newer technology emerges that helps to alleviate the waste problem, they'll be ready.

Link to Nuclear Technology—Plutonium

Early in the nineteenth century, the farthest planet known in the solar system was Uranus. The first planet to be discovered beyond Uranus was named Neptune. In 1930, a planet beyond Neptune was discovered and was named Pluto. During this time the heaviest element known was uranium. Appropriately enough, the first heavier-than-uranium element to be discovered was named *neptunium,* and the second was named *plutonium.*

Neptunium is produced when a neutron is absorbed by a U-238 nucleus. Rather than undergoing fission, the resulting nucleus, U-239, emits a beta particle and becomes neptunium-239, the first synthetic element heavier than uranium. The half-life of neptunium-239 is only 2.3 days, and so this element isn't around very long. Neptunium-239 is also a beta emitter, and very soo n becomes plutonium-239. The half-life of plutonium-239 is about 24,000 years, and so it lasts a considerable time. The isotope plutonium-239, like U-235, undergoes fission when it captures a neutron. Whereas the separation of fissionable U-235 from uranium metal is a very difficult process (because U-235 and U-238 have the same chemistry), the separation of plutonium from uranium metal is relatively easy. This is because plutonium is an element distinct from uranium and so has its own chemical properties.

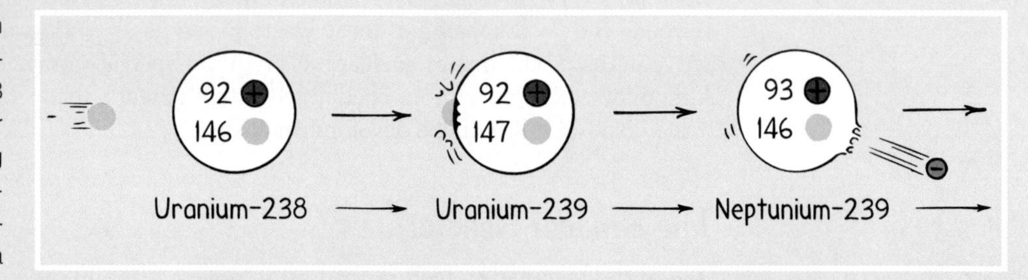

The element plutonium is chemically poisonous to humans and other animals in the same sense as are lead and arsenic. It attacks the nervous system and can cause paralysis. Death can follow if the dose is sufficiently large. Fortunately, plutonium does not remain in its elemental form for long because it rapidly combines with oxygen to form three compounds: PuO, PuO_2 and Pu_2O_3, all of which are relatively benign chemically. They will not dissolve in water or in biological systems. These plutonium compounds do *not* attack the nervous system and have been found to be biologically harmless.

Plutonium in any form, however, is radioactively toxic to humans and other animals. It is more toxic than uranium, although less toxic than radium. Plutonium emits high-energy alpha particles, which kill cells rather than simply disrupting them. Because it is damaged cells rather than dead ones that produce mutations and lead to cancer, plutonium ranks relatively low as a cancer-producing substance. The greatest danger that plutonium presents to humans is its potential for use in nuclear fission bombs. Its usefulness is in fission reactors—particularly breeder reactors.

decide to turn to breeder reactors, they may well find themselves digging up the radioactive wastes they once buried.[7]

The benefits of fission power are plentiful electricity, conservation of many billions of tons of fossil fuels that every year are literally turned to heat and smoke (which in the long run may be far more precious as sources of organic molecules than as sources of heat), and the elimination of the megatons of sulfur oxides and other poisons put into the air each year by the burning of fossil fuels.

14.10 Mass-Energy Relationship

Fig. 14.24
Work is required to pull a nucleon from an atomic nucleus. This work increases the energy and hence the mass of the nucleon once it is outside the nucleus.

In the early 1900s, Albert Einstein discovered that mass is actually "congealed" energy. He realized that mass and energy are two sides of the same coin, as stated in his celebrated equation $E = mc^2$. In this equation E stands for the energy that any mass at rest has, m stands for mass, and c is the speed of light. The quantity c^2 is the proportionality constant of energy and mass. This relationship between energy and mass is the key to understanding why and how energy is released in nuclear reactions.

Is the mass of a nucleon inside a nucleus the same as the mass of the same nucleon outside a nucleus? This question can be answered by considering the work required to separate nucleons from a nucleus. From physics we know that work, which is expended energy, is equal to the product of force and distance. Think of the enormous external force required to pull a nucleon out of a nucleus and through a distance sufficient to overcome the attractive nuclear force. Enormous force, and therefore enormous work, are required. The work done is manifest in the energy of the pulled-out nucleon: It has more energy outside the nucleus—an additional amount equal to the work required to separate it from the nucleus. This energy is manifest in a change in the nucleon's mass. Therefore the mass of a nucleon outside a nucleus is greater than the mass of the same nucleon locked inside a nucleus. For example, in the units used to measure the mass of atoms and atomic particles, a carbon-12 atom—the nucleus of which is made up of six protons and six neutrons—has a mass of exactly 12.00000 units. However, outside the nucleus a proton has a mass of 1.00728 units, and a neutron has a mass of 1.00867 units. Thus we see that the combined mass of six free protons and six free neutrons—$(6 \times 1.00728) + (6 \times 1.00867) = 12.09570$—is greater than the mass of one carbon-12 atom.

The masses of the isotopes of various elements can be very accurately measured with a mass spectrometer (Figure 14.25). This important device uses a magnetic field to deflect ions of these isotopes into circular arcs. The greater the inertia (mass) of an ion, the more it resists deflection and the greater the radius of its curved path. The magnetic force sweeps lighter ions into shorter arcs and heavier ions into larger arcs.

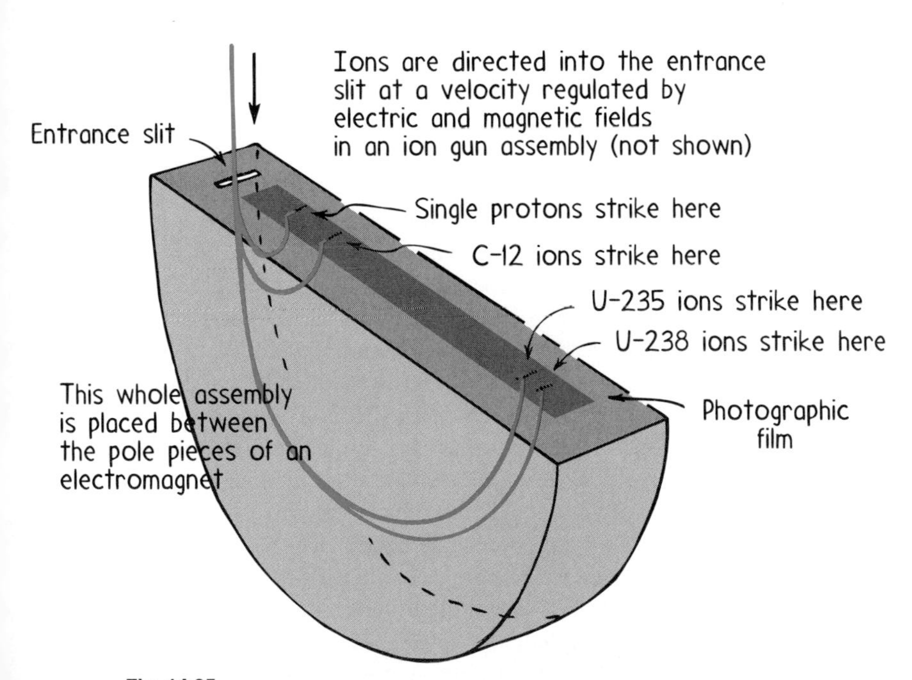

Fig. 14.25
How a mass spectrometer works. Ions directed into the semicircular "drum" are swept into semicircular paths by a strong magnetic field. Because of differences in inertia, heavier ions are swept into curves of large radii and lighter ions are swept into curves of smaller radii. The radius of any curve is directly proportional to the mass of the ion traveling along that curve.

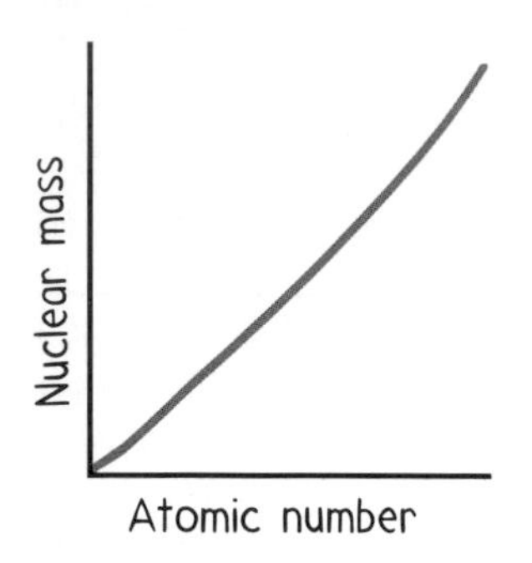

Fig. 14.26
Nuclear mass increases with increasing atomic number.

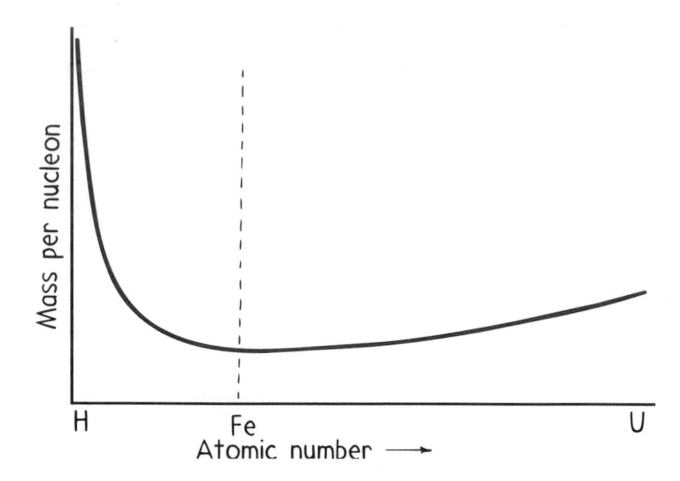

Fig. 14.27
The average mass of a nucleon depends on which nucleus it is in. A nucleon has the greatest mass when it is in the lightest nucleus—that of hydrogen. The nucleon has the least mass when in an iron nucleus. It has intermediate mass values when in nuclei heavier than that of iron.

A graph of nuclear mass as a function of atomic number is shown in Figure 14.26. The graph slopes upward with increasing atomic number, as expected, telling us that elements are more massive as atomic number increases. (The slope curves because there are proportionally more neutrons in the more massive atoms.) A more important graph results from the plot of average mass *per nucleon* for the elements hydrogen through uranium (Figure 14.27). This is perhaps the most important graph in this book, for it is the key to understanding the energy associated with nuclear processes like those that cause our Sun to burn. To obtain the average mass per nucleon, you divide the total mass of a nucleus by the number of nucleons in the nucleus. (If you divide the total mass of a roomful of people by the number of people in the room, you get the average mass per person.) The important fact we learn from Figure 14.27 is that the mass of a given nucleon changes, depending on which element the nucleon finds itself in.

The greatest mass per nucleon occurs for a proton when it is alone in a hydrogen nucleus because here it has no binding energy to pull its mass down. As we progress to elements beyond hydrogen, Figure 14.27 tells us that the mass per nucleon gets smaller and is least for a proton in an iron nucleus. Beyond iron, the trend reverses itself as protons (and neutrons) have progressively more and more mass in atoms of increasing atomic number. This continues all the way through the list of elements.

From Figure 14.27 we can see why energy is released when a uranium nucleus splits into two nuclei of lower atomic number. When the uranium nucleus splits, the masses of the two fission fragments lie about halfway between the masses of uranium and hydrogen on the horizontal scale of the graph (a little to the right of the value for iron, Fe). With a pencil, sketch a horizontal line on the graph that is parallel to the bottom black line and passes through the vertical black line and the rightmost tip of the red curve. The point where this line intersects the vertical black line is the mass/nucleon value of uranium. Now sketch a second horizontal line parallel to the one you just drew and passing through the red curve in the middle region, a little to the right of Fe. This second line represents the fission fragments of the uranium atom. Note that it intersects the vertical black line at a lower point than the first, indicating a lower mass-per-nucleon value for the fragments than for the parent uranium. In other words, the masses of nucleons in the fission fragments are less than the masses of the same nucleons when combined in the uranium nucleus. Protons and neutrons lose mass as they change from uranium nucleons to fragment nucleons. When this decrease in mass is multiplied by the speed of light squared (c^2 in Einstein's equation), the product is equal to the energy yielded by each uranium nucleus as it undergoes fission.

■ **Question**

Correct the following incorrect statement: When a heavy element such as uranium undergoes fission, there are fewer nucleons after the reaction than before.

14.11 Nuclear Fusion

As mentioned, a drawback to nuclear fission is the production of radioactive fission fragments. A more promising long-range source of nuclear energy is found in the lightest elements. In a nutshell, energy is produced as light nuclei *fuse* (which means they combine). This process is **nuclear fusion**—the opposite of nuclear fission. We see from Figure 14.27 that as we move along the list of elements from hydrogen to iron (the steepest part of the energy valley), the average mass per nucleon decreases. Thus if two small elements were to fuse, the mass of the fused nucleus would be less than the mass of the two single nuclei before fusion (14.28). Energy is gained as light nuclei fuse rather than split apart.

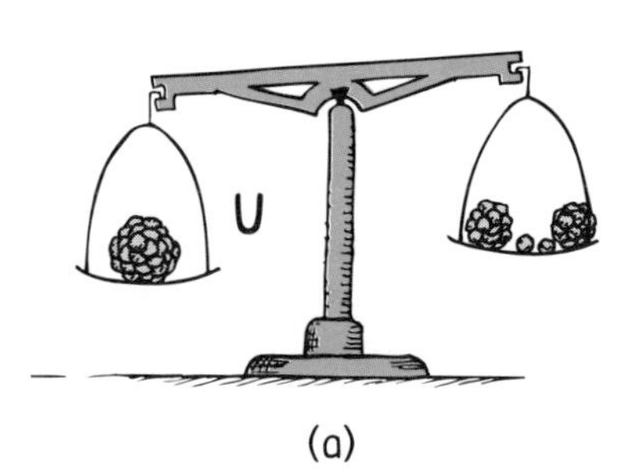

Fig. 14.28
The mass of a nucleus is not equal to the sum of the mass of its parts. (a) The fission fragments of a heavy nucleus like uranium are less massive than the uranium nucleus. (b) Two protons and two neutrons are more massive in their free states than when combined to form a helium nucleus.

Consider hydrogen fusion. For a fusion reaction to occur, the nuclei must collide at a very high speed to overcome their mutual electric repulsion. The required speeds correspond to the extremely high temperatures found in the Sun and other stars. Fusion brought about by high temperatures is called **thermonuclear fusion**. In the high temperatures of the Sun, approximately 657 million tons of hydrogen is fused to 653 million tons of helium each second. The "missing" 4 million tons of mass is discharged as radiant energy. Such reactions are, quite literally, nuclear burning.

Thermonuclear fusion is analogous to ordinary chemical combustion. In both chemical and nuclear burning, a high temperature starts the reaction; the release of energy by the reaction maintains a high-enough temperature to spread the fire. The net result of the chemical reaction is a combination of atoms into more tightly bound molecules. In nuclear reactions, the net result is more tightly bound nuclei. In both cases mass decreases as energy is given off.

■ **Answer**

When a heavy element such as uranium undergoes fission, there aren't fewer nucleons after the reaction. Instead, there's *less mass* in the same number of nucleons.

Questions

1. First we stated that nuclear energy is released when atoms split apart. Now we state that nuclear energy is released when atoms combine. Is this a contradiction? How can energy be released by opposite processes?
2. To get energy from the element iron, should iron nuclei be fissioned or fused?

Controlling Fusion

Although the fusion energy per reaction of individual hydrogen atoms is less than the energy given up by the fission of individual uranium atoms, gram for gram, fusion is several times more energy-producing than fission. This is because there are many more hydrogen atoms in a gram of hydrogen than there are heavier uranium atoms in a gram of uranium. Thus fusion holds more promise for our energy needs. Fusion reactions require temperatures of millions of degrees. There are a variety of techniques for attaining high temperatures. No matter how the temperature is produced, a technological problem is that all materials melt and vaporize at the temperatures required for fusion. The solution to this problem is to confine the reaction in a *nonmaterial container.*

Fig. 14.29
Fusion with multiple laser beams. Pellets of deuterium are rhythmically dropped into synchronized laser crossfire in this device. The resulting heat is carried off by molten lithium to produce steam.

One type of nonmaterial container is a magnetic field, which can exist at any temperature and can exert powerful forces on charged particles in motion. "Magnetic walls" provide a kind of magnetic "straight jacket" for hot gases called plasmas (Section 6.8). Magnetic compression further heats the plasma to fusion temperatures. At about a million degrees, some nuclei are moving fast enough to overcome electrical repulsion and slam together and fuse. The energy output, however, is small relative to the energy used to heat the plasma. Even at 100 million degrees, more energy must be put into the plasma than is given off by fusion. At about 350 million degrees, the fusion reactions produce enough energy to be self-sustaining. At this ignition temperature, all that is needed to produce continuous power is a steady feed of nuclei.

Although net energy production has been achieved in several fusion devices, instabilities in the plasma have thus far prevented a sustained reaction. A big problem has been devising a field system that can hold the plasma in a stable and sustained position while an ample number of nuclei fuse. A variety of magnetic confinement devices are the subject of much present-day research.

Another approach uses high-energy lasers. One proposed technique is to aim an array of laser beams at a common point and drop solid pellets of hydrogen isotopes through the synchronous cross fire (Figure 14.29). The energy of the multiple beams should crush the pellets to densities 20 times that of lead. Such a fusion could produce

---

Answers

1. Energy is released in any nuclear reaction in which the mass of the nucleons decreases. Light nuclei, such as hydrogen, have less mass after they fuse to form heavier nuclei. They therefore release energy in this reaction. Heavy nuclei, such as uranium, have less mass after they split to become lighter nuclei. For energy to be released, "Decrease Mass" is the name of the game—any game, chemical or nuclear.
2. Neither way will get you any energy because iron is at the very bottom of the curve (energy valley). If you fuse two iron nuclei, the product lies somewhere to the right of iron on the curve, which means the product has a higher mass per nucleon. If you split an iron nucleus, the products lie to the left of iron on the curve, which again means a higher mass per nucleon. Because no mass decrease occurs in either reaction, no energy is ever gained.

Fig. 14.30
Pellet chamber at Lawrence Livermore Laboratory. The laser source is Nova, the most powerful laser in the world, which directs 10 beams into the target region.

several hundred times more energy than is delivered by the laser beams that compress and ignite the pellets. Like the succession of fuel/air explosions in an automobile engine's cylinders that convert into a smooth flow of mechanical power, the successive ignition of pellets in a fusion power plant may similarly produce a steady stream of electric power.[8] The success of this technique requires precise timing, for the necessary compression must occur before a shock wave causes the pellet to disperse. High-power lasers that work reliably are yet to be developed. Break-even has not yet been achieved with laser fusion.

Other fusion schemes involve the bombardment of fuel pellets not by laser light but by beams of electrons and ions.

■ **Question**

Consider a star having a core composed of medium-mass elements and iron. Would you expect the stellar temperature to increase or decrease with the fusion of iron?

■ **Answer**

In the fusion of iron and any heavier nuclei, energy is absorbed. Thus the star tends to cool. (Interestingly, as we shall see in Chapter 29, elements beyond iron are not manufactured in normal fusion cycles in stellar sources but are manufactured when stars explode.)

---

[8] The rate of pellet fusion is 5 per second on the projected Cascade power plant, now on the drawing board at Lawrence Livermore Laboratory. (For comparison, about 20 explosions per second occur in each automobile engine cylinder in a car traveling at highway speed.) Such a plant could produce 1000 million watts of electric power, enough to supply a city of 600,000 people. Five fusion burns per second provides about the same power as 60 l of fuel oil or 70 kg of coal per second in conventional power plants.

Link to Recycling Technology—Fusion Torch

A fascinating application of the abundant energy of fusion is the *fusion torch,* a star-hot flame or high-temperature plasma into which all waste materials—whether liquid sewage or solid industrial refuse—could be dumped. At such high temperatures, the materials would be reduced to their constituent atoms and separated by a mass-spectrometer-type device into various "bins" ranging from hydrogen to uranium. In this way a single fusion plant could, in principle, not only dispose of thousands of tons of solid wastes per day but also provide a continuous supply of fresh raw material—thereby closing the cycle from use to reuse.

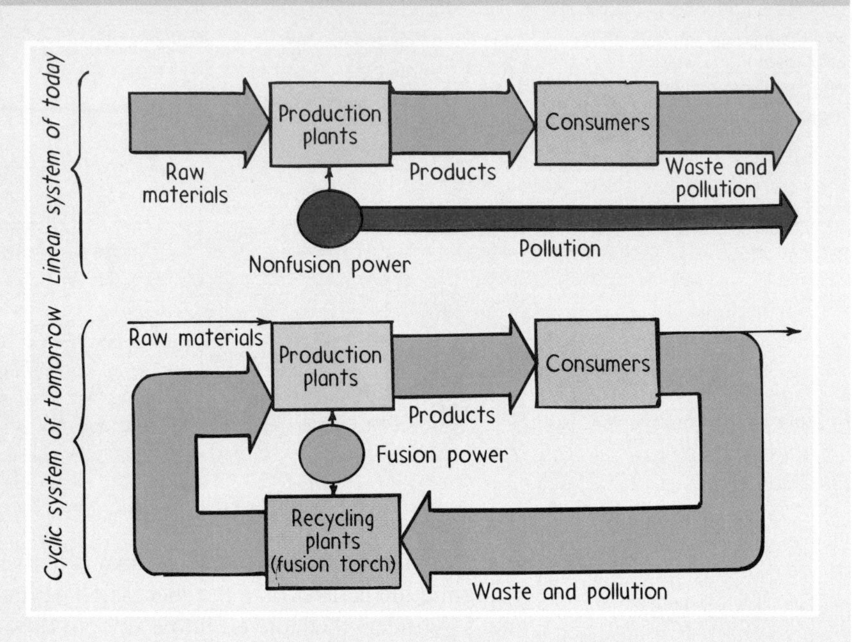

This would be a major turning point in materials economy. The following diagram (courtesy of *Scientific American* magazine) contrasts the present linear system of today with a proposed closed cycle system.

Our present concern for recycling materials will reach a grand fruition with this or a comparable achievement, for it would be recycling with a capital *R*! Rather than gut our planet further for raw materials, we'd be able to recycle our existing stock over and over again, adding new material only to replace the relatively small amounts that are lost. Fusion power can produce abundant electrical power, desalinate water, help cleanse our world of pollution and wastes, recycle our materials, and in so doing provide the setting for a better world—not in the far-off future but perhaps in the first half of the 21st century. If and when fusion power plants become a reality, they are likely to have an even more profound impact upon almost every aspect of human society than did the harnessing of electromagnetic energy in the previous century.

When we think of our continuing evolution, we can see that the universe is well suited to those who will live in the future. If people are one day to dart about the universe in the same way we jet about the world today, their supply of fuel is assured. The fuel for fusion—hydrogen—is found in every part of the universe, not only in the stars but also in the space between them. About 91 percent of the atoms in the universe are estimated to be hydrogen. For people of the future, the supply of raw materials is also assured because all the elements known to exist result from the fusing of more and more hydrogen nuclei. Simply put, if you fuse 8 deuterium nuclei, you have oxygen; 26, you have iron; and so forth. Future humans might synthesize their own elements and produce energy in the process, just as the stars have always done.

Summary of Terms

- **Alpha particle** The nucleus of a helium atom, which consists of two neutrons and two protons, ejected by certain radioactive elements.
- **Beta particle** An electron (or positron) emitted during the radioactive decay of certain nuclei.
- **Gamma ray** High-frequency electromagnetic radiation emitted by the nuclei of radioactive atoms.
- **Nucleon** A nuclear proton or neutron; the collective name for either or both.
- **Quarks** The elementary constituent particles of building blocks of nuclear matter.
- **Isotopes** Atoms whose nuclei have the same number of protons but different numbers of neutrons.
- **Half-life** The time required for half the atoms in a sample of a radioactive isotope to decay.
- **Transmutation** The conversion of an atomic nucleus of one element into an atomic nucleus of another element through a loss or gain in the number of protons.
- **Nuclear fission** The splitting of the nucleus of a heavy atom, such as uranium-235, into two main parts, accompanied by the release of much energy.
- **Chain reaction** A self-sustaining reaction in which the products of one reaction event stimulate further reaction events.
- **Critical mass** The minimum mass of fissionable material in a reactor or nuclear bomb that will sustain a chain reaction.
- **Breeder reactor** A fission reactor that is designed to breed more fissionable fuel than is put into it by converting nonfissionable isotopes to fissionable isotopes.
- **Nuclear fusion** The combination of the nuclei of light atoms to form heavier nuclei, with the release of much energy.
- **Thermonuclear fusion** Nuclear fusion produced by high temperature.

Review Questions

Alpha, Beta, and Gamma Rays

1. Compare the electric charges carried by alpha, beta, and gamma rays.
2. Why are alpha and beta rays deflected in opposite directions in a magnetic field? Why are gamma rays undeflected?
3. Which of the three rays has the greatest penetrating power?

Effects of Radiation on Humans

4. Is radioactivity in the world something relatively new? Defend your answer.
5. Where does most of the radiation you encounter originate?
6. What is a radioactive tracer?

The Nucleus

7. Name the two different nucleons.
8. How does the mass of a nucleon compare with the mass of an electron?
9. In what way is the emission of gamma rays from a nucleus similar to the emission of light from an atom?
10. Why does an alpha particle leave at high speed once it gets outside an atomic nucleus?
11. What are quarks?

Isotopes

12. Distinguish between an isotope and an ion.
13. Distinguish between atomic number and atomic mass number.
14. Distinguish between deuterium and tritium.

Half-Life

15. What is meant by radioactive half-life?
16. What is the half-life of radium-226?
17. For any radioactive isotope, what is the relationship between decay rate and half-life?
18. What is the long-range fate of all the uranium that exists in the world today?

Natural Transmutation of Elements

19. What is transmutation?
20. When thorium, atomic number 90, decays by emitting an alpha particle, what is the atomic number of the resulting nucleus?
21. When thorium decays by emitting a beta particle, what is the atomic number of the resulting nucleus?
22. What change in atomic mass is there for each of the preceding two reactions?
23. What change in atomic number occurs when a nucleus emits an alpha particle? A beta particle? A gamma ray?

Artificial Transmutation of Elements

24. When did the first successful intentional transmutation of an element occur?
25. Why are the elements beyond uranium not common in the Earth's crust?

Isotopic Dating

26. What do cosmic rays have to do with transmutation?
27. How is carbon-14 produced in the atmosphere?
28. Which is radioactive, carbon-12 or carbon-14?
29. Why is there more carbon-14 in new bones than in old bones of the same mass?
30. Why would carbon-14 dating be useless for dating old coins but not old pieces of cloth?
31. Why is lead found in all deposits of uranium ores?

32. What does the proportion of lead and uranium in a rock tell us about the age of the rock?

Nuclear Fission

33. When a nucleus undergoes fission, what role can the ejected neutrons play?
34. Why does a chain reaction not occur in uranium mines?
35. Is a chain reaction more likely in two separate pieces of uranium or in the same pieces stuck together?
36. How is a nuclear reactor similar to a conventional fossil fuel plant? How is it different?
37. What is the function of control rods in a nuclear reactor?
38. What is the effect of mixing small amounts of Pu-239 with large amounts of U-238?
39. How does a breeder reactor breed nuclear fuel?

Mass-Energy Relationship

40. Is work required to pull a nucleon out of an atomic nucleus? Does the nucleon, once outside, then have more mass than it did when it was inside the nucleus?
41. Is the mass per nucleon of a nucleus greater than, less than, or the same as the mass of a nucleon outside a nucleus?
42. How does the mass per nucleon in uranium compare with the mass per nucleon in the fission fragments of uranium?
43. If an iron nucleus split in two, would its fission fragments have more mass per nucleon or less mass per nucleon?
44. If a pair of iron nuclei were fused, would the product nucleus have more mass per nucleon than an iron nucleus or less?

Nuclear Fusion

45. When a pair of hydrogen isotopes are fused, is the mass of the product nucleus more or less than the sum of the masses of the two hydrogen nuclei?
46. What is thermonuclear fusion?
47. How is nuclear burning similar to chemical burning?
48. How do the products of fusion reactions differ from the products of fission reactions?
49. What kind of containers are used to contain multimillion-degree plasmas?
50. How is fusion accomplished with lasers?

Home Project

Some watches and clocks have luminous hands that glow continuously. In some of these, what causes the glow is traces of radioactive radium bromide mixed with zinc sulfide. (Safer clock faces use light rather than radioactive decay as a means of excitation and as a result become progressively dimmer in the dark.) If you have a glow-all-the-time type available, take it into a completely dark room and, after your eyes adjust to the dark, examine the hands with a very strong magnifying glass or the eyepiece of a microscope or telescope. You should be able to see individual tiny flashes, which together seem to be a steady source of light to the unaided eye. Each flash occurs when an alpha particle ejected by a radium nucleus strikes a molecule of zinc sulfide.

Exercises

1. What is the evidence that radioactive decay is the changing of one element to another?
2. Why is a sample of radium always a little warmer than its surroundings?
3. Some people say that all things are possible. Is it possible for a hydrogen nucleus to emit an alpha particle? Defend your answer.
4. An alpha particle has twice the electric charge of a beta particle but deflects less than a beta in a magnetic field. Why?
5. How would the paths of alpha, beta, and gamma radiations compare in an electric field?
6. Which type of radiation—alpha, beta, or gamma—results in the greatest change in mass number? In atomic number?
7. Which type of radiation—alpha, beta, or gamma—results in the least change in mass number? In atomic number?
8. When atomic nuclei are bombarded with proton "bullets," why must the protons be accelerated to high energies in order to make contact with the target nuclei?
9. Why would you expect alpha particles to be less able to penetrate materials than beta particles of the same energy?
10. Within an atomic nucleus, which interaction tends to hold it together and which interaction tends to push it apart?
11. What evidence supports the contention that the attractive nuclear interaction is stronger than the repulsive electrical interaction at short internuclear distances?
12. If a sample of radioactive isotope has a half-life of 1 year, how much of the original sample will be left at the end of the second year? The third year? The fourth year?
13. The isotope cesium-137, which has a half-life of 30 years, is a product of nuclear power plants. How long will it take for this isotope to decay to about one-sixteenth its original amount?
14. A sample of a radioactive isotope is placed near a Geiger counter that then registers 160 counts per minute. Eight hours later, the counter registers 10 counts per minute. What is the half-life of the isotope?
15. Radiation from a point source obeys the inverse-square law. If a Geiger counter 1 m from a small sample reads 360 counts per minute, what will it read 2 m from the source? 3 m from the source?
16. When the isotope bismuth-213 emits an alpha particle, which element results? Which element results if the bismuth-213 emits a beta particle?
17. When $^{226}_{88}Ra$ decays by emitting an alpha particle, what are the atomic number and atomic mass number of the resulting nucleus?
18. When $^{218}_{84}Po$ emits a beta particle, what are the atomic number and atomic mass number of the resulting nucleus?

What are they if the polonium-218 emits an alpha particle instead?

19. State the number of neutrons and protons in $^{2}_{1}H$, $^{12}_{6}C$, $^{56}_{26}Fe$, $^{197}_{79}Au$, $^{90}_{38}Sr$, and $^{238}_{92}U$.

20. How is it possible for an element to decay to an element of higher atomic number?

21. Elements heavier than uranium do not exist in any appreciable amounts in nature because they have short half-lives. Yet there are several elements lighter than uranium that have equally short half-lives but do exist in appreciable amounts in nature. How can you account for this?

22. You and a friend journey to the mountains to get closer to nature and escape such things as radioactivity. While bathing in the warmth of a natural hot spring, she wonders aloud how the spring gets its heat. What do you tell her?

23. Coal contains very tiny quantities of radioactive materials, and yet there is more environmental radiation surrounding a coal-fired power plant than a fission power plant. What does this indicate about the shielding that typically surrounds the two types of power plants?

24. When we speak of dangerous radiation exposure, are we generally speaking of alpha, beta, or gamma radiation? Discuss.

25. People who work around radioactivity wear film badges to monitor the amount of radiation that reaches their bodies. These badges consist of a small piece of photographic film enclosed in a light-proof wrapper. What kind of radiation do these devices monitor, and how can they determine the amount of radiation a body receives?

26. A friend produces a Geiger counter to check the local background radiation. It ticks. Another friend, who normally fears most that which is understood least, makes an effort to keep away from the region of the Geiger counter and looks to you for advice. What do you say?

27. Why is carbon dating currently not accurate for estimating the ages of materials older than 50,000 years?

28. The age of the Dead Sea Scrolls, which are written on parchment, was found by carbon dating. Could this technique have worked if they were carved in stone tablets? Explain.

29. Why will nuclear fission probably not be used directly for powering automobiles? How could it be used indirectly?

30. Why does a neutron make a better nuclear bullet for penetrating a nucleus than a proton or an electron?

31. Does the average distance a neutron travels through fissionable material before escaping increase or decrease when two pieces of fissionable material are assembled into one piece? Does this assembly increase or decrease the probability of an explosion?

32. U-235 releases an average of 2.5 neutrons per fission, whereas Pu-239 releases an average of 2.7 neutrons per fission. Which of these elements might you therefore expect to have the smaller critical mass?

33. Why does plutonium not occur in appreciable amounts in natural ore deposits?

34. As a supply of uranium fuel reaches the end of its usefulness, (typically 3 years), why is most of its energy coming from the fission of plutonium?

35. If a nucleus of $^{232}_{90}Th$ absorbs a neutron and the resulting nucleus undergoes two successive beta decays (emitting electrons), what is the final product nucleus?

36. Energy is released in nuclear fission because fissionable nuclei have about 0.1 percent more mass per nucleon than do the nuclei resulting from the fission. What would be the effect on the amount of energy released if the 0.1 percent figure were instead 1 percent?

37. Explain how a physicist uses the equation $E = mc^2$ combined with either the curve of Figure 14.27 or a table of nuclear masses to predict the approximate energy release of either a fission or a fusion reaction.

38. Does the mass of a nucleon after it has been pulled from an atomic nucleus depend on which nucleus it was extracted from?

39. Which process would release energy from gold, fission or fusion? From carbon? From iron?

40. If uranium were to split into three segments of equal size instead of two, would more energy or less energy be released? Defend your answer in terms of Figure 14.27.

41. Explain how radioactive decay has always warmed the Earth from the inside and nuclear fusion has always warmed the Earth from the outside.

42. What effect on the mining industry can you foresee if future generations work out a way to dispose of urban waste by a process that couples a fusion torch with a mass spectrometer?

43. Everyday life was enormously changed by the discovery of electromagnetic induction and its applications to electric motors and generators. Speculate on changes that are likely to follow the advent of successful fusion reactors.

44. Compare pollution from conventional fossil-fuel power plants and nuclear-fission power plants. Consider thermal pollution, chemical pollution, and radioactive pollution.

Problems

1. A certain radioactive substance has a half-life of 1 hour. If you start with 1 g of the material at noon, how much is left at 3:00, at 6:00, and at 10:00 P.M.?
2. Suppose you measure the intensity of radiation from carbon-14 in an ancient piece of wood to be 6% of what it would be in a freshly cut piece of wood. How old is this artifact?
3. The kiloton, a unit used to measure the energy released in an atomic explosion, is equal to 4.2×10^{12} J (approximately the energy released in the explosion of 1000 tons of TNT). Recalling that 1 kilocalorie of energy raises the temperature of 1 kg of water by 1 C° and that 4184 J is equal to 1 kilocalorie, how many kilograms of water can be heated 50 C° by a 20-kiloton atomic bomb?

Part V—Sample Exam Questions

Choose the BEST answer to each of the following.

1. What makes one element distinct from another is the number of
(a) protons in its nucleus (b) neutrons in its nucleus (c) electrons in its nucleus (d) total particles in its nucleus

2. In the atomic nucleus of a certain element are 26 protons and 28 neutrons. The atomic number of the element is
(a) 26 (b) 27 (c) 28 (d) 54 (e) none of these

3. The origin of atoms that comprise the body of a newborn baby is
(a) tissues from the mother's body (b) the father's sperm (c) ancient stars

4. The weight of matter comes mostly from its
(a) electrons (b) protons

5. The volume of matter comes mostly from its
(a) electrons (b) protons

6. Deflection of alpha particles in Rutherford's gold foil experiments indicate that atoms are
(a) positively charged (b) in constant motion (c) composed of a dense nucleus (d) smaller than the wavelength of visible light (e) none of these

7. The atomic model explaining why electrons orbit at discrete distances from the nucleus involves
(a) electron wavelengths (b) electrons acting like planets orbiting the sun (c) quantized energy levels (d) springs connecting electrons to the nucleus

8. Astronomers are able to identify the elements that make up a star by studying its
(a) Doppler effect (b) structure (c) temperature (d) spectra (e) standing waves

9. Two beams of light, a red beam and a blue beam, have the same energy. The beam with the greater number of photons is the
(a) red beam (b) blue beam (c) same in each (d) cannot say

10. A photosensitive surface is illuminated with both blue and violet light. The light that will cause the most electrons to be ejected is
(a) blue light (b) violet light (c) both eject the same number (d) not enough information is given

11. Which experiences the greatest electrical force in an electrical field?
(a) alpha particle (b) beta particle (c) gamma ray (d) none of these

12. The discovery of radioactivity was a boost to Earth scientists who were then able to know more about
(a) ages of various rocks (b) why the Earth's interior is hot (c) both (d) neither

13. The radioactive half life of a certain isotope is 1 day. At the end of 4 days the amount remaining is
(a) 1/2 (b) 1/4 (c) 1/8 (d) 1/16 (e) none of these

14. There is a greater proportion of C-14/C-12 in
(a) old bones (b) new bones (c) same in each (d) not enough information to say

15. Most of the radioactivity we personally encounter comes from
(a) fallout from past and present testing of nuclear weapons (b) nuclear power plants (c) medical X rays (d) the natural environment

16. Electrical forces within the atomic nucleus tend to
(a) hold the nucleus together (b) push the nucleus apart (c) neither of these

17. In both nuclear fission and fusion, most energy release is initially in the form of
(a) gamma radiation (b) the kinetic energy of recoiling particles (c) potential energy of newly formed nuclei (d) high temperatures (e) each of these, about equally

18. If uranium fissioned into three equal segments instead of two, the reaction would be
(a) as energetic (b) less energetic (c) more energetic (d) not enough information is given for a reasonable estimate

19. Fusing a pair of iron nuclei yields a net
(a) absorption of energy (b) release of energy (c) neither (d) actually both

20. Compared with the energy produced by fissioning a gram of uranium, the energy produced by fusing a gram of deuterium is
(a) less (b) more (c) about the same

Answers: 1a, 2a, 3c, 4b, 5a, 6c, 7a, 8d, 9b, 10d, 11a, 12c, 13d, 14b, 15d, 16b, 17b, 18c, 19a, 20(b)

PART

VI

Chemistry

Atoms of the same kind can be combined in different ways to produce totally different materials. For example, carbon atoms arranged in flat planes like playing cards that slide over one another make up graphite, used as a lubricant and as the "lead" in pencils. Arranged in a three-dimensional structure, carbon atoms form diamonds, the hardest natural substance known. Carbon atoms may also bond in this soccer-ball shape -- a "buckyball," which when crystallized can conduct electricity with zero resistance. All these and more from only carbon! Onward to chemistry!

CHAPTER

15

Elements of Chemistry

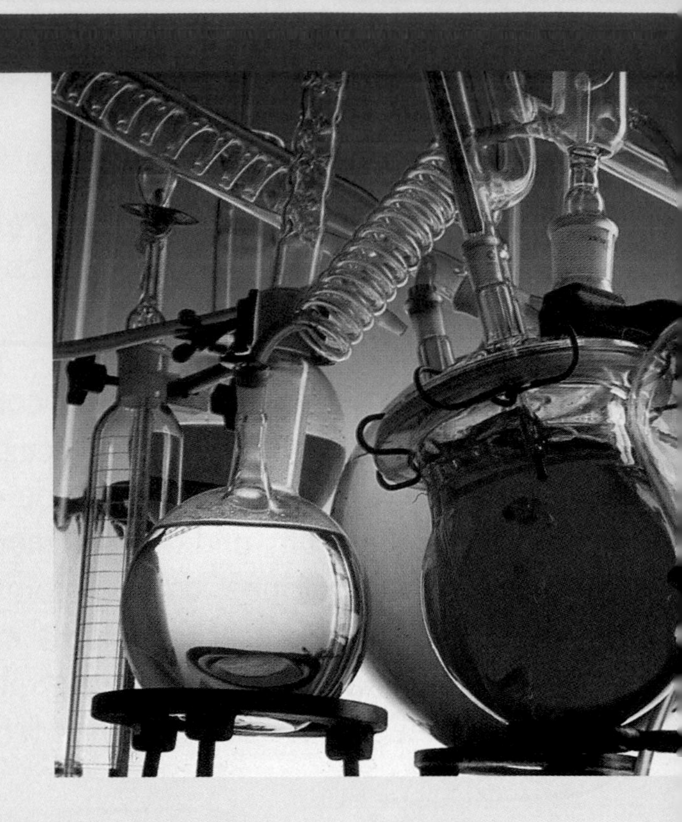

In the previous two chapters we studied the atom and its nucleus. Learning about these topics has prepared us for our study of chemistry, the part of physical science we turn to next. Chemistry is often called the central science, forming a bridge between the more fundamental science of physics and the applied sciences such as geology, meteorology, and biology. All chemical processes obey the laws of physics, and it is these law-abiding chemical processes that govern the nature of all materials that make up our universe, including rocks, weather, and living organisms.

The primary goal of this and the following chemistry chapters is to reveal the submicroscopic workings of nature and to show how knowledge of these workings translates into a deeper understanding of our macroscopic surroundings. Underlying every macroscopic or microscopic phenomenon is a submicroscopic explanation. Chemistry is the science of discovering these explanations and connecting them to our everyday world.

15.1 About Chemistry

Chemistry is the practice of thinking carefully about matter, where *matter* includes anything you can touch, taste, smell, see, or hear. When you wonder about what the Earth, sky, and oceans are made of, you're thinking chemistry. When you wonder how a puddle of water disappears in the sunlight, how a car gets its energy from gasoline, or how your body gets its energy from the food you eat, you're again within the realm of chemistry. By definition, **chemistry** is the study of matter and the transformations it can undergo. The scope of chemistry is very broad, and its concepts underlie much of our understanding of the natural universe.

The science of chemistry is central to our daily lives. Through chemistry we have learned how to manipulate matter into a wide variety of useful materials, such as synthetic fibers, fuels, medicines, fertilizers, herbicides and pesticides—all of which have dramatically altered our standards of living. Chances are, most all of the material items within your household are shaped by some human-devised chemical process.

Chemistry is a physical science, one that focuses on the study of nonliving things. More and more, however, advances in chemistry are giving rise to clearer understanding in the life sciences—biology, zoology, botany, and medicine. Indeed, many of the great advances in the life sciences today, such as genetic engineering, are applications of some very exotic chemistry.

Chemistry is central to the economy. The textile, agricultural, telecommunications, travel, timber, automotive, electronics, and metals industries, as well as many others, all depend on chemical goods. The chemical industry is the fifth largest in the United States, employing well over a million people. And the chemical industry has a remarkable safety record boasting one of the lowest rates of occupational illness and injury involving death or days away from work.

Through chemistry, we gain a deeper understanding of our macroscopic surroundings. A leaf is green because of the energy-absorbing pigment molecules it contains. Clouds are white because of the way water molecules clump together and reflect light. Ice floats on water because of how water molecules orient themselves in the solid phase. Fish don't really breathe water, they breathe the same oxygen we do dissolved in the world's oceans, lakes, and streams. Fluoride ions are added to toothpaste, making our teeth stronger and more resistant to decay. What happens to a city's wastewater? To understand these and the many other properties of matter and to deal with many of today's technological issues, we turn to the realm of chemistry.

15.2 Phases of Matter—A Molecular View

One of the most apparent ways we can describe matter is by its phase (sometimes called "state"), which may be *solid, liquid,* or *gas.*[1] A **solid** material, such as a rock, has a definite volume and shape and is not readily deformed. A **liquid**, on the other hand, has definite volume but indefinite shape. A liter of milk, for example, may take the shape of its carton or the shape of a bowl, but its volume is the same in both cases. A **gas** is diffuse, having neither definite volume nor definite shape. Any sample of gas assumes

[1]As discussed in Chapter 6 there is a fourth phase of matter—the plasma phase. In our discussion of chemistry, however, we'll focus on the more familiar solid, liquid, and gas phases.

The Impact of Materials

Throughout history and across cultures, standards of living have been closely tied to the materials available for use. Stone Age people learned how to design tools from stone, which made hunting, food preparation, and consequently life in general easier. The advent of new materials in the Bronze Age and Iron Age further raised standards of living. Later, the introduction of brick, paper, glass, and gunpowder also enhanced human capability. In the present time, the list of new materials is being extended daily.

Our present time is not an age of any one type of material. It is the rapidly growing diversity of materials that makes our time unique. Fabrication of new materials is possible because we have learned how to manipulate atoms. Evidence of the atomic age is all around us. Steel and other alloys are used to make buildings, cars, and appliances. We walk on carpets of synthetic fibers. We cook on nonstick Teflon surfaces. Food is kept sanitized by plastic wrapping. Semiconductor materials enable computers. Superconductors levitate trains. Modern ceramics boost efficiency in automobile engines. Our ability to manipulate atoms for the production of nuclear energy is also a hallmark of our atomic age.

The chemist's command of materials goes far beyond providing more "material things." The effects on life itself are direct. Nitrogen combined with hydrogen makes ammonia for fertilizers. Formulations of herbicides, pesticides, and fungicides aid in the large-scale production of foodstuffs. Formulations of other materials produce substances we use to treat ills ranging from headaches to cancer. Expanded knowledge of DNA may lead to a cure for such genetic diseases as muscular dystrophy, cystic fibrosis, and sickle-cell anemia.

The science of materials is an important one.

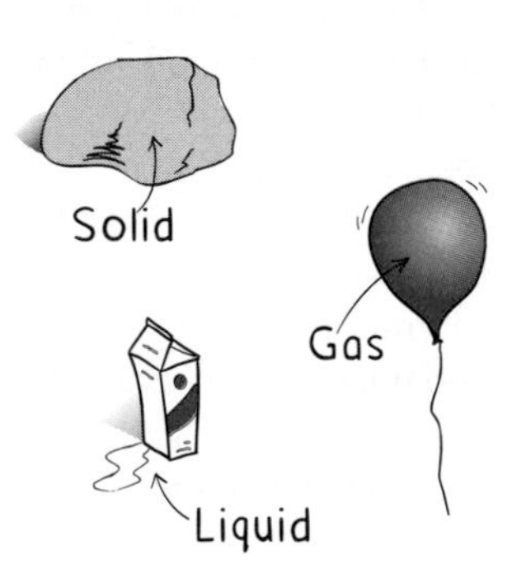

Fig. 15.1
The familiar bulk properties of a solid, liquid, and gas.

both the shape and the volume of the container it occupies. Compressed air, for example, may assume the volume and shape of a toy balloon or the volume and shape of an automobile tire. Released from its container, a gas diffuses into the atmosphere, which is a collection of various gases held to our planet only by the force of gravity.

On the submicroscopic level, the solid, liquid, and gas phases are readily distinguished. In solid matter the attractions among the submicroscopic particles (either atoms or molecules) are strong enough to hold them together in some fixed three-dimensional arrangement. The particles are able to vibrate about their fixed positions, but they cannot move past one another (Figure 15.2). Adding heat causes these vibra-

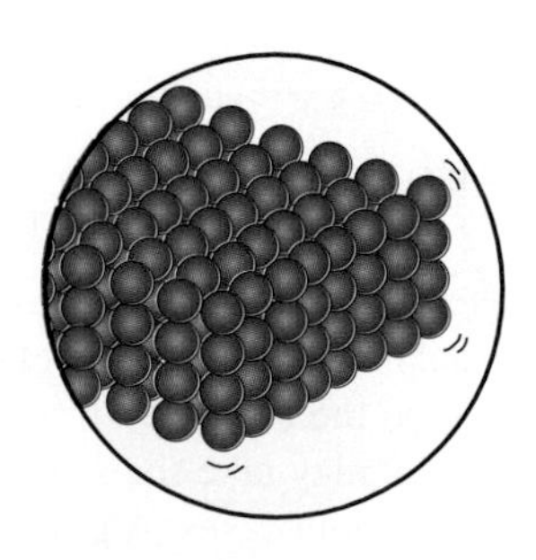

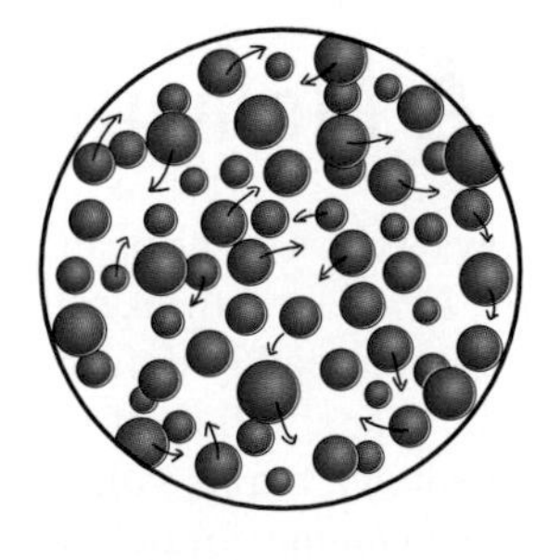

Fig. 15.2
(a) In the solid phase, the submicroscopic units vibrate about fixed positions but cannot move past one another. (b) In the liquid phase, the submicroscopic units slip past one another.

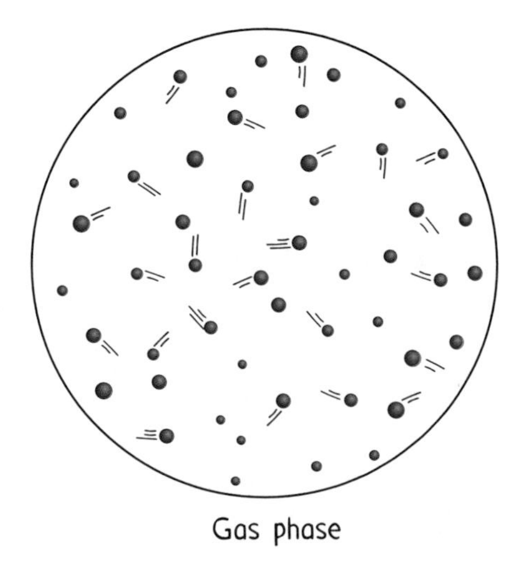

Fig. 15.3
The fast-moving submicroscopic units of the gaseous phase are separated by large average distances.

Fig. 15.4
The gaseous phase of a material occupies significantly more volume than either its solid or liquid phase. (a) Solid carbon dioxide, CO_2, is broken up with a hammer into powder form and (b) funneled into a balloon, which (c) expands as the CO_2 transforms from solid to gas.

tions to increase until, at a certain temperature, they are rapid enough to disrupt the fixed arrangement. The particles can then slip past one another and tumble around much like a bunch of marbles held within a plastic bag. This now is the liquid phase of matter, and it is the mobility of submicroscopic particles that gives rise to the liquid's fluid character—it flows and takes on the shape of its container.

A liquid can be heated so that it transforms to the gaseous phases—a phase in which the submicroscopic particles are widely separated (more than ten particle diameters is typical). Because particles of a gas are separated by large average distances, matter in the gaseous phase occupies much more volume than it does in the solid or liquid phase (Figures 15.3 and 15.4). This explains how gases are easily compressed. Enough air for an underwater diver to breathe for many minutes, for example, is readily squeezed into a tank small enough to be fitted onto the diver's back.

Although the particles of a gas move at high speeds (on the order of 1500 kilometers per hour), they do not drift very far because the particles are continually hitting one another. You can find evidence for the low drift speed of gas particles in your home when someone opens an oven door after baking: A shot of aromatic gas molecules escapes from the oven, but, sitting in the next room, you don't smell the aroma until the particles drift from the oven to your nose (Figure 15.5).

Fig. 15.5
In getting from point A to point B, the typical gas molecule travels a very circuitous route because of numerous collisions with other gas molecules. These collisions are indicated here by all the changes in direction in the line pointing from A to B. Although the gas molecule travels at very high speeds, it takes a relatively long time to cross between two distant points

15.3 Physical and Chemical Properties

Physical properties describe the physical attributes of a substance, such as color, hardness, density, texture, and phase. Each substance has its own set of characteristic physical properties. Gold, for instance, is a yellowish, soft, dense solid at room temperature, and diamond is a transparent, extremely hard, not very dense solid at room temperature. Two other physical properties are the ability to conduct electricity and heat. Gold is a good conductor of both electricity and heat. Diamond, on the other hand, is a poor conductor of electricity but an excellent conductor of heat (Figure 15.6).

(a)

(b)

Fig. 15.6
(a) Electrical contacts plated with gold conduct electricity while resisting corrosion.
(b) Drill bits and saws coated with a diamond film stay relatively cool because of diamond's ability to conduct heat away from the cutting edge.

A **physical change** is a change in some physical property of a substance. Most physical changes occur either during heating/cooling or when there is a change in pressure. During a physical change, one or more physical attributes transform, but the chemical identity of the substance remains the same. The melting of ice to liquid water is an example of a physical change, as is the boiling of liquid water to steam. In all three phases—solid, liquid, and gas—water maintains its chemical identity—it is still water with two hydrogen atoms and one oxygen atom per molecule (Figure 15.7). The only difference is in how the molecules are packed. In solids the molecules are generally packed tightly together, while in liquids they are packed more loosely. In the gaseous phase, molecules are not packed at all as they bounce around in random directions.

Chemical properties characterize the tendency of a substance to transform into a different substance. Iron, for example, changes to rust when exposed to water and oxygen. The tendency of iron to rust, therefore, is one of iron's chemical properties. Similarly, it is a chemical property of grape juice to ferment into wine and of wood to transform into carbon dioxide and water vapor upon burning. Gold has the interesting chemical property of *resisting* a change in identity, as do other relatively inert materials, such as platinum and helium.

The process of forming a new substance is called a **chemical change**. Atoms rearrange during a chemical change, switching partners as previous connections are broken and new ones are formed. It is the new arrangement of atoms that results in a material completely different from the starting material. Metallic iron, for example, consists of many iron atoms all connected to one another in one large mass. As iron rusts, iron atoms break away from one another and combine with oxygen atoms. As grape juice ferments, the atoms that make up sugars in the juice rearrange to form the intoxicating substance ethyl alcohol. As wood burns, the atoms that make up wood rearrange to form carbon dioxide and water vapor. During each of these chemical processes new substances are formed by a rearrangement of atoms (Figure 15.8).

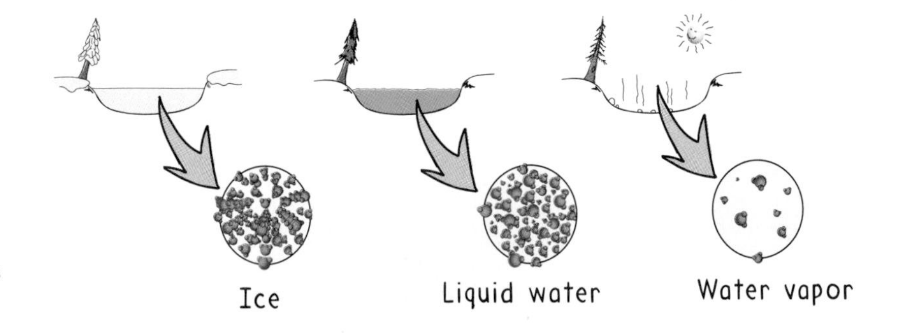

Fig. 15.7
The identity of a substance is not lost in a physical change. Ice, liquid water, and water vapor are merely different forms of the same thing— H_2O.

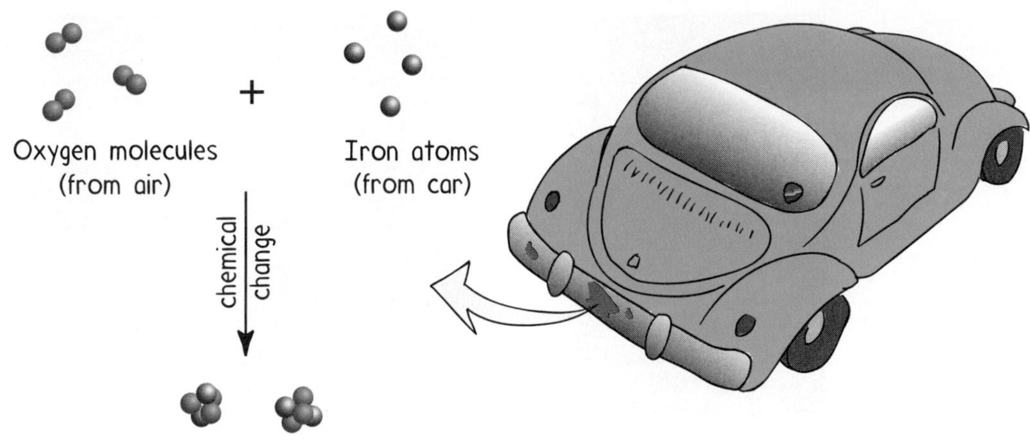

Fig. 15.8
The rusting of a car involves a chemical change in which iron atoms combine with oxygen atoms to form the material we see as rust.

In the language of chemistry, materials undergoing a chemical change are said to be "reacting." Iron, for example, *reacts* with water and oxygen to form rust, and grape juice *reacts* to form alcohol. Synonymous to chemical change, therefore, is the phrase *chemical reaction.* During a **chemical reaction**, a new material is formed by the rearrangement of atoms.

Discerning between a physical change and a chemical change can be tricky. In both cases, there is a change in physical attributes. In a physical change, the new attributes are the result of a new set of conditions imposed upon the material. For example, change water's temperature from 20°C to −5°C, and the water undergoes a physical change from liquid to solid. In a chemical change, by contrast, the new attributes are the physical properties of a fundamentally different material. Rust, for example, is a material that is neither the iron nor the oxygen from which it is produced.

The distinction between a physical and chemical change is straightforward—only in a chemical change are new chemicals formed. If you observe some change, however, identifying it as either physical or chemical can be tricky because in both cases there are changes in physical appearance. In general, if the change involves the release of large amounts of energy, then chances are good that the change is chemical. Fuels burning and explosives exploding are examples of chemical changes. Physical changes, on the other hand, are noted for the ease with which they can be reversed. One need only restore the original conditions and the substance reverts back to its original form: Frozen water melts upon warming.

Consider potassium chromate, a material whose color depends on its temperature. At room temperature, potassium chromate is a bright canary

The tree that Stephanie admires is composed of cellulose, a material made of the three elements hydrogen, oxygen, and carbon. The hydrogen and oxygen came from the water that was absorbed by the tree through its roots. The carbon was taken in by the leaves from carbon dioxide in the air. The atoms of liquid water and gaseous air rearranged into a solid tree—now there's a chemical change!

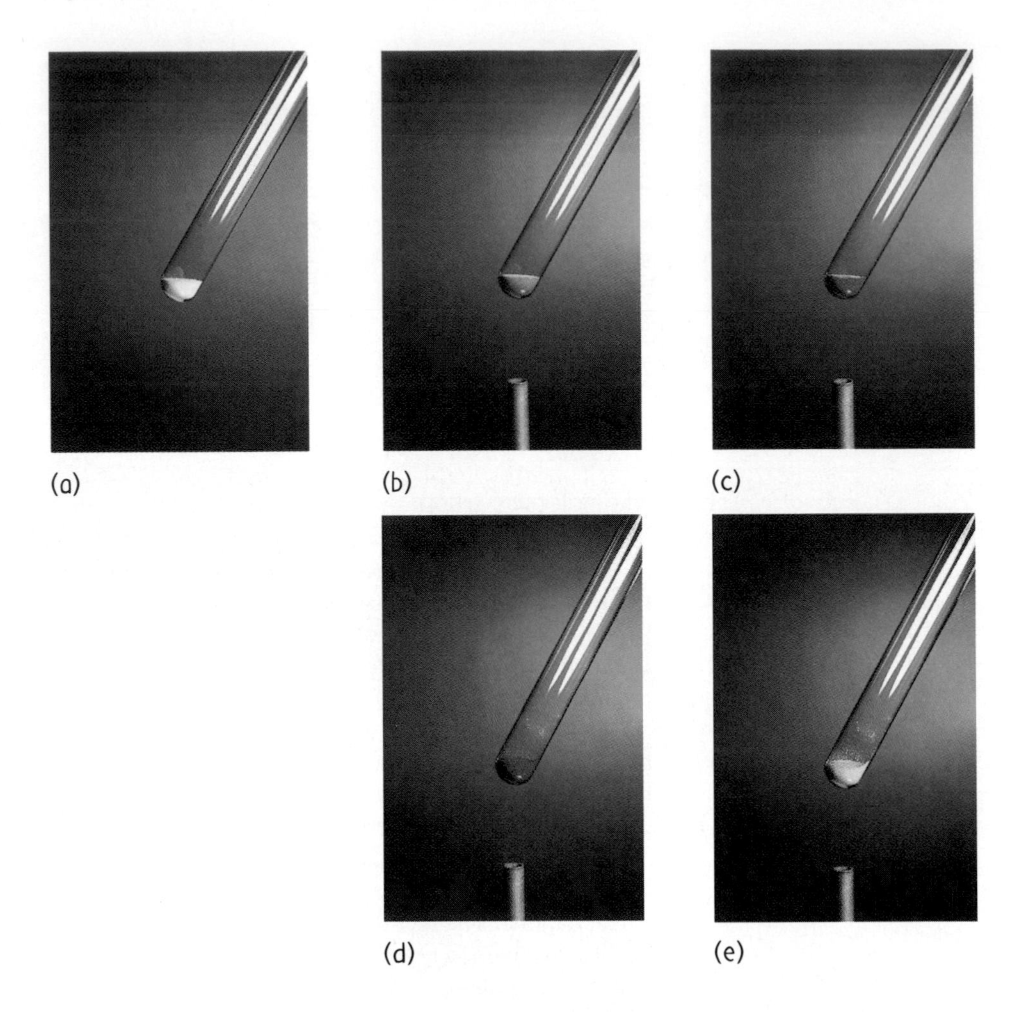

Fig. 15.9
Potassium chromate changes color as its temperature changes. This is a reversible physical change.

yellow. At higher temperatures, it is a deep reddish orange. Upon cooling, the canary color returns, suggesting that the change is physical (Figure 15.9). With a chemical change, reverting to original conditions does not restore the original attributes. Ammonium dichromate, for example, is an orange material that when heated explodes into ammonia, water vapor, and green chromium trioxide. Upon cooling, there remains no trace of orange ammonium dichromate. In its place are new substances having completely different physical properties (Figure 15.10).

Question

Your friend's kid sister has grown 2 inches in height. Is this better described as a physical change or a chemical change?

Answer

Biological growth involves the formation of numerous biomaterials, which are ultimately derived from the food we eat. The human body takes the atoms of various foods and rearranges them into the substances it needs for building itself. Therefore biological growth is better described as a chemical change.

Sure, there is a change in physical appearance, but don't confuse this change with a "physical change" in the chemical sense of the term. The physical appearance of a car is quite different after it rusts not because of physical changes but because of chemical changes. The effects of chemical change on physical appearance can be most dramatic. As another sign that biological growth involves chemical changes, consider its irreversibility: We grow older, but never do we grow younger.

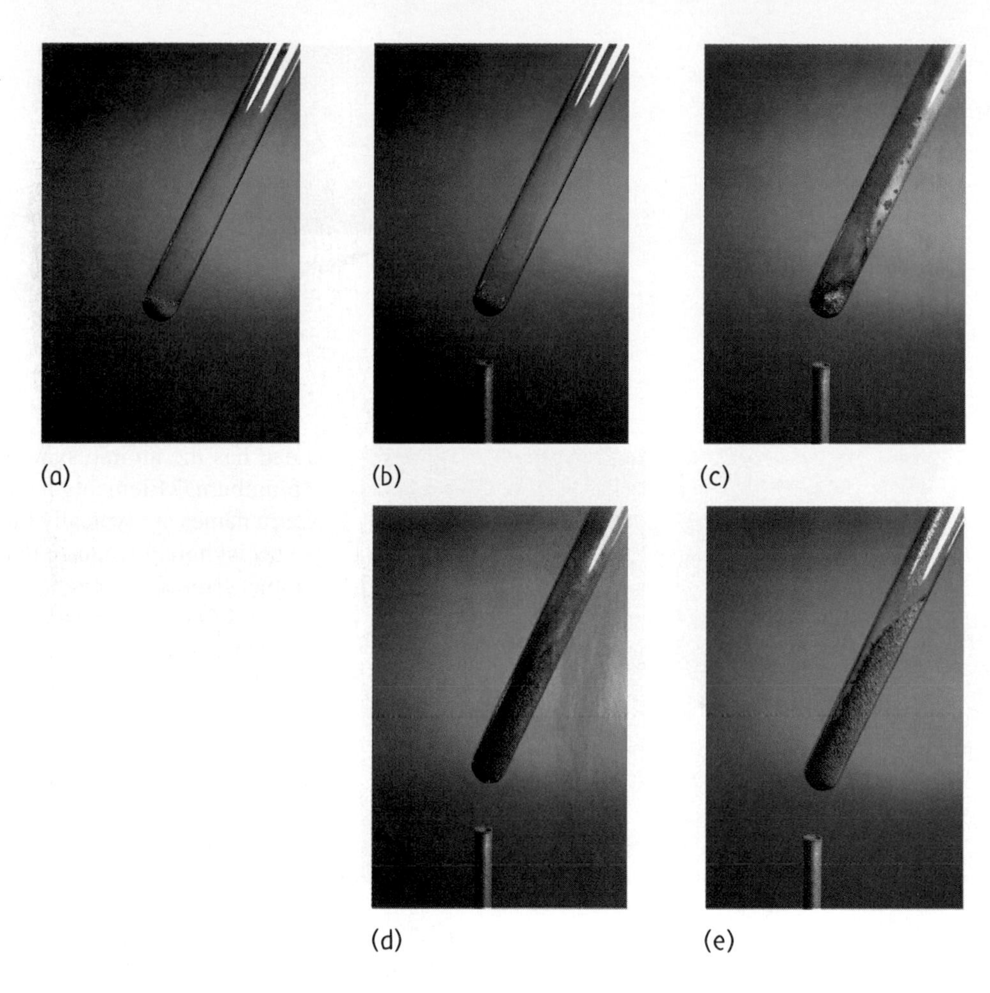

Fig. 15.10
Ammonium dichromate chemically changes to ammonia, water vapor, and solid chromium trioxide. Because this is a chemical change, it is irreversible.

15.4 Elements, Compounds, and Mixtures

Any material consisting of only one type of atom is classified as an **element**. A sample of gold, for example, consisting of only gold atoms is identified as an element. As of this writing, chemists have identified 112 elemental materials. Of these, only about 90 are natural; the remainder have been created in the laboratory. All of the known elements are listed in the *periodic table,* as mentioned in Chapter 13. We'll be exploring the periodic table and its many intricacies in much more detail in Chapter 16.

The terms *element* and *atom* are often used in a similar context. You might hear, for example, that the element hydrogen combines with the element oxygen to form water. Alternatively, you might hear that hydrogen atoms combine with oxygen atoms to form water. Generally, *element* is used in reference to microscopic or macroscopic samples, while *atom* is used when speaking of the submicroscopic. The important distinction is that elements are made of atoms and not the other way around. Technically, an *atom* is the smallest particle of an element that still retains all the chemical properties of that element (Figure 15.11).

Each element is designated by its **atomic symbol**. The atomic symbol is derived from the first letters of the element's name. For example, the atomic symbol for carbon is C, and that for chlorine is Cl. In many cases the atomic symbol is derived from the element's Latin name. Gold has the atomic symbol Au, after its Latin name "aurum."

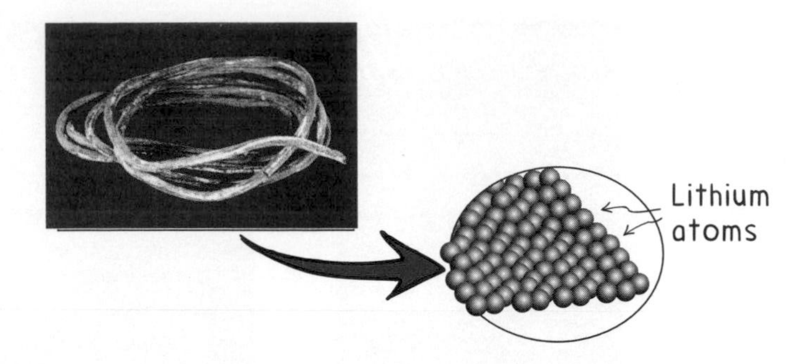

Fig. 15.11
Lithium metal is an example of an element. It consists only of lithium atoms, all of which share the same physical and chemical characteristics.

Fig. 15.12
Plumbers got their name because they once worked with lead pipes (the Latin word for lead is *plumbum,* and the atomic symbol of lead is Pb). A plumb bob gets it name from the lead that is still sometimes used as the weight.

Lead has the atomic symbol Pb, after its Latin name "plumbum." Elements having symbols derived from Latin names are typically those discovered earliest. On a special note, recognize that only the first letter of the atomic symbol is capitalized. Whereas Co is the element cobalt, CO is a combination of two elements, carbon, C, and oxygen, O.

The atoms that compose an element may combine with each other in a variety of ways. In gold, for example, the atoms are all grouped together in a single mass. In sulfur, however, the atoms are connected to one another in rings, each ring containing eight sulfur atoms. The nitrogen and oxygen we breathe contain paired atoms. How atoms are grouped in an element is shown using the **elemental formula**, which consists of the atomic symbol along with a numerical subscript to indicate the number of atoms grouped together. Because gold atoms do not group in any particular fashion, gold is not usually given an elemental formula and is simply represented using the atomic symbol, Au. Sulfur, on the other hand, is given the elemental formula S_8, and nitrogen and oxygen are given the elemental formulas N_2 and O_2, respectively.

■ **Question**

The oxygen we breathe, O_2, is converted to ozone, O_3, in the presence of electrical sparks. Is this a physical or chemical change?

■ **Answer**

When atoms regroup, the result is an entirely new substance, and that is what happens here. The oxygen we breathe, O_2, is odorless and life-giving. Ozone, O_3, on the other hand, is toxic and has a pungent smell commonly associated with electric motors. The conversion of O_2 to O_3, therefore, is a chemical change.

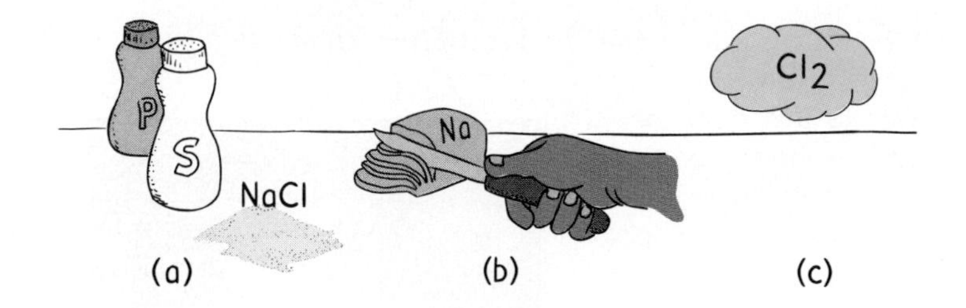

Fig. 15.13
The physical and chemical properties of (a) sodium chloride are very different from the physical and chemical properties of (b) sodium metal and (c) chlorine gas.

When atoms of *different* elements attach to one another, they make a **compound**. Sodium atoms and chlorine atoms, for example, join to make the compound sodium chloride, commonly known as table salt. Iron atoms and oxygen atoms join to make the compound iron oxide, commonly known as rust. A compound is represented by its **chemical formula**, in which the symbols for the different elements of the compound are written together. The chemical formulas for sodium chloride and iron oxide are NaCl and Fe_2O_3, respectively. Numerical subscripts indicate the ratio in which atoms combine to make the compound. By convention, subscripts of 1 are understood and omitted. So, we see that in NaCl there is one sodium atom for every one chlorine atom and in Fe_2O_3 there are two iron atoms for every three oxygen atoms.

Compounds have physical and chemical properties that are different from the prop erties of their elemental components. Sodium chloride, NaCl, is very different from either elemental sodium or elemental chlorine (Figure 15.13). Elemental sodium consists of nothing but sodium atoms, which form a soft but solid silvery metal that can be cut easily with a knife. Elemental sodium's melting temperature is 97.5°C, and it reacts violently with water. This is *not* the "sodium" you put on your popcorn. Elemental chlorine has the elemental formula Cl_2. This material is very toxic, and it was used as a lethal chemical warfare agent during World War I. Elemental chlorine's boiling temperature is −34°C, and the gas has a yellow-green color. The compound sodium chloride, however, is a translucent, brittle, colorless crystal that melts at 800°C. Sodium chloride contains the "sodium" used to season foods. It does not react with water and is an essential compound for living organisms. Sodium chloride is not sodium, nor is it chlorine; it is uniquely sodium chloride, a tasty chemical when sprinkled lightly over popcorn.

Question

Hydrogen sulfide, H_2S, is one of the smelliest compounds. Rotten eggs get their characteristic bad smell from the hydrogen sulfide they release. Can you infer from this information that elemental sulfur, S_8, is also quite smelly?

Answer
No, you cannot. In fact, elemental sulfur's odor is negligible compared with that of hydrogen sulfide. Compounds are truly different from the elements from which they are composed. Hydrogen sulfide, H_2S, is as different from elemental sulfur, S_8, as water, H_2O, is from elemental oxygen, O_2.

Link to Health—What's in a Glass of Water?

The answer to this question is more than meets the eye because tap water is anything but pure water. Depending on your location, tap water also contains a variety of compounds, such as calcium carbonate; magnesium carbonate; calcium fluoride; chlorine disinfectants; the ions of metals such as iron and potassium; trace amounts of heavy metals such as lead, mercury, and cadmium; and trace amounts of organic compounds in addition to dissolved gaseous materials such as oxygen, nitrogen, and carbon dioxide. There is no need to panic and go thirsty, however.

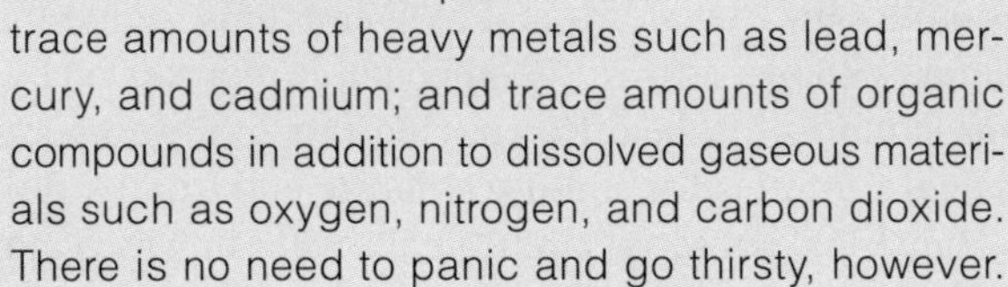

While it is surely important to minimize any toxic components of your drinking water, it is unnecessary and undesirable to remove all other substances from it. Some of the dissolved gases and minerals give water a pleasing taste, and many of the other dissolved substances promote human health: The fluoride ion protects teeth; trace amounts of chlorine destroy harmful bacteria; and as much as 10 percent of our daily requirements for iron, potassium, calcium, and magnesium is obtained from ordinary drinking water.

Most materials we encounter are mixtures: mixtures of elements, mixtures of compounds, or mixtures of elements and compounds. Stainless steel, for example, is a mixture of the elements iron, chromium, nickel, and carbon. Sparkling mineral water is a mixture of the compounds water, mineral salts, and carbon dioxide gas. Our atmosphere is a mixture of the elements nitrogen, oxygen, and argon plus small amounts of such compounds as carbon dioxide, water vapor, and various pollutants. Even the cleanest water is a mixture of various materials, including dissolved atmospheric gases, such as the oxygen that fish require to live.

The components of any mixture can be separated from one another by physical means. Consider air, which is a colorless liquid at −200°C. The boiling points of its major components, nitrogen and oxygen, differ by 13C° (−196°C and −183°C, respectively). If liquid air is warmed to −196°C, the nitrogen boils away (a physical change), leaving the oxygen. Alternatively, we can cool gaseous air to −183°C and watch the oxygen condense to a liquid (another physical change) while the nitrogen remains a gas (Figure 15.14).

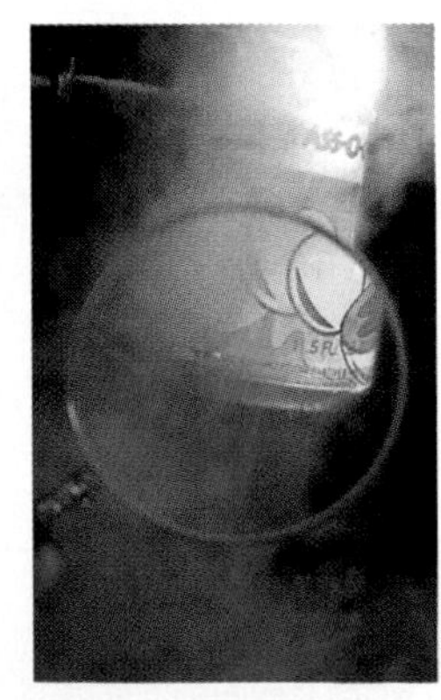

Fig. 15.14
Oxygen and nitrogen in the air can be separated from each other by holding up a soda can filled with liquid nitrogen at a temperature of about −196°C. Oxygen is a liquid at this incredibly cold temperature, and so the oxygen in the atmosphere condenses as it comes into contact with the outside of the can. Most of the nitrogen in the atmosphere stays gaseous because at −196°C nitrogen only begins to condense. The oxygen, however, begins to condense at 183°C, and so at 196°C its rate of condensation is much faster than that of the nitrogen. The resulting drips from the outside of the can are nearly pure oxygen.

Fig. 15.15
A simple distillation set-up that can be used to separate one component—water—from the mixture we call seawater.

Fig. 15.16
At the southern end of San Francisco Bay you'll find areas where the seawater has been partitioned off. These are evaporation ponds where the water is allowed to evaporate, leaving behind solid sea salts, which are then further refined for commercial sale. The remarkable color of the ponds results from suspended particles of iron oxide and other minerals, which are easily removed during refining.

As another example, consider seawater, which is a mixture of pure water and a variety of salts. One way to separate the water from seawater is to heat a sample of seawater until it boils. As the boiling water vaporizes (a physical change), it travels away from the mixture. At a separate location, the water vapor is then cooled to return it to its liquid form (another physical change). This process of collecting a vaporized substance is **distillation** (Figure 15.15). After all the water has been distilled from seawater, what remains are the dry salts. These salts, also a mixture of compounds, contain a variety of valuable materials such as sodium chloride, potassium bromide, and small amounts of various metals such as gold!

■ **Question**

Why would your fish die in an aquarium filled with only water?

■ **Answer**

Fish don't breathe water. They breathe the small amounts of oxygen, O_2, mixed in with the water. Without this O_2, your fish would promptly drown! This is why aquariums require aerators—as bubbles from the aerator rise, some of the air mixes into the water. It's the oxygen from the air dissolved in the water that the fish's gills are able to extract.

Naming Compounds

In the late 1780s, the French chemist Antoine Lavoisier pioneered the systematic naming of compounds. Prior to Lavoisier, chemical names were often meaningless and left over from alchemy: powder of algaroth, pomphlix, colcothar, turbith. If a name did indicate elemental content, it was usually misleading—butter of antimony, sugar of lead, flowers of sulfur. Working with others, Lavoisier devised a practical system in which the name of a compound reflected its elemental composition. In this system, for example, cinnabar became mercuric sulfide and saltpeter became sodium nitrate. For the chemistry student, Lavoisier's system was a blessing, for it provided an alternative to having to memorize the arbitrary names of the many known compounds. Moreover, chemists could, for the first time, predict the results of chemical reactions that had never been performed. From the old names no one could have guessed that Epsom salt was formed from magnesia and vitriol. Now, knowing from the labels on the bottles that magnesia is magnesium carbonate and that vitriol is sulfuric acid, the chemist can easily guess that the reaction product would be magnesium sulfate (Epsom salt). The reaction can now be written in words:

magnesium carbonate + sulfuric acid → magnesium sulfate

To accommodate the countless number of possible compounds, the rules for naming them have evolved since the time of Lavoisier. We need not go through the task of learning all these rules at this point. Instead, it suffices to become acquainted with how the system works for a variety of simple compounds.

Naming compounds made up of only two elements is easy. The name of the element farthest to the left in the periodic table is followed by the name of the element farthest to the right with the suffix *-ide:*

$NaCl$	Sodium chloride
LiH	Lithium hydride
CaF_2	Calcium fluoride
HCl	Hydrogen chloride
MgO	Magnesium oxide
Sr_3P_2	Strontium phosphide

For compounds that contain different numbers of the same elements, prefixes can be added to remove the ambiguity. The first four prefixes are *mono-* (one),* *di-* (two), *tri-* (three), and *tetra-* (four):

CO	Carbon monoxide
CO_2	Carbon dioxide
NO_2	Nitrogen dioxide
N_2O_4	Dinitrogen tetroxide
SO_2	Sulfur dioxide
SO_3	Sulfur trioxide

Common names are used to designate many compounds when the systematic names are tedious to use repeatedly, when they are difficult to pronounce, or because of tradition. Examples of common names are water, ammonia, and methane. The systematic names for these compounds are respectively dihydrogen oxide, H_2O; trihydrogen nitride, NH_3; and tetrahydrogen carbide, CH_4. Care should be taken in using common names, however, because they provide little to no information regarding the composition and structure of a compound—the name *water,* for example, tells us nothing of the hydrogen and oxygen of which it is composed. Common names are useful only to those who are already familiar with the compound in question.

■ **Question**

Name the compounds $CsCl_2$ and H_2O_2.

■ **Answer**

Cesium chloride; dihydrogen dioxide.

* The prefix *mono-* is typically omitted from the beginning of the name. The compound NO, for instance, is usually called nitrous oxide, not nitrous monoxide. Carbon monoxide is an exception.

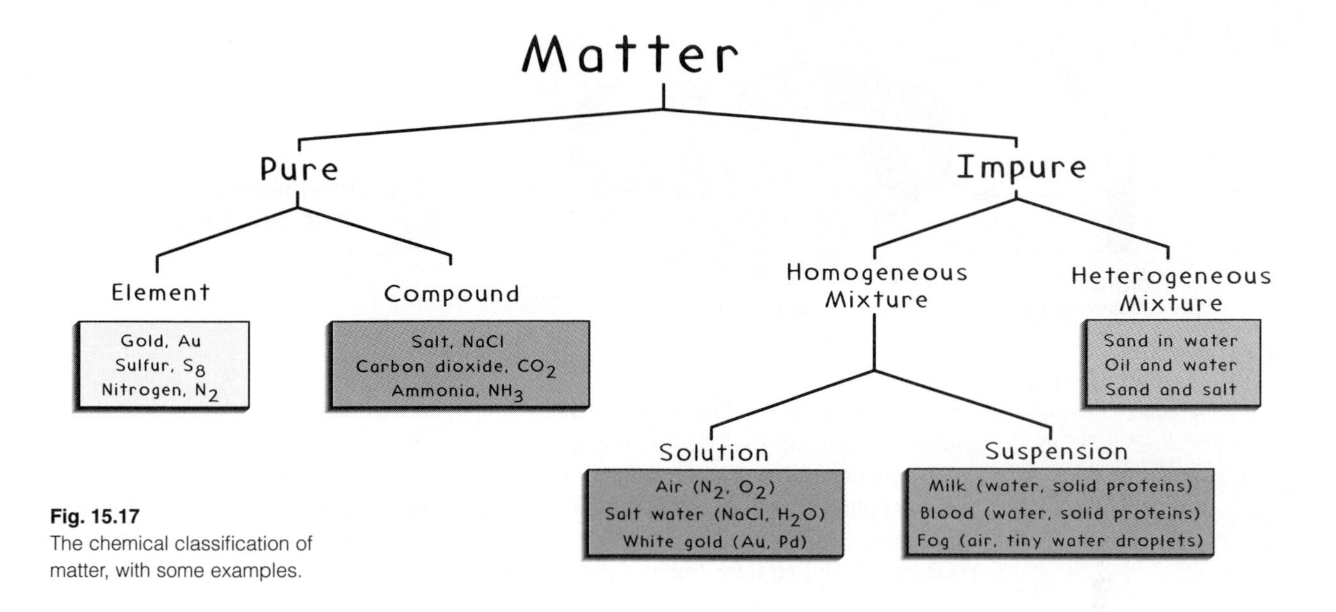

Fig. 15.17
The chemical classification of matter, with some examples.

15.5 Classification of Matter

Matter can be classified as pure or impure. If a material is **pure**, it consists of only a single element or a single compound. In pure gold, for example, there is nothing but the element gold. In pure table salt there is nothing but the compound sodium chloride. If a material is **impure**, it is a *mixture* of more than one element or compound.

The smallness of atoms and molecules makes it impractical to prepare a sample that is truly 100 percent of a single material. Samples, however, can be "purified" through any impurity-removing process, such as distillation. Comparing the purity of two samples, the purer sample is that containing the fewer impurities. Sometimes naturally occurring mixtures are labeled as pure, as in "pure" orange juice. According to a chemist's definition, however, orange juice is anything but pure, as it contains a wide variety of materials, such as water, pulp, flavorings, and sugars.

Mixtures may be **heterogeneous** or **homogeneous**. In a heterogeneous mixture the different components can be seen as individual substances, like pulp in orange juice, sand in water, or oil on water. The different components are obvious. In a homogeneous mixture, substances are mixed together so finely that they cannot be distinguished. Homogeneous mixtures have the same composition throughout—one portion of the mixture has the same ratio of substances as does any other portion. Milk from a cow naturally separates into a heterogeneous mixture of cream and milk. When milk is homogenized, though, the particles are blended into a homogeneous mixture.

A homogeneous mixture may be either a *solution* or a *suspension.* In a **solution**, all components are of the same phase. The atmosphere we breathe is an example of a gaseous solution that consists of the gaseous elements nitrogen and oxygen as well as other gaseous materials. Salt water is an example of a liquid solution because both the water and the dissolved sodium chloride are found in a single liquid phase. An example

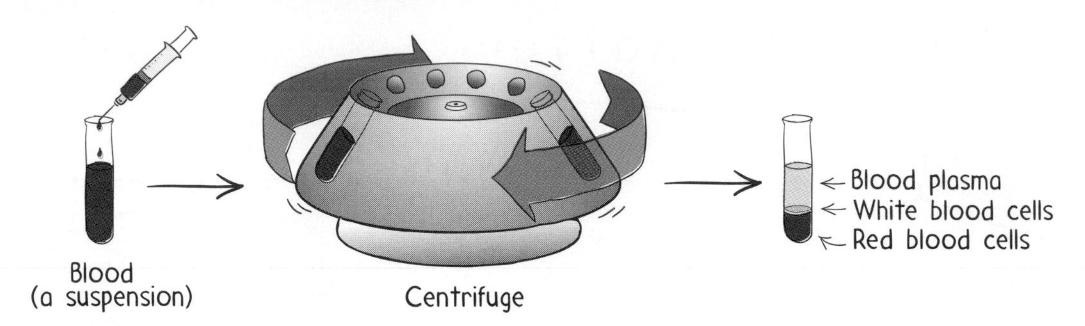

Fig. 15.18
Blood that's not used as whole blood at the blood bank is centrifuged into its useful components, which include the blood plasma (a yellowish solution), white blood cells, and red blood cells.

of a solid solution is white gold, which is a homogeneous mixture of the elements gold and palladium. We shall be discussing solutions in more detail in Chapter 18.

A **suspension** is a homogeneous mixture in which the different components are in different phases, such as solids in liquids or liquids in gases. In a suspension the different phases are finely mixed together. Milk is a suspension, for it is a homogeneous mixture of solid proteins finely dispersed throughout water. Blood is also a suspension, composed of finely dispersed solid proteins in water. Another example of a suspension is clouds, which are homogeneous mixtures of tiny water droplets in air.

The easiest way to distinguish a suspension from a solution is to apply centrifugal forces, as with a centrifuge. A centrifuge separates the components of suspensions but not those of solutions. The common solution salt water, for example, cannot be separated into salt and water in a centrifuge. Nor can air be separated into nitrogen and oxygen. However, put blood or any other suspension in a centrifuge, and the different phases are separated based upon differing densities (Figure 15.18). When blood is centrifuged, the solid blood proteins are nudged outward away from the liquid plasma, which is a solution of components not separable by the centrifuge.

■ Question

Multiple Choice: Impure water can be purified by

a. removing the impure water molecules
b. removing everything that is not water
c. breaking down the water to its simplest components
d. adding some disinfectant such as chlorine

■ Answer

The answer is b. Water, H_2O, is a compound made of the elements hydrogen and oxygen in a 2-to-1 ratio. Every H_2O molecule is exactly the same, and there's no such thing as an impure H_2O molecule. Just about anything, including you, beach balls, rubber ducks, dust particles, bacteria, and even other compounds can be found in water. When something other than water is found in water, we say that the water is not pure. It is important to see that the impurities are in the water and not part of the water, which means that it is possible to remove them by a variety of physical means, such as filtration or distillation.

Summary of Terms

- **Chemistry** The study of matter and of the transformations it can undergo.
- **Solid** Phase of matter characterized by definite volume and shape.
- **Liquid** Phase of matter characterized by definite volume but no definite shape; a liquid takes on the shape of its container.
- **Gas** Phase of matter in which molecules fill whatever space is available to them, taking neither definite shape nor definite volume.
- **Physical property** Any physical characteristic of a substance, such as color, density, and hardness.
- **Physical change** A change in which a substance changes its physical properties without changing its chemical identity.
- **Chemical property** The tendency of a substance to change chemical identity.
- **Chemical change** A change in which a substance changes its chemical identity. During a chemical change atoms are rearranged to give a new substance.
- **Chemical reaction** Synonymous to chemical change.
- **Element** A fundamental material consisting of only one type of atom.
- **Atomic symbol** An abbreviation for an element derived from the first one or two letters of the element's name.
- **Elemental formula** A notation that uses the atomic symbol and a numerical subscript to denote the composition of an element.
- **Compound** A material in which atoms of different elements are chemically held to one another.
- **Chemical formula** A notation used to denote the composition of a compound. In a chemical formula the atomic symbols for the elements making up the compound are written along with numerical subscripts that indicate their proportions.
- **Distillation** The process of recollecting a vaporized substance.
- **Pure** The state of a material that consists of only a single element or compound.
- **Impure** The state of a material that consists of more than one element or compound. A chemical mixture is impure.
- **Heterogeneous mixture** A mixture in which the components can be seen as individual substances.
- **Homogeneous mixture** A mixture composed of components so finely mixed that composition is the same throughout the mixture.
- **Solution** A homogeneous mixture in which all components are of the same phase.
- **Suspension** A homogeneous mixture in which different components are of different phases.

Review Questions

About Chemistry

1. What is chemistry?
2. Why is chemistry often called the central science?
3. Is chemistry built upon the principles of physics or those of biology?

Phases of Matter—A Molecular View

4. As liquid water evaporates, what is happening to the molecules?
5. How are the submicroscopic particles of a solid arranged differently from those of a liquid?
6. How does the arrangement of submicroscopic particles in a gas differ from the arrangements in liquids and solids?
7. Which occupies the most volume: 1 g of ice, 1 g of liquid water, or 1 g of water vapor?
8. Why are gases so easily compressed but not so with solids and liquids?
9. Why does it take so long for a gas particle to travel between two distant points?

Physical and Chemical Properties

10. What does a physical property describe?
11. What does a chemical property describe?
12. Why is it sometimes difficult to distinguish between a physical change and a chemical one?
13. What features of a chemical change allow it to be distinguished from a physical change?
14. What is a chemical reaction? What happens during one?
15. What evidence typically suggests that a chemical change has taken place rather than a physical change?

Elements, Compounds, and Mixtures

16. What is so unique about an element?
17. How many elements are there?
18. Distinguish between an atom and an element.
19. What is the difference between an element and a compound?
20. Distinguish between a compound and a mixture.
21. What is the number of atoms of each element and the total number of atoms per molecule in H_3PO_4?
22. Eat metallic sodium and you die. Inhale chlorine and you die. Mix them together, however, and you sprinkle the product on your tomatoes for better taste. What is going on?
23. What defines a material as a mixture?
24. How is it possible to separate components of a mixture?
25. How might you separate a heterogeneous mixture of sand and salt? How about iron and sand?

Classification of Matter

26. How does a chemist distinguish between a pure and an impure substance?
27. How are homogeneous and heterogeneous mixtures different?
28. What do a homogeneous and heterogeneous mixture have in common?
29. Classify the following as A, homogeneous mixture; B, heterogeneous mixture; C, element; or D, compound.
    Milk:___ Steel:___ Ocean water:___ Blood:___
    Sodium:___ Planet Earth:___
30. By what means can a solution and a suspension be distinguished from each other?

Home Projects

1. This one is for those of you who have access to a gas stove. Place a large pot of cool water on the stove and set the burner on high. What product from the combustion of the natural gas do you see condensing on the outside of the pot? Is this product really what you think it is or is it just some artificial form? How might you test its authenticity? Would you form more or less if the pot contained ice water and why? Where does this product go after the pot gets warmer? Identify the various physical and chemical changes.
2. Hydrogen peroxide, H_2O_2, is not a disinfectant. Rather, when you pour a solution of this liquid over a cut, an enzyme in your blood decomposes the hydrogen peroxide to produce oxygen. It is this oxygen at high concentrations at the site of injury that kills microorganisms. This blood enzyme is the same type of enzyme found in baker's yeast.

   Pour a packet of baker's yeast into a wide, short glass. Follow with a couple of swills of 3% hydrogen peroxide and watch the oxygen bubbles form. You can test for the presence of oxygen by holding a lighted match to the bubbles. Use tweezers to hold the match if necessary. Wear safety glasses and be sure to remove all combustibles, such as paper towels, from the area. Look for the flame to glow brighter as it bursts into one of the oxygen bubbles. Describe oxygen's physical and chemical properties.
3. What's in a glass of water? Here's a way to find out. Add water from a full glass to a sparkling clean cooking pot and boil the water to dryness. (Be sure to turn off the burner before the water is all gone.) Examine the resulting residue by scraping it with a knife. These are the solids you ingest with every glass of water you drink.

   To see the gases dissolved in your water, fill a cooking pot with water and let it stand at room temperature for several hours. Note the bubbles that adhere to the inner sides of the pot. Where did they come from? What do you suppose they contain? Could a fish breathe one of these bubbles, or are fish not equipped to breathe bubbles?

Exercises

1. Is chemistry the study of the submicroscopic, the microscopic, the macroscopic, or all three? Defend your answer.
2. In what sense is a color computer monitor or TV screen similar to our particulate view of matter? Place a drop of water on your computer screen for a closer look.
3. A covered glass of water sits for days with no drop in water level. Strictly speaking, can you say that nothing has happened? Formulate your answer from a submicroscopic point of view.
4. What is the relationship between the phase of a material and the motion of its submicroscopic particles?
5. What happens to the density of a gas as the gas is compressed?
6. Gas particles travel at speeds of up to 1500 km/h. Why, then, does it take so long for aromatic gas molecules to travel the length of a room?
7. Ozone, O_3, has three atoms per molecule, while oxygen, O_2, has only two. Does it follow that a given number of ozone molecules in the gas phase should occupy a greater volume than the same number of oxygen molecules in the gas phase?
8. The drawing on the left shows two phases of a single substance. In the middle box, draw what these particles would look like if heat were taken away. In the box on the right, draw what they would look like upon the addition of heat. If each particle were a water molecule, what would be the temperature in the box on the left?

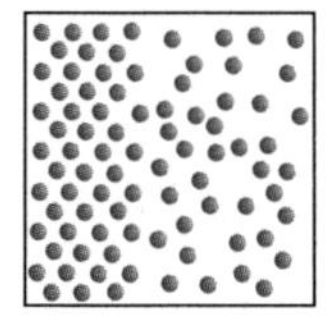

9. The central fuel tank of the space shuttle consists of two compartments, one filled with oxygen and the other with combustible hydrogen. Both of these materials are carried in the liquid phase. Why? What might you say about the temperature inside the central fuel tank?
10. Humidity is a measure of the amount of water vapor in the atmosphere. Why is humidity always very low inside your kitchen freezer?
11. Each night you measure your height just before going to bed. When you arise each morning, you measure your height again and consistently find that you are 1 inch taller than you were the night before but only as tall as you were 24 hours ago! Is what happens to your body in this instance best described as a physical change or a chemical change? Be sure to try this activity if you haven't already.
12. How is a chemical property much like one's personality?

13. Classify the following changes as physical or chemical:
(a) Grape juice turns into wine.____________
(b) Wood burns to ashes.____________
(c) Photographic film is exposed to light.____________
(d) Water begins to boil.____________
(e) A broken leg mends itself.____________
(f) Grass grows.____________
(g) An infant gains 10 pounds.____________
(h) Rock is crushed into a powder.____________

14. One set of drawings represents a physical change and the other a chemical change. Which is which?

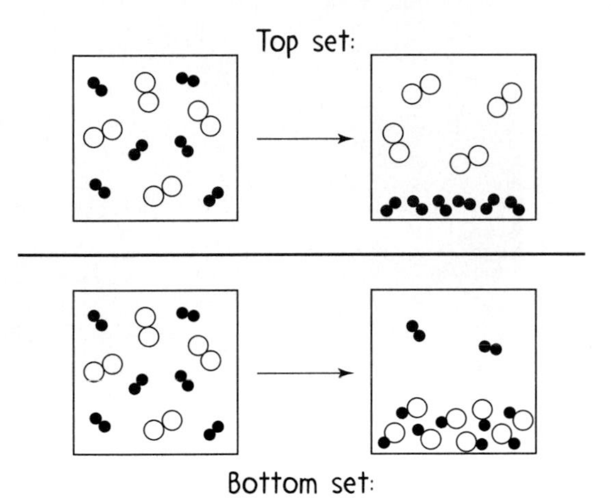

15. Is learning an example of a physical change or a chemical change?

16. Which elements are some of the oldest known? What is your evidence?

17. Why is it impossible for us to breathe water even though water is 88.88% oxygen by mass?

18. Do compounds have the same physical and chemical properties as the elements from which they are made? Give an example to support your answer.

19. The following materials are probably in daily use in your home: salt, stainless steel, water, sugar, vanilla extract, butter, pepper, aluminum, ice, milk, aspirin. Classify these materials as elements, compounds, or mixtures.

20. Why are most biological materials mixtures of chemicals?

21. The freezing point of oxygen is $-218°C$ and that of nitrogen is $-210°C$. Which is in the liquid phase and which is in the solid phase at $-214°C$?

22. What is the common name for dioxygen oxide? What about oxygen oxide?

23. Distinguish between a compound and a chemical mixture.

24. Of the three boxes, which contains an elemental material, which a compound, and which a mixture? How many types of molecules are shown in all three boxes?

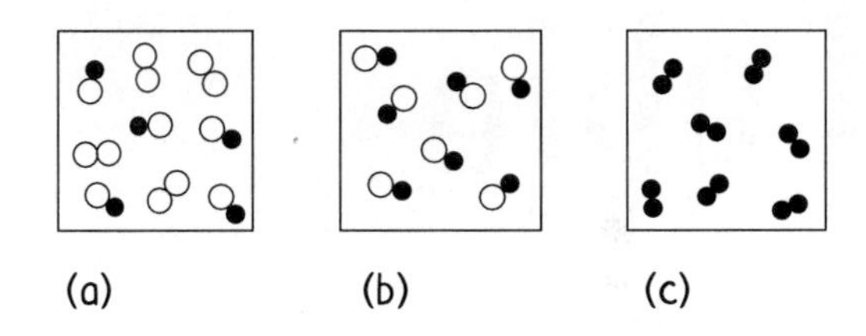

25. Why is it impossible to have a material that is 100% pure?

26. Why is half-frozen fruit punch always sweeter than the same fruit punch when melted?

27. Why does a gas occupy so much more volume than a solid or a liquid?

28. Why is biology more dependent on chemistry than chemistry is dependent on biology?

29. Explain to someone what chicken noodle soup and soil have in common—without using the phrase heterogeneous mixture.

30. There are only three abundant materials on this planet found in the liquid phase. One you burn, another burns anything in its path, and the last puts out that which is burning. What are they?

CHAPTER

16

The Periodic Table

You go to the pantry to get ingredients for a cake. You select sugar, flour, eggs, milk, baking soda plus an assortment of other ingredients, all of which you will measure and mix in some particular fashion. If you change your measurements or alter your procedure, however, you can create any of a large number of different products—pancakes, muffins, biscuits, or who knows what—instead of a cake.

Nature is a bit like a baker. The ingredients in nature's pantry are the chemical elements. Different combinations of the elements give rise to the diversity of materials in our environment. Cotton, for instance, is made of carbon, hydrogen, and oxygen, but the same elements combined in a different way make sugar. Arranged still differently, we have a potato.

In this and in subsequent chapters we explore the remarkable organization of the **periodic table** of the elements. What you will find is that the periodic table is a powerful tool—a device that one learns how to *read.* The periodic table is to a chemist what the dictionary is to a writer. Both are useful as references and should always be available. Neither, of course, need be memorized.

16.1 Organizing the Elements

A cook organizes spices in a spice rack, and chemists do the same sort of thing with the elements. The simplest way to organize the elements is to list them in a vertical column according to increasing atomic number, as shown in Table 16.1.

If various physical or chemical properties of the elements are listed alongside their atomic number, some interesting trends become apparent. First, consider the atomic sizes of the elements as measured by *atomic radius,* which is the distance from the atomic nucleus to the atom's outer "surface." For most elements, the atomic radii gradually get smaller with increasing atomic number (Table 16.1). This is understandable because the greater nuclear charge pulls electrons closer to the nucleus. At certain intervals, however, the atomic radius of an element is dramatically *greater* than that of the previous element. This occurs for the elements lithium, Li; sodium, Na; potassium, K; rubidium, Rb; cesium, Cs; and francium, Fr. We see these jumpy intervals when atomic radius is plotted against atomic number (Figure 16.1a).

A similar pattern arises when we look at the *ionization energies* of the elements. **Ionization energy** is the amount of energy required to pull an electron away from an atom to form a positively charged ion (Section 8.1). The ionization energy of atoms generally increases with increasing atomic number. At the same break points at which atomic radii increase, however, there is a dramatic decrease in ionization energy, as Figure 16.1b shows.

Table 16.1
Elements Listed by Atomic Number

Element	Atomic Number	Atomic Radius (picometers)	Ionization Energy (kJ/mole)
H	1	37	1,300
He	2	50	2,380
Li	3	152	520
Be	4	111	900
B	5	88	800
C	6	77	1,100
N	7	70	1,400
O	8	66	1,310
F	9	64	1,700
Ne	10	70	2,050
Na	11	186	480
Mg	12	160	700
Al	13	143	600
Si	14	117	800
P	15	110	1,010
S	16	104	1,000
Cl	17	99	1,280
Ar	18	94	1,540
K	19	231	400
⋮	⋮	⋮	⋮

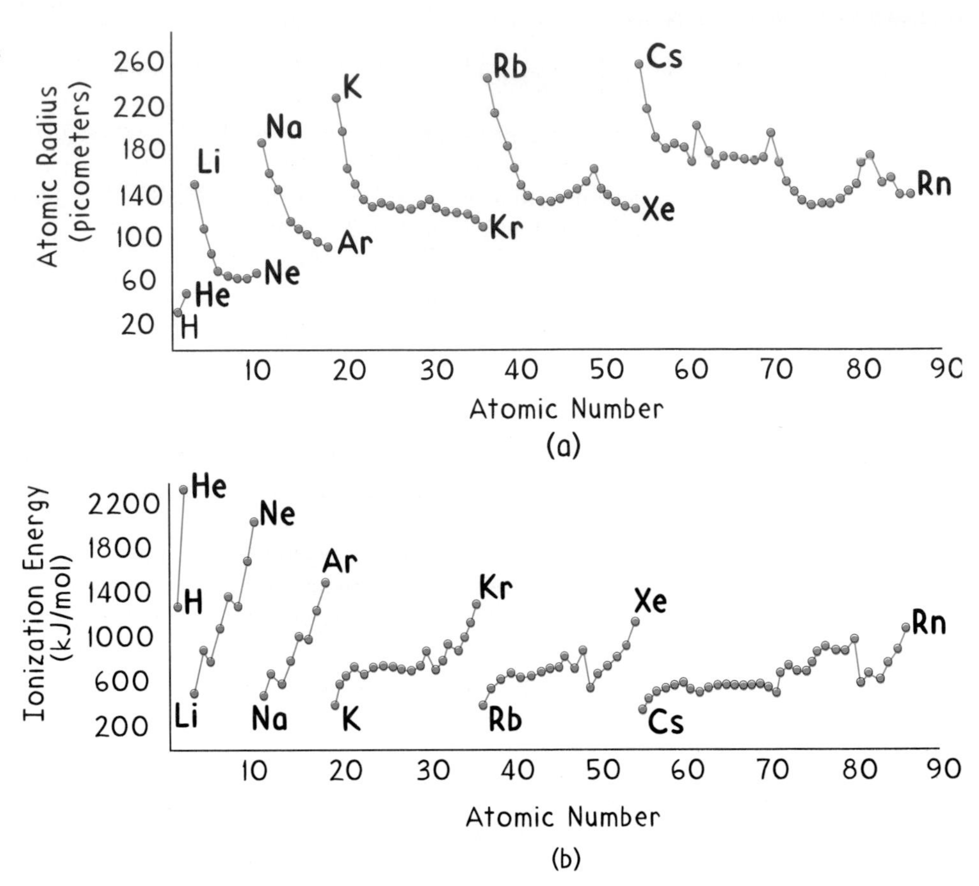

Fig. 16.1
(a) The atomic radii of elements gradually decrease with increasing atomic number but occasionally increase sharply. (b) The ionization energies of elements gradually increase with increasing atomic number but occasionally decrease sharply. Note that the discontinuities (sharp breaks) occur in the same places for both plots.

Because other properties of the elements follow the same pattern, it is natural to group the elements according to these intervals. In Table 16.1, we indicated this grouping by extra space every so often. Another way to indicate groups is to list the elements in a series of horizontal rows such that elements of the same interval appear in the same row. Each row is commonly referred to as a **period**.[1] When we place the seven known periods on top of one another, we get the following:

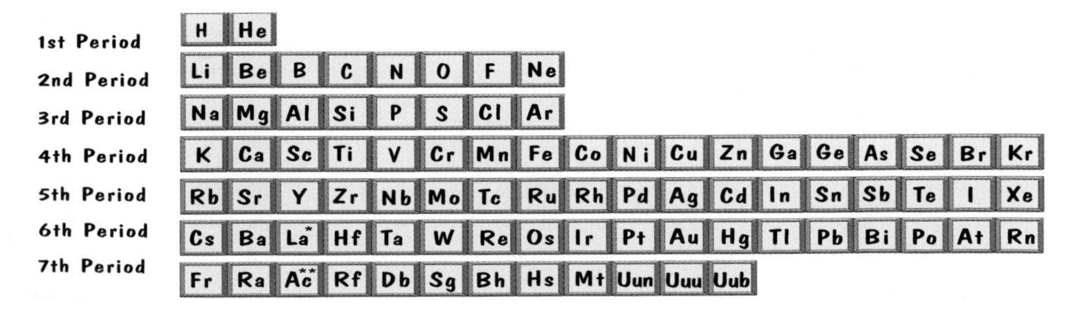

Note that different periods contain different numbers of elements.

[1] "Period" in this sense refers to the cyclic repetition of atomic properties.

Interestingly enough, this horizontal listing of the elements leads to further organization. All the first elements of the periods, for example, have similar chemical and physical properties. In other words, the elements hydrogen (in its atomic form), lithium, Li, sodium, Na, potassium, K, rubidium, Rb, cesium, Cs, and francium, Fr, all react chemically with water and are all soft metals.

With some slight modifications, we are able to generate even more organization. The second element of the first period, helium, He, has nothing in common with the second elements of the other periods, which all have very similar properties. Helium, however, is a very unreactive element, as are the elements neon, Ne, and argon, Ar. For these reasons, we slide helium rightward until it is aligned with neon and argon:

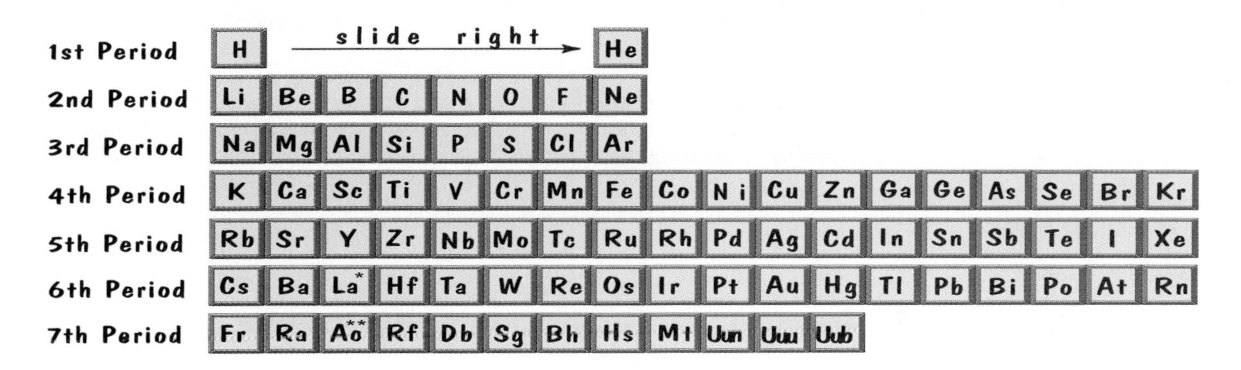

Also, for the purpose of grouping elements with similar properties above or below one another, we slide helium along with the second-period elements boron, B, through neon and the third-period elements aluminum, Al, through argon to the far right:

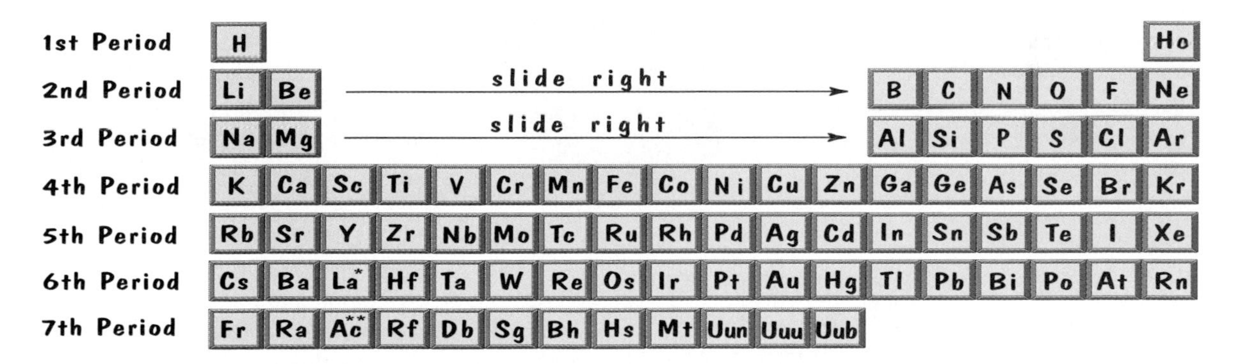

The resulting arrangement we recognize as the common form of the periodic table. The organization of elements in the table is astonishing. We find that not just a few but rather all elements listed directly above or below one another share similar physical and chemical properties. Equally astonishing are the gradual changes in the properties of the elements as you move in any direction. The atoms of elements toward the lower left-hand corner tend to be the largest, whereas those toward the upper right tend to be the smallest (Figure 16.2a). In addition, the largest atoms toward the lower left tend to have the lowest ionization energy, and the smallest atoms toward the upper right tend to have the highest ionization energies (Figure 16.2b).

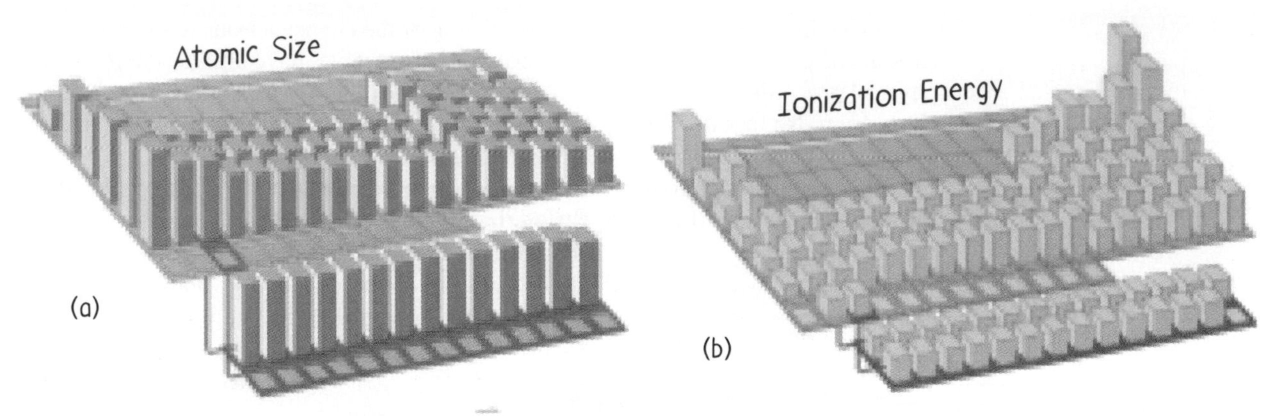

Fig. 16.2
(a) Atomic size generally increases in moving toward the lower left-hand corner of the periodic table.
(b) The amount of energy required to remove an electron from an atom is greatest for elements toward the upper right-hand corner of the periodic table.

Link to History—Dmitri Mendeleev—Father of the Periodic Table

Science is the search for patterns and regularities in nature. More than a century ago many scientists working independently had noted how the physical and chemical properties of elements tend to repeat themselves. Many charts were produced in which elements of similar properties were listed close to one another.

In 1869, a Russian chemistry professor, Dmitri Mendeleev, produced for his students a chart summarizing the known properties of the elements. Mendeleev's chart was unique in that it resembled a calendar, with the weeks in horizontal rows and each day of the week represented by a vertical column. Across each row Mendeleev placed all the elements that appeared within one period of repeating properties. Down each vertical column he placed elements of similar properties. He found, however, that in order to properly align elements within a column he had to occasionally shift the elements sideways. This left obvious gaps—blank spaces that could not be filled by any known element. Instead of looking upon these blank spaces as defects, Mendeleev boldly predicted the existence of elements that had not yet been discovered. Furthermore, it was his predictions about the properties of some of those missing elements that led to their discovery!

ПЕРИОДИЧЕСКАЯ СИСТЕМА ЭЛЕМЕНТОВ

		ГРУППЫ ЭЛЕМЕНТОВ										
		I	II	III	IV	V	VI	VII	VIII			0
1	I	H 1 1,008										He 2 4,003
2	II	Li 3 6,940	Be 4 9,02	5 B 10,82	6 C 12,010	7 N 14,008	8 O 16,000	9 F 19,00				Ne 10 20,183
3	III	Na 11 22,997	Mg 12 24,32	13 Al 26,97	14 Si 28,06	15 P 30,98	16 S 32,06	17 Cl 35,457				Ar 18 39,944
4	IV	K 19 39,096	Ca 20 40,08	Sc 21 45,10	Ti 22 47,90	V 23 50,95	Cr 24 52,01	Mn 25 54,93	Fe 26 55,85	Co 27 58,94	Ni 28 58,69	
	V	29 Cu 63,57	30 Zn 65,38	31 Ga 69,72	32 Ge 72,60	33 As 74,91	34 Se 78,96	35 Br 79,916				Kr 36 83,7
5	VI	Rb 37 85,48	Sr 38 87,63	Y 39 88,92	Zr 40 91,22	Nb 41 92,91	Mo 42 95,95	Ma 43 —	Ru 44 101,7	Rh 45 102,91	Pd 46 106,7	
	VII	47 Ag 107,88	48 Cd 112,41	49 In 114,76	50 Sn 118,70	51 Sb 121,76	52 Te 127,61	53 J 126,92				Xe 54 131,3
6	VIII	Cs 55 132,91	Ba 56 137,36	La 57 * 138,92	Hf 72 178,6	Ta 73 180,88	W 74 183,92	Re 75 186,31	Os 76 190,2	Ir 77 193,1	Pt 78 195,23	
	IX	79 Au 197,2	80 Hg 200,61	81 Tl 204,39	82 Pb 207,21	83 Bi 209,00	84 Po 210	85 —				Rn 86 222
7	X	87 —	Ra 88 226,05	Ac 89 227	Th 90 232,12	Pa 91 231	U 92 238,07					

* ЛАНТАНИДЫ 58-71

Ce 58 140,13	Pr 59 140,92	Nd 60 144,27	61 —	Sm 62 150,43	Eu 63 152,0	Gd 64 156,9
Tb 65 [illegible]	Dy 66 162,46	Ho 67 164,94	Er 68 167,2	Tu 69 169,4	Yb 70 173,04	Cp 71 174,99

Mendeleev's chart, which ultimately led to our modern periodic table, is recognized as one of the most important achievements of modern science. It helped to confirm the atomic theory of matter and laid the ground work for future understanding of atomic behavior.

■ Question

Was the great organization of the periodic table invented by humans or discovered by humans?

16.2 Metals, Nonmetals, and Metalloids

Elements are sorted in the periodic table in a number of ways. One immediately apparent way is as *metals, nonmetals,* and *metalloids.*

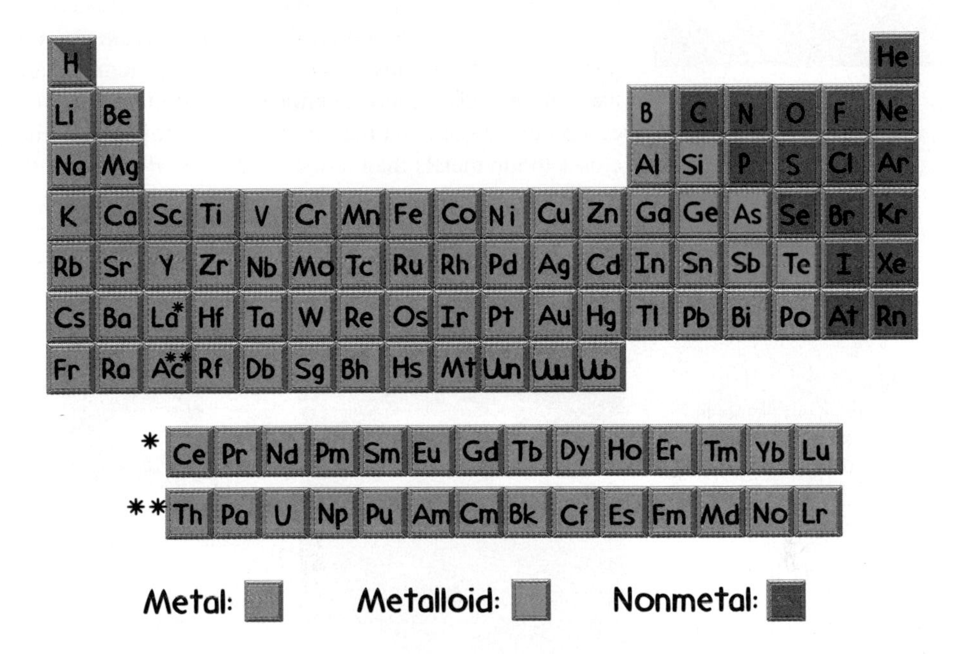

Fig. 16.3
Elements classified as metal, nonmetal, or metalloid.

Fig. 16.4
Metals bend and give when hammered—in other words, they are malleable.

About 85 percent of the known elements are **metals**, defined as those elements that are generally shiny, opaque, and good conductors of electricity and heat. Metals are also *malleable,* which means they can be hammered into different shapes or bent without breaking. They are also *ductile,* which means they can be drawn into wires. All but a few metals are solid at room temperature. The exceptions include mercury, Hg; gallium, Ga; cesium, and francium, which are all in the liquid phase at a warmish room temperature of 30°C (86°F). Another interesting exception is hydrogen, which takes on the properties of a liquid metal only at very high pressures.

The nonmetallic elements are grouped together on the right of the periodic table. In contrast to metals, **nonmetals** are very poor conductors of electricity and heat, and may also be transparent (Figure 16.6). Nonmetals are neither malleable nor ductile.

■ **Answer**

This organization was *discovered* because it exists regardless of whether or not we know about it. The natural world has order before we humans figure it out. Science is a way of discovering this order and presenting it in a form we can understand.

Fig. 16.5
Geoplanetary models suggest that hydrogen exists as a liquid metal deep beneath the surfaces of the larger planets, such as Jupiter and Saturn. These planets are composed mostly of hydrogen, and internal pressures are more than 3 million times the Earth's atmospheric pressure. At these high pressures, hydrogen is compressed to a liquid-metal phase. Back here on the Earth at our relatively low atmospheric pressures, hydrogen exists as a gas of hydrogen molecules, H_2.

Rather, they are brittle and shatter when hammered. At room temperature, some are solid (carbon, C), others are liquid (bromine, Br), and still others are gaseous (helium).

Six elements are classified as **metalloids**: boron, B; silicon, Si; germanium, Ge; arsenic, As; antimony, Sb; and tellurium, Te. Situated between the metals and the nonmetals in the periodic table, the metalloids take on a mix of both metallic and nonmetallic characteristics. For example, these elements are weak conductors of electricity, which makes them useful as the semiconductors in the integrated circuits of computers. Note from the periodic table how germanium, Ge (atomic number 32), is closer to the metals than to the nonmetals. Because of this positioning, we can deduce that germanium has more metallic properties than silicon (atomic number 14) and is a slightly better conductor of electricity. So we find that integrated circuits fabricated with germanium operate faster than those fabricated with silicon. Because silicon is much more abundant and less expensive to obtain, however, silicon computer chips remain quite common (Figure 16.7).

Fig. 16.6
Some nonmetallic elements are transparent, such as carbon in its diamond form. When hammered, it—like many other nonmetals—shatters into tiny fragments and dust. Instead, the skilled diamond cutter chisels at precise angles with extreme care to produce the familiar multifaceted gemstone.

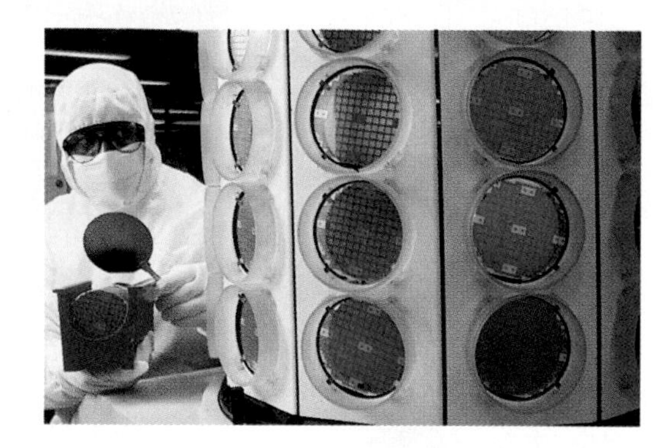

Fig. 16.7
Purified silicon cylinders are sliced into wafers for the manufacture of silicon-based integrated circuits.

16.3 Atomic Groups and Periods

Because of their similar properties, elements in the same vertical column are referred to as a **group**. In the modern periodic table, the 18 known groups are numbered from left to right (Figure 16.8).[2]

---

[2]A system of roman numerals and letters is also used to represent groups. Although more complex, this system is more descriptive of the behavior of the elements. Ask your instructor about this alternative system or, better yet, take a follow-up chemistry course!

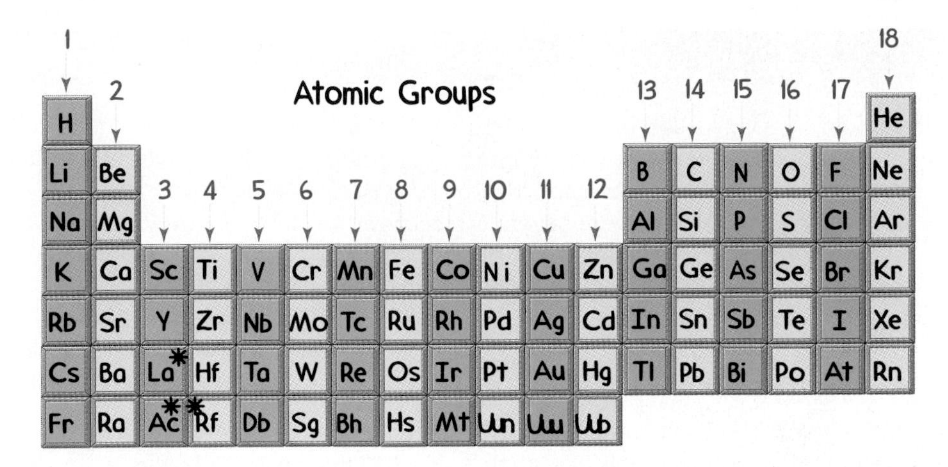

Fig. 16.8
Each vertical column is a *group*. Elements in the same group have similar chemical and physical properties.

Several groups have traditional names that describe the properties of the elements in them (Figure 16.9). Early in human history, people discovered that ashes mixed with water produces a slippery solution useful for removing grease. By the Middle Ages such mixtures were described as *alkaline,* a term derived from the Arab word for ashes, *al-qali.* Alkaline mixtures found many uses, particularly in the preparation of soaps (Figure 16.10). We now know that alkaline ashes contain compounds of group 1 elements, most notably potassium carbonate, also known as *potash.* Because of this history, group 1 elements, which are metals, are called the **alkali metals**.

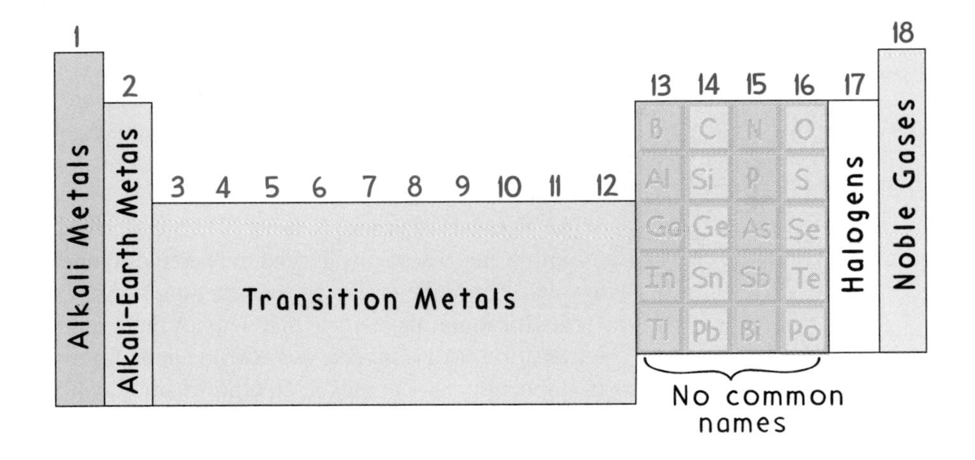

Fig. 16.9
The common names for various groups of elements.

Fig. 16.10
Ashes and water make a slippery alkaline solution once used to clean hands.

Elements of group 2 also form alkali solutions when placed in water. Furthermore, medieval alchemists noted that certain minerals (which we now know are made up of group 2 elements) do not melt or change when put in fire. These fire-resistant substances were known to the alchemists as "earth." As a holdover from these ancient times, group 2 elements are known as the **alkali-earth metals**. Over toward the right-hand side of the periodic table, elements of group 17 are known as the **halogens** ("salt-forming" in Swedish) because of their

Chlorine in Water

One of the major uses of chlorine is to disinfect water. Chlorination of public water supplies has prevented much illness and increased the quality of living for many people. There are several ways to add chlorine to water. Chlorine gas, Cl_2, may be bubbled through the water, or chemical compounds containing chlorine may be dissolved in the water. One such compound is calcium hypochlorite, $Ca(OCl)_2$, the powdery solid commonly added to swimming pools. Another compound is sodium hypochlorite, NaOCl, the active ingredient in laundry bleach. Chlorine gas, calcium hypochlorite, and sodium hypochlorite each mix with water to form the chemical compound hypochlorous acid, HOCl, which kills disease-producing microorganisms. Monitors watch the levels of chlorine in water carefully. Too little chlorine does not control waterborne diseases; too much chlorine can give the water a bad taste and even cause sickness.

Fig. 16.11
Noble gases in glass tubes glow their characteristic colors when energized with electricity. The most familiar of these is the classic neon light.

tendency to form various salts. Group 18 elements are all unreactive gases that tend not to combine with other elements. For this reason, they are called the **noble gases**, presumably because the nobility of earlier times were above interacting with common folk (Figure 16.11).

The elements of groups 3 through 12 are all metals that do not form alkaline solutions with water. These metals tend to be harder than the alkali metals and less reactive with water, hence they are used for structural purposes. Collectively they are known as the **transition metals**, a name that denotes their central position in the periodic table. The transition metals include some of the most familiar and important elements, such as iron, Fe; copper, Cu; nickel, Ni; chromium, Cr; silver, Ag; and gold, Au. They also include many lesser-known elements that are nonetheless important in modern technology. Persons with hip implants appreciate the transition metals titanium, Ti; molybdenum, Mo; and manganese, Mn; for these noncorrosive metals are used in implant devices.

■ Question

What do the transition metals of group 11 have in common with one another?

■ **Answer**
In addition to all the typical properties of metals, the group 11 elements, copper, silver, and gold, are all precious metals used for currency and jewelry. If element 111 were not so radioactive and could be isolated in large enough quantities, it, too, would likely be regarded as a precious metal.

Transitions Metals in Society

Our standard of living and our economy are completely dependent upon the production of transition metals. We get them either from natural sources or through recycling. Approximately 500 million tons of iron, 8 million tons of copper, and 750,000 tons of nickel are produced each year worldwide for the manufacture of coins, cars, appliances, bridges, buildings, and other metal-requiring commodities. Production involves the mining of naturally occurring ores and subsequent chemical processing. Many transition-metal compounds are highly colored. This makes them useful as pigments in paints and dyes. The white of this page, for example, is produced by titanium dioxide, TiO_2, and the colors of rubies, sapphires, and other gemstones are due to trace quantities of such transition metals as chromium and iron.

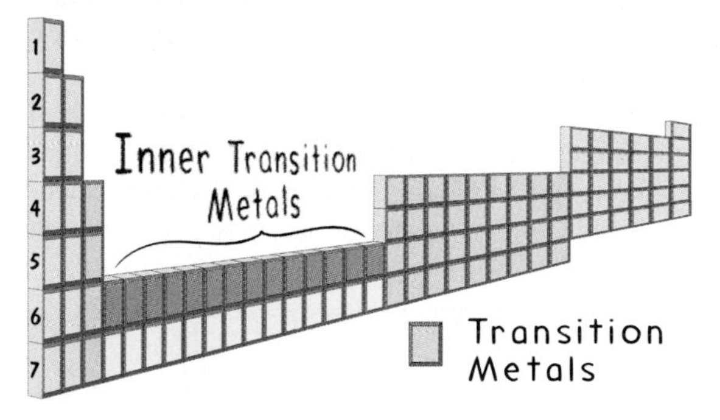

Fig. 16.12
The inner transition metals inserted between atomic groups 3 and 4 results in an unusually long periodic table whose form is not conducive to being printed on a standard sheet of paper.

In the sixth period is a subset of 14 metallic elements (atomic numbers 58–71) that are quite unlike any of the other transition elements. A similar subset (atomic numbers 90–103) is found in the seventh period. These two subsets are the **inner transition metals**. Inserting the inner transition metals into the main body of the periodic table as shown in Figure 16.12 results in a long and cumbersome periodic table. So that the periodic table can fit nicely within a standard paper size, these elements are instead commonly shown below the main body of the periodic table, as in Figure 16.13.

The inner transition metals are not assigned group numbers. Rather, the elements of each subset have properties that are so similar to one another that each subset can be viewed as a group. The sixth-period inner transition metals are called the **lanthanides** because they fall after lanthanum, La. Because of their similar physical and chemical properties, the lanthanides tend to occur mixed together

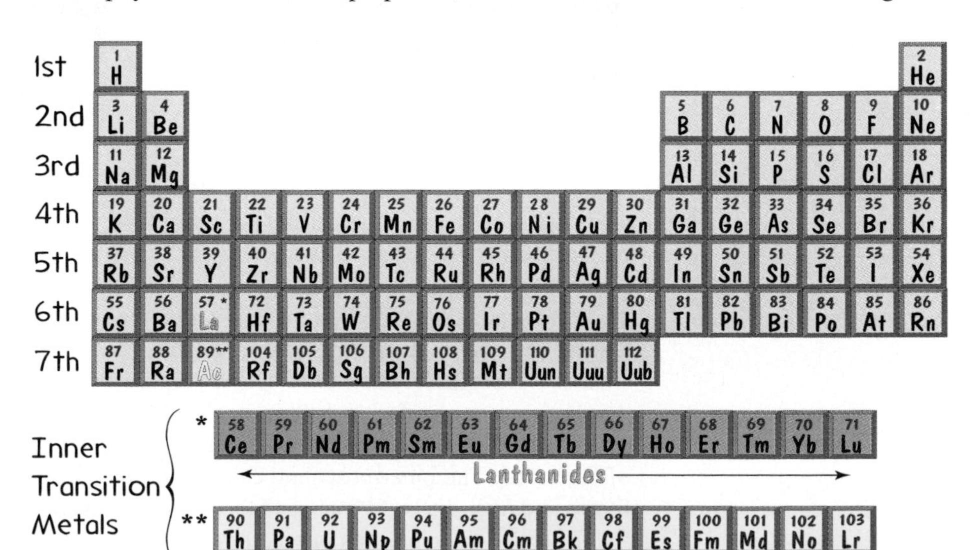

Fig. 16.13
The typical display of the inner transition metals. The sequence in the sixth period goes from lanthanum (La, 57) to cerium (Ce, 58) on through to lutetium (Lu, 71) then back to hafnium (Hf, 72). A similar jump is made in the seventh period.

in the same geologic zones. Also, because of their similarities, lanthanides are unusually difficult to purify. Recently they have been used in an increasing number of applications. Several lanthanide elements, for example, are used in the fabrication of the light-emitting diodes (LEDs) of laptop computer monitors.

The seventh-period inner transition metals are called the **actinides** because they fall after actinium, Ac. They, too, are a group of elements having similar properties and hence are not easily purified. The nuclear power industry faces this obstacle because it requires purified samples of two of the most publicized actinides: uranium, U, and plutonium, Pu. Actinides heavier than uranium are not found in nature but are synthesized in the laboratory.

16.4 Noble Gas Shells

The quality of a song depends upon the arrangement of its musical notes. In a similar fashion, the chemical and physical properties of a substance depend upon the arrangements of electrons in its atoms. To grasp how electrons arrange themselves in an atom, we turn to a conceptual model.

The model we choose is an abbreviated version known as the **noble gas shell** model. This model gets its name from the noble gas elements of group 18, which, as we shall see, are the only elements to have *shells* that are completely filled with electrons. A **shell** is defined as a region of space about the atomic nucleus within which electrons may reside. These shells may be drawn to appear similar to Bohr's planetary model, as was presented in Chapter 13. They are different, however, in that each shell represents a collection of probability clouds of similar energy. Recall from Section 13.7 that probability clouds come in a variety of shapes and sizes. The actual "look" of a noble gas shell, therefore, is most complex. For our purposes, it suffices to treat each shell as a smooth, 3-dimensional sphere (Figure 16.14).

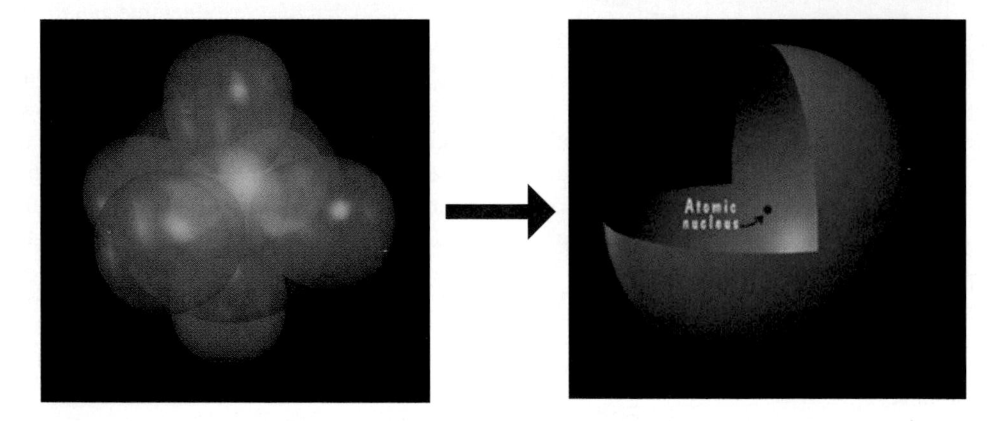

Fig. 16.14
A noble gas shell consists of a number of overlapping probability clouds that can be approximated by a single smooth sphere or circle.

We can account for the behavior of all the elements in the periodic table by using a system of seven concentric noble gas shells. Each shell can hold only a limited number of electrons. The innermost shell can hold 2; the second and third shells, 8 each; the fourth and fifth shells, 18 each; and the sixth and seventh shells, 32 each (Figure 16.15).

Because the electrons of an atom are attracted to the positive charge of the nucleus, they tend to cluster as close to the nucleus as possible. For this reason, they fill inner shells before outer shells.

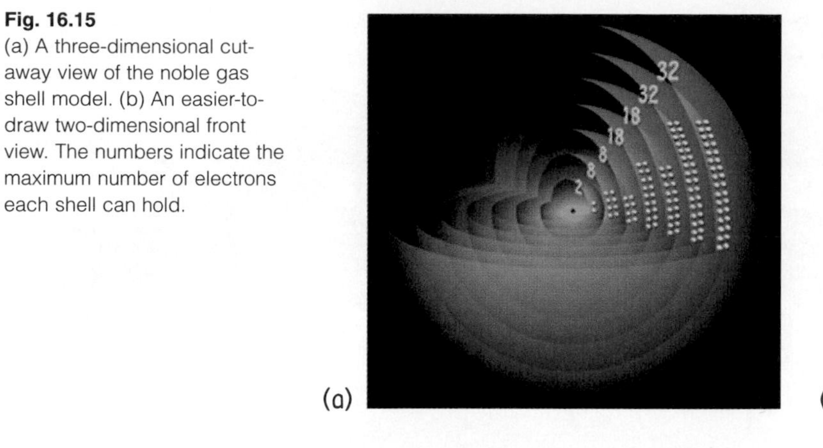

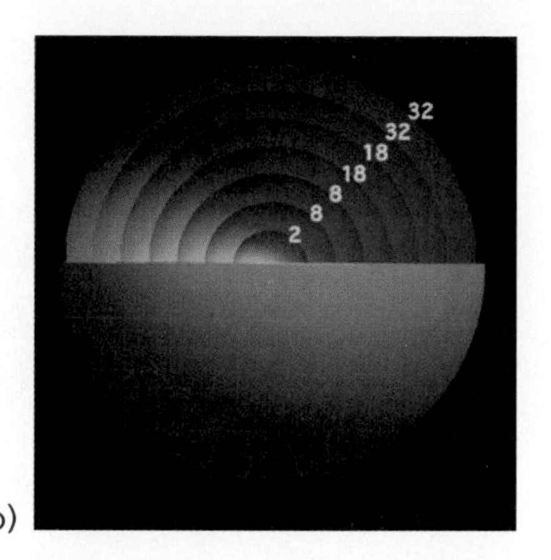

Fig. 16.15
(a) A three-dimensional cutaway view of the noble gas shell model. (b) An easier-to-draw two-dimensional front view. The numbers indicate the maximum number of electrons each shell can hold.

The seven noble gas shells account for the seven periods in the periodic table. Furthermore, the number of elements in each period is related to the shell's capacity for electrons. The first shell, for example, has a capacity for only two electrons, and so we find only two elements, hydrogen and helium, in the first period (Figure 16.16). The second and third shells each have a capacity for eight electrons, and so eight elements are found in both the second and third periods.

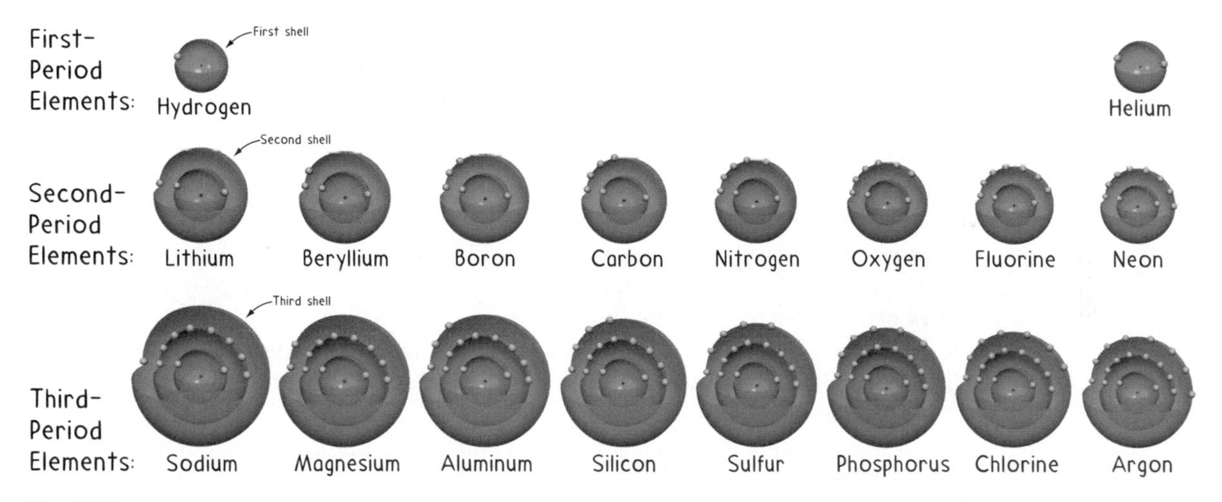

Fig. 16.16
The first three periods of the periodic table according to the noble gas shell model. Elements in the same period have electrons in the same shells. The elements in a period differ from one another by the number of electrons in the outermost shell. Electrons in each shell are shown congregated together so that they may be easily counted; a more accurate depiction would show them spread far apart and moving at high speeds.

Question

Was the noble gas shell model of the atom discovered or invented?

Answer

As we addressed in an earlier question, the organization of the periodic table was *discovered.* Here we see that models are then invented to explain this organization. The noble gas shell model is one such invention. There are others, many of them far more complex than what we are studying here. The noble gas shell model is a conceptual model that allows us to account for observed behavior. An atom doesn't actually contain a series of concentric shells; it merely behaves as though it does.

Fig. 16.17
(a) As students enter the bus, each goes to a separate double seat. (b) Two students sit together only when no more empty double seats are available—that is, when the bus is half full. Similarly, electrons pair up in a shell only after the shell is half full.

The electrons of the outermost occupied shell in any atom are directly exposed to the external environment and are the first to interact with other atoms. Most notably, they are the ones that participate in chemical bonding, as we shall be discussing in the following chapter. These electrons, therefore, are most significant. They are called the **valence electrons**, where the term *valent* refers to the "combining power" of an atom.

The electrons in any shell interact among themselves. Because of their like charge, they repel one another and, as much as is possible, keep to separate locations in the same shell. After the shell is half-filled, however, they can no longer avoid one another and then begin to pair up. A good analogy is kids piling into a school bus furnished with double seats. The students represent electrons, and the bus represents a shell. These particular students—being an unfriendly bunch—prefer to occupy double seats alone, and so as each one boards, she or he plops down in a seat and places a bookbag on the adjoining seat, as in Figure 16.17a. Once all seats are occupied, either by a student or by a bookbag, students boarding have to ask a classmate to remove a bookbag so that they can sit down (Figure 16.17b). Only then do students pair with each other. Electrons in shells are similar. When a shell is filled more than halfway, electrons are forced to pair up.

Electrons spin about a central axis much as the Earth spins about its axis. This electron spin affects the pairing of electrons. A spinning electron generates a magnetic field, and two electrons spinning in opposite directions generate oppositely aligned magnetic fields that attract each other. To indicate the orientation of an electron's spin, the electron is represented as an arrow. Two electrons coupled together spinning in opposite directions are drawn as a pair of arrows pointing in opposite directions (Figure 16.18).

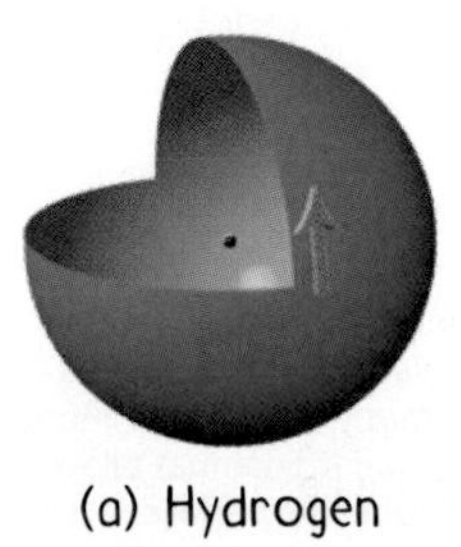

Fig. 16.18
The capacity of the first shell is two electrons. (a) The one electron of a hydrogen atom, seen here as a single arrow, is unpaired. (b) The two electrons of a helium atom, seen here as two oppositely directed arrows, are paired.

Consider the configurations of the electrons in elements of the second period (Figure 16.19). The lithium atom has three electrons: two paired nonvalence electrons in the first (innermost) shell and one unpaired valence electron in the second shell. The carbon atom has six electrons: two paired nonvalence electrons in the first shell and four unpaired valence electrons in the second. Recall that the second shell has a capacity of eight electrons, and this is why four electrons can stay unpaired. With the nitrogen atom, two valence electrons are forced to pair, which leaves only three unpaired. Likewise, oxygen has only two unpaired valence electrons, fluorine has one, and neon, which has enough electrons to fill both the first and second shells, has none.

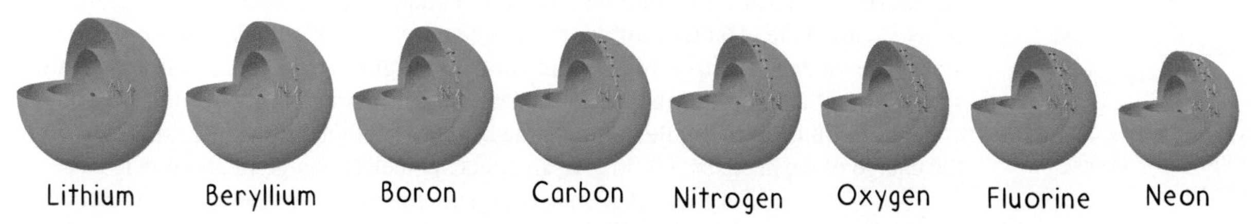

Fig. 16.19
Valence electrons begin to pair once a shell is half-filled.

The periodic table is organized such that elements within the same group have the same number of valence electrons. All alkali metals (group 1), for example, have one unpaired valence electron, and all alkali-earth metals (group 2) have two. All halogens (group 17) have one unpaired valence electron, and all noble gases (group 18) have none.

Question

How many unpaired valence electrons are there in a potassium atom, which has 19 electrons? How many in a rubidium atom, which has 37 electrons? Note the relative location of these elements in the periodic table.

Inner-Shell Shielding

Imagine you are one of two electrons in the lone shell of a helium atom. You share this shell with one other electron, but it doesn't affect your attraction to the nucleus because you both have the same "line of sight" to the nucleus. You and your neighboring electron sense a nucleus of two protons, and you are each attracted to it to the same extent (Figure 16.20). This is like two honeybees buzzing closely around a single flower. Each bee is equally attracted to the flower despite the other bee's presence.

The situation is different for atoms beyond helium, which have more than one shell of electrons. In these cases, inner-shell electrons weaken the attraction between

---

Answer
One for both potassium and rubidium. Two of potassium's electrons pair up in the first shell, eight pair up in the second shell, and another eight in the third shell. This leaves one electron unpaired in the fourth shell. Rubidium has enough electrons to fill the first four shells. This leaves one unpaired in the fifth shell. Another touch of grandeur in the periodic table.

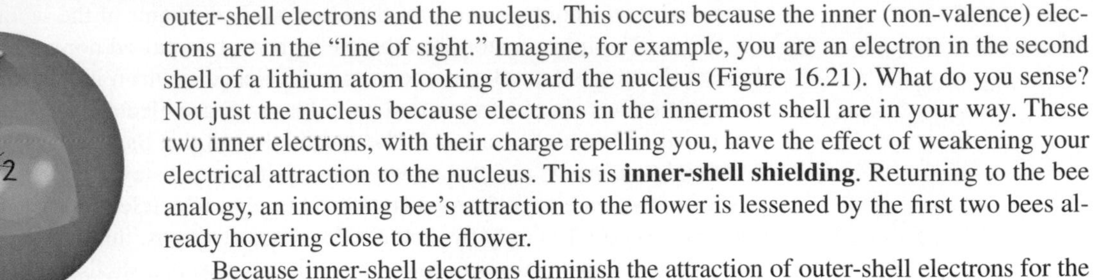

outer-shell electrons and the nucleus. This occurs because the inner (non-valence) electrons are in the "line of sight." Imagine, for example, you are an electron in the second shell of a lithium atom looking toward the nucleus (Figure 16.21). What do you sense? Not just the nucleus because electrons in the innermost shell are in your way. These two inner electrons, with their charge repelling you, have the effect of weakening your electrical attraction to the nucleus. This is **inner-shell shielding**. Returning to the bee analogy, an incoming bee's attraction to the flower is lessened by the first two bees already hovering close to the flower.

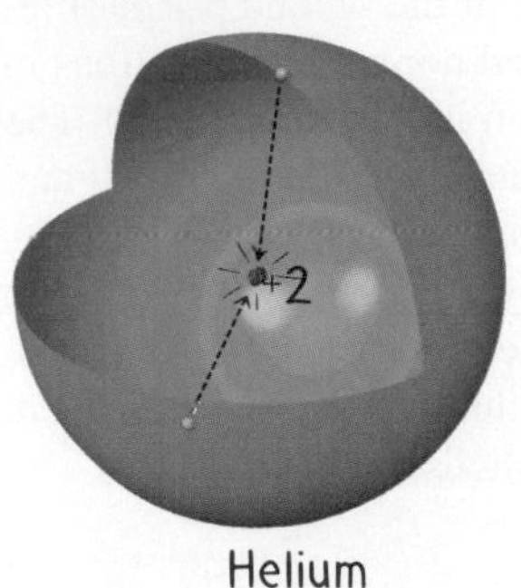

Fig. 16.20
Both electrons in a helium atom have equal exposure to the nucleus; hence they both experience the same nuclear charge.

Because inner-shell electrons diminish the attraction of outer-shell electrons for the nucleus, the nuclear charge sensed by outer-shell electrons is always less than the actual charge of the nucleus. This diminished nuclear charge experienced by outer-shell electrons is called the **effective nuclear charge** and is abbreviated Z^* (pronounced zee-star). The valence electron for lithium shown in Figure 16.21, for example, does not sense the full effect of lithium's +3 nuclear charge (there are three protons in the nucleus of lithium). Instead, the total charge of inner-shell electrons, −2, subtracts from the charge of the nucleus, +3, to give an effective nuclear charge of about +1.

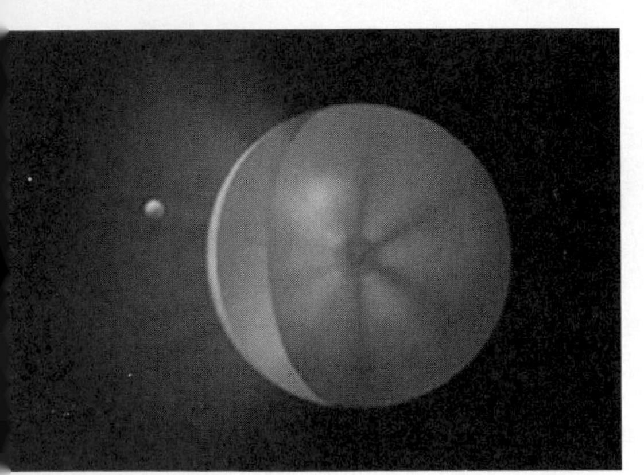

Fig. 16.21
Lithium's valence electron is shown here outside of the first shell. The light streaming from the hypothetical opening of lithium's first shell is suggestive of the nuclear charge the valence electron would experience if it were not for inner shellshielding.

For most elements, subtracting the total number of inner-shell electrons from the actual nuclear charge provides a convenient rough estimate of the effective nuclear charge. The effective nuclear charge experienced by electrons in the third shell of a chlorine atom, for example, is about $17 - 10 = +7$. Similarly, electrons in the fourth shell of a potassium atom sense an effective nuclear charge of $19 - 18 = +1$ (Figure 16.22).

Looking only at inner-shell shielding, we might expect elements in the same group to have the same effective nuclear charge. As you move down any group, however, the atoms become larger because of the increasing number of filled shells. Outer-shell electrons become farther removed from the positively charged nucleus. In accordance with the inverse-square nature of Coulomb's law (Section 8.1), therefore, the electric force of attraction between an electron and the nucleus becomes less as the distance from the nucleus increases (Figure 16.23).

As an example, consider the group 1 elements lithium, sodium, and potassium. According to inner-shell shielding alone, the electron in the outermost shell of each of these elements should experience an effec-

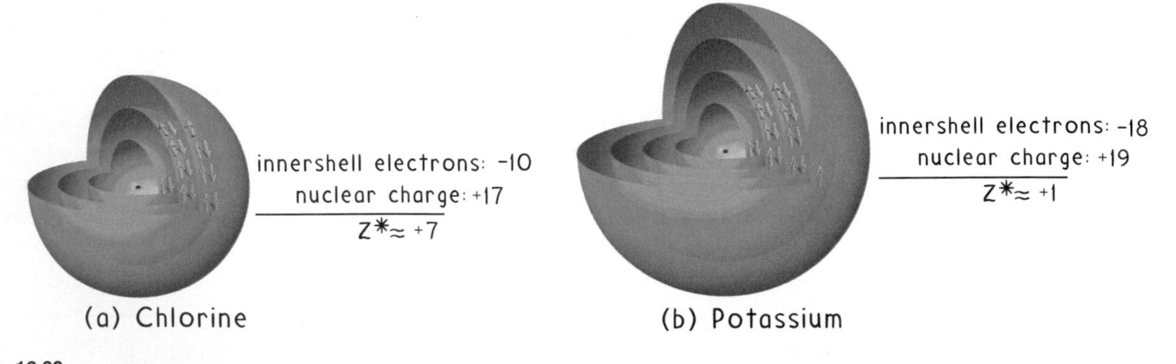

Fig. 16.22
(a) A chlorine atom consists of three shells of electrons. The 2 + 8 = 10 electrons of the first two shells shield the 7 electrons of the third shell from the +17 nucleus. The third-shell electrons therefore experience an estimated effective nuclear charge of +7. (b) For potassium, fourth-shell electrons experience an estimated effective nuclear charge of +1.

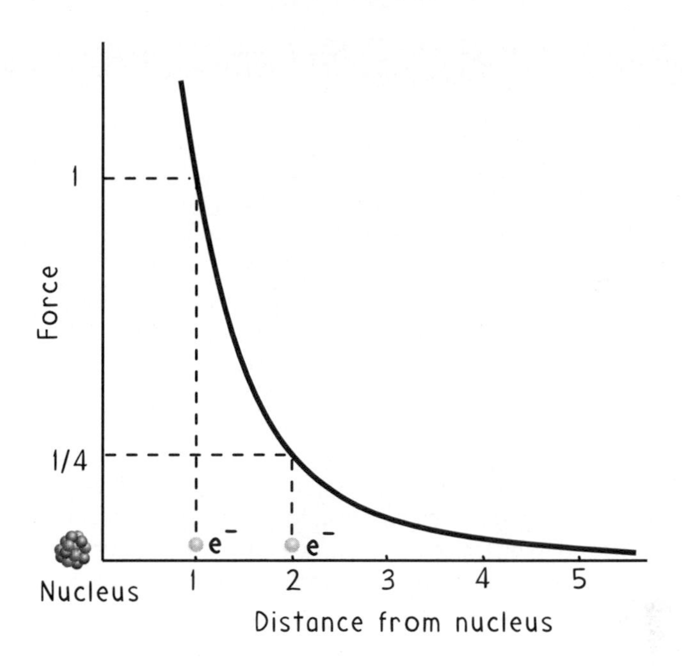

Fig. 16.23
The electric force is inversely proportional to the square of the distance from the nucleus. Double the distance and the strength of the force becomes $\frac{1}{2^2} = \frac{1}{4}$ as great. Triple the distance and the force becomes $\frac{1}{3^2} = \frac{1}{9}$ as great.

tive nuclear charge of about +1. Taking the distance factor into account, however, we judge that the effective nuclear charge in the outermost shell of lithium is greater than that of sodium, which is greater than that of potassium. Why? Because as we go from lithium to sodium to potassium, the outermost electron is farther from the nucleus (Figure 16.24).

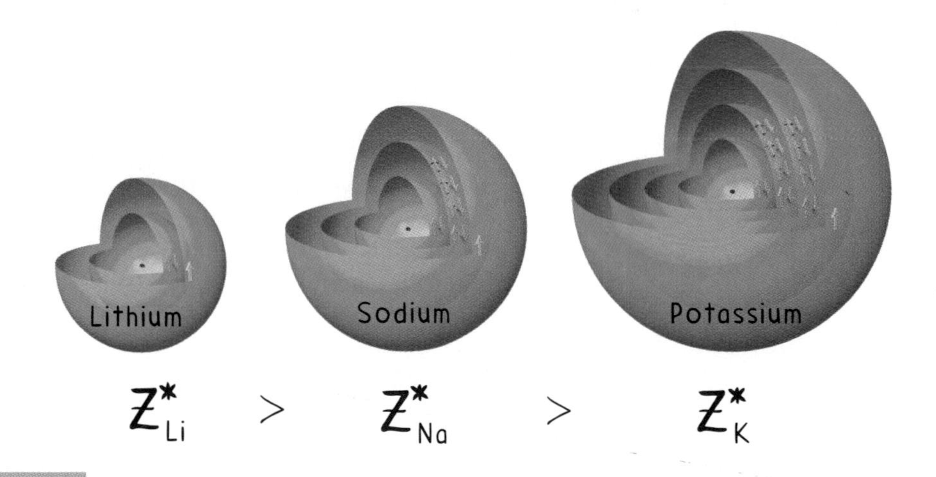

Fig. 16.24
The effective nuclear charge Z^* experienced by an outermost electron decreases with increasing distance from the nucleus. For this reason, the effective nuclear charge experienced by the outermost electron in lithium is greater than that in sodium, which is greater than that in potassium.

16.5 Periodic Trends

Various properties of elements change gradually as you move in any direction across or down the periodic table. The effective nuclear charge experienced by electrons influences such properties as atomic size and ionization energy. As you move across any period from left to right, the degree of inner-shell shielding remains the same. Conse-

Link to Relativity—Why Gold's Color

The type of light an element emits or absorbs depends on the arrangement of its electrons. Recall from Chapter 13 that the emission and absorption of light has to do with the electrons that orbit the atomic nucleus. The arrangement of electron orbits (shells) in a silver atom, for example, results in strong absorption of ultraviolet light but not light of visible frequencies. So instead of being absorbed by silver atoms, all the frequencies of visible light are reflected. Hence the characteristic silvery-white color of silver when viewed in white light.

Gold lies directly below silver in the periodic table, so we might think that gold should similarly absorb ultraviolet light and reflect visible light. Why gold doesn't look like silver in color baffled early investigators, until they realized that Einstein's special theory of relativity is involved. In the early 1900s Einstein discovered that the nature of certain things changes markedly at speeds approaching the speed of light (300,000 km/sec). At near–light speeds, for example, the momentum of a moving object becomes more than its mass multiplied by its velocity (Section 3.1). Instead, a warping of space–time dictates that the object's momentum is appreciably greater by a factor known as the Lorentz factor. This is explored more fully in Chapter 30 (Section 30.6).

The speeds of electrons in lighter atoms, including silver, are much less than the speed of light, and their momenta are closely approximated by classic methods. But in heavier atoms like gold, where greater nuclear charge tugs on electrons, the speeds of innermost electrons are about 60 percent the speed of light. According to relativity, these electrons have a greater than usual momentum. A consequence of this increased momentum is that gold's innermost electrons are drawn closer to the nucleus. This increases their ability to shield outer electrons, which expand farther away from the nucleus. The upshot is a shift in electron orbits, which in turn shifts the frequencies of light that the atom absorbs. While silver absorbs ultraviolet light, gold shifts to absorption of the violet and blue regions of the visible spectrum. The absorption of these components of white light results in reflection of red and yellow frequencies of light, which combine to produce the characteristic gold color.

So gold's color is a result of special relativity. In Section 30.4 you'll see that at 60 percent the speed of light, gold's innermost electrons experience only 52 seconds for each one of our minutes. A diamond may be forever, but the innermost electrons of gold are 8 seconds per minute slow. How about that!

quently, as the actual nuclear charge becomes greater, so does the effective nuclear charge. As you move down any group, the effective nuclear charge becomes less because of an increasing number of shells. Putting these two trends together, we find that elements at the upper right of the periodic table have the greatest effective nuclear charges in their outermost shell while those at the lower left have the weakest (Figure 16.25).

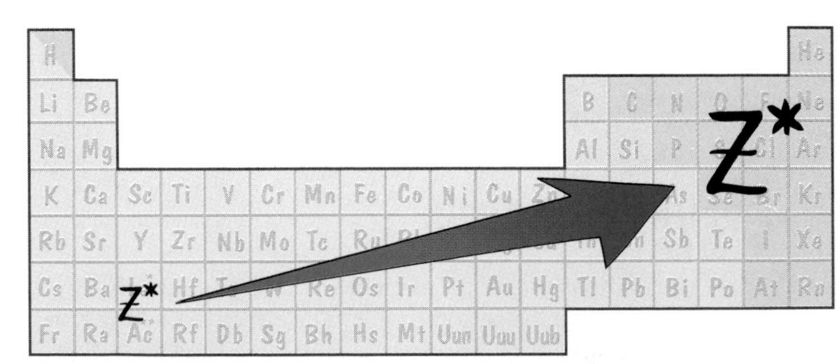

Fig. 16.25
The effective nuclear charge Z^* experienced by the outermost electrons of elements at the upper right of the periodic table is greater than that experienced by elements at the lower left.

Atomic Size

From left to right across any row of the periodic table, the atoms get *smaller.* Let's look at this trend from the point of view of effective nuclear charge. Consider lithium's second shell valence electron, which experiences an effective nuclear charge of about $+1$ (see Figure 16.19). Then look at the second shell valence electrons for neon, which experience an effective nuclear charge of about $+8$. Because the neon valence electrons experience a greater net attraction to the nucleus, they are pulled in closer to it. So neon, which is nearly three times as massive as lithium, is considerably smaller. In general, across any period from left to right the atoms become smaller because of an increase in effective nuclear charge.

Moving down a group, atoms get larger because of an increasing number of occupied shells. Whereas helium is small, having only one occupied shell, radon is large, having six occupied shells. As was shown in Figure 16.2(a) atoms at the upper right are smallest and atoms at the lower left are largest.

Ionization Energy

In the beginning of this chapter we stated that when elements are listed in order of atomic number, there are jumps in the ionization energies—the amount of energy required to remove an outer electron from an atom. The more tightly the electron is held to the nucleus, the more difficult it is to remove. Now we see that ionization energy and effective nuclear charge go hand in hand: The greater the effective nuclear charge, the higher the ionization energy. So, as we saw in Figure 16.2(b) elements at the upper right of the periodic table have the greatest ionization energies and those at the lower left have the smallest.

Question

From which would an electron be more easily removed, a sodium atom or a neon atom?

From Figure 16.16 we can see that in each group 1 element, there is one electron in the outermost occupied shell. The ionization energy for this electron is relatively low, which is to say, this lone electron is easily knocked away, resulting in the $+1$ ion. So, how easily might a second electron be pulled away from any group 1 atom? This second electron would necessarily have to come from an inner shell, where the effective nuclear charge is about eight times greater. So, while it is certainly possible for a group 1 atom to lose a second electron, this does not readily happen without the input of an unusually large amount of energy (Figure 16.26).

By the same rationale we can see why the atoms of group 2 elements all tend to form $+2$ ions. Note from Figure 16.16 how these atoms have two electrons in the outermost shell. Because of low ionization energies, both of these electrons are easily lost.

Answer

The effective nuclear charge in the outermost occupied shell of a sodium atom is about $+1$, while in the outermost occupied shell of the neon atom it is about $+8$. It is markedly easier, therefore, to pluck an electron from the sodium. Indeed, experiments show that the ionization energy of sodium is about four times less than that of neon.

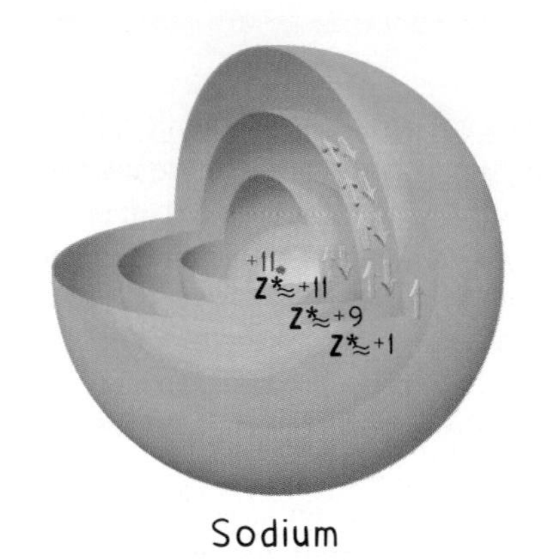

Fig. 16.26
The outermost electron of a sodium atom is relatively easy to remove. Removing a second electron, however, would mean extracting it from an inner shell, where the effective nuclear charge is much greater. As a result, sodium and the other group 1 elements lose inner electrons only under extraordinary conditions.

Group 2 atoms, therefore, commonly form the +2 ion. Only under extremely forced conditions will a third electron be lost and a +3 ion be formed.

For most of the transition metals as well as for the lanthanides and actinides, different numbers of electrons may be lost to form a variety of positively charged ions. In principle, the highest charge possible is the number of electrons in the outermost shell, which is given by the group number. In practice, the common ions formed by most metals are either +2 or +3 ions (Figure 16.27).

For the elements of groups 13 through 18, the number of electrons lost in forming ions is generally equal to or less than the group number minus ten. The elements of group 13, for example, tend to form +3 ions, while those of group 14 tend to form +4 ions. This assumes, however, that enough energy is available to remove these electrons. Recall that as you move across the periodic table the ionization energy becomes greater; hence removing electrons becomes increasingly difficult. Ultimately, there comes a point where the atoms have a greater tendency to gain electrons than to lose them. This transition occurs in moving from the metallic to the nonmetallic elements.

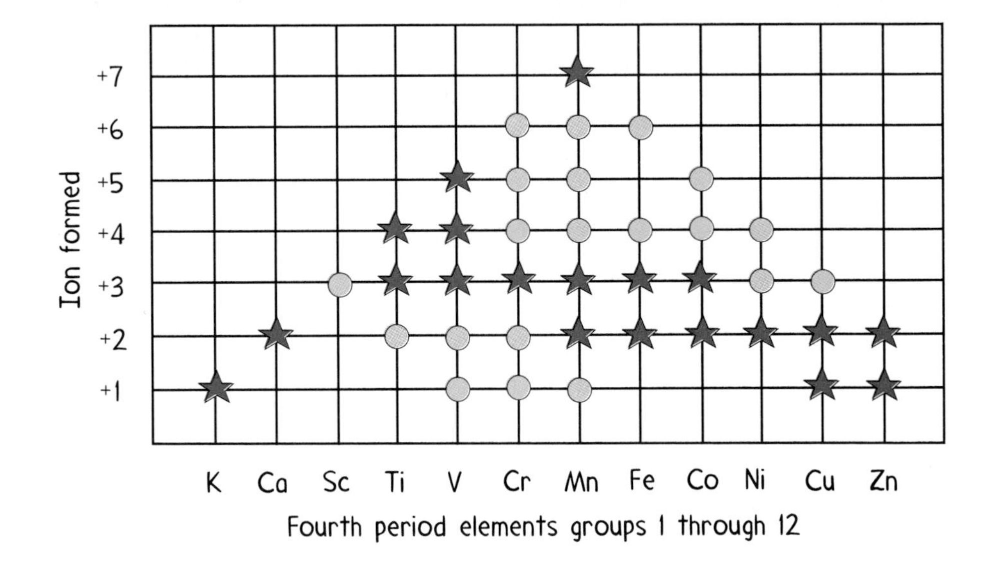

Fig. 16.27
The positive ions formed by the first 12 elements of the fourth period. The more commonly formed ions are indicated by stars; ions less commonly formed are indicated by dots. Fig 16.30 Hydrogen has an affinity for one additional electron.

Electron Affinity

Whereas ionization energy relates to the difficulty of *losing* electrons, *electron affinity,* by contrast, relates to the ease of *gaining* them. **Electron affinity** is the ability of an atom to attract additional electrons. A neutral hydrogen atom, for example, has one positive proton balanced by a single negative electron. Given the opportunity, this neutral hydrogen readily picks up a second electron to become a negatively charged ion (Figure 16.28).

Like ionization energy, electron affinity depends on how tightly the positively charged nucleus is able to hold electrons as measured by the effective nuclear charge.

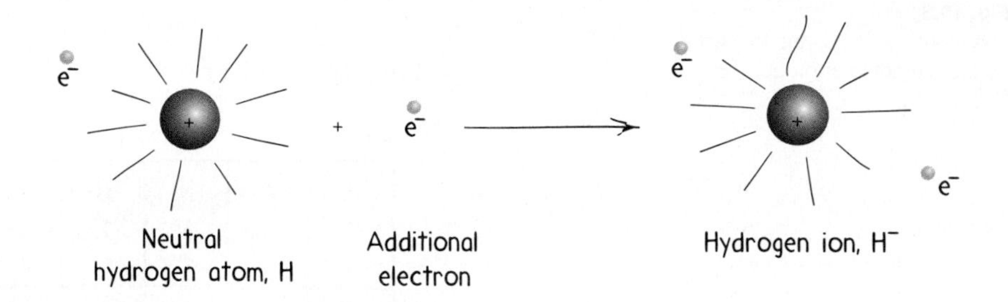

Fig. 16.28
Hydrogen has an affinity for one additional electron.

Atoms with the greatest electron affinities (and ionization energies) are the nonmetals at the upper right of the periodic table, while the metal atoms at the lower left have the smallest electron affinities (and ionization energies) (Figure 16.29).

Electron affinity depends on the effective nuclear charge and also on the number of empty spaces available in the outermost shell. For example, the effective nuclear

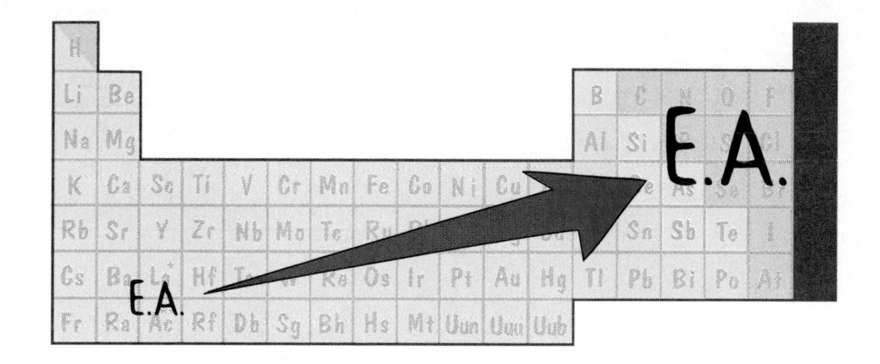

Fig. 16.29
Electron affinity (E.A.) parallels effective nuclear charge and ionization energy as a periodic trend. Those elements at the upper right have the greatest electron affinity, and those at the lower left have the smallest.

charge as sensed from the outermost shell of noble gas atoms is quite large. Because the shell is filled, however, there is no space available for an additional electron (Figure 16.30a). The noble gas atoms, therefore, tend not to gain (or lose) electrons, which is why noble gas elements are so chemically unreactive. Each group 17 atom, however, has space for an additional electron in its valence shell. These atoms, with their strong effective nuclear charges, therefore have an affinity for one additional electron, which means that they tend to form the -1 ion (Figure 16.30b). Each group 16 atom has space

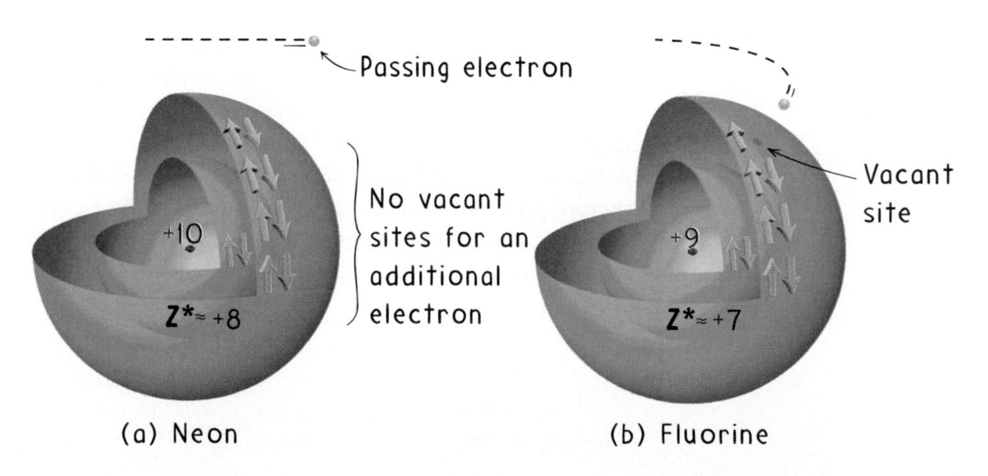

Fig. 16.30
(a) The effective nuclear charge felt from the outermost shell of a neon atom is about +8, which is quite large. This outermost shell, however, is already full, and so no additional electrons can be accommodated. (b) The effective nuclear charge felt from the outermost shell of a fluorine atom is about +7. Any passing free electron is immediately pulled into this +7 charge, thereby giving the fluorine atom one additional electron and making it a fluoride ion, F^-. Might a second additional electron be pulled into a fluorine atom?

Fig. 16.31
The most common ions formed by the elements of groups 13 through 18.

Group number

13	14	15	16	17	18
					He none
B -5 +3	C -4 +4	N -3 +5	O -2 +6	F -1 +7	Ne none
Al +3	Si -4 +2,4	P -3 +5	S -2 +6	Cl -1 +7	Ar none
Ga +1,3	Ge -4 +2,4	As -3 +5	Se -2 +6	Br -1 +7	Kr none
In +1,3	Sn -4 +2,4	Sb -3 +5	Te -2 +6	I -1 +7	Xe none
Tl +1,3	Pb +2,4	Bi -3 +5	Po -2 +6	At -1 +7	Rn none

Ge
Negative ion formed → -4
Positive ions formed → +2,4

for two additional electrons, which explains how it is that these atoms tend to form −2 ions. Similarly, group 16 atoms tend to form −3 ions, and group 14 atoms tend to form −4 ions (see Figure 16.31).

Mendeleev and other early investigators couldn't comprehend this organization. Connections were suspected, but they didn't make sense. That's because these early scientists had no suitable conceptual model of the atom's internal structure. They had no idea that an atom has a tiny positively charged nucleus that attracts negatively charged electrons. They couldn't possibly think in terms of shells, valence electrons, inner-shell shielding, and effective nuclear charge—the keys to understanding the regularities of the periodic table. This knowledge came only after enormous effort and experimentation. And today we have only to read about the results. With study we can picture processes that the early chemists couldn't even dream of and begin to understand the remarkable organization of the periodic table. Chemistry may be intricate, but it *is* understandable. The connections make sense.

Periodic Trends

(Data unavailable for nonprojected elements)

Atomic Size

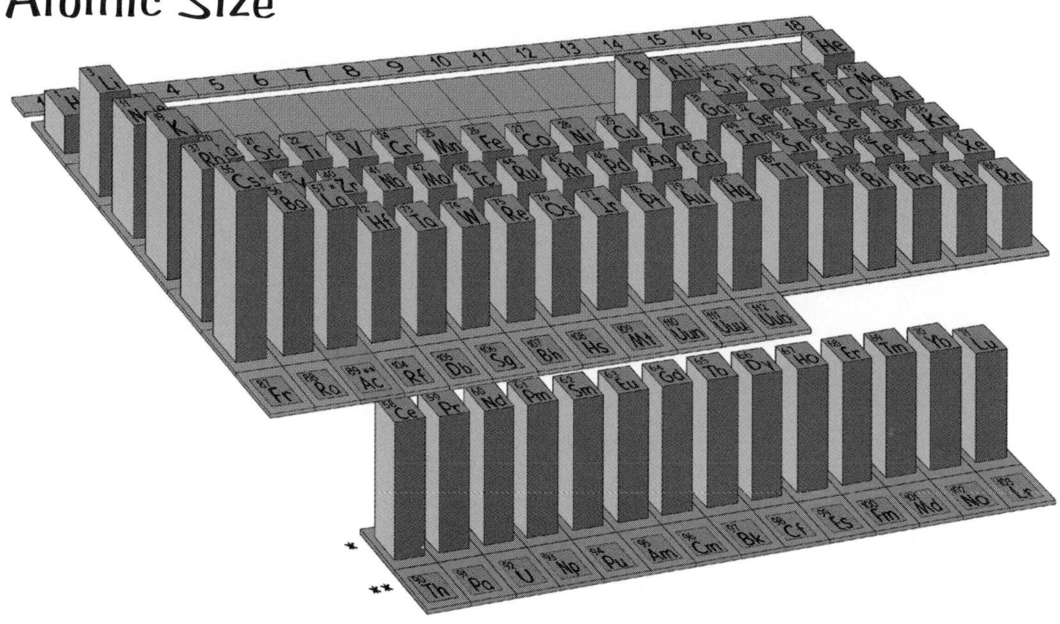

Ionization Energy

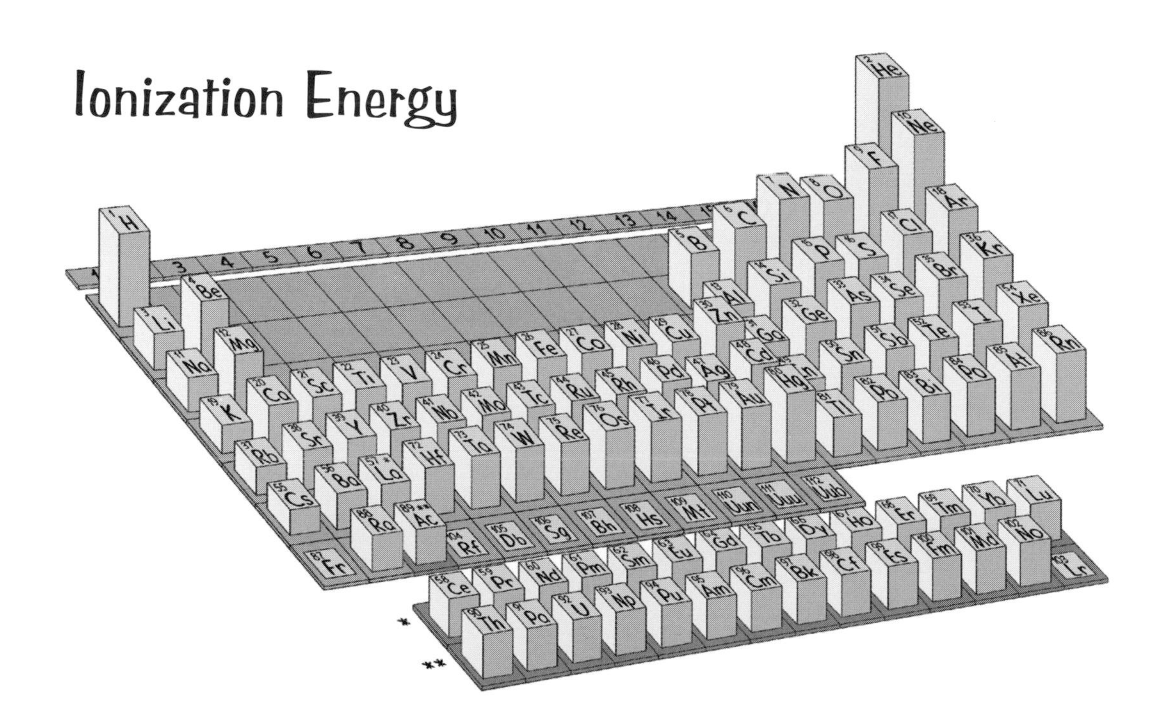

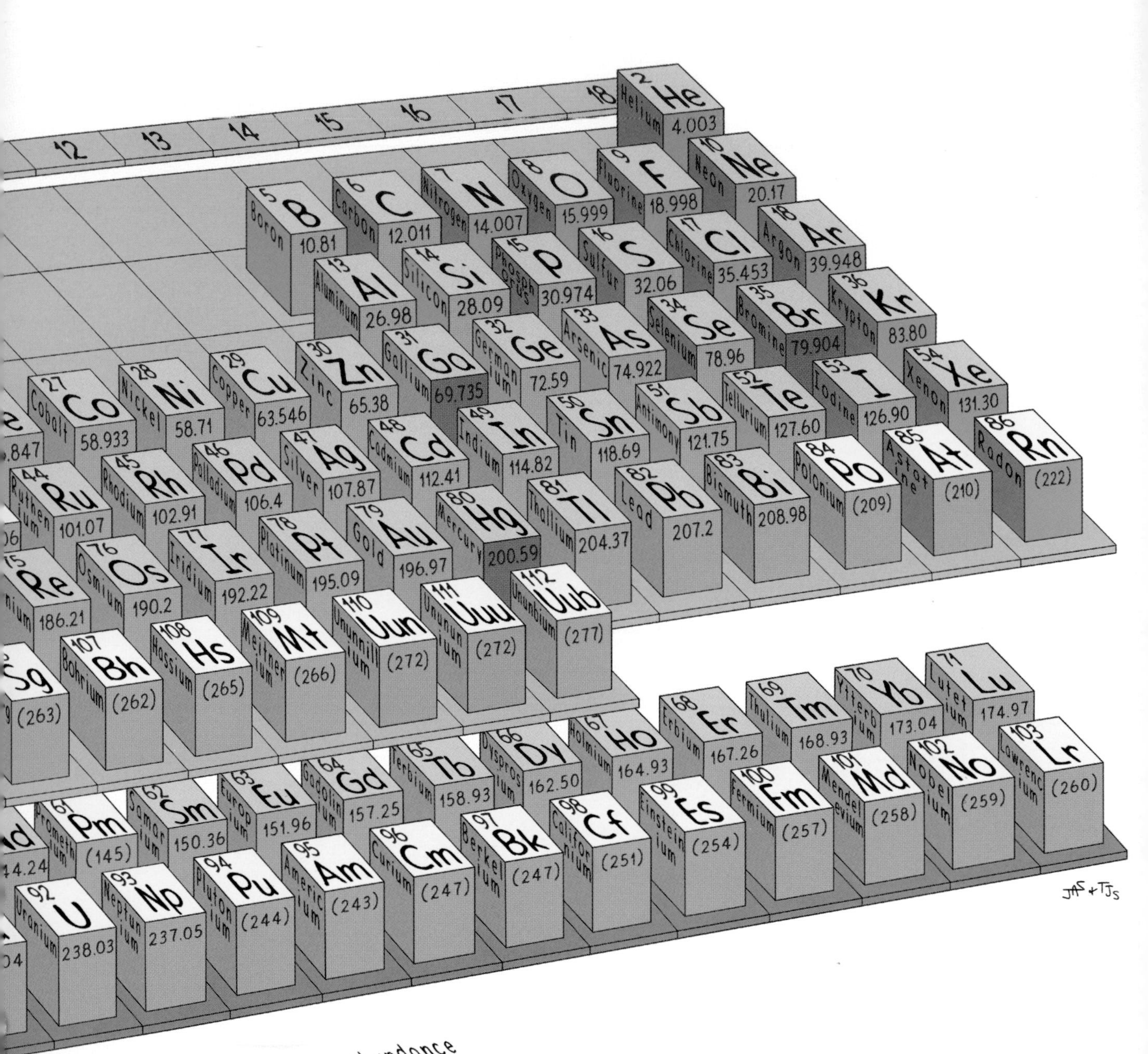

asses are averaged by isotopic abundance
earth's surface, expressed in atomic mass units
u). Atomic masses for radioactive elements shown
parentheses are the whole number nearest the most
stable isotope of that element.

The Periodic Table

Guide to the Elements

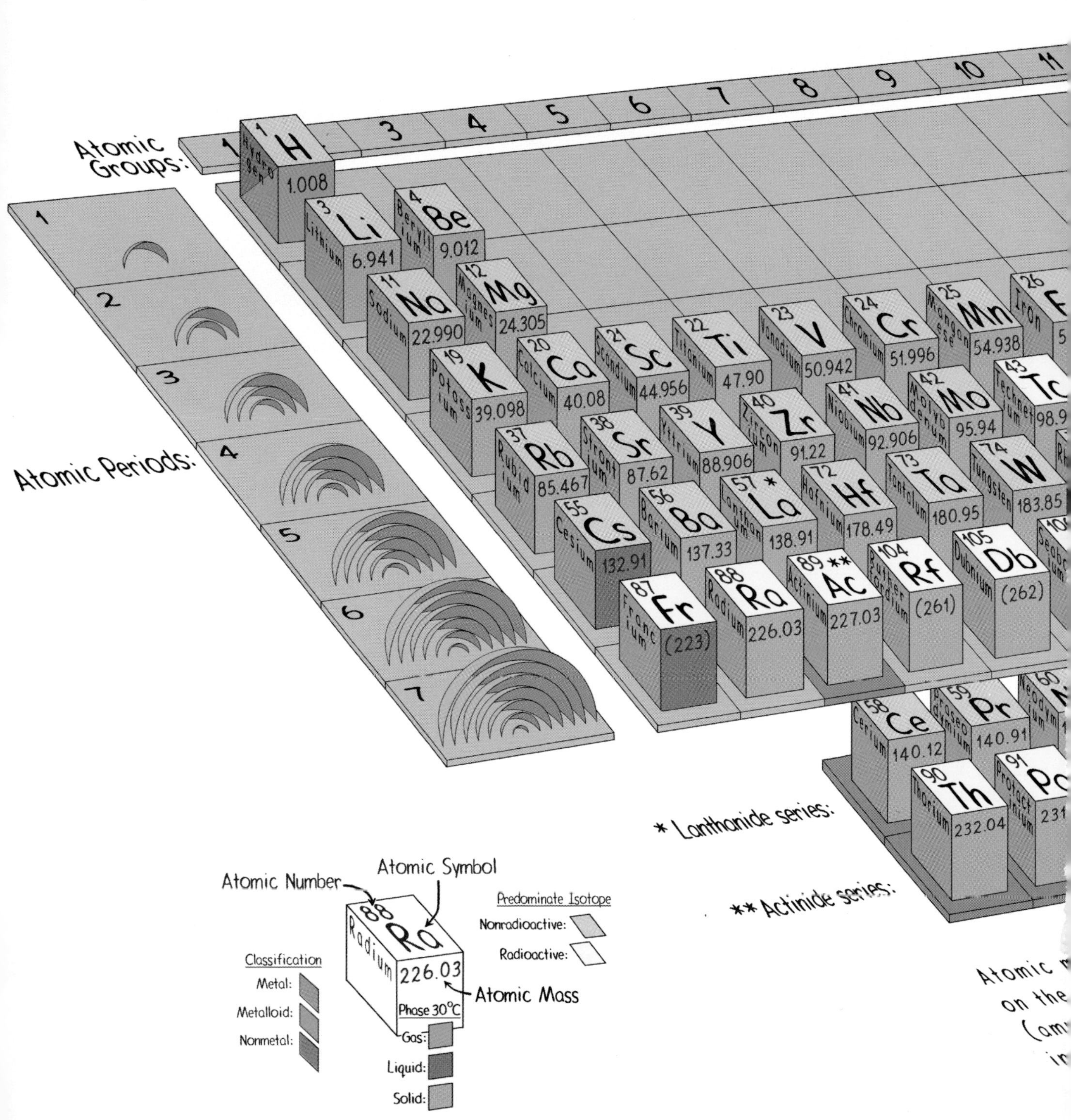

Electronegativity

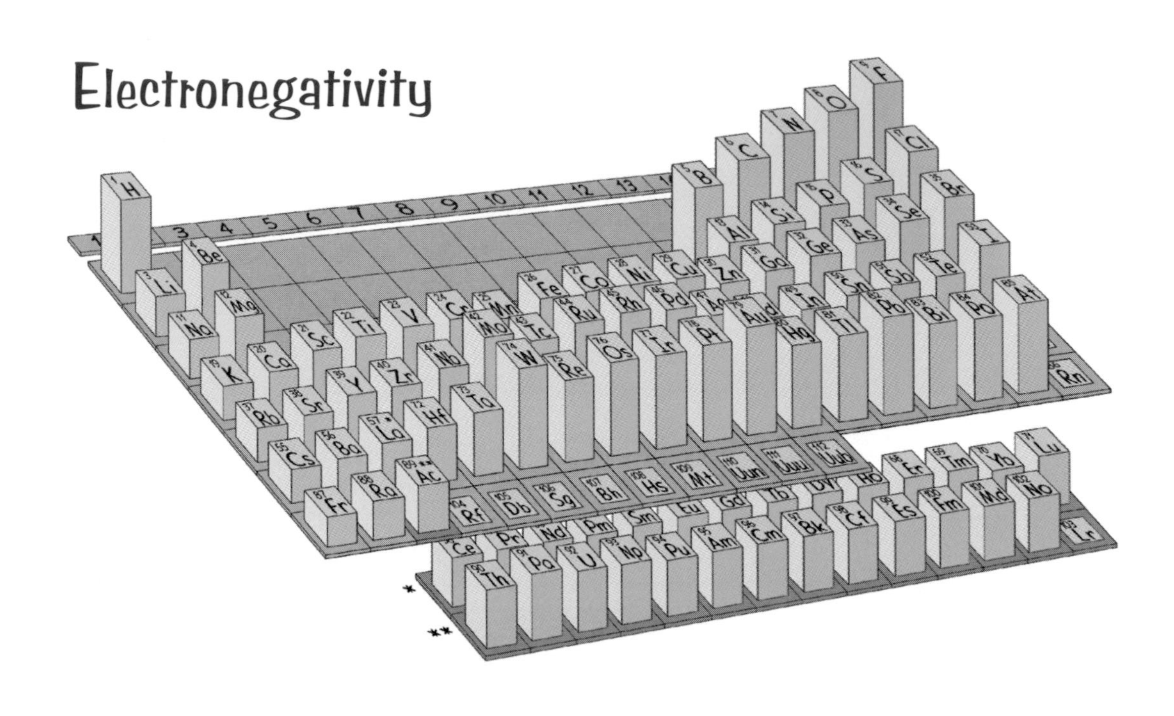

Density

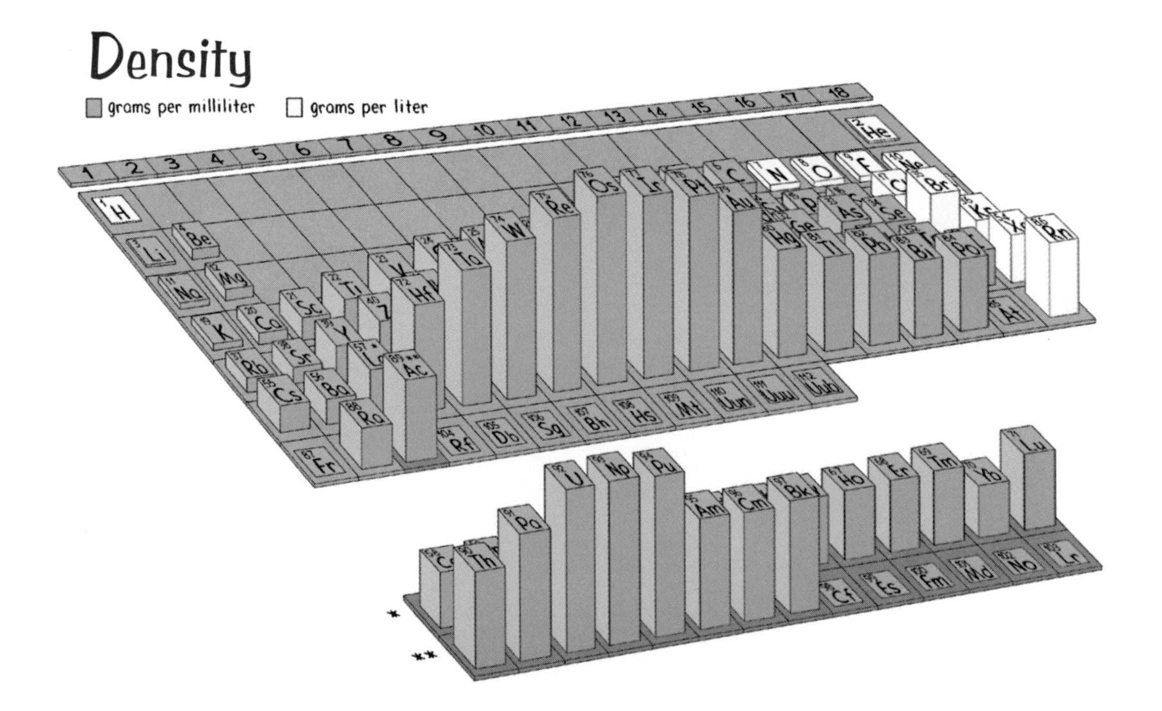

Summary of Terms

- **Periodic table** A highly organized chart listing all the known elements arranged in horizontal rows called periods and vertical columns called groups.
- **Ionization energy** The amount of energy required to pull an electron away from an atom.
- **Period** A horizontal row in the periodic table.
- **Metal** An element that is generally shiny, opaque, malleable, ductile, and a good conductor of electricity and heat.
- **Nonmetal** An element that is nonmalleable, nonductile, and a poor conductor of electricity and heat.
- **Metalloid** One of six elements—boron, silicon, germanium, arsenic, antimony, tellurium—that exhibit both metallic and nonmetallic properties.
- **Group** A vertical column in the periodic table.
- **Alkali metal** A group 1 element.
- **Alkali-earth metal** A group 2 element.
- **Halogen** A group 17 element.
- **Noble gas** A group 18 element.
- **Transition metals** The elements of groups 3 through 12.
- **Inner transition metals** Two subgroups—lanthanides and actinides—of the transition metals.
- **Lanthanides** The inner transition metals of the sixth period.
- **Actinides** The inner transition metals of the seventh period.
- **Noble gas shell** A spherical region of space about the atomic nucleus where electrons of a similar energy level may be found.
- **Shell** A region of space about the atomic nucleus within which an electron may reside.
- **Valence electron** Any electron in the outermost shell of an atom.
- **Inner-shell shielding** The tendency of inner-shell electrons to partially shield outer-shell electrons from the nuclear charge.
- **Effective nuclear charge** The nuclear charge experienced by outer-shell electrons, which is diminished by their distance from the nucleus and by the shielding effect of inner-shell electrons.
- **Electron affinity** The ability of an atom to attract additional electrons.

Review Questions

Organizing the Elements

1. What fact about the properties of elements becomes apparent as we list them in a single vertical column?
2. What is represented by each horizontal row of the periodic table?
3. Helium and beryllium are both the second element in their respective rows. Why are they not placed directly above and below each other in the periodic table?
4. Why is the position of helium shifted so that this element appears directly above neon?

Metals, Nonmetals, and Metalloids

5. Are most elements in the periodic table metallic or nonmetallic?
6. What are some properties of metals, nonmetals, and metalloids that allow us to differentiate among the three groups?
7. Where are the metalloid elements located in the periodic table?

Atomic Groups and Periods

8. How many groups are there in the periodic table?
9. Why are the group 1 elements called alkali metals?
10. Why does hydrogen take on the properties of a metal in the planet Jupiter but not on the planet Earth?
11. Why are the group 2 elements called alkali-earth metals?
12. What is the primary use of transition metals?
13. Why are the inner transition metals not listed in the main body of the periodic table?
14. Why is it difficult to purify an inner transition metal?
15. Why are the group 17 elements called halogens?

Noble Gas Shells

16. What is the relationship between the number of shells in atoms and the number of periods in the periodic table?
17. What is the relationship between the electron capacities of shells and the number of elements in each period of the periodic table?
18. What is a valence electron?
19. When do valence electrons start to pair up?
20. Place the proper number of electrons in each shell for sodium, Na (atomic number 11); rubidium, Rb (atomic number 37); krypton, Kr (atomic number 36); and chlorine, Cl (atomic number 17). Use arrows to represent electron spin.

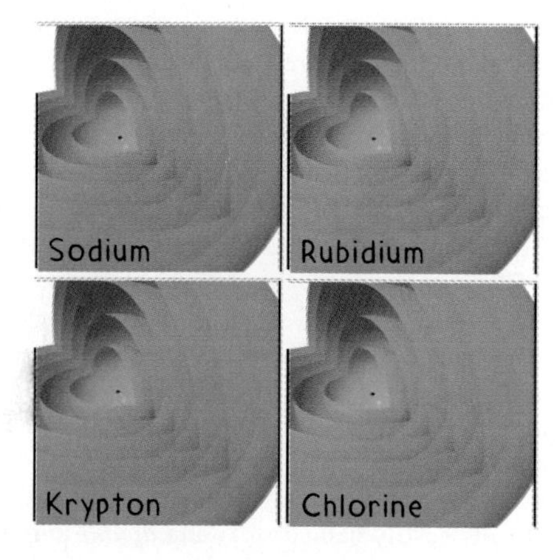

21. How many atomic shells are occupied by electrons in a gold atom, Au (atomic number 79)?
22. How many electrons are there in the outermost shell of a carbon atom?
23. The nucleus of a carbon atom has a charge of +6, but this is not the charge sensed by carbon's valence electrons. Explain.
24. What is meant by effective nuclear charge?
25. What is the approximate effective nuclear charge for a valence electron in krypton, Kr (atomic number 36)?

Periodic Trends

26. How is it possible that as atoms get more massive they become smaller?
27. How is it possible that as atoms get more massive they become larger?
28. Smaller atoms tend to have greater ionization energies. Why?
29. The neon atom has no affinity for an additional electron. Why?
30. Which of the following concepts underlies all the others: electron affinity, ionization energy, effective nuclear charge, atomic size?

Home Project

Build a periodic table with a deck of playing cards. Lay out all 52 cards in a grid of 4 horizontal rows and 13 vertical columns. For each horizontal row, use the same suit starting with the ace on the left and ending with the king on the right. Lay out clubs on the top row, followed by diamonds, hearts, and spades. To mimic the chemical periodic table, assume that the suits have weight such that any diamond is heavier than any club, any heart is heavier than any diamond, and any spade is heavier than any heart.

Bored by this activity, imagine you meander away to read the newspaper, where you see the headlines: "Three New Suits Discovered! Card Players of the World Stunned." Reading on, you find that the new suits were hastily named *squares, circles,* and *X's*. Furthermore, initial trials at a recent bridge tournament indicated that all three suits were lighter than any of the traditional suits. The squares weighed in as the lightest of all, followed by the heavier circles, and then the X's. Most peculiar was the discovery that the squares consisted only of two cards (ace and king), while both the circles and the X's were missing the cards numbered 3 through 7. Curious? OK, go ahead now and make these new cards by cutting up some paper, 3 × 5 cards, or envelopes and see how they might be organized around or within the array of cards you have already laid down.

Wanting to learn more, you turn on your TV to a 24-hour news channel, and sure enough the grand master of the bridge tournament is seen announcing the startling discovery of two additional suits hastily named *stars* and *double stars*. Tournament play showed clearly that all star and double star suits consisted only of the ace through 7. More puzzling still was the evidence suggesting that all stars were heavier than the 3 of hearts but lighter than the 4 of hearts. Similarly, all double stars appeared heavier than the 3 of spades but lighter than the 4 of spades. Curious? OK, go ahead now and make these new cards and see how they might be organized around or within the array of cards you have already laid down.

Questions

1. Why is it so unorganized to have the ace and king of squares right next to each other?
2. Why place a large gap between the 2 and 8 cards of the circles and X's?
3. Why is it so unappealing to shift the 4 through king of hearts and the 4 through king of spades over to the right to make room for the new suits of stars and double stars?
4. Why is there such a large gap between beryllium and boron in the periodic table?
5. Why do you suppose the elements of atomic numbers 58 through 71 and 90 through 103 are set apart from the main body of the periodic table?

Exercises

1. A radioactive isotope of strontium, Sr, is especially dangerous to humans because it tends to accumulate in calcium-dependent bone marrow tissues. Suggest how this fact relates to what you know about the organization of the periodic table.
2. Germanium computer chips operate faster than silicon computer chips. So how might a gallium chip compare with a germanium chip?
3. What physical property of gallium prohibits its use in the fabrication of integrated circuits?
4. If an element of atomic number 118 is ever prepared, what properties might you expect it to have?
5. Another interesting periodic trend is density. We find that osmium, Os, has the greatest density of all elements, and, with some exceptions, the closer an element is positioned to osmium, the greater its density. Based upon this trend, list the following elements in order of increasing density: copper, gold, platinum, Pt, and silver.
6. Name ten elements that you have access to as a consumer here on planet Earth.
7. How many valence electrons are there in magnesium, Mg (atomic number 12)? How many of these are unpaired?
8. How many valence electrons are there in fluorine, F (atomic number 9)? How many of these are unpaired?
9. Which experiences a greater effective nuclear charge, an electron in the outer shell of neon or an electron in the outer shell of sodium? Why?
10. An electron in the outermost shell of which element experiences the greatest effective nuclear charge:

(a) sodium (b) potassium (c) rubidium (d) cesium (e) all the same

11. List the following atoms in order of increasing atomic size: thallium, Tl; germanium, Ge; tin, Sn; phosphorus, P.
(smallest) _______ < _______ < _______ < _______ (largest)

12. Arrange the following elements in order of increasing ionization energy: tin, Sn; lead, Pb; phosphorus, P; arsenic, As.
(weakest) ______ < ______ < ______ < ______ (strongest)

13. Suppose that, for all elements, the innermost shell had an unlimited capacity for electrons. If this were true (it most definitely is not), what trend would you predict for atomic size as a function of increasing atomic number?

14. Bromine has an affinity for how many additional electrons?

15. How many electrons are there in the third shell of the sodium ion, Na^{+}? How about in the third shell of the chloride ion, Cl^{-}?

16. How are the arrangements of electrons in the neon atom and the calcium ion, Ca^{2+}, similar? How are they different?

17. Based on the noble gas shell model, which would you expect to be larger: a neutral sodium atom, Na, or a positively charged sodium ion, Na^{+}?

18. Based on the noble gas shell model, which would you expect to be larger: a neutral fluorine atom, F, or a negatively charged fluoride ion, F^{-}?

19. Use the noble gas shell model to explain why the alkali metals (group 1) tend to form $+1$ ions whereas the alkali-earth metals (group 2) tend to form $+2$ ions.

20. It is relatively easy to pull one electron away from a potassium atom but very difficult to remove a second one. Use the noble gas shell model and the concept of effective nuclear charge to explain why this is so.

21. Oxygen, O, has an affinity for two additional electrons. Its affinity for the second electron, however, is somewhat less than its affinity for the first. Explain.

22. The effective nuclear charge in the outer shell of a neon atom is relatively strong (about $+8$). Why is it that the neon atom has no affinity for an additional electron?

23. Why do nonmetals tend to have a greater electron affinity than metals?

24. While it's true that atoms that have high electron affinities also have high ionization energy, it is not always true that atoms that have high ionization energies also have high electron affinities. Why?

25. Does a shell have to contain electrons in order to exist?

26. Not mentioned in the text is that noble gas shells may be subdivided into smaller capacity shells called *subshells*. As shown in the figure below, the fourth noble gas shell, for example, can be subdivided into three subshells with capacities of 2, 10, and 6 electrons, which together make a total fourth shell capacity of 18 electrons. Examine the charts of Figure 16.2 carefully and you will find evidence for these subshells. What is this evidence? Explain your observations.

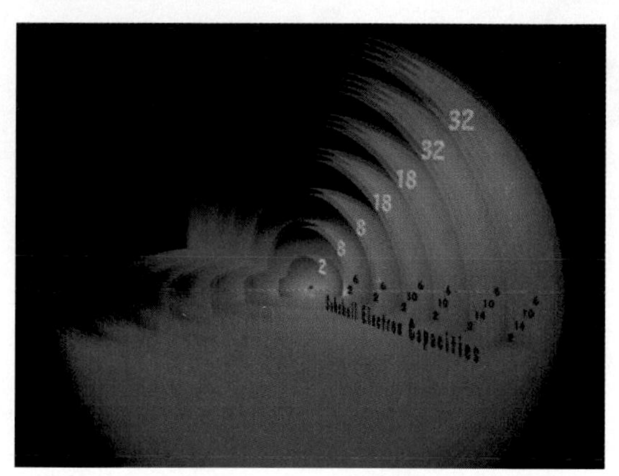

27. Why does hydrogen take on the properties of a metal in the planet Jupiter, but not the planet Earth?

28. Noble gas atoms tend not to form negatively charged ions because there is no room available for additional electrons in the valence shell. Why do they tend not to form positively charged ions?

29. Why is the ionzation energy of a sodium atom greater than that of a potassium atom?

30. Why are the inner transition elements placed below the main body of the periodic table?

CHAPTER

17

Chemical Bonding

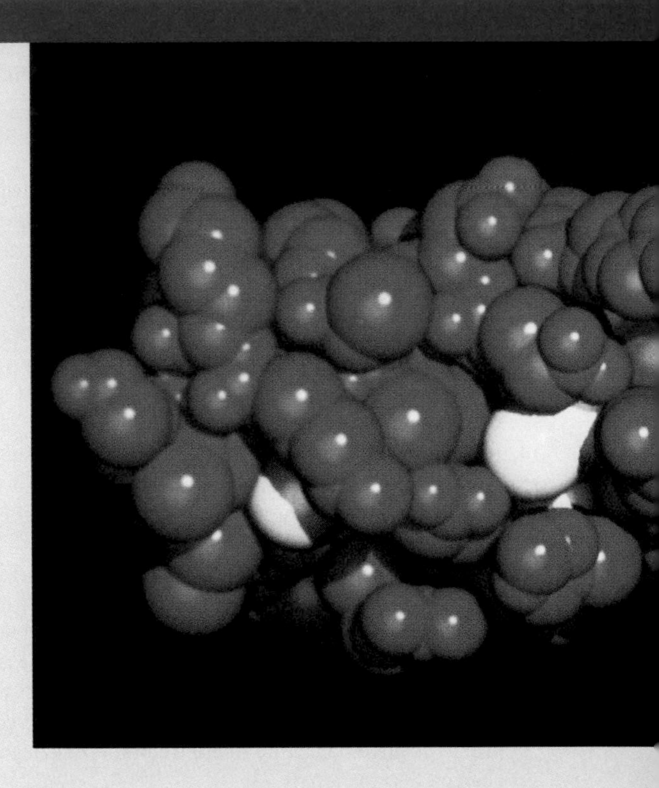

In our study of electricity in Chapter 8 we learned about Coulomb's law, which states that electric charges attract or repel each other depending on their sign and on the square of the distance between them. If the charges are of like signs, as is the case with two electrons, they repel each other; if the signs are opposite, as with an electron and a proton, the charges attract. Charged things interact electrically with other charged things. In this chapter we learn that the forces bonding atoms together are electrical. Chemical bonds are governed by Coulomb's law at the atomic level, and in this chapter we see how electrical interactions produce three types of chemical bonds: *metallic, ionic,* and *covalent.*

As we saw in the previous chapter, the periodic table is our road map to understanding chemical principles. This road map is particularly useful in understanding bonding between atoms. As illustrated in Figure 17.1, metallic bonds occur between the atoms of metallic elements; ionic bonds occur between the atoms of elements located on opposite sides of the periodic table; and covalent bonds occur between the atoms of the nonmetallic elements.

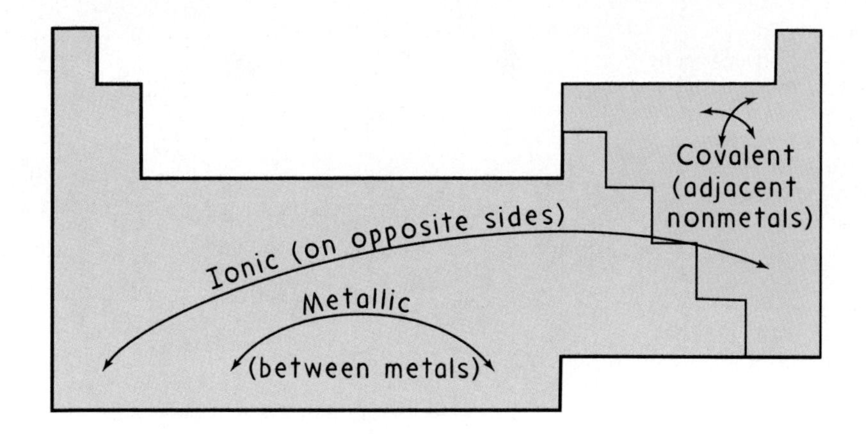

Fig. 17.1
The character of a chemical bond between atoms of two elements can be assessed using the relative position of the two elements in the periodic table.

We begin with the most conceptually straightforward type of chemical bond—the metallic bond. From there we build our way to understanding the ionic bond and then to the conceptually more intricate covalent bond.

17.1 Metals and Alloys

The outer electrons of most metal atoms tend to be weakly held to the atomic nucleus. Consequently, these outer electrons are easily dislodged, leaving behind positively charged metal ions. Recall from Section 8.1 that an *ion* is an atom that has either lost or gained one or more electrons so that the number of electrons in the atom is no longer equal to the number of protons. If electrons are gained by an atom, then electrons outnumber protons and the ion's net charge is negative. If electrons are lost, as is the case with metals, then protons outnumber electrons and the ion's net charge is positive (Figure 17.2).

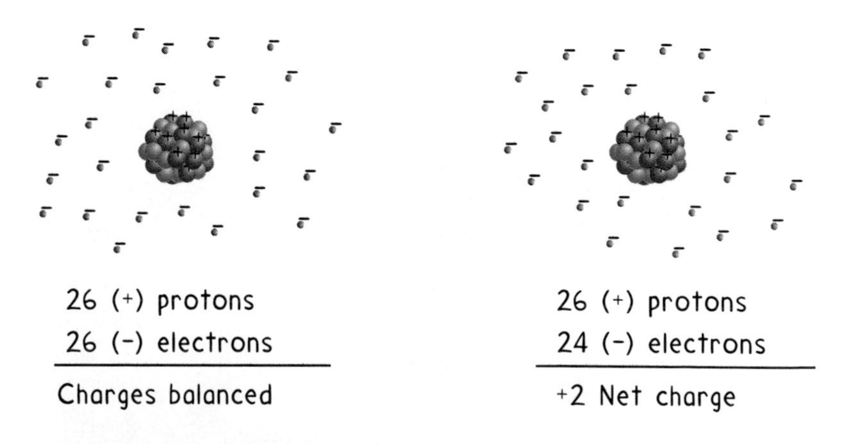

Fig. 17.2
Remove an electron from a metal atom and the result is a positively charged metal ion, which contains more protons than electrons.

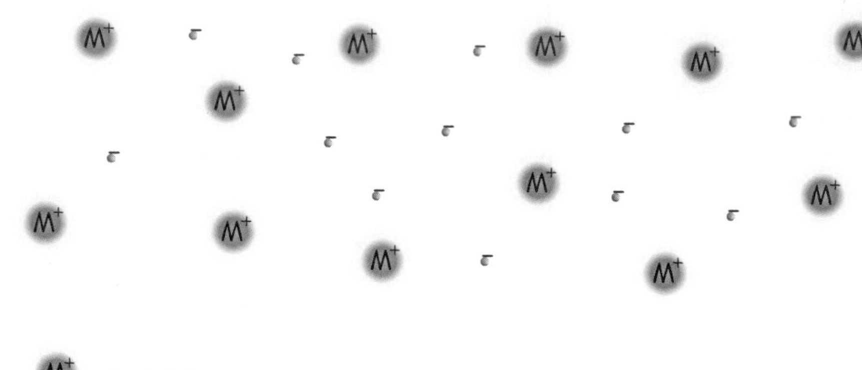

Fig. 17.3
Metal ions are held together by freely flowing electrons. These loose electrons form a kind of "electronic fluid" that flows through the lattice of positively charged metal ions.

The many electrons easily dislodged from a large grouping of metal atoms flow freely through the assembly of resulting metal ions. This "fluid" of electrons holds the positively charged metal ions together, as depicted in Figure 17.3. This flowing-electron action constitutes the **metallic bond**.

The mobility of electrons in a metal accounts for the metal's high electrical and thermal conductivities. Also, metals are opaque and shiny because the free electrons easily vibrate to the oscillations of light when it falls upon them, reflecting most of it. It is the mobility of electrons in metals that makes metals both malleable and ductile (Section 16.2).

Two or more metals can be bonded to one another by metallic bonds. This occurs, for example, as molten gold and molten palladium are blended together to form the homogeneous solution known as white gold. The quality of the white gold can be modified simply by changing the proportions of gold and palladium. White gold is an example of an **alloy**, which is any mixture composed of two or more metallic elements. By playing around with proportions, the properties of an alloy can be readily modified. Sterling silver, for example, is an alloy containing 92.5 percent silver and 7.5 percent copper. Change the proportion of silver and copper to 70 percent and 10 percent, respectively, throw in 18 percent tin and 2 percent mercury, and you have the composition of a dentist's tooth filling, which has the same rate of thermal expansion as that of human teeth (Figure 17.4).

Fig. 17.4
Your teeth expand as they are warmed by a hot beverage. If the alloy making up the fillings in your teeth doesn't expand at the same rate, your teeth may crack.

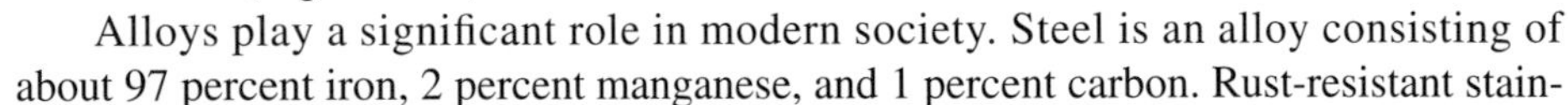

Alloys play a significant role in modern society. Steel is an alloy consisting of about 97 percent iron, 2 percent manganese, and 1 percent carbon. Rust-resistant stainless steel is an alloy of about 71 percent iron, 18 percent chromium, 10 percent nickel, and 1 percent carbon. Because pure aluminum is too soft and weak to be useful for most structural purposes, aluminum cans, roadway signs, and airplanes are fabricated using an alloy made of aluminum plus small amounts of such other metals as zinc and manganese.

Fig. 17.5
Aluminum is alloyed to give it the strength necessary for a wide variety of structural uses.

17.2 Ionic Bonds

Whereas metal atoms lose electrons to become positively charged ions, nonmetal atoms gain electrons to become negatively charged ions. A sodium atom, for example, loses one electron to become a positively charged sodium ion, $Na+$, and a chlorine atom gains one electron to become a negatively charged chloride ion, Cl^-. These two oppositely charged ions are pulled together by the electrical force as spelled out by Coulomb's law (opposites attract). This force of attraction between two oppositely charged ions is called the **ionic bond** (Figure 17.6).

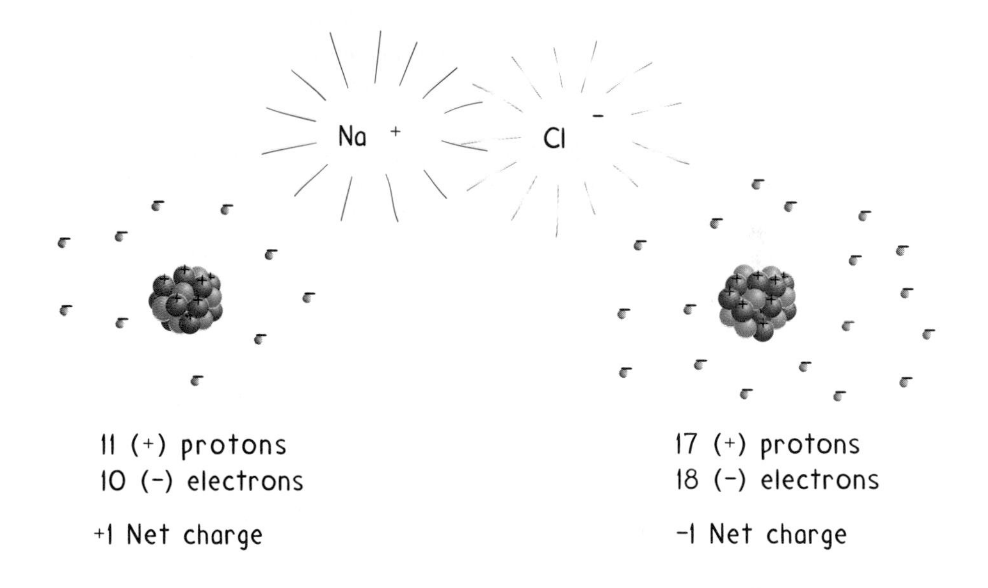

Fig. 17.6
Sodium and chloride ions are held together by an electrical force called the ionic bond. Each sodium ion possesses 11 positively charged protons balanced by only 10 negatively charged electrons. Each chloride ion possesses 17 protons balanced by 18 electrons. Together, sodium and chloride ions make up the ionic compound sodium chloride, which you know as table salt.

Any chemical compound made of ionically bonded ions is called an **ionic compound**. Sodium chloride with its oppositely charged sodium and chloride ions is a good example. As we discussed in Section 15.4, the ionic compound sodium chloride is different from the sodium and chlorine elements from which it is made. Another example of an ionic compound is potassium iodide, KI, which is added in minute quantities to commercial table salt because the iodide ion, I^-, is a mineral essential to humans. Also important to human health is the tooth-strengthening fluoride ion, F^-, which occurs naturally in the drinking water of some communities in the form of the ionic compound calcium fluoride, CaF_2. In communities where there is little calcium fluoride in the drinking water, the ionic compound sodium fluoride, NaF, is often added as a substitute. Whether your drinking water is fluoridated or not, your toothpaste most likely is.

Fig. 17.7
The relatively small size of the fluoride ion (Section 16.5) means it is able to get closer to an oppositely charged ion, which translates into a stronger ionic bond. It is the strong ionic bonding between fluoride ions and the compound of your teeth that is responsible for the ions' effectiveness in preventing tooth decay. People born before municipalities began adding fluoride ions to their water supplies in the early 1970s tend to have a greater number of fillings in their teeth. Compare mouths to see for yourself.

As mentioned in the opening of this chapter, ionic compounds typically consist of elements found on opposite sides of the periodic table. This is because elements located on the left side of the periodic table have the greatest tendency to *lose* electrons and those located on the right (with the exception of the noble gases) have the greatest tendency to *gain* electrons. In all cases the positively charged ions are derived from metallic elements, and the negatively charged ions are derived from nonmetallic elements. Although all ionic compounds consist of a metal and a nonmetal, not all bonds between metals and nonmetals are ionic. The chemical bond between the metallic germanium atom and the nonmetallic phosphorus atom, for example, is primarily *covalent* in character, not ionic, as we shall see in Section 17.3. Nevertheless, a useful rule is

that ionic compounds consist of elements located on opposite sides of the periodic table, which necessarily involves a metal ion coupled together with a nonmetal ion.

We use a superscript placed to the right of the atomic symbol to indicate the magnitude and sign of an ion's charge. For example, a sodium atom that loses one electron is written Na^{1+}, a calcium atom that loses two electrons is written Ca^{2+}, a chlorine atom that gains a single electron is written Cl^{1-}, and an oxygen atom that gains two electrons is written O^{2-}. Be sure to keep in mind that a positive superscript means that one or more electrons have been *lost,* while a negative superscript means that one or more electrons have been *gained.* By convention, the numeral 1 is often left out of superscript notation, so that the sodium ion and chloride ion, for instance, are usually written Na^{+} and Cl^{-}, respectively.

As was discussed in Section 16.5, elements are most particular about the type of ions they form. Each sodium atom, for example, typically loses only one electron. The calcium atom, on the other hand, typically loses two electrons to form the 2+ calcium ion. In most instances, elements in the same atomic group form the same kind of ion. All elements of the first vertical column, group 1, for example, form 1+ ions, while all elements of the second vertical column, group 2, form 2+ ions. Also, as was discussed in Section 16.5, various incremental trends are found across the atomic periods. The type of ions formed is another example of such a trend. Across the second and third atomic periods, for example, the pattern of ions formed is as follows: 1+, 2+, 3+, 4−, 3−, 2−, 1−, 0.

We can use this information to deduce the chemical formulas for various ionic compounds. For all ionic compounds, positive and negative charges must balance. With sodium chloride, for example, there is one sodium 1+ ion for every one chloride 1- ion. This concept of charge balance holds true even for compounds containing ions that carry multiple charges. The calcium ion, for example, carries a charge of 2+, but the fluoride ion carries a charge of only 1−. Therefore two fluoride ions are needed to balance each calcium ion. The formula for calcium fluoride is thus CaF_2 (Figure 17.8).

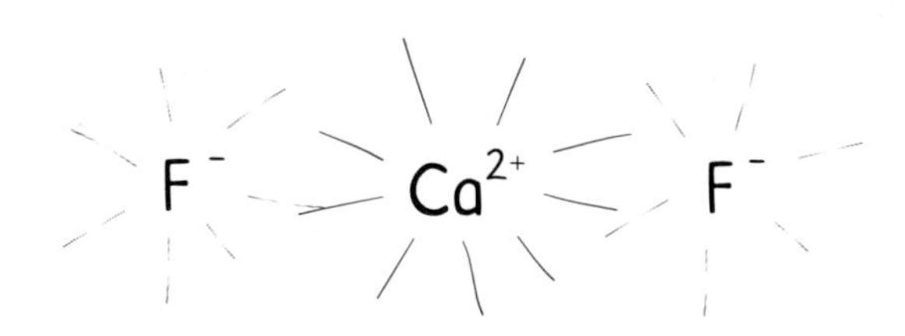

Fig. 17.8
In the ionic compound calcium fluoride, CaF_2, two fluoride ions, F^-, balance the charge of a single calcium ion, Ca^{2+}.

Similarly, the aluminum ion carries a 3+ charge, and the oxygen ion carries a 2− charge. Together, these ions make the ionic compound aluminum oxide, Al_2O_3, the main component of many gemstones. In aluminum oxide, the three oxide ions carry a total charge of 6−, which balances the total 6+ charge of the two aluminum ions (Figure 17.9).

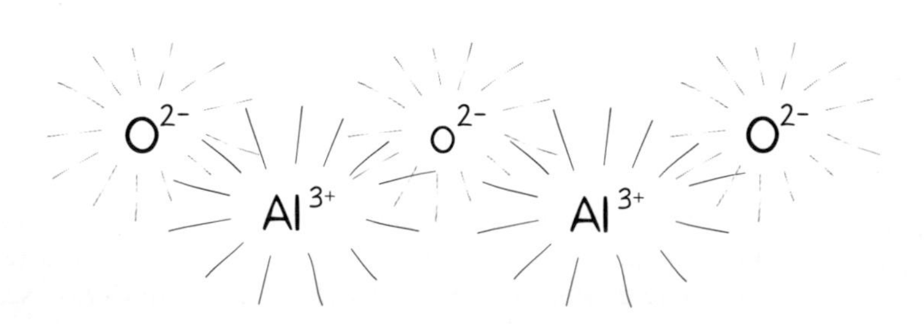

Fig. 17.9
In the ionic compound aluminum oxide, Al_2O_3, three oxide ions, O^{2-}, balance the charge of two aluminum ions, Al^{3+}

■ **Question**

What is the chemical formula for the ionic compound magnesium oxide?

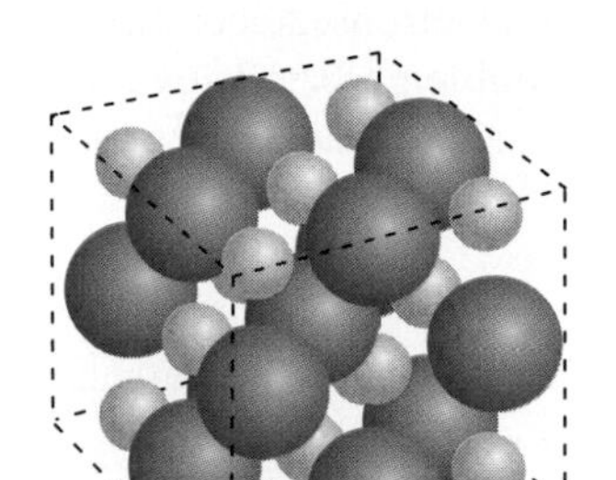

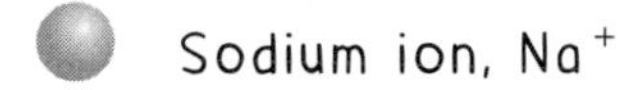

Fig. 17.10
Sodium chloride (as well as other ionic compounds) forms ionic crystals in which every internal ion is surrounded by others of opposite charge. (For simplicity, only a 3-by-3 ion lattice is shown here. A typical NaCl crystal involves millions and millions of ions.)

The arrangement of ions in an ionic compound is highly ordered. Regular patterns appear in which ions of opposite charges surround one another. For sodium chloride, each sodium ion is surrounded by six chloride ions and each chloride ion is surrounded by six sodium ions (Figure 17.10). Overall there is one sodium ion for each chloride ion, but there are no identifiable sodium chloride pairs. When the ions of an ionic compound are arranged in this type of three-dimensional array, they form an **ionic crystal**. On the atomic level, the crystalline structure of sodium chloride is cubic, which is why macroscopic salt crystals are also cubic (Figure 17.11). Smash a larger cubic salt crystal with a hammer and what do you get? Smaller cubic salt crystals! Similarly, the crystalline shapes of other ionic compounds depend on how the ions are able to pack together.

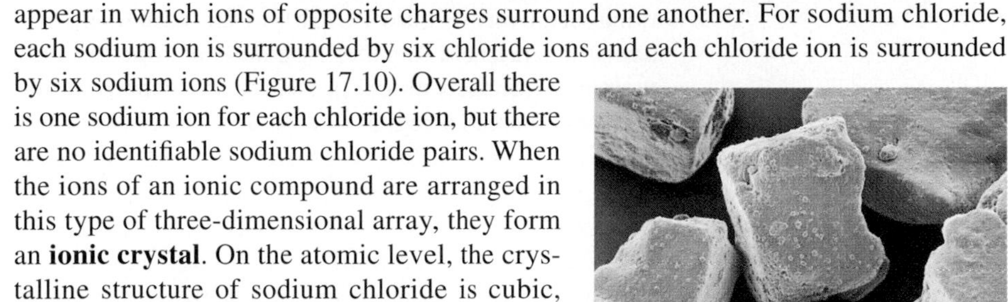

Fig. 17.11
A view of salt crystals through a microscope. The cubic shape of the crystals is a natural consequence of the cubic arrangement of sodium and chloride ions.

17.3 Covalent Bonds

Imagine two children playing together and sharing their toys. A force that keeps the children together is their mutual attraction for the toys they share. In a similar fashion, two atoms can be held together by their mutual attraction for electrons they share. In the hydrogen molecule, H_2, for example, each of the two hydrogen atoms shares its electron with the other so that they are both attracted to a total of two electrons (Figure 17.12a). This type of electrical attraction generally involves the valence electrons (Section 16.4) and is therefore called the **covalent bond**, where *co-* signifies sharing and *-valent* refers to the valence electrons being shared. Any group of atoms held together by one or more covalent bonds is a **covalent compound**. Hydrogen, H_2, is an example. By convention, two atoms covalently bonded are represented by drawing a straight line between their atomic symbols (Figure 17.12b).

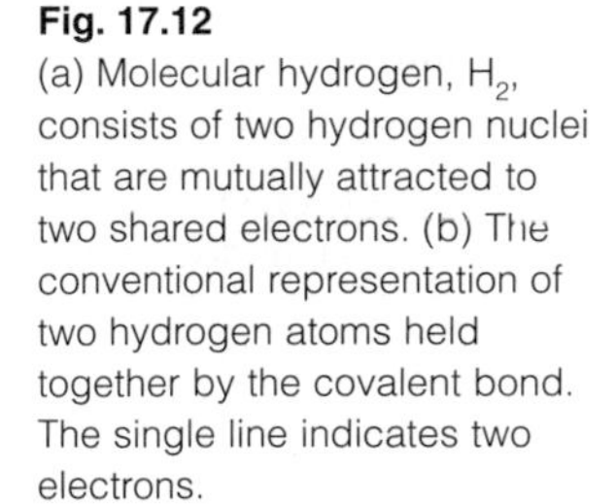

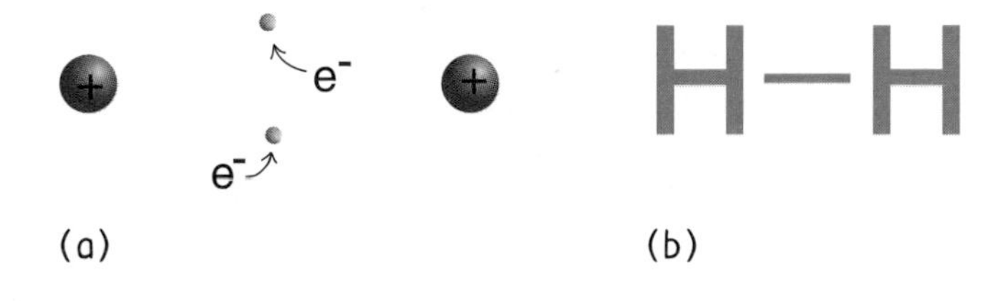

Fig. 17.12
(a) Molecular hydrogen, H_2, consists of two hydrogen nuclei that are mutually attracted to two shared electrons. (b) The conventional representation of two hydrogen atoms held together by the covalent bond. The single line indicates two electrons.

■ **Answer**

Because the magnesium atom is a group 2 atom, you know it must lose two electrons to form the Mg^{2+} ion. Oxygen, being a group 16 atom, gains two electrons to form the O^{2-} ion. These charges balance in a one-to-one ratio, and so the formula for magnesium oxide is MgO.

The principle underlying covalent bonding is *electron affinity,* which, as discussed in Section 16.5, is the ability of an atom to attract electrons. A hydrogen atom, for example, has an affinity for one additional electron. It can find this additional electron in one of two ways. The electron may be a free-roaming electron, in which case the hydrogen atom becomes the negatively charged ion H− (Figure 17.13a). Alternatively, the electron may be already bound to a second atom that has a similar affinity for additional electrons (Figure 17.13b). In this case the result is a covalent bond, where the two atoms are held together by a mutual attraction for shared electrons. According to the noble gas shell model presented in Chapter 16, the two noble gas shells overlap such that there is a region of common space within which the shared electrons reside (Figure 17.14).

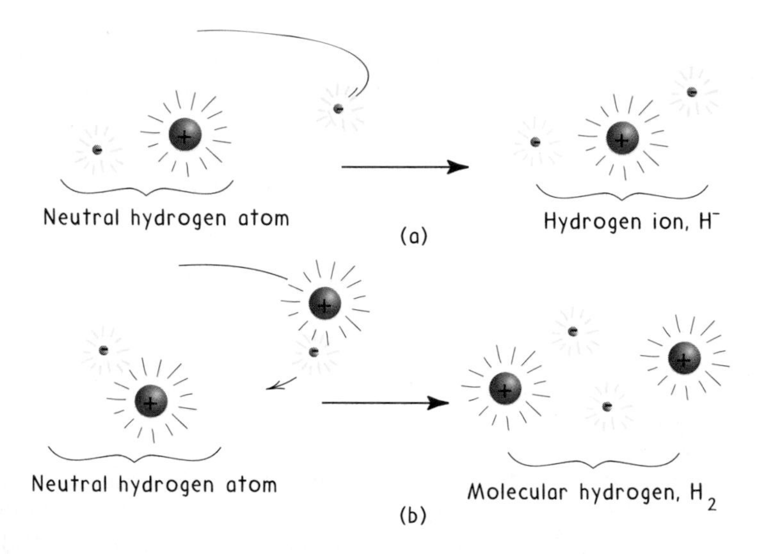

Fig. 17.13
(a) A neutral hydrogen atom can attract a free-roaming electron to become a negatively charged hydrogen ion. (b) It can also attract an electron already attached to a second nucleus. In such a case the result is a covalent bond.

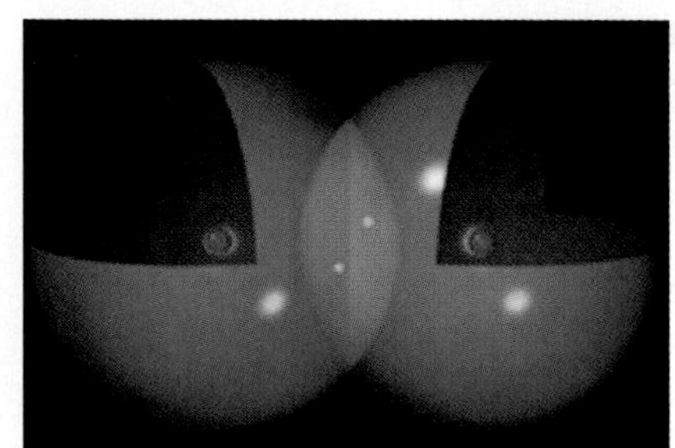

Fig. 17.14
The shells of two atoms joined in a covalent bond overlap.

Because most elements have some degree of electron affinity, most elements have some ability to bond covalently. Hydrogen and the nonmetal elements at the upper right of the periodic table, however, have the strongest electron affinities. These elements therefore form covalent bonds most readily. The noble gas elements helium through krypton are an exception. They have no electron affinity because they have no room for additional electrons in their filled shells. Thus they tend not to form covalent bonds.[1]

The number of covalent bonds an atom is able to form is equal to the number of additional electrons it is able to attract. Hydrogen has an affinity for only one additional electron, and so it is able to form only one covalent bond. Oxygen, however, has an affinity for two additional electrons. It finds two such electrons when it encounters two hydrogen atoms and reacts to form water, H_2O. In water, not only does the oxygen atom have access to two additional electrons by covalently bonding to two hydrogen atoms, but each hydrogen atom has access to an additional electron by bonding to the oxygen atom (Figure 17.15). Everybody's happy!

---

[1]Xenon is able to accommodate additional electrons by borrowing space from higher empty shells. Xenon therefore has some electron affinity and can participate in covalent bonding, as in the compound xenon difluoride, XeF_2. Similar radon compounds are hypothesized to be possible but have not been isolated owing to radon's highly radioactive character.

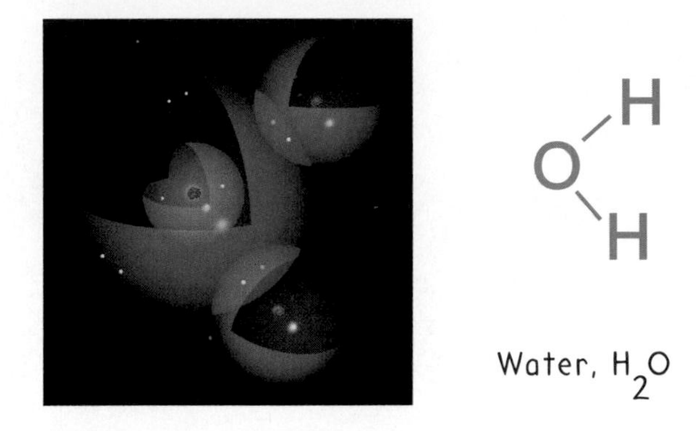

Fig. 17.15
Two valence electrons of oxygen get together with the valence electrons of two hydrogen atoms to produce water. Note how the shells overlap where the covalent bonds are formed. Oxygen's electrons are shown in yellow, and hydrogen's electrons are shown in green.

In a similar fashion, nitrogen attracts three additional electrons and is thus able to form three covalent bonds, as occurs in ammonia, NH_3. Carbon can attract four additional electrons and is thus able to form four covalent bonds, as occurs in methane, CH_4 (Figure 17.16). Note that the number of covalent bonds formed by these and other nonmetal elements parallels the type of negative ions they tend to form, as shown in Figure 16.33. This makes sense because covalent bond formation and negative ion formation are both applications of the same concept—electron affinity.

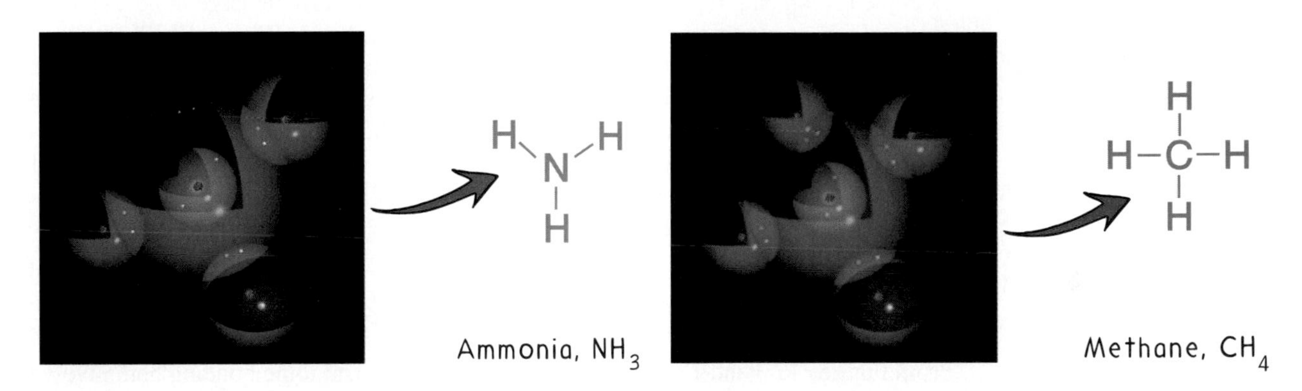

Fig. 17.16
Nitrogen attracts three additional electrons from three hydrogen atoms to form ammonia, NH_3. Carbon attracts four additional electrons from four hydrogen atoms to form methane, CH_4. In these and most other cases of covalent bond formation, the result is a filled outer noble gas shell.

It is possible to have more than one covalent bond between two atoms. For example, molecular oxygen, O_2, consists of two oxygen atoms connected by two covalent bonds, called a *double covalent bond* or just *double bond* for short (Figure 17.17). In a double bond, there are four electrons being shared, two from each atom. As another example, the covalent compound carbon dioxide, CO_2, consists of two double bonds connecting two oxygen atoms to a central carbon atom. Some atoms can form *triple covalent bonds,* in which six electrons—three from each atom—are shared. One example is molecular nitrogen, N_2. Any double or triple bond is often referred to as a *multiple covalent bond.* Multiple bonds higher than these, such as the quadruple bond, have not been observed and are not possible according to current models.

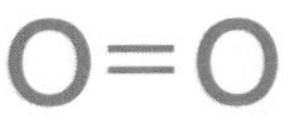

O=O — Oxygen, O_2

N≡N — Nitrogen, N_2

O=C=O — Carbon dioxide, CO_2

Fig. 17.17
Multiple covalent bonds may occur between atoms.

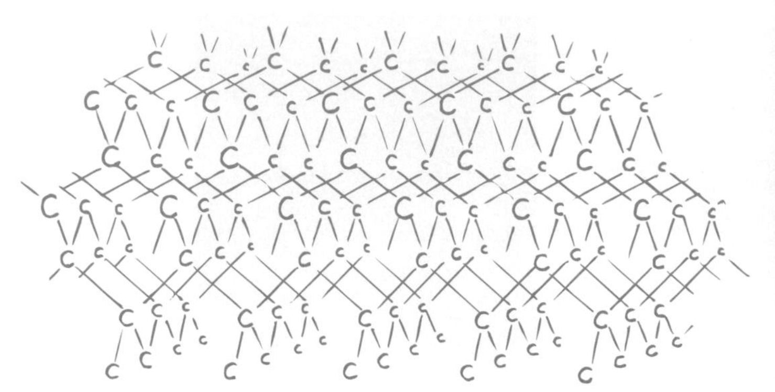

Fig. 17.18
A diamond is a large molecule consisting of many covalently bonded carbon atoms. Diamond's molecular nature results in a set of unusual properties, such as its hardness and its behavior in light.

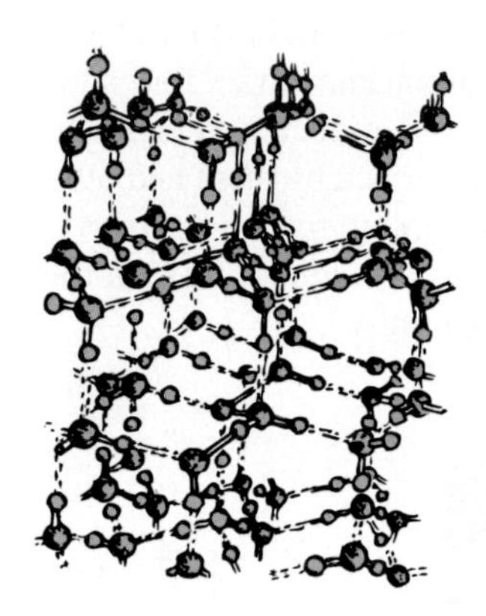

Fig. 17.19
Water molecules pack together in an orderly fashion to produce ice—a covalent crystal.

In Chapter 6 we defined a *molecule* as a group of atoms. We can now present a more accurate definition: A **molecule** is a group of atoms held tightly together by covalent bonds. Molecules, therefore, are the fundamental units of any covalent compound. The fundamental units of water, for example, are water molecules, H_2O, and the fundamental units of oxygen are oxygen molecules, O_2. Most molecules are far too small to be seen with optical microscopes. Some atoms, however, are able to bond covalently many times over to produce molecules large enough to be seen with the naked eye! An example is diamond. Each stone in a jeweler's window is a single large molecule made of many carbon atoms connected by numerous covalent bonds (Figure 17.18).

Molecules that arrange themselves in an orderly fashion form a **covalent crystal**. An example of a covalent crystal is ice, which consists of water molecules frozen into a solid three-dimensional array (Figure 17.19). Another example is sugar, which consists of many sugar molecules packed neatly together in an orderly three-dimensional fashion. Diamond is also a covalent crystal because of the ordered packing of its covalently bonded carbon atoms.

Metallic, Ionic, and Covalent Bonds Compared

From Figures 17.3 and 17.10 you can see how metallic and ionic bonding both involve whole groupings of ions. In the metallic bond, a network of positively charged metal ions are held together within a "fluid" of electrons. In the ionic bond, each ion is attracted not to one oppositely charged ion but to *all* the oppositely charged ions surrounding it. By contrast, from Figure 17.12 you can see that a covalent bond involves only the two atoms sharing electrons.

In an anthropomorphic sense, covalent bonding is much like marriage, where two individuals are—in theory, at least!—held exclusively to each other, forming a single identifiable unit: the family. Likewise, two covalently bonded atoms constitute a single identifiable unit called the *molecule.* With children and relatives added, families can be of different sizes. Similarly, some molecules consist of only a couple of atoms (N_2 and H_2, for example), and others consist of many more (a diamond on a wedding ring, for example). From a chemist's point of view, a molecule is a tightly knit *family* of atoms held together by covalent bonds. The ions of metals and ionic compounds, on the other hand, lead more of the "single life," being less particular about any individual partner.

In the following section, we delve into some of the "family dynamics" of molecules. We learn, for example, how a molecule can take on some of the charged properties of ions and how this ability plays a role in the interactions among molecules.

17.4 Covalent Bond Polarity

If the two atoms in a covalent bond are identical, their nuclei have the same positive charge and the electrons are shared *evenly.* The electrons of a hydrogen-hydrogen covalent bond, for example, experience identical nuclear charges from both sides of the bond. So the electrons are attracted to both sides equally and do not tend to accumulate to one side or the other. We can represent these electrons as being centrally located, as shown in Figure 17.20a.

H : H (a) Molecular hydrogen

H :F (b) Hydrogen fluoride

Fig. 17.20
(a) The two hydrogen nuclei in an H_2 covalent bond attract the electrons with equal forces. The electrons therefore are not pulled to one side or the other. (b) The fluorine nucleus attracts the bonding electrons with a greater force because of its greater nuclear charge; the electrons are therefore on average closer to the fluorine nucleus than to the hydrogen nucleus.

In a covalent bond between a pair of nonidentical atoms, the nuclear charges are different, and consequently the bonding electrons are shared *unevenly.* This occurs in a hydrogen-fluorine bond, for instance, where electrons are more attracted to fluorine's greater nuclear charge and are drawn to the fluorine side of the covalent bond, as shown in Figure 17.20b. The bonding electrons spend more time around fluorine, much like a bunch of bees spend more time around a sweeter flower. For this reason, the fluorine side of the bond is slightly negative and, because the bonding electrons have been drawn away from the hydrogen atom, the hydrogen side of the bond is slightly positive. This segregation of charge, called a **dipole** (pronounced die-pole), may be represented using the characters "δ−" and "δ+," which respectively read "slightly negative" and "slightly positive" (Figure 17.21a). A second way to represent a dipole is to draw a crossed arrow that points to the negatively charged side of the bond (Figure 17.21b).

Fig. 17.21
(a) In a molecule of hydrogen fluoride, the hydrogen atom is slightly positive, δ+, and the fluorine atom is slightly negative, δ−. (b) A crossed arrow that points toward the negatively charged side of the bond may also be used to represent a dipole.

A chemical bond is a tug-of-war between atoms for electrons. How well atoms are able to "tug" on bonding electrons has been measured experimentally and quantified as an atom's **electronegativity**, which ranges on a scale from 0 to 4 (Figure 17.22). Electronegativity is best thought of as an atom's "electron pulling power." The greater the electronegativity of an atom, the greater its ability to pull electrons toward itself when bonded. In hydrogen fluoride, fluorine has a greater electronegativity, or pulling power, than does hydrogen.

Electronegativity and electron affinity are very similar concepts. Both refer to the attraction an atom has for electrons. Electronegativity, however, is more specific, referring to the attraction an atom has for *bonding* electrons. Like electron affinity, electronegativity is greatest for the elements to the upper right of the periodic table and lowest for elements at the lower left of the periodic table. Noble gases are not considered in electronegativity discussions because, with only a few exceptions, they do not participate in chemical bonding.

Depending on the electronegativity of the atoms involved, a covalent bond is either *nonpolar* or *polar.* When the two atoms forming the bond have the same elec-

Electronegativities

H 2.2																	He --
Li .98	Be 1.57											B 2.04	C 2.55	N 3.04	O 3.44	F 3.98	Ne --
Na .93	Mg 1.31											Al 1.61	Si 1.9	P 2.19	S 2.58	Cl 3.16	Ar --
K .82	Ca 1.0	Sc 1.36	Ti 1.54	V 1.63	Cr 1.66	Mn 1.55	Fe 1.83	Co 1.88	Ni 1.91	Cu 1.90	Zn 1.65	Ga 1.81	Ge 2.01	As 2.18	Se 2.55	Br 2.96	Kr --
Rb .82	Sr .95	Y 1.22	Zr 1.33	Nb 1.6	Mo 2.16	Tc 1.9	Ru 2.2	Rh 2.28	Pd 2.20	Ag 1.93	Cd 1.69	In 1.78	Sn 1.96	Sb 2.05	Te 2.1	I 2.66	Xe --
Cs .79	Ba .89	La* 1.10	Hf 1.3	Ta 1.5	W 2.36	Re 1.9	Os 2.2	Ir 2.20	Pt 2.8	Au 2.54	Hg 2.00	Tl 2.04	Pb 2.33	Bi 2.02	Po 2.0	At 2.2	Rn --
Fr .7	Ra .9	Ac** 1.1	Rf --	Db --	Sg --	Bh --	Hs --	Mt --	Uun --	Uuu --	Uub --						

Fig. 17.22
The electronegativities of elements as arranged in the periodic table.

tronegativity, no dipole is formed (as is the case with H_2) and the bond is classified as **nonpolar**. When the electronegativities of the atoms differ, a dipole is formed (as is the case with HF) and the bond is classified as **polar**. The bond in H_2 is an example of a nonpolar covalent bond, and that in HF an example of a polar covalent bond.

The greater the difference in electronegativity between the two bonding atoms, the greater the dipole of the bond, which is to say the bond has a greater *polarity.* In general, the farther apart two bonding atoms are in the periodic table, the greater the difference in their electronegativities and hence the greater the polarity of the bond between them (Figure 17.23). So a chemist doesn't need a list of electronegativities to predict which bonds are more polar than others; to find out, he or she need only look at the relative positions of the atoms in the periodic table—the farther apart they are from each other, the greater the polarity of the bond between them.

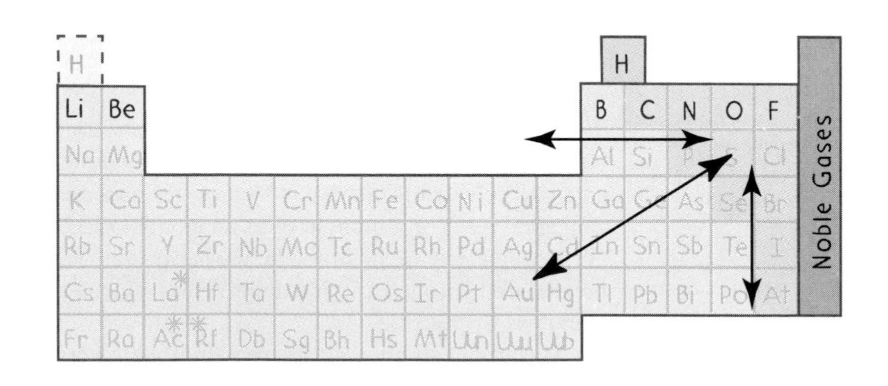

Fig. 17.23
The farther apart two elements are positioned along these three orientations, the greater the polarity of the bond between them. Hydrogen is shown directly above boron, B, and carbon, C, because its electronegativity is most similar to the electronegativities of these elements.

Question

List these bonds in order of increasing polarity:

P–F S–F Ga–F Ge–F

(F, fluorine, atomic no. 9; S, sulfur, atomic no. 16; P, phosphorus, atomic no. 15; Ge, germanium, atomic no. 32; Ga, gallium, atomic no. 31)

We can indicate the magnitude of bond polarity by the size of the arrow or δ+/δ− symbol used to depict a dipole. The larger the arrow or symbol, the greater the polarity (Figure 17.24). An alternative way to indicate bond polarity is to note the **percent ionic character** of the bond, which is determined by the difference in electronegativities (Figure 17.25). Absolutely no sharing of electrons is taken as 100 percent ionic character, and a perfect sharing is 0 percent ionic character. The percent ionic character of a polar covalent bond with its slightly charged atoms is somewhere between 0 and 100 percent.

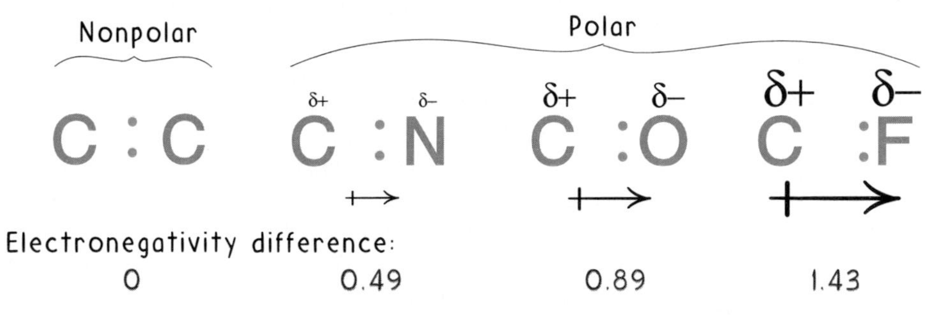

Fig. 17.24
These bonds are listed in order of increasing polarity from left to right. Which of these pairs of elements are farthest apart from each other in the periodic table?

Difference in electronegativity	Percent ionic character
0.1	0.5
0.2	1
0.3	2
0.4	4
0.5	6
0.6	9
0.7	12
0.8	15
0.9	19
1.0	22
1.1	26
1.2	30
1.3	34
1.4	39
1.5	43
1.6	47
1.7	51
1.8	55
1.9	59
2.0	63
2.1	67
2.2	70
2.3	74
2.4	76
2.5	79
2.6	82
2.7	84
2.8	86
2.9	88
3.0	89
3.1	91
3.2	92

Fig. 17.25
The percent ionic character of a chemical bond as a function of the difference in electronegativities of bonding atoms.

In actuality, there is no black-and-white distinction between ionic and covalent bonds. Rather, there is a gradual change from one to the other as the atoms that bond are found farther apart in the periodic table. Atoms positioned on opposite sides of the periodic table have great differences in electronegativity, and hence the bonds between them are predominantly ionic. Such bonds are not 100 percent ionic, however, because some sharing of electrons takes place. In contrast, covalently bonding atoms close together in the periodic table have similar electronegativities, and so their bonds are less ionic. The polar covalent bond with its uneven *sharing* of electrons and slightly *charged* atoms is a little of both (Figure 17.26).

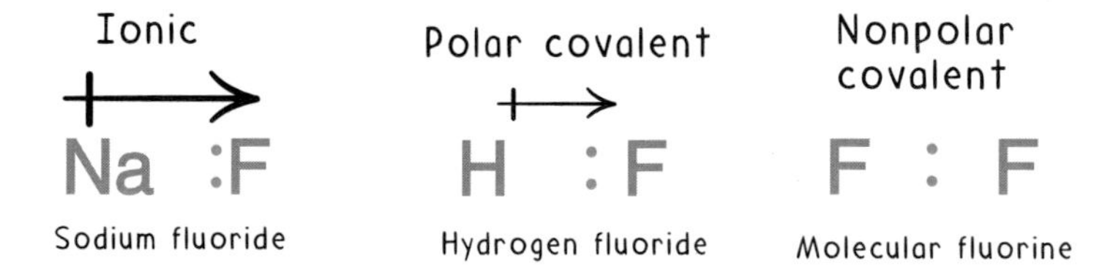

Fig. 17.26
The ionic bond and the nonpolar covalent bond represent two extremes of chemical bonding. The ionic bond involves an *exchange* of electrons, and the covalent bond involves the *sharing* of electrons. The character of a polar covalent bond falls between these two extremes, being both partly ionic and partly covalent.

Answer
If you answered this, or attempted to, before reading this answer, hooray for you! You're doing more than reading the text—you're learning chemistry. The greater the *difference* in electronegativities between two bonded atoms, the greater the polarity of the bond, and so the order of increasing polarity is S–F < P–F < Ge–F < Ga–F (S–F least polar, Ga–F most polar).

17.5 Molecular Polarity

A covalent bond may be polar or nonpolar, but what about an entire molecule? If all the bonds in the molecule are nonpolar, then the molecule as a whole is also nonpolar—as is the case with H_2, O_2, and N_2. If a molecule consists of only two atoms and the bond between them is polar, then the molecule has this same polarity—as with HF, HCl, and ClF. Complexities arise, however, when trying to assess the polarity of a molecule containing more than two atoms. Consider carbon dioxide, CO_2 (Figure 17.27). The cause of the dipole in a lone carbon-oxygen bond is oxygen's greater pull (electronegativity) on the bonding electrons. The second oxygen atom placed on the opposite side of the carbon, however, has the effect of pulling those electrons back to the carbon. The net result is an even distribution of bonding electrons around the whole molecule, with no accumulation to one side or the other. So, dipoles equal and opposite each other within the same molecule effectively cancel each other out and transform a would-be polarity into nonpolarity.

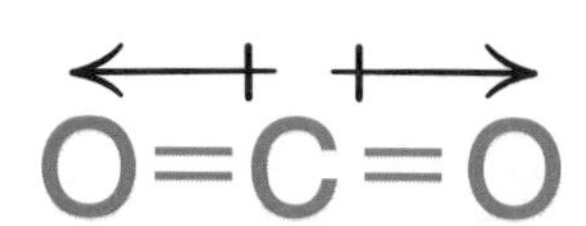

Fig. 17.27
There is no net dipole in a carbon dioxide molecule, and so the molecule is nonpolar.

Dipoles are vector quantities because they have both magnitude and direction. Thus we can graphically illustrate the nonpolarity of carbon dioxide and the polarity or nonpolarity of many other molecules by *vector addition* (Chapter 1). Consider two people pulling on a rope in opposite directions, as in Figure 17.28. If they both pull with a force of 500 newtons, the situation can be illustrated by two arrows of the same size pointing in opposite direction. The *resultant* of these two vectors—the vector sum—is 0 newtons, which is why the rope is not pulled to one side or the other. This is analogous to the case of carbon dioxide, where the two oppositely oriented oxygen atoms negate the effect of each other's pull.

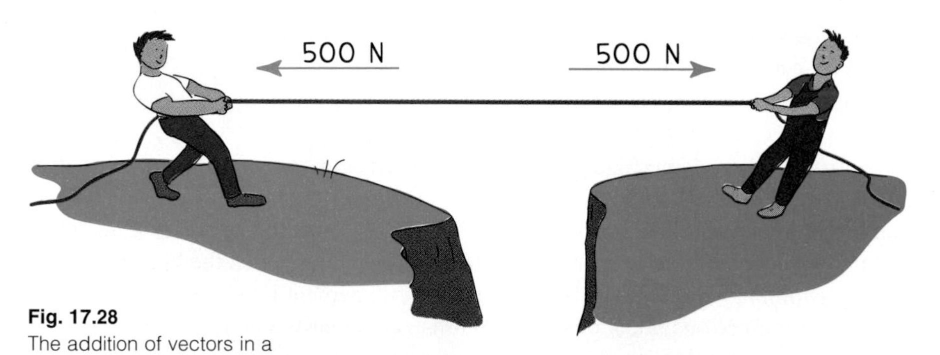

Fig. 17.28
The addition of vectors in a game of tug-of-war.

Vectors need not be aligned along a straight line in order to be added. Consider three people oriented 120 degrees apart and pulling on three ropes all tied to a central ring (Figure 17.29). If all three pull with the same force, the vector sum is zero and the ring is not pulled to any particular side. We can see how this is so by using vector addition. Recall the parallelogram rule from Section 1.2 and from the *Conceptual Physical Science Practice Book.* Figure

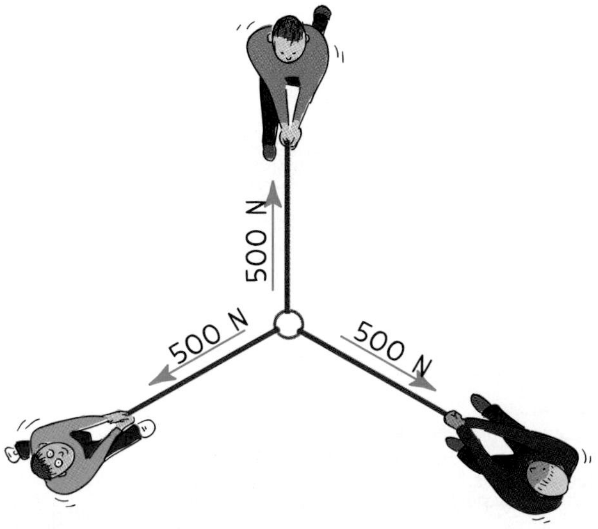

Fig. 17.29
The addition of vectors in a three-way tug-of-war.

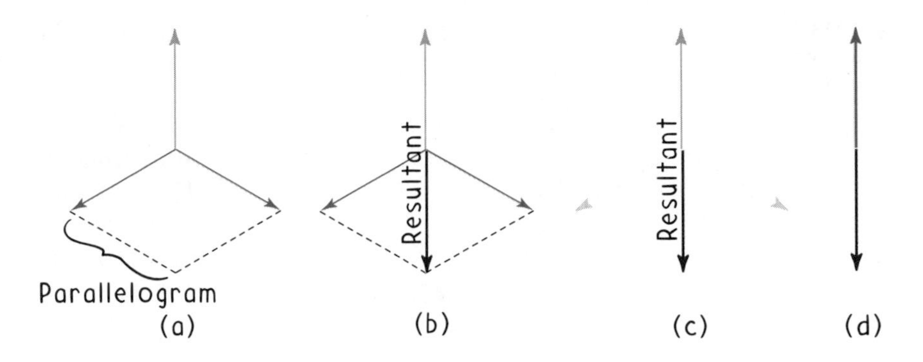

Fig. 17.30
(a) The resultant of two vectors at an angle to each other can be found by drawing a parallelogram and then (b) connecting the parallelogram's opposite corners. (c) This resultant vector, because it is the vector sum of the original two vectors, takes their place. (d) The remaining vectors are equal in magnitude and opposite in direction, and so they cancel each other.

17.30a shows a parallelogram drawn from two of the three vectors of Figure 17.29. The resultant of these two vectors is equal in both magnitude and direction to the arrow drawn across the diagonal of the parallelogram, as shown in Figure 17.30b. The original two vectors can be erased and this resultant left in their place (Figure 17.30c). We then see that this resultant is equal in magnitude to the third vector and pointing in the opposite directions, which gives for the whole system a final resultant of zero (Figure 17.30d).

The three-vector situation just described is analogous to the arrangement of the atoms in boron trifluoride, BF_3, where three fluorine atoms are oriented 120 degrees from one another around a central boron atom. Because the angles are all the same and because each fluorine atom pulls on the electrons of its boron-fluorine bond with the same force, the resultant polarity of this molecule is zero (Figure 17.31). So despite the would-be polarity of the individual boron-fluorine bonds, boron trifluoride with its particular geometry is a nonpolar molecule.

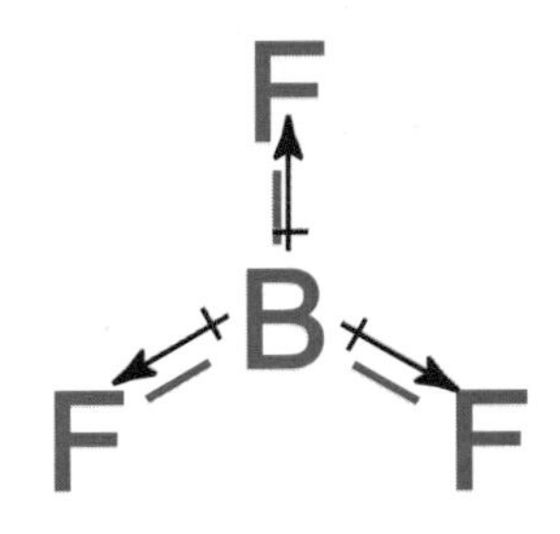

Fig. 17.31
The three dipoles of a boron trifluoride molecule, BF_3, oppose each other at 120° angles, which makes the overall molecule nonpolar.

Nonpolar molecules are very self-contained, having little attraction to other molecules. The bonding forces within a carbon dioxide molecule, for example, are many times stronger than any forces of attraction that might occur between two adjacent carbon dioxide molecules. In general, nonpolar molecules are not strongly attracted to other nonpolar molecules, which explains the low boiling temperatures of many nonpolar substances. Recall from Section 6.8 that boiling is a process wherein closely packed molecules of a substance in the liquid phase separate from one another as they transfer to a gaseous phase. The weaker the attractions between molecules of the liquid, the less thermal energy is required to liberate molecules from one another into the gaseous phase. This translates into a relatively low boiling temperature for the liquid. The boiling temperatures of liquid hydrogen, H_2, oxygen, O_2, nitrogen, N_2, carbon dioxide, CO_2, and boron trifluoride, BF_3, for example, are well below room temperature, which is why they are commonly observed as gases (Figure 17.32).

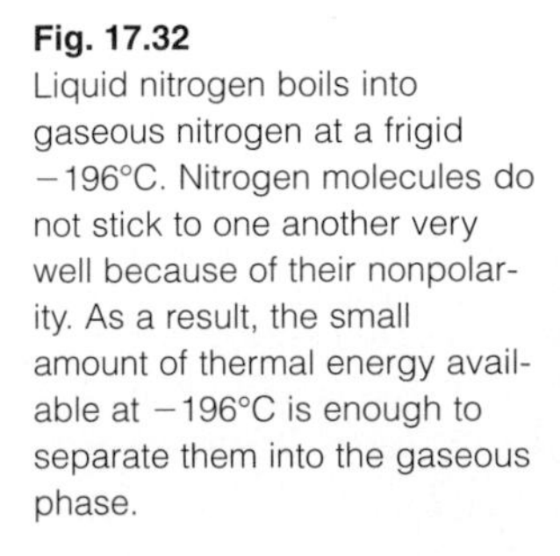

Fig. 17.32
Liquid nitrogen boils into gaseous nitrogen at a frigid −196°C. Nitrogen molecules do not stick to one another very well because of their nonpolarity. As a result, the small amount of thermal energy available at −196°C is enough to separate them into the gaseous phase.

Not all molecules are nonpolar. To understand why, reconsider the analogy of three people pulling three separate ropes attached to a central ring. If the ropes are oriented 120 degrees apart, the resultant force on the ring is zero and it is not pulled to any one side.

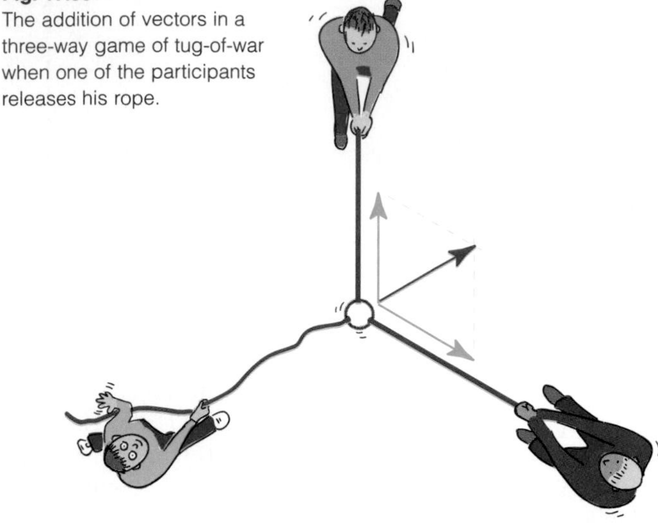

Fig. 17.33
The addition of vectors in a three-way game of tug-of-war when one of the participants releases his rope.

Imagine, however, that one person begins to ease off or perhaps even let go of his or her rope. In such a case the pulls are no longer balanced, and the central ring begins to move in the direction away from the person who is letting go (Figure 17.33). A similar situation occurs in many molecules where the various polar covalent bonds are not equal and opposite. Perhaps the most relevant example is the water molecule, H_2O. Each water molecule has a bent shape (Figure 17.34a).[2] The covalent bond between each hydrogen and the oxygen is a relatively large dipole because of the great electronegativity difference. Given the geometry of the water molecule, however, the two dipoles do not cancel each other. To assess the magnitude and direction of water's overall dipole, we can use the parallelogram method described earlier for boron trifluoride (Figure 17.34b). From this vector addition, we see that the water molecule is very polar, with the oxygen side negative and the hydrogen side positive.

Fig. 17.34
(a) The bent shape of a water molecule. (b) The individual dipoles in a water molecule add together to give a resultant polar molecule. (c) As a result, the region around the oxygen atom is slightly negative, and the region around the two hydrogens is slightly positive.

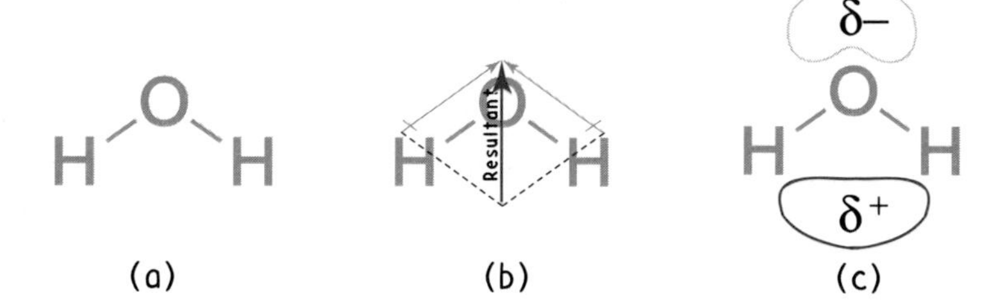

Question

Which of these molecules is polar and which is nonpolar?

F₍ ₎F structures:

F, F — C=C — F, F (F and F on the left carbon; F and F on the right carbon)

H, H — C=C — F, F (H and H on the left carbon; F and F on the right carbon)

[2]We can see via the noble gas shell model of Section 16.4 that, in the outer shell of the oxygen atom in a water molecule, there are two unbonded electron pairs. These unbonded electron pairs push the two hydrogen atoms together. Without these unbonded electron pairs, a water molecule would be as straight as a carbon dioxide molecule with the two hydrogen atoms 180° apart.

Polar molecules electrically attract one another and are relatively difficult to separate (Figure 17.35). Substances composed of polar molecules therefore typically have higher boiling temperatures than those composed of nonpolar molecules (Table 17.1). Water, for example, boils at 100°C, whereas carbon dioxide boils at −79°C. This 179 C° difference is quite dramatic when you consider that water and carbon dioxide differ by only one type of atom. Molecular polarity, to many chemists, is one of the most important and central concepts of chemistry.

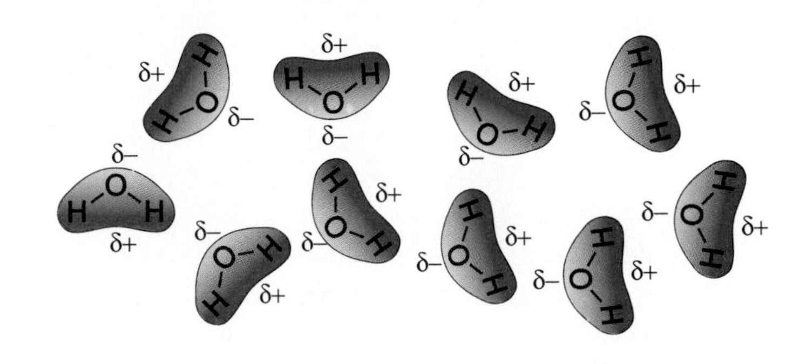

Fig. 17.35
Polar molecules stick to one another by an electrical force of attraction.

Table 17.1
Boiling Temperatures of Some Polar and Nonpolar Substances

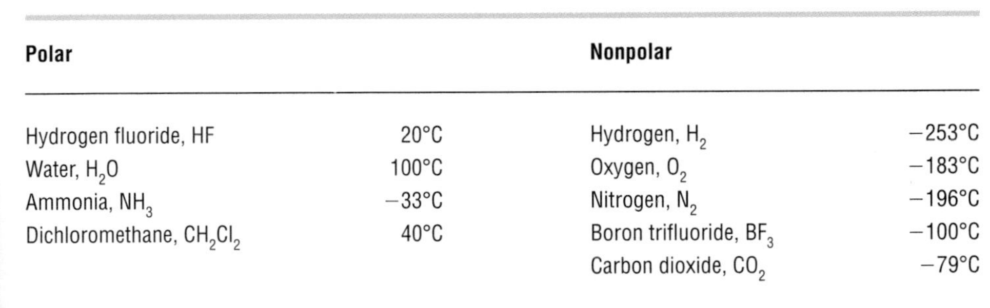

Polar		Nonpolar	
Hydrogen fluoride, HF	20°C	Hydrogen, H_2	−253°C
Water, H_2O	100°C	Oxygen, O_2	−183°C
Ammonia, NH_3	−33°C	Nitrogen, N_2	−196°C
Dichloromethane, CH_2Cl_2	40°C	Boron trifluoride, BF_3	−100°C
		Carbon dioxide, CO_2	−79°C

Fig. 17.36
Oil and water are difficult to mix, but not because they repel each other. Rather, water molecules are so attracted to themselves because of their polarity that they pull themselves together into a single phase. The nonpolar oil molecules are thus excluded and left to themselves in a separate phase.

Answer

Symmetry is often the greatest clue for determining polarity. The molecule on the left is symmetrical, and the resultant dipoles on the two sides cancel each other. The left molecule is therefore nonpolar:

The molecule on the right is less symmetrical (more "lop-sided"). Inspection of the electronegativities shows that carbon is more electronegative than hydrogen. Therefore the resultant dipole of the two H–C bonds points to the right. Because fluorine is more electronegative than carbon, the resultant dipole of the C–F bonds also points to the right. The sum of these two resultants (not shown) is an even larger resultant pointing to the right. The right structure is therefore polar, with the fluorine side slightly negative and the hydrogen side slightly positive.

Questions

1. A substance made of polar molecules tends to have a higher boiling temperature than one made of nonpolar molecules. Why?
2. Consider two substances—one made of the molecule shown on the left, the other made of the molecule shown on the right. Which will have a higher boiling temperature?

```
     Cl              Cl
     |               |
F—C—Cl          H—C—Cl
     |               |
     F               H
```

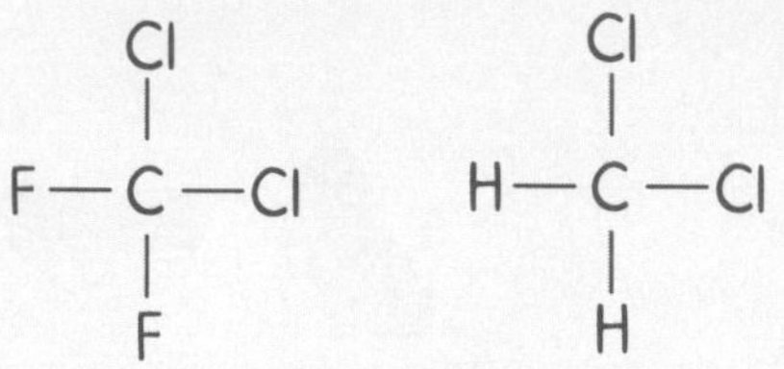

The vector addition techniques presented in this section are also applicable to molecules in their three-dimensional forms. Such a feat, however, is perhaps best explored hands-on in a molecular modeling laboratory that may accompany your course of study. For now, the important message is that the shape of a molecule determines its degree of polarity and that polarity has a great effect on macroscopic behavior. If water molecules were not bent, for example, they would be as nonpolar as carbon dioxide and the nature of water would be completely different. Water would not be a liquid at the ambient temperatures of our planet, and we, in turn, would not be here discussing the issue. Welcome to the joy of thinking molecular.

Answers

1. Polar molecules stick to one another, which means a lot of thermal energy (higher boiling temperature) must be added to separate them into the gaseous phase. Nonpolar molecules are much less "sticky," and so less thermal energy (lower boiling temperature) is needed to separate them into the gaseous phase.
2. Vector analysis of the F–C–F and Cl–C–Cl bonds in the left molecule shows two resultants that oppose and cancel each other. This molecule therefore has polarity only to the extent that the resultant of the F–C–F bond is greater than that of the Cl–C–Cl bond. (If you don't know why the F–C–F polarity is greater, check the periodic table.) Its boiling temperature is about 9°C, which means it is a gas at room temperature.

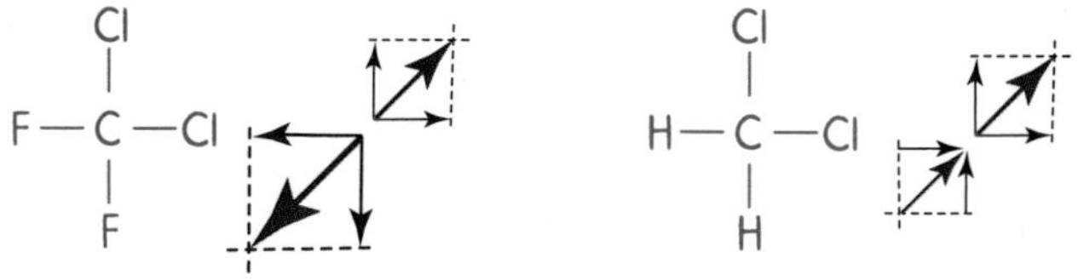

```
     Cl              Cl
     |               |
F—C—Cl          H—C—Cl
     |               |
     F               H
```

Vector analysis of the molecule on the right shows that, because carbon has a greater electronegativity than hydrogen, the direction of the H–C–H resultant vector is the same as that of the Cl–C–Cl resultant vector. Rather than canceling each other, these two resultants reinforce each other, giving rise to a molecule that is very polar overall. Its boiling temperature is 40°C, which means that it is a liquid at room temperature.

The compound on the left is a chlorofluorocarbon, a class of compounds once widely used as the refrigerant in air conditioners and refrigerators. These compounds are now banned because of evidence suggesting they play a significant role in the depletion of stratospheric ozone (Chapters 19 and 26). The compound on the right is dichloromethane. Once a popular organic solvent, its use is now limited due to its toxicity.

Link to Cooking—Heating Bonds and Molecular Polarity

Chemical bonds behave like tuning forks. They vibrate to and fro and even twist and turn in response to an input of energy. The natural frequency of a turning fork depends on the material it is made of and on its construction. Similarly, the natural frequency of a chemical bond depends on the masses of the bonded atoms and on the strength of the bond. The chemical bonds in a water molecule, for example, have a natural frequency in the range of infrared radiation, 10^{11} to 10^{14} hertz. Exposed to infrared radiation, water molecules begin to vibrate wildly. Bumping into one another, they twist, rotate, and bounce. These faster molecular motions translate into an increase in molecular kinetic energy and hence an increase in temperature.

Human skin is particularly sensitive to infrared radiation because of the significant amount of water it contains. As your skin is exposed to infrared radiation, the water molecules vibrate faster, and you feel the sensation of warmth. For this reason, infrared radiation—whether from the Sun, a campfire, or a hot stove—is known as heat radiation.

It is the polar nature of water molecules that underlies how microwave ovens work. The natural frequency of vibration of the H_2O dipole is about 10^9 to 10^{11} hertz. This frequency is in the range of microwaves, which flip water molecules to and fro, thus increasing their kinetic energy. Because food molecules aren't composed of strong dipoles, food is cooked in a microwave oven as the water it contains is heated. The flipping H_2O molecules slam into and heat the food from within. And because water retains heat so well, microwave cooking continues even after the food is removed from the oven. This is why many microwave recipes call for a wait period before serving.

It's also interesting to note that, because microwave cooking is just a fancy way of boiling, the temperature of the food never gets much hotter than that of boiling water. The browning of food, however, requires much higher temperatures. So bread won't ever toast in your microwave oven. For that you use the infrared radiation of your toaster.

Summary of Terms

- **Metallic bond** A chemical bond in which metal atoms are held together by their attraction to a common pool of electrons.
- **Alloy** A mixture of two or more metallic elements.
- **Ionic bond** The electrical force of attraction that holds ions of opposite charge together.
- **Ionic compound** Any chemical compound containing ions.
- **Ionic crystal** A group of many ions held together in an orderly three-dimensional array.
- **Covalent bond** A chemical bond in which atoms are held together by their mutual attraction for two electrons they share.
- **Covalent compound** An element or chemical compound in which atoms are held together by the covalent bond.
- **Molecule** A group of atoms held tightly together by covalent bonds.
- **Covalent crystal** A group of molecules arranged in an orderly fashion.
- **Dipole** A separation of charge.
- **Electronegativity** The ability of an atom to attract a bonding pair of electrons to itself when bonded to another atom.
- **Nonpolar** The state of a chemical bond or molecule having no dipole.
- **Polar** The state of a chemical bond or molecule having a dipole.
- **Percent ionic character** A measure of the degree of charge separation in a chemical bond as determined from the difference in electronegativities of the bonded atoms.

Review Questions

1. Which elements tend to form covalent bonds? Which tend to form ionic bonds? Which tend to form metallic bonds?

Metals and Alloys

2. Do metals more readilv gain or lose electrons?
3. Where are the bonding electrons within a metal located?
4. Why are metals malleable?
5. Which is the better statement: Metals are made out of alloys or alloys are made out of metals?

Ionic Bonds

6. Suppose an oxygen atom gains two electrons to become an oxygen ion. What is its electrical charge?
7. What is an ionic crystal?

Covalent Bonds

8. By what means are the atoms of a covalent bond held together?
9. What role does electron affinity play in covalent bond formation?
10. Which is the true statement: "Atoms are composed of molecules" or "molecules are composed of atoms"?
11. How many electrons are shared per covalent bond?
12. How many electrons are shared in a triple covalent bond?
13. Cite an example of a molecule easily seen by the naked eye.
14. How is frozen water different from liquid water on a molecular level?

Covalent Bond Polarity

15. What is a dipole?
16. Does a dipole have a net electric charge? Does it have electric properties?
17. Which is more polar, a carbon-oxygen bond or a carbon-nitrogen bond?
18. What is electronegativity?
19. An individual carbon-oxygen bond is polar. Yet carbon dioxide, CO_2, which has two carbon-oxygen bonds, is nonpolar. Why?
20. Why do nonpolar substances tend to boil at relatively low temperatures?
21. How do the outer electrons in metal atoms differ from the outer electrons in nonmetal atoms?
22. A friend says all chemical bonds are electrical. Do you agree or disagree?

Molecular Polarity

23. How can a molecule be nonpolar when it consists of atoms that have different electronegativities?
24. What two aspects of a dipole make it a vector?
25. Which would you describe as "stickier": a polar or nonpolar molecule?

Home Projects

1. Its polarity is responsible for most all of water's physical behavior. For a demonstration of water's marked polarity, rub an inflated balloon over your hair to give the balloon a static charge, and then hold it close to a thin stream of water running from a kitchen faucet. If the balloon held the opposite charge, would you see the same behavior?
2. View crystals of table salt with a magnifying glass or, better yet, a microscope if you have one available. Purchase some sodium-free salt (potassium chloride) and some flaked salt (an unusual form of sodium chloride) from the grocery store and examine these ionic crystals as well. Look at salt and sugar crystals mixed together and list how these two materials differ from each other. If you have a microscope, crush the salt crystals with a spoon and examine the resulting powder.
3. Use toothpicks and different colors of jelly beans or gum drops to build models of the molecules shown in this and other chapters. Assign a certain color to represent each type of atom and estimate the polarity—nonpolar, slightly polar, or very polar—of each molecule you build. As a general rule, nonpolar molecules are always more symmetric than polar molecules. Explain your conclusions to your instructor.

Exercises

1. How are metal atoms held together in the metallic bond?
2. Why is brass not an element?
3. Can an alloy contain a nonmetallic element?
4. A sodium atom loses an electron. Is this an example of a physical or chemical change?
5. Make an argument for why the splitting of a diamond should be viewed as a chemical change.
6. Magnesium ions carry a $+2$ charge, and chloride ions carry a -1 charge. What is the chemical formula for the ionic compound magnesium chloride?
7. Barium ions carry a $+2$ charge, and nitrogen ions carry a -3 charge. What is the chemical formula for the ionic compound barium nitride?
8. Which is easier to pull away from a sodium ion: a fluoride ion or a chloride ion? Why?
9. What is the fundamental unit of an ionic crystal?
10. How are ionic and covalent crystals different from each other?
11. What differentiates a metallic bond from a covalent bond?
12. Why might a chemist refer to the ions in a metallic or ionic bond as ambivalent as opposed to covalent?
13. Name four elements that tend to form covalent bonds.
14. Phosphine is a covalent compound of phosphorus and hydrogen. What is its chemical formula?

15. Explain the difference between electron affinity and electronegativity.

16. Which bond is most polar?
(a) H–N (b) N–C (c) C–O (d) C–C (e) O–H
(f) C–H

17. Which molecule is most polar:
(a) S=C=S (b) O=C=O (c) O=C=S

18. For each of the following diatomic molecules, circle the atom that carries the greater positive charge:
H–Cl Br–F C–O Br–Br

19. List the following bonds in order of increasing polarity:
(a) N–N (b) N–F (c) N–O (d) H–F
_______ < _______ < _______ < _______
(least polar) (most polar)

20. Which should be more polar, a sulfur-bromine (S–Br) bond or a selenium-chlorine (Se–Cl) bond?

21. List the following in order of increasing boiling temperature. Briefly explain your reasoning.

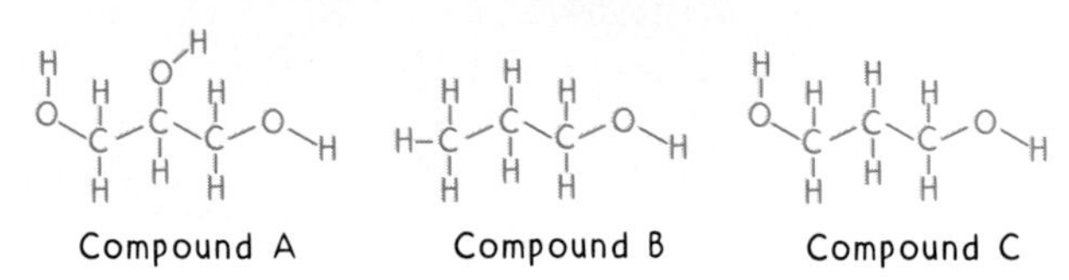

22. Water, H_2O, and methane, CH_4, have about the same mass and differ by only one type of atom. Why are their boiling temperatures so different from each other?

23. Classify the following bonds as ionic, covalent, or metallic (O, atomic no. 8; F, atomic no. 9; Na, atomic no. 11; Cl, atomic no. 17; U, atomic no. 92).
O with F _______________
Ca with Cl _______________
Na with Na _______________
U with Cl _______________

24. Circle the molecule from each pair that should have a higher boiling temperature: (atomic numbers: Cl=17; S=16; O=8; C=6; H=1)

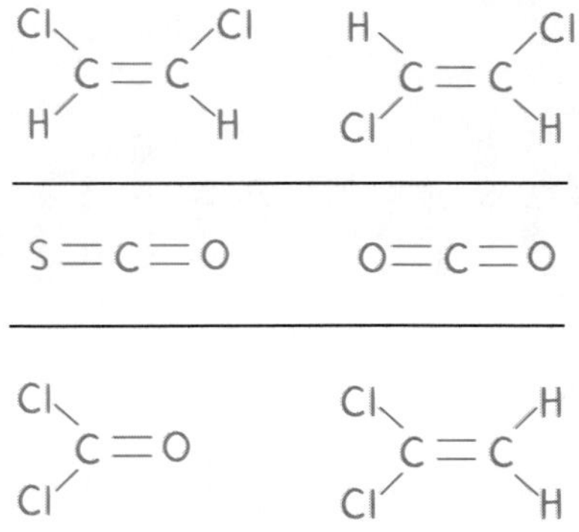

Cl(H)C=C(Cl)(H) Cl(H)C=C(Cl)(H)

S=C=O O=C=O

Cl_2C=O Cl_2C=CH_2

25. At room temperature, water is a liquid but carbon dioxide is a gas. Based on this information, which do you suppose is heavier: a water molecule or a carbon dioxide molecule? Defend your answer.

26. How is it that an electrically neutral hydrogen atom is more reactive (and dangerous in the laboratory!) than a negatively charged hydrogen ion?

27. Why can't a hydrogen atom form more than one covalent bond?

28. Which drives an atom to form a covalent bond: electron affinity or the need to have a filled outer shell? Explain.

29. Nonmetal atoms form covalent bonds, but they can also form ionic bonds. How is this possible?

30. In what sense are ionic bonds also covalent?

CHAPTER

18 Molecular Mixing

We have explored how atoms combine to form metals, ionic compounds, and molecules. We have also seen how the properties of a molecule depend on the number and types of atoms in the molecule and on their spatial orientation. With that foundation of understanding for what goes on within a molecule, we now explore the interactions that occur between molecules. Through this chapter we'll see how the interactions among molecules affect such things as a material's phase; oxygen dissolving in blood; the mixing of sugar and water; and the mixing of soap, water, and grease.

18.1 Molecular Interactions

Molecular interactions are the electrical attractions that can occur between molecules. The four types we consider in this chapter are listed in Table 18.1. They differ from one another primarily in their relative strengths, but it should be noted that all of them are more than 100 times weaker than any chemical bond.

Table 18.1
Relative Strengths of Molecular Interactions

Type of Interaction	Relative Strength
ion-dipole	Strongest
dipole-dipole	↑
dipole–induced dipole	↓
induced dipole–induced dipole	Weakest

Ion-Dipole

What happens when polar molecules, such as water, approach an ionic compound such as sodium chloride? The positive sodium ions are attracted to the negative side of the water molecules, and the negative chloride ions are attracted to the positive side of the water molecules. Such an interaction between an ion and a dipole is called an **ion-dipole interaction**.

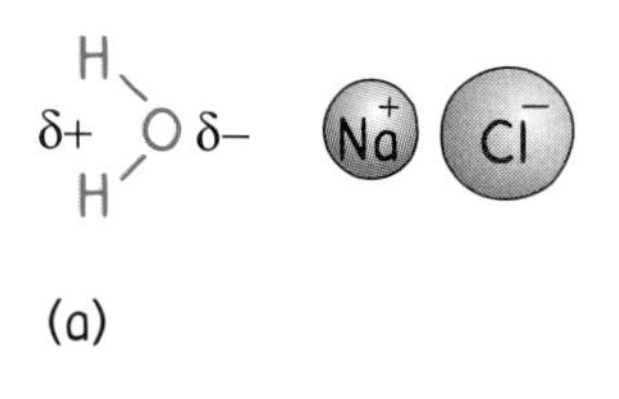

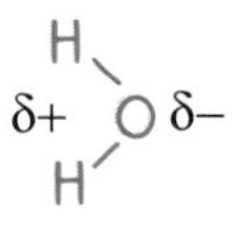

Fig. 18.1
(a) An ion-dipole interaction occurs between the negative side (oxygen side) of a water molecule and the positively charged sodium ion of sodium chloride. (b) Another ion-dipole interaction occurs between the positive side (hydrogen side) of a water molecule and the negatively charged chloride ion.

Ion-dipole interactions, although the strongest of the interactions listed in Table 18.1, are much weaker than ionic bonds. A large number of them, however, can act collectively to disrupt an ionic bond, and this is what happens with sodium chloride in water. The sodium and chloride ions are strongly held together by ionic bonds, but a multitude of ion-dipole interactions with water can break the bonds and pull the ions apart. The result is a solution of sodium chloride in water.

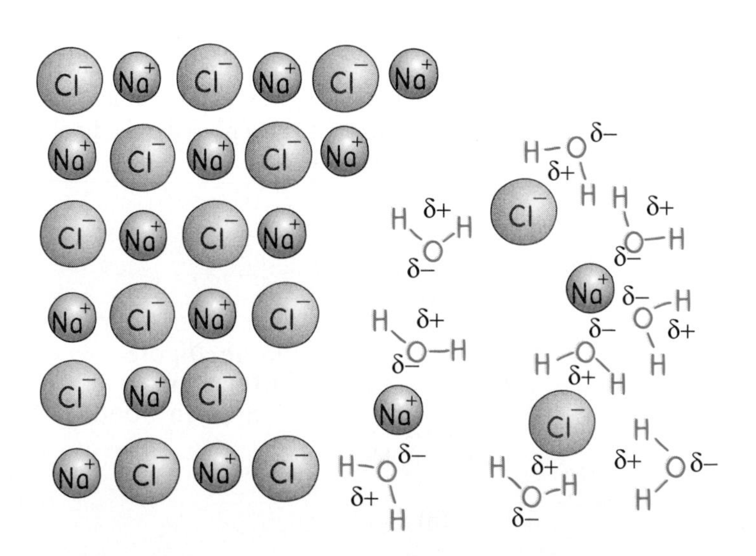

Fig. 18.2
In a water solution of sodium chloride, tightly bound sodium and chloride ions are separated from one another by the collective attractions of many water molecules.

Fig. 18.3
Seawater is a mixture of many types of ions dissolved in water by ion-dipole interactions. By weight, the primary ions are the sodium ion, Na^+ (32.2%), and the chloride ion, Cl^- (58.4%). Secondary ions include the magnesium ion, Mg^{2+} (3.9%), the sulfate ion, SO_4^{2-} (2.7%), the calcium ion, Ca^{2+} (1.2%), and the potassium ion, K^+ (1.1%).

Dipole-Dipole

When dipoles interact among themselves, we have the **dipole-dipole interaction**. If we represent dipole molecules as cigar-shaped objects having negative and positive ends, their charges align in an orderly and mutually attractive fashion as shown in Figure 18.4.

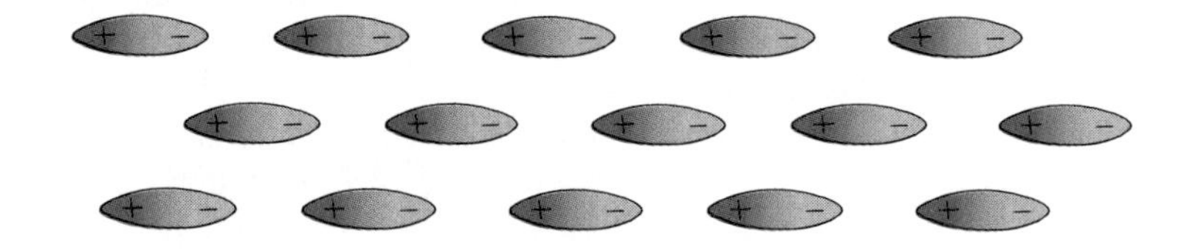

Fig. 18.4
Dipoles align with one another because of the attractive forces between them.

An unusually strong dipole-dipole interaction is the **hydrogen bond**. This interaction occurs between molecules that have hydrogen atoms covalently bonded to highly electronegative atoms, such as nitrogen, oxygen, and fluorine. A hydrogen bond occurs when the hydrogen side of one such molecule is attracted to the negatively charged side of a neighboring molecule. The strength of the hydrogen bond lies in the magnitude of the dipoles involved. The term *hydrogen bond* is somewhat of a misnomer and shouldn't be confused with a "real" chemical bond. Recall from Chapter 17 that a chemical bond may be either metallic, ionic, or covalent. A hydrogen bond, however, is none of these; it is simply a dipole-dipole interaction that happens to involve hydrogen atoms (Figure 18.5).

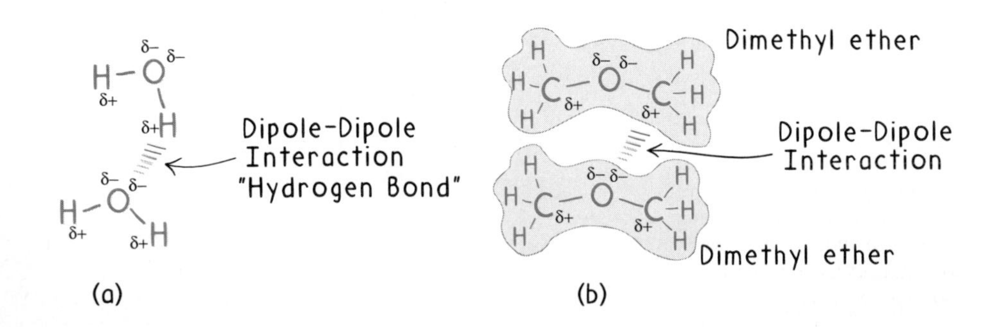

Fig. 18.5
(a) The dipole-dipole interaction between two water molecules (shown as hatched lines) is a hydrogen bond because it involves hydrogen atoms attached to highly electronegative atoms (oxygen). (b) The dipole-dipole interaction between two dimethyl ether molecules (circled in gray) does not involve any hydrogen bonding. The hydrogen atoms in dimethyl ether are attached to carbon, which has an electronegativity similar to that of hydrogen.

The hydrogen bond is responsible for many of the unusual properties of water and other similar compounds, and it is of great importance in the chemistry of large biomolecules such as DNA. Although the hydrogen bond is stronger than other dipole dipole interactions, it, like all the other molecular interactions discussed in this chapter, is much weaker than any chemical bond. Perhaps elevation to the status of "bond" owes to its significance rather than its strength.

Dipole–Induced Dipole

Electrons in many molecules are distributed evenly, and so there is no dipole, as is the case with the oxygen molecule, O_2. Such a nonpolar molecule can be induced into becoming a temporary dipole, however, when it is brought close to a water molecule or any other dipole (Figure 18.6). The negative side of the water molecule pushes the oxygen's electrons to the side of the oxygen molecule farthest from the water. The result is an uneven distribution of electrons and therefore a dipole in the oxygen molecule. Because this dipole needs to be induced, it is called an **induced dipole**. The resulting attraction between the permanent dipole (water) and the induced dipole (oxygen) is an example of the **dipole–induced dipole interaction**.

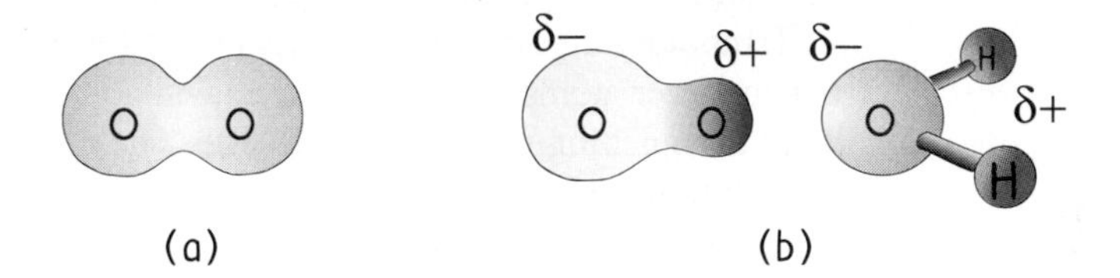

Fig. 18.6
(a) An isolated oxygen molecule, O_2, has no dipole; its electrons are distributed evenly. (b) An adjacent water molecule induces a redistribution of electrons in the oxygen molecule.

■ **Question**

How does the electron distribution in an oxygen molecule change when the hydrogen end of a water molecule is nearby?

Induced dipoles are only temporary. If the water molecule of Figure 18.6b were, say, knocked away by thermal motion, the oxygen molecule would return to its normal nonpolar state. As a consequence, dipole–induced dipole interactions are weaker than dipole-dipole interactions (Table 18.1). They are strong enough to hold relatively small quantities of oxygen dissolved in water, however. This is vital for fish and other forms of marine life, who breathe not water but the oxygen mixed in with the water.

Dipole–induced dipole interactions also occur between molecules of nonpolar carbon dioxide and water. This helps to keep carbonated beverages (which are mixtures of carbon dioxide in water) from losing their fizz too quickly after they've been opened. Dipole–induced dipole interactions are also responsible for holding plastic wraps to glass (Figure 18.7). These wraps are made of very long nonpolar molecules that are induced to have dipoles when placed in contact with the highly polar glass molecules.

Fig. 18.7
Polar induction of the normally nonpolar molecules in plastic wrap makes it stick to the highly polar glass molecules.

■ **Answer**
Because the hydrogen side of a water molecule is slightly positive, the electrons in oxygen are pulled toward it and a temporary dipole is induced in the oxygen.

■ **Question**

Distinguish between a dipole-dipole interaction and a dipole–induced dipole interaction.

Induced Dipole–Induced Dipole

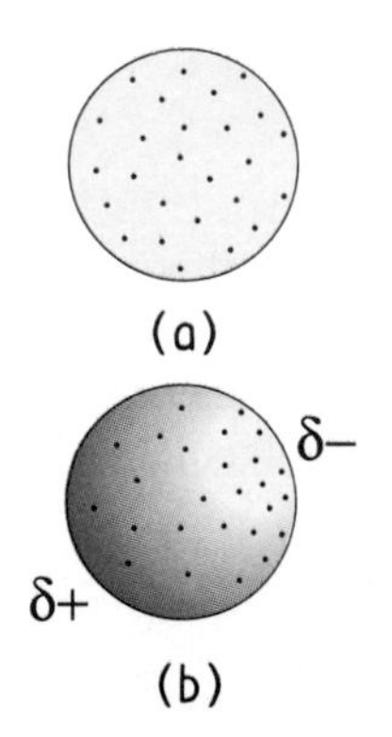

Fig. 18.8
(a) The electron distribution in a nonpolar atom is normally even. (b) The distribution of electrons at any moment, however, may be less than even, resulting in a temporary dipole.

Nonpolar atoms and molecules, on average, have a fairly even distribution of electrons. Because of the randomness of electron motion, however, at any given moment the electrons in a nonpolar atom or molecule may be bunched to one side. The consequence is a temporary dipole (Figure 18.8).

Just as a permanent dipole of a polar molecule can induce a dipole in a nonpolar molecule, a temporary dipole can do the same thing. This gives rise to the relatively weak **induced dipole–induced dipole interaction** (Figure 18.9).

Temporary dipoles are more significant for larger atoms. This is because the electrons in larger atoms have more space available for random motion and a higher likelihood of bunching together on one side. The electrons in smaller atoms are less able to bunch to one side because they are confined to a smaller space, where greater electrical repulsion tends to keep them evenly spread. So it is the larger atoms, and molecules made of larger atoms, that have the strongest induced dipole–induced dipole interactions.

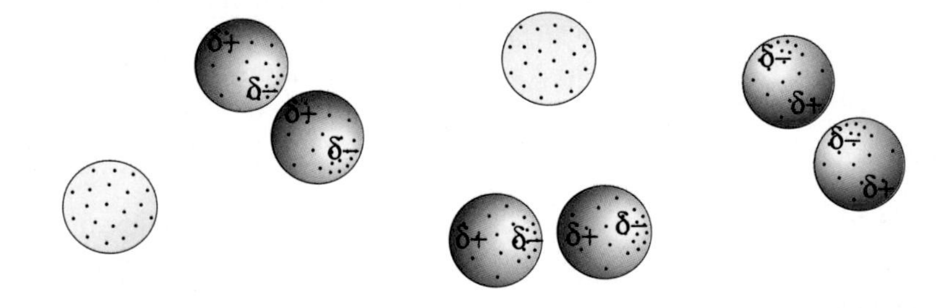

Fig. 18.9
Nonpolar atoms can be attracted to one another by induced dipole–induced dipole interactions.

That temporary dipoles are more important in larger atoms is illustrated by the different boiling temperatures of different-sized molecules. Compare the boiling temperatures of the halogens iodine, I_2; bromine, Br_2; chlorine, Cl_2; and fluorine, F_2 (Table

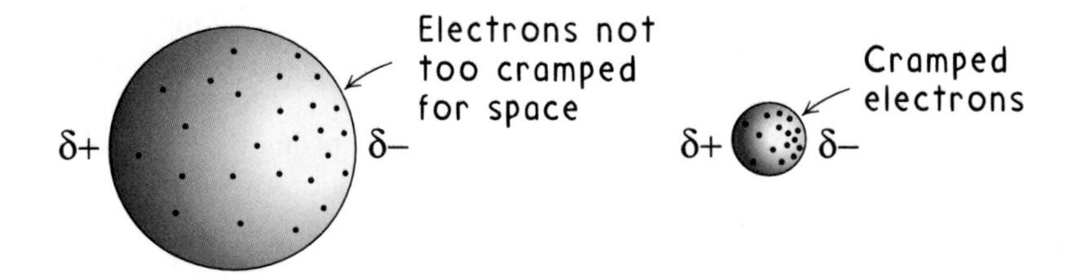

Fig. 18.10
Temporary dipoles are easier to form in larger atoms, where electrons bunched to one side are still relatively far apart from one another.

■ **Answer**
The dipole-dipole interaction is stronger and involves two permanent dipoles. The dipole–induced dipole interaction is weaker and involves a permanent dipole and a temporary one.

Table 18.2
Boiling Points of Halogens

Halogen	Atomic Radius*	Boiling Temperature	Phase at 25°C
Iodine, I_2	1.3 Å	184°C	solid
Bromine, Br_2	1.1 Å	59°C	liquid
Chlorine, Cl_2	0.98 Å	−35°C	gas
Fluorine, F_2	0.68 Å	−188°C	gas

*As measured in Angstroms, Å, where 1Å$=10^{-10}$ meter.

18.2). Iodine is a large molecule. This relatively large size allows more frequent temporary dipoles and many induced dipole–induced dipole interactions among iodine molecules. Because of these interactions, iodine molecules are relatively hard to pull apart. Hence iodine has a high boiling temperature of 184°C. At room temperature iodine's induced dipole–induced dipole interactions are strong enough to hold it together as a solid. Fluorine is the smallest halogen molecule. Temporary dipoles for fluorine are infrequent, and so induced dipole–induced dipole interactions are weak. As a consequence fluorine has a low boiling temperature of −188°C and, at room temperature fluorine molecules remain separated from one another in the gaseous phase.

Because of the very weak interactions in fluorine, it is very difficult for any other molecule to stick. This is the principle behind the Teflon nonstick surface. The chemical structure of Teflon consists of a long chain of carbon atoms chemically bonded to fluorine atoms (Figure 18.11).

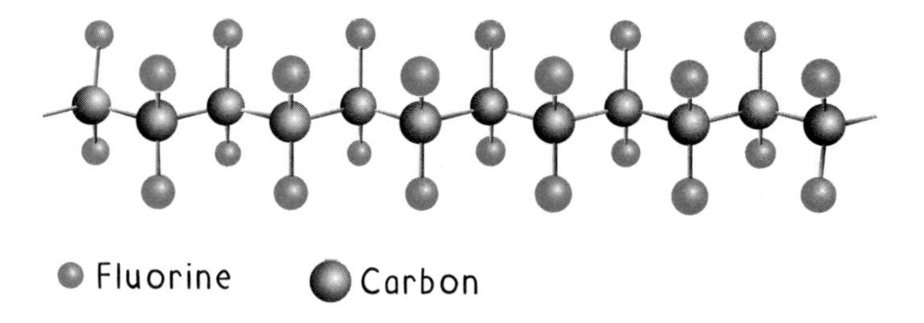

Fig. 18.11
Few things stick to Teflon because of the high proportion of fluorine atoms it contains. The structure depicted here is only a portion of the full length of the molecule.

Question

Distinguish between a dipole–induced dipole interaction and an induced dipole–induced dipole interaction.

Induced dipole–induced dipole interactions explain why, at room temperature, natural gas is a gas but gasoline is a liquid. The major component of natural gas is methane, CH_4. The chemical composition of methane is similar to that of molecules found in gasoline, such as octane, C_8H_{18}. We see in Figure 18.12 that the number of induced dipole–induced dipole interactions that can take place between two methane

---

Answer
The dipole–induced dipole interaction is stronger and involves a permanent dipole and a temporary one. The induced dipole–induced dipole interaction is weaker and involves two temporary dipoles.

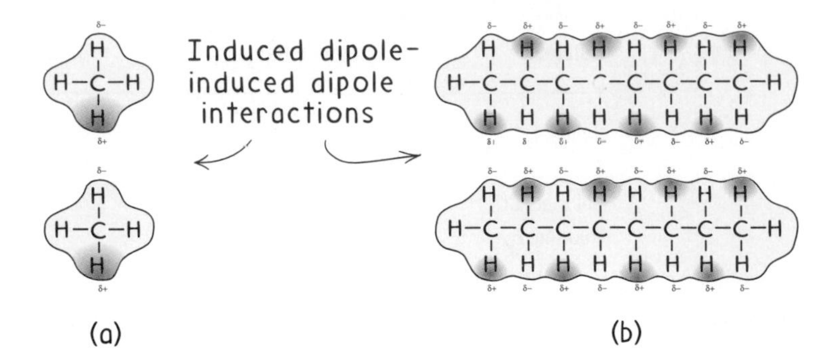

Fig. 18.12
(a) Nonpolar methane molecules, CH_4, are attracted to one another by induced dipole–induced dipole interactions. The number of these interactions is relatively low, however. (b) The nonpolar molecules found in gasoline, such as octane, C_8H_{18}, are similar to methane but larger. The number of induced dipole–induced dipole interactions among these larger nonpolar molecules is greater.

molecules is appreciably less than the number that can occur between two octane molecules. Octane molecules are larger, and therefore more induced dipole–induced dipole interactions are possible. Have you ever noticed that two small pieces of Velcro are easier to pull apart than two long pieces? Like short pieces of Velcro, methane molecules can be pulled apart with little energy. That's why methane has a low boiling temperature, −161°C, and is a gas at room temperature. Octane molecules, on the other hand, like long strips of Velcro, are relatively hard to pull apart because of the larger number of induced dipole–induced dipole interactions. The boiling temperature of octane, 125°C, is therefore much higher than that of methane and octane is a liquid at room temperature.

Questions

1. Gasoline is a gas at the temperatures of an operating car engine, whereas motor oil, which is used for lubricating the engine, is not. Both gasoline and motor oil are made of nonpolar molecules that interact by induced dipole–induced dipole interactions. Suggest how the sizes of gasoline molecules and motor oil molecules may differ.
2. Methanol, CH_3OH, which can also be used as a fuel, is not much larger than methane yet is a liquid at room temperature. Suggest why.

18.2 Solutions

What happens to table sugar as it is stirred into water? Is the sugar destroyed? We know it isn't because it sweetens the water. Does the sugar disappear because it somehow ceases to occupy space, or because it fits within the nooks and crannies of the water? Not so, for the presence of sugar does affect volume. This may not be noticeable at first, but continue to add sugar to a glass of water and you'll see that the water level rises just as it would if you were adding sand.

Answers

1. To transform a liquid into a gas requires that the molecules of the substance be separated from one another. The stronger the interactions among these molecules, the harder it is for them to be separated. Because motor oil remains a liquid at the high operating temperatures of an engine, the interactions among the motor oil molecules (including induced dipole–induced dipole) must be strong. These molecules are therefore longer than the molecules in gasoline and, like long strips of Velcro, harder to separate.
2. The polar oxygen-hydrogen covalent bonds in methanol lead to strong dipole-dipole interactions among methanol molecules. These strong interactions hold methanol molecules together as a liquid at room temperature.

What happens is that the sugar loses its crystalline form. Each crystal of sugar consists of billions upon billions of sugar molecules packed neatly together. When the sugar crystals are exposed to water, an even greater number of water molecules pull them apart as new dipole-dipole interactions form between the sugar and the water (Figure 18.13). With a little stirring the sugar molecules soon mix throughout the water. In place of sugar crystals and water we have a homogeneous mixture of sugar and water.[1]

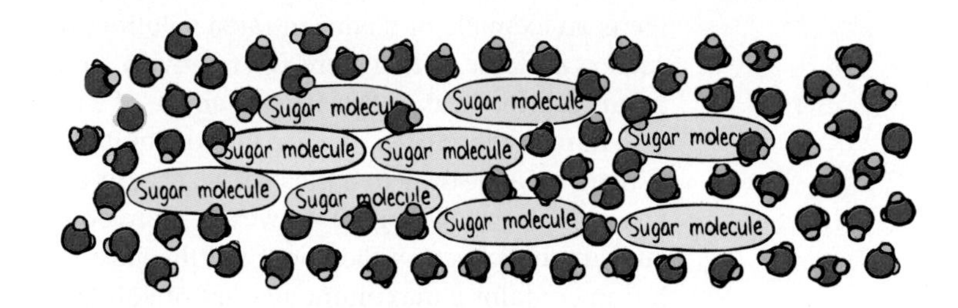

Fig. 18.13
Water molecules break the sugar molecules in a sugar crystal away from one another.

Recall from Section 15.5 that a homogeneous mixture consisting of only a single phase is called a *solution.* Sugar in water is a solution where the phase is liquid, as is copper sulfate in water (Figure 18.14). Solutions aren't always liquids, however. They can also be solid or gaseous. Gem stones are examples of solid solutions. A ruby, for example, is a solid solution of trace quantities of red chromium ions, Cr^{3+}, in transparent aluminum oxide. A blue sapphire is a solid solution of trace quantities of light green iron ions, Fe^{2+}, and blue titanium ions, Ti^{4+}, in aluminum oxide. Other important examples of solid solutions include metal alloys. Brass is a solid solution of copper and zinc, and stainless steel is a solid solution of metallic iron, chromium, nickel, and a little nonmetallic carbon. An example of a gaseous solution is the air we breathe, which is a gaseous solution of 78 percent nitrogen, 21 percent oxygen, and 1 percent other gaseous materials, including water vapor and carbon dioxide. The air we *exhale* is a gaseous solution of 75 percent nitrogen, 14 percent oxygen, 5 percent carbon dioxide, and 6 percent water vapor.

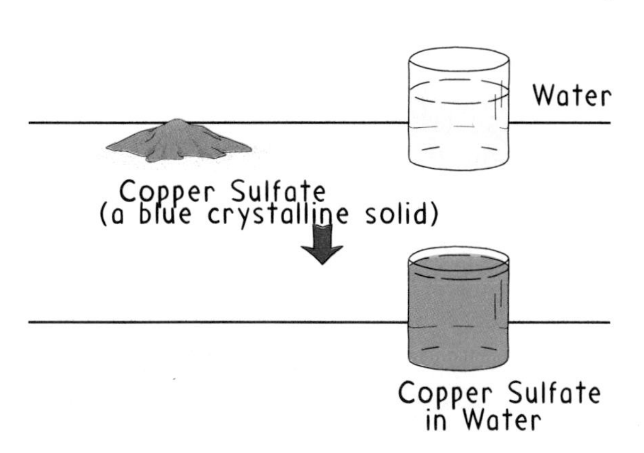

Fig. 18.14
The properties of a solution are a mixture of the properties of its components. Copper sulfate is blue, and water is colorless. The color of a solution of copper sulfate in water is light blue. Adding more copper sulfate results in a deeper blue color.

In describing solutions, it is usual to think of the component present in the largest amount as the **solvent** and the other component(s) as the **solute(s)**. For example, when a teaspoon of sugar is mixed with a liter of water, we identify the sugar as the solute and the water as the solvent. The process of mixing a solute in a solvent is called **dissolving**. To make a solution, a solute must *dissolve* in a solvent; that is, the solute and solvent must form a homogeneous mixture consisting of only one phase. Whether one material dissolves in another is a function of molecular interactions. As we explore in Section 18.3 the stronger the molecular interactions, the greater the likelihood that the two materials will form a solution.

---

[1]As discussed in Section 15.5, homogeneous means that a sample from one part is the same as a sample from any other part. For example, the sweetness of the first sip of the sugar water is the same as the sweetness of the last sip.

■ **Question**

What is the solvent of the air we breathe?

Solutions are often described by the amount of solute they contain. When there is a relatively large amount of dissolved solute, the solution is **concentrated**. Strong coffee is an example of a concentrated solution, for it contains large amounts of solutes, such as caffeine. When there is a relatively small amount of dissolved solute, the solution is **dilute**. Weak coffee is an example of a dilute solution.

In many instances, the amount of solute that can dissolve in a solvent is limited. When you add sugar to a glass of water, for example, the sugar rapidly dissolves. As you continue to add sugar, however, there comes a point when it no longer dissolves and simply collects at the bottom of the glass, even after stirring. At this point, the solution contains a maximum amount of solute—the solution is **saturated**. A solution that has *not* reached the limit of solute that will dissolve in it is said to be **unsaturated** (Figure 18.15).

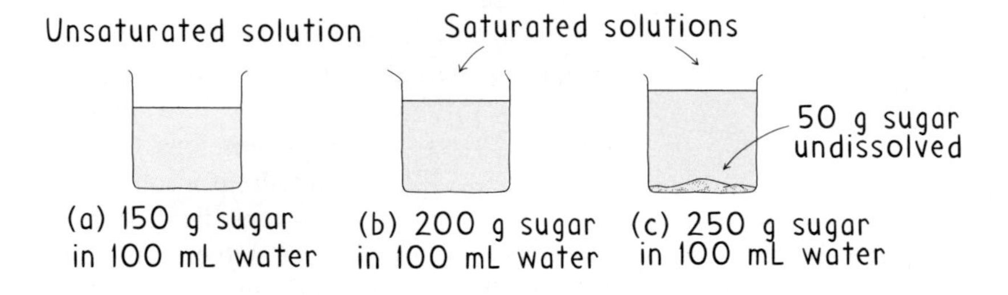

Fig. 18.15
A maximum of about 200 g of sugar dissolves in 100 mL of water at 20°C. (a) 150 g of sugar in 100 mL of water at 20°C produces an unsaturated solution. (b) 200 g of sugar in 100 mL of water at 20°C produces a saturated solution. (c) If 250 g of sugar is mixed into 100 mL of water at 20°C, 50 g remains undissolved.

The quantity of solute dissolved in a solution is described in mathematical terms by the solution's **concentration**, which is the amount of solute dissolved per volume of solution:

$$\textbf{Concentration} = \frac{\textbf{amount of solute}}{\textbf{volume of solution}}$$

For example, sugar water may have a concentration of 1 gram of sugar for every liter of solution.[2] This can be compared with concentrations of other solutions. A sugar solution containing 2 grams of sugar per liter of solution, for example, is more *concentrated,* and one containing only 0.5 gram of sugar per liter of solution is less concentrated, or more *dilute.*

We can calculate the amount of solute in a solution if we know the concentration of the solution and its volume:

$$\textbf{Amount of solute} = \textbf{concentration} \times \textbf{volume of solution}$$

■ **Answer**

Nitrogen is the solvent because it is the component present in the greatest quantity.

[2]Note carefully that the definition of concentration used here refers to liters of *solution,* not liters of solvent.

For example, if the concentration of sugar water is 1 gram per liter and the volume of solution is 1 liter, then the solution contains 1 gram. In using this relationship, notice that the unit of volume used for the concentration must be the same as the unit used for the volume of solution:

$$\textbf{Amount of solute} = \textbf{(1 gram/1 liter)} \times \textbf{(1 liter)} = \textbf{1 gram}$$

Questions

1. How many grams of sugar is dissolved in 5 L of sugar water that has a concentration of 0.5 g per liter of solution?
2. A saturated solution of sodium chloride in water has a concentration of about 300 g per liter of solution. How many grams of sodium chloride are required to make 3 L of a saturated solution?

Chemists are often more interested in the number of molecules of solute in a solution rather than in the number of grams. Molecules, however, are so very small that the number of them in any observable sample is incredibly large. To get around cumbersome numbers, a unit called the **mole** was defined. For reasons that are explained in Section 19.6, 1 mole is an incredibly large number—6.02×10^{23}. A mole of marbles, for example, which means 6.02×10^{23} marbles, is enough to cover the entire land area of the United States to a depth greater than 4 meters! Because molecules are so much smaller than marbles, there are that many sugar molecules—6.02×10^{23} of them—in only 342 grams of sugar. Therefore a solution of sugar water that has a concentration of 342 grams of sugar per liter of solution also has a concentration of 6.02×10^{23} sugar molecules per liter of solution or, by definition, a concentration of 1 mole per liter.

The most common unit of concentration used by chemists is **molarity**, which is the solution's concentration expressed in moles per liter:

$$\textbf{Molarity} = \frac{\textbf{number of moles}}{\textbf{1 liter of solution}}$$

A solution that contains 1 mole of solute per liter of solution is said to have a concentration of 1 *molar,* which is often abbreviated 1 M. Similarly, a more concentrated 2 molar solution (2 M) contains 2 moles of solute per liter of solution.

The importance of referring to the number of molecules of a solute rather than the number of grams can be illustrated by the following question: The saturated solution of sugar water in Figure 18.15b contains 200 grams of sugar and 100 grams of water. Which is the solvent: sugar or water? There are 3.5×10^{23} molecules of sugar in 200 grams of sugar but almost ten times as many molecules of water in 100 grams of water—3.3×10^{24}.[3] As defined earlier, the solvent is the component present in the

Answers

1. Multiply concentration by volume: (0.5 g/L)(5 L) = 2.5 g.
2. Multiply concentration by volume to obtain the amount of solute required: (300 g/L)(3 L) = 900 g.

[3]We'll learn how to calculate the number of molecules in a given mass in Chapter 19.

largest *amount,* but what do we mean by amount? If amount means number of molecules, then water is the solvent. If amount means mass, then sugar is the solvent. So, the answer depends on how you look at it. From a chemist's point of view, "amount" typically means the number of molecules, and so water is the solvent in this case.

Questions

1. How many moles of sugar are there in 0.5 L of a 4 M solution? How many molecules of sugar is this?
2. Does 1 L of a 1 M solution of sugar water contain 1 L of water, less than 1 L of water, or more than 1 L of water?

18.3 Solubility

The **solubility** of a solute refers to its ability to dissolve in a solvent. As we might expect, this ability depends in part on the molecular interactions that take place between the solute and the solvent. With strong molecular interactions, such as dipole-dipole, a solute may be very **soluble**—a lot can be dissolved before the solution is saturated. With weaker molecular interactions, such as induced dipole–induced dipole, a solute may be less soluble, and the solution may become saturated after only very little of the solute is dissolved.

In some instances molecular interactions are so significant that there is no limit to the amount of solute that can be dissolved in a solvent—there is no point of saturation. In these cases the two materials are said to be *infinitely soluble* in one another. This is the situation with ethanol and water, where the molecules are attracted to one another by hydrogen bonds. We can even add ethanol to water until the ethanol rather than the water may be considered the solvent. In fact, ethanol and water stick so well to each other that, even after distillation, the purest ethanol we can get is 95 percent.

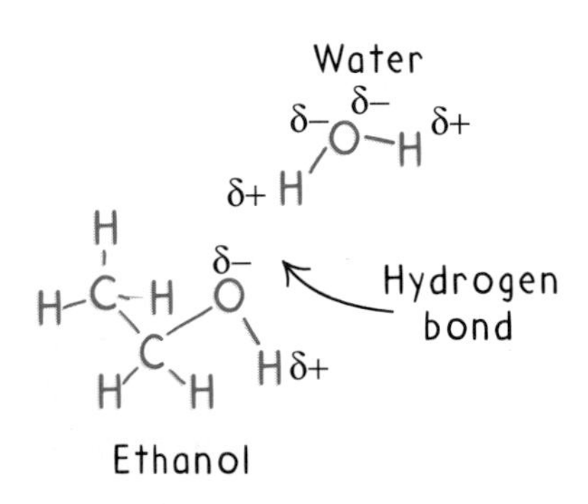

Fig. 18.16
Ethanol and water attract each other via hydrogen bonds. Any amount of ethanol mixes with any amount of water, and we say that these two liquids are infinitely soluble in each other.

As discussed in Section 18.1, sodium chloride has excellent solubility in water because of the ion-dipole interactions. Sugar also has excellent solubility in water, not because of ion-dipole interactions but because of dipole-dipole

Answers

1. First you need to understand that 4 M means 4 moles per liter. Then multiply concentration by volume to obtain amount of solute: (4 moles/L)(0.5 L) = 2 moles. Because 1 mole equals 6.02×10^{23} molecules, 2 moles equals twice this much, 12.04×10^{23} molecules.

2. The definition of molarity refers to number of liters of *solution,* not liters of solvent. When sugar is added to water, the volume of the liquid increases. So, if 1 mole of sugar is added to 1 L of water, the result is more than 1 L of solution. Therefore, 1 L of a 1 M solution requires less than 1 L of water. If you wish to prepare a solution of known concentration, it is best to add the solvent to the solute rather than the other way around. Discuss with a classmate or your instructor why this is so.

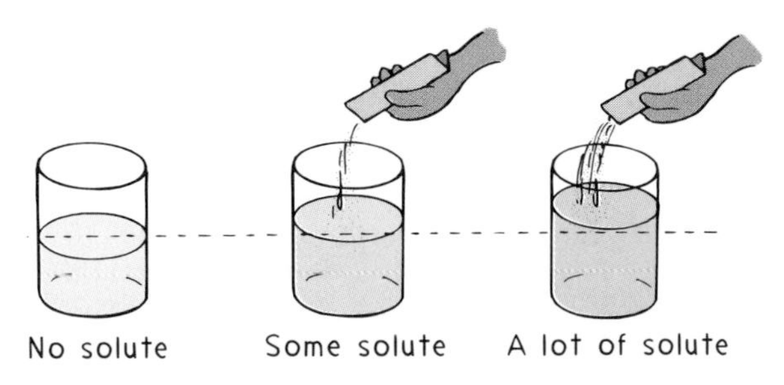

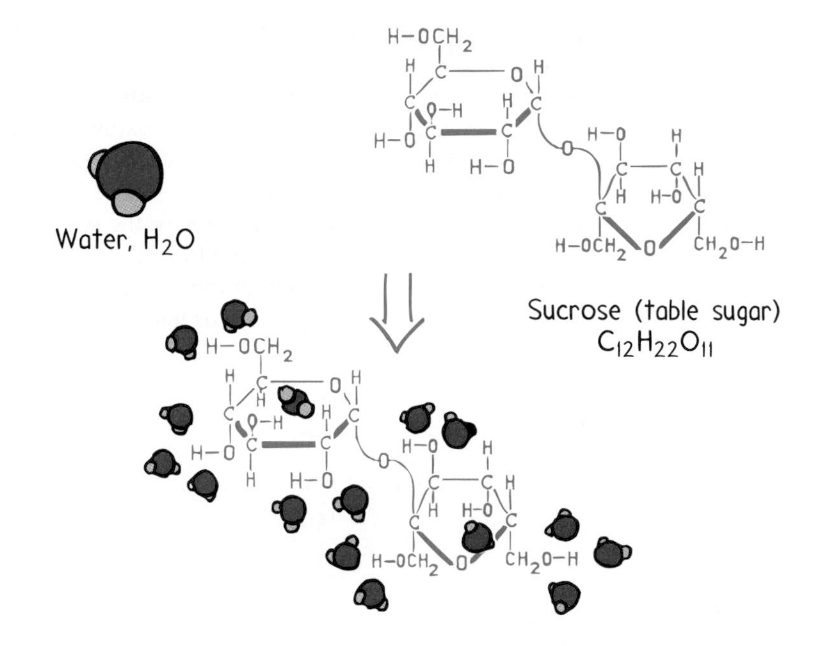

Fig. 18.17
Many water molecules are attracted to a sugar molecule by way of dipole-dipole interactions.

Figure 18.18
Glass is frosted by dissolving its outer surface in hydrofluoric acid.

Figure 18.19
Is this cup melting or dissolving?

interactions. A sugar molecule has many hydrogen-oxygen covalent bonds, as Figure 18.17 shows. Each of these bonds is polar, with the hydrogen side slightly positive and the oxygen side slightly negative. This polarity attracts many water molecules, with the result that each sugar molecule becomes surrounded by a multitude of water molecules and the sugar dissolves.

Considering the many dipole-dipole interactions that can take place between sugar and water, we may wonder why sugar is not infinitely soluble in water. Solubility has to do not only with interactions between the solute and the solvent but also with interactions among solute molecules and interactions among solvent molecules. In the case of sugar, solute-solute interactions are strong enough to make sugar a solid at room temperature and give it a relatively high melting temperature of 185°C. The fact that water molecules must work to overcome the sugar-sugar interactions and pull sugar molecules away from one another diminishes the solubility of sugar in water. This is why sugar is not infinitely soluble in water.

An example of a solute that has low solubility in water is oxygen, O_2. In contrast to sugar, which has a solubility of about 200 grams per 100 milliliters of water, only 0.004 gram of oxygen can dissolve in 100 milliliters of water. We can account for oxygen's low solubility in water by noting that the only molecular interactions that occur between oxygen and water are relatively weak dipole–induced dipole interactions. More important, however, is the fact that the stronger attraction of water molecules for one another effectively excludes oxygen molecules from intermingling.

A material that does not dissolve in a solvent to any appreciable extent is said to be **insoluble** in that solvent. There are many substances we consider to be completely insoluble in water, including sand, glass, and Styrofoam.

Just because a material is not soluble in one solvent, however, does not mean it won't dissolve in another. Sand and glass, for example, are soluble in hydrofluoric acid, HF, which is used to give glass a decorative frosted look. Styrofoam is partially soluble in acetone, a solvent used in nail-polish remover. Pour a little acetone into a Styrofoam cup and the acetone soon dissolves the bottom of the cup and leaks out.

Solubility and Temperature

We know from experience that water-soluble solids dissolve better in hot water than in cold water. A molecular explanation is that hot water molecules have greater thermal energy and as a result are able to "attack" the solid solute more vigorously. However, although the solubilities of some solid solutes—sugar, to name just one example—are greatly affected by temperature changes, the solubilities of other solid solutes, such as sodium chloride, are only mildly affected. Why this is so has to do with a number of factors, including the strength of the chemical bonds within the solid and the way the molecules of the solid are packed together.

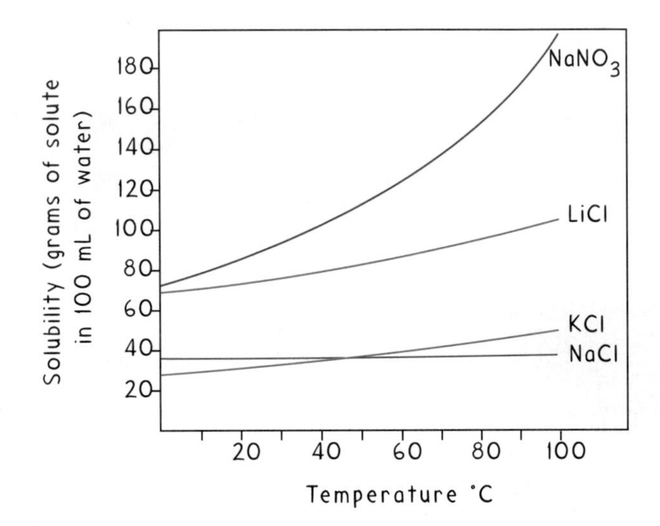

Fig. 18.20
The solubility of many water-soluble solids increases with temperature.

When a solution saturated at a high temperature is allowed to cool, some of the solute usually comes out of solution and forms what is called a **precipitate**. When this happens, the solute is said to have *precipitated.* For example, at 100°C the solubility of sodium nitrate, $NaNO_3$, in water is 180 grams per 100 milliliters of water. As we cool this solution, the solubility of $NaNO_3$ decreases, and this causes some of the dissolved $NaNO_3$ to precipitate (come out of solution). At 20°C, the solubility of $NaNO_3$ is 87 grams per 100 milliliters of water. So if we cool the 100°C solution to 20°C, 93 grams (180 grams − 87 grams) precipitates (Figure 18.21).

If a hot saturated solution is allowed to cool slowly and without disturbance, the solute may stay in solution. The result is a **supersaturated** solution. Supersaturated

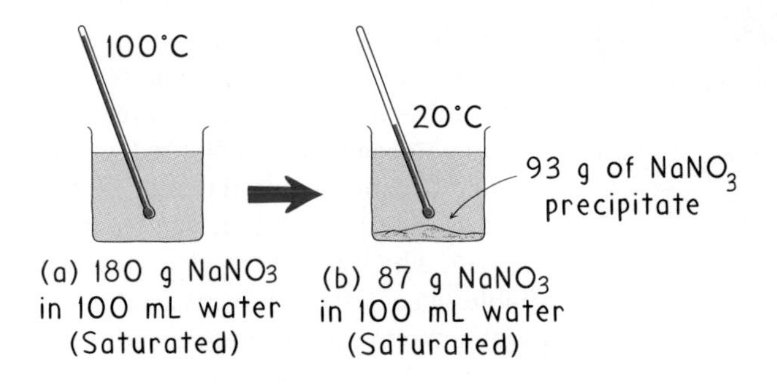

Fig. 18.21
(a) The solubility of sodium nitrate, $NaNO_3$, is 180 g per 100 mL of water at 100°C but only 87 g/100 mL at 20°C.
(b) Cooling a 100°C saturated solution of $NaNO_3$ to 20°C causes 93 g of the solute to precipitate.

Fig. 18.22
Rock candy growing out of a supersaturated solution of sugar water.

solutions of sugar water are fairly easy to make. Just dissolve as much sugar as possible in some boiling water, then allow the solution to cool. You will find that this solution is somewhat unstable, for if you disturb it—by stirring, for example—much of the sugar precipitates. It is possible to grow large sugar crystals, also known as "rock candy," out of a supersaturated solution. One good way is to tie some string to a weight, such as a nut or bolt, and lower the string into the solution *before* it cools. Support the string with a pencil such that the weight does not touch the bottom. Leave the mixture undisturbed for about a week, but check it periodically. The longer you wait, the larger the crystals.

Solubility of Gases

The solubilities of gases in liquids *decrease* with increasing temperature (Table 18.3). With an increase in temperature, the solvent has more thermal energy. This makes it more difficult for a gaseous material, such as oxygen, to stay in solution in the hot liquid solvent because the gas solute molecules are literally being kicked out by the solvent molecules. Perhaps you have noticed that warm carbonated beverages go flat faster than cold ones. Greater warmth causes the molecules of carbon dioxide gas to leave the liquid solvent at a higher rate.

The solubility of a gas in a liquid also depends on the pressure of the gas immediately above the liquid. In general, higher pressure means more of the gas dissolves. A gas at a high pressure corresponds to many, many gas particles crammed into a given volume. The pocket of space above an enclosed soda drink, for example, is crammed with carbon dioxide molecules in the gaseous phase. With nowhere else to go, many of these molecules "escape" into the liquid. Alternatively, we might say that the great pressure forces the carbon dioxide molecules into solution. When the soda bottle is opened, the "head" of highly pressured carbon dioxide escapes. With a lower pressure above the drink, the solubility of the carbon dioxide drops and the carbon dioxide molecules once squeezed into the solution begin to escape. The rate at which carbon dioxide leaves, however, is relatively slow. You can increase the rate by pouring in

Table 18.3
The Temperature-Dependent Solubility of Oxygen in Water*

Temperature, °C	Solubility of O_2 (g O_2/L H_2O)
0	0.0141
10	0.0109
20	0.0092
25	0.0083
30	0.0077
35	0.0070
40	0.0065

*All data at a pressure of 1 atmosphere.

Fig. 18.23
Marine life flourishes in cold polar waters, where the solubility of oxygen and carbon dioxide gases are greater than in warm tropical waters, which are not as fertile.

Link to Medicine—Perfluorocarbons

Oxygen gas is not very soluble in water because polar water molecules are so attracted to each other that they exclude the nonpolar oxygen molecules. Oxygen, however, is very soluble in a class of nonpolar compounds known as *perfluorocarbons,* which consist of carbon atoms bonded to many fluorine atoms:

Interestingly, a saturated solution of oxygen in liquid perfluorocarbons contains about 20 percent more oxygen than does the atmosphere we breathe. When this solution is inhaled, the lungs (animal or human) are able to absorb the oxygen in much the same way they absorb it from air, and because perfluorocarbons are as inert as Teflon, a related fluorocarbon, negative side effects of having such a fluid in the lungs are minimal.

Much research is currently being conducted on perfluorocarbons and their potential applications. For example, it is nearly impossible for babies born before 7 months gestation to breathe air. This is because their lungs have yet to develop an inner lining that prevents the lungs from collapsing due to the cohesiveness of the moisture within them. Researchers have found that premature infants can breathe oxygenated perfluorocarbons quite effectively. Adults may also benefit from inhaling perfluorocarbons because when the liquid is drained from the lungs, it carries with it much foreign matter that has accumulated over time. Have you had your lungs cleaned lately?

Another exciting application that has already been demonstrated is perfluorocarbons as a blood substitute. Among the many advantages of such an "artificial blood" are long-term storage capability and the elimination of the transmission of such diseases as hepatitis and AIDS through blood transfusions.* Artificial blood would also be a useful medium for preserving organs prior to transplantation. The material would also have applications for the treatment of both anemia and sickle cell anemia. Significantly, artificial blood would help to alleviate present and anticipated shortfalls in our blood supply. Currently less than 5 percent of the population donates blood. This percentage is dropping in the face of growing demand, which increases worldwide by about 7.5 million liters each year. Within the next 30 years, the shortfall could be most critical. There is still much research, however, that needs to be done with perfluorocarbons and their potential applications. So for both today and tomorrow, donating blood is still a very worthwhile idea.

*Because of precautionary measures taken by blood banks, our current supply of blood is most safe from these diseases. For example, the chance of dying from a blood transfusion is only about 1 in 100,000, which is less than the 2-in-100,000 chance of dying in a car accident.

granulated sugar, salt, or sand. The nooks and crannies within the grains serve as *nucleation* sites where carbon dioxide bubbles are able to form rapidly and then escape by buoyant forces. Shaking the beverage increases the surface area of the liquid-to-gas interface, making it easier for the carbon dioxide to escape the supersaturated gas/liquid solution. Upon shaking, the rate at which carbon dioxide escapes becomes so great that the beverage froths over. You also increase the rate at which carbon dioxide escapes by pouring the beverage into your mouth, which abounds in nucleation sites—the resulting sensation is called *effervescence.*

Question

Which provides more effervescence in your mouth: a just-opened warm carbonated beverage or a cold one?

Answer

The solubility of carbon dioxide in water decreases with increasing temperature. Warm carbonated beverages therefore fizzle in your mouth more than do cold ones. So, which makes you burp less?

Oxygen in Our Blood

If oxygen gas is not very soluble in water, how is it possible for blood, which is mostly water, to deliver oxygen efficiently to cells? The answer is that blood is much more than just water. It is both a solution of compounds such as sodium chloride and a suspension of blood cells. Red blood cells contain the large, oxygen-capturing molecule hemoglobin. There are about 250 million hemoglobin molecules in each red blood cell, and within each hemoglobin molecule there are four iron atoms in the +2 ionized state, Fe^{2+}. This form of iron readily combines with oxygen to become Fe^{3+} (This is the same chemistry that causes an iron nail to rust.) So, when red blood cells are passed through the blood/lung interface, the Fe^{2+} ions of hemoglobin attract available oxygen molecules. These oxygen molecules, however, never make contact with the iron because neighboring proteins in the red blood cell get in the way and hold the oxygen molecule just out of reach. Oxygen is normally a nonpolar molecule, but in the presence of the charged Fe^{2+} ion, polarity is induced. The result is a significant *ion–induced dipole* interaction. This interaction is strong enough to hold the oxygen molecule onto the hemoglobin as it gets transported to other parts of the body, but it is also weak enough so that once the hemoglobin gets to an oxygen-requiring cell, the oxygen molecule is readily released.

A hemoglobin molecule.

18.4 How Soap Works

Why are dirt and grime so difficult to remove with water alone? Most dirt and grime is made of nonpolar molecules that do not mix well with polar water molecules. They mix quite readily, however, with nonpolar organic solvents, such as turpentine. This is because organic solvent molecules and dirt and grime molecules are all nonpolar and able to interact strongly by way of induced dipole–induced dipole interactions.

Rather than washing our dirty hands and clothes with turpentine, however, we have a more pleasant alternative—soap and water. Soap works because soap molecules have both nonpolar and polar properties. A typical soap molecule has two parts: a long *nonpolar tail* consisting of carbon and hydrogen atoms and a *polar head* containing at least one ionic bond:

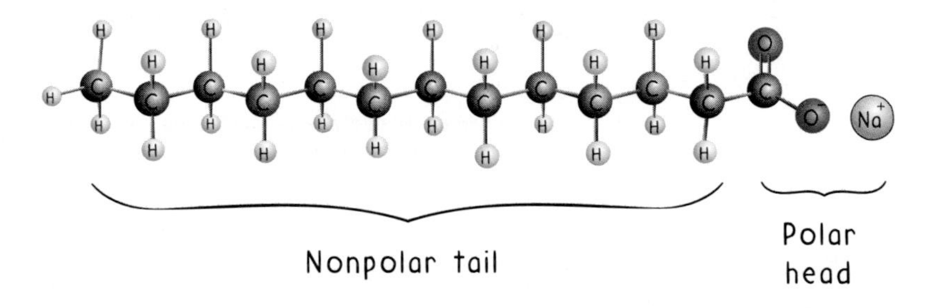

Because most of a soap molecule is nonpolar, it interacts quite well with nonpolar dirt and grime. In fact, dirt or grime quickly finds itself penetrated by three dimensions by the nonpolar tails of soap molecules. This interaction is usually enough to lift the dirt or grime away from the surface being cleaned. With the nonpolar tails faced inward toward the grime, the polar heads are all directed outward, where they are attracted to water molecules by relatively strong ion-dipole interactions. If the water is flowing, the whole enterprise of grime and soap molecules flows with it, away from your hands or clothes and down the drain.

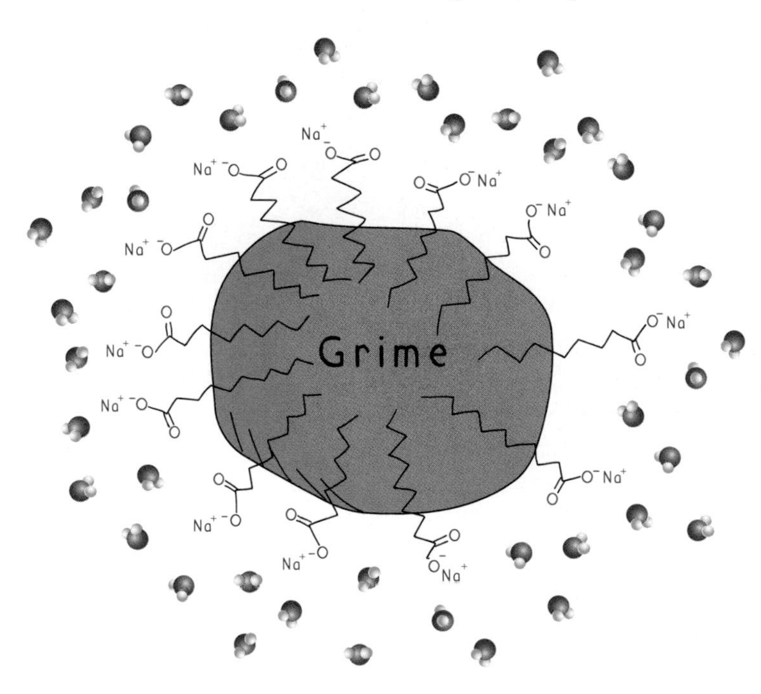

For the past several centuries, soaps have been prepared by treating animal fats, such as lard, with sodium hydroxide, NaOH, also known as caustic lye. In this reaction, which is still used today, each fat molecule is broken down into three soap molecules, and one glycerol molecule (Figure 18.24). In the 1940s, chemists began developing a class of synthetic soaps, known as detergents, which offer several advantages over soaps. The chemical structures of detergents are most similar to that of soap in that both possess a polar head attached to a nonpolar tail. The polar head, however, typically consists of either a sulfate, $—OSO_3^-$, or sulfonate, $—SO_3^-$, group and the nonpolar tail can have an assortment of

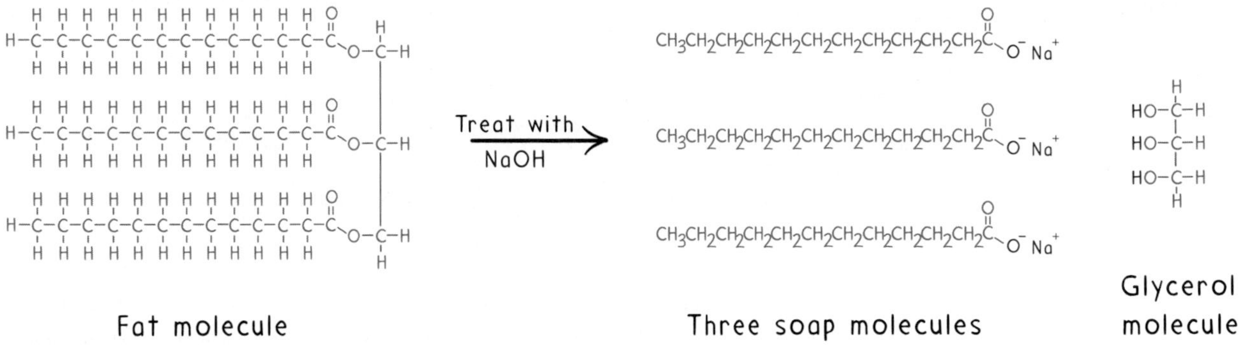

Fat molecule

Treat with NaOH

$CH_3CH_2CH_2CH_2CH_2CH_2CH_2CH_2CH_2CH_2CH_2C(=O)O^- Na^+$

$CH_3CH_2CH_2CH_2CH_2CH_2CH_2CH_2CH_2CH_2CH_2C(=O)O^- Na^+$

$CH_3CH_2CH_2CH_2CH_2CH_2CH_2CH_2CH_2CH_2CH_2C(=O)O^- Na^+$

Three soap molecules

Glycerol molecule

Fig. 18.24
Treated with caustic lye (NaOH), a single fat molecule is broken down into three soap molecules and a glycerol molecule. Atoms that originated from the NaOH are shown in purple. For aesthetics, the long carbon chains in the soap molecules have been redrawn in a chemical formula notation where each CH_2 corresponds to a carbon bonded to two hydrogen atoms as is seen in the fat molecule.

structures. One of the most common sulfate detergents is sodium lauryl sulfate, a main ingredient of many toothpastes and shampoos (Figure 18.25a). A common sulfonate detergent is sodium dodecyl benzenesulfonate, also known as a linear alkylsulfonate, or LAS (Figure 18.25b). You'll often find this compound in dishwashing liquids. Both of these detergents are biodegradable, which means that microorganisms can break down these molecules after their release into the environment

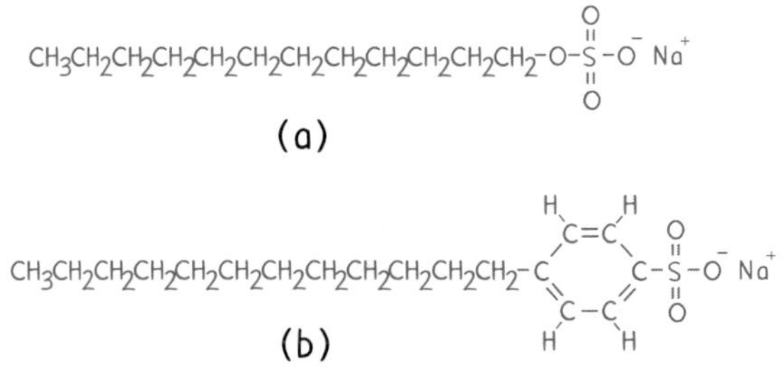

$CH_3CH_2CH_2CH_2CH_2CH_2CH_2CH_2CH_2CH_2CH_2CH_2{-}O{-}SO_2{-}O^- Na^+$

(a)

(b)

Fig. 18.25
The synthetic detergents (a) sodium lauryl sulfate, and (b) sodium dodecyl benzenesulfonate.

18.5 Hard Water and Soap Scum

Impurities are naturally found in drinking water. These include dissolved gases such as nitrogen, oxygen, and carbon dioxide, as well as dissolved solids such as salts of sodium, calcium, and magnesium. All these dissolved impurities give water its natural flavor, and without them water would be flat and tasteless.

Water containing relatively large amounts of calcium and magnesium ions is said to be hard water, and it has many undesirable qualities. For example, when hard water is heated, the calcium and magnesium ions tend to form solid compounds, which can clog hot-water heaters, boilers, and heat exchangers. You'll find these calcium and magnesium compounds coated on the inside surface of a well-used tea kettle. Don't drink hard water if you have a problem with frequent kidney stones—high intakes of calcium ions leads to kidney stone formation.

Hard water also inhibits the cleansing action of soaps and, to a lesser extent, that of detergents. The sodium ions of soap and detergent molecules carry a +1 charge, and calcium and magnesium ions carry a +2 charge (note their positions in the periodic

Table 18.4
Ratings of Water Hardness*

Very hard	Over 200 ppm calcium/magnesium
Hard	150 to 200 ppm calcium/magnesium
Moderately hard	50 to 100 ppm calcium/magnesium
Soft	0 to 50 ppm calcium/magnesium

*As measured in parts per million (ppm), which is the same as milligrams solute per liter of solution (mg/L)

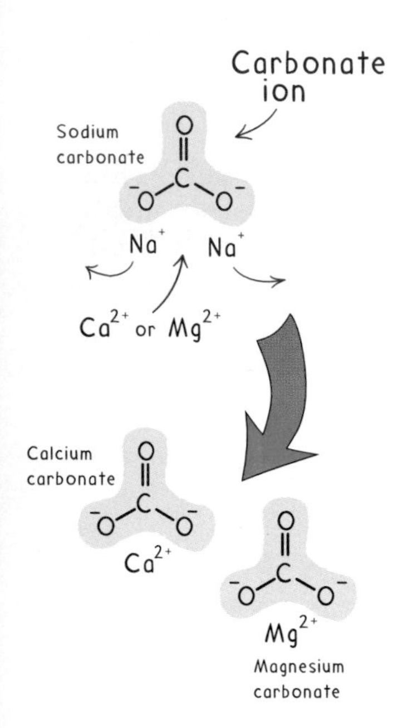

Fig. 18.26
The doubly negative carbonate ion of sodium carbonate preferentially binds with the calcium ions, Ca^{+2}, or the magnesium ion, Mg^{+2}. Sodium carbonate thus transorms into either calcium or magnesium carbonate

table). The negatively charged portion of a soap or detergent molecule's polar head is more attracted to the double positive charges of calcium and magnesium ions. Soap or detergent molecules therefore give up their sodium ions to selectively bind with calcium or magnesium ions.

$$CH_3CH_2CH_2CH_2CH_2CH_2CH_2CH_2CH_2CH_2CH_2C(=O)O^- \quad Ca^{2+} \quad {}^-O\text{–}C(=O)CH_2CH_2CH_2CH_2CH_2CH_2CH_2CH_2CH_2CH_2CH_3$$

Soap or detergent molecules bound to calcium or magnesium ions tend to be insoluble in water. As they come out of solution, they form a scum, which ends up as that ring around the bathtub. Since the soap or detergent molecules are tied up with calcium and magnesium ions, more of the cleanser must be added to maintain cleaning effectiveness. Many detergents today, however, are formulated with sodium carbonate, Na_2CO_3, popularly known as washing soda. Calcium and magnesium ions are most attracted to the carbonate ion of sodium carbonate (Figure 18.26). With the calcium and magnesium ions bound to the carbonate ion, the soap or detergent is free for cleansing. Because it removes the hard water ions, sodium carbonate is known as a water softening agent.

In a typical home water-softening system, hard water is passed through a large tank filled with tiny rubberlike beads made of a water-insoluble resin, known as an *ion-exchange resin.* The resin consists of many negatively charged ions, which capture the positively charged calcium and magnesium ions as they pass through the ion-exchange cartridge (Figure 18.27). Eventually, all the sites for calcium and magnesium on the resin are filled and the resin needs to be either discarded or recharged. The resin is recharged by flushing it with a concentrated solution of sodium chloride, NaCl. The abundant sodium ions displace the calcium and magnesium ions, thereby freeing up the binding sites on the resin for repeated use.

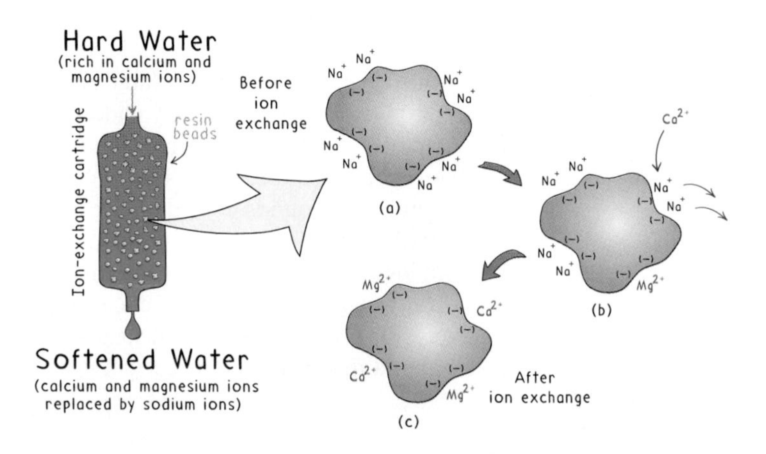

Fig. 18.27
(a) Negatively charged sites on the ion-exchange resin are occupied by sodium ions. (b) As hard water passes over the resin, sodium ions are displaced by calcium and magnesium ions. (c) After the resin becomes saturated with calcium and magnesium ions, it is no longer effective at softening water.

18.6 Surface Tension and Capillary Action

Fig. 18.28
A paper clip lies on the surface of the water, pushing the water down slightly but not sinking.

Take a metal paper clip and drop it into water. Because the weight of the paper clip is more than the buoyant force acting on it, the paper clip sinks. Now dry the clip thoroughly and very gently lay it on top of the water. If done properly, the clip floats! Why?

Water molecules on the surface are attracted to one another, and the sideways attraction between them allows the paper clip to stay afloat. Imagine a circle of people holding hands. To go into the circle, you must break through one pair of held hands. Similarly, if the paper clip is to enter the water, it must break through the grip of cohesive water molecules (Figure 18.29). The water's surface behaves as if it were a tightened elastic film upon which various light objects are able to rest, provided they do not pierce it. Dry steel needles on their sides or razor blades lying flat can be made to rest on the water in this fashion. Also, many insects that "skate" on top of streams and ponds use this means of support.

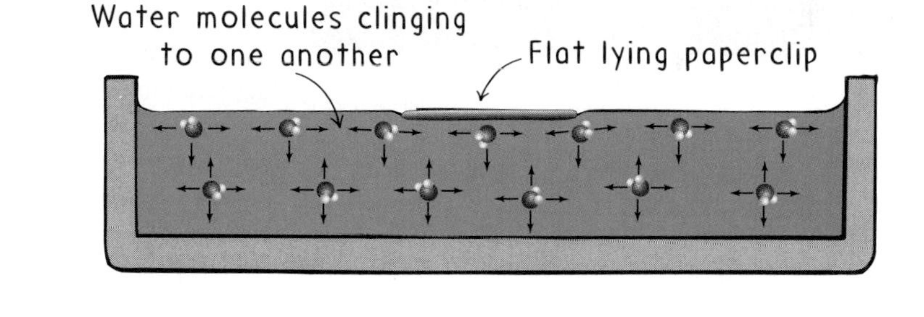

Fig. 18.29
The paper clip encounters the surface water molecules, which are clinging to one another so as to form a somewhat impenetrable barrier.

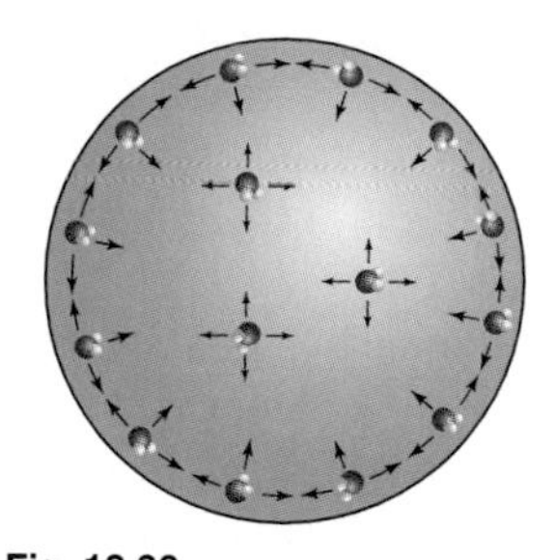

Fig. 18.30
The molecules on the surface have a net pull inward toward the center. This causes the material to reduce its surface area and form a sphere.

The **surface tension** of a liquid is defined as the energy required to break through the surface. Liquids in which there are strong molecular interactions, such as water, typically have high surface tensions. Surface tension accounts for the spherical shape of liquid drops. The surface molecules of a liquid are pulled sideways, but attractions from underneath also pull them downward into the liquid. This pulling of surface molecules into the liquid causes the surface to contract and become as small as possible. Guess which geometrical shape has the least surface for a given volume? That's right—a sphere. So we see why raindrops, drops of oil, and falling drops of molten metal are all spherical (Figure 18.30).

The surface tension in water is dramatically reduced by the addition of soap or detergent. Soap or detergent molecules tend to aggregate at the surface of water, where their nonpolar tails stick out away from the water (Figure 18.31). At the surface, these molecules interfere with the dipole-dipole interactions among water molecules, thereby reducing the surface tension, often by as much as 90 percent. Get a metal paper clip floating on the surface of some water and then carefully touch the water a few centimeters away with the corner of a bar of wet soap or a dab of liquid detergent. You will be amazed by how quickly the surface tension is destroyed.

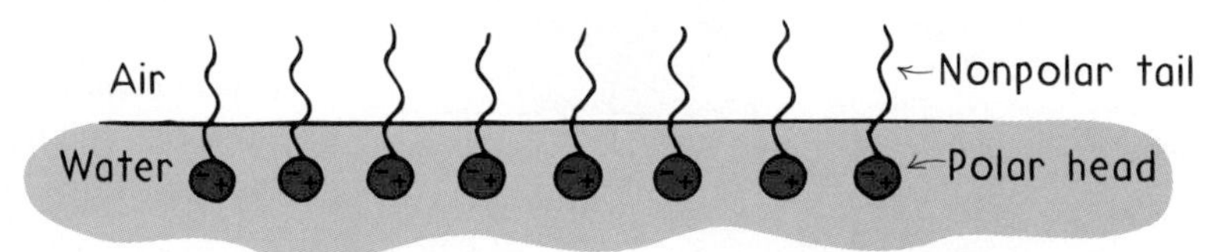

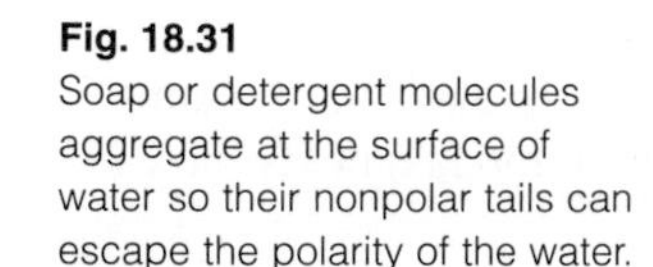

Fig. 18.31
Soap or detergent molecules aggregate at the surface of water so their nonpolar tails can escape the polarity of the water.

It is the strong surface tension of water that prevents it from wetting certain materials, such as waxy leaves, umbrellas, and freshly polished automobiles. Rather than wetting (spreading out evenly), the water simply beads as the water molecules pull themselves together by the dipole-dipole forces. This is good if the idea is to keep water away. If we want to clean, however, the idea is to get the object as wet as possible. This is another way in which soaps and detergents assist in cleaning. By destroying water's surface tension, they enhance its ability to wet. Dirty fabrics and dishes, for example, are penetrated by the water more rapidly and cleaning is more efficient.

Fig. 18.32
Water spreads evenly on a plate treated with detergent because the detergent interferes with the water's surface tension.

Dip a paper clip into water and then slowly pull it out. You'll find that for a short distance, the water is brought up with the metal (Figure 18.33). This happens because water molecules are polar and the metal is full of loose electrons, a combination that makes for relatively strong dipole–induced dipole interactions.

Molecular interactions arising between molecules of two different substances are called **adhesive forces**. When the molecular interactions are between the molecules of one substance, they are called **cohesive forces**. Because glass is made of polar molecules, there are adhesive forces between glass and water. These forces cause water to creep up the sides of glass containers, as depicted in Figure 18.35. We call the curving of water (or any other liquid) at the interface of a container a **meniscus**.

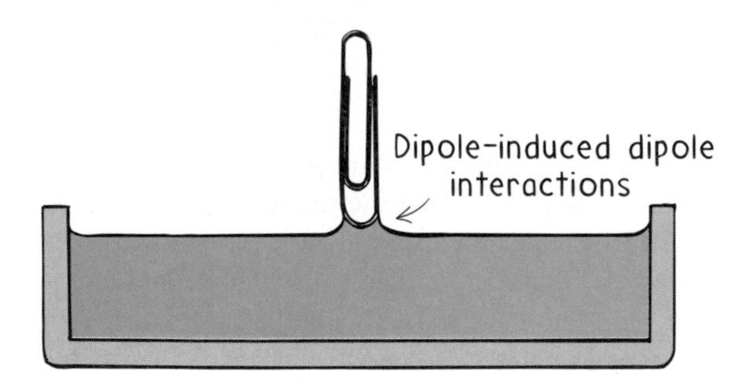

Fig. 18.33
Water molecules are attracted to the metal paper clip because of dipole–induced dipole interactions.

When a small-diameter glass tube is placed in water, adhesive forces initially cause a relatively steep meniscus (Figure 18.35a). As soon as the meniscus forms, however, the attractive cohesive forces among water molecules respond to the steepness by acting to minimize the surface area of the meniscus (Figure 18.35b). The result is that the water level in the tube rises. Adhesive forces then cause the formation of another steep meniscus (Figure 18.35c). This is followed by the action of cohesive forces, which cause the steep meniscus to be "filled in" (Figure 18.35d). This cycle is repeated until the upward adhesive force equals the weight of the raised water in the tube. This rise in the fluid due to the interplay of adhesive and cohesive forces is called **capillary action**. In a tube that has an internal diameter of about 0.5 millimeter, the water rises slightly higher than 5 centimeters. With a smaller internal diameter, the water rises much higher (Figure 18.36).

Fig. 18.34
Adhesive forces account for the observation that when you empty a glass of water, some water drops invariably cling to the glass and remain behind. Cohesive forces account for the beading together of water drops.

We see capillary action at work in many phenomena. If a paintbrush is dipped into water, the water rises up into the narrow spaces between the bristles by capillary action. Hang your hair in the bathtub, and water seeps up to your scalp in the same way. This is how oil soaks upward in a lamp wick and water in a bath towel when one end hangs in water. Dip one end of a lump of sugar in coffee, and the entire lump is quickly wet. The

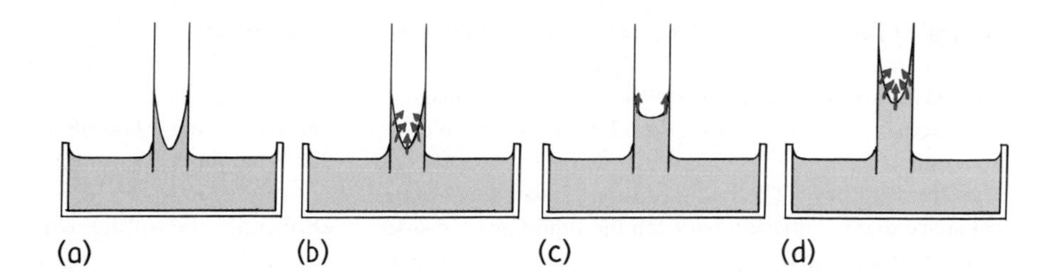

Fig. 18.35
Water is spontaneously drawn up a narrow glass tube by an interplay of adhesive and cohesive forces.

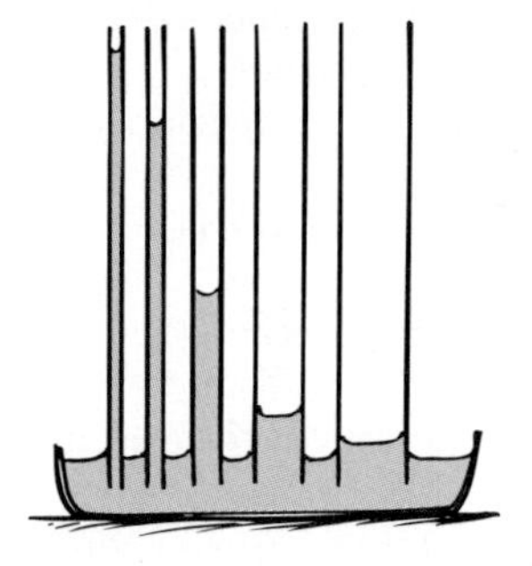

Fig. 18.36
Capillary tubes. The smaller the internal diameter of the tube, the higher the liquid rises in it

capillary action occurring between soil particles is important in bringing water to the roots of plants.

The concepts of chemistry tell us much about how the world and all its materials are put together. They also stimulate us to think critically. For example, we needn't memorize some long list of which solutes are soluble in which solvents. Instead, fairly accurate predictions about solubility can be made based upon an understanding of molecular interactions.

But where do the concepts of chemistry come from? The concepts we are learning about in these chapters were developed from the results of experiments—the nature of matter was first explored and only then were the concepts developed to explain behavior. A thorough knowledge of chemistry, therefore, is a blend of conceptual understanding and personal, hands-on experience. Indeed, much of the excitement and flavor of chemistry come from hands-on experience, which should occur in the laboratory part of your physical science course.

Summary of Terms

- **Ion-dipole interaction** The molecular interaction involving an ion and a dipole.
- **Dipole–dipole interaction** The molecular interaction involving dipoles.
- **Hydrogen bond** A strong dipole-dipole interaction that involves a hydrogen atom chemically bonded to a strongly electronegative element, such as nitrogen, oxygen, or fluorine.
- **Induced dipole** A dipole temporarily created in an otherwise nonpolar molecule. It is induced by a neighboring charge or dipole.
- **Dipole–induced dipole interaction** The molecular interaction involving a dipole and an induced dipole.
- **Induced dipole–induced dipole interaction** The molecular interaction involving only induced dipoles. This is a relatively weak molecular interaction.
- **Solvent** The component in a solution present in the largest amount.
- **Solute** Any component in a solution that is not the solvent.
- **Dissolving** The process of mixing a solute in a solvent.
- **Concentrated** A solution containing a relatively large amount of solute.
- **Dilute** A solution containing a relatively small amount of solute.
- **Saturated** A solution containing as much solute as will dissolve.
- **Unsaturated** A solution that will dissolve additional solute if added.
- **Concentration** A quantitative measure of the amount of solute in a solution.
- **Mole** A very large number equal to 6.02×10^{23}. This number is a unit commonly used when describing a number of molecules.
- **Molarity** A common unit of concentration measured by the number of moles in 1L of solution.
- **Solubility** The ability of a solute to dissolve, which depends not only on molecular interactions between the solute and the solvent, but also on the interactions among both solute molecules and among solvent molecules.
- **Soluble** Capable of dissolving in a solvent.
- **Insoluble** Not capable of dissolving to any appreciable extent in a solvent.
- **Precipitate** A solute that has come out of solution.
- **Supersaturated** A solution that contains more solute than it normally contains.
- **Surface tension** The energy required to break through the surface of a liquid.

- **Adhesive forces** Molecular interactions that arise between two different substances.
- **Cohesive forces** The attractive forces within a substance.
- **Meniscus** The curving of a liquid at the interface of its container.
- **Capillary action** The rising of liquid into a small vertical space due to adhesion between the liquid and the sides of the container and to cohesive forces within the liquid.

Review Questions

Types of Molecular Interactions

1. What distinguishes a molecular interaction from a chemical bond?
2. Which is stronger, the ion-dipole interaction or the induced dipole–induced dipole interaction?
3. Why are water molecules attracted to sodium chloride?
4. How are ion-dipole interactions able to break apart the ionic bond, which is relatively strong?
5. Are electrons distributed evenly or unevenly in a molecular dipole? Why?
6. What is a hydrogen bond?
7. How are oxygen molecules attracted to water molecules?
8. Are induced dipoles normally permanent?
9. How can nonpolar atoms induce dipoles in other nonpolar atoms?
10. According to Table 18.2, which is the larger molecule, iodine, I_2, or fluorine, F_2?
11. Why is it difficult to induce a dipole in a fluorine atom?
12. Why is the boiling point of octane, C_8H_{18}, so much higher than the boiling point of methane, CH_4?

Solutions

13. What happens to the volume of water as sugar is dissolved in it?
14. Why is a ruby gemstone considered to be a solution?
15. Distinguish between a solute and a solvent.
16. What does it mean to say that a solution is concentrated?
17. Distinguish between a saturated and an unsaturated solution.
18. How is the amount of solute in a solution calculated?
19. Is a mole a very large or very small number?
20. By what means are ethanol and water molecules attracted to each other?
21. Why does oxygen have such a low solubility in water?
22. What effect does temperature have on the solubility of a solid solute in a liquid?
23. What effect does temperature have on the solubility of a gas solute in a liquid solvent?
24. How are supersaturated solutions made?

Solubility

25. What does it mean to say that two materials are infinitely soluble in each other?
26. Why is sugar not infinitely soluble in water?
27. What molecular interaction is responsible for oxygen dissolving in water?
28. Name one solute whose solubility in water does not markedly increase with increasing temperature.
29. What is the relationship between a precipitate and a solute?
30. Why does the solubility of a gas decrease with increasing temperature?
31. What molecular interactions occur between oxygen and perfluorocarbon molecules?
32. What molecular interactions occur between oxygen and hemoglobin?

How Soap Works

33. Should soap molecules be classified as polar, nonpolar, or both?
34. Water and soap are attracted to each other by what type of molecular interaction?
35. Soap and grime are attracted to each other by what type of molecular interaction?
36. What is the difference between a soap and a detergent?

Hard Water and Soap Scum

37. What component of hard water makes it "hard"?
38. Why are soap molecules so attracted to the calcium and magnesium ions in hard water?
39. Calcium and magnesium ions are even more attracted to sodium carbonate than to soap. Why?

Surface Tension and Capillary Action

40. What is the cause of surface tension?
41. Why do liquids in which the molecules take part in strong interactions have greater surface tension than those with only weak interactions?
42. What kind of molecular interactions take place between water and metal?
43. In which does water rise higher: a narrow tube or a wide tube?
44. What determines the height that water will rise by capillarity?

Home Projects

1. Black ink is made by combining many different-colored inks, which together absorb all the frequencies of light. Because no light is reflected, the ink appears black. Use molecular interactions to separate the components of black

ink through a special technique called *paper chromatography.* Place a concentrated dot of ink at the center of a piece of porous paper, such as a paper towel, napkin, or coffee filter. Next, carefully place a drop of solvent, such as water, acetone (fingernail polish remover), rubbing alcohol, or white vinegar on top of the dot and watch the ink spread radially with the solvent. Because the various components have differing affinities for the solvent, they travel with the solvent at differing rates. Just after your drop of solvent is completely absorbed, add a second drop, then a third, and so on until the ink components have separated to your satisfaction. How they separate depends on several factors, including your choice of solvent and your technique. The ink from black felt-tip pens or water-soluble markers for overhead projectors tends to work best. It's also interesting to watch the leading edge of the moving ink under a strong magnifying glass or microscope. Check for capillary action!

2. Just because a solid dissolves in a liquid doesn't mean the solid no longer occupies space. Fill a tall glass to its brim with warm water. Carefully pour all the water into a larger container. Now add three or four heaping tablespoons of sugar or salt to the empty glass. Return half of the warm water to the glass and stir to dissolve all of the solid. Return the remaining water, and as you get close to the top ask a friend to predict whether the water level will be less than before, about the same as before, or whether the water will spill over the edge. If your friend doesn't understand the result, ask him or her what would happen if you had added a single large crystal of sugar or salt directly to the full glass of water.
3. Be sure to make those sugar crystals from a supersaturated solution of sugar water as described in Section 18.3. Interesting crystals can also be made from supersaturated solutions of Epsom salts ($MgSO_4 \cdot 7H_2O$) and alum ($KAl(SO_4)2 \cdot 12\ H_2O$), which is available in the spice sections of some grocery stores. Crystal shape directly relates to how the molecules of the substance pack together. In fact, substances are often characterized by the shape of the crystals they form. Note how different solutes give rise to differently shaped crystals.

4. Fill a large pot with water and allow to stand. Bubbles will form on the inside of the pot when the water is left to stand for several hours. Where do these bubbles come from? What is their composition?

Exercises

1. Why are ion-dipole interactions stronger than dipole-dipole interactions?
2. Two molecules are attracted to each other by the dipole-dipole interaction. Is it possible for a third molecule to join in the attraction? Is it possible for a third molecule that has no dipole to join in the attraction?
3. Fish don't breathe water, they breathe the oxygen dissolved in the water. Being that water is a polar substance and oxygen a nonpolar substance, how can there be any dissolved oxygen in the water?
4. At room temperature, chlorine, Cl_2, is a gas but bromine, Br_2, is a liquid. Explain.
5. Dipole–induced dipole forces of attraction exist between water and gasoline, and yet these two substances do not mix because water has such a strong attraction for itself. Which of the following compounds might best help make these two substances mix into a single liquid phase?

```
    H H H                       H
    | | |                       |
H-O-C-C-C-H       Na+ Cl-     H-C-H
    | | |                       |
    H H H                       H
```

6. Consider the boiling points of the following compounds and their solubilities in room-temperature water. Then briefly explain how the solubilities in water go down as the boiling points of these alcohols go up.

	$CH_3{-}O{-}H$	$CH_3CH_2CH_2CH_2{-}O{-}H$	$CH_3CH_2CH_2CH_2CH_2{-}O{-}H$
Boiling point:	65°C	117°C	138°C
Solubility:	infinite	8g/100mL	2.3g/100mL

7. The boiling point of 1,4-butanediol is 230°C. Would you expect this compound to be soluble or insoluble in room-temperature water? Explain your answer.

$$H{-}O{-}CH_2CH_2CH_2CH_2{-}O{-}H$$

1,4-butanediol

8. Are noble gases infinitely soluble in noble gases? Defend your answer.
9. To avoid a painful and potentially lethal medical condition known as the *bends,* professional deep-sea divers breathe mixtures of helium and oxygen rather than normal air, which is a mixture nitrogen and oxygen. Nitrogen is avoided because, under great pressure, it dissolves in the blood. The dissolved nitrogen then tends to accumulate in nonpolar fatty tissues, such as those found in joints and in

the outer linings of nerve cells. If a diver rises toward the surface too rapidly, bubbles of nitrogen form and lead to severe joint aches as well as damage to the nervous system. (a) Suggest why nitrogen tends to accumulate in nonpolar regions of the body. (b) Helium is used in place of nitrogen because of helium's lower solubility in blood. Suggest why helium has a lower solubility in blood.

10. Describe two ways to tell whether a sugar solution is saturated or not.

11. At 10°C, which is more concentrated—a saturated solution of sodium nitrate, $NaNO_3$, or a saturated solution of sodium chloride? (See Figure 18.21.)

12. Suggest why salt is insoluble in gasoline. Consider the molecular interactions.

13. When 50.0 mL of water is combined with 50.0 mL of ethanol, the result is 98.0 mL of solution. Why?

14. When 50.0 mL of ethanol are combined with 50.0 mL of hexane (C_6H_{14}), the result is 100.4 mL of solution. Why?

15. Deuterium oxide, D_2O, and water, H_2O, have the same chemical structure and differ only in that D_2O possesses the deuterium isotope of hydrogen, whereas water possesses the protium isotope. Deuterium oxide, also known as "heavy water," is 11% heavier than water. Might you expect its boiling temperature also to be about 11% greater? Why or why not?

16. Which would you expect to have a higher melting point: sodium chloride, NaCl, or aluminum oxide, Al_2O_3? Why?

17. Why does the humpback whale spend winters in the warm tropical waters of Hawaii and Baja California but migrate each summer to the cold polar waters of Alaska?

18. How necessary is soap for removing salt from your hands?

19. Why must a razor blade be lying flat in order for it to be held up on the surface of water by surface tension?

20. Capillary action causes water to climb up the internal walls of narrow glass tubes. Why does the water not climb so high when the glass tube is wider?

21. Mercury forms a *convex* meniscus with glass and not the *concave* meniscus formed by water (shown in Figures 18.37 and 18.38). What does this tell you about the cohesive forces within mercury versus the adhesive forces between mercury and glass? Which are stronger?

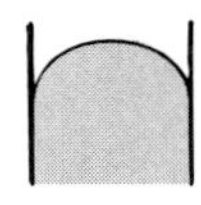

Convex meniscus

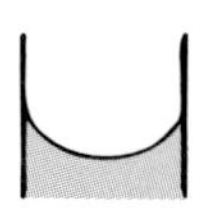

Concave meniscus

22. Would you expect the surface tension of water to increase or decrease with temperature? Defend your answer.

23. Two chemical structures are shown, one of a typical gasoline molecule and the other of a typical motor oil molecule. Which is which? Base your reasoning not on memorization but rather on what you know about molecular interactions and the various physical properties of gasoline and motor oil.

```
  H H H H H H H H H H H H H H H H
  | | | | | | | | | | | | | | | |
H-C-C-C-C-C-C-C-C-C-C-C-C-C-C-C-C-H
  | | | | | | | | | | | | | | | |
  H H H H H H H H H H H H H H H H
```

Structure A

```
  H H H H H H H H
  | | | | | | | |
H-C-C-C-C-C-C-C-C-H
  | | | | | | | |
  H H H H H H H H
```

Structure B

24. What is the boiling temperature of a single water molecule?

25. Account for the observation that ethyl alcohol, C_2H_5OH, dissolves readily in water but dimethyl ether, CH_3OCH_3, which has the same number and kinds of atoms, does not.

26. An inventor claims to have developed a new perfume that lasts a long time because it doesn't evaporate. Comment on this claim.

27. Water is 88.88% oxygen by mass. Oxygen is exactly what a fire needs to grow brighter and stronger. So why is water so effective at *stopping* a fire rather than feeding it?

28. Why are the melting temperatures of most ionic compounds far greater than the melting temperatures of most covalent compounds?

29. When you set a pot of tap water on the stove to boil, you'll often see bubbles start to form well before boiling temperature is reached. Explain.

30. Fish don't live very long in water that has just been boiled and brought back to room temperature. Suggest why.

Problems

1. How many grams of sodium chloride is needed to make 15 L of a solution that has a concentration of 3.0 g per liter of solution?

2. If 1 mole of sodium chloride is put into water to make a total of 1 L of solution, what is the molarity? What about for 2 moles in 0.5 L?

3. There is 1 mole of oxygen molecules, O_2, in every 32 g of oxygen gas. If 0.04 g of O_2 is able to dissolve in 1 L of water, how many moles is this? How many molecules?

CHAPTER

19

Chemical Reactions

When a substance undergoes a chemical change, it changes its identity. Grape juice treated with yeast ferments into wine; iron exposed to air and water transforms into rust; dynamite explodes into a variety of gaseous compounds; wood burns to ashes, and aspirin is synthesized from petroleum. During these changes, chemical bonds are broken and new ones are formed—molecules are plucked apart and their atoms reassembled in different arrangements. This shuffling of atoms is what we call a *chemical reaction* (Section 15.3). In this chapter we focus on some fundamental concepts of chemical reactions, including chemical equations used to represent the reactions, the role of energy and catalysts, the reversibility of various chemical reactions, and, lastly, how to keep quantitative tabs on chemicals as they transform.

19.1 The Chemical Equation

During a chemical reaction a new material is formed by the rearrangement of atoms. To represent a chemical reaction on paper, we turn to the *chemical equation.* A **chemical equation** shows the materials about to react, called the **reactants**, to the left of an arrow that points to the newly formed materials, called the **products**:

$$\text{Reactants} \rightarrow \text{products}$$

Typically, the reactants and products are represented by their atomic or molecular formulas, but molecular structures or simple names may be used. Phases are also often shown: (*s*) for solid, (*l*) for liquid, and (*g*) for gas. Materials dissolved in water are designated (*aq*) for aqueous. For example, a chemical reaction occurs when solid carbon, C, reacts with gaseous oxygen, O_2, to form gaseous carbon dioxide, CO_2:

$$\underbrace{C\,(s) + O_2\,(g)}_{\text{reactants}} \rightarrow \underbrace{CO_2\,(g)}_{\text{product}}$$

Numbers are placed in front of the chemical formulas to show the ratio in which reactants combine and products form. These numbers are called *coefficients.* One carbon atom, for example, reacts with one oxygen molecule to form one molecule of carbon dioxide:

$$1\,C\,(s) + 1\,O_2\,(g) \rightarrow 1\,CO_2\,(g)$$

In another chemical reaction, two hydrogen gas molecules, H_2, react with one oxygen gas molecule, O_2, to produce two molecules of water , H_2O, in the gaseous phase:

$$2\,H_2\,(g) + 1\,O_2\,(g) \rightarrow 2\,H_2O\,(g)$$

By convention, a coefficient of 1 is omitted, so that the preceding chemical equations are typically shown as

$$C\,(s) + O_2\,(g) \rightarrow CO_2\,(g)$$
$$2\,H_2\,(g) + O_2\,(g) \rightarrow 2\,H_2O\,(g)$$

One of the most important principles of chemistry is the **principle of the conservation of mass**, which states that matter is neither created nor destroyed during a chemical reaction. Rather, the atoms present at the beginning of the reaction merely switch partners to form new molecules. This means that no atoms are lost or gained during any reaction. The chemical equation must therefore be *balanced;* that is, each atom in it must appear on both sides of the arrow the same number of times. The preceding equation for the formation of carbon dioxide is balanced because each side shows one carbon and two oxygen atoms. The equation for the formation of gaseous water is also balanced—there are four hydrogen and two oxygen atoms before and after the arrow.

Note that a coefficient in front of a chemical tells us the number of times that chemical must be counted. For example, 2 H_2O indicates two water molecules, which contain four hydrogen and two oxygen atoms. An option used by many students is to draw the chemical structures instead of the formulas (Figure 19.1). How many times you draw a structure is specified by the coefficient.[1]

---

[1]Drawing accurate structures is not a necessity when trying to balance a chemical equation. It is vital, however, that you accurately represent the correct number of atoms per molecule as specified by the chemical formula.

Fig. 19.1
A conceptual presentation of the equation $2\ H_2 + O_2 \rightarrow 2\ H_2O$.

H–H H–H + O=O ⟶ H–O–H H–O–H

Question

Which of the following chemical equations are balanced?

(a) $3\ NO \rightarrow N_2O + NO_2$

(b) $SiO_2 + 4\ HF \rightarrow SiF_4 + 2\ H_2O$

(c) $4\ NH_3 + 5\ O_2 \rightarrow 4\ NO + 6\ H_2O$

Balancing Equations

Equations can be balanced by changing coefficients to produce correct proportions. It's important *not to change subscripts,* however, because to do so changes the chemical's identity—H_2O is water, but H_2O_2 is hydrogen peroxide! The following steps are a guide to balancing a variety of chemical equations:

1. Focus on balancing one element at a time. Start with the leftmost element and modify the coefficients to make this element appears on both sides of the arrow the same number of times.
2. Move to the next element and modify the coefficients to balance this element. Don't worry if you unbalance the previous element. You will come back to it in subsequent steps.
3. Continue from left to right, balancing each element individually.
4. Several left-to-right passes may be required, and so repeat steps 1–3 until all elements are balanced.

Helpful hints: Use a pencil so that you can erase coefficients as needed. Never, ever, EVER alter a subscript. Remember that coefficients must appear *before* a chemical compound, not within.

Example

Balance this equation:

$$\_\_\_\ \mathbf{Fe}\ (s) + \_\_\_\ \mathbf{O_2}\ (g) \rightarrow \_\_\_\ \mathbf{Fe_2O_3}\ (s)$$

Start with iron, which can be balanced by adding a coefficient of 2:

$$\mathbf{2\ Fe}\ (s) + \_\_\_\ \mathbf{O_2}\ (g) \rightarrow \_\_\_\ \mathbf{Fe_2O_3}\ (s)$$

The oxygen can be balanced by placing a 3 in front of the O_2 and a 2 in front of the iron oxide product, Fe_2O_3. This now provides for 6 oxygen atoms before and after the arrow:

$$\mathbf{2\ Fe}\ (s) + \mathbf{3\ O_2}\ (g) \rightarrow \mathbf{2\ Fe_2O_3}\ (s)$$

Answer
In all these equations the number of times each element appears before the arrow is equal to the number of times it appears after the arrow. So they are all balanced.

We've made one full pass through all the elements (iron and oxygen) in the equation. Start over with a second pass, beginning again with iron. Note that in balancing the oxygen, the iron was inadvertently unbalanced. This can be remedied by changing its coefficient from 2 to 4:

$$4\ Fe\ (s) + 3\ O_2\ (g) \rightarrow 2\ Fe_2O_3\ (s)$$

Proceed on to the oxygen to double-check that it is still balanced. Continue to make passes until you are confident the equation is balanced, which this last equation is.

Practicing chemists develop a knack for balancing equations. This knack involves creative energy and like most other knacks improves with experience. There are some useful tricks of the trade for balancing equations, and maybe your instructor will share some with you. For brevity, we introduce only the rudiments here. More important than techniques for balancing equations is knowing why they need to be balanced. And that is this: Atoms are neither created nor destroyed in a chemical reaction—they are simply rearranged.

Question

Find the coefficients that balance these equations:

(a) ___ H_2 (*g*) + ___ N_2 (*g*) → ___ NH_3 (*g*)

(b) ___ Cl_2 (*g*) + ___ KBr (*aq*) → ___ Br_2 (*l*) + ___ KCl (*aq*)

(c) ___ CH_4 (*g*) + ___ O_2 (*g*) → ___ CO_2 (*g*) + ___ H_2O (*l*)

19.2 Energy and Chemical Reactions

Where does the energy that lifts a rocket into space come from? Where does a campfire get the energy to glow red hot? Where does the energy in the food we eat come from? These questions all have the same answer: chemical reactions. Let's look at the intimate relationship between energy and chemical reactions.

During a chemical reaction, chemical bonds are broken and atoms rearrange themselves to form new chemical bonds. Breaking and forming chemical bonds involves changes in energy. Consider a pair of strong magnets held together by the magnetic force. To separate them requires an input of "muscle energy." Conversely, when the two separated magnets collide, they are warmed, much as a hammer and nail are warmed upon impact. This warmth is evidence of energy released. So we see that energy must be put into the system in order to pull the two magnets apart, and energy is released from the system as they come together. The same principle applies to atoms. To pull bonded atoms apart requires energy input. When atoms combine, there is energy output. Released energy is usually in the form of either heat or light.

Answer

(*a*) 3, 1, 2; (*b*) 1, 2, 1, 2; (*c*) 1, 2, 1, 2. (Remember that, by convention, 1's are typically not shown in the balanced equation.)

Table 19.1
Selected Bond Energies

Bond	Bond Energy (kJ/mole)	Bond	Bond Energy (kJ/mole)
H—H	436	C—C	414
H—C	414	Cl—Cl	243
H—N	389	N—N	159
H—O	464	O═O	498
H—F	569	O═C	803
H—Cl	431	N≡N	946
H—S	339	C≡C	837

Chemical reactions work in accord with the conservation of energy principle: The amount of energy required to pull two bonded atoms apart is the same as the amount released when they are brought together via a chemical bond. This amount of energy absorbed upon bond breaking and released upon bond formation is called the **bond energy**. Each chemical bond has its own characteristic bond energy. The hydrogen-hydrogen bond energy, for example, is 436 kilojoules per mole.[2] This means that 436 kilojoules of energy is required to pull apart 1 mole of hydrogen-hydrogen bonds. Conversely, 436 kilojoules of energy is released upon the formation of 1 mole of hydrogen-hydrogen bonds. Different bonds involving different elements have different bond energies. The oxygen-oxygen double bond has a bond energy of 498 kilojoules per mole, and the oxygen-hydrogen bond has a bond energy of 464 kilojoules per mole. These and some other bond energies are listed in Table 19.1.

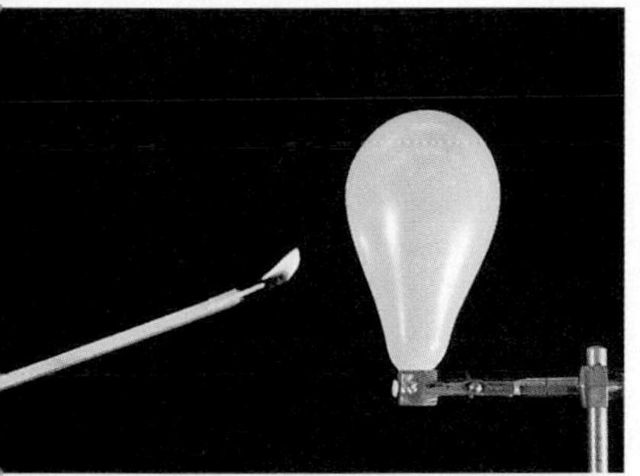

Fig. 19.2
A balloon filled with a mixture of hydrogen gas and oxygen gas is ignited to produce water and lots of energy.

The role of bond energies in a chemical reaction is nicely illustrated by the reaction of hydrogen with oxygen to form water:

H−H H−H + O=O ⟶ H−O−H H−O−H

In the reactants we see that hydrogen atoms are bonded to hydrogen atoms and oxygen atoms are bonded to oxygen atoms. In the products, however, hydrogen atoms are bonded to oxygen atoms. So, for the reaction to proceed, hydrogen-hydrogen and oxygen-oxygen bonds must first be broken. Only then can the atoms rearrange to form hydrogen-oxygen bonds.

Breaking chemical bonds requires energy—the bond energy. We see in the preceding balanced equation that two hydrogen-hydrogen bonds are broken (2 × 436 kJ/mole) for every one oxygen-oxygen bond (498 kJ/mole). This is a total of (2 × 436 kJ/mole) + 498 kJ/mole = 1370 kJ/mole (Table 19.2). Compare this amount with the energy released upon the formation of products. The formation of four hydrogen-oxygen bonds releases 4 × 464 kJ/mole = 1856 kJ/mole. So we find there is a net release of energy in this reaction—486 kJ/mole, which is 1856 kJ/mole subtract 1370 kJ/mole. We therefore show energy as a product of this reaction:

$$\mathbf{2\,H_2 + O_2 \rightarrow 2\,H_2O + energy}$$

[2]Recall from Section 18.2 that a *mole* is the quantity of any substance that contains 6.02×10^{23} particles of that substance. Atoms are so small that they are measured in bulk, not individually.

Table 19.2
Bond Energies in the Formation of Water

Amount of energy **required** to break chemical bonds in reactants

Type of bond	Number of Bonds	Bond Energy	Total
H—H	2	436 kJ/mole	872 kJ/mole
O=O	1	498 kJ/mole	498 kJ/mole
		Total energy required:	1370 kJ/mole

Amount of energy **required** upon bond formation in products

Type of bond	Number of Bonds	Energy per Bond	Total
H—O	4	464 kJ/mole	1856 kJ/mole
		Total energy released:	1856 kJ/mole

Net change in energy:

1856 kJ/mole released − 1370 kJ/mole required = 486 kJ/mole released

Fig. 19.3
The space shuttle uses exothermic chemical reactions to lift off from the Earth's surface.

A chemical reaction that produces energy is said to be *exothermic*. The formation of water from hydrogen and oxygen is a good example. For exothermic reactions, some initial amount of energy is required to start the reaction, to break apart the reactant molecules. When room-temperature hydrogen and oxygen are mixed, they do not react until ignited by a spark. Once the reaction starts, the energy generated by the reaction keeps it self-sustaining.

The amount of energy produced in an exothermic reaction depends upon the amount of reactants. For the formation of water, the preceding analysis shows that 2 moles (4 grams) of H_2 and 1 mole (32 grams) of O_2 yields 486 kilojoules. With greater quantities of these reactants, correspondingly more energy is released. In fact, the reaction of large amounts of hydrogen and oxygen provides the energy to lift the space shuttle into orbit (Figure 19.3). There are two vast compartments in the large central tank to which the orbiter is attached, one filled with liquid hydrogen and the other with liquid oxygen. Upon ignition, they mix and react chemically to produce water, which is expelled out of the rocket cones to produce thrust. Additional thrust is obtained by a pair of solid-fuel rocket boosters that contain a mixture of ammonium perchlorate, NH_4ClO_4, and powdered aluminum, Al. Upon ignition, these chemicals also react to produce chemical products that are energetically expelled out the back of the rocket. This reaction is:

$$3\ NH_4ClO_4 + 3\ Al \rightarrow Al_2O_3 + AlCl_3 + 3\ NO + 6\ H_2O + \text{energy}$$

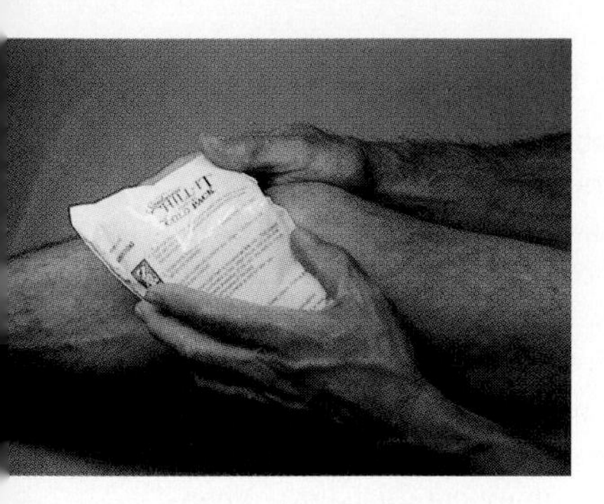

Fig. 19.4
Punch the seal of this cold pack and it quickly becomes cold to your touch—and your bruises if need be.

A chemical reaction in which the amount of energy needed to start the reaction is greater than the amount of energy released is said to be *endothermic*. Endothermic reactions require the continuous input of energy from an external source in order to proceed. The conversion of sodium chloride to its elemental components, sodium metal and chlorine gas, is an endothermic reaction. This reaction occurs only when electrical energy is put into the system:

$$\textbf{Electricity} + 2\,NaCl\,(l) \rightarrow 2\,Na\,(s) + Cl_2\,(g)$$

Some endothermic reactions are able to sustain themselves by absorbing heat from their surroundings. An example is solid ammonium nitrate, NH_4NO_3, dissolving in water and separating into ammonium, NH_4^+ (*aq*) and nitrate, NO_3^- (*aq*), ions:

$$\textbf{Heat} + NH_4NO_3\,(s) \rightarrow NH^{4+}(aq) + NO_3^-(aq)$$

As the ammonium nitrate is mixed with the water, heat is absorbed from the water, the container, and any material in contact with the container. We can sense this action by the resulting drop in temperature. Ammonium nitrate is the active component of many "cold packs" used to diminish swelling (Figure 19.4). To activate the pack, it must be punched. This breaks an inner seal and allows the ammonium nitrate to mix with water. As the ammonium nitrate dissolves, heat is absorbed and the temperature of anything in contact with the pack—including sprained ankles—decreases.

Questions

1. Use data from Table 19.1 to calculate whether the following reaction is exothermic or endothermic.

   H–H H–H + N≡N ⟶ H₂N–NH₂ (H(H)N–N(H)H)

2. Without looking at bond energies, deduce whether the following reaction is exothermic or endothermic. Should energy be written as a reactant or as a product?

   $O_2N\text{–}NO_2 \longrightarrow NO_2 + NO_2$

Answers

1. The amount of energy required to break all the chemical bonds in the reactants is (2 × 436 kJ/mole) + 946 kJ/mole = 1818 kJ/mole. The amount of energy released upon bond formation is (4 × 389 kJ/mole) + 159 kJ/mole = 1715 kJ/mole. More energy is required than is released, and so the reaction is endothermic. The net consumption of energy is 1818 kJ/mole − 1715 kJ/mole = 103 kJ/mole.
2. The nitrogen-nitrogen bond is broken during this reaction, but no new bonds are formed. Because energy is required to break a chemical bond, this reaction is endothermic, and energy should be written as a reactant: Energy + $N_2O_4 \rightarrow 2\,NO_2$. (If the reaction were to run backward, there would be one bond formed but none broken, meaning a net release of energy and an exothermic reaction: $2\,NO_2 \rightarrow N_2O_4$ + Energy.)

Catalysts

A **catalyst** is any substance that increases the rate of a chemical reaction. Iron rusts faster, for example, in the presence of sodium chloride; food molecules are more efficiently digested in the presence of vitamins; and gasoline is more thoroughly combusted in the presence of platinum, the active component of most automotive catalytic converters. In these examples, sodium chloride, vitamins, and platinum are catalysts.

Rather than being consumed during a chemical reaction, a catalyst is regenerated. This means that a single catalyst molecule or atom can be recycled to facilitate the formation of many thousands of product molecules. A catalyst need therefore be present only in tiny amounts. This is clearly an advantage because many catalysts are expensive. After the reaction is complete, a catalyst can be recovered and used for future reactions.

The chemical industry depends on catalysts because they lower the costs of manufacturing. Without catalysts, the price of gasoline would be much higher, as would be the price of rubber, carpets, plastics, pharmaceuticals, automobile parts, clothing, all food grown with chemical fertilizers, and most other chemical products. If it were not for catalysts, many chemical reactions would not even be feasible. For example, carbon dioxide, water, and sunlight do not normally react to form glucose, which is an important food source. However, this reaction proceeds efficiently in green plants thanks to catalysts in their chlorophyll. The action of catalysts can be beneficial.

Not so beneficial is the role of catalysts in the layer of ozone, O_3, high in the Earth's atmosphere. Ozone, which absorbs most of the ultraviolet radiation that impinges on the Earth, is destroyed by chlorine atoms, which act as a catalyst to turn ozone into diatomic oxygen, O_2. Initially, chlorine monoxide, ClO, and diatomic oxygen are formed by the following reaction:

Cl	+	O_3	→	ClO	+	O_2
atomic chlorine		ozone		chlorine monoxide		diatomic oxygen

Chlorine monoxide then reacts with atomic oxygen, which is abundant in the upper atmosphere, to re-form the chlorine atom:

ClO	+	O	→	Cl	+	O_2
chlorine monoxide		atomic oxygen		atomic chlorine		diatomic oxygen

The net result is the destruction of ozone and the regeneration of chlorine, which then catalyzes more ozone destruction.

Atomic chlorine is generated in the upper atmosphere from a number of sources, including a class of human-made compounds known as chlorofluorocarbons (CFCs), whose atmospheric concentrations have increased 500 percent in the last 20 years. Atomic chlorine, however, is not the only catalyst favoring the destruction of ozone: Sulfur dioxide, SO_2 (from volcanoes and the burning of coal and petroleum); methyl iodide, CH_3I (released by the ocean and carried to the upper atmosphere by thundercloud updrafts); and nitrogen oxides, N_2O, NO, and NO_2 (products of soil microbes, fossil fuel combustion, and lightning strikes) also catalyze the destruction of ozone. In short, there are many causes of ozone destruction, both natural and human. Ozone does re-form, but because of increased concentrations of atomic chlorine in the atmosphere these days, the present rate of depletion is greater than the rate of formation.

Stratospheric Ozone and CFCs

Ozone is a gas formed of three oxygen atoms, O_3. Some ozone is formed from auto emissions and is an urban air pollutant (See Section 24.6). Ozone is also formed naturally at altitudes from 20 to 30 kilometers, within a zone of the atmosphere known as the stratosphere. At these altitudes high-energy ultraviolet (UV) radiation breaks diatomic oxygen into atomic oxygen, which then reacts with additional O_2 to form ozone:

$$O_2 + UV \rightarrow 2\,O$$
$$2\,O + 2\,O_2 \rightarrow 2\,O_3$$
$$\textbf{Net reaction: } 3\,O_2 + UV \rightarrow 2\,O_3$$

This synthesis of ozone is of great benefit to life on the Earth because ozone absorbs high-energy UV radiation, which, if it were to reach the Earth's surface, would cause immediate harm to living tissue.

When it absorbs UV radiation, ozone fragments into molecular and atomic oxygen. These fragments eventually come back together to re-form ozone. Because chemical bonds are formed when ozone reforms, energy is released as heat:

$$O_3 + UV \rightarrow O_2 + O$$
$$O_2 + O \rightarrow O_3 + \text{heat}$$
$$\textbf{Net reaction: } O_3 + UV \rightarrow O_3 + \text{heat}$$

So harmful UV radiation is transformed by ozone into a not-so-harmful slight heating of the stratosphere. Note in the preceding equation that ozone is not lost by this transformation, which means it can continue to shield UV from the Earth's surface indefinitely. Very beneficial.

The concentration of ozone in the stratosphere is quite small. If it were subjected to the atmospheric pressure at the Earth's surface, it would comprise a layer of gas only 3 millimeters thick! Nevertheless, this ozone layer absorbs more than 95 percent of the UV radiation that comes to our planet from the Sun. It is the safety blanket of planet Earth.

In the early 1970s, professors Mario Molina of the Massachusetts Institute of Technology, F. Sherwood Rowland of the University of California Irvine, and Paul J. Crutzen of the Max Planck Institute in Germany recognized the potential threat to stratospheric ozone posed by chlorofluorocarbons. Because CFCs are inert gases, they were once commonly used in air conditioners and aerosols.

Professors Mario Molina, F. Sherwood Rowland, and Paul Crutzen at a press conference before receiving the 1995 Nobel Prize in Chemistry for their work in atmospheric chemistry, particularly concerning the formation and decomposition of stratospheric ozone.

Estimates are that CFCs are so stable they remain in the atmosphere from 80 to 120 years, and they are now spread throughout the Earth's atmosphere. Even in the most remote location, there are no less than 25 quadrillion CFC molecules in every liter of air you breathe.* Estimates are that one chlorine atom causes the destruction of at least 100,000

Two of the most common CFCs, also known as "freons," were CFC-11, trichlorofluoromethane, and CFC-12, dichlorodifluoromethane. At the height of CFC production in 1988, some 1.13 million tons was produced worldwide.

(continued)

*This figure assumes an average concentration of one part per billion. Concentrations in urban areas are usually much higher.

Stratospheric Ozone and CFCs—continued

ozone molecules in the year or two before it forms a hydrogen chloride molecule, HCl, and is carried away by atmospheric moisture.

The fragility of stratospheric ozone came to world focus in 1985 with the discovery of a seasonal depletion of stratospheric ozone over the Antarctic continent—the so-called ozone hole. That atomic chlorine plays an active role in the destruction of Antarctic ozone is clearly seen in the profile of chlorine monoxide and ozone concentrations relative to latitude. Subsequent studies have shown that the cold, still, sunless Antarctic winters favor the formation of stratospheric ice crystals upon which airborne compounds containing chlorine atoms accumulate. Chemical reactions on and within the ice crystals lead to the formation of Cl_2. In October of the Antarctic spring, sunlight returns and fragments the molecular chlorine into vast amounts of ozone-depleting atomic chlorine:

$$Cl_2 + \text{sunlight} \rightarrow 2\ Cl$$

There has been an unprecedented level of international cooperation toward banning ozone-destroying chemicals. The first major step was the signing of the 1987 *Montreal Protocol on Substances that Deplete the Ozone Layer* which called for the reduction of CFC production to one-half of 1986 levels by 1998. By the time this treaty was signed, however, scientists had confirmed that chlorine from CFCs is a significant cause of the recurring Antarctic ozone hole. An important piece of evidence was the unusually high levels of fluorine compounds detected in the Antarctic stratosphere. Whereas chlorine compounds come from a number of natural sources, fluorine compounds in nature are relatively rare. So the fluorine found in the Antarctic stratosphere was, in a sense, the "smoking gun" showing CFC involvement. In addition to the fluorine evidence, minor ozone depletion around the north pole and mid-latitudes was also beginning to appear. The protocol was soon amended to initiate a more rapid phase-out of CFCs and related compounds. A complete ban of CFCs was established in January 1996. Even with the Montreal Protocol in place and the continued cooperation of all signed nations, however, the ozone-destroying actions of CFCs will be with us for some time. Atmospheric CFCs are not expected to drop back to pre-ozone-hole levels until sometime in the 22nd century.

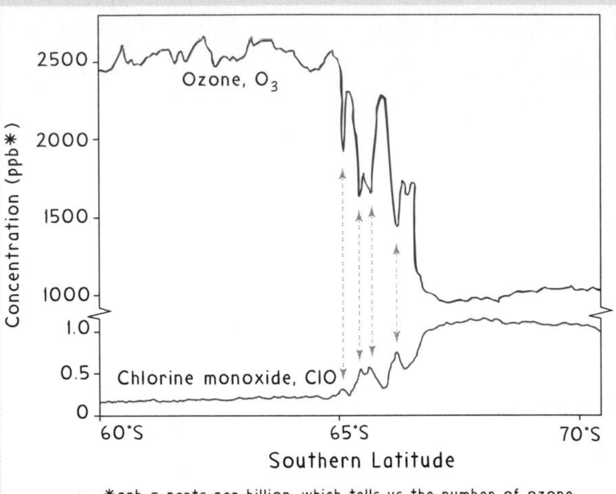

Data collected during flights across the southern polar region of our planet show the close relationship between stratospheric levels of chlorine monoxide and ozone. As chlorine monoxide levels increase, ozone levels decrease. Likewise, as chlorine monoxide levels decrease, ozone levels increase. Also note how only relatively small concentration changes of chlorine monoxide appear to be related to large concentration changes of ozone. This is consistent with chlorine monoxide behaving as a catalyst.

A satellite image of the southern hemisphere showing relative concentrations of stratospheric ozone.

19.3 Chemical Equilibrium

The arrow of a chemical equation indicates the direction of the reaction. In the equation for the formation of water, for example, the arrow indicates that hydrogen and oxygen combine to form water:

$$2\,H_2 + 1\,O_2 \rightarrow 2\,H_2O$$

In principle, apply a strong electric current to the ocean for an incredibly long time, and all the water will be converted into gaseous hydrogen and oxygen. The first strike of a match, however, would initiate a massive explosion as the two gases exothermically convert back into water.

The reverse reaction is also possible; that is, oxygen and hydrogen can be formed from water:

$$2\,H_2 + 1\,O_2 \rightarrow 2\,H_2O$$

Most chemical reactions are *reversible,* but the extent of the reverse reaction depends very much on conditions. Under usual conditions, for example, water does not readily convert to oxygen and hydrogen. When electricity passes through water, however, the water does decompose to O_2 and H_2.

Nitrogen dioxide, NO_2, a brown gas and one of the toxic components of air pollution, provides a good example of a reaction that is fairly easy to reverse. At room temperature, nitrogen dioxide molecules pair off to become colorless dinitrogen tetroxide, N_2O_4:

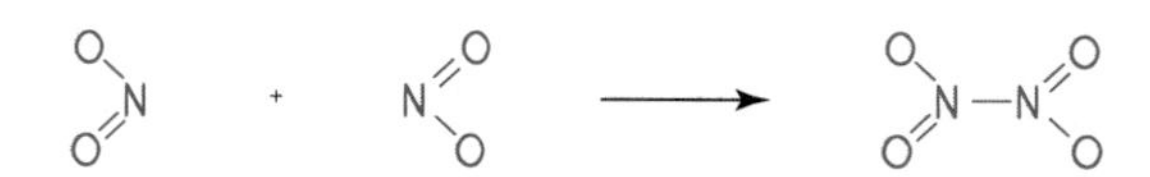

Once dinitrogen tetroxide molecules are formed, however, they break apart to re-form nitrogen dioxide:

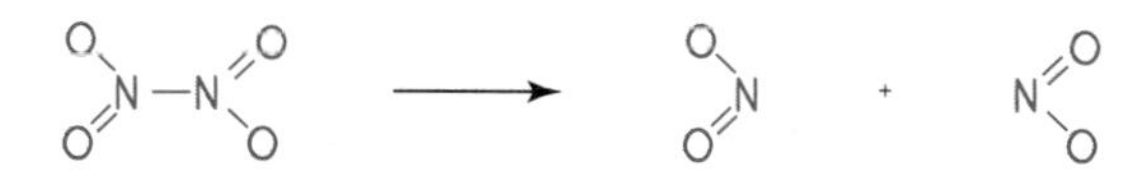

Of course, once the nitrogen dioxide molecules are re-formed, they can get together to re-form the dinitrogen tetroxide. The net result is two competing reactions, which we depict using double arrows:

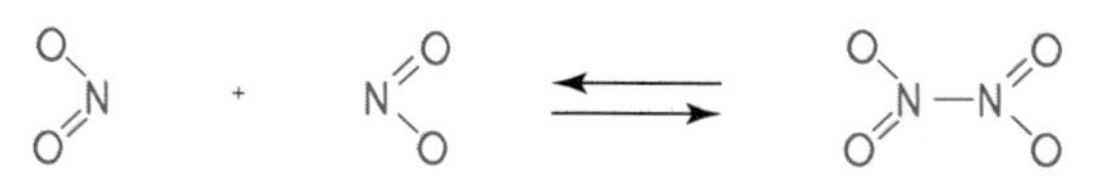

Initially, we may start with pure nitrogen dioxide. In time, however, the percentage of nitrogen dioxide decreases and the percentage of dinitrogen tetroxide increases. This shift in amounts present continues to a point where the two reactions balance each other and the percentages of NO_2 and N_2O_4 settle to constant values (at room temperature, about 31 percent NO_2 and 69 percent N_2O_4). At this point the forward and reverse

Fig. 19.6
The NO_2 and N_2O_4 in this flask are in equilibrium at 16 percent NO_2 and 84% N_2O_4. Although the nonchanging brown color seems to indicate no activity in the flask, individual molecules are continuously converting back and forth with a balance between forward and reverse reactions.

reactions have achieved what is called **chemical equilibrium**, where the rate at which products are formed is equal to the rate at which they are converted back into reactants (Figure 19.6).

Chemical equilibrium is like a store in which there is a steady flow of customer traffic. Think of people inside the store as product molecules and those outside as reactant molecules. The steady customer flow means that the number of people entering the store in any given time interval equals the number of people leaving in the time interval. As a result, the number of people inside the store never changes. The individual people in the store change as Ms. X and Mr. Y enter and Mr. A and Ms. B leave, but the total number inside remains constant. In the same way, the number of N_2O_4 molecules in the flask of Figure 19.6 stays constant because the rate at which N_2O_4 molecules convert to NO_2 molecules equals the rate at which N_2O_4 molecules are formed from NO_2 molecules.

Chemical equilibrium does not mean that the amounts of reactants and products are equal. The equilibrium point in our store analogy might be ten people in the store and 100 outside. For the NO_2/N_2O_4 reaction an equilibrium point may be reached when the NO_2:N_2O_4 ratio is 31:69. The important thing to remember is that chemical equilibrium is a dynamic state. Although the concentrations of products and reactants remain constant, individual molecules are continuously changing back and forth.

If the store holds a big sale, the equilibrium of customers is affected. At the beginning of the sale, the rate at which people enter the store is greater than the rate at which they leave. Hence the number of people inside the store increases. There is a limit to the number of customers the store can hold, however. Eventually a door attendant is hired to allow people to enter only at the rate at which other people leave. Thus a new equilibrium is established. In this case, however, there are now more people in the store at any given moment than there were before the sale. Likewise, chemical equilibria are affected by conditions. For example, when dinitrogen tetroxide is exposed to energy, it tends to form nitrogen dioxide. Increasing the temperature of a flask of nitrogen dioxide and dinitrogen tetroxide increases the rate of nitrogen dioxide formation until a new equilibrium is reached. At this higher-temperature equilibrium, nitrogen dioxide predominates and the overall color of the reaction mixture is dark brown. This is one of the reasons air pollution becomes particularly noticeable on warm days; the higher the temperature, the greater the concentration of brown nitrogen dioxide.

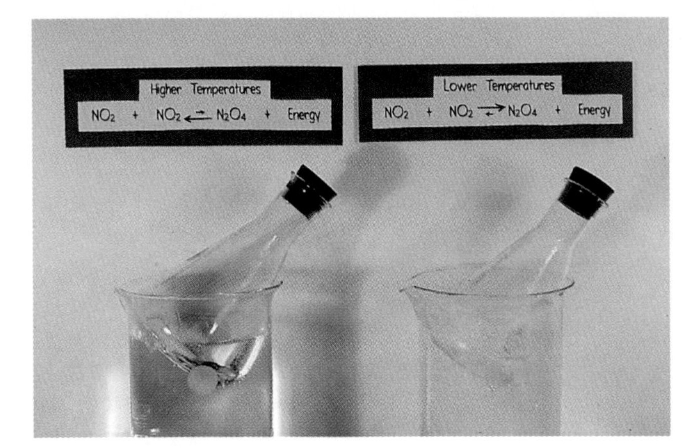

Fig. 19.7
Higher temperatures favor formation of brown nitrogen dioxide, and the mixture in the flask is dark brown. Lower temperatures favor formation of colorless dinitrogen tetroxide, and the mixture is a much lighter brown.

Pressure as well as temperature can affect the chemical equilibrium of reacting gases. As we saw in Section 5.6, we can increase the pressure of a gas by decreasing the volume of its container. This decrease in volume brings the gas molecules closer together, which has the effect of increasing the number of collisions among them (Figure 19.8).

Take a container filled with an equilibrium mixture of NO_2 gas and N_2O_4 gas and squeeze it to a smaller volume. The rate at which the molecules collide with one another increases. This is advantageous for the formation of N_2O_4, which requires the coming together (collision) of two NO_2 molecules. Under higher pressures, there

■ Question

After a mixture of nitrogen dioxide and dinitrogen tetroxide initially at room temperature is brought to 100°C and kept at that temperature for a few minutes, which is greater: the rate of nitrogen dioxide formation or the rate of dinitrogen tetroxide formation?

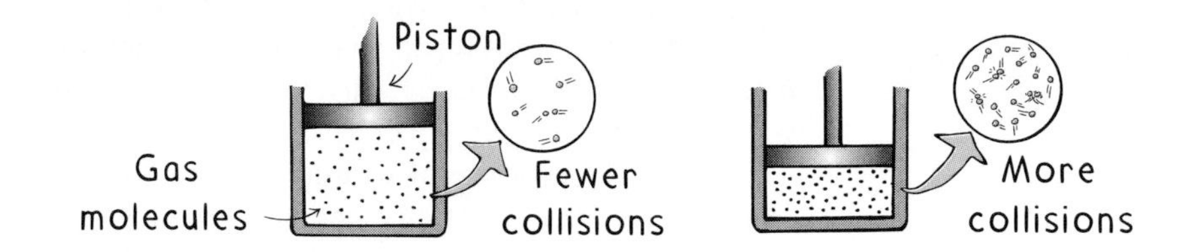

Fig. 19.8
Gas particles at higher pressures are closer together, which increases the rate at which they collide with one another.

fore, the rate of N_2O_4 formation becomes greater than the rate of NO_2 formation and the gaseous mixture becomes lighter in color, as shown in Figure 19.9. As more and more N_2O_4 is formed at high pressure, however, the supply of NO_2 dwindles, reducing the number of NO_2 collisions, which decreases the rate of N_2O_4 formation. Eventually a new equilibrium is reached where the rates of NO_2 formation and N_2O_4 formation are once again the same. The relative concentration of N_2O_4 at this new equilibrium, however, remains greater than it is at lower pressures. At twice the pressure, for example, the relative proportions of NO_2:N_2O_4 changes to 24:76.

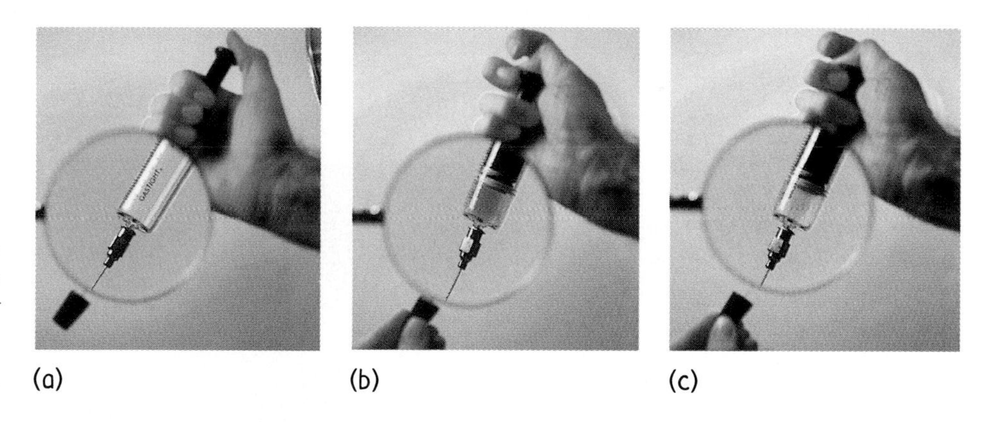

Fig. 19.9
(a) A syringe containing an equilibrium mixture of nitrogen dioxide and dinitrogen tetroxide. (b) Compressing the gas initially shows a deepening of the brown color as brown NO_2 molecules are bunched together. (c) The higher proportion of NO_2 molecules, however, favors a greater number of collisions among these molecules, which upon colliding convert back to the colorless N_2O_4 molecules. This is evidenced by a rapid lightening of the brown color soon after the NO_2 molecules have been compressed.

The same sort of thing happens when pressure is applied to a mixture of gaseous nitrogen and hydrogen, which react to form gaseous ammonia, NH_3. As can be seen in Figure 19.10, the formation of ammonia requires the collision of four molecules (one

---

■ **Answer**

Neither! After a few minutes a new equilibrium is achieved. It's like the store holding a sale. Initially, equilibrium is lost as more customers pile in. Eventually, however, the store becomes so packed that the rate at which people come in matches the rate at which they go out. Similarly, with the application of heat, the rate of NO_2 formation is initially greater than the rate of N_2O_4 formation. As the N_2O_4 becomes depleted, however, the rate of NO_2 formation decreases to the point where the forward and reverse reactions become matched as they were before.

nitrogen and three hydrogen), whereas the formation of nitrogen and hydrogen requires the collision of only two ammonia molecules. Because an increase in pressure favors the reaction requiring the greater number of collisions, higher pressures favor the formation of ammonia over the formation of nitrogen and hydrogen. This is a practical matter. Ammonia is a valuable commodity, especially as a fertilizer. Very little ammonia, however, is formed from nitrogen and hydrogen at atmospheric pressures. The efficient production of ammonia at high pressures was first realized by the German chemist Fritz Haber during World War I. His discovery led to an agricultural revolution because before then nearly all fertilizers came from natural sources, which were already becoming scarce. With Haber's discovery, fertilizer (ammonia) could now be produced economically using air, which is 78 percent nitrogen, as one of the starting materials. Thus came the promise of feeding a much larger human population. Today, in the United States alone, this high-pressure technique is used to produce about 36 billion pounds of ammonia every year.

$$\mathrm{N{\equiv}N} \quad \mathrm{H{-}H} \quad \mathrm{H{-}H} \quad \mathrm{H{-}H} \rightleftarrows \mathrm{NH_3} \quad \mathrm{NH_3}$$

Fig. 19.10
Four gas molecules combine to produce gaseous ammonia, NH_3, while only two gas molecules combine to form gaseous N_2 and H_2.

Question

Would an increase in pressure favor the formation of gaseous sulfur trioxide, SO_3, from gaseous sulfur dioxide, SO_2, and oxygen, O_2?

$$2\,SO_2\,(g) + O_2\,(g) \rightleftarrows 2\,SO_3$$

Link to the Environment—Equilibrium in Nature

The concept of equilibrium is not limited to chemical reactions. Recall we introduced it back in Chapter 2, when we discussed mechanical equilibrium. In general, equilibrium is the balance between opposing forces or actions. Rain falling on the Earth is in equilibrium with the evaporation of water from oceans, lakes, and rivers. Populations of reindeer are in equilibrium with the number of wolves who eat them. Wolves primarily kill and eat weaker reindeer, which leaves healthier reindeer to multiply. With their population in check, the reindeer are also less likely to exhaust the scarce food supply during the winter months. Here, as with any other ecological system, the survival of a species depends on a balance of forces. Remove the wolves and the reindeer population is no longer in check; remove the reindeer and the wolves starve. Planet Earth is one large system of related equilibria, where all forces and actions affect one another. Insofar as we humans participate in global forces and actions, we must be aware of the concept of equilibrium. Technically speaking, it's not planet Earth that needs protection, for it will be here whatever we humans do or don't do. It is the many dynamic equilibria of the biosphere that require our utmost attention.

The tendency of reindeer to overpopulate is balanced by the nutritional needs of the wolf.

19.4 Relative Masses of Atoms and Molecules

In any chemical reaction, a specific number of atoms or molecules react to form a specific number of product molecules. For example, when carbon and oxygen combine to form carbon dioxide, they always combine in the ratio of one carbon atom to one oxygen molecule (Figure 19.11). A chemist wanting to carry out this reaction in the laboratory would be wasting chemicals and money if she were to add, say, four carbon atoms to the reaction flask for every one oxygen molecule. The excess carbon atoms would just be wasted.

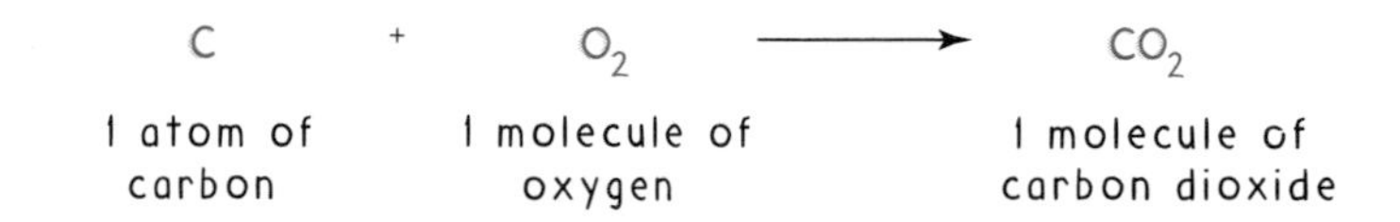

Fig. 19.11
The balanced chemical equation for the formation of carbon dioxide from carbon and molecular oxygen.

How is it possible to measure out a specific number of atoms or molecules? Because atoms and molecules are so small, we don't count them individually. Instead, we use a scale that measures the masses of bulk quantities. Some special thinking, however, is required in order to measure out a specific number of atoms or molecules. Say, for example, we needed the same number of carbon atoms as oxygen molecules. Measuring equal masses of the two materials would not provide equal numbers. You know that 1 kilogram of golf balls contains fewer balls than 1 kilogram of Ping-Pong balls (Figure 19.12). Likewise, because different atoms and molecules have different masses, there are different numbers of them in, say, a 1-gram sample of each. In fact, there are more lighter carbon atoms in a gram of carbon than there are heavier oxygen molecules in a gram of oxygen. So, clearly, weighing equal masses won't give us equal numbers.

Fig. 19.12
The same number of golf balls and Ping-Pong balls have different masses. Clearly, each golf ball weighs more than each Ping-Pong ball.

If we know the *relative masses* of different materials, however, we can measure equal numbers. Golf balls, for example, are about 40 times more massive than Ping-Pong balls. In other words, the number of golf balls in 40 kilograms of golf balls equals the number of Ping-Pong balls in 1 kilogram of Ping-Pong balls. Hence, if we want equal numbers of the two types of balls, we should measure out a mass of golf balls that is 40 times the mass we measure out for the Ping-Pong balls. Similarly, because we know the relative masses of carbon and molecular oxygen (from the periodic table), we can measure out equal numbers of these two species. Because carbon is three-eighths as massive as molecular oxygen ($12.011 \div 31.999 = \frac{12}{32} = \frac{3}{8}$), we weigh out only three-eighths as much carbon. For example, 3 grams of carbon contains the same number of particles as 8 grams of molecular oxygen (3 is three-eighths of 8). Alternatively, we might measure out 12 grams of carbon and 32 grams of molecular oxygen (12 is three-eighths of 32). In both cases we would have as many carbon atoms as oxygen molecules (Figure 19.13).

Answer

Yes. An increase in pressure favors a greater number of collisions, which has a greater benefit for the formation of SO_3. As we shall see in the next chapter, SO_2 is an industrial pollutant. In the atmosphere it transforms into SO_3 as shown above. The SO_3 then reacts with water to form sulfuric acid, H_2SO_4—a prime component of acid rain.

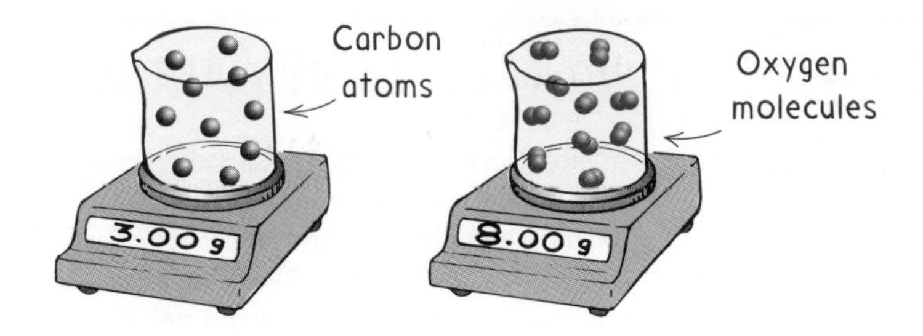

Fig. 19.13
To have the same number of carbon atoms as oxygen molecules requires weighing out three-eighths as much carbon as oxygen.

According to the balanced equation shown in Figure 19.11, the number of carbon dioxide molecules formed is the same as the number of carbon atoms and oxygen molecules that react. One carbon atom combines with one oxygen molecule to produce one carbon dioxide molecule. Likewise, a billion carbon atoms combine with a billion oxygen molecules to produce a billion carbon dioxide molecules. Similarly, we might say that 6.02×10^{23} carbon atoms combine with 6.02×10^{23} oxygen molecules to produce 6.02×10^{23} carbon dioxide molecules. Because 6.02×10^{23} is the same as 1 *mole* (Section 18.2), we can more conveniently say that 1 mole of carbon atoms combines with 1 mole of oxygen molecules to produce 1 mole of carbon dioxide molecules. We exaggerate our examples here to emphasize the point that we are referring to a *count* of atoms and molecules.

As Figure 19.11 shows, in any reaction between carbon and oxygen to form carbon dioxide, the numbers of atoms or molecules of all the species are the same. Although the counts are the same, the masses of these species are not. For example, 3 grams of carbon (which calculates to 1.51×10^{23} atoms) combines with 8 grams of molecular oxygen (1.51×10^{23} molecules) to form 11 grams of carbon dioxide (1.51×10^{23} molecules). So we see that the mass of the carbon dioxide produced is the sum of the masses of carbon and oxygen that reacted (3 grams + 8 grams = 11 grams).

■ **Question**

Reacting 3 g of carbon with 8 g of oxygen results in 11 g of carbon dioxide. Would reacting 6 g of carbon with 8 g of oxygen also result in 11 g of carbon dioxide?

Atomic and Formula Masses

With modern techniques the absolute masses of individual atoms are measured to a high degree of accuracy. The hydrogen atom, for example, has a mass of 1.674×10^{-24} gram, and the oxygen atom has a mass of 2.657×10^{-23} gram.[3] These numbers are unwieldy and can be simplified by a change of units. Just as a grocer calls 12 eggs a dozen, and a printer calls 500 sheets of paper a ream, a chemist handles the masses of atoms in

---

■ **Answer**
Yes, with 3 g of carbon left over. Recall that carbon and oxygen react in a 1:1 ratio, and, as we've been discussing, 3 g carbon contains just as many particles as 8 g of oxygen. For all 6 g of carbon to react, we would need a total of 16 g of oxygen (6 g $\times \frac{8}{3}$ = 16 g), which is twice the amount available.

---

[3]These are average values based on the different isotopes of these elements and their abundances.

atomic mass units (amu). For several reasons, one of which is nuclear stability, the carbon-12 isotope was chosen as a standard. The atomic mass unit is defined as exactly one-twelfth the mass of the carbon-12 isotope:

$$\textbf{Mass of carbon-12 atom} = 1.993 \times 10^{-23}\ \textbf{gram}$$

$$\downarrow$$

$$\textbf{divide by 12}$$

$$\downarrow$$

$$1\ \textbf{amu} = 1.661 \times 10^{-24}\ \textbf{gram}$$

Thus we can say that the mass of the hydrogen atom is 1.01 amu (rather than 1.674×10^{-24} gram), and that of the oxygen atom is 16.00 amu (rather than 2.657×10^{-23} gram).

Both the gram and the atomic mass unit are units of mass—they differ only in scale. The gram is used in measuring macroscopic quantities. A 1-gram sample is visible because it consists of billions upon billions of atoms. The atomic mass unit, on the other hand, is used in measuring the mass of individual atoms or molecules.

The mass of an element in atomic mass units is its *atomic mass,* a concept introduced in Chapter 13. As noted there, the atomic mass of each element is written in the periodic table directly below the atomic symbol. With these values, the relative masses of any two elements can be calculated. For example, by dividing the atomic mass of oxygen (16.00 amu) by the atomic mass of hydrogen (1.01 amu), we find that oxygen atoms are 15.84 times more massive than hydrogen atoms.

■ **Question**

Which is greater: 1.01 amu of hydrogen or 1.01 g of hydrogen?

The **formula mass** of a substance is the sum of the atomic masses of elements in its chemical formula. For example, the formula mass of molecular oxygen, O_2, is the atomic mass of oxygen (16.00 amu) times 2:

16.00 amu	×	2	=	32.00 amu
atomic mass of O				formula mass of O_2

Similarly, the formula mass of carbon dioxide, CO_2, is the atomic mass of carbon plus 2 times the atomic mass of oxygen:

Carbon (12.01 amu)	×	1	=	12.01 amu
Oxygen (16.00 amu)	×	2	=	32.00 amu
				44.01 amu
				formula mass of CO_2

By knowing the formula masses of different compounds, we can determine their relative masses. From the preceding data, for example, we find that carbon dioxide is $44.01 \div 32.00 = 1.375$ times as massive as molecular oxygen (about eleven-eighths).

■ **Answer**

1.01 amu is the mass of a single hydrogen atom, which is far less than the mass of all the hydrogen atoms in a 1.01-g sample of hydrogen.

By the same analysis we see how carbon is 12.01 ÷ 32.00 = 0.375 (about three-eighths) the mass of molecular oxygen. Atomic masses give us a handle on how *atoms* should be measured out so that they are in correct proportions. Formula masses are useful because we can determine what proportion, by mass, *compounds* should be mixed so that the number of molecules of each are in a desired ratio.

So we see that the rules of cooking and chemistry are similar in that they both require the measuring of ingredients. Just as a cook looks to a recipe to find the necessary quantities measured by the cup or the tablespoon, a chemist looks to the periodic table to find the necessary quantities measured by a count of atoms or molecules.

Questions

1. Which has the greater number of molecules: 32.00 g of molecular oxygen or 44.01 g of carbon dioxide?
2. Which has the greater mass, 12.04×10^{23} molecules of molecular hydrogen or 12.04×10^{23} molecules of water?

19.5 Avogadro's Number and the Mole

We are now in a position to learn the origin of that unusually large number we have been referring to: the mole, which is 6.02×10^{23}. Why this particular value and what does it mean? Read this section carefully to find out.

If you were given 100 grams of jelly beans and each jelly bean had a mass of 2 grams, how many jelly beans would you have? Mathematically, we find the answer by taking the total mass of jelly beans and dividing by the mass of a single jelly bean:

$$\frac{\textbf{Total mass of jelly beans}}{\textbf{Mass of a single jelly bean}} = \frac{\textbf{100 grams}}{\textbf{2 grams per jelly bean}} = \textbf{50 jelly beans}$$

Similarly, if you wish to know the number of atoms in a particular sample of an element, you divide the total mass of the sample by the mass of a single atom. For example, the number of carbon-12 atoms in a sample weighing 3.986×10^{-20} gram, where each carbon atoms has a mass of 1.993×10^{-23} gram, is

$$\frac{\textbf{Total mass of carbon-12}}{\textbf{Mass of single carbon-12 atom}} = \frac{\mathbf{3.986 \times 10^{-20}}\ \textbf{gram}}{\mathbf{1.993 \times 10^{-23}}\ \textbf{gram per atom}} = \textbf{2000 atoms}$$

Of course, a 3.986×10^{-20} gram of carbon-12 is a speck too small for us to see—it consists of only 2000 atoms! Let's consider a larger mass of carbon-12, but not just

Answers

1. There are the same number of molecules in 32.00 g of molecular oxygen as there are in 44.01 g of carbon dioxide because carbon dioxide molecules are 44.01 ÷ 32.00 times heavier than oxygen molecules.
2. The water has the greater mass, just as a bunch of golf balls has more mass than the same number of Ping-Pong balls. The water has about 9 times as much mass because each H_2O molecule (16 + 1 + 1 = 18 amu) is about 9 times as massive as each H_2 molecule (1 + 1 = 2 amu): 18 amu ÷ 2 amu = 9. The big numbers don't change anything: 12.04×10^{23} molecules of water have a greater mass than 12.04×10^{23} molecules of molecular hydrogen.

Fig. 19.14
Express the atomic mass of any element in *grams*, and that many grams contains 6.02×10^{23} atoms.

any mass. Let's choose 12 grams, the same number as carbon's atomic mass number. Our question, then, is how many atoms of carbon-12 are there in a 12.0000-gram sample? Our answer is found when we divide 12.0000 grams by 1.993×10^{-23} gram per atom, which gives us the astronomically large number 6.02×10^{23} atoms. Take a moment to appreciate the enormity of this number:

$$6.02 \times 10^{23} = 602{,}000{,}000{,}000{,}000{,}000{,}000{,}000$$

This many grains of wheat would occupy 20 million cubic kilometers (about the volume of the Arctic ocean). Atoms are so small, however, that this many atoms of carbon-12 fit within a 12-gram sample (about the size of five sugar cubes if the carbon is in the form of graphite).

When the atomic mass of any element is expressed in grams, the number of atoms in that sample equals 6.02×10^{23}. A 4.00-gram sample of helium, for example (He, atomic mass = 4.00), contains 6.02×10^{23} helium atoms; a 55.85-gram sample of iron (Fe, atomic mass = 55.85) contains 6.02×10^{23} iron atoms; and a 196.97-gram sample of gold (Au, atomic mass = 196.97) contains 6.02×10^{23} gold atoms. So, look to the periodic table and read any atomic mass in grams. The number of atoms in a sample of this mass is always 6.02×10^{23} (Figure 19.14). Again, for emphasis, the atomic mass of an element shown in the periodic table is the mass of 6.02×10^{23} atoms of that element.

Question

How many atoms are there in a 6.94-g sample of lithium (atomic mass 6.94)?

When the formula mass of a compound is expressed in grams, we get the same number. There are 6.02×10^{23} molecules of molecular oxygen in 32.00 grams of molecular oxygen (formula mass = 32.00 amu), just as there are 6.02×10^{23} molecules of carbon dioxide in 44.01 grams of carbon dioxide (formula mass = 44.01 amu).

Question

How many molecules are there in a 18.00-g sample of water (formula mass 18.00)?

How do we arrive at this same number whenever any atomic or formula mass is expressed in grams? It so happens that this large number is the ratio of *relative* mass to *actual* mass. The relative mass is the atomic or molecular formula expressed in grams. The actual mass is the mass of an individual atom or molecule expressed in grams. The

---

Answer
In 6.94 g of lithium there are 6.02×10^{23} lithium atoms.

Answer
6.02×10^{23} molecules.

key to understanding this tricky concept is to recognize that the bigger the atom or molecule, the bigger its relative mass *and* the bigger its actual mass. The carbon-12 isotope, for example, has a relative mass of 12.0000 grams and an actual mass of 1.993×10^{-23} gram. The oxygen-16 isotope has a greater relative mass of 15.9998 grams and a correspondingly larger actual mass of 2.658×10^{-23} gram. In all cases the ratio of the two (12.0000 grams divided by 1.993×10^{-23} gram and 15.9998 divided by 2.658×10^{-23} gram) is 6.02×10^{23}. This is similar to the ratio of a circle's circumference to diameter. This ratio is the constant value π (3.14159265 . . .), and it is always the same no matter what the size of the circle (Figure 19.15).

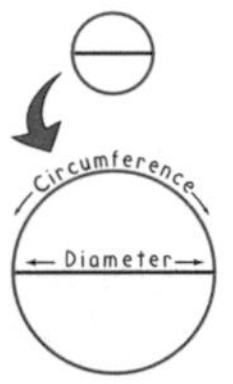

$$\frac{C}{D} = \pi \qquad \frac{\text{RELATIVE MASS}}{\text{ACTUAL MASS}} = 6.02 \times 10^{23}$$

Circumference
Diameter

$$\frac{C}{D} = \pi \qquad \frac{\text{RELATIVE MASS}}{\text{ACTUAL MASS}} = 6.02 \times 10^{23}$$

Fig. 19.15
As the circumference of a circle gets bigger, so does the diameter. The ratio of the two is always equal to the constant π. In a similar manner, the ratio of an atom or molecule's relative mass to its actual mass is always equal to the constant value 6.02×10^{23}.

The existence of this special number for the ratio of relative mass to actual mass was recognized in the late 1800s. Its value of 6.02×10^{23}, however, could not be determined until the early 1900s, when techniques for measuring the mass of a single atom were developed. Today, this number is known as **Avogadro's number**, in honor of Amadeo Avogadro, who in the early 1800s was one of the pioneers in the development of atomic theory.[4]

As discussed in Section 19.4, we refer to 6.02×10^{23} atoms or molecules of a substance as 1 *mole* of that substance. Thus the atomic or formula mass of a substance in grams is equal to 1 mole of that substance. So we say that 12.0000 grams of carbon-12 (12.0000 amu) is 1 mole of carbon-12, which equals 6.02×10^{23} atoms of carbon-12. And 32.00 grams of molecular oxygen (32.00 amu) is 1 mole of molecular oxygen, which equals 6.02×10^{23} molecules of oxygen. Thus atomic and formula masses are often given using the unit *grams per mole.* The atomic mass of carbon-12, for example, is 12.0000 grams per mole, and that of molecular oxygen is 32.00 grams per mole.

■ **Question**

What is the mass in grams of 1 mole of water (H_2O = 18 amu)?

So 1 mole of any substance always contains the same number of particles, 6.02×10^{23}. The mole, therefore, is an ideal unit when dealing with chemical reactions. For example, 1 mole of carbon (12.01 grams) reacts with 1 mole of molecular oxygen (32.00 grams) to give 1 mole of carbon dioxide (44.01 grams). In many instances, the ratio in which chemicals react is not one to one. This is most easily depicted using the mole unit. For example, 2 moles of molecular hydrogen, H_2 (4.04 grams), react with 1 mole of molecular oxygen, O_2 (32.00 grams), to give 2 moles of water, H_2O (36.04 grams), (Figure 19.16).

■ **Answer**
From the formula mass expressed in grams per mole, we see there are 18 grams of water in 1 mole of water.

[4]Avogadro died in 1856, some 50 years before this honor was bestowed upon him.

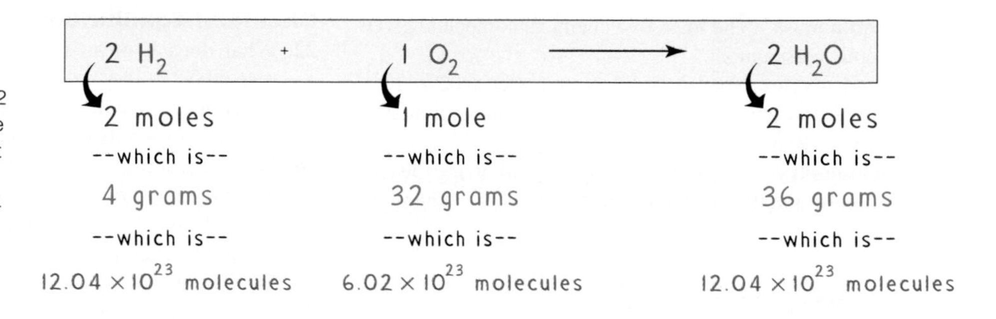

Fig. 19.16
Two moles of molecular hydrogen, H_2, reacts with 1 mole of molecular oxygen, O_2, to give 2 moles of water, H_2O. This is the same as saying 4 g of H_2 react with 32 g of O_2 to give 36 g of H_2O, or equivalently, that 12.04 $\times$ 10^{23} molecules of H_2 react with 6.02 $\times$ 10^{23} molecules of O_2 to give 12.04 $\times$ 10^{23} molecules of H_2O.

Questions

1. How many moles of carbon dioxide, CO_2, are produced from the reaction of 1 mole (16 g) of methane, CH_4, and 2 moles (64 g) of molecular oxygen, O_2? How many grams of CO_2 is this?

$$\mathbf{CH_4 + 2\ O_2 \rightarrow CO_2 + 2\ H_2O}$$

2. How many grams of calcium oxide, CaO, reacts with 64.058 g of sulfur dioxide, SO_2, in the formation of calcium sulfite, $CaSO_3$? How many grams of $CaSO_3$ is formed in this reaction?

Formula masses:

CaO = 56.079 g/mole

SO_2 = 64.058 g/mole

$CaSO_3$ = 120.137 g/mole

Balanced equation:

$CaO + SO_2 \rightarrow CaSO_3$

The quantitative material covered in these last two sections is known as **stoichiometry**, the science of calculating the amount of reactants or products in any chemical reaction. Stoichiometric calculations can be quite complex, requiring a very systematic approach, which is a focus of the typical introductory general chemistry course. For our purposes, however, the message is not that you must become an expert at juggling numbers. Rather, your awareness of stoichiometry and how it serves as a powerful tool is sufficient. Consider the premise of stoichiometry—that the substances around us are composed of molecules, which are in turn composed of atoms. In the laboratory, stoichiometry is never-failing in its predictions. The success of stoichiometry, therefore, is yet another strong piece of evidence for the accuracy of the atomic model upon which chemistry is based.

Summary of Terms

- **Chemical equation** A representation of a chemical reaction showing the relative numbers of reactants and products.
- **Reactants** The starting material for a chemical reaction. It appears before the arrow in the chemical equation.
- **Products** The new material formed by a chemical reaction. It appears after the arrow in the chemical equation.
- **Principle of the conservation of mass** A principle stating that matter is neither created nor destroyed in a chemical reaction, as far as we are able to detect.
- **Bond energy** The amount of energy absorbed upon bond breaking and released upon bond formation. Each chemical bond has its own characteristic bond energy.
- **Catalyst** Any substance that serves to increase the rate of a chemical reaction.
- **Chemical equilibrium** A dynamic state in which the rate of the forward chemical reaction is equal to the rate of the reverse chemical reaction. At chemical equilibrium the concentrations of reactants and products remain constant.
- **Atomic mass unit** A very small unit of mass used for atoms and molecules. One atomic mass unit (amu) is equal to $\frac{1}{12}$ the mass of the carbon-12 atom, or 1.661×10^{-24} g.

- **Formula mass** The mass of a chemical compound given in atomic mass units.
- **Avogadro's number** A very large number: 6.02×10^{23}. This is the number of atoms in the atomic mass of an element when expressed in grams.
- **Stoichiometry** An aspect of chemistry involving the calculation of quantities of substances involved in chemical reactions.

Review Questions

The Chemical Equation

1. What is the purpose of coefficients in chemical equations?
2. What is meant by saying a chemical equation is balanced?
3. How many chromium atoms are indicated on the right hand side of the following balanced chemical equation? How many oxygen atoms?

$$4\ Cr\ (s) + 3\ O_2\ (g) \rightarrow 2\ Cr_2O_3\ (g)$$

4. Why are the letters (*s*), (*l*), (*g*), and (*aq*) often included in a chemical reaction?
5. Why is it important that a chemical equation be balanced?
6. Why is it important never to change the subscript when balancing a chemical equation?
7. Why are coefficients always shown ***before*** a chemical compound?
8. Which of the following chemical equations are balanced?
   (a) $Mg\ (s) + 2\ HCl\ (aq) \rightarrow MgCl_2\ (aq) + H_2\ (g)$
   (b) $3\ Al\ (s) + 3\ Br_2\ (l) \rightarrow Al_2Br_3\ (s)$
   (c) $2\ HgO\ (s) \rightarrow 2\ Hg\ (l) + O_2\ (g)$

Energy and Chemical Reactions

9. If it takes 436 kJ of energy to break a bond, how many kilojoules are released when the bond is formed?
10. Why is energy required to break apart a chemical bond?
11. What's the difference between an exothermic reaction and an endothermic one?
12. In what sense does an exothermic reaction consume energy?
13. What effect does a catalyst have on a chemical reaction?
14. What effect does a chemical reaction have on a catalyst?
15. How is ozone produced?
16. Why is stratospheric ozone so important to life on the Earth?
17. Is there any chemical difference between stratospheric ozone and the ozone found in air pollution?
18. CFCs are not the only chemical compounds that catalyze the destruction of ozone. What are some others?
19. When was the potential harm of CFCs first recognized?
20. During what time of the year is the Antarctic ozone hole most apparent?
21. How close are you to a CFC molecule right now? Explain.

Chemical Equilibrium

22. What does it mean to say that a chemical reaction is reversible?
23. What does it mean to say that the condition of chemical equilibrium is a dynamic situation?
24. When a reaction is at chemical equilibrium, what is true about the forward and reverse reactions?
25. Does chemical equilibrium mean that the number of product and reactant molecules are the same?
26. What is "equal" in a chemical equilibrium?
27. What effect does increasing pressure have on the number of collisions among molecules in a gas?

Relative Masses of Atoms and Molecules

28. What device is commonly used to measure out atoms and molecules?
29. Why don't equal masses of golf balls and Ping-Pong balls contain the same number of balls?
30. Why don't equal masses of carbon atoms and oxygen molecules contain the same number of particles?
31. Is the atomic mass unit a very large or very small quantity?
32. How does formula mass differ from atomic mass?
33. What is the mass of a sodium atom in atomic mass units?
34. What is the relationship between the atomic mass unit and the gram?

Avogadro's Number and the Mole

35. Could Avogadro's number of golf balls fit in your physical-science classroom?
36. If you had 1 mole of marbles, how many marbles would you have?
37. If you had 2 moles of pennies, how many pennies would you have?
38. Avogadro's number is the ratio of what two values?
39. How many moles of water are there in 18 g of water?
40. How many molecules of water are there in 18 g of water?

Home Projects

1. An example of an endothermic process is as near as the salt in your cupboard. Add lukewarm water to two plastic cups. Transfer the liquid back and forth between containers to ensure equal temperatures, ending up with the same amount in each. Dissolve a lot of table salt in one of the cups (stirring helps). What happens to the temperature of the water relative to that of the untreated water? (Hold the cups up to your cheeks to tell.) Heat energy can be used to break chemical bonds. What type of chemical bonds are being broken here?
2. Holding your hand over a sink, pour some water from the faucet into your cupped palm. Pour an equal amount of isopropyl rubbing alcohol into the water in your cupped

palm. Is this an exothermic or endothermic process? What's going on at the molecular level? Rinse away the excess rubbing alcohol and avoid the fumes.

Exercises

1. Is the following chemical equation balanced?

$$2\ C_4H_{10}\ (g) + 13\ O_2\ (g) \rightarrow 8\ CO_2\ (g) + 10\ H_2O\ (l)$$

2. Some chemical equations can be difficult to balance (unless your instructor provides you with a successful methodology). For a challenge, try this one:

$$\_\_\_\ Na_2SO_3 + \_\_\_\ S_8 \rightarrow \_\_\_\ Na_2S_2O_3$$

3. Balance the following chemical equations:
   (a) ___ Fe (s) + ___ S (s) → ___ Fe_2S_3 (s)
   (b) ___ P_4 (s) + ___ H_2 (g) → ___ PH_3 (g)
   (c) ___ NO (g) + ___ Cl_2 (g) → ___ NOCl (g)
4. Use the bond energies given in Table 19.1 to deduce whether the following reactions are exothermic or endothermic:
   $H_2 + Cl_2 \rightarrow 2\ HCl$
   $2\ H\text{–}C{\equiv}C\text{–}H + 3\ O_2 \rightarrow 4\ CO_2 + 2\ H_2O$
5. Why do multiple bonds generally have higher bond energies than single bonds?
6. Note in Table 19.1 how bond energies increases going from H–N to H–O to H–F. Explain this trend based upon the atomic sizes of these atoms as deduced from their positions in the periodic table.
7. Many people hear about atmospheric ozone depletion and wonder why we don't simply replace that which has been destroyed. Knowing about CFCs and about how catalysts work, explain how this would not be a lasting solution.
8. As stated in the text, there are about 25 quadrillion (25,000,000,000,000,000 or 2.5×10^{16}) CFC molecules in every liter of air you breathe. What information would you need to know if you wanted to calculate the amount of CFCs in our air as a percentage of air?
9. Is the synthesis of ozone from oxygen an example of an exothermic or endothermic reaction?
10. Do CFCs catalyze the destruction of ozone?
11. Other compounds besides CFCs result in the destruction of ozone. Why have CFCs gotten so much of the public attention?
12. Is the endothermic reaction of ammonium nitrate discussed in Section 19.2 an example of a chemical or physical change?
13. Increasing temperature favors the formation of nitrogen dioxide, NO_2, from dinitrogen tetroxide, N_2O_4. However, no matter how high the temperature gets, the N_2O_4 never completely disappears. Explain.
14. Why is the production of ammonia more efficient at high pressures?
15. The plunger on a sealed syringe containing a mixture of NO_2 and N_2O_4 at equilibrium is pulled outward. What happens to the rate of N_2O_4 formation? Why?
16. Is there equilibrium in a growing population?
17. You are given two samples of matter, 10 g each. If the number of atoms is the same in both, what must be true of the two samples?
18. Two atomic mass units equals how many grams?
19. What's the mass of a single oxygen atom, O, in atomic mass units?
20. What's the mass of a single oxygen atom, O, in grams?
21. What's the mass of a single molecule of water, H_2O, in atomic mass units?
22. What's the mass of a single molecule of water, H_2O, in grams?
23. Is it possible to have a 14-amu sample of oxygen?
24. Why are the relative masses of atoms and molecules more convenient to work with than their absolute masses?
25. What is the formula mass of ammonia, NH_3?
26. Which has more atoms: 17.04 g of ammonia, NH_3, or 72.92 g of hydrogen chloride, HCl?
27. Which has the greatest number of molecules?
   (a) 32 g of nitrogen, N_2 (b) 32 g of oxygen, O_2
   (c) 32 g of methane, CH_4 (d) 38 g of fluorine, F_2
28. Which has the greatest number of atoms?
   (a) 28 g of nitrogen, N_2 (b) 32 g of oxygen, O_2
   (c) 16 g of methane, CH_4 (d) 38 g of fluorine, F_2
29. Hydrogen and oxygen always react in a 1:8 ratio by mass to form water. Early investigators took this to mean that oxygen was 8 times as massive as hydrogen. What did these early investigators assume about water's chemical formula?
30. What are the formula masses (in amu) of water, H_2O; propene, C_3H_6; and 2-propanol, C_3H_8O?

Problems

1. How many grams of water, H_2O, and propene, C_3H_6, can be formed from the reaction of 6.0 g of 2-propanol, C_3H_8? (*Hint:* Find their formula masses.)
2. How many molecules of aspirin (formula mass 180 amu) are there in a 0.250-g sample?

2-Propanol → Propene + Water

3. Small samples of oxygen gas needed in the laboratory can be generated by any number of simple chemical reactions, such as

$$2\ KClO_3\ (s) \rightarrow 2\ KCl\ (s) + 3\ O_2\ (g)$$

What mass of oxygen (in grams) is produced when 122.55 g of $KClO_3$ (formula mass = 122.55 amu) takes part in this reaction?

CHAPTER

20

Acid, Base, and Redox Reactions

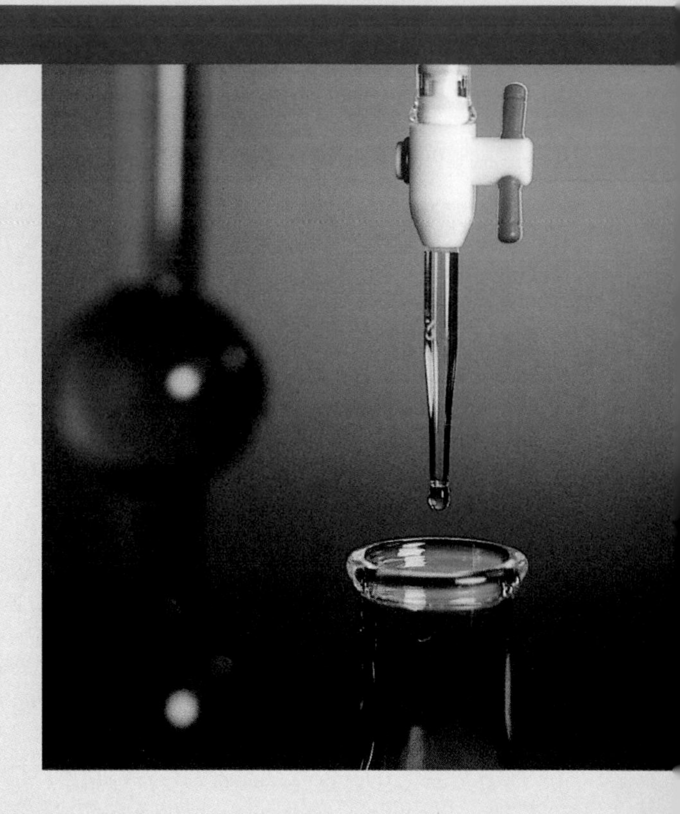

We begin this chapter with **acid-base reactions**, which involve the transfer of hydrogen ions, H^+, from one molecule to another. Knowledge of acid-base reactions is relevant to our everyday world. Many foods, such as citrus fruits, are acidic, and many drugs, such as caffeine, nicotine, and morphine, are basic (alkaline). Blood in our bodies undergoes acid-base reactions. Rain is acidified by carbon dioxide and sulfur dioxide in the atmosphere. When rain falls into the ocean, it is neutralized because ocean water is basic as a result of dissolved calcium carbonate, which is also the active ingredient in many over-the-counter antacids.

We end the chapter with **oxidation-reduction reactions** (redox), which involve the transfer of electrons. A common example of an oxidation-reduction reaction is *combustion.* Rapid combustion of fuels at a power plant provides electrical energy. Slower combustion of food in our bodies provides energy for life. Oxidation-reduction reactions produce the energy in batteries and are responsible for the images that develop on photographic film, the corrosion of metals, and the electroplating of metals.

20.1 Acids and Bases Defined

The term *acid* comes from the Latin word *acidus,* which means "sour." The sour taste of vinegar and citrus fruits is due to the presence of acids. The term *base* comes from the early observation that certain substances react with acids to form salts, which are discussed later in this section. These substances were seen as the foundation, or "base," upon which the salts were able to form. The term *base* is still used today to describe a general class of substances that react with acids. The bitterness of many foods, such as unsweetened chocolate, is due to the bases they contain. When wet, many bases have a slippery feel as they react with skin oils to produce a thin slippery solution of soap. Solutions containing bases are often called *alkaline.* As was discussed in Section 16.3, this term is derived from the Arabic word for ashes (*al-qali*), which are slippery when wet because they contain the base potassium carbonate, K_2CO_3.

Fig. 20.1
Citrus fruits contain many types of acids, including ascorbic acid, $C_8H_8O_6$, which is vitamin C.

The first person to recognize the essential nature of acids and bases was Svante Arrhenius. Before the turn of the 20th century, Arrhenius postulated that **acids** are substances that produce hydrogen ions, H^+, when dissolved in water, and **bases** are substances that produce hydroxide ions, HO^-, when dissolved in water.[1] For example, when hydrogen chloride dissolves in water, it breaks apart into hydrogen and chloride ions:

$$\mathbf{HCl}\ (g) \rightarrow \mathbf{H^+}\ (aq) + \mathbf{Cl^-}\ (aq)$$

Because hydrogen ions are produced by this process, hydrogen chloride is considered an acid. Sodium hydroxide, NaOH, on the other hand, dissolves in water to produce hydroxide ions:

$$\mathbf{NaOH}\ (s) \rightarrow \mathbf{Na+}\ (aq) + \mathbf{HO^-}\ (aq)$$

According to Arrhenius, therefore, sodium hydroxide is a base.

Fig. 20.2
These supermarket products all contain bases.

The Arrhenius definition of acids and bases is somewhat limited. There are many substances—ammonia, NH_3, for example—that are bases but do not have hydroxide ions as part of their formula. In 1923, a more general definition of acids and bases was suggested by the Danish chemist Johannes Brønsted and the English chemist Thomas Lowry. In the Brønsted-Lowry definition, an acid is a H^+ donor and a base is a H^+ acceptor.[2] Consider what happens when hydrogen chloride is mixed into water:

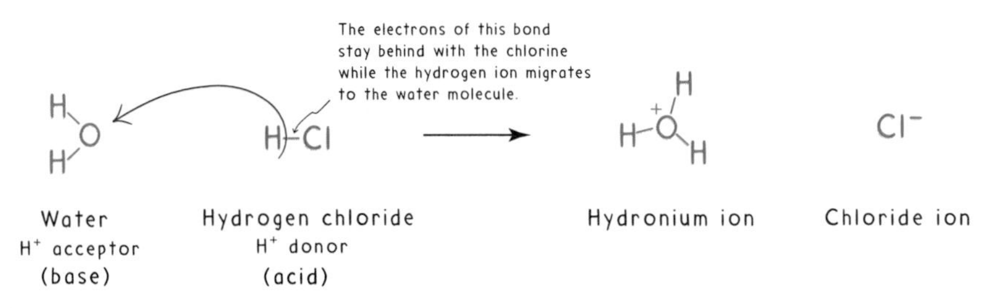

---

[1]The hydroxide ion consists of a hydrogen atom covalently bonded to an oxygen atom: H—O⁻. The single negative charge of this *polyatomic* ("many atoms") ion resides on the oxygen atom, which has an additional electron.

[2]The acronym BAAD is useful for remembering this definition: **B**ase **A**ccepts hydrogen ion, **A**cid **D**onates.

Hydrogen chloride donates a hydrogen ion to an accepting water molecule. In this case hydrogen chloride behaves as an acid and water behaves as a base. The products of this reaction are a chloride ion and a **hydronium ion**, H_3O^+, which is a water molecule with an extra hydrogen.[3]

The Brønsted-Lowry definition accounts for ammonia's behaving as a base: It *accepts* a hydrogen ion from water, which behaves as an acid. This results in formation of an ammonium ion and a hydroxide ion:

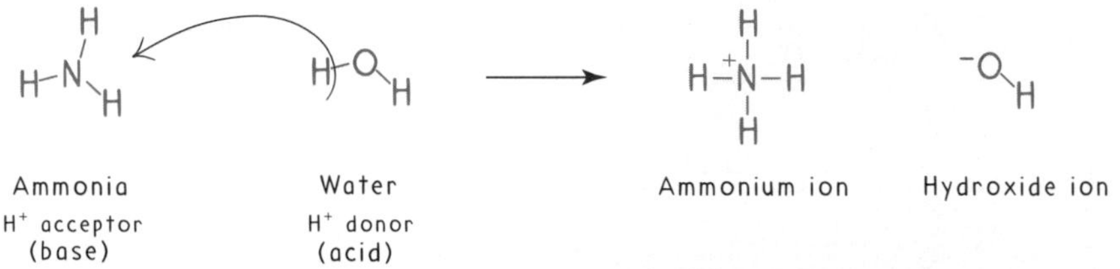

Ammonia, H^+ acceptor (base) + Water, H^+ donor (acid) → Ammonium ion + Hydroxide ion

An important aspect of the Brønsted-Lowry definition is that it recognizes acid/base as *behavior*. We say hydrogen chloride *behaves* as an acid when mixed with water, which in this reaction *behaves* as a base. Similarly, ammonia *behaves* as a base when mixed with water, which in this case *behaves* as an acid. Because acid/base is seen as behavior, there is really no contradiction when a chemical, such as water, behaves as a base in one instance but as an acid in another instance. By analogy, consider yourself. You are who you are, but your behavior changes depending upon whom you are with. Likewise, it is a chemical property of water to behave as a base (accept H^+) when mixed with hydrogen chloride and as an acid (donate H^+) when mixed with ammonia.

The products of an acid-base reaction can also behave as acids or bases. The ammonium ion, for example, may donate the hydrogen ion back to the hydroxide ion to re-form ammonia and water:

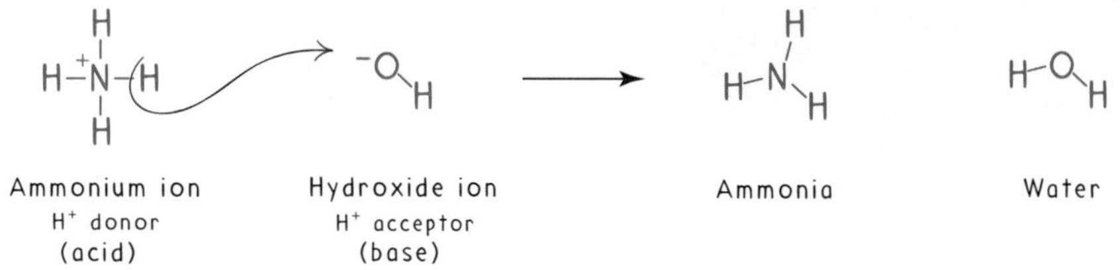

Ammonium ion, H^+ donor (acid) + Hydroxide ion, H^+ acceptor (base) → Ammonia + Water

Forward and reverse acid-base reactions eventually reach chemical equilibrium. These two reactions can therefore be represented as

NH_4^+	+	HO^-	$\rightleftharpoons$	NH_3	+	H_2O
acid		**base**		**base**		**acid**
(H^+ donor)		**(H^+ acceptor)**		**(H^+ acceptor)**		**(H^+ donor)**

[3]The single positive charge of the hydronium ion resides on the oxygen atom, which is deficient one electron.

When the equation is viewed from left to right, the ammonium ion behaves as an acid because it donates a hydrogen ion to the hydroxide ion, which therefore acts as a base. Viewed in the reverse direction, the equation says that water donates a hydrogen ion to ammonia. In this case water behaves as an acid and ammonia behaves as a base.

Questions

In the following equations, identify the acid or base behavior of each participant:

1. $H_3O^+ + Cl^- \rightleftharpoons H_2O + HCl$

   ____ ____ ____ ____

2. $NH_2^- + H_3O^+ \rightleftharpoons NH_3 + H_2O$

   ____ ____ ____ ____

Salt: The Ionic Product of an Acid and a Base

In everyday language, the word *salt* implies table salt, the common food seasoning sodium chloride. In the language of chemistry, however, **salt** is a general term meaning any ionic compound formed from the reaction of an acid and a base. The reaction between hydrogen chloride and sodium hydroxide, for example, yields the salt sodium chloride plus water:

HCl	+	NaOH	→	Na^+Cl^-	+	H_2O
hydrogen chloride		sodium hydroxide		sodium chloride (a salt)		water

Likewise, the reaction between hydrogen chloride and potassium hydroxide yields the salt potassium chloride plus water:

HCl	+	KOH	→	K^+Cl^-	+	H_2O
hydrogen chloride		potassium hydroxide		potassium chloride (a salt)		water

Fig. 20.3
Table salt substitutes contain potassium chloride in lieu of sodium chloride. Because their "salty" tastes are not quite identical, half-and-half mixtures of sodium chloride and potassium chloride are also sold.

The two preceding reactions are examples of **neutralization**, defined as any reaction in which an acid and a base combine to form a salt. During a neutralization reaction, the salt is formed from a rearrangement of atoms, as Figure 20.4 shows.

Salts are generally far less corrosive than the acids and bases from which they are formed. Hydrogen chloride, for example, is a remarkably corrosive acid, a property that makes it useful for cleaning toilet bowls and etching through metal surfaces. Simi-

---

Answers

1. The H_3O^+ transforms into a water molecule. This means the H_3O^+ loses a hydrogen ion, which is donated to the Cl^-. The H_3O^+ therefore behaves as an acid, and the Cl^- behaves as a base. In the reverse direction, we see the H_2O gaining a hydrogen ion (behaves as a base) to become H_3O^+. It gets this hydrogen ion from the HCl, which in donating behaves as an acid.
2. The NH_2^- gains a hydrogen ion to become NH_3. In accepting the hydrogen ion, it behaves as a base. It gets the hydrogen ion from the H_3O^+, which in donating behaves as an acid. In the reverse direction, NH_3 loses a hydrogen ion to become NH_2^- and thus behaves as an acid. The recipient of the hydrogen ion is the H_2O, which thus behaves as a base as it transforms into H_3O^+.

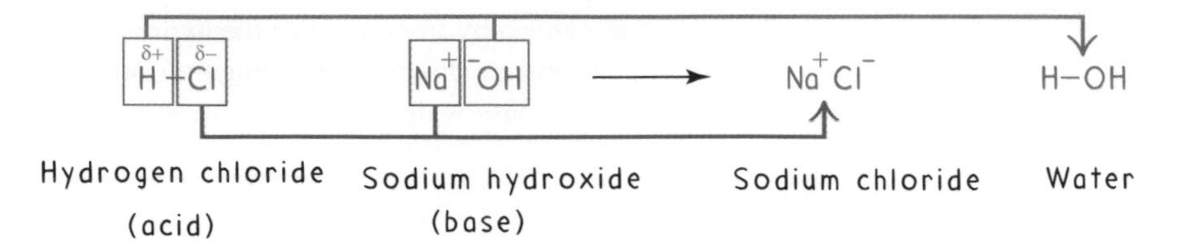

Fig. 20.4
During the neutralization reaction between hydrogen chloride and sodium hydroxide, the compounds change partners. The H^+ and HO^- ions come together to form water, and the Na^+ and Cl^- ions come together to form sodium chloride.

larly, sodium hydroxide is a very corrosive base, useful for unclogging drains stopped up with hair. Mixing hydrogen chloride and sodium hydroxide in equal portions, however, produces an aqueous solution of sodium chloride—salt water, which is somewhat corrosive (as people who live near the ocean and own a car know) but nowhere as destructive as either starting material.

As can be seen in Figure 20.4, the positive ion in any salt comes from the base in the acid-base reaction, and the negative ion comes from the acid. We thus characterize a salt as being related to the acid or base from which it was formed. Sodium chloride or potassium chloride, for instance, can both be described as salts of HCl. Alternatively, sodium chloride can be described as a salt of NaOH, and potassium chloride can be described as a salt of KOH.

Consider the drug cocaine. In the presence of hydrogen chloride, cocaine behaves as a base to form a salt called the "cocaine hydrogen chloride salt":

Cocaine (base) + H–Cl Hydrogen chloride (acid) ⟶ Cocaine hydrogen chloride salt

The salt form of cocaine is soluble in water and so is absorbed through the moist membranes of nasal passages or mouth. The nonsalt form, also known as free-base cocaine or crack cocaine, is a nonpolar material that vaporizes easily when heated. Its vapors are inhaled directly into the lungs, resulting in dangerously high concentrations of cocaine in the bloodstream. We shall return to the actions of various drugs in Chapter 21.

20.2 Acid Strength

Acids are characterized as being either strong or weak.[4] The stronger an acid, the greater its ability to donate a hydrogen ion. Hydrogen chloride is a strong acid because

[4]Bases can also be characterized as strong or weak, but for ease of discussion we focus primarily on acids.

it very willingly donates its hydrogen ion to water, forming chloride and hydronium ions. Because HCl is such a strong acid, nearly all of it is converted to these ions. At equilibrium, only a very few non-ionized hydrogen chloride molecules remain:

$$\mathrm{HCl} + \mathrm{H_2O} \rightleftharpoons \mathrm{Cl^-} + \mathrm{H_3O^+}$$

Hydrogen chloride (strong acid) | Water | Chloride ion | Hydronium ion

An example of a weak acid is acetic acid, CH_3CO_2H, which has much less tendency to donate a hydrogen ion to water. When dissolved in water, only a small portion of acetic acid is converted to ions. This ionization occurs as the polar O—H bond is broken (the C—H bonds are unaffected by the water because they are nonpolar). The majority of acetic acid molecules remain intact in their non-ionized form:

$$\mathrm{CH_3COOH} + \mathrm{H_2O} \rightleftharpoons \mathrm{CH_3COO^-} + \mathrm{H_3O^+}$$

Acetic acid (weak acid) | Water | Acetate ion | Hydronium ion

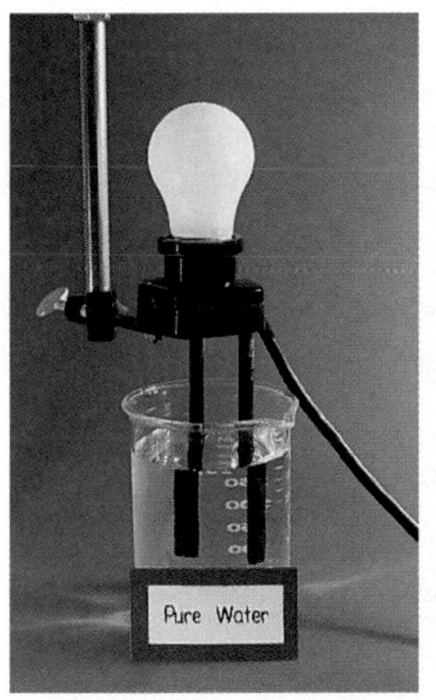

Fig. 20.6
Left circuit: Because HCl is a strong acid, nearly all of its molecules break apart in water, giving a high concentration of ions, which are able to conduct an electric current that lights the bulb. Right circuit: Acetic acid, CH_3COOH, is a weak acid, and in water only a small portion of its molecules break up into ions. Because fewer ions are generated, only a weak current exists and the bulb is dimmer.

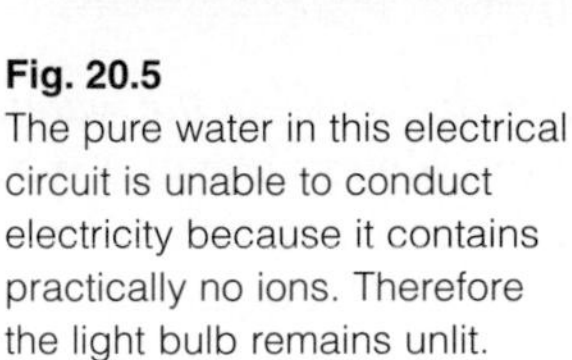

Fig. 20.5
The pure water in this electrical circuit is unable to conduct electricity because it contains practically no ions. Therefore the light bulb remains unlit.

Whether an acid behaves as a strong or weak acid can be determined by measuring the solution's ability to conduct an electric current. Electric charge does not flow very well through pure water because in pure water there are practically no ions to conduct the electricity (Figure 20.5). When a strong acid is dissolved in water, many ions are generated, which allows for a large electric current, indicated by the brightly lit bulb in the left circuit in Figure 20.6. A weak acid dissolved in water releases a relatively small number of ions, which allows for a smaller electric current and the dimmer bulb in the right circuit in Figure 20.6.

Water as an Acid and a Base

A substance that behaves as either an acid or a base is said to be an **amphoteric substance**. Water is a good example because of its ability to accept and donate hydrogen ions. Through its amphoteric nature, water even has the ability to react with itself. In behaving as an acid, a water molecule donates a hydrogen ion to a neighboring water molecule, which in accepting the hydrogen ion is behaving as a base. This reaction produces a hydroxide ion and a hydronium ion:

Water (very weak acid) + Water ⇌ Hydroxide ion + Hydronium ion

Water, however, is a very, very weak acid. This means that in pure water there are very, very few ions, as evidenced by the unlit light bulb in Figure 20.5.

20.3 Acidic, Basic, or Neutral

Associated with pure water and with any solution that contains water is an interesting rule pertaining to the relative concentrations of hydronium and hydroxide ions. At equilibrium, the concentration of hydronium ions in any aqueous solution multiplied by the concentration of hydroxide ions always equals a constant K_w that is a very small number:

$$\textbf{Concentration } H_3O^+ \times \textbf{concentration } HO^- = K_w = 0.00000000000001$$

For concentration we use units of molarity (Section 18.2), which is implied by abbreviating the preceding equation using brackets:

$$[H_3O^+] \times [HO^-] = K_w = 0.00000000000001$$

The brackets tell us that this equation is read "the molarity of H_3O^+ times the molarity of HO^- equals K_w." Omitting the multiplication sign and writing in scientific notation, we have

$$[H_3O^+][HO^-] = K_w = 1.0 \times 10^{-14}$$

The constant value of K_w is significant. *No matter what is dissolved in the water,* the product of the hydronium and hydroxide ion concentrations always equals 1.0×10^{-14}. This means that if the concentration of H_3O^+ goes up, the concentration of HO^- must come down—so that the product of the two remains 1.0×10^{-14}. For example, if HCl gas is dissolved in water, the hydronium ion concentration increases as the HCl breaks up into H^+ and Cl^-, and the water molecules accept the H^+ to become H_3O^+. The hydroxide ion concentration must therefore decrease so that the product $[H_3O^+][HO^-]$ stays constant. The hydroxide ion concentration goes down because hydroxide ions are neutralized by the additional hydronium ions (Figure 20.7a). In a similar manner, adding a base to water increases the hydroxide ion concentration. The response is a decrease in the hydronium ion concentration as hydronium ions become neutralized by the added hydroxide ions (Figure 20.7b).

Fig. 20.7
(a) Neutral water contains as many hydronium ions as hydroxide ions. (b) Hydronium ions produced from the addition of HCl neutralize hydroxide ions in water, thereby decreasing the hydroxide ion concentration. (c) Added hydroxide ions neutralize water's hydronium ions, thereby decreasing the hydronium ion concentration.

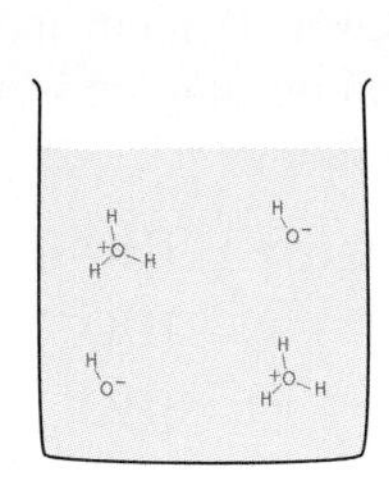

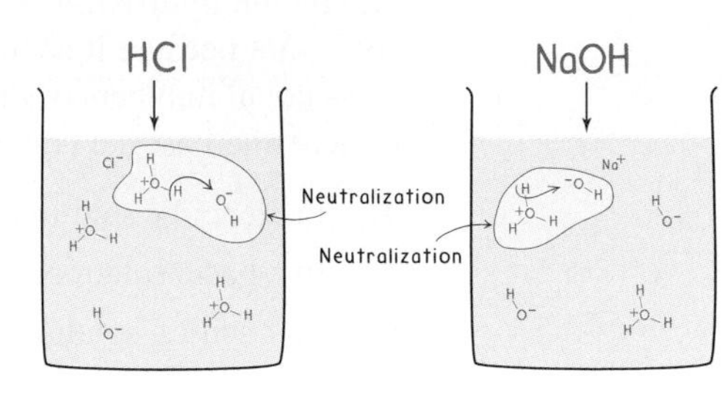

(a) Neutral water
Equal number of H_3O^+ and ^-OH

(b) Acidic water
More H_3O^+ than ^-OH after HCl addition

(c) Basic water
More ^-OH than H_3O^+ after NaOH addition

Example

What is the concentration of hydronium ions in a solution if the concentration of hydroxide ions is 1.0×10^{-9} M?

Questions

1. What is the concentration of hydronium ions in a solution if the concentration of hydroxide ions is 1.0×10^{-5} M?
2. In pure water the hydroxide ion concentration is 1.0×10^-7 M. What is the hydronium ion concentration in pure water?

A solution can be described as acidic, basic, or neutral. An **acidic solution** is one in which the hydronium ion concentration is greater than the hydroxide ion concentration. An acidic solution is made by adding an acid, such as hydrogen chloride, to water. The effect of this is to increase the concentration of hydronium ions, which necessarily decreases the concentration of hydroxide ions. A **basic solution** is one in which the hydroxide ion concentration is greater than the hydronium ion concentration. A basic so-

---

Answer

The hydronium ion concentration multiplied by the hydroxide ion concentration always equals 1.0×10^{-14}. Because the units of H_3O^+ and HO^- concentration are both given as M (for molarity), the units of K_w are M^2:

$$[H_3O^+][1.0 \times 10^{-9}\ M] = 1.0 \times 10^{-14}\ M^2$$

We solve for the hydronium ion concentration by dividing each side by 1.0×10^{-9} M:

$$[H_3O^+] = \frac{1.0 \times 10^{-14}\ M^2}{1.0 \times x10^{-9}\ M} = 1.0 \times 10^{-5}\ M$$

Answers

1. 1.0×10^{-9} M.
2. 1.0×10^{-7} M.

lution is made by adding a base, such as sodium hydroxide, to water. This increases the concentration of hydroxide ions, which necessarily decreases the concentration of hydronium ions. A **neutral solution** is one in which the hydronium ion concentration equals the hydroxide ion concentration. Pure water is an example of a neutral solution—not because it contains so few hydronium and hydroxide ions but because it contains equal numbers of them. A neutral solution is also obtained when equal quantities of acid and base are combined. In short,

1. In an acidic solution, $[H_3O^+] > [HO^-]$.
2. In a basic solution, $[H_3O^+] < [HO^-]$.
3. In a neutral solution, $[H_3O^+] = [HO^-]$.

■ **Question**

How does adding ammonia, NH_3, to water make a basic solution when there are no hydroxide ions in the formula for ammonia?

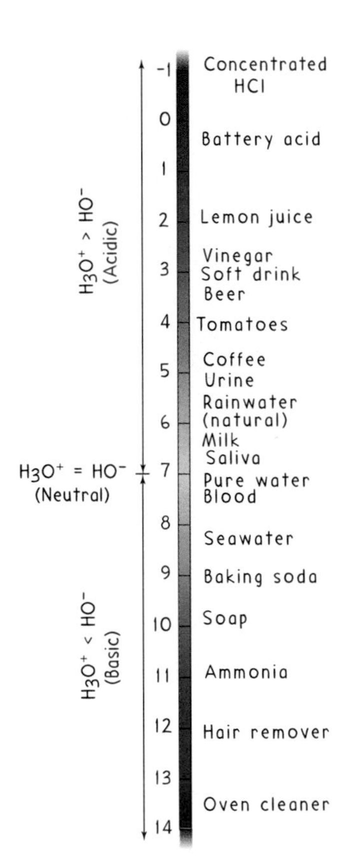

Fig. 20.8
Some typical pH values.

20.4 The pH Scale

The *pH scale* is a method for expressing the acidity of a solution. Mathematically, **pH** is equal to the negative logarithm of the hydronium ion concentration:[5]

$$\mathbf{pH = -\log[H_3O^+]}$$

When a solution is neutral, the hydronium ion and hydroxide ion concentrations both equal 1.0×10^{-7} M. To find the pH of such a solution, we first take the log of this value, which equals -7 (see box). We then change the sign, which gives a positive value. Hence, in a neutral solution, where the hydronium ion concentration equals 1.0×10^{-7} M, the pH = 7. Acidic solutions have pH values that are less than 7; the more strongly acidic they are, the lower the pH. Basic solutions have pH values greater than 7; the more strongly basic they are, the higher the pH.

■ **Answer**

Ammonia increases the hydroxide ion concentration by reacting with water:

$$NH_3 + H_2O \rightarrow NH_4^+ + HO^-$$

This increase in hydroxide ion concentration has the effect of lowering the hydronium ion concentration. With the hydroxide ion concentration now greater than the hydronium ion concentration, the solution is basic.

[5] The "p" in pH is a mathematical notation indicating the negative logarithm. The "H" indicates the hydrogen ion, H^+, which can be used interchangeably with the hydronium ion, H_3O^+. In some references, you'll see pH = $-\log[H^+]$. The hydrogen and hydronium ions can be used interchangeably because they are both indicators of the number of extra protons found in the aqueous solution. Recall that H^+ is simply a free proton and H_3O^+ is the result of that free proton's bonding to a water molecule.

Link to Mathematics—Logarithms

The logarithm of a number can be conveniently found on any scientific calculator. Briefly, to find the logarithm of a number, often abbreviated "log," you are finding the exponent to which the number 10 must be raised to equal this number. The log of 100, for example, is 2 because 10 raised to the second power, 10^2, equals 100. Similarly, the log of 1000 is 3 because 10 raised to the third power, 10^3, equals 1000.

Quiz: What is the log of 10,000?

Answer: The number 10,000 is 10^4, the log of which is 4.

Any positive number, including a very small one, has a log value. The log of 0.0001, which is 10^{-4} in scientific notation, for example, is -4 (the power to which 10 is raised to equal this number).

20.5 Acid Rain and Basic Oceans

Rain water is naturally acidic. One source of this acidity is carbon dioxide, CO_2, the same gas that gives fizz to soda drinks. There are some 670 billion tons of CO_2 in the atmosphere, most of it from such natural sources as volcanoes and decaying organic matter but a growing amount from human activities. Water in the atmosphere reacts with carbon dioxide to form *carbonic acid:*

CO_2	+	H_2O	$\rightleftharpoons$	H_2CO_3
carbon dioxide		**water**		**carbonic acid**

Carbonic acid, as its name implies, behaves as an acid and lowers the pH of water. The concentration of CO_2 in the atmosphere brings the pH of rain water to about 5.6—noticeably below the neutral pH value of 7. Because of local fluctuations, the normal pH of rain water varies between 5 and 7.

By convention, *acid rain* is rain having a pH less than 5. It is created when airborne pollutants, such as sulfur dioxide, SO_2, are absorbed by atmospheric moisture. Sulfur dioxide is readily converted to sulfur trioxide, which reacts with water to form *sulfuric acid:*

$2\ SO_2$	+	O_2	$\rightleftharpoons$	$2\ SO_3$
sulfur dioxide		**oxygen**		**sulfur trioxide**
SO_3	+	H_2O	$\rightleftharpoons$	H_2SO_4
sulfur trioxide		**water**		**sulfuric acid**

Each year about 20 million tons of SO_2 is released into the atmosphere by the combustion of sulfur-containing coal and oil. Sulfuric acid is a much stronger acid than carbonic acid. Rain laced with sulfuric acid eventually corrodes metals, paints, and other exposed substances. Each year the damages cost billions of dollars. Also, many rivers and lakes receiving acid rain become less capable of sustaining life. Much vegetation that receives acid rain doesn't survive. This is particularly evident in regions around heavy industry.

The environmental impact of acid rain depends on the local geology. In certain regions, such as the midwestern United States, the ground contains significant quantities of the alkaline compound calcium carbonate, which was deposited when these lands were submerged under oceans many years ago. Acid rain pouring into these regions is often neutralized by the calcium carbonate before any damage is incurred. The grounds of the northeastern United States and many other regions, however, contain very little calcium carbonate and are composed primarily of chemically less reactive materials, such as granite. In these regions, the effect of acid rain on lakes and rivers accumulates. One demonstrated solution is to raise the pH of these acidified water systems by adding calcium carbonate—a process known as *liming*. The cost of transporting the calcium carbonate coupled with the need to monitor treated water systems closely limits liming to only a small fraction of the vast number of water systems already affected. Furthermore, as acid rain continues to pour into these regions, the need to lime also continues. Ultimately, the effects of adding neutralizing agents to the water systems may eventually prove to be just as damaging as the acid rain.

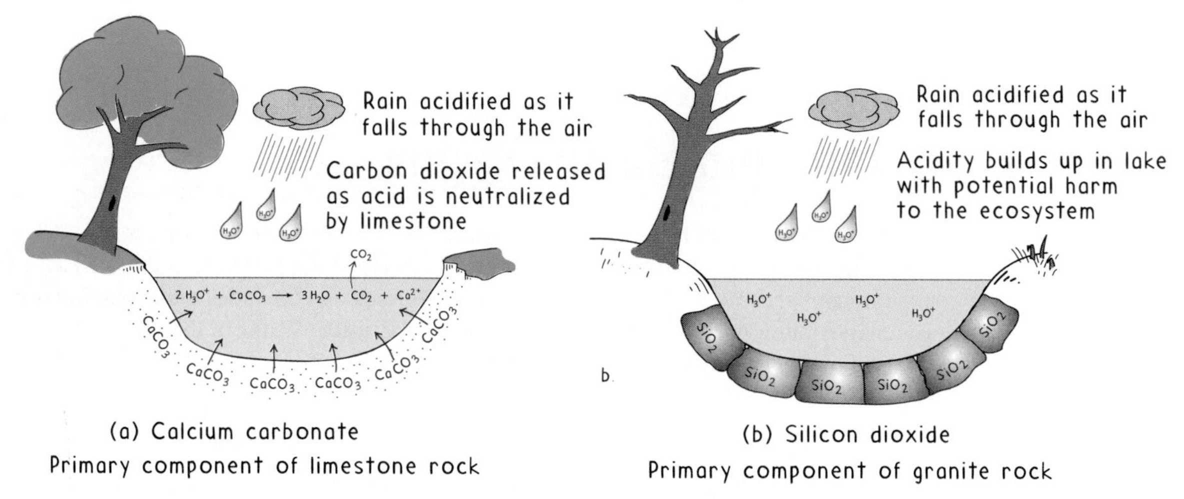

Fig. 20.9
(a) The damaging effects of acid rain do not appear in bodies of fresh water lined with calcium carbonate, which acts to neutralize any acidity. (b) Lakes and rivers lined with inert materials are not so protected.

A longer-term solution to acid rain is to prevent most of the generated sulfur dioxide and other pollutants from entering the atmosphere in the first place. Toward this end, smokestacks have been designed or retrofitted to minimize the quantities of pollutants released. Though costly, the positive effects of these adjustments have been demonstrated. An ultimate long-term solution, however, would be a shift from fossil fuels to alternative, clean-air energy sources such as nuclear or solar energy.

It should come as no surprise that the amount of carbon dioxide put into the atmosphere by human activities is growing. What is surprising, however, is that studies indicate that atmospheric CO_2 concentrations are not increasing proportionately. A likely explanation has to do with the oceans. When atmospheric CO_2 dissolves in any body of water—a raindrop, a lake, or the ocean—it reacts to form carbonic acid. In fresh water, this carbonic acid transforms through the reverse reaction back into water and carbon dioxide, and the carbon dioxide is released back into the atmosphere. Carbonic acid in the ocean, however, is quickly neutralized by dissolved alkaline substances such as calcium carbonate (the ocean is alkaline, pH around 8.2). The products of this neutralization eventually end up on the ocean floor as insoluble precipitates (Figure 20.10). Thus carbonic acid neutralization in the ocean prevents CO_2 from being released back into the atmosphere. The ocean is therefore a genuine carbon dioxide sink because most of the CO_2 that goes in doesn't come out. So, pushing more CO_2 into our atmosphere means pushing more of it into our vast oceans. This is another of the many ways our oceans moderate our environment.

Nevertheless, concentrations of atmospheric CO_2 *are* increasing, even though not as much as would happen without ocean absorption. Carbon dioxide is being produced

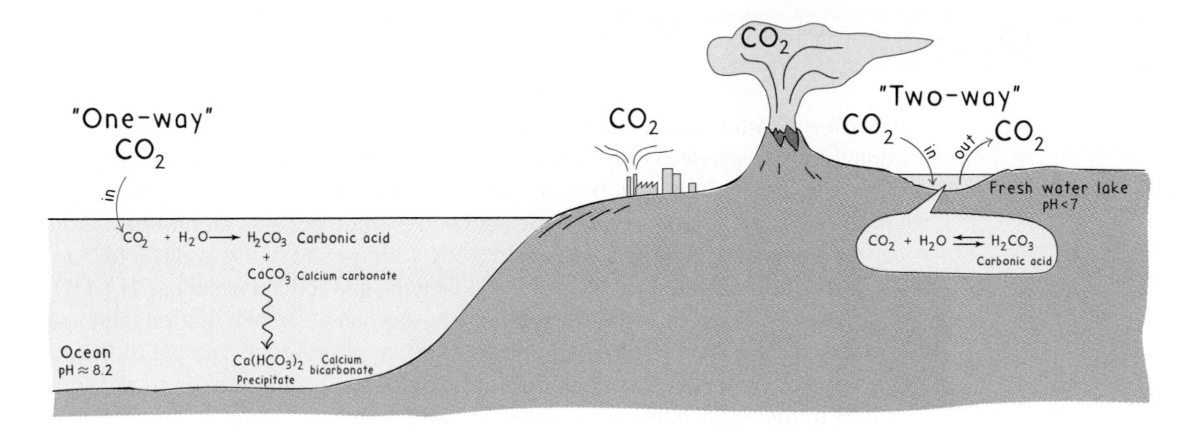

Fig. 20.10
Carbon dioxide forms carbonic acid upon entering any body of water. In fresh water, this reaction is reversible and the carbon dioxide is released back into the atmosphere. In the alkaline ocean, however, the carbonic acid is neutralized to compounds such as calcium bicarbonate, $Ca(HCO_3)_2$, which precipitate to the ocean floor. Most of the atmospheric carbon dioxide that enters our oceans stays there.

faster than the ocean can absorb it, and this may alter the Earth's environment. Carbon dioxide is a *greenhouse gas,* which means it helps keep the surface of the Earth warm by preventing infrared heat from escaping into outer space. Without greenhouse gases in the atmosphere, the Earth's surface would average a frigid −18°C. With increasing concentrations of CO_2 in the atmosphere, we can expect higher average temperatures, which may significantly alter global weather patterns as well as raise the average sea level due to partial melting of the vast polar ice caps and thermal expansion of seawater as it warms. We shall explore global warming in greater detail in Chapter 26.

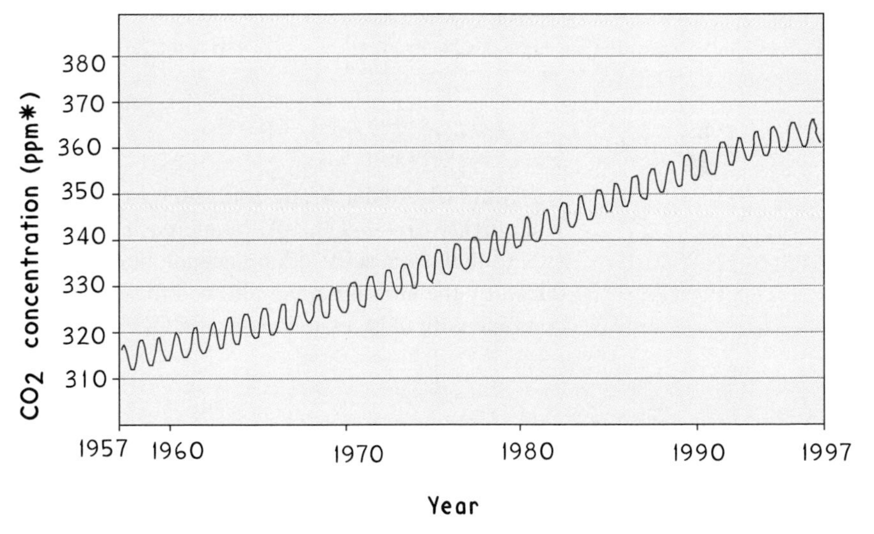

Fig. 20.11
Researchers at the Mauna Loa Weather Observatory in Hawaii have recorded increasing concentrations of atmospheric carbon dioxide since they began collecting data in the 1950s. The oscillations of this graph reflect seasonal changes in the levels of CO_2.

So we find the pH of rain depends, in great part, on the concentration of atmospheric CO_2, which depends on the pH of the oceans. These equilibria are in equilibrium with global temperatures, which naturally connect to the countless equilibria of all living systems on Earth. How true it is—all the parts are intricately connected down to the level of atoms and molecules!

20.6 Buffers

A **buffer solution** is any solution that resists changes in pH. Buffer solutions work by containing at least two components. One component serves to neutralize any incoming base, while the second component serves to neutralize any incoming acid. As an example, consider a mixture of acetic acid and sodium acetate. Good buffer solutions can be prepared by mixing a weak acid along with the salt of this weak acid. An example would be a mixture of acetic acid, CH_3COOH, and sodium acetate, $CH_3COO^-Na^+$. To see how this particular buffer solution resists changes in pH, first imagine adding a strong base, such as sodium hydroxide, NaOH, to plain water. The pH of the solution quickly *increases* because the concentration of hydroxide ions is increasing. Add NaOH to the acetic acid/sodium acetate buffer solution, however, and the HO^- ions produced by the NaOH are never allowed to stay in solution to raise the pH because they combine with H^+ ions from the acetic acid to form water:

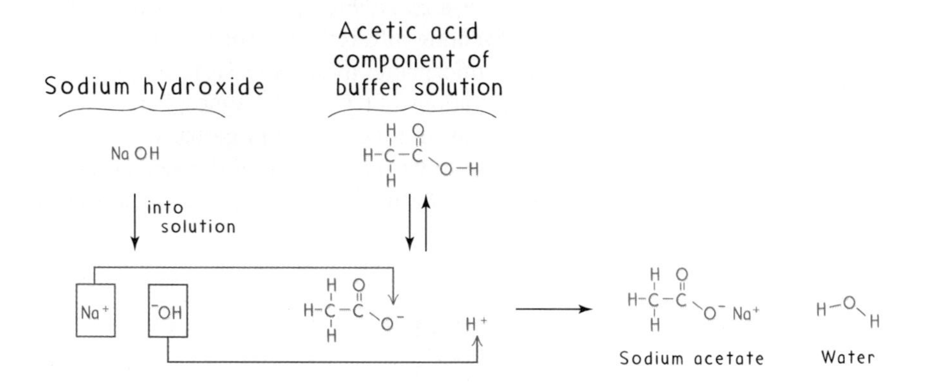

Similarly, add a strong acid, such as hydrogen chloride, HCl, to plain water and you quickly *decrease* the pH by increasing the concentration of hydronium ions. Add HCl to the acetic acid/sodium acetate buffer solution, however, and the H^+ ions produced by the HCl are never allowed to stay in solution to lower the pH because they combine with the acetate ions (CH_3COO^-) of sodium acetate to form acetic acid:

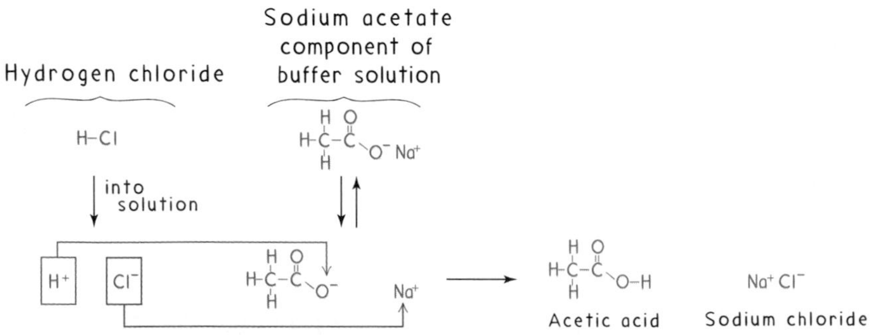

Sodium acetate
component of
buffer solution
Hydrogen chloride
H-Cl
into
solution
H^+ Cl^- Na^+
Acetic acid
Na^+Cl^-
Sodium chloride

So in a buffer solution, strong bases and acids are neutralized by the various components. This does not mean that the pH remains unchanged, however. When NaOH is added to this buffer system, sodium acetate is produced. Because sodium acetate behaves as a weak base (accepts hydrogen ions, but not very well), there is a slight increase in the pH. When HCl is added, acetic acid is produced. Because acetic acid behaves as a weak acid, there is a slight decrease in pH. Buffer solutions therefore resist only *large* changes in pH when a strong acid or base is added.

Question

Why must a buffer solution contain two dissolved components?

There are many different types of buffer systems useful for maintaining particular pH values. The acetic acid/sodium acetate system is good for maintaining a pH of around 4.8. Buffer solutions containing equal mixtures of a weak base and a salt of that weak base may be used to maintain alkaline pH values. For example, a buffer solution of ammonia, NH_3 (a weak base), and ammonium chloride, $NH_4{}^+Cl^-$ (the hydrogen chloride salt of ammonia), is useful for maintaining a pH of about 9.3.

Blood consists of several buffer systems that work together to maintain a narrow pH range between 7.35 and 7.45. A pH value above or below these levels can be lethal, primarily because cellular proteins become *denatured,* which is what happens when vinegar is added to milk or a raw egg hits the frying pan.

The major buffer system of the blood involves carbonic acid and its salt sodium bicarbonate (Figure 20.12). An advantage of this particular buffer system is that the levels of the buffering components can be modified by altering one's breathing pattern. This ability to adjust concentrations helps the body control blood pH. Recall that carbonic acid is produced when carbon dioxide reacts with water (Section 20.5). Holding one's breath or slowing down the breathing rate allows a build up of CO_2 in the bloodstream because it continues to enter the bloodstream from the many cells in which it is produced.[6] The result is an increase in carbonic acid levels and a *decrease* in pH. Hyperventilating, on the other hand, which means greatly increasing one's breathing rate, forces CO_2 out of the bloodstream. This causes a decrease in carbonic acid levels and hence an *increase* in blood pH (Figure 20.13).

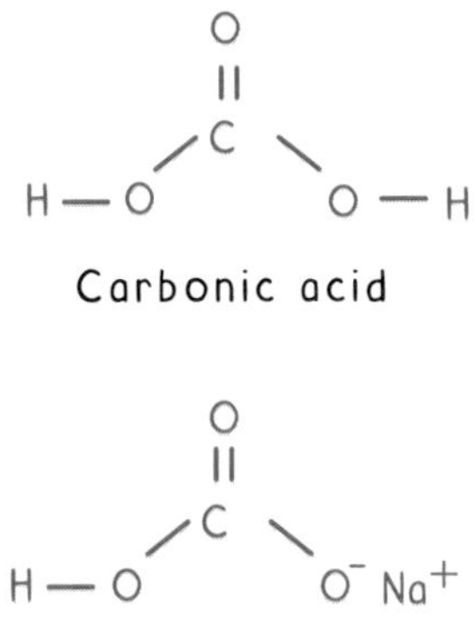

Carbonic acid

Sodium bicarbonate

Fig. 20.12
Carbonic acid, H_2CO_3, and sodium bicarbonate, $HCO_3{}^-Na^+$.

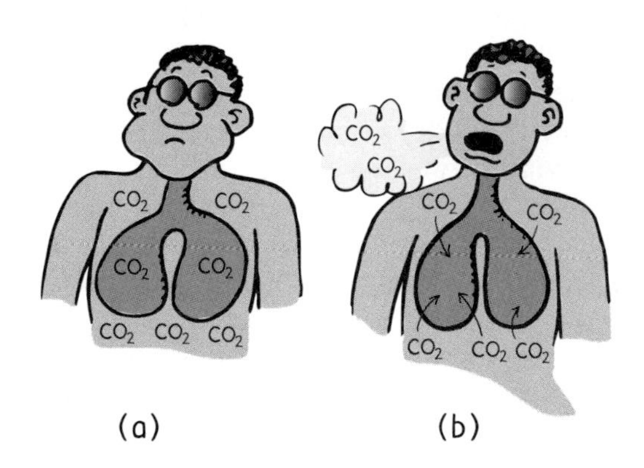

Fig. 20.13
(a) Hold your breath and CO_2 builds up in your blood stream. The increase in CO_2 increases the amount of carbonic acid in the blood, which has the effect of lowering your blood pH. (b) Hyperventilate and the amount of CO_2 in your bloodstream decreases. This decrease in CO_2 decreases the amount of carbonic acid in the blood, which has the effect of increasing your blood pH.

Answer
One component neutralizes any incoming acid, and the second neutralizes any incoming base.

[6]Carbon dioxide is a by-product of several chemical reactions that take place in all the cells of the body.

To protect the brain from high pH values, cerebral blood vessels constrict when the blood pH begins to rise. This reduces the flow of high-pH blood to the brain but also causes the lightheadedness commonly associated with hyperventilation. Sometimes an individual going through a traumatic experience cannot stop hyperventilating. In such a circumstance, breathing into a paper bag or cupped hands is a useful way to avoid passing out. By this method, carbon dioxide initially exhaled is reinhaled to help maintain adequate levels of carbonic acid in the blood. As blood pH returns to normal, blood flow to the brain is restored and the sense of lightheadedness disappears.

20.7 Oxidation-Reduction

Fig. 20.14
In the exothermic formation of sodium chloride, sodium metal is oxidized by chlorine gas and chlorine gas is reduced by sodium metal.

Whereas an acid-base reaction involves the transfer of a hydrogen ion from one reactant to another, an oxidation-reduction reaction involves the transfer of electrons. **Oxidation** is the process whereby a reactant loses one or more electrons. **Reduction** is the opposite process whereby a reactant gains one or more electrons. Oxidation and reduction are complementary processes. They always occur together; you cannot have one without the other. Electrons that are lost by one chemical don't simply disappear; they are gained by another.

Different atoms have different oxidation and reduction tendencies. Some lose electrons more readily, while others gain electrons more readily. The tendency to do one or the other is a function of how strongly the atomic nucleus is able to hold onto electrons. The greater the effective nuclear charge (Sections 16.4 and 16.5), the greater an atom's tendency to *gain* electrons. The weaker the effective nuclear charge, the greater the atom's tendency to *lose* electrons. Because atoms to the upper right of the periodic table have the strongest effective nuclear charges (noble gases excluded), these atoms have the greatest tendency to gain electrons. Those atoms to the lower left of the periodic table, having the weakest effective nuclear charges, have the greatest tendency to lose electrons.

When an atom loses an electron, we say it has been *oxidized*. When an atom gains an electron, we say it has been *reduced*.[7] We can look at a material composed of atoms that tend to oxidize and see that by losing electrons this material causes other materials to pick up the electrons and become reduced. Because of this tendency, a material that loses its electrons (becomes oxidized) is referred to as a *reducing agent*. Similarly, a material that accepts electrons (becomes reduced) is referred to as an *oxidizing agent* (Figure 20.15).

An example of an oxidation-reduction reaction is that between elemental sodium and chlorine to form sodium chloride:

$$2\,Na\,(s) + Cl_2\,(g) \rightleftharpoons 2\,Na^+Cl^-\,(s)$$

Figure 20.15
Reduction and oxidation are complementary processes.

[7]**L**oss of **e**lectrons is **o**xidation, and **g**ain of **e**lectrons is **r**eduction. As a mnemonic: **Leo** the lion with laryngitis went "**ger**."

Note that the sodium atom changes from an electrically neutral state to a positively charged ion. Each sodium atom loses an electron and is therefore oxidized:

$$Na \rightarrow Na^{+} + e^{-} \quad \textbf{Oxidation}$$

The chlorine atom changes from an electrically neutral state to negatively charged ions. Each chlorine atom gains an electron and is therefore reduced:

$$Cl + e^{-} \rightarrow Cl^{-} \quad \textbf{Reduction}$$

When sodium and chlorine are combined, the net result is that sodium atoms transfer electrons to chlorine atoms. Because sodium causes the reduction of chlorine, we refer to the sodium as a reducing agent. Because the chlorine causes the oxidation of sodium, the chlorine is referred to as an oxidizing agent.

Questions

True or false:

1. Reducing agents are oxidized in an oxidation-reduction reaction.
2. Oxidizing agents are reduced in an oxidation-reduction reaction.

Link to Photography

Lay some wax paper on the back of an open camera. Hold the shutter open using the B-stop, then focus. Voila! you have an image. But close the shutter and the image is gone. This is the same image that forms on the photographic film inside a camera. The photographic film differs, however, in that it is able to retain the image after the shutter has closed. How does it do that? The answer is with oxidation-reduction chemistry.

A camera can be used to focus an image on some wax paper as well as it does on photographic film.

Unexposed black-and-white photographic film is a strip of plastic typically coated with silver bromide, $Ag^{+}Br^{-}$, an unusual salt that becomes susceptible to reduction after being exposed to light. The image of a picture is focused on silver bromide particles. Because lighter regions of the image activate more silver bromide particles than darker regions do, the image is encoded. The film is then "developed" by immersing it in a solution of reducing agent, such as hydroquinone. This, of course, must be done in the dark so as not to activate additional silver bromide. Red light, however, can be used because the low frequency of red light does not activate the silver bromide. Hydroquinone loses electrons to silver ions according to the equation

$C_6H_4(OH)_2$	+	$2\,Ag^{+}Br^{-}$	$\rightleftharpoons$	$C_6H_4O_2$	+	$2\,Ag$	+	$2\,H^{+}Br^{-}$
hydroquinone hydrogen		**silver bromide**		**quinone**				**metallic silver bromide**

box continues

Answers

Both statements are true, in accord with one of the great conservation principles in science—the *conservation of charge.* The electrical charge lost by one substance is gained by another substance. Charge is simply transferred in these reactions.

Link to Photography (continued)

Note that two positive silver ions reduce to a neutral state while two hydrogens from the hydroquinone, subsequently bonded to bromide ions, oxidize to a positive state. The opaque metallic silver adheres to the film. Unreduced silver bromide is then washed away by a "hypo"—a solution of sodium thiosulfate, $Na_2S_2O_3$. Hence we have a *negative* , which is dark in regions of brightness and light in regions of darkness. You can see the high reflection of the metallic silver in the dark regions of the negative by looking at the film while holding it at an oblique angle.

(a) A photographic negative of Ian and Reece Suchocki. (b) The positive print.

Project light through a negative, and the image cast is the original *positive.* This positive image can be captured on film or on specially coated paper by repeating the same procedure that gave you the negative.

Photography has come a long way since it was invented in the mid-1800s. Color photographic film is coated with a variety of chemicals that respond to light of different frequencies. There are more oxidation-reduction reactions involved in the developing of a color photograph, but the basic principle is the same—the selective reduction of only those chemicals exposed to light.

Question: Does silver bromide, AgBr, behave as a reducing agent or an oxidizing agent?

Answer: Because it causes the hydrogen of the hydroquinone to become oxidized, AgBr behaves as an oxidizing agent. It is reduced to metallic silver, Ag.

20.8 Electrochemistry

Electrochemistry concerns the relationship between electrical energy and chemical change. It typically involves either the production of an electric current via an oxidation-reduction reaction or the use of an electric current to produce a chemical change.

To understand how an oxidation-reduction reaction can generate an electric current, consider what happens when an oxidizing agent is placed in direct contact with a reducing agent: Electrons flow from the reducing agent to the oxidizing agent. This flow of electrons is an electric current that can be harnessed.

Iron, Fe, is a better reducing agent than the copper ion, Cu^{2+}. So when a piece of iron metal and some copper ions are placed in contact with each other, electrons tend to flow from the iron to the copper ions. The result is oxidation of the iron and reduction of the copper ions (Figure 20.16).

The iron and copper ions need not be in physical contact with each other for electrons to flow between them. Instead, they can be separated by a conducting wire through which the electrons may flow (Figure 20.17). Then if electrons were to flow from the iron to the copper ions, the resulting electric current in the wire could be attached to some useful device, such as a light bulb.

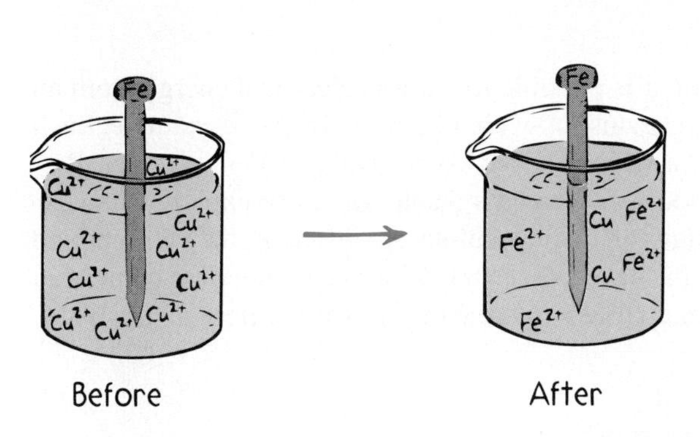

Fig. 20.16
An iron nail, Fe, placed in a solution of copper ions, Cu^{2+}, oxidizes to Fe^{2+} ions, which dissolve in the water. Copper ions, meanwhile, are reduced to metallic copper, Cu, which coats the nail. For simplicity, the negatively charged ions that balance these positively charged ions in solution are not depicted.

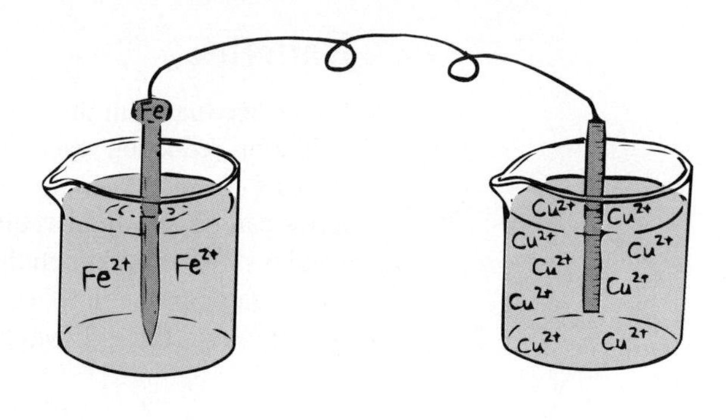

Fig. 20.17
An iron nail, Fe, is placed in water and connected to a solution of copper ions, Cu^{2+}, by way of a conducting wire. Nothing happens because this arrangement does not make a complete electric circuit.

But alas, an electric current is not sustained by this arrangement. Such a current would result in an impossible buildup of electric charge in each container. The container of copper ions would build up a negative charge as electrons accumulated on this side. This negative charge would prevent any more electrons from flowing over to this side because electrons are repelled by negative charges.

The answer to this problem is to complete the electric circuit by allowing ions to migrate between the solutions. This is accomplished with a *salt bridge,* a U-shaped tube filled with a salt, such as sodium nitrate, $NaNO_3$, and enclosed by semiporous plugs. A salt bridge allows ions to pass freely between the two solutions, and this free passage of ions permits the flow of electrons through the conducting wire (Figure 20.18).

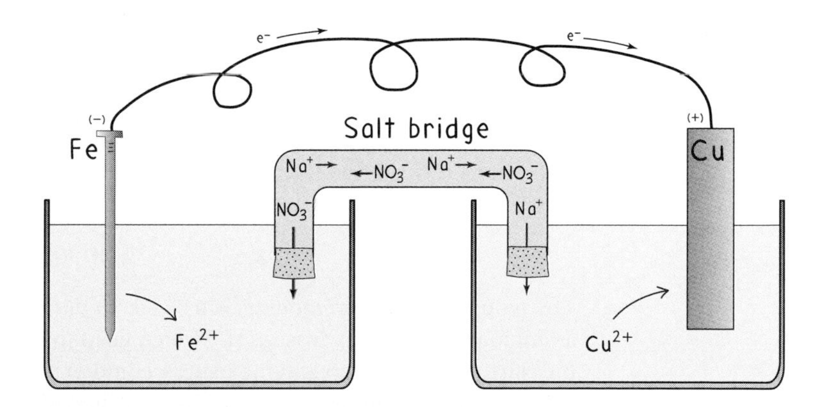

Fig. 20.18
The salt bridge completes the circuit. In the left container, iron atoms in the nail release electrons that pass through the wire to the Cu^{2+} ions in the right container. As the iron loses electrons, Fe^{2+} ions form and dissolve in the solution around the nail. Oppositely charged NO_3- ions from the salt bridge enter the solution around the nail to balance the charges of the Fe^{2+} ions. Meanwhile, Na^+ ions from the salt bridge (not shown) enter the right container to replace the Cu^{2+} that are lost as they pick up electrons to become metallic copper.

Batteries

So we see that with the proper setup it is possible to harness electrical energy from an oxidation-reduction reaction. The apparatus shown in Figure 20.18 is one example. Such devices are called *voltaic cells,* and a self-contained voltaic cell is called a *battery.* Batteries can be classified as either disposable or rechargeable. Here, we explore some examples of each. Although their designs and compositions are different, they all function by the same principle: Two materials that oxidize and reduce each other are connected by a medium through which ions may travel to balance an external flow of electrons.

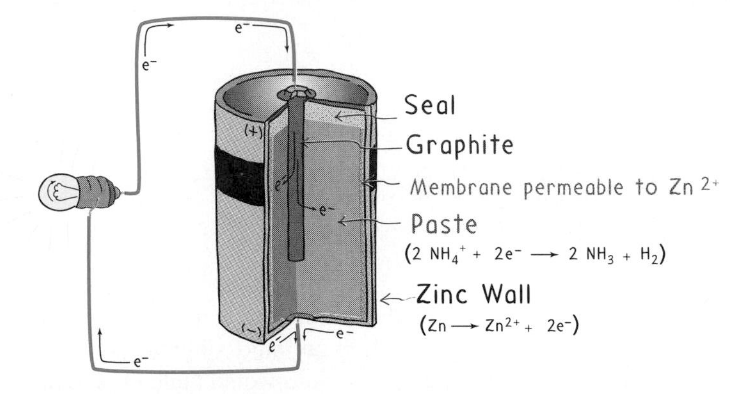

Fig. 20.19
A common dry-cell battery, with a graphite rod immersed in a paste of ammonium chloride, manganese oxide, and zinc chloride.

The common dry-cell battery was invented in 1866 and is still used today as a disposable energy source for flashlights, toys, and the like. The basic design consists of a zinc metal cup filled with a thick paste of ammonium chloride, NH_4Cl, zinc chloride, $ZnCl_2$, and manganese oxide, MnO_2. Immersed into this paste is a porous graphite *electrode* that projects to the top of the battery. An **electrode** is a conducting material that establishes electrical contact with the nonmetallic part of a circuit, which in this case is the paste (Figure 20.19).

Electrons are generated from the walls of the battery as zinc atoms oxidize to zinc ions:

$$\mathbf{Zn\ (s) \rightleftharpoons Zn^{2+}\ (aq) + 2\ e^- \quad Oxidation}$$

The zinc ions migrate into the paste, while the electrons removed from the zinc atoms as the ions formed travel through an external circuit and return through the graphite, where they are accepted by ammonium ions in the paste:

$$\mathbf{2\ NH_4^+\ (aq) + 2\ e^- \rightarrow 2\ NH_3\ (g) + H_2\ (g) \quad Reduction}$$

The reduction of the ammonium ions leads to the formation of two gaseous products—ammonia, NH_3, and hydrogen, H_2, which need to be absorbed. If they aren't absorbed, pressure could build up and the battery could explode. Absorption is accomplished by the remaining components of the paste. Zinc chloride reacts with ammonia to form solid zinc ammonium chloride, $Zn(NH_3)_2Cl_2$, and manganese dioxide reacts with hydrogen to form solid dimanganese trioxide, Mn_2O_3, plus water:

$$\mathbf{ZnCl_2\ (aq) + 2\ NH_3\ (g) \rightarrow Zn(NH_3)_2Cl_2\ (s)}$$

$$\mathbf{2\ MnO_2\ (s) + H_2\ (g) \rightarrow Mn_2O_3\ (s) + H_2O\ (l)}$$

The life span of a dry cell battery is relatively short. Oxidation of the zinc wall causes the wall to deteriorate and eventually the contents leak out. Even while the battery is not operating, the zinc corrodes as it reacts with ammonium ions. This latter

Fig. 20.20
Alkaline batteries offer greater performance but they are expensive.

Fig. 20.21
Lead storage batteries are most reliable; however, their weight is a major disadvantage. Much research is currently under way to find lighter-weight batteries, especially for use in electric vehicles.

mode of zinc corrosion can be inhibited by storing the battery in a refrigerator, which can significantly increase the battery's life.

More expensive alkaline batteries avoid many of the problems of the dry-cell battery by operating in a strongly alkaline medium. In the presence of hydroxide ions, the zinc oxidizes to insoluble zinc oxide, ZnO:

$$\mathrm{Zn}\ (s) + 2\ \mathrm{HO^-}\ (aq) \rightarrow \mathrm{ZnO}\ (s) + \mathrm{H_2O}\ (l) + 2\ e^- \quad \textbf{Oxidation}$$

Meanwhile, manganese dioxide is reduced to dimanganese trioxide, Mn_2O_3:

$$2\ \mathrm{MnO_2}\ (s) + \mathrm{H_2O}\ (l) + 2\ e^- \rightarrow \mathrm{Mn_2O_3}\ (s) + 2\ \mathrm{HO^-}\ (aq) \quad \textbf{Reduction}$$

Note how these reactions avoid the use of the zinc-corroding ammonium ion, NH_4^+, and avoid the formation of any gaseous products. Furthermore, these alternative chemical reactions are better suited to maintaining a given voltage during longer periods of operation.

The smaller mercury and lithium batteries used for calculators and cameras are variations of the alkaline battery. In the mercury battery, mercuric oxide, HgO, is reduced rather than manganese dioxide. Manufacturers are phasing out these batteries because of the environmental hazard posed by mercury, which is poisonous. In the lithium battery, lithium metal is used as the source of electrons rather than zinc. Not only is lithium able to maintain a higher voltage than zinc, it is about 13 times less dense, which allows for a lighter battery.

Disposable batteries have relatively short lives because electron-producing chemicals are consumed. The main feature of the rechargeable battery is the reversibility of oxidation and reduction reactions. In your car's *lead storage battery,* for example, electrical energy is produced as lead oxide, lead, and sulfuric acid are consumed to form lead sulfate and water:

$$\mathrm{PbO_2} + \mathrm{Pb} + 2\mathrm{H_2SO_4} \rightarrow 2\ \mathrm{PbSO_4} + 2\ \mathrm{H_2O} + \text{electrical energy}$$

This reaction can be reversed by supplying electrical energy. This is the task of the car's alternator, which is powered by the engine:

$$\text{Electrical energy} + 2\ \mathrm{PbSO_4} + 2\ \mathrm{H_2O} \rightarrow \mathrm{PbO_2} + \mathrm{Pb} + 2\ \mathrm{H_2SO_4}$$

So running the motor maintains concentrations of lead oxide, lead, and sulfuric acid in the car battery. With the engine turned off, these reactants stand ready to supply electrical power as needed to start the engine, operate the emergency blinkers, or play the radio.

Question

Your car lights were left on while you were shopping and now your car battery is dead. Has the pH of the battery fluid increased or decreased?

Many small rechargeable batteries are made of compounds of nickel and cadmium (ni-cad batteries). As with the lead storage battery, ni-cad reactants are replenished by supplying electrical energy from some external source, such as an electrical wall outlet. Like mercury batteries, ni-cad batteries pose an environmental hazard because cadmium is toxic to humans and other life forms. Alternative metal hydride batteries are rapidly gaining a place in the market.

Answer
As the battery supplied electrical energy to the headlights, sulfuric acid was consumed. With less sulfuric acid in the battery fluid, the pH has increased. Recall from Section 20.4 that a higher pH means less acidity.

Fuel Cells

A *fuel cell* is a device that continuously changes a fuel's chemical energy into electrical energy. Fuel cells are by far the most efficient means of generating electricity.

The hydrogen-oxygen fuel cell has two compartments, one for hydrogen and the other for oxygen, separated by a set of porous electrodes (Figure 20.22). Hydrogen is oxidized (loses electrons) upon contact with the hydroxide ions in the hydrogen-facing electrode. The free electrons flow through an external circuit and provide electrical power before meeting up with oxygen at the oxygen-facing electrode. The oxygen readily picks up the electrons (becomes reduced) and reacts with water to form hydroxide ions. To complete the circuit, these hydroxide ions migrate across the porous electrodes and through an ionic medium of potassium hydroxide to meet up with additional hydrogen at the hydrogen-facing electrode. The hydrogen and hydroxide ions react to produce steam as well as electrons. The energy of the steam may be used for industrial purposes or even to generate more electricity using the steam turbine. Furthermore, the water that condenses from the steam is pure water, most suitable for drinking!

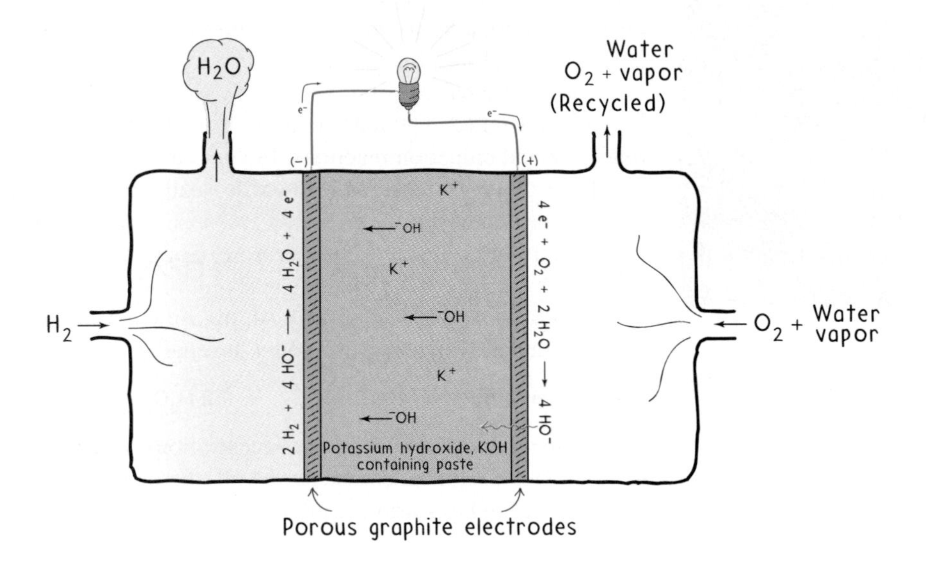

Fig. 20.22
The hydrogen-oxygen fuel cell.

Fuel cells are similar to batteries except they don't run down so long as fuel is supplied. Furthermore, they are much lighter than batteries and can produce water. This gives fuel cells some very important applications. The space shuttle uses hydrogen-oxygen fuel cells to meet its electrical needs. On board, the fuel cells also produce more than 100 gallons of drinking water for the astronauts during a typical week-long mission. Back on the Earth, electric motor vehicles may be equipped with hydrogen-oxygen fuel cells. Because these vehicles rely on a fuel rather than a set of charged batteries, driving range is limited only by the availability of a network of hydrogen stations. Ideally the hydrogen will be generated by the *electrolysis* of water, using electric energy from nonpolluting renewable sources, such as solar energy. In the meantime, a fuel-cell system that extracts hydrogen from gasoline has recently been developed. The advantage here, of course, is that the gasoline infrastructure already exists. Furthermore, electric cars running off of this system get at least twice as many miles to the liter or gallon as today's most efficient cars.

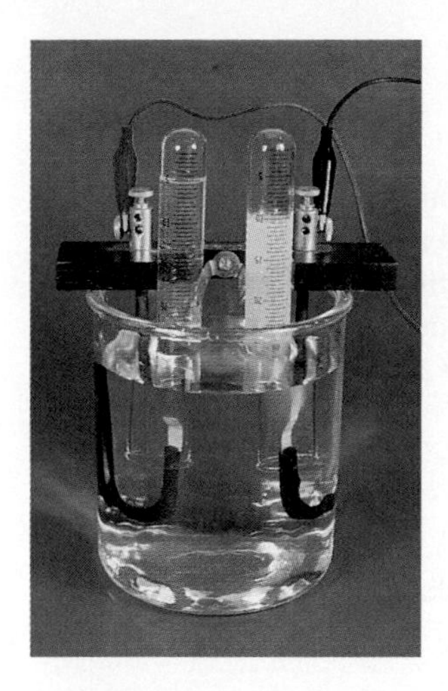

Fig. 20.23
The electrolysis of water produces hydrogen gas and oxygen gas in a 2:1 ratio by volume. Why must ions be dissolved in the water for this electrolysis to work?

So why aren't fuel cells widely used today? The internal combustion engine works really well, having the advantage of many years of development. These engines are also well entrenched in our society, and so change comes slowly. Fuel cells and many other alternative energy systems are still under development, however. A changeover to these alternative systems will occur only when there is the demand, which will come only after the systems have proved themselves more economical than the status quo.

Electrolysis

Electrolysis is the use of electrical energy to produce chemical change. The recharging of a car battery is an example of electrolysis. Another is the process of passing an electric current through water, which can cause the water to break down into its elemental components:

$$\text{Electrical energy} + 2\,H_2O \rightarrow 2\,H_2\,(g) + O_2\,(g)$$

The purification of metals from ores is done by electrolysis. Foremost is aluminum, the third most abundant element in the Earth's crust. Aluminum occurs naturally bonded to oxygen in an ore called *bauxite*.[8] Aluminum metal wasn't known until about 1827, when it was first prepared by reacting bauxite with hydrochloric acid. This gave the aluminum ion, Al^{3+}, which was reduced to aluminum metal using sodium metal:

$$Al^{3+} + 3\,Na \rightarrow Al + 3\,Na^{+}$$

Aluminum thus became an expensive rarity. In 1855, pieces were exhibited in Paris with the crown jewels of France. The price of aluminum at that time was about $100,000 per pound. Then, in 1886, two men working independently, Charles Hall in the United States and Paul Heroult in France, almost simultaneously discovered an electrolytic process for producing aluminum directly from bauxite. This greatly facilitated mass production, and by 1890 the price of aluminum dropped to about $2 per pound. Today, about 16,000 kilowatt-hours of energy is required to produce each ton of aluminum from its ore. This is an enormous amount of energy, considering that worldwide production of aluminum is about 16 million tons annually. Recycling aluminum, on the other hand, consumes only about 700 kilowatt-hours for every ton. Recycling aluminum not only reduces litter but also helps reduce the load on electric companies, which in turn reduces air pollution.

Fig. 20.24
For a nerve-wracking experience, bite a piece of aluminum foil with a tooth filled with dental amalgam. (If you don't have one, hooray for you! You'll need to ask a less fortunate friend what this is like.) Aluminum is a better reducing agent than dental amalgam and therefore gives electrons to the amalgam. The slight current that results produces a jolt of . . . ouch . . . pain.

Question

Is the reaction that takes place in a hydrogen-oxygen fuel cell an example of electrolysis?

Answer
No. During electrolysis, electrical energy is used to produce chemical change. In the hydrogen-oxygen fuel cell, chemical change is used to produce electrical energy.

[8]Bauxite is named after the place of its discovery, Les Baux, France.

20.9 Oxidizing Power of Oxygen

Look at the upper right of the periodic table and you will find one of the most common oxidizing agents—oxygen. In fact, if you haven't guessed already, the term *oxidation* is derived from this element. Oxygen is able to pluck electrons from many other types of elements, typically, those that lie to its lower left in the periodic table. Two common types of oxidation-reduction reactions involving oxygen as the oxidizing agent include *corrosion* and *combustion.*

■ **Question**

Oxygen is a good oxidizing agent, but so is chlorine. What does this say about their relative positions on the periodic table?

Corrosion

Fig. 20.25
Structures made of iron are most susceptible to the destructive powers of corrosion.

Metals deteriorate—a destructive process that costs billions of dollars annually. This is *corrosion.* About one-quarter of steel production in the United States goes into replacing corroded iron. The corrosion of iron—rusting—occurs when iron reacts with oxygen and water to form iron oxide trihydrate—the naturally occurring reddish-brown substance you know as rust:

$$\underset{\text{iron}}{4\,Fe} + \underset{\text{oxygen}}{3\,O_2} + \underset{\text{water}}{6\,H_2O} \rightarrow \underset{\text{rust}}{2\,Fe_2O_3{\cdot}3\,H_2O}$$

We can better understand rusting by considering this equation in steps. First, iron loses electrons to form the Fe^{2+} ion . Second, oxygen accepts these electrons and then reacts with water to form hydroxide ions, HO^-. In a subsequent step, iron and hydroxide ions combine to form an intermediate, iron hydroxide, $Fe(OH)_2$:

$$2\,Fe \rightarrow 2\,Fe^{2+} + 4\,e^- \quad \text{Oxidation}$$

$$O_2 + 4\,e^- + 2\,H_2O \rightarrow 4\,HO^-$$

$$2\,Fe^{2+} + 4\,HO^- \rightarrow \underset{\text{iron hydroxide}}{2\,Fe(OH)_2}$$

Through a series of similar steps, iron hydroxide is further oxidized by oxygen to form rust:

$$\underset{\text{iron hydroxide}}{4\,Fe(OH)_2} + \underset{\text{oxygen}}{O_2} + \underset{\text{water}}{2\,H_2O} \rightarrow \underset{\text{rust}}{2\,Fe_2O_3{\cdot}3\,H_2O}$$

Another common metal oxidized by oxygen is aluminum. The product of aluminum oxidation is aluminum oxide, Al_2O_3, which is water-insoluble. Because of its

---

■ **Answer**
Chlorine and oxygen must lie in the same area of the periodic table. Both have strong effective nuclear charges, and both are strong oxidizing agents. Chlorine's oxidizing powers are what make it useful as the active component of bleach.

Fig. 20.26
Galvanized nails are protected from rusting by the sacrificial oxidation of zinc.

water insolubility, aluminum oxide forms a protective coat that shields aluminum from further oxidation. This coat is so thin that it's transparent, which is why aluminum maintains its metallic shine.

A protective water-insoluble oxidized coat is the principle underlying galvanized nails. Zinc has a greater tendency to oxidize than does iron. For this reason, many iron articles, such as nails, are "galvanized" by coating them with a thin layer of zinc. The zinc oxidizes to zinc oxide, an inert, insoluble substance that protects the inner iron from rusting.

Another way to protect iron and other metals from oxidation is *electroplating* with a corrosive-resistant metal, such as chromium, platinum, or gold. *Electroplating* is the operation of coating one metal with another by electrolysis. This is done by connecting the object to be electroplated to a negative battery terminal and then submerging the object into a solution of desirable metal ions, such as chromium ions, Cr^{2+} (Fig-ure 20.27). The positive terminal of the battery is connected to an electrode made of the desirable metal. The circuit is completed when this electrode is submerged into the solution. Dissolved metal ions are attracted to the negatively charged object, where they pick up electrons and are deposited as metal atoms. The ions in solution are replenished by the forced oxidation of the metal at the positively charged electrode. The net effect is that desirable metal atoms from the positively charged electrode are deposited on the negatively charged object.

Iron oxidation can also be inhibited by alloying the iron with metals that resist oxidation, such as chromium and nickel. This combination yields stainless steel (Section 17.2).

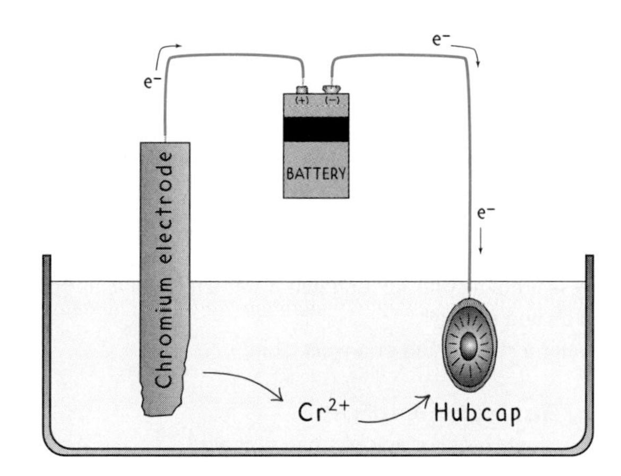

Fig. 20.27
A hubcap is submerged in a solution of chromium ions, Cr^{2+}. As electrons flow, metal atoms are transferred from a positively charged electrode to the negatively charged hubcap.

Combustion

Combustion is an oxidation-reduction reaction between a chemical compound and molecular oxygen. Combustion reactions characteristically release energy (in other words, they are exothermic). An example of slow combustion is the rusting of iron. Its exothermic nature can be witnessed by sealing moist steel wool in a glass jar. Heat from the rusting iron vaporizes the water, which condenses onto the inner walls of the container. A more violent combustion reaction is the formation of water from hydrogen and oxygen. As discussed in Section 19.2, the energy from this reaction is used to power rockets into outer space. More common examples of combustion include the burning of wood or fossil fuels such as gasoline and heating oil. The combustion of these and other organic chemicals forms carbon dioxide and water. Consider, for example, the combustion of methane, the major component of natural gas:

$$\underset{\text{methane}}{CH_4} + \underset{\text{oxygen}}{2\,O_2} \rightarrow \underset{\text{carbon dioxide}}{CO_2} + \underset{\text{water}}{2\,H_2O} + \text{energy}$$

During combustion it is always the molecular oxygen that gets reduced (gains electrons) because electrons in an oxygen atom are in a much lower energy state than electrons in the fuel being combusted. The energy state in the oxygen atom is lower because the effective nuclear charge the electrons experience is stronger in oxygen than in carbon (Section 16.4). The transferring of electrons to oxygen is therefore a downhill, energy-releasing process. Typically, the energy is released in the form molecular kinetic energy (heat), or even light, which we see as a flame.

It is the oxidizing power of oxygen that makes it so vital to life. The oxygen we inhale is held to the iron of hemoglobin by this very characteristic (see the Chapter 18 box on *Oxygen in Our Blood*). The proteins of hemoglobin, however, prevent a complete reduction of oxygen so that it still has its oxidizing powers when it reaches the cells. Once inside the cell, it readily accepts electrons from food molecules, which are transformed (oxidized) into carbon dioxide and water. The carbon dioxide is exhaled, and the water is either excreted or lost through perspiration. The energy these reactions yield is retained and utilized by the body for all its living processes.

Summary of Terms

- **Acid-base reaction** A reaction involving the transfer of a hydrogen ion, H^+, from one reactant to another.
- **Oxidation-reduction reaction** A reaction involving the transfer of one or more electrons from one reactant to another.
- **Acid** A substance that produces or donates hydrogen ions in solution.
- **Base** A substance that produces hydroxide ions in solution or accepts hydrogen ions.
- **Hydronium Ion** A water molecule after accepting a hydrogen ion, H_3O^+.
- **Salt** An ionic compound formed from the reaction of an acid and a base.
- **Neutralization** A reaction in which an acid and base combine to form a salt plus water.
- **Amphoteric substance** A substance that can behave as either an acid or a base.
- **Acidic solution** A solution in which the hydronium ion concentration is greater than the hydroxide ion concentration.
- **Basic solution** A solution in which the hydroxide ion concentration is greater than the hydronium ion concentration.
- **Neutral solution** A solution in which the hydronium ion concentration is equal to the hydroxide ion concentration.
- **pH** A measure of the acidity of a solution. The pH is equal to the negative logarithm of the hydronium ion concentration. At 25°C, neutral solutions have a pH of 7, acidic solutions have a pH less than 7, and basic solutions have a pH greater than 7.
- **Buffer solution** A solution of either a weak acid and one of its salts or a weak base and one of its salts that resists large change in pH.
- **Oxidation** The process whereby a reactant loses one or more electrons.
- **Reduction** The process whereby a reactant gains one or more electrons.
- **Electrochemistry** A branch of chemistry concerned with the relationship between electrical energy and chemical change.
- **Electrode** A conducting material used to establish electrical contact between metallic and nonmetallic parts of an electric circuit.
- **Electrolysis** The use of electrical energy to produce chemical change.

Review Questions

Acids and Bases Defined

1. What is the Arrhenius definition of an acid? Of a base?
2. How do the Arrhenius definitions of acid and base differ from the Brønsted-Lowry definitions?
3. When an acid is dissolved in water, what ion does the water form?
4. When a molecule loses a hydrogen ion, is it behaving as an acid or as a base?
5. Does a salt always contain sodium ions?
6. What two types of substances are necessarily involved in a neutralization reaction?

Acid Strength

7. What does it mean to say that an acid is strong in aqueous solution?
8. Why does a solution of a strong acid conduct electricity better than a solution of a weak acid?
9. Why is it not a contradiction that water can behave as both an acid and a base?
10. Is water a strong acid or a weak acid?

Acidic, Basic, or Neutral

11. Is K_w a very large or a very small number?
12. As the concentration of H_3O^+ ions in an aqueous solution increases, what happens to the concentration of HO^- ions?
13. How do we characterize solutions as acidic, basic, or neutral, in terms of the relative concentrations of hydronium and hydroxide ions?

The pH Scale

14. What does the pH of a solution indicate?
15. As the hydronium ion concentration of a solution increases, does the pH of the solution increase or decrease?

Acid Rain and Basic Oceans

16. What is the product of the reaction between carbon dioxide and water?
17. In terms of pH, what is the definition of acid rain?
18. What does sulfur dioxide have to do with acid rain?

19. What are some sources of sulfur dioxide in our atmosphere?
20. How does one "lime" a lake?
21. Why aren't atmospheric levels of carbon dioxide rising as rapidly as might be expected based upon the increase output of carbon dioxide by human activities?

Buffers

22. What is a buffer solution?
23. A strong acid, such as HCl, added to water causes a large drop in the pH of the water. The same acid added to a buffer solution causes only a very small decrease in the pH of the solution. Why?
24. Do buffer solutions prevent or inhibit changes in pH?
25. Why is it so important that the pH of our blood be maintained within a narrow range of values?
26. Holding your breath causes the pH of your blood to decrease. Why?

Oxidation-Reduction

27. What elements in the periodic table have the greatest tendency to behave as oxidizing agents?
28. Write an equation showing a potassium atom (atomic symbol K) being oxidized.
29. Write an equation showing a bromine atom (atomic symbol Br) being reduced.
30. What is the difference between an oxidizing agent and a reducing agent?
31. What happens to a reducing agent as it reduces something?

Electrochemistry

32. What is electrochemistry?
33. What is the purpose of the manganese dioxide in a dry-cell battery?
34. What chemical reaction is forced to occur while a car battery is being recharged?
35. Why doesn't a fuel cell run down the way a battery does?
36. What is electrolysis and how does it differ from the underlying principle of a battery?

Oxidizing Power of Oxygen

37. Why is oxygen such a good oxidizing agent?
38. What do the oxidation of zinc and the oxidation of aluminum have in common?
39. What is electroplating and how is it accomplished?
40. Is corrosion an example of combustion or is combustion an example of corrosion?

Home Projects

1. Put a teaspoon of baking soda (a base) in a tall drinking glass. Follow with 3 or 4 capfuls of vinegar. Add the vinegar slowly so that the bubbles don't overflow the glass. After the bubbles subside, light a match and hold it adjacent to the lip of the glass. Extinguish the match by pouring the gaseous contents of the glass over the it. The gas is carbon dioxide, which is heavier than air.
2. The pH of a solution can be approximated with a *pH indicator,* which is any chemical whose color changes with pH. Many pH indicators are found in plants; a good example is red cabbage. Shred about a quarter of a head of red cabbage and then boil it in 2 cups of water for about 5 minutes. Strain the cabbage and collect the broth, which contains the pH indicator. Red cabbage indicator is red at low pH values (1–4), a light purple at neutral pH values (7), green at moderately alkaline pH values (8–11), and yellow at very alkaline pH values (13). Add small amounts of the cabbage broth to various solutions, such as white vinegar, rain water, ammonia, baking soda, and bleach, to estimate their pH.
3. You can see the electrolysis of water by immersing the top end of a 9-V battery into a solution of salt water. The bubbles that form contain hydrogen, which results from the decomposition of water. Why does this activity work bet ter with salt water than with pure water? Why does this activity quickly ruin your battery? (Remember to dispose of the battery properly after performing this home project.)

Exercises

1. Suggest why people once washed their hands with ashes.
2. This chapter places a greater emphasis on which definition of acids and bases: the Arrhenius definition or the Brønsted-Lowry definition?
3. What is the relationship between the hydroxide ion and a water molecule?
4. To what subatomic particle is the hydrogen ion equivalent?
5. An acid and a base react to form a salt, which consists of positive and negative ions. Which forms the positive ions: the acid or the base? Which forms the negative ions?
6. Water is formed from the reaction of an acid and a base. Why is it not classified as a salt?
7. What happens to the corrosive properties of an acid and a base after they neutralize each other? Why?
8. In the following equations identify the acid or base behavior of each participant:

(a) $H_2PO_4 \quad + \quad H_2O \quad \rightleftharpoons \quad H_3O^+ \quad + \quad HPO_4^-$

$\_\_\_\_\_\_ \quad \_\_\_\_\_\_ \quad \_\_\_\_\_\_ \quad \_\_\_\_\_\_$

(b) $HSO_4^- \quad + \quad H_2O \quad \rightleftharpoons \quad H_3O^+ \quad + \quad SO_4^{2-}$

$\_\_\_\_\_\_ \quad \_\_\_\_\_\_ \quad \_\_\_\_\_\_ \quad \_\_\_\_\_\_$

(c) $HSO_4^- \quad + \quad H_2O \quad \rightleftharpoons \quad HO^- \quad + \quad H_2SO_4$

$\_\_\_\_\_\_ \quad \_\_\_\_\_\_ \quad \_\_\_\_\_\_ \quad \_\_\_\_\_\_$

(d) $O^{2-} \quad + \quad H_2O \quad \rightleftharpoons \quad HO^- \quad + \quad HO^-$

$\_\_\_\_\_\_ \quad \_\_\_\_\_\_ \quad \_\_\_\_\_\_ \quad \_\_\_\_\_\_$

9. Which should conduct electricity better, a solution saturated with a strong acid or one saturated with a weak acid? Defend your answer.
10. Why is HF a stronger acid than HCl?

11. What does the value of K_w say about the extent of the ionization of pure water?
12. The value of K_w increases with increasing temperature. Does this indicate that the acid-base reaction of two water molecules is exothermic or endothermic? Explain.
13. Because the value of K_w increases with increasing temperature, the pH of neutral water also changes with increasing temperature. Which has a higher pH, hot pure water or cold pure water?
14. Why do we use the pH scale to indicate the acidity of a solution rather than simply stating the concentration of hydronium ions?
15. Along with the pH scale, there is the pOH scale, which indicates the level of "basicity" in a solution. The definition is $pOH = -\log[HO^-]$. What is the sum of the pH and the pOH of a solution always equal to?
16. When the hydronium ion concentration equals 1 mole per liter, what is the pH of the solution? Is the solution acidic or basic?
17. When the hydronium ion concentration equals 10 moles per liter, what is the pH of the solution? Is the solution acidic or basic?
18. Lakes lying in granite basins, such as those in the northeastern United States, tend to become acidified by acid rain more readily than lakes lying in limestone basins, such as those found in the midwestern United States. Suggest why this is so. (Granite is an inert rock composed primarily of silicon dioxide, SiO_2, whereas limestone is composed primarily of calcium carbonate, $CaCO_3$.)
19. Cutting back on the pollutants that cause acid rain is one solution to the problem of acidified lakes. Suggest another.
20. How might warmer oceans accelerate global warming?
21. Could a single chemical capable of behaving as an acid and a base be used to make a buffer solution consisting of only one component? Defend your answer.
22. Hydrogen chloride is added to a buffer solution of ammonia, NH_3, and ammonium chloride, $NH_4^+Cl^-$. What is the effect on the concentration of ammonia? On the concentration of ammonium chloride?
23. Sodium hydroxide is added to a buffer solution of ammonia, NH_3, and ammonium chloride, $NH_4^+Cl^-$. What is the effect on the concentration of ammonia? On the concentration of ammonium chloride?
24. At what point does a buffer solution cease to moderate changes in pH?
25. An overdose of aspirin (chemical name: acetyl salicylic acid) causes the body to hyperventilate. How so?
26. Does an oxidizing agent donate or accept electrons? Answer the same question for a reducing agent.
27. What correlation might you expect between the electron affinity of an element (Section 16.5) and its ability to behave as an oxidizing agent? How about its ability to behave as a reducing agent?
28. What correlation might you expect to find between the ionization energy of an element (Section 16.5) and its ability to behave as an oxidizing agent? How about its ability to behave as a reducing agent?
29. Based upon relative positions in the periodic table, which might you expect to behave as a stronger oxidizing agent: chlorine or fluorine? Why?
30. Iron is a better reducing agent than the copper ion Cu^{2+}. So, in which direction do electrons flow when an iron nail is submerged in a solution of Cu^{2+}?
31. What is the purpose of the salt bridge depicted in Figure 20.18?
32. Is sodium metal oxidized or reduced when used in the production of aluminum?
33. Why is the formation of iron hydroxide, $Fe(OH)_2$ from Fe^{2+} and HO^- not considered an oxidation-reduction reaction?
34. Consider the oxidation-reduction reaction

$$Mg(s) + Cu^{2+}(aq) \rightleftharpoons Mg^{2+}(aq) + Cu(s)$$

Sketch a voltaic cell that uses this reaction. Which species is reduced? Which species is oxidized?
35. Jewelry is often manufactured by electroplating an expensive metal such as gold over a cheaper metal. Draw a sketch showing how such a process may be set up.
36. Some car batteries require the periodic addition of water. Does adding the water to the battery increase or decrease its electric power potential? Explain.
37. Which weighs more: a heavy-duty dry-cell battery or a less expensive regular-duty dry-cell battery? Why?
38. The oxidation of iron to rust can be a problem, but the oxidation of aluminum to aluminum oxide is not. Why?
39. How many electrons transfer from iron atoms to oxygen atoms in the formation of two molecules of iron hydroxide, $Fe(OH)_2$, from iron metal?
40. Why are combustion reactions generally exothermic?

Problems

1. What is the hydroxide ion concentration in an aqueous solution when the hydronium ion concentration is 1×10^{-10} mole per liter?
2. When a hydronium ion concentration equals 1×10^{-10} mole per liter, what is the pH of the solution? Is the solution acidic or basic?
3. When a hydronium ion concentration equals 1×10^{-4} mole per liter, what is the pH of the solution? Is the solution acidic or basic?
4. What is the hydroxide ion concentration in an aqueous solution having a pH of 5?
5. The value of K_w depends on temperature. At 25°C, K_w is 1.0×10^{-14}, as discussed in the text. At 45°C, it is 4.0×10^{-14}. What is the pH of neutral water at 45°C?

CHAPTER

21

Organic Chemistry

Carbon is a unique element. Its ability to connect with other carbon atoms through strong and stable covalent bonds sets it apart from other elements. Whereas no more than three oxygen atoms can bond together, as occurs in ozone, O_3, carbon atoms can form long chains well over 100,000 carbon atoms in length. In addition to forming chains, carbon atoms can link to form a multitude of unusual structures—much like a child's set of Tinker Toys. Life itself is based upon the remarkable chemistry of carbon. Reflecting this fact, the branch of chemistry that studies carbon-containing compounds has come to be known as **organic chemistry**. Likewise, carbon-containing compounds have come to be known as **organic compounds**. Organic compounds, however, need not come from living or once-living organisms. Many are purely *synthetic,* which means they are solely the product of chemical reactions carried out in the laboratory.

In the periodic table, carbon is the king of versatility. It is able to form strong covalent bonds not only with itself but also with most other nonmetal atoms. More than 13 million known organic compounds exist, with more than 100,000 new ones added to the list each year. By contrast, there are only 200,000 to 300,000 known inorganic compounds. Because the growth in the number of organic compounds is so great, it's important to have a basic understanding of them—what they can and can't do, and what implications they have for medicine and the biological sciences. We begin with the simplest organic molecules—those consisting of only carbon and hydrogen, and we conclude with the more complex organic polymers.

21.1 Hydrocarbons

Compounds that contain only carbon and hydrogen are **hydrocarbons**. Although they all contain only two types of atoms, hydrocarbons exist in tremendous variety. They include natural gas, gasoline, motor and heating oil, and many plastics.

Hydrocarbons differ from one another in the numbers of carbon and hydrogen atoms per molecule (Table 21.1). Natural gas, for example, is mostly methane, which is a hydrocarbon having only one carbon atom per molecule. Gasoline, on the other hand, is a mixture of hydrocarbons containing between five and ten carbons per molecule. Motor and heating oil are mixtures of hydrocarbons containing 10–16 carbon atoms per molecule. The hydrocarbon polyethylene contains thousands of carbon atoms per molecule.

Hydrocarbons also differ from one another in the way the carbon atoms connect to each other. Consider the three hydrocarbons *n*-pentane, *iso*-pentane, and *neo*-pentane (Figure 21.1). These hydrocarbons all have the same molecular formula, C_5H_{12}, but they are structurally different because their carbon atoms are connected differently. The carbon framework of *n*-pentane is a chain of five consecutive carbon atoms. In *iso*-

Table 21.1
Properties of Hydrocarbons

Molecular Formula	Name	Melting Point (°C)	Boiling Point (°C)	Phase at Room Temperature
CH_4	methane	−184	−161	Gas
C_2H_6	ethane	−183	−88	
C_3H_8	propane	−188	−42	
C_4H_{10}	*n*-butane	−138	−0.5	
C_5H_{12}	*n*-pentane	−130	36	Liquid
C_6H_{14}	*n*-hexane	−94	69	
C_7H_{16}	*n*-heptane	−91	98	
C_8H_{18}	*n*-octane	−57	126	
$C_{16}H_{34}$	*n*-hexadecane	18	288	
$C_{17}H_{36}$	*n*-heptadecane	23	303	Solid
$C_{18}H_{38}$	*n*-octadecane	28	317	
$C_{2000}H_{4002}$*	polyethylene	136	dec.**	

*The number of carbon atoms per molecule in polyethylene varies greatly but is generally a large number not more than 100,000.

**Decomposes before boiling.

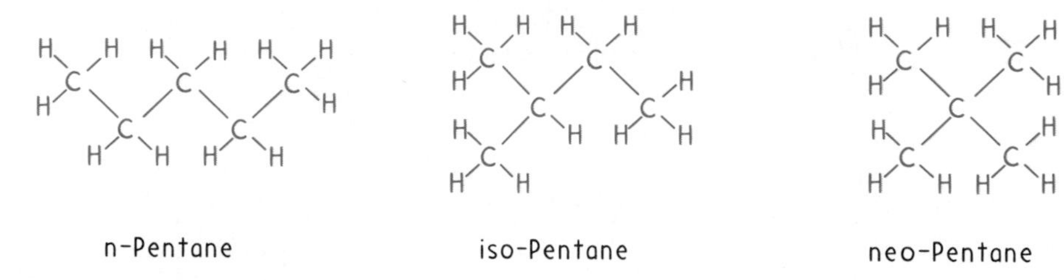

Fig. 21.1
The three hydrocarbons *n*-pentane, *iso*-pentane, and *neo*-pentane are different structurally, but all three have the same molecular formula: C_5H_{12}.

pentane, the carbon chain branches. In *neo*-pentane, a central carbon atom is bonded to four surrounding carbon atoms (Figure 21.2). Molecules that have the same molecular formula but differ in their chemical structures are **structural isomers**. Structural isomers have different chemical and physical properties. For example, *n*-pentane has a boiling point of 36°C, *iso*-pentane's boiling point is 30°C, and *neo*-pentane's is 10°C.

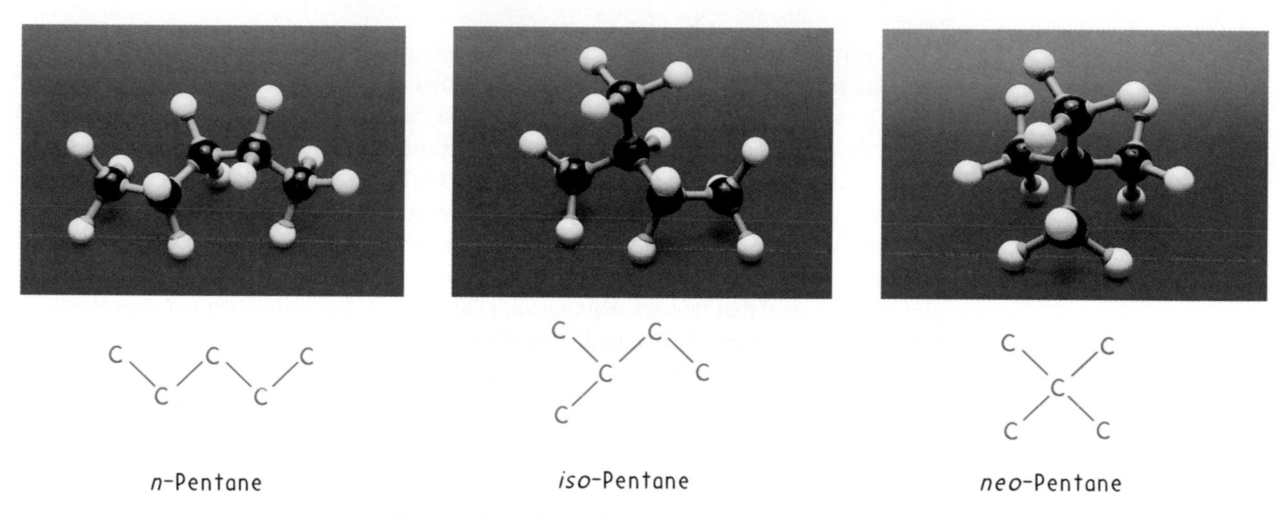

Fig. 21.2
We can see the different structural features of *n*-pentane, *iso*-pentane, and *neo*-pentane more clearly using molecular models or by ignoring the hydrogen atoms.

The number of possible structural isomers for a chemical formula increases rapidly as the number of carbon atoms increases. There are 3 structural isomers of pentane, C_5H_{12}; 5 of hexane, C_6H_{14}; nine of heptane, C_7H_{16}; 18 of octane, C_8H_{18}; 75 of decane, $C_{10}H_{22}$; and a whopping 366,319 of $C_{20}H_{42}$!

Because the carbon-carbon single bond can rotate, a structural isomer may be found in a number of different spatial orientations called **conformations**. This is analogous to your arm. Flex your wrist, elbow, and shoulder joints, and you'll find your arm passing through a range of different conformations. Likewise, organic molecules can twist and turn about their carbon-carbon single bonds to be found in a range of different conformations. The following structures, for example, are all different conformations of the same molecule—*n*-pentane:

In a sample of liquid *n*-pentane, the many individual molecules are found in all the different conformations possible—not unlike a bucket of worms.

Question

How many structural isomers are shown here?

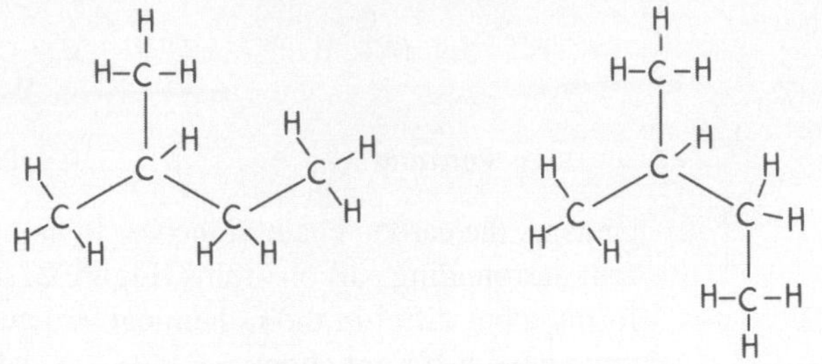

Hydrocarbons are obtained primarily from either coal or petroleum. Most of the coal and petroleum that exists today was formed between 280 and 395 million years ago. The formation of coal begins with the slow bacterial decay of plants under water, in the absence of air. As decay takes place, oxygen is gradually removed from plant fibers, and a residue composed largely of hydrocarbons remains.

Beds of coal are the remains of ancient swamps. Swamps have produced coal ever since the appearance of plant life. Seldom, however, were conditions as favorable as they were more than 280 million years ago. At that time there were extensive swamps close to sea level that periodically became submerged. Partially decayed vegetation in the swamps was trapped beneath layers of marine sediments.

The origin of petroleum (also called crude oil) is more obscure. Because fossils are not preserved in fluids, little record of the organisms responsible for petroleum formation is found. Also, oil can migrate long distances after formation, and so it is unreasonable to surmise anything about the origin of a body of petroleum based upon the local fossil records and geology. However, most geologists believe that petroleum forms in much the same way that coal forms—through the decay of organic matter with no oxygen present. Both plant and animal matter probably contribute to the formation of petroleum.

About 17 million barrels of petroleum is consumed each day in the United States. Approximately 8 million barrels of this total is converted to gasoline, and another 8 million barrels is converted to such fuels as heating oil, diesel fuel, jet engine fuel, and oil for electric power plants. The remaining 1 million barrels used daily provides raw material for the production of organic chemicals and polymers as discussed in Section 21.7. Thus only one-seventeenth of the hydrocarbons consumed daily go into useful materials; most ends up as heat and smoke.

Petroleum is a complex mixture of various hydrocarbons and other compounds. To obtain useful materials at a reasonable cost, the components must be separated effi-

---

Answer

Both these structures represent two possible conformations for a single structural isomer, that of *iso*-pentane. Although these structures have different spatial orientations, close examination shows that the bonding of the atoms in each structure is the same. In both cases, for example, there is no chain more than four carbon atoms long. They differ in that the central carbon-carbon bond has rotated, which has the effect of moving the upper-right carbon atom to down below.

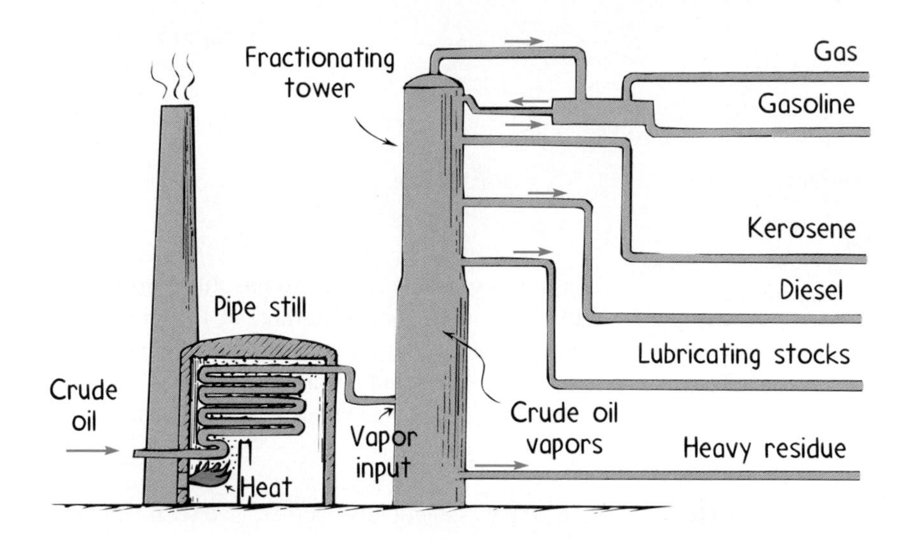

Fig. 21.3
A schematic for the fractional distillation of petroleum. The petroleum is volatilized in the pipe still and then sent to a fractionating tower, which is warmer at the bottom and cooler at the top. Hydrocarbons that have low boiling points, such as natural gas and gasoline, can travel to the top of the tower before condensing. Hydrocarbons that have higher boiling points, such as lubricating stocks, condense at lower heights. Pipes drain the various liquid hydrocarbon fractions from the tower.

ciently. Because most of the components are volatile, *fractional distillation* is used, a method in which the components are separated into fractions according to boiling temperature.

Special refinery processes are used to increase the production of high-demand hydrocarbons such as gasoline. For example, large-molecule hydrocarbon oils are broken down into smaller, less viscous gasoline hydrocarbons through *catalytic cracking,* a process in which hydrocarbon vapor is passed over a silica and aluminum oxide ($SiO_2 \cdot Al_2O_3$) catalyst at 450–550°C.

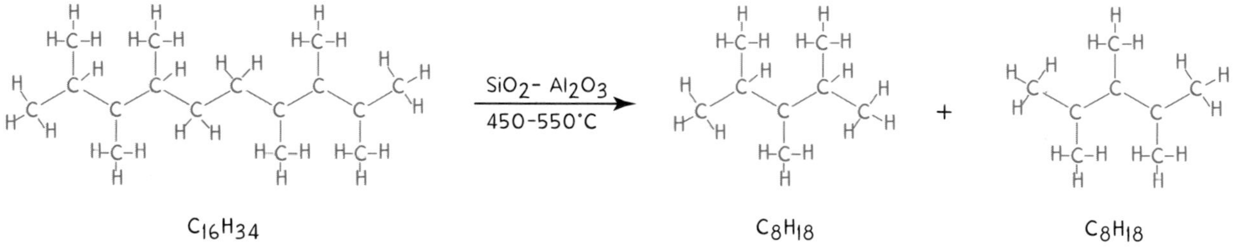

$$C_{16}H_{34} \xrightarrow[450\text{-}550^{\circ}C]{SiO_2 - Al_2O_3} C_8H_{18} + C_8H_{18}$$

Fig. 21.4
Larger hydrocarbons can be broken down into smaller ones by catalytic cracking.

A second procedure to increase the yield of gasoline from petroleum combines lightweight hydrocarbons and treats them with acids to form heavier gasoline hydrocarbons.

$$C_4H_8 + C_4H_{10} \xrightarrow[H_2SO_4]{HF} C_8H_{18}$$

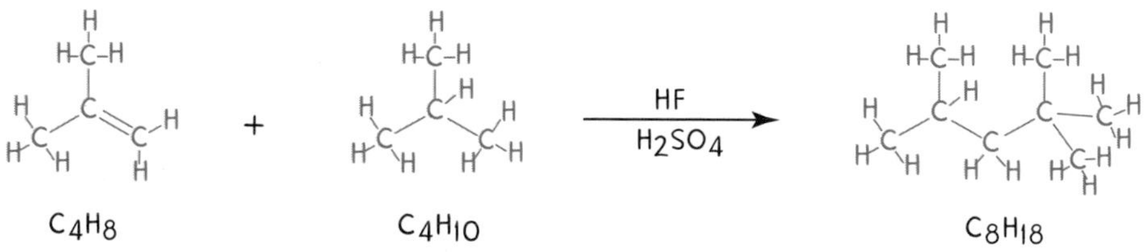

Fig. 21.5
Through the use of an acid catalyst, lightweight hydrocarbons can be combined to form gasoline hydrocarbons.

Some hydrocarbons burn more efficiently than others in a car engine. In general, the more branching in a gasoline hydrocarbon, the better it burns. *iso*-Octane, for example, burns quite well; *n*-hexane, which has no branching, causes an engine to fire irregularly (engine "knock") (Figure 21.6). These two compounds are used as standards

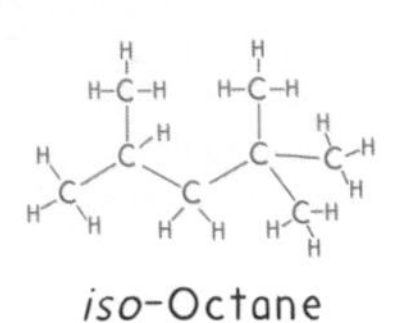

iso-Octane

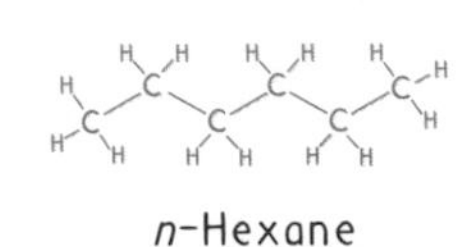

n-Hexane

Fig. 21.6
iso-Octane has more branching in its structure and burns smoothly. *n*-Hexane, with no branching, burns irregularly to produce "knock" in car engines.

Fig. 21.7
The higher the octane number, the greater the degree of hydrocarbon branching.

in assigning octane ratings to gasoline. *iso*-Octane is arbitrarily assigned an octane number of 100, and *n*-hexane is assigned 0. The antiknock performance of a particular gasoline is compared with that of various mixtures of *iso*-octane and *n*-hexane, and an octane number is assigned.

■ **Question**

Of the three structural isomers shown in Figure 21.1, which should have the highest octane rating?

Construction or Combustion?

A great number of commercial items are made from petroleum-based materials. Here are a few: plastics, synthetic fibers, cosmetics, chewing gum, wax, flavorings, dyes, asphalt, paints, ceramics, explosives, and pharmaceuticals (even aspirin is produced from oil).

Dmitri Mendeleev, the nineteenth-century Russian chemist who proposed the modern periodic table, was one of the first to recognize the value of petroleum as a raw material for industry. He cautioned, however, that burning petroleum as a fuel would be like burning money. Today, his warning holds special urgency as we find the world's supply of petroleum limited and no match for the growing needs of our growing human population. Many petroleum experts predict a sharp decline in petroleum production as the world's finite reserves dwindle. Within a century, oil production is expected to fall back to the meager level of the early 1900s.

Burning petroleum for its energy content may be worse than burning money. Money only *represents* wealth; petroleum *is* wealth. It is rich with usable energy, and it is the most versatile of all building materials. There are alternative sources of energy, such as solar, hydroelectric, and nuclear, that may be developed to substitute for petroleum, but there are no alternatives to petroleum or recycled petroleum products for the fabrication of new petroleum-based products. The continued burning of petroleum means that the price of all petroleum-based products will gradually increase. Eventually these now-pervasive valuables will become less affordable, and the growing human population may have to turn to some other as yet unknown resource for its building needs.

21.2 Unsaturated Hydrocarbons

Hydrocarbons come in two classes: *saturated* and *unsaturated.* In a **saturated hydrocarbon**, there are no multiple bonds between carbon atoms. Instead, every carbon atom is bonded to four neighboring atoms by four single bonds. The term *saturated* comes from the idea that the hydrocarbon has as many hydrogen atoms as is possible—it is *saturated* with hydrogens. Consider the saturated hydrocarbon *n*-butane (Figure 21.8a).

(a) *n*-Butane (b) 2-Butene

Fig. 21.8
(a) The saturated hydrocarbon *n*-butane is *saturated* with as many hydrogen atoms as possible—three for each terminal carbon and two for each internal carbon. (b) Because of the double bond, the unsaturated hydrocarbon 2-butene has two fewer hydrogen atoms than *n*-butane.

With only single bonds, the number of hydrogens attached to the carbons is ten. If there is a double bond between two carbons, as with 2-butene in Figure 21.8b, the number of hydrogen atoms drops to eight. Look back at the previous section and you'll see that all the hydrocarbons discussed are saturated.

The 2-butene of Figure 21.8b is an example of an **unsaturated hydrocarbon,** one that contains at least one multiple covalent bond, such as a double bond or a triple bond. The multiple-bonded carbon atoms are said to be *unsaturated* with hydrogen, and the entire molecule is classified as an unsaturated hydrocarbon. Besides 2-butene, another example of an unsaturated hydrocarbon is benzene, C_6H_6, which has three double bonds contained within a flat hexagonal ring (Figure 21.9).

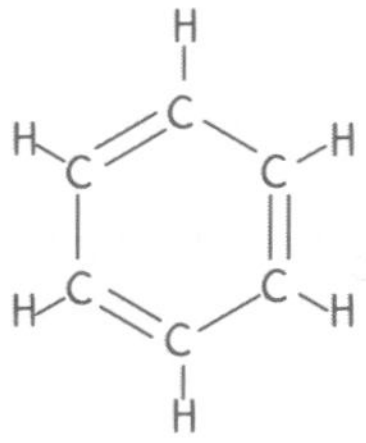

Benzene

Fig. 21.9
Benzene, C_6H_6, an unsaturated hydrocarbon with its carbon atoms joined in a ring.

Many organic compounds contain benzene rings, and many of these are fragrant. So by convention, any organic molecule containing the benzene ring is classified as an **aromatic compound.**[1] Toluene, a common solvent and paint thinner, is one example. Toluene gives airplane glue its distinctive yet toxic odor. Some aromatic compounds, such as naphthalene, contain two or more benzene rings fused together. Naphthalene gives rise to the smell of moth balls. Most moth balls sold today, however, are made of the less toxic 1,4-dichlorobenzene.

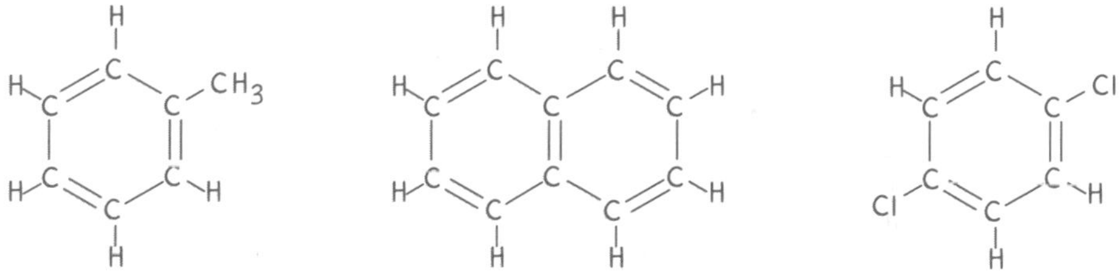

Toluene Napthalene 1,4-Dichlorobenzene

Fig. 21.10
The structures for three odoriferous ("aromatic") benzene-ring-containing organic compounds: toluene, naphthalene, and 1,4-dichlorobenzene.

The carbon-carbon multiple bond is chemically more reactive than the carbon-carbon single bond. Unsaturated hydrocarbons therefore tend to burn better as fuels than do saturated hydrocarbons of a similar carbon framework. For this reason automotive fuel additives commonly contain mixtures of aromatic compounds such as toluene. Consider also the organic compound acetylene, H_2C_2, which is an unsaturated hydrocarbon because of the triple bond between its two carbon atoms. A confined flame of acetylene burning in oxygen is hot enough to melt iron, which makes it a choice fuel for welding.

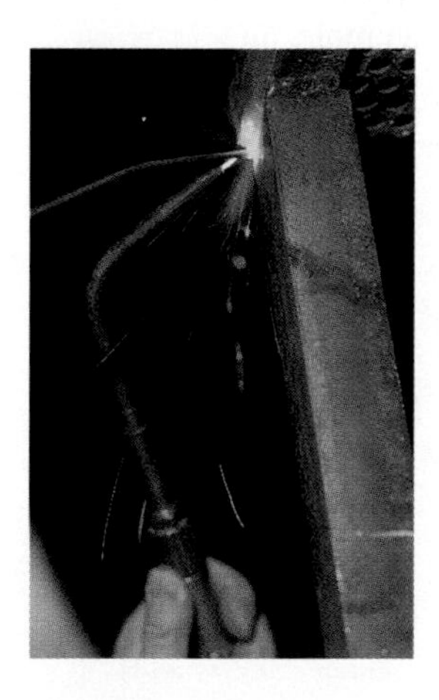

Fig. 21.11
The unsaturated hydrocarbon acetylene, H—C≡C—H, burned in this torch produces a flame hot enough to melt iron.

Answer

neo-Pentane, because it has the greatest amount of branching in its carbon framework:

Isomer	Octane Rating
n-Pentane	61.7
iso-Pentane	92.3
neo-Pentane	116

[1]Even if it is not especially fragrant.

21.3 Noncarbon Atoms in Organic Molecules

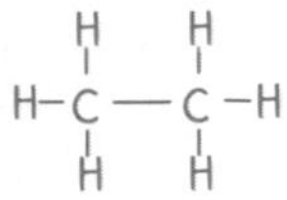

Ethane

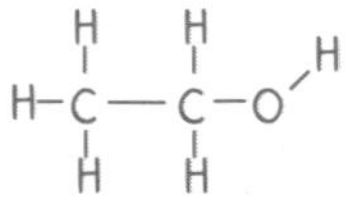

Ethanol

Fig. 21.12
The oxygen atom accounts for the remarkably different physical and chemical properties of ethane and ethanol.

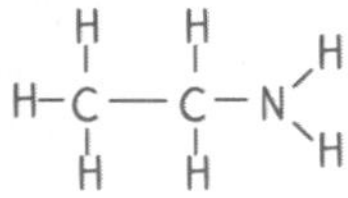

Ethylamine

Fig. 21.13
Heteroatoms give character to organic molecules. Ethanol is drinkable, but ethylamine is not.

Carbon atoms can bond to each other and to hydrogen in many ways, which results in an incredibly large number of hydrocarbons. But carbon can bond to other elements as well. This ability further increases the number of possible organic molecules. Adding even a single atom other than carbon or hydrogen to a hydrocarbon may change its physical and chemical properties markedly. For example, consider the change that occurs when an oxygen atom is added to the hydrocarbon ethane, C_2H_6, as shown in Figure 21.12. Ethane which has a melting point of $-183°C$ and a boiling point of $-88°C$, is a gas at room temperature, and does not dissolve in water. Addition of an oxygen atom to the ethane structure produces the molecule ethanol, C_2H_6O. Ethanol has a melting point of $-130°C$ and a boiling point of 78°C, is a liquid at room temperature, is infinitely soluble in water, and is the active ingredient of alcoholic beverages.

Organic chemists call any noncarbon or nonhydrogen atom in an organic molecule a **heteroatom**.[2] As the example just cited shows, adding a heteroatom to a hydrocarbon structure can produce an enormous chemical difference. For instance, hydrocarbons are relatively inert. The fact that they persist for millions of years as coal and petroleum suggests this. Their inert structures serve as three-dimensional carbon frameworks to which various heteroatoms can be added. We have seen that adding an oxygen atom to an ethane molecule produces ethanol. Adding a nitrogen atom (and one more hydrogen atom) produces a substance called *ethylamine,* a corrosive, pungent, potentially lethal gas (Figure 21.13).

Because the properties of organic molecules are largely determined by heteroatoms, organic molecules are classified according to which heteroatoms they contain and how the heteroatoms are attached to the carbon framework. Examples of organic compounds so classified are alcohols, phenols, ethers, amines, ketones, aldehydes, amides, carboxylic acids, and esters. The essential heteroatom-containing structural feature used to identify each one of these classes of organic compounds is called a **functional group** (Table 21.2). Through the next several sections we shall be examining these various functional groups to see how they affect the properties of organic molecules.

Table 21.2
Functional Groups in Organic Molecules

General Structure	Name	Class of Organic Molecule	General Structure	Name	Class of Organic Molecule
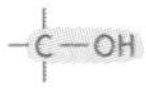 –C–OH	hydroxyl group	alcohols	 C–C(=O)–C	ketone group	ketones
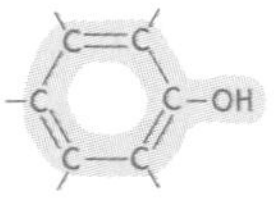 (benzene ring)–OH	phenolic group	phenols	 –C(=O)–H	aldehyde group	aldehydes
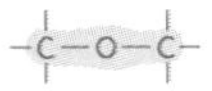 –C–O–C–	ether group	ethers	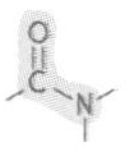 –C(=O)–N–	amide group	amides
 –C–N<	amine group	amines	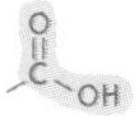 –C(=O)–OH	carboxyl group	carboxylic acids
 C=O	carbonyl group	common to ketones, aldehydes, amides, carboxylic acids, and esters	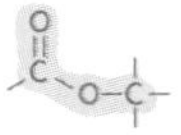 –C(=O)–O–C–	ester group	esters

[2]*Hetero* means "different." In organic chemistry a heteroatom is any atom different from carbon or hydrogen.

21.4 Alcohols, Phenols, Ethers, and Amines

Alcohols are a class of organic molecules in which a *hydroxyl group* is bonded to a saturated carbon. As can be seen in Table 21.2, the hydroxyl group consists of an oxygen bonded to a hydrogen, OH. Because of the polarity of the oxygen-hydrogen bond, smaller alcohols are often soluble in water. Some common alcohols are listed in Table 21.3.

Table 21.3
Properties of Some Simple Alcohols

Scientific Name	Common Name	Structure	Melting Point (°C)	Boiling Point (°C)
methanol	methyl alcohol	$CH_3—OH$	−97	65
ethanol	ethyl alcohol	$CH_3—CH_2—OH$	−115	78
2-propanol	isopropyl alcohol	$CH_3—CH(OH)—CH_3$	−126	97

Where:

$CH_3—$ = H—C(H)(H)— $—CH_2—$ = —C(H)(H)— $—CH—$ = —C(H)—

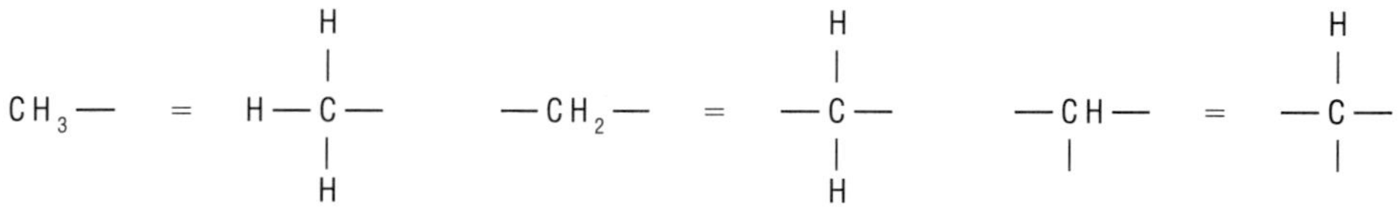

More than 11 billion pounds of methanol, CH_3OH, is produced annually in the United States. Most of it is used for making formaldehyde and acetic acid, which are important starting materials for the production of various polymers. In addition, methanol is used as a solvent and as an octane booster and anti-icing agent in gasoline. Methanol is sometimes called wood alcohol because it can be obtained from the distillation of wood. Methanol should never be ingested because once in the body it is metabolized to formaldehyde and formic acid. Formaldehyde is harmful to the eyes and can lead to blindness; it is familiar to biology students as the chemical used to preserve dead lab specimens. Formic acid, the active ingredient of an ant bite, can lower the pH of the blood to dangerous levels. Ingesting only about 15 milliliters of methanol may lead to blindness, and about 30 milliliters can cause death.

Ethanol is one of the oldest chemicals cultivated by humans. It is the "alcohol" of alcoholic beverages. Ethanol is prepared for drink by fermenting sugars from various plants. Ethanol is also widely used as an industrial solvent, which for many years was also made by fermentation. Today, industrial-grade ethanol is more cheaply manufactured from petroleum byproducts, such as ethene (Figure 21.14).

Fermentation can lead to concentrations of ethanol no greater than about 12 percent. At this concentration the ethanol producing yeast begins to die. Higher concentra-

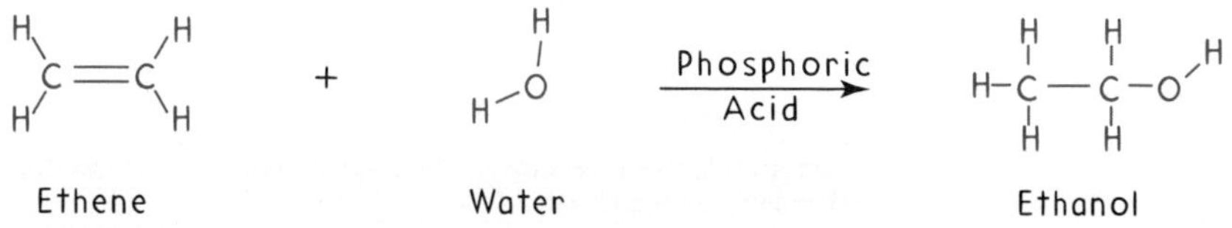

$$H_2C{=}CH_2 + H{-}O{-}H \xrightarrow{\text{Phosphoric Acid}} H{-}CH_2{-}CH_2{-}O{-}H$$

Ethene   Water   Ethanol

Fig. 21.14
Ethanol can be synthesized from the unsaturated hydrocarbon ethene with phosphoric acid as a catalyst.

tions are obtained by distilling the fermented brew. This technique began in the 1200s with the development of glass, which facilitated the design of distillation devices. In the United States, the alcohol content of strong alcoholic beverages is measured as *proof,* which is twice the percent ethanol. An 86-proof whiskey, for example, is 43 percent alcohol. The term evolved from a crude method once employed to test alcohol content. Gun powder was wetted with a beverage of suspect alcohol content. If the beverage was primarily water, the gun powder would not ignite. But if the beverage contained significant percentages of ethanol, the gun powder would explode, thus providing "proof" of the beverage's worth. Sadly, more than 15,000 people die in the United States each year from alcohol-related traffic accidents, accounting for about 40 percent of all traffic fatalities. Furthermore, the intoxicating effects of ethanol relate to many other social problems, such as violent crime and alcoholism.

A third well-known alcohol is isopropyl, or "rubbing," alcohol, also called 2-propanol. Although 2-propanol has a relatively high boiling point, it readily evaporates, leading to a pronounced cooling effect when applied to skin—an effect once used to reduce fevers.[3] Also, 2-propanol is used to swab skin prior to needle injection because it readily dissolves grime and disinfects.

Phenols are similar to alcohols in that they also contain a hydroxyl group. The hydroxyl group of a phenol, however, is attached directly to a benzene ring. The simplest phenol, shown in Figure 21.15, bears the same name: phenol. In 1867, after learning of Louis Pasteur's discovery of bacteria, Joseph Lister discovered the antiseptic value of phenol. Solutions of phenol applied to surgical instruments and to incisions greatly increased surgery survival rates. Phenol was the first purposefully used antibacterial solution, or *antiseptic.* Because phenol damages healthy tissue, a number of alternative

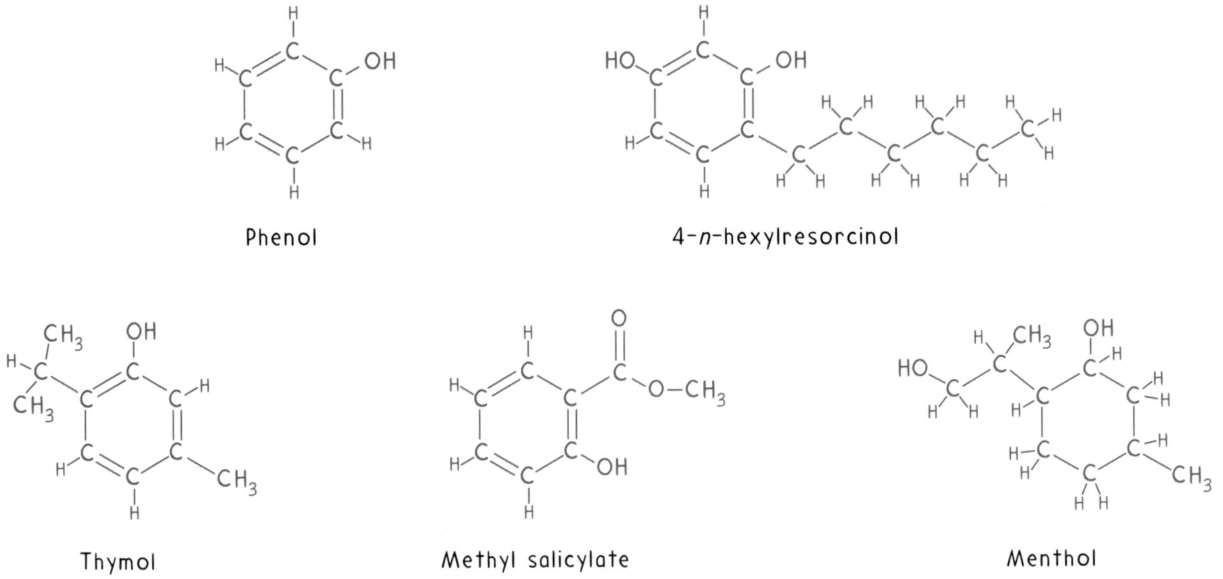

Phenol

4-*n*-hexylresorcinol

Thymol

Methyl salicylate

Menthol

Fig. 21.15
Phenols contain a hydroxyl group attached to a benzene ring (highlighted in green). Which one of these structures is not a phenol?

[3]Isopropyl alcohol is very toxic if ingested. A cool washcloth wetted with water is nearly as effective in lowering fever and far safer.

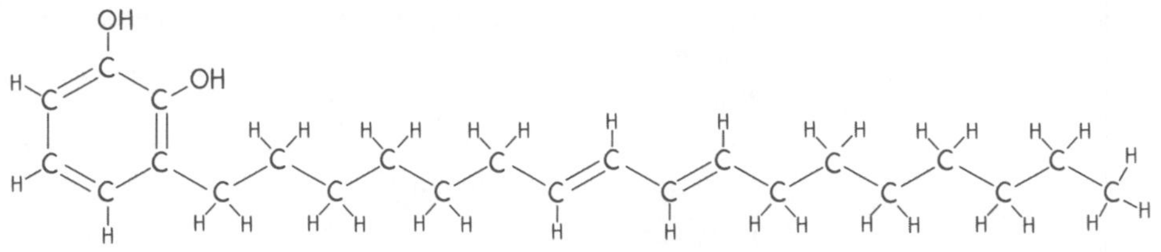

Fig. 21.16
Tetrahydrourushiol is the active component of poison oak.is not a phenol?

Tetrahydrourushiol

phenols with antiseptic properties have been introduced. The phenol 4-*n*-hexylresorcinol is commonly used in throat lozenges and mouthwashes. This compound has even greater antiseptic properties than phenol, and yet it is mild to tissue. Listerine brand mouthwash contains the antiseptic phenols thymol and methyl salicylate, as well as menthol, which is not a phenol but an alcohol added for its minty flavor and throat-soothing properties.

The skin-irritating component of poison ivy and similar plants is the phenol tetrahydrourushiol, illustrated in Figure 21.16. Its long hydrocarbon tail embeds itself within the skin, where the molecule initiates an allergic response. Scratching the itch spreads tetrahydrourushiol molecules over a greater surface area, causing the zone of irritation to grow.

Ethers are a class of organic compounds structurally related to alcohols. The oxygen atom in ethers, however, is not part of a hydroxyl group. Instead, it is bonded to two carbon atoms, as we see in Figure 21.17. Although ethanol and dimethyl ether have the same chemical formula, C_2H_6O, their physical properties are vastly different. Whereas ethanol mixes with water and boils at 78°C, dimethyl ether is immiscible with water and boils at −25°C. These differences are due to the fact that ethers lack the polar hydroxyl group of alcohols. Ethers are immiscible with water because, without the hydroxyl group, they are unable to form strong dipole-dipole attractions with water. Furthermore, without the polar hydroxyl group, the molecular interactions among ether molecules are relatively weak. Therefore ether molecules are easy to pull apart from one another. This is why ethers have relatively low boiling points and evaporate so readily.

Diethyl ether, with a boiling point of 35°C, was one of the first anesthetics. The anesthetic properties of diethyl ether were discovered in the early 1800s and revolutionized the practice of surgery. Because of its high volatility at room temperature, diethyl ether can be rapidly administered to the bloodstream by way of inhalation. Because it has low solubility in water, however, it quickly leaves the bloodstream once introduced. Because of these physical properties, a surgical patient can be brought in and out of anesthesia (a state of unconsciousness) at will simply by regulating the gases breathed. Modern-day gaseous anesthetics, which have fewer side effects such as nausea and headaches, work by the same principle.

Amines are a class of organic compounds that contain nitrogen bonded to saturated carbon atoms. The compounds shown in Table 21.4 are examples. The nitrogen of an amine may be connected to one, two, or three carbons. Because the polarity of the nitrogen-hydrogen or the nitrogen-carbon bond is not as great as the polarity of the

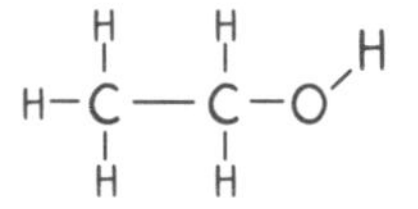

Ethanol
soluble in water
boiling temperature = 78°C

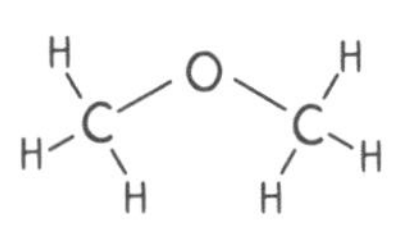

Dimethyl ether
insoluble in water
boiling temperature = -25°C

Fig. 21.17
The oxygen of alcohols is bonded to one carbon atom and one hydrogen atom. The oxygen of ethers is bonded to two carbon atoms. Alcohols and ethers of similar molecular mass have vastly different physical properties.

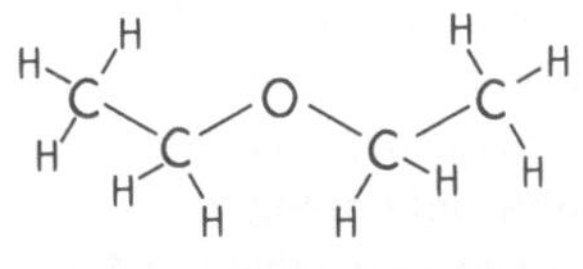

Diethyl ether

Fig. 21.18
Diethyl ether is the technical name for the "ether" that historically was used as an anesthetic.

Table 21.4
Properties of Some Simple Amines

Scientific Name	Structure (°C)	Melting Point (°C)	Boiling Point
ethylamine	$CH_3CH_2—NH_2$	−81	17
diethylamine	$CH_3CH_2—NH—CH_2CH_3$	−50	55
triethylamine	$CH_3CH_2—N(CH_2CH_3)—CH_2CH_3$	−7	89

oxygen-hydrogen bond, amines are typically less soluble in water than alcohols. Also, their boiling points are typically somewhat less than those of alcohols of similar molecular mass.

One of the most notable physical properties of many lightweight amines is their offensive odor. Two appropriately named amines, putrescine and cadaverine, for example, are responsible for the odor of decaying flesh.

$H_2N—CH_2CH_2CH_2CH_2—NH_2$

Putrescine
(1,4-butanediamine)

$H_2N—CH_2CH_2CH_2CH_2CH_2—NH_2$

Cadaverine
(1,5-pentanediamine)

Organic amines typically are alkaline because the nitrogen atom readily accepts a hydrogen ion (Figure 21.19). In general, any organic molecule containing a nitrogen tends to be slightly alkaline.

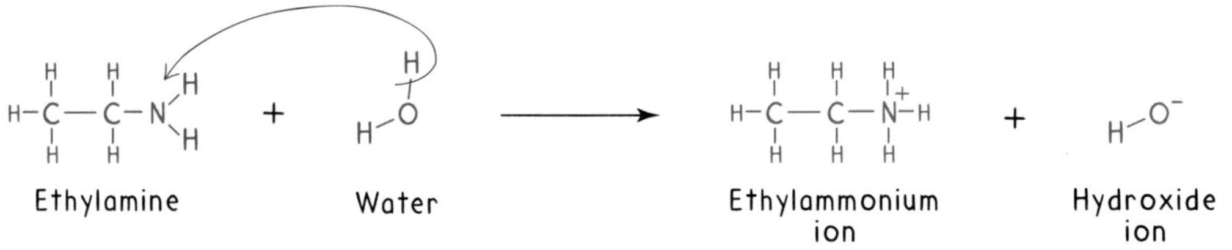

$$CH_3CH_2NH_2 + H_2O \longrightarrow CH_3CH_2NH_3^+ + OH^-$$

Ethylamine + Water → Ethylammonium ion + Hydroxide ion

Fig. 21.19
Ethylamine acts as a base and accepts a hydrogen ion from water to become the ethylammonium ion. This generates the hydroxide ion, which increases the pH of the solution.

Molecules found in nature that are alkaline because they contain one or more nitrogen atoms are often called *alkaloids* (Figure 21.20). Many have medicinal value. There is great interest in isolating these compounds from the natural materials in which they occur, which may be of plant or marine origin. Alkaloids are basic, and so they react with acids to form neutral salts that are usually quite soluble in water (Figure 21.21). An efficient way to isolate alkaloids, therefore, is to expose alkaloid-containing material, which is usually insoluble in water, to an aqueous solution of acid. By this method, all the alkaloids present convert to a water-soluble salt form. Alkaloid salts are then carried away by water. Once the alkaloids are isolated, other chemical separation techniques can be used to purify them.

Nature is usually a step ahead of us in forming the salts of alkaloids. Frequently, natural salts are made using certain organic acids called *tannic acids.* Tannic acid salts are usually soluble only in hot water. This is why we use hot water when we brew coffee beans or tea leaves for their caffeine, which is found in these plant materials primarily as the tannic acid salt.

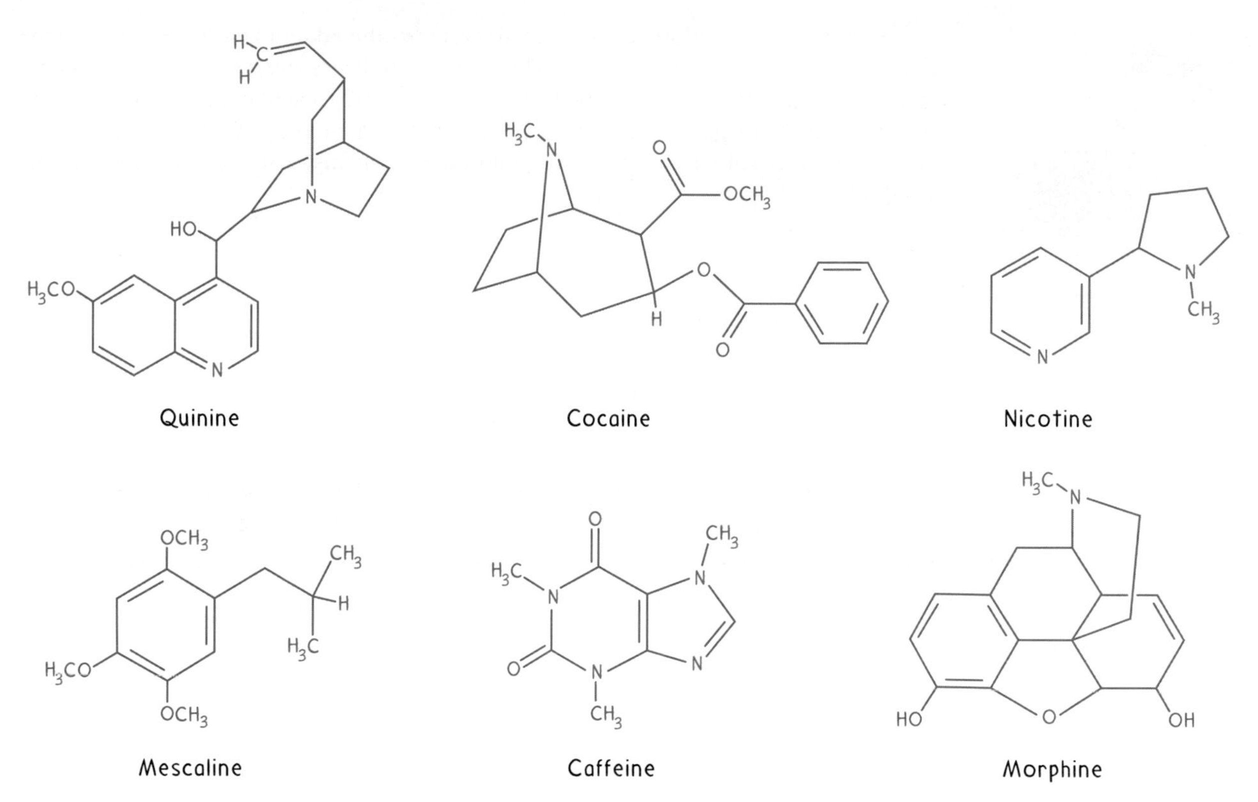

Fig. 21.20
Alkaloids are a class of naturally occurring compounds that are alkaline because of the presence of nitrogen atoms. An alkaloid molecule may also contain other heteroatoms. In the structures drawn here, each corner represents a carbon atom. Most of the hydrogen atoms have been excluded to make it easier to view the great variety of carbon frameworks. We'll continue to use this common method of drawing organic structures in subsequent illustrations. Watch carefully.

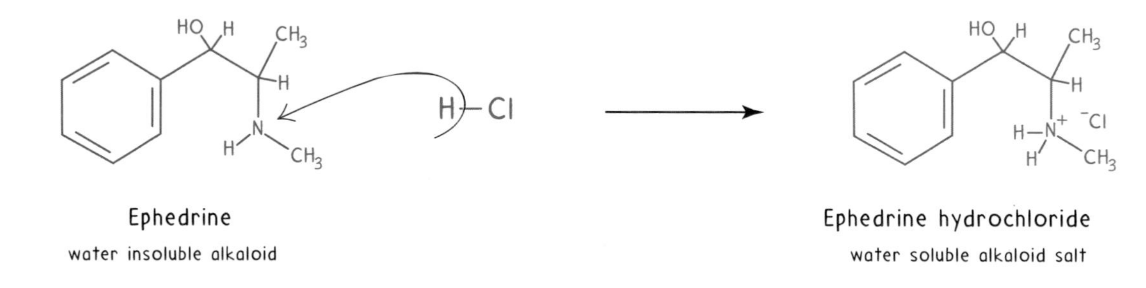

Fig. 21.21
Ephedrine, an alkaloid that is not soluble in water, reacts with hydrochloric acid, HCl, to form the hydrogen chloride salt of ephedrine, a substance that is soluble in water and formulated for prescription use. As we saw in Section 20.1, the nonsalt form is referred to as a *free base*.

Humans have developed a remarkable taste for the alkaloid caffeine, but caffeine by itself is not very soluble in water. The beverage industry overcomes this by making beverages acidic. Read the label on your next can of caffeinated cola soft drink. You'll find that it contains phosphoric acid. The phosphoric acid guarantees that the caffeine stays in its water-soluble salt form. The phosphoric acid also enhances the flavor of the soft drink.

Fig. 21.22
Tannins are responsible for the brown stains left behind in coffee mugs or on a coffee drinker's teeth. Because tannins are acidic, they can be readily removed with an alkaline cleanser. For the coffee mug, use a little bleach. For your teeth, use baking soda.

Question

What is the significance of heteroatoms in an organic molecule?

21.5 Carbonyl-Containing Organic Molecules

The remaining major classes of organic compounds are the ketones, aldehydes, amides, carboxylic acids, and esters. All these compounds have in common a special group called the **carbonyl group**, which, as shown in Table 21.2, consists of a carbon atom double-bonded to an oxygen atom.

Ketones and Aldehydes

A **ketone** is a carbonyl-containing organic molecule in which the carbonyl carbon is bonded to two carbon atoms. A familiar example of a ketone is *acetone,* often used in nail polish remover (Figure 21.23a). In an **aldehyde,** the carbonyl carbon is bonded either to one carbon atom and one hydrogen atom, as in Figure 21.23b, or, in the special case of formaldehyde, to two hydrogen atoms.

O, C, H_3C, CH_3

(a) Acetone

O, H_3C, C, C, H, H, H

(b) Propionaldehyde

Fig. 21.23
(a) When the carbon of a carbonyl group is bonded to two carbon atoms, the result is a ketone. An example is acetone. (b) When the carbon of a carbonyl group is bonded to at least one hydrogen atom, the result is an aldehyde. An example is propionaldehyde.

Ketones and aldehydes are quite similar structurally. With aldehydes, however, the carbonyl group must always come at the end of a carbon chain. If it were in the middle, the carbonyl would be surrounded by two carbons and the molecule would be a ketone.

Answer
Heteroatoms largely determine the "personality" of an organic compound. The varied properties of all the different classes of organic compounds discussed in this and future sections, for example, are a consequence of the heteroatoms they contain.

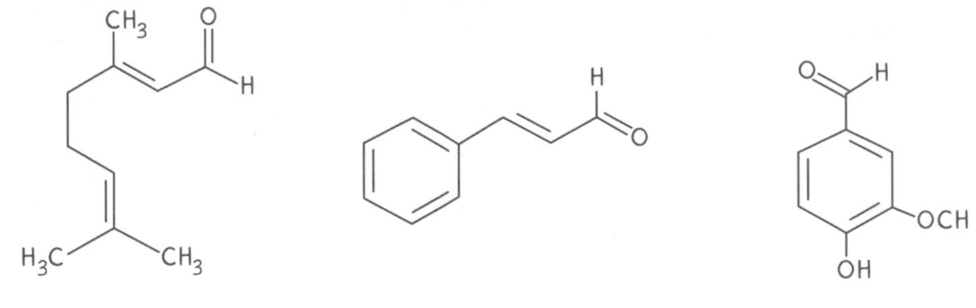

Fig. 21.24
Aldehydes are responsible for many familiar fragrances. As in Figure 21.20, each corner represents a carbon atom and most of the hydrogen atoms are not drawn.

Citral

Cinnamonaldehyde

Vanillin

Many aldehydes are particularly fragrant. A number of flowers, for example, owe their pleasant odor to the presence of simple aldehydes. The smell of lemons is due to the aldehyde *citral.* The smells of cinnamon, vanilla, and almond are due to the aldehydes *cinnamaldehyde, vanillin,* and *benzaldehyde,* respectively.

Amides, Carboxylic Acids, and Esters

An **amide** is a carbonyl-containing organic molecule in which the carbonyl carbon is bonded to a nitrogen atom (Figure 21.25). The active ingredient of most mosquito repellents is an amide whose chemical name is *N,N*-diethyl-*m*-toluamide but is commercially known as DEET. This compound is actually not an insecticide. Rather, it causes certain insects, especially mosquitoes, to lose their sense of direction, which effectively protects DEET wearers from being bitten.

A **carboxylic acid** is a carbonyl-containing organic molecule in which the carbonyl carbon is bonded to a hydroxyl group (Figure 21.26). An example is *salicylic acid,* found in the bark of the willow tree. Once brewed for its antipyretic (fever-reducing) effect, salicylic acid became an important analgesic (painkiller), but it also causes nausea and severe stomach upset. In 1899, Friederich Bayer and Company, in Germany, introduced a chemically modified version of this compound that has fewer side effects. This compound was acetylsalicylic acid, the chemical name for aspirin.

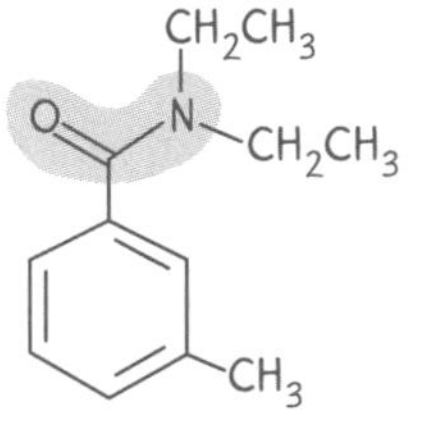

N,N-Diethyl-m-toluamide (DEET)

Fig. 21.25
Amides contain the amide group, highlighted in blue, in which a nitrogen is bonded to a carbonyl carbon.

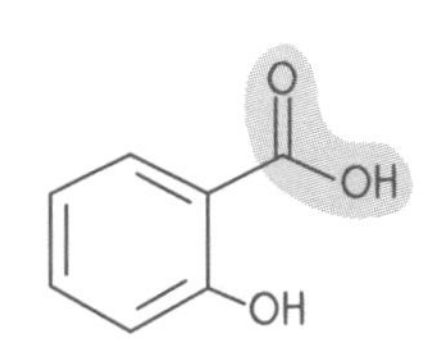

Salicyclic acid

Fig. 21.26
The carboxyl group, highlighted in blue, consists of a hydroxyl group bonded to a carbonyl group.

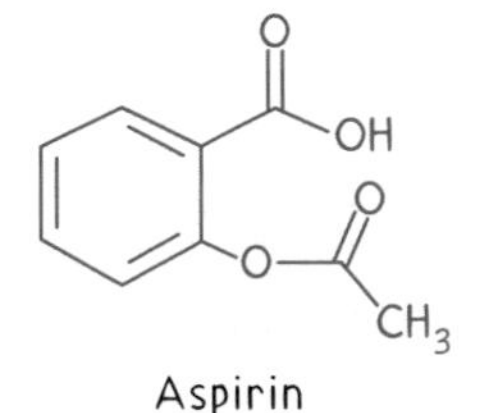

Aspirin
(Acetylsalicylic acid)

Fig. 21.27
Aspirin was originally synthesized from the naturally occurring salicylic acid found in willow trees.

Table 21.5
Some Esters and Their Flavors

Structure	Name	Flavor
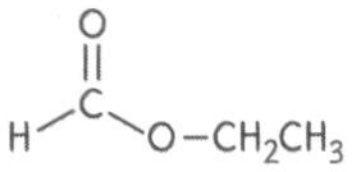 $H{-}C({=}O){-}O{-}CH_2CH_3$	Ethyl formate	Rum 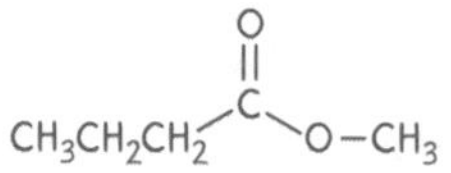
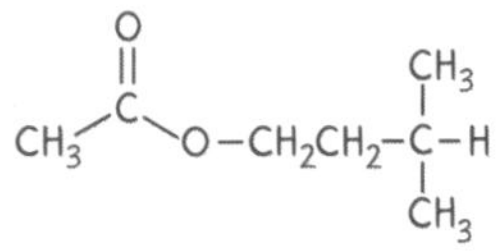 $CH_3{-}C({=}O){-}O{-}CH_2CH_2{-}C(CH_3)_2{-}H$	Isopenyl acetate	Banana 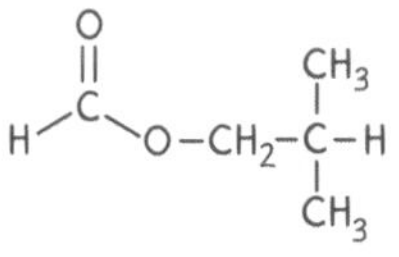
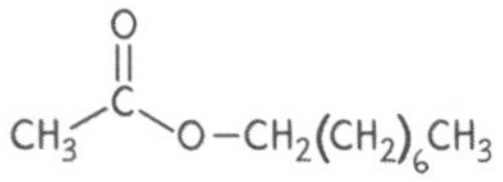 $CH_3{-}C({=}O){-}O{-}CH_2(CH_2)_6CH_3$	Octyl acetate	Orange 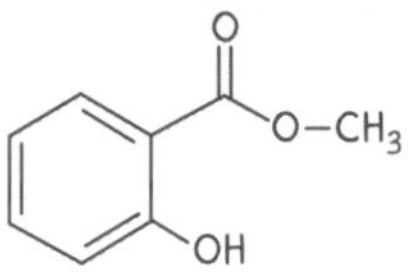
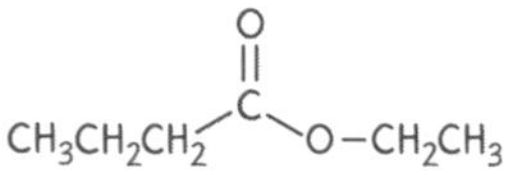 $CH_3CH_2CH_2{-}C({=}O){-}O{-}CH_2CH_3$	Ethyl butyrate	Pineapple
$CH_3CH_2CH_2{-}C({=}O){-}O{-}CH_3$	Methyl butyrate	Apple
$H{-}C({=}O){-}O{-}CH_2{-}C(CH_3)_2{-}H$	Isobutyl formate	Raspberry
$C_6H_4(OH){-}C({=}O){-}O{-}CH_3$	Methyl salicylate	Wintergreen

An **ester** molecule is similar to a carboxylic acid except that in the ester the hydrogen of the hydroxyl group is replaced by a carbon (Table 21.2). Like aldehydes, many simple esters have notable fragrances and are used as flavorings (Table 21.5).

■ Question

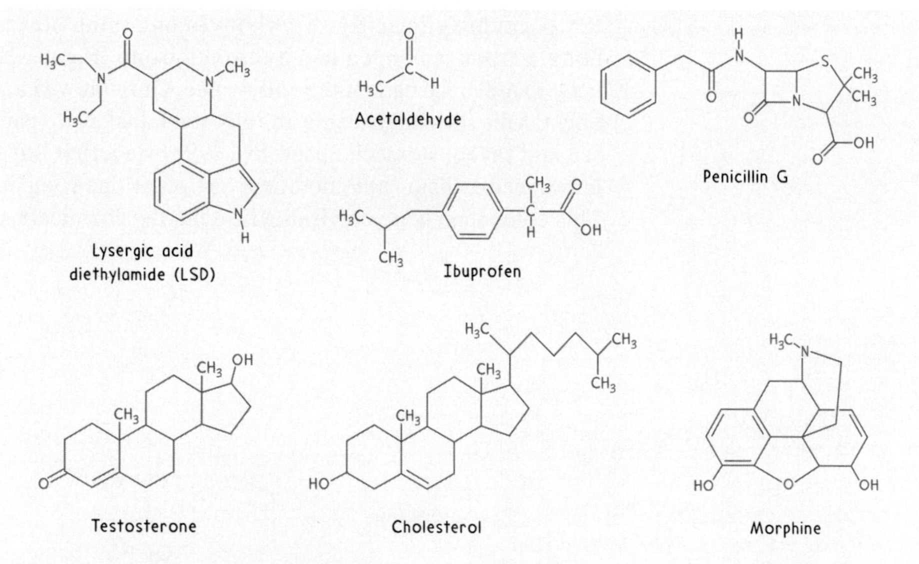

These organic molecules may be one or more of the following: alcohol, phenol, ether, amine, ketone, aldehyde, amide, carboxylic acid, or ester. Identify each. (You can ignore the sulfur group in penicillin G.)

■ **Answer**

LSD: amide, amine; *acetaldehyde:* aldehyde; *ibuprofin:* carboxylic acid; *penicillin G:* amide, carboxylic acid; *testosterone:* alcohol, ketone; *cholesterol:* alcohol; *morphine:* alcohol, phenol, ether, amine.

Link to Medicine—Drugs

A drug may be loosely defined as any chemical that elicits a biological effect. Most drugs used today are organic compounds, and they come from different origins. Many drugs come directly from terrestrial or marine plants or animals. Others are natural products that have been chemically modified to increase potency or decrease harmful side effects. There is also a growing number of drugs that are produced only in the laboratory.

Drug	Biological Effect	Origin
Caffeine	nerve stimulant	natural product
Reserpine	reduces hypertension	
Vincristrine	anticancer agent	
Penicillin	antibiotic	
Morphine	analgesic	
Prednisone	antirheumatic	chemical derivative of a natural product
Ampicillin	antibiotic	
LSD	hallucinogen	
Chloroquinine	antimalarial	
Ethynodiol diacetate	contraceptive	
Valium	antidepressant	laboratory
Benadryl	antihistamine	
Allobarbital	sedative-hypnotic	
Phencyclidine	veterinary anesthetic	
Methadone	analgesic	

When you pop an aspirin pill for a headache, how does the aspirin know to go to your head rather than your big toe? The answer is, it doesn't! After aspirin dissolves in your stomach, it gets absorbed into your bloodstream, which distributes the drug over your whole body—from head to toe. Consequently, aspirin is good for headaches, muscle aches, backaches, and toe aches.

One effect of aspirin is to alleviate pain. However, aspirin can also cause ringing in your ears and can even inhibit your blood from clotting. Like most other drugs, aspirin has more than one effect on the body. Much effort in the pharmaceutical industry is directed toward modifying the chemical structure of drugs so that their effects are more specifically aimed at treating ailments.

An interesting example of this approach is cancer chemotherapy. Many anticancer drugs work by killing cells that are in the process of dividing. This selectively kills cancer cells because, unlike normal cells, cancer cells are nearly always dividing. This treatment has had much success; however, it's not perfect. One problem is that even normal cells divide occasionally. Some cells, such as those in your intestines and hair follicles, divide quite frequently. Anticancer drugs therefore take their toll on normal cells, too. As researchers focus on other differences between cancer cells and normal cells, however, it is hoped that more-specific anticancer drugs may be found. For example, it now appears that many cancer cells have unusual outer surfaces. Anticancer drugs that are able to selectively stick to this outer surface might help to increase the success of cancer chemotherapy manyfold.

Many drugs work like a lock and key. There are different types of receptor sites in the body, each acting like a lock in a door. When a drug molecule fits into one of these sites, the way a key fits into a lock, a particular biological effect, such as a nerve impulse or change in cellular morphology, is triggered. To fit into a particular receptor site, a drug molecule must have the proper shape, just as a key must have properly shaped notches to fit the lock.

According to this model, the problem with nonspecific drugs is that they fit into too many receptor sites. Hence, like a skeleton key, they "unlock" a variety of biological effects. Knowing the precise shape of a target receptor site, such as one on the surface of a cancer cell, allows chemists to design molecules that have an optimum fit and a specific biological effect.

21.6 Modified Natural Polymers

Polymers are exceedingly long organic molecules that consist of repeating molecular units called **monomers** (Figure 21.28).[4] Monomers have relatively simple structures consisting of anywhere from 4 to 100 atoms per molecule. When chained together, they can form polymers consisting of hundreds of thousands of atoms per molecule. These large molecules are still too small to be seen with the unaided eye. They are, however, giants in the world of the submicroscopic—if a typical polymer molecule were as thick as a kite string, it would be as much as a kilometer in length.

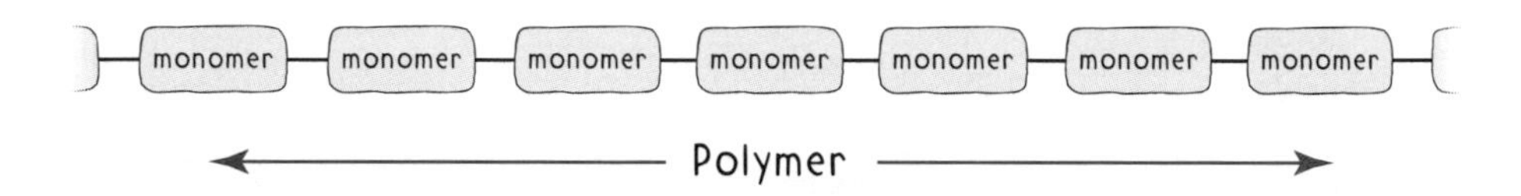

Fig. 21.28
A polymer is a long molecule comprising many smaller monomer molecules linked together.

Many of the molecules that make up living tissue are polymers. This includes the cellulose of plants, the complex carbohydrates of starchy foods, protein molecules, and DNA. We leave a discussion of these important molecules to a follow-up course in the life sciences. In this and the following section, we instead focus on the human-made polymers that have had a profound effect on our standard of living. This includes natural polymers that have been chemically modified to increase their utility to humans and polymers that are solely a product of human invention.

The first polymers to be manufactured were modified versions of naturally occurring polymers. *Vulcanized rubber* is an important example. This material is derived from natural rubber, which is a semisolid, elastic, natural polymer excreted by various plants, such as the rubber trees of Malaysia and Indonesia. The fundamental chemical unit of natural rubber is polyisoprene, which is produced by plants from monomeric isoprene molecules (Figure 21.30).

Fig. 21.29
Indonesian rubber trees ready for harvest.

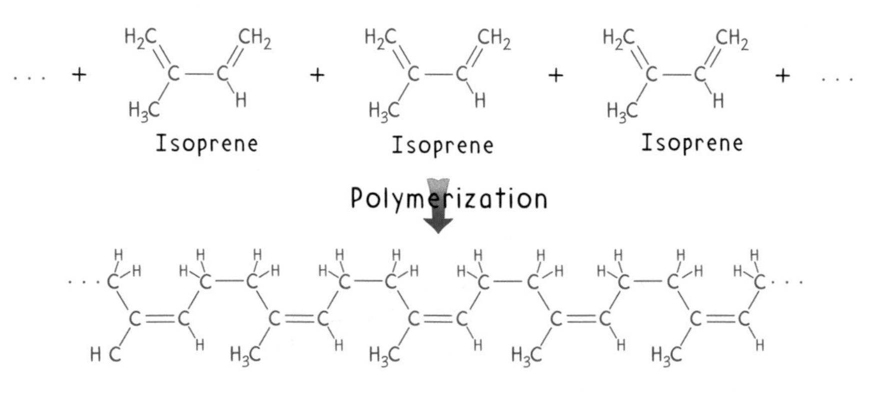

Fig. 21.30
Isoprene molecules (monomers) react with one another to form the polymer polyisoprene, the main component of natural rubber.

[4]The term *polymer* is derived from the Greek *poly-* ("many") and *-mer* ("parts"); *monomer* comes from the Greek *mono-* ("one").

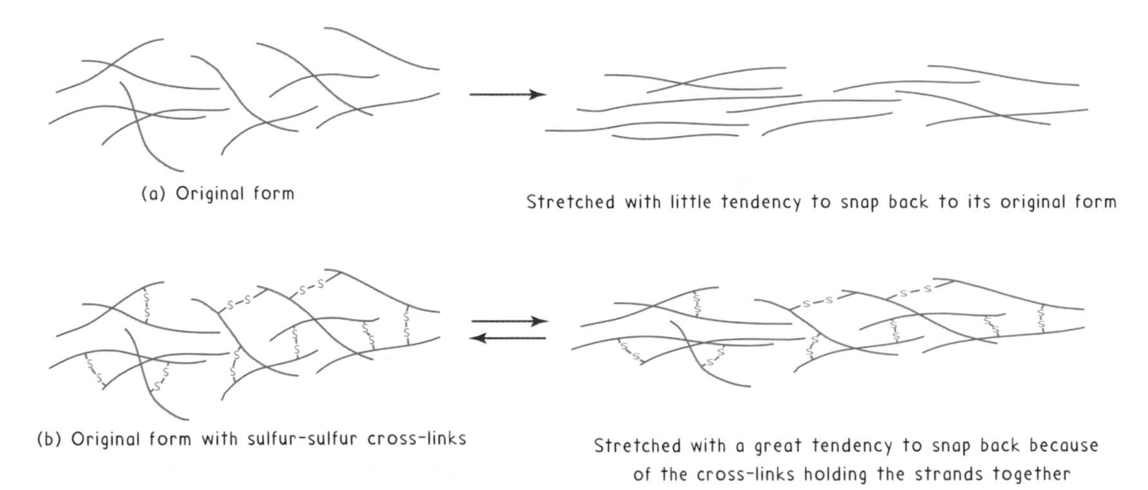

Fig. 21.31
(a) When stretched, individual polyisoprene strands slip past one another. (b) Vulcanized rubber is both harder and more elastic than natural rubber because of sulfur cross-links between polyisoprene strands. When stretched, the sulfur cross-links hold the strands together, allowing the rubber to return to its original shape.

In the 1700s, natural rubber was noted for its ability to rub off pencil marks, which is the origin of the term *rubber.* Natural rubber has few other uses, however, because at warmer temperatures it turns gooey and at colder temperatures it becomes brittle and subject to breaking. In 1839, an American inventor, Charles Goodyear, discovered the process of *rubber vulcanization,* in which natural rubber is heated in the presence of sulfur. The product, vulcanized rubber, is harder than natural rubber and also retains its elastic properties over a wide range of temperatures. This is the result of sulfur *cross-linking* between polymer chains (Figure 21.31). Vulcanized rubber has found innumerable applications, from tires to rain gear, and has grown into a multibillion dollar industry. Goodyear unfortunately reaped very few rewards from his discovery. He was a man of ill-health who died in jail while serving time for debts he was unable to pay. The present-day Goodyear Corporation was founded not by Charles Goodyear, but by others who sought to pay tribute to his name—15 years after he died.

Fig. 21.32
To help quench our ever-growing thirst for vulcanized rubber, natural rubber (polyisoprene) is now synthesized from petroleum distillates. Synthetic polymers that mimic vulcanized rubber are also pro-

Some of the first modified natural polymers of the 20th century were derivatives of cellulose, which is the basic structural component of all plant life. *Cellulose nitrate,* for example, is produced by treating cellulose with nitric acid (Figure 21.33). When mixed with ethyl alcohol, cellulose nitrate transforms into a moldable material

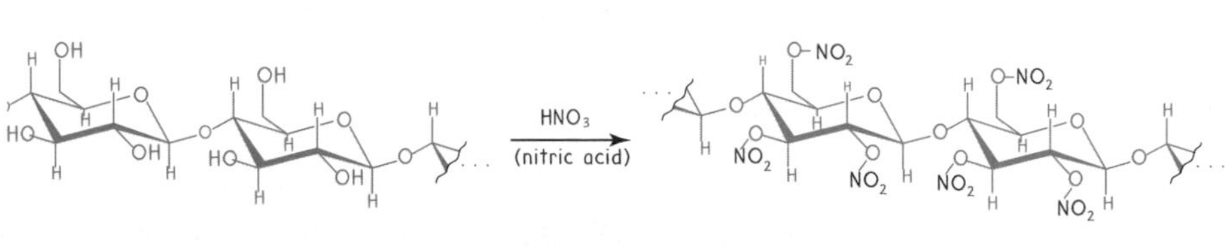

Cellulose $\xrightarrow[\text{(nitric acid)}]{HNO_3}$ Cellulose nitrate

Fig. 21.33
Treatment of cellulose with nitric acid results in cellulose nitrate.

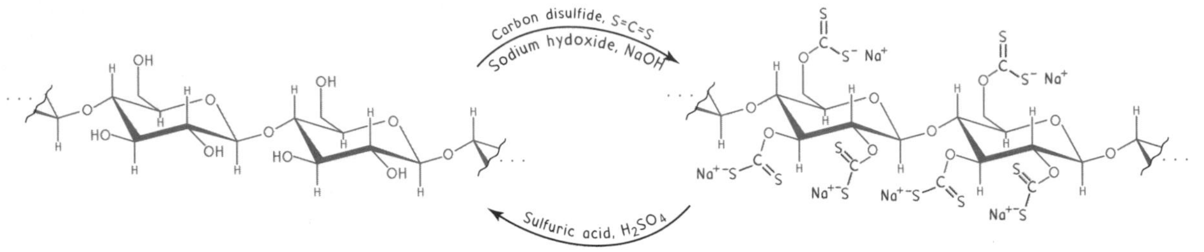

Carbon disulfide, S=C=S
Sodium hydoxide, NaOH
Sulfuric acid, H_2SO_4

Cellulose
(water insoluble)

Cellulose xanthate
(water soluble)

Fig. 21.34
(a) Cellulose xanthate is produced by reacting cellulose with carbon disulfide in the presence of sodium hydroxide. (b) After fibers have been spun, the xanthate attachment is hydrolyzed by sulfuric acid to produce rayon, which is a tightly bound form of celluslose.

that hardens once the alcohol evaporates. The resulting product, known as *celluloid,* was used in the 1920s for making a wide variety of products including billiard balls, combs, and collar stays. Thin films of celluloid were also used then as the backing for photographic film. Cellulose nitrate, however, is very flammable, and many movie theater fires started in projection booths where movie film and hot lamps were in close proximity. Stronger and more flame-resistant cellulose derivatives, such as cellulose acetate, have since replaced cellulose nitrate for most of these applications. Cellulose nitrate, however, is still used in the manufacture of Ping-Pong balls and magicians' flash paper. It also serves as a propellant for firearms, cannons, and rockets.

Fig. 21.35
Molten polymers can be blown into transparent plastic films.

Fig. 21.36
Cellophane wrapping is manufactured from cellulose.

A water-soluble cellulose derivative known as cellulose xanthate can be made by reacting cellulose with carbon disulfide in the presence of sodium hydroxide (Figure 21.34a). Thick solutions of cellulose xanthate can be drawn into thin fibers by extruding the solution through a mesh of tiny holes. Acid is then used to catalyze the removal of the xanthate attachment to yield *rayon* (Figure 21.34b). Although rayon and cellulose have the same chemical composition, rayon is stronger because the extruding process produces a better alignment of polymer chains that allows greater intermolecular attractions. Cellulose xanthate solutions are also pushed through a thin slit into an acidic solution, a process that produces the packaging film *cellophane.*

Paints

Paints consist of three fundamental components: pigment, binder, and solvent. *Pigments* are chemical compounds that give paints their color. Most paints start with a base of titanium dioxide, TiO_2, a bright-white pigment that has excellent covering ability. (Lead carbonates were once used for this purpose but have since been banned because of their toxicity.) A desired color is obtained by adding compounds such as cadmium sulfide, CdS (yellow), chromium trioxide, Cr_2O_3 (green), iron oxide, Fe_2O_3 (red), or carbon graphite, C (black). *Binders* are organic chemicals that polymerize and thereby fix the pigment to the surface. Historically, food oils such as linseed oil were used for this purpose. These oil paints, however, never really dry. Instead, they remain soft until the unsaturated carbon chains polymerize by cross-linking with one another upon exposure to atmospheric oxygen (Figure 21.37). Weather-resistant synthetic analogs of linseed oil that require less drying time are now widely used.

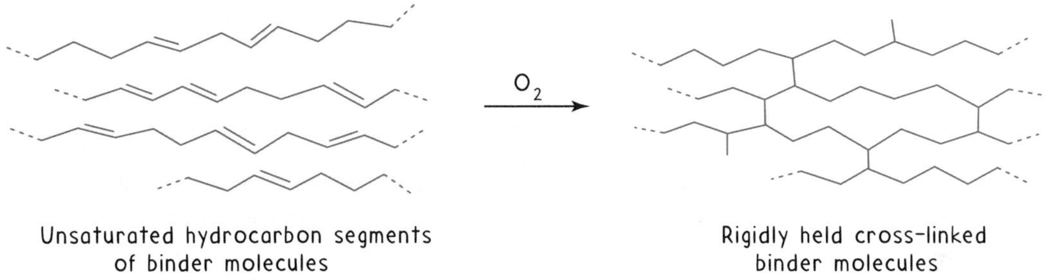

O_2

Unsaturated hydrocarbon segments of binder molecules

Rigidly held cross-linked binder molecules

Fig. 21.37
Atmospheric oxygen catalyzes the formation of covalent bonds among the unsaturated hydrocarbon chains of adjacent binder molecules. As these cross-links form, the paint hardens and becomes fixed to the painted surface.

Solvents allow paints to be spread easily over a surface. Oil-based paints use hydrocarbon solvents, such as turpentine. Water-based paints use water as the solvent even though the binders and pigments are insoluble in water. The trick is to create an emulsion using detergentlike molecules (Section 18.4). Within the emulsion, organic binders and pigments are corralled into tiny suspended globules, as illustrated in Figure 21.38.[5] The binders of a water-based paint are only partially polymerized. As water evaporates from the paint, the tiny suspended globules come together, allowing a completion of polymerization. The end result is a water-insoluble covering. A great advantage of water-based paints, aside from the easy clean-up, is that air-polluting organic solvents are avoided.

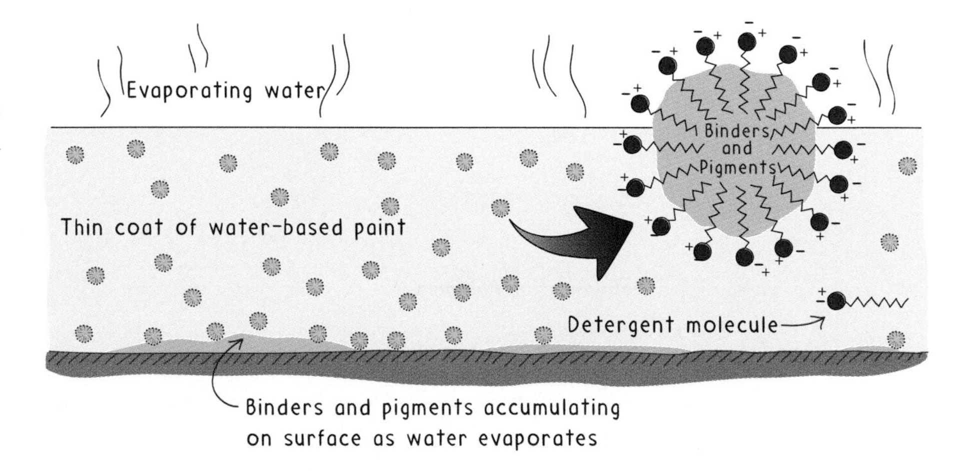

Fig. 21.38
Water-based paints are an emulsion of partially polymerized binders and pigments. As the water evaporates, binders and pigments accumulate on the painted surface.

[5]An emulsion is a type of suspension in which mutually insoluble liquids are dispersed as tiny droplets. Oil and vinegar when shaken together vigorously form a temporary emulsion.

21.7 Synthetic Polymers

A **synthetic polymer** is any polymer not found in nature. Solely the product of human design, synthetic polymers pervade modern living. They have been available since about the time of World War II, when the supply of natural rubber from Malaysia and Indonesia was cut off by Japan. In response, chemical researchers of the Allies looked intensively for alternatives. Their efforts resulted in much more than a mere replacement for natural rubber, for many new polymers with remarkable properties were discovered. Ultimately this led to the establishment of the plastics industry, which has grown enormously ever since. In the United States synthetic polymers have outstripped steel as the most widely used material.

Polymers are classified according to how they are formed. **Addition polymers** form by the joining of monomer units. For this to happen, each monomer must contain

Table 21.6
Addition and Condensation Polymers

Addition Polymers	Repeating Unit	Common Uses	Recycling Code
Polyethylene (PE)	$\cdots CH_2{-}CH_2 \cdots$	Plastic bags, bottles	2 HDPE, 4 LDPE
Polypropylene (PP)	$\cdots CH_2{-}CH(CH_3) \cdots$	Indoor-outdoor carpets	5 PP
Polystyrene (PS)	$\cdots CH_2{-}CH(C_6H_5) \cdots$	Plastic utensils, insulation	6 PS
Polyvinyl chloride (PVC)	$\cdots CH_2{-}CHCl \cdots$	Shower curtains, tubing	3 V
Polyvinylidene chloride (Saran)	$\cdots CH_2{-}CCl_2 \cdots$	Plastic wrap	--
Polytetrafluoroethylene (Teflon)	$\cdots CF_2{-}CF_2 \cdots$	Non-stick coating	--
Polyacrylonitrile (Orlon)	$\cdots CH_2{-}CH(C{\equiv}N) \cdots$	Yarn, paints	--
Polymethyl methacrylate (Lucite, Plexiglas)	$\cdots CH_2{-}C(CH_3)(C(=O)OCH_3) \cdots$	Windows, bowling balls	--
Polyvinyl acetate (PVA)	$\cdots CH_2{-}CH(O{-}C(=O){-}CH_3) \cdots$	Adhesives, chewing gum	--
Condensation Polymers			
Nylon	$\cdots C(=O){-}(CH_2)_4{-}C(=O){-}NH{-}(CH_2CH_2)_3{-}NH \cdots$	Carpeting, clothing	--
Polyethylene terephthalate (Dacron, Mylar)	$\cdots C(=O){-}C_6H_4{-}C(=O){-}O{-}CH_2CH_2{-}O \cdots$	Clothing, plastic bottles	1 PET
Melamine-Formaldehyde resin (Melmac, Formica)	$\cdots NH{-}C_3N_3(NH{-}C{=}O \cdots){-}NH{-}C(=O) \cdots$	Dishes, countertops	--

at least one multiple bond. Polymerization occurs when one of the bonds in the multiple bond opens up to form a new bond with a neighboring monomer molecule, as we saw in the formation of polyisoprene in Figure 21.30. During this process, no atoms are lost so that the total mass of the polymer is equal to the sum of the masses of all the monomers. Table 21.6 shows some examples of addition polymers.

For some polymers, it is not necessary that the monomers contain multiple bonds. Instead, as we'll be showing in this section, the joining of monomer units is accompanied by the loss of small molecules, such as water or hydrochloric acid. These are the **condensation polymers**, which include Nylon, Dacron polyester, and various hard-setting resins (Table 21.6).

Addition Polymers

The most common polymer is the addition polymer polyethylene. Nearly 12 million tons of polyethylene is produced annually in the United States; that's about 90 pounds per U.S. citizen. It is synthesized from ethylene, an unsaturated hydrocarbon produced in large quantities from petroleum. Polyethylene is the simplest polymer, consisting of straight hydrocarbon chains.

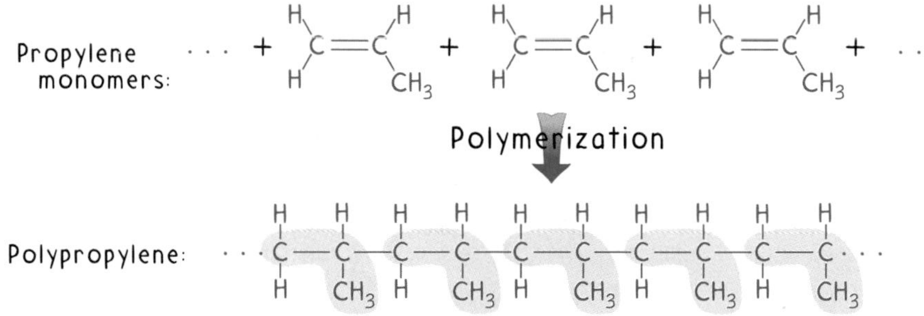

Propylene monomers: $\cdots + H_2C{=}CH(CH_3) + H_2C{=}CH(CH_3) + H_2C{=}CH(CH_3) + \cdots$

Polymerization

Polypropylene: $\cdots -CH_2-CH(CH_3)-CH_2-CH(CH_3)-CH_2-CH(CH_3)-CH_2-CH(CH_3)-CH_2-CH(CH_3)- \cdots$

Two principal forms of polyethylene are produced by using different catalysts and reaction conditions. High-density polyethylene (HDPE) consists of long, straight chains packed closely together (Figure 21.39a). The tightness of these chains makes HDPE a relatively rigid and tough plastic useful for such things as bottles and milk jugs. Low-density polyethylene (LDPE) has many branching side chains, an architecture that prevents polymer strands from packing close together (Figure 21.39b). This makes LDPE a more bendable plastic than HDPE and gives it a lower melting point. While HDPE holds its shape in boiling water, LDPE deforms. Low-density polyethylene is most useful for such items as plastic bags, photographic film, and electric wire insulation.

(a) Molecular strands of HDPE

(b) Molecular strands of LDPE

Fig. 21.39
(a) The polyethylene strands of HDPE are able to pack close together, much like strands of uncooked spaghetti. (b) The polyethylene strands of LDPE are branched, which prevents a close-packing among the molecules.

Other widely used addition polymers consist of a polyethylene backbone with the periodic placement of side groups, which are incorporated by selecting the proper monomer. A methyl group can be substituted on every other carbon, for example, by using propylene as the monomer. This yields polypropylene, a tough plastic material useful for pipes, hard-shell suitcases, and appliance parts. Fibers of polypropylene are also used for upholstery and indoor-outdoor carpets.

Propylene monomers: $\cdots + H_2C{=}CH(CH_3) + H_2C{=}CH(CH_3) + H_2C{=}CH(CH_3) + \cdots$

Polymerization

Polypropylene: $\cdots -CH_2-CH(CH_3)-CH_2-CH(CH_3)-CH_2-CH(CH_3)-CH_2-CH(CH_3)-CH_2-CH(CH_3)- \cdots$

Adding a benzene ring to every other carbon yields polystyrene, which is manufactured from styrene. Transparent plastic cups are made of polystyrene, as are thousands of other throwaway household items. Blowing gas into liquid polystyrene generates Styrofoam, widely used for coffee cups, packing material, and insulation.

Styrene monomers: $\cdots + H_2C{=}CH(C_6H_5) + H_2C{=}CH(C_6H_5) + H_2C{=}CH(C_6H_5) + \cdots$

Polymerization

Polystyrene: $\cdots -CH_2-CH(C_6H_5)-CH_2-CH(C_6H_5)-CH_2-CH(C_6H_5)-CH_2-CH(C_6H_5)-CH_2-CH(C_6H_5)- \cdots$

Fig. 21.40
PVC is tough and easily molded, which is why it is used to fabricate many household items.

A chlorine atom on every other carbon of a polyethylene backbone yields polyvinyl chloride (PVC), which is manufactured from vinyl chloride. PVC is both tough and easily molded. Floor tiles, shower curtains, and pipes are most often made of PVC.

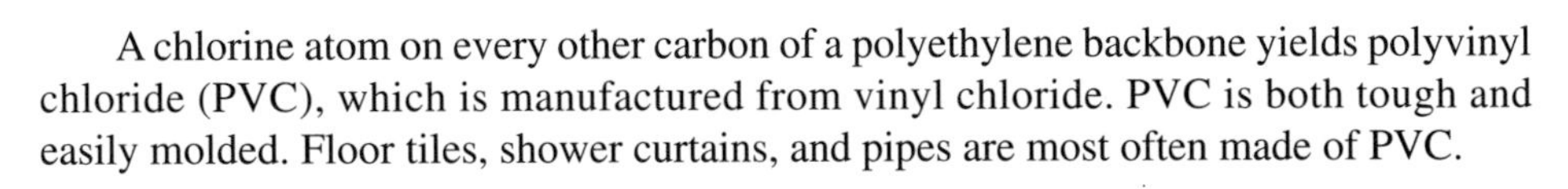

Adding a second and third chlorine atom to every other carbon atom of PVC produces a polymer having a lot of cling capacity. This is polyvinylidene chloride (trade name: Saran), used as plastic wrap for food. The large chlorine atoms in this polymer help it stick to surfaces such as glass by induced dipole chemical interactions.

Polyvinylidene chloride: (Saran) $\cdots -CH_2-CCl_2-CH_2-CCl_2-CH_2-CCl_2-CH_2-CCl_2-CH_2-CCl_2- \cdots$

Replacing all the hydrogens of polyethylene with fluorine gives Teflon, which is synthesized from tetrafluoroethylene. In contrast to the chlorine-containing Saran, fluorine-containing Teflon has a nonstick surface because the fluorine atoms tend not to experience any intermolecular attractions (Section 18.1). Also, carbon-fluorine bonds are unusually strong, which means that Teflon can be heated to high temperatures before decomposing. These properties make Teflon an ideal coating for cooking surfaces. It is also relatively inert, which is why many corrosive chemicals are shipped or stored in Teflon containers.

Polytetrafluoroethylene: (Teflon)

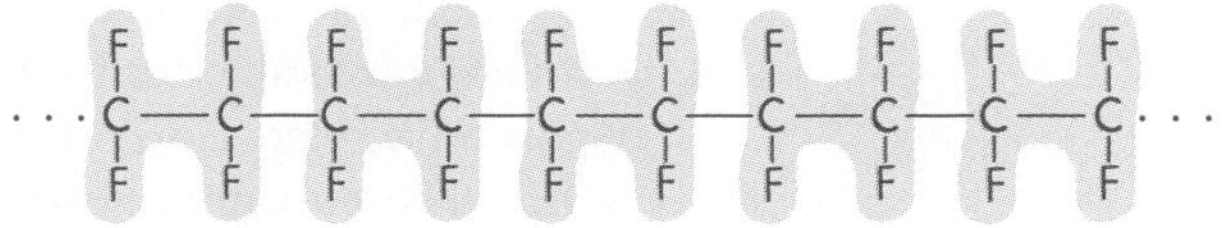

$$\cdots -CF_2-CF_2-CF_2-CF_2-CF_2-CF_2-CF_2-CF_2-CF_2-CF_2- \cdots$$

Question

What do all monomers that form addition polymers have in common?

Condensation Polymers

Monomers that form condensation polymers have at least two reactive ends. As two monomers come together, a reactive end of one bonds with a reactive end of the other. When this happens over and over with huge numbers of monomers, the result is a long polymeric chain. Bond formation between the monomers of condensation polymers generally occurs with the release of water or some other small molecule.

One of the first synthetic polymers to be marketed was the condensation polymer *66 polyamide,* discovered in 1937 by DuPont chemist Wallace Carothers. This polymer is composed of two different monomers, which classifies it as a **copolymer**. One monomer is adipic acid, which contains two carboxylic acid groups. The second monomer is hexamethylenediamine, which contains two amine groups. The ends of the adipic acid and hexamethylamine can be made to react with each other with the loss of a water molecule.

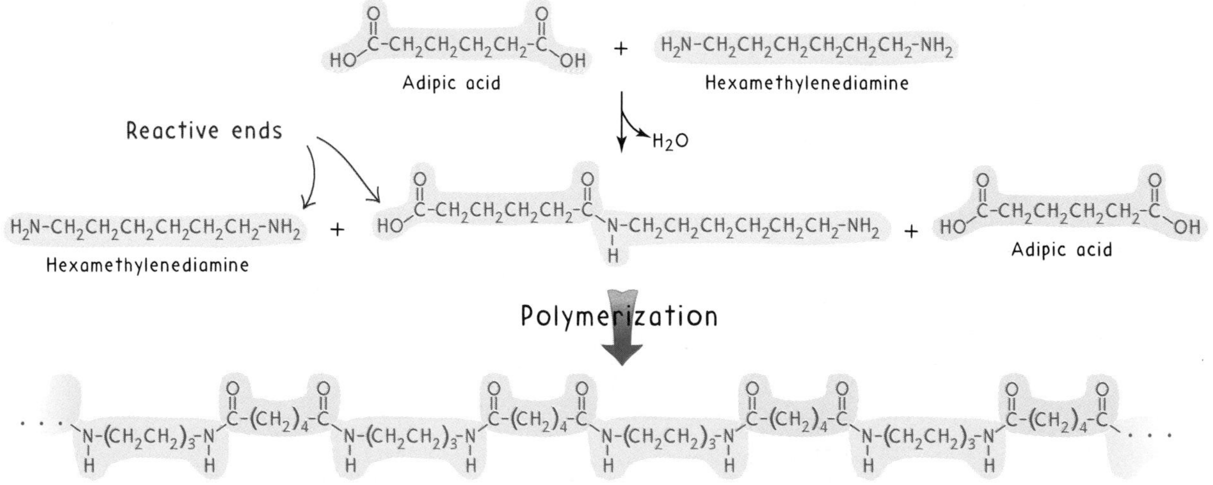

$$HO-C(=O)-CH_2CH_2CH_2CH_2-C(=O)-OH + H_2N-CH_2CH_2CH_2CH_2CH_2CH_2-NH_2$$

Adipic acid + Hexamethylenediamine

$$\downarrow -H_2O$$

Reactive ends

$$H_2N-CH_2CH_2CH_2CH_2CH_2CH_2-NH_2 + HO-C(=O)-CH_2CH_2CH_2CH_2-C(=O)-NH-CH_2CH_2CH_2CH_2CH_2CH_2-NH_2 + HO-C(=O)-CH_2CH_2CH_2CH_2-C(=O)-OH$$

Hexamethylenediamine + ... + Adipic acid

Polymerization

$$\cdots -NH-(CH_2CH_2)_3-NH-C(=O)-(CH_2)_4-C(=O)-NH-(CH_2CH_2)_3-NH-C(=O)-(CH_2)_4-C(=O)-NH-(CH_2CH_2)_3-NH-C(=O)-(CH_2)_4-C(=O)-NH-(CH_2CH_2)_3-NH-C(=O)-(CH_2)_4-C(=O)- \cdots$$

6,6-Polyamide (Nylon)

Answer

A multiple covalent bond between two carbon atoms.

After two monomers have joined, reactive ends still remain for further reactions, which leads to a growing polymer chain. The "66" associated with this polymer's name is a measure of the number of carbon atoms in each monomer—six in the adipic acid and six in the hexamethylenediamine.

Soon after its invention, DuPont sought a market for 66 polyamide, which they found remarkable for its ability to form strong, silklike fibers. Stockings made of 66 polyamide were comfortable and resistant to the tears and runs common to silk stockings. The polyamide stockings were also easy to manufacture and could be sold at prices much cheaper than silk stockings. Tremendous demand for polyamide stockings ensued, and commercial success was immediate. To reflect the benefits of the polyamide stockings, the folks at DuPont considered marketing their polymer under the trade name "norun." That didn't fly, and so they spelled it backwards to come up with "nuron." When that didn't fly either, they changed the second and third letters to "y" and "l" to come up with what we all now know as "nylon." Aside from ladies' hosiery, nylon also finds great use in the manufacture of ropes, parachutes, clothing, and carpets.

Question

Would 6-aminohexanoic acid be suitable for forming a condensation polymer?

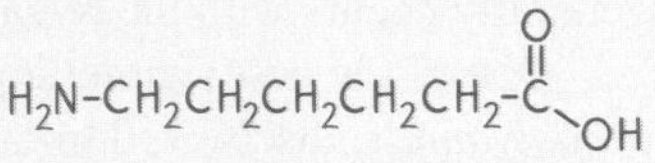

$$H_2N\text{-}CH_2CH_2CH_2CH_2CH_2\text{-}C(=O)OH$$

6-Aminohexanoic acid

Another widely used condensation polymer is polyethylene terephthalate (PET), formed from the copolymerization of ethylene glycol and terephthalic acid:

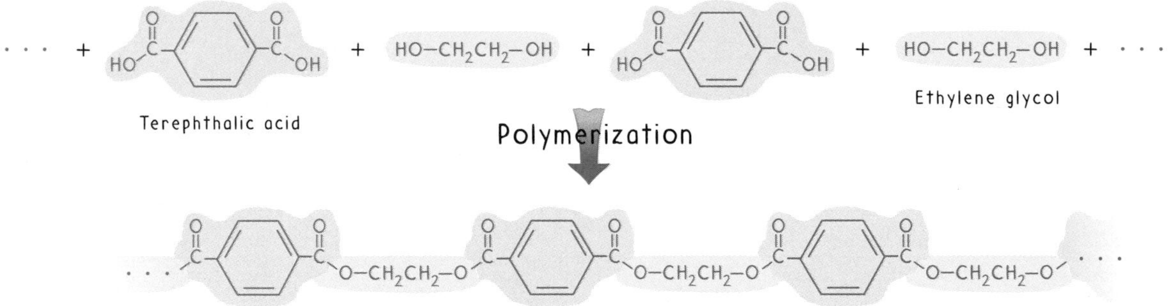

$$\cdots + HO\text{-}C(=O)\text{-}C_6H_4\text{-}C(=O)\text{-}OH + HO\text{-}CH_2CH_2\text{-}OH + HO\text{-}C(=O)\text{-}C_6H_4\text{-}C(=O)\text{-}OH + HO\text{-}CH_2CH_2\text{-}OH + \cdots$$

Terephthalic acid

Ethylene glycol

Polymerization

$$\cdots\text{-}C(=O)\text{-}C_6H_4\text{-}C(=O)\text{-}O\text{-}CH_2CH_2\text{-}O\text{-}C(=O)\text{-}C_6H_4\text{-}C(=O)\text{-}O\text{-}CH_2CH_2\text{-}O\text{-}C(=O)\text{-}C_6H_4\text{-}C(=O)\text{-}O\text{-}CH_2CH_2\text{-}O\text{-}\cdots$$

Polyethylene terephthalate (PET)

Answer

Yes, because the molecule has the same two reactive ends found in the starting materials used to make nylon. The only difference is that here both types of reactive ends are on the same molecule. Many of these monomers combine to form a polymer known as nylon-6, which after the loss of water has the

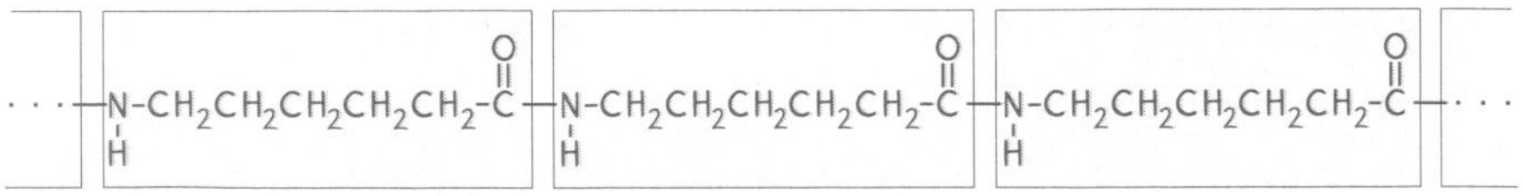

$$\cdots\text{-}[N(H)\text{-}CH_2CH_2CH_2CH_2CH_2\text{-}C(=O)]\text{-}[N(H)\text{-}CH_2CH_2CH_2CH_2CH_2\text{-}C(=O)]\text{-}[N(H)\text{-}CH_2CH_2CH_2CH_2CH_2\text{-}C(=O)]\text{-}\cdots$$

Nylon-6

Plastic soda bottles are made from this polymer. Also, PET fibers are sold as Dacron polyester, which is used in clothing and stuffing for pillows and sleeping bags. Thin films of PET are called Mylar and can be coated with metal particles to make magnetic recording tape or those metallic-looking balloons you see for sale at most grocery store check-out counters.

Monomers that contain three reactive functional groups can also form polymer chains. These chains become interlocked in a very rigid three-dimensional network that lends considerable strength and durability to the polymer. Once formed, these condensation polymers cannot be remelted or reshaped, which makes them hard-set, or *thermoset,* polymers. A good example is the thermoset polymer formed from the reaction of formaldehyde with melamine:

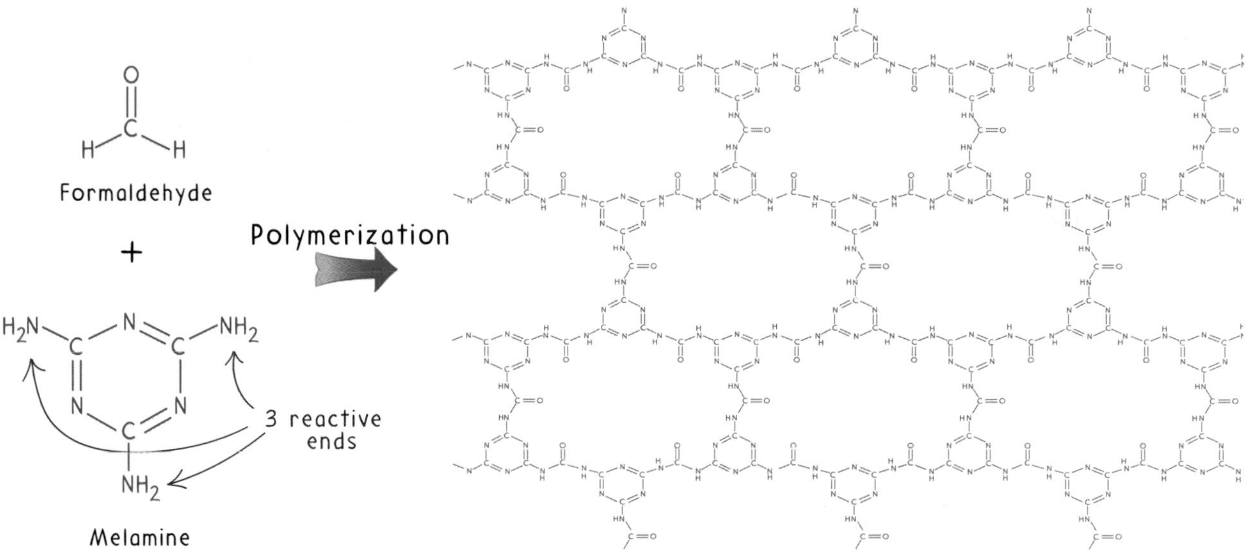

Formaldehyde

+

Melamine

3 reactive ends

Polymerization

Melmac

Hard plastic dishes (Melmac) and countertops (Formica) are made of this material. A similar polymer, Bakelite, made from phenol and formaldehyde, is used to bind plywood and particle board. Epoxy resins and polyacrylonitriles (super glue) are other examples where strength is obtained by three-dimensional cross-linking.

Thermoplastics

By definition, a *plastic* is any material capable of being molded—even clay. With the exception of the thermosetting polymers, all the synthetic polymers we have discussed can be molded only upon heating. These polymers are therefore appropriately called *thermoplastics.* Thermoplastic polymers have both crystalline and amorphous regions. In crystalline regions, chains are aligned and held together by molecular interactions, lending strength to the polymer. In amorphous regions, chains are more randomly oriented, lending flexibility to the polymer (Figure 21.41). The molecular interactions in crystalline regions are relatively weak, which means they are easily disrupted when heat is applied. At some elevated temperature, called the *glass transition temperature,* the polymer becomes fluid enough to be shaped by pressure, as in a mold. As the polymer cools, the molecular interactions take hold once again and the thermoplastic retains its molded form. Alternatively, molten thermoplastic may be either passed through

Fig. 21.41
A schematic of how the molecules of a thermoplastic align in crystalline regions but are all tangled together in amorphous regions.

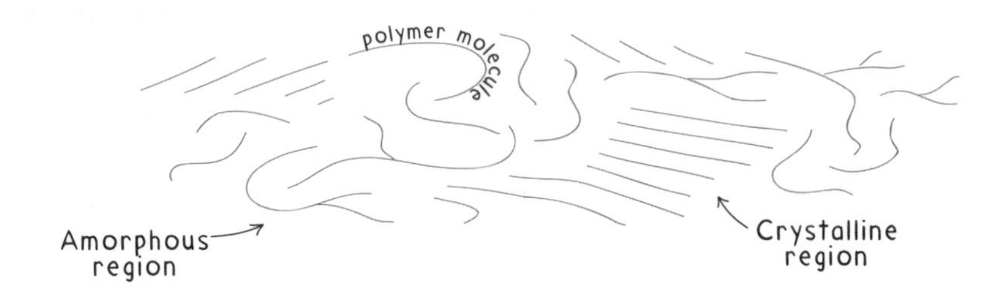

a series of tiny holes to create fibers or extruded through thin slits to create films. Thermoplastics can be recast into different shapes many times, which makes them easy to recycle.

Question

Which would you expect to have a higher glass transition temperature: a thermoplastic in which there are more crystalline regions or one in which there are more amorphous regions?

When chemicals known as *plasticizers* are added to thermoplastics, softening can be accomplished without increasing temperature. The highly branched nature of plasticizers inhibits the formation of crystalline regions that otherwise cause the polymer to harden and become brittle. Plasticizers are particularly useful for PVC, which is naturally brittle at room temperature. Plasticizer molecules enable the polymer to remain amorphous over long periods of time. Ultimately, however, they gradually escape, which accounts for such things as brittle dashboards and cracked seat covers. Many vinyl-care products simply replenish the plasticizers that have been lost over time. Toxic polychlorinated biphenyls (PCBs) were once used as plasticizers, but these have since been replaced by derivatives of phthalic acid, which are far less harmful to us and the environment.

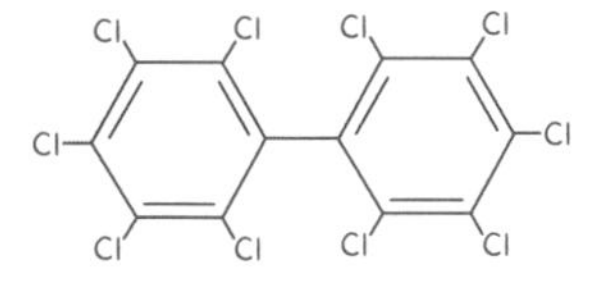

Polychlorobiphenyl (PCB)

Diethyl phthalate (DEP)

The plastics industry has grown remarkably over the past 50 years. Annual production of plastics in United States has grown from about 3 billion pounds in 1950 to nearly 100 billion pounds in the year 2000. Today, it is a challenge to find any consumer item that does *not* contain a plastic of one sort or another. And the applications of plastics and the polymers they contain continue to grow. In the future, watch for new kinds of polymers having a wide range of remarkable properties. We already have polymers that conduct electricity, replace body parts, and are stronger but much lighter than steel. Imagine synthetic polymers that mimic photosynthesis by transforming solar energy to chemical energy, or synthetic polymers that mimic biochemical enzymes, or synthetic polymers that efficiently separate fresh water from the oceans. These are not dreams. They are realities already demonstrated in the laboratory. Polymers hold a clear promise for the future. Let's work to ensure that the petroleum starting materials from which we fabricate most polymers are not exhausted before this promise is realized.

Answer
The one with more crystalline regions, because of the greater molecular interactions between aligned polymer chains.

Summary of Terms

- **Organic chemistry** The study of carbon compounds.
- **Organic compound** Any compound featuring the element carbon covalently bonded to a variety of nonmetal atoms including itself.
- **Hydrocarbon** A compound containing only carbon and hydrogen atoms.
- **Structural isomers** Molecules that have the same molecular formula but different chemical structures.
- **Conformation** The spatial orientation of a molecule, which changes as the single bonds in the molecule rotate.
- **Saturated hydrocarbon** A hydrocarbon containing no multiple covalent bonds; the carbon atoms are "saturated" with hydrogen atoms.
- **Unsaturated hydrocarbon** A hydrocarbon containing at least one multiple covalent bond.
- **Aromatic compound** Any organic molecule containing a benzene ring.
- **Heteroatom** Any atom other than carbon or hydrogen in an organic molecule.
- **Functional group** The essential heteroatom-containing structural feature found in all members of a class of compounds.
- **Alcohols** A class of organic molecules that contain a hydroxyl group bonded to a saturated carbon.
- **Phenols** A class of organic molecules that contain a hydroxyl group bonded to a benzene ring.
- **Ethers** A class of organic molecules containing an oxygen atom bonded to two saturated carbon atoms.
- **Amines** A class of organic molecules containing the element nitrogen bonded to saturated carbon atoms.
- **Carbonyl group** A carbon atom double-bonded to an oxygen atom, C=O. The carbonyl group is found in ketones, aldehydes, amides, carboxylic acids, and esters.
- **Ketones** A class of organic molecules containing a carbonyl group, the carbon of which is bonded to two carbon atoms.
- **Aldehydes** A class of organic molecules containing a carbonyl group, the carbon of which is bonded to one carbon atom and one hydrogen atom.
- **Amides** A class of organic molecules containing a carbonyl group, the carbon of which is bonded to one carbon atom and one nitrogen atom.
- **Carboxylic acids** A class of organic molecules containing a carbonyl group, the carbon of which is bonded to one carbon atom and one hydroxyl group.
- **Esters** A class of organic molecules containing a carbonyl group, the carbon of which is bonded to one carbon atom and one oxygen atom that is also bonded to a carbon atom.
- **Polymer** A long molecule made of many repeating parts.
- **Monomer** The small molecular unit from which a polymer is formed.
- **Synthetic polymer** A polymer not found in nature.
- **Addition polymer** A polymer formed simply by joining monomer units.
- **Condensation polymer** A polymer formed by joining monomers with the concomitant loss of a small molecule, such as water.
- **Copolymer** A polymer composed of at least two different types of monomers.

Review Questions

Hydrocarbons

1. What are some examples of hydrocarbons?
2. What are some uses of hydrocarbons?
3. How do two structural isomers differ from each other?
4. How are two structural isomers similar to each other?
5. How is the formation of coal different from the formation of petroleum?
6. What physical property of hydrocarbons is used for fractional distillation?
7. How is it possible to increase the yield of light hydrocarbons from petroleum?
8. What types of hydrocarbons are more abundant in higher-octane gasoline?

Unsaturated Hydrocarbons

9. What is the difference between a saturated and an unsaturated hydrocarbon?
10. How many multiple bonds must a hydrocarbon contain to be classified as unsaturated?
11. To how many atoms is a saturated carbon atom bonded?
12. Aromatic compounds contain what kind of ring system?

Noncarbon Atoms in Organic Molecules

13. What is a heteroatom?
14. Why do heteroatoms make such a difference in the physical and chemical properties of an organic molecule?
15. Which molecule should have the higher boiling point and why?

$CH_3CH_2CH_2CH_3$ $\qquad$ $CH_3CH_2CH_2CH_2{-}OH$

Alcohols, Phenols, Ethers, and Amines

16. Why are small alcohols soluble in water?
17. What distinguishes an alcohol from a phenol?
18. What distinguishes an alcohol from an ether?
19. Why do ethers typically have lower boiling points than alcohols?
20. What heteroatom is characteristic of an amine?
21. Do amines tend to be acidic, neutral, or basic?
22. Where might one find an alkaloid?
23. What are some examples of alkaloids?

Carbonyl-Containing Organic Molecules

24. Which elements make up the carbonyl group?

25. How are ketones and aldehydes related to each other? How are they different from each other?
26. What is one commercially useful property of aldehydes?
27. How are amides and carboxylic acids related to each other? How are they different from each other?
28. From what naturally occurring compound is aspirin prepared?
29. Identify each of the following molecules as hydrocarbon, alcohol, or carboxylic acid:

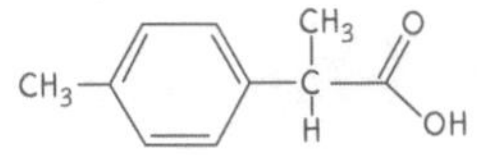

$CH_3CH_2CH_2CH_3$ $CH_3CH_2CH_2CH_2{-}OH$

Modified Natural Polymers

30. How does natural rubber differ from vulcanized rubber?
31. How is cellulose nitrate softened into a moldable material?
32. What are the three basic components of a paint?
33. What's the difference between cellophane and celluloid?

Synthetic Polymers

34. How did World War II boost the development of synthetic polymers?
35. What happens to the multiple bond of a monomer participating in the formation of an addition polymer?
36. What is released in the formation of a condensation polymer?
37. Why is Saran a stickier plastic than polyethylene?
38. What is a copolymer?
39. What happens to a thermoplastic at its glass transition temperature?
40. What are plasticizers used for?

Home Projects

1. Write your name across a sheet of paper using your finger coated with petroleum jelly, which is a mixture of nonpolar hydrocarbons. Add drops of food coloring to and around the letters. Watch the behavior of the liquid carefully. Rinse off the paper and explain your observations in terms of molecular interactions. What is the effect on the back side of the paper? Why?
2. The chemical composition of a polymer has a significant effect on its macroscopic properties. To see this for yourself, place a drop of water on a new plastic sandwich bag. Tilt the bag vertically so that the drop races off. Observe the behavior of the water carefully. Now race a drop of water off a freshly pulled strip of plastic food wrap, such as Saran wrap. How does the behavior of water on the food wrap compare with the behavior of water on the sandwich bag? Most brands of sandwich bags are made of polyethylene terephthalate, and most brands of food wrap are made of polyvinylidene chloride. Look carefully at the chemical compositions of these polymers shown in Section 21.7. Which consists of larger atoms? Which might be involved in stronger dipole–induced dipole interactions with the water? Need help with these questions? Refer back to Sections 16.5 and 18.1.

Exercises

1. Which contains more hydrogen atoms, a five-carbon saturated hydrocarbon molecule or a five-carbon unsaturated hydrocarbon molecule?
2. Why does the melting temperature of hydrocarbons increase as the number of carbon atoms per molecule increases?
3. Draw all the structural isomers for hydrocarbons having the molecular formula C_4H_{10}.
4. Draw all the structural isomers for hydrocarbons having the molecular formula C_6H_{14}.
5. The temperatures in a fractionating tower at an oil refinery are important, but so are the pressures. Where might the pressure within the fractionating tower be greatest, at the bottom or at the top? Defend your answer.
6. Identify the following functional groups in this organic molecule: amide, ester, ketone, ether, alcohol, aldehyde, amine: 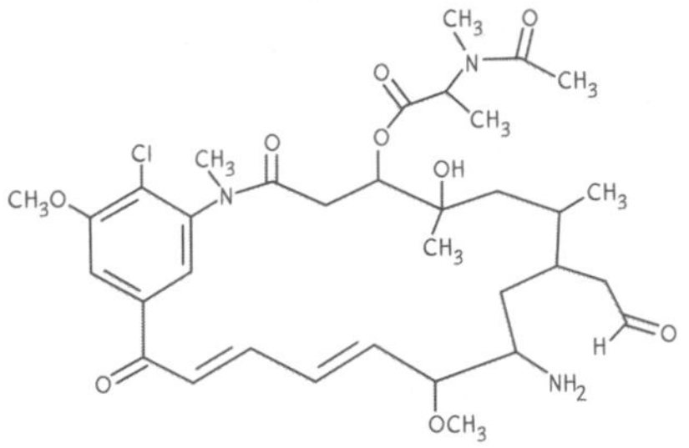
7. Why might a large alcohol be insoluble in water?
8. What is the percent by volume of water in 80-proof vodka?
9. Is ingesting methanol directly or indirectly harmful to one's eyes? Explain.
10. Briefly describe how to remove the caffeine from a cola drink. (*Hint:* The free base of caffeine is soluble in the organic solvent diethyl ether. Also, diethyl ether and water are immiscible.) 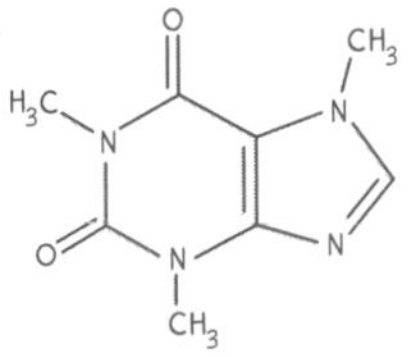
11. Draw all the structural isomers for amines having the molecular formula C_3H_9N.

12. Explain why caprylic acid, $CH_3(CH_2)_6COOH$, is soluble in 5% aqueous NaOH but caprylaldehyde, $CH_3(CH_2)_6CHO$, is not.
13. If you saw the label phenylephrine·HCl on a decongestant, would you worry that consuming it exposes you to the strong acid HCl? Explain.

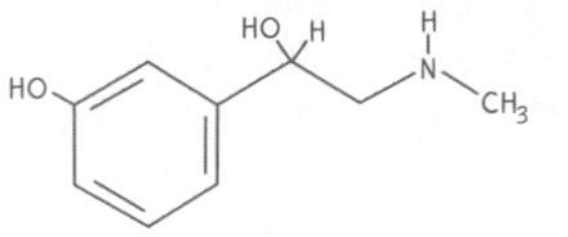

Phenylephrine

14. In water, the following molecule tends to act as (a) an acid; (b) a base; (c) neither, for it is neutral; or (d) both an acid and a base?

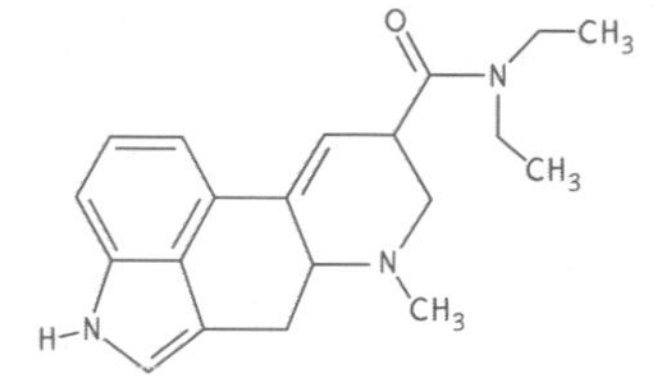

Lysergic acid diethylamide

15. Examine the structure of vanillin in Figure 21.24 and suggest why this molecule has antiseptic properties.
16. Suggest an explanation for why aspirin has a sour taste.
17. Why are organic chemicals so suitable for making drugs?
18. As noted in the text, the compound 6-aminohexanoic acid is used for forming the condensation polymer nylon-6. Polymerization is not always successful, however, because of a competing side reaction. What is this side reaction? Would polymerization be favored in a dilute or concentrated solution of this monomer? Why?
19. Would you expect polypropylene to be more or less dense than low-density polyethylene? Why?
20. Many polymers emit toxic fumes when burning. One produces hydrogen cyanide, HCN. Which one? Which one produces toxic hydrogen chloride, HCl, gas? (See Table 21.6.)
21. One solution to the buildup of plastics in landfills is to burn the plastic instead of burying it. What would be some of the advantages and disadvantages of this practice?
22. Most chewing gums use a polymer with a glass transition temperature of about 5°C, or 38°F. Suggest how you might remove chewing gum from clothing using this information.
23. Which would you expect to be more viscous, a polymer made of long molecular strands or one made of short molecular stands? Why?
24. What effect does a plasticizer have on the glass transition temperature of a polymer?
25. Which of the following chemicals might work best as a plasticizer?

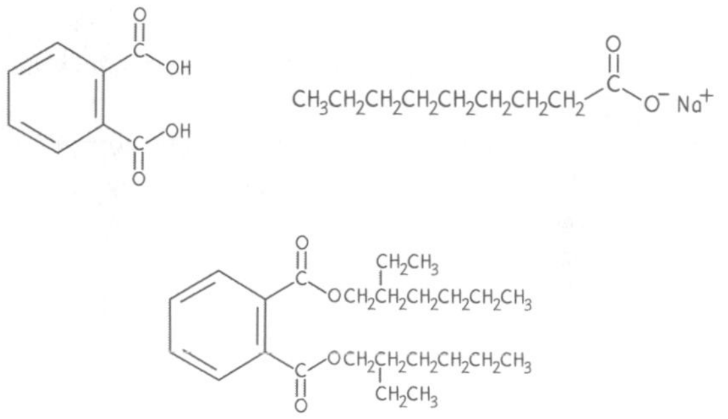

26. Are the structures of dibutyl phthalate and sodium decanoate (shown in the preceding exercise) rigid, or are each of the long carbon chains able to flop around like a cooked strand of spaghetti?
27. Hydrocarbons release a lot of energy when ignited. Where does this energy come from?
28. What type of polymer would be best to use in the manufacture of stain-free carpets?
29. Draw the structure for the monomer starting material for polyvinylidene chloride.
30. Polyethylene food containers stain easily from the pigment-containing oils in such foods as spaghetti sauce because the nonpolar oils are absorbed into the polymer. Polyvinyl chloride food containers, however, have less of a tendency to stain. Why?

Part VI — Sample Exam Questions

Choose the BEST answer to each of the following.

1. Chemical and physical changes
   (a) are virtually indistinguishable on the molecular level
   (b) differ in that only during a chemical change do the atoms of a substance change their identity
   (c) differ in that only during a physical change do the molecules of a substance maintain their identity
   (d) differ in that physical changes usually involve a greater input or output of energy than do chemical changes
2. Solutions and suspensions
   (a) are both examples of hetergeneous mixtures
   (b) are both examples of homogenous mixtures
   (c) cannot be separated into separate components by physical means
   (d) may be either pure or impure
3. Which figure best illustrates the surface of a liquid near its boiling temperature?

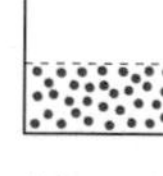
(a)

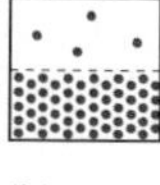
(b)

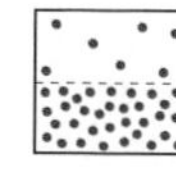
(c)

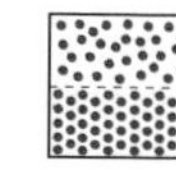
(d)

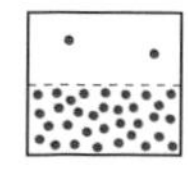
(e)

4. The inner transition elements are placed below the main body of the periodic table because
(a) these elements do not have as many applications as the elements shown in the main body
(b) they are the heaviest of all the known elements
(c) they are not true metals
(d) such positioning makes the periodic table fit better on a sheet of paper

5. The ionization energies of elements to the left side of the periodic table are so relatively small because
(a) of significant inner shell shielding
(b) these elements have weak nuclear charges
(c) there are too many electrons in the outermost shells of these elements
(d) False! These elements tend to have high ionization energies compared to elements to the right side of the periodic table

6. What type of bonds generally involve metal atoms?
(a) metallic bonds (b) covalent bonds
(c) ionic bonds (d) two of the above

7. Electron affinity is not directly related to which of the following types of chemical bonding?
(a) metallic bonds (b) covalent bonds
(c) ionic bonds (d) two of the above

8. Water is considered a polar compound because
(a) it is found in its frozen state in both the arctic and antarctic regions of our planet
(b) it has a strong attraction to magnets
(c) one side is slightly negative while the other side is slightly positive
(d) each molecule consists of fewer than 10 atoms

9. Which type of molecular interaction is considered the strongest?
(a) ion-dipole
(b) dipole-dipole
(c) dipole-induced dipole
(d) induced dipole-induced dipole

10. Only so much sugar dissolves in water because
(a) sugar molecules have such a strong attraction to themselves
(b) water molecules have such a strong attraction to themselves
(c) it is infinitely soluble
(d) water can only get so hot before it begins to boil

11. A razor blade must be lying flat in order to be held up on the surface of water by surface tension because
(a) the greater the surface area interacting with the water, the greater the surface tension
(b) the sharp edge of the razor blade disrupts surface tension by slicing through individual water molecules
(c) a vertically placed razor blade falls sideways and sinks after splashing into the surface
(d) this minimizes the pressure that the blade exerts on the surface of the liquid

12. Is the following equation balanced or unbalanced?
$2\ Fe + 2\ Na_2CrO_4 + 2\ H_2O \longrightarrow Fe_2O_3 + Cr_2O_3 + 4\ NaOH$
(a) balanced (b) unbalanced

13. Energy is required to break apart a chemical bond to overcome
(a) gravitational forces of attraction
(b) nuclear forces of attraction
(c) electrical forces of attraction
(d) It is not! Energy is actually released when a bond is broken

14. How many molecules are there in 34 grams of ammonia, NH_3?
(a) 6.02×10^{23} (b) 12.04×10^{23}
(c) 3.01×10^{23} (d) 24.08×10^{23}

15. The pH of a solution is less than zero
(a) when the concentration of hydronium ions is greater than 10 M
(b) when the concentration of hydronium ions is less than 10 M
(c) only after all of the hydroxide ions within a solution have been removed
(d) Such a solution is not possible because the lower limit of pH is zero

16. What do rusting and combustion have in common?
(a) Neither involve the reduction of a substance
(b) They are the reverse processes and so they share nothing in common
(c) They are both impeded by the presence of water
(d) They are both examples of oxidation

17. Electrolysis is important to the national economy because it
(a) provides a means of producing electricity
(b) allows the convenient production of chemicals, especially metals
(c) promotes clean air
(d) all of the above

18. How many structural isomers are shown?

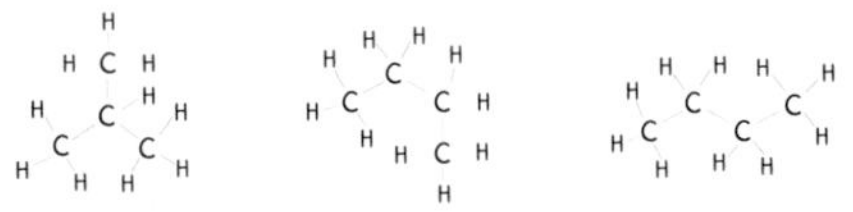

(a) 0 (b) 1 (c) 2 (d) 3

19. Heteroatoms make a difference in the physical and chemical properties of an organic molecule because
(a) they add extra mass to the hydrocarbon structure
(b) each heteroatom has its own characteristic chemistry
(c) they can enhance the polarity of the organic molecule
(d) all of the above

20. The difference between an addition polymer and a condensation polymer is that
(a) addition polymers tend to be longer
(b) condensation polymers form with the loss of a small molecule, such as HCl
(c) addition polymers are synthetic
(d) condensation polymers tend to be more highly branched

Answers: 1c, 2b, 3c, 4d, 5a, 6d, 7c, 8a, 9a, 10a, 11d, 12a, 13c, 14b, 15a, 16d, 17b, 18c, 19d, 20b

PART

Earth Science VII

This hunk of igneous rock fascinates me. Before being erupted by volcanic action, it was part of the Earth's interior, kept hot by radioactive decay. Volcanoes intrigue me too. Over the vast span of geologic time they've spewed out not only molten rock, but gases and water vapor that formed much of our atmosphere. Upon condensation, the water vapor, together with cometary debris from outer space, formed our oceans. Ah... physical science!

CHAPTER

22

Rocks and Minerals

When we apply physics and chemistry to Planet Earth, we have Earth science, the topic we turn to next. Earth science includes geology—the study of the Earth's natural materials and processes. It also includes the study of the Earth's atmosphere, oceans, and weather. Our study of Earth science begins with geology and an investigation, here in Chapter 22, of the rocks and minerals directly beneath our feet. In Chapter 23, we'll look at the continents. We'll see that they are not permanent land masses but rather like drifting ice flows, amalgams of lands repeatedly broken up, dispersed, and then crunched together into new shapes. Then we'll examine water and its effect on the Earth's surface in Chapter 24. In Chapter 25, we'll probe geologic time and emphasize an underlying theme in Earth science—that change is ever-present and ongoing. Just as an insect with a life span of a few-hours has no concept of the growth rate of the plant on which it lives, it is difficult for us to imagine this planet's dynamic nature because of the immense amount of time involved. In Chapter 26, we'll expand our study of

Fig. 22.1
The erosion that cut through these layers of the Grand Canyon continues, and one day the Grand Canyon will appear as a wide-open stretch similar to the Great Plains that now dominate central North America.

Fig. 22.2
The three main types of rock. (a) Basalt and granite are igneous rocks. (b) Sandstone and limestone are sedimentary rocks. (c) Marble and slate are metamorphic rocks.

Earth Science with an exploration of atmospheric-oceanic interactions, and then in Chapter 27 we'll conclude with a look at our planet's weather.

The Earth's crust is relatively thin, less than 1 percent of the Earth's radius—about the thickness of the skin of a tomato relative to the tomato's radius. In this very thin layer are found the materials of the Earth's surface—the minerals and the aggregates of minerals we know as rocks.

22.1 Rocks

Rocks help us decipher the Earth's past and understand the processes that have shaped our planet. The Earth is not static but is instead continually regenerating and rearranging itself. As land and rock are formed in one area, they are destroyed in another. This reworking of rock is summed up in the theory of plate tectonics, which we discuss in Chapter 23. According to plate tectonics, the Earth's surface is broken into several large, rigid plates that move in conjunction with convection currents that operate in the Earth's interior. The boundaries of these plates are places of intense geological activity—earthquakes, volcanoes, and young mountain ranges all tend to be concentrated along the edges of plates. For this reason, plate boundaries are where many rocks are either created, changed, and/or destroyed.

The rocks of the Earth's surface are classified into three types according to their origin: *igneous, sedimentary,* and *metamorphic.*

Igneous rocks are formed by the cooling and crystallization of hot, molten rock material called magma. The word *igneous* means "formed by fire." Igneous rocks make up about 95 percent of the Earth's crust. Basalt and granite are common igneous rocks.

Sedimentary rocks are formed from weathered material (sediments) carried by water, wind, or ice. Sedimentary rocks are the most common rocks in the uppermost part of the Earth's crust. They cover more than two-thirds of the Earth's surface. Sandstone, shale, and limestone are common sedimentary rocks.

Metamorphic rocks are formed from preexisting rocks (igneous, sedimentary, or metamorphic) that are transformed by high temperature, high pressure, or both—without melting. The word *metamorphic* means "changed in form." Marble and slate are common metamorphic rocks.

Because all three types of rock are composed of one or more minerals, we now focus on minerals and their properties.

22.2 Minerals

Minerals are the building blocks of rocks. Geologists define a **mineral** as a naturally formed, generally inorganic, crystalline solid composed of an ordered array of atoms and having a specific chemical composition. Minerals differ from one another in their combination and proportion of elements and/or in the internal arrangement of their atoms.

A few minerals, such as gold, copper, and iron, are composed of single elements. Most minerals, however, are compounds of more than one element. Chemical composition and internal crystal structure provide the basis for the classification of minerals. Although sophisticated instruments can be used to determine a mineral's composition and crystal structure, minerals are most often identified by their easily observable physical properties. The physical properties include crystal form, hardness, cleavage, luster, color, streak, and specific gravity.

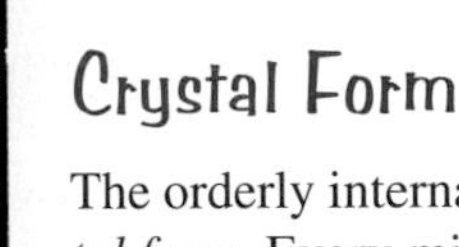

(a)

(b)

Fig. 22.3
The unique crystal form of each mineral is the external expression of the mineral's internal arrangement of atoms. (a) The intergrown cubes of pyrite. (b) The six-sided prisms of quartz.

Crystal Form

The orderly internal arrangement of atoms in a crystal is expressed in its shape, or *crystal form.* Every mineral has its own characteristic crystal form. Some minerals are easily identified by their unique crystal form. The mineral pyrite, for example, commonly forms as intergrown cubes, while quartz commonly forms as six-sided prisms that terminate in a point (Figure 22.3). Unfortunately, well-shaped crystals are rare in nature because minerals typically grow in cramped spaces. Even though crystal form is usually imperfect, most minerals can still be identified by their most-often-seen crystal growth pattern. For example, asbestos minerals often resemble narrow threadlike fibers, and the mineral hematite often assumes a globular form that resembles a bunch of grapes (Figure 22.4).

Two or more minerals that contain the same elements in the same proportions but have different crystal structures are called **polymorphs** (many forms) of each other. Graphite and diamond are examples of polymorphs, for they both consist only of carbon atoms—yet they exhibit vastly different properties (Figure 22.5). Because formation of polymorphs depends on particular temperatures and pressures, they are good indicators of the geological conditions at their sites of formation.

(a)

(b)

Fig. 22.4
Well-shaped crystal forms do not develop when growing occurs in a confined space. Nevertheless, the distinctive growth patterns of many minerals are apparent. (a) The narrow, threadlike fibers of the asbestos group minerals. (b) The grape-cluster shape of hematite.

■ Question

In what way are the minerals calcite ($CaCO_3$) and aragonite ($CaCO_3$) similar? How are they different?

■ Answer
Every mineral is unique in its chemical composition and/or internal crystal structure. Because calcite and aragonite have the same chemical composition, $CaCO_3$, they must be different from one another in their crystal structures. Because they contain the same elements in the same proportions but have different crystal structures, calcite and aragonite are polymorphs.

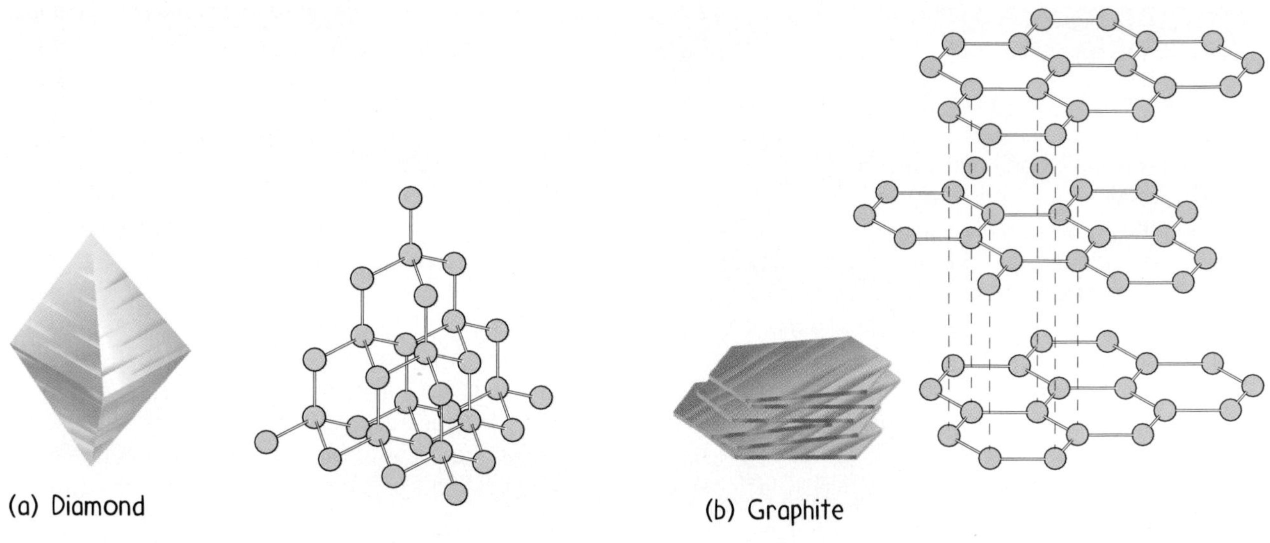

Fig. 22.5
The polymorphs graphite and diamond are pure carbon. (a) Diamond, the hardest substance known, has a tightly packed, symmetric structure. (b) Graphite has an open, layered structure. When rubbed between your fingers, individual graphite molecules glide over one another like cards in a stack, giving it a slippery feel, hence its use as a lubricant. It also glides easily when stroked onto paper and leaves an opaque tracing, hence its use in pencils (it is also much less toxic than lead).

Table 22.1
Mohs Scale of Hardness

Mineral	Scale Number	Common Objects
Diamond	10	Jewelry
Corundum	9	Machine tools
Topaz	8	Jewelry
Quartz	7	Steel file
Feldspar	6	Window glass
Apatite	5	Pocket knife
Fluorite	4	
Calcite	3	Copper wire or coin
Gypsum	2	Fingernail
Talc	1	

Hardness

Just as a diamond will scratch glass, a quartz crystal will scratch a feldspar crystal because quartz is harder than feldspar. The resistance of a mineral to being scratched or its ability to scratch other minerals is a measure of its hardness. The varying degrees of hardness are represented by the **Mohs scale of hardness** (Table 22.1).

A mineral's hardness depends on the strength of its chemical bonds—the stronger the bonds, the harder the mineral. Factors that influence bond strength—ionic charge, atom/ion size, and packing—influence hardness. The greater the ionic charge, the greater the attraction between ions, and the stronger the bond. Small atoms or ions can generally pack closer together than large atoms or ions because their electrons are closer to their nucleus. Closely packed atoms/ions have a smaller distance between one another, and form stronger bonds than do minerals in which the atoms/ions are not so closely packed. Gold, because of its large atomic size, is relatively soft (hardness less than 3), whereas diamond, with its small carbon atoms and tightly packed structure, is relatively hard (hardness = 10).

Cleavage and Fracture

Cleavage is the tendency of a mineral to break along planes of weakness. Planes of weakness are determined by crystal structure and chemical bond strength. In general, minerals that have strong bonds between planar (flat) crystal surfaces show poor cleavage, and minerals that have weak bonds between planar crystal surfaces show more developed cleavage. If a mineral can break along one surface, it can also break along all parallel planes, for all those potential breaks represent the same cleavage direction.

Asbestos: Friend and Foe

When the word *asbestos* is mentioned, people tend to think of lung disease and/or removal problems, but it hasn't always been that way. The first known reference to asbestos goes back to the time of Aristotle when this material was discovered to have fireproof qualities. Since then, the incombustability and low heat conductivity of asbestos, plus its fibrous, flexible nature, have prompted humans to use it in many ways. It has been woven into fabrics (as theater curtains and fireproof suits), and utilized in building materials (fireproof insulation) and as a flame retardant in plaster, ceiling, and floor tile. It has also been used in automobile brake shoes and clutch facings, air and water filters, cigarette tips, military gas masks, and toothpaste! In the 1970s, the commercial use of asbestos reached an all-time high, but then it was found to be linked to lung disease. The fibrous nature that makes asbestos so flexible also allows easy penetration into bodily tissues, particularly the lungs. The history of asbestos is one of bitter paradox because the unique qualities that allowed it to save lives have also been found to endanger lives.

Asbestos is not a single mineral but rather a family of silicate minerals known for their fibrous structure. There are six types of asbestos minerals, but only two are of commercial importance—chrysotile and crocidolite. The asbestos mineral chrysotile accounts for 95 percent of asbestos production worldwide, and crocidolite accounts for the remaining 5 percent. Chrysotile has a sheet silicate structure that makes it soft and flexible. Because of its softness, chrysotile is easily broken down in the body, producing no apparent damage. This form of asbestos, leached from the ground, is present in many reservoirs of quite-safe drinking water. Recent scientific medical evidence indicates that people exposed for long periods to moderate amounts of chrysotile show no lung ailments.

Crocidolite is a different story, however, for this type of asbestos has a double-chain silicate structure that makes it strong and stiff, and thus more dangerous in the body. People exposed either to high levels of crocidolite or to moderate levels over a prolonged period of time have been found to develop lung disease. Thus it is crocidolite that is the principal culprit in asbestos-related lung diseases. Despite this knowledge, many reports on asbestos health hazards fail to make a distinction between the various types.

This failure to distinguish between harmless and dangerous asbestos has contributed to a public view that any asbestos mineral is fatal. The widely embraced and emotionally volatile premise that "one fiber can kill" has made asbestos the most feared contaminant on the Earth. It is by far the most expensive pollutant in terms of regulation and removal. The removal of asbestos-containing materials from schools, hospitals, and other public buildings has cost billions of dollars over the past 20 years. With only 5 percent of asbestos minerals posing a health problem, however, many scientists question the practice of eliminating all forms of asbestos, proposing that a more responsible method of remediation would distinguish among the different types. As with electricity in the 1800s, gasoline-powered vehicles in the early 1900s, and radioactivity in the late 1900s, public fears about asbestos will likely persist for some time before informed common sense prevails. Then we may view asbestos as both friend (chrysotile) and foe (crocidolite).

Some minerals, such as muscovite and calcite, have very distinct cleavage. Muscovite, for example, has perfect cleavage in one direction and breaks apart to form thin, flat sheets (Figure 22.6a). Calcite has perfect cleavage in three directions and breaks to produce rhombohedral faces that intersect at 75-degree angles (Figure 22.6b). Garnet, whose crystal structure has strong bonds in all directions, shows no cleavage.

A break that is not along a cleavage plane is a *fracture*. When a mineral fractures with a smooth, curved surface resembling broken glass, the fracture is *conchoidal*. Quartz and olivine display conchoidal fractures when broken. Some minerals, such as hematite and serpentine, break into splinters or fibers, but most fracture irregularly. The degree and type of cleavage or fracture are useful guides for identifying minerals.

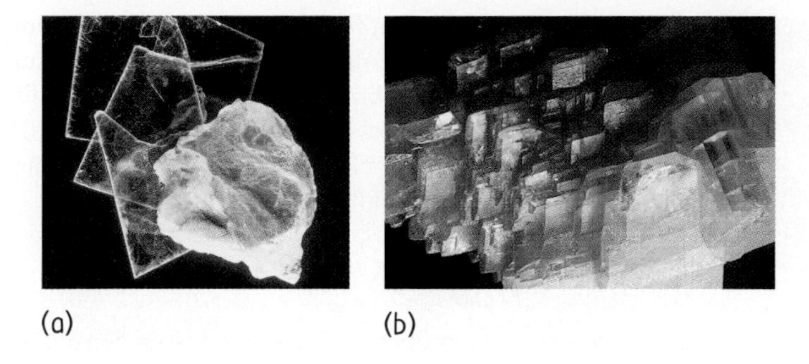

Fig. 22.6
A mineral's cleavage is very useful in its identification. (a) Muscovite (mica) has perfect cleavage in one direction. (b) Calcite has perfect cleavage in three directions.

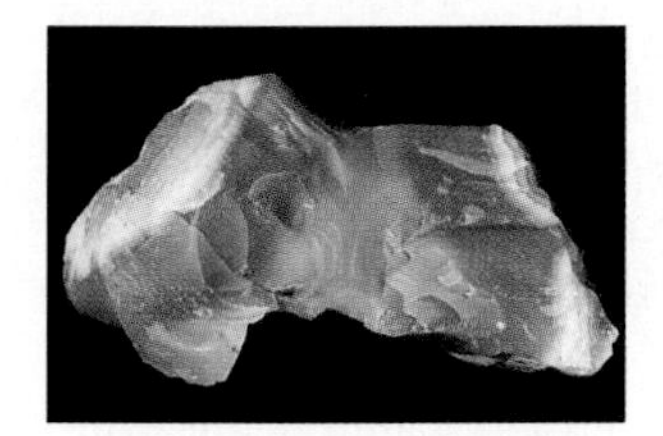

Fig. 22.7
Quartz does not exhibit cleavage. When it breaks, it instead develops a conchoidal fracture—a curved, smooth surface that resembles broken glass.

Question

When pieces of calcite and fluorite are scraped together, which scratches which?

Luster

The *luster* of a mineral is the appearance of its surface as it reflects light. Luster is independent of color; minerals of the same color may have different lusters, and minerals of the same luster may have different colors. Mineral lusters are listed in Table 22.2.

Color and Streak

Although color is an obvious feature of a mineral, it is not a very reliable means of identification. Some minerals—copper and turquoise are two examples—have a distinctive color, but the majority of minerals may occur in a variety of colors or be colorless. Chemical impurities in a mineral affect color. For example, the common mineral quartz, SiO_2, can be found in a variety of colors, depending on slight impurities. It can be clear and colorless (no impurities), milky white from minute fluid inclusions, rose-colored from small amounts of titanium, violet from small amounts of iron, or smoky gray to black from a radiation-damaged crystal lattice. The color of the mineral corundum, Al_2O_3, is commonly white or grayish, but impurities give us rubies and sapphires.

Streak, the name given to the color of a mineral in its powdered form, is an important characteristic for identifying minerals that have a metallic or semimetallic luster. When rubbed across an unglazed porcelain plate, all minerals leave behind a thin layer of powder—a streak. Although different samples of the same mineral often vary in color, the color of the streak is constant

Table 22.2
Mineral Lusters

Mineral Lusters	Appearance
Metallic	Strong reflection; polished or dull
Vitreous	Bright, glassy
Resinous	Resins, waxy
Greasy	Oily glass, also may feel greasy
Pearly	Pearly iridescence
Silky	Sheen of silk
Adamantine	Diamond, brilliant

Answer
Looking at Table 22.1, we see that fluorite is harder than calcite. So fluorite scratches calcite.

(a)

(b)

Fig. 22.8
The mineral corundum (Al_2O_3) comes in a variety of colors as a result of chemical impurities. The addition of small amounts of chromium in place of aluminum produces the red gemstone *ruby*, and with the addition of small amounts of iron and titanium, the result is the blue gemstone *sapphire*.

(Figure 22.9). Minerals that have a metallic luster generally leave a dark streak that may be different from the color of the mineral. For example, the mineral hematite is normally reddish-brown to black but always streaks reddish-brown. Magnetite is normally iron-black but streaks black. Limonite is normally yellowish-brown to dark brown and always streaks yellowish-brown. Minerals that have a nonmetallic luster generally leave behind an undiagnostic white streak.

Specific Gravity

An obvious physical property of a mineral is its density. In practical terms, density is how heavy a mineral feels for its size. The standard measure of density is *specific gravity*—the ratio of the weight of a certain volume of a substance to the weight of an equal volume of water. For example, if 1 cubic centimeter of a mineral weighs three times as much as an equal volume of water, its specific gravity is 3. The specific gravities of some minerals are shown in Table 22.3.

Table 22.3
Specific Gravity of Various Minerals

Mineral	SG	Mineral	SG
Borax	1.7	Chromite	4.6
Quartz	2.65	Pyrite	5.0
Talc	2.8	Hematite	5.26
Mica	3.0	Silver	10.5
Olivine	3.6	Gold	19.3

Gold's particularly high specific gravity of 19.3 is nicely taken advantage of by miners panning for gold. Fine gold pieces hidden in a mixture of sediments settle to the bottom of the pan when the mixture is swirled in water. Water and less dense materials spill out upon swirling. After a succession of dousings and swirls, only the substance with the highest specific gravity remains—gold!

Question

Why are there no units for specific gravity?

Fig. 22.9
The streak test can be used to identify minerals that have a metallic or semimetallic luster.

Chemical Properties of Minerals

Sometimes chemical properties are used to identify a mineral. Two simple chemical tests for identifying minerals are the taste test and the fizz test. The taste test is commonly used to identify the mineral halite, NaCl (common table salt), which has a distinctive salty taste. The fizz test on carbonate minerals is also common. Carbonate minerals effervesce (fizz) in dilute hydrochloric acid, giving off bubbles of carbon dioxide gas produced by a chemical reaction between carbonate minerals and HCl (Figure 22.10). Some carbonate minerals react more readily with HCl than others.

Answer
Specific gravity is a ratio of densities. Density units divided by density units cancel out. For example, the density of the mineral hematite, Fe_2O_3, is 5.26 g/cm^3, and that of water is 1.0 g/cm^3. Therefore the specific gravity of hematite is (5.26 g/cm^3) ÷ (1.0 g/cm^3) = 5.26.

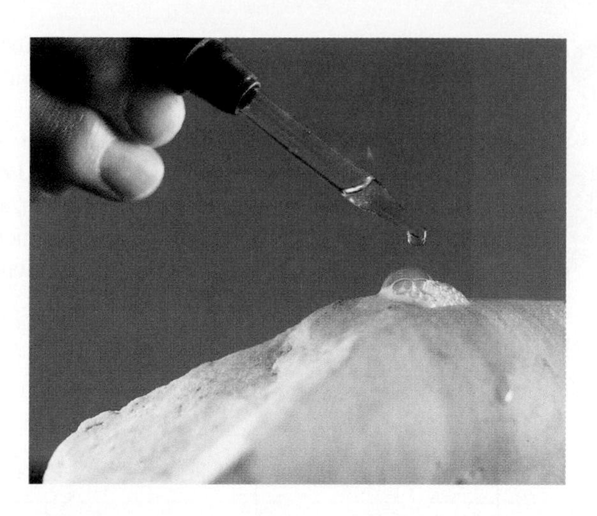

Fig. 22.10
Bubbles of carbon dioxide effervesce when carbonate minerals make contact with HCl.

22.3 Building Blocks of Rock-Forming Minerals

Of the 112 known elements, 88 occur naturally in the Earth's crust. These 88 elements combine to make up the more than 3400 types of minerals. Of this huge number of minerals, only about two dozen are abundant, and these are composed predominantly of eight elements (Table 22.4). These eight elements represent about 98 percent of the mass of the Earth's crust. Almost half of this mass is the element oxygen, which is found in such common minerals as the *silicates, oxides,* and *carbonates;* other rock-forming minerals are *sulfides* and *sulfates.* With few exceptions, all rock-forming minerals are members of these five groups.

The Silicates

After oxygen, the second most abundant element in the Earth's crust is silicon. The tendency of silicon to bond with oxygen is so strong that silicon is never found in na-

Table 22.4
Most Common Chemical Elements in the Earth's Crust

Element	Symbol	Percent by Mass	Percent by Volume
Oxygen	O	46.60	93.8
Silicon	Si	27.72	0.9
Aluminum	Al	8.13	0.5
Iron	Fe	5.00	0.4
Calcium	Ca	3.63	1.0
Sodium	Na	2.83	1.3
Potassium	K	2.59	1.8
Magnesium	Mg	2.09	0.3
Total		98.59	100.0

SOURCE: *Principles of Geochemistry* by Brian Mason and Carleton B. Moore. Copyright 1982 by John Wiley & Sons, Inc.

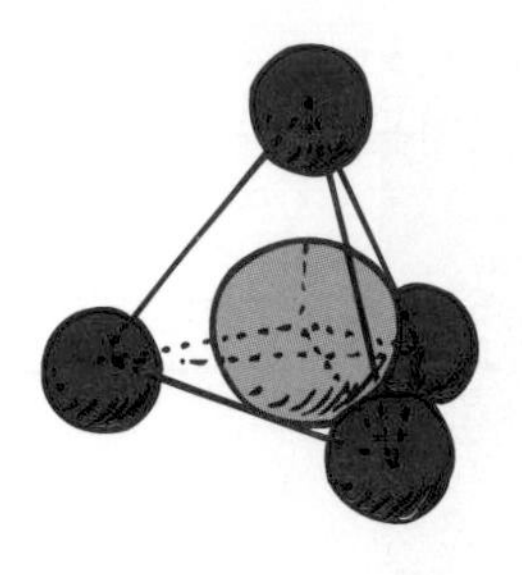

Fig. 22.11
The silicon-oxygen tetrahedron is four oxygen atoms surrounding a central silicon atom.

ture as a pure element; it is always combined with oxygen. Oxygen and silicon combine to form the most common mineral group, the *silicates.* The silicate known as quartz, the second most common mineral in the Earth's crust, is composed only of oxygen and silicon, but most other silicates usually contain other elements in addition to oxygen and silicon. Feldspars, the most common and abundant minerals, for example, are silicates that also contain aluminum, sodium, potassium, and/or calcium.

All silicates have the same fundamental structure, the silicon-oxygen tetrahedron (Figure 22.11). Stability is achieved when tetrahedra form polymers by electrically linking to one another (Figure 22.12).

Silcate Mineral		Typical Formula	Cleavage	Silicate Structure
Olivine		$(Mg.Fe)_2SiO_4$	None	Single tetrahedron
Pyroxene		$(Mg.Fe)SiO_3$	Two planes at right angles	Chains 
Amphibole		$(Ca_2Mg_5)Si_8O_{22}(OH)_2$	Two planes at 60° and 120°	Double chains 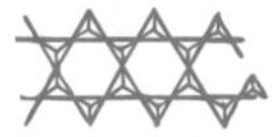
Micas	Muscovite	$KAl_3Si_3O_{10}(OH)_2$	One plane	Sheets 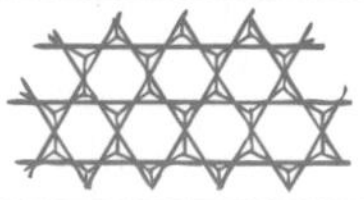
	Biotite	$K(Mg.Fe)_3Si_3O_{10}(OH)_2$		
Feldspars	Orthoclase	$KAlSi_3O_8$	Two planes at 90°	Three-dimensional networks 
	Plagioclase	$(Ca.Na)AlSi_3O_8$		
Quartz		SiO_2	None	

Fig. 22.12
As silicate tetrahedra link to one another, they polymerize to form chains, sheets, and various network patterns. The complexity of the silicate structure increases down the chart.

The Oxides

In minerals known as the oxides, oxygen is combined with one or more metals. These include iron (hematite and magnetite), chromium (chromite), manganese (pyrolusite), tin (cassiterite), and uranium (uraninite). Oxides are of great economic importance because they make up the ores necessary for industrial and technological manufacture. Most oxides are ionically bonded, their structures varying with the size of the metallic cations.

The Carbonates

Fig. 22.13
The fundamental carbonate ion structure, CO_3^{2-}: a central carbon atom bonded to three oxygen atoms.

The carbonate minerals are much simpler in structure than the silicates. Carbonate structure is triangular, with a central carbon atom bonded to three oxygen atoms, CO_3^{2-}(Figure 22.13). Two common carbonate minerals are calcite, which is the chemical compound calcium carbonate, $CaCO_3$, and dolomite, which is a mixture of calcium carbonate and magnesium carbonate, $CaMg(CO_3)_2$. Calcite and dolomite are the chief minerals that make up the group of rocks called limestone.

The Sulfides and Sulfates

As the names imply, the sulfide and sulfate minerals have sulfur as a main constituent. In sulfides, negatively charged sulfide ions combine with metallic elements; thus most sulfide minerals look like metals. In fact, the sulfides form an important class of minerals that includes the majority of ore minerals. The most common sulfide mineral is pyrite (fool's gold), FeS_2.

In sulfates, sulfur is present as a sulfate ion, SO_4^{2-}, a tetrahedron made up of one sulfur atom and four oxygen atoms. One of the most abundant minerals of the sulfate group is gypsum, a calcium sulfate ($CaSO_4 \cdot 2\ H_2O$).

22.4 Igneous Rocks

Most of the Earth's crust—about 95 percent—is composed of a wide variety of igneous rock. On the continents, the most common igneous rocks are granite and andesite. On the ocean floor, basalt is predominant. All igneous rock originated as magma—molten rock from the Earth's interior.

Magma and the Evolution of Igneous Rocks

Just as there is water and ice, there is magma and rock. Magma that cools and solidifies becomes rock, and rock that is heated melts to become magma. Just as ice melts at the same temperature that water freezes, the temperature at which a solid mineral melts is the same temperature at which the same mineral in molten form solidifies. So when we discuss the melting temperature of a mineral, we imply that the magma form of that mineral solidifies at the same temperature.

For rock to melt, the temperature needs to be very high—usually above 750°C for granitic rocks and above 1000°C for basaltic rocks. The Earth's temperature increases with depth—about 30°C for each kilometer (Figure 22.14). Thus all rocks melt if they are deep enough inside the Earth.

A rock's melting point is affected by two factors: the pressure placed on the rock and the amount of water trapped in the rock. Inside the Earth, pressure increases with depth as a result of the increased load of rock above. In general, as pressure on rock increases, melting point increases. (This is similar to increased atmospheric pressure causing an increase in the boiling point of water.) The water content of a rock also affects its melting point. Rocks with a high water content have a lower melting point because water dissolves in the magma. Rocks with a low water content have a higher melting point and therefore require higher temperatures to melt.

Keep in mind that rock is composed of various minerals. As rock is heated, the first minerals to melt are those with the lowest melting points. Thus the melting of rock into magma occurs over a broad temperature range. When conditions are such that all minerals within a rock can melt, the composition of the resulting magma is the same as the composition of the original rock. Most often, however, melting is not complete; we have **partial melting**. This is analogous to the partial distillation of crude oil (Chapter 20), where the various component hydrocarbons in the oil depend upon its temperature. Similarly, the magma resulting from partial melting is derived only from those minerals that have melted, the ones having the lowest melting points. All this results in magmas of many different compositions and—because these magmas cool to form igneous rock—many different igneous rocks.

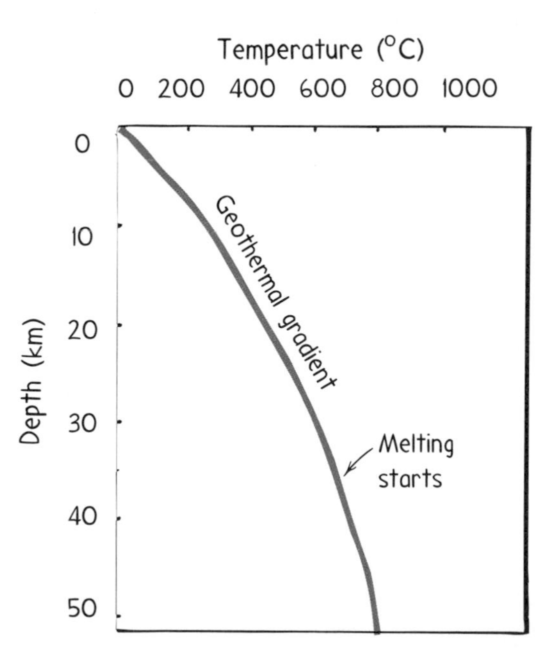

Fig. 22.14
Temperature inside the Earth increases about 30° C for each kilometer of depth. This increase of temperature with depth is known as the *geothermal gradient.* At sufficient depth, temperatures become hot enough to melt rock.

Magmas are classified by the amount of *silica* they contain. Silica is a chemical compound composed of one silicon atom and two oxygen atoms—SiO_2.[1] Minerals with a high silica content tend to have relatively low melting temperatures. Partial melting thus produces magmas with a higher silica content than the parent rock because the high silica content minerals are the first to melt, and the lower silica content minerals are left unmelted. There are three major types of magma—*basaltic, andesitic,* and *granitic. Basaltic magma* is about 50 percent silica. Basaltic magma that has solidified is the dark igneous rock—basalt—that makes up the Hawaiian Islands. *Andesitic magma,* composed of 60 percent silica, is produced from partial melting of basaltic oceanic crust. The rock andesite, produced from andesitic magma, gets its name from the Andes Mountains in South America, where it is very common. When water-rich andesitic rocks undergo partial melting, *granitic magma,* composed of about 70 percent silica, is produced. This magma, when solidified, forms granite and other granitic rocks. In the plate tectonic model, discussed in Chapter 23, basaltic magma is typically found where plates move apart from one another. Andesitic and granitic magmas form where plates collide and one plate plunges beneath the other and undergoes partial melting. Of all igneous rocks in the crust, oceanic and continental combined, approximately 80 percent forms from basaltic magma, 10 percent from andesitic magma, and 10 percent from granitic magma.

Igneous Rock Crystallization

Igneous rocks form from the cooling and crystallization of magma. The process of crystallization is very similar to the process of partial melting but in the reverse order: Solid crystals form out of a liquid mixture. Minerals that have the highest melting points—which means those that have the lowest silica content—crystallize first, followed by minerals with lower melting points (those containing larger amounts of silica). As crystallization proceeds, the composition of the liquid in which the crystals form changes continuously. It becomes *depleted* in the constituents of minerals that have already crystallized and *enriched* in the constituents of minerals yet to crystallize. This process is called **fractional crystallization**, and it enriches the remaining liquid in silica.

Sometimes newly formed crystals are prevented, either because they settle to the bottom of a magma chamber or because magma moves out of a chamber, from reacting with the remaining magma. When this happens, the composition of the rock formed will be different from the composition of the magma. When this doesn't happen—in other words, when newly formed crystals stay in contact with the remaining magma—the resulting solid rock will have the same bulk composition as the original magma. Thus fractional crystallization allows a single magma to generate several different igneous rocks.

■ **Question**

How are partial melting and fractional crystallization similar?

■ **Answer**

Both produce temperature-dependent materials but by opposite processes. Partial melting (solid to liquid) produces *magmas* of various compositions that depend on the melting temperatures of the minerals making up the rock that is melting. Fractional crystallization (liquid to solid) produces *crystals* of various compositions that depend on the solidification temperatures of the minerals that form from the magma that is solidifying (which is the same as the melting temperature). In both processes, materials separate as a function of temperature.

---

[1]Silica is not to be confused with quartz, a mineral having the same chemical formula.

Igneous Rock Formation at the Earth's Surface

Igneous rocks may form either at or below the Earth's surface. Igneous rocks that form at the Earth's surface are called **extrusive** rocks. Igneous rocks that form beneath the Earth's surface are called **intrusive** rocks.

Magma that moves upward from inside the Earth and extrudes onto the surface is called **lava**. The term *lava* refers to both the molten material itself and to rocks that form from it. Lava may be extruded through cracks and fractures in the Earth's surface, or through a central vent—a **volcano**. Although eruptions from a volcano are more familiar, the outpourings of magma through fissures are much more common.

Fig. 22.15
Fissure eruptions along the mountainous ridges of the ocean floors have produced enormous amounts of basalt—enough to build the crust of the Earth's entire present seafloor in the past 200 million years. The most common of these submarine lava flows are *pillow basalts.*

Fissure Eruptions Most fissure eruptions occur as outpourings of basaltic lava on the ocean floor. The hot magma that comes out in such an eruption is quickly cooled by cold seawater. The final form of the cooled lava depends on how far it gets from the fissure before it solidifies. When solidification occurs close to the fissure, the lava is very hot and not very viscous[2] immediately prior to solidification. As a result, the solid rock produced resembles a stack of thin layers in which each layer may be only 20 centimeters or so thick. When solidification takes place farther from the fissure, the lava is cooler and more viscous immediately prior to solidification. In this case, instead of forming thin layers, the cooling lava forms a thicker, distinct shape that resembles a pillow. Pillow basalts are the most common submarine extrusive rock.

Fissure eruptions also occur on land. Lava outpourings known as *flood basalts* have flooded large areas that, after solidification, have created extensive lava plains or have piled up to form lava plateaus. The Columbia Plateau in the Pacific Northwest is the result of an extensive flood basalt (Figure 22.16), as is the Deccan Plateau in India.

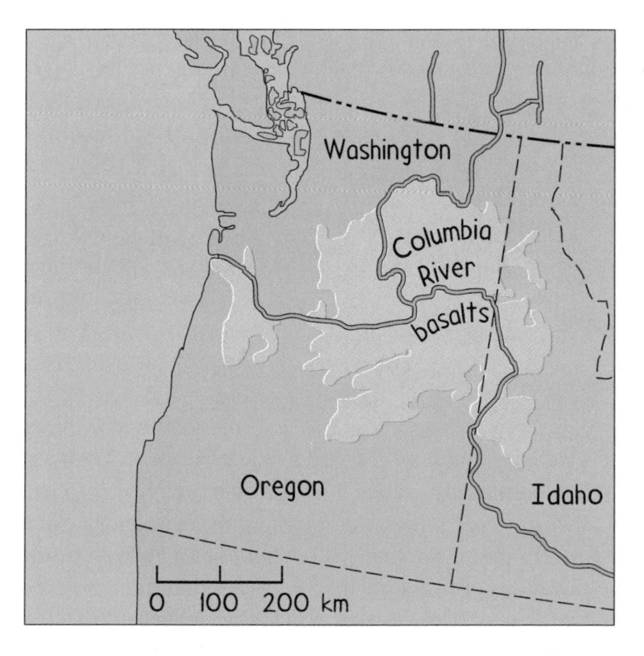

Fig. 22.16
The flood basalts that produced the Columbia Plateau covered more than 200,000 km^2 of the preexisting land surface.

Volcanoes Volcanoes come in a variety of shapes and sizes (Figure 22.17). Those built by a steady supply of very fluid basaltic lava have a broad, gently sloping cone that resembles a shield. These are *shield volcanoes,* built from the accumulation of successive flows that pour out in all directions to cool as thin, gently sloping sheets. Some of the largest volcanoes in the world are shield volcanoes. The enormous size of Mauna Loa in Hawaii is the result of the accumulation of individual lava flows,

[2]*Viscosity,* a measure of internal resistance to flow, is directly related to a magma's silica content. Basaltic magmas, with their low silica contents, have a low viscosity and tend to be quite fluid, while granitic magmas have a high viscosity and flow so slowly that movement is often (or would be) difficult to detect. The same is true for lava—molten rock at the Earth's surface. Therefore, the flow behavior of magma and lava is very dependent on silica content.

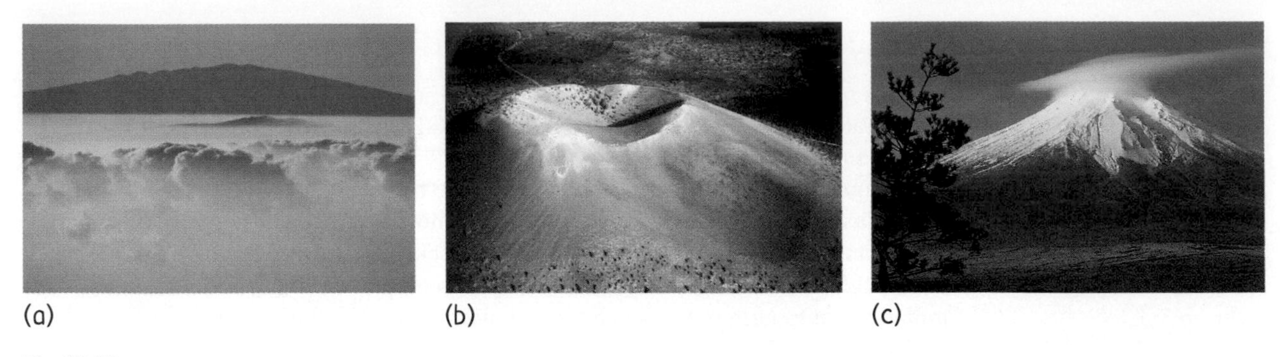

Fig. 22.17
The three types of volcanoes. (a) Shield volcanoes, such as Mauna Loa, have broad, gentle slopes that average between 1° and 10° (from the horizontal). (b) Cinder cones, such as Sunset Crater, generally have smooth steep slopes of 25° to 40° and bowl-shaped summit craters. (c) Composite cones, such as picturesque Mount Fujiyama, are also very steep. On average, the slope of a composite cone starts out at 30° at the summit and gradually flattens to 10° at the base.

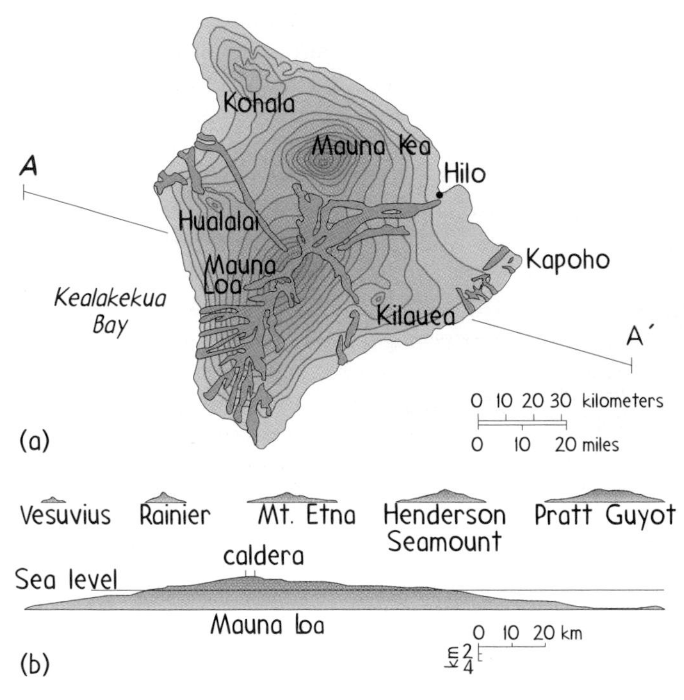

Fig. 22.18
(a) Mauna Loa, a shield volcano on the Island of Hawaii, is the largest volcano on Earth. (b) When compared to other large volcanoes, its immense size and volume is dramatic.

each only a few meters thick. Mauna Loa is the largest volcano on Earth (and from the ocean floor, it is also the largest mountain on Earth!), projecting 4145 meters above sea level and more than 9750 meters above the deep ocean floor.

Solidified basalt flows have several different appearances depending on the fluidity, or viscosity, of the molten lava. Freely flowing lava often forms a thin, smooth outer skin that often wrinkles as the lava advances away from the vent. These flows are known by the Hawaiian name *pahoehoe* (pronounced pa-hoy-hoy) and, while molten and after solidifying, resemble the twisting braids in ropes (Figure 22.19a). Another common type of basaltic lava has a rough, jagged surface with dangerously sharp edges and spiny projections. The Hawaiian name *aa* (pronounced ah-ah) is given to this type of flow. Because they are generally cooler and more viscous than pahoehoe flows, aa flows have the appearance of an advancing mass of lava rubble (Figure 22.19b). Movement downslope results in cooling and the escape of gas, which increases the viscosity and may convert a pahoehoe flow to an aa flow. So smooth, ropelike pahoehoe commonly grades into aa as the lava flow progresses downslope.

Cinder cones are common in many volcanically active areas. They are very steep and rarely rise more than 300 meters or so above ground level. They are formed from the piling up of ash, cinders, and rocks that have been explosively erupted from a single vent. As debris showers down, the larger fragments pile up near the summit of the developing cone to form a symmetrical, steep sided cone around the vent. The finer particles fall farther from the vent to form gentle slopes at the base. Two well-known examples of cinder cones are Sunset Crater in Arizona (Figure 22.17b) and Parícutin in Mexico.

Link to Mythology

According to Greco-Roman mythology, volcanic activity can be traced to Vulcan, the Roman god of volcanic fire and metal working. The word *volcano* comes from the island of Vulcano off the coast of southern Italy, which in ancient times was believed to be the metal workshop of Vulcan. The people surrounding Vulcano believed that the lava fragments and glowing ash that erupted from the island were a result of Vulcan's work as he forged thunderbolts for Jupiter, king of the gods, and weapons for Mars, god of war. A similar story is told by the people of Polynesia, who attribute volcanic activity to Pele, goddess of volcanoes.

In ancient times, people used humanized stories and myths to describe and understand nature. Today we use science to make sense of the events that shape our natural environment.

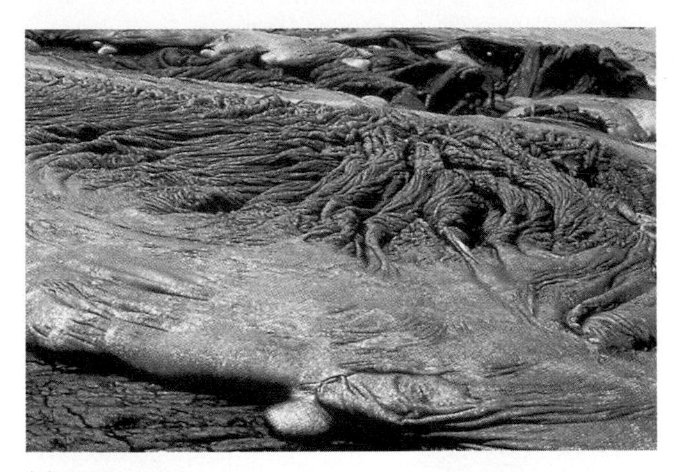

(a)

(b)

Fig. 22.19
(a) Smooth, flowing *pahoehoe* lava is characterized by a twisting rope-like appearance. (b) Jagged *aa* lava moves as an advancing mass of rubble.

When a volcano erupts both lava and ash, a *composite cone* of alternating layers of lava, ash, and mud is produced. The layers build up to form a volcano with a steep-sided summit and gently sloping lower flanks. Mount Fujiyama (Figure 22.17c) is a classic example of a majestic composite cone.

Composite cones tend to erupt explosively because they usually have very viscous magmas and lavas. Viscous magma traps gases, thereby increasing the pressure in the volcano. We can compare the gases in a magma to gases in a bottle of carbonated soda. If we cover the top of the bottle and shake vigorously, the gases separate from the soda and form bubbles. When we remove the cover, pressure is released and gases and liquid explode from the bottle. The gases in magma behave in much the same way. In a volcanic blast, as pressure and temperature exceed confinement, the whole mass of viscous magma and overlying rock explodes into dust and rubble, which when mixed with volcanic ash, can expand and engulf everything in its path. Examples of such volcanic activity are Mount Vesuvius in 79 A.D., Mount Pelee in 1902, Mount St. Helens in 1980, and Mount Pinatubo in 1991.

A *crater* is commonly formed above the central vent of an erupting volcano (Figure 22.20). During eruption, the upwelling lava overflows the crater walls and then sinks back into the vent as the eruption subsides. The walls of a crater often collapse after an eruption, enlarging the area of the central vent.

A crater can grow to more than a kilometer in diameter. A very large crater is referred to as a *caldera.* Calderas range from 5 to 30 kilometers in diameter. Most calderas are formed when the central part of a composite cone collapses into the partially emptied

Blast from Our Recent Past: Mount St. Helens

Imagine the energy released by dropping 30,000 atomic bombs at a rate of one per second over a period of several hours. Such was the explosive energy released when Mount St. Helens erupted on May 18, 1980. The eruption did not come as a surprise. For months prior to the eruption, earthquake activity beneath the volcano caused by the upward movement of magma signaled that the mountain's 123 years of dormancy was over.

To scientists, the dangers of Mount St. Helens were well known. Evidence in the local geologic record showed that over the past 4500 years, Mount St. Helens had erupted more often and more violently than any other volcano in the contiguous United States.

And so it was on May 18, 1980, when the sleeping giant awoke. Within seconds of a magnitude-5 earthquake, the entire northern flank of the mountain fell apart in a massive landslide. Explosions ripped through the sliding debris, blasting gases, ash, and rock across the land at hurricane speeds. With more than 300 meters of its summit gone, the volcano was like a gigantic pressure cooker suddenly uncorked. Vertical explosions of rapidly expanding ash thrust 25 kilometers into the sky (about twice as high as a commercial jet flies). Blown eastward by strong winds, the ash cloud cast an eerie darkness across the state of Washington. In a matter of hours, the ash covered the landscape; 540 million tons of ash was deposited over an area of more than 22,000 square miles. In two weeks time, the ash circled the Earth.

As ash particles filled the skies, mud flows scoured the land. Roaring clouds of superheated gas, steam, and ash, traveling at speeds in excess of 350 miles per hour, obliterated everything for miles to the north of the volcano. The glowing debris blasted from the volcano quickly melted Mount St. Helens' cover of ice and snow, sending torrents of scalding mud flows down the North and South forks of the Toutle River. When it was over, 150 square miles of forest lay in ruins and about 57 people had perished.

Even in the face of such violence, the rebirth of life often occurs quickly and dramatically. Although devastating, the 1980 eruption of Mount St. Helens provides an example of the Earth's cyclical nature. The blast virtually obliterated everything in its path but very soon afterwards, life returned to the blast area and now is anchoring itself in the desolate volcanic landscape; pockets of plant and animal life have gained a foothold. Like pioneers moving west, life slowly returns to the mountain.

Fig. 22.20
Craters form in the central vent of volcanoes. This crater in Mount St. Helens shows a rising steam plume and a lava dome.

space in the interior of the volcano where magma used to be (the magma chamber), but a few have been formed by explosive eruptions in which the top of a volcano was blown out. The volcanic eruption 7000 years ago of Mount Mazama in Oregon is one of these few catastrophic events. The eruption blasted ash throughout the northwestern United States. After the eruption, most of the cone collapsed into the emptied magma chamber. The caldera, filled with rainwater, resulted in Oregon's famous Crater Lake which is 9 kilometers wide and 590 meters deep.

Yellowstone National Park, located in one of the most seismically active regions of the Rocky Mountains, is a "hotspot" in the Earth's crust. Situated in a caldera 70 kilometers long and 45 kilometers wide, Yellowstone Park is the remnant of an ancient volcano that violently erupted about 600,000 years ago. Most of the hot springs, bubbling muds, steaming pools, and spouting geysers for which the park is famous lie within the caldera. Heat from the enormous reservoir of molten rock that produced the massive eruption still remains not too far below the Earth's surface, sustaining the present thermal activity. Although no eruption has occurred in historical time, the molten rocks remain so close to the surface that the possibility of an eruption in the future cannot be disregarded. Yellowstone is an

Fig. 22.21
Crater Lake in Oregon is a remnant of the eruption of Mount Mazama 7000 years ago.

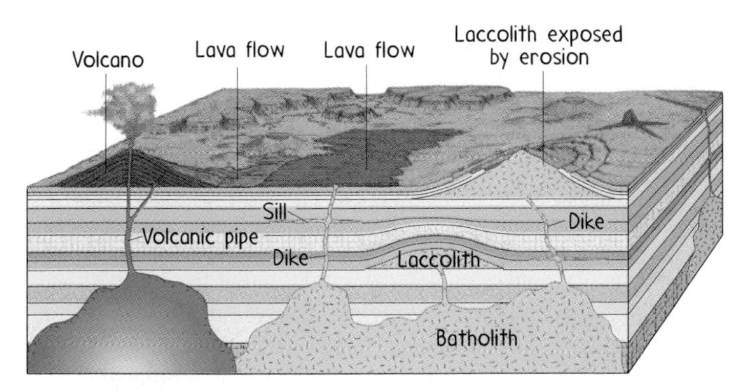

Fig. 22.22
Intrusive igneous features in cross-sectional view.

example of geology in action, where the eruption of scalding water may be but a precursor to more violent activity in the near future.

Igneous Rock Formation Beneath the Earth's Surface

As mentioned earlier, igneous rocks that form beneath the Earth's surface are called *intrusive* rocks. Large intrusive igneous rock bodies are called **plutons**. They occur in a great variety of shapes and sizes, ranging from slabs to massive, nondescript blobs (Figure 22.22). As you might have guessed, intrusive rocks can only be studied after the processes of uplift and erosion have exposed them.

A common pluton is a *dike*—formed by the intrusion of magma into fractures that cut across the layering of existing rock. Dikes, the ancient channel ways for rising magma, are closely associated with volcanic vents. A spectacular example of this association is the radiating dikes around the eroded volcanic neck at Shiprock, New Mexico (Figure 22.23). At Shiprock, the volcanic neck and the dikes are more resistant to erosion than the rock they intruded, and so erosion has preferentially removed the surrounding material, leaving the volcanic rocks behind as a peak (the volcanic neck) and wall-like ridges (the dikes). If a dike is more prone to erosion than the surrounding rock, it may leave a trench or ditch at the surface.

Another pluton is a *sill*—formed by the intrusion of magma into fractures that are parallel to the layering of existing rock. Most sills are formed by the intrusion of very fluid basaltic magmas at shallow depth. Because sills form at shallow depths, they often resemble buried lava flows. A variation of a sill is a *laccolith,* a body of rock that is created when viscous magma rising upward in the Earth's crust encounters a more resistant layer that forces the magma to spread, forming a mushroom shape. Unlike sills, laccoliths push the overlying layers upward in domelike fashion.

Batholiths are the largest of the plutons and are defined as having more than 100 square kilometers of surface exposure. A batholith is usually not generated by a single intrusion. Instead, numerous intrusive events over millions of years account for the development of a massive batholith. Batholiths form the cores of many

Fig. 22.23
Shiprock New Mexico. Radiating dikes surround the eroded remains of a volcanic vent.

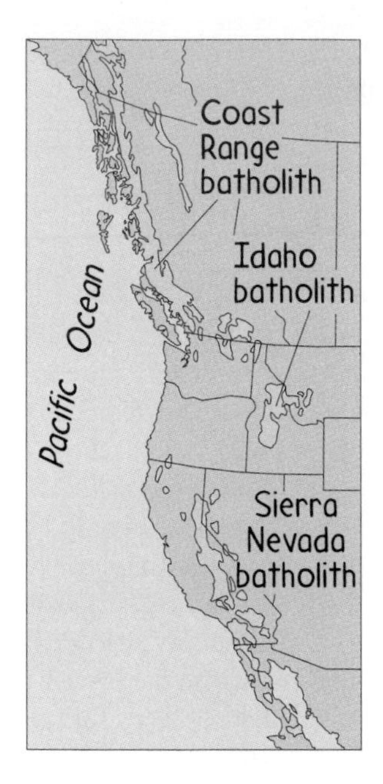

Fig. 22.24
(a) Some of the largest batholiths in North America include the Coast Range batholith and the Sierra Nevada batholith.

major mountain systems of the world. Many mountains are actually the exposed cores of the batholiths of larger mountains that have long since eroded. Some of the largest batholiths in North America include the Coast Range batholith and the Sierra Nevada batholith. These batholiths continue to push upward, increasing the height of the mountain range (Figure 22.24).

Question

1. Why is it incorrect to say that igneous rocks may form from the intrusion of lava?
2. Is it correct to say that igneous rocks may form from the extrusion of lava?

22.5 Sedimentary Rocks

Sedimentary rocks are the most common rocks in the uppermost portion of the crust. They cover two-thirds of the Earth's surface and form a thin and extensive blanket over igneous and metamorphic rocks. Because sedimentary rocks are the remains of the preexisting rocks that they cover, they provide information about geological events that occurred at the Earth's surface.

Sedimentation

While the process of volcanism continually generates new rock material at the Earth's surface, the opposing force of *weathering* breaks down and decomposes surface rock. There are two kinds of weathering, *mechanical* and *chemical,* and both produce *sediment.* **Mechanical weathering** physically breaks rocks into smaller and smaller pieces. **Chemical weathering**, which involves reactions with water, decomposes rock into smaller pieces. As rock is weathered, it erodes. **Erosion** is the process by which weathered rock particles are removed and transported away by water, wind, or ice.

Sediments composed of small fragments of preexisting rocks are called *clastic* sediments, whereas those produced by chemical precipitation are called chemical sediments. When clastic particles are first produced, they are normally quite angular and jagged. During transportation, especially by water, the particles are continuously abraded by impacts with other particles, which decreases their size and rounds off their sharp edges. When transportation stops, deposition and sedimentation begin. As wind dies down, dust settles; as water currents subside, sand settles; as glaciers melt, the small rocks embedded in the ice get left behind. Erosion, transportation, deposition, and sedimentation follow a downhill path in response to gravity.

Answers

1. The terminology in the statement is wrong. To begin with, the term intrusion refers to solidification that occurs in the Earth's interior and therefore has nothing to do with lava. Secondly, lava is not a synonym for magma, but is the term for magma that has been extruded at the Earth's surface in molten form. There is no flowing lava beneath the Earth's surface.
2. Yes, once magma is extruded from the Earth it is called lava, which when solidified becomes igneous rock.

(a)

(b)

Fig. 22.25
(a) Well-sorted sediment grains.
(b) Poorly-sorted sediment grains.

The larger a sediment particle, the stronger a water current must be to carry it. Typically, water currents get weaker as they move away from their source area, and so larger particles being transported by moving water are the first to be deposited, while smaller particles are able to stay with the flow. Particles therefore tend to be sorted according to size as they are deposited. A deposit that contains particles of very similar sizes is said to be well sorted, while a deposit that contains particles of various sizes is poorly sorted. During transportation, sediment is continuously sorted and abraded, and so the size, shape, and sorting of particles in a sedimentary deposit provide clues to the time and distance of transport and to the depositional environment. In general, poorly sorted, angular grains of various shapes imply a short transportation distance, whereas well-sorted, well-rounded grains imply a greater transportation distance. We can also get clues to the method of transportation. Glacial deposits, for example, tend to be very poorly sorted and angular, whereas wind-blown deposits tend to be very well sorted and fine-grained.

As deposited sediment accumulates, it begins to transform into sedimentary rock. The transformation of sediments to sedimentary rock occurs in two ways—compaction and cementation. As the weight of overlying sediments presses down upon deeper layers, sediment grains are squeezed and compacted together. This compaction squeezes water out of the spaces (called *pores*) between sediment particles. This water often contains compounds such as silica, calcite, and hematite (iron oxide) in solution. These compounds, which are chemically precipitated from solution, partially fill the pore spaces with mineral matter and thereby act as cementing agents. Thus calcite, silica, and iron oxide make up the most common cementing agents. Silica cement, the most durable, produces some of the hardest and most resistant sedimentary rocks. When iron oxide acts as a cementing agent, it produces the red or orange stain of many sedimentary rocks. The colors of Bryce Canyon provide a picturesque example of iron oxide stain.

Fig. 22.26
The red and orange at Bryce Canyon are caused by the presence of iron oxide.

(a)

(b)

(c)

Fig. 22.27
Sedimentary rocks. (a) Shale, composed of fine sized particles (this sample contains a fossil brachiopod); (b) sandstone, composed of medium-sized particles; and (c) conglomerate, made up of a wide variety of large, rounded particles.

Clastic Sediments

Clastic sedimentary rocks are classified by particle size (Table 22.5). The most abundant clastic sedimentary rocks are *shale,* composed of very fine particles too small to be visible with a magnifying hand lens, *sandstone,* composed of medium sized particles such as those found in typical beach sand, and *conglomerate,* composed of a variety of larger particles, ranging from pebbles to boulders.

Shale is formed by the compaction of silt and clay-sized particles. It is finely laminated and has the ability to split into thin layers, or flakes, parallel to depositional layers. The extremely fine grain size suggests that particle deposition occurred in relatively quiet waters, such as deep ocean basins, flood plains, deltas, lakes, or lagoons. The color of shale ranges from gray to black and from red to brown to green, indicating its environment of formation. Gray to black shale indicates buried organic matter, which can be preserved only in an oxygen-deficient, swampy environment. Black shale is commercially important, for it is the principal source rock for petroleum. Red to brown shale indicates ferric oxide (red) or hydroxide (brown). The absence of these colors show the characteristic green color of shale.

Sandstone is the second most abundant clastic sedimentary rock and can be classified into three types. When quartz is the primary mineral, the rock is simply called *quartz sandstone.* Quartz sandstone is composed of well-sorted, well-rounded quartz grains. Sandstone that contains considerable amounts of the mineral feldspar is called *arkose.* The grains in arkose tend to be poorly rounded and not as well sorted as those in quartz sandstone. Sandstone composed of quartz, feldspar, and angular rock fragments is called *graywacke.* Sandstones form in a variety of environments, including desert dunes, beaches, marine sand bars, river channels, and alluvial and submarine fans.

Conglomerates are composed of gravels and rounded rock fragments. The rock fragments are usually large enough for easy identification, thus readily providing useful information about the areas from which the sediments were eroded. Larger rock fragments must have been transported by currents strong enough to carry them. Because these strong currents rounded the rock fragments, the roundness of their edges and corners are good guides to the distance they traveled. Conglomerates are often found in river channels and in rapidly eroding coastlines.

Clastic Sedimentary Environments

The most dominant feature of sedimentary rocks is the way the particles of sediment are laid down, layer upon layer. These layers are referred to as *beds.* Varying in both thickness and area, each bed represents one episode of deposition. For example, flooding in a particular year might produce a layer of sediment adjacent to a river. A flood any time after that produces an overlying layer. The deposition of clastic sediments occurs in many different environments, including alluvial, desert, delta, and shoreline environments.

The term *alluvial* refers to unconsolidated gravels, sand, and clay that have been deposited by streams. When a fast-flowing mountain stream leaves its narrow valley and abruptly emerges onto a broad, relatively flat plain, the speed of the flow drops and the stream dumps its load of sediment. These deposits are generally fan-shaped *alluvial fans* (Figure 22.28). The steep, upper slopes of the fan are dominated by boulders, cobbles, and gravels, while the base of the fan and the alluvial plain are made up of sand, silt, and mud. With each episode of deposition, the alluvial fan grows upward and outward.

As streams continue their downhill path, they meander back and forth across a river valley, depositing sediments as they go (Figure 22.29). The meandering move-

Table 22.5
Particle Size Classification of Clastic Sediments and Sedimentary Rocks

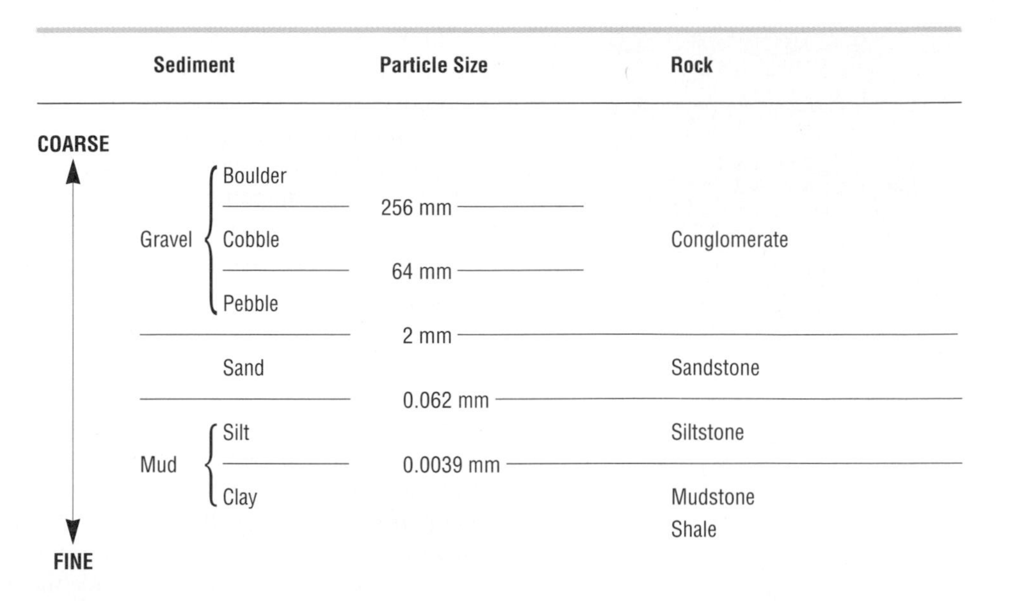

COARSE ↑

Sediment		Particle Size	Rock
Gravel	Boulder		
		256 mm	
	Cobble		Conglomerate
		64 mm	
	Pebble		
		2 mm	
	Sand		Sandstone
		0.062 mm	
Mud	Silt		Siltstone
		0.0039 mm	
	Clay		Mudstone Shale

↓ FINE

Fig. 22.28
An alluvial fan in Death Valley, California.

ment creates a wide belt of almost flat land—a floodplain. As the name implies, it is this section of the river valley that becomes flooded with water when a river overflows its banks. When flooding occurs, sands and gravels are deposited on the river banks, and these deposits create natural levees that confine the river, while silt and clay particles spread out over the floodplain.

The arid desert environment is characterized by angular hills, shear canyon walls, and sand dunes. Because such an environment lacks moisture, mechanical weathering predominates. Interestingly enough, although the desert lacks moisture, water is the main cause of erosion and transportation of sediments. Rare as water is in the desert, when a heavy rain falls, the water does not have time to soak into the ground and causes flash floods. These flash floods transport and then deposit great quantities of debris and sediment, forming alluvial fans at the bases of mountain slopes and depositing alluvium on the floors of wide valleys and basins. Evidence of alternating wet and dry conditions

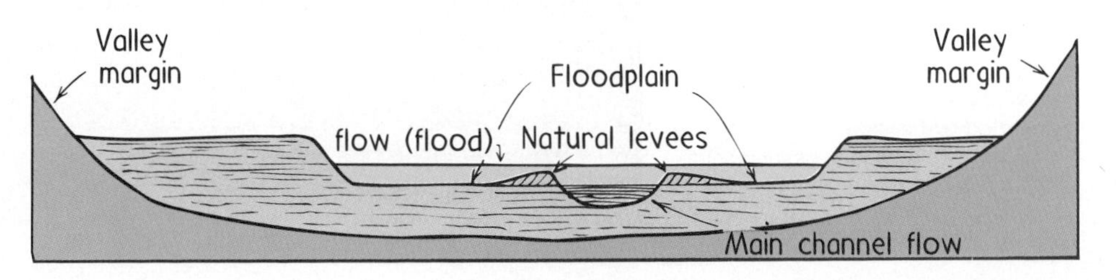

Fig. 22.29
Cross section of an alluvial valley. A floodplain is created when a river overflows its banks. Sands and gravels settle out first and act as natural levees to confine the river. Because the finer silt and clay particles are able to flow as a suspended load, they move beyond the levees and settle on the floodplain.

Fig. 22.30
Desert erosion on a cliff face. A desert has many extremes—scorching daytime heat, chilling night air, and strong winds. Mechanical weathering physically breaks down the rocks to smaller and smaller pieces.

can be seen in the unique sedimentary feature called *mudcracks,* found in desert basins (Figure 22.31). When exposed to air, mud dries out and shrinks, producing cracks. Mudcracks are also associated with shallow lakes and tidal flats.

Wind also transports and deposits sediments. If you've ever been in a wind storm or at the beach on a windy day, you may have felt the sandblasting effect of the wind. Once in the air, particles of sediment can be carried great distances by the wind. Red dust from the Sahara is found on glaciers in the Swiss Alps, and fine quartz from central Asia has been detected on the Hawaiian Islands! In the desert, winds move over surfaces of dry sand, picking up the small, more easily transported particles but leaving the large, harder-to-move particles behind. The small particles bounce across the desert floor, knocking more particles into the air, to form ripple marks, which are actually tiny sand dunes. Ripple marks can also be formed by the movement of sand grains in water currents, as seen in shallow streams or under the waves at beaches.

Sand dunes are formed when air flow is obstructed by an obstacle, such as a rock or clump of vegetation. As the wind sweeps over and around the obstacle, wind speed drops and, as a result, sand grains settle in the wind shadow. As more sand settles, mounds form and further impede the flow of air. With more sand and more wind, the mound becomes a dune. As a dune grows, the whole mound starts moving downwind as sand grains on the windward slope move up and over the crest of the dune to fall on the leeward slope (Figure 22.33). Over time, this continuous process moves the entire dune. A unique feature of this action can be seen in the *cross-bedding* found on the leeward side of the dune. The direction of cross-bedding indicates the direction of the wind (or water current) that deposited the sediments. Cross-bedding is also a common feature in river deltas and certain stream-channel deposits.

Another common depositional environment for clastic sediments occurs where a river ends as it enters a sea, bay, or lake. A river that encounters a standing body of water dumps its load of sediment as its speed decreases. The dumped sediment is referred to as a *delta.* A delta environment resembles an alluvial fan

Fig. 22.31
Mudcracks, a feature unique to sedimentary rocks, are evidence of alternating wet and dry conditions in the desert.

Fig. 22.32
Generated by blowing winds, ripple marks are narrow ridges of sand separated by wider troughs. They are small, elongated sand dunes.

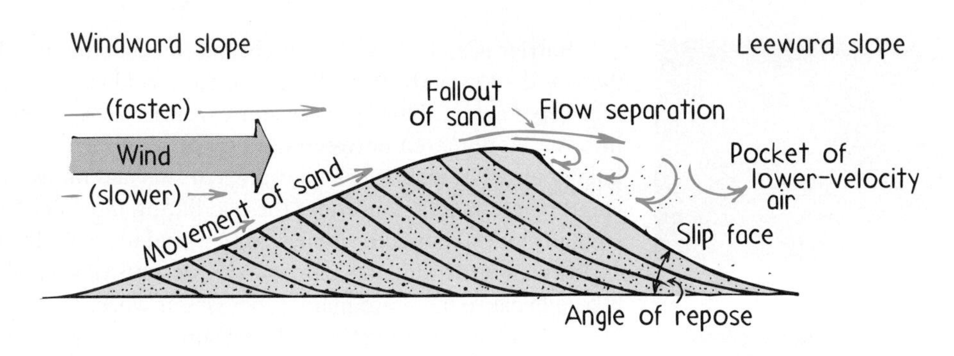

Fig. 22.33
Formation of a sand dune. When air flow is obstructed, air speed drops and as a result sand grains settle in the wind shadow. With more wind, more sand settles, and a dune is formed. As a dune grows, sand grains on the windward slope move up and over the crest to fall on the leeward slope, which results in motion of the whole mound downwind.

Fig. 22.34
Cross-bedding in an ancient sand dune. Can you tell the direction of windflow?

in two respects. Both are fan-shaped, and both are formed by a stream's inability to transport sediment indefinitely. As a delta forms, coarser material settles first, then medium, and finer material farther out.

Deltas begin to form underwater, but the addition of incoming sediment eventually causes the delta to emerge as new land. The stream channel becomes choked with sediment causing the formation of smaller branches off the main river (Figure 22.35). Some of the world's greatest rivers have massive deltas at their mouth. Millions of years ago, the mouth of the Mississippi River was where Cairo, Illinois, is today. Since that time the delta has extended 1600 kilometers south to the city of New Orleans. Less than 5000 years ago the site of New Orleans was underwater in the Gulf of Mexico!

Clastic sedimentary environments also occur in coastal areas. Shoreline environments are dominated by beaches and barrier islands. Winds blowing across the ocean surface generate waves; as the waves approach shallow water near land, they become higher and steeper until they finally collapse, or break. This is the *surf zone,* where wave activity moves sediment back and forth, shoreward and seaward. Sandy beaches are the result of the turbulent motion of the surf zone.

Fig. 22.35
Satellite image of the Mississippi Delta. Note how smaller streams are formed branching off from the main river.

(a)

(b)

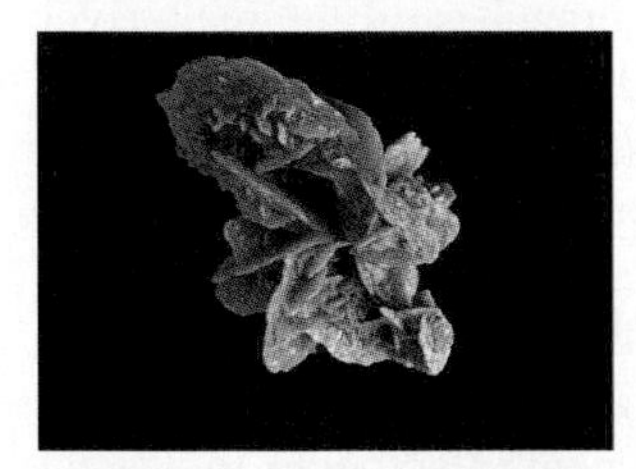
(c)

(d)

Fig. 22.36
Chemical sedimentary rocks: (a) limestone, (b) dolomite, (c) gypsum, (d) and halite.

Barrier islands, low offshore islands of sand that parallel the coast, build up along the world's lowland coasts. Ridges of dune sand are built up on the island from successive wave action. During large storms, surf washes over the lowlands, making inlets into the lagoon area between the barrier island and the shore. The lagoon area is a quieter environment. It has finer-grained silts and muds that feature cross-bedding and small ripple marks caused by the oscillating motion of the lagoon water. On shore, smooth stones, rounded pebbles, and/or sand make up the beaches.

Glacial environments can contain elements of all these clastic environments. This is true because there are many processes at work, driven by water, wind, and ice. For example, glacial meltwater forms streams that may terminate as deltas in lakes, bays, or seas. Glacial deposits have some unique features, however. When glacial ice melts, it drops a poorly sorted, heterogeneous load of boulders, pebbles, sand, and clay. A wide range of particle sizes is the hallmark that differentiates glacial sediment from the much better sorted material deposited by streams and winds.

Chemical Sediments

Chemical sediments are formed by the precipitation of minerals from water. The process can occur directly, as the result of an inorganic reaction, or indirectly, as the result of a biochemical reaction. Chemical sediments fall into two categories: *carbonates* and *evaporites.* Carbonates are the best example of rocks formed by biochemical reactions because most of them form from the shells of dead marine organisms. Evaporites are good examples of rocks formed by inorganic processes because they form as a result of the evaporation of water.

Carbonates are minerals and rocks composed mostly of calcite, $CaCO_3$, or dolomite, $CaMg(CO_3)_2$. They make up about 10 percent of the total volume of sedimentary rocks. Although most carbonates are formed biologically as a result of shell growth, some form without the aid of organisms. Cave dripstones, such as stalactites and stalagmites, provide an interesting example of calcium carbonate precipitating inorganically from dripping water (Figure 22.37). The common rock *limestone,* the most abundant carbonate rock, is formed predominantly as a result of biologic activity. Many organisms living in the sea extract calcium carbonate from water to build hard, protective shells. When these organisms die, their shells accumulate on the sea floor, eventually solidifying into limestone. Because of compaction and the high solubility of calcium carbonate, the original textures and structures of the sea shells are often obliterated. Closely related to limestone, *dolostone,* composed predominantly of dolomite, results from the replacement of calcium by magnesium.

Evaporites are minerals and rocks that are precipitated when a restricted body of seawater or the water of a saline lake evaporates. Examples are gypsum, anhydrite, and halite. These names apply both to individual minerals and to rocks made of a single type of evaporite mineral. As evaporation proceeds, the least soluble minerals, such as gypsum (used for the making of plaster of Paris) precipitate first, followed by the more soluble minerals, such as anhydrite and then halite (common table salt). Although carbonates make up the bulk of chemical sediments, evaporites make up a small but significant portion.

Chemical Sedimentary Environments

Warm climates favor carbonate deposition because carbonates are less soluble in warm water than in cold. Carbonate depositional environments include coral reefs and carbonate platforms.

Fig. 22.37
Calcium carbonate precipitating from dripping water in a cave forms icicle-shaped stalactites hanging down from the ceiling and cone-shaped stalagmites protruding upward from the ground.

Coral reefs are made up of actively growing, individual coral organisms, each of which is only a few millimeters across. The organisms secrete calcium carbonate as they grow. When we see a piece of coral or a coral reef, it is the calcium carbonate that we are seeing, not the organism. Most, but not all, corals form colonies, and it is the colony-forming corals that are the major reef-builders. Reefs grow outward with time, but they can also grow upward as new corals cement themselves to the dead coral below. Reefs can only survive in shallow water because a major food source for coral is photosynthetic algae, which need light to live. The coral and algae live in mutual support. The coral provides protection for the algae, and the algae provides oxygen and nutrients for the coral. Coral also require warm, clear, and relatively sediment-free water to flourish.

A coral reef is usually divided into several sections: a wave-resistant front, a flat reef platform, a shallow protected lagoon, and an island (or small patches of reefs) (Figure 22.38). Each has its own characteristics. Because the coral reef is partially destroyed by wave action as it grows, carbonate particles adjacent to the reef range in size from blocks several meters across to fine mud. Because ancient coral reefs, composed of alternating layers of porous material and impermeable muds, have the potential to act as traps for oil and gas, they are economically important. An example of ancient coral reefs can be found in the Guadalupe Mountains of western Texas (Figure 22.39).

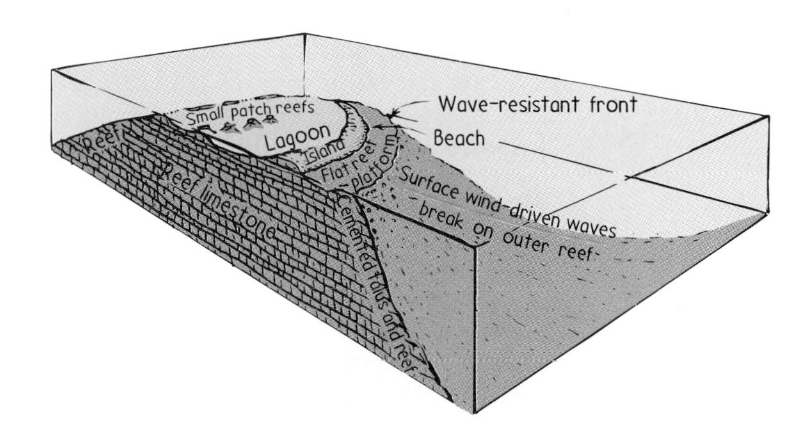

Fig. 22.38
The sections of a typical coral reef.

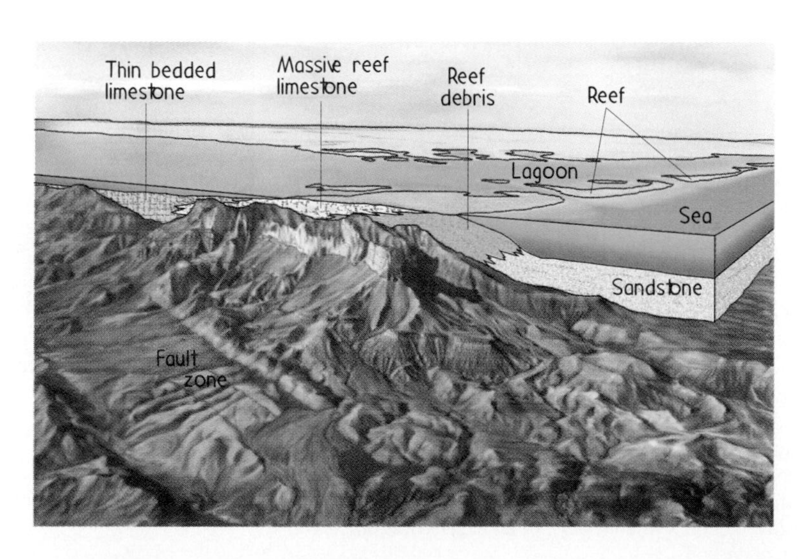

Fig. 22.39
Ancient coral reefs in the Guadalupe Mountains of western Texas.

(a)

(b)

Fig. 22.40
(a) The white-sand beaches of California are composed of silicate minerals and are classified as inorganic. (b) The white-sand beaches of Hawaii are composed of carbonate minerals—the sediment remains of tiny shells—and are classified as organic.

Carbonate platforms are much larger than coral reefs, but their existence is also due to organisms. Carbonate platforms are the graveyards of calcium-secreting organisms and are formed in shallow waters either close to or attached to continents. They account for the largest portion of carbonate sediment produced in the ocean.

Sand-sized fragments from coral reefs and carbonate platforms make up the white-sand beaches in many island areas, such as Hawaii (Figure 22.40). Look carefully at the sand in such tropical beaches, and you'll see it is predominantly composed of shell fragments. In contrast, the sand on the beaches of the continents is predominantly composed of silicate minerals. So, whereas sand in Hawaii is organic in origin, beach sand along the American western coast is largely inorganic.

Evaporite deposits require a dry climate conducive to the evaporation of lake water or seawater. As the water dries out, evaporite minerals precipitate. Modern-day as well as ancient evaporites are found in desert basins, tidal flats, and restricted sea basins. Vast carbonate and evaporite deposits on the continents are evidence that expansive, shallow seas have periodically covered the land surfaces in the past.

Fossils

Because sedimentary rocks are formed at the Earth's surface, they often contain remains of preexisting life forms—fossils—that provide important information for interpreting the Earth's geologic past. As we shall see in Chapter 25, fossils play an important role as time indicators and in the correlation of rocks from different places of similar age. Some fossils consist of whole organisms, but most are just their parts. Other fossils are simply an impression, made in the rock before it hardened. Plants commonly leave their impression as a thin film of carbon. There are many methods of fossilization (Figure 22.41).

(a)

(b)

(c)

(d)

Fig. 22.41
Some of the many methods of fossilization.
(a) Permineralization occurs when the porous remains of an organism become filled with water that is rich in dissolved minerals—like petrified wood.
(b) Impression is made by an organism buried quickly, before it can decompose, thereby preserving its impression.
(c) Replacement occurs when the remains of organisms are replaced by a mineral. Pyrite has replaced the original shell in this specimen.
(d) Carbonization occurs when a plant or an organism is preserved as a thin film of carbon.

Fossil Fuels

Most of the organic matter from animals and plants of the past was quickly decomposed by bacteria and converted to nutrients that became available for use by new life. Material that escaped bacterial decay was either stored as sparsely distributed organic matter or converted to petroleum or coal.

Although coal, oil, gas, and oil shale are all fossils in the sense that they are the remains of past organisms, these remains have been so changed over time that the form and even the composition of the accumulated organisms are beyond recognition.

The source of oil and gas is fossil organic matter integrated throughout buried sediments. When buried organic-rich sediment is heated over a sufficient period of time, chemical changes take place that create oil. Under pressure of the overlying sediments, the minute droplets are squeezed out of the source rocks and into overlying porous rocks that become reservoirs.

Oil shale is formed when algae produce an ooze rich in hydrocarbons on the bottom of a lake or sea. Just as in the metamorphism of rocks, deeper burial results in higher temperatures that generate gas rather than oil.

Coal is formed from plants that do not completely decay but are so altered that the original structure is destroyed. Coal, petroleum, and natural gas are the primary fuels of our modern economy.

22.6 Metamorphic Rocks

Igneous or sedimentary rocks may undergo change—**metamorphism**—if they are heated and/or compressed for long periods of time. Potter's clay, for example, is soft and pliable at room temperature but when heated turns to a hard ceramic. Similarly, limestone subjected to enough heat and pressure becomes marble, and shale is metamorphosed to slate. Rocks may also be drastically stretched or compressed. It is important to note that no minerals are melted during metamorphism. Change instead occurs by means of recrystallization of preexisting minerals and mechanical deformation of rock.

Recrystallization often occurs when rocks subjected to high temperatures and pressures go through a change in mineral assemblage—usually accompanied by the loss of H_2O or CO_2. Although temperature and pressure are often high enough to cause a metamorphic reaction, fluids enclosed in pore spaces act as a catalyst to aid the reaction. When the rock is subjected to increased temperature and pressure, the amount of pore space decreases and the fluid is squeezed out. The fluid can then readily react with the surrounding rock. In general, the more fluid, the faster the reaction.

Mechanical deformation, which may or may not involve elevated temperatures, occurs when a rock is subjected to stress. Surface rocks that become deeply buried and undergo increased pressure may flow plastically, bending into intricate folds. Or pressure may deform and flatten the rock, or shear it, breaking it and grinding it into fragments.

Types of Metamorphism

Metamorphism can be *dynamic, contact,* or *regional.* Each type is characterized by differences in mechanical deformation and recrystallization.

Dynamic metamorphism primarily involves mechanical deformation. For example, the shearing and grinding that take place in a fault zone crush, flatten, and elongate preexisting crystals to produce broken and distorted textures.

Contact metamorphism occurs when a body of rock is intruded by magma (Figure 22.42). The high temperature of the magma produces a zone of alteration that surrounds the intrusion. The alteration is greatest at the *contact,* which is the surface that separates the solidified intrusive rock from the surrounding rock. The width of the altered zone may range from a few centimeters to several hundred meters. Around a small intrusive body, such as a dike, the altered zone is very narrow and resembles "baked" rock, but with a larger intrusive body, such as a batholith, the altered zone may be 100 meters thick or more. One of the most common changes is an increase in grain size due to recrystallization. The grain size is greatest at the contact, and decreases with distance from that point. The water content of the rock also changes with distance from the contact. At the contact, where temperature is high, water content is low. So we find dry, high-temperature minerals such as garnet and pyroxene at the contact. Farther away, we find water-rich, low-temperature minerals such as muscovite and chlorite.

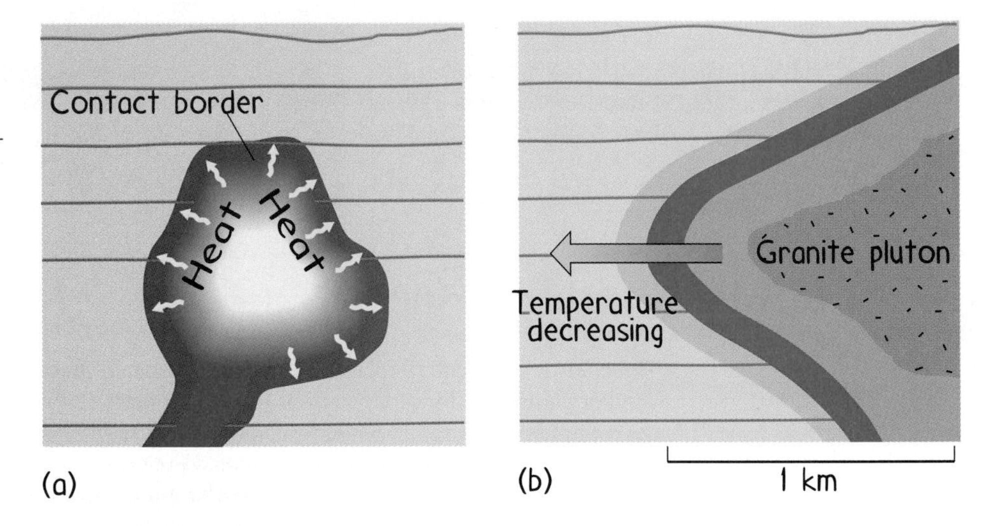

Fig. 22.42
(a) Contact metamorphism is the result of rising molten magma that intrudes a rock body. (b) Surrounding the solidified intrusive rock is a zone of alteration. Alteration is greatest at the contact and decreases away from the contact area.

Regional metamorphism is the alteration of rock by both thermal and mechanical means over a whole region, but it is more than just dynamic metamorphism added to contact metamorphism. For example, the intrusion of magma is not the only cause of the elevated temperatures associated with regional metamorphism. The deep burial of large tracts of rock results in heating simply because the Earth's interior is hotter than its surface. Regionally metamorphosed rocks are found in all the major mountain belts

Fig. 22.43
Folded rock layers in the Rocky Mountains.

of the world. During the process of mountain-building, the Earth's crust is severely compressed into a mass of highly deformed rock. This deformation can be seen in the folded and faulted rock layers in many mountain ranges. The effects of regional metamorphism are most pronounced in the cores of deformed mountains. Rocks develop distinctly foliated and layered textures and zonal sequences of minerals and textures. Because of the large-scale nature of regional metamorphism, these zones tend to be broad and extensive. Regions of structural deformation are the hunting grounds of gem prospectors, for the heat and pressure that accompany these changes produce beautiful minerals.

Metamorphic rocks are defined by their texture and the minerals they contain. For classification and identification, metamorphic rocks can be divided into two groups: *foliated* and *nonfoliated.*

Foliated Metamorphic Rocks

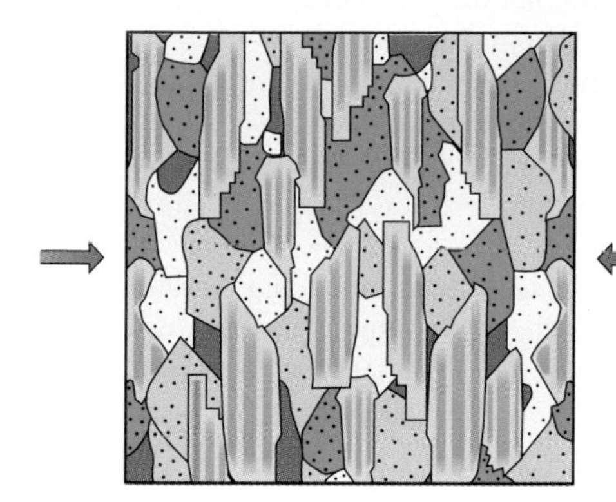

Fig. 22.44
As compressive forces squeeze platy and sheet structured minerals, the grains align themselves perpendicular to the main direction of force. Arrows indicate direction of compressive force.

As rock is subjected to an increase in temperature and pressure, some of its minerals realign themselves in parallel planes, with the face of each plane perpendicular to the main direction of the compressive force. This leads to a layered texture (much different from the layering seen in sedimentary rocks!), called *foliation,* that is a prominent textural feature of regionally metamorphosed rocks. For example, sheet-structured minerals such as the micas start to grow and orient themselves so that their sheets are perpendicular to the direction of maximum pressure (Figure 22.44). The resulting rock, which now has parallel flakes, or plates, of mica, is said to be foliated. The most common foliated metamorphic rocks—slate, schist, and gneiss—are derived from fine-grained sedimentary rocks that have the appropriate chemical composition for micas to form.

Slate is the lowest-grade foliated metamorphic rock, meaning that it is formed under relatively low temperatures and pressure. Metamorphosed shale, slate is a very fine-grained foliated rock composed of minute mica flakes. The most noteworthy characteristic of slate is its excellent rock cleavage, which allows it to be split into thin slabs. The best pool tables and chalk boards are made from slate quarried in metamorphic terrains where slaty cleavage is well developed. Slate is also used as roof and floor tile.

Schist is one of the most distinctive metamorphic rocks. It forms under higher temperature and pressure conditions than slate, which causes the constituent minerals to grow large enough to be identified with the naked eye. Schists typically contain 50 percent platy minerals, most commonly muscovite and biotite. The larger mica flakes give the rock a shiny surface that is quite striking. Schists are named according to the major minerals in the rock (biotite schist, staurolite-garnet schist, and so on).

(a)

(b)

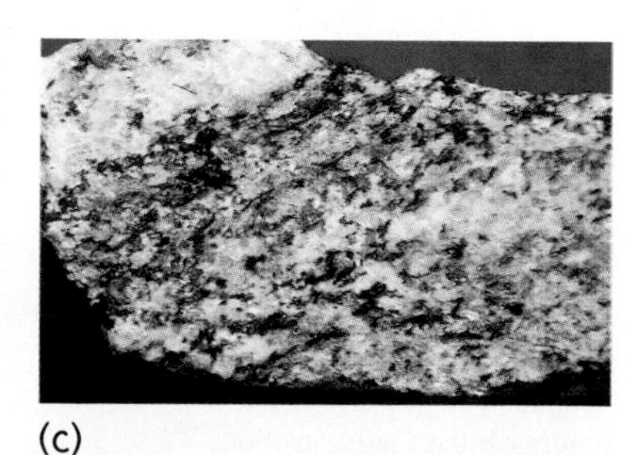

(c)

Fig. 22.45
Common foliated metamorphic rocks: (a) slate, (b) schist, (c) gneiss.

Gneiss (pronounced "nice") is a foliated metamorphic rock that contains alternating layers of dark platy minerals and lighter granular minerals, which imparts a characteristic banded appearance. This change in texture is caused by still greater temperature and pressure conditions than those for schists. The most common granular minerals found in gneisses are quartz and feldspar, which are also the most common granular minerals in granite. In fact, some gneisses are actually metamorphosed granites.

(a)

(b)

Fig. 22.46
Nonfoliated metamorphic rocks: (a) marble, (b) quartzite.

Nonfoliated Metamorphic Rocks

Nonfoliated metamorphic rocks can form either in areas of increased temperature and pressure, or in areas where only the temperature increased. Under high pressure, foliation cannot occur if the chemical constituents necessary for the growth of platy minerals, such as micas, are absent. If these chemical constituents are present, but the pressure is not high enough, such as in contact metamorphism, foliation cannot develop. Two common nonfoliated rocks that generally do not have the potential for platy minerals to grow are marble and quartzite.

Marble (Figure 22.46a) is a coarse, crystalline, metamorphosed limestone or dolostone. Pure marble is white and virtually 100 percent calcite. Because of its color and relative softness (hardness 3) marble is a popular building stone. Often the limestone from which marble forms contains impurities that produce various colors. Thus marble can range from pink to gray, green, or even black.

Quartzite (Figure 22.46b) is metamorphosed quartz sandstone and therefore very hard (hardness 7). The recrystallization of quartzite is so complete that the rock splits across the original quartz grains when broken, rather than between them. Although pure quartzite is white, it commonly contains impurities and thus can be a variety of colors, such as pink, green, and light gray.

22.7 The Rock Cycle

We have seen that the igneous, sedimentary, and metamorphic rocks of the Earth's crust have different origins. Although formed by different processes, the three rock types are related. This interrelationship is graphically portrayed in the model of the rock cycle (Figure 22.47). By following the different pathways in the model, we can determine the origin of the three basic rock types and the various geologic processes that transform one rock type into another. The figure helps to summarize this chapter.

Magma (molten rock)
Cooling and solidification (Crystallization)
Igneous ("fire-formed") Rock
Weathering, transportation, and deposition
Sediment
Cementation and compaction (Lithification)
Sedimentary ("settling") Rock
Heat and pressure (Metamorphism)
Metamorphic ("changed form") Rock
Melting
Heat and pressure
Weathering, transportation, and deposition
Weathering, transportation, and deposition

Fig. 22.47
The rock cycle: Igneous rock subjected to heat and pressure far below the Earth's surface may become metamorphic rock. Or metamorphic or sedimentary rocks at the Earth's surface may decompose to become sediment that becomes new sedimentary rock. Whatever the variety of routes, molten rock rises from the depths of the Earth, cools, and solidifies to form a crust that over eons is reworked by shifting and erosion, only to eventually return to become magma in the Earth's interior.

We have seen that igneous rock is formed when the molten magma beneath the Earth's crust cools and crystallizes. Magma can crystallize into many different kinds of igneous rock. Although most of the Earth's crust is either igneous or derived from rock that was initially igneous, the rock we see at the surface is mainly sedimentary.

We have seen that sedimentary rock is the result of the decomposition and disintegration of igneous (or other) rocks by weathering and erosion. When sedimentary rock is buried deep within the Earth or is involved in mountain building, we have seen that great pressures and heat can transform it into metamorphic rock. When subjected to still greater heat and pressure, metamorphic rock melts and turns to magma, which eventually solidifies as igneous rock to complete the rock cycle.

The rock cycle varies in its routes. Igneous rock, for example, may be subjected to heat and pressure far below the Earth's surface to become metamorphic rock. Or metamorphic or sedimentary rocks at the Earth's surface may decompose to become sediment that becomes new sedimentary rock. Cycles within cycles occur in the dynamics of the Earth's crust. Whatever the route, molten rock rises from the depths of the Earth, cools, and solidifies to form a crust that over eons is reworked by shifting and erosion, only to eventually return to the interior, where it once again becomes magma.

Summary of Terms

- **Igneous rocks** Rocks formed by the cooling and crystallization of hot, molten rock material called magma.
- **Sedimentary rocks** Rocks formed from the accumulation of weathered material (sediments) carried by water, wind, or ice.
- **Metamorphic rocks** Rocks formed from preexisting rocks that have been changed or transformed by high temperature, high pressure, or both.
- **Mineral** A naturally formed, inorganic solid composed of an ordered array of atoms chemically bonded to form a particular crystalline structure.
- **Polymorph** Minerals that have the same chemical composition but different crystal structures.
- **Mohs scale of hardness** A ranking of the relative hardness of minerals.
- **Magma** Molten rock from the Earth's interior.
- **Partial melting** The incomplete melting of rocks, resulting in magmas of different compositions.
- **Fractional crystallization** The process and sequence by which different minerals in a magma crystallize as the magma cools.
- **Extrusive rocks** Igneous rocks that form at the Earth's surface.
- **Intrusive rocks** Rocks that crystallize below the Earth's surface.
- **Lava** Magma once it reaches the Earth's surface.
- **Volcano** A central vent through which lava, gases, and ash erupt and flow.
- **Pluton** A large intrusive body formed below the Earth's surface.
- **Mechanical weathering** The breakdown of rocks on the Earth's surface by physical means.
- **Chemical weathering** The breakdown of rocks on the Earth's surface by chemical means.
- **Erosion** The process by which rock particles are transported away by water, wind, or ice.
- **Metamorphism** The changing of one kind of rock into another kind as a result of high temperature, high pressure, or both.
- **Recrystallization** Metamorphism caused by high temperatures.
- **Mechanical deformation** Metamorphism caused by stress, such as increased pressure.
- **Rock cycle** A sequence of events involving the formation, destruction, alteration, and reformation of rocks as a result of the generation and movement of magma, the weathering, erosion, transportation, and deposition of sediment, and the metamorphism of preexisting rocks.

Review Questions

Rocks

1. Name the three major types of rocks, and cite the conditions of their origin.

Minerals

2. A rock may be defined as an aggregate of one or more minerals. What, then, is a mineral?
3. What physical properties are used in the identification of minerals?
4. All minerals are defined by an orderly internal arrangement of their constituent atoms—the crystal form. Yet, most mineral samples do not display their crystal form. Why?
5. What is a polymorph?
6. What are the two classifications for mineral luster?
7. Although color is an obvious feature of a mineral, it is not a very reliable means of identification. Why?
8. Will the mineral topaz scratch quartz, or will quartz scratch topaz? Why?
9. The minerals calcite, halite, and gypsum are all nonmetallic, light and softer than glass, and all have three directions of cleavage. In what ways can they be distinguished from one another?

10. Silver has a density of 10.5 g/cm^3. What is its specific gravity?
11. What is the relationship of density to specific gravity?
12. What are two common chemical tests for identifying some minerals?

Building Blocks of Rock Forming Minerals

13. What is the most abundant element in the Earth's crust? What is the second most abundant element?
14. What is the most abundant mineral in the Earth's crust? What is the second most abundant mineral?
15. Name the five most common mineral groups found in most rocks.

Igneous Rocks

16. What are the most common igneous rocks and where do they generally occur?
17. What percentage of the Earth's surface is composed of igneous rocks?
18. What is meant by *partial melting*?
19. What are the three main types of magma? Relate the different magmas to silica content.
20. What two reasons make silica an important element for the different magmas?
21. What is meant by *fractional crystallization* and what are its consequences?
22. How are partial melting and fractional crystallization similar?
23. Because sills form at shallow depths, they are often confused with buried lava flows. In what ways does a sill differ from a buried lava flow?
24. In what way does a sill differ from a dike?
25. Where are lava flows most common?
26. What are the three major types of volcanoes?
27. What is viscosity?
28. Which type of volcano produces the most violent eruptions? Which type produces the quietest eruptions?
29. What does Yellowstone National Park have to do with volcanic activity?

Sedimentary Rocks

30. How does weathering produce sediment? Distinguish between weathering and erosion.
31. What does roundness tell us about sediment grains?
32. What can we say about a rock that is composed of various sizes of sediments in a disorganized pattern?
33. Relate the size and shape of sand grains to the way they were probably transported.
34. What can we say about a rock that is composed of very angular sediments?
35. What is a clastic sedimentary rock?
36. What are the three most common clastic sedimentary rocks?
37. Give two examples of sedimentary rocks that provide information about past geologic events at the Earth's surface.
38. What is the most dominant feature of a sedimentary rock environment?
39. What is alluvium? Where do we find it?
40. Deserts are generally dry areas. Why is water still a major factor of erosion in the desert environment?
41. Name two environments where cross-bedding occurs. What information do cross-beds provide?
42. What is a delta?
43. Are all beaches sandy? Why or why not?
44. What is a chemical sediment?
45. What are three common chemical sedimentary rocks?
46. When water evaporates from a body of water, what type of sediment is left behind?
47. What is the most common chemical sediment?
48. Why do coral building organisms require clear, shallow water?
49. The deposition of carbonate sediments is more prevalent in warm-water environments. What does this tell us about carbonate deposition near Dallas, Texas?
50. What is a fossil? How are they used in the study of geology?

Metamorphic Rocks

51. What is metamorphism? What are the agents of metamorphism?
52. What are the two processes by which rock is changed?
53. What changes are characteristic of contact metamorphism?
54. What changes are characteristic of dynamic metamorphism?
55. What changes are characteristic of regional metamorphism?
56. Distinguish between foliated and nonfoliated metamorphic rocks.
57. How does gneiss differ from granite?

The Rock Cycle

58. Explain the different cycles of rock formation.

Home Projects

1. Look at some crystals of table salt under a microscope or magnifying glass and observe their generally cubic shapes. There's no machine at the salt factory specifically designed to give these cubic shapes, as opposed to spherical or triangular. The cubic shape occurs naturally and is a reflection of how the atoms of salt are organized—cubically. Smash a few of these salt cubes and then look at them carefully. What you'll see are smaller salt cubes! Use the cleavage properties of crystals to explain these results.
2. A physical property of any material is its density—its mass per volume. We know that lead is denser than zinc and copper. Pennies fabricated after 1982 contain both copper and zinc. Being that zinc is less dense than copper, post-1982 pennies are less dense—hence, less massive. Dig into your penny collection and find 20 pre-1982 and 20

post-1982 pennies. Measure their masses on a sensitive scale, such as a home postage scale. Alternatively, hold the pennies in opposite hands to see if you can feel the difference in their masses. How few pennies can you hold and still feel the difference? Try holding single pennies on your left and right index fingers. Can you tell the difference with your eyes closed? Try this with a friend.

Exercises

1. Clearly distinguish between physical and chemical properties, and give examples of each.
2. Name at least one other national park, beside Yellowstone National Park, that was formed by volcanic or plutonic activity.
3. What type of rock is formed when magma rises slowly and solidifies before reaching the Earth's surface? Explain the process.
4. Why are minerals in volcanic rocks usually smaller than minerals in plutonic rock?
5. Which has the greater density—magma that rises slowly, or magma that rises quickly?
6. What physical property of magma, other than density, causes one magma to rise through the Earth's crust more slowly than another?
7. In what parts of the Earth's crust do we find the two most common igneous rocks in the Earth's crust, basalt and granite?
8. How do the densities of granitic and basaltic rock relate to continental land and oceanic crust?
9. What is the evidence for saying whether rocks that surround a batholith are older than, younger than, or equal in age to the batholith?
10. Are the Hawaiian Islands primarily made up of igneous, sedimentary, or metamorphic rock?
11. Can metamorphic rocks exist on an island of purely volcanic origin? Explain.
12. What accounts for the differences in lava composition of the two volcanoes that have erupted in recent times, Mauna Loa in Hawaii, and Mount St. Helens in Washington?
13. What mainly determines the viscosity of a magma?
14. What mainly determines a rock's initial melting temperature?
15. When the numerous craters formed by meteor impact on the Moon are compared with the relative absence of them on the Earth, a geologist comments that the Moon wears no makeup. What is meant by this?
16. Does the rock cycle apply to rocks on the Moon? Defend your answer.
17. What composes clastic sedimentary rocks, and what rocks are the most abundant of them?
18. In a conglomerate rock, why are pebbles of granite very common and pebbles of marble relatively uncommon?
19. In the formation of a river delta, why is it that coarser material is deposited first, followed by medium and finer material farther out? What type of bedding and gradation of sediments results from this sequence? Defend your answer.
20. What causes the formation of branches off the main channel of a river delta?
21. Dig a hole in an alluvial fan and the size of the rocks change with depth. Are rocks larger or smaller in the bottom layers? Why?
22. What rock feature does a geologist look for in a sedimentary rock to determine the distance the rock has traveled from its place of origin?
23. What textural features does a geologist look for in a sedimentary rock to determine the conditions of its place of origin?
24. On a beach where sand has a granitic source area, what dominant minerals likely make up the sand?
25. What feature of clastic sedimentary rock enables the flow of oil after it has been formed?
26. Of the rocks (a) granite (b) sandstone (c) limestone (d) halite, which are the first to weather in a humid climate? Which are the last to weather? Defend your answers.
27. How do chemical sediments produce rock? Name two rock types that form by chemical sedimentation.
28. Which type of rock is most sought by petroleum prospectors—igneous, sedimentary, or metamorphic? Why?
29. What kind of weathering is imposed on a rock when it is smashed into small pieces? When dissolved in acid?
30. What type of rock is made of previously existing rock whose formation does not involve melting?
31. What properties of slate make it good roofing material?
32. In metamorphosed rocks there is a sequence of mineral assemblages that indicate pressure and temperature conditions during formation. Metamorphosed rocks also show a sequence of texture characteristics corresponding to the different conditions. Which more likely determines metamorphic grade—foliation or mineralogy? Why?
33. Name two minerals that can give a metamorphic rock its foliation.
34. How is foliation different than sedimentary layering?
35. Can metamorphism, caused solely by elevated temperature occur without the presence of magma? Why or why not?
36. Each of the following statements describes one or more characteristics of a particular metamorphic rock. For each statement, name the metamorphic rock being described.
    (a) Foliated rock derived from granite.
    (b) Hard, nonfoliated, monomineral rock, formed under high to moderate metamorphism.
    (c) Foliated rock possessing excellent rock cleavage. Generally used in making blackboards.
    (d) Nonfoliated rock composed of carbonate minerals.
    (e) Foliated rock containing about 50% platy minerals; named according to the major minerals in the rock.
37. Why is asbestos in drinking water not particularly harmful to humans, whereas asbestos particles in air is very harmful?
38. Make an argument that the removal of asbestos products such as ceiling tile and pipe coverings, may be more hazardous to humans than simply covering them up.

CHAPTER

23

The Earth's Internal Properties

When you were a child, did you ever think about digging a hole to China? The idea is intriguing, for why go all the way around the Earth if you can take a shortcut straight through? Digging such a hole, unfortunately, is not a very realistic possibility. If such a hole were possible, it might be more valuable as a source of geological information than as a means of travel—at least to a geologist.

Because they can't dig straight through the Earth, scientists look for clues about the Earth's interior in surface rocks and in data from earthquakes and volcanic eruptions—the external expressions of the Earth's internal processes. If we plot the locations of modern earthquakes and volcanic eruptions on a map of the world, we see a distinct pattern (Figure 23.1). The great majority of these events occur in narrow zones, with very few occurring outside these zones. In science, natural processes are sought to explain observations, and in this chapter we apply the scientific method to try and understand how the Earth's internal properties cause rocks to fold and/or break, and how they dictate where earthquakes and volcanoes are located.

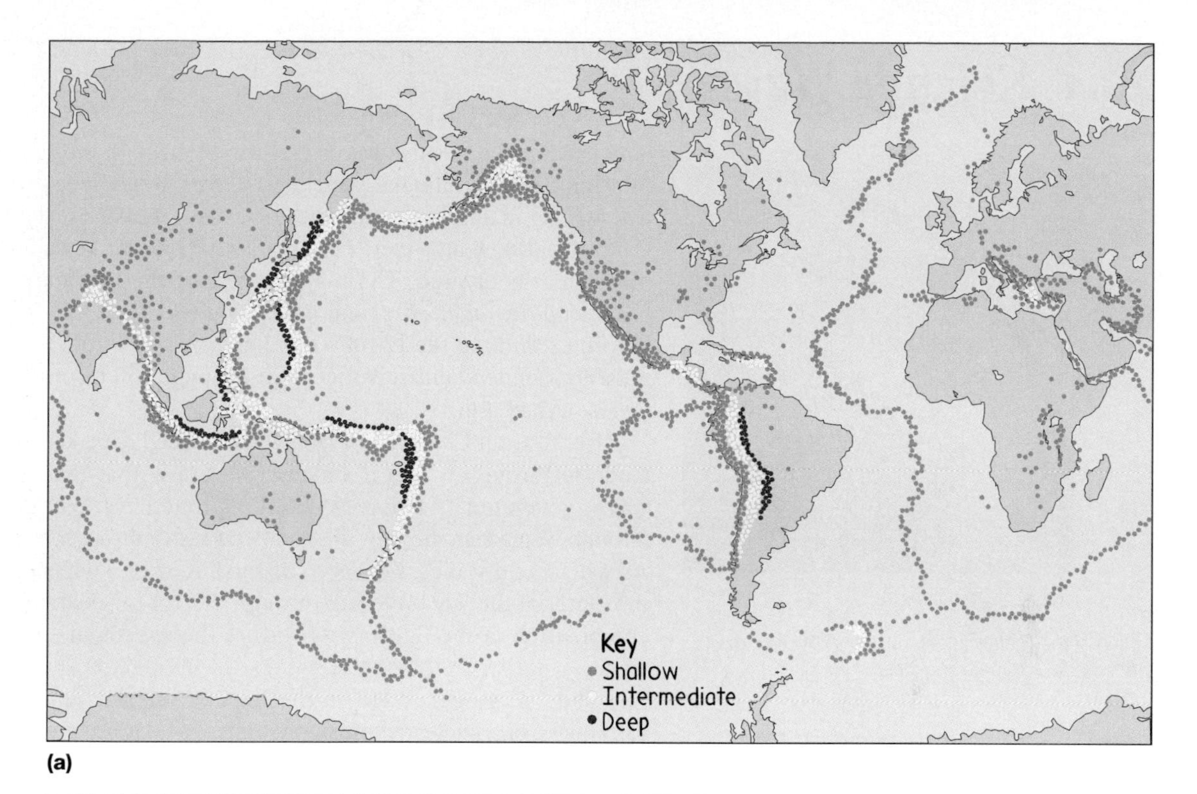

(a)

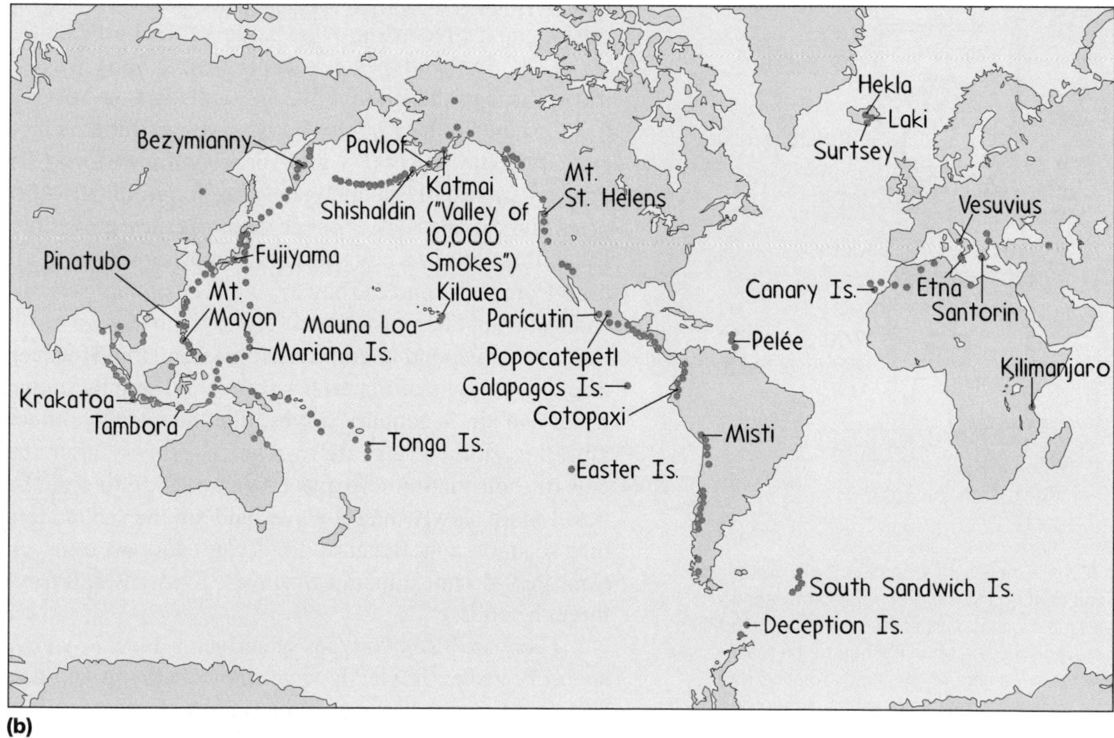

(b)

Fig. 23.1
(a) The global distribution of earthquakes.
(b) The global distribution of recent volcanic centers. (*Source:* NOAA)

23.1 Seismic Waves

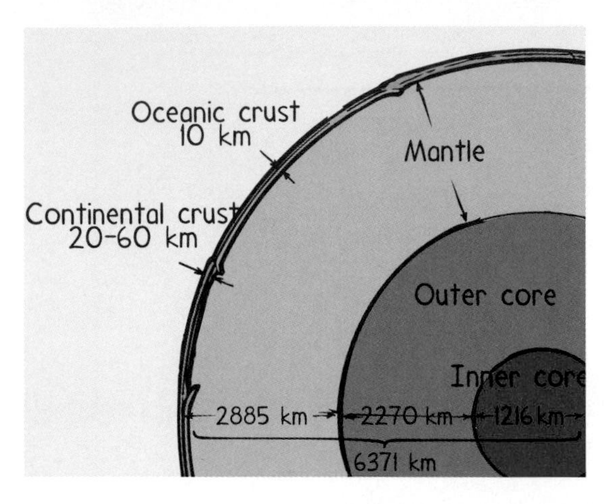

Fig. 23.2
Cross section of the Earth's interior showing the four major layers and their approximate thickness.

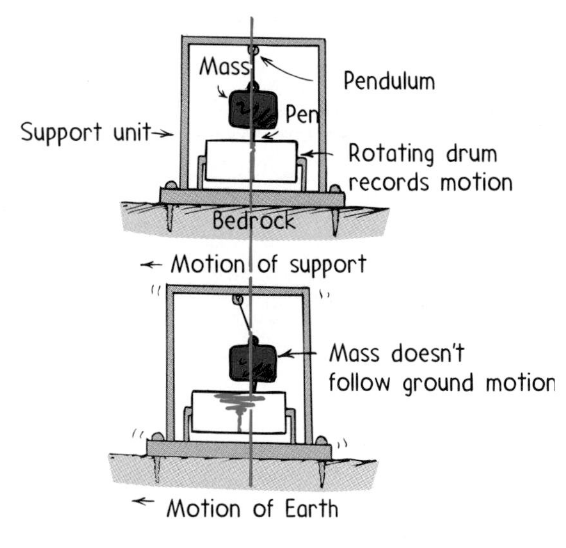

Fig. 23.3
Diagram of a seismograph. When the Earth moves, the support unit attached to the ground also moves but, because of inertia, the mass at the end of the pendulum tends to stay in place. A pen attached to the mass marks the relative displacement on the slowly rotating drum beneath. In this way, ground movement is recorded.

Any earthquake creates waves that travel through the Earth's interior. Such earthquake-generated waves are called *seismic waves.* The manner in which these waves travel has provided Earth scientists with a view into the Earth's interior, revealing a planet that is layered. The major layers of the Earth are the *crust, mantle, outer core,* and *inner core* (Figure 23.2). Thus, before examining the Earth's composition and structure, we must first understand seismic-wave propagation through the layers of the Earth.

Recall from Chapter 11 that a wave's speed depends on the medium through which it travels. We know that the sound waves generated by two rocks clicking together travel faster through water than through air, and even faster through a solid. As with sound waves, the speed of seismic waves depends on the material they're traveling through. So for clues about the composition of the Earth, we measure the speeds of seismic waves.

The energy generated in the Earth's interior during an earthquake or a nuclear explosion radiates in all directions. This energy travels in the form of seismic waves to the Earth's surface. Evidence of the waves is gathered on a seismograph, and seismographic records provide a map of the Earth's interior.

There are two types of seismic waves: **body waves**, which travel through the Earth's interior, and **surface waves**, which travel on the Earth's surface. Body waves are further classified as either **primary waves** (P-waves) or **secondary waves** (S-waves). Primary waves, like sound waves, are longitudinal—they compress and expand rock or other media as they move through it. Like vibrations in a bell, primary waves move out in all directions from their source. They are the fastest of all seismic waves and so the first to register on a seismograph. Because solids and fluids both respond to compression/expansion, P-waves travel through any type of material—from solid granite to magma to water and air. Secondary waves, like the waves produced by a vibrating violin string, are transverse—they vibrate the particles of their medium up and down and side to side. S-waves travel more slowly than P-waves and are the second to register on a seismograph. Because fluids can't support transverse motion, they do not transmit S-waves. S-waves can travel only through solids.

There are also two types of surface waves: *Rayleigh waves* and *Love waves.* Rayleigh waves move in an up-and-down motion, and Love waves move in a side-to-side, whiplike motion. Both of these surface-wave types travel at lower speeds than P- and S-waves and so are the last to register on a seismograph.

Seismic waves are reflected by surfaces in the Earth's interior and are refracted where wave speed changes. Thus in their detective work geologists study these reflections and refractions as well as the speeds of the various types of seismic waves. It was research with seismic waves that led geologists to conclude that the Earth's interior is layered (Figure 23.5).

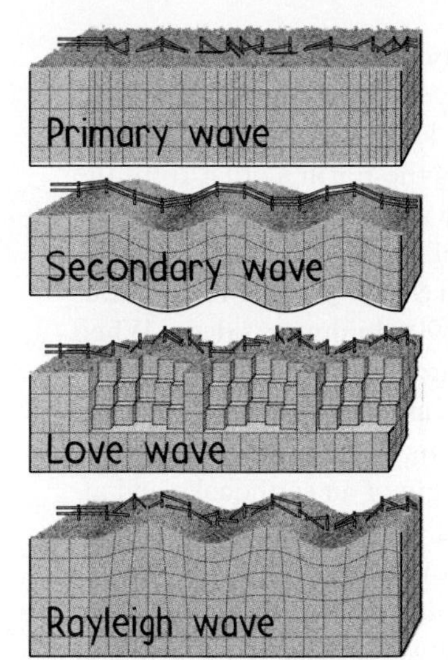

Fig. 23.4
Block diagrams show the effects of seismic waves. The yellow portion on the left side of each diagram represents the undisturbed area. (a) Primary body waves alternately compress and expand the Earth's crust, as shown by the different spacing between the vertical lines, similar to the action of a spring. (b) Secondary body waves cause the crust to oscillate up and down and side to side. (c) Love surface waves whip back and forth in horizontal motion. (d) Rayleigh surface waves operate much like secondary body waves but affect only the surface of the Earth.

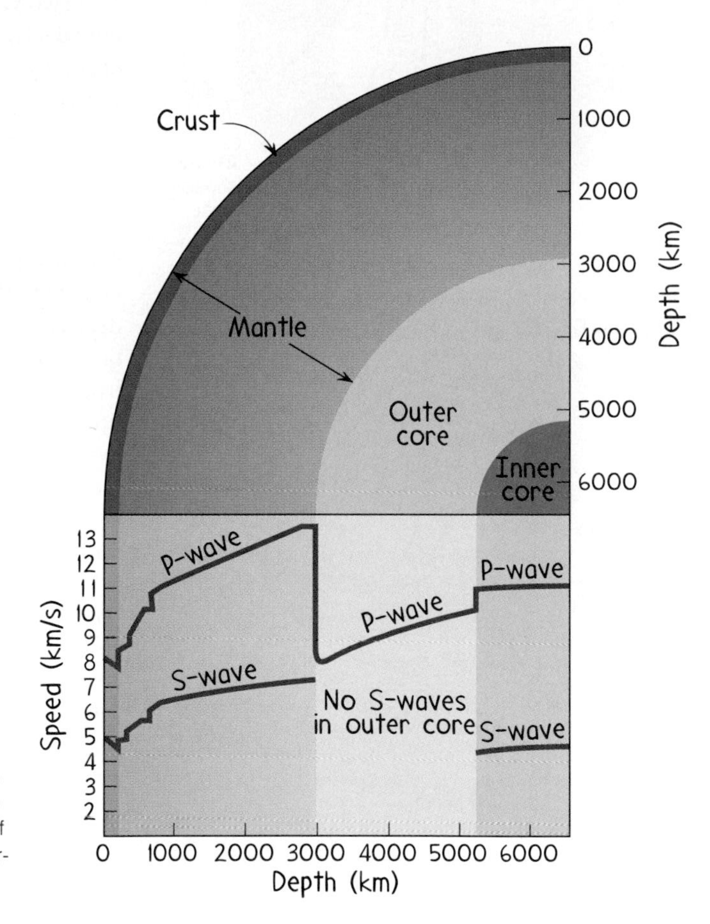

Fig. 23.5
Cross section of the Earth's internal layers, showing the increases and decreases of P- and S-waves in the different layers.

23.2 Earth's Internal Layers

In 1909, the Croatian seismologist Andrija Mohorovičić presented the first convincing evidence that the Earth's "innards" are layered. Studying seismographic data from a recent earthquake, he discovered that the seismic waves generated by the quake suddenly picked up speed at a certain depth below the surface. Knowing that the speeds of these waves depend on the density of the material they pass through, Mohorovičić concluded that the speed increase he observed was due to variation in the density of the Earth.

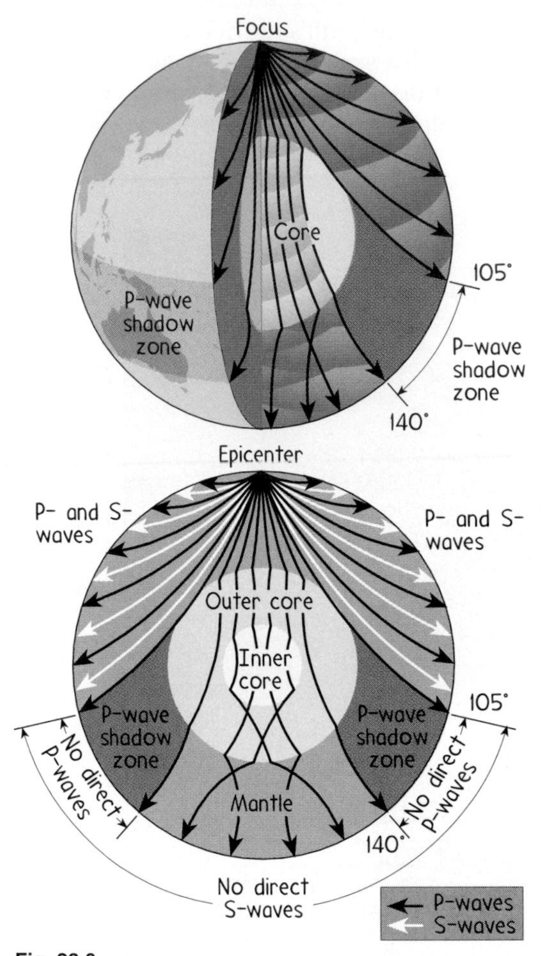

Fig. 23.6
Cut-away and cross-sectional diagrams showing the change in wave paths at the major internal boundaries and the P-wave shadow. The P-wave shadow between 105° and 140° from an earthquake's epicenter is caused by the refraction of the P-waves at the core-mantle boundary. Note that any location that is more than 105° from an earthquake's epicenter does not receive S-waves because the liquid outer core does not transmit S-waves.

The seismographic data had literally drawn a map of the upper boundary of the Earth's mantle, a layer of dense rock underlying the less dense crust. This boundary, known as the **Mohorovičić discontinuity** ("Moho"), separates the Earth's crust from the rocks of different composition in the mantle below.

Two years after the discovery of the Moho, the mantle-core boundary was detected. Both P- and S-waves are strongly influenced by a pronounced boundary 2900 kilometers deep. When P-waves reach that boundary, they are reflected and refracted so strongly that the boundary actually casts a P-wave shadow over part of the Earth (Figure 23.6). This wave shadow develops 105–140 degrees of arc from the origin of an earthquake and allows no direct penetration of seismic waves. Because the boundary is so pronounced, we infer that it marks a very significant change in the density of the materials present. Both the overall density[1] of the Earth and the speed at which seismic waves travel through the core suggest that the core is composed of metallic iron, a material that is much denser than the silicates that make up the mantle. This boundary casts an S-wave shadow that is even more extensive than the P-wave shadow, suggesting that S-waves are unable to pass through the core. Because S-waves, being transverse, cannot travel through liquids, we infer that the outer portion of the core is liquid.

In 1936, the discovery that P-waves are reflected from a boundary within the core showed there was another layer. The P-waves passing through the inner portion of the core had greater speed than those passing through the outer core. Therefore, the inner core must be solid. Do you suppose these layers in the Earth's interior influence the geologic changes our planet experiences? The answer is yes, as you will now see.

Question

What evidence supports the theory that the Earth's inner core is solid and the outer core liquid?

Answer
The differences between P- and S-wave propagation through the Earth's interior. As these waves encounter the boundary at 2900 km, a very pronounced wave shadow develops. P-waves are both reflected and refracted at the boundary, but S-waves are only reflected. S-waves cannot travel through liquids, implying a liquid outer core. As P-waves propagate through the outer core, there is a depth at which there is a sudden increase in speed. Knowing that waves travel faster in solids, we infer a solid inner core.

---

[1]The density of rocks at the Earth's surface is typically 2.7-3.0 g/cm^3, whereas the average density of the Earth as a whole is 5.5 g/cm^3. Thus surface rocks are not representative of the planet's interior. To account for the Earth's high average density, the density of the core must be at least 10 g/cm^3. This and other reasons suggest that the core is composed of iron and nickel, the most abundant of the heavier elements.

The Core

The **core** is composed mostly of iron and nickel. In the inner core, the iron and nickel are solid. Although the inner core is very hot, intense pressure from the weight of the rest of the Earth prevents the material of the inner core from melting (much as a pressure cooker prevents high-temperature water from boiling). Because less weight is exerted on the outer core, the pressure is less here, with the result that the iron and nickel are liquid. The molten outer core flows at the low rate of several kilometers per year. This flow is evident far outside the Earth's surface as it generates the electric current that powers the Earth's magnetic field. This magnetic field is not stable but has changed throughout geologic time. Recall from Chapter 10 that there have been times when the Earth's magnetic field has diminished to zero, only to build up again with the poles reversed. These magnetic pole reversals probably result from changes in the direction of outer core flow.

■ **Question**

Iron's normal melting point is 1535°C, yet the Earth's inner core temperature is about 4300°C. Why doesn't the solid inner core melt?

The Dynamic Mantle

Surrounding the core of the planet is the **mantle**, a rocky layer some 3000 kilometers thick. Composed of hot, iron-rich silicate rocks, the mantle behaves like plastic, which means that it responds in a semifluid manner. The upper portion of the mantle, which extends from the crust-mantle boundary down to a depth of about 350 kilometers, has two zones, as Figure 23.7 shows. The lower part of the upper mantle, called the **asthenosphere**, is especially plastic. Thermal convection currents in the asthenosphere contribute to its gradual flow. The constant flowing movements in the asthenosphere greatly affect the surface features of our planet.

Situated above the asthenosphere is the **lithosphere**. The lithosphere is about 100 km thick and includes the entire crust and the uppermost portion of the mantle (Figure 23.7). Unlike the asthenosphere, the lithosphere is relatively rigid and brittle and resists deformation, instead of flowing. The lithosphere is, in a sense, floating on top of the asthenosphere like a raft on a pond and is carried along by the motions of the material in the asthenosphere. The motions in the asthenosphere are not uniform, and because of this, as we

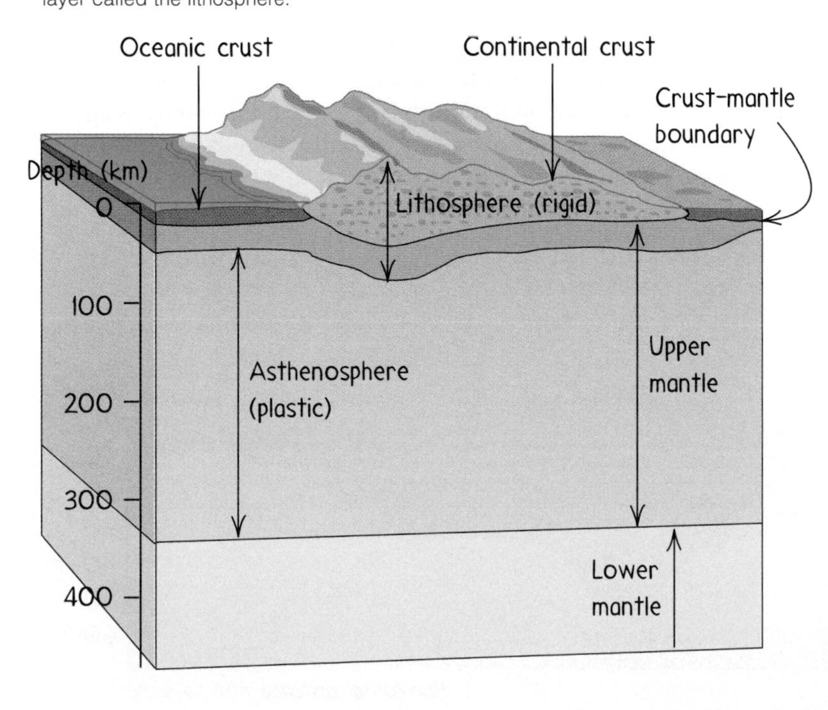

Fig. 23.7
The bottom portion of the Earth's upper mantle is the plastic asthenosphere. The top portion of the upper mantle plus the crust form the rigid layer called the lithosphere.

■ **Answer**
The intense pressure from the weight of the Earth above crushes atoms so tightly that even the high temperature cannot budge them. In this way melting is prevented.

shall see later in this chapter, the brittle lithosphere is broken into many individual pieces called *plates.*

The lithospheric plates are continuously in motion. They float on the circulating asthenosphere. Although asthenosphere currents move at a leisurely pace—taking hundreds of millions of years to complete one loop—they are powerful enough to move continents and reshape many of our surface features. The movement of the lithospheric plates cause earthquakes, volcanic activity, and the deformation of large masses of rock that create mountains.

The Crustal Surface

The uppermost portion of the lithosphere, the portion on which we live, is the **crust**. The density, composition, and thickness of the crust vary markedly from the deep ocean basins to the lofty continental plateaus. The crust of the ocean basins is compact. It's only about 10 kilometers thick and is composed of dense basaltic rocks. The part of the crust we know as continents is between 20 and 60 kilometers thick. Continental masses are composed of granitic rocks, which are less dense than basaltic rock.

If continental crust is so much thicker than oceanic crust, why are the ocean basins underwater and the continents high and dry? The answer is found in their density differences and buoyancy (Chapter 5). The less dense continental crust always floats higher than the more dense oceanic crust, even if the continental crust has more mass. This is the principle of **isostasy**. Either type of crust reaches an equilibrium height with respect to the mantle when the upward-acting buoyant force, provided by the mantle, equals the weight of the crust. Let us consider two blocks of crust of equal mass, one composed of oceanic crust and the other continental. The mantle exerts the same buoyant force on both blocks, but the continental block floats higher because it is less dense. If we add more crust to the continental block, its upper surface will be higher than before but it will also sink farther into the mantle because of the additional weight. So, the higher the crust floats above the mantle the deeper its "root" must be to support it.

Blocks of crust come to equilibrium with each other. The weight of the mantle material displaced by a block of floating crust equals the weight of the block. Where does the displaced mantle material go? Because the pressure in the mantle from the overlying crust must be the same at any given depth, the displaced material simply flows away from whatever is pressing down on it until the pressure in the mantle is equalized. This means that the pressure in the mantle underneath the thick but less dense continental crust is about the same as the pressure in the mantle beneath the thin but denser oceanic crust. To achieve this state of balance, the vertical positions of the different blocks of crust adjust until the pressure in the mantle is balanced.

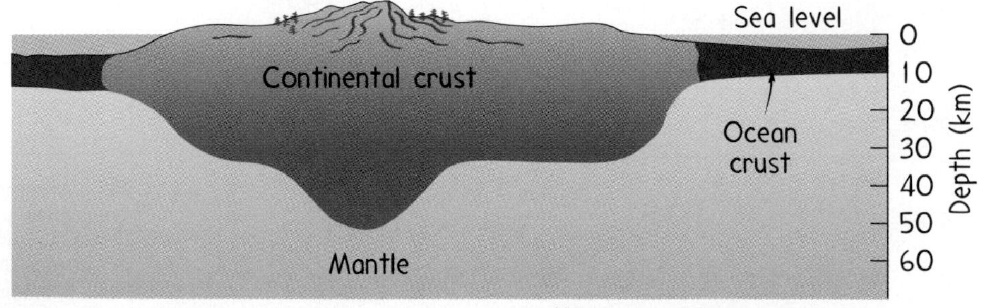

Fig. 23.8
Continental crust is thicker and less dense than oceanic crust. Due to isostasy, continental crust floats higher on the mantle than oceanic portions. The pressure in the mantle at any given depth must be the same beneath both the continental crust and the oceanic crust.

■ **Question** If you wished to drill the shortest hole to the mantle, would you drill in western Colorado or in Florida?

23.3 The Theory of Continental Drift

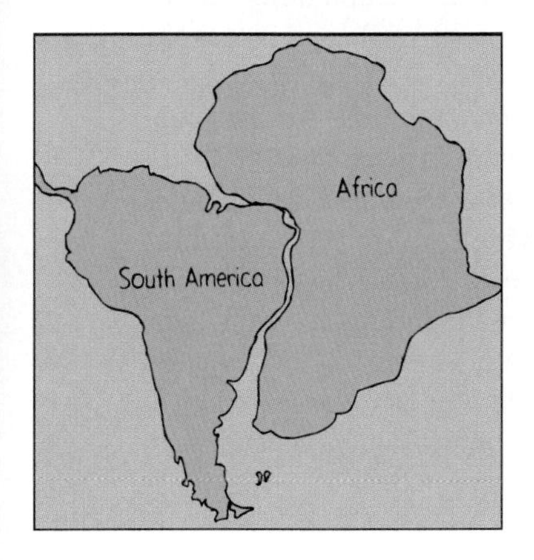

Fig. 23.9
When you align the shorelines of South America and Africa, the continents fit together like pieces in a jigsaw puzzle.

Scientists of the early 20th century believed that oceans and continents were geographically fixed. They regarded the surface of the planet as a static skin spread over a molten, gradually cooling interior. They believed that the cooling of the planet resulted in its contraction, which caused the outer skin to contort and wrinkle into mountains and valleys.

Many people had noticed, however, that the eastern shorelines of South America and the western shoreline of Africa seemed to fit together like a jigsaw puzzle (Figure 23.9). One Earth scientist who took this observation seriously was Alfred Wegener, who saw the Earth as a dynamic planet with the continents in constant motion. He believed that all the continents had once been joined together in one great supercontinent he called **Pangaea**, meaning "all land." His hypothesis was that Pangaea had fractured into a number of pieces, and that South America and Africa had indeed once been joined together as part of a larger land mass.

Wegener supported his hypothesis with impressive geological, biological, and climatological evidence. He proposed that the geological boundary of each continent lay not at its shoreline but at the edge of its *continental shelf* (the gently sloping platform between the shoreline and the steep slope that leads to the deep ocean floor). When Wegener fit Africa and South America together along their continental shelves, the fit was even better than it was at the shorelines (Figure 23.10). Furthermore, rocks on different continents that are brought into juxtaposition when the continental shelves are matched up are virtually identical. In addition, many of the mountain systems in Africa and South America show strong evidence of a previous connection. Similarly, fossils of identical land-dwelling animals are found in South America and Africa but nowhere else. And fossils of identical trees are found in South America, India, Australia, and Antarctica.[2]

Even stronger evidence for a supercontinent comes from paleoclimatic (ancient climate) data. More than 300 million years ago, a huge continental ice sheet covered parts of South America, southern Africa, India, and southern Australia. The ice sheet left evi-

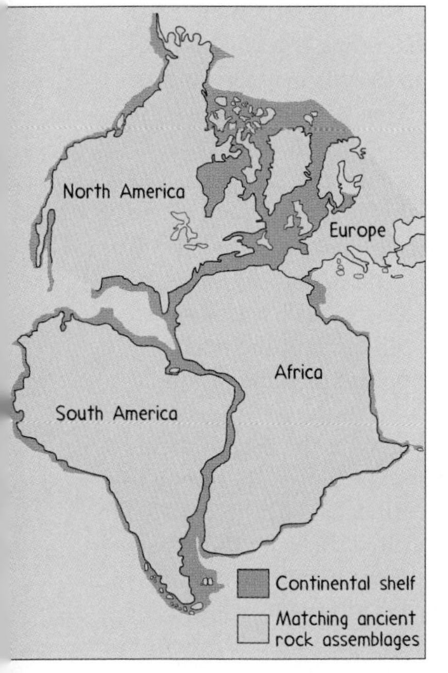

Fig. 23.10
The jigsaw-puzzle fit between continents is even better at the continental shelves than at the shorelines.

■ **Answer**
Put the question another way: If you wanted to drill the shortest hole through ice to the water below, would you drill atop an iceberg or through a slab of ice that hardly extends out of the water? You would drill your hole in the slab, of course, and likewise you should drill through the thinner crust of nonmountainous Florida. In mountainous western Colorado, the crust is much thicker. (If you really want the shortest hole, drill through the ocean floor—exactly what scientists have done in Project MoHole, in the East Pacific Ocean.)

[2]One fossil plant assemblage that offers especially strong support for Wegener's idea is the *Glossopteris flora,* which is named after the dominant gymnosperm tree found in the prehistoric southern temperate forests of South America, India, Australia, and Antarctica. Because the seeds from these trees were too large to be distributed by air, the wide distribution of this flora supports Wegener's theory that the continents were once joined together.

dence of its existence in thousands of well-preserved glacial striations that reveal the directions of ice flow. If these continents were in their present positions, the ice sheet would have had to cover the entire Southern Hemisphere, and in some places, cross the equator! If the ice sheet was that extensive, the world climate would have been very cold. But there is no evidence of glaciation in the Northern Hemisphere at that time. In fact, the time of glaciation in the Southern Hemisphere was a time of subtropical climate in the Northern Hemisphere. To account for this enigmatic distribution of paleoclimates, Wegener proposed that Pangaea had been in existence 300 million years ago, with South Africa located over the Earth's south pole. This reconstruction brings all the glaciated regions into close proximity in the vicinity of the south pole and places the modern northern continents nearer the tropics.

Wegener described continental drift in *The Origin of Continents and Oceans,* published in 1915. Although he used evidence from different scientific disciplines, his well-founded hypothesis was ridiculed by the community of Earth scientists. Antagonists complained that Wegener failed to provide a suitable driving force to account for the continental movements. (Wegener wrongly proposed that the tidal influence of the Moon could produce the needed force. He also proposed that the continents broke through the Earth's crust like ice breakers cutting through ice.) Without a convincing explanation for his theory, the scientific community of the early part of this century was not ready to believe that the continents had drifted to their present position. It is only recently, with new-found discoveries, that Wegener's concept has become accepted.

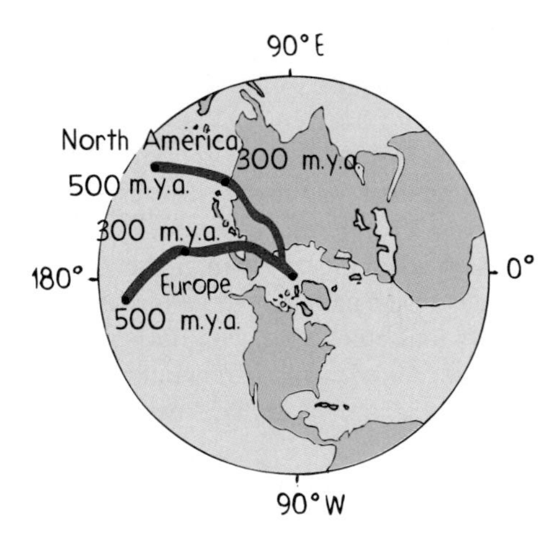

Fig. 23.11
The path of the magnetic north pole during the last 500 million years. (The unit m.y.a. stands for "millions of years ago.") The lower red line is derived from evidence collected in Europe, and the upper red line is derived from evidence collected in North America. One would expect that these two lines would overlie each other. Thus either the magnetic pole wanders erratically, or the continents have moved. But how could the pole be in more than one place at the same time? This question strongly suggests that the continents have indeed moved relative to each other.

A Scientific Revolution

One of the first key discoveries in support of continental drift came about through studies of the Earth's magnetic field. We know from Chapter 9 that the Earth is a huge magnet and that its magnetic north and south pole are near the geographic poles. Because certain minerals align themselves with the magnetic field when a rock is formed, rocks have a preserved imprint of the changes in the Earth's magnetism over the eons of geologic time. This magnetism from the geologic past is known as **paleomagnetism**.

Three essential bits of information are contained in the preserved magnetic record: (1) the polarity of the Earth's magnetic field at the time the rock was formed, (2) the direction to the magnetic pole from the rock's location at the time the rock was formed, and (3) the magnetic latitude of the rock's location at the time the rock was formed. Once the magnetic latitude of a rock and the direction of the magnetic poles are known, the position of the magnetic pole at the time of formation can be determined. During the 1950s, a plot of the position of the magnetic north pole through time revealed that, over the past 500 million years, the position of the pole had wandered extensively throughout the world (Figure 23.11)! It seemed that either the magnetic poles migrated through time or the continents had drifted. Because the apparent path of polar movement varied from continent to continent, it was more plausible that the continents had moved. Thus the hypothesis of continental drift was revived, but a mechanism to explain how the movement occurred was still lacking.

The 1950s were a time of extensive and detailed mapping of ocean floors. Topographic features revealed huge mountain ranges running down the middle of the Atlantic, Pacific, and Indian Oceans; a major rift

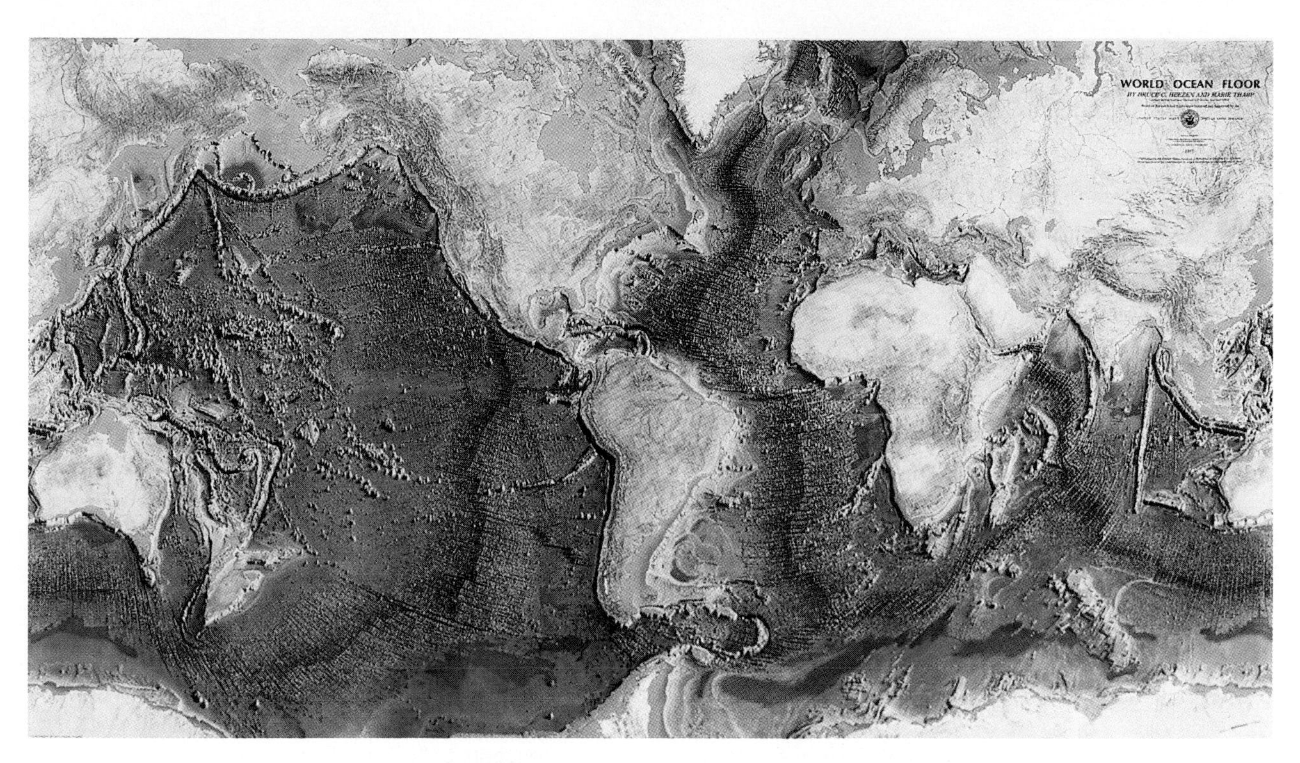

Fig. 23.12
A detailed map of the ocean floor reveals enormous mountain ranges (brown regions) in the middle of the oceans and deep ocean trenches near some continental landmasses.

valley along each crest; and deep ocean trenches near some of the continental landmasses, particularly around the edges of the Pacific (Figure 23.12). So, some of the deepest parts of the ocean are actually near some of the continents, and out in the middle of the oceans the water is relatively shallow because of the underwater mountains. Volcanism and high thermal energies were found to be generated at the ridge systems. With this new information, H. H. Hess, an American geologist, presented the hypothesis of **seafloor spreading**. Hess proposed that the seafloor is not permanent but is constantly being renewed. He theorized that the ocean ridges are located above upwelling convection cells in the mantle. As rising material from the mantle oozes upward, new lithosphere is formed. The old lithosphere is simultaneously destroyed in the deep ocean trenches near the edges of continents. Thus in a conveyor belt fashion new lithosphere forms at a spreading center, and older lithosphere is pushed from the ridge crest to be eventually recycled back into the mantle at a deep ocean trench (Figure 23.13).

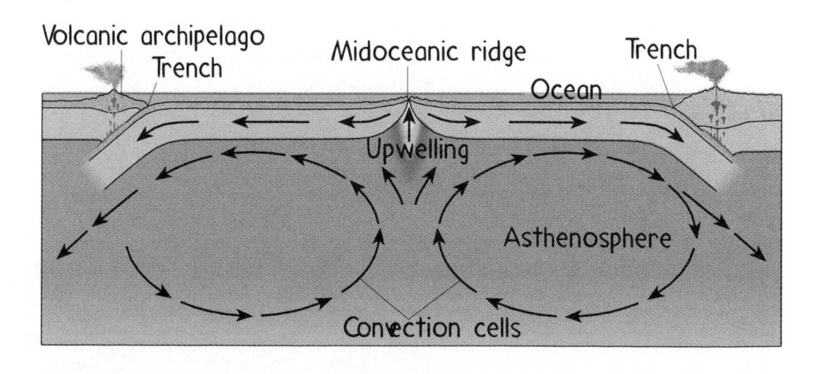

Fig. 23.13
In conveyor-belt fashion, new lithosphere is formed at the midocean ridges (also called "spreading centers") as old lithosphere is recycled back into the asthenosphere at a deep ocean trench.

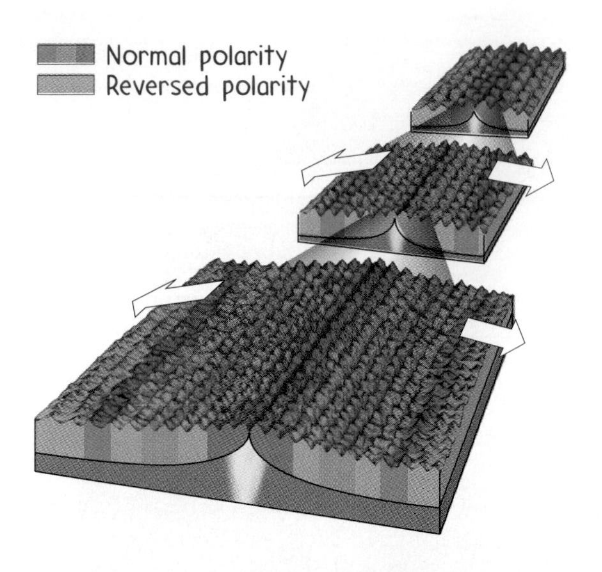

Fig. 23.14
As new material is extruded at an oceanic ridge (spreading center), it is magnetized according to the existing magnetic field. Magnetic surveys show alternating strips of normal and reversed polarity paralleling both sides of the rift area. Like a very slow magnetic tape recording, the magnetic history of the Earth is thus recorded in the spreading ocean floors.

Support for this theory came from paleomagnetic analysis of the ocean floor. As new basalt is extruded at an oceanic ridge, it is magnetized according to the existing magnetic field. The magnetic surveys of the ocean's floor showed alternating strips of normal and reversed polarity, paralleling either side of the rift areas (Figure 23.14). As in a very slow magnetic tape recording, the magnetic history of the Earth is thus recorded in spreading ocean floors, forming a continuous record of the movement of the seafloors. Since the dates of pole reversal can be determined, the magnetic pattern of the spreading seafloor documents both the age of the seafloor and the rate at which it spreads. Thus the oceanic crust was found to be thin and young near the central ridge region and progressively thicker and older away from the ridge.

The theory of seafloor spreading provided a mechanism for continental drift. The time was right for the revolutionary concepts to follow. The tide of scientific opinion had indeed switched in favor of a mobile Earth.

■ **Question**

Why was Wegener's theory of continental drift not taken more seriously in the early part of this century?

23.4 The Theory of Plate Tectonics

The framework that allows us to understand how and why the various features of the Earth constantly change is called the theory of plate tectonics. It describes the forces within the Earth that give rise to continents, ocean basins, mountain ranges, earthquake belts, and other large-scale features of the Earth's surface. The **theory of plate tectonics** holds that the Earth's outer shell, the lithosphere, is divided into eight relatively large plates and a number of smaller ones (Figure 23.15). These lithosphere plates ride

■ **Answer**

Wegener failed to produce a suitable driving mechanism to support his theory. Even if he had postulated the role of the convective interior, though, we can only speculate about how quickly the scientific community would have accepted his hypothesis. Scientists, like all other human beings, tend to identify with the ideas that characterize their time. Do advances in knowledge, scientific or otherwise, occur because they are accepted by the status quo or because holders of the status quo eventually die off? Knowledge that is radical and unacceptable to the old guard is often easily accepted by newcomers who use it to push the knowledge frontier further. Hooray for the young (and the young-at-heart)!

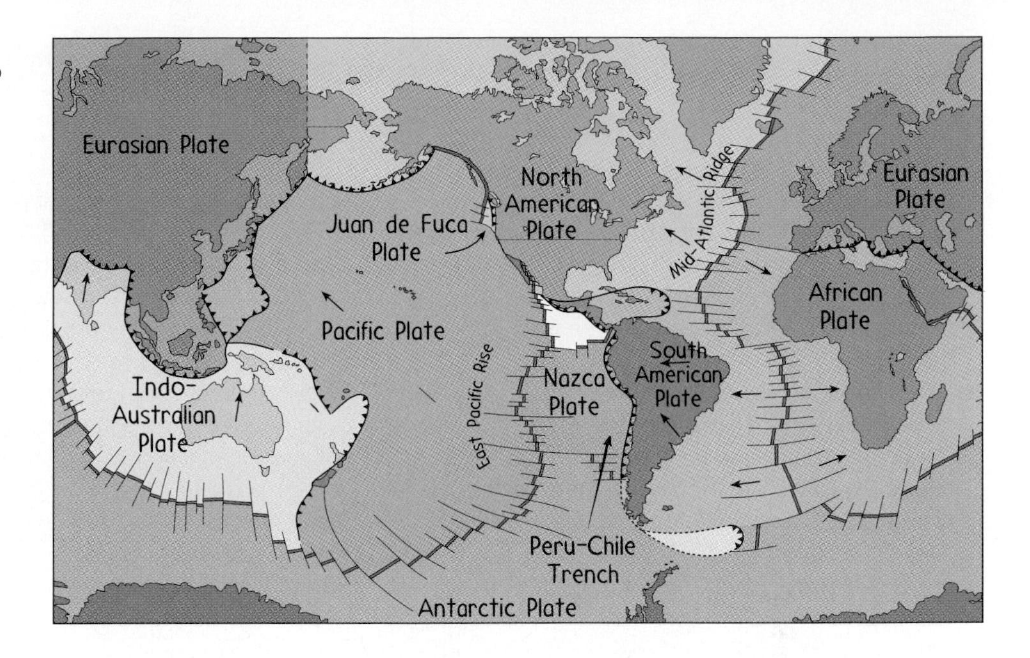

Fig. 23.15
The lithosphere is divided into eight large plates and a number of smaller ones.

atop the plastic asthenosphere below. Because each plate moves as a distinct unit in relation to other plates, the interiors of the plates are generally geologically quiet. All major interactions between plates are manifested along plate boundaries. Thus most of the Earth's seismic activity, volcanism, and mountain building occur along these dynamic margins. In fact, the creation and destruction of lithosphere described in Figure 23.13 take place at such margins.

Before we examine plate boundaries in detail, we must first discuss how plate motion deforms the Earth's crust—the realm of *Structural Geology.*

Structural Geology

Convection in the Earth's mantle causes the overlying lithospheric plates to be in slow but constant motion, which creates stress within the plates. Rocks subjected to stress begin to deform into intricate and broad *folds,* and if enough stress is applied, rocks break and then move along *faults.* Faults range all the way from small and virtually unnoticeable to large and loaded with the potential to devastate.

Folds When a rock is subjected to compressive stress it begins to buckle and fold. To see what these terms mean, suppose you had a throw rug on your floor, with a friend standing on one end. If you push the rug toward your friend while keeping it on the floor, the rug begins to tilt away from your hand, and a series of ripples, or **folds**, develop. This is what happens to the Earth's crust when it is subjected to compressive stress.

We know from Chapter 22 that sediments settling from water in an ocean or bay are deposited in horizontal layers, with the layer at the bottom deposited first. This bottom layer is therefore the oldest in the sequence of deposited layers. Each new layer is deposited on top of the previous layer, with the youngest at the top. As originally flat sedimentary rock layers are subjected to compressive stress, they tilt and become folded, just as the throw rug became tilted and folded. Each high point and low point in a series of folds is an axis, which you can imagine as a plane extending downward into the Earth,

Hot Spots and Lasers—A Measurement of Tectonic Plate Motion

Motion is relative. Whenever we discuss the motion of something, all we can do is describe its motion relative to something else. We call this something else—this place from which motion is observed and measured—a reference frame. Living in a world where everything is in motion, how can we measure rates of plate motion? What do we choose for our reference frame?

The Canadian geophysicist J. Tuzo Wilson suggested that one measure of Pacific seafloor movement can be found in the ages of the volcanic Hawaiian Islands. Wilson postulated that the islands are the tips of huge volcanoes that formed as the Pacific seafloor moved over a fixed *hot spot*—a magma source rising from the Earth's interior. The concept of a fixed hot spot provides a stationary reference point on the surface of the Earth against which plate motion can be determined. There are about 100 such reference hot spots around the world, and they are used together to determine rates of plate movements.

There is, however, some debate as to whether the hot spots are truly stationary. A fascinating and more precise way to measure Earth movement is by a laser beam reflecting off "mirrors" in outer space. Lasers are used to detect the broad movements of tectonic plates. Laser pulses beamed to the reference point in space from a pair of ground stations located on opposite sides of a plate boundary or a fault start timers that run until the reflected pulses are received back at the stations. A computer combines this elapsed time with the known position of the reference point to determine the exact position of the ground station. Any movement of the ground station is registered as a change in the elapsed time.

This type of measurement can also be done with radio signals. Radio telescopes are keyed to a reference point in outer space—a quasar or a satellite—from which relative positions of points on Earth can be plotted. Local movements along faults can be measured by lasers as well. For this case, the mirror is placed on the opposite side of the fault from the laser, instead of in outer space. Neato!

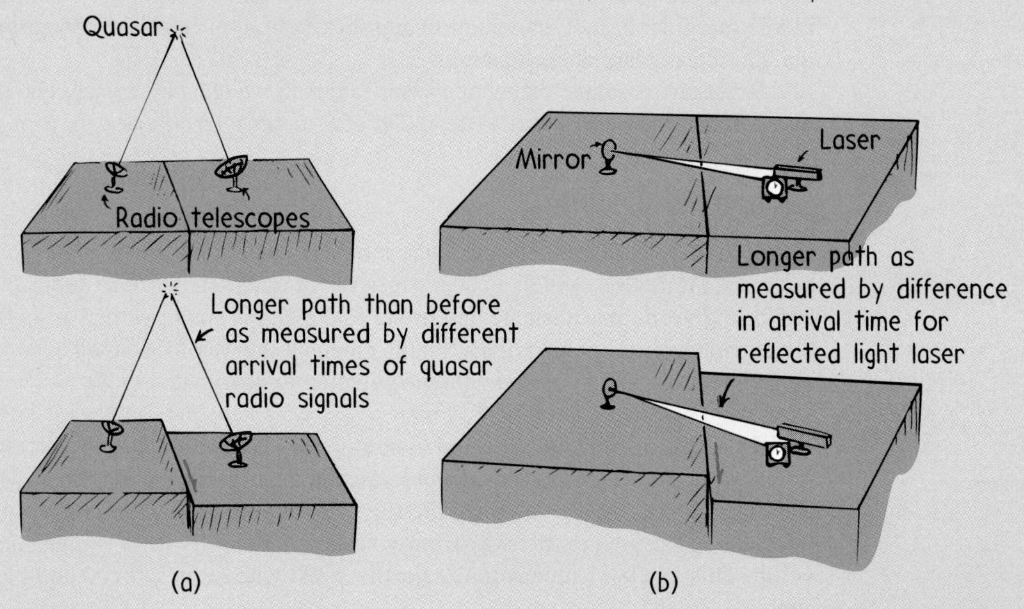

Earth movements are measured by radio telescopes or lasers that are keyed to a stationary reference point. (a) Broad movements in the range of 1 cm per year are detected using satellites or quasars as reference points. (b) Fault movements are measured by laser beams shot from opposite sides of a fault. The laser flashes a beam off a reflector; the bounce-back time is recorded. Because of light's constant speed, the time of bounce-back will change with Earth movement.

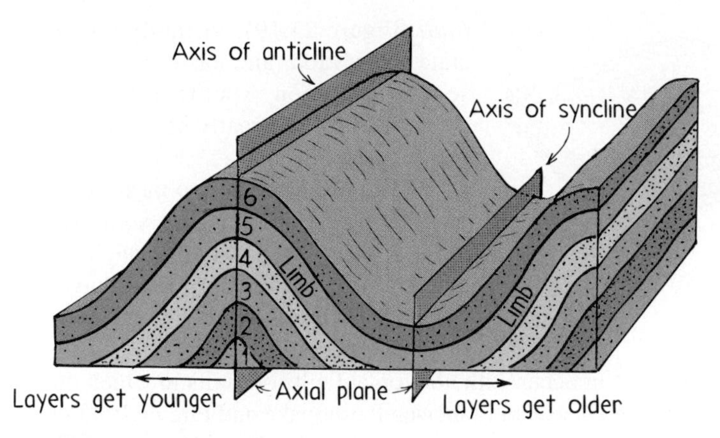

Fig. 23.16
Anticline and syncline folds. Layer 1 is the oldest rock, and layer 6 is the youngest. The limbs of an anticline tilt away from the axis of the fold (a marble would roll away from the axis), and the rock layers are oldest at the core of the fold. The limbs of a syncline tilt toward the axis of the fold (a marble would roll toward the axis), and the rocks are youngest at the core.

as Figure 23.16 shows. When the layers tilt toward the fold axis, so that if you put a marble on the rock it would roll toward the axis, the fold is called a **syncline**. The rocks at the center, or *core,* of a syncline are the youngest, and as you move horizontally away from the axis, the rocks get older and older. If the fold layers tilt away from the axis, so that if you put a marble on the rock it would roll away from the axis, the fold is called an **anticline**. The rocks in an anticline are oldest at the core, and as you move horizontally away from the axis, the rocks get younger. Another way to think about it is that anticlines fold upward and synclines fold downward.

Question

Why are rocks at the core of a syncline younger than those farther out from the core while the opposite is true for an anticline?

Faults When compressional stress overcomes the strength of rock, the rock fractures into two parts. If one part then moves relative to the other part, the fracture is called a **fault**.

Note the angle of the fault in Figure 23.17 (the thick black line in the top drawing) with respect to the horizontal ground surface. Imagine you could pull the block diagram apart at the fault, as is done in the lower drawing. The half containing the fault surface where someone could stand is the *footwall* block. The fault surface of the other half is inclined to make standing impossible; this is the *hanging wall* block. These terms were coined by miners because one could hang a lamp on a hanging wall and could stand on a footwall.

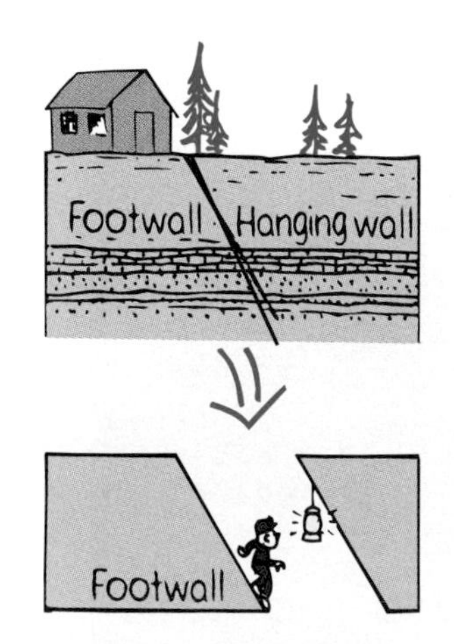

Fig. 23.17
The terms *footwall* and *hanging wall* were coined by miners because one could hang a lamp on a hanging wall and stand on a footwall.

Now we can discuss what happens in a zone of compressional faulting. Once a fault has been created by compressional forces, the forces cause rocks in the hanging wall to be pushed upward along the fault plane relative to rocks in the footwall, as Figure 23.18 shows, forming a *reverse fault.* Sometimes, the angle between the fault plane and a horizontal plane is quite low, and the reverse fault is referred to as a *thrust fault.* The Rocky Mountain foreland, the Canadian Rockies, and the Appalachian Mountains, to name a few, were formed in part by reverse and thrust faulting.

In addition to compression, stress can also occur by tension. Whereas compression pushes, tension pulls. Tension causes rocks in a hanging wall to drop downward along the fault plane relative to those in the adjacent footwall, producing a *normal*

Answer
Think of the rug example. Assume the top surface of the rug is younger than the lower surface. When you push the rug it can (1) fold upward (like the letter "A"), or (2) fold downward (like a sag). In the first case, the bottom surface makes up the core—an anticline. In the second case, the top surface makes up the core—a syncline. Makes sense!

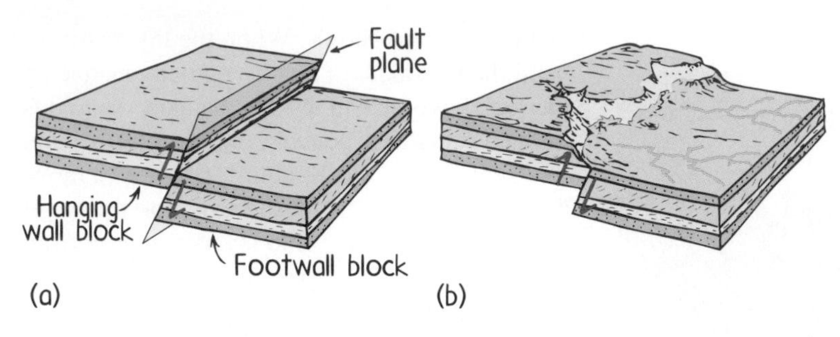

Fig. 23.18
A reverse fault. In a zone of compressional faulting, rocks in the hanging wall are pushed up relative to rocks in the footwall. (a) A reverse fault before erosion; (b) the same reverse fault after erosion.

fault (Figure 23.19). Virtually the entire state of Nevada, and eastern California, southern Oregon, southern Idaho, and western Utah are greatly affected by normal faulting.

The faults described so far have dominantly vertical motion, but some of the world's most famous faults, such as the San Andreas fault in California, have virtually no vertical motion; their relative motion is horizontal.

Devastating earthquakes can occur with horizontal faults or faults in which movement is primarily vertical. The Great San Francisco Earthquake and Fire of 1906 registered near 8.3 on the Richter scale (a measure of ground motion caused by earthquakes). It caused 700 deaths and extensive fire damage. The Loma Prieta earthquake near Santa Cruz, California, in 1989, registered 7.1 on the Richter scale. It sadly caused 62 deaths and more than $6 billion in damage. Still larger and more catastrophic earthquakes have occurred along reverse faults. The 1964 earthquake in Anchorage, Alaska, registered 8.5 on the Richter scale, and caused 131 deaths and $300 million in damage.[3] One of the greatest tragedies of recent times was the Mexico City earthquake of 1985. It measured 8.5 on the Richter scale and caused 7000 deaths!

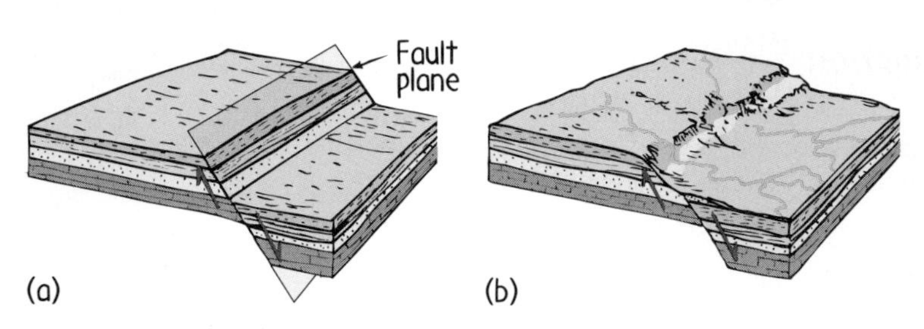

Fig. 23.19
A normal fault. In a zone of tensional faulting, rocks in the hanging wall drop down relative to those in the footwall, forming a normal fault. (a) A normal fault before erosion; (b) the same normal fault after erosion.

Understanding earthquakes is obviously of major importance to society. Table 23.1 lists some of the world's most notable earthquakes according to their impact on society. The 1906 San Francisco earthquake is notable because of its damage to the city and its inhabitants. But interestingly enough, in the winter of 1811–1812 a much greater earthquake on the New Madrid fault in Missouri changed the landscape beyond recog-

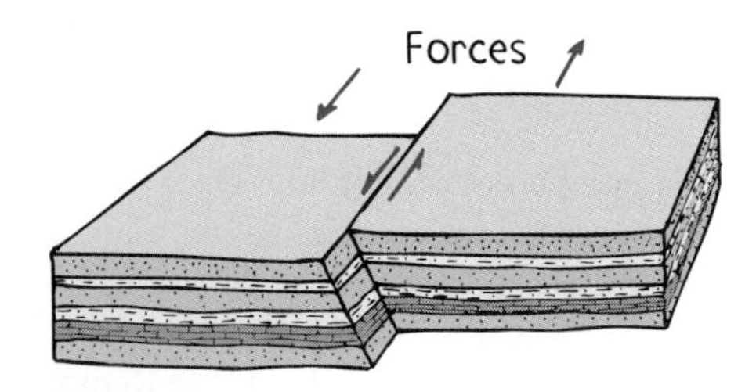

Fig. 23.20
The relative movement of a fault having horizontal offset.

Fig. 23.21
Offset orchard rows in an orange grove that straddles the right-lateral (an object on the opposite side of the fault moves to the right) San Andreas Fault. The rows in the background have moved to the right relative to the rows in the foreground.

[3]The death toll was largely due to great seismic sea waves, or tsunamis. A tsunami is generated from the displacement of water as a result of an earthquake, submarine landslide, or an underwater volcanic eruption.

Table 23.1
Some notable earthquakes

Year	Location	Magnitude	Estimated Deaths	Comments
1556	Shensei, China		830,000	Possibly the greatest natural disaster
1811	New Madrid, Missouri		few	
1906	San Francisco, California	8.25	700	Fires caused extensive damage
1908	Messina, Italy	7.5	120,000	
1920	Kansu, China	8.5	180,000	
1923	Tokyo, Japan	8.2	150,000	Fire caused extensive destruction
1960	Southern Chile	8.7	5700	The largest earthquake ever recorded
1964	Anchorage, Alaska	8.5	131	More than $300 million in damage
1970	Peru	7.8	66,000	Great rockslide
1971	San Fernando, California	6.5	65	More than $5 billion in damage
1975	Liaoning, China	7.5	few	First major earthquake to be predicted
1976	Tangshan, China	7.6	500,000	
1985	Mexico City, Mexico	8.5	7,000	
1989	San Francisco, California	7.1	62	More than $6 billion in damage
1994	Northridge, California	6.7	57	More than $25 billion in damage
1995	Kobe, Japan	7.2	5,500	Between $95 to $147 billion in damage

Table 23.2
Richter Magnitude

Magnitude	Number per year	Maximum Mercalli Intensity	Characteristic Effects
<3.4	800,000	I	Recorded only by seismographs.
3.4–4.4	30,000	II and III	Felt by some people in the area.
4.4–4.8	4,800	IV	Felt by many people in the area.
4.8–5.4	1,400	V	Felt by everyone in the area.
5.4–6.0	500	VI and VII	Slight building damage.
6.0–7.0	100	VIII and IX	Much building damage.
7.0–7.4	15	X	Serious damage, bridges twisted, walls fractured.
7.4–8.0	4	XI	Great damage, buildings collapse.
>8.0	one every 5–10 years	XII	Total damage, waves seen on ground, objects thrown in air.

Source: From B. Gutenberg, 1950

nition as it shifted the direction and course of the Mississippi River. Fortunately, because the quake occurred in a remote region where there were few settlers, human casualties were few. Going much further back in time, the presently inactive Appalachians and parts of the Rocky Mountains were once zones of intense earthquake activity, much like the places listed in Table 23.1 that have recently endured the awesome power of the quaking Earth.

Plate Boundaries Moving plates interact at their adjacent boundaries. Two types of plate boundaries mark the edges of lithospheric formation and destruction.

Earthquake Measurements—Mercalli and Richter Scales

Every year hundreds of thousands of earthquakes occur. Although most are small and go undetected, the danger of large earthquakes certainly exists. Earthquake-prone regions experience large earthquakes about every 50–100 years.

The Mercalli scale measures quake intensity in terms of the effects the quake produces on the local environment. The scale ranges from an intensity of I, barely detectable, to an intensity of XII, total destruction. The Mercalli intensity at any location depends (1) on how far that location is from where the earthquake occurred, and (2) on the nature of the subsurface materials (for example, sedimentary or artificial fill versus bedrock) at the location.

Based purely on observation, the Mercalli scale, though a valuable yardstick, does not provide a precise measurement of quake size. For this reason seismologists developed a more precise way to estimate the energy released in an earthquake. The Richter scale, which is a magnitude scale, measures quake severity in terms of the amount of energy released and the amount of ground-shaking at a standard distance from the location of the quake. The Richter magnitude scale, which measures the amplitude of seismic waves recorded by a seismograph, is logarithmic. This means that each increase of 1 unit on the scale is equivalent to an increase of 10 in the amplitude of ground-shaking. Because seismic waves occur over a range of frequencies, this tenfold increase in amplitude translates into 30 times more energy released! For example, the 1985 Mexico City earthquake, which had a magnitude of 8.5, released about 30 times as much energy, and had 10 times more ground-shaking than the 1908 Messina earthquake, which had a magnitude of 7.5. The magnitude 8.5 Mexico City quake released 900 times as much energy as and produced 100 times more ground shaking than the 1971 magnitude 6.5 quake in San Fernando, California.

The Mercalli Scale of Intensity

I Not felt except by a very few under especially favorable circumstances.

II Felt only by a few persons at rest, especially on upper floors of buildings.

III Felt quite noticeably indoors, especially on upper floors of buildings, but many people do not recognize this as an earthquake.

IV Most people feel it indoors, a few outdoors. Dishes, windows, doors rattle.

V Felt by nearly everyone. Disturbances of trees, poles, and other tall objects.

VI Felt by all; many frightened and run outdoors. Some heavy furniture moved; a few instances of fallen plaster or damaged chimneys. Damage slight.

VII Everybody runs outdoors. Damage negligible in buildings of good design and construction; slight to moderate in well-built structures; considerable in poorly built structures.

VIII Damage slight in specially designed structures; considerable in ordinary substantial buildings with partial collapse; great in poorly built structures (fall of chimneys, factory stacks, columns, monuments, walls).

IX Damage considerable in specially designed structures. Buildings shifted off foundations. Ground conspicuously cracked.

X Some structures destroyed. Most masonry and frame structures destroyed with foundations. Ground badly cracked.

XI Few, if any, structures (masonry) remain standing. Bridges destroyed. Broad fissures in ground.

XII Damage total. Waves seen on ground surfaces. Objects thrown up in air.

Source: U.S. Coast Guard and Geodetic Survey

Divergent Boundaries Where two plates are moving apart, as in Figure 23.22a, tensional forces act to stretch the lithosphere and generate a spreading center. Hot, molten rock from the Earth's asthenosphere buoyantly rises up to form new lithosphere. The lithosphere near the spreading edge is thin and has a relatively low density due to heat and expansion of the rising magma. As it moves away from the spreading center, new lithosphere cools, contracts, and becomes more dense.

The Mid-Atlantic Ridge is a spreading center that has been producing the Atlantic Ocean floor for 160 million years. The spreading is accompanied by almost continuous

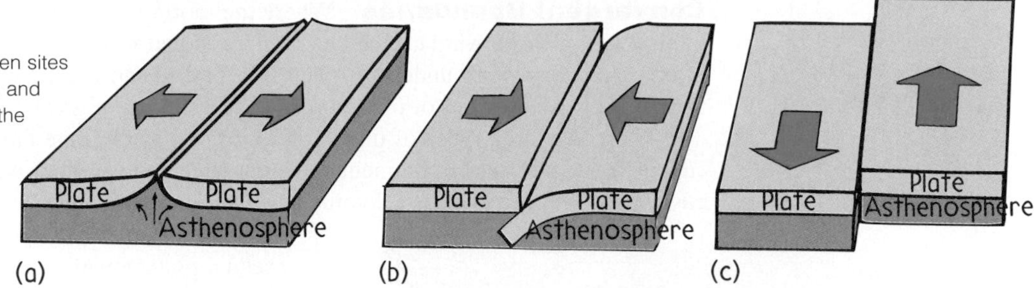

Fig. 23.22
Plate boundaries are often sites of lithospheric formation and destruction. Named for the movement they accommodate, three types of plate boundaries are (a) divergent, (b) convergent, and (c) transform-fault boundaries.

earthquake activity that, except by scientists, goes largely unnoticed because of the low magnitude and harmlessness to humans. With production of lithosphere at the ridges, the ocean floor grows and the continents drift apart. The lithosphere is younger near the central ridge region and progressively older away from the ridge. The spreading rate has been slow, about 2 centimeters each year (about as fast as fingernails grow). This rate over 160 million years adds up to 3200 kilometers—the width of the Atlantic Ocean. So, it has taken only 160 million years for a mere fracture in an ancient continent to turn into the Atlantic Ocean!

Spreading centers are not restricted to the ocean floors, but also develop on land. Hot, rising molten material in the Earth's interior beneath continental landmasses causes upwarping of the Earth's crust. Gaps in the crust are produced, and large slabs of rock sink and slide down into these gaps. The large down-faulted valleys generated by this process are either **rifts** or **rift valleys** (Figure 23.23). The Great Rift Valley of East Africa is a prime example of such a feature and, if spreading continues, may be the beginning of a new ocean basin.

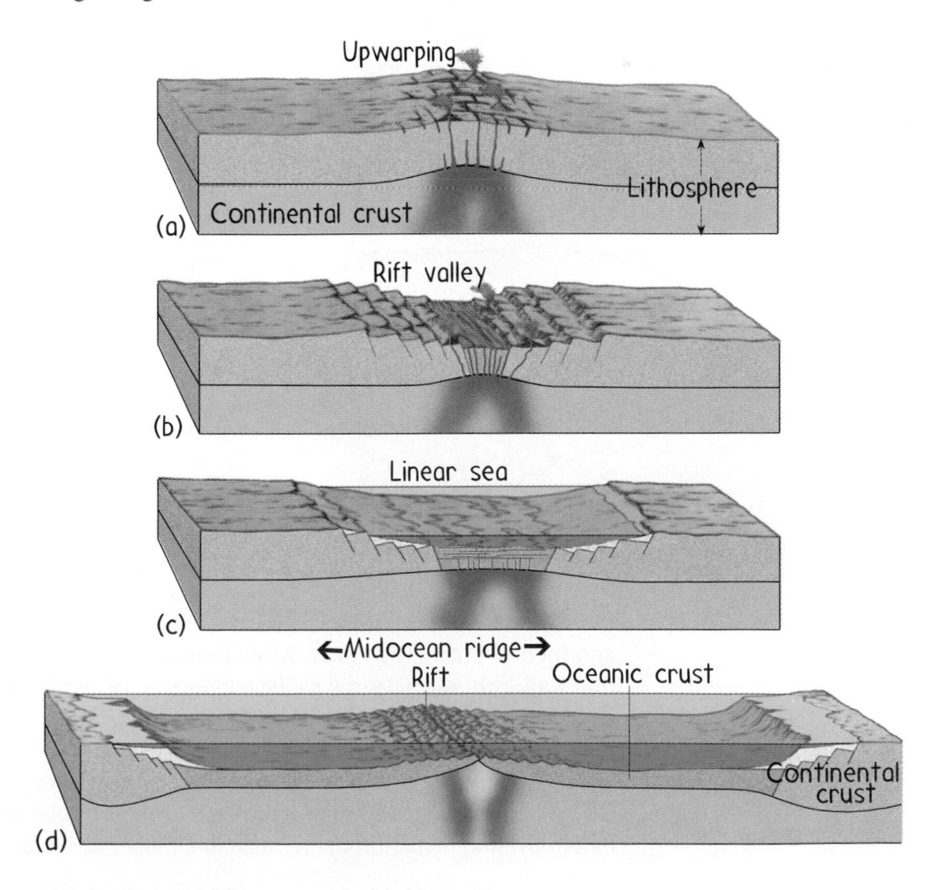

Fig. 23.23
Formation of a rift valley and its transformation into an ocean basin. (a) Rising magma uplifts continental crust, causing the surface to crack. (b) Rift valley forms as crust is pulled apart. It is in this stage that we find Africa's Great Rift Valley today. (The two sides of the valley move away from each other because they happen to be located above mantle convection cells that have the same circulation pattern as the cells in Fig. 23.13.) (c) Water from the ocean drains in as the rift drops below sea level, forming a linear sea, so called because it is usually long and narrow. (d) Over millions of years, the rift widens to become an ocean basin.

Convergent Boundaries Where the motion created by convection cells acts to push two plates toward each other, compressional forces either push the lithosphere of one plate downward under the other plate or else shorten the lithosphere by folding and faulting. The regions of plate collisions are regions of great mountain building.

There are three types of plate collisions: (1) Each plate has an oceanic leading edge, (2) one plate has a continental leading edge and the other has an oceanic leading edge, and (3) each plate has a continental leading edge (Figure 23.24).

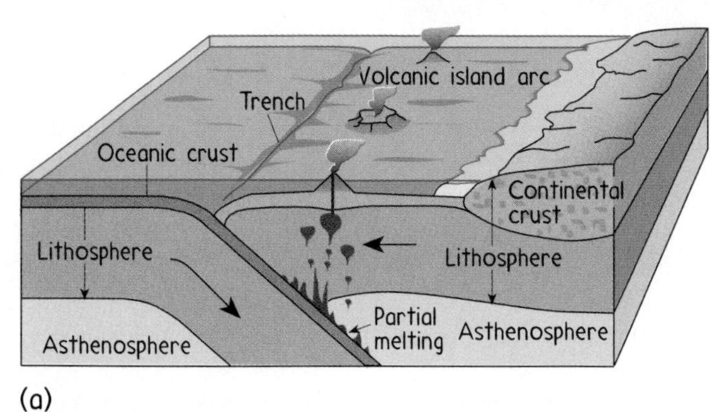

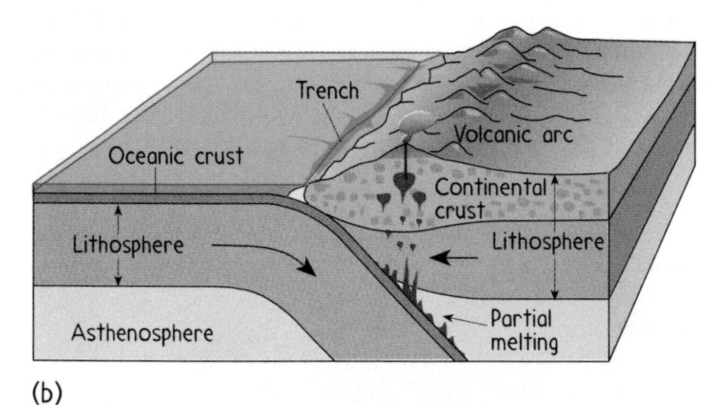

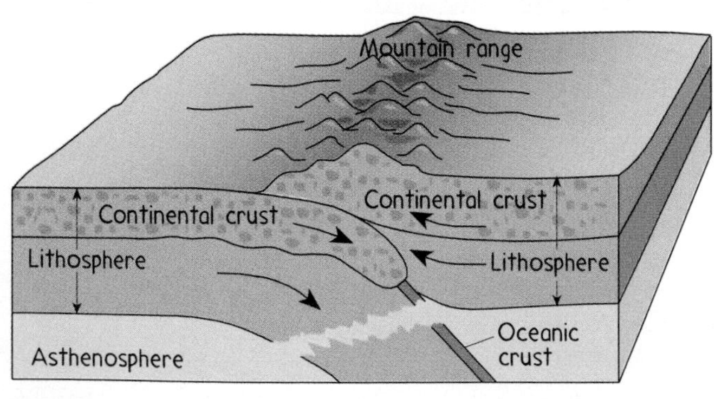

Fig. 23.24
The three types of convergent margins: (a) oceanic-oceanic, (b) oceanic-continental, and (c) continental-continental.

Convergence (collision) between two oceanic plates results in *subduction,* a process in which one plate bends and descends beneath the other to produce a deep ocean trench. The deepest trench known is the Marianas Trench in the western Pacific Ocean, where the seafloor is as much as 11 kilometers below sea level. Partial melting of the subducted crust produces andesitic magma that buoyantly migrates upward, eventually erupting onto the ocean floor to produce a series of volcanoes that constitute a volcanic *island arc.* The size and elevation of the arc increase over time because of continued volcanic activity. Just offshore from the island arc, sediments scraped off the descending plate pile up in the trench. As subduction continues, the pile of sediment becomes folded and faulted until it is a thick deformed wedge. Such island arcs have formed the Aleutian, Marianas, and Tonga Islands and the island arc systems of the Alaskan Peninsula, the Philippines, and Japan.

Deep ocean trenches mark the active subduction zones that border island arcs. Earthquakes occur along the subduction zones as the downgoing plate grinds against the overriding plate. The earthquakes get deeper and deeper in the direction of subduction (Figure 23.25).

When an oceanic plate and a continental plate converge, the denser oceanic plate is subducted beneath the lighter continental plate, and a deep ocean trench is formed. The descending oceanic crust carries a mixture of oceanic basaltic rock and continental sediment. As with oceanic-oceanic convergence, andesitic magma rises upward in oceanic-continental convergence. Some of this magma crystallizes below the surface, and some reaches the surface where it erupts to form a chain of volcanic mountains on the overriding continental plate. The Andes Mountains of western South America, which formed in this way, continue to grow as the subduction of the Nazca Plate beneath the continent of South America along the Peru-Chile trench causes sediments from the Nazca Plate to be scraped off onto the granitic roots of the Andes. This scraped material adds thickness and buoyancy to the mountains so that they rise upward more rapidly than they are eroded by wind and rain. Remnants of the original volcanic chain are the exposed batholiths and metamorphic terrains that flank the Andes on the western coast of South

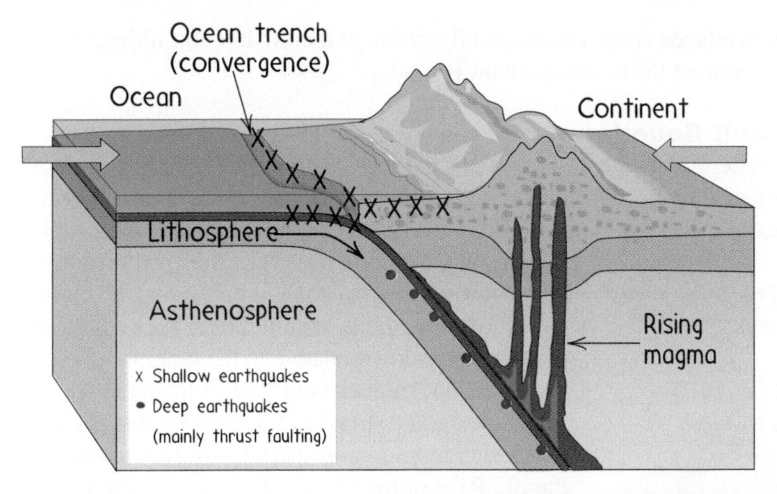

Fig. 23.25
Earthquakes at a subduction zone get deeper and deeper in the direction of subduction.

America. In the western United States, examples of such volcanic activity are found in the Sierra Nevada, an ancient volcanic range, and the Cascade Range, which is currently active. The Sierra Nevada was produced by the subduction of the ancient Farallon Plate beneath the North American Plate. The Sierra Nevada batholith is a remnant of the original volcanic range, while the Coast Range has remnants of the sediment that accumulated in the trench. The Cascade Range, produced from the subduction of the Juan de Fuca Plate (a remnant of the Farallon Plate) beneath the North American Plate, includes the volcanoes Mounts Rainier, Shasta, and St. Helens. The eruption of Mount St. Helens gives testimony that the Cascade Range is still quite active.

Earthquake activity, similar to that in oceanic-oceanic convergence, is also characteristic of oceanic-continental convergence.

■ Question

Erosion wears mountains down, and yet the Andes Mountains grow taller each year. Why?

The collision between two continental landmasses is always preceded by oceanic-continental convergence. Because continental crust, being light and buoyant, does not undergo any appreciable subduction, convergence between two continental plates is a head-on collision (Figure 23.24c). Compressional forces cause the plates to buckle and fold upon each other, making the crust very thick. Intensely compressed and metamorphosed rock defines the zone where the plates meet. In contrast with convergence involving either two oceanic plates or one continental and one oceanic plate, volcanic activity is not characteristic of continental-continental collisions, but earthquakes are.

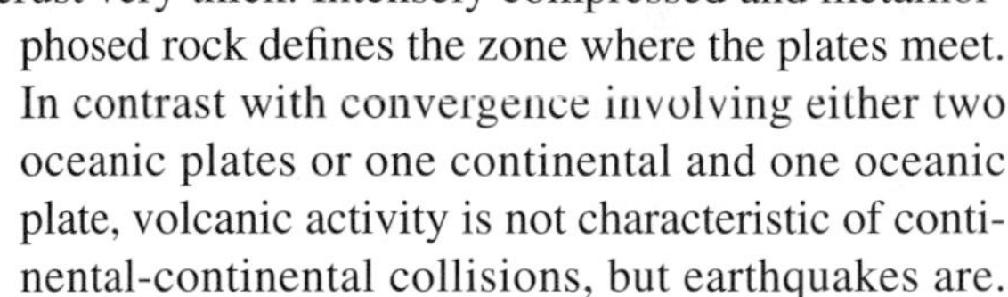

The collision between continental plates has produced some of the most famous mountain ranges, one majestic example being the snow-capped Himalayas, the highest mountain range in the world. This chain of towering peaks is still being thrust upward as India continues crunching against Asia (Figure 23.26). The European Alps were formed in a similar fashion when fragments of the African Plate collided with the Eurasian Plate 80 million years ago. Relentless pressure between the two plates continues and is slowly closing up the Mediterranean Sea. In America, the Appalachian

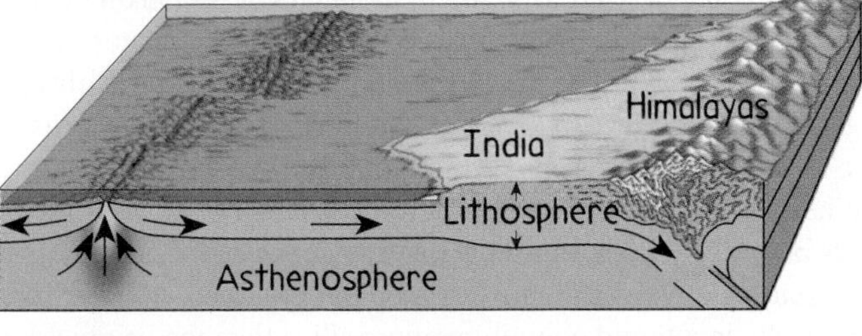

Fig. 23.26
The continent-to-continent collision of India with Asia produced—and is still producing—the Himalayas.

■ **Answer**

Subduction is still occurring at the Peru-Chile trench causing the uplift of the Andes Mountains. Because the rate of uplift exceeds the erosion rate, the mountains continue to grow.

Mountains were produced from a continental-continental collision that ultimately resulted in the formation of the supercontinent Pangaea.

Transform-Fault Boundaries A **transform fault** is a plate boundary that occurs where two plates are neither colliding nor pulling apart but rather sliding horizontally past each other (Figure 23.27). The fault "transforms" the motion from one ridge segment to the other. Because contact forces are neither tensional nor compressional, there is no creation or destruction of the lithosphere. A transform fault is a zone of horizontal accommodation of movement, with neither side of the fault moving up or down relative to the other.

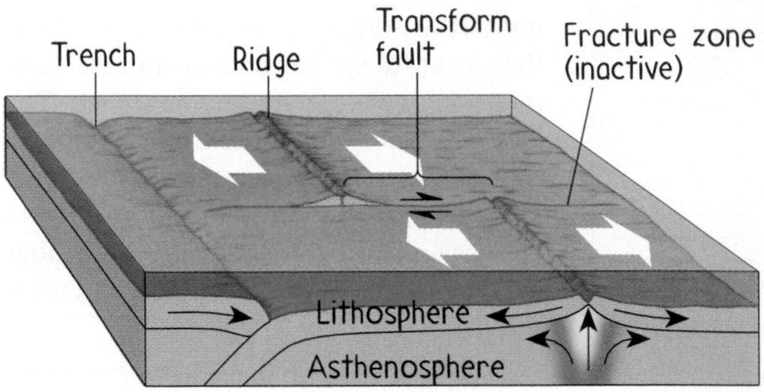

The San Andreas fault, one of the most famous transform faults, stretches for 1500 kilometers from Cape Mendocino in northern California to the East Pacific Rise in the Gulf of California. The Pacific Plate is moving northwest at a rate of about 5.0 centimeters per year relative to the North American Plate. The San Andreas fault accommodates about 70 percent of this motion, or about 3.5 centimeters per year. The remaining motion is taken up by other faults. Grinding and crushing take place as the two plates move past each other. When sections of the plates become locked, stress builds up until the friction is relieved in the form of an earthquake. On April 18, 1906 the Pacific Plate lurched about 6 meters northward over a 434 kilometer stretch of the fault, releasing the built-up stress, and causing the catastrophic San Francisco earthquake.

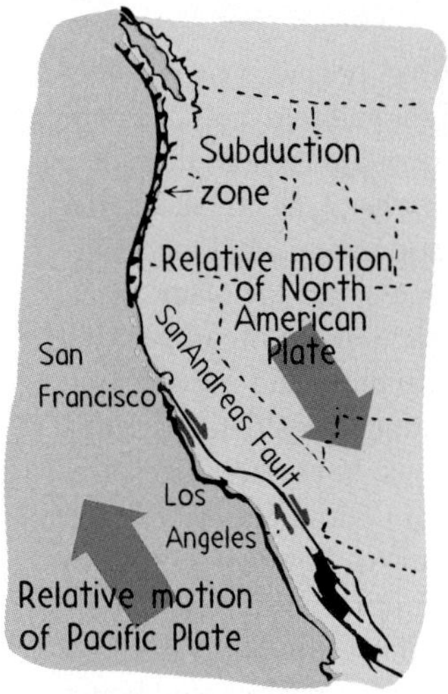

Fig. 23.27
Transform faults allow two plates to slide past one another where ridge segments are offset.

Fig. 23.28
The San Andreas Fault shows horizontal offset, where the Pacific Plate slides northward past the adjacent North American Plate.

The Theory That Explains Much

Before the theory of plate tectonics was put forth, processes such as mountain building, folding, and faulting were poorly understood. Without a mechanism to explain a lithosphere that shifts, scientists did the best they could to explain their observations. Plate tectonics offers explanations as to the where and why of many geologic processes. Indeed, it can be thought of as a unifying theory because it links causes and effects.

Why are the Appalachian Mountains located where they are? What about the Sierra Nevada? The Rocky Mountains? The Alps? The plate tectonic model gives an answer: All the mountain-building events take place near convergent plate boundaries.

We can also relate the formation of the three different rock types to the plate tectonic theory. Although we simplify the discussion of rock formation for the sake of clarity, all rocks are tied to plate interaction in some manner or another.

The intense heat and pressure caused by subduction and continental collisions result in the metamorphism of preexisting rocks. Here is where regional metamorphism occurs. As the subducting slab is heated, it eventually begins to melt. Recall from Chapter 22 that when rock melts, the liquid contains more silica than the parent rock (partial melting). Thus andesite forms from basaltic material (subducted ocean floor), and we find belts of composite cone volcanoes, like those in the Andes Mountains and the Cascade Range.

What about granite? Where does most of that come from? As the ocean floor melts, it can produce large volumes of andesitic magma. This magma doesn't all erupt at once, but rather accumulates in the Earth's crust. As the magma bodies cool, they undergo fractional crystallization, which results in a liquid containing more silica than the original liquid. When this silica-enriched magma cools, it forms granite. Where in nature has this occurred? The Sierra Nevada is a largely granitic mountain range in California that formed in such a manner. The large batholiths of granitic rocks are the "roots," or solidified magma bodies, from a once-extensive volcanic belt formed as a result of subduction.

We can also explain the creation of basaltic rocks using the plate tectonic model. Where plates are diverging, such as at a midocean ridge, mantle material partially melts to form new lithosphere that is capped by oceanic crust—basalt.

What about sedimentary rocks? As mountains grow as a result of plate collisions, they also begin to weather and erode. The clastic sediments produced are transported downslope, where they accumulate, layer upon layer, eventually becoming sedimentary rock.

Virtually all earthquake and volcanic activity can be tied directly to plate tectonics. These energetic responses to plate interactions are almost always found where plates interact; earthquakes are found at all types of plate boundaries, and volcanoes are concentrated where plates either collide or pull apart.

So we see that the tectonic interaction between lithospheric plates, which occurs primarily at their boundaries, provides an explanation to the origin of mountain chains, the development and destruction of ocean basins, the three types of rocks found on the Earth, and the global distribution of earthquakes and volcanoes. The internal activities that change the Earth's surface do so in a cyclical manner. In Chapter 25, we shall see the effects of plate tectonic interaction through time. The study of geology uses processes that occur today to understand what may have occurred in the past. This concept is commonly stated as "the present is the key to the past." But what has happened in the past provides clues as to what may happen in the future. The Earth is indeed a dynamic planet.

Summary of Terms

- **Body wave** A seismic wave that travels through the Earth's interior.
- **Surface wave** A seismic wave that travels along the Earth's surface.
- **Primary waves** A longitudinal body wave; travels through solids, liquids, and gases and is the fastest seismic wave.
- **Secondary waves** A transverse body wave; cannot travel through liquids and so does not travel through the Earth's outer core.
- **Mohorovičić discontinuity (Moho)** The crust-mantle boundary, marking the depth at which the speed of P-waves traveling toward the Earth's center increases.
- **Core** The central layer in the Earth's interior, divided into an outer liquid core and an inner solid core.
- **Mantle** The middle layer in the Earth's interior, between crust and core.
- **Asthenosphere** A subdivision of the upper mantle situated below the lithosphere, a zone of plastic, easily deformed rock.
- **Lithosphere** The entire crust plus the portion of the mantle above the asthenosphere.
- **Crust** The Earth's outermost layer.
- **Isostasy** The principle that dictates how high the crust stands above the mantle.
- **Pangaea** A single, large landmass that existed in the geologic past and was composed of all the present-day continents.
- **Paleomagnetism** The study of natural magnetization in a rock to determine the intensity and direction of the Earth's magnetic field at the time of the rock's formation.
- **Seafloor spreading** The divergence of two oceanic plates to form a rift in the seafloor.
- **Theory of plate tectonics** The idea that the Earth's lithosphere is broken into pieces (plates) that move over the asthenosphere; boundaries between plates are where most earthquakes and volcanoes occur and where lithosphere is created and recycled.
- **Fold** A series of ripples in the crust that result from compressional deformation of the lithosphere.
- **Syncline** A fold in strata that has relatively young rocks at its core, with rock age increasing as you move horizontally away from the fold core.
- **Anticline** A fold in strata that has relatively old rocks at its core, with rock age decreasing as you move horizontally away from the fold core.
- **Fault** A fracture along which visible displacement can be detected on one side relative to the other.
- **Rift (rift valley)** A long, narrow trough that forms as a result of divergence of two plates.
- **Transform fault** A plate boundary formed by two plates that are sliding horizontally past each other.

Review Questions

Seismic Waves

1. P-waves and S-waves move through the Earth's interior in two ways. What is the difference in their mode of propagation?

2. Can S-waves travel through liquids? Explain.
3. Name the two types of surface waves and describe the motion of each.

The Earth's Internal Layers

4. What was Andrija Mohorovičić's major contribution to geology?
5. List the different properties of seismic waves and how they contributed to the discovery of the Earth's internal boundaries.
6. What does the wave shadow that develops 105–140° from the origin of an earthquake tell us about the Earth's composition?
7. What is the evidence for the solidity of the Earth's inner core?
8. Even though the inner and outer cores are both composed predominantly of iron and nickel, the inner core is solid and the outer core is liquid. Why?
9. What is the evidence that the Earth's outer core is liquid?
10. Describe the asthenosphere and the lithosphere. In what way are they different from each other?
11. What convectional movement is responsible for the motion of lithospheric plates?
12. How does continental crust differ from oceanic crust?
13. Why does continental crust float higher on the Earth's surface than oceanic crust?
14. Define *isostasy* and relate it to continental and oceanic crust.

The Theory of Continental Drift

15. What key evidence did Alfred Wegener use to support his idea of continental drift?
16. How do the glacial striations found in parts of South America, southern Africa, India, and southern Australia support the concept of a supercontinent?
17. What was the stated reason for the scientific community rejecting Wegener's idea of continental drift?

A Scientific Revolution

18. What information can be learned from a rock's magnetic record?
19. What role did paleomagnetism play in supporting continental drift?
20. Where are the deepest parts of the ocean?
21. What major discovery at the bottom of the ocean was made by H. H. Hess?
22. How is the ocean floor similar to a gigantic slow-moving tape recorder?
23. What does the Earth's crust have in common with a conveyor belt?
24. In what way does seafloor spreading support continental drift?

The Theory of Plate Tectonics

25. Name and describe the three types of plate boundaries.
26. What are folds?
27. Are folded rocks the result of compressional or tensional forces?
28. Distinguish between anticlines and synclines.
29. There are three major types of faults: thrust, reverse, and normal. What is the difference between them, and how can geologists differentiate between them in the field?
30. Which kinds of fault result primarily from tension in the Earth's crust? Primarily from compression?
31. What type of fault is associated with the 1964 earthquake in Alaska?
32. How old is the Atlantic Ocean thought to be? For how many years has magma been extruding in the mid-Atlantic?
33. What is a rift?
34. What kind of boundary separates the South American Plate from the African Plate?
35. Describe the three types of plate collision.
36. What is the driving force for mountain building in the coastal western United States? In the Andes?
37. The Appalachian Mountains were produced at what type of plate boundary?
38. What clues could we use to recognize the boundaries between ancient plates no longer in existence?
39. What is a transform fault?
40. What kind of plate boundary separates the North American Plate from the Pacific Plate?

Home Project

Look for a very old window, and note the lens effect in the bottom part of the glass. Glass has both solid and liquid properties; in fact, it is often thought of as a very viscous liquid. Over many years, its downward flow due to gravity is evidenced by the increased thickness near the bottom of the pane. Tie this observation of "plastic" behavior into our discussion of plate tectonics. Which parts of the Earth behave plastically? Which parts are rigid? What do these ideas have to do with plate tectonics?

Exercises

1. Compare the relative speeds of primary and secondary seismic waves, and relate speeds of travel to the medium in which the waves travel.
2. Explain how seismic waves indicate whether regions within the Earth are solid or liquid.
3. How do seismic waves indicate layering of materials in the Earth's interior?
4. What is the evidence for the Earth's central core being solid?
5. What is the evidence for believing the part of the Earth beneath the crust is semi-liquid?
6. Speculate on why the lithosphere is rigid and the asthenosphere plastic, even though they are both part of the mantle.

7. If the Earth's mantle is composed of rock, how can we say that the crust floats on the mantle?
8. Why is the Earth's crust thicker beneath a mountain range?
9. What is the principle of isostasy, and what evidence supports it?
10. Which extends farther into the asthenosphere, the continental or the oceanic crust? Why?
11. How does erosion and wearing away of a mountain affect the depth to which the crust extends into the asthenosphere?
12. Describe how the different paths of polar wandering helped establish that continents move over geologic time.
13. Describe how the presence of faults and folds supports the idea that lithospheric plates are in motion.
14. Why are most earthquakes generated near plate boundaries?
15. Most volcanoes are located near plate boundaries. Why are the volcanic Hawaiian Islands located in the interior of a plate?
16. Why do mountains tend to form in long thin ranges?
17. What type of plate boundary most likely forms mountains? Explain.
18. Relate the formation of metamorphic rocks to plate tectonics. Would you expect to find metamorphic rocks at all three types of plate boundaries? Why or why not?
19. Cite one line of evidence that suggests subduction once occurred off the coast of California.
20. Distinguish between continental drift and plate tectonics.
21. Are the present ocean basins a permanent feature on our planet? Discuss why or why not.
22. Are the present continents a permanent feature on our planet? Discuss why of why not.
23. Why is it that the most ancient rocks are found on the continents, and not on the ocean floor?
24. Upon crystallization, certain minerals (the most important being magnetite) align themselves in the direction of the surrounding magnetic field, providing a magnetic fossil imprint. How does the Earth's magnetic record support the theory of continental drift?
25. How is the theory of seafloor spreading supported by paleomagnetic data?
26. What is meant by magnetic pole reversals? What useful information do they tell us about the Earth's history?
27. What kind of boundary is associated with seafloor spreading?
28. Using a photocopy of Figure 23.15, mark the different boundaries of plate interaction. Draw arrows showing direction of plate movement for convergent, divergent and transform fault boundaries.
29. Earthquakes, the result of sudden motion in the Earth caused by abrupt release of slowly accumulated stress, cause rock to fracture or fault. Relate such faulting to horizontal movement of plates. Where does this type of movement occur?
30. Does the fact that the mantle is beneath the crust necessarily mean that the mantle is denser than the crust? Explain.
31. How does the age distribution of the Hawaiian Islands chain relate to the direction of Pacific Plate movement?
32. What is a very likely candidate for the existence of the Earth's magnetic field?
33. Lithospheric material is continuously created and destroyed. Where does this creation and destruction take place? Do the rates of the two processes balance each other?
34. Subduction is the process of one lithospheric plate descending beneath another. Why does the oceanic portion of the lithosphere undergo subduction while the continental portion does not?
35. What geologic features are explained by plate tectonics?
36. In 1964, a large tsunami struck the Hawaiian Islands without warning, devastating the coastal town of Hilo, Hawaii. Since that time a tsunami warning station has been established for the coastal areas of the Pacific. Why do you think these stations are located around the Pacific rim?
37. What type of volcano would you expect to find at continental subduction margins such as the Andes Mountains and the Cascade Range?
38. How did the Himalayan mountains originate? The San Andreas fault? The Andes Mountains?
39. What is the major source of energy responsible for earthquakes in southern California?
40. Where is the world's longest mountain range located?

Problems

1. If the mid-Atlantic ocean is spreading at 2 cm per year, how many years has it taken for it to reach its present width of 5000 km?
2. The San Andreas Fault separates the northwest moving Pacific Plate, on which Los Angeles sits, from the North American Plate, on which San Francisco sits. If the plates slide past one another at a rate of 3.5 cm per year, how long will it take the two cities to form one large city? (The distance between Los Angeles and San Francisco is 600 km.)
3. The weight of ocean floor bearing down upon the lithosphere is increased by the weight of ocean water. Relative to the weight of the 10-km-thick basaltic ocean crust (specific gravity 3), how much weight does the 3-km-deep ocean (specific gravity 1) contribute? Express your answer as a percent of the crust's weight.
4. The Richter scale is logarithmic, meaning that each increase of 1 on the Richter scale corresponds to an increase of 10 in the amplitude of the seismic waves created by an earthquake. An earthquake that measures 8 on the Richter scale has how many times more ground shaking than a quake that measures 6 on the Richter scale?

CHAPTER

24

Water and Surface Processes

A view of the Earth from space shows our planet to be a vast expanse of water interrupted here and there by islandlike continents. About 70 percent of the Earth's surface is covered with water, and water plays an important role in just about every natural process on the Earth's surface.

Water is vital to life and has been since the first life forms evolved. Our bodies are more than 65 percent water by weight and require water to function. Water is used not only for drinking but also for agriculture, industry, sanitation, and transportation. Where does the water we use come from and, more important, will it last?

Slightly more than 97 percent of all the Earth's water is in the oceans, and a little more than 2 percent is frozen in the polar icecaps and glaciers. The remainder, less than 1 percent, consists of water vapor in the atmosphere, water in the ground, and water in rivers and lakes.

This chapter begins with the Earth's freshwater supply—the water frozen in glaciers and ice caps and that found either underground or in rivers and lakes—and concludes with the great reserves of water in the Earth's oceans.

Fig. 24.1
Distribution of the Earth's water supply.

24.1 The Hydrologic Cycle

The Earth's water is constantly circulating, powered by the heat of the Sun and the force of gravity. As the Sun's energy evaporates ocean water, a cycle begins (Figure 24.2). Water molecules move from the Earth's surface to become part of the atmosphere. The resulting moist air may be transported over great distances by winds. Some of the water molecules condense to form clouds and then precipitate as rain or snow. Precipitation falling on the ocean completes the cycle from ocean back to ocean.

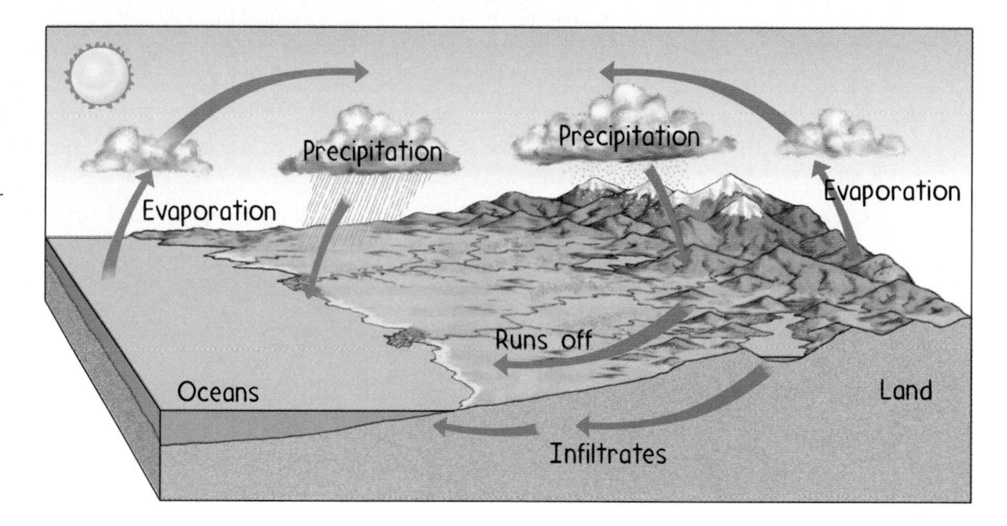

Fig. 24.2
The hydrologic cycle. Water evaporated at the Earth's surface enters the atmosphere as water vapor, condenses into clouds, precipitates as rain or snow, and falls back to the surface, only to evaporate again and go through the cycle yet another time.

Completion of the cycle is more complex when precipitation falls on land, for water may drain to streams, then to rivers, and then journey back into the ocean. Or it may percolate into the ground, or evaporate back into the atmosphere before reaching the ocean. Also, water falling on land may become part of a snow pack or glacier. Although snow or ice may lock water up for many years, it eventually melts or evaporates and returns to the cycle. This natural circulation of water from the oceans to the air, to the ground, then to the oceans and then back to the atmosphere is called the **hydrologic cycle**.[1]

The total amount of water vapor in the atmosphere remains relatively constant. Therefore, evaporation and precipitation balance each other. Being that most of the Earth's surface area is ocean, it makes sense that evaporation and precipitation are greatest over the oceans. In fact, 85 percent of the atmosphere's water vapor is water evaporated from the ocean, and 75 percent of the atmosphere's water vapor is precipitated back to the oceans. On the continents, precipitation exceeds evaporation. Of the atmosphere's water vapor, 15 percent is water evaporated from the continents, and 25 percent is precipitated back to the land. Balance is maintained between the amount of water taken up into the atmosphere (85% + 15%) and the amount precipitated out of the atmosphere (75% + 25%).

---

[1] This key concept is another conservation principle, of the sort we first saw in Chapter 3 when we studied conservation of momentum and energy. In Chapter 9, we saw it as conservation of electric charge; in Chapter 16, as conservation of nucleons; and in Chapter 19, as conservation of atoms. So now we learn that the amount of water on Earth is conserved. A lack of it in one place means an abundance someplace else, which in most instances is the ocean.

The rain or snow that reaches the continents is the Earth's only natural supply of fresh water. More than three-quarters of the Earth's fresh water is in the polar ice caps and glaciers. It is surprising to note that, of the fresh water not locked in glacial ice, most is not in lakes and rivers but rather is beneath the Earth's surface. As rain falls and sinks into the ground, it percolates downward to fill the open pore spaces between sediment grains. This water is now called **groundwater**.

Questions

1. What fraction of the Earth's water supply is fresh water?
2. The volume of water evaporated from all the land surface of the Earth is 60,000 km^3 per year, but the volume precipitated over the land surface is 96,000 km^3 per year. Being that the volume that precipitates each year is 36,000 km^3 more than the volume that evaporates, why isn't all the land flooded?

24.2 Groundwater

Except in polar regions, the water in lakes, ponds, rivers, streams, springs, and puddles is the only fresh water that meets our eye, but all these water reservoirs together hold only about 1.5 percent of the Earth's non-ice fresh water. The other 98.5 percent resides in porous regions beneath the Earth's surface as groundwater.

Have you ever noticed how during a rainstorm sandy ground soaks up rain like a sponge? The water literally disappears into the ground. The nature of the surface material influences the amount of water that penetrates the ground. Some soils, like sand, readily soak up water. Other soils, like clay, impede the infiltration of water and cause runoff. Rocky surfaces with little or no soil are the poorest absorbers of water, with penetration only through cracks in the rock.

The amount of water that can be contained underground at a given location depends on the porosity of the soil or rock at that location. **Porosity** is the volume of open space, or voids, in a soil or rock sample compared to the total volume of solids plus voids. Porosity depends on the size and shape of the soil or rock particles and on how tightly these particles are packed. For example, a soil composed of rounded particles all about the same size has a higher porosity than a soil composed of rounded particles of many sizes. This is because the smaller particles fill up the pores formed by the larger particles, thereby reducing the overall porosity. In addition, because of their irregular shape, angular particles can fill in pores created by other particles.

The ability of a material to transmit fluid is its **permeability**. If the spaces between particles are extremely small and poorly connected (as is the case with flattened clay

---

Answers

1. Less than 3%, as you can see by adding the freshwater values in Figure 24.1: 2.14% + 0.61% + 0.009% + 0.005% = 2.764%.
2. The excess water works its way back to the oceans. Excess water to the oceans does not cause sea level to rise because evaporation (85%) exceeds precipitation (75%) by 10%. Balance is maintained between the amount of water evaporated and precipitated over the oceans (85% − 75% = 10%) and the amount precipitated and evaporated over land (25% − 15% = 10%).

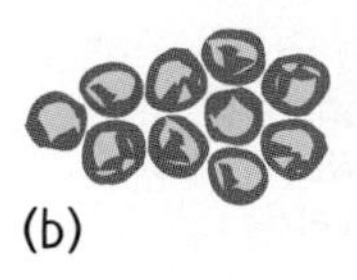

Fig. 24.3
(a) The sediment particles in clay are small, flat, and tightly packed. Because the sediment particles are flat, the many small pore spaces are poorly connected. Thus clays have high porosity but low permeability. (b) Sediment particles in sand or gravel are relatively uniform in size and shape, with large and well-connected pore spaces. This allows water to flow freely. So sands and gravels can have both high porosity and high permeability.

particles), water may barely move at all. So although clay may have a large total pore volume, it is difficult for water to move through the pores. In other words, clay is practically impermeable. In contrast, in sand and gravel, because of the well-connected large, open pore spaces, water moves freely from one pore space to the next. Thus sand and gravels are highly porous and highly permeable (Figure 24.3).

Porosity and permeability of surface and subsurface material are very important to the storage and movement of groundwater.

The Water Table

If we were to drill a hole into the ground, we would find that wetness varies with depth. Just below the surface, we encounter an *unsaturated zone,* or *zone of aeration,* where pore spaces are filled mainly with air. As we descend farther, we enter the *zone of saturation,* where water percolating down from the surface has filled all pore spaces. The upper boundary of this saturated zone is called the **water table**.

The depth of the water table beneath the Earth's surface varies with precipitation and climate. It ranges from zero in marshes and swamps to hundreds of meters in some parts of the deserts. The water table also tends to rise and fall with the contours of the surface topography (Figure 24.4). At lakes and perennial streams, the water table is above the land surface.

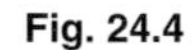

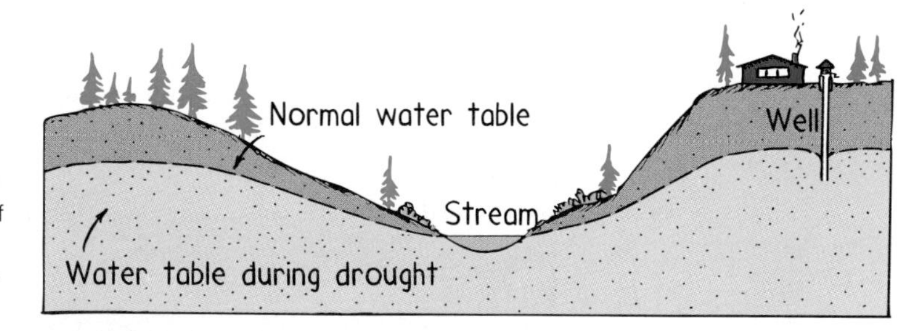

Fig. 24.4
The water table roughly parallels the ground surface. In times of drought, the water table falls, reducing stream flow and drying up wells. The water table also falls if the rate at which water is pumped out of a well exceeds the rate at which the groundwater is replaced.

Aquifers and Springs

Any water-bearing underground layer through which groundwater can flow is an *aquifer.* These reservoirs underlie the land in many places and contain an enormous amount of water—approximately 35 times the total volume of water found in freshwater lakes, rivers, and streams. More than half the land area in the United States is underlain by aquifers. One such aquifer is the Ogallala aquifer, which stretches from South Dakota to Texas and from Colorado to Arkansas!

The flow of groundwater in an aquifer can be complicated by low-permeability beds of rock or soil that hinder or prevent water movement.[2] Sometimes an aquifer becomes confined between two low-permeability layers as if in a tunnel (Figure 24.5). If the confined portion of the aquifer is at a lower elevation than the unconfined part in the *recharge area* (the area where water enters the aquifer), the confined groundwater is under pressure from the water above and flows out of the ground at any opening in the aquifer. This is an **artesian system**. If the opening is natural, it is an *artesian spring.* If the opening has been drilled, it is an *artesian well.*

---

[2]Geologists call these beds *aquicludes* or *aquitards.*

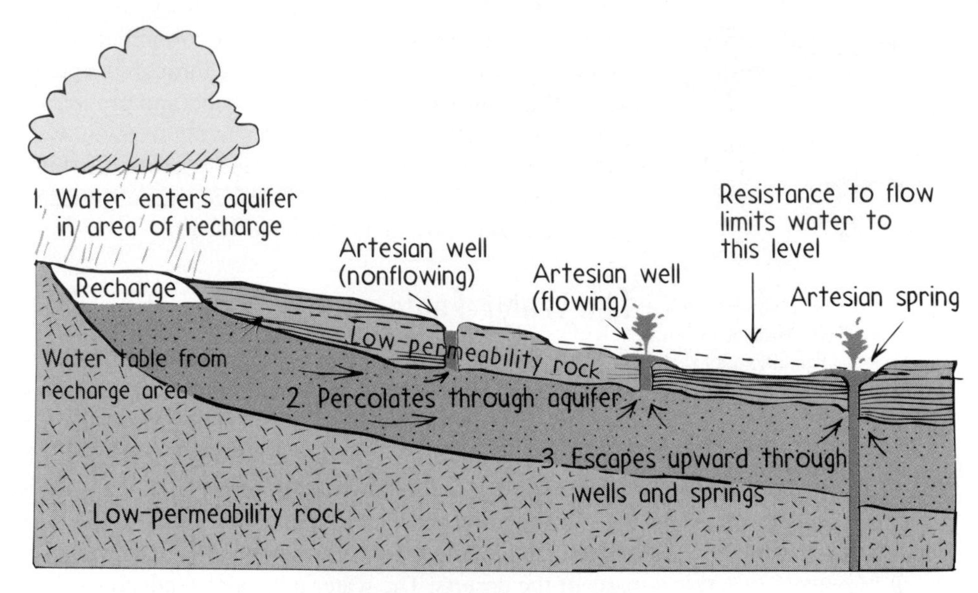

Fig. 24.5
An artesian system is formed when groundwater in an aquifer confined between layers of low-permeability rock rises to the surface through any opening that taps the aquifer. Water flows freely if the water table height in the recharge area is greater than the height of the opening (flowing artesian well and artesian spring). If the height of the opening is greater than the height of the water table in the recharge area, the water does not flow (nonflowing artesian well).

A low-permeability layer can also intercept water above the main water table; when this happens, a perched water table is created.

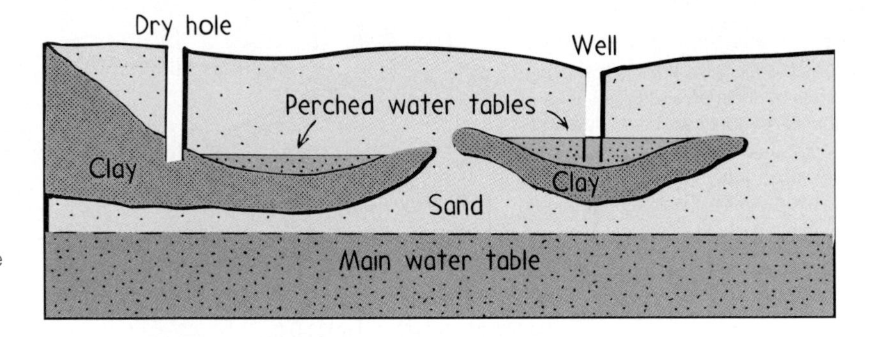

Fig. 24.6
A perched water table is separated from the main water table by a low-permeability layer, clay in this case.

When the water table intersects the land surface, groundwater emerges from an aquifer as either a spring, stream, or lake. Springs can generally be found where the water table intersects the surface abruptly, such as on a hillside or coastal cliff. Because water tends to leak out of the ground through cracks and breaks in a rock, springs are often associated with faults. In fact, field geologists can often locate faults by looking for springs.

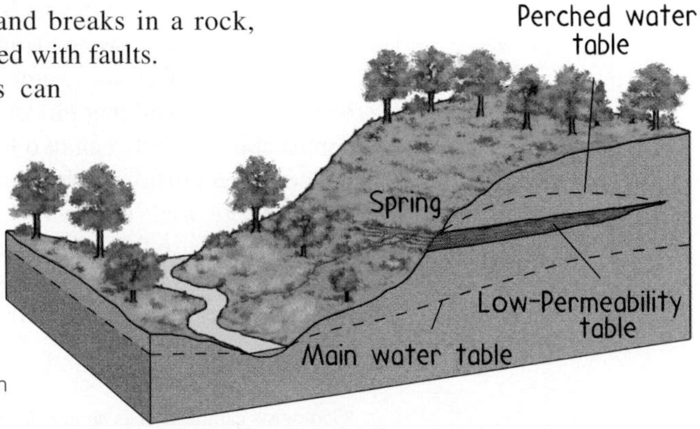

Fig. 24.7
When the water table intersects the land surface, groundwater is released. From the perched water table, water is released via a spring; from the main water table, water is released via a stream.

Questions

1. What is an aquifer?
2. What principal condition produces an artesian system?

Groundwater Movement

The rate of groundwater flow through an aquifer is directly proportional to the aquifer's permeability, but there's another factor that affects flow rate—the hydraulic gradient. To understand how the hydraulic gradient affects flow rate, you need to know what hydraulic head is. *Hydraulic head* is the height to which water rises in a well. In an unconfined aquifer, that height equals the height of the water table. *Hydraulic gradient* is the difference in hydraulic head between two points divided by the horizontal distance between those points (Figure 24.8).[3]

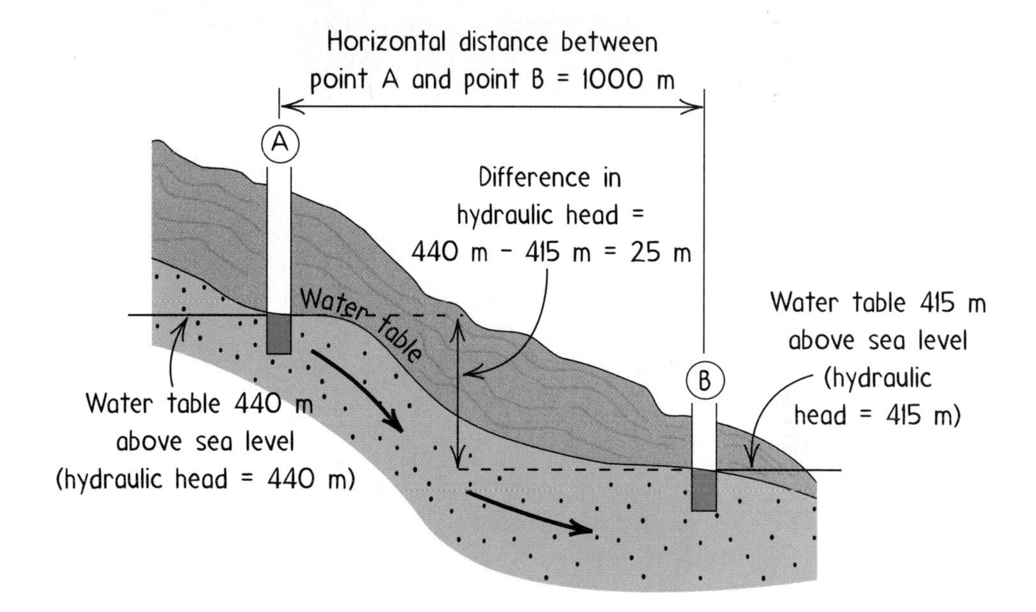

Fig. 24.8
The hydraulic gradient is the difference in hydraulic head between any two locations divided by the horizontal distance between the locations. In this example, we have (440 m − 415 m)/1000 m.

Answers

1. An aquifer is a body of rock or sediment through which groundwater moves easily .
2. An artesian system forms when an aquifer confined between two beds of low permeability and under sufficient pressure is tapped into, either naturally or by a human-made well, and water can rise above the top of the aquifer.

[3]The rate of groundwater flow in relation to hydraulic head is stated in the formula called *Darcy's law,* developed in 1856 by the French engineer Henri Darcy:

$$q = K \frac{\Delta h}{\Delta l}$$

where q is the flow rate; K is the measure of permeability, or *hydraulic conductivity;* Δh is the change in hydraulic head; and Δl is the change in horizontal distance.

The flow of groundwater depends on several geological conditions. Most water flows through pore spaces in rock. This movement is influenced by the force of gravity. Groundwater flows "downhill" underground, but the path it takes depends on differences in hydraulic head: Water flows from high head to low head. In unconfined aquifers, the water table tends to be a subdued replica of the topography, as mentioned earlier. Therefore, hydraulic head, which equals the water table height in unconfined aquifers, is generally higher beneath hills and lower beneath stream valleys, as Figure 24.9 shows. So, responding to the force of gravity, water moves from regions where the water table is high to regions where the water table is low.

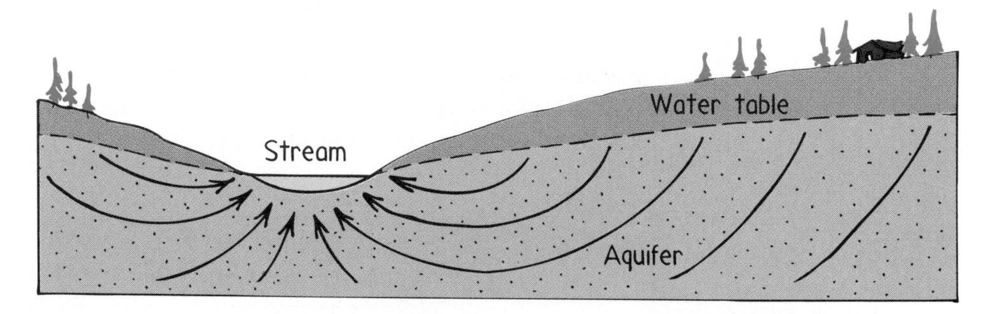

Fig. 24.9
Groundwater flows from a high-hydraulic-head area, such as beneath a hill, to a low-hydraulic-head area, such as beneath a stream valley. The curved arrows indicate flow, which show the stream is fed from below.

The speed of groundwater movement is generally very low. The more permeable the aquifer, the faster the flow; the greater the hydraulic gradient, the faster the flow. The speed and route of groundwater flow can be measured by introducing dye into a well and noting the time it takes to travel to the next well. In most aquifers, groundwater speed is only a few centimeters per day, enough to keep underground reservoirs full.

Mining Groundwater

The most common way to remove groundwater is by drilling an opening into an aquifer—a well. In an unconfined aquifer a well fills with water up to the water table level. As water is withdrawn, the water table immediately surrounding the well is lowered, creating a depression in the water table. The size of the depression depends on the amount of water withdrawn from the well. Generally, domestic wells don't cause problems, but if the well is used for irrigation or industrial purposes, the depression can be quite steep and wide.

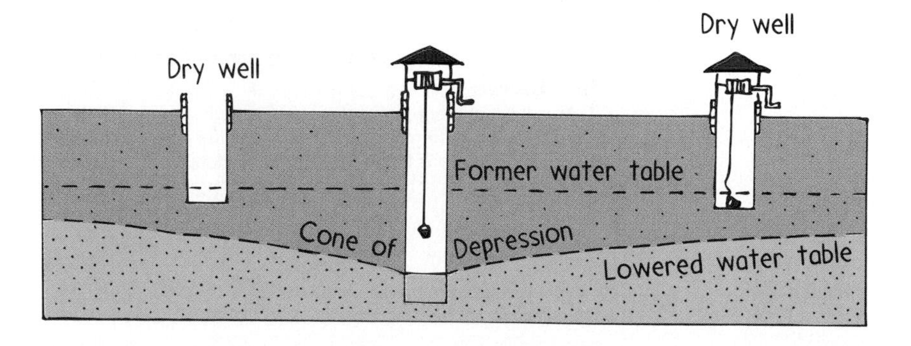

Fig. 24.10
The water table depression around a pumping well takes the shape of a cone. This cone is almost imperceptible for low-volume domestic wells but can be very steep and wide for wells used for irrigation or industrial purposes. Excessive pumping may lower the water table and cause shallow wells to dry up.

Although the reservoir of groundwater is large, when the pumping rate exceeds the water recharge rate, there can be a problem. Recall that the supply of fresh water depends on rainfall. As populations grow, the demand for water grows. In areas such as

Table 24.1
Water Resource Residence Times

Location	Average Residence Time
Atmosphere	1–2 weeks
Ocean	
Shallow depths	100–150 years
Deep depths	30,000–40,000 years
Continents	
Rivers	2–3 weeks
Lakes	10–100 years
Shallow groundwater	up to 100s of years
Deep groundwater	up to 1000s of years
Glaciers	10,000–20,000 years

the Pacific Northwest, the climate is wet and the rate at which water is being taken from the ground is balanced by the recharge rate. In dry areas, though, such as southern California and the High Plains, water recharge is very slow. Such areas are very dependent on the water reserves in the ground. The Ogallala aquifer, located under the High Plains, has supplied water to this thirsty agricultural region for more than a hundred years. Withdrawal has so greatly exceeded recharge that it will take thousands of years for the water table to recover to its original level. In certain parts of Texas, the water table has dropped so low that water has effectively become a nonrenewable resource and is literally being "mined". Likewise for many of the world's drier areas.

Recharge of an aquifer depends on the amount of rainfall and the nature of the subsurface material. It's also related to the length of time water spends in one place. Table 24.1 provides an illustration of *residence time,* the *average* time a water molecule spends in a region. As residence time varies, the time required to complete the hydrologic cycle varies. Water that resides in polar ice and glaciers has both a long residence time and a long time before completing the hydrologic cycle. The residence time of thousands of years for deep groundwater means, for all practical purposes, that this water is nonrenewable. Much of it was accumulated thousands of years ago, perhaps under wetter conditions. Just as coal and petroleum are called "fossil" fuels, such deep groundwater is called "fossil" water. Water is being recognized more and more as the most precious of our natural resources.

Fig. 24.11
The Leaning Tower of Pisa. Construction began about 1173 and was suspended when builders realized the slightly more than 2-m-deep foundation was inadequate. Work was later resumed, however, and the 60-m tower was completed 200 years later. Deviation from the vertical is about 4.6 m. The tower's foundation has been recently stabilized by groundwater withdrawal management, and so the tower should remain stable for years to come.

Land Subsidence

In areas where groundwater withdrawal has been extreme, the land surface is lowered—it *subsides.* Land subsidence is most pronounced in areas underlain by a thick sequence of poorly consolidated sediments, forming an interbedded system of aquifers and low-permeability layers. A large portion of these systems consist of highly compressible, water-bearing clay layers of very low permeability. As water is pumped from the aquifers, water slowly leaks out of the clay layers to replenish the aquifers, which usually continue to be pumped. As the clays become dewatered, they compact, and then the land subsides. Probably the most well-known example of land subsidence is the Leaning Tower of Pisa in Italy, built on the unconsolidated floodplain sediments of the Arno River. Over the years, as groundwater has been withdrawn to supply the growing city, the tilt of the tower has increased. Another region where land subsidence is evident is Mexico City, built in the middle of an ancient shallow lake. The withdrawal of groundwater beneath this city now finds many "street-level" buildings at basement level. Some areas have subsided by as much as 6–7 meters. In the United States, extensive groundwater withdrawal for irrigation in the San Joaquin Valley of California has caused the water table to drop 75 meters in 20 years, lowering the land surface by as much as 9 meters. Because corrective steps have been taken and water for irrigation is now provided by canals, the aquifer is slowly recharging.

■ **Question**

Why is land subsidence most evident in regions where the underlying geology is interbedded systems of aquifers and low-permeability layers?

24.3 Groundwater and Topography

The vast carbonate deposits that underlie millions of square kilometers of the Earth's surface provide storage areas for groundwater. The effect of groundwater on limestone and other carbonate rocks is unique, with some interesting results. Rainwater naturally reacts with carbon dioxide in the air and soil to produce carbonic acid. When this slightly acidic rainwater comes in contact with carbonate rocks, the carbonic acid partially dissolves the rocks into calcium, which is then carried away in solution. As groundwater steadily dissolves the limestone and other carbonate rocks, it creates unusual erosional features. It is in carbonate rocks that we find the only true underground rivers—in other rocks and soils underground water is found only in pore spaces, not in large, open channels.

Caverns and Caves

The dissolving action of subterranean water has carved out magnificent underground caverns and caves (a cavern is simply a large cave). As rainwater (enriched in carbonic acid) infiltrates into limestone, it moves through tiny fractures in the rock. Groundwater flow in carbonate aquifers is dominated by fracture flow. As groundwater flows naturally toward an outlet, usually a stream, as shown in Figure 24.12a, the slightly acidic water dissolves the surrounding limestone, causing the fractures to become larger. This eventually creates an underground channel. The stream that the groundwater is flowing toward also dissolves and erodes its stream channel, causing the water level in the stream and the aquifer (in other words, the water table) to drop. This lowering of the water table drains the previously filled channels in the rock, forming caves. The dropping water table also allows water in the main cave channel to seep downward to begin a new level (Figure 24.12c).

Dripping water, rich in dissolved calcium carbonate, trickles down from the cave ceiling, creating icicle-shaped stalactites as water evaporates and carbonate precipitates. Some solution drips off the end of the stalactites to build corresponding cone-shaped stalagmites on the floor. As caves enlarge and develop into interconnecting chambers, they are called caverns. One of the most impressive caverns in the United States is Carlsbad Caverns in southeastern New Mexico. The cavern descends to a maximum depth of 486 meters. Other famous caves and caverns include Mammoth Cave in Kentucky, Adelsberg Cave in Austria, and Good Luck Cave in Borneo.

■ **Answer**

Clay layers become dewatered and compact as water is pumped from the adjacent aquifers. Compaction of the clay layers and of the aquifers causes the land to subside.

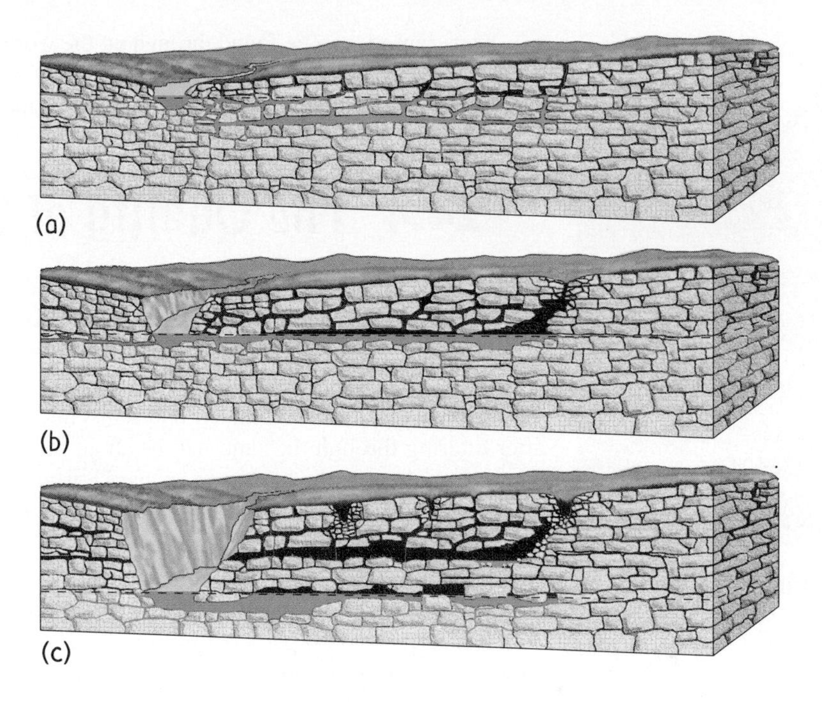

Fig. 24.12
The formation of a cave begins with a layer of carbonate rock, mildly acidic groundwater, and an enormous span of time. (a) Groundwater makes its way toward a stream. (b) As the stream valley deepens because of erosion, the water table is lowered. The carbonate rock is eaten away as acidified water erodes and enlarges the existing fractures into small caves. (c) Further deepening of the stream valley causes the water table to drop even lower; water in the cave seeps downward, leaving an empty cave above a lowered groundwater level.

Fig. 24.13
Cave dripstone formations at Carlsbad Cavern near Hobbs, New Mexico.

Sinkholes

Sinkholes are funnel-shaped cavities open to the sky. They are formed in much the same way as caves—as groundwater dissolves carbonate rock. Some sinkholes are caves whose roofs have collapsed. Some sinkholes are formed by drought conditions or excessive groundwater pumping.

When sinkholes, caves, and caverns define the land surface, the terrain is called *karst topography,* after the Karst region of Yugoslavia, where pronounced weathering and erosion of highly soluble carbonate rocks characterize the landscape. The drainage pattern in this type of landscape is very irregular; streams and rivers disappear into the ground and reappear as springs. Some karst areas appear as soft, rolling hills with large depressions that dot the landscape; the depressions are old sinkholes now covered with vegetation (Figure 24.14). In general, karst areas have sharp, rugged surfaces and thin to almost nonexistent soil as a result of high runoff and dissolution of surface material.

Fig. 24.14
Karst topography covered by vegetation makes up the rolling hills in south central Kentucky.

Fig. 24.15
The karst landscape of China has been an inspiration to classical Chinese brush artists for centuries.

Karst regions can be found throughout the world: in the Mediterranean basin; in sections of the Alps and the Pyrenees; in southern China; and in Kentucky, Missouri, and Tennessee. The beauty of southern China's karst landscape is depicted in Figure 24.15.

24.4 The Quality of Water

The quality of the water we drink as well as of the water in lakes, streams, rivers, and oceans is a crucial factor in the quality of our lives. Rainwater is used as the standard of water purity. Most of our water supply is of good quality, as good as rainwater. There are important variations though.

If rain falls through clean air and soaks into a bed of quartz sand, the water quality after filtering through the sand will be about the same as before filtration. On the other hand, groundwater in a karst area is "hard" from dissolved calcium bicarbonates. These dissolved substances can affect the taste of the water as well as its utility. The quality of groundwater depends very much on the type of soil and rock through which it flows.

The amount of dissolved substances in drinking water is generally very small. Good-quality water averages 150 parts per million for total dissolved substances, with an upper limit of 1000 parts per million. The taste of water depends on the type of dissolved substances. Water containing 1000 parts per million of dissolved calcium tastes fine, but water containing 200 parts per million of sodium chloride tastes salty. Several other dissolved substances, many introduced by human activities, have a strong effect on the quality of water; some are beneficial to health, while others can be quite dangerous. Added fluoride, for example, helps reduce tooth decay. Zeolite minerals in water filters softens hard water by chemically exchanging calcium in solution for sodium. In contrast, lead and arsenic, two naturally occurring minerals, can make water unsafe to drink even if present in minute amounts. Bacteria from sewage is also a cause of contaminated water.

Fig. 24.16
In hard water, dissolved calcium prevents soap from developing a sudsy lather and also from rinsing clean. Bathtub rings are more prevalent in areas that have hard water.

Question

How is hard water softened?

Water-Supply Contamination

The primary source of water supply contamination is human activity. As rivers and streams receive discharge from factories, sewage, and chemical spills, surface water is polluted. Because surface waters are linked to groundwaters, what affects one can easily affect the other. The groundwater supply can be adversely affected by a variety of sources, including septic tank disposal systems and sewage treatment facilities, agricultural fertilizers, municipal landfills, toxic-waste and hazardous-waste landfills, leaking underground storage tanks, and chemical and petroleum product spills.

The most common source of groundwater contamination is sewage. Sewage includes drainage from septic tanks, inadequate or broken sewer lines, and barnyard

Answer
Hard water is softened by removal of dissolved calcium. This is accomplished by passing it through a zeolite filter, which absorbs calcium ions while releasing an equivalent number of sodium ions. Unlike hard water, soft water allows the formation of soap suds that can readily be rinsed off.

wastes. Sewage water contains bacteria that, if untreated, can cause waterborne diseases such as typhoid, cholera, and infectious hepatitis. Sewage contamination can be treated naturally. If the contaminated water travels through sediments that contain large pores, such as gravel, or through rocks having large openings, such as cavernous limestone, it can travel long distances in short periods of time without much change. On the other hand, if it travels through sediments that contain small pores, such as sand, the flow takes longer but the water can be purified within short distances. The sand separates the bacteria from the water and chemical reactions remove many contaminants. Sand is a good filter for bacteria and viruses and is often used in sewage treatment plants to purify the water.

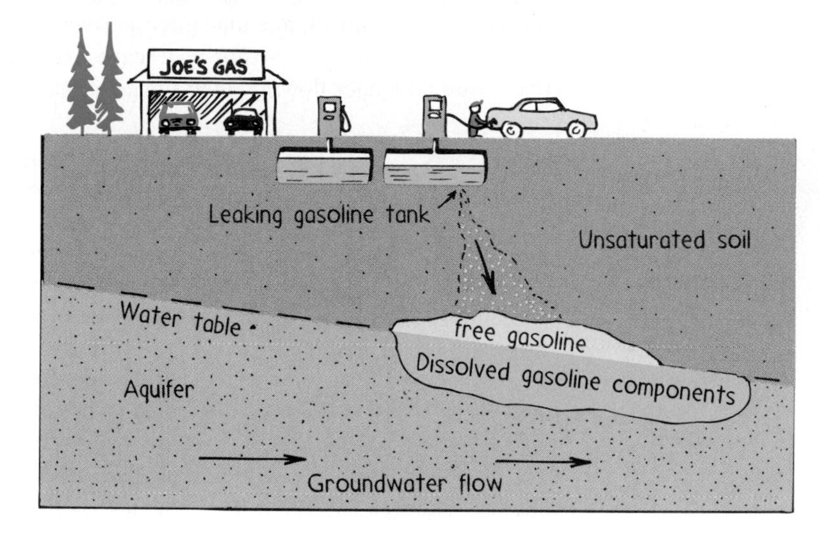

Fig. 24.17
A contaminant plume spreads in the direction of groundwater flow. In the case of a petroleum product, the specific gravity of the contaminant is less than that of water and so the contaminant floats on the water table. The concentration of contaminant decreases with increasing distance from the source.

Agricultural areas where a lot of nitrate fertilizers are used also contribute to groundwater contamination. Recall from Chapter 19 that nitrates are very soluble in water. Nitrate fertilizer spread over the land is used by plants, but the excess percolates down into the groundwater as a contaminant. Nitrate levels in groundwater have to be closely monitored because these compounds in amounts as small as 15 parts per million are toxic to humans.

As a population grows, so does its garbage. The most common means of waste and refuse disposal is burial in a landfill. Even radioactive, toxic, and hazardous wastes are buried. The location of underground storage sites is tricky to decide. For a site to be considered safe, it must be located where waste products and their containers cannot be affected chemically by water, physically by Earth movements, or accidentally by people. Precipitation infiltrating the site may dissolve a variety of compounds from the solid waste. The resulting liquid, known as **leachate**, can move vertically downward from the landfill to the water table and contaminate the groundwater. When leachate mixes with groundwater, it forms a plume that spreads in the direction of the flowing groundwater. Corrective steps can be taken. To reduce the chances of groundwater contamination, the landfill can be capped with layers of compacted clay soil or a synthetic membrane to prevent generation of leachate. It can also be lined with the same material, plus a collection system to catch any draining leachate and prevent its distribution.

Groundwater contamination also occurs as a result of spills and leaks of toxic and hazardous chemicals. These discharges can be sudden, as in a train or tanker truck accident, or as a result of slow leakage from a holding container. If the contaminant is soluble in water, it dissolves in the groundwater and flows along with it. If less dense than water, it floats on the water table (Figure 24.17). If more dense, it sinks through the water to a low-permeability layer.

24.5 Surface Water and Drainage Systems

The Grand Canyon is testimony to the mighty power of the Colorado River. For millions of years, the river has been carving out the canyon walls, cutting deeper and deeper into the rock as it makes its way to the ocean.

Streams—by which we mean all flowing water, from the Mississippi River to the shallowest woodland creek—are dynamic systems that impact both the surface of the land and the people who live on that land. Streams carve out and alter the landscape. They also provide energy, irrigation, and a means of transportation for people. What better place to set up a home than a fertile river valley? The force of a stream is indeed a powerful instrument of erosion as it leaves its mark on the land surface. Even where streams are no longer flowing, their impact remains.

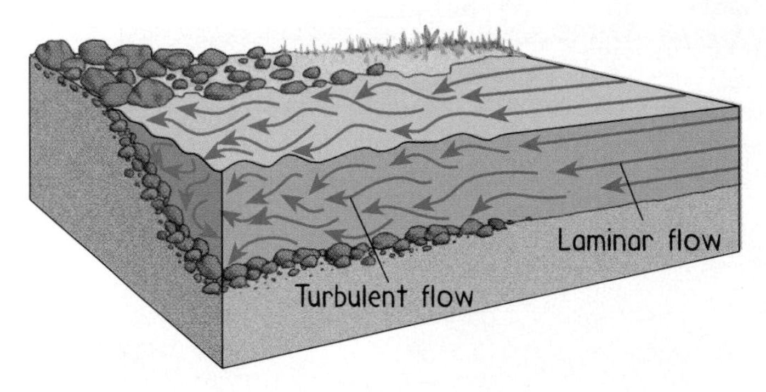

Fig. 24.18
Laminar flow is slow and steady, with no mixing of sediment in the channel. Turbulent flow is fast and jumbled, stirring up everything in the flow.

Stream Flow Geometry

As it falls on land, rain begins a complex journey back to the ocean. Some of it percolates into the ground, some evaporates back into the atmosphere, and some runs off into streams.

Streams come in a variety of forms—straight or curved, fast or slow. At their headwaters (the place where a stream originates), stream channels are narrow and water flows through deep, V-shaped mountain valleys. Farther downstream, channels widen so that water flows into and along broad, low valleys.

The flow pattern of moving water is of two types—turbulent and laminar (Figure 24.18). When water moves erratically downstream, stirring everything it comes in contact with, the flow is **turbulent**. When water flows steadily downstream with no mixing of sediment, the flow is **laminar**. In general, slow, shallow flows tend to be laminar and faster moving flows tend to be turbulent. Whether a flow is laminar or turbulent depends on the nature and geometry of the stream channel and the speed of the flow.

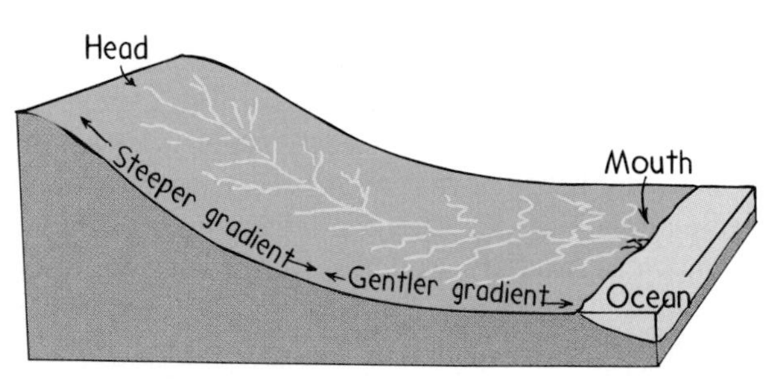

Fig. 24.19
The long profile of a stream. At a stream's headwaters, the gradient is high, the channels narrow and shallow, and the stream flow rapid. As the stream progresses downslope, the gradient decreases, the channel widens, and discharge increases.

There are three variables that influence stream flow speed—stream gradient, channel geometry, and discharge. The *gradient* of a stream is the ratio of the vertical drop to the horizontal distance for that drop. If we look at a long profile of a stream (Figure 24.19), we see that the gradient is steep near the stream's head and gentler, almost horizontal, near its mouth. Because of gravity, stream speed tends to be greater where the gradient is steep, but downstream, discharge and channel geometry also influence stream speed.

Discharge is the volume of water that passes a given point in a channel in a given amount of time.[4] It is directly related to the cross-sectional area of the channel and to the *average* stream speed.

$$\underset{\text{(m}^3\text{/s)}}{\underset{\textbf{Discharge}}{Q}} = \underset{\text{(m}^2\text{)}}{\underset{\textbf{cross-sectional area}}{A}} \times \underset{\text{(m/s)}}{\underset{\textbf{average speed}}{v}}$$

[4]Discharge can be determined by multiplying the channel's dimensions by the stream's speed:

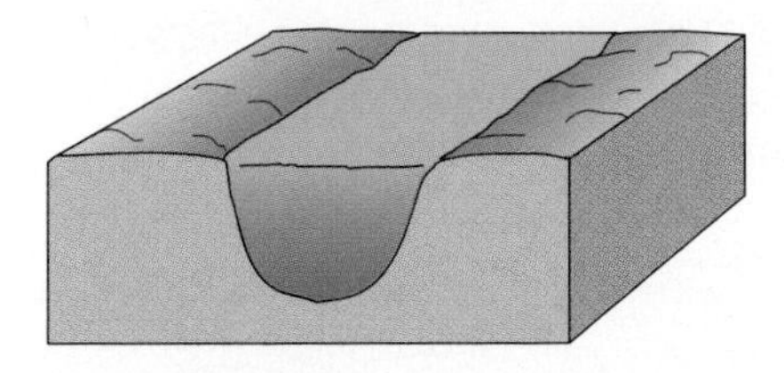

(a) Rounded, deep channel

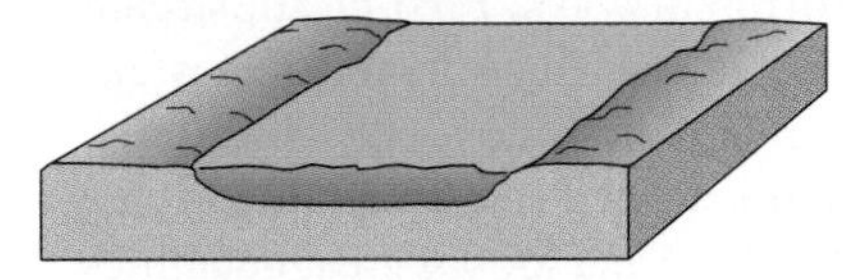

(b) Wide, shallow channel

Fig. 24.20
(a) In a rounded, deep channel, the water flow speed is relatively high because there is relatively less water in contact with the channel. (b) Wide, shallow channels tend to have slower flows because more water is in contact with the channel.

Channel geometry, the shape and dimensions of a stream channel, greatly influences stream speed. Consider two streams that have the same discharge. If one of the streams is larger, in other words it has a larger cross-sectional area, it will have a slower stream speed than a smaller stream.

The shape of the channel also affects stream speed. Now consider two streams that have the same cross-sectional area. Water flowing in the channel touches the channel bottom and sides. Friction between the water and channel slows water speed. The greater the contact area, the greater the friction. The cross-sectional shape of a channel determines the amount of water in contact with the channel and hence the amount of frictional drag (Figure 24.20). If the stream channel is rounded and deep, as opposed to flat-bottomed and relatively shallow, the stream speed will be faster because there is less channel contact.

When either channel geometry or discharge changes, so does stream speed. In the headwaters, the gradient is high, channels are narrow, and stream speed is high. In fact, these upland sections of a stream are often called "rapids." As the stream progresses downslope, the gradient gradually decreases, the channel typically widens, and, as other streams feed into the main one, discharge increases. The water's speed, which depends on all three of these variables, may or may not increase but it does change. Intuitively, we might expect speed to be directly proportional to discharge. In other words, if we double discharge, we expect speed to double. This relationship does not always hold, however, because channels typically become wider as discharge increases, which increases the cross sectional area. So we have a condition where higher discharge tends to increase the speed while at the same time the wider channel tends to decrease the speed. As a result, the net increase in speed is often less than the higher discharge would lead us to expect. In fact, water speed may even decrease downstream.

So far we have been talking about only those variations in stream speed that occur as we move downstream. Water speed also varies at a given location within a channel, however. Flow speed is lower along the stream bed, where water is in contact with the channel, and highest near the water surface. In a large stream flowing in a straight channel, the maximum flow speed is found midchannel (Figure 24.21b), while in a stream running through a bent, looping channel, the maximum flow speed is found toward the outside of each bend (Figure 24.21a and c).

Fig. 24.21
In a stream that bends (a and c), maximum flow speed is toward the outside of each bend and slightly below the surface. In a straight-channel stream (b), maximum speed is midchannel and near the water's surface. Erosion of the stream channel occurs where stream speed is greatest (cut bank); deposition occurs where stream flow slows (point bar).

Question

Stream discharge generally increases downstream as more and more tributary streams feed into the main channel. Does stream speed always increase downstream?

Erosion and Transport of Sediment

Moving water erodes stream channels in several different ways. First of all, stream water contains many dissolved substances that chemically weather the rocks they encounter. Another powerful mechanism for erosion is hydraulic action—the sheer force of running water. Swiftly flowing streams and streams at flood stage have great erosive power as they break up and loosen great quantities of sediment and rock. The most powerful type of erosion, however, is abrasion, where sediments and particles physically scour a channel, much like sandpaper on wood. When powered by turbulent spiraling water, rock particles can rotate like drill bits to carve out deep potholes. The faster the current, the greater the turbulence and the greater the erosion.

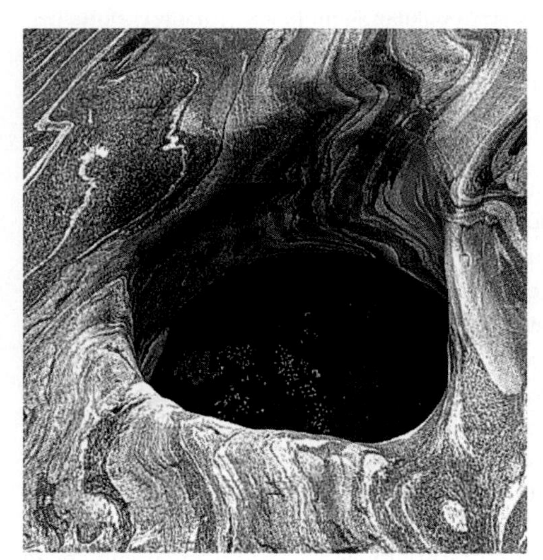

Fig. 24.22
When powered by turbulent circular currents of water, rock particles rotate like drill bits and carve out deep potholes.

Erosion is only the beginning of the story, however, for streams carry much more than just water—they transport great quantities of sediment. In general, laminar flows can lift and carry only the very smallest and lightest particles. A turbulent flow, however, depending on its speed, can move and carry a range of particle sizes—from the smallest particles of clay to large pebbles and cobbles. A turbulent current gathers and moves particles downstream mainly by lifting them into the flow or by rolling and sliding them along the channel bottom. The smaller, finer particles are easily lifted into the flow and remain suspended to make the water murky. As we would expect, the faster the current, the larger the particles that can be carried; and the larger the volume of water, the greater the sediment load. So, streams that have a higher discharge can carry larger volumes of sediment, and streams in which the water is moving fast can carry larger sizes of sediment. The continuous abrasion of sediment in the stream channel leads to an overall decrease in particle size as we progress downstream. At the river's mouth, only finer particles of sand, silt, and clay remain. As we shall soon see, these tiny particles are deposited to form deltas when the stream loses speed as it enters the sea.

Question

Which is more effective in transporting sediment, laminar or turbulent flow? Why?

Answer
Not always! If the cross-sectional area of the stream channel increases enough to compensate for the additional water, the speed may not change or may even decrease.

Answer
Turbulent flow, because in turbulent flow the water motion is irregular and sediments have a greater tendency to remain in suspension. In laminar flow, water moves steadily in a straight-line path with no mixing of sediment in the channel.

Fig. 24.23
At a stream's headwaters, high gradients contribute to fast-moving rapids. When there is an abrupt change in gradient, we see a beautiful cascading waterfall.

Stream Valleys and Floodplains

As rainfall hits the ground, it loosens soil and washes it away. As more and more rain falls and the ground continues to lose soil, a gully forms. Once water and soil particles funnel into such a gully, a stream channel is created. This erosive action may be extremely rapid, as in the erosion of unconsolidated sediments, or very slow, as in the erosion of solid rock. Its erosive power enables a stream to widen and deepen its channel, transport sediment, and, in time, create a valley. In high mountain areas, the erosive action of a stream cuts down into the underlying rock to form a narrow V-shaped valley. Because the valley is narrow, the stream channel takes up the whole valley bottom. Fast-moving rapids and beautiful waterfalls are characteristic of V-shaped mountain stream valleys.

As a stream flows downhill and its gradient becomes gentler, the focus of its energy changes from eroding downward (deepening the channel) to eroding laterally in a side-to-side motion. As a result of this lateral action, the stream develops a more sinuous form (Figure 24.24b). As the stream bends and curves, the flow speed in the channel shifts with the maximum velocity toward the outside of each bend, as we learned earlier (Figure 24.21). Rapidly moving water is very effective in eroding material from the outside of the bend, creating a steep bank called a *cut bank.* Material eroded from the cut bank is transported downstream, where it may eventually be deposited in areas where stream speed decreases. Sandy *point bars* form on the insides of bends by this process. As the stream continues to modify its profile and channel, it also widens the stream's valley. Farther downstream, sinuosity increases and the stream develops a meander pattern that winds back and forth further widening the valley into a broad, low-lying area called a *floodplain* (Figure 24.24b and c).

Floodplains are so named because they are the sites of periodic flooding. In a flood, as discharge and flow speed increase, so does the stream's capacity to carry sediment. Thus when a stream overflows its banks, sediment-rich water spills out onto the floodplain. Because the speed of the water quickly decreases as it spreads out over the floodplain, a progression of coarse to fine particles are deposited. As expected, larger, coarse-grained sediments are deposited along the edges of the channel and smaller, fine-grained sediments are deposited farther away from the stream channel on the floodplain. The coarse materials deposited close to the stream channel act as *natural levees* that help to confine future floodwaters (Figures 24.24c and 24.25).

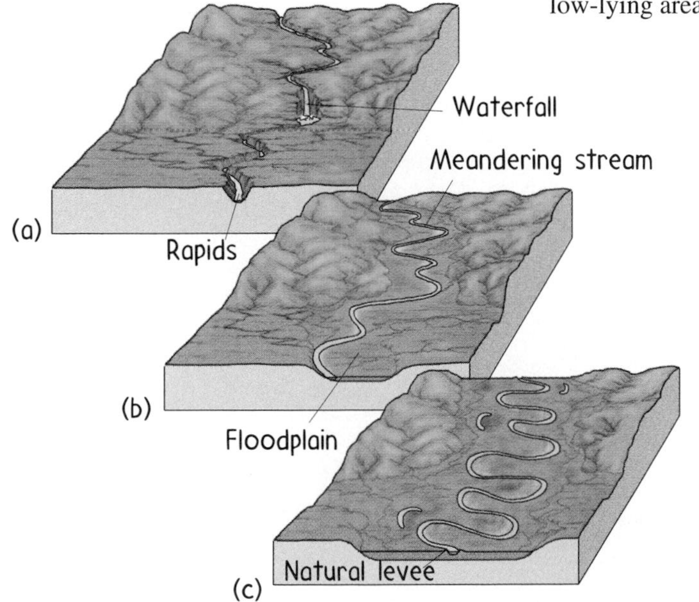

Fig. 24.24
The evolution of a stream valley and development of a floodplain. (a) At the headwaters, the V-shaped stream valley is characterized by steep gradients and fast-moving water that cuts down into the stream channel. Features in this area include cascading rapids and waterfalls. (b) Downstream, with reduced gradient, the stream focuses its erosive action in a side-to-side sinuous manner, thereby widening the stream valley. (c) Farther downstream, meandering increases and further widens the stream valley to form a large floodplain.

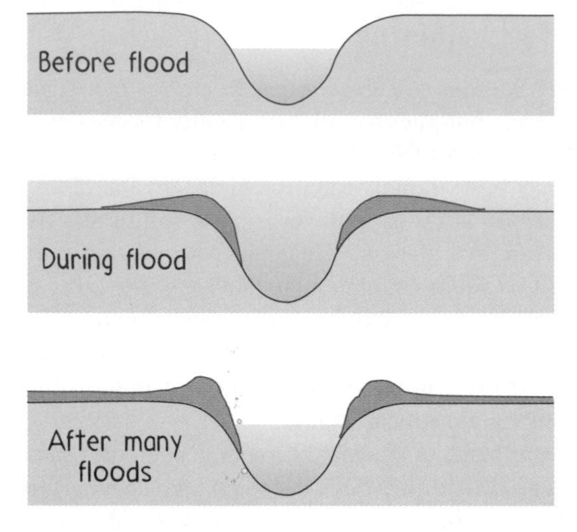

Fig. 24.25
In a flood, increased discharge and flow speed enable a stream to carry not only a large sediment load but also larger particles. Larger, coarse-grained sediment deposited close to the stream channel forms natural levees that confine the stream between flood stages. Successive floods increase the height of the levees and may even raise the overall elevation of the channel bed so that it is higher than the surrounding floodplain.

Question

Floodplains are often prime agricultural areas. Why would people want to work and live in areas so prone to flooding?

Drainage Networks

A stream is one small segment of a much larger system—a *drainage basin,* defined as the total area that contributes water to a given stream. Drainage basins are separated from one another by *divides,* lines tracing out the highest ground between streams. A drainage basin can cover a vast area or be as small as 1 square kilometer. A divide can be either very long, if it separates two enormous drainage basins, or a mere ridge separating two small gullies. The *Continental Divide,* a continuous line running north to south down the length of North America, separates the Pacific basin on the west from the Atlantic basin on the east. Water west of the Divide eventually flows to the Pacific Ocean, and water east of it flows to the Atlantic Ocean (Figure 24.26).

Streams merge with other streams as they flow downhill, becoming larger and larger. The entire assembly of streams draining a region is called a *drainage network* and can be characterized by the branching pattern formed by the streams (Figure 24.27). Because streams erode the land surface and hence erode the rocks and rock material on the land, a drainage pattern is greatly influenced by the rock and rock material eroded. The name for the most common type of drainage pattern—*dendritic*—is from the Greek word for tree, dendron; dendritic drainage patterns are typically devel-

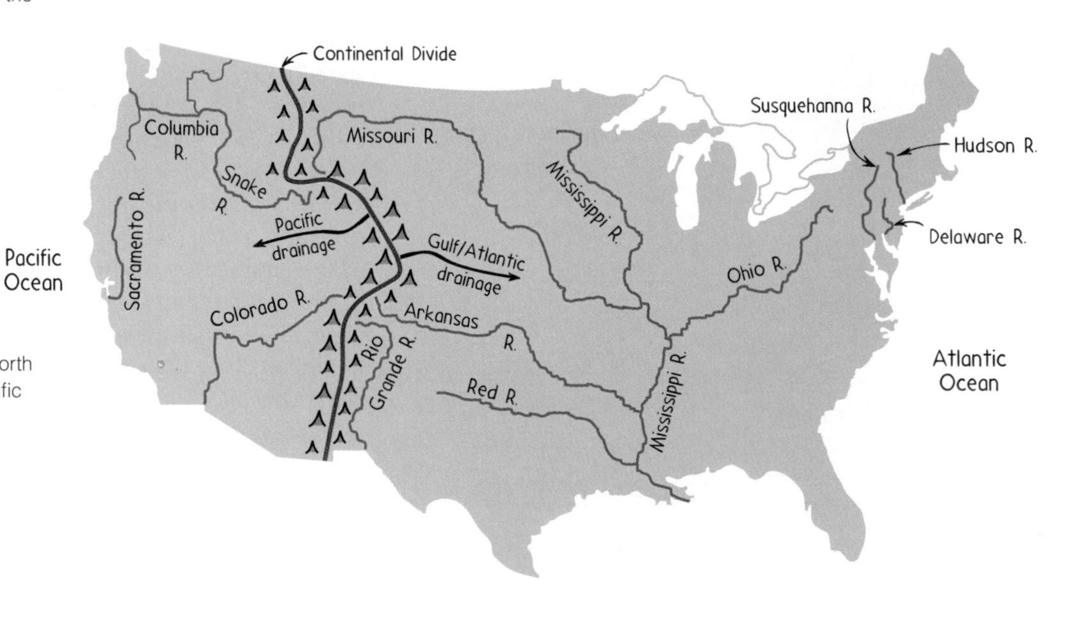

Fig. 24.26
The Continental Divide in North America separates the Pacific basin on the west from the Atlantic basin on the east.

Answer
People live and work in floodplain areas because such plains are next to rivers that provide the residents easy access to water, food, and a means of transportation. Also, because of periodic inundation of floodwaters, floodplain soils are often extremely fertile and thus serve as prime farmland.

oped in areas where the surface materials have a uniform resistance to erosion. *Rectangular* drainage patterns develop on surfaces that are fractured and jointed. A variation of rectangular drainage is the *trellis* drainage pattern, where stream channels formed in easily eroded rock material are bounded by more resistant rock material. *Radial* drainage patterns develop in areas that have a central high point, such as a volcano.

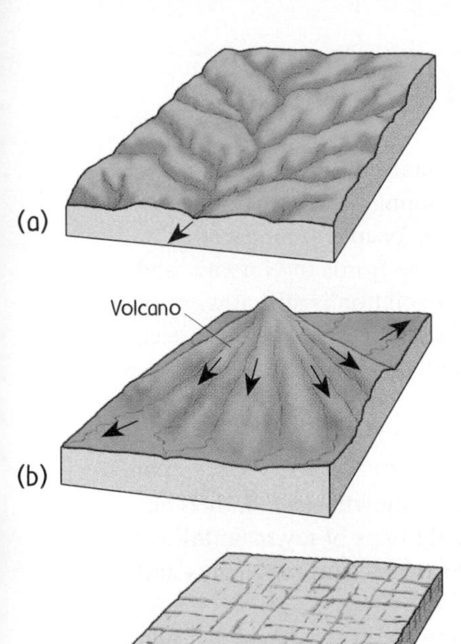

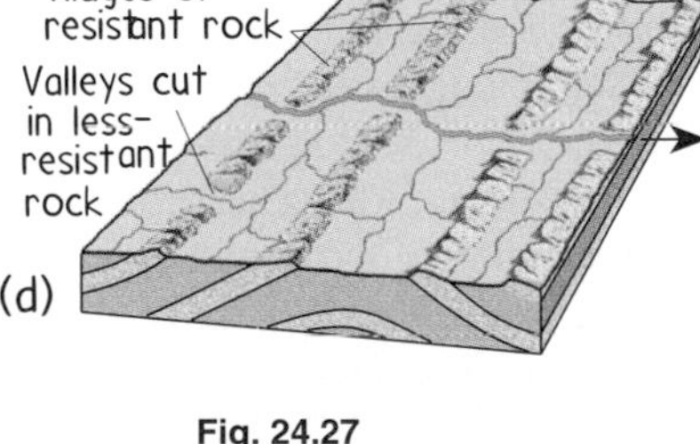

Fig. 24.27
Different drainage patterns develop according to surface material and surface structure: (a) dendritic, (b) radial, (c) rectangular, (d) trellis.

Deltas: The End of the Line for a River

As a stream flows into a standing body of water—for example, a bay—the moving water gradually loses its forward momentum, decreasing the stream's capacity to transport sediment. As the ground beneath the stream's mouth and the ground immediately offshore become filled with sediment, the depositional platform called a **delta** begins to form. Sediment is deposited in order of decreasing weight, with heavy, coarse particles settling at and near the shoreline and light, fine particles settling farther offshore. The continuous supply of sediment thickens the delta so that it builds itself upward and outward as an extension of land out into the bay. Once the delta has formed, stream water arriving at the point that used to be the original, pre-delta mouth of the stream still has some distance to travel over the newly formed land before reaching the bay. The running stream water now creates channels in the delta. These channels are called distributaries, and through them water flows to the bay.

As the delta continues to grow in the bayward direction, the distance water in the distributaries has to travel to reach the bay becomes so great that the stream shifts its course and begins to cut new, shorter pathways to the bay. In this way the distributaries take on the appearance of branching fingers. When the fingers get too long, the process begins again. As streams continue to flow to the sea and as successive beds are deposited one on top of the other, the delta builds itself outward (Figure 24.28). Thus delta environments are areas where new land is continuously created.

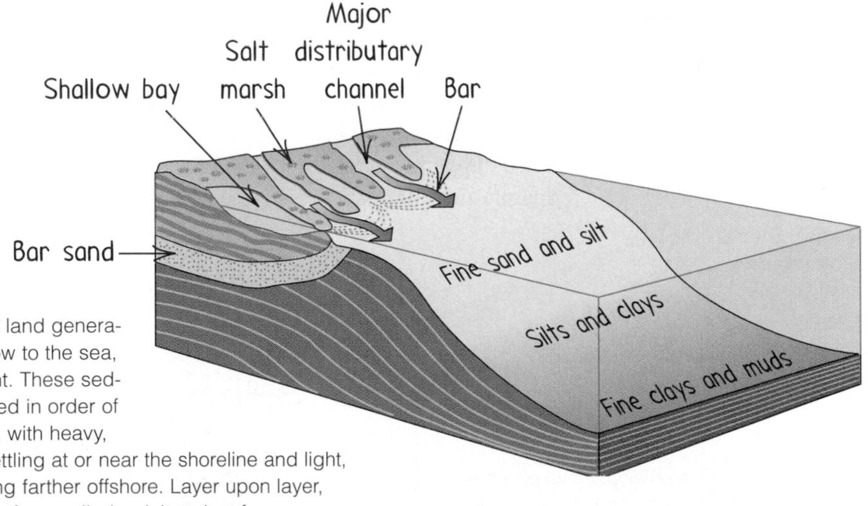

Fig. 24.28
Deltas are areas of land generation. As streams flow to the sea, they carry sediment. These sediments are deposited in order of decreasing weight, with heavy, coarse particles settling at or near the shoreline and light, fine particles settling farther offshore. Layer upon layer, the depositional platform called a delta takes form.

■ **Question**

What is the ultimate destination of all water flow and hence the eventual site of deposition of most sediments?

■ **Answer**
Water flows eventually to the ocean, and sediments to the ocean floor.

24.6 Glaciers and Glaciation

The mightiest rivers on the Earth are frozen solid and normally flow a sluggish few centimeters per day. These great icy currents are **glaciers**, among the most spectacular and powerful agents of erosion. Glaciation has given us the beautiful landscapes of Tibet, Nepal, and Bhutan in Asia; the Alps of Switzerland; the fjords of Norway; and Yosemite Valley and the Great Lakes in North America. Glaciation is still at work in many regions of the world, its agents being small alpine glaciers in mountainous areas, large alpine ice fields, and the huge Arctic and Antarctic continental ice sheets.

Geologist Bob Abrams observes the grandeur of the Juneau Ice Field, Alaska.

Glacier Formation and Movement

The ice of a glacier is formed from recrystallized snow. After snowflakes fall, their accumulation slowly changes the individual flakes to rounded lumps of icy material. As more snow falls, the pressure exerted on the bottom layers of icy snow compacts and recrystallizes it into glacial ice.

This ice does not become a true glacier, however, until it moves under its own weight. This occurs when the ice mass reaches a critical thickness of approximately 50 meters and the pressure exerted by the overlying material causes the ice at the base to deform plastically and flow downslope. This plastic deformation can be likened to what happens to a deck of playing cards. When the deck is pushed from one end, as in Figure 24.29, individual cards slide past one another shifting the entire deck. Plastic deformation of the ice in a glacier is greatest at the base, where pressure is greatest.

Fig. 24.29
When a deck of cards is pushed from one side, the individual playing cards slide past one another, thus shifting the whole deck.

Plastic flow from the slippage of ice crystals is not the only component in glacial movement. The melting point of ice decreases as pressure increases. When melted ice—*meltwater*—forms at the base of the glacier, the process called *basal sliding,* comes into play.[5] This second mechanism of glacial movement results in the sliding of the entire glacier downslope, with the meltwater acting as a lubricant. The net speed of the glacial ice increases from the base up, reaching its greatest values at the glacier's surface (Figure 24.30). Overlying layers are carried along in piggyback fashion as plastic deformation takes place in the lowest layers.

The uppermost portion of the glacier, carried along both by basal sliding and by internal plastic deformation, behaves like a rigid, brittle mass that may fracture. Huge, gap-

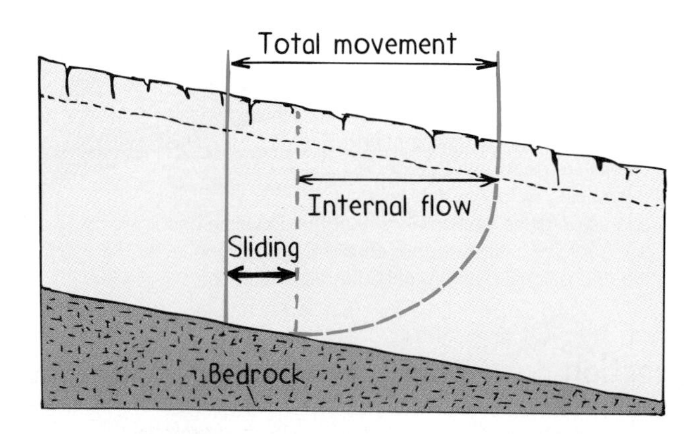

Fig. 24.30
Cross section of a glacier. Glacial movement has two components: Internal flow and sliding resulting from lubrication by meltwater. Movement is slowest at the base because of frictional drag and fastest at the surface. The upper parts of the glacier are carried along in piggyback fashion by plastic flow within the ice.

[5]Meltwater may result from the pressure of the overlying ice (the melting point of ice decreases as pressure increases), from the internal heat of the Earth, or from the generation of heat from frictional drag as the glacier moves. No matter what caused it to form, however, meltwater contributes to movement of the glacier.

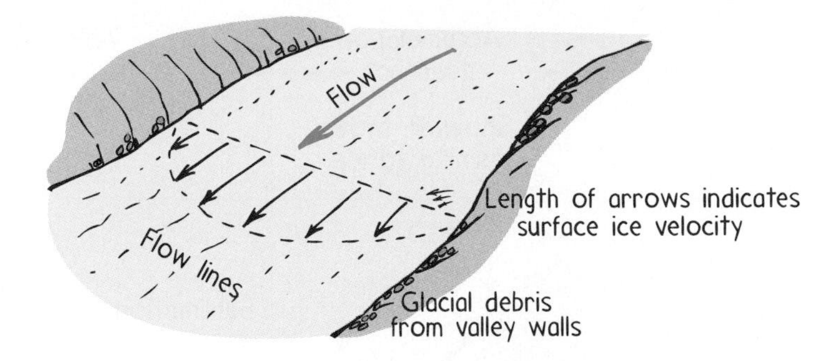

Fig. 24.31
Top view of a glacier. Movement is fastest at the center and gradually decreases along the edges because of friction.

ing cracks called *crevasses* may develop in this surface ice. These can extend to great depths and can therefore be quite dangerous for people attempting to cross a glacier.

Glacier speed is measured by placing a line of markers across the ice and recording their changes in position over a period of time, ranging from days to years. Ice is found to move fastest in the center and slower toward the edges because of frictional drag (Figure 24.31). Average speed varies from glacier to glacier and can range from a few centimeters to a few meters per day. Some glaciers experience surges, periods of much more rapid movement. These surges are likely caused by periodic melting of the base and sudden redistribution of mass. The flow rate in these relatively brief surges can be 100 times faster than the normal rate. Viewed by air, flow bands of rock debris and ice normally have a parallel pattern, but during a surge the flow bands become intricately folded.

(a)

(b)

Fig. 24.32
Glacial flows: (a) normal flow (b) surge flow.

Glacial Mass Balance

From season to season, and over longer periods of time, the mass of a glacier changes. Typically, a glacier grows in the winter as snow accumulates on its surface. The amount of snow added to a glacier annually is aptly termed **accumulation**.

As ice accumulates and begins to flow downhill, it may move to an altitude where temperatures are warmer. Then some ice melts and the glacier loses some of its mass. A glacier may also lose mass as it moves downslope to a shoreline, where ice may break off, or *calve,* to form icebergs that float away to sea. Melting and calving are the two primary mechanisms by which glaciers lose mass. Although less noticeable, glaciers may also lose mass as the ice *sublimates* to water vapor. By whatever means, the total amount of ice lost annually is called **ablation** (Figure 24.33).

When accumulation equals ablation, the size of the glacier remains constant. For example, in a mountain glacier, accumulation occurs with winter snowfall in the farther-back, higher-elevation parts of the glacier, and ablation occurs in the lower portions, where spring and summer meltings are greatest. When accumulation rate and ablation rate are equal, the melting of the lower portions is offset by the downslope flow of ice from higher portions. As a result the location of the front edge of the glacier does not change. When accumulation exceeds ablation, the glacier advances—it grows. When ablation exceeds accumulation, the glacier retreats—it shrinks. Naturally, in all these cases, the ice of the glacier is always flowing downslope.

Question

Under what conditions does the front of a glacier remain at the same location from year to year?

Answer
The front of a glacier remains at the same location when the rate of growth for the year equals the rate of shrinking. In the spring, as ice at the glaciers front melts away, the glacier retreats upslope. At the same time, the increased mass from the past winter's accumulation causes the glacier to move forward. When the rate at which this forward movement matches the rate of melting, the location of the front edge remains constant.

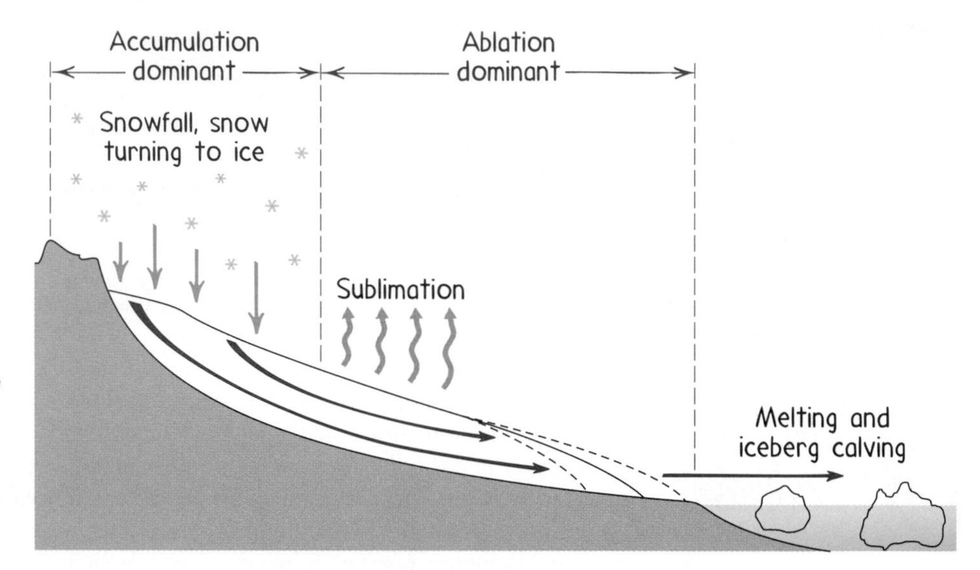

Fig. 24.33
Accumulation on a glacier takes place as snow falls on the glacier and turns to ice at high elevations. Ablation takes place as ice melts and/or calves into icebergs at lower, warmer elevations, or as ice is lost through sublimation.

Glacial Erosion and Erosional Landforms

Fig. 24.34
Striations mark the presence of a former glacier.

In many ways, a glacier is like a plow as it scrapes and plucks up rock and sediment. It is also like a sled as it carries its heavy load to distant places. As it moves across the Earth's surface, a glacier loosens and lifts up blocks of rock, incorporating them into the ice. The large rock fragments carried at the bottom of a glacier scrape the underlying bedrock and leave long, parallel scratches aligned in the direction of ice flow (Figure 24.34). These are called *striations.*

The two main types of glaciers, *alpine* and *continental,* have different erosional effects and produce different landforms. Alpine glaciers develop in mountainous areas and are often confined to individual valleys, while continental glaciers cover much larger areas. An alpine glacier erodes mostly near its head as it cuts deep into bedrock to form a steep-sided, bowl-shaped depression called a *cirque.* After the glacier has melted away, portions of the cirque may become filled with water, forming a small lake called a *tarn.* When two glaciers carve out cirques on opposite sides of a divide, the cutting action steepens the divide to form a jagged, linear ridge called an *arête.* Sometimes alpine glaciers form simultaneously on several sides of a mountain, eating away at it to leave a spectacular jagged peak called a *horn.* As an alpine glacier moves down from a cirque, it carves out and accentuates existing features. For instance, it deepens, widens, and straightens the floor of a V-shaped valley into a characteristic U-shape. Because it contains a much greater amount of ice, the main glacial valley is eroded more deeply than the tributary valleys. After the glaciers have melted away, a tributary valley may lie suspended above the floor of the main valley and is called a *hanging valley.* Beautiful waterfalls are associated with hanging valleys. The erosional features of alpine glaciation are depicted in Figure 24.37.

Fig. 24.35
The Matterhorn—named for its characteristic "horn" feature.

Continental glaciers also erode the land surface, but to less dramatic effect. Continental glaciers are not confined to valleys but spread over all the land surface, smoothing and rounding the underlying topography. Although striations are produced by both types of glaciers, they have played a larger role in studies of continental glaciers. Because continental glaciers scour very large tracts of land, they tend to lack obvious valleys. By mapping striations on land once covered by continental glaciers, geologist can decipher the flow direction of the ice. The direction of ice flow can also be deciphered by small, asymmetrical hills called *roches moutonnées* (Figure 24.38). In the direction from which the ice was coming, the hill's slope is smooth and striated from the abrasion of ice on bedrock. On the downflow side, the slope is rough and steep because the ice plucked away rock fragments from joints and cracks in the bedrock.

Fig. 24.36
Hanging valleys are a spectacular feature found in areas that have been shaped by alpine glacial erosion. Bridalveil Falls in Yosemite National Park spills out of a hanging valley into the larger valley that was once occupied by the main glacier.

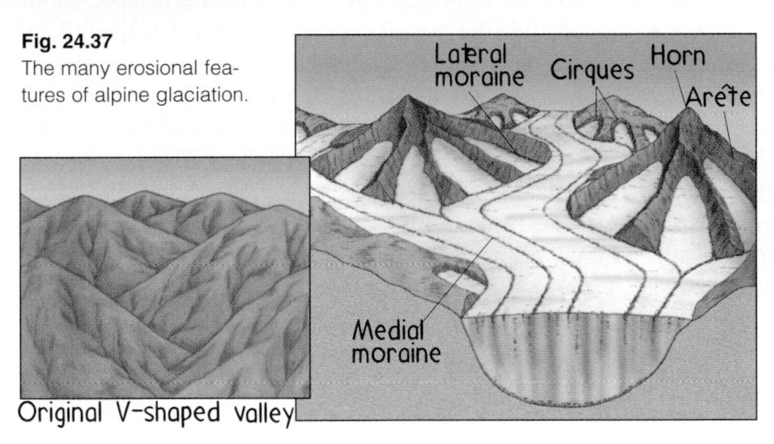

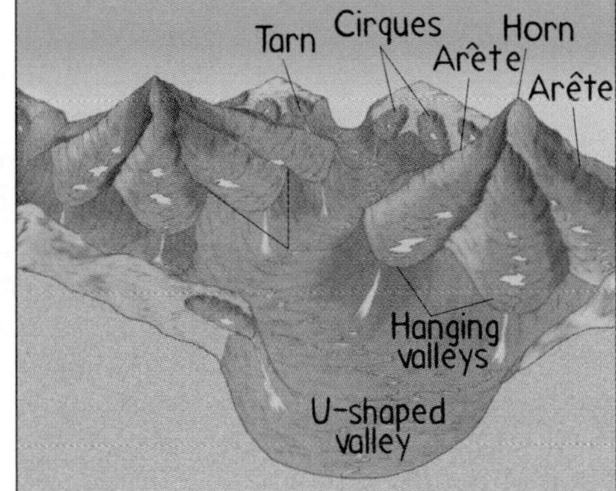

Fig. 24.37
The many erosional features of alpine glaciation.

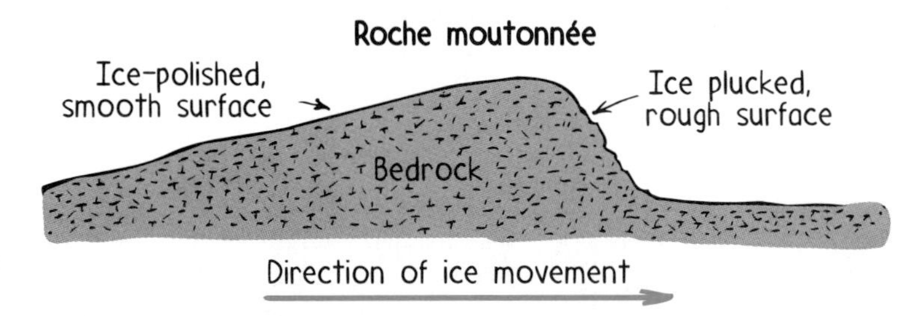

Fig. 24.38
Small asymmetrical hills called roches moutonnées show the direction of continental glacial movement. On the side of the hill facing the approaching glacier, the slope is smooth and gentle. On the side facing away from the approaching glacier, the slope is rough and steep with a plucked appearance.

Glacial Sedimentation and Depositional Landforms

As a glacier advances across the land, it acquires and transports great quantities of debris. When the glacier retreats, this debris is left behind because it is melted out of the ice. Because a glacier abrades and picks up everything in its path, glacial deposits are characteristically composed of unsorted rock fragments in a variety of shapes and sizes. Glacial deposits are collectively called **drift**, a term that dates back to the nineteenth century, when it was conjectured that all such debris had been "drifted in" by the great Biblical Flood.

Drift is deposited in two principal ways. When glacial sediment is released into meltwater, it is carried and deposited like any other waterborne sediment; thus it is well sorted. This type of drift is called *outwash.* Material deposited directly by melting ice—an unsorted mixture of clayey and bouldery rock debris—is called *till.* Many of the old stone walls and fences of New England are found in areas where the surface material is glacial till. Settlers who tried to farm this land had to remove all the larger boulders before they could plow, and piled them along the edges of their fields. A common constituent of till is an *erratic,* a rock of unknown origin. Glacial erratics are of different rock types than the bedrock on which they are found; thus they provide proof of a glacier's ability to transport heavy loads for great distances. If a bedrock outcrop that matches the rock type of the erratic can be found, then the distance and direction of glacial transport can be estimated.

The most common landform created of till is the *moraine,* a ridge-shaped landform that marks the boundaries of ice flow. There are many types of moraines. As glaciers erode material along their sides, lateral (side) moraines are formed. When two lateral moraines of adjoining glaciers merge, a medial moraine forms in the middle of the now larger flow. Of all the different types of moraines, probably the most important is the *terminal moraine,* as it marks the farthest point of a glacier's advance. Another distinctive landform consisting of till is the *drumlin,* an elongated hill shaped like the back of a whale. Formed by continental glaciation and aligned in the direction of ice flow, drumlins have a steep, blunt end in the direction from which the ice came and a tapered gentle slope on the down flow side (Figure 24.39). The most famous drumlin is Bunker Hill in Massachusetts.

Deposits of outwash from glacial meltwaters take a variety of forms. *Kames* (small hills of sand and gravel) and *eskers* (long, winding ridges of sand and gravel) were formed by meltwater flowing on, within, beneath, or away from the margins of continental glaciers (Figure 24.40). Deposited in separate layers, kames are often exploited as a commercial source of sand and gravel.

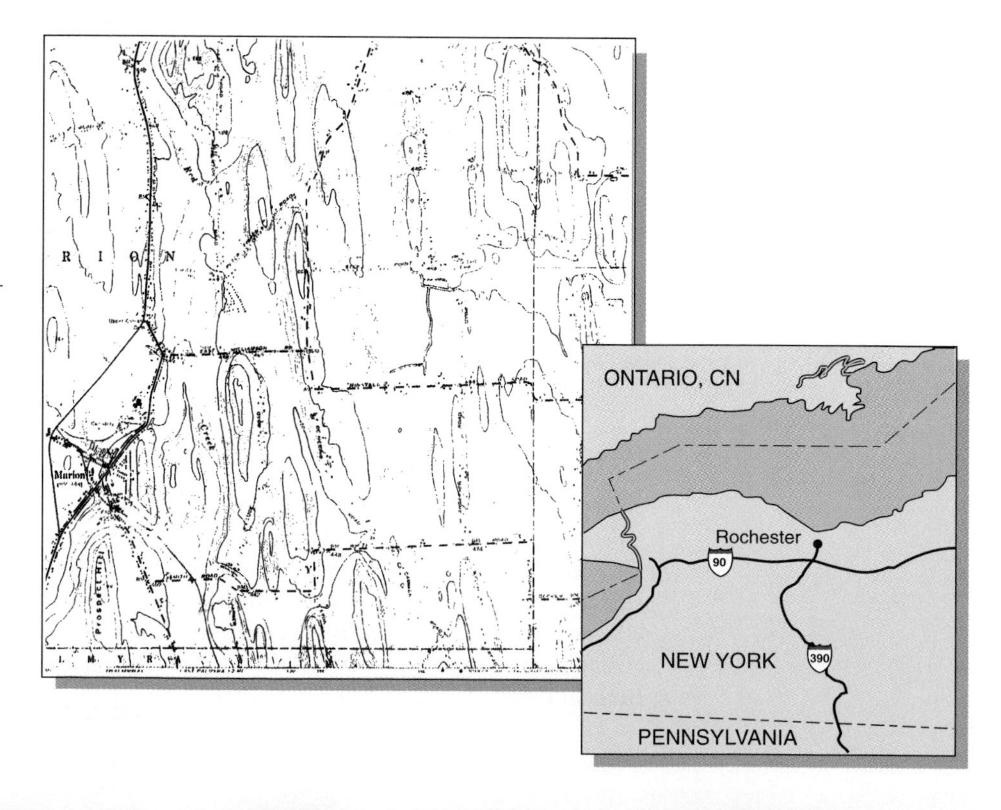

Fig. 24.39
Topographic map showing numerous oval-shaped drumlins in upstate New York. Drumlins are steep and blunt on the side that faced the approaching glacier but tapered and gently sloping on the downflow side. Looking at the map, can you tell the direction of continental ice flow?

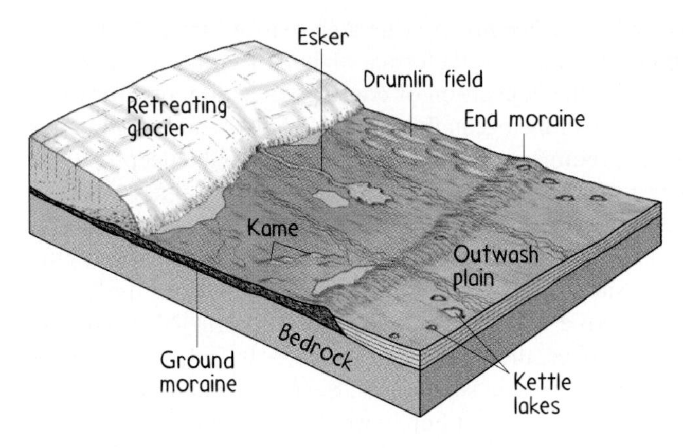

Fig. 24.40
Kames and eskers are features of outwash glacial deposits. The drumlin field is a glacial till deposit.

As continental glaciers retreated northward, large blocks of stagnant ice were sometimes left behind, often becoming incorporated in morainal deposits or partially buried by outwash. As the chunks of ice melted, drift material caved in around them creating large dish-shaped hollows called *kettles.* Such kettles dot the surface of the northern United States by the tens of thousands. Today, filled with water, these kettle lakes make up the "10,000 Lakes" of Minnesota.

Many of the world's lakes, small and large, are the products of glacial action. In addition to creating kettles, glaciers deepened valleys and deposited sediments that acted as dams, blocking stream drainage within some valleys and creating lakes. The Five Finger Lakes in upstate New York and the Great Lakes of North America are all products of glacial action.

24.7 The Oceans

If we could ever drain the water from the Earth's oceans, we'd see, as we learned in Chapter 23, enormous mountain ranges in the middle of the ocean basins and deep trenches bordering many of the continents. These features of the ocean bottom are very pronounced. In fact, land rises on average about 840 meters above sea level while the ocean bottom averages about 3800 meters below sea level. If we were to compare the height of Mount Everest in the Himalayas, a majestic 8848 meters above sea level, to the Marianas Trench in the Pacific Ocean, an unfathomable 11,035 meters below sea level we would see that the oceans are much deeper than land mountains are high!

The **continental margin** is the boundary between the continents and the ocean. As Figure 24.41 shows, it consists of a *continental shelf* (the submerged upper portion of the margin), a *continental slope* (the break point where the shelf steepens as it descends

Fig. 24.41
Profile of the continental margin going from land to the deep ocean bottom. The vertical dimension is exaggerated for clarity.

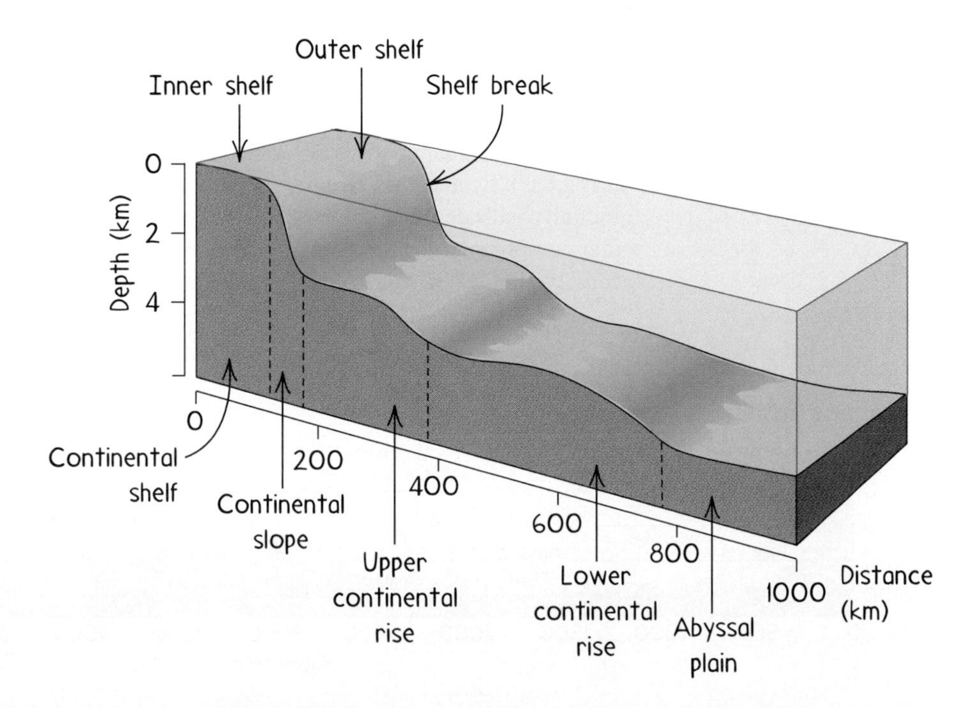

Fig. 24.43
Space-shuttle photograph of circular rings of coral known as atolls. Atolls are often built on undersea volcanoes.

to depths of 2–3 kilometers), and a *continental rise* (the area seaward from the base of the slope to the deep ocean floor). Similar to delta formation but on a much larger level, continental margins are formed by the deposition of continental sediments. Between continental margins, the topography of the ocean floor varies from the midocean ridges that encircle the globe to the flat, sediment-covered ocean bottom to the deep seafloor trenches near continental margins (Figure 24.42).

As we learned in Chapter 23, the midocean ridges are sites of ongoing volcanic activity; hence they are the sites where new oceanic crust is formed. In some places the oceanic ridge reaches the water surface and forms volcanic islands. One of the largest of these islands is Iceland, an active volcanic island centered on the Mid-Atlantic Ridge. Away from the midocean ridges, the floor of the deep oceans is a landscape of hills, plateaus, sediment-floored basins, and seamounts. Seamounts and volcanic islands are formed along a midocean ridge or at hot spots. They form either alone, in clusters, or in chains. As underwater seamounts and dormant volcanic islands are subjected to surface waves and erosion, they flatten and subside farther beneath the water. In warm areas these eroded islands are colonized by corals and algae thus forming circular coral reefs known as *atolls* (Figure 24.43).

Ocean Waves

Ocean waves come in a variety of sizes and shapes, from tiny ripples to the gigantic waves in hurricanes, to the huge tides as we saw in Chapter 4. Water waves, like all other waves, begin by some disturbance. The most common disturbing force is the wind. Blow on a bowl filled with water and you'll see a succession of small ripples moving across the water surface. The generation of ripples in the ocean is similar. As wind speed increases, the ripples grow to full-sized waves; as stronger winds blow, larger waves are created. As waves travel from their origin, they develop into regular patterns of smooth, rounded waves called *swells*—the mature undulations of the open ocean.

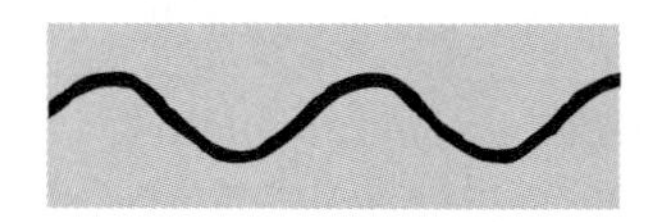

Fig. 24.44
Ocean waves have characteristics of simple sine waves.

Recall from our study of waves in Chapter 10 that wave motion can be described in terms of a sine curve (Figure 24.44), and that it is the *disturbance* rather than the medium that is carried by a wave. The wave form of an ocean wave travels across the ocean, while the water making up the wave remains for the most part in one place.

However, because they are both transverse and circular, ocean waves are more complicated than the simple transverse waves discussed in Chapter 10. As a water wave passes a given point, the water particles at that point move in circular paths. This circular motion can be seen by observing the behavior of a floating piece of wood on the ocean surface. The wood sways to and fro while bobbing up and down, tracing a circle during each cycle. This circular motion occurs near the water surface and decreases gradually with depth (Figure 24.45). At a depth of about one-half wavelength, the circular component of the wave is negligible. For this reason, we can say with reasonable accuracy that water waves occur mainly at the surface.

Fig. 24.42
Map of the ocean floor showing variation in topography.
(a) Atlantic profile
(b) Pacific profile

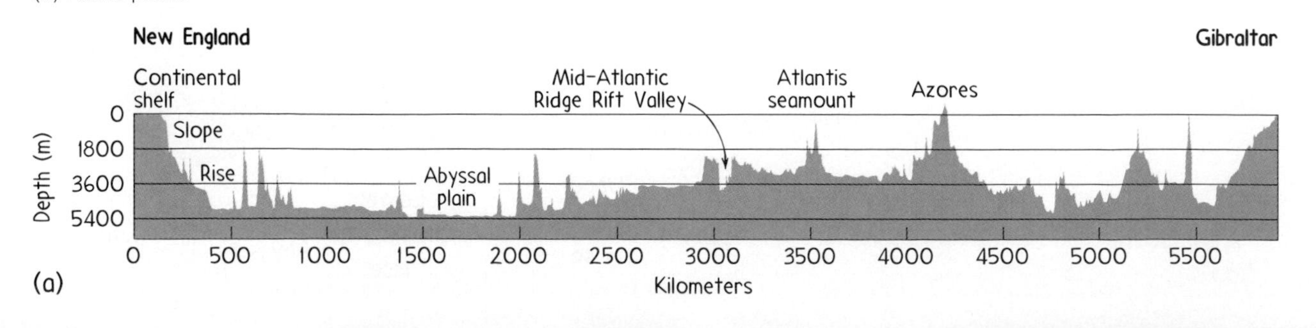

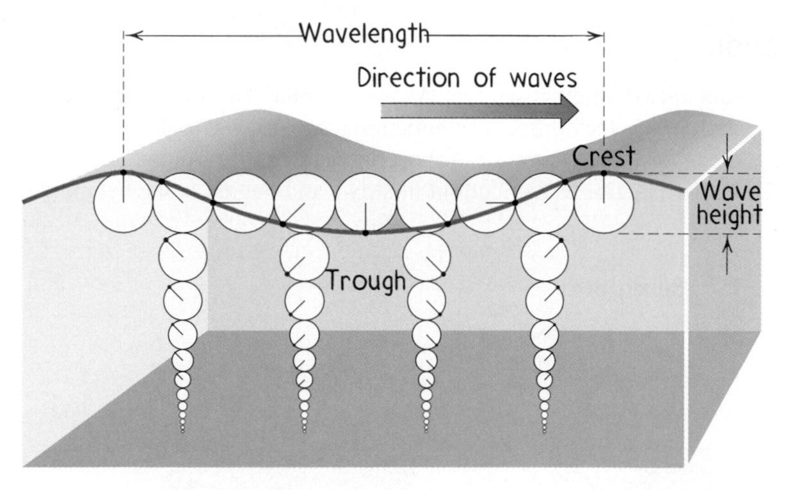

Fig. 24.45
Movement of water particles with the passage of a wave. The particles move in a circular orbit. Orbital motion is greatest at the surface and gradually decreases with depth. At depths greater than half a wavelength, orbital motion is negligible.

When a wave approaches the shore, where water depth decreases, circular motion is interrupted by the ocean bottom. When water depth approaches half a wave's wavelength, the bottom of the circular path flattens, slowing the wave. The wave period remains unchanged because the swells from deeper waters continue to advance. As a result, incoming waves gain on leading slower waves and the distance between waves decreases. This concentration of water in a narrower zone produces higher, steeper waves. When wave height steepens to the point where water can no longer support itself, the wave overturns, breaking with a crash. This breakwater is called *surf*—the area of wave activity between the line of breakers and the shore (Figure 24.46). As we would expect, the surf area is an area of continuous erosion.

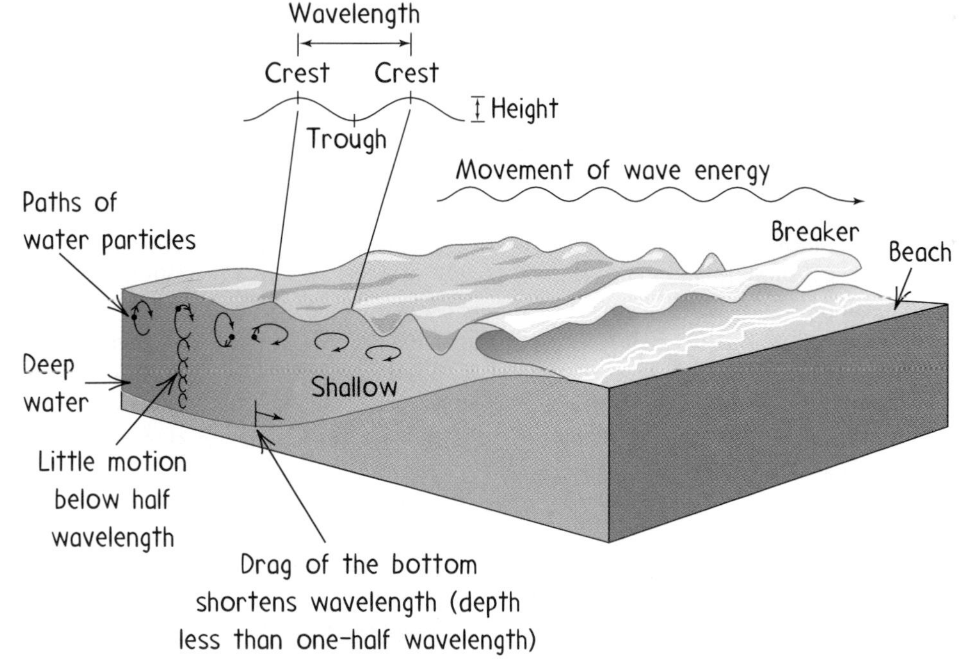

Fig. 24.46
Waves change form as they travel from deep water through shallow water to shore. In deep water, orbital motion is circular. In shallow water orbital motion becomes elliptical as a result of contact with the bottom. This change decreases the wave speed. As incoming waves continue to advance, the distance between waves decreases, causing wave height to increase. When waves reach a critical height, they break and crash into the surf zone.

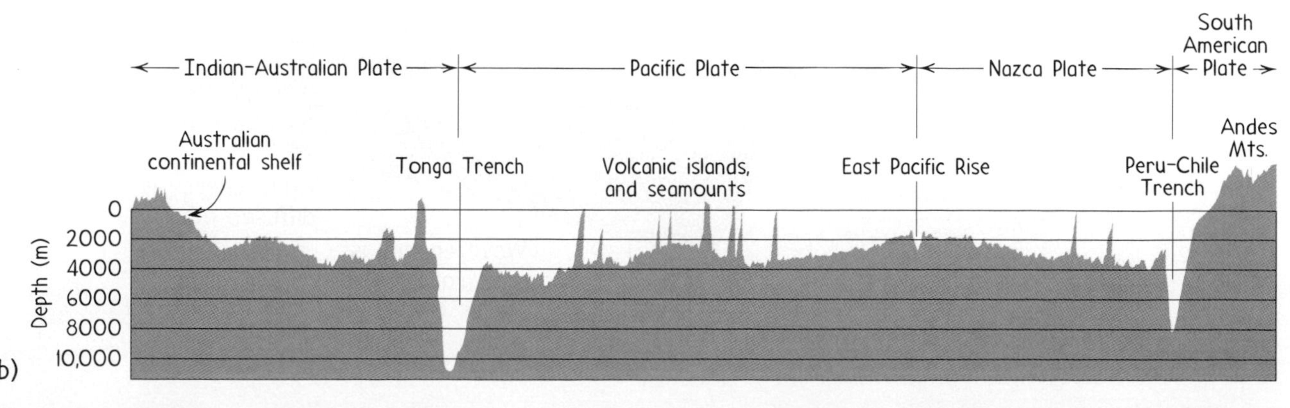

Wave Refraction

Wave refraction occurs in two different settings. As waves enter shallow water, their direction of approach changes as they refract to a direction more parallel to the shore. This refraction occurs whenever waves approach a shoreline at an angle. As the portion of the wave nearest the shore begins to feel bottom, it slows and begins to lag behind the portions still in deeper water. As the next portion of the wave feels bottom, it too slows. Thus in a continuous fashion, the line of the wave crest bends as it moves into shallower water, becoming more nearly parallel to the coastline (Figure 24.47). This bending is not always complete and some waves approach the shoreline at a small angle. The oblique approach of waves causes a *longshore current* that flows parallel to shore.

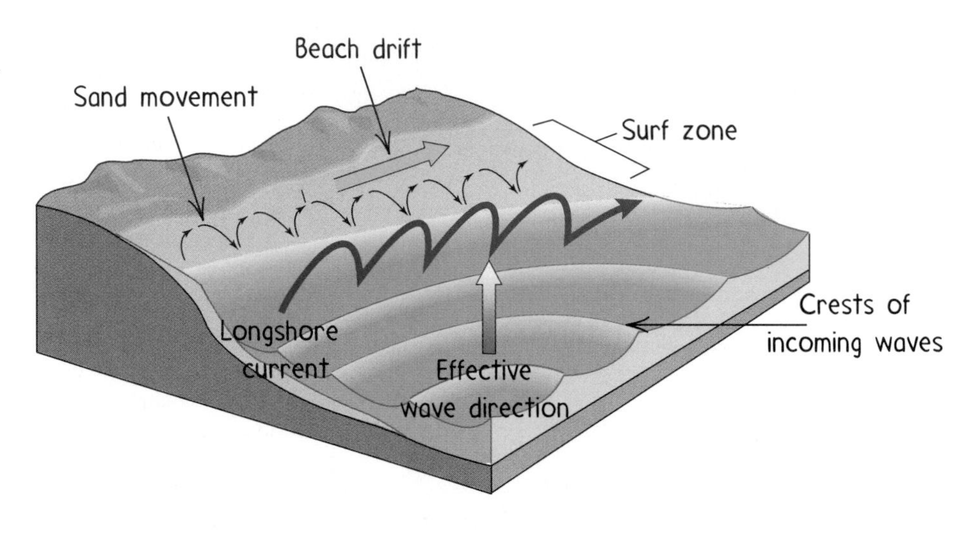

Fig. 24.47
When waves approach a shoreline, they refract (bend) so that the crests of the approaching waves become more parallel to the shore as they move into shallower water. Because the overall direction of wave movement is oblique to the shore, a longshore current forms, which causes water and sand to move parallel to the shoreline.

Geologic features are best viewed from an airplane. Next time you are flying in an airplane, request a window seat and enjoy the geology below.

Wave refraction also greatly impacts irregular shorelines, mainly those where there are protruding headlands and small bays. Refraction causes wave energy to be unevenly distributed. It is concentrated at headland areas, where shorelines project into the water, and diluted in adjacent bays (Figure 24.48). Over a long period of time, wave refraction acts to straighten out shorelines.

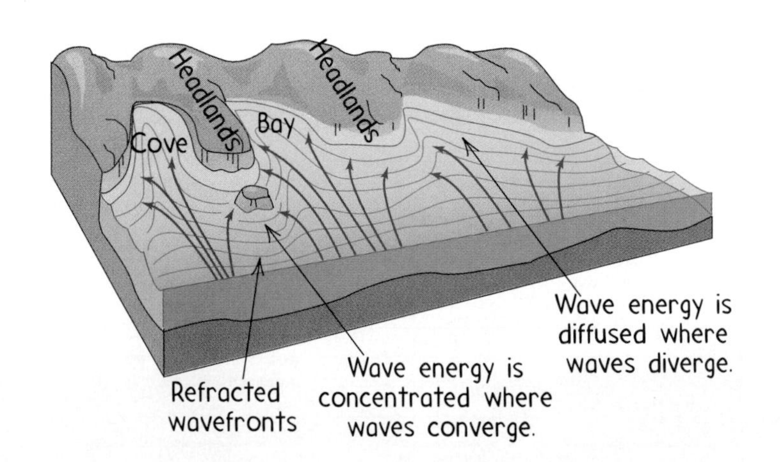

Fig. 24.48
Wave refraction ultimately results in coastal straightening. On irregular coastlines, wave energy is concentrated as it converges on headlands and diffused as it diverges in coves and bays.

24.8 Shorelines

Wind is the driving force for wave action, and waves provide the energy that shapes our shorelines. Because the amount of surf at a shoreline varies with time and because the rocks at any shoreline have different degrees of resistance to erosion, surf can form many different erosional features. Soft rocks and highly fractured rocks erode fastest, whereas hard rocks and unfractured ones erode more slowly.

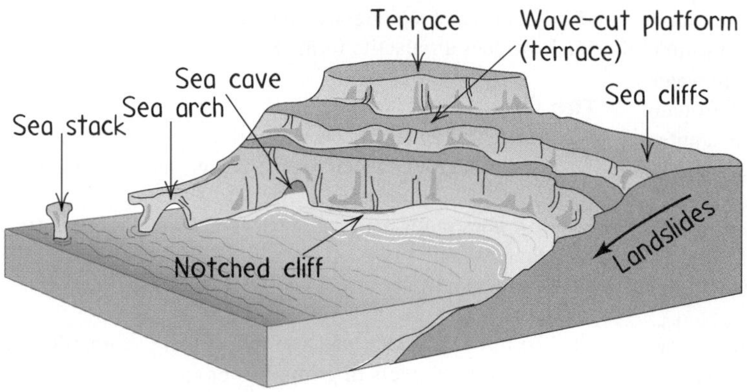

Along shorelines consisting of resistant rock, the pounding surf cuts into and notches the base of the land. As erosion proceeds, the notch deepens and the rocks above begin to jut out over the empty space at the base. As the overhanging rocks fall into the surf, the cliff progressively retreats. In time, waves cut into the cliff to form a relatively flat surface known as a *wave cut platform.*

Fig. 24.49
Characteristic coastal erosional landforms.

Along some rocky coastlines, *sea caves* form along with cliffs. *Sea arches* can form if two caves, usually on opposite sides of a headland, unite. When an arch collapses, an isolated remnant called a *sea stack* is left behind. In time, wave action also erodes away the sea stack.

Fig. 24.50
Spit forming on the Western Fjords in Iceland.

Rock particles worn from the coast must sooner or later be deposited, and beaches are the most common shoreline depositional feature. Beaches tend to be elongated by longshore currents that form when waves approach the shoreline at an oblique angle. For example, waves approaching a north-south coastline from the northwest would cause a longshore current southward down the beach. These currents move sand down the length of the coast and may cause the formation of *spits.* Spits begin as submerged ridges of sand. As sand accumulates, the spit rises above the surface and projects from the coast into open water as a continuation of the beach, frequently as a fingerlike piece of land (Figure 24.50).

When sand ridges form parallel to the coast, they eventually grow into *barrier islands.* Barrier islands form where ridges of sand break the surface of the water for a long enough period of time that vegetation begins to take hold. Vegetation allows the new barrier island to become resistant to surf and storm erosion. With continued safeguards, the barrier island grows even more. Separated from the coast by tidal flats or shallow lagoons, barrier islands form a barricade between the coast and the open ocean. The Gulf Coast of the United States and much of the eastern shore south of New York City have abundant barrier islands. Since the lagoons separating these narrow islands from the shore are zones of relatively quiet water, small boats often use the lagoons as a "freeway" between Florida and New York, thus avoiding the potentially rough waters of the open Atlantic.

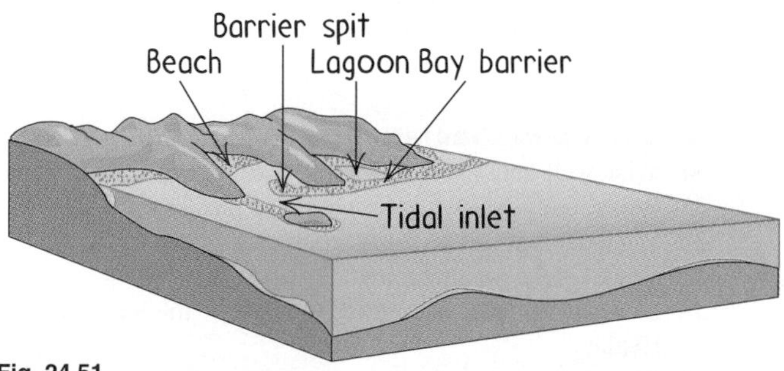

Fig. 24.51
Characteristic coastal depositional landforms.

Summary of Terms

- **Hydrologic cycle** The natural circulation of water from ocean to atmosphere to ground, back to ocean.
- **Groundwater** Subsurface water in the zone of saturation.
- **Porosity** The volume of open space in a rock or sediment compared to the total volume of solids plus open space.
- **Permeability** The ability of a material to transmit fluid.
- **Water table** The upper boundary of the zone of saturation, the area where every pore space is filled with water.
- **Artesian system** A system in which confined groundwater under pressure can rise above the level of an aquifer.
- **Leachate** A solution formed by water that has percolated through soil containing water-soluble substances.
- **Turbulent flow** Water flowing erratically in a jumbled manner, stirring up everything it touches.
- **Laminar flow** Water flowing smoothly in a straight line with no mixing of sediment.
- **Delta** A flat-topped accumulation of sediments, commonly forming a triangular or fan-shaped plain, deposited where a stream flows into a standing body of water.
- **Glacier** A large mass of ice formed by the compaction and recrystallization of snow, moving downslope under its own weight.
- **Accumulation** The amount of snow added to a glacier in a year.
- **Ablation** The amount of ice a glacier loses in a year.
- **Drift** Generic term for all glacial deposits. Waterborne glacial deposits are known as *outwash.* Material deposited directly by melting ice is *till.*
- **Continental margin** The boundary between continental land and deep ocean basins, consisting of continental shelf, continental slope, and continental rise.

Review Questions

Groundwater

1. Distinguish between *porosity* and *permeability.*
2. Name two rock types that can have high porosity but low permeability.
3. A kitchen table is usually flat, but the water table is generally not flat. Why?
4. Compare and contrast the zone of aeration with the zone of saturation.
5. What type of soils allow for the greatest infiltration of rainfall?
6. How does an aquiclude differ from an aquifer?
7. Can an aquifer be composed of igneous rocks? Explain.
8. What is an artesian system and how is it formed?
9. What factors affect the rate of groundwater movement?
10. What are the consequences of overpumping groundwater?
11. What two factors contribute to the ability of an aquifer to recharge?
12. What is meant by the statement "In parts of Texas the groundwater is now literally mined?"

Groundwater and Topography

13. How does groundwater affect carbonate rocks?
14. Carbonate rocks are mainly formed in marine environments. Why do we find abundant carbonate deposits on continental land?
15. What is karst topography? Where is it found?
16. How does a stalactite form? How does a stalagmite form?

The Quality of Water

17. Which aquifer would be most effective in purifying contaminated groundwater: coarse gravel, sand, or cavernous limestone? Defend your answer.
18. How does rainwater become acidic? How does this affect groundwater?
19. Does groundwater affect surface water? If so, why? If not, why not?
20. List three ways our water supply is being contaminated.
21. High nitrate levels in groundwater can come from what source?
22. How does leachate form?

Surface Water and Drainage Systems

23. Rivers are important to human culture. Why?
24. What is meant by stream gradient, and how does it affect stream velocity?
25. What is the greater transporter of sediment, a laminar flow or a turbulent flow? Why?
26. What are the consequences when (a) the discharge of a stream increases, and (b) the speed of a stream increases?
27. How does the shape of a stream channel affect flow?
28. Name three ways the movement of water erodes the stream channel. Which creates potholes?
29. What factors are responsible for the formation of a stream valley?
30. Under what conditions do sinuous, meandering rivers form along a floodplain?
31. What type of stream flow do we generally find in high mountainous regions?
32. What is a continental divide?
33. What is the significance of the Continental Divide in North America with respect to water flow to the Atlantic and Pacific Oceans?

Glaciers and Glaciation

34. What well-known landscapes have been carved by glaciers?
35. What conditions are necessary for a glacier to form?
36. What distinguishes a huge block of ice from a glacier?
37. What are the main features of glacial flow?
38. Does all the ice in a glacier move at the same speed? Explain.
39. Why do crevasses form on the surface of glaciers?
40. Under what conditions does a glacier front advance?
41. Under what conditions does a glacier front retreat?

42. Under what conditions does a glacier front remain stationary?
43. What is a glacial surge?
44. What are striations? What is their significance?
45. How do glacially deposited rocks differ from river deposited rocks?
46. What erosional features might you find in an area of alpine glaciation?
47. What features might you find as a glacial deposit?

The Oceans

48. Is Mount Everest higher than the oceans are deep?
49. How is wavelength related to wave speed? How is wave speed affected by water depth?
50. Why do waves become taller as they approach shore?
51. Why are headland areas prime areas for erosion?
52. What is wave refraction? Why does it happen in ocean waves?

Shorelines

53. Describe how a sea stack forms.
54. How does the refraction of incoming waves help to straighten out a jagged coastline?
55. What types of land features are associated with transport of sand from a longshore current?

Home Project

Water is "hard" if it contains dissolved calcium and magnesium. To test the hardness of the water in your area, collect four water samples from four different local sources—a nearby pond or well, a stream, your kitchen faucet, and bottled distilled water. Add a drop of liquid soap to each water sample and shake. The bottle with the most suds should be the softest water. Record your observations.

Exercises

1. Look at a map of any part of the world and you'll see the locations of older cities where rivers occur, or have occurred. What is your explanation?
2. The oceans are salt water, yet evaporation over the ocean surface produces clouds that precipitate fresh water. Why no salt?
3. How much of the Earth's supply of water is fresh water, and where is most of it located?
4. Where does most rainfall on Earth finally end up before becoming rain again?
5. Compare the recycling of water in an orbiting space shuttle to recycling of water on Earth.
6. In a confined aquifer the water in a well can rise above the top of the aquifer. What is this system called?
7. How high can water rise in a well supplied by an unconfined aquifer?
8. Are natural groundwater recharge rates greatest on steep, rocky slopes or on gentle, sandy slopes? Defend your answer.
9. How is the local hydrologic cycle affected by the practice of drawing drinking water from a river, then returning sewage to the same river?
10. Why is pollution of groundwater a greater environmental hazard than pollution of surface waters?
11. Some metals can be extremely dangerous to water supplies. Aluminum has been linked to Alzheimer's and Parkinson's disease; cadmium is known to cause liver damage; and lead affects the circulatory, reproductive, nervous, and kidney systems. What are the likely ways these metals can get into our water supply?
12. When a water supply becomes overly rich in nitrogen and phosphorus, plant life thrives to the point of destruction. The overgrowth of algae causes unsightly scum, unpleasant odors, and robs the water supply of dissolved oxygen. What are the sources of this type of pollution and how does it affect other aquatic life?
13. What are the major factors that determine the length of time a well will produce water?
14. By what means can Earth scientists predict the discharge of a stream after a rainstorm?
15. During a 20-year period, withdrawal of groundwater for irrigation purposes in the San Joaquin Valley of California caused the water table to drop by 75 m. Pumping has been greatly reduced, the aquifer is slowly recharging, and water for irrigation is now provided by canals that bring water from the Sierra Nevada Mountains. Speculate as to whether the problem of water supply is now solved in this region.
16. Extend your speculation in the previous question to other regions, and defend your conclusions.
17. Removal of groundwater can cause subsidence. If removal of groundwater is stopped, will the land likely rise again to its original level? Defend your answer.
18. What is a sinkhole? What factors contribute to its formation?
19. As a population increases so does the amount of garbage produced by that population. In many areas the way to deal with increasing wastes is by burial in a landfill or underground storage facilities. What principal factors must be considered in the planning and building of such sites?
20. Recall from Chapter 22 that the Mississippi Delta has moved south from near Cairo, Illinois, to its present location in Louisiana. Other than the length of time, why has the delta moved so far?
21. Which of the three agents of transportation—wind, water, or ice—transports the largest boulders? Why?
22. Which of the three agents of transportation is limited to transporting small size rocks? Why?

23. With regard to residence time, how may the process of cleaning up groundwater contamination differ from cleaning up surface contamination in a lake?

24. What effect does a dam have on the water table in the vicinity of the dam?

25. Once a dam is constructed and filled, what effect does increased evaporation of water behind the dam have on the volume of water that flows through the dam and downstream from the dam?

26. What effect does a dam have on erosional activity downstream from the dam?

27. What effect does the accumulation of sediments behind a dam have on its capacity for storing water?

28. In an aquifer, if the hydraulic head next to a stream is lower than the water level in the stream, does groundwater flow into the stream or does stream water flow into the ground? Explain.

29. As runoff into streams increases, what variable of stream flow (as discussed in the text) increases?

30. How is a glacier formed?

31. In what way does a glaciated mountain valley differ from a nonglaciated mountain valley?

32. The Earth's periods of glaciation have had a major impact on the surface features of our planet. What places, other than the examples cited in this chapter owe their striking features to glaciers?

33. How does "frictional drag" play a role in the external movement of a glacier? How about the internal movement?

34. If a volcano erupted onto a glacier, speculate about the long-term differences in resulting surface features when the molten rock melts the ice all the way to the ground surface, and when the molten rock solidifies on the ice surface.

35. The color of deep glacial ice is a brilliant greenish blue. Suggest a reason for this difference in color compared with ordinary ice.

36. As waves approach shallow water, those with longer wavelengths slow down before those with shorter wavelengths. Why?

37. Suppose a breakwater is built offshore and perpendicular to the shore. How will this structure affect the longshore current and its transport of sand? Explain.

38. Can a stream erode land that lies below sea level? Explain.

CHAPTER

25 A Brief History of the Earth

The Earth is some 4.6 billion years old. This vast span of time, called *geologic time,* is difficult to comprehend. If we were to compress all of geologic time into a single year, so that our planet formed from matter surrounding the Sun on January 1st, the oldest known Earth rocks would appear in early March. Simple bacterial life would appear in the sea at the end of March, and more complex plants and animals would not emerge until late October or early November. Dinosaurs would rule the Earth in mid-December and disappear by December 26. *Homo sapiens* would appear at 11:50 P.M. on the evening of December 31, and all of recorded human history would occur in the last minute of New Year's Eve!

The Earth's history is recorded in the rocks of its crust. Scientists use an assumption called *uniformitarianism* to relate what we know about present-day processes to past events—the present is the key to the past. Simply put, uniformitarianism states that the natural laws we know about today have been constant over the geologic past. The rock record is like a long and detailed diary, containing the history of Earth-shaping events. The book is incomplete, however. Many pages, especially in the early part, are missing, and many others are tattered, torn, and difficult to read. But enough pages are preserved to give an account of the remarkable events of the Earth's four and a half billion years of history.

25.1 Relative Dating

A sequence of rock layers composed of sedimentary rocks and/or lava flows provides immediate evidence of their relative ages. Simply put, lower layers were formed before top layers. Perhaps the world's most spectacular display of the rock record is the Grand Canyon of the Colorado River in Arizona. The many layers of rock exposed in the canyon walls and the thickness of these layers are testimony to great geologic activity over millions of years. The conditions under which the sedimentary layers were deposited varied widely, changing from season to season and year to year. Some layers reveal climatic cycles that span centuries, others indicate regions submerged beneath a shallow sea, while still others show periods of tremendous increase in rainfall accompanied by gradual uplift of the entire area. Millions of years after the top layer was deposited, abrasive erosion from the modern and ancestral Colorado River cut through all these accumulated layers of sedimentary rock like a notched knife into a layer cake, forming the canyon!

Fig. 25.1
The lowermost layers of the Grand Canyon are older than the uppermost layers—the principle of superposition.

In the Grand Canyon and elsewhere, Earth scientists use five common-sense principles to discern the nature and sequence of geological events and the relative ages of rocks:

1. **Original horizontality** Layers of sediment are deposited evenly, with each new layer laid down nearly horizontally over older sediment. Layers that are inclined at any angle—from very slight to very steep—indicate they were moved into that position by crustal disturbances after deposition.
2. **Superposition** In an undeformed sequence of sedimentary rocks, each layer is older than the one above and younger than the one below. Like the layers of a huge wedding cake, the rock record was formed from the bottom layer to the top. Upper layers are younger than lower layers.
3. **Cross-cutting** An igneous intrusion or fault that cuts through preexisting rock is younger than the rock through which it cuts (Figure 25.2).

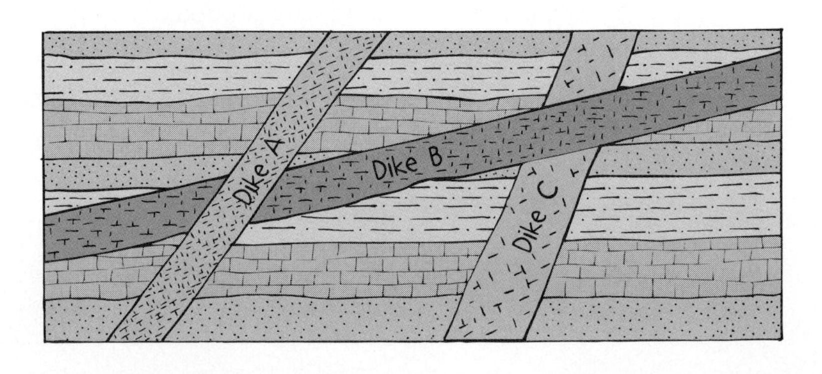

Fig. 25.2
Dikes cutting into a rock body are younger than the rock they cut into. In the diagram, dike A cuts into dike B, and dike B cuts into dike C. From the principle of cross-cutting relationships, A is the youngest dike, B the next youngest, and C the oldest of the three. The horizontal layers, which are cut by all three dikes, are all older than C.

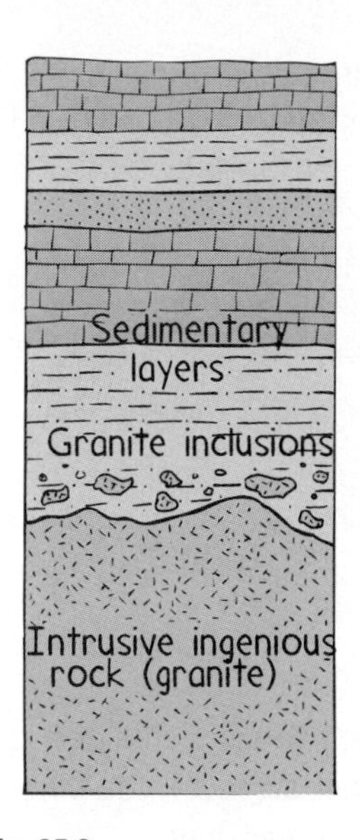

Fig. 25.3
The rocks locked in the sedimentary layer existed before the sedimentary layer formed—the principle of inclusion.

4. **Inclusion** Inclusions are pieces of one rock type contained within another. Any inclusion is older than the rock containing it, just as pieces of rock that make up a slab of concrete were formed before the concrete was formed (Figure 25.3).
5. **Faunal succession** The evolution of life is recorded in the rock record in the form of fossils. Fossil organisms succeed one another in a definite, irreversible, determinable order. Fossils provide a great tool for correlating rocks of similar age in different regions because any time period can be uniquely recognized by its fossil content. Once a period is established, the fossils that rocks contain can be used to identify rocks of the same age in widely separated regions of the Earth.

It always comes as a surprise to find the fossil of a sea animal known to be extinct for millions of years encased in rock at an elevation of a kilometer or so above sea level. Such fossils are evidence that many of today's land surfaces were yesterday's sea bottoms. Finding fossils is a delight to casual and experienced fossil hunters alike.

Although most rock layers were deposited without interruption, nowhere is there a continuous sequence from the Earth's formation to the present time. Weathering and erosion, crustal uplifts, and other geologic processes interrupt the normal sequence of deposition, creating breaks or gaps in the rock record as Figure 25.5 shows. These gaps, called **unconformities**, are detected by observing the relationships of strata and fossils.

The most easily recognized of all unconformities is an **angular unconformity**—tilted or folded sedimentary rocks overlain by younger, relatively horizontal rock layers. An angular unconformity forms when older, previously horizontal rock layers are uplifted and tilted by the Earth's movements (Figure 25.6). During and after the uplift, erosion wears down the tilted layers so that rocks at the surface are eroded to a more or less even plane. After the period of erosion is over, more layers are deposited over the tilted ones, and these younger layers are horizontal. The angular unconformity is the "surface" that separates the tilted layers from the horizontal layers. It represents the long interval of time during which uplift and erosion took place. The part of the rock record representing this long interval is now missing because of erosion, and the unconformity is the evidence that remains.

Fig. 25.4
Hunting for fossils can be a lot of fun. Finding one is a delightful experience, as the author indicates.

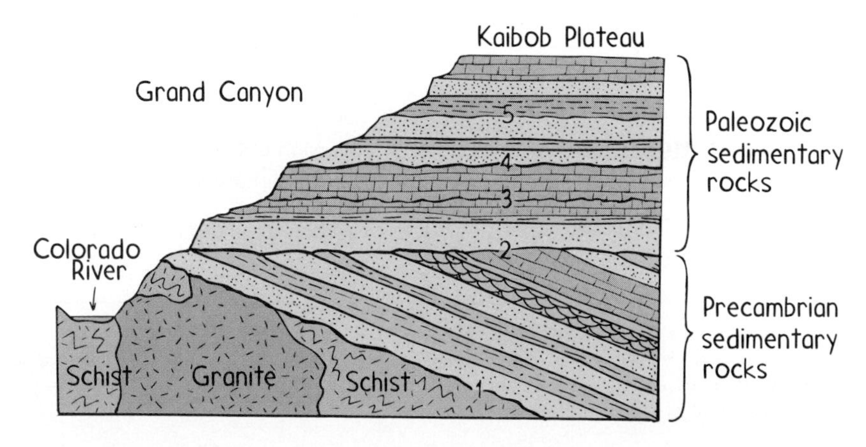

Fig. 25.5
The age of the Grand Canyon can be deciphered by its sequence of rock layers. As in other places, the sequence is not continuous and there are time gaps. (1) A nonconformity separating older metamorphic rocks from sedimentary layers. (2) An angular unconformity separating older tilted layers from horizontal layers. Time gaps are also represented between horizontal sedimentary layers. The unconformities (3)–(5) are difficult to identify and often require a good eye and a knowledge of fossils.

Fig. 25.6
Sequence of events in the formation of an angular unconformity.

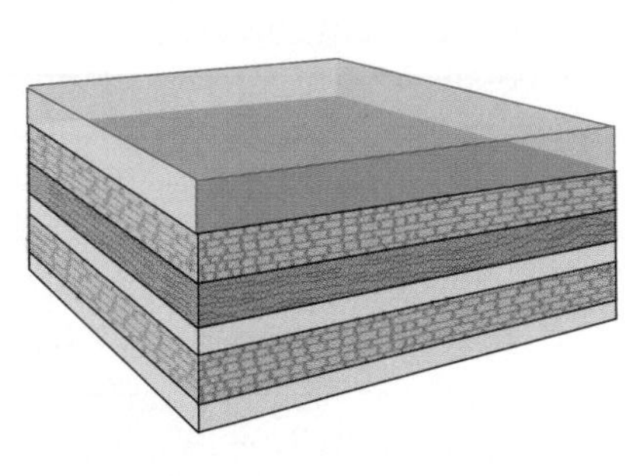

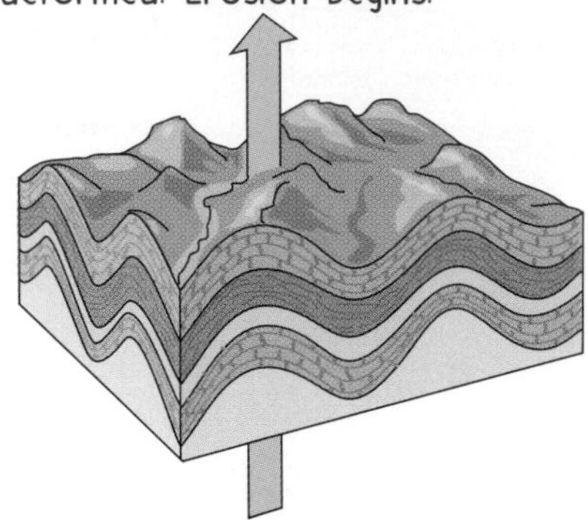

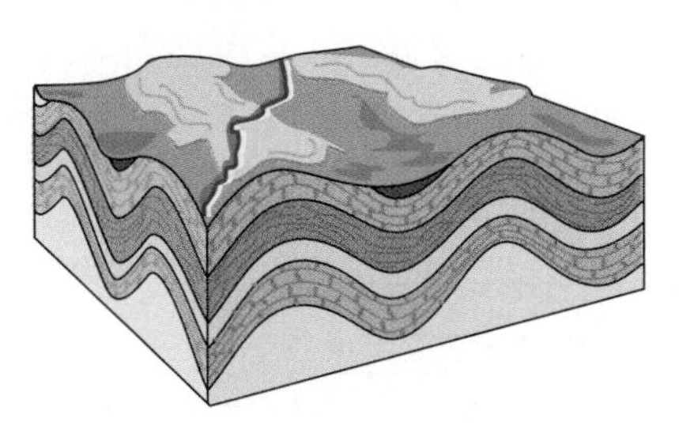

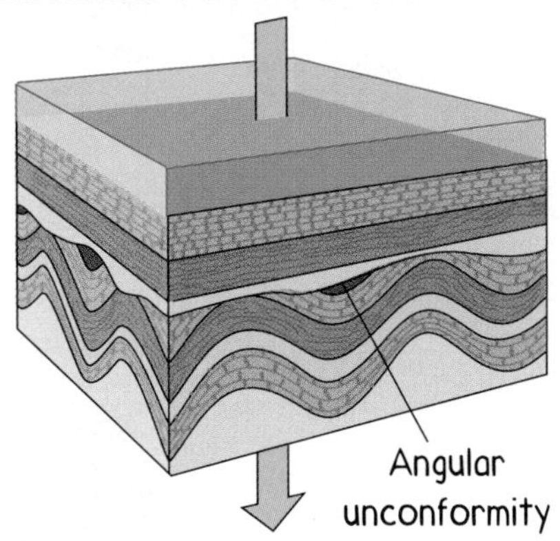

When overlying sedimentary rocks are found on an eroded surface of metamorphic or intrusive igneous rocks, the unconformity is called a **nonconformity**. The older intrusive igneous or metamorphic rocks formed deep beneath the Earth's surface but were present at the Earth's surface when the overlying sedimentary rocks were deposited on top of them. This type of unconformity indicates that a great deal of uplift and erosion occurred before the sedimentary layers were deposited, with a large stretch of time "missing" from the rock record. Such time gaps are often quite difficult to identify.

Question

If a granitic intrusion cuts into or across sedimentary layers, which is older: the granite or the sedimentary layers?

Answer
The intrusion is new rock in the making. Therefore the sedimentary layers are older than the intrusions that cut into them.

25.2 Radiometric Dating

Relative dating indicates which parts of the Earth's crust are older or younger, but it doesn't tell us the actual age of rock—the amount of time that has passed since the rock was formed. The actual age of a rock can be estimated by **radiometric dating**, which entails measuring the ratio of radioactive isotopes to their decay products.

Table 25-1
Isotopes Most Commonly Used for Radiometric Dating

Radioactive Parent	Stable Daughter Product	Currently Accepted Half-life Value
Uranium-238	lead-206	4.5 billion years
Uranium-235	lead-207	704 million years
Potassium-40	argon-40	1.3 billion years
Carbon-14	nitrogen-14	5730 years

Recall from Chapter 14 that atoms of the same element that contain different numbers of neutrons are *isotopes*. And recall our discussion of isotopic dating in Section 14.8. Some of the common radioactive isotopes frequently used for dating and estimates of geologic time are shown in Table 25.1.

In a uranium-bearing rock, the naturally occurring radioactive isotopes uranium-238 and uranium-235 decay to the lead isotopes lead-206 and lead-207, respectively, but not to the common isotope of lead, lead-208. Therefore, any lead-206 and lead-207 found in a rock today were at one time uranium. So the age of a sample containing equal numbers of uranium-235 and lead-207 atoms is one uranium-235 half-life, or 704 million years old. (We assume all the lead-207 used to be uranium-235.) If uranium ore contains only a relatively small amount of lead-207, it is relatively young.

Radiometric dating with uranium-238 is useful for very old rocks. In fact, scientists have used this isotope to estimate that the oldest rocks on the Earth are 3.8 billion years old. Old, potassium-rich rocks (such as those containing micas and feldspars) are often dated by measuring the amounts of parent potassium-40 and daughter argon-40.

Radiometric dating is based on the assumption that once a mineral has crystallized, any daughter product found within it originates only from the decay of the unstable parent—there was no daughter product present initially. Another important assumption is that there is no "leakage" of parent or daughter products into or out of the mineral. If a mineral is, for instance, reheated by metamorphism, its "time clock" is reset. If a potassium-bearing mineral is reheated, some or all of the gaseous daughter product argon-40 might diffuse out of the crystalline structure of the mineral, resetting the time clock for that sample. This can complicate age estimation. Happily, cross checking by different radiometric methods can increase accuracy. Radiometric dating in general is subject to some uncertainty due to the detailed analytical procedures required and the random nature of radioactive decay.

Fig. 25.7
Amount of parent material versus number of half-lives.

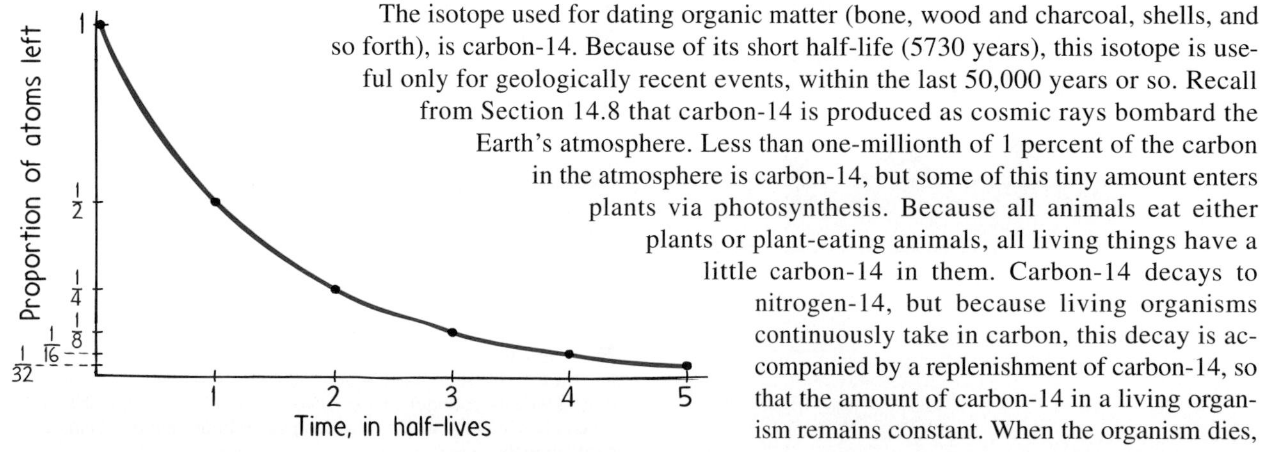

The isotope used for dating organic matter (bone, wood and charcoal, shells, and so forth), is carbon-14. Because of its short half-life (5730 years), this isotope is useful only for geologically recent events, within the last 50,000 years or so. Recall from Section 14.8 that carbon-14 is produced as cosmic rays bombard the Earth's atmosphere. Less than one-millionth of 1 percent of the carbon in the atmosphere is carbon-14, but some of this tiny amount enters plants via photosynthesis. Because all animals eat either plants or plant-eating animals, all living things have a little carbon-14 in them. Carbon-14 decays to nitrogen-14, but because living organisms continuously take in carbon, this decay is accompanied by a replenishment of carbon-14, so that the amount of carbon-14 in a living organism remains constant. When the organism dies,

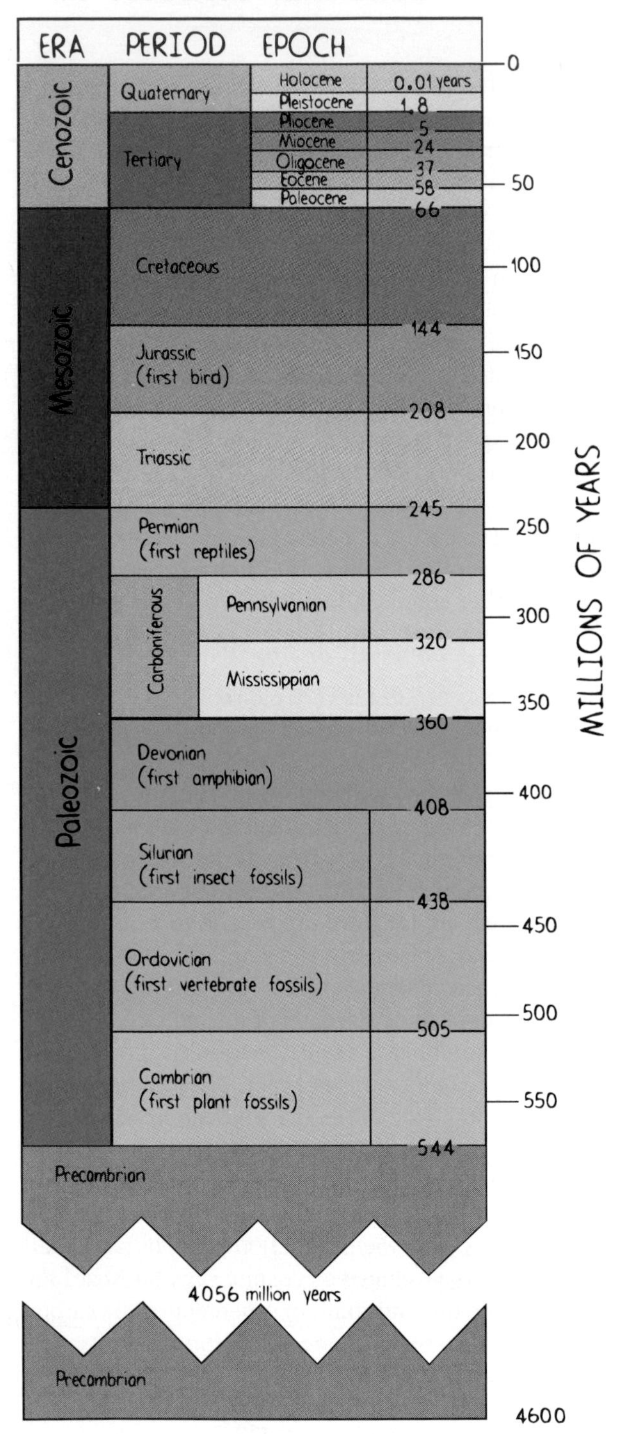

the replenishment stops, and so carbon-14 dating tells us the amount of time that has elapsed since the time of the organism's death.

The largest uncertainties related to carbon dating are the fluctuation of carbon-14 production in the atmosphere (due to fluctuations in cosmic ray intensity and variations in the strength of the Earth's magnetic field) and the amount of carbon dioxide present in the atmosphere. Changes in global climate affect the amount of carbon dioxide in the atmosphere. The onset of the industrial revolution, for example, has increased the amount of "old" carbon in the atmosphere, which has reduced the proportion of carbon-14.

Using both relative and radiometric dating, scientists learn the sequence of events and how long ago each occurred. Although radiometric dating gives us the age of minerals and/or organic matter, it cannot directly give the age of sedimentary rocks. But radiometric dating gives the maximum age of the sedimentary rock in which the datable material is found—the rock can be no older than the age of the datable material within it. So, if we know the maximum age of an overlying and an underlying rock layer, we can bracket the age of a layer in between by using the principle of superposition. Cross-cutting relationships can also be used to bracket the age of sedimentary rocks. An igneous intrusion that can be radiometrically dated provides the minimum age for the rock through which it intrudes—the intruded rock must be older than the intrusion.

The geologic time scale was developed through the use of relative dating, and specific dates were applied to it via radiometric dating. By convention the geologic time scale is divided into three eras, the **Paleozoic** (time of ancient life), the **Mesozoic** (time of middle life), and the **Cenozoic** (time of recent life). Table 25.2 shows each era further divided into periods, which are still further divided into epochs. The largest span of time, the time period preceding the Paleozoic, is known as the **Precambrian** (the time of hidden life).

Question

Could carbon-14 be used for dating rocks from Precambrian time?

Answer

No. Carbon-14 has a half-life of 5730 years and can be used to date only relatively younger rocks. Any carbon-14 in Precambrian carbonaceous material would have long since been reduced to insignificant amounts.

25.3 The Precambrian Time

The Precambrian era ranges from about 4.6 billion years ago, when the Earth formed, to about 544 million years ago, when abundant macroscopic life appeared. The Precambrian—the time of which we know the least—makes up 85 percent of Earth's history! Most of the rocks that formed in this early part of the Earth's history have been eroded away, metamorphosed, or recycled into the Earth's interior. Relatively few fossils are known from Precambrian rocks because organisms didn't have the easily fossilized hard parts that later organisms developed.

The beginning of the Precambrian era is thought to have been a time of considerable volcanic activity and frequent meteorite[1] bombardment. Imagine the Earth as it was at that time: an oceanless planet covered with countless volcanoes belching forth gases and steam from its scorching interior. Huge holes and gashes left by falling meteorites scarred its surface. There was intense convection in the mantle, and severe heat escaping from the interior left the surface of the Earth's early crust in turmoil. The earliest crustal formations were short-lived, ever-changing small lithospheric plates. About 4 billion years ago, heat dissipated, large meteorite impacts decreased, and crustal blocks began to survive. All were completely devoid of life, however.

Fig. 25.8
Primitive stromatolites found in western Australia are dated as old as 3.5 billion years. They are very similar in structure to the present-day stromatolites pictured here. Although the first stromatolites evolved in an anaerobic (oxygen-poor) environment, in time they developed the ability to use sunlight to convert carbon dioxide to food, generating oxygen as a waste product. With the production of oxygen, anaerobic life forms became poisoned while stromatolites flourished. Stromatolites thus changed the Earth's history as its atmosphere became oxygen-rich.

Gases brought to the surface by volcanic processes created both a primitive atmosphere and an ocean. The first atmosphere was rich in water vapor but *anaerobic*—very poor in free oxygen. The first simple organisms for which fossils have been found were plants. The fossils, known as stromatolites, are the remains of mats of wavy layers of algae that lived in shallow seas.

During the mid-Precambrian era, organisms such as stromatolites and blue-green algae developed a simple version of photosynthesis. Photosynthetic organisms require CO_2 to utilize the Sun's energy. They keep the carbon and expel oxygen. With the release of free oxygen, a primitive ozone layer began to develop above the Earth's surface. The ozone layer reduced the amount of harmful ultraviolet radiation reaching the Earth. This protection and the accumulation of free oxygen in the Earth's atmosphere permitted the emergence of new life.

The primitive blue-green algae and bacteria that lived during this time were composed of cells without a nucleus. Reproduction was by simple cell division. The first evidence of nucleated single-celled organisms (green algae) occurs in rocks dated at 1.3–1.5 billion years ago. The discovery of multicellular plants and animals, dated at approximately 700 million years ago, shows evidence of major evolutionary changes that began during the latter half of the Precambrian and greatly increased toward the end of the Precambrian. Some rocks in southern Australia contain diverse fossils of soft-bodied animals, ranging from jellyfish to wormlike forms. This area provides us with the first evidence of an animal community from shallow marine waters.

Precambrian Tectonics

Evidence from folded and faulted rocks and radiometric ages indicates that the first continental crust movements took place 2.5 billion years ago. Continents then began to form from the accretion of smaller landmasses. Speculation holds that about a billion

---

[1]A meteorite is any solid rock object from interplanetary space that has fallen to the Earth's surface without being vaporized during its passage through the atmosphere. We shall learn more about these objects in Chapter 28 when we study the formation of the solar system.

and a half years ago Siberia fused to the western edge of North America while Europe was converging with the eastern region of North America. Other continents were converging from the south to form the first documented supercontinent (long before Pangaea). Large-scale rifting in the central North American crust began about 1 billion years ago, resulting in extensive flood basalts but not in the breakup of North America.

25.4 The Paleozoic Era

Better known than the Precambrian is the Paleozoic era, which was very short in comparison. The Paleozoic began about 544 million years ago and lasted about 300 million years, during which time sea levels rose and fell several times worldwide, allowing shallow seas to cover the continents and marine life to flourish. Changing sea levels greatly influenced the progression and diversification of life forms—from marine invertebrates to fishes, amphibians, and reptiles. The beginning of the Paleozoic is marked by the development of shells, the preservation of which as fossils is why so much more is known about the Paleozoic than the Precambrian. The Paleozoic era is divided into six major periods, each characterized by profound changes in life forms as well as by major tectonic changes.

Fig. 25.9
The trilobite was the dominant fossil of the Cambrian period.

Cambrian Period

The Cambrian period marks the base of the Paleozoic era. Almost all major groups of marine organisms came into existence during this time, as evidenced by abundant fossils. A most important event in the Cambrian was the development of organisms having the ability to secrete calcium carbonate and calcium phosphate for the formation of outer skeletons, or shells. This ability helped organisms become less vulnerable to predators and provided protection against ultraviolet rays, allowing the organisms to move into shallower habitats. In addition, the support provided by a skeleton allowed organisms to grow larger.

The fossil record of the Cambrian is dominated by the skeletons of shallow marine organisms. A variety of these organisms flourished, including the *trilobite,* the armored "cockroaches" of the Cambrian sea.

Ordovician Period

Fossil records show that the Ordovician period was a time of great diversity and abundant marine life. By the end of the Ordovician, all major groups of animals that could be preserved as fossils had appeared. Although isolated bonelike fragments have been identified from the late Cambrian, the Ordovician period marks the earliest unquestionable appearance of vertebrates—the jawless fish known as the *agnatha.* The end of the Ordovician brought many extinctions, likely a result of widespread glaciation. Tropical shallow-water marine groups were the most affected, while high-latitude and deep-water organisms were relatively unaffected.

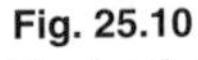

Fig. 25.10
The hagfish is a descendent of the *agnatha,* a primitive jawless fish that made its debut in the Cambrian and flourished in the Ordovician.

Silurian Period

During the Silurian period much of the North American continent was at or above sea level. Thick gypsum and other evaporites accumulated in the vanishing seas. The Silurian brought the emergence of terrestrial life, the earliest known being terrestrial plants that had a well-developed circulatory system (vascular plants). These plants were closely tied to their water origins and inhabited only low wetlands. As plants moved ashore, so did other terrestrial organisms. Air-breathing scorpions and millipedes were common terrestrial animals during this period.

Fig. 25.11
Life in the Devonian sea. In the front center, a nautiloid, which is related to the modern squid, is attacking a trilobite. The colorful organisms on the left are corals.

Devonian Period

By the Devonian period, known as the "age of fishes," many dramatic changes had taken place. Vascular plants were spreading over land surfaces. Lowland forests of seed ferns, scale trees, and true ferns flourished. In the seas, the armored fishes diversified into many new groups. Some, such as the shark and bony fishes, are still present today. In the bony fish group, the lobe-finned fishes are of particular interest because of their development of internal nostrils, which enabled some species to breathe air. Today, the lungfishes and the *coelacanth,* a "living fossil," have such internal nostrils and breathe in a similar way.[2] Another important characteristic of the lobe-finned fishes is that their fins were lobed and muscular with jointed appendages that enabled the animals to walk. Animal life moved to land. Descended from the lobe-finned fishes, the first amphibians made their appearance during the late Devonian. The advent of amphibians was of enormous importance on the evolutionary chain of air-breathing vertebrate land animals. Amphibians, although able to live on land, need to return to water to lay their eggs.

Fig. 25.12
Warm, moist climatic conditions contributed to the lush vegetation and swampy coal forests of the Carboniferous period.

Carboniferous Period

The Carboniferous period encompasses both the Mississippian and the Pennsylvanian periods.[3] Warm, moist climatic conditions contributed to lush vegetation and dense swampy forests. These swamps were the source of the extensive coal beds that now lie under North America, Europe, and northern China. In the Carboniferous period, insects underwent rapid evolution that led to such diverse forms as giant cockroaches and dragonflies having wingspans of 80 cm. The evolution of the first reptiles, the group known as *anapsids,* took place with the development

---

[2]The coelacanth was thought to have become extinct after the Mesozoic. However, in 1938, the first living specimen was caught off the coast of East Africa. Since then other specimens have been discovered in the Madagascar area. The coelacanth is now considered a "living fossil."

[3]The term *Carboniferous period* originated in England but is now used around the world. In North America, the terms *Mississippian period,* named for the Mississippi River valley, and *Pennsylvanian period,* named for the state of Pennsylvania, refer to localities where rocks of these periods are well exposed. Whatever the name, rocks from this time period are known for their coal beds.

of the amniote egg, a porous shell containing a membrane that provided a completely self-contained environment for an embryo. The shell protected the embryo from drying out, allowing animals to complete the transition, begun by amphibians in the Devonian period, from aquatic environments to land. Thanks to the amniote egg, reptiles do not need to lay their eggs in water the way amphibians do.

Permian Period

The evolution of reptiles continued in the Permian period. The reptiles must have been well suited to their environment, for they ruled the Earth for 200 million years. (By comparison, modern humans have inhabited the Earth for less than 100,000 years.) Joining the anapsids from the Carboniferous period, the other two major groups of reptiles appeared during Permian time: the *diapsids* and the *synapsids*. The synapsids, which include ancestors of the earliest mammals, dominated the Permian. The most famous of the synapsids are the fin-backed *pelycosaurs* (Figure 25.13), whose fins may have served to regulate body temperature. The diapsids were less conspicuous in the Permian period, but it was this group that gave rise to the dinosaurs early in the Mesozoic era.

Fig. 25.13
The fin-backed pelycosaurs, a famous member of the reptile group *synapsid*.

The end of the Permian period saw one of the greatest extinctions of animals in the Earth's history. Marine invertebrates were affected much more than terrestrial life. Half of all invertebrate marine families and possibly 90 percent of all species disappeared from the sea. The cause of the extinction is not well understood. One hypothesis is that worldwide global cooling resulted in glaciation and an accompanying lowering of sea level. Climatic extremes ranging from glaciers to deserts are well recorded in the rocks of this time. The duration of low sea level, about 20–25 million years, undoubtedly put much stress on the environments of marine organisms. Yet this alone cannot account for the magnitude of their extinction. Whatever happened took a less drastic toll on terrestrial life. Terrestrial life, although affected, continued to evolve and expanded rapidly as new land habitats appeared, likely due to a lowered sea level. As we shall see in the next section, one likely explanation for the Permian crisis is the tectonic activity that accompanied the formation of Pangaea.

Paleozoic Tectonics

There is more than one interpretation of early Paleozoic tectonics. Science is an ongoing process, and as new evidence is gathered, old models are updated or new ones proposed. In the following discussion, we describe one model of the tectonic development of North America and western Europe, illustrated in Figure 25.14, and the effect that plate movements had on sea level.

Central to the development of this model is the idea that as tectonic plates move away from a spreading oceanic ridge, the newly produced oceanic lithosphere begins to cool and contract. If seafloor spreading is relatively slow, the seafloor as a whole becomes cooler and more contracted, and thus denser. As a result, the seafloor "stands" lower on the asthenosphere (recall isostasy from Section 23.2), and ocean basins are deeper. If seafloor spreading is relatively rapid, the oceanic lithosphere as a whole becomes warmer and less dense (less contracted) and "stands" higher on the asthenosphere. This results in shallower ocean basins, a condition that forces seawater (assuming seawater volume remains constant) onto low portions of the continents. The result is a rise in sea level and the formation of shallow seas on top of continental crust.

The breakup of the Precambrian supercontinent began about 600 million years ago (latest Precambrian) and continued into the Cambrian period (earliest Paleozoic era). This was a time of active seafloor spreading, with the North American and European Plates diverging from each other (Figure 25.14a). This activity opened up new ocean basins, which, because they stood relatively high on the asthenosphere, resulted in the first of several major worldwide rises in Paleozoic sea level. In the mid-Ordovician, the direction of plate motions changed (Figure 25.14b). Eastward subduction of the seafloor adjacent to North America beneath an offshore volcanic arc resulted in the collision of that arc with the North American continental margin by the end of the Ordovician. In Silurian-Devonian time, an ancient microcontinent known as Avalonia (thought possibly to be a displaced piece of Africa) collided with North America. Europe then collided with North America, which now included Avalonia, to form the large landmass known as *Laurasia* (Figure 25.14d).

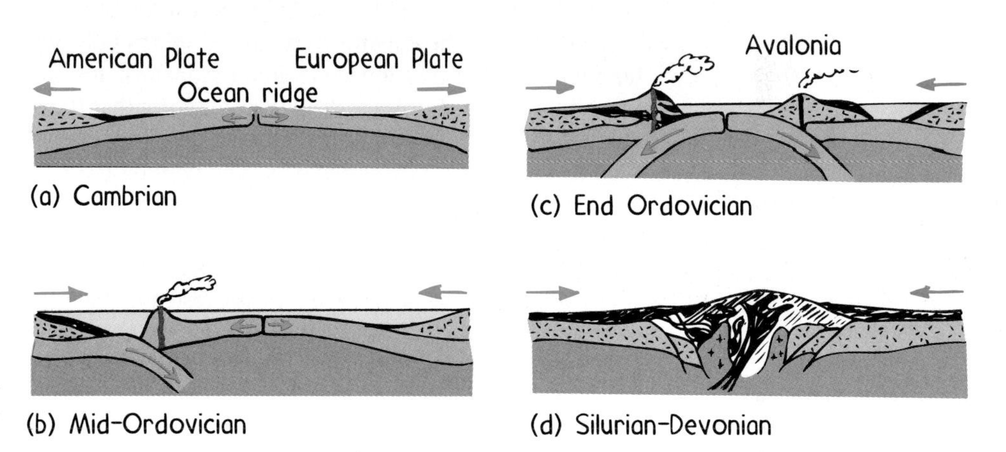

Fig. 25.14
Sequence of tectonic events of the early to mid-Paleozoic for North America and western Europe: (a) continued divergence of American and European Plates after the breakup of the Precambrian supercontinent; (b) Proto-Atlantic ocean begins to close as American and European Plates converge, causing the formation of a subduction zone and a volcanic arc off the North American margin; (c) continued plate convergence, subduction, and the approach of Avalonia; (d) Avalonia and European Plate collide with North American Plate, forming Laurasia.

The Southern Hemisphere was home to the large landmass known as *Gondwanaland,* which contained the present-day continents of Africa, Australia, Antarctica, South America, New Zealand, India, and southeastern Asia. During the Ordovician, large ice sheets developed over much of present-day Africa, which was then located at the south pole (Figure 25.15). By the Silurian, western Africa and South America had shifted to the south pole.

Later in the Paleozoic (Devonian to Permian periods), the collision of Siberia with eastern Europe (which was still part of Laurasia) and of Gondwanaland with Laurasia resulted in the supercontinent of *Pangaea* (Figure 25.16). Mountain-building activity continued and was widespread throughout the Appalachian Mountains in North America, the Hercynian and Caledonian Mountains in Europe, and the Ural Mountains in Russia. Disturbances affected not only continental margins but also inner regions. The ancestral Rocky Mountains, for example, were intracontinental byproducts of the dramatic collision.

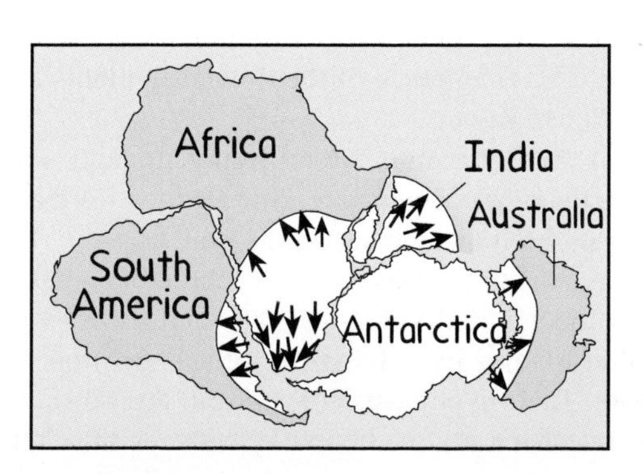

Fig. 25.15
During the Ordovician period, Gondwanaland was situated in the Southern Hemisphere, with Africa over the south pole. Arrows depict direction of glacial movement. Recall that glacial striations provide clues for the positioning of the continents.

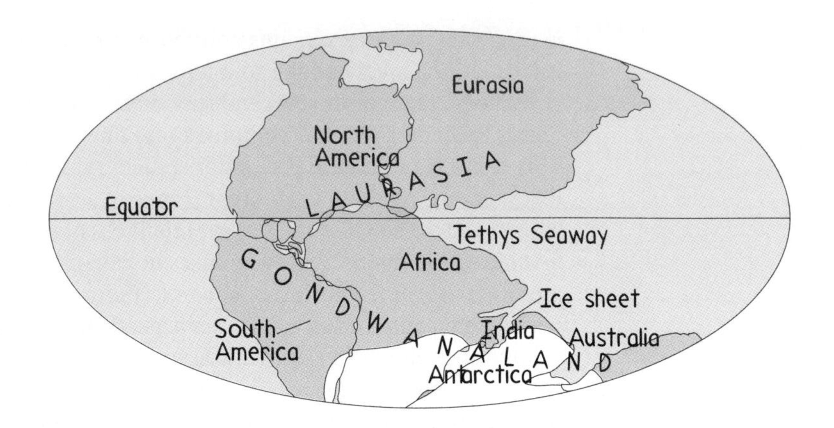

Fig. 25.16
With the dramatic collision of Gondwanaland with Laurasia in the late Paleozoic, the supercontinent Pangaea was formed.

With Pangaea now formed, the repositioned landmasses had a tropical seaway between them, known as the Tethys Sea. This paleogeographical arrangement contributed greatly to the different climatic belts. North America, Europe, and North Africa were located close to the equator in the trade-winds belt, where the climate ranged from humid uplands to drier lowlands. In the Appalachians, the climate was monsoonal with seasonal rainfall. Geologic evidence suggests that the ancient Appalachians were at least as high as the present-day Rockies or Andes Mountains, and possibly even as high as the present-day Himalayas. The high mountains blocked incoming moisture, casting a rain shadow across all of central North America and parts of western North America. The mid-African region was also very dry. The southern climate of Gondwanaland was dominated by widespread glaciation due to the close proximity to the south pole. Paleomagnetic evidence suggests that Pangaea was drifting as a unit across the south pole, accounting for the shift in centers of glaciation. Being very large, Pangaea greatly influenced the climate belts and the evolution of land life.

25.5 The Mesozoic Era

The Mesozoic era, known as "the age of reptiles," is made up of three periods: *Triassic, Jurassic,* and *Cretaceous.* Reptiles that survived the Permian extinction at the end of the Paleozoic era evolved to become the rulers of the world. The most significant event of the Mesozoic era was the rise of the dinosaurs. Mammals evolved from reptiles early in the Mesozoic but were relatively small and insignificant compared to the dinosaurs.

Land plants greatly diversified during the Mesozoic era. True pines and redwoods appeared and rapidly spread throughout the land. Flowering plants arose in the Cretaceous period and diversified so quickly that by the end of the period they were the dominant flora. The emergence of the flowering plants also accelerated the evolution and specialization of insects.

The end of the Cretaceous, 66 million years ago, was a time of great extinction. The dinosaurs, flying reptiles, and marine reptiles were completely wiped out, as were many nonreptiles, both on land and in the seas.

The cause of the extinction is still a source of some debate among scientists. Perhaps the best-documented hypothesis was put forth by the father-son Alvarez team. Luis and Walter Alvarez hypothesize that the extinction was caused by the impact of a large meteorite. Their hypothesis comes from their discovery of an abundance of iridium in sediments that mark the boundary between the Cretaceous and Tertiary periods.

In general, the composition of large meteorites is similar to the composition of the Earth, including similar concentrations of iridium. Recall from Chapter 23 that the Earth is layered (core, mantle, and crust). Because most of the Earth's iridium is deep in its interior, the concentration of iridium in a meteorite is higher than the iridium concentration in the Earth's crust. Yet all over the world the concentration of iridium in sediments at the Cretaceous-Tertiary boundary is much greater than in sediments above or below the boundary. The Cretaceous-Tertiary boundary layer was deposited about 66 million years ago—the time of the great dinosaur extinction.

The Alvarez team postulated that a large meteorite hit the Earth with such force that a gigantic light-blocking cloud of dust developed—a "nuclear winter"—that lasted for months, perhaps even longer. The dust cloud stopped photosynthesis, terminated the food supply, and chilled the Earth. Nuclear winter subsided when the dust settled, depositing the layer of iridium-enriched sediment discovered by the Alvarezes. Other killing mechanisms associated with an impact of this magnitude include acid rain, tsunamis, wildfires, and a delayed greenhouse effect. According to recent research, the site of the impact crater is located in the Mexican Yucatan peninsula.

An alternative to the Alvarez hypothesis suggests that the iridium layer may have been generated from massive volcanic eruptions. The ash and debris from these eruptions also could have blocked out the Sun. A third possibility is that large-scale volcanic eruptions could have been caused by the impact of an extraterrestrial object.

The Cretaceous extinction marked the close of the Mesozoic era.

Mesozoic Tectonics

The Mesozoic era witnessed the initial breakup of Pangaea (Figure 25.17). The breakup began at the end of the Triassic period with the eruption of extensive basalt flows associated with two major rift zones. The northern rift zone initiated the separation of North America from Gondwanaland, thus forming the central Atlantic ocean basin. The southern rift zone developed into a triple junction—a point where three lithospheric plates meet. During the Jurassic period, India started on a northward journey from the triple junction, while South America/Africa separated from Australia/Antarctica. The south Atlantic Ocean formed after the split of South America and Africa during the Cretaceous period. A major fault developed between Africa and Europe, resulting in the fragmentation of the western continental shelf. As the continental shelf broke up, numerous microcontinents were created. The breakup of Pangaea occurred during the entire Mesozoic era, and the plate movements that initiated the breakup continue today. Of all the former continental unions begun in Paleozoic time, only that of Europe and Asia has survived to the present time.

Subduction of the Farallon Plate (Section 23.4) and tectonic accretions to the North American continent began no later than the Triassic period. This activity produced deformation and widespread volcanism in both the North American and the Andean mountain belts. Granitic batholiths of the Andes and the Sierra Nevadas are the remnants left behind from the numerous volcanic arcs that rimmed the eastern Pacific basin. Successive volcanic arc collisions with the continental margin triggered mountain-building activity across California and Nevada.

Probably due to the breakup of Pangaea, a worldwide rise in sea level occurred during the Cretaceous period. Over one-third of the present land area was submerged. With such an expanse of water covering the land, mild temperate to subtropical conditions dominated most continental areas. Mild ocean temperatures were spread worldwide, with waters in midlatitude regions averaging 25°C and those in north polar regions averaging 10°C.

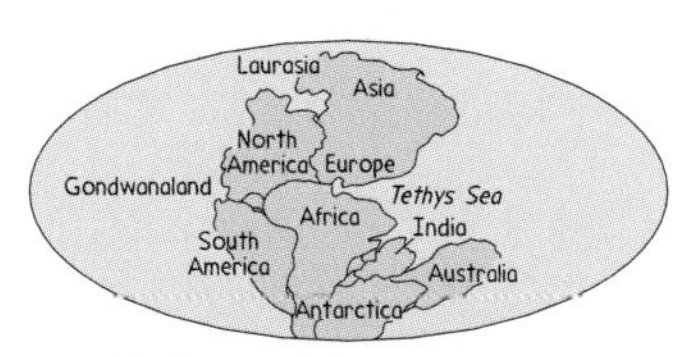

200 million years ago
Mesozoic Era

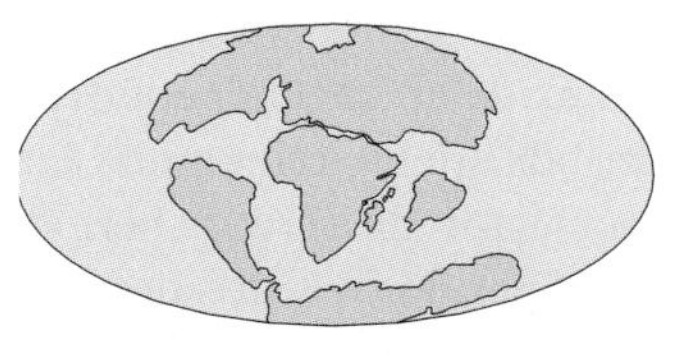

65 million years ago
Cenozoic Era

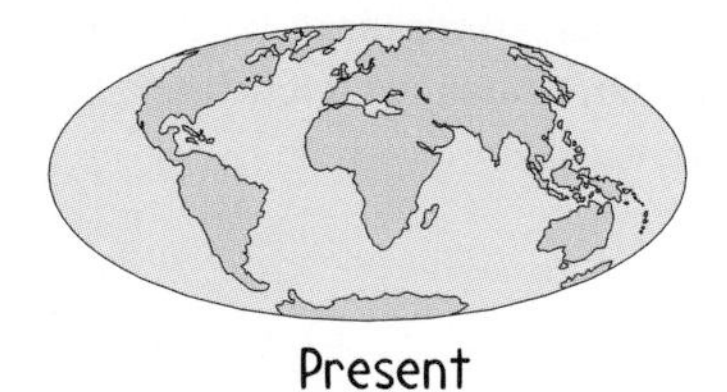

Present

Fig. 25.17
Stages in the breakup of Pangaea.

Link to Biology—Life Can Thrive in *Very* Inhospitable Places

Recent reports of possible fossil evidence for primitive life on Mars certainly spark the imagination. Here on the Earth, we know that life can be found in every nook and cranny of what we term the *biosphere.* We know that simple life forms thrive even in hot springs on land and in the deep ocean near spreading ridges. Whereas we used to think of life as confined to the biosphere, however, we now find life some 3000 meters beneath the Earth's surface—in solid rock! Clearly, there is no photosynthesis at this depth—nor is there free oxygen. And it is very hot: 45–85°C. The bacteria found here, named *Bacillus infernus,* survive under pressures exceeding 3000 pounds per square inch.

Bacteria need to "breathe" just as we do. The breathing is actually an oxidation/reduction reaction between oxygen (the oxidizing agent) and organic carbon (the reducing agent) that gives the bacterium energy much in the same way we get energy from eating and breathing air. Not all bacteria use oxygen, however. It has been known for some time that near-surface bacteria, such as those found in lakes or shallow aquifers, can use such other oxidants as iron, manganese, nitrate, and sulfate to oxidize organic carbon and gain energy. These bacteria can also utilize a variety of compounds that contain organic carbon, such as gasoline and many industrial solvents. This fact has been a boon to people who are attempting to clean up groundwater aquifers contaminated with such compounds.

But it is only very recently that scientists have discovered bacteria that utilize iron surviving at depths as great as 3000 meters. These bacteria were recovered in 1992 and 1994 from deep holes drilled in sedimentary rock in Colorado and Virginia. The colonies appear to be more than 100 million years old (Jurassic period).

The predecessors of these hearty little creatures could have survived on the surface of the very hot, oxygen-devoid, primitive Earth. Because a byproduct of their metabolic process is carbon dioxide, related creatures may even have played a pivotal role in the formation of the early atmosphere. They eat iron oxide (rust), and the mineral magnetite is formed in the process. Thus bacteria such as these may have been responsible for the formation of the Earth's very old and poorly understood banded iron deposits.

Amazingly, subsurface life, which includes *Bacillus infernus,* may add up to more weight than all the living things on the Earth's surface combined.

25.6 The Cenozoic Era

The Cenozoic era, known as the "age of the mammals," is made up of two periods—*Tertiary* and *Quaternary.* From oldest to youngest, the periods are broken up into the *Paleocene, Eocene, Oligocene, Miocene,* and *Pliocene* epochs for the Tertiary period; and the *Pleistocene* and *Holocene* epochs for the Quaternary period. We are currently in the Holocene epoch.

After the mass extinctions at the end of the Mesozoic era, many environmental niches were left vacant, allowing the relatively rapid evolution of mammals in habitats formerly occupied by their extinct predecessors.[4] Flying bats, some large land mammals, and marine animals such as whales and dolphins began to occupy niches left vacant by the demise of many of the Mesozoic reptiles.

---

[4]Biologists refer to this phenomenon as *adaptive radiation* because organisms begin to adapt to new environments and radiate, or diversify, away from a smaller set of ancestors.

Link to Anthropology—Early Humans

The earliest known hominid fossils are 3.8 million years old. Of particular interest is the nearly complete skeletal fossil of Lucy, a 20-year-old female of the *Australopithecus afarensis* species. This species was partially bipedal but probably not as erect as the later *Homo erectus.* Lucy (named after the popular Beetles song, "Lucy in the Sky with Diamonds") had a large, overhung jaw with canine and incisor teeth, and an undetermined brain size. Later species of *Australopithecus* became more upright as bipedalism increased and a slightly larger brain developed. *Homo erectus,* dated at 1.6 million years old, had a fully erect posture. With a brain size that surpassed 1000 cubic centimeters, evidence of complex behavior, and the development of social patterns, it is quite likely that *Homo erectus* evolved into the modern human species, *Homo sapiens.* Remnants of the first *Homo sapiens* were found in Africa and Europe in middle Pleistocene rocks. These early *Homo sapiens* are called *Neanderthal* people. They're characterized by heavy eyebrow ridges, a pronounced chin, and a stocky body. The brain size averaged 1300 cubic centimeters. The Neanderthal people were cave dwellers. Evidence of their well-developed society is indicated by their fabrication of stone tools, knowledge of fire, and burial of the dead.

Homo sapiens first appeared 90,000 years ago in South Africa, 50,000 years ago in the Middle East, and 35,000 years ago in Europe. These early humans, known as *Cro-Magnon,* are characterized by a high, vertical forehead, a short skull, and a rounded face. Cro-Magnon brain size averaged 1300 cubic centimeters, but may have been as large as 1400 cubic centimeters. Neanderthal and Cro-Magnon humans coexisted for about 10,000 years, but Cro-Magnon humans dominated and spread rapidly throughout the world.

As humans moved into new areas, many characteristics evolved to facilitate adaptation to particular environments. For example, different skin color probably evolved in response to the ultraviolet light intensity at different latitudes. Pigmented skin acts as an effective ultraviolet filter, protecting delicate tissue and preventing the overproduction of vitamin D. As humans moved into higher latitudes, the ultraviolet light grew less intense, likely insufficient to promote the synthesis of vitamin D that occurs in the skin. So selective pressures operated toward lighter skin.

No later than 10,000 years ago, all major groups of modern humans had appeared and occupied their primary distribution areas.

Although modern humans appeared more than 90,000 years ago, investigators today believe that it wasn't until about 25,000 years ago that humans reached North America. With the lowering of sea level due to glaciation, the Bering land bridge allowed the entrance of humans from Asia to the North American continent. The same route was likely used earlier by the woolly mammoth, reindeer, and other late Cenozoic animals.

Climates cooled during much of the Cenozoic era, culminating in the widespread glaciation that characterized the Pleistocene. Although this *ice age* continues today, there have been many alternations between glacial and interglacial conditions (see box). During the glacial episodes, as much as one-third of the present land area has been covered by great thicknesses of ice, depressing the land by its weight and altering the courses of many streams and rivers. The glaciers eroded and scratched the land in some places and deposited huge moraines in others, leaving behind abundant evidence of the extent of their former existence.

The Cenozoic era also brought about the advent of humans. The extensive glaciation of the Pleistocene caused sea level to drop because a great deal of water was tied up in glaciers. Even though the distribution of landmasses was essentially the same as it is today, the lowered sea level resulted in "land bridge" connections between landmasses that are now separated by water. One of these land bridges existed across the present-day Bering Strait and provided the route for the human migration from Asia to North America. The expansion of humans, not only into North America but throughout

Link to Global Thermodynamics—Is It Cold Outside?

Yes it is, relatively speaking. For 90 percent of the Earth's history, there were no glaciers of continental magnitude anywhere. In fact, because such glaciers exist today, mainly in the polar ice caps and Greenland, we are technically now in an *ice age.* Because continental-scale glaciers are currently restricted to the polar regions, we are in what is known as an *interglacial* period of an ice age.

Ice ages have occurred five times over the course of Earth's history. The first one we find evidence of occurred more than 2 billion years ago. Another began about 840 million years ago and lasted an incredible 240 million years! There were two ice ages during the Paleozoic era, but none in the Mesozoic. For the first 50 million years or so of the Cenozoic era, there were also no ice ages. The present ice age actually began 8–10 million years ago, but the extensive glaciation that characterized the Pleistocene epoch began about 1 million years ago .

So what causes ice ages? There most likely is no single explanation, but most scientists agree that global-scale cooling leading to ice ages is caused by the right combination of three things: (1) the arrangement of continents around the globe, (2) the amount of sunlight reflected back into space, and (3) the geometry of the Earth's rotation on its axis and revolution around the Sun.

The arrangement of continents greatly influences ocean and atmospheric currents, which are the main mechanisms for redistributing ocean and atmospheric thermal energy around the globe. Continents grouped together in one location are easier to warm, as equatorial waters flow with less obstruction toward the poles. For continents spread out around the globe, as they are today, circulation "cells" are smaller and heat redistribution is more local and less efficient.

When sea level is lower, for whatever reason, more land area is exposed. This increased amount of land area tends to increase the amount of sunlight reflected back into space, which results in cooler temperatures globally. Cloud cover and/or dust in the atmosphere also causes sunlight to be reflected back into space, reducing the absorption of solar radiation.

The *Milankovitch effect* refers to a combination of factors that affects the distribution of solar radiation over the Earth's surface during the year: (1) variations in the angle at which the Earth's rotation axis is tilted (currently about 23.5 degrees), (2) the wobbling of the Earth's rotation axis, and (3) variations in the eccentricity ("ovalness") of the Earth's orbit around the Sun. Certain combinations of these factors, which recur periodically, lead to reduced solar radiation at high northern latitudes during the summer. If the reduction is great enough, then all the snow from the preceding winter does not melt and, if these conditions persist over many years, continental-scale glaciers eventually form. The periodic nature of the Milankovitch effect may be the primary cause of glacial-interglacial cycles. According to this theory, the first two processes—arrangement of continents and amount of reflected radiation—make the Earth cold enough, and the third—the Milankovitch effect—causes the climate to teeter-totter between glacial and interglacial periods.

So what's next? Large-scale glaciation or global warming? Only further research can provide hope of finding an answer. Anything less is pure speculation, at best.

the world, coincided with a period of extinction that occurred during the Pleistocene. The Pleistocene extinctions primarily involved large terrestrial mammals; marine animals were for the most part unaffected. In North America, many large mammals became extinct after humans arrived, and in Africa, mammalian extinctions can be related to the appearance of the Stone Age hunters.

The cause of the Pleistocene extinction is a much-debated issue. The extreme climatic variation that existed at the time could have been partly responsible. However, even though large-scale glaciation was occurring in some regions, the climate in many areas was relatively mild, leading some scientists to believe that harsh climate was not the only factor in the extinctions. The Pleistocene extinctions were probably the result of a combination of numerous factors. Further research may elucidate the causes.

Cenozoic Tectonics

Structural disturbances of great magnitude occurred rapidly throughout the world during the Tertiary period. In the early Tertiary period, there was a spreading center off the western margin of North America, with the Pacific Plate on the west and the Farallon Plate on the east (Figure 25.18 and 25.19). As the Farallon Plate was subducted beneath North America, the spreading center approached the North American continental

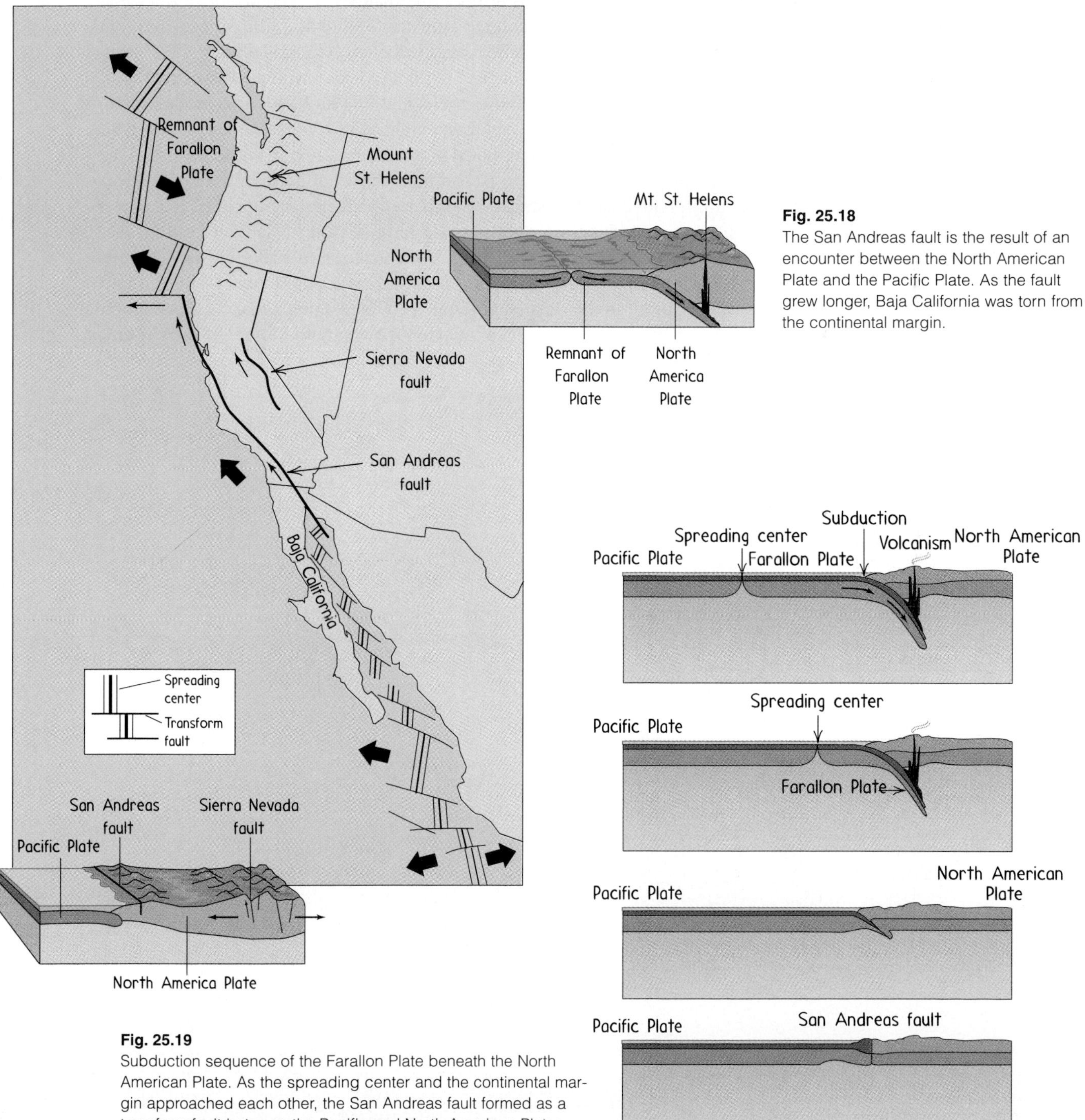

Fig. 25.18
The San Andreas fault is the result of an encounter between the North American Plate and the Pacific Plate. As the fault grew longer, Baja California was torn from the continental margin.

Fig. 25.19
Subduction sequence of the Farallon Plate beneath the North American Plate. As the spreading center and the continental margin approached each other, the San Andreas fault formed as a transform fault between the Pacific and North American Plates.

margin. The collision between the westward moving North American Plate and the Pacific ridge system occurred about 30 million years ago, giving birth to the San Andreas Fault. Baja California was torn away from the Mexican mainland, and the Gulf of California was created. Because the plates are still moving, western California and Baja California will eventually either become completely detached from the mainland or will find themselves joined to western Canada.

The Hawaiian Island/Emperor Seamount chain (Figure 25.20) gives evidence of another significant Tertiary structural disturbance: the change in direction of the Pacific Plate. The bend in the chain of islands occurred between 30 and 40 million years ago (mid-Tertiary) when plate motion changed from nearly due north to northwesterly. The change in direction corresponds with the collision of northern Mexico with the Pacific ridge.

The collision between the North American and Pacific Plates also produced compressional forces within the North American continent, forces that led to events that characterize our present era. Compressional mountain building in western North America culminated in the regional upwarping and final phases of uplift of the Rocky Mountains. This uplift rejuvenated all streams and rivers by steepening the gradients. The reinforced streams and rivers brought about a long period of canyon cutting. Also, widespread flood basalts spilled out from deep fissures to make up the Columbia Plateau in Washington and the Craters of the Moon in Idaho. Crustal extension produced normal faulting in the Basin and Range province, encompassing all of Nevada and parts of California and Utah. The late Cenozoic saw the uplift of the Colorado

Fig. 25.20
The Hawaiian Island/Emperor Seamount Chain. The bend in the chain shows the change in direction of the Pacific Plate as a result of the collision of northern Mexico with the Pacific ridge. The red numbers indicate the age (in millions of years) of the individual islands and seamounts.

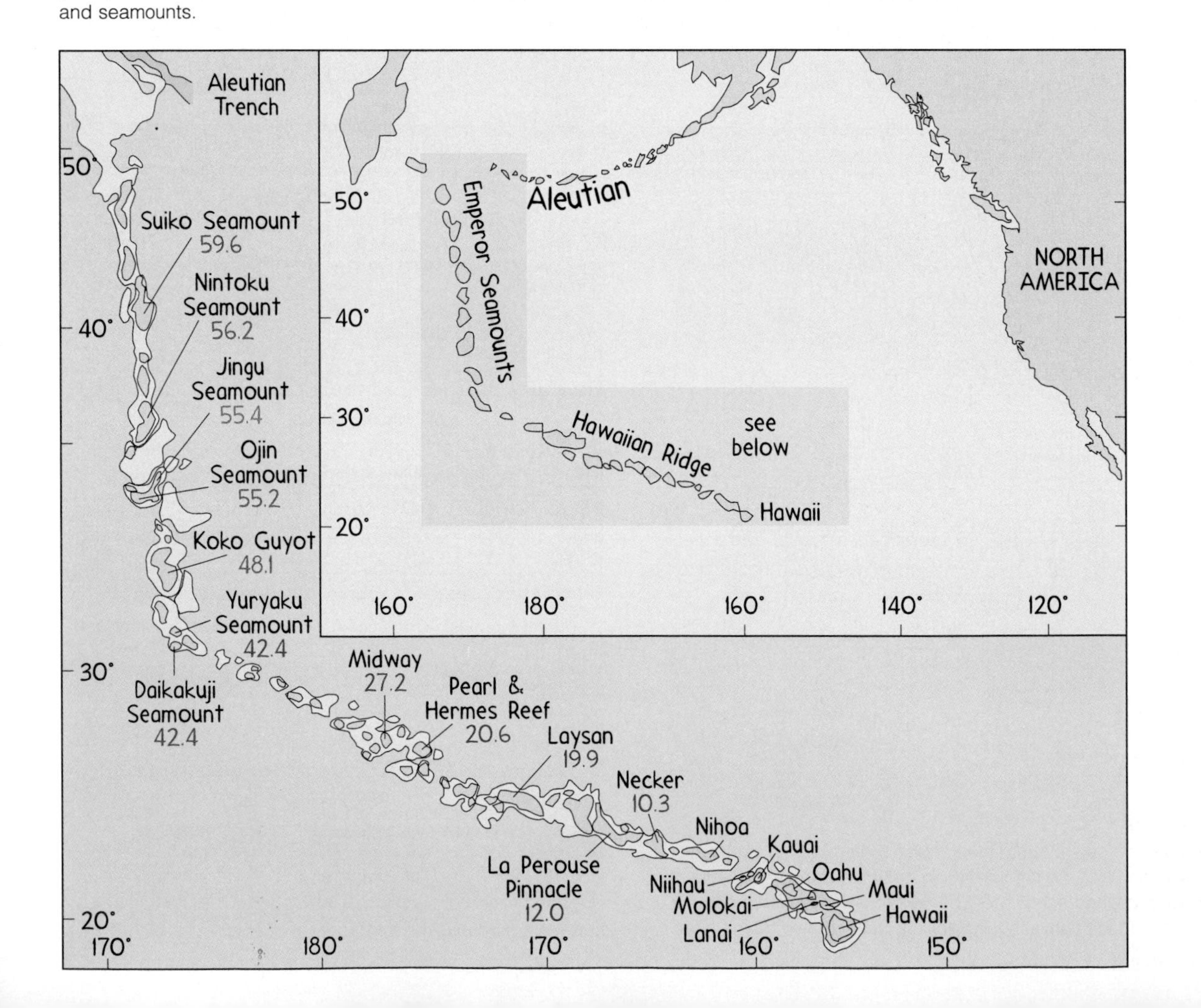

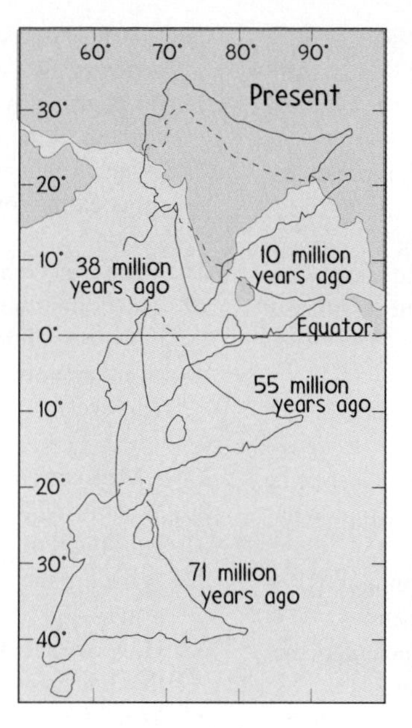

Fig. 25.21
The formation of the Himalayas was a result of the collision of India with Asia. As this was a continent-to-continent collision, the Himalayas have an unusually thick accumulation of continental lithosphere. Like the bottom of an iceberg, the mountains run deeper than they are high.

Plateau and the resulting downcutting of the Colorado River, which formed the Grand Canyon.

Considerable tectonic activity also occurred in Eurasia during the Tertiary period. Recall the formation of microcontinents after the breakup of Pangaea. These microcontinents were transported northeastward and eventually collided with southern Eurasia. The culmination of these collisions occurred in the mid-Cenozoic when Afro-Arabia collided with Europe to produce the Alps and India collided with Asia to produce the Himalayas. The leading edge of the Indian Plate was forced partially under Asia, generating an unusually thick accumulation of continental lithosphere. Due to isostasy, the thick lithosphere gave additional uplift to the Himalayas.

Human Geologic Force

Although the "human age" amounts to only a brief 0.05 percent of geologic time, we are almost certainly the most clever and adaptable organism to have evolved on the planet. All life forms alter their environment. Humans do it more, as we manipulate it to meet our needs. We have but to look at the irrigation systems of Mesopotamia, the cultivation of the Nile, the plowing of the prairies in the Great Plains, the invention of machines to further utilize the land, and the dams and locks on the Mississippi, Missouri, and Colorado rivers to illustrate the human role in geologic changes. These geologic changes also include such problems as deterioration of the ozone layer, hydrocarbon pollution, and global warming. Because we have the capacity to affect geologic change, it is imperative that we take care of our terrestrial home. It's the only one we've got!

Summary of Terms

- **Original horizontality** Layers of sediment are deposited evenly, with each new layer laid down almost horizontally over the older sediment.
- **Superposition** In an undeformed sequence of sedimentary rocks, each bed or layer is older than the one above and younger than the one below.
- **Cross-cutting** When an igneous intrusion or fault cuts through other rocks, the intrusion or fault is younger than the rock it cuts.
- **Inclusions** Any inclusion (pieces of one rock type contained within another) is older than the rock containing it.
- **Faunal succession** Fossil organisms succeed one another in a definite, irreversible, determinable order.
- **Unconformity** A break or gap in the geologic record, caused by an interruption in the sequence of deposition or by erosion of preexisting rock.
- **Angular unconformity** An unconformity in which older tilted strata are overlain by horizontal younger beds.
- **Nonconformity** Overlying sedimentary rocks on an eroded surface of intrusive igneous or metamorphic rocks.
- **Radiometric dating** A method of calculating the age of geologic materials based on the nuclear decay of naturally occurring radioactive isotopes.
- **Paleozoic** The time of ancient life, from 544 million years ago to 245 million years ago.
- **Mesozoic** The time of middle life, from 245 million years ago to about 66 million years ago.
- **Cenozoic** The time of recent life, began 66 million years ago and still going on.
- **Precambrian** The time of hidden life, which began about 4.6 billion years ago when the Earth formed, lasted until about 544 million years ago (beginning of Paleozoic), and makes up 85% of the Earth's history.

Review Questions

Relative Dating

1. Does the concept of relative dating state that rocks are formed in a definite sequence?
2. What five principles are used in relative dating?
3. When a granitic dike is found in a bed of sandstone, what can be said about the age of the dike and the age of the sandstone? What is this principle called?
4. What is a fossil?
5. How are fossils used in determining geologic time?
6. In what type of rock do we find fossils?
7. In a sequence of sedimentary rock layers, the oldest layer is on the bottom and the youngest layer at the top. What principle does this observation fit?
8. Why aren't rock formations found with a continuous sequence from the beginning of time to the present?
9. Explain how fossils of fishes and other marine organisms occur at high elevations such as the Himalayas.
10. In an undeformed sequence of rocks, fossil X is found in a limestone layer at the bottom of the formation, and fossil Y is found in a shale layer at the top of the formation. What can we say about the ages of fossils X and Y?

Radiometric Dating

11. What is the definition of half-life?
12. What are the half-lives of uranium-238, potassium-40, and carbon-14?
13. What isotope is best for dating very old rocks?
14. What isotope is commonly used for dating sediments or organic material from the Pleistocene?

The Precambrian Time

15. Which of the geologic time units spans the greatest length of time?
16. How old is the Earth?

The Paleozoic Era

17. Which major events in Earth history begin and end each of the three eras of the geologic time scale?
18. The Paleozoic era experienced several fluctuations in sea level. What effect did this have on life forms?
19. Name the periods for the Paleozoic era.
20. What is the Silurian period best known for?
21. The Devonian is known as "the age of fishes." What were some of the Devonian life forms?
22. What is the significance of the development of internal nostrils in the lobe-finned fishes?
23. Why do many geologists consider the lobe-finned fishes especially significant?
24. During what time period were coal deposits laid down? Why was this period unique?
25. In what area of the United States do we find rich coal deposits?
26. What group evolved from the amphibians with the development of the amniote egg?
27. What circumstances are thought to have contributed to the great extinctions that took place at the end of the Permian period?
28. What was Gondwanaland?
29. The collision of Gondwanaland and Laurasia resulted in the supercontinent of Pangaea. What effect did this collision have on the land features? On climatic regions? On sea level?

The Mesozoic Era

30. What is the Mesozoic era known as?
31. Dinosaurs ruled the Earth during which periods?
32. What are the most likely causes of the Cretaceous extinction?
33. How does iridium relate to the time of the extinction of the dinosaurs?
34. What effect did the breakup of Pangaea have on sea level?
35. What Pangaean landmass survives to this day?

The Cenozoic Era

36. What life forms characterized the Tertiary period?
37. What life forms characterized the Quaternary period?
38. Which epochs make up the Tertiary? The Quaternary?
39. Explain how the Alps were formed.
40. What route did humans use to enter the Western Hemisphere?
41. What process is thought to be responsible for the folding and volcanism of the western margins of North and South America during the Cenozoic era?
42. What geological event resulted in the bending of the Hawaiian Island/Emperor Seamount chain?
43. How did tectonic activity in the Cenozoic era effect the Grand Canyon?
44. What role did tectonic activity play in the formation of the San Andreas fault?
45. How did the Pleistocene glaciation effect the land surface?
46. How was the Gulf of California formed?

Exercises

1. Suppose you encounter an outcrop of sedimentary rock overlaid by a basalt flow. A fault displaces the bedding of the sedimentary rock but does not intersect the basalt flow. Relate the fault to the ages of the two rock types in the formation.
2. If a sedimentary rock contains inclusions of metamorphic rock, which rock is older?

3. Refer to the following figure. Using the principles of relative dating, determine the relative ages of the rock bodies and other lettered features. Start with the question: What was there first?

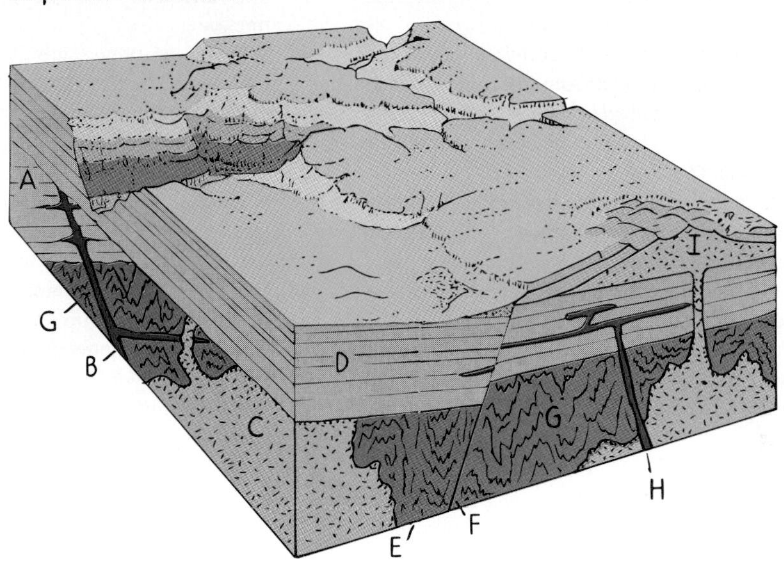

4. Before the discovery of radioactivity, how did geologists estimate the age of rock layers?
5. Which isotopes are most appropriate for dating rocks from the following ages: (a) the early Precambrian; (b) the Mesozoic; (c) the late Pleistocene?
6. Has the amount of uranium in the Earth increased over geologic time? Has the amount of lead increased? Explain.
7. In dating a mineral, what is meant by "resetting of the mineral's time clock"?
8. If we divide a number by 2, and then divide the result by 2, and so on indefinitely, the answer will never be zero. Why then is carbon dating useful only for materials that are no older than about 50,000 years? (*Hint:* What is the half-life of carbon-14?)
9. Granitic pebbles within a conglomerate have a radiometric age determination of 300 million years. What can you say about the age of the conglomerate? Nearby, an outcrop of the same conglomerate is intruded by a dike having a radiometric age of 200 million years. What can we say about the age of the conglomerate?
10. Suppose that in an undeformed sequence of rocks you find a trilobite embedded in shale layers at the bottom of the formation and fossil leaves embedded in shale at the top of the formation. From your observation, what can you tell about the ages of the formation?
11. In a sequence of sedimentary rock layers the youngest layer is found at the bottom and the oldest layer at the top. What does this type of layering signify?
12. Geologists often refer to the "Cambrian Explosion" when describing the record of the early Paleozoic. What is meant by this phrase?
13. What is the difference between a nonconformity and an angular unconformity?
14. If fine muds are laid down at a rate of 1 cm per 1000 years, how long would it take to accumulate a sequence 1 km thick?
15. What key developments in life occurred during the Precambrian era?
16. What factors are believed to have contributed to the generation of free oxygen during the early Precambrian? In what way did the increase in oxygen affect our planet?
17. The earliest stromatolites were anaerobic. Are the modern-day stromatolites anaerobic? Explain.
18. What evidence do we have of Precambrian life?
19. Why can we find Paleozoic marine sedimentary rocks, such as limestone and dolomite, widely distributed in the continental interiors?
20. Coal beds are formed from the accumulation of plant material that becomes trapped in swamp floors. Yet coal deposits are present in the continent of Antarctica, where no swamps or vegetation exists. How can this be?

21. Did humans and the dinosaurs exist at the same time? What is your evidence?
22. If large meteorites and the Earth as a whole have the same concentration of iridium, why does the discovery of an iridium-rich sedimentary layer suggest that the impact of an extraterrestrial object occurred at the end of the Cretaceous period?
23. How have recent humans affected geological processes?
24. A radiometric date is determined from mica that has been removed from a rock. What does the date signify if the mica is found in granite? What does the date signify if the mica is found in schist?
25. During the Earth's long history, life has emerged and life has perished. Briefly discuss the emergence of life and the extinction of life for each era.
26. Do you think the breakup of Pangaea affected the Cretaceous extinction? Explain.
27. Some of the Cenozoic life forms grew to great proportions. What are some of the possible causes for their great size?
28. Why does sea level rise when the rate of seafloor spreading increases?
29. In what ways could sea level be lowered? What effect might this have on existing life forms?
30. What could cause a rise in sea level? Is this likely to happen in the future? Why or why not?
31. If sea level were to rise today, what land areas would be most affected? What life forms would be in danger of extinction?
32. What circumstances are likely to lead to the formation of continental-scale glaciers? What is a likely cause of glacial-interglacial cycles? Are we currently in an ice age? Explain.
33. When humans spread to new areas, they changed as they adapted to their new environments. What were some of these changes? Are the effects of these changes still seen today?
34. What is the distinguishing difference between humans and other life forms from the past to the present? Discuss the possibility that humans will one day become extinct.
35. What would you say to convince a friend with no scientific background that the Earth is as old as it is?

CHAPTER

The Atmosphere, the Oceans, and Their Interactions 26

The view of the Earth from space shows our planet to be distinct shades of blue and silver. The blue is due to the water in the oceans and the silver to clouds in the atmosphere. How our atmosphere and oceans came to be and how their properties and functions affect us are the subject of this chapter. We begin with the evolution of the atmosphere and the oceans, and explore important features of both. We then investigate the transfer of heat between these two fluid shells and how this transfer affects the Earth's climate. We conclude with a look at the mechanisms that influence atmospheric and oceanic circulation patterns and their relationship to changes in the Earth's climate.

Seventy percent of the Earth's surface is covered by water. The remaining 30 percent is land, most of which is located in the Northern Hemisphere (Figure 26.1). Although the many oceans are named for their various locations, they are really one big continuous ocean. As we learned in Section 24.1, the oceans are the

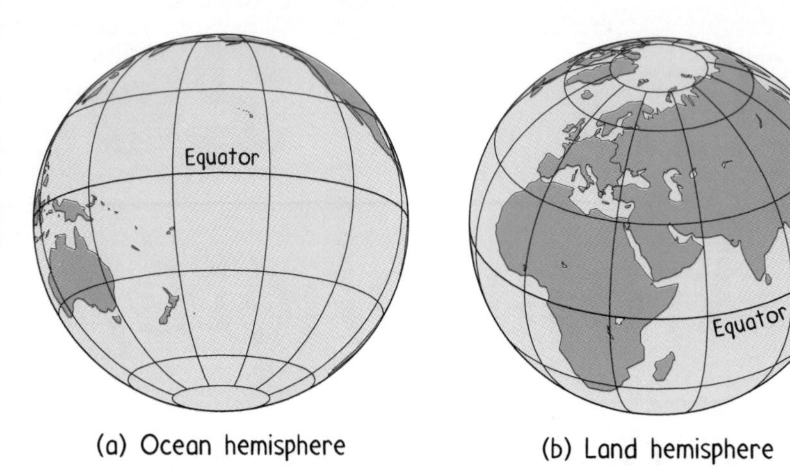

Fig. 26.1
Most of the Earth's surface is covered by water. We can divide the Earth into (a) an ocean-dominated hemisphere, and (b) a land-dominated hemisphere.

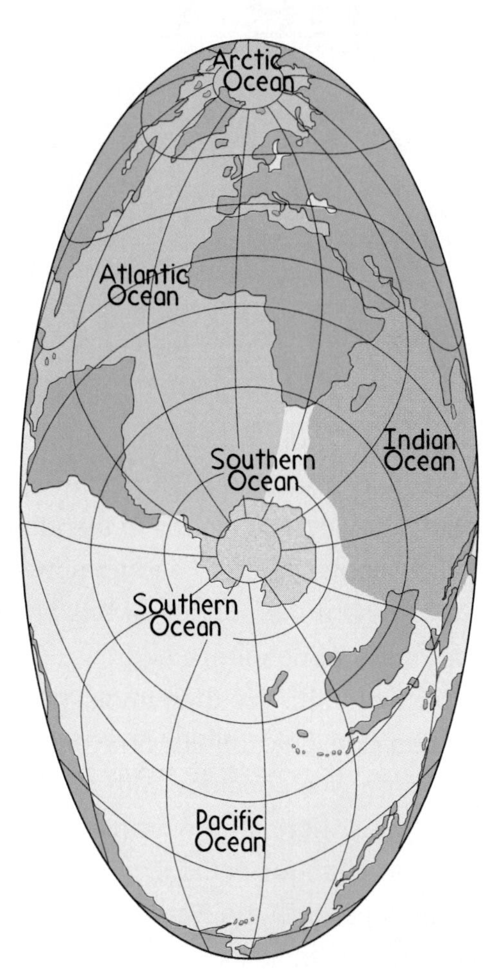

Fig. 26.2
When a map is centered over Antarctica, the expanse of the world ocean can be seen. In terms of size and volume, the Pacific Ocean accounts for more than half of the world ocean and is thus the largest ocean. In fact, the Atlantic and Indian Oceans combined would easily fit into the space occupied by the Pacific.

reservoirs from which water evaporates into the atmosphere to later precipitate as rain and snow. The oceans also play a direct role in moderating the Earth's temperature and climate. Recall from Chapter 6 that water has a high specific heat—which means it is slow to heat up or cool down and it transfers large amounts of thermal energy when its temperature changes. This property of water accounts for the moderate temperatures on lands bordering the oceans. The moderating influence of the oceans can be seen when we look at seasonal temperature variations for two cities at the same latitude: coastal San Francisco and continental Wichita (Figure 26.3). Whereas temperatures in San Francisco tend to have small seasonal variations, temperatures in Wichita show strong seasonal fluctuations—cold winters and hot summers. As we shall see, the interconnections between the oceans and the atmosphere have many far reaching consequences.

26.1 Evolution of the Earth's Atmosphere and Oceans

The Earth probably had an atmosphere before the Sun was fully formed. This primitive atmosphere was possibly composed principally of hydrogen and helium, the two most abundant gases in the universe, along with a few simple compounds such as ammonia and methane. Then when the temperature and pressure in the contracting center of the still-forming Sun became high enough to ignite thermonuclear reactions, our Sun was born. The blast from the Sun's formation must have produced a strong outflow of charged particles—an outflow strong enough to sweep the Earth of its earliest atmosphere.

The next stage in the formation of the atmosphere likely occurred when gases trapped in the Earth's hot interior escaped through volcanoes and fissures at the Earth's

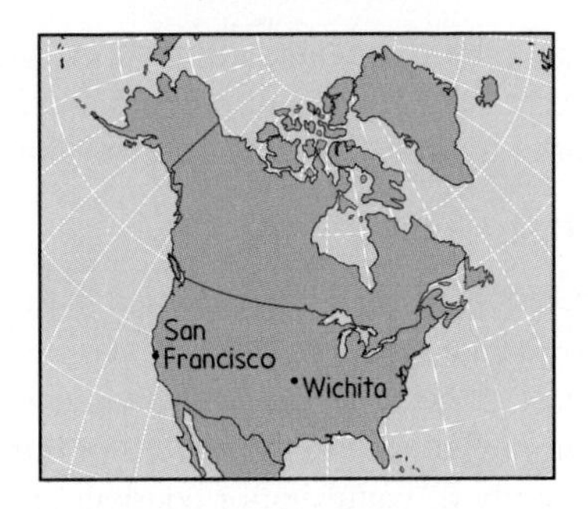

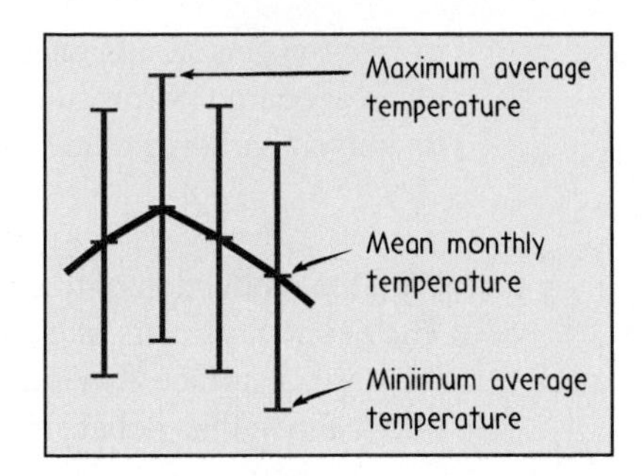

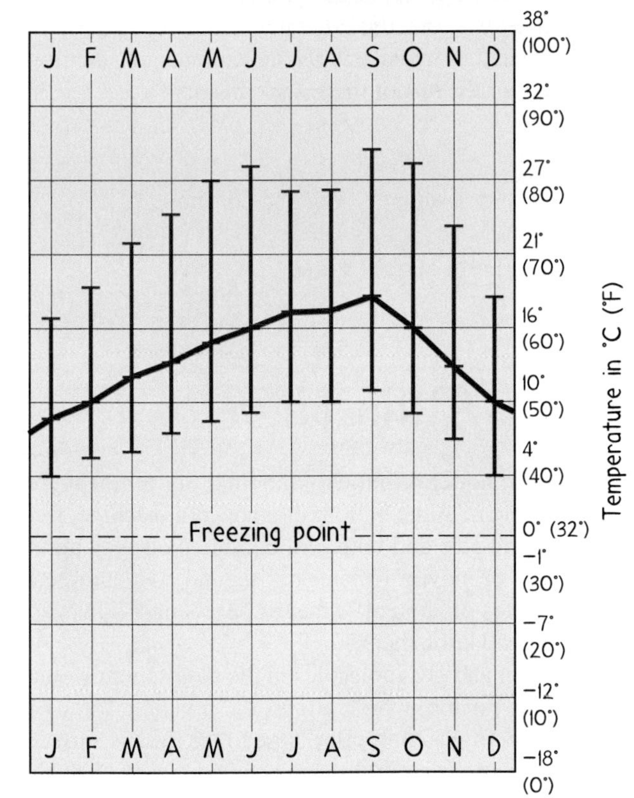

Station: San Francisco, California
Latitude/longtitude: 37°37′ N, 122°23′ W
Average annual temperature: 14°C (57.2°F)
Total annual precipitation: 47.5 cm (18.7 in.)
Elevation: 5 m (16.4 ft)
Population: 750,000
Annual temperature range: 9°C (16.2°F)

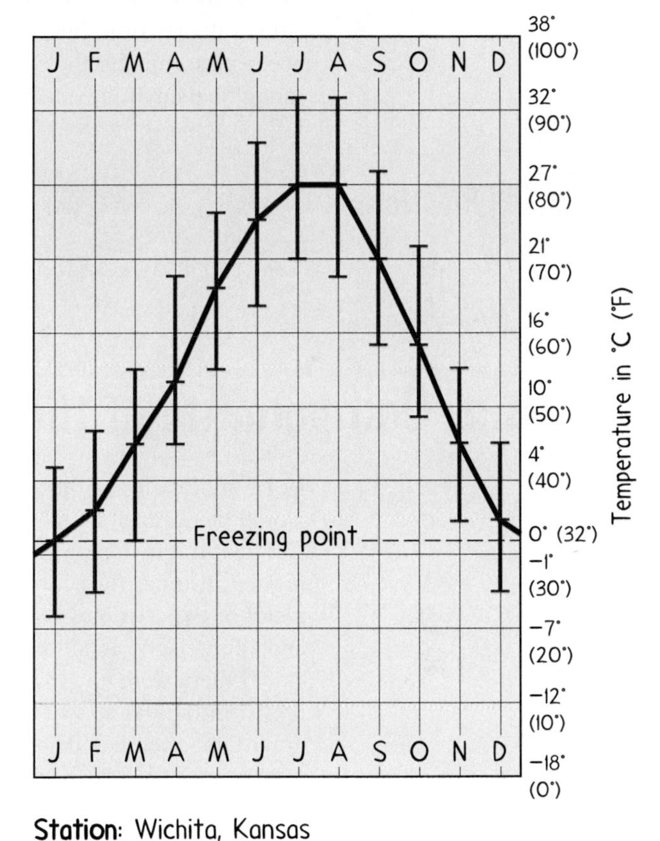

Station: Wichita, Kansas
Latitude/longtitude: 37°39′ N, 97°25′ W
Average annual temperature: 13.7°C (56.6°F)
Total annual precipitation: 72.2 cm (28.4 in.)
Elevation: 402.6 m (1321 ft)
Population: 280,000
Annual temperature range: 27°C (48.6°F)

Fig. 26.3
Comparison of seasonal temperature ranges for coastal San Francisco, California, and continental Wichita, Kansas.

surface. The gases spewed out by these early eruptions were probably much like the gases found in the volcanic eruptions of today—about 85 percent water vapor, 10 percent carbon dioxide, and 5 percent nitrogen, by mass. The early atmosphere had no free oxygen and therefore was inhospitable to the type of life we have today. The production of free oxygen did not occur until the primitive plants known as stromatolites and green algae appeared. Stromatolites and green algae, like all green plants, use photosynthesis to convert carbon dioxide and water to hydrocarbon and free oxygen.

$$CO_2 + H_2O + \text{light} \rightarrow CH_2O + O_2$$

With the production of free oxygen, an ozone (O_3) layer formed in the upper atmosphere. The ozone layer acts like a filter to reduce the amount of ultraviolet radiation reaching the Earth's surface. The surface therefore became more hospitable to life. The evolution of oxygen in this global envelope was a vital step in the history of Earth and its life.

As the Earth cooled, the rich supply of water vapor condensed to form the oceans. Comet debris from interplanetary space also contributed water to the oceans. These oceans, essential to the evolution of life and ultimately to the development of the present global environment, have remained for the rest of the Earth's history.

Questions

1. Why are the hottest climates on the Earth typically found in continental interiors?
2. Did the ozone layer exist before the Earth acquired green plants?

26.2 Components of the Earth's Atmosphere

If gas molecules in the atmosphere were not continuously moving, our atmosphere would lie dormant on the ground like popcorn at the bottom of a popcorn machine. But add heat to the popcorn, or to atmospheric gas, and both will bumble their way up to higher altitudes. Popcorn attains speeds of maybe 1 meter per second and can rise a meter or two. Air molecules move at speeds of about 450 meters per second, however, and a few rise to an altitude of more than 50 kilometers.

If there were no gravity, both popcorn and gas molecules in the atmosphere would fly off into outer space. Gases are compressible, which allows the invisible force of gravity to squeeze and hold a great number of gas molecules close to the Earth's surface (where gravity is strongest). Thus the density of air molecules is greatest at the Earth's surface and gradually decreases with height. Because air molecules have weight, they exert pressure, known as *atmospheric pressure* or simply *air pressure*. Like density, air pressure also decreases with increasing height above the Earth's surface.

Table 26.1 shows that the Earth's present-day atmosphere is a mixture of various gases—primarily nitrogen and oxygen with small percentages of water vapor, argon, and carbon dioxide[1], and minute traces of other elements and compounds.

Answers

1. The large heat capacity of water tends to keep coastal areas from experiencing extreme temperatures. Therefore very hot climates are usually some distance away from the ocean.
2. No. The formation of ozone, O_3, was preceded by the introduction of free oxygen, which came from photosynthesizing plants.

---

[1]Carbon dioxide is a minor constituent in the Earth's atmosphere because much of the carbon dioxide gases spewed into the air from the Earth's interior readily dissolve in ocean water. After carbon dioxide is dissolved in our oceans, it undergoes various chemical reactions, most of which lead to the formation of carbonate precipitates such as limestone.

Table 26.1
Composition of the Atmosphere

Permanent Gases			**Variable Gases**		
Gas	**Symbol**	**Percent by Volume**	**Gas**	**Symbol**	**Percent by Volume**
Nitrogen	N_2	78%	water vapor	H_2O	0 to 4
Oxygen	O_2	21%	carbon dioxide	CO_2	0.035
Argon	Ar	0.9%	ozone	O_3	0.000004*
Neon	Ne	0.0018%	carbon monoxide	CO	0.00002*
Helium	He	0.0005%	sulfur dioxide	SO_2	0.000001*
Methane	CH_4	0.0001%	nitrogen dioxide	NO_2	0.000001*
Hydrogen	H_2	0.00005%	particles (dust, pollen)		0.00001*

* Average value in polluted air.

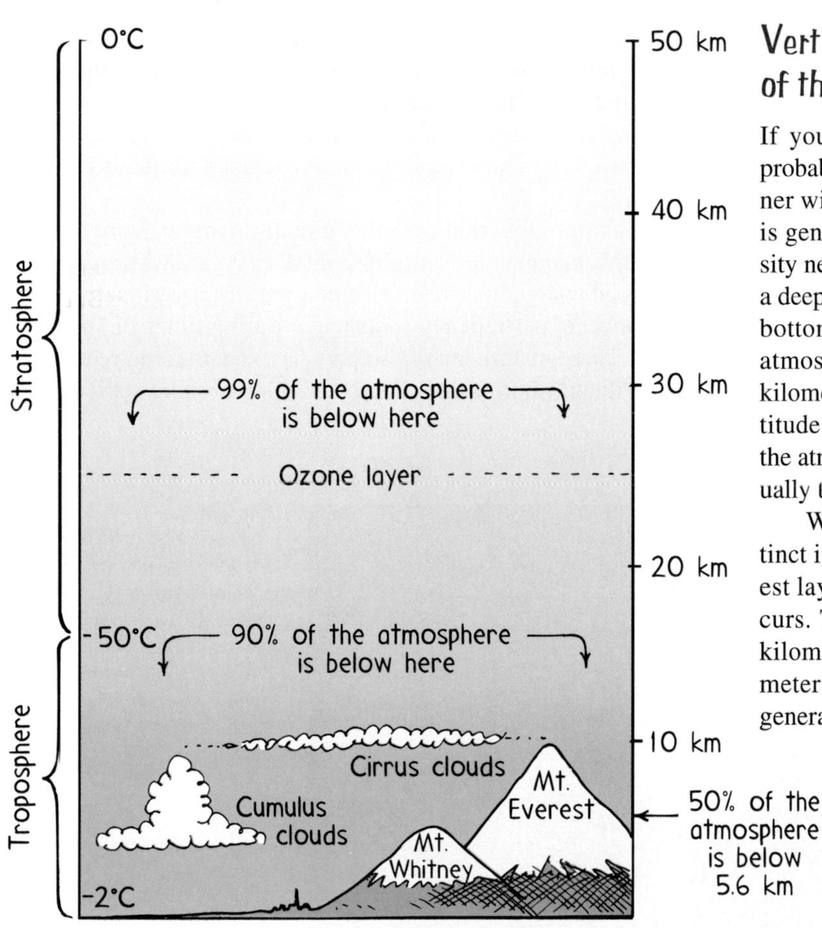

Fig. 26.4
The two lowest atmospheric layers, the troposphere and stratosphere.

Vertical Structure of the Atmosphere

If you have ever gone mountain climbing you probably noticed that the air grows cooler and thinner with increasing elevation. At sea level, the air is generally warmer and denser. The greater density near the Earth's surface is due to gravity. Like a deep pile of feathers, the density is greatest at the bottom and least at the top. More than half the atmosphere's mass lies below an altitude of 5.6 kilometers, and about 99 percent lies below an altitude of 30 kilometers. Unlike a pile of feathers, the atmosphere doesn't have a distinct top. It gradually thins to the near vacuum of outer space.

We classify the atmosphere in layers, each distinct in its characteristics (Figure 26.4). The lowest layer, the **troposphere**, is where weather occurs. The troposphere extends to a height of 16 kilometers over the equatorial region and 8 kilometers over the polar regions. Commercial jets generally fly at the top of the troposphere to minimize the buffeting and jostling caused by weather disturbances. Even though the troposphere is the thinnest atmosphere layer, it contains 90 percent of the atmosphere's mass and essentially all of the atmosphere's water vapor and clouds. Temperature in the troposphere decreases steadily (6°C per kilometer) with increasing altitude. At the top of the troposphere, temperature averages about –50°C.

Above the troposphere is the **stratosphere**, which reaches a height of 50 kilometers. Ultraviolet radiation from the Sun is absorbed by the ozone layer in the stratosphere. Because of this absorption of energy, temperature in the stratosphere rises from about −50°C at the bottom to about 0°C at the top.

(a)

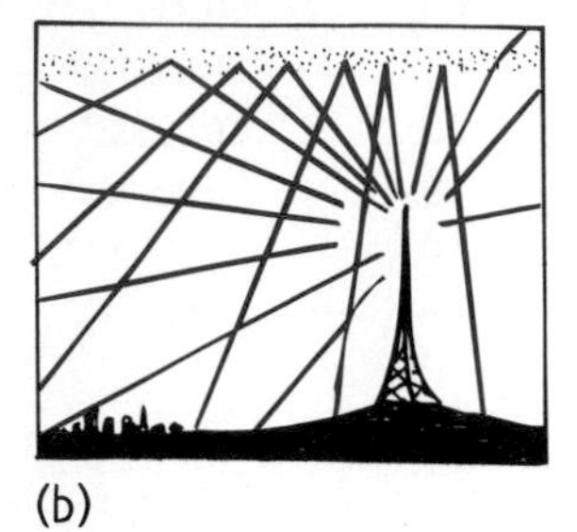

(b)

Fig. 26.5
(a) During the day, solar radiation increases ionization and energized ions move throughout the ionosphere. This scattered distribution of ions hinders AM radio-wave reception, because most of the AM radio-waves are greatly weakened by collisions with the ions as they travel through the ionosphere. (b) At night, ionization decreases and the remaining ions tend to concentrate in a thick band at the top of the ionosphere. With fewer ionic collisions in the lower ionosphere, AM radio-waves are able to travel to the upper ionosphere where, because of the concentrated band of ions, they are reflected back to the Earth's surface. That's why long distance AM radio reception is better at night.

Above the stratosphere, the **mesosphere** extends upward to about 80 kilometers. The gases that make up the mesosphere absorb very little of the Sun's radiation. As a result, temperature decreases from about 0°C at the bottom of the layer to about −90°C at the top.

The situation is just the opposite in the layer above the mesosphere, the **thermosphere**. Extending upward to 500 kilometers, this layer contains very little air. What air there is absorbs enough solar radiation to bring about a 2000°C temperature! Because of the low air density, however, this extreme temperature has little significance. Very little heat would be transferred to a slowly moving body in this region.

The **ionosphere** is an ion-rich region within the thermosphere and uppermost mesosphere. The ions in it are produced from the interaction between high-frequency solar radiation and atmospheric atoms. The incoming solar rays strip electrons from nitrogen and oxygen atoms, producing a large concentration of free electrons and positive ions in this layer. The degree of ionization in the ionosphere depends on air density and on the amount of solar radiation. Because ions are more prevalent where air density is low and solar radiation high, the number of ions increases with altitude and is greatest in the upper part of the ionosphere.

AM radio waves transmitted from the Earth's surface are reflected back to the Earth by the ionosphere, and this reflection is greatest at night (Figure 26.5). At night, the ions in the ionosphere tend to congregate in a dense band in the upper part of the ionosphere. As Figure 26.5b shows, this dense band of ions causes many AM waves to be reflected back to the Earth.[2] Thus AM radio reception can be extended for thousands of kilometers.

Ions in the ionosphere cast a faint glow that prevents moonless nights from becoming stark black. Near the Earth's magnetic poles, fiery light displays called *auroras* occur as the solar wind (high-speed charged particles ejected by the Sun) agitates the ionosphere. These auroral displays are particularly spectacular during times of solar disturbances such as solar flares. These disturbances also play havoc with radio reception, and so when the aurora is brilliant, short-wave radio operators hear more static on their receivers.

Fig. 26.6
The aurora borealis over Alaska is created by solar-charged particles that strike the upper atmosphere and light up the sky (just as similar particles on a smaller scale light up a fluorescent lamp).

Fig. 26.7
Short-wave radio operators hear more static on their receivers during solar disturbances such as sunspots and solar flares.

[2]Shorter wavelengths penetrate the ionosphere rather than being weakened or reflected in this layer. AM radio waves have a long wavelength, making them unable to penetrate the ionosphere and therefore prime candidates for ionosphere reflection. Waves of shorter wavelengths, such as FM radio waves, short-wave radio waves, and television waves are able to penetrate right through the ionosphere and travel out toward space.

Finally, above 500 kilometers, in the **exosphere**, the thinning atmosphere gradually yields to the radiation belts and magnetic fields of interplanetary space.

Questions

1. Is long-distance AM radio reception more likely in the daytime or at night?
2. Why do commercial airliners tend to fly at the top of the troposphere?

26.3 Solar Energy

Why are the Earth's equatorial regions always warmer and the polar regions always colder? The temperature of the Earth's surface depends very much on the energy per unit of surface area received from the Sun each day, which depends on the angle between the Sun's rays and the Earth's surface. This can be seen by holding a flashlight vertically over a table, and shining the light directly down on the flat surface (Figure 26.8a). The light produces a bright circle. Now tip the flashlight at various angles to the horizontal and notice that the circle elongates into ellipses, spreading the same amount of energy over more area and therefore decreasing the intensity of the light. Likewise for sunlight on the Earth's surface. High noon in equatorial regions is akin to the vertically held flashlight; high noon at higher latitudes is akin to the flashlight held at an angle.

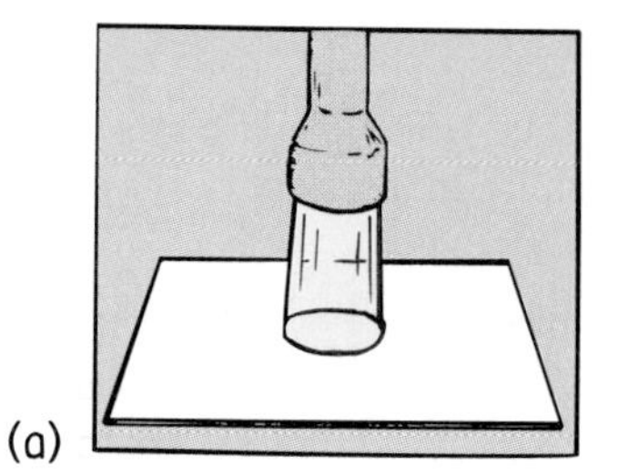

(a)

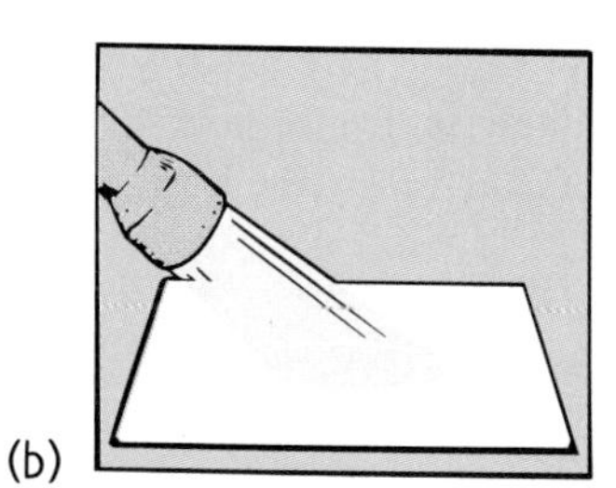

(b)

Fig. 26.8
(a) When the flashlight is held directly above at a right angle to the surface, the beam of light produces a bright circle.
(b) When the light is shone at an angle, the light beam is dispersed over a larger area and is therefore less intense.

The Seasons

The northern United States and Canada, both temperate regions, have distinct summer and winter seasons because of variations in the angle at which the Sun's rays strike the Earth's surface. Figure 26.9 shows the tilt of the Earth and how the corresponding different spreadings of solar radiation produce the yearly cycle of seasons. When the Sun's rays are closest to perpendicular at any spot on the Earth, that region experiences summer. Six months later the rays fall upon the same region more obliquely, and we have winter. In between are fall and spring.

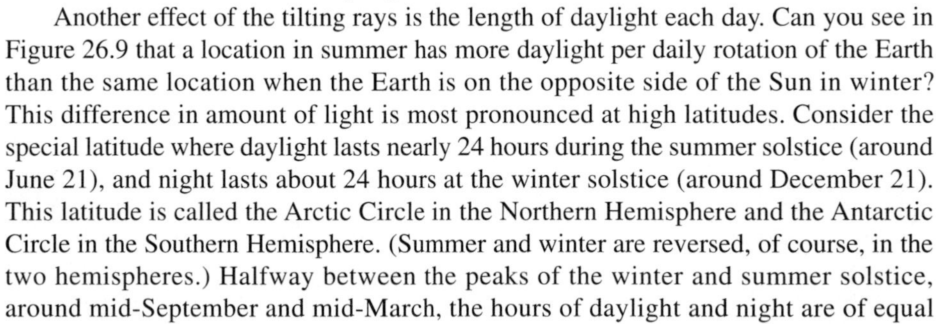

Another effect of the tilting rays is the length of daylight each day. Can you see in Figure 26.9 that a location in summer has more daylight per daily rotation of the Earth than the same location when the Earth is on the opposite side of the Sun in winter? This difference in amount of light is most pronounced at high latitudes. Consider the special latitude where daylight lasts nearly 24 hours during the summer solstice (around June 21), and night lasts about 24 hours at the winter solstice (around December 21). This latitude is called the Arctic Circle in the Northern Hemisphere and the Antarctic Circle in the Southern Hemisphere. (Summer and winter are reversed, of course, in the two hemispheres.) Halfway between the peaks of the winter and summer solstice, around mid-September and mid-March, the hours of daylight and night are of equal

---

Answers

1. At night. Ionization is greatest in the upper levels of the ionosphere where air density is lowest and solar energy highest. During the day, AM radio signals are unable to penetrate beyond the lower ionosphere because of high air density and frequent collisions between ions and other molecules. At night, when there are fewer ion collisions, AM radio waves are able to travel to higher altitudes where they can be reflected. This increases the overall distance AM radio waves can travel.
2. Because the captain wants to have a smooth ride! Most weather disturbances don't extend past the top of the troposphere.

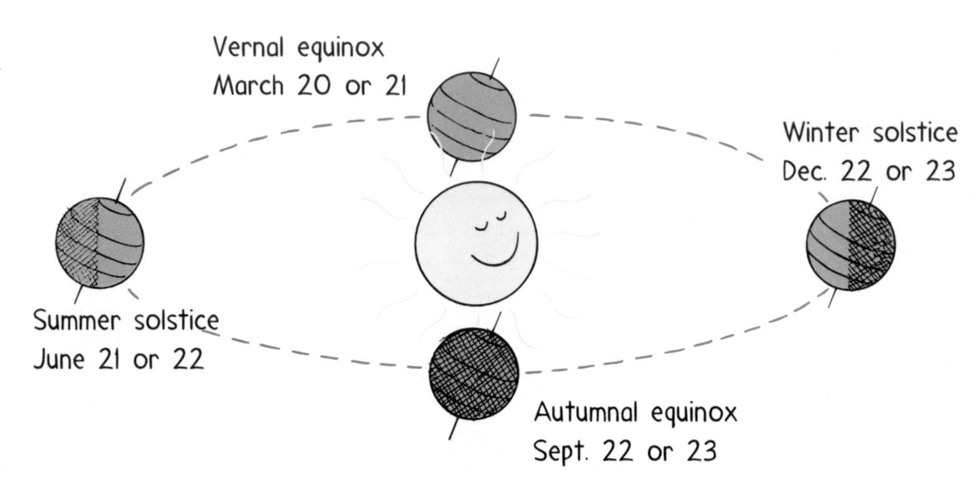

Fig. 26.9
The tilt of the Earth and the corresponding different spreading of solar radiation produce the yearly cycle of seasons.

length. These are called the equinoxes (Latin for "equal nights"). The equal hours of day and night during the equinoxes are not restricted to high latitudes, they occur all over the world.

As you travel north of the Arctic Circle (or south of the Antarctic Circle), the greater are the number of summer days with the Sun always above the horizon and the greater are the number of winter days with the Sun always below the horizon! At the poles there is a full six months of continuous sunlight followed by a full six months of continuous night! These 24-hour-long "days" in the polar regions are never very bright because the Sun is never very far above the horizon. Likewise, the 24-hour-long "nights" aren't all that dark because the Sun never sinks very far below the horizon. The polar regions can be eerie places.

The Solar Constant

We feel the Sun's energy when we step from the shade into the sunshine. The warmth we feel isn't so much because the Sun is hot, for its surface temperature of 6000°C is no hotter than the flames of some welding torches. We are warmed mainly because the Sun is so *big*. As a result, it emits enormous amounts of radiant energy.

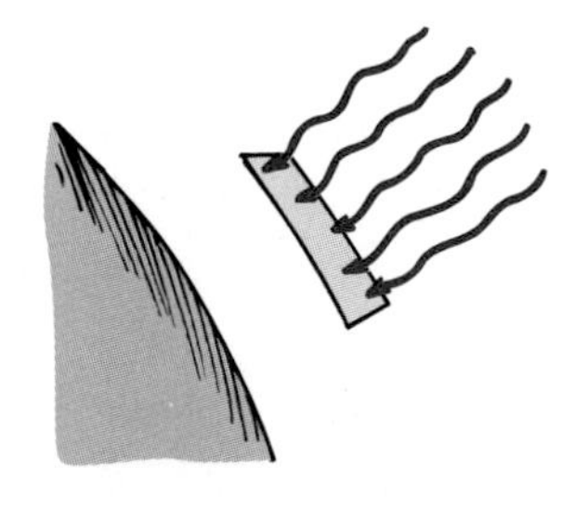

Fig. 26.10
Over each square meter of area perpendicular to the Sun's rays at the top of the atmosphere, the Sun emits 1400 J of radiant energy each second. Hence the solar constant is 1.4 $kJ/s/m^2$, or 1.4 kW/m^2.

The amount of solar energy received by the Earth is measured by a quantity known as the **solar constant**. To see how this constant is defined, imagine a square pane of glass measuring 1 meter on a side and floating at the top of the atmosphere, as shown in Figure 26.10. Scientists have determined that the amount of solar energy striking this imaginary square meter of glass perpendicularly is 1400 joules each second. Recall from Section 3.9 that power is defined as the amount of energy per unit of time, and so we can express the value 1400 joules per second per square meter in the power unit of 1.4 kilowatts per square meter. This latter quantity is the definition of the solar constant. Note that it refers to the amount of energy *striking the top of the atmosphere*. Once sunlight enters the atmosphere, however, it is scattered by fine dust and air molecules, reflected by cloud cover, and absorbed by atmospheric gases. For this reason, the solar intensity reaching the ground is much lower than the solar constant value.

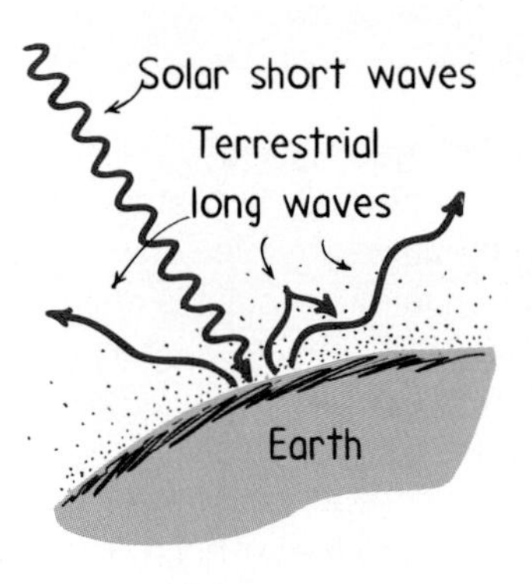

Fig. 26.11
The hot Sun emits short waves, and the cool Earth re-emits long waves. Radiation emitted from the Earth is terrestrial radiation.

Terrestrial Radiation

Solar radiation covers a wide spectrum of wavelengths, primarily in the visible short-wavelength part of the spectrum. The Earth absorbs this energy, and in turn, reradiates part of it back to space in the infrared long-wavelength part of the spectrum. As we learned in Section 7.3, we call this reradiated energy *terrestrial radiation* because it is emitted from the Earth's surface.

It is terrestrial radiation rather than solar radiation that directly warms the lower atmosphere, which explains why air close to the ground is appreciably warmer than air at higher elevations. The temperature of the Earth's surface depends on the amount of solar radiation coming in compared with the amount of terrestrial radiation going out. In direct sunlight, the net effect is warming because the Earth's surface absorbs more energy from the Sun than it emits. At night, the net effect is cooling because the Earth's surface emits more energy than it absorbs. Cloud cover blocks either incoming solar radiation or outgoing terrestrial radiation. Can you now see why cloudy days are cooler than sunny days and cloudy nights warmer than clear nights?

The Greenhouse Effect and Global Warming

The Earth absorbs short-wavelength radiation from the Sun and reradiates it as long-wavelength terrestrial radiation. Incoming short wavelength solar radiation easily penetrates the atmosphere to reach and warm the Earth's surface, but outgoing long wavelength terrestrial radiation cannot penetrate the atmosphere to escape into space. Instead, atmospheric gases (mainly water vapor and carbon dioxide) absorb the long-wave terrestrial radiation. As a result, this long-wave radiation ends up keeping the Earth's surface warmer than it would be if the atmosphere were not present. This process is very nice, for the Earth would be a frigid $-18°C$ otherwise. Our present environmental concern, however, is that increased levels of carbon dioxide and other gases in the atmosphere may make the Earth *too* warm.

Similar to the panes of glass in a greenhouse, atmospheric gases trap long-wave terrestrial radiation, thereby warming the lower atmosphere. This warming of the lower atmosphere is called the **greenhouse effect** and plays a significant role in global warming. Gases released primarily by volcanic eruptions, and also by the burning of fossil fuels (coal, oil, and gas) and from agricultural and manufacturing industries, add carbon dioxide and other greenhouse gases to the atmosphere, altering its composition. This compositional change, in turn, affects atmospheric absorption of both solar and terrestrial energy.

Of all the greenhouse gases, water vapor plays the largest role in confining the Earth's heat. As part of the Earth's natural water cycle, water vapor levels have remained relatively constant throughout time. Like water vapor, carbon dioxide occurs naturally in the Earth's atmosphere. Unlike the case with water vapor, however, carbon dioxide levels are on the rise. Since the Industrial Revolution of the 1800s, atmospheric carbon dioxide levels have been steadily increasing and within the next century will probably double pre-industrial levels. This increase may account for the warming of the Earth's surface by about 0.6°C since 1850. Some scientists and policy makers believe that further warming will likely occur if carbon dioxide emissions are not held in check. Other gases, such as methane, nitrous oxides, and CFCs, are also on the increase and as such, they too may play a role in changing the Earth's atmosphere.

The effects of warming the Earth's surface are not known. One concern is that warming will cause polar ice caps to melt, raising sea level and flooding low-lying coastal lands. Warming would also likely change rainfall patterns, seriously affecting

Global Warming

There are 670 billion tons of CO_2 in the atmosphere. This amounts to about 0.035 percent of the atmosphere, or about 350 parts per million of atmospheric gases. Be glad this is so. Without carbon dioxide, plants could not photosynthesize the biomaterials essential to almost all life forms on our planet. Optimum plant growth occurs at a temperature of 25°C and a CO_2 level of at least 800 parts per million (which is why CO_2 generators are used to elevate the CO_2 level in the air of commercial greenhouses). Concentrations of 5 parts per million CO_2 barely support some tropical grasses, while most other plants require a minimum of 50 parts per million.

Human use of fossil fuels has added to the amount of CO_2 in the atmosphere, but atmospheric concentrations of CO_2 haven't increased proportionately, mainly because of oceanic absorbtion (Section 20.5). Increased amounts of CO_2 accelerate plant growth and also allow plants to grow in drier regions. Animal life, which depends upon plants, also flourishes. So why such an alarm over increased levels of CO_2 in the atmosphere?

Because increased CO_2 concentrations heighten the greenhouse effect; that is, sunlight gets through the atmosphere but infrared heat waves emitted by the Earth do not. Interestingly, the gas most responsible for the greenhouse effect is not CO_2 but rather H_2O! There is little concern about water vapor, however, because levels have been fairly constant in recent times. Increased amounts of CO_2 mean a higher average temperature for the planet, which means more melting of polar ice caps and higher sea levels. As the relatively slight changes in ocean temperature caused by El Niño have shown, we're talking about a different world climate. Does "different" mean better or worse? Although the Earth is currently in a long-term warming trend, it has been much warmer during the past 3000 years—without environmental calamities. The Earth's average temperature goes through cycles, altered by differences in sunlight and a host of factors other than CO_2 levels. Change one thing and you change another. So between the extremes of no CO_2 (a world without plants and animals as we know them) and exceedingly high CO_2 as occurs on the planet Venus (a world without life as we know it) lies a favorable balance.

What the favorable balance is and how the world climate will change are both questions open to debate.

agricultural industries. Grain-growing regions of North America and Asia might shift northward as local climates warmed and growing seasons lengthened. On the other hand, deserts in the interior of continents might spread to cover much larger areas. We don't know. What we do know is that the Earth has experienced warmer and colder periods in times past.

Questions

1. What does it mean to say the greenhouse effect is like a one-way valve?
2. Which gas in the atmosphere is the greatest contributor to the greenhouse effect?
3. What is the primary contributor to carbon dioxide in the Earth's atmosphere?

Answers

1. The transparent material—atmosphere for the Earth and glass for the florist's greenhouse—passes only incoming short waves and blocks outgoing long waves. In other words, radiation travels only one way.
2. Water vapor.
3. Volcanic eruptions. Interestingly, the volcanic eruption of Mount Pinatubo in 1991 spewed more chlorine into the atmosphere than all the leakage of CFCs in the world combined.

26.4 Components of the Earth's Oceans

Interaction between the oceans and the atmosphere greatly influences the Earth's climate. The oceans provide a sink for excess atmospheric carbon dioxide. Carbon dioxide readily dissolves in the ocean, where it undergoes various chemical reactions, most of which lead to the formation of carbonate precipitates such as limestone.

Salts make up 99.7 percent of the ocean's dissolved materials. The amount of dissolved salts in seawater is measured as **salinity**—the mass of salts dissolved in 1000 grams of seawater. Although salinity varies from one part of the ocean to another, the overall composition of seawater is fairly uniform—a mixture of about 96.5 percent water and 3.5 percent salt. Variation is, of course, influenced by factors that increase or decrease supplies of fresh water. Fresh water enters the ocean in three ways: Runoff from streams and rivers, precipitation, and melting of glacial ice. Fresh water leaves the ocean in two ways: Evaporation and formation of ice. Overall balance is maintained when evaporation is offset by precipitation/runoff and ice formation is offset by ice melting.

Physical Structure of the Ocean

Like the atmosphere, the ocean can be divided into several vertical layers. If you scuba dive or snorkel, you have probably noticed an increase in water pressure as you swim to lower depths. The deeper you descend, the greater the water pressure. The pressure is simply the weight of the water above you pushing against you. Another factor that generally changes as you descend is temperature. Deeper waters are cooler. So, in addition to variations in salinity, seawater also varies in temperature and pressure. These variations are best illustrated when we look at the ocean's vertical structure (Figure 26.12).

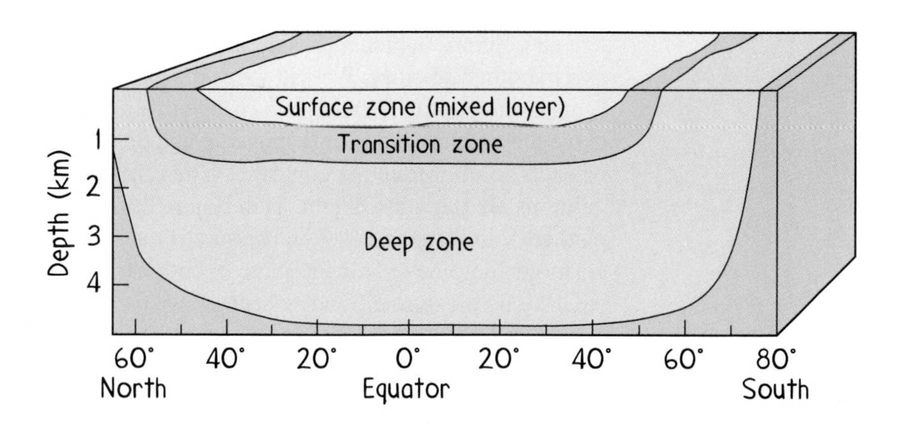

Fig. 26.12
The ocean's vertical structure. In the surface zone, water is well mixed as it moves vertically in response to temperature and density changes, and horizontally in response to wind. Water in the transition zone moves along density surfaces. Water in the deep zone is density-driven as it circulates from cold polar regions to warmer equatorial regions.

The top layer, the *surface zone,* represents only 2 percent of the ocean's volume. It is the zone with which we are most familiar. Seawater in the surface zone is well mixed as it moves vertically in response to changes in the overlying atmosphere. As expected, surface zone water temperatures are warmer in equatorial areas and colder toward the poles. Because cold water is denser than warm water, cold seawater sinks below warmer seawater. Salinity also affects density: The greater the salinity, the greater the density. As expected, salinity is greatest in the centers of ocean basins where there is no dilution by river discharge.

Below the surface zone is the *transition zone*—the area between the surface and the deep ocean. The exact depth of this zone changes due to the factors that influence seawater density—temperature and salinity. In the open ocean where salinity is fairly constant, the depth of the transition zone is controlled by temperature. In coastal areas, where fresh water enters the ocean, transition-zone depth is controlled by salinity. Seawater in the transition zone moves horizontally according to density.

Below the transition zone we encounter the deep zone, a fairly uniform zone that accounts for 80 percent of the ocean's volume. Except near high latitudes, the deep zone is isolated from contact with the atmosphere (Figure 26.12). In fact, most deep water forms at or near the surface in the high latitudes, where exposure to the atmosphere makes surface water as cold as the water below. With no temperature difference, surface water and deep water share similar characteristics so that they are both considered to be part of the deep zone.

Question

Would you expect the pressure 100 m deep in the equatorial Pacific to be the same as it is 100 m deep in the northern Pacific?

26.5 Driving Forces of Air Motion

We know that as warm air rises, it expands and cools. As the air rises, cooler air sinks to occupy the region left vacant by the rising warm air. Such motion constitutes a convection cycle and thermal circulation of the air—in other words, a *convection current.* As convection currents stir the atmosphere, the result is *wind*—defined as air in nearly horizontal motion. Wind is generated in response to pressure differences in the atmosphere, which are largely the result of temperature differences.

To see how air pressure can vary from one location to another, let's look at equivalent air columns (which means the same number of molecules in the two columns) above two hypothetical cities. We will focus the discussion by making three important simplyfying assumptions: 1) there is no change in air density with height, 2) air cannot enter or leave the column, it can only move up and down within each column, and 3) the width of each column remains constant. When the cities are at the same temperature, the columns are the same height, as in Figure 26.13a. At any elevation—say the elevation marked X in the drawing—the pressure is the same in both cases because the number of air molecules above X is the same in both cases. This is true for any elevation, and so we have no pressure difference because we have no temperature difference.

Now suppose the temperature in city 1 drops while that in city 2 goes up. In this case, we get the situation shown in Figure 26.13b. The cooling air over city 1 contracts and becomes denser, and the warming air over city 2 expands and gets less dense. Now the same number of air molecules are in a short column over city 1 and in a much taller column over city 2. Although the ground-level pressure is still the same in the two cities (because equal numbers of air molecules press down on both), we now see differences in pressure at elevations above ground level. Consider elevation Y, halfway up the column above city 1. Because Y is at the midway point of the column, the top half of the column must contain 50 percent of the air molecules. This same elevation Y

Answer

Probably not. If salinity is the same at both locations (a good assumption if we are away from coastal areas), the pressure will be slightly higher in the cold, northern Pacific. Remember, cold water is denser than warm water, and so a volume of cold water weighs more than an equal volume of warm water. Pressure is the weight per unit of area.

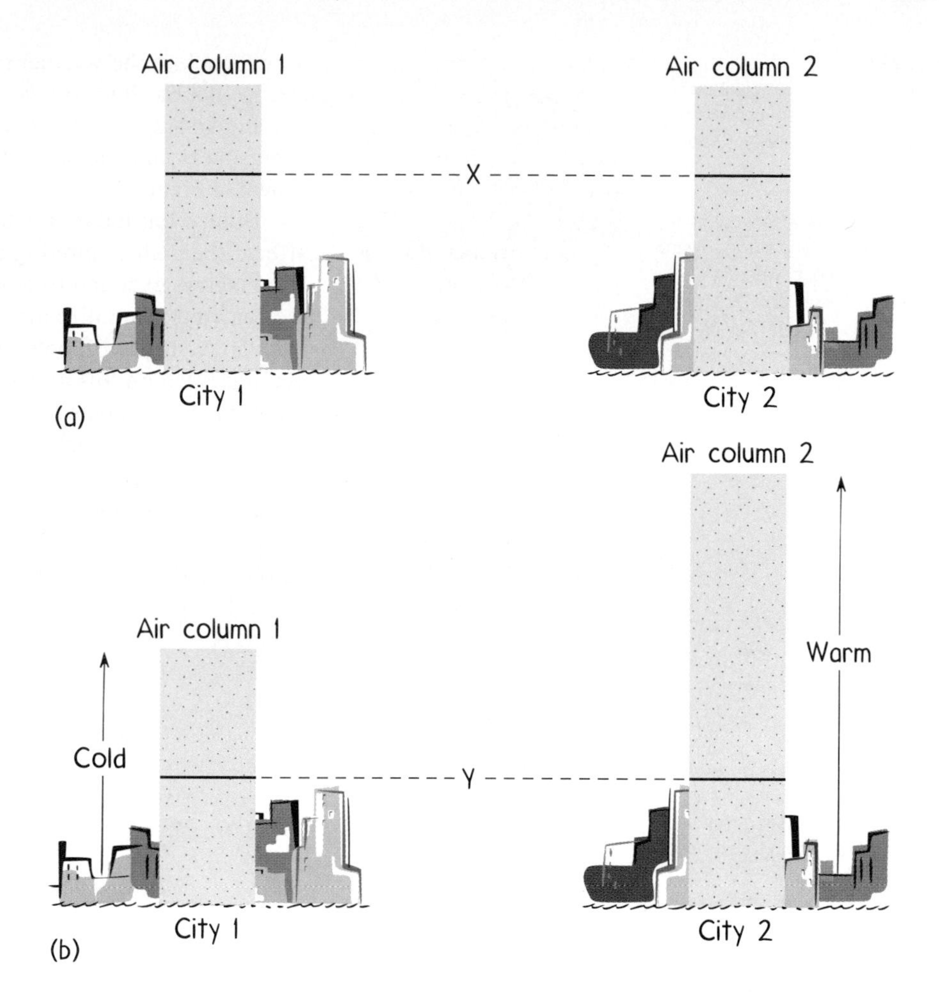

Fig. 26.13
(a) Same-temperature air columns over two cities. Note that air pressure at any elevation, such as elevation X shown here, is the same for both cities. (b) When city 1 is cold and city 2 warm, the air column over city 2 is taller due to expansion, which means for an equal given elevation, pressure is greater in the taller column. This produces a pressure-gradient between the two cities. So wind blows from city 2 to city 1.

must be less than midway up the column over city 2, and so the part of the column above this elevation must contain more than 50 percent of the air molecules. Because there are more molecules bearing down on elevation Y above city 2 than above city 1, the air pressure at Y must be greater above city 2 than above city 1. In other words, a difference in temperature has led to a difference in pressure. For values of Y that are 1 kilometer or more above the ground, warm air is high-pressure air and cold air is low-pressure air.

So what does this pressure difference at elevations of 1 kilometer or more have to do with wind at the Earth's surface. Differences in pressure cause the air to move and hence the wind to blow. Now let's disregard assumption number 2 and allow air to flow from one column to another. In Figure 26.13b, the air at elevation Y moves horizontally from the warm column (city 2) to the cold column (city 1). In other words, a wind blowing from city 2 toward city 1 is created at elevation Y. Once the wind starts blowing, there will be more air in the cold column than in the warm column, increasing the ground-level pressure (surface pressure) in city 1. Thus cold days are associated with high surface pressure, and warm days are associated with low surface pressure. A difference in pressure between two different locations is called a *pressure gradient,* and forces caused by changes in pressure are called **pressure-gradient forces**.

The underlying cause of general air circulation is unequal heating of the Earth's surface. On a global level, equatorial regions receive optimum radiant energy from the Sun and as a result have higher average temperatures than other regions. As air heated by the hot ground or ocean at the equator rises, it moves toward the polar regions, cooling gradually in the upper atmosphere. This cooled air then sinks at the poles and is

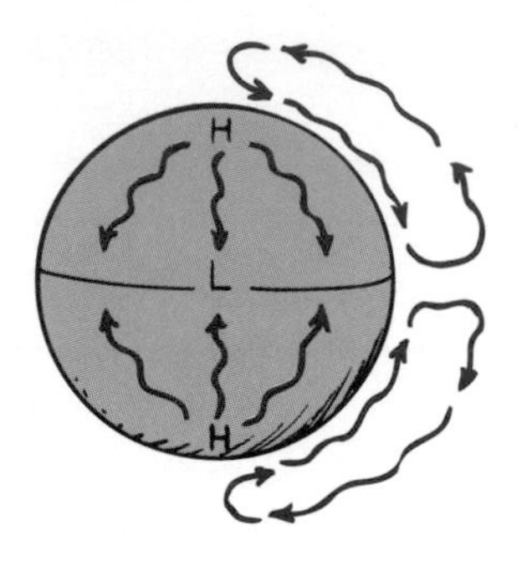

Fig. 26.14
If the Earth were simply a nonrotating sphere, air circulation would be in a single Northern Hemisphere cell and a single Southern Hemisphere cell. In each cell, heated air would rise at the equator and move toward the polar regions, where it would cool, sink, and be drawn back to the warmer regions of the equator.

drawn back to the warmer regions of the equator. If we assume the Earth to be a nonrotating sphere, the effect is one simple single-cell circulation pattern in the Northern Hemisphere and another in the Southern Hemisphere, as shown in Figure 26.14.

But the Earth does rotate, and this movement greatly affects the path of moving air.[3] Think of the Earth as a large merry-go-round rotating in a counterclockwise direction (the same direction the Earth spins as viewed from above the north pole). Pretend you and a friend are playing catch on this merry-go-round. When you throw the ball to your friend, the direction the ball appears to travel is affected by the circular movement of the merry-go-round. Although the ball travels in a straight-line path, it appears to curve to the right, as shown in Figure 26.15. (The ball travels straight, but your friend never catches it because the movement of the merry-go-round causes his position to change.) This apparent curving is similar to what happens on the Earth. As the Earth spins, all free-moving objects—air and water, aircraft and ballistic missiles, and even snowballs to a small extent—appear to deviate from their straight-line paths as the Earth rotates under them. This apparent deflection due to the rotation of the Earth is called the **Coriolis effect**.

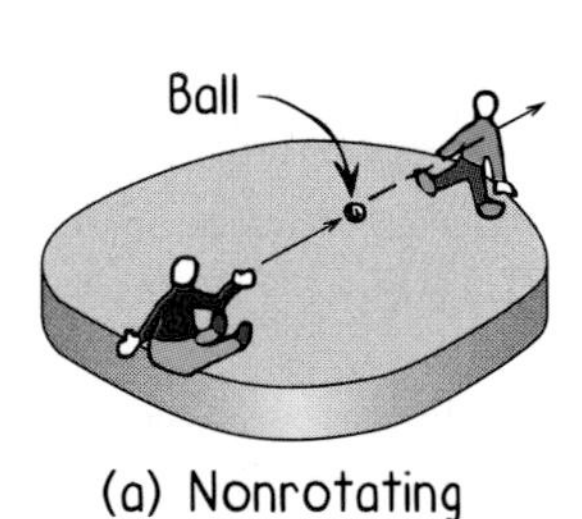

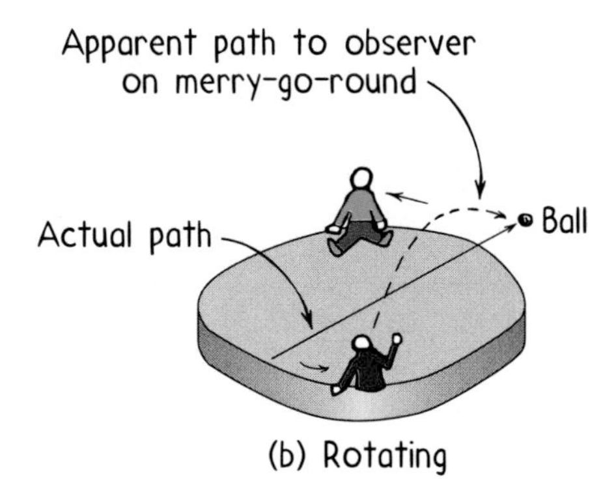

Fig. 26.15
(a) On the nonrotating merry-go-round, a thrown ball travels in a straight line.
(b) On the counterclockwise rotating merry-go-round, the ball moves in a straight line. However, because the merry-go-round is rotating, the ball appears to deflect to the right of its intended path.

A significant result of the Coriolis effect is the apparent deflection of winds toward the right in the Northern Hemisphere and toward the left in the Southern Hemisphere (Figure 26.16). The magnitude of the Coriolis effect varies according to the speed of the wind. The faster the speed, the greater the deflection. Latitude also influences the degree of deflection. Deflection is greatest at the poles and decreases to zero at the equator (Figure 26.17). As we shall soon see, the Coriolis effect is nonexistent for equatorial winds.

Air moving close to the Earth's surface encounters a *frictional force.* The rougher the surface, the greater the friction and so the greater the drag. Because surface friction

---

[3]As air circulates over the oceans, it causes surface water to drift along with it. A severe storm in 1990 has given scientists an unusual tool for studying the currents of the Pacific Ocean. Five cargo containers of athletic shoes were washed overboard from freighters that ran into stormy seas en route from South Korea to the Pacific Northwest. Since then, in a confirmation of theories about currents in the Northwest Pacific, thousands of sneakers, hiking boots, children's sandals and other shoes have been picked up along beaches from British Columbia to Oregon and as far into the Mid-Pacific as Hawaii. Despite months at sea, most shoes have been wearable after washing. The problem, though, is that the shoes were not tied together. Beachcombers formed "swap meets" to search for mates of found shoes!

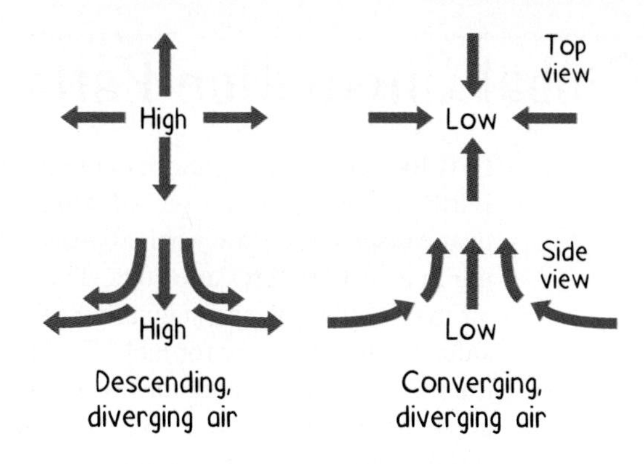

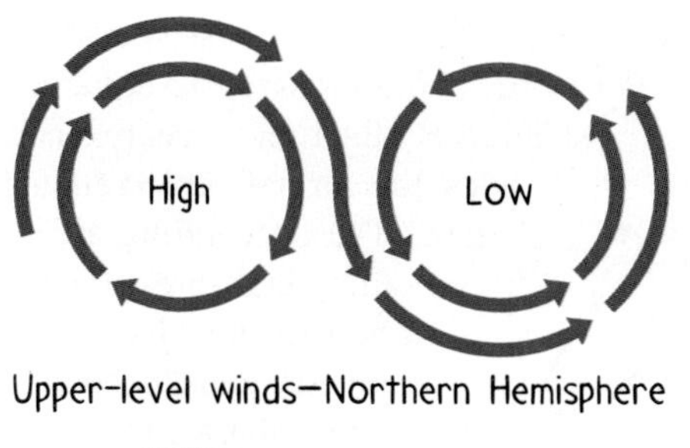

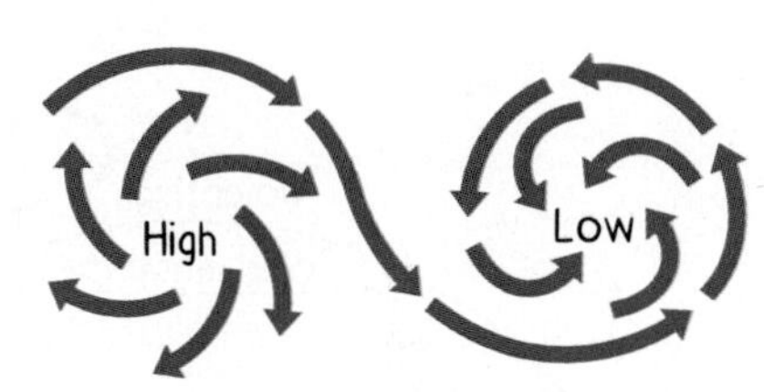

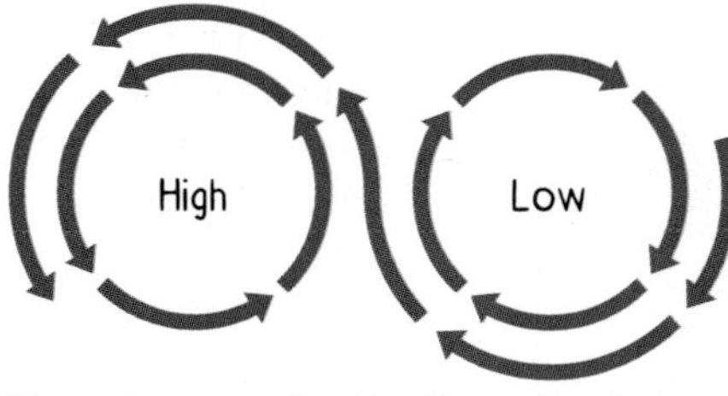

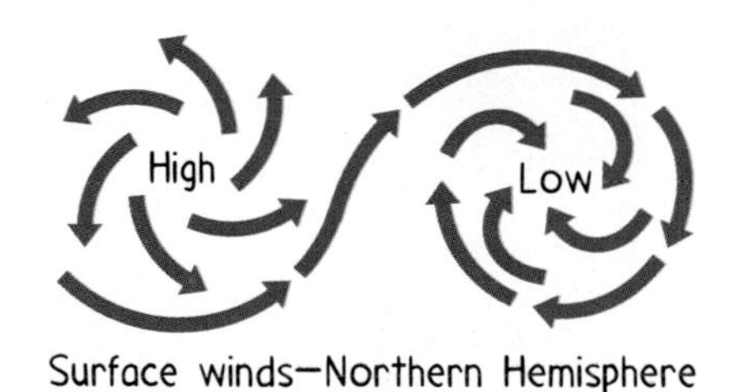

Fig. 26.16
The Coriolis effect—the apparent deflection of winds from straight-line paths by the Earth's rotation, is a principal force in the production of wind. It is, however, not the only force. (a) First of all, air moves due to pressure differences—the pressure gradient force. (b) Once the air is moving, it is effected by the Earth's rotation—the Coriolis effect. (c) As air moves close to the ground it slows due to frictional force.

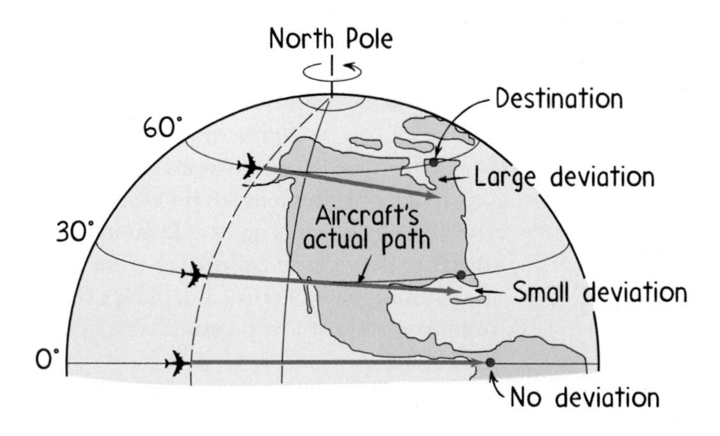

Fig. 26.17
Latitude influences the apparent deflection resulting from the Coriolis effect. A free-moving object heading east (or west) appears to deviate from its straight-line path as the Earth rotates beneath it. Deflection is greatest at the poles and decreases to zero at the equator.

reduces wind speed, it reduces the effect of the Coriolis force causing winds in the Northern Hemisphere to spiral out clockwise from a high-pressure region and spiral counterclockwise into a low-pressure region (top part of Figure 26.16c). In the Southern Hemisphere, these circulation patterns are reversed (bottom part of Figure 26.16c).

26.6 Global Circulation Patterns

Cell-like circulation patterns are responsible for the redistribution of heat across the Earth's surface and our global winds (Figure 26.18). Beginning at the equator, direct heat causes air to flow vertically upward with very little horizontal movement, resulting in a vast low-pressure zone. This rising motion creates a narrow, windless realm of air that is still, hot, and stagnant. Seamen of long ago cursed the equatorial seas as their ships floated listlessly for lack of wind, and referred to the area as the *doldrums.* When the moist air from the doldrums rises, it cools and releases torrents of rain. Over land areas these frequent rains give rise to the tropical rain forests that characterize the equatorial region.

The air of the sweltering doldrums rises to the boundary between the troposphere and stratosphere, where it divides and spreads out either to the north or south. (Very little wind crosses the equator into the neighboring hemisphere.) By the time it has reached about 30° N and 30° S latitudes, this air has cooled enough to descend toward the surface. The descending air is compressed and warms. A resulting high-pressure zone girdles the Earth, creating a belt of hot, dry surface air. On land, these high-pressure zones account for the world's great deserts—the Sahara in Africa, the Arabian Desert in the Middle East, the Mojave Desert in the United States, and the great Victoria Desert in Australia. At sea, the hot, descending air produces very weak winds. According to legend, early sailing ships were frequently stalled at these latitudes north and south. As food and water supplies dwindled, horses on board were either eaten or cast overboard to conserve fresh water. As a result, this region is now known as the *horse latitudes.* The thermal convection cycle that starts at the equator is completed when air flowing southward from the horse latitudes in the Northern Hemisphere and northward in the Southern Hemisphere is deflected westward to produce the *trade winds.* Air that flows northward from the horse latitudes in the Northern Hemisphere and southward in the Southern Hemisphere is deflected eastward to produce the prevailing *westerlies.*[4]

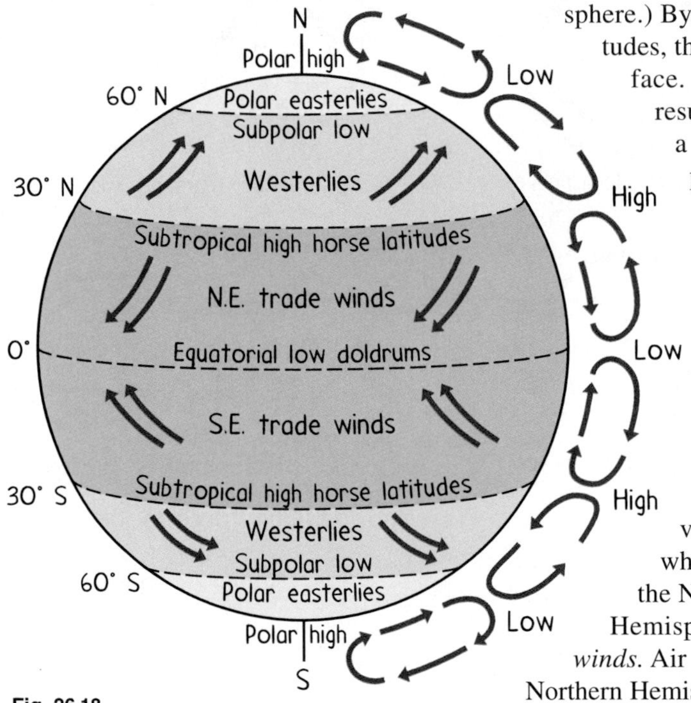

Fig. 26.18
Global winds are the result of several cell-like circulation patterns, brought about by unequal heating of the Earth's surface and compounded by effects of the Earth's rotation.

In the polar regions, frigid air continuously sinks, pushing the surface air outward. The Coriolis effect is quite evident in the polar regions as the wind deflects to the west to create the *polar easterlies* (Figure 26.18). The cool dry polar air meets the warm moist air of the westerlies at latitudes 60° N and 60° S. This boundary, called the *polar front,* is a zone of low pressure where contrasting air masses converge, often resulting in storms.

The midlatitudes are noted for their unpredictable weather. Although the winds tend to be westerlies, they are often quite changeable as the temperature and pressure

[4]Meteorologists refer to wind direction as the direction from which the winds come. In the case of the westerlies, the wind comes from the west and moves toward the east.

differences between the subtropical and polar air masses at the polar front produce powerful winds. As air moves from regions of high pressure, where air is denser, toward regions of low pressure, the result is a cyclone effect.

Irregularities in the Earth's surface also influence wind behavior. Mountains, valleys, deserts, forests, and great bodies of water all play a part in determining which way the wind blows.

Upper Atmospheric Circulation

In the upper troposphere, "rivers" of rapidly moving air meander around the Earth at altitudes of 9–14 kilometers. These high-speed winds are the *jet streams.* With wind speeds averaging between 95 and 190 kilometers per hour, the jet streams play an essential role in the global transfer of thermal energy from the equator to the poles.

The two most important jet streams, the *polar jet stream* and the *subtropical jet stream* form in both the Northern and Southern Hemispheres in response to temperature and pressure gradients. The formation of polar jet streams is a result of a temperature gradient at the polar front, where cool polar air meets warm tropical air. This temperature gradient causes a steep pressure gradient that intensifies the wind speed. During the winter, the polar jet is strong and extensive as it migrates to lower latitudes, bringing strong winter storms and blizzards to the United States. In summer, the jet stream is weaker and migrates to higher latitudes.

The subtropical jet stream is generated as warm air is carried from the equator to the poles, producing a sharp temperature gradient along the boundary (subtropical front). Once again a pressure gradient caused by the temperature gradient generates strong winds.

The subtropical jet stream above Southeast Asia, India, and Africa merits special mention. The formation of this jet stream is related to the warming of the air above the Tibetan highlands. During the summer, the air above the continental highlands is warmer than the air above the ocean to the south. Thus temperature and pressure gradients generate strong on-shore winds that contribute to the region's *monsoon* (rainy) climate. During winter, the winds change direction to produce a dry season.

This cycle of winds characterizes the climates of much of Southeast Asia. The predictable rain-bearing summer wind from the sea that moves over the heated land is called the *summer monsoon;* the prevailing wind from land to sea in winter is called the *winter monsoon.*

Fig. 26.19
Winds over Southeast Asia. (a) During the summer months, air over the oceans is cooler than the air over land. The summer monsoon brings heavy rains as the winds blow from sea to land. (b) During the winter months, air over continents is cooler than air over oceans. The winter monsoon generally has clear skies and winds that blow from land to sea.

Questions

1. What are the main causes of the trade winds, jet streams, monsoons, and ultimately their bearing on the world's climates?
2. In the midlatitudes, airlines allot shorter flight times for planes traveling west to east and longer flight times for planes traveling east to west. Why are eastbound planes faster?

Answers

1. Simply enough, the unequal heating of the Earth's surface, coupled with the Earth's rotation.
2. The upper-level westerly moving winds of the jet stream account for faster-moving eastbound aircraft. As the jet stream moves from west to east it carries along everything in its path. To save time and fuel, air pilots seek the jet stream when traveling west to east and avoid it when traveling east to west.

Oceanic Circulation

The forces that drive the winds also impact the movement of seawater. In the open ocean, the major movement of seawater results from two types of currents: Wind-driven surface currents and density-driven deep-water currents. Near coastlines, water movement is affected not only by surface and deep-water currents but also by coastal boundaries.

Surface Currents As winds blow across the ocean, frictional forces set surface waters into motion. If distances are short, the surface waters move in the same direction as the wind. For longer distances, however, other factors come into play. One such factor is the deflective Coriolis effect which causes surface waters to spiral in a circular whirl pattern called a **gyre**. The circular motion is clockwise in the Northern Hemisphere and counterclockwise in the Southern Hemisphere.

In the tropics, the trade winds drive equatorial ocean currents westward. When the westward flow is obstructed by a continental shoreline, the current diverges, with some flow going north and some going south. At temperate latitudes, the prevailing westerlies take over to drive the surface currents eastward. In the Northern Hemisphere, huge gyres are created as eastward-moving water encounters land boundaries. In the Southern Hemisphere, with fewer land obstructions, eastward flow is able to encircle the globe (Figure 26.20).

An important consequence of these large gyres is the transport of heat from equatorial regions to higher latitudes. In the North Atlantic Ocean, for example, warm equatorial water flows westward into and around the Gulf of Mexico then northward along the eastern coast of the United States. This warm-water current is called the *Gulf*

Fig. 26.20
Circulation of the ocean's surface waters. The names of the major currents are indicated.

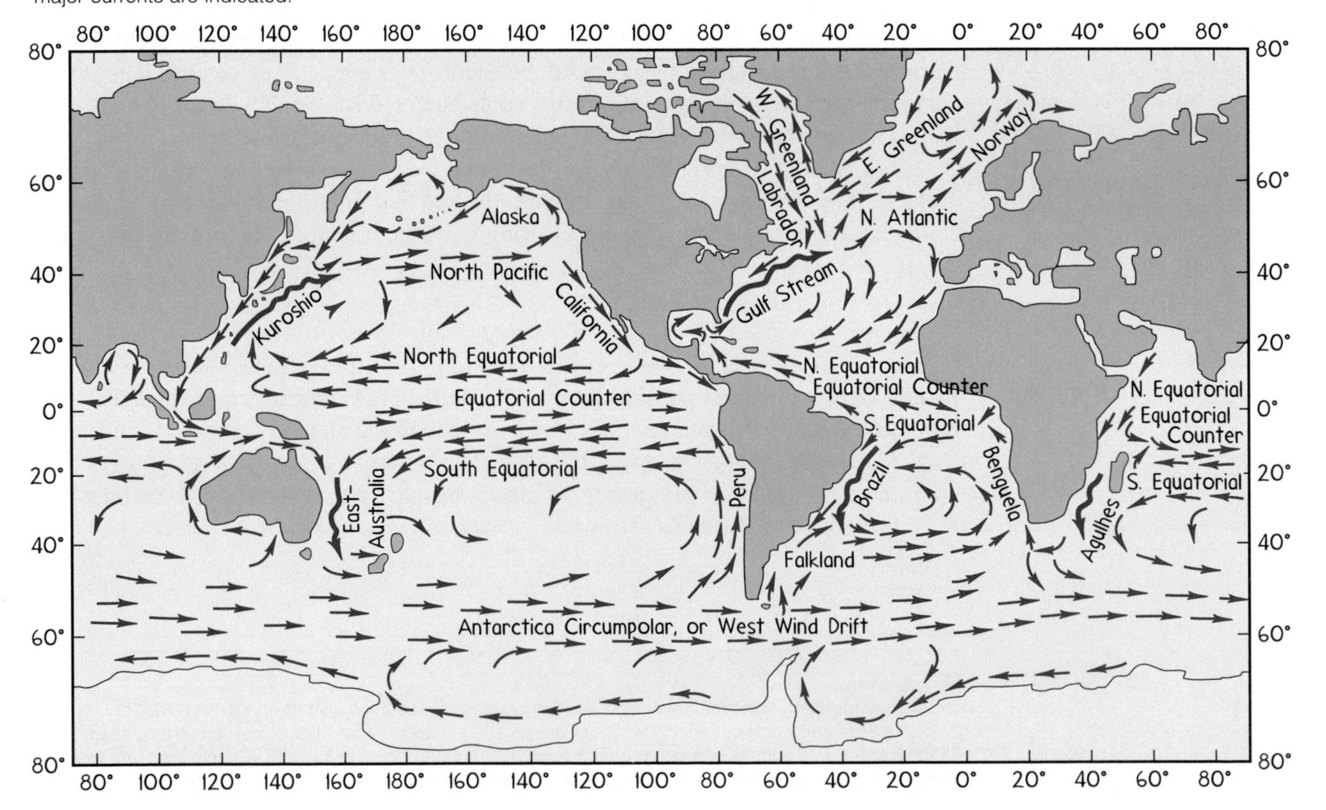

The El Niño Condition

When weather is measured over time, an *average* weather pattern can be seen. The consistent behavior of weather over time is referred to as climate. There are periods when the *average* weather pattern departs from its norm, thus disrupting the expected climate. A prime example of such a disruption is the El Niño condition.

Under normal conditions, weather patterns in the Pacific are controlled by warm, high-pressure systems located in both hemispheres near the equator. These high-pressure systems cause the trade winds to blow westward along the equator, dragging the warm equatorial surface waters along with them. As warm surface waters move westward, deeper, colder waters to the east rise upward to occupy the space left vacant by the warm surface water. The upwelling cold waters, rich in nutrients, attract a variety of sea life. Upwelling of these cold waters has been especially important to fishing interests along the coast of South America, where people earn their living catching anchovies that come to feed in the nutrient-rich waters.

Fishing is not always good, however. Each year in October the trade winds slacken, reversing the normal westward flow of warm tropical surface waters. As the warm surface waters drift eastward, upwelling decreases and so does the fishing industry. People along the South American coast refer to this occurrence as El Niño because it begins each year around the traditional December celebration of Christ's birth (Christ is El Niño, The Child, in Spanish). Under normal conditions, the trade winds pick up again in early spring, the surface waters are again blown westward across the ocean, and everything returns to normal.

There are some years, however, when the trade winds fail to strengthen and the warm surface waters remain off the coast of South America for a year or longer. During these abnormal conditions, upwelling of cold water ceases, and South American fishing industries fail. Although a small El Niño occurs each year, it is this extended El Niño that is referred to as the **El Niño condition.**

The El Niño condition influences climate on both sides of the tropical Pacific Ocean. Under normal conditions, upwelling cold water on the eastern side of the Pacific coincides with dry cool air, high pressures, and clear skies. On the western side of the Pacific, surface waters—warmed and fueled by their long journey across the ocean—warm the surrounding air. As the warm moist air rises, low pressures and storms develop on the warm western side of the Pacific.

During an extended El Niño condition, the pattern is reversed. Warm water, rising warm moist air, low pressures, and storms are found on the eastern side of the Pacific rather than the western side. This exchange of pressure systems and weather patterns between east and west during an El Niño condition is sometimes referred to as the *southern oscillation.*

Stream. As the Gulf Stream flows northward along the North American coast, the prevailing westerlies steer the warm current eastward toward Europe (Figure 26.20). Great Britain and Norway benefit from the warm waters in the Gulf Stream, for lands at this northern latitude would be much colder without being bathed in warm water from the Gulf of Mexico. As the warm current encounters Europe, it is diverted southward toward the equator where it is once again picked up by the trade winds to move westward into the Gulf of Mexico and once again become part of the Gulf Stream.

Oceanic circulation in the North Pacific is similar to that in the North Atlantic. The Pacific counterpart of the Gulf Stream is the warm, northward-flowing current known as the *Kuroshio.* In the Southern Hemisphere, surface oceanic circulation (with the exception of the Antarctica Circumpolar Current) is similar except that the gyres move counterclockwise.

Deep-Water Currents Surface waters are driven by winds, but deeper waters are driven by gravity. In essence, deep water flows because dense water sinks. Although

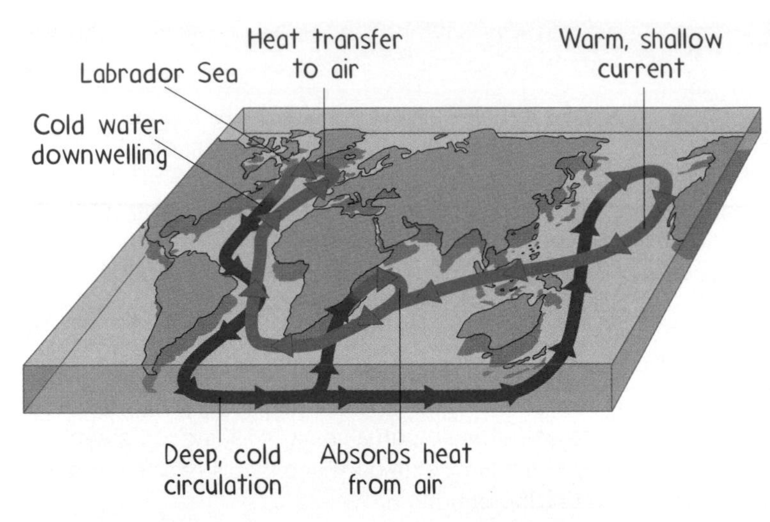

Fig. 26.21
Deep-ocean currents act like a conveyor belt, transporting cold water from the North Atlantic to the equator and onto the Antarctic. From the Antarctic, water flows eastward then northward into the Pacific and Indian Oceans.

deep water flows more slowly than surface water, the volume of deep-water flow can be likened to a large global conveyor belt (Figure 26.21).

In the high latitudes, where seawater in the deep zone interacts with seawater in the surface zone, vertical mixing helps to set up a very slow worldwide, north-south circulation pattern. To understand how this pattern develops, we need to look at what happens when seawater begins to freeze.

First of all, seawater does not freeze easily. When it does, however, only the water freezes, and the salt is left behind. Thus the seawater that does not freeze experiences an increase in salinity, which in turn brings about an increase in density. The cold, denser, saltier seawater sinks, which means a pattern of vertical movement is set up. There is also horizontal movement as the dense water that sinks in the polar regions flows along the bottom to the deeper parts of the ocean floor.

Thus conveyor-belt circulation begins in the North Atlantic as dense, cold, salty seawater around Greenland and Iceland sinks and flows along the ocean bottom toward the equator then onto the Antarctic Ocean. Once near Antarctica, the water flows eastward around the continent, then northward into the Pacific and Indian Oceans. Thus deep-water currents flow in a north-south circulation pattern.

Summary of Terms

- **Troposphere** The atmospheric layer closest to the Earth's surface, 16 km high over the equator and 8 km high over the poles, and containing 90% of the atmosphere's mass and essentially all its water vapor and clouds.
- **Stratosphere** The second atmospheric layer above the Earth's surface, extending from the top of the troposphere up to 50 km.
- **Mesosphere** The third atmospheric layer above the Earth's surface, extending from the top of the stratosphere to 80 km.
- **Thermosphere** The fourth atmospheric layer above the Earth's surface, extending from the top of the mesosphere to 500 km.
- **Ionosphere** An electrified region within the thermosphere and uppermost mesosphere where fairly large concentrations of ions and free electrons exist.
- **Exosphere** The fifth atmospheric layer above the Earth's surface, extending from the thermosphere upward and out into interplanetary space.
- **Solar constant** The 1400 J/m^2 received from the Sun each second at the top of the Earth's atmosphere. Expressed in terms of power, it is 1.4 kW/m^2.
- **Greenhouse effect** Warming caused by short-wavelength radiant energy from the Sun that easily enters the atmosphere and is absorbed by the Earth. This energy is then reradiated at longer wavelengths that cannot easily escape the Earth's atmosphere. Thus the Earth's atmosphere and surface are warmed.
- **Salinity** The mass of salts dissolved in 1000 g of seawater.
- **Pressure-gradient force** The force that moves air from a region of high-pressure to an adjacent region of low-pressure air.
- **Coriolis effect** The apparent deflection from a straight-line path observed in any body moving near the Earth's surface, caused by the Earth's rotation.
- **Gyre** Circular or spiral whirl pattern, usually applied to very large current systems in the open ocean.

Review Questions

Evolution of the Earth's Atmosphere and Oceans

1. What were the main components of the Earth's first atmosphere? What happened to this atmosphere?
2. The Earth's present atmosphere likely developed from gases that escaped from the interior of the Earth during

volcanic eruptions. What were the three principal atmospheric gases produced by these eruptions?

3. Explain the importance of photosynthesis in the evolution of the atmosphere.
4. Explain the connection between life on Earth and the ozone layer.

Components of the Earth's Atmosphere

5. Why doesn't gravity flatten the atmosphere against the Earth's surface?
6. What elements make up today's atmosphere?
7. Being that our atmosphere developed as a result of volcanic eruptions, why aren't there higher traces of atmospheric carbon dioxide, one of the principal volcanic gases?
8. In which atmospheric layer does all our weather occur?
9. Does temperature increase or decrease as one moves upward in the troposphere? As one moves upward in the stratosphere?
10. Why is the thermosphere so much hotter than the mesosphere?
11. What is the source of the ions that give the ionosphere its name?
12. Why can you often pick up AM stations hundreds of kilometers away at night, but not at all during the day?
13. What causes the fiery displays of light called the auroras?

Solar Energy

14. What does the angle at which the Sun strikes the Earth have to do with the temperate and polar regions?
15. What does the tilt of the Earth have to do with the change of seasons?
16. If it is winter and January in Chicago, what are the corresponding season and month in Sydney, Australia?
17. Why are the hours of daylight equal all around the world on the two equinoxes?
18. Do all parts of the Earth receive the same amount of solar energy? Explain.
19. How does radiation emitted from the Earth differ from that emitted by the Sun?
20. How is the atmosphere near the Earth's surface heated from below?

Components of the Earth's Oceans

21. How does the density of seawater vary with changes in temperature? How does density change with salinity?

Forces Driving Air Motion

22. What are the three main driving forces of air motion?
23. What is the underlying cause of air motion?
24. In what direction does the Earth spin—west to east or east to west?
25. How does the Coriolis effect determine the general path of air circulation?
26. What does the Coriolis effect do to winds? To ocean currents?
27. As a volume of seawater freezes, the salinity of the surrounding water increases. Explain.

Global Circulation Patterns

28. Why do winds come predominately from the west in temperate latitudes?
29. What is the characteristic climate of the doldrums and why does it occur?
30. Why are most of the world's deserts found in the area known as the horse latitudes?
31. What are the trade winds?
32. What is the name given to the area where cold polar air meets the warm air of the temperate zone?
33. Why are eastbound aircraft flights usually faster than westbound ones?
34. Why do the temperate zones have unpredictable weather?
35. What are the jet streams and how do they form?
36. In summer, Southeast Asia, India, and Africa experience heavy flooding. Why?
37. What factors set surface ocean currents into motion?
38. How does the Coriolis effect influence the movement of surface waters?
39. Explain the circulation pattern of the Gulf Stream.
40. Explain why most of the bottom water of the oceans forms in the North Atlantic and near Antarctica.

Exercises

1. It being true that a gas fills all the space available to it, why doesn't the atmosphere go off into space?
2. How does the density of air in a deep mine compare with the density of air at sea level? Explain.
3. Why do your ears pop when you ascend to higher altitudes? Explain.
4. In a still room, smoke from a cigar sometimes rises and then settles in the air before reaching the ceiling. Explain.
5. If the composition of the upper atmosphere were changed so that a greater amount of terrestrial radiation escaped, what effect would this have on the Earth's climate? How about if the atmosphere reduced the amount of terrestrial radiation that could escape?
6. As we humans consume more energy, the overall temperature of the Earth tends to rise. Regardless of the increase in energy consumption, however, the temperature does not rise indefinitely. By what process is an indefinite rise prevented? Explain.
7. The Earth is closest to the Sun in January, but January is cold in the Northern Hemisphere. Why?
8. How do the total number of hours of sunlight in a year compare for tropical regions and polar regions of the Earth? Why are polar regions so much colder?

9. How do the wavelengths of radiant energy vary with the temperature of the radiating source? How does this affect solar and terrestrial radiation?
10. How is global warming affected by the relative transparencies of the atmosphere to long- and short-wavelength electromagnetic radiation?
11. Why is it important that mountain climbers wear sunglasses and use sunblock even when the temperature is below freezing?
12. If there were no water on the Earth's surface, would weather occur? Defend your answer.
13. As the world's population increases, the amount of carbon dioxide emissions from fossil fuel combustion also increases. Yet the amount of carbon dioxide emitted is greater than the amount found in the atmosphere. Where might some of the excess atmospheric carbon dioxide be going?
14. The amount of chlorine expelled into the atmosphere by Mount Pinatubo in the Philippines during its eruption was greater than the chlorine present in all the chemical plants of the world. What are the consequences to the environment of this amount of chlorine in the atmosphere?
15. Ozone is a component of automobile exhaust. Does the pollution from automobiles help to alleviate the ozone hole problem above the south pole? Defend your answer.
16. If the Earth were not spinning, what direction would the surface winds blow where you live? What direction does it blow on the spinning Earth at 15° S latitude and why?
17. What is the relationship between global atmospheric circulation and ocean currents?
    Relate oceanic gyres to patterns of subtropical high pressure.
18. Relate the jet stream to upper-air circulation. How does this circulation pattern relate to airline schedules from New York to San Francisco and the return trip to New York?
19. What is EI Niño? Why do surface temperatures along the eastern ocean margins rise during EI Niño? Why do fishing industries suffer?
20. How and why do weather patterns change during the EI Niño condition?
21. How might global warming affect the frequency of EI Niños?
22. Why are temperature fluctuations greater over land than water?
23. Why is the ocean usually blue-green?
24. How does the ocean influence weather on land?
25. Because seawater does not freeze easily, sea ice never gets very thick. This being true, where do large icebergs come from?
26. What happens to the salinity of seawater when evaporation at the ocean surface exceeds precipitation? When precipitation exceeds evaporation? Explain
27. Water denser than surrounding water sinks. With respect to the densities of deeper water, how far does it sink?
28. What effect does the formation of sea ice in polar regions have on the density of seawater? Explain.
29. The Mediterranean Sea is highly salty. What can you say about the rates of evaporation at the surface of the Mediterranean compared with precipitation over the Mediterranean?
30. What role does the Sun play in the circulation of ocean currents?
31. Which receives more solar energy over the course of a year—tropical regions or temperate regions? How does this affect ocean salinity?
32. In tropical regions, solar energy exceeds terrestrial radiation. What effect does this have on the salinity of oceans in tropical regions?
33. Most industrial pollutants spewed into the air in the United States end up in the North Sea where they are absorbed. Most pollutants spewed into the air in China end up in the icy waters of northern Canada. What does this tell you about prevailing wind directions in the Northern Hemisphere?
34. What happens to the water level in a glass of water when a floating ice cube in the glass melts? Similarly, what happens to the water level in the Great Lakes when floating chunks of ice melt?
35. What happens to the water level in a glass of salt water when a floating ice cube of freshwater melts? (Hint: the effect is much more pronounced for greater differences in density, say an ice cube that melts when floating in mercury.) Is there a rise in ocean level when floating icebergs melt? If so, is it significant?
36. Why is there more concern about melting of polar ice caps than there is about melting icebergs?

Problem

1. Fill a tall vessel with a 1-meter deep column of water. When you heat the water it will expand, rising by 0.0007 centimeters for each Celsius degree increase in temperature. (a) If the same water were 1 km deep and its temperature increased by 10 C°, how high would its surface rise by thermal expansion? (b) Why does this variation not occur in 1 km deep lakes that experience 10 C° degree differences in surface temperature?

CHAPTER 27

Weather

The Earth and its atmosphere make up a dynamic system that is always in a state of change. Although changes on the Earth's surface take place over long periods of time, changes in the atmosphere can occur in minutes. We need only look upward to the sky to see changes in the wind, cloud cover, and in the weather.

This chapter examines factors that influence the weather—atmospheric moisture content, temperature, air pressure, and the arrangement of land features and water. As we saw in Chapter 26, water on the Earth's surface and in the atmosphere greatly influences climate. We therefore begin our discussion of weather by looking at atmospheric moisture, and at how the amount of moisture in the air influences atmospheric stability. Then we discuss the development of different air masses and resulting weather patterns. We conclude with a look at the violent weather forces that impact our planet's surface.

27.1 Atmospheric Moisture

As we saw in previous chapters, water plays a dynamic role both on the Earth's surface and in its atmosphere. Any phase change in water is accompanied by an increase or decrease in the thermal energy of the water. Ice melts or water freezes, and water evaporates or water vapor condenses. All these phase changes occur with water both in the ocean and in the atmosphere.

There is always some water vapor in the air. A measure of the amount of this water vapor is called **humidity** (mass of water per volume of air). Weather reports often use **relative humidity**—the ratio of the amount of water vapor currently in the air at a given temperature to the largest amount of water vapor the air can contain at that temperature. A relative humidity of 50 percent, for example, means the water content in the air is half the amount it could be at that temperature.[1]

Air that contains as much vapor as it can possibly contain is *saturated.* Saturation occurs when the air temperature drops and water vapor molecules in the air begin condensing. Recall from Section 18.1 that water molecules are electric dipoles and so tend to stick together. Because of their normally high average speeds in air, however, most water molecules do not stick together when they collide. Instead, these fast-moving molecules bounce back when they collide and so remain in the gaseous phase. Some water molecules move more slowly than average, though, and these slow ones are more likely to stick to each other upon collision (Figure 27.1). Therefore, slow-moving molecules are the ones that condense and form droplets of water in saturated air. Because lower air temperatures are characterized by slower-moving molecules, saturation and condensation are more likely to occur in cool air than in warm air. Warm air can contain more water vapor than cold air.

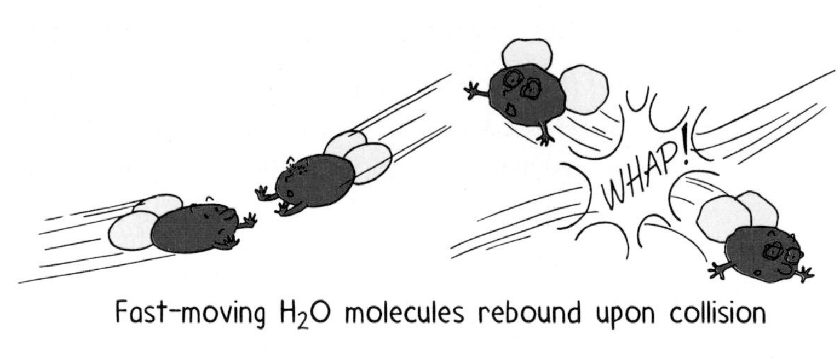

Fig. 27.1
Condensation of water molecules.

[1]Relative humidity is a good indicator of comfort. For most people, conditions are ideal when the temperature is about 20°C and the relative humidity 50-60%. When the relative humidity is too high, moist air feels "muggy" as condensation counteracts the evaporation of perspiration. Cold air that has a high relative humidity feels colder than dry air of the same temperature because of increased conduction of heat from the body. When the relative humidity is high, hot weather feels hotter, and cold weather feels colder.

Fig. 27.2
San Francisco is well known for its summer fog.

As air rises, it expands because it moves into a region of lower air pressure. As we learned in Section 7.2, when air expands, it cools. As air molecules cool, they slow down and then are more apt to stick together. If there are larger and slower-moving particles or ions present in the air, water vapor condenses on these particles, and we have a cloud. As the size of the cloud droplets grow, they fall as rain, sleet, or snow. This is *precipitation.*

When condensation in the air occurs near the Earth's surface, we call it *dew, frost,* or *fog.* On cool, clear nights, objects near the ground cool more rapidly than the surrounding air. As the air cools below a certain temperature, called the *dew point,* the air's ability to hold water vapor decreases and water from the now-saturated air condenses on any available surface. This may be a twig, a blade of grass, or the windshield of a car, and we have early-morning dew. When the dew point is at or below freezing, we have frost. When a large mass of air cools and reaches its saturation point (dew point), the relative humidity approaches 100 percent producing a cloud near the ground—fog.

Question

What is the major difference between fog and a cloud?

27.2 Air Behavior and Atmospheric Stability

Air is a mixture of molecules moving haphazardly and colliding with one another like ricocheting marbles. Moving molecules exert a small push on whatever they hit. This force (push) spreads over the area on which it is exerted as pressure. Because we are talking about air molecules, we use the term *air pressure.* The higher the kinetic energy of the air molecules (in other words, the faster they are moving), the greater the air pressure and the temperature. And the greater the number of air molecules that collide (in other words, the denser the air), the greater the air pressure. How air behaves depends on its pressure, temperature, and density.

A key concept in explaining air behavior is energy—especially thermal energy. According to the *first law of thermodynamics* (Section 7.4), when heat flows to or from a system, the system gains or loses an amount of thermal energy equal to the amount of heat transferred. There is no net loss in energy, and no net gain in energy. When heat is added to an air mass, the thermal energy of the air increases, which means its temperature, its pressure, or both increase. Heat can be added by solar radiation, by moisture condensation, or by contact with warm ground. When heat is subtracted from an air mass, the temperature of the air falls. Heat can be subtracted by radiation to space, by evaporation of rain falling through dry air, or by contact with cold surfaces.

Answer
Altitude.

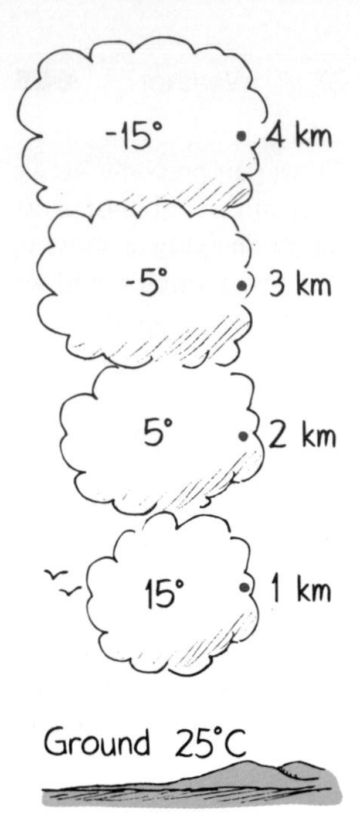

Fig. 27.3
The temperature of a parcel of dry air that expands adiabatically changes by about 10°C for each kilometer of elevation.

There are many atmospheric processes, usually involving time scales of a day or less, in which the amount of thermal energy added or subtracted is very small—so small that the process is nearly *adiabatic* (no thermal energy enters or leaves the system; see Section 7.4). Then the change in temperature is due only to pressure changes. The adiabatic form of the first law of thermodynamics is

Temperature change ~ pressure change

Adiabatic processes in the atmosphere are characteristic of large bodies of air. To illustrate how bodies of air behave, let us imagine a body of air enclosed in a very thin plastic garment bag—an air *parcel.* Like a free-floating balloon, the parcel can expand and contract freely without heat exchange with air outside the parcel. Recall from Section 26.2 that at higher elevations the air thins and has fewer air molecules. Air pressure always decreases with increasing height. So, as an air parcel flows up the side of a mountain, air pressure within the parcel decreases, allowing it to expand and cool.

With adiabatic expansion, the temperature of a dry air parcel decreases some 10°C for each kilometer it rises (Figure 27.3). This rate of cooling for dry air is called the *dry adiabatic rate.* Air flowing up and over tall mountains or rising in thunderstorms may change elevation by several kilometers. Thus when a dry air parcel at ground level at a comfortable 25°C rises 6 kilometers, its temperature drops to a frigid −35°C. On the other hand, if air at a typical temperature of −20°C at 6 kilometers descends to the ground, its temperature rises to a whopping 40°C.

Adiabatic processes are not restricted to dry air. As rising air cools, its capacity for containing water vapor decreases, increasing the relative humidity of the rising air. If the air cools to its dew point, the water vapor condenses and a cloud forms. Because condensation releases energy, the surrounding air is warmed. This added heat offsets the cooling due to expansion, making the air cool at a lesser rate—a *moist adiabatic rate.* Although the moist adiabatic rate varies according to temperature and moisture content, on average a moist air parcel cools 6°C for every kilometer it rises.

Atmospheric Stability

Stable air is air that resists upward movement. When a parcel of rising air is cooler than its surroundings, it is also denser than its surroundings and so tends to sink. This air is stable. Stable air that is forced to rise spreads out horizontally. When clouds develop, they too spread out into thin horizontal layers having flat tops and bottoms.

On the other hand, when a rising air parcel is warmer than its surroundings, it is less dense and continues to move upward until its temperature equals the temperature of its surroundings. In this case the air is unstable, for it favors upward movement. When the rising air is dry it cools at the dry adiabatic rate and a shallow layer of hot surface air results. When the rising air is moist, billowy towering clouds develop.

A dramatic example of adiabatic warming is the *Chinook*—a dry wind that blows down from the Rocky Mountains across the Great Plains. Cold air moving down a mountain slope is com-

Fig. 27.4
Chinooks—which are warm, dry winds—occur when high-altitude air descends and is adiabatically warmed.

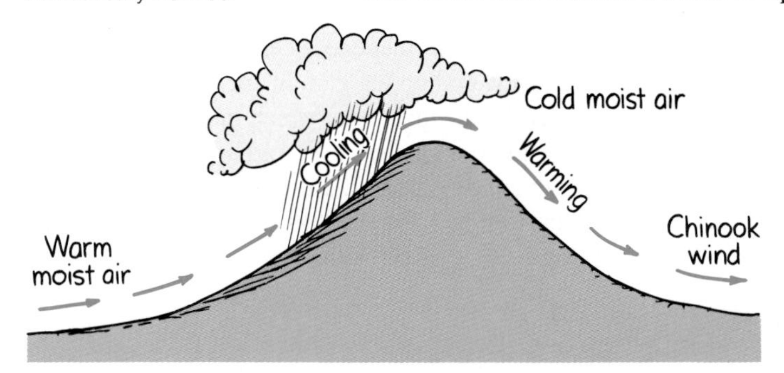

[2]When you're flying at high altitudes where outside air temperature is typically −35°C, you're quite comfortable in your warm cabin—but not because of heaters. The process of compressing outside air to a cabin pressure nearly that of sea level normally heats the air to a roasting 55°C (131°F). So air conditioners must be used to extract heat from the pressurized air.

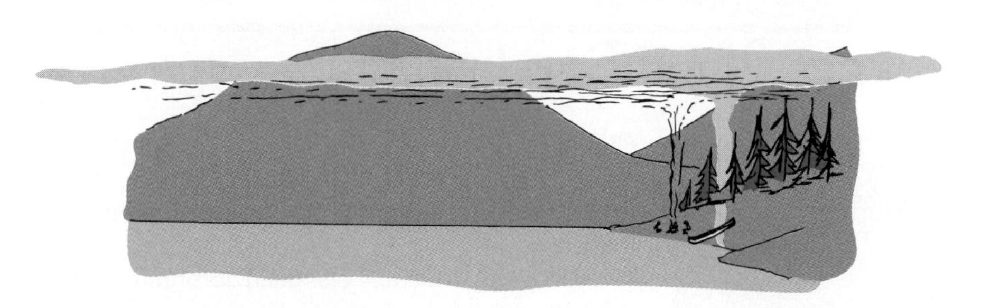

Fig. 27.5
The layer of campfire smoke over the lake indicates a temperature inversion. The air above the smoke is warmer than the smoke, and the air below is cooler.

pressed as it moves to lower elevations (where the pressure is greater than at higher elevations) and becomes appreciably warmer. The effect of expansion or compression of gases is quite impressive.[2]

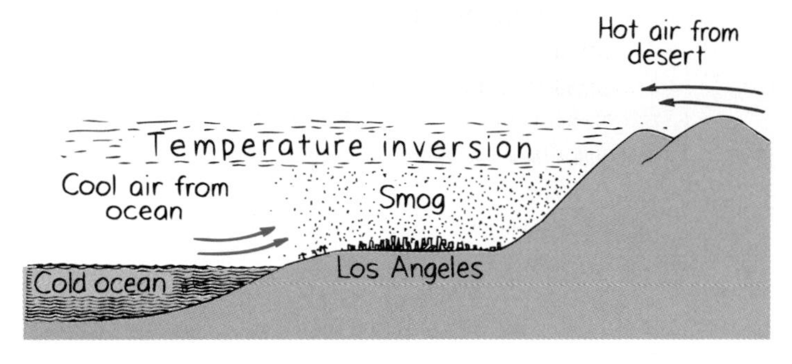

Fig. 27.6
Smog in Los Angeles is trapped by the mountains and a temperature inversion caused by warm air from the Mohave Desert overlying cool air from the Pacific Ocean.

A rising parcel of air continues to rise as long as it is warmer and less dense than the surrounding air. If it gets cooler and denser than its surroundings, it sinks. Under some conditions, large parcels of cold air sink and remain at a low elevation. This results in air above that is warmer. When the upper regions of the atmosphere are warmer than the lower regions, we have a **temperature inversion**. Unless rising warm air is less dense than this upper layer of warm air, the rising air rises no farther. Evidence of this is commonly seen over a cold lake when visible gas and other small particles such as smoke spread out in a flat layer above the lake rather than rising and dissipating higher in the atmosphere (Figure 27.5).

The smog of Los Angeles is trapped by such an inversion, caused by low-level cold air from the ocean capped by a layer of hot air moving westward over the mountains from the hot Mojave Desert. The west-facing side of the mountains helps hold the trapped air (Figure 27.6). The mountains on the edge of Denver play a similar role in trapping smog beneath a temperature inversion.

Questions

1. If a parcel of dry air initially at 0°C expands adiabatically while flowing upward alongside a mountain, what is its temperature when it has risen 2 km? 5 km?
2. What happens to the air temperature in a valley when dry, cold air blowing across the mountains descends into the valley?
3. Imagine a gigantic dry-cleaner's garment bag full of dry, −10°C air floating 6 km above the ground like a balloon with a string hanging from it. If you yank it suddenly to the ground, what will its temperature be?

Answers

1. The air cools at the dry adiabatic rate of 10°C for each kilometer it rises. When the parcel rises to an elevation of 2 km, its temperature is −20°C. At an elevation of 5 km, its temperature is −50°C.
2. The air is adiabatically compressed, and so its temperature increases. Residents of some valley towns in the Rocky Mountains, such as Salida, Colorado, benefit from this adiabatic compression and enjoy "banana belt" weather in midwinter.
3. If the bag of air is pulled down quickly and heat conduction is negligible, the air is adiabatically compressed by the atmosphere and its temperature rises to a piping hot 50°C.

27.3 Cloud Development

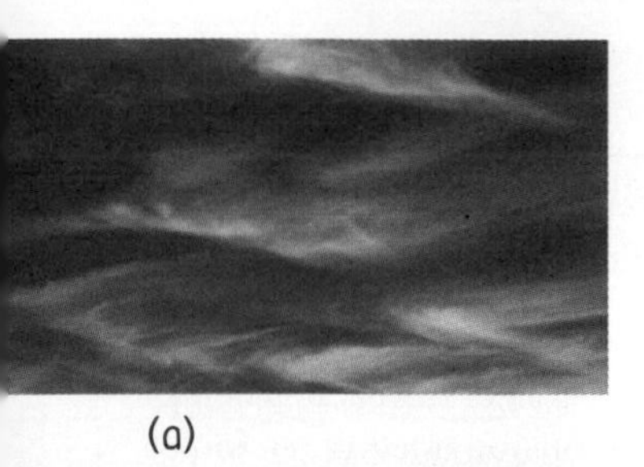

(a)

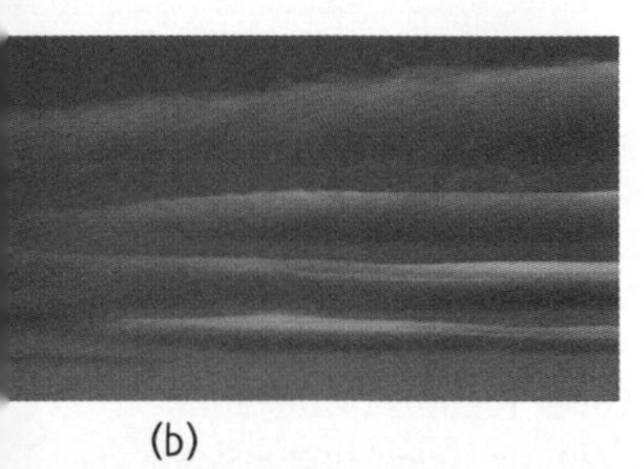

(b)

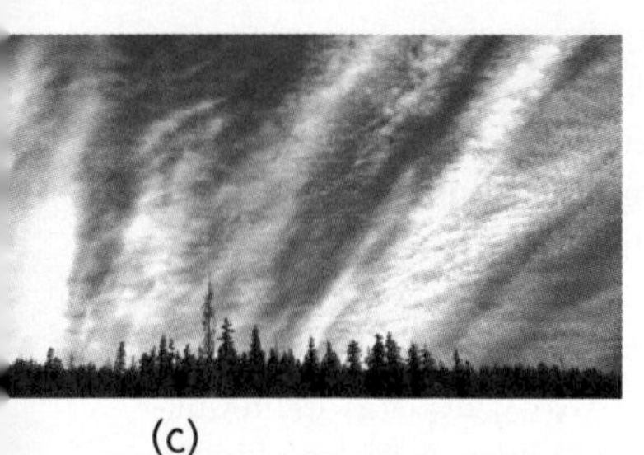

(c)

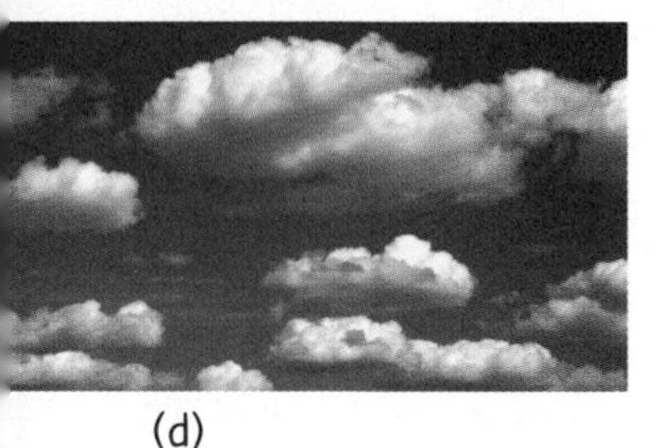

(d)

Fig. 27.7
The four cloud groups and ten cloud forms. (a) High clouds: cirrus, cirrostratus, cirrocumulus. (b) Middle clouds: altostratus, altocumulus. (c) Low clouds: stratus, stratocumulus, nimbostratus. (d) Clouds with vertical development: cumulus, cumulonimbus.

Clouds are an aggregation of suspended water droplets and ice crystals, formed from rising warm, moist air. As the warm air rises, it cools and therefore becomes less able to contain water vapor. As the water vapor condenses into tiny droplets, clouds are formed.

Clouds are generally classified according to altitude and shape. There are ten principal cloud forms, each of which belongs to one of four major groups (Table 27.1).

Table 27.1
The Four Major Cloud Groups and Ten Cloud Forms

1. High clouds (above 6000 m)	**3.** Low clouds (below 2000 m)
Cirrus	Stratus
Cirrostratus	Stratocumulus
Cirrocumulus	Nimbostratus
2. Middle clouds (2000–6000 m)	**4.** Clouds having vertical development
Altostratus	Cumulus
Altocumulus	Cumulonimbus

High Clouds

High clouds generally form above 6000 meters. High clouds are denoted by the prefix *cirro-*. The air at this elevation is quite cold and dry, and so clouds this high are made up almost entirely of ice crystals. The most common high clouds are thin, wispy *cirrus* clouds, blown by high winds into the well-known wispy “mare’s tail” or “artist’s brush.” Cirrus clouds usually indicate fair weather, but may also indicate approaching rain.

Cirrocumulus clouds are the familiar rounded white puffs, in patches, seldom covering more than a small portion of the sky. Small ripples and a wavy appearance make the cirrocumulus clouds resemble fish scales, hence they are often said to make up a *mackerel sky.*

Cirrostratus clouds are thin and sheetlike, and often cover the whole sky. The ice crystals in these clouds refract light and produce a halo around the Sun or the Moon. When cirrostratus clouds thicken, they give the sky a white, glary appearance—an indication of coming rain or snow.

Middle Clouds

Middle clouds form between 2000 and 6000 meters and are denoted by the prefix *alto-*. These clouds are made up of water droplets and, when temperature allows, ice crystals. *Altostratus* clouds are gray to blue-gray, and often cover the sky for hundreds of square kilometers. Thin altostratus clouds are often confused with thick cirrostratus clouds. Altostratus clouds are often so thick they diffuse incoming sunlight so that objects on the ground don’t produce shadows. Because altostratus clouds often form before a storm, look on the ground and if you don’t see your shadow, cancel that picnic.

Altocumulus clouds appear as gray, puffy masses in parallel waves or bands. The individual puffs are much larger than those found in cirrocumulus, and the color is also much darker. The appearance of these clouds on a warm, humid summer morning often indicates thunderstorms by late afternoon.

Low Clouds

Low clouds ranging from the surface up to 2000 meters are called *stratus* clouds. They are almost always made up of water droplets but in cold weather may also contain ice crystals and snow. *Stratus* clouds are uniformly gray and often cover the whole sky. They are very common in winter and as a consequence give rise to the sky's "hazy shade of winter." They resemble a high fog that doesn't touch the ground. Although stratus clouds are not directly associated with falling precipitation, they sometimes generate a light drizzle or mist.

Stratocumulus clouds either form a low, lumpy layer that grows in horizontal rows or patches or, with weak rising motion, appear as rounded masses. The color is generally light to dark gray. They are often confused with altocumulus clouds but can be distinguished by their size; hold your hand at arm's length and point toward the cloud in question. An altocumulus cloud commonly appears to be the size of your thumbnail; a stratocumulus cloud appears to be about the size of your fist. Precipitation of rain or snow does not usually fall from stratocumulus clouds.

Nimbostratus clouds are dark and foreboding. They are a wet-looking cloud layer associated with light to moderate rain or snow.

Clouds Having Vertical Development

Vertical cloud development is caused by rising air currents. These are *cumulus* clouds—the most familiar of the many cloud types. Cumulus clouds resemble pieces of floating cotton, with sharp outlines and a flat base. They are white to light gray and generally occur about 1000 meters above the ground. The tops of cumulus clouds are often in the form of rising towers, denoting the limit of the rising air. These are the clouds childhood daydreams are made of. Remember the horses, dragons, and magic palaces you saw in them?

When cumulus clouds turn dark and are accompanied by precipitation, they are referred to as *cumulonimbus* clouds and indicate a coming storm. As we shall see, cumulonimbus clouds often become thunderheads.

Although water vapor is less dense than air, a water droplet in air is considerably denser than air, and the gravitational force pulling it down is much greater than any buoyant force supplied by the surrounding air. So why don't all the water droplets in clouds fall to the ground? The answer has to do with updrafts. A typical cumulus cloud has an updraft speed of at least 1 meter per second, more than the maximum speed of typical falling droplets and more than enough to support the droplets (preventing them from falling to the ground). Without updrafts, the droplets drift so slowly out of the bottom of the cloud and evaporate so quickly that they have no chance to reach the ground. They are replaced by new droplets forming above.[3]

Raindrops are huge compared with typical cloud droplets. Their maximum speed is greater than the speed of most updrafts and they evaporate slowly enough that they can easily reach the ground.

[3]A 20 μm diameter droplet (μm = micrometer), typical in thin clouds or fog, has a maximum speed of about 0.2 m/s. A droplet having a diameter of 100 μm, found in some dense clouds, has a maximum speed of about 1 m/s. A typical raindrop, which has a diameter of 2000 μm (2 mm), has a maximum speed of nearly 10 m/s.

27.4 Air Masses, Fronts, and Storms

An *air mass* is a volume of air much larger than the parcels of air we have discussed. Various air masses cover large portions of the Earth's surface, and each portion imparts its unique characteristics to the air it touches. For this reason, the characteristics of an air mass depend on where it forms. An air mass formed over water in the tropics is different from one formed over land in the polar regions. Air masses are divided into six general categories according to what type of land or water they form over and the latitude where formation occurs (Table 27.2 and Figure 27.8). The surface type is designated by a lowercase letter (m for maritime, c for continental), and the source region is designated by an uppercase letter (A for Arctic, P for polar, T for tropical.)

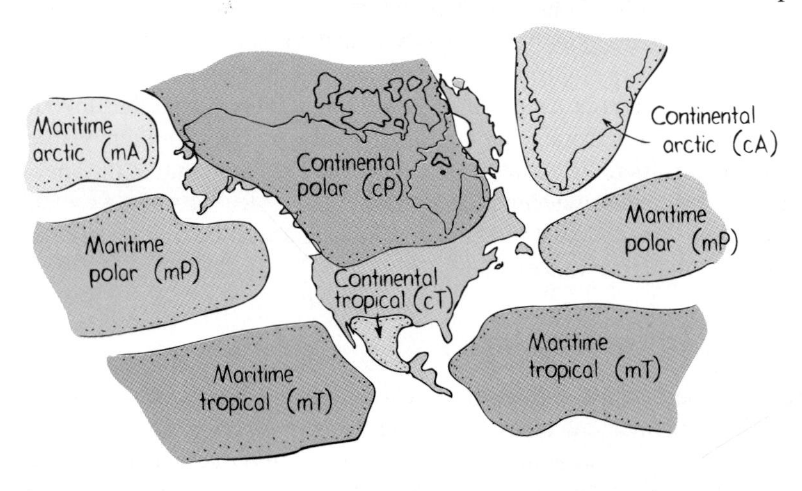

Fig. 27.8
Typical source regions of air masses for North America.

Each air mass has its own characteristics. Continental polar and continental Arctic air masses generally produce very cold, dry weather in winter and cool, pleasant weather in summer. Maritime polar and maritime Arctic air masses, picking up moisture as they travel across the oceans, generally bring cool, moist weather to a region. Continental tropical air masses are generally responsible for the hot, dry weather of summer, and warm, humid conditions are due to maritime tropical air masses. Where two air masses having different properties meet, we find a variety of weather.

Atmospheric Lifting Mechanisms

Clouds are great indicators of weather. For clouds to form, air must be lifted. The three principal lifting mechanisms in the atmosphere are convectional lifting, orographic lifting, and frontal lifting.

Table 27.2
Classification of Air Masses and Their Characteristics

Typical Source Region	Classification	Symbol	Characteristics
Arctic	maritime Arctic	mA	cool, moist, unstable
Greenland	continental Arctic	cA	cold, dry, stable
North Atlantic, Pacific Oceans	maritime Polar	mP	cool, moist, unstable
Alaska, Canada	continental Polar	cP	cold, dry stable
Caribbean Sea, Gulf of Mexico	maritime Tropical	mT	warm, moist; usually unstable
Mexico, Southwest United States	continental tropical	cT	hot, dry, stable—aloft; unstable—surface

Convectional Lifting

The Earth's surface is heated unequally. Some areas are better absorbers of solar radiation than others and so heat up more quickly. The air that touches these surface "hot spots" becomes warmer than surrounding air and forms thermal bubbles that rise, expand, and cool. This rising of air is accompanied by the sinking of cooler air aloft. This circulatory motion produces **convectional lifting**.

If cooling occurs close to the air's saturation temperature, the condensing moisture forms a cumulus cloud. Air movement within the cumulus cloud moves in a cycle: Warm air rises, cool air descends. Because descending cool air inhibits the expansion of warm air beneath it, small cumulus clouds usually have a great deal of blue sky between them (Figure 27.9).

Fig. 27.9
Cumulus clouds are often found as individual towering white clouds separated from each other by expanses of blue sky.

Cumulus clouds often remain in the same space as they form, dissipating and reforming many times. As they grow, they shade the ground from the Sun. This slows surface heating and inhibits the upward convection of warm air. Without a continuous supply of rising air, the cloud begins to dissipate. Once the cloud is gone, the ground reheats, allowing the air above it to warm and rise. Thus convectional lifting begins again, and another cumulus cloud begins at the same place.

Orographic Lifting

An air mass that is pushed upward over an obstacle such as a mountain range undergoes **orographic lifting**. The rising air cools, and if it is humid, clouds form. The type of cloud formed depends on the air's stability and moisture content. If the air is stable, a layer of stratus clouds may form; if the air is unstable, cumulus clouds may form. As the air mass moves down the other side of the mountain (the leeward slope), it warms. This descending air is dry because most of its moisture was removed in the form of clouds and precipitation on the windward (upslope) side of the mountain. Because the dry leeward sides of mountain ranges are sheltered from rain and moisture, they are often referred to as regions of *rain shadow* (Figure 27.10).[4]

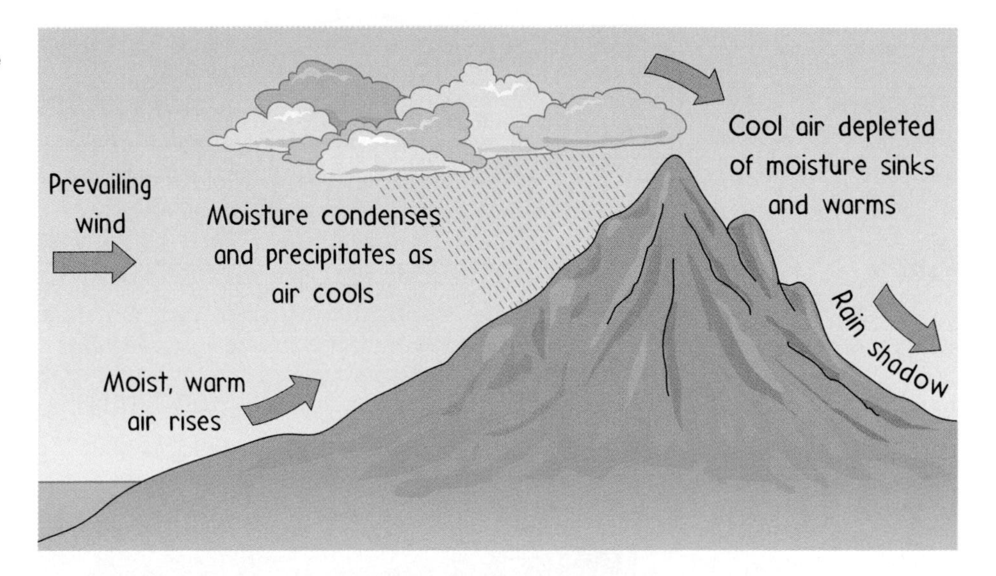

Fig. 27.10
A mountain range may produce a rain shadow on its leeward slope. As warm, moist air rises on the windward slope, the air cools and precipitation develops. By the time it reaches the leeward slope, the air is depleted of moisture, so that the leeward side is dry. It lies in a rain shadow.

[4]Just as a "regular" shadow is a place where no light falls because some obstacle blocks the light, a rain shadow is a place where no rain falls because some obstacle (such as a mountain) blocks precipitation.

Frontal Lifting

The contact zone between two different air masses is called a **front**. When two air masses make contact, differences in temperature, moisture, and pressure can cause one air mass to ride over the other. When this occurs, we have **frontal lifting**. If a cold air mass moves into an area occupied by a warm air mass, the contact zone between them is called a *cold front,* and if warm air moves into an area occupied by cold air, the zone of contact is called a *warm front.* If neither of the air masses is moving, the contact zone is called a *stationary front.* Fronts are usually accompanied by wind, clouds, rain, and storms.

Meteorologists and other observers of the sky can often tell when a cold front is approaching by observing high cirrus clouds, a shift in wind direction, a drop in temperature, and a drop in air pressure. As cold air moves into a warm air mass, forming a cold front, the warm air is forced upward. As it rises, it cools, and water vapor condenses into a series of cumulonimbus clouds. The advancing wall of clouds at the front develops into thunderstorms with heavy showers and gusty winds. After the front passes, the air cools and sinks, pressure rises, and rain ceases. Except for a few fair-weather cumulus clouds, the skies clear and we have the calm after the storm.

When warm air moves into a cold air mass, forming a warm front, the less-dense warmer air gradually rides up and over the colder, denser air. The approach of a warm front, although less obvious and more gradual than the approach of a cold front, is also

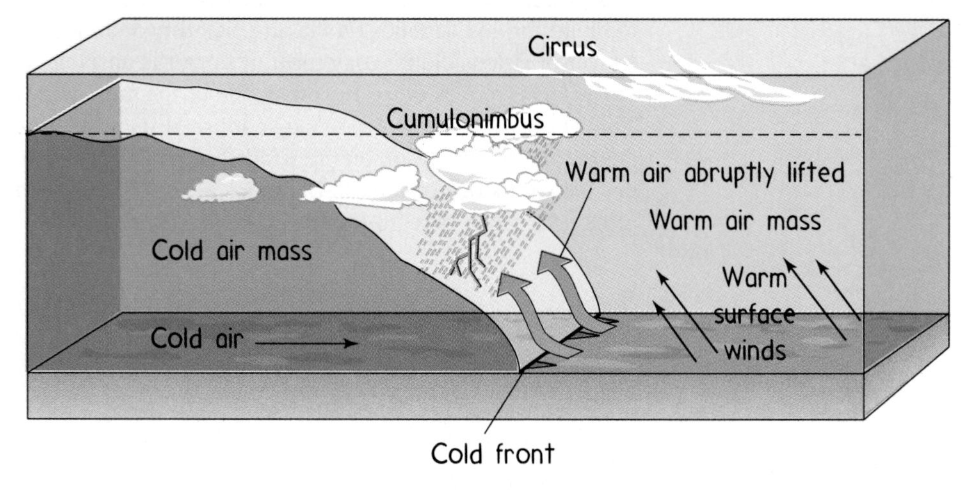

Fig. 27.11
A cold front occurs when a cold air mass moves into a warm air mass. The cold air forces the warm air upward, where it condenses to form clouds. If the warmer air is moist and unstable, heavy rainfall and gusty winds develop.

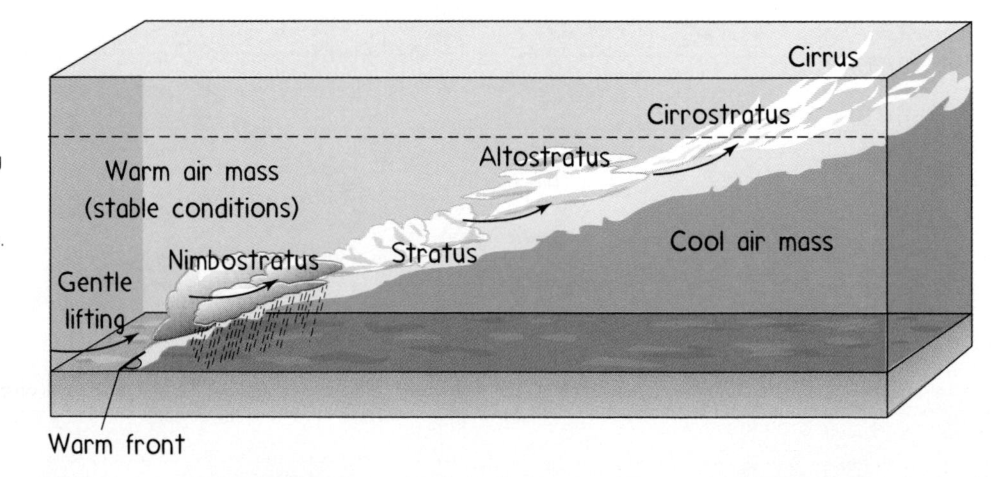

Fig. 27.12
A warm front occurs when a warm air mass moves into a cold air mass. The less-dense warmer air rides up and over the colder, denser air, resulting in widespread cloudiness and light to moderate precipitation that can cover great distances.

indicated by cirrus clouds. Ahead of the front, the cirrus clouds lower and thicken into altocumulus and altostratus clouds that turn the sky an overcast gray. Moving still closer to the front, light to moderate rain or snow develops, and winds become brisk. At the front, air gradually warms, and the rain or snow turns to drizzle. Behind the front, the air is warm and the clouds scatter.

Middle-Latitude Cyclones

The convergence of contrasting air masses along frontal zones can lead to the development of a **cyclone**—a migrating center of low-pressure with converging air masses spiraling inward. In the Northern Hemisphere, the spiral moves in a counterclockwise direction, and in the Southern Hemisphere, motion is clockwise. Recall from Section 26.5 that such spiraling (cyclonic) motion is generated by the pressure-gradient force, the Coriolis effect, and surface friction. To better understand the development of cyclonic motion refer to Figure 27.13. In the middle latitudes, as cold polar air from the north moves southward, it converges with warm, northward-moving air from the south; the paths of the air masses are in opposite directions. As the two air masses meet, a trough of low pressure develops between them. The cold air flows in under the less-dense warm air, forcing the warm air upward. The rising warm air cools adiabatically, and we have clouds, precipitation, and storms.

In the United States, cyclones generally develop along the eastern slopes of the Rocky Mountains, the Great Basin, the Gulf of Mexico, and east of the Carolinas in the Atlantic Ocean. Although surface conditions in these different areas greatly influence the development of cyclones, the polar jet stream also plays a role. Recall from Section 26.6 that the polar jet stream varies seasonally and is strongest during the winter. As surface air moves upward into the polar jet stream, wind speeds intensify, bringing strong winter storms and blizzards to the United States.

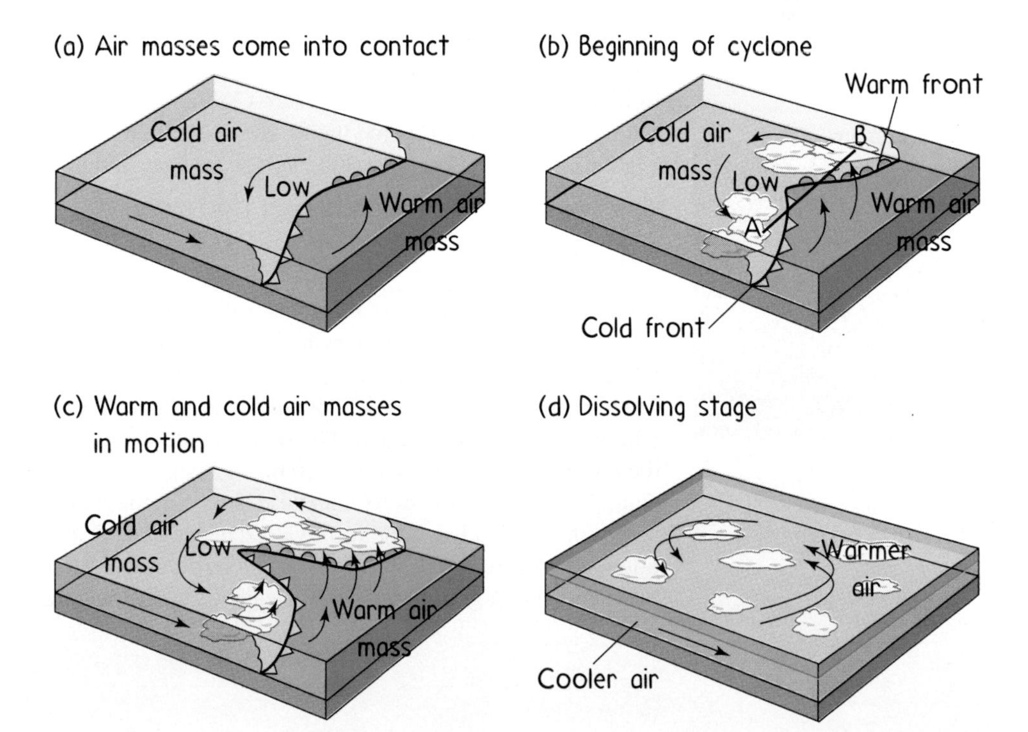

Fig. 27.13 Stages in a middle-latitude cyclone.

27.5 Violent Weather

The three types of lifting mechanisms bring about a variety of weather conditions. As contrasting air masses converge, the weather resulting from their contact depends on the conditions in their source regions. Weather changes can occur slowly or very quickly. The most rapid changes, and the most violent ones occur with three major types of storms: Thunderstorms, tornadoes, and hurricanes.

Fig. 27.14
The mature stage of a thunderstorm cloud appears as a towering cumulonimbus cloud that reaches up to about 12 km. Strong horizontal winds and icy temperatures flatten and distend the cloud's crown into a characteristic anvil shape.

Thunderstorms

A thunderstorm cycle begins with humid air rising, cooling, and condensing into a single cumulus cloud. This cloud builds and grows upward as long as it is fed by an updraft of rising warm air from below. Particles of precipitation grow larger and heavier within the cloud until they eventually begin to fall as rain. The falling rain drags some of the cool air along with it, creating a downdraft, and so the chilled air becomes colder and denser than the air around it. Together, the rising warm updraft and the sinking chilled downdraft make up a storm cell within the cloud. This is the mature stage, where the thunderstorm cloud appears as a lonely giant—dark and brooding in the sky. It typically has a base several kilometers in diameter and can tower to altitudes up to 12 kilometers. At these high altitudes, horizontal winds and lower temperatures flatten and distend the thunderhead crown into a characteristic anvil shape (Figure 27.14). After the thunderstorm dissipates, it leaves behind the cirrus anvil as a reminder of its once mighty presence.

At any given time, there are about 1800 thunderstorms in progress in the Earth's atmosphere. Wherever thunderstorms occur, there is lightning and thunder. As water droplets in the cloud bump into and rub against one another, the cloud becomes electrically charged—usually positively charged at the top and negatively charged at the base. As electrical stress between the oppositely charged regions builds up, the charge becomes great enough that electrical energy is released and passed to other points of opposite charge, which quite often means the ground. The electrical energy flow from cloud to ground is lightning. As lightning heats up the air, the air expands and we hear lightning's noisy companion, thunder. Lightning strikes the Earth some 100 times every second, with some bolts having an electric potential of as much as 100 million volts. Lightning claims more than 200 victims per year in the United States alone.

Fig. 27.15
Time exposure of cloud-to-ground lightning during an intense thunderstorm.

Tornadoes

A revolving object, such as a whirling ball on a string, speeds up when pulled toward its axis of revolution, thus conserving its angular momentum. Similarly, winds slowly rotating over a large area speed up when the radius of rotation decreases. This increase in speed produces a *tornado,* which is a funnel-shaped cloud that extends downward from a large cumulonimbus cloud. The funnel cloud is called a tornado only after it touches the ground. The winds of a tornado travel at speeds of up to 800 kilometers per hour in a counterclockwise direction (clockwise in the Southern Hemisphere). As a tornado moves across the land, advancing at forward speeds from 45 to 95 kilometers per hour, it may bounce and skip, rising briefly from the ground and then touching down again. A tornado acts like a gigantic vacuum cleaner, picking up everything in its path. It wreaks havoc not only by suction but also by the battering power of its whirling winds. In its wake, an explosive trail of flying dirt and debris is left behind (Figure 27.16).

Fig. 27.16
Like a gigantic vacuum cleaner, the strong wind of a tornado can pick up and obliterate everything in its path.

Tornadoes occur in many parts of the world. In the flat central plains of the United States, a tornado zone extends from northern Texas through Oklahoma, Kansas, and Missouri, where more than 300 tornadoes touch down each year—a zone known as Tornado Alley. Tornadoes are so frequent in this part of the country that some homes are built with underground shelters (do you recall the *Wizard of Oz*?). The power of a tornado is terrifying and devastating.

Hurricanes

In the steamy tropics where the Sun warms the oceans, the transfer of thermal energy to the atmosphere by evaporation and conduction is so thorough that air and water temperatures are about equal. The high humidity prevalent in this part of the world favors the development of cumulus clouds and afternoon thunderstorms. Most of the individual storms are not severe. However, as the moisture content and thermal energy of the air increase and surface winds converge, a strong vertical wind shear can cause the rising warm, moist air to tilt inward and spiral. This condition produces a more violent storm—a *hurricane*—with wind speeds up to nearly 300 kilometers per hour. Gaining energy from its source area, a hurricane grows as more air rises and increasing winds rotate around a central relatively calm, low-pressure zone—the *eye* of the storm.

Would storms occur if all parts of the Earth's surface were heated evenly?

Answer
No. The principal factor in the formation of storms is contact between warm air and cool air.

27.6 Weather Forecasting

Fig. 27.17
On August 24, 1992, Hurricane Andrew struck the southern peninsula of Florida with gusts measured at 200 km per hour. The storm was catastrophic, with 45 deaths and more than 180,000 people left homeless. Without the timely hurricane watch issued by the National Weather Service, casualty losses could have been much more severe.

Forecasting hurricanes and other storms is a prime concern of meteorologists. Weather forecasting is in part a matter of determining air-mass characteristics, predicting how and why the characteristics might change, and in what direction an air mass might move. In the case of hurricanes and tornadoes, such predictions are life-saving. Meteorologists have a long and remarkable record of reducing property loss and saving many lives.

All of our Earth's weather occurs in the troposphere. Forecasting weather changes requires first knowing present weather conditions over a large area. The data needed include temperature, air pressure, humidity, type of clouds, level of precipitation, and wind direction and speed. This is a great deal of data, especially for measurements throughout a large region. A network of observation stations has been set up with more than 10,000 land-based stations, hundreds of ship-based stations, and several satellites that provide surface-weather information to the World Meteorological Organization (WMO).

The WMO is an international organization that shares and exchanges weather information. The system that enables this information exchange is the Global Telecommunications System. Atmospheric data are fed into supercomputers that carry out billions of calculations to generate models of real atmospheric conditions. These models are then analyzed so that weather maps and charts can be prepared, and the prediction of the weather on a global scale can begin. Weather charts and predictions are then distributed to public and private agencies worldwide. In the United States, the information is used and adapted for regional and local weather forecasts. These forecast reports are then broadcast to the general public by radio or television.

There are several methods of weather forecasting. Probably the simplest weather forecast is based on the *persistence method*—the continuity of a weather pattern. For example, if it is raining and local atmospheric conditions are likely to persist, then rain will most likely occur tomorrow. Or, because surface-weather systems tend to move in the same direction and at the same speed, a forecast might be based on the trend of the weather pattern—the *trend method.* For example, if a cold front is moving eastward at an average speed of 20 kilometers per hour, it can be expected to affect the weather 80 kilometers away in four hours.

Another method of forecasting is the *analog method,* which compares present features of the weather (cloud cover, wind, temperature, humidity, and so on) with weather conditions produced by such features in the past. If the features are similar enough, forecasters assume that present weather conditions will be the same as past weather conditions. Say, for example, the average weather condition for San Francisco during the summer months is fog. Then, if features that produced summer fog in the past occur in the present, forecasters can predict foggy weather conditions for San Francisco for any particular day in July with a high chance of being correct.

We often hear about the probability of a weather condition—for example, that the probability of rain is 70 percent. This is an expression of chance, meaning that there is a 70 percent chance that rain will fall somewhere in the forecast area.

Another forecast we often hear about is the extended forecast. This forecast is based upon weather features that develop in certain areas. If a continental polar air mass is approaching, for instance, we can expect cold, dry weather. If a maritime polar air mass is approaching, we can expect cold, moist weather. All these methods of weather prediction are based on the statistical analysis of weather information.

Weather forecasting involves great quantities of data from all over the world. Before the 1960s, most of these data were assembled, analyzed, and plotted on weather

Weather Maps

The weather forecaster's primary tool is the surface weather map or chart. A weather map is essentially a representation of the frontal systems and the high-pressure and low-pressure systems that overlie the areas outlined in the map. Symbols on such a map are a shorthand notation to represent data gathered from various observation stations. These symbols are called weather codes.

This shorthand notation compiles 18 items of data into a very small area called a *station model.* The circle at the center describes the overall appearance of the sky. Jutting from the circle is a wind arrow, its tail in the direction from which the wind comes and its feathers indicating wind speed. The other 15 weather elements are in standard position around the circle.

continued

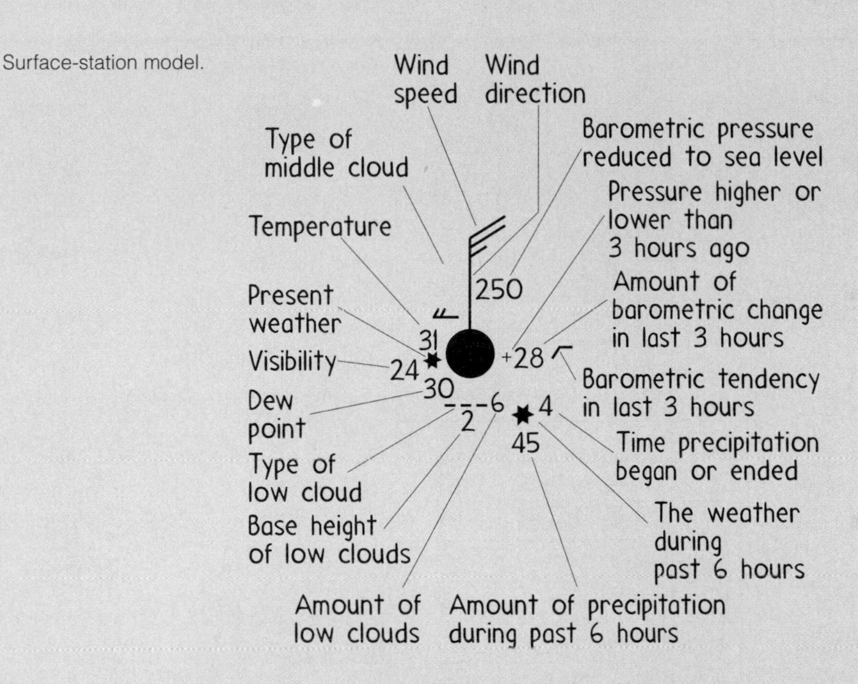

Surface-station model.

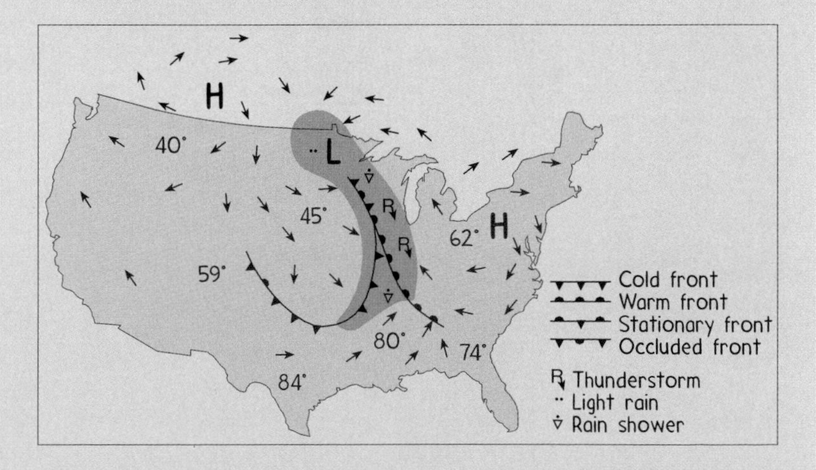

Weather maps show atmospheric conditions. As warm air rises, it expands and chills. As it chills, the water vapor molecules condense to form clouds. Because air moves from a high-pressure region to a low-pressure region, low-pressure zones are accompanied by cloud cover. In a high-pressure zone, air generally sinks. Because sinking air does not usually produce clouds, we find clear skies and fair weather.

Weather Maps *(continued)*

A weather map is covered with lines—*isobars*—that connect points of equal pressure. As air moves from a high-pressure region to a low-pressure region, it rises and cools and the moisture in it condenses into clouds. In the vicinity of the low (L on map), we see an extensive cloud cover. In the vicinity of the high (H on map), we see clear skies. In a high-pressure region, air sinks and warms adiabatically. Because sinking air does not produce clouds, we find clear skies and fair weather. The heavy lines on a weather map represent fronts. Because fronts generally mean a change in the weather, they are of great importance on weather maps.

Weather Symbols

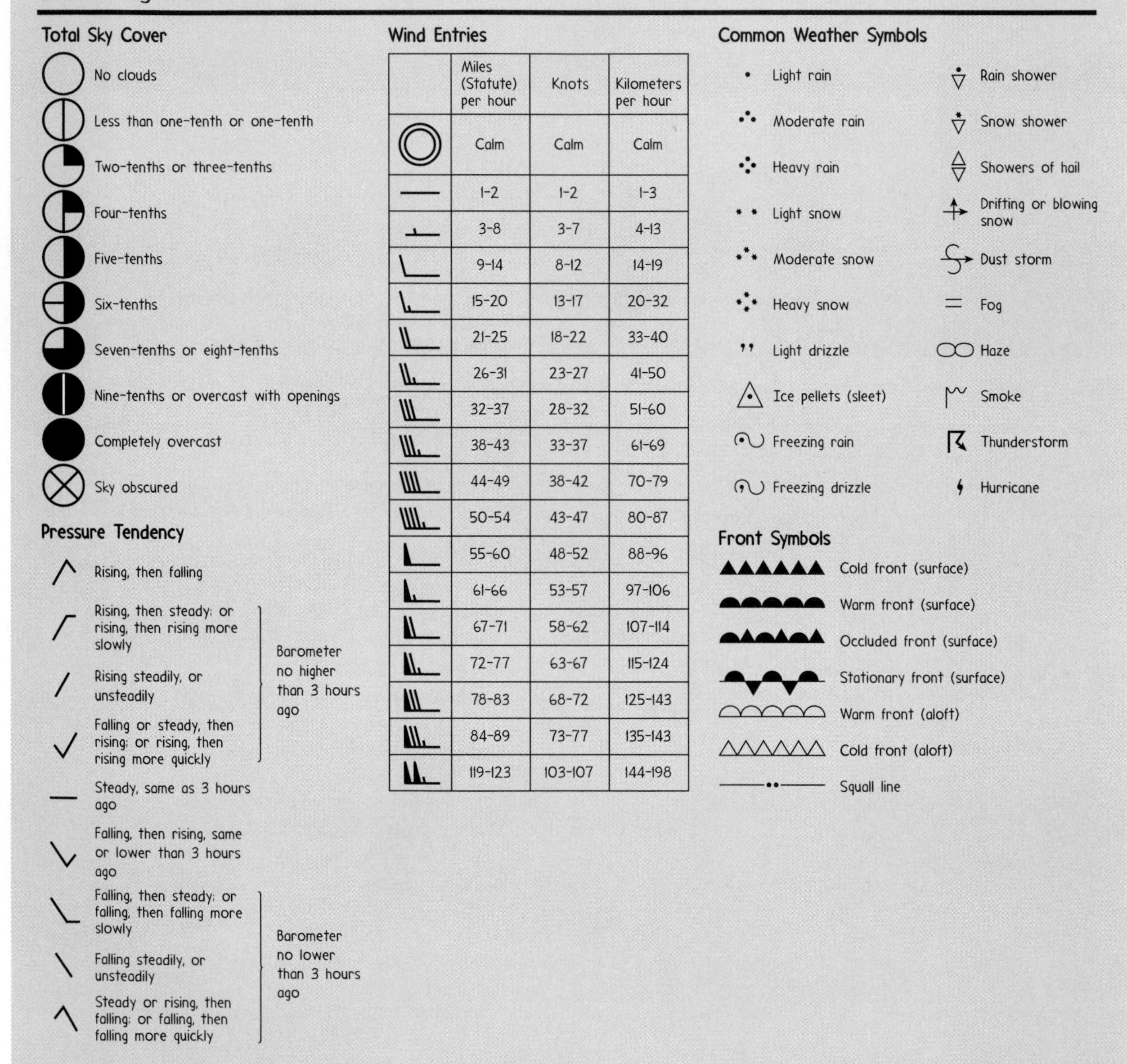

	Miles (Statute) per hour	Knots	Kilometers per hour
	Calm	Calm	Calm
	1-2	1-2	1-3
	3-8	3-7	4-13
	9-14	8-12	14-19
	15-20	13-17	20-32
	21-25	18-22	33-40
	26-31	23-27	41-50
	32-37	28-32	51-60
	38-43	33-37	61-69
	44-49	38-42	70-79
	50-54	43-47	80-87
	55-60	48-52	88-96
	61-66	53-57	97-106
	67-71	58-62	107-114
	72-77	63-67	115-124
	78-83	68-72	125-143
	84-89	73-77	135-143
	119-123	103-107	144-198

maps and charts by hand. This took thousands of calculations, a large work force, and long hours. Now, with computers, great quantities of data from around the world can be processed in a matter of minutes. Computers not only plot and analyze data, they also predict the weather. Meteorologists design atmospheric models using mathematical equations that describe how atmospheric temperature, pressure, and moisture content will change with time. These equations are programmed into the computer and measurements of temperature, air pressure, humidity, and wind direction are fed into the equations. Each equation is solved for an increment of time and for a large number of locations. The results of these equations are used by the computer to draw a chart of the projected weather conditions. The weather forecaster then uses these projections as a guide to predicting the weather. Even so, the many variables involved are not exactly predictable, and so it may unexpectedly rain on your parade!

Summary of Terms

- **Humidity** A measure of the amount of water vapor in the air.
- **Relative humidity** The amount of water vapor in the air at a given temperature expressed as a percentage of the maximum amount of water vapor the air can hold at that temperature.
- **Temperature inversion** A condition in which the upper regions of the atmosphere are warmer than the lower regions.
- **Clouds** The condensation of water droplets above the Earth's surface.
- **Convectional lifting** An air-circulation pattern in which air warmed by the ground rises while cooler air aloft sinks.
- **Orographic lifting** The lifting of air over a topographic barrier such as a mountain.
- **Front** The contact zone between two different air masses.
- **Frontal lifting** The lifting that occurs as two air masses converge.

Review Questions

Atmospheric Moisture

1. Distinguish between humidity and relative humidity.
2. Why does relative humidity increase at night?
3. As air temperature decreases, does relative humidity increase, decrease, or stay the same?
4. What factors are responsible for condensation?
5. Does condensation occur more readily at high temperatures or low temperatures? Explain.
6. What does saturation point have to do with dew point?
7. Distinguish between dew and frost.
8. What happens to the water vapor in saturated air as the air cools?
9. When water vapor condenses to liquid water, is heat absorbed or released?
10. Distinguish between condensation and precipitation.

Air Behavior and Atmospheric Stability

11. Explain why warm air rises and cools as it expands.
12. Does a rising parcel of air get warmer, get cooler, or stay the same temperature?
13. In what ways can the thermal energy of air be increased?
14. In what ways can the thermal energy of air be decreased?
15. What is an adiabatic process?
16. What is a temperature inversion? Give examples of where these inversions may occur.
17. Explain how a convection cycle is generated.
18. What happens to the air pressure of an air parcel as it flows up the side of a mountain?
19. Which cloud form is associated with a stable air mass.

Cloud Development

20. Explain how clouds form.
21. Name the cloud form associated with (a) the hazy shade of winter, (b) mackerel sky, (c) floating cotton, and (d) snowfall.
22. Name the cloud group for (a) altocumulus, (b) cirrostratus, (c) nimbostratus, and (d) cumulus clouds.
23. Rain or snow is most likely to be produced by which of the following cloud forms? a) cirrostratus, b) nimbostratus, c) altocumulus, d) stratocumulus
24. Are clouds having vertical development characteristic of stable air, stationary air, unstable air, or dry air?
25. Which type of clouds can become thunderheads?

Air Masses, Fronts, and Storms

26. What are the characteristics of continental polar air, continental tropical air, maritime polar air, and maritime tropical air?
27. What are the three main atmospheric lifting mechanisms?
28. Explain how convectional lifting plays a role in the formation of cumulus clouds?
29. Does a rain shadow occur on the windward side of a mountain range or on the leeward side? Explain.
30. Under what conditions does orographic precipitation occur?
31. What is a front?
32. Differentiate between a cold front and a warm front.

33. List four regions in North America where cyclones tend to develop.
34. How does the polar jet stream influence cyclone formation?
35. In which direction do the winds in a Northern Hemisphere cyclone spiral—clockwise or counterclockwise? Does the air move toward the center of the spiral or away from the center?
36. In which direction do the winds in a Southern Hemisphere cyclone spiral—clockwise or counterclockwise? Does the air move toward the center of the spiral or away from the center?

Violent Weather

37. What cloud form is associated with thunderstorms?
38. How do downdrafts form in thunderstorms?
39. Briefly describe how thunder and lightning develop.
40. Where is the highest frequency of tornadoes in the United States?
41. When and where are hurricanes most likely to develop? Explain.

Weather Forecasting

42. In which atmospheric layer does all our weather occur?
43. What information must be known to predict the weather?
44. What are three methods of weather forecasting?
45. The accuracy of weather forecasts depends on great quantities of data and thousands of calculations. If the number of data points were decreased, would accuracy also decrease?

Exercises

1. Distinguish between weather and climate.
2. As a given area on the Moon's surface rotates from sunlight into darkness, the surface temperature goes from searing hot to bitter cold. But when an area on the Earth rotates from sunlight into darkness, its surface temperature decreases only slightly. Why?
3. Why do clouds tend to form above mountain peaks?
4. Why does warm, moist air blowing over cold water result in fog?
5. Why does dew form on the ground during clear, calm summer nights?
6. Why does a July day in the Gulf of Mexico generally feel appreciably hotter than a July day in Arizona?
7. Would you expect a glass of water to evaporate more quickly on a windy, warm, dry summer day or a calm, cold, dry, winter day? Defend your answer.
8. Why does surface temperature increase on a clear, calm night as a low cloud cover moves overhead?
9. During a summer visit to Cancun Mexico, you stay in an air conditioned room. Getting ready to leave your room for the beach, you put on your sunglasses. The minute you step outside your sunglasses fog up. Why?
10. After a day of skiing in the Rocky Mountains, you decide to go indoors to get a warm cup of cocoa. As you enter the ski lodge, your eyeglasses fog up. Why?
11. Is it possible for the temperature of an air mass to change if no thermal energy is added or subtracted? Defend your answer.
12. Which produces precipitation, a rising moist air mass or a descending moist air mass? Why?
13. Why is it necessary for an air mass to rise if it is to produce precipitation?
14. Why does a drop in air pressure indicate the coming of cloudy weather and a rise in air pressure indicate clear weather?
15. Why do air temperatures tend to be higher on the eastern side of the Appalachian Mountains than on the western side, even though both regions are subjected to winter cP air masses?
16. As an air mass moves first upslope and then downslope over a mountain, what happens to the air's temperature and moisture content?
17. What conditions produce a seabreeze? A land breeze?
18. The sky is overcast, and it is raining. What type of cloud is above you, nimbostratus or cumulonimbus? Defend your answer.
19. How can a layer of altostratus clouds change into altocumulus clouds?
20. What accounts for the large spaces of blue sky between cumulus clouds?
21. Why don't cumulus clouds form over cool water?
22. Antarctica is covered by glaciers and large ice sheets. Is the snowfall in Antarctica therefore heavy or light? Why?
23. What is the difference between rainfall that accompanies the passage of a warm front and rainfall that accompanies the passage of a cold front?
24. How do fronts cause clouds and precipitation?
25. Explain why freezing rain is more commonly associated with warm fronts than with cold fronts.
26. How does a rain-shadow desert form?
27. Why are clouds that form over water more efficient in producing precipitation than clouds that form over land?
28. How is it possible for a layer of air to be convectively unstable and absolutely stable at the same time?
29. Sinking air warms, and yet the downdrafts in a thunderstorm are cold. Why?
30. What factors contribute to making the central part of the United States the tornado capital of the world?
31. Tornadoes form in the regions of a strong updraft, and yet they descend from the base of a cloud. Explain.
32. What is the source of the enormous amount of energy released by a hurricane?
33. Why are hurricanes more likely to occur on the eastern coast of the United States than on the western coast?
34. How does a tsunami differ from high waves generated by hurricanes?
35. Weather forecasts are as good as the available data collected—in most cases the more data, the better the forecast. Even though many thousands of weather observations are taken worldwide each day, there are still many regions where data collection is weak. In your own words, state how weather prediction can become more accurate.

Part VII — Sample Exam Questions

Choose the BEST answer to each of the following.

1. When different minerals are composed of the same elements but have different crystal structures the minerals are called
(a) polygamous (b) polymorphs
(c) polygonal (d) polymers

2. The process by which a single magma can generate several different igneous rocks is
(a) partial melting
(b) partial crystallization
(c) fractional crystallization
(d) partial distillation

3. The silica content of magma greatly affects its viscosity Magma with high silica content has a
(a) high viscosity and flows quickly
(b) high viscosity and flows slowly
(c) low viscosity and flows quickly
(d) low viscosity and flows slowly

4. Fluids in a metamorphic reaction
(a) speed up the process
(b) neutralize the process
(c) slow down the process
(d) have no effect on metamorphic process

5. Convergent boundaries are areas of
(a) compressional forces that generate a spreading center
(b) tensional forces that generate a spreading center
(c) crustal formation
(d) plate collision

6. The evidence to support the concept that the inner core is solid and the outer core is liquid comes from
(a) the refraction of seismic waves as they encounter dif ferent mediums
(b) the wave shadow effect of P- and S-waves as they encounter the solid and liquid core
(c) the wave shadow effect of P- and S-waves and the increase in velocity of P-waves as it encounters the solid inner core

7. Suppose we have two blocks of rock—one basalt, one granite—each rock block has the same mass and shape. Which rock mass would stand higher on the asthenosphere?
(a) basalt
(b) granite
(c) both rocks would stand the same height

8. If seafloor spreading creates new lithosphere, does the size of the Earth change?
(a) Yes, the Earth is getting larger
(b) No, the extra crust is crumpled up to form mountains.
(c) No, older crust is recycled back into the asthenosphere
(d) Yes, the Earth is shrinking because it is cooling

9. A subsurface layer that holds and transmits water is a/an
(a) aquifer (b) hydrometer
(c) water table (d) aquiclude

10. The volume of water that flows past a given point in a channel during a specified time is termed
(a) load (b) gradient
(c) discharge (d) runoff

11. Most of the world's water is in
(a) icecaps (b) glaciers
(c) lakes (d) rivers
(e) oceans

12. Formation of the supercontinent of Pangaea caused
(a) a worldwide rise in sea level
(b) a worldwide drop in sea level
(c) no change in sea level
(d) none of the above

13. Radiometric dating is based on the
(a) decay of Uranium-238 to Lead-206
(b) sequence of rocks and the relative position of one layer to another
(c) proportions of radioactive isotopes and their decay products
(d) half life of radiometric atoms

14. In a folded sequence of rocks we find older rocks at the axis of the fold and younger rocks away from the fold axis. The fold
(a) is called a syncline (b) is called an anticline
(c) is plunging (d) has been tilted

15. The wind blows in response to
(a) pressure differences
(b) the Earth's rotation
(c) temperature differences
(d) both a and c
(e) all of the above

16. Evaporation of raindrops in the atmosphere
(a) warms the air
(b) cools the air
(c) does not happen, raindrops always reach the Earth's surface
(d) is greatest above polar icecaps

17. A rising parcel of air
(a) warms
(b) cools
(c) stays the same temperature
(d) none of the above

18. Whenever water condenses
(a) heat is absorbed (b) heat is released
(c) frost forms (d) temperature drops

19. What happens to relative humidity as air temperature decreases?
(a) it increases (b) it decreases
(c) it stays the same (d) it drops

Answers: 1b, 2c, 3b, 4a, 5d, 6c, 7b, 8c, 9a, 10c, 11e, 12a, 13c, 14b, 15d, 16b, 17b, 18b, 19a

PART

Astronomy VIII

We began this book with the *physics* of the everyday world, then progressed to the microscopic realm, the *chemistry* of molecules. Then to the *geology* of the whole earth -- big by comparison. Now we conclude our study of *Conceptual Physical Science* by studying the very, very big -- the universe! We are of the stars, in matter and energy. Atoms form in stars, then disperse when the stars explode to become the material that becomes us. Most of our energy comes from the nearest star, the sun. So we are literally stardust -- the conscious part of nature looking at itself. Onward to *astronomy*!

CHAPTER

28

The Solar System

For thousands of years humans have looked into the night sky and wondered about the stars. With only the unaided eye, they neither saw nor dreamed that the stars are greater in number than all the grains of sand on all the beaches of the world! Nor did they realize that the Sun is a star—simply the nearest star of all in the universe. Probably the most fascinating was the Moon, which when full was perceived as a flat, circular disk rather than as the three-dimensional sphere we now know it to be. The fact that we see only one face of the Moon further supported the idea that the Moon is a disk. Why do we see only one side of the Moon? Why does the Moon go through phases of full to thin crescent but the Sun remains round?

Fig. 28.1
Both the Moon and the spherical fireworks display look two-dimensional because they are far from our eyes.

These and other questions about the night sky have intrigued humans for millennia. Explanations for what goes on "up there" are as numerous as the cultures of those who wondered about the cosmos. We'll not treat the false explanations of those who came before us. Instead, we begin with what we know today about the Moon and then study how the solar system came to be. After a brief tour of the planets, we conclude the chapter with other bodies in the solar system—asteroids, meteoroids, and comets. In the following chapter we'll learn about stars, both ordinary stars like the Sun and exotic black holes, and how they came to be. The final chapter deals with the bigger picture—the universe.

28.1 The Moon

On July 20, 1969, Neil Armstrong set foot on the Moon. To date, 12 people have stood on the Moon. We know more about the Moon than any other celestial body. From nearly 400 kilograms of rock and soil samples brought back from lunar landings, we know the Moon's age, its composition, and a lot about its history. From its low density (3.34 grams per cubic centimeter), we know the Moon cannot have a substantial iron core. From its extremely weak magnetic field (less than 0.0001 that of Earth's), we know it cannot have a large molten core. But there is so much we don't know about the Moon—how it formed, for example. Did it split off from the Earth while the Earth was forming? Did the Earth and Moon condense from the same material as the solar system formed? Did the Moon form somewhere else, then fall into the Earth's gravitational grip? Did it form from debris ejected from the Earth when some other body slammed into the Earth? Or perhaps the Earth and Moon are the result of a collision and merger of two very large planets in the making—a hypothesis presently gaining the favor of many astronomers. How the Moon is a much-debated subject.

Fig. 28.2
Edwin E. Aldrin Jr., one of the three *Apollo 11* astronauts, stands on the dusty lunar surface. Old Glory is rigged to appear flapping in the wind, for the Moon is too small to have an atmosphere.

The Moon is small, its diameter about the distance from San Francisco to New York City. It began with a molten surface, cooled too rapidly for plate motion like the Earth's, formed an igneous crust thicker than the Earth's, and underwent intense meteoroid bombardment early in its evolution. A little more than 3 billion years ago, basins formed by bombardment and volcanic activity filled with lava to produce a surface that has undergone very little change since. The Moon is too small to have an atmosphere, and so without weather the only eroding agents have been meteoroid impacts.

We know less about the surface of the far side of the Moon, which is always directed away from the Earth. But why does only one side of our Moon face the Earth?

Fig. 28.3
The Earth and Moon as photographed in 1977 from the *Voyager 1* spacecraft on its way to Jupiter and Saturn. (Source: NASA)

One Side of the Moon

The first humans to see the back side of the Moon were Russian cosmonauts who orbited the Moon in 1968. From the Earth we always see the same side of the Moon. Even naked-eye observations tell us this is true. The familiar facial features of the "man in the moon" are always turned toward us on Earth. Because the same side always faces us, the Moon appears to be without axial spin, or rotation. But with respect to the stars, the Moon clearly rotates—although quite slowly, about once every 27 days. This rotational rate matches the rate at which the Moon revolves about the Earth and explains why the same side of the Moon is always facing the Earth (Figure 28.4). If the Moon rotated faster or slower, all of its faces would reveal themselves. This matching of axial rotation rate and orbital revolution rate is not a coincidence. Let's see why.

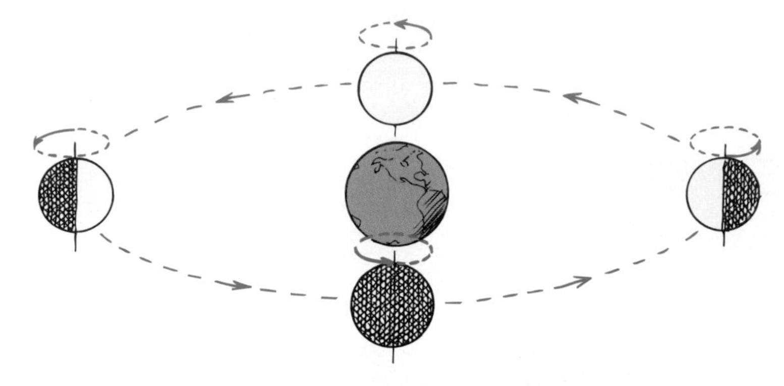

Fig. 28.4
The Moon rotates about its own polar axis just as often as it circles the Earth: once every 28 days. So as the Moon circles the Earth, it rotates so that the same side (shown yellow) always face the Earth. In each of the four successive positions shown here, the Moon has rotated one-quarter of a turn.

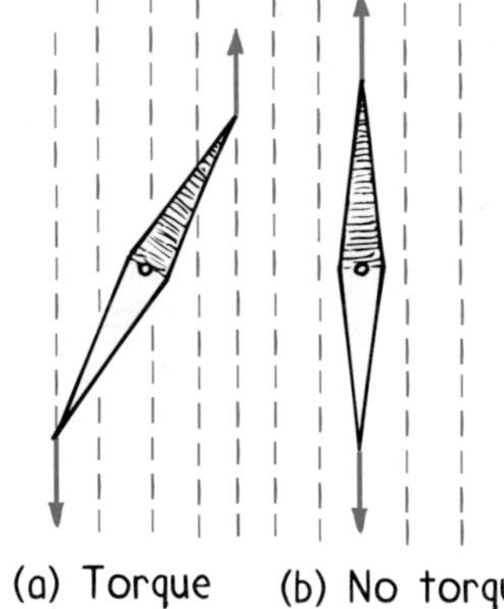

Fig. 28.5
(a) When the compass needle is not aligned with the magnetic field (dashed lines), the forces represented by the blue arrows at either end produce a pair of torques that rotate the needle. (b) When the needle is aligned with the magnetic field, the forces no longer produce torques.

We can better understand this phenomenon if we first understand why a compass needle aligns with a magnetic field. This aligning involves a *torque*—a "turning force with leverage" (like that produced by a child at the end of a see-saw)—that tends to produce rotation. The compass needle in Figure 28.5a rotates because of a pair of torques produced by off-axis forces. The needle rotates counterclockwise until it attains the alignment shown in (b), parallel with the magnetic field. Then the forces are no longer off-axis, and the needle is stable. In a similar manner, the Moon aligns with the Earth's gravitational field.

We know that gravity weakens with distance (the inverse-square law, Chapter 4), and for this reason the side of the Moon nearest the Earth is pulled toward the Earth with more force than the side farthest from the Earth. This elongates the Moon and makes it slightly football-shaped. (Recall from our treatment of ocean tides in Chapter 4 that the Earth is similarly elongated.) The greater pull on the near side means that the half of the Moon closest to the Earth is "heavier," making the Moon's center of gravity (CG, center of "weight") closer to us than its center of mass (CM, geometrical center). To put it another way, the elongated Moon's center of gravity is slightly displaced from its center of mass. Whenever the Moon's long axis is not lined up with the Earth's center (Figure 28.6), the Earth exerts a small torque on the Moon. This torque tends to twist the Moon toward aligning with the Earth's gravitational field, just as the torque in Figure 28.5 aligns the compass needle with the magnetic field. We say there is a *gravity lock* between the Earth and the Moon. This gravity lock acts to some degree on all astronomical bodies close to each other. The degree to which it acts on the Moon is enough to produce the result we see—namely, that the same side of the Moon always faces us. Each month the Moon makes one complete rotation (spin) as it revolves around the Earth.

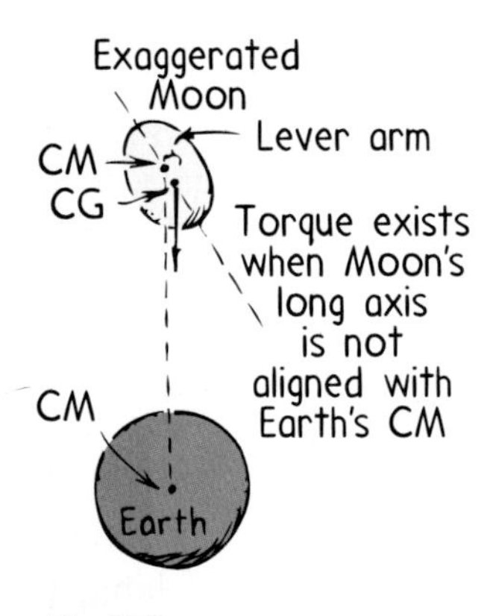

Fig. 28.6
When the long axis of the Moon is not aligned with the Earth's gravitational field, the Earth exerts a torque that tends to rotate the Moon into alignment.

■ **Question**

A friend says the Moon does not spin about its polar axis and cites as evidence the fact that the same side of the Moon always faces the Earth. What do you say?

Phases of the Moon

At any given moment rays of sunlight illuminate one-half of the Moon's surface. Because the Moon both rotates on an axis and revolves around the Earth, we have the Moon's **phases**, changes in its visible shape that occur in monthly cycles (Figure 28.7). The first half of the Moon cycle begins with the new moon (totally dark; we see nothing) and climaxes with the full moon. The *new moon* phase occurs when the Sun, Moon, and Earth are lined up, with the Moon in the middle, as shown in position 1 in Figure 28.8. The new moon is not visible to us in this phase because the side of the Moon facing the Earth is dark; the Moon is in a position such that it cannot reflect any of the Sun's rays toward the Earth. The Moon's illuminated side is the side we cannot see.

Fig. 28.7
The Moon in its various phases.

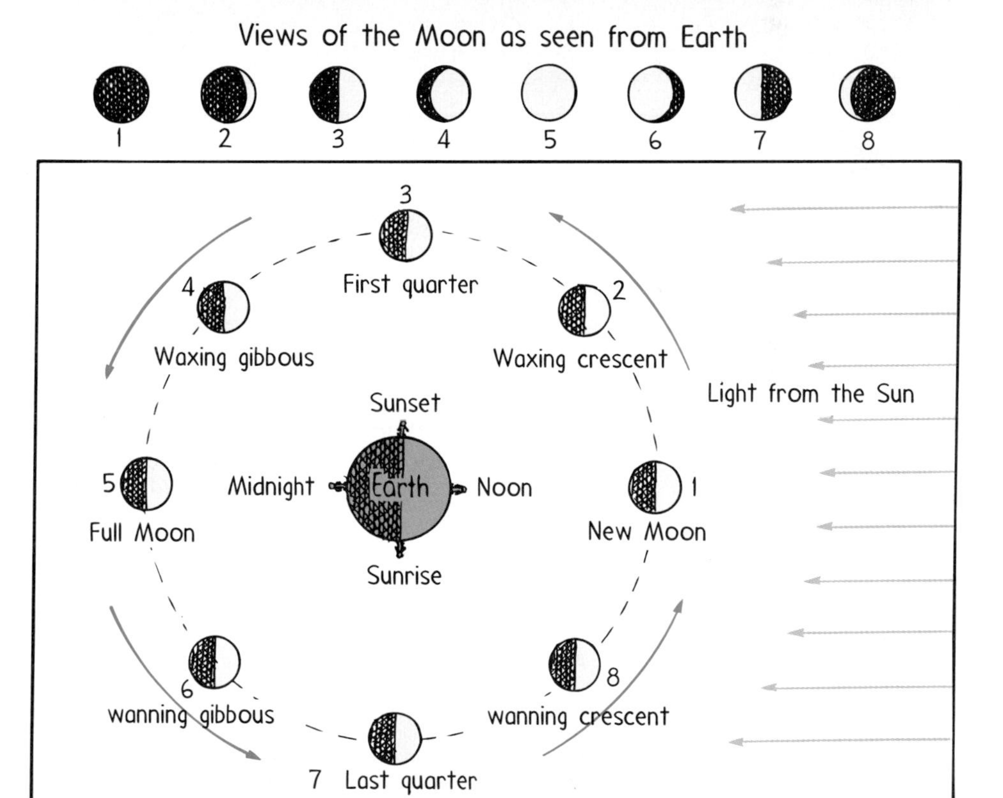

Fig. 28.8
Sunlight illuminates only one-half of the Moon. As the Moon orbits the Earth, we see varying amounts of its sunlit side. One lunar phase cycle takes 29½ days.

Answer

Help your friend distinguish between apparent spin and actual spin. With respect to the Earth, the Moon has no apparent spin. But if we broaden our point of view and look at things from the stars, we see that the Moon does spin—but just as slowly as it revolves about the Earth. If your friend were correct and the Moon did not spin, then different parts of it would face the Earth throughout the month. Every two weeks opposite sides of the Moon would face us. The fact that only one side of the Moon ever faces the Earth is evidence that the Moon *does* rotate, not that it doesn't. The key concept is that its rotation rate is the same as its revolution rate.

During the next seven days, progressively more and more of the Moon's side exposed to our view becomes illuminated, as indicated by position 2 in Figure 28.8. The Moon is going though its *waxing crescent* phase.[1] At the *first quarter,* the angle between the Sun, Moon, and Earth is 90 degrees. At this point we see half the sunlit portion of the Moon (position 3).

During the next week, more and more of the sunlit part is exposed to us as the Moon goes though its *waxing gibbous* phase (position 4). We see a *full moon* when the side of the Moon that faces us is completely illuminated—which means when Sun, Earth, and Moon are lined up with the Earth in the middle.

The cycle reverses during the following two weeks as we see less and less of the sunlit side while the Moon continues in its orbit. This movement produces the *waning gibbous*, *last quarter*, and *waning crescent* phases. The time elapsed during one complete cycle is about $29\frac{1}{2}$ days.[2]

Questions

1. Can a full moon be seen at noon? Can a new moon be seen at midnight?
2. Astronomers prefer to view the stars when the Moon is absent from the night sky. When, and how often, is the night sky moonless?

28.2 Eclipses

Although the Sun is 400 times larger in diameter than the Moon, it is also 400 times farther away from us. So from the Earth, the Sun and Moon subtend the same angle (0.5°) and appear the same size in the sky. It is this coincidence that allows us to see solar eclipses. Both the Earth and the Moon cast a shadow when sunlight shines on them. When either body crosses into the shadow cast by the other, an eclipse occurs.

A **solar eclipse** occurs when the Moon's shadow falls on the Earth. Because of the large size of the Sun, rays of sunlight taper to provide an umbra and a surrounding penumbra (Figure 28.9). An observer in the umbra part of the shadow experiences darkness during the day—a total eclipse, *totality*. Totality begins when the Sun disappears behind the Moon and ends when the Sun reappears on the other edge of the Moon.

Answers

1. Inspection of Figure 28.8 shows that a noontime observer is on the wrong side of the Earth to see the full moon. Likewise, an observer at midnight is out of sight of a new moon. The new moon is overhead in the daytime, not the nighttime.
2. At the time of the new moon and during the week on either side of the new moon, the night sky does not show the Moon. Unless an astronomer wishes to study the Moon, these dark nights are the best time for viewing other objects. Astronomers usually view the night skies during two-week periods every two weeks.

---

[1] *Waxing* means increasing; *waning* means shrinking. *Crescent* means less than half; *gibbous* means more than half.

[2] The Moon actually orbits the Earth once each 27.3 days with respect to the stars. The 29.5-day cycle is with respect to the Sun and is due to the motion of the Earth-Moon system as it revolves about the Sun.

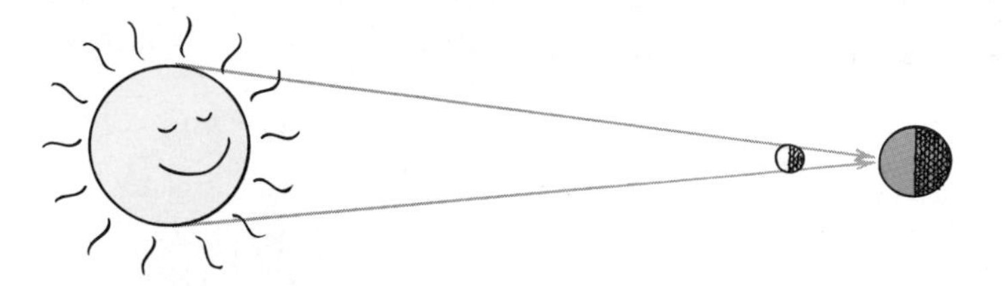

Fig. 28.9
A solar eclipse occurs when the Moon is directly between the Sun and the Earth and the Moon's shadow is cast on the Earth. Because of the small size of the Moon and tapering of the solar rays, a solar eclipse occurs only on a small part of the Earth.

The average time of totality is 2 or 3 minutes, with a maximum of 7.5 minutes. The duration of a solar eclipse is always brief because of the Moon's motion. An observer in the penumbra experiences not totality but rather a partial eclipse and can still see part of the Sun the whole time the eclipse is taking place.

It is interesting to note that the darkness of totality is not complete because of the bright corona that surrounds the Sun (Figure 28.10). The corona is seen only during an eclipse, when the light from the Sun is blocked. We don't normally see it because it is overwhelmed by the Sun's brightness (similar to the way we don't see stars in the daytime).

Fig. 28.10
Eclipsed view of the Sun, showing the corona, a pearly white halo of solar gases that extends several million kilometers beyond the Sun's surface.

The lining up of the Earth, Moon, and Sun produces a **lunar eclipse** when the Moon passes into the shadow of the Earth (Figure 28.11). Usually a lunar eclipse either precedes or follows a solar eclipse by two weeks. Just as all solar eclipses involve a new moon, all lunar eclipses involve a full moon. A lunar eclipse may be partial or total. All observers on the dark side of the Earth see a lunar eclipse at the same time. Interestingly, when the Moon is fully eclipsed, it is still visible and reddish (see box).

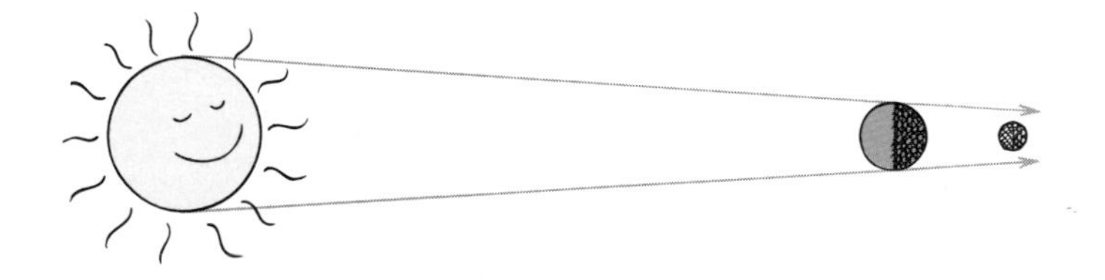

Fig. 28.11
A lunar eclipse occurs when the Earth is directly between the Moon and the Sun and the Earth's shadow is cast on the Moon.

Appearance of the Moon During a Lunar Eclipse

A fully eclipsed Moon is quite visible. This is because the Earth's atmosphere acts as a lens and refracts light into the shadow region—enough to faintly illuminate the Moon. More interestingly, an eclipsed Moon is reddish. To understand why, recall the reason for red sunsets from Chapter 11: The atmosphere scatters high-frequency light from sunlight and the low frequencies that aren't scattered make the sunlight reddish (which produces the redness of sunsets). Beams of sunlight through the air travel the longest distance from the Sun to our eyes when the Sun is on the horizon—at sunset or sunrise. The longer "filtering path" at that time makes the light red.

The next time you view a sunset (or sunrise), quickly move your head to one side so that the light that would have met your eye instead misses and continues to the horizon behind you (Figure 28.12). If nothing is in the way, the light will continue through the atmosphere and refract into space. If the light was reddish when it passed you, it will get even redder as it travels farther through the atmosphere before continuing into space. This is the light that shines on the eclipsed Moon—hence its reddish color (Figure 28.13). So poetically enough, the redness of an eclipsed Moon is the result of red light from all the sunsets and sunrises going on around the world.

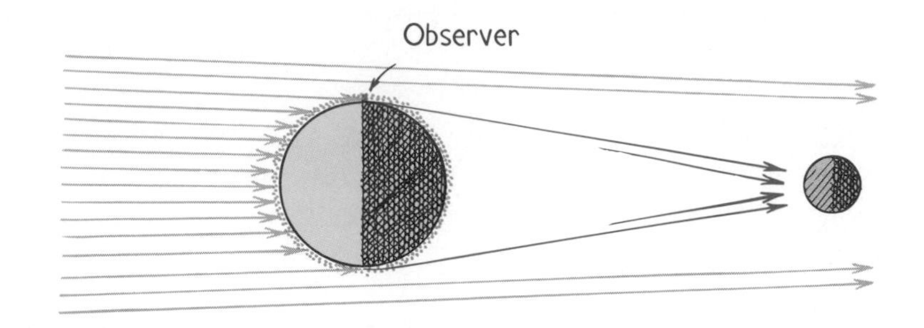

Fig. 28.12
When the Sun is low in the sky, the long path through the atmosphere to the observer filters high frequencies to make the sunlight reddish. Hence the red sunset. Light that continues past the observer travels through twice as much atmosphere and is even redder when it shines on the eclipsed Moon.

Fig. 28.13
The fully eclipsed Moon is often reddish because the Earth's atmosphere acts like a lens and refracts light from sunsets and sunrises all around the world onto the otherwise dark Moon.

Why are eclipses relatively rare? This has to do with the different orbital planes of the Earth and Moon. The Earth revolves around the Sun in a flat planar orbit. The Moon similarly revolves about the Earth in a flat planar orbit. But the two planes of revolution are slightly tipped relative to each other—a 5.2 degree tilt (Figure 28.14). If the planes weren't tipped, eclipses would occur monthly. Because of the tip, eclipses occur only when the Moon intersects the Earth-Sun plane at the time of a three-body alignment (Figure 28.15). This occurs some two times per year, which is why there are at least two solar eclipses per year (visible only from certain locations on the Earth). Sometimes there are as many as seven solar and lunar eclipses in a year.

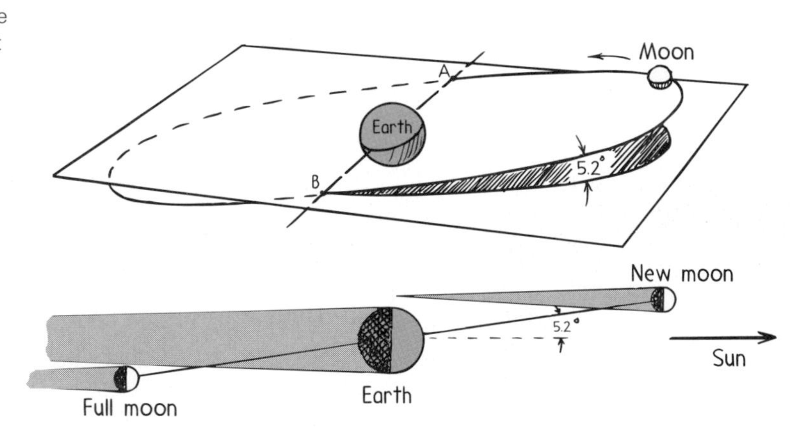

Fig. 28.14
The Moon orbits the Earth in a plane that is tipped 5.2° with respect to the plane of the Earth's orbit around the Sun. A solar or lunar eclipse occurs only when the Moon intersects the Earth-Sun plane (points A and B) at the precise time of a three-body alignment.

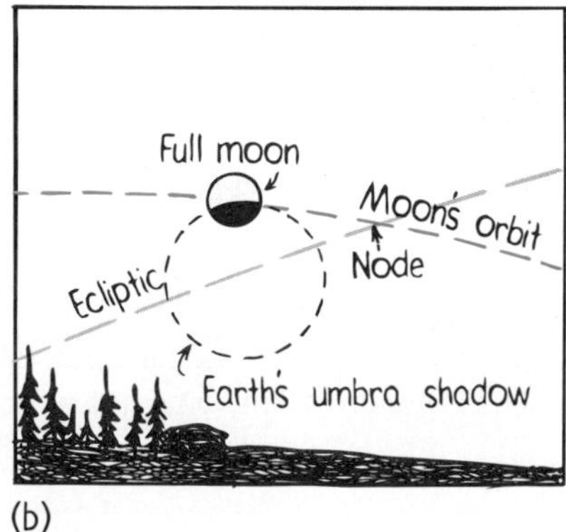

Fig. 28.15
An eclipse can occur only when the Earth and Moon are near a node, (points A and B in Figure 28.22), which is a point where the tipped orbital planes intersect. (a) Partial solar eclipse; (b) partial lunar eclipse.

Questions

1. Does a solar eclipse occur at the time of a full moon or a new moon?
2. Does a lunar eclipse occur at the time of a full moon or a new moon?
3. Why are lunar eclipses more commonly seen than solar eclipses?

28.3 The Sun

Fig. 28.16
Every second, 4.5 million tons of solar mass is converted to radiant energy in the Sun. The Sun is so massive, however, that in a million years only one ten-millionth of its mass is consumed.

Beyond the Moon is the Sun—our nearest star. Ancients who worshipped the Sun seem to have realized that it is the source of all life on the Earth. We are able to see, hear, touch, laugh, and love only because every second $4\frac{1}{2}$ million tons of mass in the Sun is converted to radiant energy, a tiny fraction of which hits the Earth.

This is the energy of thermonuclear fusion taking place in the interior of the Sun, where hydrogen nuclei are being crushed together to form helium (Section 14.11). The solar conversion of hydrogen to helium has been going on since the Sun formed nearly 5 billion years ago and is expected to continue at this rate for another 5 billion years.

The parts of the Sun visible to us are its surface and its atmosphere. The Sun's tenuous surface is neither solid nor liquid nor gas but a glowing 5800-kelvin plasma, probably no more than 500 kilometers thick. At nearly 6000 kelvins, the Sun's surface is well above the temperature required to vaporize any known material. This transparent solar surface is the *photosphere* (sphere of light). On the surface are relatively cool regions that are created by strong magnetic fields and show up as **sunspots** when viewed from the Earth. These can be seen by the unaided eye through protective filters or when the Sun is low enough on the horizon not to hurt the eyes. Sunspots are typically twice the size of the Earth, appear to move around because of the Sun's rotation, and last about a week or so. Often they cluster in groups (Figure 28.17).

The layer of the Sun's atmosphere just above the photosphere is a transparent, 10,000-kilometer-thick shell of plasma called the *chromosphere* (sphere of color), seen during an eclipse as a pinkish glow surrounding the eclipsed Sun. Beyond the chromosphere are streamers and filaments of outward-moving, high-temperature plasma curved by the Sun's magnetic field. This outermost region of the Sun's atmosphere is the *corona*, extending out several million kilometers until it merges into a hurricane of high-speed protons and electrons—the *solar wind.* It is the solar wind that powers the aurora borealis on Earth and produces the tails of comets.

The Sun spins on its axis, but slowly. Interestingly, different latitudes of the Sun spin at different rates. Equatorial regions spin once in 25 days, but higher latitudes take

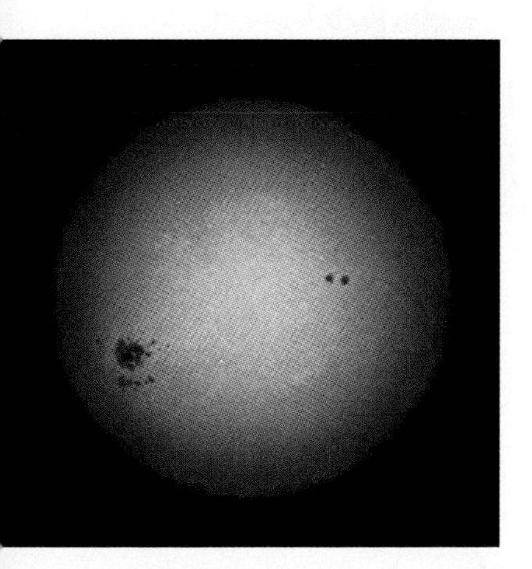

Fig. 28.17
Sunspots on the solar surface are relatively cool regions. We say *relatively* cool because they are hotter than 4000 K. They look dark only by contrast with their 5800 K surroundings.

Answers

1. A solar eclipse occurs at the time of a new moon, when the Moon is directly in front of the Sun. Then the shadow of the Moon falls on part of the Earth. Because of the tilt in the Moon's orbit around the Earth, most of the time the shadow misses the Earth.
2. A lunar eclipse occurs at the time of a full moon, when the Moon and Sun are aligned on opposite sides of the Earth. Then the shadow of the Earth falls on the full moon. Because of the tilt in the Moon's orbit around the Earth, most of the time the shadow misses the Moon.
3. Relatively few people witness solar eclipses because the shadow of the small Moon tapers to a very small part of the Earth's surface. During a lunar eclipse, the similarly tapered shadow of the large Earth completely covers the Moon, so that everybody on the dark side of the Earth can see the eclipse. That's why nearly all your friends have seen a total lunar eclipse, but relatively few of them have ever witnessed a total solar eclipse.

up to 36 days per rotation. This differential rotation means the surface near the equator pulls ahead of the surface farther north or south.

Formation of the Solar System

How the Sun formed is still a matter of conjecture, but it is generally believed to have originated from the gravitational contraction of a huge amount of interstellar matter nearly 5 billion years ago. The universe began primarily as hydrogen, with some helium. By the time the solar system formed some 10 billion or so years later, however, it contained small amounts of all the elements known today. All elements beyond hydrogen and helium had to be formed in the cores of stars that formed before our Sun did. When these stars underwent their death throes, their heavier elements were spewed into the interstellar mix, providing material for new stars.

As interstellar matter moves randomly in space, pockets of gas are believed to condense and dissolve repeatedly, as wisps of fog similarly form and disperse in air. Pressure waves originating in exploding stars and passing through the interstellar medium compress the gas pockets even more. Sometimes these temporary clumps become permanent because the atoms and molecules in them are held together by mutual gravity. A clump adds mass to itself by attracting neighboring atoms and molecules. Because of gravity, it falls in upon itself. Gravitational potential energy becomes thermal energy, and the temperature at the clump's center rises. Continued gravitational contraction and rising temperatures contribute to a concentration of hot matter toward the center. Like an ice skater drawing arms inward when going into a spin, the contraction is accompanied by an increase in angular speed. As the collapsing matter spins faster, it flattens into a disk shape (Figure 28.18). A disk has more surface area per volume than its configuration before flattening and consequently radiates more of its energy into space. So this flattening results in cooling. In the formation of our solar system, this decreasing temperature probably was accompanied by the condensation of matter in swirling eddies—the birthplaces of the planets.

Temperatures near the center of the disk would have been too high for matter to solidify. Farther out though, material likely solidified to become the four inner planets: Mercury, Venus, Earth, and Mars. Farther from the hot center, condensations of larger amounts of more volatile matter, mostly hydrogen, formed the gigantic outer planets: Jupiter, Saturn, Uranus, and Neptune. Small and distant Pluto, as we shall see, is an exception to this creation hypothesis.

So we have a solar system, our home in the universe. Of the countless generations who have wondered before us, only in our lifetime have we begun to understand what and where we are in the universe.

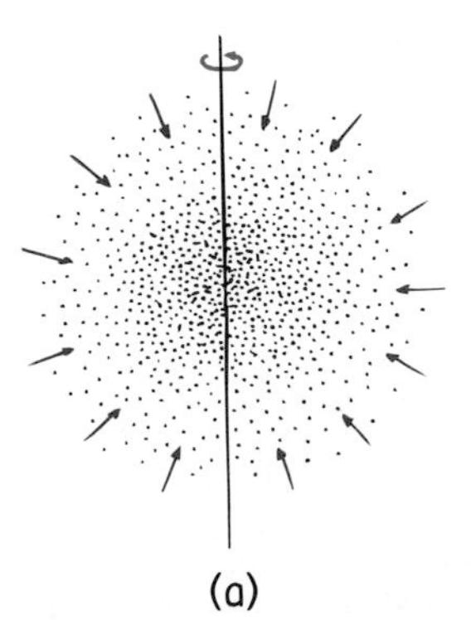

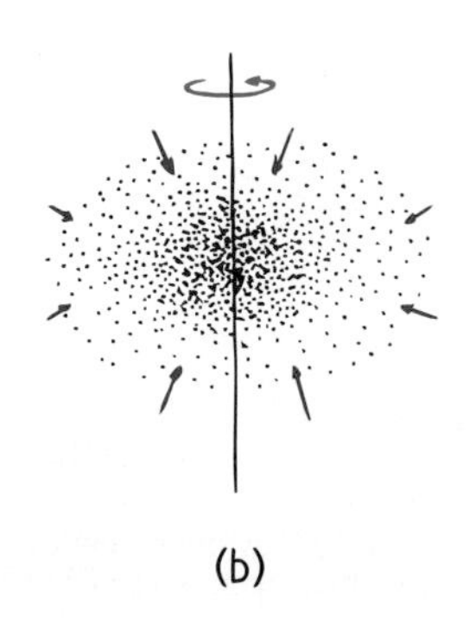

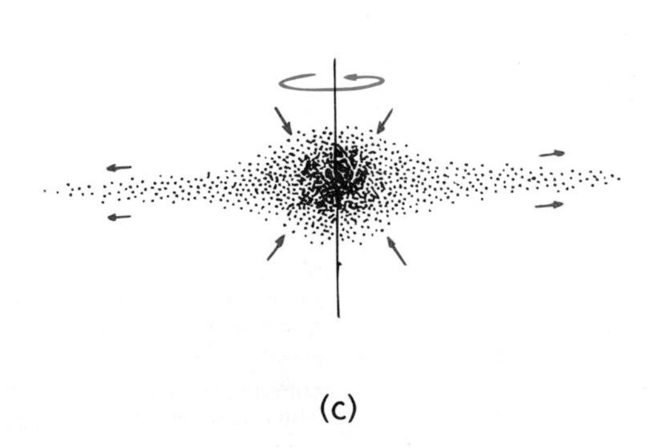

Fig. 28.18
(a) A slowly rotating pocket of interstellar gas contracts as a result of the mutual gravitation between all the particles in it. (b) The law of conservation of angular momentum accounts for the pocket's speeding up. (c) The increased momentum of individual particles and clusters of particles causes them to move in wider paths about the rotational axis, producing an overall disk shape.

Kepler's Laws of Planetary Motion

Long before the invention of the telescope, the Danish astronomer Tycho Brahe spent his lifetime making accurate observations of the positions of the planets. Upon his death in 1601, his charts and record books were passed on to his gifted assistant Johannes Kepler. At this time Kepler and other astronomers assumed that the planets move in perfect circles. Kepler performed the enormous task of transforming Brahe's observations into a diagram showing the motions of the planets as seen by a stationary observer *outside* the solar system. After years of effort, he had to give up his belief in perfectly circular orbits because he found the paths to be ellipses.

Kepler also found that the planets do not go around the Sun at a uniform speed. Instead, each planet moves faster when it is near the Sun and more slowly when it is farther away. He found a planet's changes in speed were such that an imaginary line or spoke joining the Sun and any planet sweeps out equal areas of space in equal time intervals. The triangular area swept out during a month when a planet is far from the Sun, represented by triangle *ASB* in the sketch, is equal to the triangular area swept out during a month when the planet is closer to the Sun (triangle *CSD*).

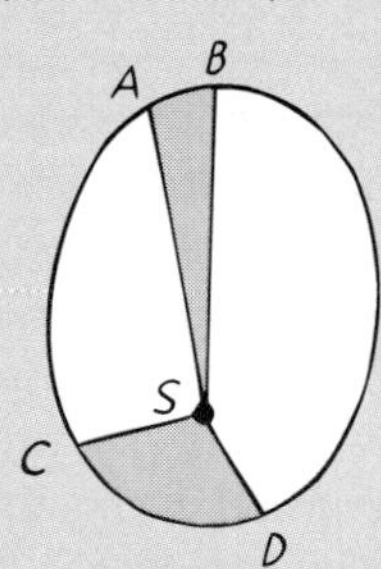

Ten years later, Kepler discovered a third law. He had spent these years searching for a connection between the size of a planet's orbit and its *period* (the time the planet takes to make one complete revolution around the Sun). From Brahe's data Kepler found that the square of the period of any planet is proportional to the cube of the planet's average distance from the Sun. His discovery was that the ratio R^3/T^2 is the same for all planets, where R is the average radius of the planet's orbit and T is the planet's period measured in Earth days.

Kepler's laws of planetary motion are as follows:

Law 1: Each planet moves in an elliptical orbit having the Sun at one focus.

Law 2: An imaginary line from the Sun to any planet sweeps out equal areas in equal time intervals.

Law 3: The square of any planet's period is proportional to the cube of the planet's average distance from the Sun. ($T^2 \sim R^3$ for all planets.)

Kepler's laws apply not only to planets but to moons or any artificial satellite in orbit around any body. Except for Pluto (which Kepler had no knowledge of) and Mercury, the elliptical orbits of the planets are very nearly circular. Only the precise measurements of Brahe showed the slight differences.

Kepler preceded Newton and had no idea why the planets moved in elliptical paths and no general explanation for the mathematical relationships he discovered. He was familiar with Galileo's ideas about inertia and accelerated motion, but he failed to apply them to his own work. As a consequence, Kepler thought there must be some forces pushing the planets and keeping them moving. He never appreciated the concept of inertia—that the planets, once moving, would continue to move without any forces pushing them. And he didn't realize that the forces that really do act on the planets are directed not along their direction of motion but toward the Sun. Kepler's work was soon to be an important stepping stone for Isaac Newton in his development of the law of universal gravitation.

■ 

What is the evidence that our Sun is a second- or third-generation star?

■ **Answer**

Heavy elements are normally fused from lighter elements only in star cores. When stars explode, they contribute heavy elements to the primordial (original) hydrogen-helium mix. The fact that there is an abundance of heavy elements in the solar system is evidence of heavy elements in the mixture from which the solar system formed, and therefore evidence of previous stars. So the Sun is a newcomer. There are many stars twice as old as the Sun in our galaxy. These older stars have lower abundances of heavy elements.

28.4 The Planets

The ancients distinguished between the planets and the stars because of the different apparent motions of the two types of objects. The stars remain relatively fixed in their patterns in the sky, but the planets wander. The planets were called the *wanderers*. Today we know that planets are relatively cool bodies orbiting the Sun. Planets emit no visible light of their own and, like the Moon, simply reflect sunlight.

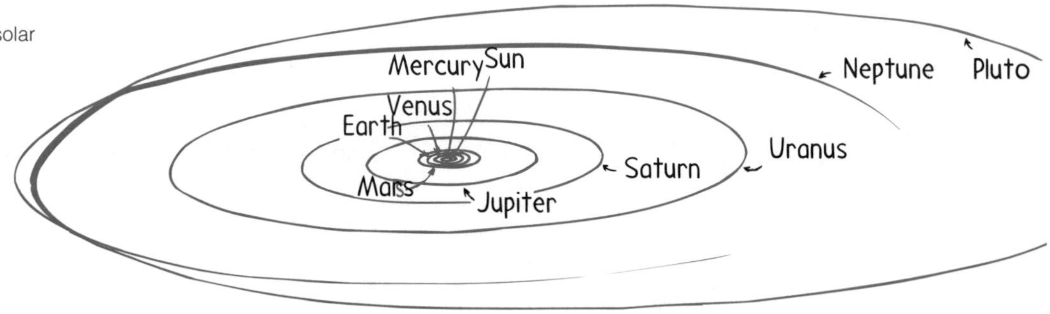

Fig. 28.19
Scale drawing of the solar system, showing the four inner planets crowded around the Sun and the five outer planets orbiting at greater distances.

To informed minds a century and more ago, knowledge of the planets was very meager. Detailed knowledge of planets today is enormous. How human knowledge advanced from knowing planets as no more than starlike spots that crossed the nighttime skies is a fascinating detective story—made possible first by careful observation with the unaided eye, then with telescopes, and more recently with satellite probes launched from Earth. For brevity, we leave this fascinating story to the recommended reading, and here offer only a brief description of the planets. We divide the planets into two groups: the inner planets and the outer planets.

Inner Planets

The four planets closest to the Sun are relatively close together. These are Mercury, Venus, Earth, and Mars, which are small, dense worlds each having a sparse atmosphere. Each inner planet has a solid, mineral-containing crust and Earthlike composition, which is why they are collectively called the *terrestrial planets*. The only terrestrial planets to have natural satellites are the Earth and Mars, a fact that enabled scientists to calculate their masses before the era of space probes. It is interesting to note that the mass of a celestial body cannot be found unless it has a satellite. The masses of Mercury and Venus were known only after these planets were circled by space probes in the early 1960s. Let's visit the inner planets in order.

Mercury Mercury, somewhat larger than the Moon and similar in appearance, is the planet closest to the Sun. Because of its closeness, it is the fastest planet, taking only 88 Earth-days to make one revolution. Thus one "year" on Mercury lasts only 88 Earth-days. Mercury rotates only three times for each two revolutions around the Sun, and so Mercury's "daytime" is both very long and very hot, as high as 430°C.

Table 28.1

	Mean Distance from Sun (Earth-distances, Au)	Orbital Period (years)	Diameter		Mass		Average Density (g/cm³)
			(km)	(Earth = 1)	(g)	(Earth = 1)	
Sun			1,392,000	109.1	1.99×10^{33}	3.3×10^{5}	1.41
Mercury	0.39	0.24	4,880	0.38	3.3×10^{26}	0.06	5.4
Venus	0.72	0.62	12,100	0.95	4.9×10^{27}	0.81	5.2
Earth	1.00	1.00	12,760	1.00	6.0×10^{27}	1.00	5.5
Mars	1.52	1.88	6,800	0.53	6.4×10^{26}	0.11	3.9
Jupiter	5.20	11.86	142,800	11.19	1.90×10^{30}	317.73	1.3
Saturn	9.54	29.46	120,700	9.44	5.7×10^{29}	95.15	0.7
Uranus	19.18	84.0	50,800	3.98	8.7×10^{28}	14.65	1.3
Neptune	30.06	164.79	49,600	3.81	1.0×10^{29}	17.23	1.7
Pluto	39.44	247.70	2,300	0.18	10^{25}	0.002	1.9

Because of its smallness and weak gravitational field, Mercury holds very little atmosphere. Its atmosphere is about a trillionth as dense as the Earth's atmosphere. Because there is no insulating blanket of atmosphere and no winds to carry heat from one region to another, nighttime is very cold, about −170°C.

Venus Venus is the second planet from the Sun. It is brighter than Mercury in the sky and is easily seen near the Sun at either sunup or sunset. Because Venus is often the first starlike object to appear after the Sun goes down, it is often called the evening "star" (first to be seen as the sky darkens) during March and April or a morning "star" (last to disappear as the sky lightens) during September and October. Nearby Mercury, although less bright, is also seen as an evening and morning "star." Both are seen near the Sun at either sunup or sunset.

Compared with the other planets, Venus most closely resembles the Earth with respect to size, density, and distance from the Sun. A major difference between the two planets is the very dense atmosphere and opaque cloud cover on Venus, causing a

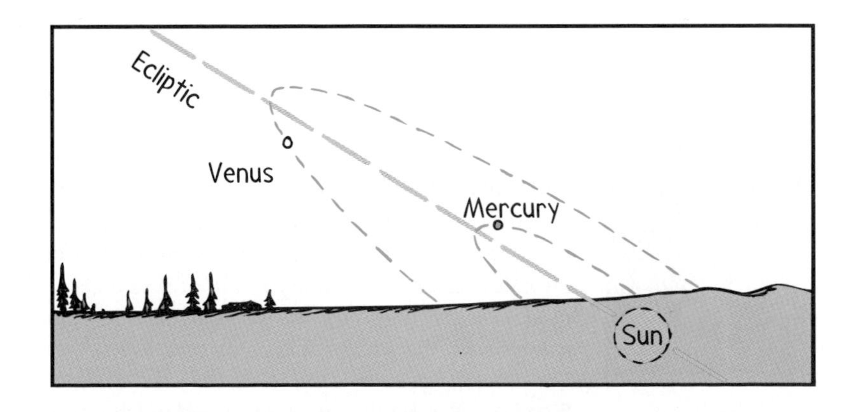

Fig. 28.20
Because the orbits of Mercury and Venus lie inside the orbit of the Earth, they are always near the Sun. Near sunset (or sunrise), they are visible in the sky as evening "stars" (or morning "stars").

Determining the Mass of a Planet

When the first artificial satellite, *Sputnik,* was launched by the Russians in 1958, United States President Dwight Eisenhower was surprised to learn from physicists that there was no way to determine its mass. No amount of observation would help. One would have to poke it with a probe of known mass, measure its rebound speed, and then use momentum conservation to calculate *Sputnik's* mass. Or, quite far fetched, one would have to send a smaller satellite in orbit around *Sputnik* and measure the smaller satellite's period of revolution. Clearly *Sputnik's* mass was too small to employ this latter method, but it is precisely how scientists measure the masses of larger celestial bodies. In fact, it is their only way of measuring such masses. Unless another body orbits a planet or star, and unless the average radius and period of the orbit are known, astronomers can only guess about the mass of that planet or star.

If a planet does have a satellite, the planet's mass can be calculated. It is found to be directly proportional to the cube of the average distance between the planet and its satellite and inversely proportional to the square of the satellite's period. In symbols, $M \sim R^3/T^2$. Find this out for yourself by equating the centripetal force[3] on a satellite, mv^2/R, to the gravitational force, $G(mM/R^2)$, between the satellite (mass m) and its planet (mass M). The speed of the satellite v is equal to the orbital distance $2\pi R$ per period T. After substituting $2\pi R/T$ for v and canceling the mass m of the satellite, you should get the planet's mass $M = 4\pi^2/G(R^3/T^2)$. Try it and see.

Hooray, this also gives you Kepler's third law!

[3]*Centripetal force* is the name given to any force that pulls an object into a circular path. In our present example, centripetal force is the force of gravity pulling the satellite toward the planet. In other situations it may be the force supplied by a string on a whirling object or the electrical force exerted on a circularly moving charge. Any object of mass m moving at constant speed v in a circle of radius R has an acceleration v^2/R directed toward the center of the circle. From Newton's second law, $F = ma$, the centripetal force on this object is therefore $F = mv^2/R$.

greenhouse effect that makes it an unbearably hot place(460°C)—too hot for oceans. Another difference between Venus and the Earth is in how the two planets rotate. Venus takes 243 Earth days to make one full rotation about its central axis. Being that Venus makes one revolution around the Sun in only 225 Earth-days, we can say that on Venus a day lasts longer than a year! Venus rotates in a direction opposite the direction of the Earth's rotation. An observer hovering about the solar system who sees the Earth rotating clockwise sees Venus rotating counterclockwise.

Because Venus has been regarded as almost an Earth twin, early speculations were that its surface is a steamy swamp inhabited by unfamiliar creatures. Speculations of life there have been severely dimmed in recent years by data from 17 probes that have landed on Venus's surface and 18 flyby spacecraft (notably *Pioneer Venus* in 1978 and *Magellan* in 1993). Venus has been very active volcanically and is an extremely harsh place. However, evidence gathered by satellite probes suggests that the surface temperature and atmosphere of Venus were once very much like those of the Earth. Whereas most of the Earth's carbon dioxide is locked up in limestone formations and oceans, with very little in the atmosphere, the greater amount of sunlight on Venus has produced the opposite result on that planet. Much carbon dioxide was released into its atmosphere, which increased the greenhouse effect, which released even more carbon dioxide, which now makes up about 95 percent of the atmosphere. Venus serves as a model for the Earth, for we wonder if a small temperature rise here on the Earth could trigger a similar irreversible chain reaction.

Fig. 28.21
Earth, the blue planet.

Earth The Earth is well described in Chapters 22–27, so our treatment here is brief. In answer to the question of what it must be like to live in outer space, we must not forget that we in fact *are* in outer space—gathered together on a hospitable planet in an inhospitable universe. Planet Earth is our haven and deserves our greatest respect.

Ours is the blue planet, with more water surface than land. Not too close to the Sun and not too far away, with an average surface temperature delicately between that of freezing and boiling water and a just-dense-enough atmosphere to keep the oceans liquid. The insulating properties of our atmosphere and our relatively high rotational rate ensure only a brief and small lowering of temperature on the side of the Earth away from the Sun. So temperature extremes of day and night are conducive to life as we know it. Considering the harshness of most of the universe, the Earth is a nice place to live. Our activities ought to be consistent with keeping it that way.

Fig. 28.22
The rover Sojourner on the surface of Mars before setting out to explore the red planet.

Mars Our nearest neighbor, Venus, is too close to the Sun and too hot for human habitation, and our next nearest neighbor, Mars, is farther from the Sun than the Earth is but not too cold for human habitation. Mars captures our fancy as another world, perhaps one with life. This is because of the similarities between the Earth and Mars: Mars is a little more than half the Earth's size, its mass is about $\frac{1}{9}$ that of the Earth, and it has a core, mantle, and crust as well as a thin, nearly cloudless atmosphere. Its axis tilts from the perpendicular to its orbital plane, a configuration that gives it polar ice caps and seasons that are nearly twice as long as Earth's because Mars takes nearly 2 Earth-years to orbit the Sun. When Mars is closest to the Earth, which happens once every 15–17 years, its bright, ruddy color outshines the brightest stars.

The Martian atmosphere is about 95 percent carbon dioxide, with only about 0.15 percent oxygen. So bring your own air supply if you plan to visit there. Also bring warm clothing, for its surface temperature at the equator goes from a comfortable 30°C in the day to a chilly −130°C at night. Night is only slightly longer than on the Earth, for the Martian day lasts 24 hours and 37.4 minutes. Never mind your raincoat, for there is far too little water vapor in the atmosphere for rain. Even the ice at the poles is primarily carbon dioxide. And never mind your rubbers, for the low atmospheric pressure won't allow any puddles or lakes.

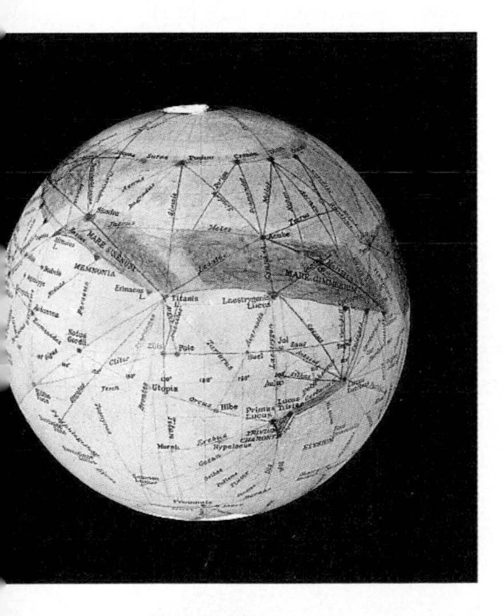

Fig. 28.23
Martian canals mapped by astronomer Percival Lowell in the late 1800s. The canals proved to be optical illusions produced by the brain's ability to assemble vague markings into a coherent image (the same ability that enables us to see TV images rather than swarms of incoherent dots).

Investigators speculate that water may have been more abundant in the Martian past, abundant enough to carve some of the channels seen on the Martian surface. The largest of these channels intrigued early investigators, who saw them through their telescopes as canals, reinforcing speculation of a Martian civilization. Except for questionable traces of life in Martian meteorites found in Antarctica, landings on Mars show that it has no life at the surface and no canals. The 1997 *Pathfinder* mission shows it to be a very dry place and windy, too. Because the Martian atmosphere has a very low density, unequal heating produces Martian winds about 10 times stronger than winds on the Earth.

Mars has two small moons—Phobos, the inner, and Deimos, the outer. Both are potato-shaped and have cratered surfaces. Phobos orbits in the same easterly direction that Mars rotates (like our Moon), at a distance of almost 6000 kilometers in a period of 7.5 hours. From Mars it appears about half the size of our Moon. But because Phobos revolves about Mars much faster than Mars rotates, it rises on the western horizon and sets on the eastern horizon 5.5 hours later. It appears to be "going backward."

Deimos is about half the size of Phobos and orbits Mars in 30.3 hours at a distance of 20,000 kilometers. Its orbital period is somewhat longer than the rotational period of Mars, and so it rises on the eastern horizon and sets on the western horizon nearly three days later. It appears small, subtending the same angle that a 25-cent piece subtends when it is 37 meters distant.

Outer Planets

The more widely spaced outer planets beyond Mars are considerably different from the inner planets—in size, in composition, and in the way they were formed. Jupiter, Saturn, Uranus, and Neptune are gigantic, gaseous, low-density worlds. They are typified by Jupiter and are called *Jovian* planets. All have ring systems, Saturn's being the most prominent. Beyond these giants is outermost Pluto, the most different among planets, being neither terrestrial nor Jovian. We consider the outer planets in turn.

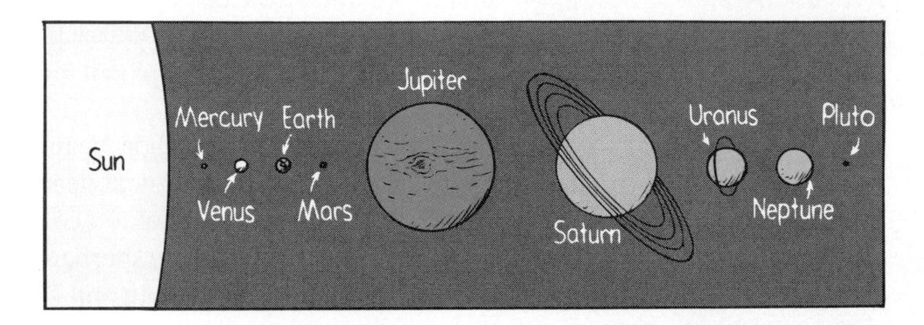

Fig. 28.24 Relative sizes of the Sun and planets. The Sun's diameter is 13 times that of Jupiter.

Jupiter The largest of all the planets is Jupiter, whose yellow light in the night sky outshines the stars. In prespacecraft years, Jupiter was thought of as a failed star, for its composition is closer to that of the Sun than to that of the terrestrial planets. Jupiter is more liquid than gaseous or solid. It rotates rapidly about its axis in about 10 hours, a speed that produces a flattening that makes the equatorial diameter about 6 percent greater than the polar diameter. As with the Sun, all parts do not rotate in unison. Equatorial regions complete a revolution several minutes before adjacent higher and lower latitude regions. Jupiter doesn't have a hard surface crust that an astronaut could walk on. And if there were a place to stand, atmospheric pressure would be a crushing millionfold the atmospheric pressure of the Earth. Jupiter's atmosphere is about 82 percent hydrogen; 17 percent helium; and 1 percent methane, ammonia, and other molecules.

The average diameter of Jupiter is about 11 times greater than the Earth's, which means Jupiter's volume is more than a thousand times the Earth's. Jupiter's mass ex-

Fig. 28.25 Jupiter, with its moons Io and Europa, as seen from the *Voyager I* spacecraft in February 1979. The great red spot, lower left, is a cyclonic weather pattern of high winds and turbulence larger than the Earth. (Source: NASA)

Fig. 28.26
Saturn surrounded by its famous rings, believed to be composed of rocks and chunks of ice.

ceeds the combined masses of all the other planets. Due to its low density, however—about one-fourth the Earth's—Jupiter's mass is barely more than 300 times the Earth's. Jupiter's core is a solid sphere about 20 times as massive as the whole Earth and composed of iron, nickel, and other minerals.

More than half of Jupiter's volume is an ocean of liquid hydrogen. Beneath the hydrogen ocean lies an inner layer of hydrogen compressed into a liquid metallic state, permeated by abundant conduction electrons that flow to produce an enormous magnetic field. The strong magnetic field about Jupiter captures high-energy particles and produces radiation belts 400 million times as energetic as the Earth's Van Allen radiation belts. Radiation levels surrounding Jupiter are the highest ever recorded in space.

Surface temperatures are about the same day and night. Jupiter radiates about twice as much heat as it receives from the Sun. The excess heat likely comes from internal heat generated long ago by gravitational contraction at the time the planet formed. When forming planets contract, gravitational potential energy is converted to thermal energy.

If you're planning to visit Jupiter, choose one of its moons instead. At least 16 of them, along with a faint ring, orbit Jupiter. Among the four largest moons, discovered by Galileo in 1610, Io and Europa are about the size of our Moon, and Ganymede and Callisto are about as large as Mercury. The most interesting of Jupiter's moons seems to be Io, which has more volcanic activity than any other body in the solar system.

Saturn Because its rings are clearly visible with binoculars, Saturn is one of the most remarkable objects in the sky. It is brighter than all but two stars and is second among the planets in mass and size. Saturn is twice as far from us as Jupiter. Its mean diameter, not counting its ring system, is nearly 10 times that of the Earth and its mass nearly 100 times greater. It is composed primarily of hydrogen and helium and has the lowest density of any planet, 0.7 times that of water. Saturn would easily float in a bathtub if the bathtub were big enough. Because of Saturn's low density and its 10.2-hour rapid rotation, it has more polar flattening than any other planet, about 11 percent. Like Jupiter, Saturn radiates about twice as much heat as it receives from the Sun.

Saturn's rings, likely only a few kilometers thick, lie in a plane coincident with Saturn's equator. Four major rings have been known for many years, and spacecraft missions have detected others composed of hundreds of ringlets. The rings are composed of chunks of frozen water and rocks, believed to be the remnants of a moon that never formed or one ripped apart by tidal forces. All the rocks and other bits of matter that make up the rings travel about Saturn in independent orbits.[4] Inner parts of the ring travel faster than outer parts, just as any satellite near a planet travels faster than a more distant satellite.

Saturn has some 23 moons beyond its rings. The largest is Titan, 1.6 times larger than our Moon and even larger than Mercury. It revolves once each 16 days and has a methane atmosphere with atmospheric pressure likely greater than the Earth's. Its surface temperature is a cold −170°C. So bring a heavy coat and breathing gear if you plan to visit Titan. If that doesn't work out, try another of Saturn's large moons, Iapetus. One side of Iapetus is very bright and the other very dark. Try the region between these two extremes.

---

[4]The entire ring system lies within a critical distance called the *Roche limit,* equal to about 2.4 Saturn-radii. The Roche limit, named after the nineteenth-century French mathematician Edouard Roche, is the distance where gravitational attraction by a planet on two adjacent orbiting particles is larger than the attraction of the two particles to each other. If our Moon were within 2.4 Earth-radii, its expected fate would be to be torn apart to form a ring system about the Earth.

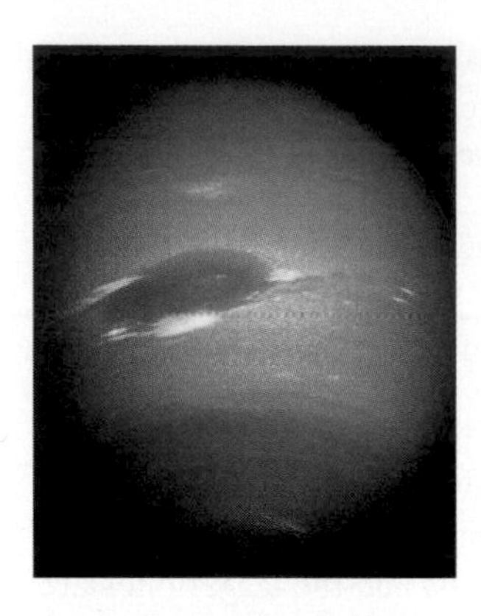

Fig. 28.27 Cyclonic disturbances on Neptune produce a great dark spot, which is larger than the Earth and similar to Jupiter's great red spot. (Source: NASA)

Uranus Uranus is twice as far from the Earth as Saturn is and barely perceptible to the naked eye. Uranus was unknown to ancient astronomers. It has a diameter four times larger than the Earth's and a density slightly greater than that of water. Put Uranus in a huge bathtub and it would sink. The most unusual feature of Uranus is its tilt. Its axis is tilted 98 degrees to the perpendicular of its orbital plane, so it lies on its side. Unlike Jupiter and Saturn, it appears to have no internal source of heat. It is a cold place.

Uranus has at least 17 moons, in addition to a complicated faint ring system.

Neptune All planets are held in the solar system by their gravitational interaction with the Sun. But the planets interact with each other and with everything else as well. When one planet is near another, the pull between them slightly disturbs their orbits. This disturbance is called a *perturbation*. Early in the nineteenth century, unexplained perturbations were observed for Uranus. Either the law of gravitation was failing at this great distance from the Sun or an unknown eighth planet was perturbing Uranus. An Englishman and a Frenchman, J. C. Adams and Urbain Leverrier, independently calculated where an eighth planet should be. At about the same time, both sent letters to their respective observatories with instructions to search a certain area of the sky. The request by Adams was delayed by misunderstandings at Greenwich, but Leverrier's request to the director of the Berlin observatory was heeded immediately. The planet Neptune was discovered one-half hour later.[5]

Neptune's diameter is about 3.9 times the Earth's, its mass is 17 times greater, and its mean density is about a third that of the Earth. Its atmosphere is mainly hydrogen and helium, with some methane and ammonia. Like Jupiter and Saturn, it emits about 2.5 times as much heat as it receives from the Sun.

Neptune has at least eight moons, in addition to a ring system. The largest moon is Triton, which orbits Neptune in 5.9 days in a direction opposite the planet's eastward rotation. Triton is 0.75 the Moon's diameter, with a mass double the Moon's. It has bright polar caps and geysers of liquid nitrogen. A smaller moon, Nereid, takes nearly a year to orbit Neptune in a highly elongated elliptical path.

Pluto The relative positions of stars on photographs do not change. That is, star images taken on any particular night will be in the same relative positions as images taken nights later. But the images of planets will be in different places. Careful examination of such photographic plates resulted in the 1930 discovery of Pluto at the Lowell Observatory in Arizona. Because Pluto takes 248 years to make a single revolution about the Sun, no one will see it in its discovered position again until 2178!

Whereas most of the planetary orbits in the solar system are nearly circular, Pluto's is the most elliptical and most steeply inclined to the planetary plane. Pluto's orbit is so eccentric that it is presently closer to the Sun than Neptune. Its most recent perihelion (closest distance to the Sun) occurred in 1989, which found it more than 100 million kilometers closer to the Sun than Neptune was at that time. After March 4, 1999, it will move farther than Neptune from the Sun.

Pluto is smaller than our Moon. It has a diameter about one-fifth the Earth's diameter and a mass of about 0.002 that of the Earth. Pluto has bright polar caps that are likely frozen methane. Its rotational period is 6.4 days, and it has a moon named Charon that is about 5–10 percent the mass of Pluto. Because Charon also has a period of 6.4 days, it appears motionless in Pluto's sky.

---

[5]Recent studies of Galileo's notebooks show that Galileo saw Neptune in December of 1612 and again in January 1613. He was interested in Jupiter at the time and so merely plotted Neptune as a background star.

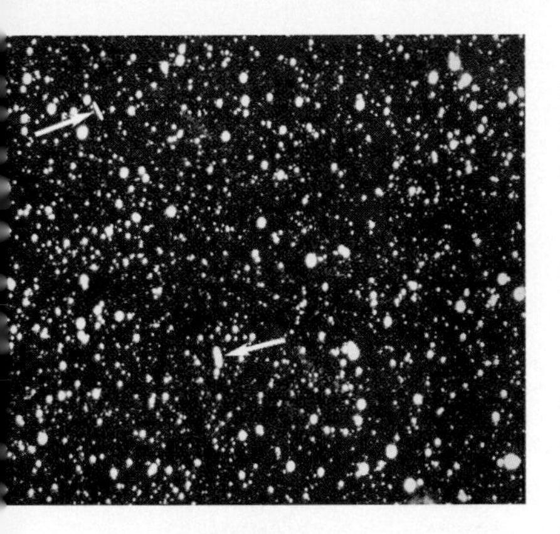

Fig. 28.28
Asteroids leave blurred trails on time-exposure photographs of the stars. The images of two asteroids are indicated by the two white arrows.

Not surprisingly, less is known about Pluto than about the other planets. Its uniqueness suggests that it and Charon may be remnants of some long ago collision. Their distance from the Sun is so great that from Pluto's surface the Sun would appear as an ordinary star among stars. Pluto must be a very cold place.

28.5 Asteroids, Meteoroids, and Comets

There is a large gap between Mars and Jupiter. In this gap is the *asteroid belt*, populated by tens of thousands of small rocky bodies called **asteroids** that orbit the Sun. The smallest asteroids are irregular in shape, like boulders, and the larger ones are roughly spherical. Asteroids vary in size from grains of sand to rocky hunks hundreds of kilometers in diameter. The largest is Ceres, which is often called a *minor planet* and has a diameter of 750 kilometers. Asteroids are thought to be material that failed to become a planet during the formation of the solar system. If the planet had formed, it would have been small, for the combined mass of all the asteroids is considerably less than the mass of the Earth's Moon.

Although many asteroids neatly circle the Sun, others do not. Collisions among asteroids are common, sending some of them helter-skelter. Some stray toward Planet Earth. Asteroids smaller than a few hundred kilometers across are called **meteoroids**. A **meteor** is a meteoroid that strikes the Earth's atmosphere, usually at an altitude of about 80 kilometers. A meteor is heated white-hot by friction with the atmosphere and is seen from the Earth as a flash of light—a "falling star." Most meteors we see are very small meteoroids, about the size of a grain of sand. Any meteor that survives its fiery decent through the atmosphere and reaches the ground is called a **meteorite** once it lands.

Fig. 28.29
A meteor is produced when a meteoroid enters the Earth's atmosphere, usually about 80 km high. Most are grains of sand, seen as "falling stars."

Most meteorites are small and strike the Earth with no more energy than a falling hailstone. Some are big, though, and evidence of their impact is seen as craters. If the Earth were without weather and other eroding elements, its surface would likely be as cratered as the Moon's. Most impact craters on the Earth were either eroded or covered by geologic processes long ago. More recent impacts, however, leave telltale marks (Figure 28.30). Perhaps the most dramatic impact of all was one we have a record of—the impact near the Yucatan Peninsula in Mexico 65 million years ago, discussed in Chapter 25. The effects of that impact are thought to be responsible for the extinction of dinosaurs and half the living species in the Cretaceous period.

Fig. 28.30
The Barringer Crater in Arizona, made 25,000 years ago by an iron meteorite having a diameter of about 50 m. The crater is 1.2 km across and 200 m deep.

Question

A school project is to visit a science museum and view both a meteor and a meteoroid. Is this a reasonable project?

Answer

Any extraterrestrial rocks you see in a museum are, by definition, meteor*ites*. So unless the science museum is in outer space, you'll never view a meteoroid from there, for meteoroids exist only outside the Earth's atmosphere. And the only meteor you can possibly see in a museum is one streaking across the sky through a window at night. Look at it this way: The same body has three names—meteoroid when it's far from the Earth, meteor when it's traveling through the Earth's atmosphere, meteorite once it hits the ground.

Fig. 28.31
When the Earth crosses the orbit of a comet, we see a *meteor shower.*

Fig. 28.32
The two tails of Comet West. A comet is always named after the person who first sees it. Guess who was the first person to see this comet?

A **comet** is a chunk of dust and ice that orbits the Sun and becomes partly vaporized as it passes near the Sun. Meteoroids come not only from the asteroid belt but from comets as well. As a matter of fact, most of the meteors we see are small particles of comet debris (Figure 28.31). Unlike meteors that shoot briefly across the sky, a comet moves slowly and gracefully to display one of nature's most beautiful astronomical spectacles (Figure 28.32).

Whereas asteroids travel between the planets in roughly circular orbits, the orbits of comets are highly elliptical, extending far beyond Pluto's orbit. As a comet approaches the Sun, solar heat vaporizes the ices. Escaping vapors glow to produce a fuzzy, luminous ball called a *coma*, typically a million kilometers in diameter. Within the bright coma is the solid part of the comet, the *nucleus*, typically a chunk of ice, dust, and other materials measuring a few kilometers across.

The solar wind (Section 28.4) and radiation pressure blow luminous vapors from the coma outward, away from the Sun, into a long flowing *tail.* A comet's tail can extend over 100 million kilometers. Most often the Sun produces two tails on a comet: an *ion tail* and a *dust tail* (Figure 28.32). The ions are largely the remnants of water vapor, too massive to be affected by the pressure of sunlight. They flow with the high-speed solar wind, directly away from the Sun. The dust tail is composed of micron-sized dust particles large enough to be affected by radiation pressure. The lower-speed dust tail curves, much as a water stream curves from the nozzle of a moving hose.

Fig. 28.33
Comet Hale-Bopp in 1997.

The density of the material in a comet's tails is quite low—less than the density achieved in typical industrial vacuums. So compared with the Earth's atmosphere, the tails of a comet are "nothing at all." When a tail crosses the Earth directly, except for meteor showers high in the atmosphere, nothing at all changes at the Earth's surface. The incidence of a comet nucleus, however, is a different story. "Meteor" craters are formed by the impact of comets as well as meteors. Only the impact debris indicates the difference.

Comets are plentiful. There is almost always a comet in the sky, but most are too faint to be seen without a good telescope. About half a dozen new comets are discovered each year, many by amateur astronomers. Most comets have no visible tails, for their supply of ice is eventually exhausted. After about 100–1000 passes around the Sun, a comet is pretty much burned out.

Link to Space Technology—Comet Chasing

In the early 1980s, co-author Paul Hewitt convinced his bright student Tenny Lim to pursue a career in science and engineering. Tenny, very good with her hands and her mind, had just graduated in dental technology. Nevertheless, she began her studies all over again, spending four years rather than two years at City College of San Francisco, graduating and going on to Cal Poly at San Luis Obispo. A photograph of Tenny in which she shows the relationship between the potential energy of a drawn bow and the subsequent kinetic energy of an arrow has been in Hewitt physics textbooks ever since and is in this book on page 62.

Today Tenny is part of a team of engineers and scientists at the Jet Propulsion Laboratory in Pasadena working on a fascinating project: to land a spacecraft onto the comet Tempel 1 to find out what the nucleus is made of. The project, Deep Space 4/Champollion, is scheduled for launch in April–May 2003. This will be the first landing of scientific instruments on the surface of a comet nucleus. Samples will be collected and returned to a carrier spacecraft and then back to the Earth in 2010. In this photograph we see Tenny with an arrow again—this time a penetrator being tested for impact into concrete blocks having properties that simulate cometlike materials. The penetrator is part of a deployable harpoon that will anchor the spacecraft to the surface to permit drilling operations and other relevant science measurements. We're very proud of Tenny Lim!

Summary of Terms

- **Moon phases** The cycles of change of the face of the Moon as seen from the Earth: new (dark) to waxing to full to waning and back to new.
- **Solar eclipse** The phenomenon whereby the shadow of the Moon falls on the Earth, producing a region of darkness in the daytime.
- **Lunar eclipse** The phenomenon whereby the shadow of the Earth falls on the Moon, producing relative darkness of the full moon.
- **Sunspots** Temporary relatively cool, dark regions on the Sun's surface.
- **Asteroid** A small, rocky fragment that orbits the Sun.
- **Meteoroid** A very small asteroid.
- **Meteor** A meteoroid once it enters the Earth's atmosphere; a "shooting star."
- **Meteorite** A meteoroid that survives passage through the Earth's atmosphere and hits the ground.
- **Comet** A body composed of ice and dust that orbits the Sun, usually in a very eccentric orbit, and which has one or more luminous tails when it is close to the Sun.

Review Questions

1. How numerous are the stars compared with the number of grains of sand on all the deserts and beaches of the Earth?
2. Why do the Sun and Moon look to be disks rather than spheres?

The Moon

3. How does the Moon's rate of rotation about its own axis compare with its rate of revolution around the Earth?
4. Why does the same side of the Moon always face the Earth?
5. What is the relationship between the action of a magnetic compass aligning with a magnetic field, and the Moon's orientation to the Earth?
6. What is meant by gravity lock?
7. What Moon-Earth-Sun relative positions account for the various phases of the Moon?

Eclipses

8. In what alignment of Sun, Moon, and Earth does a solar eclipse occur?
9. Why is totality during a solar eclipse not altogether dark?
10. In what alignment of Sun, Moon, and Earth does a lunar eclipse occur?
11. Why is totality during a lunar eclipse not altogether dark?
12. What two conditions are necessary for either a solar or lunar eclipse?

The Sun

13. What happens to the amount of the Sun's mass as it "burns?"

14. Distinguish between the Sun's photosphere and its chromosphere.
15. What is the solar wind?
16. How does the rotation of the Sun differ from the rotation of a solid body?
17. How old is the Sun?
18. Where are elements heavier than hydrogen and helium formed?

The Planets

19. Why did the ancients call planets "wanderers"?
20. Into what two major groups are the planets divided?
21. Name the inner planets. Why are they called the terrestrial planets?
22. Why are days on Mercury very hot and nights very cold?
23. Why are Mercury and Venus seen as evening or morning "stars"?
24. Why is Earth called the blue planet?
25. What predominant gas makes up the Martian atmosphere?
26. Name the Jovian planets. Why are they so named?
27. What is the major difference between the terrestrial and Jovian planets?
28. Why was Jupiter once thought to be a failed star?
29. What surface feature do Jupiter and the Sun have in common?
30. Why does Jupiter bulge at the equator?
31. What is thought to produce the strong magnetic fields of Jupiter?
32. What distinguishes the rings of Saturn from the rings of the other Jovian planets?
33. Which move faster, Saturn's inner rings or the outer ones?
34. How did Uranus lead to the discovery of Neptune?
35. By what investigative method was Pluto discovered?
36. Why is Pluto sometimes closer to the Earth than to Neptune?

Asteroids, Meteorites, and Comets

37. Distinguish between an asteroid, a meteoroid, and a comet.
38. Where, as far as we can tell, do most asteroids reside?
39. Distinguish between a meteoroid, a meteor, and a meteorite.
40. What is a falling star?
41. Why do the tails of comets point away from the Sun?
42. Why do most comets have two tails? How do the tails differ from each other?
43. What would be the consequence of a comet's tail sweeping across the Earth?
44. A meteor is visible only once, but a comet may be visible at regular intervals throughout its lifetime. Why?
45. Why does a comet eventually burn out?

Exercises

1. Which is larger, the radius of the Sun or the Moon-Earth distance? (See the inside back cover.)
2. What is the main advantages of placing a telescope in orbit?
3. Why would it be advantageous to have a telescope mounted on the Moon?
4. Why does the Moon lack an atmosphere? Defend your answer.
5. Is the fact that we see only one side of the Moon evidence that the Moon rotates or that it doesn't rotate? Defend your answer.
6. Is it necessary that the Moon be pulled into an oblong shape for it to be gravity-locked to the Earth? (If its material resisted stretching, would its center of gravity and center of mass still be at different locations?)
7. Photograph A shows the Moon partially lit by the Sun. Photograph B shows a Ping-Pong ball in sunlight. Compare the positions of the Sun in the sky when each photograph was taken. Do the photographs support or refute the claim that they were taken on the same day? Defend your answer.

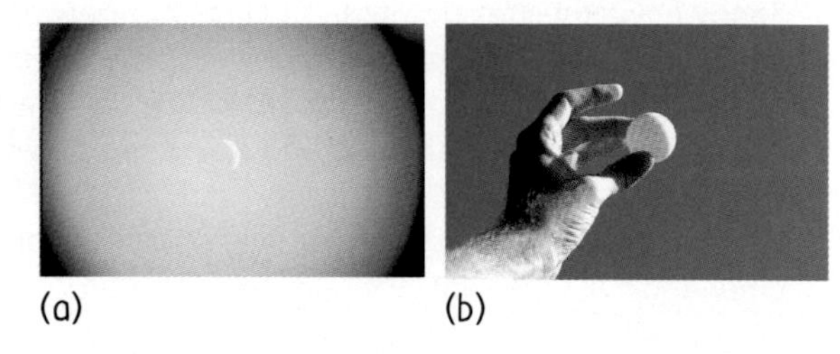

(a) (b)

8. Distinguish between a waning moon and a waxing moon, and between gibbous and crescent phases.
9. Do star astronomers work during the full moon part of the month or during the new moon part of the month? Why?
10. Why are there a lot more craters on the surface of the Moon than on the surface of the Earth?
11. Nearly everybody has seen a lunar eclipse, but relatively few people have seen a solar eclipse. Why?
12. If the Sun's diameter were only 10 times the diameter of the Moon and if the Sun-Earth distance were only 10 times greater than the Earth-Moon distance, what differences would occur for (a) solar eclipses and (b) lunar eclipses?
13. Because of the Earth's shadow, a partially eclipsed Moon looks like a cookie with a bite taken out of it. Explain with a sketch how the curvature of the bite indicates the size of the Earth relative to the size of the Moon. How does the tapering of the Sun's rays affect the curvature of the bite?
14. What energy processes make the Sun shine? In what sense can it be said that gravity is the prime source of solar energy?
15. A television screen is light gray when not illuminated. This being so, how is the blackness of sunspots similar to the blackness in images on a television screen?
16. When a contracting hot ball of gas spins into a disk shape, it cools. Why? Defend your answer.
17. The greenhouse effect is very pronounced on Venus but doesn't exist on Mercury. Why?

18. What is the cause of winds on Mars (and on almost every other planet, too)?

19. Explain how the amount of sunlight Venus receives has caused its atmosphere to be very different from the Earth's atmosphere.

20. How much of Jupiter's surface is ocean?

21. Jupiter's mass is more than 100 times the mass of the Earth, and yet your weight on the surface of Jupiter would be only about three times your weight on the Earth. Why? (*Hint:* Go back to the equation for gravitational force in Chapter 4 and take into account Jupiter's larger radius.)

22. It is said that Saturn would float in a bathtub (if the tub were big enough) because of the planet's low density. Would it float in a water-filled bathtub in *any* gravitational field or only in the gravitational field near the surface of the Earth? Defend your answer.

23. A friend suggests that the rings of Saturn are probably solid flat disks (with a hole for Saturn in the center). Then your friend looks to you for confirmation or refutation—with your usual well-thought-out explanation. What is your response?

24. Could the rings of Saturn be composed of a concentric series of thin rings and be consistent with your answer to Exercise 23?

25. Why are the seasons on Uranus different from the seasons on any other planet?

26. What were the similar historical circumstances that link the names of the planets Neptune and Pluto with the elements Neptunium and Plutonium?

27. We call Pluto the ninth planet, but sometimes it is the eighth planet and Neptune the ninth. Explain.

28. Why are meteorites so much more easily found on Antarctica than on other continents?

29. Chances are about 50-50 that in any night sky there is at least one visible comet that has not been discovered. This keeps amateur astronomers busy looking night after night, for the discoverer of a comet gets the honor of having it named for him or her. With this high probability of comets in the sky, why aren't more of them found?

30. In terms of the conservation of energy, describe why comets eventually burn out.

Problems

1. Perform the derivation described in the box on finding the mass of a planet and show by a series of algebraic steps that a planet's mass $M = 4\pi^2/G(R^3/T^2)$.

2. Find the appropriate data for the variables in Problem 1 and calculate the mass of Planet Earth.

3. Show that the results of Problem 1 lead to Kepler's third law.

CHAPTER

29

The Stars

The roots of astronomy reach back to prehistoric times when humans first noted star patterns in the night sky. The earliest astronomers divided the night sky into groups of stars called *constellations.* The names of the constellations are mainly a carryover from the names assigned by early Greek, Babylonian, and Egyptian astronomers.

The grouping of stars and the significance given to them varied from culture to culture. In some cultures, the constellations stimulated storytelling and the creation of great myths. In others, the constellations honored great heroes like Hercules and Orion or served as navigational aids for travelers and sailors. In still other cultures, constellations provided a guide for planting and harvesting crops—for they were seen to move periodically in the sky, in concert with the seasons. Charts of this periodic movement became some of the first calendars.

Fig. 29.1
The constellations and Taurus represent figures from astrology.

Stars were thought to be points of light on a great revolving celestial sphere having the Earth at its center. Positions of the sphere were believed to affect earthly events and so were carefully measured. Keen observations and logical reasoning gave birth to both astrology and later, astronomy. So we find that astronomy and astrology—science and pseudoscience—

share the same roots. Today we know that the Earth orbits the Sun, with its night side always turned away from the Sun. We see in Figure 29.2 why the background of stars varies in the night sky throughout the year.[1]

When we look at the stars on a moonless night, we might guess we see many thousands or even millions of them, but the unaided eye sees at most about 3000 stars, horizon to horizon. We see many more stars with a telescope, of course, but disappointingly,

Fig. 29.2
The night side of the Earth always faces away from the Sun. As the Earth circles the Sun, different parts of the universe are seen in the nighttime sky. Here the circle, representing 1 year, is divided into 12 parts—the monthly constellations. To Earth observers, the stars seen in the night sky appear to move in a yearly cycle.

stars appear as point sources with or without magnification. Telescopes can show the details of the Moon or planets but no details of stars. This is because stars are really far away. Many of the brightest are 10–1000 light-years distant.[2] Because of their great distance, they all appear equally remote, as on the celestial sphere imagined by the ancients. This chapter is about stars—how they are born, how they live, and how they die.[3]

29.1 Birth of Stars

Interstellar space is not empty but contains faint amounts of elements, primarily hydrogen and, to a lesser degree, a wide variety of other molecules ranging in complexity from ammonia to ethyl alcohol. Among these atoms and molecules are specks of interstellar dust composed of carbon and silicates, sometimes coated with frozen ices of water, carbon dioxide, methane, and ammonia. These interstellar dust particles may

[1]Place a lamp on a table in the middle of your room and move counterclockwise around the table while keeping your back to the lamp at all times. You'll see different parts of the room as you walk. Likewise, an observer on the night side of the Earth sees different parts of the sky as the Earth orbits about the Sun. Looking at Figure 28.2, can you see why the background stars during a midday solar eclipse are of constellations normally seen six months earlier or later?

[2]One light-year is the distance light travels in 1 year, about 9.5×10^{12} km. Another unit of distance popular with astronomers is the *parsec,* which is the same as 3.26 light-years.

[3]This chapter presents a brief "this is how it is" treatment of astronomy. For an expanded "this is how we know this is how it is" treatment, refer to the suggested readings at the end of the book.

The Big Dipper and the North Star

Perhaps the most easily recognized star group in the Northern Hemisphere is the Big Dipper (Figure 29.3). Because of its great distance, it seems to form a plane, but the seven stars actually lie at quite different distances from us. The Big Dipper and the larger groups that make up constellations, of course, would take on entirely different patterns if viewed from other locations in the universe. Because of the variety of speed and directions of stars, the familiar patterns of all groups are temporary. We see in Figure 29.4 how the Big Dipper looked from Earth in the past and how it is projected to look in the future.

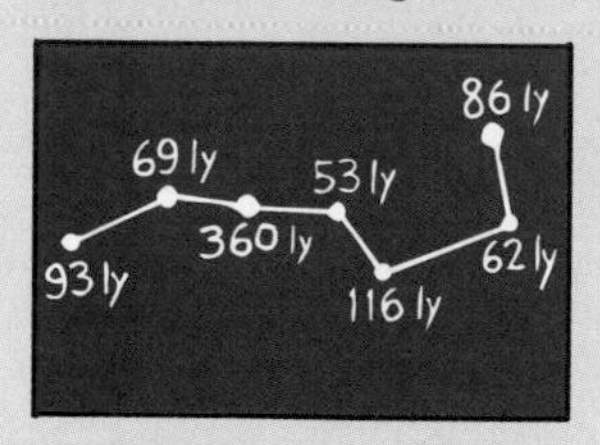

Fig. 29.3
The familiar Big Dipper. The size of the dots represents the apparent brightness of the stars, which are not all the same distance from the Earth. Their distances are noted in light-years (ly).

The Big Dipper is useful for locating the North Star (Polaris), which happens to lie almost exactly on the Earth's rotational axis. The North Star is easily located by drawing a line through the two stars in the end of the bowl of the Big Dipper and extending the line upward about five times the distance between these two stars (Figure 29.5). Because it lies very close to the extended Earth's rotational axis, the North Star appears to remain stationary as the Earth rotates. All the surrounding stars appear to move in circles around the North Star, as evidenced in time-exposure photographs (Figure 29.6).

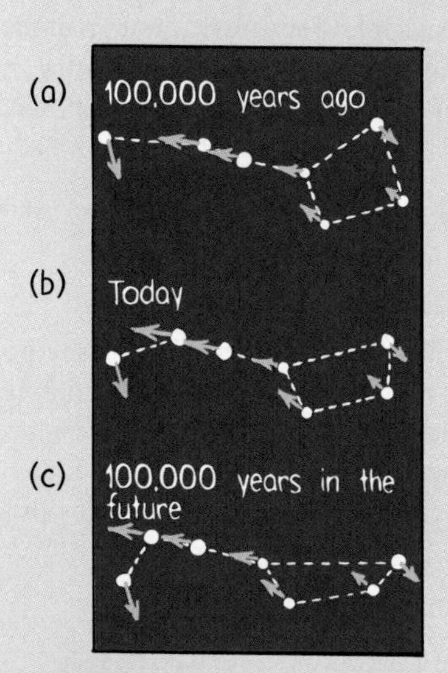

Fig. 29.4
The pattern of the Big Dipper is temporary: (a) its pattern 100,000 years ago; (b) as it appears at present; (c) what it will look like 100,000 years from now.

For centuries, one test of good eyesight has been to see if you can see which star in the Big Dipper is actually two closely spaced stars—the next-to-last star in the handle. Although these two stars appear to be closely spaced, a star and a companion star, they are actually quite far apart in space. They look close because they happen to lie approximately along the same line of sight for an observer on the Earth. Interestingly, the brighter star, Mizar, *is* actually a pair of stars—a **binary**—the first optical binary to be observed by telescope. No amount of good eyesight will show the double star Mizar without the aid of a good-sized telescope.

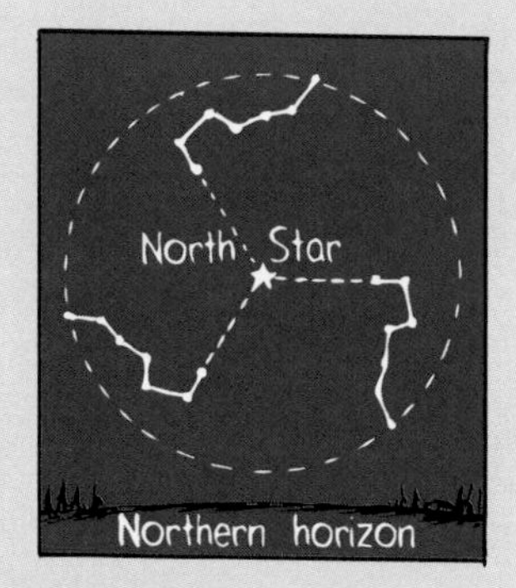

Fig. 29.5
The pair of stars in the end of the Big Dipper's bowl point to the North Star. The Earth rotates about its axis and therefore about the North Star, so over a 24-hour period the Big Dipper (and other surrounding star groups) makes a complete revolution.

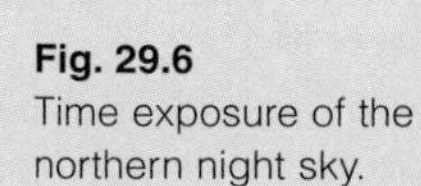

Fig. 29.6
Time exposure of the northern night sky.

play the same condensing role in star formation that similar dust particles play in cloud formation. The density of all this interstellar material is a million times lower than the densities of the highest vacuums achieved in earthbound laboratories.

To make a star, begin with a huge cloud of low-temperature interstellar material. The density of such a cloud will not be perfectly uniform. Regions of slightly greater gas density will have slightly more mass and a slightly greater gravitational field. Therefore they will more strongly attract neighboring particles. This attracting of neighbors increases the mass and gravitational field of the region, which then attracts still more particles. In time we have an aggregation of matter many times the mass of the Sun spread out over a volume many times larger than the solar system. This is a forming star—a **protostar**.

Mutual gravitation between the gaseous particles in a protostar results in an overall contraction of this huge ball of gas. The density at the center increases dramatically as matter is scrunched together, with an accompanying rise in pressure and temperature. When the central temperature reaches about 10 million kelvins, some of the hydrogen nuclei fuse to helium nuclei. As we learned in Chapter 14, this thermonuclear reaction converting hydrogen to helium releases an enormous amount of radiant and thermal energy. This ignition of nuclear fuel marks the change from protostar to star. Outward-moving radiant energy and the gas it pushes with it exert an outward pressure on the contracting matter, ultimately becoming strong enough to stop the contraction. Radiation and gas pressures balance gravitational pressure, resulting in a full-fledged star.

The material composing the star depends on how old the universe was when the star formed. The very first protostars had only primordial hydrogen and some helium to work with. Stars run their life cycles and then, like living things, return their materials to the overall environment. Elements heavier than hydrogen and helium are manufactured in star cores, and when the stars run their life courses, they spew these heavier elements into the interstellar mix. So the protostars that followed the earliest ones were enriched with heavier elements. The heavy elements that make up the Sun and its planets are testimony that many stars lived and died before the solar system came to be. All atoms on the Earth heavier than helium were once part of another star. So we are quite literally made of stardust.

■ **Question**

What do thermonuclear fusion and gravitational contraction have to do with the physical size of a star?

29.2 Life of Stars

The life span of a star depends in good part on the rate at which it burns its fuel. Astronomers are most familiar with stars like our Sun, a hydrogen burner. Our Sun has an expected life span of some 10 billion years. Hydrogen fusion in stars more massive

---

■ **Answer:**
Thermonuclear fusion tends to blow the star outward, and gravitation tends to pull it inward. The outward thermonuclear expansion and inward gravitational contraction produce an equilibrium that accounts for the star's size.

Fig. 29.7
The Sun is about 5 billion years old, having gone through about half its expected life span of 10 billion years.

than the Sun occurs at a more furious rate; these stars are very bright and have a relatively short life. In low-mass stars, hydrogen fusion occurs at a rate much slower than the rate in the Sun, and the stars are dimmer and live longer.

Surprisingly, about half the stars seen in the sky do not live alone but are actually two stars revolving around a common center just as the Earth and the Moon revolve around each other. These double stars are binary stars. By observing how the two stars in a binary revolve around their common center, we can calculate their masses. Recall from the box Determining the Mass of a Planet in Chapter 28 that the only way astronomers can calculate the masses of planets is by measuring the average distance and period of a satellite about a planet. This is also true of stars. The only way to determine the mass of a star is to find it in a binary system (Sun excluded, for its planets provide this information). The size and orbital period of binary stars depend on the masses of the stars.[4] So binaries provide astronomers with a basic means to determine how much matter is contained in stars.

There is speculation that our Sun does not live a solitary life but is part of a binary system. If it is, its partner is very small and very distant. This partner star is thought to travel in a very large elliptical orbit, as much as 3 light-years away from the Sun. At its closest approach, which would occur every 26–30 million years, it would pass near the fringes of the outermost planets. Its gravity would perturb billions of comets into the inner solar system, all within a few million years. This star has been named *Nemesis,* after the goddess of divine retribution, because some speculators credit it with triggering the meteorite impact that may have led to the demise of the dinosaurs 65 million years ago. This deadly companion has not been found and may not exist.

With or without a companion star, our Sun certainly does not live alone. It has us and at least eight other planets for company. The Sun contains 99 percent of all the mass in the solar system but only about 2 percent of the solar system's angular momentum.[5] Hence it has a slow spin, and 98 percent of the solar system's angular momentum is in the orbiting planets. Measurements of the spin rates of thousands of stars show that hot, massive stars are spinning at rates 100 times that of the Sun and cooler, less massive stars spin slowly like our Sun. Perhaps the massive stars spin rapidly because they alone possess angular momentum, while the less massive stars have formed planetary systems that absorb most of the system's angular momentum.

If other stars with as little mass as our Sun have planetary systems, today's telescopes cannot see the planets because of their nonluminosity and great distance. For many years it has seemed a fair speculation that some are located, like the Earth, at a distance from their star that is not too hot and not too cold—at a location that would support life. This idea is appealing. Based upon recent "hard science modeling," however, a growing number of astronomers contend that the range of distances from a star for life-supporting conditions as we know them is *much* tinier than previously thought.

---

[4]Recall from the box in Chapter 28 that the mass of a planet is found from the ratio of (satellite distance)^3 to (satellite period)^2: R^3/T^2 (in accord with Kepler's third law). Being that satellite mass is negligible relative to a planet's mass, the center of rotation is at the center of the planet. In much the same way, the *sum of the masses* of two stars is proportional to R^3/T^2, where R is the average distance between the stars' centers and T the period of revolution. Individual star masses are found by the stars' relative distances from the center of revolution, which is located somewhere between the stars. A star four times as massive as its companion, for example, is four times closer to the center. Equal masses are equidistant from the center.

[5]Whereas linear momentum, as studied in Chapter 3, is inertia × velocity, angular momentum is *rotational inertia × rotational velocity.* The conservation of angular momentum states that angular momentum is conserved during internal processes. Just as a spinning figure skater spins faster when he draws in his arms, a rotating ball of gas spins faster when it contracts (a decreased rotational inertia is compensated by an increased rotational velocity). And just as the spinning skater slows when he extends his arm, the Sun slowed as its planets formed.

The H-R Diagram

One of the most important tools of astronomers is the **Hertzsprung-Russell diagram**, or **H-R diagram**, developed in the early 1900s by Danish astronomer Ejnar Hertzsprung and American astronomer Henry Norris Russell. The H-R diagram is a plot of stellar variables equivalent to brightness versus temperature. The temperature of a star is evident by its color. Cooler stars are red, medium-temperature stars are white, and hotter stars are bluish-white. The graph shown here is a typical H-R diagram. Each dot represents a star whose absolute magnitude of brightness, or *intrinsic luminosity,* has been determined. Bright stars are near the top of the diagram, and dim stars are toward the bottom. Hot stars are toward the left, and cool stars toward the right.

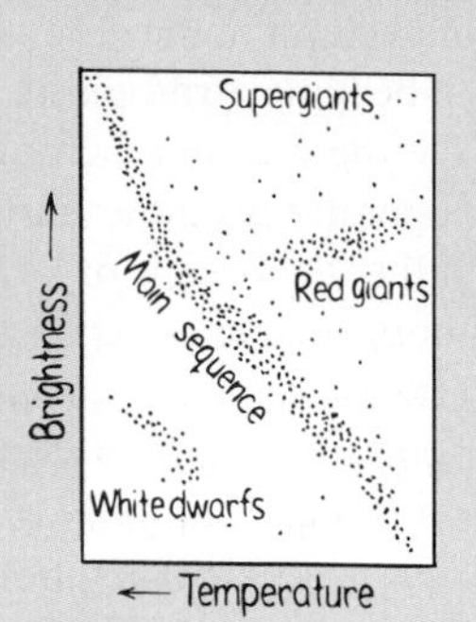

The H-R diagram shows several distinct regions of stars. The band that stretches diagonally across the diagram represents a majority of stars seen in the night sky. This band is called the *main sequence* and extends from hot, bright, bluish stars in the upper left to cool, dim, reddish stars in the lower right. About 90 percent of all stars, including our Sun, lie in the main sequence of stars.

Toward the upper right of the diagram are the group of stars called *red giants.* They are cooler stars and appear reddish in the night sky. Above these are a few rare stars, the supergiants, larger and brighter than the red giants. Toward the lower left are the dim, hot stars known as white dwarfs, which cannot be seen by the unaided eye.

When a spectroscope is attached to a telescope, stars are seen to have a variety of spectral patterns that indicate their elemental makeup and temperatures. Stars are arranged into various *spectral types* according to their spectra. When brightness versus spectral type is plotted, the same H-R pattern results. The H-R diagram dramatically shows the fundamental fact that different types of stars exist. These stars represent different stages of stellar evolution.

Nevertheless, a large NASA SETI (Search for Extraterrestrial Intelligence) program is presently in progress. Our own civilization is so young that there has hardly been enough time for it to have come to the attention of others. The most conspicuous evidence of life on the Earth—radio, TV, and radar broadcast—has by now reached some 65 light-years into space, a distance encompassing only a few hundred stars.[6]

In a similar way, most of the starlight from far-away stars has not reached the Earth yet. That's how far away most of the universe is. And the light from these far-away stars that does reach us is mostly Doppler-shifted below the visible part of the spectrum and is thus invisible to us. Hence the night sky is black instead of ablaze with starlight!

29.3 Death of Stars

All luminous stars "burn" nuclear fuel. A star's life begins when it ignites its nuclear fuel, and it ends when its nuclear fires go out. The first ignition in a star core is the fusion of hydrogen to helium, a step that may last from a few million to a few hundred billion years, depending on the star's mass. In the old age of an average-mass star like

[6]On the positive side, one verified contact will prove that intelligent life exists elsewhere. On the negative side, we can never prove that extraterrestrial life *doesn't* exist. However intense our search, life could always be "just around the next corner." If after centuries of listening and looking we find no sign of extraterrestrial intelligence, we might then be justified in assuming that we are alone. If we are the sole inhabitants of the galaxy, then our present concern for tending Planet Earth should extend to being the guardians of the galaxy.

our Sun, the burned-out hydrogen core that has been converted to helium contracts because of gravity, and this contraction raises its temperature. The increase in temperature ignites both the helium in the core and the unfused hydrogen outside the core, and the star expands to become a **red giant**. Our Sun will reach this stage about 5 billion years from now. On the way to reaching this stage, the swelling and more luminous Sun will cause temperatures on the Earth to rise, first stripping the Earth of its atmosphere and then boiling the oceans dry. Ouch!

The cores of both solar-mass stars and those having mass lower than the Sun are not hot enough to fuse carbon, and lacking a source of nuclear energy, they shrink. As they shrink, their outer layers are sometimes ejected and form expanding shells that look like smoke rings and eventually disperse and mix with the interstellar material. This expanding shell is a **planetary nebula** (Figure 29.8). The shrunken core that is left blazes white-hot and is a **white dwarf**. Here matter is so compressed that a teaspoonful of it weighs tons.

Fig. 29.8
The planetary ring nebula in the constellation Lyra, which can be seen through a modest telescope.

Because the nuclear fires of a white dwarf have burned out, it is not actually a star anymore. It's more accurate to call it a *stellar remnant.* It may continue to radiate energy and change from white to yellow and then to red, until it slowly but ultimately fades to a cold, black lump of matter—a **black dwarf**. The density of a black dwarf is enormous. Into a volume no more than that of an average-size planet is concentrated a mass hundreds of thousands times greater than that of the Earth. The black dwarf has a density comparable to that of a battleship squeezed into a pint jar!

There is another possible fate for a white dwarf, if it is part of a binary and if its partner is close enough. Because its great mass means the white dwarf exerts a gravitational pull on nearby objects, the white dwarf may pull hydrogen from its companion star and deposit this material on its own surface as a very dense hydrogen layer. Continued compacting increases the temperature of this layer, which ignites to embroil the white dwarf's surface in a thermonuclear holocaust that we see as a **nova**. (A nova is an event, not a stellar object. Because the explosion causes the white dwarf to brighten significantly, astronomers who first observed this event thought they had sighted a new star and so named it *stellar nova,* Latin for "new star.") After a while, a nova subsides until enough matter again accumulates to repeat the event. A given nova flares up at irregular intervals that may range from decades to hundreds of thousands of years.

The Bigger They Are, the Harder They Fall

How a star evolves depends on its mass. Low-mass and medium-mass stars become white dwarfs, but more massive stars don't. When a star having a mass much greater than the Sun's mass contracts, more heat is generated than in the contraction of a smaller star. This heat keeps the star from shrinking to a white dwarf because carbon nuclei in the core fuse and liberate energy while synthesizing heavier elements such as neon and magnesium. Gas pressure and radiation pressure halt further gravitational contraction until all the carbon is fused. Then the core contracts again to produce even greater temperatures and a new fusion series that produces even heavier elements. The fusion cycles repeat until the element iron is formed. The fusion of elements beyond iron requires energy rather than liberating energy (recall from Figure 14.27 that the mass of nucleons increases and absorbs energy as elements beyond iron are fused). With no energy coming from the iron core, the center of the star collapses without rekindling. The star begins its final collapse.

The collapse is catastrophic. When the core density is so great that all the nuclei are compressed against one another, the collapse momentarily comes to a halt. The collapsed star, compressed like a spring, rebounds violently in a great explosion, hurling

Link to Pseudoscience—Astrology

There is more than one way to view the cosmos and its processes. Astronomy is one way and astrology is another. Astrology is a belief system that began more than 2000 years ago in Babylonia. It has survived nearly unchanged since the second century A.D., when some revisions were made by the Egyptians and the Greeks, who believed that their gods moved heavenly bodies to influence the lives of people on the Earth.

Astrology today holds that the position of the Earth in its orbit around the Sun at the time of a person's birth, as well as the relative positions of the planets at that moment, has some influence over the person's life. The stars and planets are said to affect such personal things as character, marriage, friendships, wealth, and death.

The question is often raised as to whether the force of gravity exerted by these celestial bodies is a legitimate factor in human affairs. After all, the ocean tides are the result of the Moon's and Sun's gravitational pull, and the gravitational pull of one planet on another perturbs both their orbits. Because slight variations in gravity produce these effects, might not slight variations in the planetary positions at the time of birth affect a newborn? If the influence of stars and planets is in their gravity, then credence must also be given to the effect of the gravitational pull between the newborn and the Earth, for this pull is enormously greater than the combined pull of all the planets, even when they are all lined up in a row (as occasionally happens). And the gravitational influence of the hospital building on the newborn would surely exceed that of the distant planets! So planetary gravitation cannot be an underlying agent for astrology.

Astrology must therefore look to another realm for its basis, for all attempts to find physical explanations to support it have failed. Astrology is not a science, for it doesn't change with new information as science does, nor are its predictions borne out by fact. So the realm of astrology may be spiritual, a religion of sorts. Or it may be a primitive psychology where the stars serve as a point of departure for musings about personality and personal decisions. Or astrology may be in the realm of numerology or phrenology—rigid and empty superstitions that prevail because of their focus on what is very important to each of us—ourselves.

A common position is that astrology is a harmless belief—a little fun at minimum harm. But is it harmless when believers are led to think their personalities are fixed by the stars at birth, that weak people will remain weak, that sad people will remain sad, that one's fate is dictated by the stars? We must also question the harm dealt people whose astrological signs are deemed incompatible with the signs of others. Astrology teaches, in a nutshell, that people are hostages to the stars, and to contend that this is a harmless belief is questionable.

into space the elements previously manufactured over billions of years. The entire episode can be over in a few minutes. It is during this brief time that the heavy elements beyond iron are synthesized, as protons and neutrons mash with other nuclei to produce elements such as silver, gold, and uranium. Because the time available for making these heavy elements is so brief, they are not as abundant as iron and the lighter elements.

Such a stellar explosion is a **supernova**, one of nature's most spectacular cataclysms. Supernovae are fiery cauldrons that generate the elements essential to life, for all the elements beyond iron that make up our bodies originated in far-off, long-ago supernovae. A supernova flares up to millions of times its former brightness. In 1054 A.D., Chinese astronomers recorded their observation of a star so bright it could be seen by day as well as by night. This was a supernova, its glowing plasma remnants now making up the spectacular Crab nebula (Figure 29.9). The 1987 supernova (see box) was less spectacular but afforded astronomers an exciting first-hand look at one of these seldom-seen events.

Fig. 29.9
The Crab nebula, the remnant of a supernova that was seen from the Earth in 1054 A.D.

The inner part of a supernova star implodes to form a core compressed to *neutron density,* which means protons and electrons have been compressed together to form a core of neutrons just a few kilometers wide. This superdense, central remnant of a supernova survives as a **neutron star**. In accord with the law of conservation of angular momentum, these tiny bodies, which have densities hundreds of millions times that of white dwarfs, spin at fantastic speeds. Neutron stars explain **pulsars**, discovered in 1967 as rapidly varying sources of low-frequency radio emission. As a pulsar spins, the beam of radiation it emits sweeps around the sky, and if the beam sweeps over the Earth, we detect pulses. Of the approximately 300 known pulsars, only a few have been found emitting X-ray or visible light. One is in the center of the Crab nebula (Figure 29.10). It has one of the highest rotational speeds of any pulsar studied, rotating more than 30 times per second. This is a relatively young pulsar, and it is theorized that X-ray and optical radiation is emitted only during a pulsar's early history.

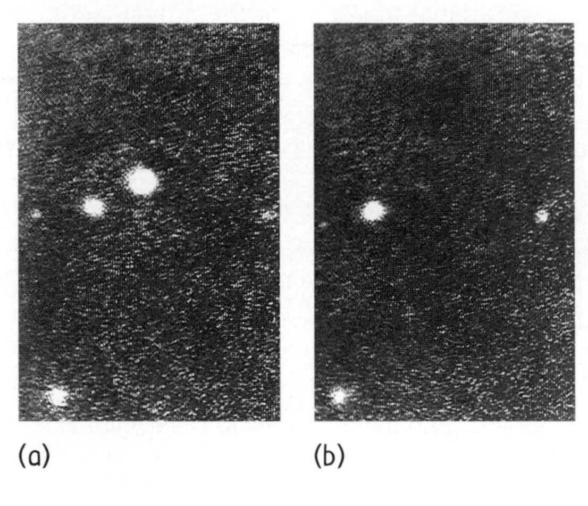

Fig. 29.10
The pulsar in the Crab nebula rotates like a searchlight, beaming light and X rays toward the Earth about 30 times a second, blinking on and off. (a) pulsar on, (b) pulsar off.

Dying stars that have cores greater than 3–5 solar masses collapse so violently that no physical forces are strong enough to inhibit continued contraction. The bigger they are, the harder they fall! The enormous gravitational field about the imploding concentration of mass makes explosion impossible. When the star has collapsed down to only 18 kilometers in diameter, collapse is unrestrained, and the star disappears from the observable universe. A most peculiar *black hole* is all that is left.

29.4 Black Holes

A very massive star that undergoes gravitational collapse leaves a **black hole**. The collapsed star is black because the gravitational force at its surface is so enormous that light cannot escape.

We can see why gravity is so great in the vicinity of a black hole by considering the change in the gravitational field at the surface of any star that collapses. In accord with Newton's law of gravity, any object at the surface of a star, whether it be a large object or simply a particle, has weight that depends both on its mass and on the mass of the star. But more important, the object's weight also depends on the distance between it and the center of the star. When the star collapses, the distance between the object on the surface and the star's center decreases. The object's weight increases without any change in its mass. And how much is this increase? That depends on the amount of collapse. If a star collapses to half its original size, then in accord with the inverse-square law, the weight of an object at its surface quadruples (Figure 29.11). If it collapses to a

The 1987 Supernova

Johannes Kepler is credited for spotting, in 1604, the last supernova before the invention of the telescope. Since then a dozen generations of astronomers have lived and died without ever witnessing such a stellar explosion. Astronomers have contented themselves with studying the remnants of explosions that occurred before their time. This generation is more fortunate, for on February 24, 1987, Ian Shelton, graduate-school dropout and resident observer at the University of Toronto telescope in the Andes Mountains of northern Chile, stumbled upon a curious blotch in a photograph he had just made of the Large Magellanic Cloud (which is a dwarf companion to our own Milky Way Galaxy). He stepped outside and confirmed his finding firsthand. He saw a pinpoint of light less bright than others in the Magellanic Cloud that hadn't been there the night before. This was Supernova 1987A.

The 1987A Supernova, seen in the southern sky in the Large Magellanic Cloud.

During the first few days after Shelton's sighting, the supernova's brightness increased a thousandfold, much more swiftly than expected, and leveled off dimmer than expected. Although initially very hot and blue, it turned red very quickly, a change that occurred as the shell of debris raced outward at 80,000,000 kilometers per hour and cooled. In March, the supernova brightened again, powered by the decay of radioactive elements in the stellar remains. Enormous quantities of nickel and cobalt forged in the detonation were decaying into iron—enough to construct 20,000 Earths. The burst reached its peak toward the end of May, when it was billions of times more luminous than our Sun. This colossal event ran its course during the summer, and a year later the visible fireworks were completely over. To the unaided eye Supernova 1987A had faded into oblivion in the southern sky.

Centuries from now astronomers will still be studying the wispy, incandescent filaments of Supernova 1987A, just as observers today meticulously examine the remains of the "new stars" observed centuries ago by Kepler and others. Theories rise and fall with each new measurement of the ever-expanding star remnants.

How awe inspiring it is that men and women on a small, obscure planet are able to investigate spots of light in the night sky and from their examinations arrive at a magnificent description of creation. This was well put in 1948 during the opening of the famous Hale telescope: "In the last analysis, the mind which encompasses the universe is more marvelous than the universe which encompasses the mind."

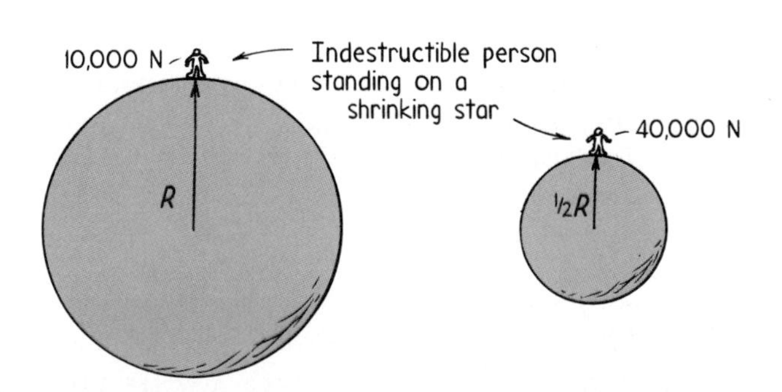

Fig. 29.11
If a star collapses to half its radius and there is no mass change, gravitation at its surface increases fourfold (inverse-square law). If the star collapses to one-tenth its radius, gravitation at its surface increases a hundredfold.

tenth its original size, the weight at the surface is 100 times as much. Gravitation at the surface continues to increase as collapse continues.

Along with the increase in gravitational field, the escape speed from the surface of the collapsing star increases. If our Sun were to collapse to a radius of 3 kilometers, the escape speed from its surface would exceed the speed of light, and so nothing—not even light—could escape![7] The Sun would be invisible. It would be a black hole.

The Sun in fact has too little mass to experience such a collapse. But when some stars having a core mass greater than four times the mass of the Sun reach the end of their nuclear resources, they do collapse. And unless rotation is high enough, the collapse continues until the stars reach infinite densities. Gravitation near the surfaces of these shrunken stars is so enormous that light cannot escape from them. They have crushed themselves out of visible existence. The results are completely invisible black holes.

A black hole is no more massive than the star from which it formed. For this reason, the gravitational field at a distance from the center equal to the star's original radius is no different after collapse than before. But distances closer than one original radius are nothing less than the *collapse of space,* with a surrounding warp into which anything that passes too close—light, dust, or a spaceship—is drawn. Astronauts could enter the fringes of this warp and with a powerful spaceship still escape. Below a certain distance, however, they could not, and they would disappear from the observable universe.

Questions

1. What determines whether a star becomes a white dwarf, a neutron star, or a black hole?
2. If the Sun somehow suddenly collapsed to a black hole, what change would occur in the orbital speed of Planet Earth?

Black Hole Geometry

If we shine a beam of light past a black hole, gravity is intense enough to noticeably deflect (bend) the beam (Figure 29.12). If the beam passes very far from the hole, where gravity is not as strong, the beam deflects only slightly. The closer it is to the black hole, the more the beam bends. If a beam passes the black hole at precisely the right

Answers

1. The mass of a star is the principal factor that determines its fate. Any star that has a mass either equal to or less than the mass of the Sun becomes a white dwarf. Any star that has a mass equal to 2 or 3 solar masses becomes a neutron star. Any star that has a mass equal to 3–5 solar masses ultimately becomes a black hole.

2. None. This is best understood classically; nothing in Newton's law of gravitation, $F = G\frac{(mM)}{d^2}$, changes. The fact that the Sun is compressed doesn't change its mass M or its distance, d, from the Earth. Because the Earth's mass m and G don't change either, the force F holding the Earth in its orbit does not change.

[7]In Chapter 30, we'll see that light, like massive things, is affected by gravity. Just as we fail to see the gravity-caused curvature of a high-speed bullet when viewed along short segments, we also fail to see the gravity-caused curvature of even higher-speed light.

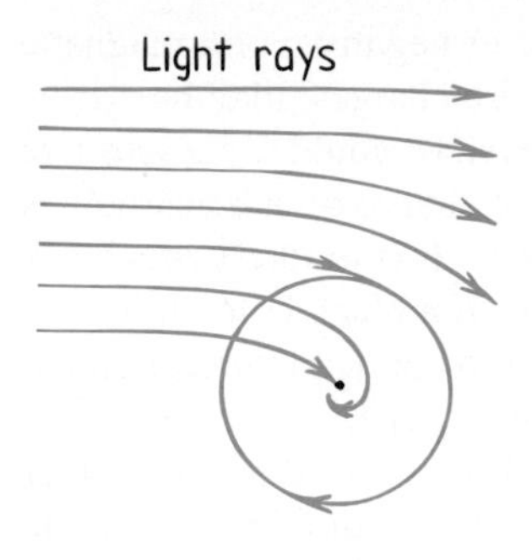

Fig. 29.12
Light rays deflected by the gravitational field around a black hole. Light passing far away is bent only slightly, light passing closer can be captured into circular orbit, and light passing closer still is sucked into the hole.

distance, the beam goes into *circular orbit* about the hole. The three-dimensional projection of this precise distance above the black hole is called the *photon sphere*. The spherical surface that constitutes the photon sphere is very thin, and the slightest variation in the path of a beam traveling on the sphere sends the beam either inside or outside the shell, either spiraling into the hole or traveling back off into space.

An indestructible astronaut in a powerful enough spaceship could venture through the photon sphere of a black hole and come out again. While inside the photon sphere, she could send beams of light back into the outside universe (Figure 29.13). If she directed her flashlight sideways and toward the black hole, the light would quickly spiral into the hole, but light directed vertically and at angles close to the vertical would escape. As she gets closer and closer to the black hole, however, she finds she must shine the light beam closer and closer to the vertical for it to escape. Moving closer still, our astronaut would find a particular distance where *no* light can escape. No matter what direction the flashlight points, all the beams are deflected into the black hole. Our unfortunate astronaut has passed within the **event horizon**. Once inside the event horizon, she can no longer communicate with the outside universe; neither light waves, radio waves, nor any matter can escape from inside the event horizon. Our astronaut would have performed her last experiment in the universe as we conceive it.

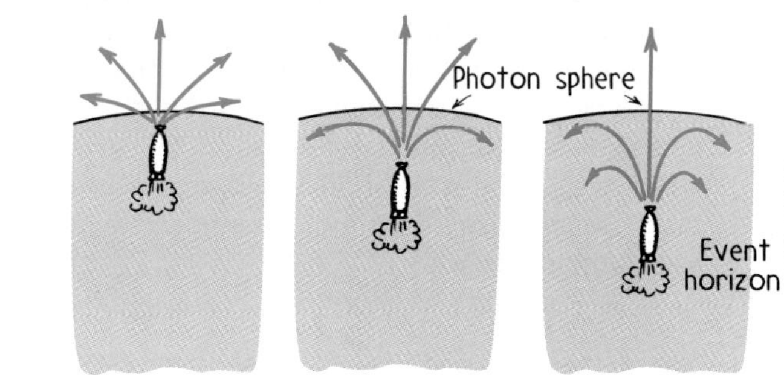

Fig. 29.13
Just beneath the photon sphere, an astronaut could still shine light to the outside. But as she gets closer to the black hole, only light directed nearer the vertical gets out, until finally even vertically directed light is trapped. This hole-light distance is the event horizon.

The event horizon surrounding a black hole is often called the *surface* of the black hole, the diameter of which depends on the mass of the hole. For example, a black hole resulting from the collapse of a star 10 times as massive as the Sun has an event horizon diameter of about 60 kilometers. The radii of event horizons for black holes of various masses are shown in Table 29.1.

Mass of Black Hole	Radius of Event Horizon
1 Earth mass	0.8 centimeter
1 Jupiter mass	2.8 meters
1 solar mass	3 kilometers
2 solar masses	6 kilometers
3 solar masses	9 kilometers
5 solar masses	15 kilometers
10 solar masses	30 kilometers
50 solar masses	148 kilometers
100 solar masses	296 kilometers
1000 solar masses	2961 kilometers

Table 29.1
Estimated Radii of Event Horizons for Nonrotating Black Holes of Various Masses

A collapsing star that contracts within its own event horizon still has substantial size. There are no forces known that can stop the continued contraction, however, and the star quickly shrinks until finally it is crushed, presumably to the size of a pinhead, then to the size of a microbe, and finally to a realm of size smaller than ever measured by humans. At this point, according to theory, there is infinite density. This point is the **black hole singularity**.

The complete description of the simplest kind of black hole is quite straightforward. It has only one property—mass. And this can be precisely determined outside the event horizon, for example, by a physicist who measures how much the trajectory of a rocket probe is deflected when in the vicinity of the black hole.

If the black hole has an electric charge (either positive or negative) or a magnetic "charge" (either north pole or south pole), the effects of these charges, like the effects of mass, also extend beyond the event horizon. A distant physicist could use a sensitive apparatus to detect these charges. So, in addition to mass, a charge on a black hole is not lost to the universe and is an additional physical property. It is unlikely that black holes having appreciable electric charges exist, however, for if a black hole did have a substantial charge, its electric field would soon tear apart atoms in nearby space and in a very short time become neutralized by particles of opposite charge.

More important is spin, for most stars are rotating and therefore possess angular momentum. If a black hole is formed from a rotating star, the surrounding space and time are dragged along with the spinning hole as it forms. (The connection between space, time, and gravity we explore in Chapter 30.) An observer in a spaceship far from the hole would notice a gradual pulling of the ship in the direction of hole rotation. The closer it is to the rotating hole, the faster the ship is pulled around. Thus we see that a black hole's angular momentum is a third property that extends beyond the event horizon.[8]

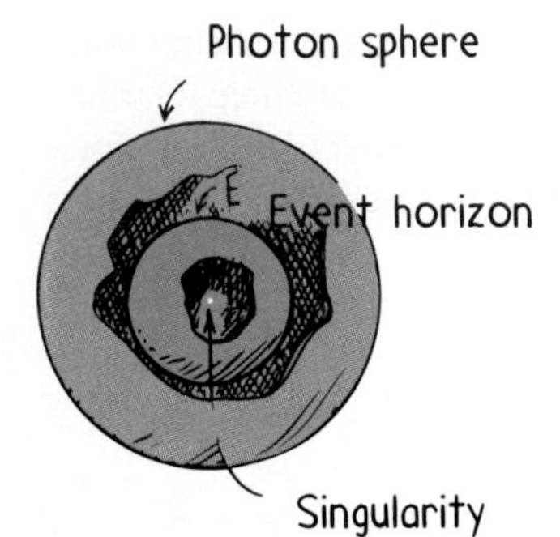

Fig. 29.14
Structure of a simple, ideal black hole (uncharged and nonrotating).

Black holes, then, can have only three properties: mass, charge, and angular momentum. Whereas a complete description of a star involves all sorts of hairy things such as chemical composition, varying pressure, densities, and, temperatures at different depths, and so on, no such complications are involved in black holes. Physicists put it simply by stating, "*Black holes have no hair*!"

Contrary to stories about black holes, they're nonaggressive and don't reach out and swallow innocents at a distance. The gravitational fields around them are no stronger than the gravitational fields about the stars before collapse—except at distances smaller than the original star radius. Being close to a black hole is what future astronauts will be wary of.

29.5 Galaxies

A **galaxy** is a large assemblage of stars, planetary nebulae, and interstellar gas and dust. Galaxies are the breeding grounds of stars. Our own star, the Sun, is an ordinary star among 200 billion others in an ordinary galaxy known as the *Milky Way Galaxy.* With unaided eyes we see our galaxy as a faint band of light across the sky, a band called the Milky Way. The early Greeks called this band the "milky circle," and the Romans called it the "milky road" or "milky way." The latter name has stuck.

Most astronomers believe that 10–15 billion years ago galaxies formed from huge clouds of primordial gas pulled together by gravity, similar to our description in Chapter 28 of the way the solar system formed. Formation begins with gravitational attrac-

[8]Science types speculate that material drawn into a black hole may reappear elsewhere through a "white hole." What vanishes in one place might be spewed out at another. The downside to this speculation is that everything that falls into a black hole is reduced to elementary particles, and everything would emerge as particles. If black holes are a gate from one realm to another, and a temporary gate at that due to fluctuations, they are a gate through which nothing can pass intact. An upside presents itself, however, if the black hole is spinning fast enough to form a *ring,* astronauts approaching this ring along its rotational axis might travel unharmed through the ring and perhaps enter another universe! The merits of these intriguing speculations are pretty much confined to science fiction stories, but they're fun to think about!

Observations of Black Holes

Finding black holes is very difficult. One way is to look for a binary in which a luminous star appears to orbit about an invisible companion. If the two members of the binary are close to each other, matter ejected by the luminous member and accelerating into the neighboring invisible black hole should emit X-rays. The first convincing candidate for a black hole, discovered by astronomers in 1972, was the X-ray-emitting star Cygnus X-1. This star is about 9 or 10 solar masses and forms a binary with a blue supergiant. Material streaming into the supposed black hole (the X-ray emitter Cygnus X-1) from the supergiant is emitting X radiation. (This is depicted in the artist's representation of a black hole stealing matter from a companion star.)

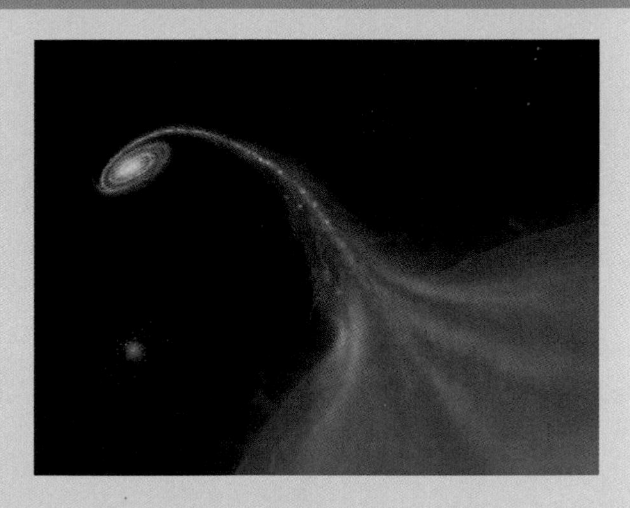

Similar radiation patterns have also been found in Circinus X-1, another binary system, and more recently in V404 Cygni. Observations by the NASA satellite *Copernicus* strongly suggest that a star in the constellation Scorpius is a black hole. This star, V861 Sco, is relatively close to us, a distance of 5000 light-years, which is slightly less than the distance from us to Cygnus X-1. The X-ray source called LMC X-3 in the Large Magellanic Cloud—a dwarf companion to our own Milky Way Galaxy—is very likely a black hole. Other massive black holes of 100–1000 solar masses are thought to exist at the centers of certain star clusters.

Fig. 29.15
A wide-angle photograph of the Milky Way, from the constellation Cassiopeia on the left to the constellation Sagittarius on the right. The dark lanes and blotches are interstellar gas and dust obscuring the light of background stars.

tion between distant particles, then contraction followed by an increased rotational rate, then in some cases flattening to a disk due to this rotation. A most striking feature of our galaxy is the spiral arms that wind outward through the disk. These arms are swarms of hot, blue stars; clouds of dust and gas; and clusters of young stars.

Masses of galaxies range from about a millionth the mass of our galaxy to some 50 times more. Galaxies are calculated to have much more mass than has been detected. This undetected mass is known as *dark matter*. The nature of this dark matter is still a question.

The millions of galaxies visible on long-exposure photographs can be separated into three main classes: *elliptical, spiral,* and *irregular*.

Elliptical galaxies are the most common galaxies in the universe. Because most of them are relatively dim, they are more difficult to see than other types of galaxies. An exception is the gigantic elliptical galaxy M87 (Figure 29.16). Elliptical galaxies contain little gas and dust and cannot make new stars.

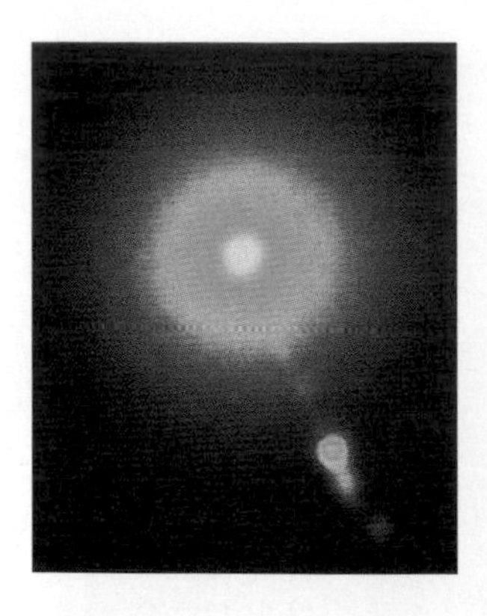

Irregular galaxies are normally small and faint. They are difficult to detect. They are without obvious spiral arms or dense centers and contain large clouds of gas and dust mixed with both young and old stars. The irregular galaxy first described by the navigator on Magellan's voyage around the world in 1521 is our nearest neighboring galaxy—the *Magellanic Clouds.* This galaxy consists of two clouds, called the Large Magellanic Cloud and the Small Magellanic Cloud. The Large Magellanic Cloud is dotted with hot young stars having a combined mass of some 20 billion solar masses, and the Small Magellanic Cloud contains stars having a combined mass of about 2 billion solar masses (Figure 29.17). The combined mass is small for a galaxy. Irregular galaxies are probably as common as spiral galaxies. Spiral and irregular galaxies contain large amounts of gas and dust and are still forming stars.

Fig. 29.16
The huge elliptical galaxy M87, one of the most luminous galaxies in the sky, is located near the center of the Virgo cluster, some 50 million light-years from the Earth. It is about 40 times more massive than our Milky Way Galaxy.

(a)

(b)

Fig. 29.17
(a) The Large Magellanic Cloud and (b) neighboring Small Magellanic Cloud make up an irregular galaxy that is only about 150,000 light-years from the Earth. The Magellanic Clouds are our closest galactic neighbors and likely orbit the Milky Way Galaxy.

Spiral galaxies are among the most beautiful arrangements of stars in the heavens. They are bright with the light of newly formed stars. The brightness of most spiral galaxies makes them easy to see at great distances. How their spiral arms form is still being investigated. Perhaps differential rotation of the galaxy stretches star-forming regions into elongated arches of stars and planetary nebulae. Or maybe the arms are created by density waves that sweep around the galaxy. We know that about two-thirds of all known galaxies are spirals, although they make up probably 15–20 percent of all galaxies. We do not see the greater number of fainter elliptical galaxies thought to exist.

Fig. 29.18
Spiral galaxy M83 in the southern constellation Centaurus, about 12 million light-years from the Earth.

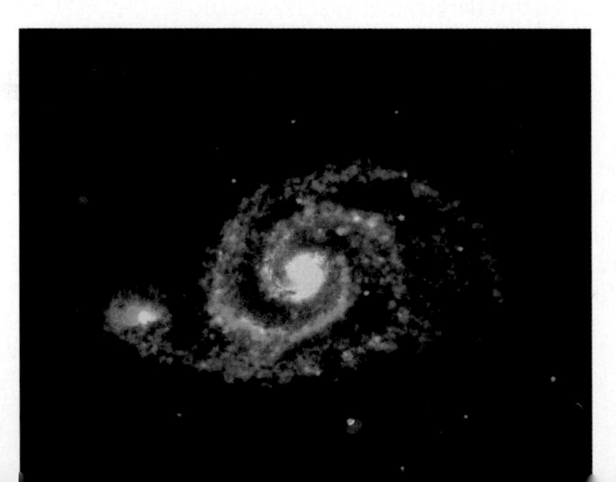

We know what it's like to live in a spiral galaxy, for our Milky Way Galaxy is a typical one. When we look at the Milky Way that crosses the night sky, we are looking through the disk of the galaxy. Interstellar dust obscures our view of most of the visible light that lies along the plane of this disk. Most of what we know of our galaxy is via infrared and radio telescopes. Infrared and radio observations reveal many details of the galactic nucleus, but astronomers are still puzzled by the processes occurring there. The nucleus seems to be crowded with stars and hot dust, and at the very center is thought to be a massive black hole that has a mass of a million Suns. Don't go too near the center of our galaxy.

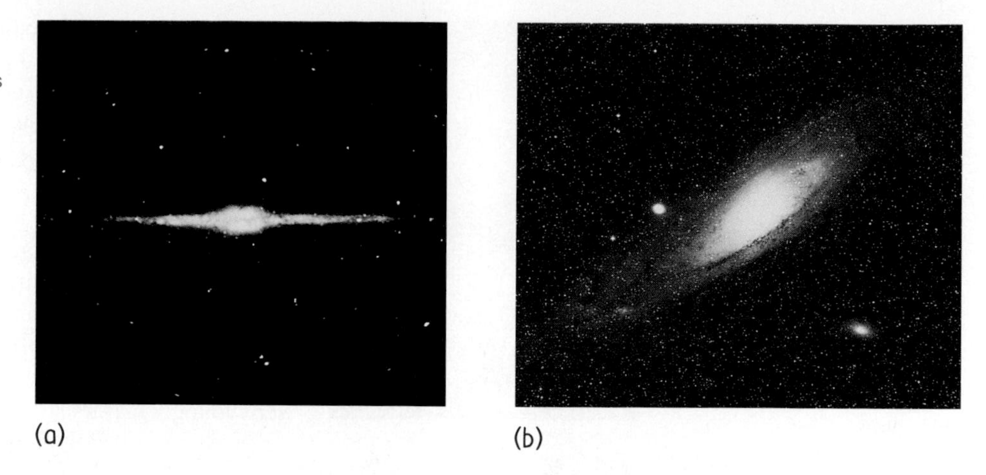

Fig. 29.19
(a) An edge-on view of our Milky Way Galaxy, which makes four rotations every billion years. Our Sun is about $\frac{5}{8}$ the distance from the galaxy's center to its outer rim.
(b) The great planetary nebula in Andromeda, a spiral galaxy about 2.3 million light-years from the Earth.

Galaxies collide. The stars in a galaxy are normally so far apart, however, that physical collisions of individual stars is a highly unlikely event. But interstellar gases and dust collide violently, with matter stripped from one galaxy and deposited in another. These collisions also prompt the formation of new stars. Low-speed collisions can result in the merger of two galaxies. There is evidence that the Milky Way Galaxy may be presently consuming the Magellanic Clouds. At high velocities, colliding galaxies can distort each other through tidal forces and create tails and bridges. The collisions of spiral galaxies are thought to form huge elliptical galaxies. Many large elliptical galaxies are believed to contain the merged remains of several spiral galaxies. Galaxies are cannibals. Spiral galaxies are survivors that have experienced either no collisions or very few collisions with large galaxies since their formation.

Galaxies are not the largest things in the universe. Galaxies come in **clusters**. Galaxy clusters appear to be part of even larger clusters, the **superclusters**. And it doesn't stop there; superclusters seem to be part of a network of filaments surrounding empty voids. Comprehension of the universe at this scale becomes mind-boggling.

29.6 Quasars

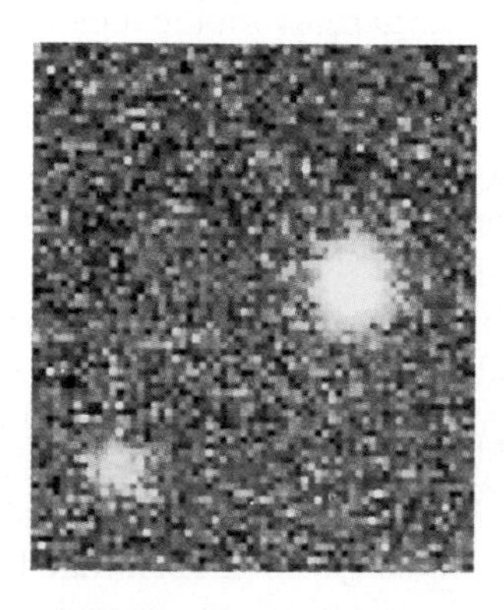

Fig. 29.20
Colorized image of the quasar BR 1202-07, the brightest observed quasar to date.

Galaxies are not the brightest parts of the universe. Brighter still are **quasars**. The energy output of these objects is enormous—hundreds of times that of the entire Milky Way Galaxy. Quasars were first thought to be relatively faint ordinary stars in our own galaxy, but in 1960, astronomers found that they emit radio waves. Radio waves are frequently observed coming from galaxies, but no star had been observed to emit strong radio signals. Further investigation revealed these "radio stars" have a pattern of spectral lines that could not be deciphered. These objects became known as "quasi-stellar sources," soon shortened to *quasars*.

The unusual spectra of quasars turned out to be a normal spectrum with an extremely large and unprecedented red shift (Section 10.10), which indicates enormous recessional speeds—some more than 90 percent the speed of light! Clearly the objects couldn't be stars in our own galaxy, for we would have noticed a change in their positions against the background of the fixed stars, and quasars had been observed for years as faint stars with no noticeable change in position. At first some investigators thought that the red shift was not a Doppler shift but a gravitational red shift characteristic of a small body that has enormous mass and a correspondingly enormous gravitational field. But spectra from quasars revealed emission lines from normal atoms with normally

Fig. 29.21
What do the atoms in Melissa's body have in common with the stars she contemplates?

orbiting electrons that wouldn't exist in a neutron star, black hole, or any other body having gravitation intense enough to produce such a large red shift.

Quasars are now thought to be young, active spiral galaxies. Because they appear to be as much as 15.5 billion light-years distant from the Earth, this places them back at the beginnings of the universe. They may be gigantic black holes that pull enormous amounts of material toward them, with resulting collisions that liberate immense energies. Current findings indicate that quasars are the brilliant cores of very distant spiral galaxies that we see as they were when they were young. Quasars still puzzle astronomers.

Summary of Terms

- **Protostar** The aggregation of matter that goes into and precedes the formation of a star.
- **Binary** A system of two stars that orbit about a common center of mass.
- **H-R diagram** (Hertzsprung-Russell diagram) A plot of brightness versus temperature for stars.
- **Red giant** A cool giant star above main sequence on the H-R diagram.
- **Planetary nebula** An expanding shell of gas ejected from a low-mass star during the latter stages of its evolution.
- **White dwarf** A dying star that has collapsed to the size of the Earth and is slowly cooling off; located at lower left of the H-R diagram.
- **Black dwarf** The presumed final stage of a white dwarf.
- **Nova** The sudden brightening of a white dwarf caused by the explosion of accumulated hydrogen gas on its surface.
- **Supernova** A stellar explosion, caused either by transfer of matter to a white dwarf or by gravitational collapse of a massive star, where enormous quantities of matter are emitted.
- **Neutron star** A small, dense star composed of tightly packed neutrons formed by the welding together of protons and electrons.
- **Pulsar** A neutron star that spins rapidly and in doing so sends out short, precisely timed bursts of electromagnetic radiation.
- **Black hole** The remains of a giant star that has collapsed into itself, so dense and of such intense gravitational field that light cannot escape.
- **Event horizon** The boundary region of a black hole from which no radiation may escape.
- **Black hole singularity** The point of zero radius and infinite density into which the matter of a black hole is compressed.
- **Galaxy** A large assembly of stars, numbering in the hundreds of millions to hundreds of billions, together with gas, dust, and other materials, all held together by the forces of mutual gravitation.
- **Cluster** A grouping of more than one galaxy.
- **Supercluster** A grouping of an enormous number of galaxies.
- **Quasar** (quasi-stellar object) A small, powerful source of energy believed to be the active core of very distant galaxies.

Review Questions

1. What are constellations?
2. Why does an observer at a given location see one set of constellations in the winter and a different set of constellations in the summer?

Birth of Stars

3. What is a protostar?
4. What process changes a protostar to a full-fledged star?
5. What are the outward forces that act on a star?
6. What are the inward forces that act on a star?
7. What do the outward and inward forces acting on a star have to do with its size?
8. Today the Earth contains many elements heavier than helium. Where did these elements originate?

Life of Stars

9. Compare the lifetimes of high- and low-mass stars.
10. How common are binaries in the universe?
11. What measurements of binaries provide data for calculating their masses?
12. What is the companion star to the hypothetical star *Nemesis*?
13. Where is most of the solar system's angular momentum?
14. What is the goal of SETI programs?

Death of Stars

15. What event marks the birth of a star, and what event marks its death?
16. When will our Sun reach the red-giant stage?
17. What is the relationship between a planetary nebula and a white dwarf?

18. What is the relationship between a white dwarf and a black dwarf?
19. What is the relationship between a white dwarf and a nova?
20. Is a white dwarf a former low-mass star or a former high-mass star?
21. What is the relationship between the heavy elements we find on the Earth today and supernovae?
22. When was the last supernova seen?
23. What is the relationship between a neutron star and a pulsar?

Black Holes

24. What is the relationship between an ordinary star and a black hole?
25. How far would the Sun have to collapse so that its light couldn't escape?
26. How does the mass of a star before collapse compare with the mass of the black hole it becomes?
27. Compare the circular orbit of light around a black hole with the circular orbit of a satellite about Planet Earth.
28. What is the relationship between the photon sphere and event horizon of a black hole?
29. What is the relationship between the event horizon and the surface of a black hole?
30. What is a black hole singularity?
31. What are the three possible properties of a black hole?
32. Are white holes facts or speculations?
33. Being that black holes are invisible, what is the evidence for their existence?

Galaxies

34. What type of galaxy is the Milky Way Galaxy?
35. What are the three types of galaxies?
36. What are the consequences of galaxies colliding?

Quasars

37. Which are brighter, galaxies or quasars?
38. What speeds do the red shifts of quasars indicate?
39. Are quasars thought to be relatively close to the Earth or relatively distant? Defend your answer.

Exercises

1. Why do we not see stars in the daytime?
2. Make a sketch similar to Figure 29.2 to show that a constellation seen in the background of a solar eclipse is one that will be seen six months later in the night sky.
3. Thomas Carlyle wrote, "Why did not somebody teach me the constellations and make me at home in the starry heavens, which are always overhead and which I don't half know to this day?" What beside the names of the constellations did he not know?
4. We see the constellations as distinct groups of stars. Discuss why they would look entirely different from some other location in the universe, far distant from the Earth.
5. The Big Dipper is sometimes right side up (can hold water) and sometimes upside down (cannot hold water). What length of time is required for the Dipper to change from one position to the other?
6. In what sense are we all made of stardust?
7. How is the gold in your sweetheart's ring evidence of ancient stars that ran their life cycles long before the solar system came into being?
8. Would you expect metals to be more abundant in old stars or new stars? Defend your answer.
9. Why is there a lower limit on the mass of a star?
10. What ordinarily keeps a star from collapsing?
11. How does the energy of a protostar differ from the energy that powers a star?
12. Why don't nuclear fusion reactions occur on the outer layers of stars?
13. Why are high-mass stars generally shorter lived than low-mass stars?
14. What is meant by the statement "The bigger they are, the harder they fall" with respect to stellar evolution?
15. Why won't the Sun be able to fuse carbon nuclei in its core?
16. Some stars contain fewer heavier elements than our Sun does. What does this indicate about the age of such stars relative to the age of our Sun?
17. Which has the highest surface temperature: red star, white star, or blue star?
18. The law of conservation of angular momentum states that the angular momentum (rotational inertia $\times$ rotational speed) of a star is the same before and after any event that does not involve external influences. How does this conservation principle explain why neutron stars spin so rapidly?
19. What does the spin rate of a star have to do with whether or not the star has a system of planets?
20. In what way is a black hole blacker than black ink?
21. If you were to fall into a black hole, you'd likely die as a result of tidal forces. Explain.
22. A black hole is no more massive than the star from which it collapsed. Why, then, is gravitation so intense near a black hole?
23. What happens to the radial distance of a black hole's photon sphere as more and more mass falls into the hole?
24. The band of light we call the Milky Way is far more prominent in July than in December. Suggest a reason.
25. Are there galaxies other than the Milky Way Galaxy visible to the unaided eye? Discuss.
26. How can collisions of galaxies affect their shapes?
27. What does it mean to say that galaxies are cannibals?
28. Quasars are the most distant objects we know of in the universe. Why do we therefore say their existence goes back to the earliest times in the universe?
29. In any direction one looks in the night sky, there is a star. This being true, why isn't the night sky ablaze with light?
30. The Andromeda Galaxy shows a blue shift in its spectra. What does this tell us about its motion relative to the Earth and the rest of the solar system?

CHAPTER

30

Relativity and the Universe

We began this book with the physics concepts of motion, force, and energy in the everyday world. In the chemistry chapters we studied these concepts in the micro world, and in the geology chapters they were applied to the processes of change on Planet Earth. Then we expanded discussion of these concepts to our solar system and the stars beyond. We conclude with the widest view of all—the universe—the whole show.

Except for an occasional exploding star here and there, earlier astronomers thought of the universe as an unchanging place. But today's scientific understanding describes a violently changing universe—one that had its beginnings in the violent fireball we call the Big Bang and that continues to change as space stretches out, carrying the galaxies with it. In this chapter we consider the evolving universe and its "fabric"—spacetime. We'll learn about the relationship between space and time developed early in this century by Albert Einstein in both his *special theory of relativity* and his *general theory of relativity.* Our treatment of the universe starts with its beginning.

30.1 The Big Bang

Compelling evidence suggests that the universe began 10–15 billion years ago, when a primordial explosion called the **Big Bang** occurred. Theorists generally believe that within the first three minutes after the Big Bang, great quantities of hydrogen and helium were created, spewing apart at great speeds. About 3 million years later, huge clouds of this matter, stretching 500 million light-years across, began to condense. After about 200 million years, these condensations formed the first galaxies—the birthplace of the stars and of elements heavier than hydrogen and helium. The universe today is the remnant of the Big Bang.

The concept of the Big Bang came into focus in the 1930s, after the American astronomer Edwin P. Hubble, for whom the Hubble Telescope is named, showed that the universe is expanding. Further findings implied that the cosmos was once concentrated in a very small, hot place at a definite time. The concept of the Big Bang holds that this special time was the *beginning of time.*

The space formed by the Big Bang was filled by intense, extremely energetic high-frequency radiation called the **primeval fireball**. Radiation from the dying embers of the primeval fireball now permeates all of space in the form of microwaves, which continuously stretch out more and more as the universe expands. These microwaves were inadvertently discovered in 1964–65 by Arno A. Penzias and Robert W. Wilson of Bell Laboratories while they were trying to rid their radio antenna of microwave noise. The microwave background radiation they discovered was predicted by Big Bang theory. Then in 1992, measurements of minuscule variations in the background radiation vindicated another prediction of Big Bang theory—that only such variations could account for the accumulation of matter to form galaxies.

That the universe is still expanding today is evident in a Doppler red shift in the light we receive from its galaxies. Recall from Section 10.10 that sound and light waves received by an observer are stretched out when a source recedes and compressed when a source approaches. The visible light we see from distant galaxies is stretched out, which means it is red-shifted and therefore indicates a receding light source. In other words, the red shift shows an increasing distance between us and other galaxies in the universe. This does not, however, place our Milky Way Galaxy in a central position. To see why not, consider a balloon with ants on it: As the balloon is inflated, every ant sees every other ant getting farther away, which certainly doesn't suggest a central position for each ant. In an expanding universe, any observer sees all other galaxies receding.

Fig. 30.1
Every ant on the expanding balloon sees every other ant getting farther away.

Red-shifted galaxy light shows us not only that the universe is expanding but also that it is expanding ever more slowly. Toss a rock skyward and it slows because of its gravitational interaction with the Earth. Similarly, every bit of matter blown apart in the primordial explosion is attracted by gravity toward every other bit of matter, resulting in a continuous slowing down of the expansion. Evidence for this slowdown is a greater red shift for the galaxies that are farthest from us. When we look at the stars and far-away galaxies, we are looking backward in time because it takes a considerable length of time for the light from those galaxies to reach all the way to the Earth. The galaxies farthest away are the ones we are seeing as they were longest ago. Because the degree of red shift indicates how quickly a light source is receding, the greater red shift of the farther galaxies shows that the universe was expanding faster in the beginning. The expansion of the universe is slowing with time.

Our analogies of the expanding balloon and the tossed rock give a visual picture of expansion and its slowing. An important point these analogies don't make is that the expansion of the universe is an expansion of space itself. The acts of blowing up a balloon or tossing a rock skyward are done in a surrounding space in some specified time—the balloon is blown against space, and the rock tossed into space. There is a space "waiting" for each event. Similarly with time. There was a time before, during, and after the balloon was blown and the rock tossed. But not so with the universe. The fundamental difference between the Big Bang and ordinary explosions is that with the Big Bang there was no space for the explosion to go into—space itself was exploding. The universe does not "exist" in space or in time; rather, both space and time "exist" within the universe. Without the universe, there would be no space and time. Space and time are in the universe, and not the other way around. The Big Bang was the expansion of space itself at the beginning of time.

To gain a perspective of the universe, we need to examine the relationship between space and time.

30.2 Spacetime

When we look at the stars, we realize we are looking backward in time. Some of the stars we see may have died long ago. The fact that we measure their distances in light-years indicates that space and time may be bound together. Einstein showed that space and time are indeed very intimately bound together.

The space we live in is three-dimensional. Because this is so, we can specify the location of any point in space with only three numbers. For example, imagine standing in one corner of a room in which a tiny spider hangs from the ceiling by a practically invisible web. We can specify the spider's location in the three-dimensional space of the room with only three numbers. The first is the number of meters to the spider measured along the line joining the floor and one of the walls adjacent to the spider. The second is the number of meters to the spider measured along the line joining the floor and the other wall adjacent to the spider. And the third is the number of meters the spider lies above the floor—measured along the vertical line joining the two walls at the corner formed by their intersection. Physicists speak of these three lines as the *coordinate axes* of a reference frame (Figure 30.2). Three numbers—the distances along the x- axis, the y- axis, and the z- axis—specify the position of any point in space.

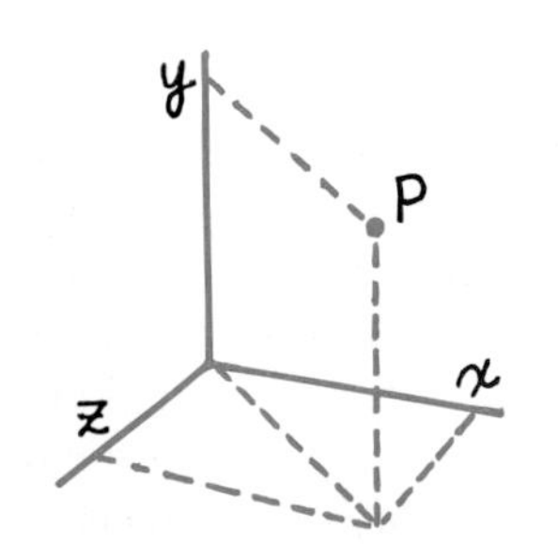

Fig. 30.2
Point *P* can be specified with three numbers: the distance along the *x-axis, y-axis,* and *z-axis.*

In addition to location, we also specify size with three dimensions. A box, for example, is described by its length, width, and height. But the three dimensions do not give a complete picture of the box because there is a fourth dimension—time. The box was not always a box of given length, width, and height. It became as a box only at a certain point in time, on the day it was made. Nor will it always be a box. At any moment it may be crushed, burned, or destroyed. So the three dimensions of space are a valid description of the box only during a certain period of time. We cannot speak meaningfully about space without implying time. Things exist in **spacetime**. Each atom, each object, each person, each planet, each star, each galaxy exists in "the spacetime continuum."[1]

[1] Points in spacetime may be quantized points in a four-dimensional spacetime lattice. From the sizes of elementary particles and minimum separation distances between colliding particles, there seems to be an elemental unit of distance (4.05×10^{-35} m), and the lifetimes of all known elementary particles are consistent with being an integral number of "chronons" (about 1.35×10^{-43} s).

30.3 Special Relativity

Einstein's concepts about space and time are part of a larger picture, a revolutionary one that predicts that motion through space causes time to slow down, that objects in motion are shorter and more massive than the same objects at rest, and that mass is actually congealed energy. These are the ideas of the **special theory of relativity**, developed by Einstein in 1905 and based upon two postulates. The first can be stated

Observers can never detect their *uniform* motion except relative to other objects.[2]

We've all noticed that from inside a car at a traffic light, we see a nearby car move only to find out it is at rest and we are moving—or vice versa. We can tell which is moving by the background of trees or other objects. Imagine, however, we are in a spaceship in interstellar space, and another spaceship coasts by at constant velocity. Which spaceship is moving? Without a background, all we can say is that the ships are moving relative to each other. And even with a background, how could we say the background wasn't moving? These are Newtonian ideas. Einstein thought about them and added his conclusion: that there is no experiment you can perform to decide which ship is moving and which is not. This means there is no such thing as absolute rest—all motion is relative.

If there is no experiment that can be performed to detect absolute motion through space, the laws of physics must be the same on both spaceships. The more general form of the first postulate is therefore

All laws of nature are the same in all uniformly moving reference frames.

On a jet airplane going 700 kilometers per hour, for example, coffee pours as it does when the plane is at rest; we swing a pendulum and it swings as it would if the plane were sitting on the runway.

According to the first postulate, any measurements of the speed of light give the same value in all uniformly moving reference frames. This constancy of the speed of light is the second postulate of special relativity:

The speed of light in free space has the same measured value for all observers, regardless of the motion of the source or the motion of the observer. The speed of light is a constant.

Every measurement of the speed of light, and there have been many, has confirmed the second postulate. Move away from a tossed baseball as you are waiting to catch it and you'll catch it at a slower speed. Do this for light and you have a different story. Imagine you're in a high-speed rocket moving away from a light source at nearly the speed of light. Good old common sense tells you that the light that catches up to you and passes you is slower than if you weren't moving. But not so. While you move, you in effect stretch out the space between you and the light source. But Einstein says you can't do that without also stretching out time. How much stretch? Just enough so that when you divide the space traveled by the time taken, your value for the speed of light is the same as if you weren't moving at all.

---

[2]By uniform motion, we mean nonaccelerated motion.

Clockwatching on a Trolleycar Ride

Imagine you are Einstein at the turn of the century, riding in a trolleycar moving away from a huge clock in the village square. The clock reads 12 noon. To say it reads 12 noon is to say that light that carries the information "12 noon" is reflected by the clock and travels toward you along your line of sight. If you suddenly move your head to the side, the light carrying the information, instead of meeting your eye, continues past, presumably out into space. Out there, an observer who *later* receives the light says, "Oh, it's 12 noon on the Earth now." But from your point of view it's now later than that. You and the distant observer see 12 noon at different times. You ponder this idea. If the trolleycar traveled as fast as the light, then the trolleycar would keep up with the light's information that says "12 noon." Traveling at the speed of light, then, tells you it's always 12 noon at the village square. In other words, time at the village square is frozen!

If the trolleycar is not moving, you see the village-square clock move into the future at the rate of 60 seconds per minute; if you move at the speed of light, you see seconds on the clock taking infinite time. These are the two extremes. What's in between? How would the advance of the clock's hands be viewed as you move at speeds less than the speed of light?

A little thought will show that you will receive the message "1 o'clock" anywhere from 60 minutes to an infinity of time after you receive the message "12 noon," depending on what your speed is between the extremes of zero and the speed of light. From your high-speed (but less than c) frame of reference, you see all events taking place in the reference frame of the clock (which is the Earth) as happening in slow motion. If you reverse direction and travel at high speed back toward the clock, you'll see all events taking place in the clock's reference frame as being speeded up. When you return and are once again sitting in the square, will the effects of going and coming compensate each other? Amazingly, no! Time will be stretched. The wristwatch you were wearing the whole time and the village clock will disagree. This is time dilation.

30.4 Time Dilation

Let's examine the notion that time can be stretched. Imagine that we are somehow able to observe a flash of light bouncing to and fro between a pair of parallel mirrors, like a ball bouncing to and fro between a floor and ceiling. If the distance between the mirrors is fixed, then the arrangement constitutes a *light clock* because the back-and-forth trips of the flash take equal time intervals (Figure 30.3). Suppose this light clock is inside a transparent high-speed spaceship. An observer who travels along with the ship and watches the light clock (Figure 30.4a) sees the flash reflecting straight up and down between the two mirrors, just as it would if the spaceship were at rest. This observer sees no unusual effects. Note that, because the observer is in the ship and so moving along with it, there is no relative motion between the observer and the light clock; we say that the observer and the clock share the same reference frame in spacetime.

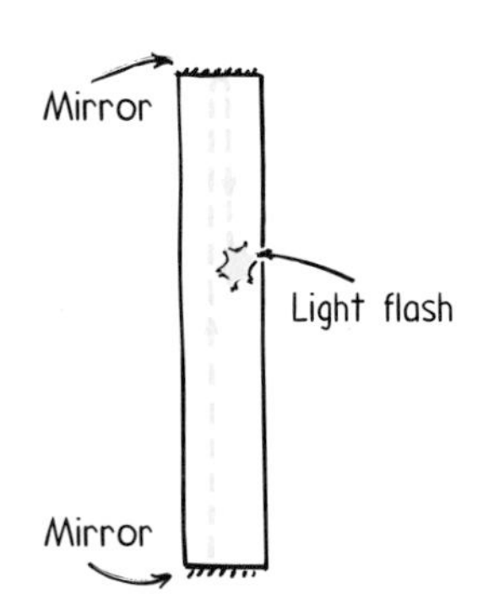

Fig. 30.3
A light clock. A parcel of light bounces up and down between parallel mirrors and "ticks off" equal intervals of time with each trip.

Now suppose we are standing on the ground as the spaceship whizzes by us at high speed—say, half the speed of light. Things are quite different from our reference frame, for we don't see the light path as being simple up-and-down motion. Because each flash moves horizontally while it moves vertically between the two mirrors, we see the flash follow a diagonal path. Notice in Figure 30.4b that from our Earthbound frame of reference the flash travels a *longer distance* as it makes one round-trip between the mirrors, considerably longer than the distance it travels in the reference frame of the observer riding along with the ship. Because the speed of light is the same in all reference frames (Einstein's second postulate), the flash must travel for a correspond-

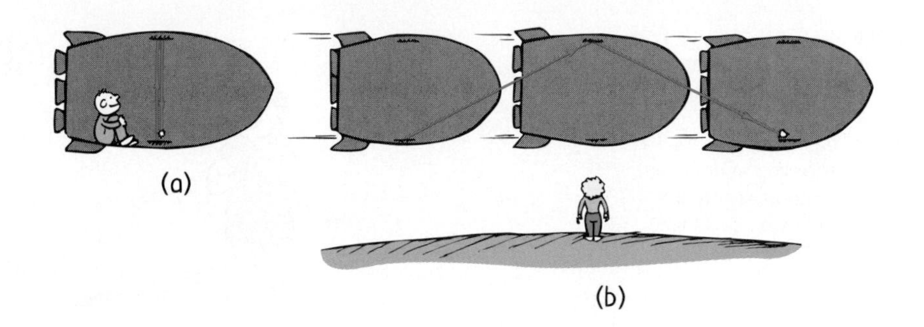

Fig. 30.4
(a) An observer moving with the spaceship observes the light flash moving vertically between the mirrors of the light clock.
(b) An observer who sees the moving ship pass by observes the flash moving along a diagonal path.

ingly longer time between the mirrors in our frame than in the reference frame of the on-board observer. This follows from the definition of speed—distance divided by time. *The longer diagonal distance must be divided by a correspondingly longer time interval to yield an unvarying value for the speed of light.* This stretching out of time is called **time dilation**.

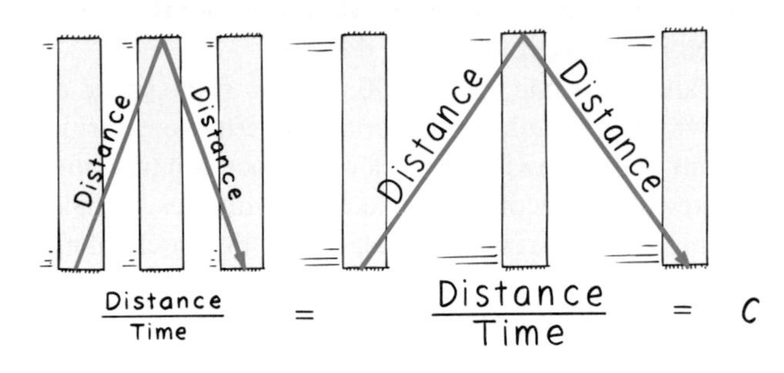

Fig. 30.5
The longer distance covered by the light flash in following the longer diagonal path on the right must be divided by a correspondingly longer time interval to yield an unvarying value for the speed of light.

We have considered a light clock in our example, but the same is true for any kind of clock. All clocks run more slowly when moving than when at rest. Time dilation has to do not with the mechanics of clocks but with the nature of time itself.

The relationship of time dilation for different frames of reference in spacetime can be derived from Figure 30.5 with simple geometry and algebra.[3] The relationship between the time t_0 (call it the *proper time*) in the observer's own frame of reference and t measured in another frame of reference (call it the *relative time*) is

$$t = \frac{t_0}{\sqrt{1-(v^2/c^2)}}$$

where v represents the relative velocity between observer and observed and c is the speed of light. Because no material object can travel at or beyond the speed of light, the

[3]The light clock is shown in three successive positions in the figure. The diagonal lines represent the path of the light flash as it starts from the lower mirror at position 1, moves to the upper mirror at position 2, and then back to the lower mirror at position 3. Distances on the diagram are marked ct, vt, and ct_0, which follows from the fact that distance traveled equals speed multiplied by time.

The symbol t_0 represents the time it takes the flash to move between mirrors as measured in a frame of reference fixed to the light clock. This is the time it takes for straight up-and-down motion. The speed of light is c, and the path of light is seen to move a vertical distance ct_0. This distance between mirrors is at right angles to the motion of the light clock and is the same in both reference frames.

The symbol t represents the time it takes the flash to move from one mirror to the other as measured from a frame of reference in which the light clock moves with speed v. Because the speed of the flash is c and the time it takes to go from position 1 to position 2 is t, the diagonal distance traveled is ct. During this time t, the clock (which travels horizontally at speed v) moves a horizontal distance vt from position 1 to position 2.

As the figure shows, these three distances make up a right triangle in which ct is the hypotenuse and ct_0 and vt are legs. A well-known theorem of geometry, the Pythagorean theorem, states that the square of the hypotenuse is equal to the sum of the squares of the two legs. If we apply this formula to the figure, we obtain

$$c^2t^2 = c^2t_0^2 + v^2t^2$$

$$c^2t^2 - v^2t^2 = c^2t_0^2$$

$$t^2[1 - (v^2/c^2)] = t_0^2$$

$$t^2 = \frac{t_0^2}{1 - (v^2/c^2)}$$

$$t = \frac{t_0}{\sqrt{1 - (v^2/c^2)}}$$

Fig. 30.6
When we see the rocket at rest, we see it traveling at the maximum rate in time: 24 hours per day. When we see the rocket traveling at the maximum rate through space (the speed of light), we see its time standing still.

ratio v/c is always less than 1; likewise for v^2/c^2. For $v = 0$, this ratio is zero, and for everyday speeds, where v is negligibly small relative to c, it's practically zero. In both these cases, $1 - (v^2/c^2)$ has a value of 1, as has $\sqrt{1 - v^2/c^2}$, and so we find $t = t_o$, which means the time intervals appear the same in both reference frames. For higher speeds, v/c is between 0 and 1, and $1 - (v^2/c^2)$ is less than 1; likewise for $\sqrt{1 - (v^2/c^2)}$. In these cases, t_o divided by a value less than 1 produces a value for t that is greater than t_o. In other words, the time interval t has been elongated, stretched, dilated.

To consider some numerical values, assume that v is 50 percent of the speed of light. Then we substitute $0.5c$ for v in the time-dilation equation and after some arithmetic find that $t = 1.15t_o$. This means that if we view a clock on a spaceship traveling past us at half the speed of light, we see the second hand take 1.15 minutes to make one revolution. If the same spaceship is at rest relative to us, we see the clock take 1 minute to make the same single revolution. If the spaceship passes us at 87 percent the speed of light, we find $t = 2t_o$, and we measure events on the spaceship taking twice the usual time intervals—hands on the ship clock turn only half as fast as those on our wristwatch. Put another way, events on the ship are seen in slow motion. At 99.5 percent the speed of light, $t = 10t_o$, we see the second hand of the spaceship's clock take 10 minutes to sweep through a revolution requiring 1 minute on our wristwatch.

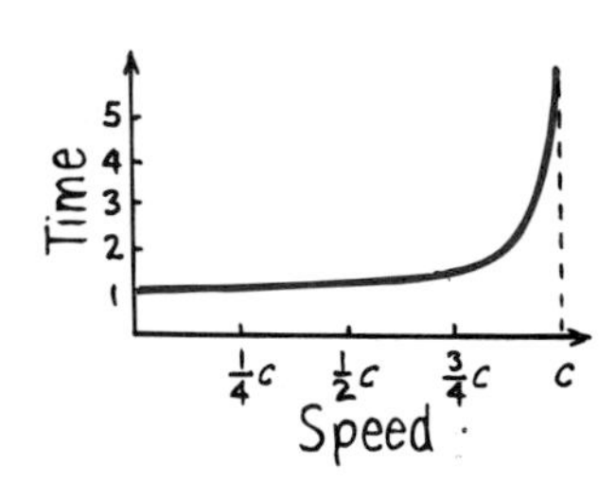

Fig. 30.7
The graph shows how 1 s on a stationary clock is stretched out when measured on a moving clock. Note that the stretching becomes significant only at speeds near the speed of light.

So, at 99.5 percent of c, the moving clock runs at one-tenth the rate of our wristwatch; it ticks only 6 seconds in the same time interval that our wristwatch ticks 60 seconds. At 87 percent of c, the moving clock ticks at half rate and shows 30 seconds to our 60 seconds. At 50 percent of c, the moving clock ticks $\frac{1}{1.15}$ as fast and ticks 52 seconds to our 60 seconds. Moving clocks run slow.

Nothing is unusual about a moving clock itself; it is simply ticking to the rhythm of a different time. The faster a clock moves, the slower it runs as viewed by an observer not moving with it. If it were possible to make a clock fly by us at the speed of light, the clock would appear to be not running at all because we measure the interval between ticks to be infinite. Time would be frozen and the clock would be ageless! If we could move with such an imaginary clock, however, the clock would not show any slowing of time at all. To us as we moved along with it the clock would be operating normally. This is because there would be no relative motion between observer and observed. The v in the time dilation equation would then be zero, and $t = t_o$; we and the clock would share the same reference frame in spacetime.

If a person whizzing past us checked a clock in our reference frame, she would find it to be running as slowly as we find hers to be. We each see the other's clock running slow. There is really no contradiction here, for it is physically impossible for two observers in relative motion to refer to one and the same realm of spacetime. All measurements made in one realm of spacetime need not agree with all measurements made in another realm of spacetime. The one measurement that all observers *always* agree on, however, is the speed of light.

Fig. 30.8
From the Earth frame of reference, light takes 25,000 years to travel from the center of our Milky Way Galaxy to our solar system. From the frame of reference of a high-speed spaceship, the trip takes less time. From the frame of reference of light itself, the trip takes no time. There is no time in a speed-of-light frame of reference.

Time dilation has been confirmed in the laboratory innumerable times with particle accelerators. The lifetimes of fast-moving radioactive particles increase as the speed goes up, and the amount of increase is just what Einstein's equation predicts.

Time dilation has also been confirmed for not-so-fast motion. In 1971, to test Einstein's theory, four cesium-beam atomic clocks were twice flown on regularly scheduled commercial jet flights around the world, once eastward and once westward. The clocks indicated different times after their round-trips. Relative to the atomic time scale of the U.S. Naval Observatory, the observed time differences, in billionths of a second, were in accord with Einstein's prediction. Now, with atomic clocks orbiting the Earth as part of the global positioning system, adjustments for the effects of time dilation are essential in order to use signals for the clocks to pinpoint locations on the Earth.

This all seems very strange to us only because it is not our common experience to deal with measurements made at speeds close to the speed of light or atomic-clock-type measurements at ordinary speeds. Due to this inexperience, the theory of relativity does not make common sense. But common sense, according to Einstein, is that layer of prejudices laid down in the mind prior to the age of 18. If we spent our youth zapping through the universe in high-speed spaceships, we would probably be quite comfortable with the results of relativity.

■ Questions

1. If you are moving in a spaceship at a high speed relative to the Earth, do you notice a difference in your pulse rate? In the pulse rate of the people back on the Earth?
2. Will observers A and B agree on measurements of time if A moves at half the speed of light relative to B? If both A and B move together at half the speed of light relative to the Earth?
3. Does time dilation mean that time really passes more slowly in moving systems or that it only seems to pass more slowly?

Adding Velocities

Most people know that if you walk at 1 kilometer per hour along the aisle of a train that moves at 60 kilometers per hour, your speed relative to the ground is 61 kilometers per hour if you walk in the same direction as the moving train and 59 kilometers per hour if you walk in the opposite direction. What most people know is *almost* correct. Taking special relativity into account, these speeds are *very nearly* 61 kilometers per hour and 59 kilometers per hour.

■ Answers

1. There is no relative speed between you and your pulse because the two share the same frame of reference. Therefore you notice no relativistic effects in your pulse. There is, however, a relativistic effect between you and people back on the Earth. You find their pulse rate slower than normal (and they find your pulse rate slower than normal). Relativity effects are always attributed to the other guy.
2. When A and B move relative to each other, each observes a slowing of time in the other's frame of reference. So they do not agree on measurements of time. When they are moving in unison, they share the same frame of reference and agree on measurements of time. They see each other's time as passing normally, and they each see events on the Earth in the same slow motion.
3. The slowing of time in moving systems is not merely an illusion resulting from motion. Time really does pass more slowly in a moving system relative to one at relative rest. (This is dramatically shown in "The Twin Trip" in the *Conceptual Physical Science Practice Book.*)

For everyday objects in uniform (nonaccelerating) motion, we ordinarily combine velocities by the simple rule

$$V = v_1 + v_2$$

But this rule does not apply to light, which always has the same velocity c. Strictly speaking, the preceding rule is an approximation of the relativistic rule for adding velocities. We'll not treat the long derivation but simply state the rule:

$$V = \frac{v_1 + v_2}{1 + \frac{v_1 v_2}{c^2}}$$

The numerator of this formula makes sense. But this simple sum of two velocities is altered by the second term in the denominator, which is significant only when both v_1 and v_2 are not much less than c.

As an example, consider a spaceship that is moving away from you at a velocity of $0.5c$ and fires a rocket. This rocket moves in the same direction as the ship at a speed of $0.5c$ relative to the ship. How fast does the rocket move relative to you? The nonrelativistic rule says that the rocket moves at the speed of light in your reference frame. But in fact,

$$V = \frac{0.5c + 0.5c}{1 + \frac{0.25c^2}{c^2}} = \frac{c}{1.25} = 0.8c$$

which illustrates another consequence of relativity: No material object can travel as fast as or faster than light.

Suppose the spaceship instead fires a pulse of laser light in its direction of travel. How fast does the pulse move in your frame of reference?

$$V = \frac{0.5c + c}{1 + \frac{0.5c^2}{c^2}} = \frac{1.5c}{1.5} = c$$

No matter what the relative velocities between two frames, light moving at c in one frame is seen to be moving at c in any other frame. Try chasing light—you will never catch it.

Space Travel

One of the old arguments against the possibility of human interstellar travel was that our life span is too short. It was argued, for example, that the star nearest the Earth (after the Sun), Alpha Centauri, is 4 light-years away, and a round-trip even at the speed of light would require 8 years. And even a speed-of-light voyage to the center of our galaxy, 25,000 light-years distant, would require 25,000 years. But these arguments fail to take into account time dilation. Time for a person on the Earth and time for a person in a high-speed rocket ship are not the same.

A person's heart beats to the rhythm of the realm of spacetime in which that heart finds itself. And one realm of spacetime seems the same as any other to the heart but not to an observer who stands outside the heart's frame of reference. For example, astronauts traveling at 99 percent of c could go to the star Procyon (10.4 light-years distant) and back in 21 Earth-years. Because of time dilation, however, it would seem to the astronauts that only 3 years had gone by. This is what all their clocks would tell them—and biologically they would be only 3 years older. It would be the space officials greeting them on their return who would be 21 years older!

Century Hopping

Let's speculate for a moment about a possible time in the future when the prohibitive problems of energy supplies and of radiation have been overcome and space travel is a routine experience. People will have the option of taking a trip and returning to any future century of their choosing. For example, one might depart from the Earth in a high-speed ship in the year 2100, travel for 5 years or so, and return in the year 2500. One could live among the Earthlings of that period for a while and depart again to try out the year 3000 for style.

People could keep jumping into the future with some expense of their own time—but they could not trip into the past. They could never return to the same era on the Earth that they bid farewell to. Time, as we know it, travels one way—forward. Here on the Earth we move constantly into the future at the steady rate of 24 hours per day. A deep-space astronaut leaving on a deep-space voyage must live with the fact that, upon return, much more time will have elapsed on the Earth than the astronaut has subjectively and physically experienced during voyage. The credo of all star travelers, whatever their physiological condition, will be permanent farewell.

At higher speeds the results are even more impressive. At a rocket speed of 99.99 percent of *c,* travelers could travel a distance of slightly more than 70 light-years in a single year of their own time; at 99.999 percent of *c,* this distance would be pushed appreciably farther than 200 light-years. A 5-year trip for them would take them farther than light travels in 1000 Earth-years!

Present technology does not permit such journeys. Getting enough propulsive energy and shielding against radiation are both prohibitive problems. Spaceships traveling at relativistic speeds would require billions of times the energy used to put a space shuttle into orbit. Even some kind of interstellar ramjet that scooped up interstellar hydrogen gas for burning in a fusion reactor would have to overcome the enormous retarding effect of scooping up the hydrogen at high speeds. And the space travelers would encounter interstellar particles just as if they had a large particle accelerator pointed at them. No way of shielding such intense particle bombardment for prolonged periods of time is presently known. The practicalities of such space journeys are enormously prohibitive. So for the time being, interstellar space travel must be relegated to science fiction.

30.5 Length Contraction

As objects move through spacetime, space as well as time changes. In a nutshell, the objects get shorter when they move by us at relativistic speeds. This **length contraction** was first proposed by the physicist George F. FitzGerald and mathematically expressed by another physicist, Hendrick A. Lorentz. Whereas these physicists hypothesized that matter contracts, Einstein saw that what contracts is space itself. Nevertheless, because Einstein's formula is the same as Lorentz's, we call the effect the *Lorentz contraction:*

$$L = L_o\sqrt{1 - (v^2/c^2)}$$

where v is the relative velocity between observed and observer, c is the speed of light, L is the measured length of the moving object, and L_o (called the *proper length*) is the measured length of the object at rest.

Suppose an object is at rest so that $v = 0$. When we substitute $v = 0$ in the Lorentz equation, we find $L = L_0$, as we would expect. When we substitute various high values of v in the Lorentz equation, we begin to see the L term get shorter and shorter. At 87 percent of c, an object would be contracted to half its original length. At 99.5 percent of c, it would contract to one-tenth its original length. If the object were somehow able to move at c, its length would be zero. This is one of the reasons we say that the speed of light is the upper limit for the speed of any moving object. A ditty popular with science types is

There was a young fencer named Fisk,
Whose thrust was exceedingly brisk.
So fast was his action
The Lorentz contraction
Reduced his rapier to a disk.

As Figure 30.9 indicates, contraction takes place only in the direction of motion. If an object is moving horizontally, no contraction takes place vertically.

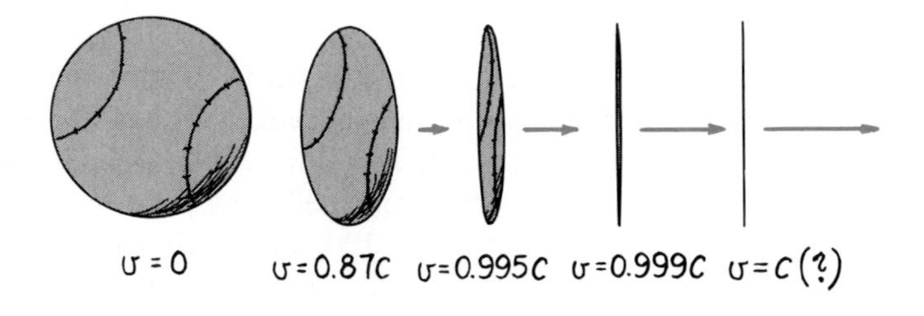

Fig. 30.9
The Lorentz contraction. As speed increases, length in the direction of motion decreases. Lengths in the direction perpendicular to the motion do not change.

Length contraction should be of considerable interest to space voyagers. The center of our Milky Way Galaxy is 25,000 light-years away. Does this mean that if we traveled in that direction at the speed of light it would take 25,000 years to get there? From an Earth frame of reference, yes, but to the space voyagers, decidedly not! At the speed of light, the 25,000 light-year distance would be contracted to no distance at all. Space voyagers would arrive there instantly![4]

Space voyagers will have to travel at speeds less than the speed of light. Nevertheless, at very high speeds, distances such as those from one galaxy to another will be shortened. If space voyagers are ever able to boost themselves to relativistic speeds, perhaps no part of the universe will be inaccessible to them.

Fig. 30.10
The meter stick is measured to be half as long when traveling at 87% of the speed of light relative to the observer.

Question

A rectangular billboard in space has the dimensions 10 m × 20 m. How fast and in what direction with respect to the billboard would a space traveler have to pass for the billboard to appear square?

Answer
For the billboard to be seen as a square, the 20-m side must contract to half that length—10 m. As we have learned, a speed of 87% of c produces a contraction of half. So the space traveler would have to travel at $0.87c$ in a direction parallel to the longer side of the board.

[4]Did songwriter Leon Russell have this in mind when he sang, "I'll love you in a place where there's no space and time; I'll love you forever, you're a friend of mine?"

30.6 Relativistic Momentum

Recall our study of momentum in Chapter 3. We learned that the change in momentum of an object is equal to the impulse Ft applied to it: $Ft = \Delta mv$ or $Ft = \Delta p$, where $p = mv$. Apply more impulse to an object that is free to move and the object acquires more momentum. Double the impulse and the momentum doubles. Apply 10 times the impulse and the object gains 10 times as much momentum. Does this mean that momentum can increase without any limit? The answer is yes. Does this mean that speed can also increase without any limit? The answer is no! As previously said, nature's speed limit for material objects is c.

To Newton, infinite momentum meant infinite speed. But not so in relativity. Einstein showed that a new definition of momentum is required:

$$p = \frac{mv}{\sqrt{1 - (v^2/c^2)}}$$

where again we see the familiar term in the denominator, the Lorentz factor. This generalized definition of momentum is valid in all uniformly moving (nonaccelerating) reference frames. *Relativistic momentum* is larger than mv by a factor of $1/\sqrt{1-(v^2/c^2)}$. For everyday speeds much less than c, this factor is nearly equal to 1, and so p is nearly equal to mv. Put another way, Newton's definition of momentum is valid at low speed.

At higher speeds the Lorentz factor grows dramatically, and so does relativistic momentum. As speed approaches c, momentum approaches infinity! No matter how close to c an object is pushed, it still requires infinite impulse to give it the last bit of speed needed to reach c—clearly impossible. Hence we see that no body that has mass can be pushed to the speed of light, much less beyond it.

Today, subatomic particles are routinely pushed to nearly the speed of light in particle accelerators. The momenta of such particles may be thousands of times more than the Newtonian expression mv predicts. The particles behave as if their mass increases with speed. In this case we find

$$m = \frac{m_0}{\sqrt{1-(v^2/c^2)}}$$

Here m represents the measured mass of an object pushed to any speed v. The symbol m_0 is the *rest mass,* the mass the object has at rest. Again, v represents the relative velocity between observer and observed.

Einstein initially favored this interpretation but later changed his mind in order to keep mass a constant—which means it is a property of matter that is the same in all frames of reference. So spacetime changes with speed, but mass doesn't.

The increased momentum of a high-speed particle is evident in the increased "stiffness" of its trajectory. The more momentum it has, the stiffer its trajectory and the harder it is to deflect. We see this when a beam of electrons is directed into a magnetic field. Charged particles moving in a magnetic field experience a force that deflects them from their normal paths. For small momentum, the path curves sharply. For large momentum, there is greater stiffness and the path curves only a little (Figure 30.11). Even though one particle may be moving only a little faster than another one—say 99.9 percent of the speed of light instead of 99 percent—its momentum is considerably greater and it follows a straighter path in the magnetic field. This stiffness must be compensated for in circular accelerators like cyclotrons and synchrotrons, where momentum dictates the radius of curvature. Physicists working with subatomic particles

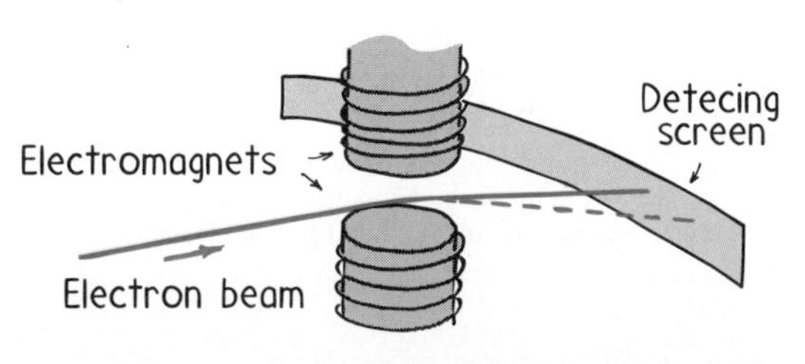

Fig. 30.11
If the momentum of the electrons were equal to the Newtonian value mv, the beam would follow the dashed line. But because the relativistic momentum is greater than the normal value, the beam follows the "stiffer" trajectory shown by the solid line.

at atomic accelerators verify every day the correctness of the relativistic definition of momentum and the speed limit imposed by nature.

So we see that as the speed of an object approaches the speed of light, its momentum approaches infinity—which means there is no way that the speed of light can be reached. There is, however, at least one thing that reaches the speed of light—light itself! But a photon of light is massless, and the equations that apply to it are different. Light travels always at the same speed. So interestingly, a material particle can never be brought to the speed of light, and light can never be brought to rest.

30.7 Mass, Energy, and $E = mc^2$

Einstein linked not only space and time but also mass and energy. A piece of matter, even at rest and not interacting with anything else, has an "energy of being." This is called its *rest energy.* Einstein concluded that it takes energy to make mass and that energy is released if mass disappears. The amount of energy E is related to the amount of mass m by the most celebrated equation of the 20th century:[5]

$$E = mc^2$$

The c^2 is the conversion factor between energy units and mass units.[6] Because of the large magnitude of c, a small mass corresponds to an enormous quantity of energy.

So when gravitation crunches mass in the Sun and other stars and ignites thermonuclear fusion, the energy that emerges is accompanied by a corresponding lowering of mass—but only a tiny bit, about one part in a thousand. The helium nucleus produced by the fusion of a pair of deuterium nuclei is about one-thousandth less massive than the two deuterium nuclei. Sunlight, then, is this small amount of mass transformed by thermonuclear fusion into radiant energy.

The relationship $E = mc^2$ is not limited to nuclear processes. It holds for all interactions of mass and energy, including all chemical reactions. When you strike a match, for example, phosphorous atoms in the match head rearrange themselves and combine with oxygen in the air to form new molecules. The resulting molecules have very slightly less mass than the separate phosphorus and oxygen molecules. For all reactions that give off energy, there is a corresponding decrease in mass. For chemical reactions the decrease in mass is about one part in a billion. For nuclear reactions, about one part in a thousand. The difference between chemical and nuclear reactions is mainly one of scale.

The transformation can go the other way. Radiant energy can be transformed into mass. The first direct proof of this was found in 1932 in a photograph emulsion used in high-altitude balloons to catch cosmic rays. C. D. Anderson, an American physicist, found that a photon of gamma radiation that had entered the emulsion had transformed into a pair of particles. This constituted direct conversion of radiant energy to mass.

---

[5] If you take a follow-up physics or astronomy course, you will likely encounter this equation in your next textbook as $E_o = mc^2$. This version emphasizes that it is an object's rest energy E_o, not its total energy, that is equal to its mass. Rest energy, the energy an object has just by existing, is nonzero even if the object's kinetic and potential energy *are* zero. That nonzero energy equals the object's mass.

[6] When c is in meters per second and m is in kilograms, E is in joules. If the equivalence of mass and energy had been understood long ago when physics concepts were first being formulated, there would probably be no separate units for mass and energy. Furthermore, with a redefinition of space and time units, c could equal 1 and then $E = mc^2$ would simply be $E = m$.

Later experiments showed that the particles he observed were most likely an electron and its *antiparticle,* called a *positron.* As we learned in Section 14.6, a positron has the same mass and spin as an electron but opposite charge. In other words, it is positively charged. When an antiparticle is formed, it doesn't last long. It soon encounters a particle that is *its* antiparticle, and the two annihilate each other, sending out a pair of gamma rays in the process.

The equation $E = mc^2$ is more than a formula for the conversion of mass into pure energy, or vice versa. It states that energy and mass *are the same thing.* Mass is simply congealed energy. If you want to know how much energy is in a system, measure its mass. For an object at rest, its energy is its mass. It is energy itself that is hard to shake.

Question

Can we look at the equation $E = mc^2$ another way and say that matter transforms into pure energy when it is traveling at the speed of light squared?

30.8 General Relativity

The special theory of relativity is about uniform motion, which is why it is called special. Einstein's conviction that the laws of nature should be expressed in the same form in every frame of reference, accelerated as well as nonaccelerated, was the primary motivation that led him to the **general theory of relativity**—a new theory of gravitation, where gravity causes space to become curved and time to slow down.

Einstein was led to this new theory of gravity by thinking about observers in accelerated motion. Long before there were real spaceships, Einstein could imagine himself in a vehicle far away from gravitational influences. In such a spaceship either at rest or in uniform motion relative to the distant stars, he and everything within the ship would float freely; there would be no up and no down. But if rocket motors were turned on and the ship accelerated, things would be different; phenomena similar to gravity would be observed. The wall adjacent to the rocket motors would push up against any occupants and become the floor, while the opposite wall would become the ceiling. Occupants in the ship would be able to stand on the floor and even jump up and down. If the acceleration of the spaceship were equal to g, the occupants could well be convinced the ship was not accelerating at all but rather was at rest on the surface of the Earth.

Fig. 30.12
(a) Everything is weightless on the inside of a nonaccelerating spaceship far away from gravitational influences. (b) When the spaceship accelerates, an occupant inside feels "gravity."

Answer

No, no, no! As matter is propelled faster and faster, its mass increases rather than decreases. As its speed approaches c, its mass approaches infinity. At the same time, its energy approaches infinity. It has more mass and more energy. Matter cannot be made to move at the speed of light, let alone the speed of light squared (which is not a speed!). The equation $E = mc^2$ simply means that energy and mass are two sides of the same coin.

Einstein concluded that gravity and motion through spacetime are related, a conclusion now called the *principle of equivalence:*

Local observations made in an accelerated frame of reference cannot be distinguished from observations made in a Newtonian gravitational field.

To examine this new "gravity" in an accelerating spaceship, let's consider the consequence of dropping two balls inside the spaceship, one of wood and the other of lead. When the balls are released, they continue to move upward side by side with the velocity the ship had at the moment of release. If the ship were moving at *constant velocity* (zero acceleration), the balls would remain suspended in the same place because they and the ship move the same distance in any given time interval. But because the ship is accelerating, the floor moves upward faster than the balls, with the result that the floor soon catches up with the balls (Figure 30.13). Both balls, regardless of their mass, meet the floor at the same time. Remembering Galileo's demonstration at the Leaning Tower of Pisa, occupants of the ship might be prone to attribute their observations to the force of gravity.

Fig. 30.13
To an observer inside the accelerating ship, a lead ball and a wood ball appear to fall together when released.

This equivalence of an accelerated reference frame and gravity would be interesting but not revolutionary if it applied only to mechanical phenomena, but Einstein went further and stated that the principle holds for *all* natural phenomena, including optical and electromagnetic ones.

Just as a tossed ball curves in a gravitational field, so does a light beam. To understand this, first consider a ball thrown sideways in a stationary spaceship in a gravity-free region. The ball follows a straight-line path relative both to an observer inside the ship and to a stationary observer outside the spaceship. But if the ship is accelerating, the floor overtakes the ball just as in our previous example. An observer outside the ship still sees a straight-line path, but to an observer in the accelerating ship, the path is curved; it is a parabola (Figure 30.15). The same holds true for a beam of light (Figure 30.16). The only difference is the curvatures of both. If the ball were somehow thrown at the speed of light, the two curvatures would be the same.

According to Newton, tossed balls curve because of a force of gravity. According to Einstein, tossed balls and light curves not because of any force but because the spacetime in which they travel is curved.

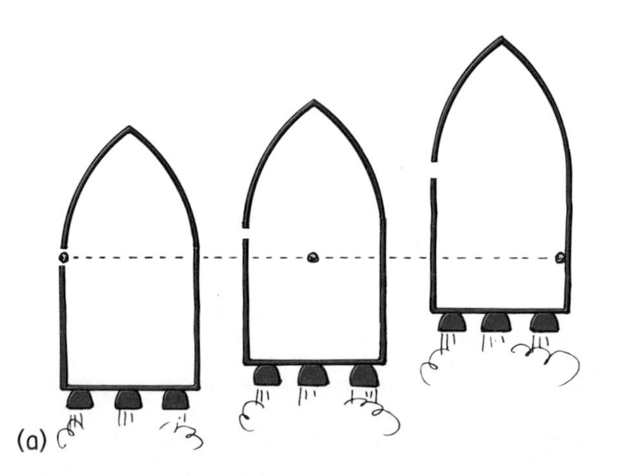

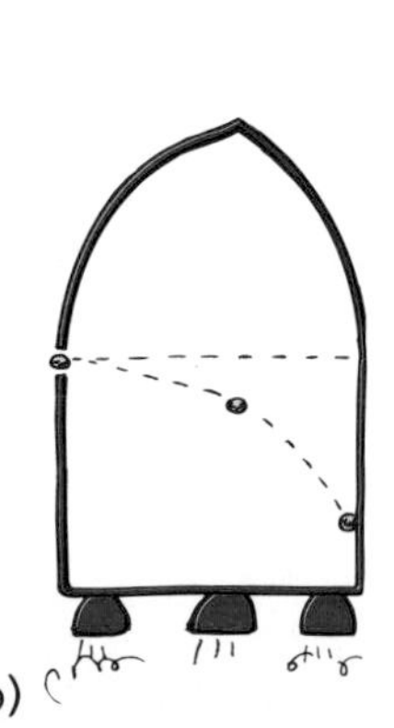

Fig. 30.14
(a) An outside observer sees a horizontally thrown ball travel in a straight line. Because the ship is moving upward while the ball travels horizontally, the ball strikes the wall below a point opposite the window.
(b) To an inside observer, the ball bends as if in a gravitational field.

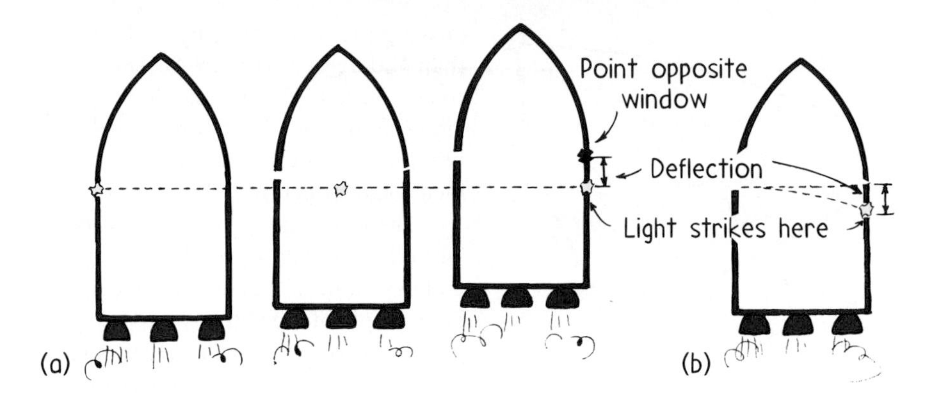

Fig. 30.15
(a) An outside observer sees light travel horizontally in a straight line, but like the ball in Figure 30.14, it strikes the wall slightly below a point opposite the window. (b) To an inside observer, the light bends as if responding to a gravitational field.

Fig. 30.16
The trajectory of a flashlight beam is identical to the trajectory of a baseball "thrown" at the speed of light. Both paths curve equally in a uniform gravitational field.

Question

Whoa! We learned previously that the pull of gravity is an interaction between masses. And we learned that light has no mass. Now we say that light can be bent by gravity. Isn't this a contradiction?

Gravity, Space, and a New Geometry

Einstein perceived a gravitational field as a geometrical warping of four-dimensional spacetime. Four-dimensional geometry is altogether different from the three-dimensional Euclidean geometry taught in high school. The laws of Euclidean geometry are not valid when applied to objects in the presence of strong gravitational fields.

The familiar rules of Euclidean geometry pertain to various figures you can draw on a flat surface: The ratio of the circumference of a circle to its diameter is equal to π; all the angles in a triangle add up to 180 degrees; the shortest distance between two points is a straight line. Draw these figures on a curved surface— on a sphere, say, or a saddle-shaped object—and the Euclidean rules no longer hold (Figure 30.17). When you measure the sum of the angles of a triangle in space, you call the space flat if the sum is equal to 180 degrees, spherelike or positively curved if the sum is larger than 180 degrees, and saddlelike or negatively curved if it is less than 180 degrees.

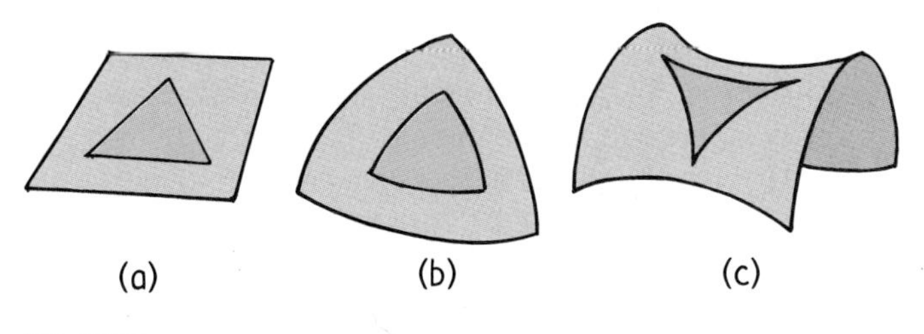

Fig. 30.17
The sum of the angles of a triangle depends on which kind of surface the triangle is drawn on. (a) On a flat surface, the sum is 180°. (b) On a spherical surface, the sum is greater than 180°. (c) On a saddle-shaped surface, the sum is less than 180°.

Of course, the lines forming the triangles in Figure 30.17 are not all "straight" from a three-dimensional point of view, but they are the "straightest," or *shortest,* distances between two points if we are confined to the curved surface. These lines of shortest distance are called *geodesic lines* or simply **geodesics**.

The path of a light beam follows a geodesic. Suppose three experimenters—one on the Earth, one on Venus, and one on Mars—measure the angles of a triangle formed by light beams traveling between these three planets. The light beams bend when

---

Answer
There is no contradiction when the mass-energy equivalence is understood. It's true that light has no mass, but it is not "energyless." The fact that gravity deflects light is evidence that gravity pulls on the energy of light. Energy indeed is equivalent to mass!

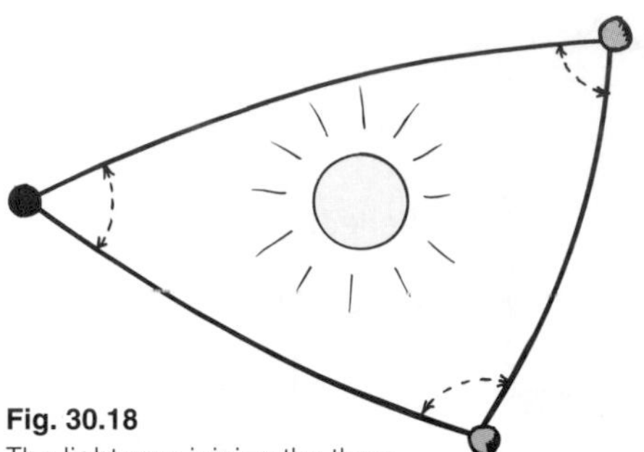

Fig. 30.18
The light rays joining the three planets form a triangle. Since light passing near the Sun bends, the sum of the angles of the resulting triangle is greater than 180°.

passing the Sun, resulting in the sum of the three angles being larger than 180 degrees (Figure 30.18). So the space around the Sun is positively curved. The planets that orbit the Sun travel along four-dimensional geodesics in this positively curved spacetime. Freely falling objects, satellites, and light rays all travel along geodesics in four-dimensional spacetime.

The whole universe may have an overall curvature. If it is negatively curved, it is open-ended like the saddle in Figure 30.17c and extends without limit. If it is positively curved, it closes in on itself. One familiar example of a positively curved space is the surface of the Earth. Our planet forms a closed curvature, so that if you travel along a geodesic, you come back to your starting point. Similarly, if the universe is positively curved, it is closed, so that if you could look infinitely into space through an ideal telescope, you would see the back of your own head! (This is assuming that you waited a long enough time or that light traveled infinitely fast.)

General relativity, then, calls for a new geometry: Rather than space simply being a region of nothingness, space is a flexible medium that can bend and twist. How it bends and twists describes a gravitational field. General relativity is a geometry of curved four-dimensional spacetime.[7] The mathematics of this geometry is too formidable to present here. The essence, however, is that the presence of mass produces the curvature, or *warping,* of spacetime. By the same token, a curvature of spacetime reveals itself as mass. Instead of visualizing gravitational forces between masses, we abandon altogether the notion of force and instead think of masses responding in their motion to the warping of the spacetime they inhabit. It is the bumps, depressions, and warpings of geometrical spacetime that *are* the phenomena of gravity.

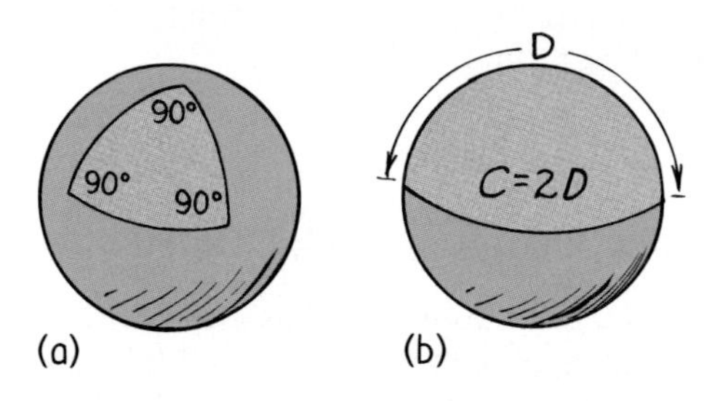

Fig. 30.19
The geometry of the curved surface of the Earth differs from the Euclidean geometry of flat space. (a) The sum of the angles for an equilateral triangle in which each side equals $\frac{1}{4}$ the Earth's circumference is clearly greater than 180°. (b) The Earth's circumference is only twice its diameter instead of π = 3.14 times its diameter. Euclidean geometry is also invalid in curved space.

We cannot visualize the four-dimensional bumps and depressions in spacetime because we are three-dimensional beings. We can get a glimpse of this warping by considering a simplified analogy in two dimensions: a heavy ball resting on the middle of a waterbed. The more massive the ball, the greater it dents the two-dimensional surface. A marble rolled across a small-dented waterbed may trace an elliptical curve and orbit the ball. The planets that orbit the Sun similarly travel along four-dimensional geodesics in the warped spacetime about the Sun.

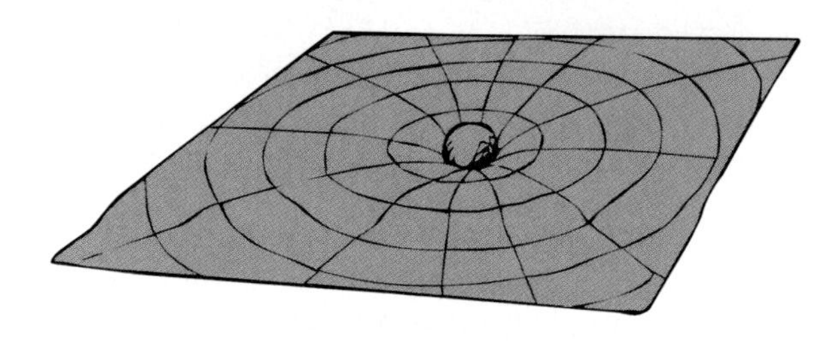

Fig. 30.20
A two-dimensional analogy of four-dimensional warped spacetime. Spacetime near a star is curved in a way similar to the surface of a waterbed when a heavy ball rests on it.

Tests of General Relativity

Using four-dimensional field equations, Einstein recalculated the orbits of the planets about the Sun. Beyond the planets, space is almost flat and objects travel along nearly

[7]Don't be discouraged if you cannot visualize four-dimensional spacetime. Einstein himself often told his friends, "Don't try. I can't do it either." Perhaps we are not too different from the great thinkers of Galileo's time who couldn't imagine a moving Earth!

Gravitational Waves

Every object has mass and therefore warps the surrounding spacetime. When an object moves, the surrounding warp moves in order to readjust to the new position. These readjustments produce ripples in the overall geometry of spacetime. This is similar to moving a ball that rests on the surface of a waterbed. A disturbance ripples across the waterbed surface in waves; if we move a more massive ball, then we get a greater disturbance and the production of even stronger waves. Similarly for spacetime in the universe: Ripples travel outward from a gravitational source at the speed of light and are *gravitational waves.*

Any accelerating object produces a gravitational wave. In general, the more massive the object and the greater its acceleration, the stronger the resulting gravitational wave. But even the strongest waves produced by ordinary astronomical events are extremely weak—the weakest known in nature. For example, the gravitational waves emitted by a vibrating electric charge are a trillion trillion trillion times weaker than the electromagnetic waves emitted by the same charge. Detecting gravitational waves is enormously difficult, and no confirmed detection has occurred to date. A new generation of wave detectors, soon to be built, is expected to detect gravitational waves from supernovae, where as much as 0.1 percent of their mass may be radiated away as gravitational waves.

As weak as they are, gravitational waves are everywhere. Shake your hand back and forth: You have just produced a gravitational wave. It is not very strong, but it exists.

Fig. 30.21
A precessing elliptical orbit.

straight-line paths. Near the Sun, however, planets and comets travel along curved paths because of the curvature of space. With only one exception, Einstein's calculations gave the same results as those obtained using Newton's law of gravity. The exception had to do with orbital precession (Figure 30.21).

Precession in the orbits of planets caused by other planets was well known. Since the early 1800s astronomers had measured a precession of Mercury's orbit—about 574 seconds of arc per century. Perturbations by the other planets were found to account for all but 43 seconds of arc per century. Even after all known corrections due to possible perturbations by other planets had been applied, the calculations of physicists and astronomers failed to account for the extra 43 seconds of arc in Mercury's precession. Before Einstein, the most plausible explanations were that either Venus was extra massive or else a never-discovered other planet (called Vulcan) was pulling on Mercury. Then came the explanation of Einstein, whose equations showed a precession in addition to that caused by planetary perturbation—precession due to the warpage of spacetime. Einstein's equations applied to Mercury accounted for the extra 43 seconds of arc per century!

As a second test of his theory, Einstein predicted that measurements of starlight passing close to the Sun would be deflected by an angle of 1.75 seconds of arc—large enough to be measured. This deflection of starlight can be observed during an eclipse of the Sun. (Measuring this deflection has become a standard practice at every total eclipse since the first measurements were made during the total eclipse of 1919.) A photograph taken of the darkened sky around the eclipsed Sun reveals the presence of nearby bright stars. The positions of the stars are compared with those in other photographs of the same area taken at other times with the same telescope. In every instance, the deflection of starlight has supported Einstein's prediction (Figure 30.22)!

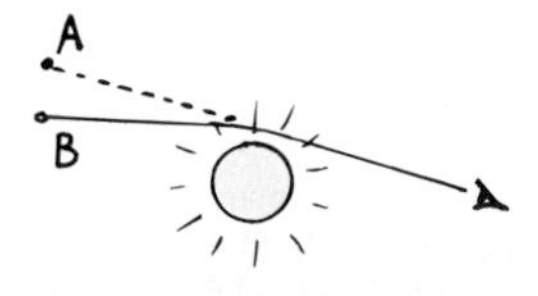

Fig. 30.22
Starlight bends as it passes close to the Sun. Point A shows the apparent position of the star emitting the light; point B shows the true position.

Question

Why do we not notice the bending of light by gravity in our everyday environment?

Fig. 30.23
If you move from a distant point down to the surface of the Earth, you move in the direction that the gravitational force acts—toward a location where clocks run more slowly. A clock at the surface of the Earth runs slower then a clock farther away.

Einstein made a third prediction—that gravity causes clocks to run slow. He predicted that clocks on the first floor of a building should tick slightly slower than clocks at the top floor, which are farther from the Earth and therefore in a weaker gravitational field (actually weaker gravitational *potential*). From the top to the bottom of the tallest skyscraper, the difference is very small—only a few millionths of a second per decade—because the difference in the Earth's gravitational potential at the bottom and top of the skyscraper is very small. For larger differences, like those at the surface of the Sun compared with the surface of the Earth, the clock-slowing effect is more pronounced. A clock at the surface of the Sun should run measurably slower than a clock at the surface of the Earth. Einstein suggested a way to measure this.

Every atom emits light at a specific frequency characteristic of the vibrational rate of the electrons in the atom. Every atom is therefore a "clock," and a slowing down of atomic vibration indicates the slowing down of the clock. An atom on the Sun, where gravitation is strong, should emit light of a lower frequency (slower vibration) than the light emitted by the same kind of atom on the Earth. Because red light is at the low-frequency end of the visible spectrum, a lowering of frequency shifts the color of the emitted light toward the red. This effect is the **gravitational red shift**. Although it is weak in the Sun, it is stronger in more compact stars because they have a greater surface gravity. An experiment confirming Einstein's prediction was performed in 1960. The gravitational red shift was detected using high-frequency gamma rays sent between the top and bottom floors of a laboratory building at Harvard University.[8] Incredibly precise measurements confirmed the gravitational slowing of time.

The gravitational red shift can be understood in terms of the gravitational force acting on photons. As it leaves the surface of a star, a photon is "retarded" by the star's gravity. It loses energy (but not speed). Being that a photon's frequency is proportional to its energy, its frequency decreases as its energy decreases. When we observe the photon, we see that it has a lower frequency than it would have if it had been emitted by a nearer source. Its time has been slowed, just like the ticking of a clock is slowed.

So measurements of time depend not only on relative motion, as we learned in special relativity, but also on the relative gravitational potentials of the regions in which the events are taking place. It is important to note the relativistic nature of time in both special relativity and general relativity. In both theories, there is no way you can ex-

---

Answer

The reason is not because the curvature of the Earth's spacetime is relatively slight, for the curvature is enough to bend a tossed baseball 5 m below a Euclidean straight line in its first second of motion. The curvature would bend a beam of light 5 m also—if the light remained in the same gravitational field for 1 s (like bouncing to and fro between parallel mirrors). We easily notice the curve of the tossed ball, and we barely notice the curve of a high-speed bullet, but we don't notice the curve of light because light travels so fast.

---

[8]In the late 1950s, shortly after Einstein's death, the German physicist Rudolph Mössbauer discovered a way to use atomic nuclei as atomic clocks. His method is called the *Mössbauer effect*, and for it he was awarded the Nobel Prize. In 1960, Professors Pound and Rebka at Harvard University used the Mossbauer effect to perform the confirming experiment.

Newton's and Einstein's Gravity Compared

When Einstein formulated his new theory of gravitation, he realized that if his theory was valid, his field equations must reduce to Newtonian equations for gravitation in the weak-field limit. He showed that Newton's law of gravitation is a special case of the broader theory of relativity. Newton's law of gravitation remains an accurate description of most of the interactions between bodies in the solar system and beyond. From Newton's law, one can calculate the orbits of comets and asteroids and even predict the existence of undiscovered planets.

Even today, when computing the trajectories of space probes throughout the solar system and beyond, scientists and engineers use only Newtonian theory. This is because the gravitational field of these bodies is very weak, and from the viewpoint of general relativity, the surrounding spacetime is essentially flat. But for regions of more intense gravitation, where spacetime is more appreciably curved, Newtonian theory cannot adequately account for various phenomena—like the precession of Mercury's orbit close to the Sun and, in the case of stronger fields, the gravitational red shift and other apparent distortions of spacetime. These distortions reach their limit in the case of a star that collapses to a black hole, where spacetime completely folds over on itself. Only Einsteinian gravitation reaches into this domain.

tend the duration of your own experience. Others moving at different speeds or in different gravitational fields may attribute a greater longevity to you, but your longevity is seen from *their* frame of reference—never your own. As said before, changes in time and other relativistic effects are always attributed to "the other guy."

We saw in Chapter 13 that Newtonian physics is linked at one end with quantum theory, whose domain is the very light and very small—tiny particles and atoms. And now we have seen that Newtonian physics is linked at the other end with relativity theory, whose domain is the very massive and very large.

30.9 Our Expanding Universe

We live in an expanding universe. Most cosmologists support the *standard model*—the theory of an expanding universe of constant mass-energy emanating from the Big Bang some 15 billion years ago. What will be the end result of this expansion? Will the universe expand forever, or will it finally slow down, stop, and then fall in on itself again? And if it does fall in on itself, does it reexplode, start a new cycle, and keep repeating this cycle ad infinitum?

Although current evidence suggests the universe will continue expanding and eventually thin out, some cosmologists subscribe to the *oscillating theory of the universe*. You know that if you throw a rock skyward, two things can happen. The first thing is that it can go up, stop, and come back down. However, if you throw it faster than 11.2 kilometers per second (escape speed for the Earth), the second thing happens: It continues its upward motion forever and never comes down again. The same thing may happen on a universal scale. If the expansion rate of the universe is less than the escape speed for the universe, the expansion will run its course, momentarily come to a stop, and then fall back in on itself. The time needed for this cycle has been estimated to be somewhat less than 100 billion years. Being now about 15 billion years on its way out, the universe should continue its outward expansion for 35 billion years and then momentarily come to rest. The universe would then be at its maximum extent.

Galaxies would be farthest apart and show no Doppler shift in their light. Then the contraction would begin. After 35 billion years of contracting, the universe would again be the size it is now, but galaxies would all show blue shifts instead of red shifts. Then after 15 billion more years would come the Big Crunch, as the universe collapsed into its own black hole—perhaps to gush out into a new universe for another 100 or so billion years. Very intriguing.

Whether the universe oscillates or whether it expands indefinitely depends on its mass density. If the mass of the universe is less than a critical value, the present expansion rate exceeds escape speed and the universe will expand indefinitely. If the mass is greater than this critical value, then the present expansion rate is less than escape speed and the expanding will someday come to a halt. Then the universe will contract to complete one cycle of its possible oscillation. Present indications are that the mass density of the universe is too low for oscillation. A search for dark matter or other mass to bring it up to critical has thus far been unsuccessful. We need to know more before we can say whether the mass density of the universe is below or above the critical value. We have much to find out.

We don't see the world the same way the ancient Egyptians, Greeks, and Chinese did. It is unlikely that people in the future will see the universe as we do. Our view of the universe may be wrong, but it is most likely less wrong than the views of others before us. Our view today stems from the findings of Copernicus and Galileo—findings that were opposed on the grounds that they diminished the importance of humans in the universe. The idea of importance then was in being able to rise above nature—in being apart from nature. We have expanded our vision since then by enormous effort, painstaking observation, and an unrelenting desire to comprehend our surroundings. Seen from today's understanding of the universe, we find our importance not in being apart from nature but rather in being very much a part of it. We are the part of nature that is becoming more and more conscious of itself.

Summary of Terms

- **Big Bang** The primordial explosion of space at the beginning of time.
- **Primeval fireball** The burst of energy during the Big Bang.
- **Spacetime** The four-dimensional continuum in which all things exist; three dimensions of space and one of time.
- **Special theory of relativity** The first of Einstein's theories of relativity, which discusses the effects of uniform motion on space, time, energy, and mass.
- **Time dilation** The slowing of time as a result of speed.
- **Length contraction** The contraction of objects in their direction of motion as a result of speed.
- **General theory of relativity** The second of Einstein's theories of relativity, which discusses the effects of gravity on space and time.
- **Geodesic** The shortest path between two points in various models of space.
- **Gravitational red shift** The lengthening of the waves of electromagnetic radiation escaping from a massive object.

Review Questions

The Big Bang

1. An ordinary explosion expands in space and in time. How does the Big Bang differ?
2. What is responsible for the slowing down of the rate at which the universe is expanding?

Spacetime

3. What are the four dimensions of spacetime?
4. Under what conditions can you and a friend share the same realm of spacetime?
5. What is special about the ratio of the distance traveled by a parcel of light and the time the light takes to travel this distance?

Special Relativity

6. Cite three examples of Einstein's first postulate.
7. Cite three examples of Einstein's second postulate.

Time Dilation

8. What do we call the phenomenon by which time stretches out when something is traveling at relativistic speeds?

9. Time is required for light to travel along a path from one point to another. If this path is seen to be longer because of motion, what happens to the time it takes for light to travel this longer path?
10. How do measurements of time differ for events in a frame of reference that is moving at 50% of the speed of light relative to us?
11. How do measurements of time differ for events in a frame of reference that is moving at 99.5% of the speed of light relative to us?
12. What is the evidence for time dilation?
13. What are the present-day obstacles to interstellar space travel?
14. When two velocities v_1 and v_2 are both much less than the speed of light, is the value v_1v_2/c^2 large or small?
15. What is the maximum value of v_1v_2/c^2 in an extreme situation? What is its smallest value?
16. Is the relativistic rule for adding velocities consistent with the fact that light can have only one speed in all uniformly moving reference frames? Can the speed of light ever be more than c?

Length Contraction

17. How long would a meter stick appear if it were traveling horizontally at 99.5% of the speed of light?
18. How long would the meter stick in the previous question appear if it were traveling with its length perpendicular to its direction of motion? (Why is your answer different from the previous answer?)
19. If you were traveling in a high-speed rocket ship, would meter sticks on board appear to you to be contracted? Defend your answer.

Relativistic Momentum

20. What is the momentum of an object that has been pushed to the speed of light?
21. When a beam of charged particles moves through a magnetic field, what is the evidence that their momentum is greater than the value mv?

Mass, Energy, and $E = mc^2$

22. What does $E = mc^2$ mean?
23. What is the evidence for $E = mc^2$ in cosmic ray investigations?
24. What is an antiparticle?

General Relativity

25. What is the principal difference between the special theory of relativity and the general theory of relativity?
26. Compare the bending of the paths of baseballs and photons by a gravitational field.
27. What does it mean to say that space is curved?
28. What is a geodesic?
29. Of all the planets, why is Mercury the best candidate for finding evidence of the relationship between gravitation and space?
30. What is the evidence for light bending near the Sun?
31. Which runs faster, a clock at the top of the Sears Tower in Chicago or a clock on the shore of Lake Michigan?
32. What is the effect of strong gravitation on measurements of time?
33. Does Einstein's theory of gravitation invalidate Newton's theory of gravitation? Explain.

Our Expanding Universe

34. What is the estimated period for the oscillating universe?
35. What condition is necessary for the universe to stop expanding and start contracting?

Exercises

1. Why are the long-wavelength microwaves that permeate the universe considered evidence of the Big Bang?
2. What is meant by saying that the Big Bang is the explosion of space at the beginning of time? Why do we say explosion *of* space rather than explosion *in* space?
3. Faraway galaxies show a red shift in their spectra. If the universe oscillates and the galaxies one day come to a halt, what shift (if any) will their spectra show then? What shift will appear if the galaxies come toward us?
4. As something expands, it spreads its energy over a greater area and it cools. How does this apply to the universe?
5. In Chapter 11, we learned that light travels more slowly in glass than in air. Does this contradict the special theory of relativity?
6. Suppose a particular clock accurately shows time passing half as fast in a particular frame of reference as in our own. What is the velocity of this frame of reference relative to us?
7. If we see somebody's clock running slow due to relative motion, do they see our clock running slow also? Or do they see our clock running fast? Explain.
8. When you drive down the highway, you are moving through space. What else are you moving through?
9. (a) Suppose the parallel mirrors of a light clock are 150,000 km apart. In the frame of reference of the clock, how much time is required for a pulse of light to make a round-trip between the mirrors? (b) How much time does a round-trip take if the parallel mirrors are 150 km apart? (c) Would your answers to these two questions be different if your measurements were made from a frame of reference that moves relative to the light clock? Explain.
10. There being an upper limit on the speed of a particle, does it follow that there is also an upper limit on its momentum? On its kinetic energy? Explain.
11. Light travels a certain distance in, say, 20,000 years. How is it possible that an astronaut could travel more slowly than light but still travel 20,000 light-years in a 20-year trip?
12. Could a human being who has a life expectancy of 70 years possibly make a round-trip journey to a part of the universe thousands of light-years distant? Explain.

13. A twin who makes a long trip at relativistic speeds returns younger than her stay-at-home twin brother. Could she return before her twin brother was born? Defend your answer.
14. Is it possible for a child to be biologically older than its parents? Explain.
15. If you were in a rocket ship traveling away from the Earth at a speed close to the speed of light, what changes would you note in your momentum? In your volume? Explain.
16. If you were on the Earth monitoring a person in a rocket ship traveling away from the Earth at a speed close to the speed of light, what changes would you note in his momentum? In his volume? Explain.
17. Is this ditty, popular with relativity types, consistent or inconsistent with relativity theory? Defend your answer.

    There was a young lady named Bright
    Who traveled much faster than light.
    She departed one day
    In an Einsteinian way
    And returned on the previous night.

18. As a meter stick moves past you at nearly the speed of light, your measurements show its momentum to be twice its classical momentum and its length to be 1 m. In what direction is the stick pointing?
19. In the preceding exercise, if the stick is moving in a direction along its length, how long will it appear to you?
20. The linear accelerator at Stanford University shoots particles along a path that is 2 miles long. Yet the path "appears" to be less than a meter long to the electrons that travel in it. Explain.
21. Muons are elementary particles that are formed high in the atmosphere by the interactions between cosmic rays and gases in the upper atmosphere. Muons are radioactive and have average lifetimes of about two-millionths of a second. Even though they travel at almost the speed of light, they are so high that very few should be detected at sea level—at least according to classical physics. Laboratory measurements, however, show that muons do reach the Earth's surface and in great proportions. What is the explanation?
22. When we look out into the universe, we see into the past. John Dobson, founder of the San Francisco Sidewalk Astronomers, says that we cannot even see the backs of our own hands *now*—in fact, we can't see anything *now*. Do you agree? Explain.
23. How does the idea that matter is frozen energy relate to $E = mc^2$?
24. An astronaut is provided a "gravity" when the ship's engines are activated to accelerate the ship. This requires the use of fuel. Is there a way to accelerate and provide "gravity" without the sustained use of fuel? Explain.
25. How would the number of pushups you could perform at the Earth's surface compare with the number you could perform in an elevator accelerating at g far from the Earth's gravitational field?
26. In his famous novel *Journey to the Moon,* Jules Verne stated that occupants in a spaceship would shift their orientation from up to down when the ship crossed the point where the Moon's gravitation became greater than the Earth's. Is this correct? Defend your answer.
27. Two persons are standing at different points on the Earth's equator. What happens to the separation distance between them as they both walk north at the same rate? And just for fun, where in the world is a step in every direction a step south?
28. We readily note the bending of light by reflection and refraction, but why is it we do not ordinarily notice the bending of light by gravity?
29. Why do we say that light travels in straight lines? Is it strictly accurate to say that a laser beam provides a perfectly straight line for purposes of surveying? Explain.
30. Light changes its energy when it "falls" in a gravitational field. This change in energy is not evidenced by a change in speed, however. What is the evidence for this change?
31. Splitting hairs, should a person who worries about growing old live at the top or at the bottom of a tall apartment building?
32. Why does the strength of the gravitational attraction between the Sun and Mercury vary? Would it vary if the orbit of Mercury were perfectly circular?

Problems

1. A passenger on an interplanetary express bus traveling at $v = 0.99c$ takes a 5-minute catnap by her watch. How long does the nap last from your vantage point on a fixed planet?
2. The relativistic addition of velocities is given by

$$V = \frac{v_1 + v_2}{1 + \frac{v_1 + v_2}{c^2}}$$

Substitute small values of v_1 and v_2 in this equation and show that for everyday velocities V is practically equal to $v_1 + v_2$.

3. The starship *Enterprise,* passing the Earth at 80% of the speed of light, sends a drone ship forward at half the speed of light relative to the *Enterprise.* What is the drone's speed relative to the Earth?
4. At the end of 1 s, a horizontally fired bullet in a gravitational field of 1g has dropped a vertical distance of 4.9 m from its otherwise straight-line path. By what distance would a beam of light drop from its otherwise straight-line path if it traveled in the same field for 1 s? For 2 s?

Part VIII — Sample Exam Questions

Choose the BEST answer to each of the following.

1. The Big Bang is considered to have been the beginning of
(a) space (b) time
(c) both (d) neither

2. When a lunar eclipse occurs, the phase of the Moon must be
(a) full (b) one-quarter
(c) one-half (d) new
(e) none of these

3. When a solar eclipse occurs, the phase of the Moon must be
(a) full (b) one-quarter
(c) one-half (d) new
(e) none of these

4. The fact that one side of the Moon always faces Earth is evidence that the Moon
(a) rotates
(b) doesn't rotate
(c) reflects rather than emits light
(d) is distorted in shape

5. Jupiter has about 300 times the mass of the Earth, yet an astronaut's weight on Jupiter's surface would be only about 3 times Earth weight-and not 300 times because of Jupiter's large
(a) gravity (b) mass
(c) radius (d) magnetic fields
(e) all of these

6. The tails of comets point away from the Sun because of
(a) radiation pressure
(b) magnetic effects
(c) gravitational interactions
(d) centripetal acceleration
(e) none of these

7. If the Sun collapsed to become a black hole, the Earth would
(a) likely be sucked into it
(b) continue in its present orbit
(c) fly off in a straight-line path
(d) be pulled apart by tidal forces
(e) none of these

8. Most of the angular momentum in our solar system is in the
(a) Sun
(b) planets
(c) either, depending on the seasons

9. A protostar becomes a full fledged star when
(a) it begins emitting light
(b) fusion occurs
(c) gravitational pressure squeezes it to its tightest configuration
(d) gravitational pressure equals radiation pressure

10. Stars with the longest lives are usually stars with
(a) small mass
(b) large mass
(c) characteristics independent of mass

11. Nuclear fusion reactions in a star occur mainly in its
(a) core
(b) `outer shell
(c) surrounding atmosphere of hot gases

12. The mass of a star is most easily determined when it is part of
(a) the Milky Way
(b) a binary
(c) a major constellation
(d) a star cluster of known mass
(e) none of these

13. Astronomers know the least about
(a) red giant stars (b) neutron stars
(c) black holes (d) pulsars
(e) quasars

14. Whether or not the expanding universe will eventually collapse has to do with the amount of its
(a) gravitational red shift
(b) gravitational blue shift
(c) mass
(d) angular momentum
(e) energy

15. Since there is an upper limit on the speed of a particle, there is also an upper limit on its
(a) momentum (b) kinetic energy
(c) temperature (d) all of these
(e) none of these

16. When a light source approaches you, your measurements of it show an increase in its
(a) speed (b) wavelength
(c) frequency (d) all of these
(e) none of these

17. Relativistic equations for time, length, and mass hold true for
(a) speeds near that speed of light
(b) everyday low speeds
(c) both
(d) neither

18. Compared with special relativity, general relativity is more concerned with
(a) acceleration (b) gravitation
(c) space-time geometry (d) all of these
(e) none of these

19. If the orbit of Mercury were perfectly circular, its rate of precession would be
(a) larger (b) smaller
(c) no different (d) zero

20. Which equation is the triumph of the theory of special relativity?
(a) $E = ma^2$ (b) $E = mb^2$
(c) $E = mc^2$ (d) $E = md^2$
(e) $E = me^2$

Answers: 1c, 2a, 3d, 4a, 5c, 6a, 7b, 8a, 9b, 10a, 11a, 12b, 13e, 14c, 15e, 16c, 17c, 18d, 19d, 20c

APPENDIX

Systems of Measurement

Two major systems of measurement prevail in the world today: the *United States Customary System* (USCS, formerly called the British system of units), used in the United States of America and in Burma, and the *Système International* (SI) (known also as the international system and as the metric system), used everywhere else. Each system has its own standards of length, mass, and time. The units of length, mass, and time and sometimes called the *fundamental units* because, once they are selected, other quantities can be measured in terms of them.

United States Customary System

Based on the British Imperial System, the USCS is familiar to everyone in the United States. It uses the foot as the unit of length, the pound as the unit of weight or force, and the second as the unit of time. The USCS is presently being replaced by the international system—rapidly in science and technology (all 1988 Department of Defense contracts) and some sports (track and swimming), but so slowly in other areas and in some specialties it seems the change may never come. For example, we will continue to buy seats on the 50-yard line. Camera film is in millimeters but computer disks are in inches.

For measuring time, there is no difference between the two systems except that in pure SI the only unit is the second (s, not sec) with prefixes; but in general, minute, hour, day, year, and so on, with two or more lettered abbreviations (h, not hr), are accepted in the USCS.

Systéme International

During the 1960 International Conference on Weights and Measures held in Paris, the SI units were defined and given status. Table A.1 shows SI units and their symbols. SI is based on the *metric system,* originated by French scientists after the French revolution in 1791. The orderliness of this system makes it useful for scientific work, and it is used by scientists all over the world. The metric system branches into two systems of units. In one of these the unit of length is the meter, the unit of mass is the kilogram, and the unit of time is the second. This is called the *meter-kilogram-second* (mks) system and is preferred in physics. The other branch is the *centimeter-gram-second* (cgs) system, which because of its smaller values is favored in chemistry. The cgs and mks units are related to each other as follows: 100 centimeters equal 1 meter; 1000 grams equal 1 kilogram. Table A.2 shows several units of length related to each other.

Table A.1
SI units

Quantity	Unit	Symbol
Length	meter	m
Mass	kilogram	kg
Time	second	s
Force	newton	N
Energy	joule	J
Current	ampere	A
Temperature	kelvin	K

One major advantage of the metric system is that it uses the decimal system, where all units are related to smaller or larger units by dividing or multiplying by 10. The prefixes shown in Table A.3 are commonly used to show the relationship among units.

Table A.2
Table Conversions Between Different Units of Length

Unit of Length	Kilometer	Meter	Centimeter	Inch	Foot	Mile
1 kilometer	= 1	1000	100,000	39,370	3280.84	0.62140
1 meter	= 0.00100	1	100	39.370	3.28084	6.21×10^{-4}
1 centimeter	= 1.0×10^{-5}	0.0100	1	0.39370	0.032808	6.21×10^{-6}
1 inch	= 2.54×10^{-5}	0.02540	2.5400	1	0.08333	1.58×10^{-5}
1 foot	= 3.05×10^{-4}	0.30480	30.480	12	1	1.89×10^{-4}
1 mile	= 1.60934	1609.34	160,934	63,360	5280	1

Table A.3
Some Prefixes

Prefix	Definition
micro-	One-millionth: a microsecond is one-millionth of a second
milli-	One-thousandth: a milligram is one-thousandth of a gram
centi-	One-hundredth: a centimeter is one-hundredth of a meter
kilo-	One thousand: a kilogram is 1000 grams
mega-	One million: a megahertz is 1 million hertz

Meter

The standard of length of the metric system orginally was defined in terms of the distance from the north pole to the equator. This distance was thought at the time to be close to 10,000 kilometers. One ten-millionth of this, the meter, was carefully determined and marked off by means of scratches on a bar of platinum-iridium alloy. This bar is kept at the International Bureau of Weights and Measures in France. The standard meter in France has since been calibrated in terms of the wavelength of light—it is 1,650,763.73 times the wavelength of orange light emitted by the atoms of the gas krypton-86. The meter is now defined as being the length of the path traveled by light in a vacuum during a time interval of 1/299,792,458 of a second.

Figure A.1
The standard kilogram

Kilogram

The standard unit of mass, the kilogram, is a block of platinum, also preserved at the International Bureau of Weights and Measures located in France (Figure A.1). The kilogram equals 1000 grams. A gram is the mass of 1 cubic centimeter (cc) of water at a temperature of 4°C. (The standard pound is defined in terms of the standard kilogram; the mass of an object that weighs 1 pound is equal to 0.4536 kilogram.)

Second

The official unit of time for both the USCS and the SI is the second. Until 1956, it was defined in terms of the mean solar day, which was divided into 24 hours. Each hour was divided into 60 minutes and each minute into 60 seconds. Thus there were 86,400 seconds per day, and the second was defined as 1/86,400 of the mean solar day. This proved unsatisfactory because the rate of rotation of the Earth is gradually becoming slower. In 1956, the mean solar day of the year 1900 was chosen as the standard on which to base the second. In 1964, the second was officially defined as the time taken by a cesium-133 atom to make 9,192,631,770 vibrations.

Newton

One newton is the force required to accelerate 1 kilogram at 1 meter per second per second. This unit is named after Sir Isaac Newton.

Joule

One joule is equal to the amount of work done by a force of 1 newton acting over a distance of 1 meter. In 1948, the joule was adopted as the unit of energy by the International Conference on Weights and Measures. Therefore, the specific heat of water at 15°C is now given as 4185.5 joules per kilogram Celsius degree. This figure is always associated with the mechanical equivalent of heat—4.1855 joules per calorie.

Ampere

The ampere is defined as the intensity of the constant electric current that, when maintained in two parallel conductors of infinite length and negligible cross section and placed 1 meter apart in a vacuum, would produce between them a force equal to 2×10^{-7} newton per meter length. In our treatment of electric current in this text, we have used the not-so-official but easier-to-comprehend definition of the ampere as being the rate of flow of 1 coulomb of charge per second, where 1 coulomb is the charge of 6.25×10^{18} electrons.

Kelvin

The fundamental unit of temperature is named after the scientist William Thomson, Lord Kelvin. The kelvin is defined to be 1/273.15 the thermodynamic temperature of the triple point of water (the fixed point at which ice, liquid water, and water vapor coexist in equilibrium). This definition was adopted in 1968 when it was decided to change the name *degree Kelvin* (°K) to *kelvin* (K). The temperature of melting ice at atmospheric pressure is 273.15 K. The temperature at which the vapor pressure of pure water is equal to standard atmospheric pressure is 373.15 K (the temperature of boiling water at standard atmospheric pressure).

Area

The unit of area is a square that has a standard unit of length as a side. In the USCS, it is a square with sides that are each 1 foot in length, called 1 square foot and written 1 ft^2. In the international system, it is a square with sides that are 1 meter in length, which makes a unit of area of 1 m^2. In the cgs system it is 1 cm^2. The area of a given surface is specified by the number of square feet, square meters, or square centimeters that would fit into it. The area of a rectangle equals the base times the height. The area of a circle is equal to πr^2, where p $\pi = 3.14$ and r is the radius of the circle. Formulas for the surface areas of other objects can be found in geometry textbooks.

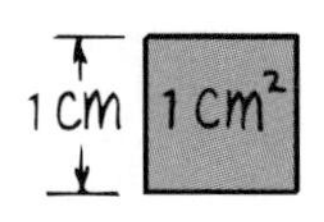

Figure A.2
Unit square.

Volume

The volume of an object refers to the space it occupies. The unit of volume is the space taken up by a cube that has a standard unit of length for its edge. In the USCS one unit of volume is the space occupied by a cube 1 foot on an edge and is called 1 cubic foot, written 1 ft^3. In the metric system it is the space occupied by a cube with sides of 1 meter (SI) or 1 centimeter (cgs). It is written 1 m^3 or 1 cm^3 (or cc). The volume of a given space is specified by the number of cubic feet, cubic meters, or cubic centimeters that will fill it.

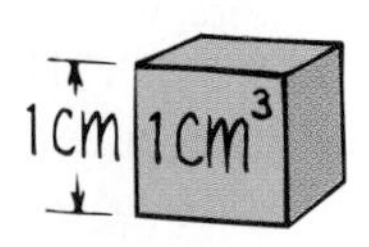

Figure A.3
Unit volume.

In the USCS, volumes can also be measured in quarts, gallons, and cubic inches as well as in cubic feet. There are 1728 ($12 \times 12 \times 12$) cubic inches in 1 ft^3. A U.S. gallon is a volume of 231 in^3. Four quarts equal 1 gallon. In the SI volumes are also measured in liters. A liter is equal to 1000 cm^3.

Exponential Growth and Doubling Time*

One of the most important things we seem unable to perceive is the process of exponential growth. We think we understand how compound interest works, but we can't get it through our heads that a fine piece of tissue paper folded upon itself 50 times (if that were possible) would be more than 20 million kilometers thick. If we could, we could "see" why our income buys only half of what it did 4 years ago, why the price of everything has doubled in the same time, why populations and pollution proliferate out of control.[†]

When a quantity such as money in the bank, population, or the rate of consumption of a resource steadily grows at a fixed percent per year, we say the growth is exponential. Money in the bank may grow at 8 percent per year; electric power generating capacity in the United States grew at about 7 percent per year for the first three-quarters of the century. The important thing about exponential growth is that the time required for the growing quantity to double in size (increase by 100 percent) is also constant. For example, if the population of a growing city takes 12 years to double from 10,000 to 20,000 inhabitants and its growth remains steady, in the next 12 years the population will double to 40,000, and in the next 10 years to 80,000, and so on.

There is an important relationship between the percent growth rate and its *doubling time,* the time it takes to double a quantity:[††]

$$\textbf{Doubling time} = \frac{\textbf{69.3}}{\textbf{percent growth per unit time}} \approx \frac{\textbf{70}}{\textbf{\%}}$$

So to estimate the doubling time for a steadily growing quantity, we simply divide the number 70 by the percentage growth rate. For example, the 7 percent growth rate of electric power generating capacity in the United States means that in the past the capacity has doubled every 10 years [70%/(7%/ year) = 10 years]. A 2 percent growth rate for world population means the population of the world doubles every 35 years [70%/(2%/year) = 35 years]. A city planning commission that accepts what seems like a modest 3.5 percent growth rate may not realize that this means that doubling will occur in 70/3.5 or 20 years; that's double capacity for such things as water supply, sewage-treatment plants, and other municipal services every 20 years.

---

*This appendix is drawn from material by University of Colorado physics professor Albert A. Bartlett, who strongly asserts, "The greatest shortcoming of the human race is man's inability to understand the exponential function." See Professor Bartlett's still-timely article, "Forgotten Fundamentals in the Energy Crisis" (*American Journal of Physics,* September 1978) or his revised version (*Journal of Geological Education,* January 1980).

[†]K. C. Cole, *Sympathetic Vibrations* (New York: Morrow, 1984).

[††]For Exponential decay we speak about half-life, the time required for a quantity to reduce to half its value. This case is treated in chapter 14.

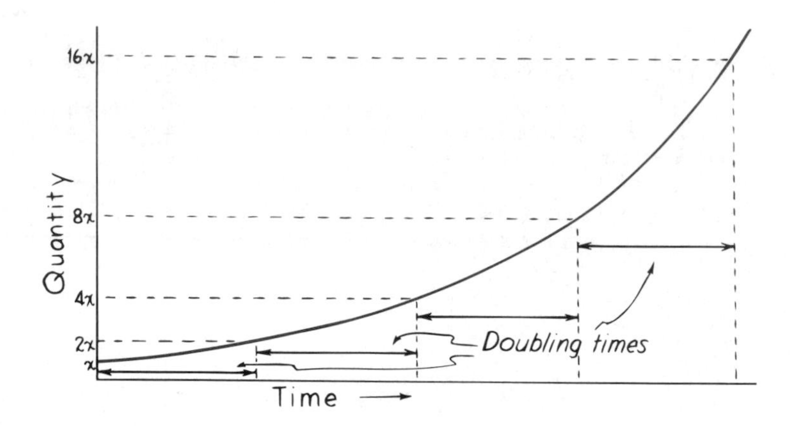

Figure B.1
An exponential curve. Notice that each of the successive equal time intervals noted on the horizontal scale corresponds to a doubling of the quantity indicated on the vertical scale. Such an interval is called the doubling time.

What happens when you put steady growth in a finite environment? Consider the growth of bacteria that grow by division, so that one bacterium becomes two, the two divide to become four, the four divide to become eight, and so on. Suppose the division time for a certain strain of bacteria is 1 minute. This is then steady growth—the number of bacteria grows exponentially with a doubling time of 1 minute. Further, suppose that one bacterium is put in a bottle at 11:00 A.M. and that growth continues steadily until the bottle becomes full of bacteria at 12 noon. Consider seriously the following question.

Question

When was the bottle half-full?

Figure B.2

It is startling to note that at 2 minutes before noon the bottle was only $\frac{1}{4}$ full. Table B.1 summarizes the amount of space left in the bottle in the last few minutes before noon. If you were an average bacterium in the bottle, at which time would you first realize that you were running out of space? For example, would you sense there was a serious problem at 11:55 A.M., when the bottle was only 3% filled ($\frac{1}{32}$) and had 97% of open space (just yearning for development)? The point here is that there isn't much time between the moment that the effects of growth become noticeable and the time when they become overwhelming.

Table B.1
The Last Minutes in the Bottle

Time	Part Full (%)	Part Empty
11:54 A.M.	$\frac{1}{64}$ (1.5%)	$\frac{63}{64}$
11:55 A.M.	$\frac{1}{32}$ (3%)	$\frac{31}{32}$
11:56 A.M.	$\frac{1}{16}$ (6%)	$\frac{15}{16}$
11:57 A.M.	$\frac{1}{8}$ (12%)	$\frac{7}{8}$
11:58 A.M.	$\frac{1}{4}$ (25%)	$\frac{3}{4}$
11:59 A.M.	$\frac{1}{2}$ (50%)	$\frac{1}{2}$
12:00 noon	full (100%)	none

Suppose that at 11:58 A.M. some farsighted bacteria see that they are running out of space and launch a full-scale search for new bottles. Luckily, at 11.59 A.M. they discover three new empty bottles, three times as much much space as they had ever known. This quadruples the total resource space ever known to the bacteria, for they now

---

Answer
11:59 A.M.; the bacteria will double in number every minute!

have a total of four bottles, whereas before the discovery they had only one. Further suppose that thanks to their technological proficiency, they are able to migrate to their new habitats without difficulty. Surely, it seems to most of the bacteria that their problem is solved—and just in time.

■ Question

If the bacteria growth continues at the unchanged rate, what time will it be when the three new bottles are filled to capacity?

We see from Table II.2 that quadrupling the resource extends the life of the resource by only two doubling times. In our example the resource is space—but it could as well be coal, oil, uranium, or any nonrenewable resource.

Continued growth and continued doubling lead to enormous numbers. In two doubling times, a quantity will double twice ($2^2 = 4$; quadruple) in size; in three doubling times, its size will increase eightfold ($2^3 = 8$); in four doubling times, it will increase sixteenfold ($2^4 = 16$); and so on.

Figure B.3
A single grain of wheat placed on the first square of the chessboard is doubled on the second square, this number is doubled on the third, and so on, presumably for all 64 squares. Note that each square contains more grain than all the preceding squares combined. Does enough wheat exist in the world to fill all 64 squares in this manner?

Table B.2
Effects of the Discovery of Three New Bottles

Time	Effect
11:58 A.M.	Bottle 1 is $\frac{1}{4}$ full
11:59 A.M.	Bottle 1 is $\frac{1}{2}$ full
12:00 noon	Bottle 1 is full
12:01 P.M.	Bottles 1 and 2 are both full
12:02 P.M.	Bottles 1, 2, 3, and 4 are all full

Table B.3
Filling the Squares on the Chessboard

Square Number	Grains on Square	Total Grains Thus Far
1	1	1
2	2	3
3	4	7
4	8	15
5	16	31
6	32	63
7	64	127
⋮	⋮	⋮
64	2^{63}	$2^{64} - 1$

This is best illustrated by the story of the court mathematician in India who years ago invented the game of chess for his king. The king was so pleased with the game that he offered to repay the mathematician, whose request seemed modest enough. The mathematician requested a single grain of wheat on the first square of the chessboard, two grains on the second square, four on the third square, and so on, doubling the number of grains on each succeeding square until all squares had been used. At this rate there would be 2^{63} grains of wheat on the 64th square. The king soon saw that he could not fill this "modest" request, which amounted to more wheat than had been harvested in the entire history of the Earth!

It is interesting and important to note that the number or grains on any square is one grain more than the total of all grains on the preceding squares. This is true anywhere on the board. Note from Table B.3 that when eight grains are placed on the fourth square, the eight is one more than the total of seven grains that were already on the board. Or the 32 grains placed on the sixth square is one more than the total of 31

■ **Answer**
12:02 P.M.!

grains that were already on the board. We see that in one doubling time we use more than all that had been used in all the preceding growth!

So if we speak of doubling energy consumption in the next however many years, bear in mind that this means in these years we will consume more energy than has heretofore been consumed during the entire preceding period of steady growth. And if power generation continues to use predominantly fossil fuels, then except for some improvements in efficiency, we would burn up in the next doubling time a greater amount of coal, oil, and natural gas than has already been consumed by previous power generation, and except for improvements in pollution control, we can expect to discharge even more toxic wastes into the environment than the millions upon millions of tons already discharged over all the previous years of industrial civilization. We would also expect more human-made calories of heat to be absorbed by the Earth's ecosystem than have been absorbed in the entire past! At the previous 7% annual growth rate in energy production, all this would occur in one doubling time of a single decade. If over the coming years the annual growth rate remains at half this value, 3.5 percent, then all this would take place in a doubling time of two decades. Clearly this cannot continue!

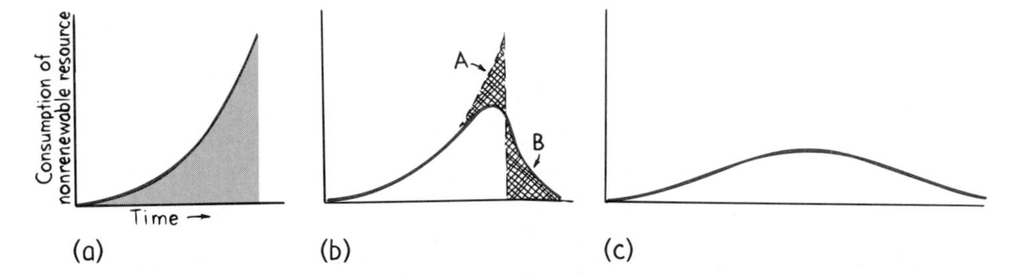

Figure B.4
(a) If the exponential rate of consumption for a nonrenewable resource continues until it is depleted, consumption falls abruptly to zero. The shaded area under this curve represents the total supply of the resource. (b) In practice, the rate of consumption levels off and then falls less abruptly to zero. Note that the crosshatched area A is equal to the crosshatched area B. Why? (c) At lower consumption rates, the same resource lasts a longer time.

The consumption of a nonrenewable resource cannot grow exponentially for an indefinite period, because the resource is finite and its supply finally expires. The most drastic way this could happen is shown in Figure B.4a, where the rate of consumption, such as barrels of oil per year, is plotted against time, say in years. In such a graph the area under the curve represents the supply of the resource. We see that when the supply is exhausted, the consumption ceases altogether. This sudden change is rarely the case, for the rate of extracting the supply falls as it becomes more scarce. This is shown in Figure B.4b. Note that the area under the curve is equal to the area under the curve in a. Why? Because the total supply is the same in both cases. The principal difference is the time taken to finally extinguish the supply. History shows that the rate of production of a nonrenewable resource rises and falls in a nearly symmetric manner, as shown in c. The time during which production rates rise is approximately equal to the time during which these rates fall to zero or near zero. If we fit the data for U.S. oil production in the contiguous 48 states to such a curve, we find that we are just to the right of the peak. This suggests that one-half of the recoverable petroleum that was ever in the ground in the U.S. has already been used and that in the future the domestic petroleum rate of production can only decrease. The U.S. production curve peaked in 1970, and by 1979 nearly half the U.S. consumption was imported. Each year we consume more oil than during the previous year.

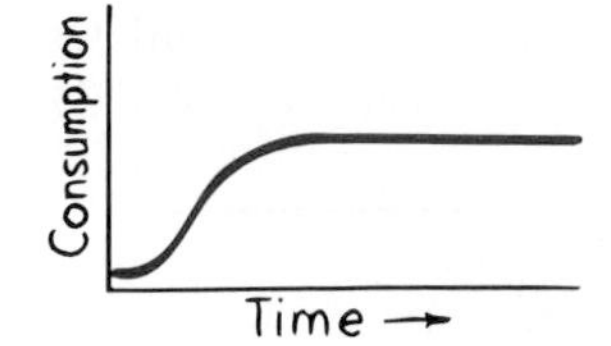

Figure B.5
A curve showing the rate of consumption of a renewable resource such as agricultural or forest products, where a steady rate of production and consumption can be maintained for a long period, provided this production is not dependent upon the use of a nonrenewable resource that is waning in supply.

Production rates for all nonrenewable resources decrease sooner or later. Only production rates for renewable resources, such as agriculture or forest products, can be maintained at steady levels for long periods of time (Figure II.5), provided such production does not depend on waning nonrenewable resources such as petroleum. Much of today's agriculture is so petroleum-dependent that it can be said that modern agriculture is simply the process whereby land is used to convert petroleum into food. The implications of petroleum scarcity go far beyond rationing of gasoline for cars or fuel oil for home heating.

The consequences of unchecked exponential growth are staggering. It is important to ask: Is growth really good? In answering this question, bear in mind that human growth is an early phase of life that continues normally through adolescence. Physical growth stops when physical maturity is reached. What do we say of growth that continues in the period of physical maturity? We say that such growth is obesity—or worse, cancer.

Questions to Ponder

1. According to a French riddle, a lily pond starts with a single leaf. Each day the number of leaves doubles, until the pond is completely covered by leaves on the 30th day. On what day was the pond half covered? One-quarter covered?
2. In an economy that has a steady inflation rate of 7% per year, in how many years does a dollar lose half its value?
3. At a steady inflation rate of 7%, what will be the price every 10 years for the next 50 years for a theater ticket that now costs $20? For a coat that now costs $200? For a car that now costs $20,000? For a home that now costs $200,000?
4. If the sewage treatment plant of a city is just adequate for the city's current population, how many sewage treatment plants will be necessary 42 years later if the city grows steadily at 5% annually?
5. In 1996 the population growth rate for the United States was 1.0%, for Mexico 1.9%, and for Rwanda the highest growth rate in the world, 16.5% (taking into account births, deaths, and immigration). If these rates were steady, how long would it take for the population in each of these countries to double?
6. If world population doubles in 40 years and world food production also doubles in 40 years, how many people then will be starving each year compared to now?
7. A continued world population growth rate of 1.9% per year would produce a density of one person per square meter in 550 years. True or false: World population growth rate will sooner or later be zero.
8. Suppose you get a prospective employer to agree to hire your services for wages of a single penny for the first day, 2 pennies for the second day, and double each day thereafter providing the employer keeps to the agreement for a month. What will be your total wages for the month?
9. In the preceding exercise, how will your wages for only the 30th day compare to your total wages for the previous 29 days?
10. If fusion power were harnessed today, the abundant energy resulting would probably sustain and even further encourage our present appetite for continued growth and in a relatively few doubling times produce an appreciable fraction of the solar power input to the Earth. Make an argument that the current delay in harnessing fusion is a blessing for the human race.

APPENDIX

Vectors

Vectors and Scalars

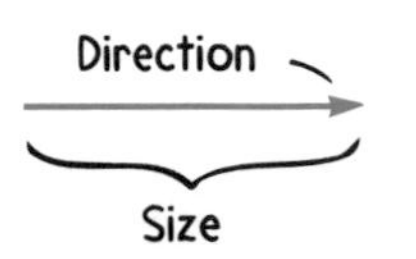

Figure C.1

A *vector* quantity is a directed quantity—one that must be specified not only by magnitude (size) but by direction as well. Recall from Chapter 3 that velocity is a vector quantity. Other examples are force, acceleration, and momentum. In contrast, a *scalar* quantity can be specified by magnitude alone. Some examples of scalar quantities are speed, time, temperature, and energy.

Vector quantities may be represented by arrows. The length of the arrow tells you the magnitude of the vector quantity, and the arrowhead tells you the direction of the vector quantity. Such an arrow drawn to scale and pointing appropriately is called a *vector.*

Adding Vectors

Vectors that add together are called *component vectors. The sum of component vectors is called a resultant.*

To add two vectors, make a parallelogram with two component vectors acting as two of the adjacent sides (Figure C.2). (Here our parallelogram is a rectangle.) Then draw a diagonal from the origin of the vector pair; this is the resultant (Figure C.3).

Caution: Do not try to mix vectors! We cannot add apples and oranges, so velocity vectors combine only with velocity vectors, force vectors combine only with force vectors, and acceleration vectors combine only with acceleration vectors—each on its own vector diagram. If you ever show different kinds of vectors on the same diagram, use different colors or some other method of distinguishing the different kinds of vectors.

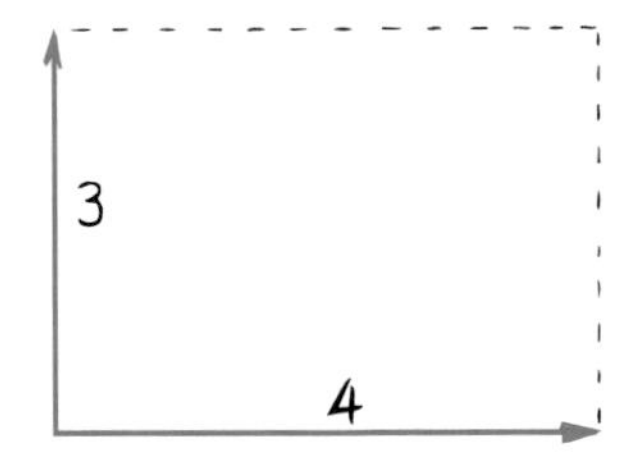

Figure C.2

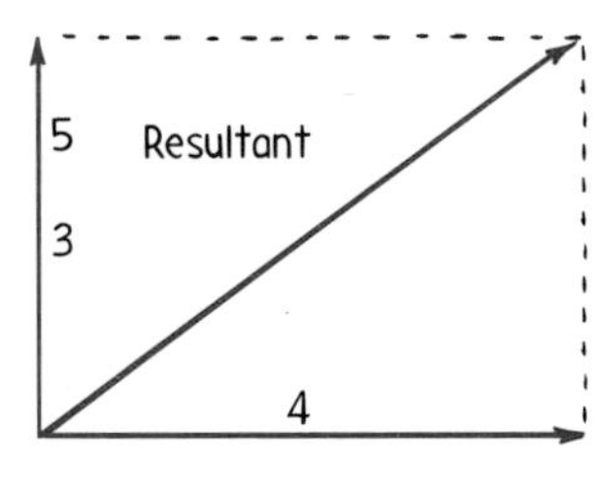

Figure C.3

Finding Components of Vectors

Recall from Chapter 3 that to find a pair of perpendicular components for a vector, first draw a dashed line through the tail of the vector (in the direction of one of the desired components). Second, draw another dashed line through the tail end of the vector at right angles to the first dashed line. Third, make a rectangle whose diagonal is the given vector. Draw in the two components. Here we let **F** stand for "total force," **U** stand for "upward force," and **S** stand for "sideways force."

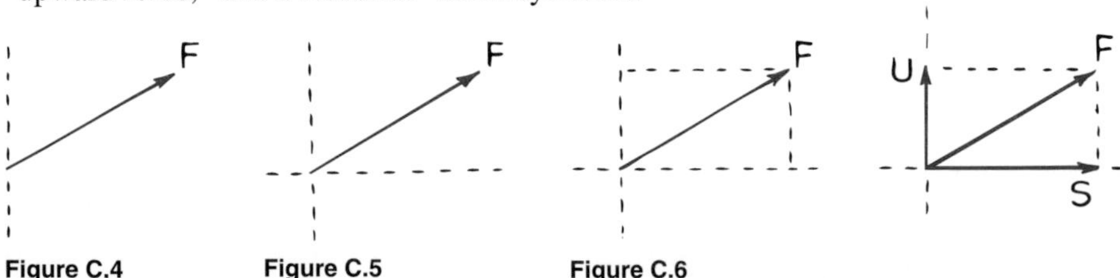

Figure C.4 Figure C.5 Figure C.6

Examples

1. A man pushing a lawnmower applies a force that pushes the machine forward and also against the ground. In Figure C.7, **F** represents the force applied by the man. We can separate this force into two components. The vector **D** represents the downward component, and **S** is the sideways component, the force that moves the lawnmower forward. If we know the magnitude and direction of the vector **F**, we can estimate the magnitude of the components from the vector diagram.

Figure C.7

2. Would it be easier to push or pull a wheelbarrow over a step? Figure C.8 shows a vector diagram for each case. When you push a wheelbarrow, part of the force is directed downward, which makes it harder to get over the step. When you pull, however, part of the pulling force is directed upward, which helps to lift the wheel over the step. Note that the vector diagram suggests that pushing the wheelbarrow may not get it over the step at all. Do you see that the height of the step, the radius of the wheel, and the angle of the applied force determine whether the wheelbarrow can be pushed over the step? We see how vectors help us analyze a situation so that we can see just what the problem is!

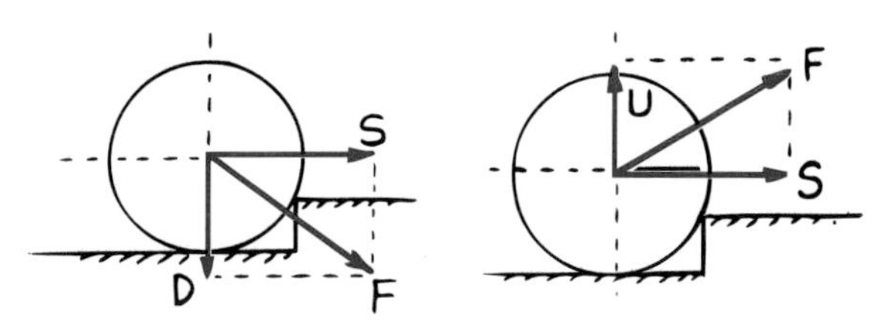

Figure C.8

3. If we consider the components of the weight of an object rolling down an incline, we can see why its speed depends on the angle. Note that the steeper the incline, the greater the component **S** becomes and the faster the object rolls. When the incline is vertical, **S** becomes equal to the weight, and the object attains maximum acceleration, 9.8 m/s^2.

   There are two more force vectors that are not shown: the normal force **N**, which is equal and oppositely directed to **D**, and the friction force **f**, acting at the barrel-plane contact.

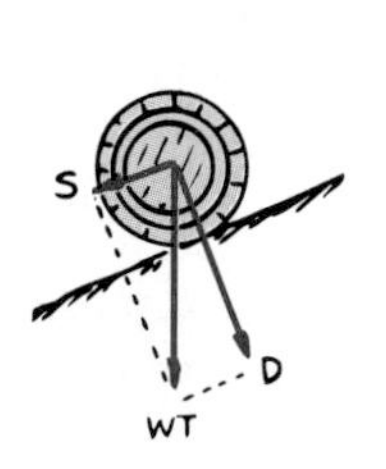

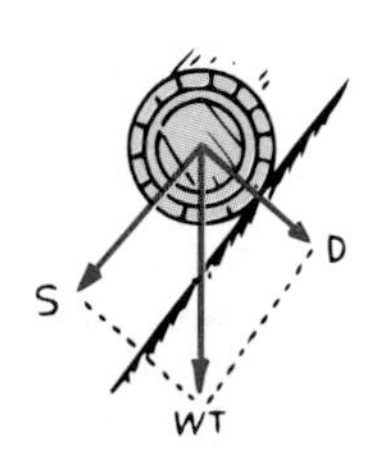

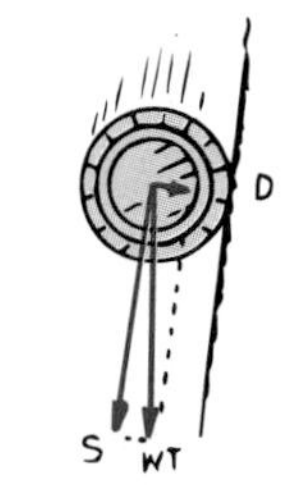

Figure C.9

4. When moving air strikes the underside of an airplane wing, the force of air impact against the wing may be represented by a single vector perpendicular to the plane of the wing (Figure III.10). We represent the force vector as acting midway along the lower wing surface, where the dot is, and pointing above the wing to show the direction of the resulting wind impact force. This force can be broken up into two components, one sideways and the other up. The upward component, **U**, is called *lift.* The sideways component, **S**, is called *drag.* If the aircraft is to fly at constant velocity at constant altitude, then lift must equal the weight of the aircraft and the thrust of the plane's engines must equal drag. The magnitude of lift (and drag) can be altered by changing the speed of the airplane or by changing the angle (called *angle of attack*) between the wing and the horizontal.

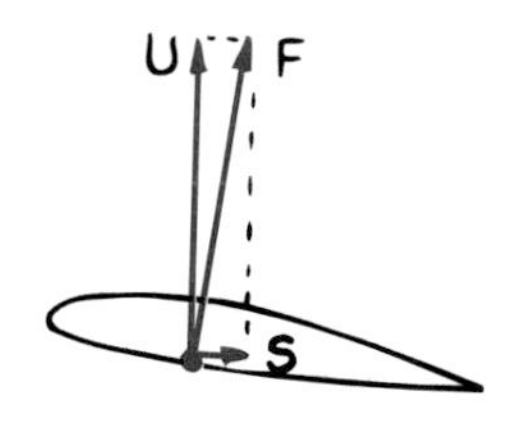

Figure C.10

5. Consider the satellite moving clockwise in Figure C.11. Everywhere in its orbital path, gravitational force **F** pulls it toward the center of the host planet. At position A we see **F** separated into two components: **f**, which is tangent to the path of the projectile, and **f'**, which is perpendicular to the path. The relative magnitudes of these components in comparison to the magnitude of **F** can be seen in the imaginary rectangle they compose: **f** and **f'** are the sides, and **F** is the diagonal. We see that component **f** is along the orbital path but against the direction of motion of the satellite. This force component reduces the speed of the satellite. The

other component f', changes the direction of the satellite's motion and pulls it away from its tendency to go in a straight line. So the path of the satellite curves. The satellite loses speed until it reaches position B. At this farthest point from the planet (apogee), the gravitational force is somewhat weaker but perpendicular to the satellite's motion, and component **f** has reduced to zero. Component **f'**, on the other hand, has increased and is now fully merged to become **F**. Speed at this point is not enough for circular orbit, and the satellite begins to fall toward the planet. It picks up speed because the component **f** reappears and is in the direction of motion as shown in position C. The satellite picks up speed until it whips around to position D (perigee), where once again the direction of motion is perpendicular to the gravitational force, **f'** blends to full **F**, and **f** is nonexistent. The speed is in excess of that needed for circular orbit at this distance, and it overshoots to repeat the cycle. Its loss in speed in going from D to B equals its gain in speed from B to D. Kepler discovered that planetary paths are elliptical, but never knew why. Do you?

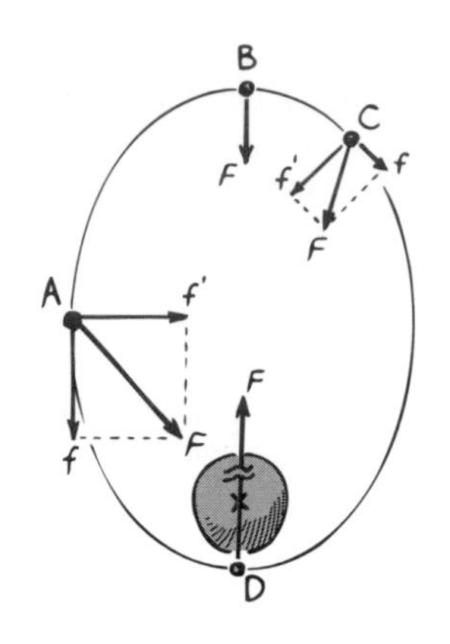

Figure C.11

6. Refer to the Polaroids held by Ludmila back in Chapter 11, in Figure 11.43. In the first picture (a), we see that light is transmitted through the pair of Polaroids because their axes are aligned. The emerging light can be represented as a vector aligned with the polarization axes of the Polaroids. When the Polaroids are crossed (b), no light emerges because light passing through the first Polaroid is perpendicular to the polarization axes of the second Polaroid, with no components along its axis. In the third picture (c), we see that light is transmitted when a third Polaroid is sandwiched at an angle between the crossed Polaroids. The explanation for this is shown in Figure C.12.

Figure C.12

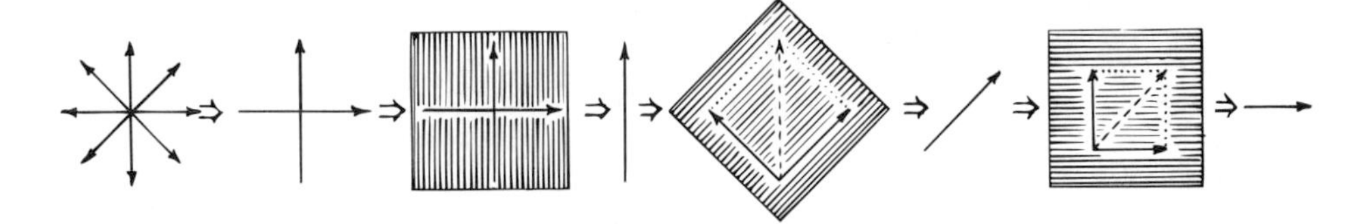

Sailboats

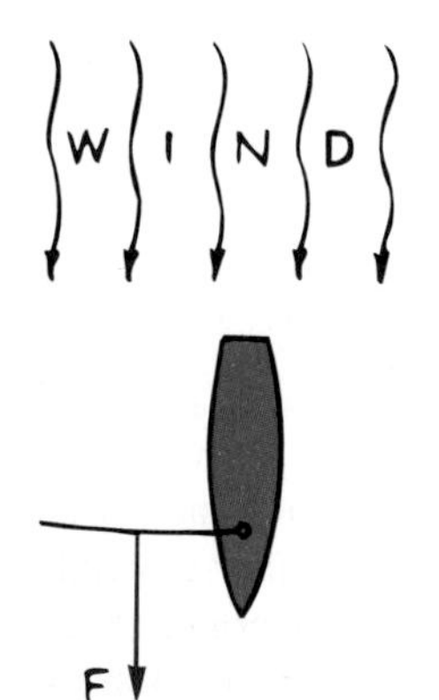

Figure C.13

Sailors have always known that a sailboat can sail downwind, in the direction of the wind. Sailors have not always known, however, that a sailboat can sail upwind, against the wind. One reason for this has to do with a feature that is common only to recent sailboats—a finlike keel that extends deep beneath the bottom of the boat to ensure that the boat will knife through the water only in a forward (or backward) direction. Without a keel, a sailboat could be blown sideways.

Figure C.13 shows a sailboat sailing directly downwind. The force of wind impact against the sail accelerates the boat. Even if the drag of the water and all other resistance forces are negligible, the maximum speed of the boat is the wind speed. This is because the wind will not make impact against the sail if the boat is moving as fast as the wind. The wind would have no speed relative to the boat and the sail would simply sag. With no force, there is no acceleration. The force vector in Figure C.13 *decreases* as the boat travels faster. The force vector is maximum when the boat is at rest and the full impact of the wind fills the sail, and is minimum when the boat travels as fast as the wind. If the boat is somehow propelled to a speed faster than the wind (by way of a motor, for example), then air resistance against the front side of the sail will produce an oppositely directed force vector. This will slow the boat down. Hence the boat when driven only by the wind cannot exceed wind speed.

If the sail is oriented at an angle, as shown in Figure C.14, the boat will move forward, but with less acceleration. There are two reasons for this:

1. The force on the sail is less because the sail does not intercept as much wind in this angular position.
2. The direction of the wind impact force on the sail is not in the direction of the boat's motion, but is perpendicular to the surface of the sail. Generally speaking,

whenever any fluid (liquid or gas) interacts with a smooth surface, the force of interaction is perpendicular to the smooth surface.* The boat does not move in the same direction as the perpendicular force on the sail, but is constrained to move in a forward (or backward) direction by its keel.

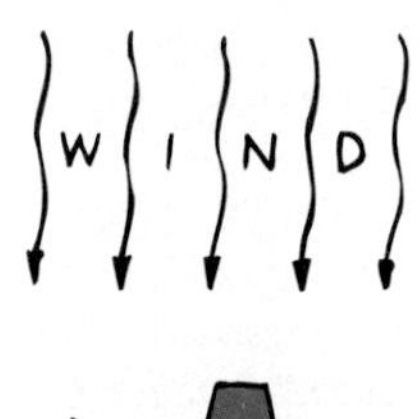

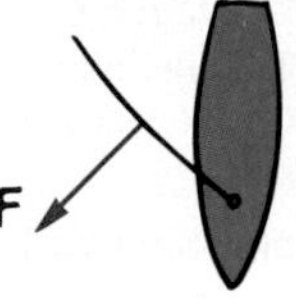

Figure C.14

We can better understand the motion of the boat by resolving the force of wind impact, **F**, into perpendicular components. The important component is that which is parallel to the keel, which we label **K**, and the other component is perpendicular to the keel, which we label **T**. It is the component **K**, as shown in Figure C.15 that is responsible for the forward motion of the boat. Component **T** is a useless force that tends to tip the boat over and move it sideways. This component force is offset by the deep keel. Again, maximum speed of the boat can be no greater than wind speed.

Many sailboats sailing in directions other than exactly downwind Figure (C.16) with their sails properly oriented can exceed wind speed. In the case of a sailboat cutting across the wind, the wind may continue to make impact with the sail even after the boat exceeds wind speed. A surfer, in a similar way, exceeds the velocity of the propelling wave by angling his surfboard across the wave. Greater angles to the propelling medium (wind for the boat, water wave for the surfboard) result in greater speeds. A sailcraft can sail faster cutting across the wind than it can sailing downwind.

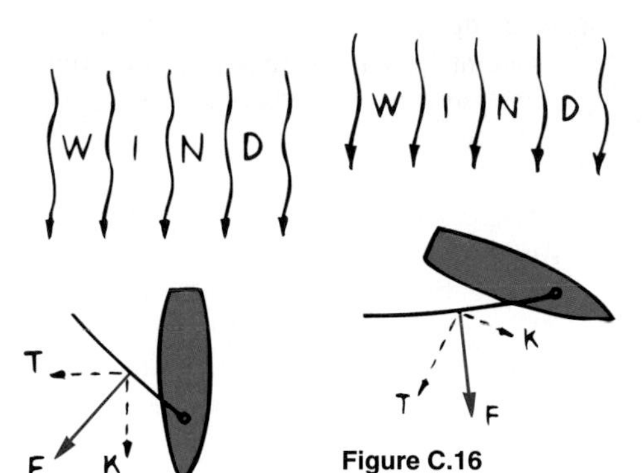

Figure C.15

Figure C.16

As strange as it may seem, maximum speed for most sailcraft is attained by cutting into (against) the wind, that is, by angling the sailcraft in a direction upwind! Although a sailboat cannot sail directly upwind, it can reach a destination upwind by angling back and forth in a zigzag fashion. This is called *tacking.* Suppose the boat and sail are as shown in Figure C.17. Component **K** will push the boat along in a forward direction, angling into the wind. In the position shown, the boat can sail faster than the speed of the wind. This is because as the boat travels faster, the impact of wind is increased. This is similar to running in a rain that comes down at an angle. When you run into the direction of the downpour, the drops strike you harder and more frequently, but when you run away from the direction of the downpour, the drops don't strike you as hard or as frequently. In the same way, a boat sailing upwind experiences greater wind impact force, while a boat sailing downwind experiences a decreased wind impact force. In any case the boat reaches its terminal speed when opposing forces cancel the force of wind impact. The opposing forces consist mainly of water resistance against the hull of the boat. The hulls of racing boats are shaped to minimize this resistive force, which is the principal deterrent to high speeds.

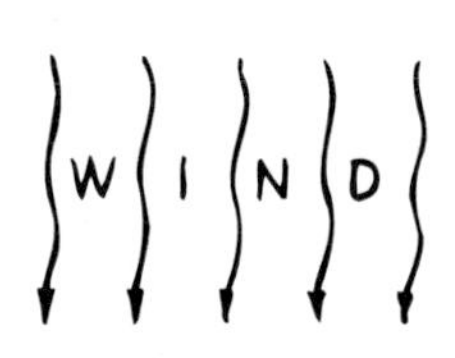

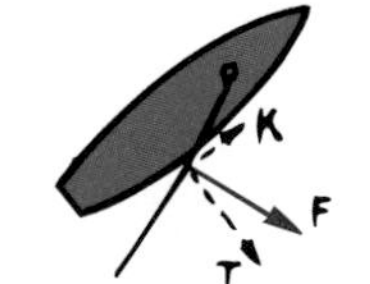

Figure C.17

Iceboats (sailcraft equipped with runners for traveling on ice) encounter no water resistance and can travel at several times the speed of the wind when they tack upwind. Although ice friction is nearly absent, an iceboat does not accelerate without limits. The terminal velocity of a sailcraft is determined not only by opposing friction forces but also by the change in relative wind direction. When the boat's orientation and speed are such that the wind seems to shift in direction, so the wind moves parallel to the sail rather than into it, forward acceleration ceases—at least in the case of a flat sail. In practice, sails are curved and produce an airfoil that is as important to sailcraft as it is to aircraft. The effects are discussed in Chapter 5.

---

*You can do a simple exercise to see that this is so. Try bouncing a coin off another on a smooth surface, as shown. Note that the struck coin moves at right angles (perpendicular) to the contact edge. Note also that it makes no difference whether the projected coin moves along path A or path B. See your instructor for a more rigorous explanation, which involves momentum conservation.

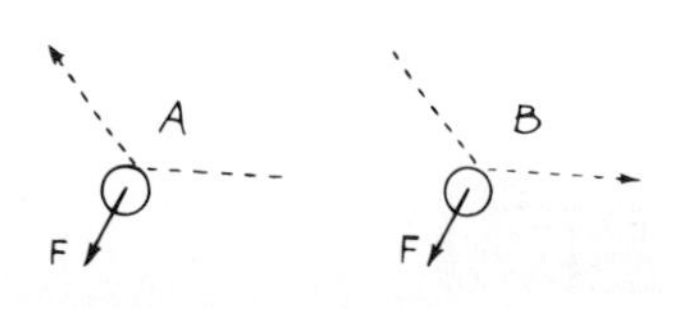

Suggested Reading

Prologue

Adams, Cecil. *The Straight Dope.* Chicago: Chicago Review Press, 1984.

Burke, James. *Connections.* Little Brown, 1995.

Burke, James. *The Pinball Effect.* Boston: Little Brown, 1996.

Cole, K. C. *Sympathetic Vibrations: Reflections of Physics as a Way of Life.* New York: Morrow, 1984.

Cole, K. C. *The Universe and the Teacup.* New York: Morrow, 1998.

Feynman, Richard P. *Surely You're Joking, Mr. Feynman.* New York: Norton, 1986.

Feynman, Richard P. *What Do You Care What Other People Think.* New York: Norton, 1989.

Gamow, George *One Two Three . . . Infinity: Facts and Speculations of Science.* New York: Viking, 1962.

Gleick, James. *Genius—The Life and Science of Richard Feynman.* Boston: Little, Brown, 1992.

Gonick, Larry, and Huffman, Art. *The Cartoon Guide to Physics.* New York: HarperCollins, 1991.

Hawking, Stephen. *A Brief History of Time.* New York: Bantam Books, 1988.

Morrison, Philip. *Powers of Ten.* New York: W. H. Freeman, 1982.

Rutherford, F. James, and Ahlgren, A. *Science for All Americans.* New York: Oxford, 1990.

Sagan, Carl. *Cosmos.* New York: Bantam Books, 1979.

Sagan, Carl. *The Demon Haunted World: Science As a Candle in the Dark.* Random House, 1997.

Sagan, Carl. *Billions and Billions: Thoughts on Life and Death at the Brink of the Millennium.* Random House, 1997.

http://www.cpsurf
Conceptual Physics on the web!

http://www.conceptualphysics.com
Conceptual Physics with Hewitt

http://maxwell.uhh.hawaii.edu/physinfo.html

Part 1

Brancazio, P. J. *Sport Science.* New York: Simon & Schuster, 1984.

Feynman, Richard P. *Lectures on Physics, Part 1,* Reading, MA: Addison-Wesley, 1963.

Feynman, Richard P. "The Great Conservation Principles," and "The Law of Gravitation," *The Character of Physical Law,* Cambridge: MIT, 1967.

Hoffman, Banish. "Introducing Vectors," *About Vectors.* New York: Dover, 1975.

Rothman, Milton A. *A Physicist's Guide to Skepticism.* Buffalo, New York: Prometheus Books, 1988.

Wood, Elizabeth A. "Frames of Reference," *Science from Your Airplane Window.* New York: Dover, 1975.

Part 2

Bridgman, P. W. *The Nature of Thermodynamics.* New York: Harper and Brothers, 1961.

Jones, Edwin R., Jr. "Fahrenheit and Celsius: A History," *The Physics Teacher,* pp. 594–595, Nov. 1980.

Rogers, Eric. *Physics for the Inquiring Mind.* Princeton, NJ.: Princeton University Press, 1960.

Wilson, S. S. "Sadi Carnot," *Scientific American,* Aug. 1981, p. 134.

Part 3

Feynman, Richard P. *Lectures on Physics, Part 2.* Reading, MA: Addison-Wesley, 1963.

Kondo, Herbert. "Michael Faraday," *Scientific American,* October 1953, p. 90.

Walker, Jearl. "The Amateur Scientist: The Secret of a Microwave Oven's Rapid Cooking Action Is Disclosed," *Scientific American,* February 1987, p. 134.

Williams, E.R. "The Electrification of Thunderstorms," *Scientific American,* Nov. 1988, p. 88.

Part 4

Greenler, Robert. *Rainbows, Halos, and Glories.* New York: Cambridge University Press, 1980.

Falk, D. S., Brill, D. R., and Stork, D. *Seeing the Light: Optics in Nature.* New York: Harper & Row, 1985.

Murphy, Pat, and Doherty, Paul. *The Colors of Nature.* San Francisco, Exploratorium, 1996.

Part 5

Cole, K. C. *Sympathetic Vibrations: Reflections of Physics as a Way of Life.* New York: Morrow, 1984.

Fermi, Laura. *Atoms in the Family; My Life with Enrico Fermi.* Chicago: University of Chicago Press, 1954.

Feynman, Richard P. *Lectures on Physics, Part 3.* Reading, MA: Addison-Wesley, 1963.

Gamow, George. *Thirty Years that Shook Physics.* New York: Dover, 1985.

Pagels, Heinz. R. *The Cosmic Code: Quantum Physics as the Language of Nature.* New York: Simon & Schuster, 1982.

Pflaum, Rosalynd. *Grand Obsession: Madame Curie and Her World.* New York: Doubleday, 1989.

Trefil, James. *Atoms to Quarks.* New York: Scribner's, 1980.

Waldrop, M. Mitchell, "The Shroud of Turin: An Answer Is at Hand." *Science,* Sept. 30, 1988, pp. 1750–1751.

http://www.achilles.net/~jtalbot/

A superb site for learning about the spectral patterns of stars and how they are used to study the universe.

http://www.iaea.or.at/worldatom

The web site for the International Atomic Energy Agency (IAEA), which monitors most all issues related to nuclear technology. This is a good starting point for exploring applications of many of the concepts discussed in this chapter.

Part 6

Atkins, P.W. *Molecules*. New York: W. H. Freeman, 1987.

Breslow, Ronald. *Chemistry, Today and Tomorrow: The Central, Useful, and Creative Science.*

Bohren, Craig F. *Clouds in a Glass of Beer*. New York: Wiley Science Editions, 1987.

Canby, Thomas Y. "Materials Reshaping Our Lives," *National Geographic*, Dec. 1989, p. 746.

Hoffmann, Roald. *The Same and Not the Same*. New York: Columbia University Press, 1997.

Poirer, Hean-Pierre, translated by Rebecca Balinski. *Lavoisier: Chemist, Biologist, Economist.* Philadelphia: University of Pennsylvania Press, 1997.

Salem, Lionel. *Marvels of the Molecule*. New York: VCH Publishers, 1987.

Salzberg, Hugh W. *From Caveman to Chemist—Circumstances and Achievements.* Washington, D.C., American Chemical Society, 1991.

Shakhashiri, Bassam Z. *Chemical Demonstrations—A Handbook for Teachers of Chemistry,*

http://www.chemcenter.org

Maintained by the American Chemical Society, this site is an excellent starting point for searching out any chemistry-related information, such as current events or the status of a particular avenue of research.

http://www.rsc.org/lap/rsccom/wcc/wccindex.htm

The home page for the Women Chemists Committee of The Royal Society of Chemistry (UK.). This organization actively promotes the entry and re-entry of women to the profession of chemistry, as well as collecting and disseminating information about women in chemistry.

http://www.csc.fi/lul/chem/graphics.html

A virtual art gallery of molecular animations, et. al, maintained by Finland's Center for Scientific Computing.

http://www.chemsoc.org

Chemistry news updates, an online chemistry magazine, and more at this site run by the Royal Society of Chemistry. There are a large number of periodic tables posted on the Web. Two notable addresses are:

http://wild-turkey.mit.edu/Chemicool/]

http://www.shef.ac.uk/~chem/web-elements/

Part 7

Abbey, Edward. *Desert Solitaire*. New York: Ballantine Books, 1968.

Ahrens, C. Donald. *Meteorology Today.* St. Paul MN: West, 1985.

Austin, Mary. *The Land of Little Rain*. New York: Penguin Books, 1988.

Bolt, Bruce. *Earthquakes*, Revised Edition. New York: W.H. Freeman, 1993.

Gould, Stephen Jay. *Wonderful Life—The Burgess Shale and the Nature of History.* New York: Norton, 1989.

Gross, M. Grant and Elizabeth. *Oceanography: A View of the Earth*. New Jersey: Prentice Hall, 1996.

Leopold, Aldo. *A Sand County Almanac*. New York: Ballantine Books, 1966.

McPhee, John. *Basin and Range*. New York: Farrar, Straus & Giroux, 1981.

McPhee, John. *The Control of Nature*. New York: Farrar, Straus & Giroux, 1989.

McPhee, John. *Assembling California*. New York: Farrar, Straus & Giroux, 1993.

Press, Frank, and Siever, Raymond. *Understanding Earth*. New York: W.H. Freeman, 1994.

Ray, Dixie Lee. *Environmental Overkill*. New York: HarperCollins, 1993.

Reisner, Marc. *Cadillac Desert*. New York: Penguin Books, 1993.

Roadside Geology Series. Missoula, Montana: Mountain Press.

Stegner, Wallace. *Beyond the Hundredth Meridian*. New York: Penguin Books, 1992.

Stowe, Keith. *Exploring Ocean Science*. New York: John Wiley, 1996.

Wenkam, Robert. *The Edge of Fire*. San Francisco: Sierra Club Books, 1987.

http://pangea.stanford.edu/GES/link.html

http://www.usgs.gov

http://water.usgs.gov/public/education.html

http://geology.er.usgs.gov/eastern/inquiries.html

Part 8

Einstein, Albert. *The Meaning of Relativity*. Princeton, NJ.: Princeton University Press, 1950.

Einstein, Albert. *Relativity: The Special and General Theory*. New York: Crown, 1961. (Orig. pub. 1916.)

Epstein, Lewis C. *Relativity Visualized*. San Francisco: Insight Press, 1983.

Ferris, Timothy. *The Whole Shebang*. Simon & Schuster, 1997.

Guth, Alan, and Lightman, Alan P. *The Inflationary Universe: The Quest for a New Theory of Cosmic Origins,* Perseus Press, 1997.

Sheaffer, Robert. *The UFO Verdict: Examining the Evidence*. Buffalo, NY: Prometheus Books, 1981.

Taylor, Edwin F., and Wheeler, John A. *Spacetime Physics*. San Francisco: W. H. Freeman, 1966.

Zeilek, Michael. *Conceptual Astronomy*. New York: John Wiley & Sons, 1992.

http://maxwell.uhh.hawaii.edu/uhhilo/physastr.html

Glossary

Ablation The amount of snow mass a glacier loses in a year.

Absolute zero The lowest possible temperature; the temperature at which all particles have their minimum kinetic energy.

Acceleration The rate at which velocity changes with time; the change in velocity may be in magnitude or direction or both.

Accumulation The amount of snow added to a glacier in a year.

Acid A substance that produces or donates hydrogen ions in solution.

Acid-base reaction A reaction involving the transfer of a hydrogen ion, H+, from one reactant to another.

Acidic solution A solution in which the hydronium ion concentration is greater than the hydroxide ion concentration.

Actinides The inner transition metals of the seventh period.

Addition polymer A polymer formed simply by the joining together of monomer units.

Additive primary colors The three colors—red, blue, and green—that when added in certain proportions produce any other color in the visible-light part of the electromagnetic spectrum.

Adhesive forces Molecular interactions that arise between two different substances.

Adiabatic process A process, usually of expansion or compression, wherein no heat enters or leaves a system.

Alcohols A class of organic molecules that contain a hydroxyl group bonded to a saturated carbon.

Aldehydes A class of organic molecules containing a carbonyl group the carbon of which is bonded to one carbon atom and one hydrogen atom.

Alkali-earth metal A group 2 element.

Alkali metal A group 1 element.

Alloy A mixture of two or more metallic elements.

Alpha particle The nucleus of a helium atom, which consists of two neutrons and two protons, ejected by certain radioactive elements.

Alternating current (ac) Electrically charged particles that repeatedly reverse direction, vibrating about relatively fixed positions. In the United States the vibrational rate is 60 Hz.

Amides A class of organic molecules containing a carbonyl group, the carbon of which is bonded to one carbon atom and one nitrogen atom.

Amines A class of organic molecules containing the element nitrogen bonded to saturated carbon atoms.

Ampere The unit of electric current; 1 ampere = 1 coulomb per second (the flow of 6.25×10^{18} electrons per second); 1 A = 1 C/s.

Amphoteric substance A substance that can behave as either an acid or a base.

Amplitude For a wave or vibration, the maximum displacement on either side of the equilibrium (midpoint) position.

Angular unconformity An unconformity in which older tilted strata are overlain by horizontal younger strata.

Anticline A fold in strata that has relatively old rocks at its core, with rock age decreasing as you move horizontally away from the core fold.

Apparent weight The force you exert against a supporting floor or a weighing scale, wherein you are as heavy as you feel.

Archimedes' principle An immersed body is buoyed up by a force equal to the weight of the fluid it displaces.

Aromatic compound Any organic molecule containing a benzene ring.

Artesian system A system in which groundwater under pressure rises above the level of an aquifer.

Asteroid A small, rocky fragment that orbits the Sun.

Asthenosphere A subdivision of the upper mantle situated below the lithosphere; a zone of plastic, easily deformed rock.

Atomic mass number The number associated with an atom that is the same as the number of protons plus neutrons in its nucleus.

Atomic mass unit A very small unit of mass used for atoms and molecules. One atomic mass unit (amu) is equal to $\frac{1}{12}$ the mass of the carbon-12 atom, or 1.661×10^{-24} grams.

Atomic nucleus The core of an atom, consisting of two basic subatomic particles—protons and neutrons.

Atomic number The number that designates the identity of an element, which is the number of protons in the nucleus of an atom; in a neutral atom, the atomic number is also the number of electrons in the atom.

Atomic orbital The region of space where electrons of a given energy are likely to be located.

Atmospheric pressure The pressure exerted against bodies immersed in the atmosphere resulting from the weight of air pressing down from above. At sea level, atmospheric pressure is about 101 kPa.

Atomic spectra The range of frequencies of light emitted by atoms.

Average speed Total distance traveled divided by time.

Atomic symbol An abbreviation for an element derived from the first one or two letters of the element's name.

Avogadro's number The number of atoms in the atomic mass of an element when expressed in grams. A very large number: 6.02×10^{23}.

Barometer Any device that measures atmospheric pressure.

Base A substance that produces hydroxide ions in solution or accepts hydrogen ions.

Basic solution A solution in which the hydroxide ion concentration is greater than the hydronium ion concentration.

Beats A series of alternate reinforcements and cancellations produced by the interference of two waves of slightly different frequency, heard as a throbbing effect in sound waves.

Bernoulli's principle When the speed of a fluid increases, pressure in the fluid decreases.

Beta particle An electron (or positron) emitted during the radioactive decay of certain nuclei.

Big Bang The primordial explosion of space at the beginning of time.

Binary A system of two stars that orbit about a common center of mass.

Black dwarf The presumed final stage of a white dwarf.

Black hole The remains of a giant star that has collapsed into itself, so dense and of such intense gravitational field that light cannot escape.

Black hole singularity The point of zero radius and infinite density into which the matter of a black hole is compressed.

Body wave A seismic wave that travels through the Earth's interior.

Boiling Rapid evaporation that takes place within a liquid as well as at its surface.

Bond energy The amount of energy absorbed upon bond breaking and released upon bond formation. Each chemical bond has its own characteristic bond energy.

Bow wave The V-shaped wave made by an object moving across a liquid surface at a speed greater than the wave speed.

Boyle's law The product of pressure and volume is constant for a given mass of confined gas so long as temperature remains unchanged:

$$P_1V_1 = P_2V_2$$

Breeder reactor A fission reactor that is designed to breed more fissionable fuel than is put into it by converting non-fissionable isotopes to fissionable isotopes.

Buffer solution A solution of either a weak acid and one of its salts or a weak base and one of its salts that resists large change in pH.

Buoyant force The net upward force that a fluid exerts on an immersed object.

Calorie A unit of thermal energy, or heat. One calorie is the thermal energy required to raise the temperature of one gram of water 1 Celsius degree (1 cal = 4.184 J). One Calorie (with a capital *C*) is equal to one thousand calories and is the unit used in describing the energy available from food.

Capillary action The rising of liquid into a small vertical space due to adhesion between the liquid and the sides of the container and to cohesive forces within the liquid.

Carbonyl group A carbon atom double-bonded to an oxygen atom, $C=O$. The carbonyl group is found in ketones, aldehydes, amides, carboxylic acids, and esters.

Carboxylic acids A class of organic molecules containing a carbonyl group the carbon of which is bonded to one carbon atom and one hydroxyl group.

Catalyst Any substance that serves to increase the rate of a chemical reaction.

Cenozoic The time of recent life. This period of time began 66 million years ago and is still current.

Chain reaction A self-sustaining reaction in which the products of one reaction event stimulate further reaction events.

Chemical change A change in which a substance changes its chemical identity. During a chemical change atoms are rearranged to give a new substance. Also referred to as a chemical reaction.

Chemical equation A representation of a chemical reaction showing the relative numbers of reactants and products.

Chemical equilibrium A dynamic state in which the rate of the forward chemical reaction is equal to the rate of the reverse chemical reaction. At chemical equilibrium the concentrations of reactants and products remain constant.

Chemical formula A notation used to denote the composition of a compound. In a chemical formula the atomic symbols for the elements making up the compound are written along with numerical subscripts that indicate their proportions.

Chemical property The tendency of a substance to change chemical identity.

Chemical reaction Synonymous with *chemical change*.

Chemical weathering The breakdown of rocks on the Earth's surface by chemical means.

Chemistry The study of matter and of the transformations it can undergo.

Clouds The condensation of water droplets above the Earth's surface.

Cluster A grouping of more than one galaxy.

Cohesive forces The attractive forces within a substance.

Comet A body composed of ice and dust that orbits the Sun, usually in a very eccentric orbit, and which has one or more luminous tails when it is close to the Sun.

Complementary colors Any two colors that when added together produce white light.

Compound A material in which atoms of different elements are chemically held to one another.

Concentrated A solution containing a relatively large amount of solute.

Concentration A quantitative measure of the amount of solute in a solution.

Condensation The change of phase from gaseous to liquid.

Condensation polymer A polymer formed by the joining of monomers with the concomitant loss of a small molecule, such as water.

Conduction The transfer of heat energy by collisions between the particles in a substance (especially a solid).

Conformation The spatial orientation of a molecule, which changes as the single bonds in the molecule rotate.

Conservation of energy Energy cannot be created or destroyed; it may be transformed from one form into another, but the total amount of energy never changes. In an ideal machine, where no energy is transformed into heat, $\text{work}_{\text{input}} = \text{work}_{\text{output}}$ and $(Fd)_{\text{input}} = (Fd)_{\text{output}}$.

Conservation of momentum When no external net force acts on an object or a system of objects, no change of momentum takes place. Hence, the momentum before an event involving only internal forces is equal to the momentum after the event:

$$mv_{\text{(before event)}} = mv_{\text{(after event)}}$$

Continental margin The boundary between continental land and deep ocean basins, consisting of continental shelf, continental slope, and continental rise.

Convection The transfer of heat energy in a gas or liquid by means of currents in the heated fluid. The fluid moves, carrying energy with it.

Convectional lifting An air-circulation pattern in which air warmed by the ground rises while cooler air aloft sinks.

Converging lens A lens that is thicker in the middle than at the edges and refracts parallel rays passing through it to a focus.

Copolymer A polymer composed of at least two different types of monomers.

Core The central layer in the Earth's interior, divided into an outer liquid core and an inner solid core.

Coriolis effect The apparent deflection from a straight-line path observed in any body moving near the Earth's surface, caused by the Earth's rotation

Coulomb The SI unit of electrical charge. One coulomb (symbol C) is equal in magnitude to the total charge of 6.25×10^{18} electrons.

Coulomb's law For any two electrically charged bodies, the relationship among the electric force the bodies exert on each other, the charge on the two bodies, and the distance between them:

$$F = k\frac{q_1 q_2}{d^2}$$

If the charges are alike in sign, the force is repelling; if the charges are unlike, the force is attractive.

Covalent bond A chemical bond in which atoms are held together by their mutual attraction for two electrons they share.

Covalent compound An element or chemical compound in which atoms are held together by the covalent bond.

Covalent crystal A group of molecules arranged in an orderly fashion.

Critical angle The minimum angle of incidence inside a medium at which a light ray is totally reflected.

Critical mass The minimum mass of fissionable material in a reactor or nuclear bomb that will sustain a chain reaction.

Cross-cutting When an igneous intrusion or fault cuts through other rocks, the intrusion or fault is younger than the rock it cuts.

Crust The Earth's outermost layer.

Delta A flat-topped accumulation of sediments deposited where a stream flows into a standing body of water.

Density The amount of matter per unit volume:

$$\text{Density} = \frac{\text{mass}}{\text{volume}}$$

Weight density is weight per unit volume.

Diffraction The bending of light as it passes around an obstacle or through a narrow slit, causing the light to spread and to produce light and dark fringes.

Diffuse reflection Reflection in many directions from an irregular surface.

Dilute A solution containing a relatively small amount of solute.

Dipole A separation of charge.

Dipole-dipole interaction The molecular interaction involving dipoles.

Dipole-induced dipole interaction The molecular interaction involving a dipole and an induced dipole.

Direct current (dc) Electrically charged particles flowing in one direction only.

Dissolving The process of mixing a solute in a solvent.

Distillation The process of recollecting a vaporized substance.

Diverging lens A lens that is thinner in the middle than at the edges, causing parallel rays passing through it to diverge as if from a point.

Doppler effect The change in frequency of wave motion resulting from motion of the wave source or receiver.

Drift Generic term for all glacial deposits. Waterborne glacial deposits are known as *outwash*. Material deposited directly by melting ice is *till*.

Effective nuclear charge The nuclear charge experienced by outer-shell electrons, which is diminished by their distance from the nucleus and by the shielding effect of inner-shell electrons.

Efficiency The percent of the work put into a machine that is converted into useful work output.

Elastic collision A collision in which colliding objects rebound without lasting deformation or the generation of heat.

Electric current The flow of electric charge that transports energy from one place to another. Measured in amperes.

Electric field Defined as force per unit charge, it can be considered an "aura" surrounding charged objects and is a storehouse of electric energy. About a charged point, the strength of the electric field decreases with distance according to the inverse-square law.

Electric motor A device that uses a current-carrying coil forced to rotate in a magnetic field to convert electrical energy to mechanical energy.

Electric potential The electric potential energy per amount of charge, measured in volts:

$$\text{Electric potential} = \frac{\text{energy}}{\text{charge}}$$

Electrical potential energy The energy a charge possesses by virtue of its location in an electric field.

Electric power The rate of energy transfer, or the rate of doing work; the amount of energy transferred per unit time, which electrically can be measured by

$$\text{Power} = \text{current} \times \text{voltage}$$

Measured in watts (or kilowatts), where $1\ \text{A} \times 1\ \text{V} = 1\ \text{W}$.

Electrical resistance The property of a material that resists the flow of charged particles through it; measured in ohms (W).

Electrically polarized Term applied to an atom or molecule in which the charges are aligned so that one side has a slight excess of positive charge and the other side a slight excess of negative charge.

Electrochemistry A branch of chemistry concerned with the relationship between electrical energy and chemical change.

Electrode A conducting material used to establish electrical contact between metallic and nonmetallic parts of an electric circuit.

Electrolysis The use of electrical energy to produce chemical change.

Electromagnet A magnet whose field is produced by an electric current. Electromagnets are usually in the form of a wire coil with a piece of iron inside the coil.

Electromagnetic induction The induction of voltage when a magnetic field changes with time. If the magnetic field within a closed loop changes in any way, a voltage is induced in the loop:

$$\text{Voltage induced} \sim \frac{\text{number of loops} \times \text{mag. field change}}{\text{change in time}}$$

This is a statement of Faraday's law. The induction of voltage is the result of a more fundamental phenomenon: the induction of an electric *field*, as defined for the more general case below.

Electromagnetic wave A wave emitted by vibrating electrical charges (often electrons) and composed of vibrating electric and magnetic fields that regenerate one another.

Electromagnetic spectrum The range of electromagnetic waves extending in frequency from radio waves to gamma rays.

Electrons The negatively charged particles in an atom.

Electron affinity The ability of an atom to attract additional electrons.

Electronegativity The ability of an atom to attract a bonding pair of electrons to itself when bonded to another atom.

Element A fundamental material consisting of only one type of atom.

Elemental formula A notation that uses the atomic symbol and a numerical subscript to denote the composition of an element.

Ellipse An oval path. The sum of the distances from any point on the path to two points inside called foci is a constant.

Energy The property of a system that enables it to do work.

Entropy A measure of the disorder of a system. Whenever energy freely transforms from one form to another, the direction of transformation is toward a state of greater disorder and therefore greater entropy.

Erosion The process by which rock particles are transported away by water, wind, or ice.

Escape speed The speed that a projectile, space probe, or similar object must reach in order to escape the gravitational influence of the Earth or celestial body to which it is attracted.

Esters A class of organic molecules containing a carbonyl group the carbon of which is bonded to one carbon atom and one oxygen atom that is also bonded to a carbon atom.

Ethers A class of organic molecules containing an oxygen atom bonded to two saturated carbon atoms.

Evaporation The change of phase from liquid to gaseous.

Event horizon The boundary region of a black hole from which no radiation may escape.

Exosphere The fifth atmospheric layer above the Earth's surface, extending from the thermosphere upward and out into interplanetary space.

Extrusive rocks Igneous rocks that form at the Earth's surface.

Faraday's law An electric field is induced in any region of space in which a magnetic field is changing with time. The magnitude of the induced electric field is proportional to the rate at which the magnetic field changes. The direction of the induced field is at right angles to the direction of the changing magnetic field.

Fault A fracture along which visible displacement can be detected on one side relative to the other.

Faunal succession Fossil organisms succeed one another in a definite, irreversible, determinable order.

First law of thermodynamics A restatement of the law of energy conservation, usually as it applies to systems involving changes in temperature: The heat added to a system equals the increase in the thermal energy of the system plus the external work the system does on its environment.

Fold A series of ripples that result from compressional deformation of the lithosphere.

Force Any influence that can cause an object to be accelerated, measured in newtons in the metric system and in pounds in the British system.

Forced vibration The setting up of vibrations in an object by a vibrating force.

Formula mass The mass of a chemical compound given in amu.

Fractional crystallization The process and sequence by which the different minerals in a magma crystallize at different temperatures as the magma cools.

Free fall A state of fall under the influence of only gravity—free from air resistance.

Freezing The change of phase from liquid to solid.

Frequency For a vibrating body or medium, the number of vibrations per unit time. For a wave, the number of crests that pass a particular point per unit time.

Friction The resistive force that opposes the motion or attempted motion of an object past another with which it is in contact, or through a fluid.

Front The contact zone between two air masses.

Frontal lifting The lifting that occurs as two air masses converge.

Functional group The essential heteroatom-containing structural feature found in all members of a class of compounds.

Fundamental frequency The lowest frequency of vibration of a musical note.

Galaxy A large assembly of stars, numbering in the hundreds of millions to hundreds of billions, together with gas, dust, and other materials, all held together by the forces of mutual gravitation.

Gamma ray High-frequency electromagnetic radiation emitted by the nuclei of radioactive atoms.

Gas Phase of matter in which molecules fill whatever space is available to them, taking neither definite shape nor definite volume.

General theory of relativity The second of Einstein's theories of relativity, which discusses the effects of gravity on space and time.

Generator An electromagnetic induction device that produces electric current by rotating a coil within a stationary magnetic field; converts mechanical energy to electrical energy.

Geodesic The shortest path between two points in various models of space.

Glacier A large mass of ice formed by the compaction and recrystallization of snow, moving downslope under its own weight.

Gravitational red shift The lengthening of the waves of electromagnetic radiation escaping from a massive object.

Greenhouse effect Warming caused by short-wavelength radiant energy from the Sun that easily enters the atmosphere and is absorbed by the Earth. This energy is then reradiated at longer wavelengths that cannot easily escape the Earth's atmosphere. Thus, the Earth's atmosphere and surface are warmed.

Groundwater Subsurface water in the zone of saturation.

Group A vertical column in the periodic table.

Gyre Circular or spiral whirl pattern, usually applied to very large current systems in the open ocean.

Half-life The time required for half the atoms in a sample of a radioactive isotope to decay.

Halogen A group 17 element.

Harmonic A frequency that is an integer multiple of the fundamental.

Heat The thermal energy that flows from an object at higher temperature to one at a lower temperature, commonly measured in calories or joules.

Heat engine A device that uses heat as input and supplies mechanical work as output or one that uses work as input and moves heat "uphill" from a cooler to a warmer place.

Heat of fusion The amount of thermal energy required to change a substance from solid to liquid or from liquid to solid.

Heat of vaporization The amount of thermal energy required to change a substance from liquid to gas or from gas to liquid.

Hertz The unit of frequency. One hertz (symbol Hz) equals one vibration per second.

Heteroatom Any atom other than carbon or hydrogen in an organic molecule.

Heterogeneous mixture A mixture in which the components can be seen as individual substances.

Homogeneous mixture A mixture composed of components so finely mixed that composition is the same throughout the mixture.

H-R Diagram (Hertzsprung-Russell diagram) A plot of brightness versus temperature for stars.

Humidity A measure of the amount of water vapor in the air.

Hydrocarbon A compound containing only carbon and hydrogen atoms.

Hydrogen bond A strong dipole-dipole interaction that involves a hydrogen atom chemically bonded to a strongly electronegative element, such as nitrogen, oxygen, or fluorine.

Hydrologic cycle The natural circulation of water from ocean to atmosphere to ground back to ocean.

Hydronium ion A water molecule after accepting a hydrogen ion, H_3O^+.

Igneous rocks Rocks formed by the cooling and crystallization of hot, molten rock material called magma.

Impulse The product of the force acting on an object and the time during which it acts.

Impure The state of a material that consists of more than one element or compound. A chemical mixture is impure.

Inclusion Any inclusion (pieces of one rock type contained within another) is older than the rock containing it.

Induced dipole A dipole temporarily created in an otherwise nonpolar molecule. It is induced by a neighboring charge or dipole.

Induced dipole-induced dipole interaction The molecular interaction involving only induced dipoles. This is a relatively weak molecular interaction.

Inelastic collision A collision in which the colliding objects become distorted, generate heat, and possibly stick together.

Inertia The tendency of things to resist changes in motion.

Inner-shell shielding The tendency of inner-shell electrons to partially shield outer-shell electrons from the nuclear charge.

Inner transition metals Two subgroups – lanthanides and actinides – of the transition metals.

Insoluble Not capable of dissolving to any appreciable extent in a solvent.

Instantaneous speed Speed at any given instant.

Interference The result of superposing two or more waves of the same wavelength. Constructive interference results from crest-to-crest reinforcement; destructive interference results from crest-to-trough cancellation.

Intrusive rocks Rocks that crystallize below the Earth's surface.

Inverse-square law A law relating the intensity of an effect to the inverse square of the distance from the cause:

$$\text{Intensity} \sim 1/\text{distance}^2$$

Ion-dipole interaction The molecular interaction involving an ion and a dipole.

Ionic bond The electrical force of attraction that holds ions of opposite charge together.

Ionic compound Any chemical compound containing ions.

Ionic crystal A group of many ions held together in an orderly three-dimensional array.

Ionization energy The amount of energy required to pull an electron away from an atom.

Ionosphere An electrified region within the thermosphere and uppermost mesosphere where fairly large concentrations of ions and free electrons exist.

Isostasy The principle that dictates how high the crust stands above the mantle.

Isotopes Atoms whose nuclei have the same number of protons but different numbers of neutrons.

Ketones A class of organic molecules containing a carbonyl group the carbon of which is bonded to two carbon atoms.

Kilogram The fundamental SI unit of mass. One kilogram (symbol kg) is this amount of mass in 1 liter (l) of water at 4°C.

Kinetic energy Energy of motion, described by the relationship

$$\text{Kinetic energy} = \frac{1}{2}mv^2$$

Laminar flow Water flowing smoothly in a straight line with no mixing of sediment.

Lanthanides The inner transition metals of the sixth period.

Lava Magma once it reaches the Earth's surface.

Law of action and reaction (Newton's third law) Whenever one object exerts a force on a second object, the second object exerts an equal and opposite force on the first.

Law of inertia (Newton's First Law) Every material object continues in its state of rest, or of uniform motion in a straight line, unless it is compelled to change that state by forces impressed upon it.

Law of reflection The angle of an reflection equals the angle of incidence. The reflected and incident rays lie in a plane that is normal to the reflecting surface.

Law of universal gravitation Every body in the universe attracts every other body with a force that, for two bodies, is directly proportional to the product of their masses and inversely proportional to the square of the distance separating them:

$$F = \frac{Gm_1m_2}{d^2}$$

Leachate A solution formed by water that has percolated through soil containing water-soluble substances.

Length contraction The contraction of objects in their direction of motion as a result of speed.

Liquid Phase of matter characterized by definite volume but no definite shape; a liquid takes on the shape of its container.

Lithosphere The entire crust plus the portion of the mantle above the asthenosphere.

Longitudinal wave A wave in which the medium vibrates in a direction parallel (longitudinal) to the direction in which the wave travels. Sound is an example of a longitudinal wave.

Lunar eclipse The phenomenon whereby the shadow of the Earth falls on the Moon, producing relative darkness of the full Moon.

Magma Molten rock from the Earth's interior.

Magnetic domains Clustered regions of aligned magnetic atoms. When these regions are aligned with one another, the substance containing them is a magnet.

Magnetic field The region of magnetic influence around either a magnetic pole or a moving charged particle.

Magnetic force (1) Between magnets, the attraction of unlike magnetic poles for each other and the repulsion between like magnetic poles. (2) Between a magnetic field and a moving charged particle, a deflecting force due to the motion of the particle.

Mantle The middle layer in the Earth's interior, between crust and core.

Mass The quantity of matter in an object. More specifically, it is the measurement of the inertia or sluggishness that an object exhibits in response to any effort made to start it, stop it, deflect it, or change in any way its state of motion.

Mechanical weathering The breakdown of rocks on the Earth's surface by physical means.

Mesozoic The time of middle life, from 245 million years ago to about 66 million years age.

Metallic bond A chemical bond in which metal atoms are held together by their attraction to a common pool of electrons.

Maxwell's counterpart to Faraday's law A magnetic field is induced in any region of space in which an electric field is changing with time. The magnitude of the induced magnetic field is proportional to the rate at which the electric field changes. The direction of the induced magnetic field is at right angles to the changing electric field.

Mechanical deformation Metamorphism caused by stress, such as increased pressure.

Mechanical equilibrium The state of an object or system of objects for which any impressed forces cancel to zero and no acceleration occurs.

Melting The change of phase from solid to liquid.

Meniscus The curving of a liquid at the interface of its container.

Mesosphere The third atmospheric layer above the Earth's surface, extending from the top of the stratosphere to 80 km.

Metal An element that is generally shiny, opaque, malleable, ductile, and a good conductor of electricity and heat.

Metalloid One of six elements—boron, silicon, germanium, arsenic, antimony, tellurium—that exhibit both metallic and nonmetallic proerties.

Metamorphic rocks Rocks formed from pre-existing rocks that have been transformed by high temperature, high pressure or both.

Metamorphism The changing of one kind of rock into another kind as a result of high temperature, high pressure, or both.

Meteor A meteoroid once it enters the Earth's atmosphere; a "shooting star."

Meteorite A meteoroid that survives passage through the Earth's atmosphere and hits the ground.

Meteoroid A very small asteroid.

Mineral A naturally formed, inorganic solid composed of an ordered array of atoms chemically bonded to form a particular crystalline structure.

Mohorovicic discontinuity (Moho) The crust-mantle boundary, marking the depth at which the speed of P-waves traveling toward the Earth's center increases.

Mohs scale of hardness A ranking of the relative hardness of minerals.

Molarity A common unit of concentration measured by the number of moles in one liter of solution.

Mole A very large number equal to 6.02×10^{23}. This number is a unit commonly used when describing a number of molecules.

Molecule A group of atoms held tightly together by covalent bonds.

Momentum The product of the mass of an object and its velocity.

Monomer The small molecular unit from which a polymer is formed.

Moon phases The cycles of change of the face of the Moon as seen from the Earth: new (dark) to waxing to full to waning and back to new.

Natural frequency A frequency at which an elastic object naturally tends to vibrate, so that minimum energy is required to produce a forced vibration or to continue vibration at that frequency.

Neutral solution A solution in which the hydronium ion concentration is equal to the hydroxide ion concentration.

Neutralization A reaction in which an acid and base combine to form a salt plus water.

Neutrons The electrically neutral particles in an atomic nucleus.

Neutron star A small, dense star composed of tightly packed neutrons formed by the welding together of protons and electrons.

Newton The SI unit of force. One newton (symbol N) is the force that will give an object of mass 1 kg an acceleration of $1 m/s^2$.

Newton's second law The acceleration of an object is directly proportional to the net force acting on the object, is in the direction of the net force, and is inversely proportional to the mass of the object.

Noble gas A group 18 element.

Noble gas shell A spherical region of space about the atomic nucleus where electrons of a similar energy level may be found.

Nonconformity Overlying sedimentary rocks on an eroded surface of intrusive igneous or metamorphic rocks.

Nonmetal An element that is nonmalleable, nonductile, and a poor conductor of electricity and heat.

Nonpolar The state of a chemical bond or molecule having no dipole.

Nova The sudden brightening of a white dwarf caused by the explosion of accumulated hydrogen gas on its surface.

Nuclear fission The splitting of the nucleus of a heavy atom, such as uranium-235, into two main parts, accompanied by the release of much energy.

Nuclear fusion The combination of the nuclei of light atoms to form heavier nuclei, with the release of much energy.

Nucleon A nuclear proton or neutron; the collective name for either or both.

Ohm's law The current in a circuit varies in direct proportion to the voltage across the circuit and inversely with the circuit's resistance:

$$\text{Current} = \frac{\text{voltage}}{\text{resistance}}$$

A potential difference of 1 V across a resistance of 1 W produces a current of 1 A.

Opaque The term applied to materials through which light cannot pass.

Organic chemistry The study of carbon compounds.

Organic compound Any compound featuring the element carbon covalently bonded to a variety of nonmetal atoms including itself.

Original horizontality Layers of sediment are deposited evenly, with each new layer laid down almost horizontally over the older sediment.

Orographic lifting The lifting of air over a topographic barrier such as a mountain.

Oxidation The process whereby a reactant loses one or more electrons.

Oxidation-reduction reaction A reaction involving the transfer of one or more electrons from one reactant to another.

Paleomagnetism The study of natural magnetization in a rock in order to determine the intensity and direction of the Earth's magnetic field at the time of the rock's formation.

Paleozoic The time of ancient life, from 544 million years ago to 245 million years ago.

Pangaea A single, large landmass that existed in the geologic past and was composed of all the present-day continents.

Parallel circuit An electric circuit in which electrical devices are connected in such a way that the same voltage acts across each one and any single one completes the circuit independently of all the others.

Partial melting The incomplete melting of rocks, resulting in magmas of different compositions.

Percent ionic character A measure of the degree of charge separation in a chemical bond as determined from the difference in electronegativities of the bonded atoms.

Period The time required for a vibration or a wave to make a complete cycle; equal to 1/frequency.

Period A horizontal row in the periodic table.

Periodic table A highly organized chart listing all the known elements arranged in horizonatal rows called periods and vertical columns called groups.

Permeability The ability of a material to transmit fluid.

pH A measure of the acidity of a solution. The pH is equal to the negative logarithm of the hydronium ion concentration. At 25°C, neutral solutions have a pH of 7, acidic solutions have a pH less than 7, and basic solutions have a pH greater than 7.

Phenols A class of organic molecules that contain a hydroxyl group bonded to a benzene ring.

Photoelectric effect The emission of electrons from a metal surface when light shines on it.

Photon A particle of light, or the basic packet of electromagnetic radiation.

Physical change A change in which a substance changes its physical properties without changing its chemical identity.

Physical property Any physical characteristic of a substance, such as color, density, and hardness.

Planetary nebula An expanding shell of gas ejected from a low-mass star during the latter stages of its evolution.

Pluton A large intrusive body formed below the Earth's surface.

Polar The state of a chemical bond or molecule having a dipole.

Polarization The alignment of the transverse electric vectors that make up electromagnetic radiation. Such waves of aligned vibrations are said to be *polarized*.

Polymer A long molecule made of many repeating parts.

Polymorph Minerals that have the same chemical composition but different crystal structures.

Porosity The volume of open space in a rock or sediment relative to the total volume of solids plus open space.

Potential difference (synonymous with *voltage difference*) The difference in electric potential between two points, measured in volts. Electrical potential difference can be compared with the difference in water pressure between two containers of water: when connected by a pipe, water flows from the container having the higher pressure to the one having the lower pressure—until the two pressures are equal. Similarly, when two points having different electric potential are connected by a conductor, charge flows from the point having the greater potential to the one having the lesser potential so long as a potential difference exists.

Potential energy The stored energy that a body possesses because of its position.

Power The time rate of work:

$$\text{Power} = \frac{\text{work}}{\text{time}}$$

Precambrian The time of hidden life, which began about 4.6 billion years ago when the Earth formed, lasted until about 544 million years ago (beginning of Paleozoic), and makes up 85 percent of the Earth's history.

Precipitate A solute that has come out of solution.

Pressure The ratio of force to the area over which that force is distributed:

$$\text{Pressure} = \frac{\text{force}}{\text{area}}$$

$$\text{Liquid pressure} = \text{weight density} \times \text{depth}$$

Pressure-Gradient force The force that moves air from a region of high-pressure air to an adjacent region of low-pressure air.

Primary waves A longitudinal body wave; travels through solids, liquids, and gases and is the fastest seismic wave.

Primeval fireball The burst of energy during the Big Bang.

Principle of flotation A floating object displaces a weight of fluid equal to its own weight.

Principle of the conservation of mass A principle stating that matter is neither created nor destroyed in a chemical reaction, as far as we are able to detect.

Probability cloud The pattern of electron positions plotted over a period of time that shows the likelihood of an electron's position at a given time.

Products The new material formed by a chemical reaction. It appears after the arrow in the chemical equation.

Projectile Any object that moves through the air or space under the influence of gravity.

Protons The positively charged particles in an atomic nucleus.

Protostar The aggregation of matter that goes into and precedes the formation of a star.

Pulsar A neutron star that spins rapidly and in doing so sends out short, precisely timed bursts of electromagnetic radiation.

Pure The state of a material that consists of only a single elemental or compound.

Quantum mechanics The field of wave mechanics where atomic and subatomic particles are treated as waves.

Quarks The elementary constituent particles of building blocks of nuclear matter.

Quasar (Quasi-stellar object) A small, powerful source of energy believed to be the active core of very distant galaxies.

Radiometric dating A method of calculating the age of geologic materials based on the nuclear decay of naturally occurring radioactive isotopes.

Radiation The transfer of energy by means of electromagnetic waves.

Reactants The starting material for a chemical reaction. It appears before the arrow in the chemical equation.

Real image An image formed by light rays that converge at the location of the image.

Recrystallization Metamorphism caused by high temperatures.

Red giant A cool giant star above main sequence on the H-R diagram.

Reduction The process whereby a reactant gains one or more electrons.

Reflection The return of light rays from a surface in such a way that the angle at which a given ray is returned is equal to the angle at which it strikes the surface.

Refraction The bending of a wave as it passes either through a nonuniform medium or from one medium to another, caused by differences in wave speed.

Relationship between impulse and momentum Impulse is equal to the change in the momentum of the object that the impulse acts on. In symbol notation:

$$Ft = \Delta mv$$

Relative humidity The amount of water vapor in the air at a given temperature expressed as a percentage of the maximum amount of water vapor the air can hold at that temperature.

Resonance The response of a body when a forcing frequency matches its natural frequency.

Resultant The single vector that results when two or more vectors are combined.

Rift (rift valley) A long, narrow trough that forms as a result of divergence of two plates.

Rock cycle A sequence of events involving the formation, destruction, alteration, and reformation of rocks as a result of the generation and movement of magma, the weathering, erosion, transportation, and deposition of sediment, and the metamorphism of preexisting rocks.

Salinity The mass of salts dissolved in 1000 grams of seawater.

Salt An ionic compound formed from the reaction of an acid and a base.

Saturated A solution containing as much solute as will dissolve.

Saturated hydrocarbon A hydrocarbon containing no multiple covalent bonds; the carbon atoms are "saturated" with hydrogen atoms.

Sea floor spreading The divergence of two oceanic plates to form a rift in the sea floor.

Secondary waves A transverse body wave; cannot travel through liquids and so does not travel through the Earth's outer core.

Second law of thermodynamics Heat never spontaneously flows from a cold object to a hot one. Also, no machine can be completely efficient in converting energy to work; some input energy is dissipated as waste heat at lower temperature. And finally, all systems tend to become more and more disordered as time goes by.

Sedimentary rocks Rocks formed from the accumulation of weathered material (sediments) carried by water, wind, or ice.

Series circuit An electric circuit in which electrical devices are connected in such a way that the same electric current exists in all of them.

Shell A region of space about the atomic nucleus within which an electron may reside.

Shock wave The cone-shaped wave made by an object moving at supersonic speed through a fluid.

Solar constant The 1400 joules per square meter received from the Sun each second at the top of the Earth's atmosphere. Expressed in terms of power, it is 1.4 kilowatts per square meter.

Solar eclipse The phenomenon whereby the shadow of the Moon falls on the Earth, producing a region of darkness in the daytime.

Solid Phase of matter characterized by definite volume and shape.

Solubility The ability of a solute to dissolve, which depends not only on molecular interactions between the solute and the solvent, but upon the interactions among both solute molecules and among solvent molecules.

Soluble Capable of dissolving in a solvent.

Solute Any component in a solution that is not the solvent.

Solution A homogeneous mixture in which all components are of the same phase.

Solvent The component in a solution present in the largest amount.

Sonic boom The loud sound resulting from the incidence of a shock wave.

Sound wave A longitudinal vibratory disturbance that travels in a medium and can be heard by the human ear when in the approximate frequency range 20 - 20,000 hertz.

Spacetime The four-dimensional continuum in which all things exist; three dimensions of space and one of time.

Special theory of relativity The first of Einstein's theories of relativity, which discusses the effects of uniform motion on space, time, energy, and mass.

Specific heat capacity The quantity of heat per unit of mass required to raise the temperature of a substance by 1 Celsius degree.

Speed Distance traveled divided by time.

Standing wave A stationary wave pattern formed in a medium when two sets of identical waves pass through the medium in opposite directions.

Stoichiometry An aspect of chemistry involving the calculation of quantities of substances involved in chemical reactions.

Stratosphere The second atmospheric layer above the Earth's surface, extending from the top of the troposphere up to 50 kilometers.

Structural isomers Molecules that have the same molecular formula but different chemical structures.

Sublimation The change of phase from solid to gaseous, skipping the liquid phase.

Subtractive primary colors The three colors of absorbing pigments—magenta, yellow, and cyan—that when mixed in certain proportions reflect any other color in the visible-light part of the electromagnetic spectrum.

Sunspots Temporary relatively cool, dark regions on the Sun's surface.

Supercluster A grouping of an enormous number of galaxies.

Supernova A stellar explosion, caused either by transfer of matter to a white dwarf or by gravitational collapse of a massive star, where enormous quantities of matter are emitted.

Superposition In an undeformed sequence of sedimentary rocks, each bed or layer is older than the one above and younger than the one below.

Supersaturated A solution that contains more solute than it normally contains.

Surface tension The energy required to break through the surface of a liquid.

Surface wave A seismic wave that travels along the Earth's surface.

Suspension A homogeneous mixture in which different components are of different phases.

Syncline A fold in strata that has relatively young rocks at its core, with rock age increasing as you move horizontally away from the core fold.

Synthetic polymer A polymer not found in nature.

Temperature A measure of the hotness of an object, related to the average kinetic energy per molecule in the object, measured in degrees Celsius, degrees Fahrenheit, or kelvins.

Temperature inversion A condition in which the upper regions of the atmosphere are warmer than the lower regions.

Terminal speed The constant speed of a falling object where acceleration terminates because air resistance balances the weight. When direction is specified, we speak of terminal velocity.

Theory of plate tectonics The theory that the Earth's lithosphere is broken up into pieces (plates) that move over the asthenosphere; boundaries between plates are where most earthquakes and volcanoes occur and where lithosphere is created and recycled.

Thermal energy (*internal energy*) The total energy (kinetic plus potential) of the particles that make up a substance.

Thermodynamics The study of heat and its transformation to mechanical energy.

Thermonuclear fusion Nuclear fusion produced by high temperature.

Thermosphere The fourth atmospheric layer above the Earth's surface, extending from the top of the mesosphere to 500 km.

Time dilation The slowing of time as a result of speed.

Total internal reflection The total reflection of light traveling within a denser medium when it strikes the boundary with a less dense medium at an angle greater than the critical angle.

Transform fault A plate boundary formed by two plates that are sliding horizontally past each other.

Transformer A device for transferring electric power from one coil of wire to another by means of electromagnetic induction, for the purpose of transforming one value of voltage to another.

Transition metals The elements of groups 3 through 12.

Transmutation The conversion of an atomic nucleus of one element into an atomic nucleus of another element through a loss or gain in the number of protons.

Transparent The term applied to materials through which light can pass in straight lines.

Transverse wave A wave in which the medium vibrates in a direction perpendicular (transverse) to the direction in which the wave travels. Light is an example of a transverse wave.

Troposphere The atmospheric layer closest to the Earth's surface, 16 kilometers high over the equator and 8 kilometers high over the poles, and containing 90 percent of the atmosphere's mass and essentially all its water vapor and clouds.

Turbulent flow Water flowing erratically in a jumbled manner, stirring up everything it touches.

Uncertainty principle It is not possible to measure exactly both the position and the momentum of a particle at the same time, and it is not possible to measure exactly both the energy of a particle and the time at which it has that energy.

Unconformity A break or gap in the geologic record, caused by an interruption in the sequence of deposition or by erosion of pre-existing rock.

Unsaturated A solution that will dissolve additional solute if added.

Unsaturated hydrocarbon A hydrocarbon containing at least one multiple covalent bond.

Valence electron Any electron in the outermost shell of an atom.

Velocity The speed of an object and specification of its direction of motion.

Velocity vector An arrow drawn to scale that represents the magnitude and direction of a given velocity.

Virtual image An image formed by light rays that do not converge at the location of the image.

Volcano A central vent through which lava, gases, and ash erupt and flow.

Volt The unit of electric potential, a potential of 1 volt equals 1 joule of energy per 1 coulomb of charge; 1 V = 1 J / C.

Water table The upper boundary of the zone of saturation, the area where every pore space is filled with water.

Wave A disturbance or vibration propagated from point to point in a medium or in space.

Wave speed The speed with which waves pass a particular point:

$$\text{Wave speed} = \text{frequency} \times \text{wavelength}$$

Wavelength The distance between successive crests, troughs, or identical parts of a wave.

Weight The gravitational force exerted on an object by the nearest most-massive body (locally, by the Earth).

White dwarf A dying star that has collapsed to the size of the Earth and is slowly cooling off; located at lower left of the H-R diagram.

Work The product of the force and the distance through which the force moves:

$$W = Fd$$

Work-Energy theorem The work done on an object is equal to the energy gained by the object.

$$\text{Work} = \Delta E$$

Photo Credits

Front Matter
p. i: G. Brad Lewis; p. xvi: Charles A. Spiegel.

Prologue
p. 1: Roger Ressmeyer/Starlight-Corbis.

Part I
p. 11: Meidor Hu.

Chapter 1
p. 12: Felix St. Claire-Renard/The Image Bank; p. 13: Terry Murphy/Animals Animals; p. 13: The Granger Collection, New York; p. 15: Erich Lessing/Art Resource, New York; p. 25: Al Tielemans/Duomo.

Chapter 2
p. 30: Larry Sergeant/Comstock; p. 35: National Portrait Gallery, London; p. 43: Fundamental Photographs; p. 48 top: Mark Romanelli/The Image Bank; p. 48 middle: John A. Suchocki.

Chapter 3
p. 53: Superstock; p. 53: Terje Rakke/The Image Bank; p. 57: Meidor Hu; p. 59: Paul G. Hewitt; p. 61: Hubrich/The Image Bank; p. 62: Paul G. Hewitt; p. 66: Meidor Hu; p. 68: NASA.

Chapter 4
p. 76: P. Wallick/H. Armstrong Roberts; p. 82: Richard Megna/Fundamental Photographs; p. 85: NASA; p. 87: Richard Megna/Fundamental Photographs; p. 92: Noren Trotman/Sports Photo Masters; p. 96: Diane Schiumo/Fundamental Photographs; p. 100: NASA.

Chapter 5
p. 105: Alan Becker/The Image Bank; p. 108: Jack Hancock; p. 113: Milo Patterson; p. 116: The Granger Collection, New York; p. 119: David Hewitt; p. 123 top: Paul G. Hewitt; p. 123 bottom: William Waterfall/The Stock Market; p. 128: Margaret Ellenstein.

Part II
p. 133: Olan Mills.

Chapter 6
p. 134: G. Brad Lewis; p. 142: Meidor Hu; p. 146: Paul G. Hewitt; p. 147 left: Paul G. Hewitt; p. 147 right: Lynn M. Stone/The Image Bank; p. 148: Jim Wallace.

Chapter 7
p. 159: Jim Wallace; p. 160: Tracy Suchocki; p. 161: Milt and Joan Mann/Cameramann International; p. 162 top: Nancy Rogers; p. 162 bottom: Paul G. Hewitt; p. 166: Robert D. Carey; p. 167: David Cavagnaro; p. 170: Meidor Hu; p. 174: John Suchocki; p. 178: Paul G. Hewitt.

Part III
p. 183: Greg Holmes.

Chapter 8
p. 184: Steven Hunt/The Image Bank; p. 190: Palmer Physical Laboratory; p. 191: Paul G. Hewitt; p. 194: Zig Leszczynski/Animals/Animals; p. 197: Jon Lightfoot.

Chapter 9
p. 211: T. S. Florian/Photo Network/PNI; p. 213: Richard Megna/Fundamental Photographs; p. 214: Richard Megna/Fundamental Photographs; p. 215: Meidor Hu; p. 217 top: Richard Megna/Fundamental Photogrpahs; p. 217 bottom left: John A. Suchocki; p. 217 bottom right: Courtesy Magplane Technology; p. 225: Paul G. Hewitt; p. 229: Jon Lightfoot.

Part IV
p. 231: Paul G. Hewitt.

Chapter 10
p. 232: Hermann Schlenker/Photo Researchers; p. 234: Gabe Palmer/Mug Shots/The Stock Market; p. 237: John A. Suchocki; p. 239: Terence McCarthy/San Francisco Symphony; p. 240: Leslie A. Hewitt; p. 241: Eric Meola/The Image Bank; p. 243: AP/Worldwide; p. 244: Education Development Center; p. 245: Paul G. Hewitt; p. 250: U.S. Navy; p. 251: Palm Press./Courtesy Estate of Dr. Harold Edgerton; p. 254: Meidor Hu.

Chapter 11
p. 258: Dr. Jeremy Burgess/SPL/Photo Researchers; p. 263: Paul G. Hewitt; p. 264: Meidor Hu; p. 267: Dave Vasques; p. 268: John A. Suchocki; p. 270: Meidor Hu; p. 272 left: Camerique/H. Armstrong Roberts; p. 272 right: Don King/The Image Bank; p. 273 top: Education Development Center; p. 273 bottom: Ken Kay/Fundamental Photographs; p. 280: Diane Schiumo/Fundamental Photographs; p. 281: Paul G. Hewitt.

Chapter 12
p. 285: Nina Leen/Life, Time Warner; p. 287: Paul G. Hewitt; p. 288 middle: Roger Ressmeyer/Starlight; p. 288 bottom: Courtesy Institute of Paper; p. 291 middle: Ted Mahieu; p. 291 bottom: Robert Greenler; p. 297: Will and Deni McIntyre/Photo Researchers; p. 298: Ronald B. Fitzgerald; p. 301: Paul G. Hewitt; p. 306: Courtesy Albert Rose; p. 309 top: Barbara Thomas; p. 309 bottom: Camerique/H. Armstrong Roberts; p. 310: Milo Patterson.

Part V
p. 313: John A. Suchocki.

Chapter 13
p. 314: Courtesy IBM Research; p. 315: Paul G. Hewitt; p. 316: Courtesy Lawrence Berkeley Laboratory; p. 322: Courtesy Sargent-Welch Scientific Company; p. 324: H. Rather, "Elektroninterferenzen" *Handbuch der Physik*, Vol. 32, 1957, Springer-Verlag, Berlin, Heidelberg, New York; p. 325 left: H. Armstrong Roberts; p. 325 right: Dr. Tony Brain/SPL/Photo Researchers.

Chapter 14
p. 332: Breck P. Kent/Earth Scenes; p. 339: Courtesy Bicran, Newberry, Ohio; p. 343: Courtesy Lawrence Berkeley Laboratory; p. 348: Roger Ressmeyer/Starlight; p. 355: Lawrence Livermore National Laboratory.

Part VI
p. 361: John A. Suchocki.

Chapter 15
p. 362: Mauritius/Phototake; p. 365: John A. Suchocki; p. 366: Jon Lightfoot; p. 368: John A. Suchocki; p. 367: Joan Lucas; p. 368: John A. Suchocki: p. 369: John A. Suchocki; p. 370 top: John A. Suchocki; p. 370 middle left: Gabe Palacio/Aurora/PNI; p. 370 right: Frank Wing/Stock, Boston/PNI; p. 372 top: Paul Hewitt; p. 372 top: David

Hewitt; p.372 bottom: John A. Suchocki; p. 373 left: Richard Megna/Fundamental Photographs; p. 373 right: USGS.

Chapter 16

p. 380: John A. Suchocki; p. 384: The Granger Collection, New York; p. 385: Michael Thomas/Stock South/PNI; p. 386 top: Courtesy of JPL/NASA; p 386 middle left: Andy Caulfield/The Image Bank; p.386 middle right: Steve Dunwell/The Image Bank; p. 387: John A. Suchocki; p 388: National Museum of Natural History/ Smithsonian Institute; p. 389: Superstock.

Chapter 17

p. 404: Leonard Lessin/Peter Arnold, Inc.; p. 406: Michael McVay/Tony Stone Worldwide; p. 409: Jason Burns/Ace/Phototake; p. 412: Julie Houck/Stock, Boston/PNI; p. 417: Stephen Frisch/Stock, Boston/PNI; p. 419: Jon Lightfoot.

Chapter 18

p. 424: Diane Schiumo/Fundamental Photographs; p. 426: Stephen Frink/Allstock/PNI; p. 427: John A. Suchocki; p. 435 top: Kip Peticolas/Fundamental Photographs; bottom: John A. Suchocki; p. 437: F. Stuart Westmorland/AllStock/PNI; p. 438: Leland C. Clark, Jr., Ph.D./Professor of Biological Sciences; p. 443: Jon Lightfoot; p. 444: Jon Lightfoot; p. 447: John A. Suchocki.

Chapter 19

p. 449: John A. Suchocki; p. 453: John A. Suchocki; p. 454: NASA; p. 455: Jon Lightfoot; p. 457: AP/Worldwide; p. 458: Phototake/NASA; p. 459: Jon Lightfoot; p. 460: John A. Suchocki; p. 461: John A. Suchocki; p. 462: Joseph van Os/The Image Bank; p. 463: Jon Lightfoot.

Chapter 20

p. 472: John A. Suchocki; p. 473 top: Jon Lightfoot; center: John A. Suchocki; p. 475: Jon Lightfoot; p. 477: John A. Suchocki; p. 486: Yoav Levy/PhotoTake; p. 487: John A. Suchocki; p. 488: John A. Suchocki; p. 491 top: Jon Lightfoot; center: Coco McCoy/Rainbow/PNI; p. 493: John A. Suchocki; p. 494: Mark Ludak/Impact Visuals/PNI.

Chapter 21

p. 499: John A. Suchocki; p. 501: Joel Gordon; p. 504: Jon Lightfoot; p. 505: Superstock, Inc.; p. 513: Jon Lightfoot; p. 516: KevinMorris/AllStock/PNI; p. 517: Martin Page/Ace Photo Agency/Phototake/PNI; p. 518 center: Keith Wood/Tony Stone Worldwide; bottom: Jon Lightfoot; p. 522: Jon Lightfoot.

Part VII

p. 531: John A. Suchocki.

Chapter 22

p. 532: Susan Middleton, 1986/California Academy of Sciences; p. 533 top: Harold W. Hoffman/Photo Researchers; bottom left: Susan Middleton, 1986/California Academy of Sciences; p. 534 top right: Lee Boltin; bottom left: (a) E. R. Degginger, (b) Breck P. Kent/Earth Scenes; p. 537 top left: Paul Silverman/Fundamental Photographs; p. top right: Gary Ratherford/Photo Researchers; center: E. R. Degginger; p. 538 top: E. R. Degginger; bottom: Leslie A. Hewitt; p. 539: Runk/Schoenberger/Grant Heilman; p. 543: Fred Grassle/Woods Hole Oceanographic Institution; p. 544 top left: Paul Dix/Impact Visuals/PNI; top center: USGS; top right: W. H. Hodge/Peter Arnold, Inc.; p. 545 center: E.R. Degginger/Earth Scenes; bottom: E.R. Degginger; p. 546: USGS; p. 547 top: Breck P. Kent/Earth Scenes; bottom: Jim Wark/Peter Arnold, Inc.; p. 549 top and center: Dr. Jeremy Burgess/ Science Photo Library/Photo Researchers; p. 549 bottom: Grant Heilman/Grant Heilman; p. 550: Paul Silverman/Fundamental Photographs; p. 551: E.R. Degginger; p. 552 top: Adrienne T. Gibson/ Earth Scenes; center: Breck P. Kent/Earth Scenes; bottom: Steve Lissau/Rainbow; p. 553 center: Richard Weiss/Peter Arnold, Inc.; bottom: Courtesy NASA/E.R. Degginger; p. 554: (a) Lee Boltin, (b) Paul Silverman/Fundamental Photographs; (c) E. R. Degginger/Earth Scenes, (d) E. R. Degginger; p. 555: E.R. Degginger; p. 556 top: Bob Abrams; center Meidor Hu; bottom: (a) A. J. Cunningham/Visuals Unlimited, (b) Alex Kerstich/Visuals Unlimited, (c) Paul Silverman/ Fundamental Photographs, (d) Cabisco/Visuals Unlimited; p. 558: Grant Heilman/Grant Heilman; p. 559: Biologica/Phototake; p. 560 top: Jeffrey Scovil; center: Joyce Photo/Photo Reserachers.

Chapter 23

p. 564: Juan Guzman/Life Magazine/Time Warner; p. 573: "Marie Tharp; p. 578: John S. Shelton.

Chapter 24

p. 588: E. R. Degginger; p. 595: Joseph Burke/Rainbow; p. 597 center: David Hiser/Photographers/PNI; bottom: Paul G. Hewitt; p. 598: Karen Su/AllStock,/PNI; p. 602: John Lemker/Earth Scenes; p. 603: E. R. Degginger; p. 606 top: Leslie A. Hewitt; center: Leslie A. Hewitt; p. 607 top: USGS anonymous; center: E. R. Degginger; p. 608 center: W. H. Hodge/Peter Arnold, Inc.; bottom: E. R. Degginger; p. 609: E. R. Degginger; p. 612: NASA/Grant Heilman; p. 615: Bob Krist/Corbis.

Chapter 25

p. 619: Charles R. Knight/Field Museum of Natural History, Chicago; p. 620: C. Bruce Forster/AllStock/PNI; p. 621: Bob Abrams; p. 625: Roland Seitre/Pter Arnold, Inc.; p. 626 center: Alex Kerstich/Visuals Unlimited; bottom: Tom McHugh/Photo Researchers; p. 627: Courtest, Department of Library Services/American Museum of Natural History; p. 628: Philip James Corwin/Corbis.

Chapter 26

p. 641: Don King/The Image Bank; p. 646: E. R. Degginger.

Chapter 27

p. 663: Michael Townsend/Tony Stone Worldwide; p. 665: Mike J. Howell/Stock, Boston; p. 668: (a) William Johnson/Stock, Boston/PNI, (b) R. L. Kaylin/AllStock/PNI, (c) Joy Spurr/Bruce Coleman/PNI, (d) Charles Mauzy/AllStock/PNI; p. 671: Jim Harrison/Stock, Boston/PNI; p. 674 top: Thomas Nes/The Stock Market; bottom: William Bickel/University of Arizona; p. 675: E. R. Degginger; p. 676: Tony Arruza/Bruce Coleman/PNI.

Part VIII

p. 683: Paul G. Hewitt.

Chapter 28

p. 684: NASA; p. 685: NASA; p. 687: Lick Observatory; p. 689: Roger Ressmeyer/Starlight; p. 690: Dennis di Cicco; p. 691: Mark Martin/NASA/Photo Researchers; p. 696: NASA; p. 697 center: NASA/JPL/Phototake; bottom: Lowell Observatory; p. 698: JPL/NASA; p. 699: NASA: p. 700: JPL/NASA; p. 701 top: Yerkes Observatory; center: Dennis Milon/SPL/Photo Researchers; bottom left: John Sanford/SPl/Photo Researchers; p. 702 center: Reverand Ronald Royer/SPL/Photo Researchers; bottom: A. Behrend/Phototake; p. 703: Paul G. Hewitt; p. 704: John A. Suchocki.

Chapter 29

p. 706 top: Paul G. Hewitt; p. 706 bottom: Paolo Ragazzini/Corbis; p. 708: Roger Ressmeyer/Starlight; p. 712: Hale Observatory/Photo Researchers; p. 714 top: Ronald Royer/Photo Researchers; center: Lick Observatory; p. 715: Royal Obseravatory, Edinburgh; p. 719 top: Julian Baum/New Scientist/JPL/Photo Researchers; bottom: Ronald Royer/SPL/Photo Researchers; p. 720 top: Dr. Steve Gull/SPL/Photo Researchers; center left: National Optical Astronomy Observatories; center right: Royal Observatory, Edimburgh/SPL/Photo Researchers; bottom: NRAO/SPL/Photo Researchers; p. 721 top: NASA/SS/Photo Researchers; center: Hale Observatory/Photo Researchers; bottom: Royal Greenwich Observatory/SPL/Photo Researchers; p. 722: Dennis McNelis.

Chapter 30

p. 724: NASA; p. 731: NOAO/Phil Degginger.

Index

Table of the Elements

Period	Name	Atomic Symbol	Atomic Number	Atomic Mass
1st	Hydrogen	H	1	1.008
	Helium	He	2	4.003
2nd	Lithium	Li	3	6.941
	Beryllium	Be	4	9.012
	Boron	B	5	10.81
	Carbon	C	6	12.011
	Nitrogen	N	7	14.007
	Oxygen	O	8	15.999
	Fluorine	F	9	18.998
	Neon	Ne	10	20.17
3rd	Sodium	Na	11	22.990
	Magnesium	Mg	12	24.305
	Aluminum	Al	13	26.98
	Silicon	Si	14	28.09
	Phosphorus	P	15	30.974
	Sulfur	S	16	32.06
	Chlorine	Cl	17	35.453
	Argon	Ar	18	39.948
4th	Potassium	K	19	39.098
	Calcium	Ca	20	40.08
	Scandium	Sc	21	44.956
	Titanium	Ti	22	47.90
	Vanadium	V	23	50.942
	Chromium	Cr	24	51.996
	Manganese	Mn	25	54.938
	Iron	Fe	26	55.847
	Cobalt	Co	27	58.933
	Nickel	Ni	28	58.71
	Copper	Cu	29	63.546
	Zinc	Zn	30	65.38
	Gallium	Ga	31	69.735
	Germanium	Ge	32	72.59
	Arsenic	As	33	74.922
	Selenium	Se	34	78.96
	Bromine	Br	35	79.904
	Krypton	Kr	36	83.80
5th	Rubidium	Rb	37	85.467
	Strontium	Sr	38	87.62
	Yttrium	Y	39	88.906
	Zirconium	Zr	40	91.22
	Niobium	Nb	41	92.906
	Molybdenum	Mo	42	95.94
	Technetium	Tc	43	98.906
	Ruthenium	Ru	44	101.07
	Rhodium	Rh	45	102.91
	Palladium	Pd	46	106.4
	Silver	Ag	47	107.87
	Cadmium	Cd	48	112.41
	Indium	In	49	114.82
	Tin	Sn	50	118.69
	Antimony	Sb	51	121.75
	Tellurium	Te	52	127.60
	Iodine	I	53	126.90
	Xenon	Xe	54	131.30
6th	Cesium	Cs	55	132.91
	Barium	Ba	56	137.33
	Lanthanum	La	57	138.91
	Cerium	Ce	58	140.12
	Praseodymium	Pr	59	140.91
	Neodymium	Nd	60	144.24
	Promethium	Pm	61	(145)
	Samarium	Sm	62	150.36
	Europium	Eu	63	151.96
	Gadolinium	Gd	64	157.25
	Terbium	Tb	65	158.93
	Dysprosium	Dy	66	162.50
	Holmium	Ho	67	164.93
	Erbium	Er	68	167.26
	Thulium	Tm	69	168.93
	Ytterbium	Yb	70	173.04
	Lutetium	Lu	71	174.97
	Hafnium	Hf	72	178.49
	Tantalum	Ta	73	180.95
	Tungsten	W	74	183.85
	Rhenium	Re	75	186.21
	Osmium	Os	76	190.2
	Iridium	Ir	77	192.22
	Platinum	Pt	78	195.09
	Gold	Au	79	196.97
	Mercury	Hg	80	200.59
	Thallium	Tl	81	204.37
	Lead	Pb	82	207.2
	Bismuth	Bi	83	208.98
	Polonium	Po	84	(209)
	Astatine	At	85	(210)
	Radon	Rn	86	(222)
7th	Francium	Fr	87	(223)
	Radium	Ra	88	226.03
	Actinium	Ac	89	227.03
	Thorium	Th	90	232.04
	Protactinium	Pa	91	231.04
	Uranium	U	92	238.03
	Neptunium	Np	93	237.05
	Plutonium	Pu	94	(244)
	Americium	Am	95	(243)
	Curium	Cm	96	(247)
	Berkelium	Bk	97	(247)
	Californium	Cf	98	(251)
	Einsteinium	Es	99	(254)
	Fermium	Fm	100	(257)
	Mendelevium	Md	101	(258)
	Nobelium	No	102	(259)
	Lawrencium	Lr	103	(260)
	Unnilquadium	Unq	104	(261)
	Unnilpentium	Unp	105	(262)
	Unnilhexium	Unh	106	(263)
	Unnilseptium	Uns	107	(262)
	Unniloctium	Uno	108	(265)
	Unnilennium	Une	109	(266)

Atomic masses are averaged by isotopic abundance in the earth's surface, expressed in atomic mass units (amu). Atomic masses for radioactive elements shown in parentheses are the whole number nearest the most stable isotope of that element.